PRINCIPLES OF
ELECTRICAL ENGINEERING

[For B.E., B.Tech. Students of Jawahar Lal Nehru Technical University, Hyderabad (A.P.), Anna University, Chennai (T.N.) and Biju Patnaik Technical University, Cuttack (Orissa)]

V.K. MEHTA

ROHIT MEHTA

S. CHAND
PUBLISHING
empowering minds

S. CHAND & COMPANY PVT. LTD.

(AN ISO 9001 : 2008 COMPANY)

RAM NAGAR, NEW DELHI-110055

S. CHAND & COMPANY PVT. LTD.

S. CHAND PUBLISHING
empowering minds

(An ISO 9001 : 2008 Company)

Head Office: 7361, RAM NAGAR, NEW DELHI - 110 055

Phone: 23672080-81-82, 9899107446, 9911310888 Fax: 91-11-23677446

www.schandpublishing.com; e-mail: helpdesk@schandpublishing.com

Branches

Ahmedabad	:	Ph: 27541965, 27542369, ahmedabad@schandpublishing.com
Bengaluru	:	Ph: 22268048, 22354008, bangalore@schandpublishing.com
Bhopal	:	Ph: 4274723, 4209587, bhopal@schandpublishing.com
Chandigarh	:	Ph: 2725443, 2725446, chandigarh@schandpublishing.com
Chennai	:	Ph. 28410027, 28410058, chennai@schandpublishing.com
Coimbatore	:	Ph: 2323620, 4217136, coimbatore@schandpublishing.com (Marketing Office)
Cuttack	:	Ph: 2332580; 2332581, cuttack@schandpublishing.com
Dehradun	:	Ph: 2711101, 2710861, dehradun@schandpublishing.com
Guwahati	:	Ph: 2738811, 2735640, guwahati@schandpublishing.com
Hyderabad	:	Ph: 27550194, 27550195, hyderabad@schandpublishing.com
Jaipur	:	Ph: 2219175, 2219176, jaipur@schandpublishing.com
Jalandhar	:	Ph: 2401630, 5000630, jalandhar@schandpublishing.com
Kochi	:	Ph: 2378740, 2378207-08, cochin@schandpublishing.com
Kolkata	:	Ph: 22367459, 22373914, kolkata@schandpublishing.com
Lucknow	:	Ph: 4026791, 4065646 lucknow@schandpublishing.com
Mumbai	:	Ph: 22690881, 22610885, mumbai@schandpublishing.com
Nagpur	:	Ph: 6451311, 2720523, 2777666, nagpur@schandpublishing.com
Patna	:	Ph: 2300489, 2302100, patna@schandpublishing.com
Pune	:	Ph: 64017298, pune@schandpublishing.com
Raipur	:	Ph: 2443142, raipur@schandpublishing.com (Marketing Office)
Ranchi	:	Ph: 2361178, ranchi@schandpublishing.com
Siliguri	:	Ph: 2520750, siliguri@schandpublishing.com (Marketing Office)
Visakhapatnam	:	Ph: 2782609 visakhapatnam@schandpublishing.com (Marketing Office)

First Edition 2003

Subsequent Editions and Reprints 1990, 92, 93, 94, 96, 97, 98, 99, 2000 (Twice), 2001 (Twice), 2002, 2003 (Twice), 2004, 2005, 2006, 2007, 2008, 2009, 2010, 2011, 2012, 2013, 2014 (Thrice),

Reprint 2016

ISBN : 978-81-219-2271-5 **Code :** 1010A 261

This book is for Sale in India & other SAARC Countries only.

PRINTED IN INDIA

By Nirja Publishers & Printers Pvt. Ltd., 54/3/2, Jindal Paddy Compound, Kashipur Road, Rudrapur-263153, Uttarakhand and published by S. Chand & Company Pvt. Ltd., 7361, Ram Nagar, New Delhi -110 055.

PREFACE TO THE SECOND EDITION

Electrical Engineering has become an important science today. More and more people are taking a serious interest in this fascinating subject either as a career or as an absorbing hobby. The rapid growth in the field of electrical engineering has created problems in engineering education, especially in the selection and arrangement of material for training the engineers of tomorrow. The present text "Principles of Electrical Engineering" is designed in such a fashion that it will meet the growing needs of the students preparing for **B.E./B. Tech. students of J.N.T.U., Hyderabad (A.P.), Anna University, Chennai (T.N.) and Biju Patnaik Technical University, Cuttack (Orissa)** and also useful for other Engineering Examinations. It also meets the needs of those readers who want to gain a sound understanding of the principles of electrical engineering.

Three outstanding features are claimed for this book. First is the styling — the authors have used the same lecture style that proved successful in their previous texts. Second the book is easy to read and stimulating in its direct personal approach. Third, the emphasis is on concepts as well as mathematical treatment. It is hoped that these features will help the readers to understand the theoretical and practical aspects of electrical engineering.

The authors wish to thank their colleagues and friends who have contributed many valuable suggestions regarding the scope and content sequence of the book. Authors are also indebted to M/s S. Chand & Co. Ltd., New Delhi for bringing out the book in a short time and pricing it moderately.

Errors might have crept in despite utmost care to avoid them. Authors shall be grateful if these are pointed out along with other suggestions for the improvement of the book.

AUTHORS

SYLLABUS

JAWAHARLAL NEHRU TECHNOLOGICAL UNIVERSITY
HYDERABAD

1 Year B. Tech CSE T P C
2 + 1* 0 4

BASIC ELECTRICAL ENGINEERING

UNIT - I : Introduction to Electrical Engineering : Essence of electricity, Conductors, semiconductors and insulators (elementary treatment only); *Electric field;* electric current, *potential* and *potential difference,* electromotive force, electric power, ohm's law, basic circuit components, electromagnetism related laws.

Magnetic field due to electric current flow, force on a current carrying conductor placed in a magnetic field, Faradays laws of electromagnetic induction. Types of induced EMF's, Kirchhoff's laws. Simple problems.

UNIT - II : Network Analysis : Basic definitions, types of elements, types of sources resistive networks, inductive networks, capacitive networks, series parallel circuits, star delta and delta star transformation, Network theorems - Superposition, Thevenins's, Maximum power transfer theorems and simple problems.

UNIT - III : Magnetic Circuits : Basic definitions, analogy between electric and magnetic circuits, magnetization characteristics of Ferro magnetic materials, self inductance and mutual inductance, energy in linear magnetic systems, coils connected in series, attracting force of electromagnets.

UNIT - IV : Alternating Quantities : Principle of ac voltages, waveforms and basic definitions, relationship between frequency, speed and number of poles, root mean square and average values of alternating currents and voltage, form factor and peak factor, phasor representation of alternating quantities, the J operator and phasor algebra, analysis of ac circuits with single basic network element, single phase series circuits, single phase parallel circuits, single phase series parallel circuits, power in ac circuits.

UNIT - V : Transformers : Principles of operation, Constructional Details, Ideal Transformer and Practical Transformer, Losses, Transformer Test, Efficiency and Regulation Calculations (All the above topics are only elementary treatment and simple problems).

UNIT - VI : Direct current machines : Principle of operation of dc machines, armature windings, e.m.f. equation in a dc machine, Torque production in a dc machine, Operation of a dc machine as a generator, operation of a dc machine as a motor.

UNIT - VII : A.C. Machines : Three phase induction motor, principle of operation, slip and rotor frequency torque (simple problems). Synchronous Machines : Principle of operation, EMF equation (Simple problems on EMF). Synchronous. motor principle and operation (Elementary treatment only)

UNIT - VIII : Basic Instruments : Introduction, classification of instruments, operating principles, essential features of measuring instruments, Moving coil permanent magnet (PMMC) instruments, Moving Iron of Ammeters and Voltmeters (elementary Treatment only)

SYLLABUS

ANNA UNIVERSITY
BE-MECHANICAL ENGINEERING
IInd Semester

EE 151	ELECTRICAL ENGINEERING

1. ELECTRICAL CIRCUITS 9

Ohms Law – Kirchoffs Laws – Steady stage solution of D.C. circuits – Introduction to AC circuits – Waveforms and RMS value power and power factor, single phase and 3 phase balanced circuits.

2. ELECTRICAL MACHINES 15

Principles of operation and characteristics of D.C. machines Transformers (single phase and three phase) – Sychronous Machines 3 phase and single phase induction motors (op principles).

3. ELECTRICAL MEASUREMENTS 6

Moving coil and moving iron instruments (Ammeter and Voltmeter) Dynamometer type watt meters and energy meters (op-principles).

ANNA UNIVERSITY
BE-CIVIL ENGINEERING
IInd Semester

EE 151	ELECTRICAL ENGINEERING

1. ELECTRICAL CIRCUITS 9

Ohms Law – Kirchoffs Laws – Steady stage solution of D.C. circuits – Introduction to AC circuits – Waveforms and RMS value power and power factor, single phase and 3 phase balanced circuits.

2. ELECTRICAL MACHINES 15

Principles of operation characteristics of D.C. machines Transformers (single phase and three phase) – Sychronous Machines – 3 phase and single phase Induction motors (op principles).

3. ELECTRICAL MEASUREMENTS 6

Moving coil and moving iron instruments (Ammeter and Voltmeter) Dynamometer type watt meters and energy meters (op-principles).

4. PRACTICAL 30

SYLLABUS

BIJU PATNAIK TECHNICAL UNIVERSITY, CUTTACK (ORISSA)

BENG 1102 – BASIC ELECTRICAL ENGINEERING (3–1–0)

This is a foundation course aimed at explaining the basic and underlying principles of Electrical circuits, Electro-mechanical devices used for energy conversion, and Generation, Transmission, Distribution & Utilization of electrical energy.

MODULE - I

(13 Lectures)

1. *Introduction* : Electricity, Electric Charge, Electric Current, Electric Potential, Resistance, Ideal and Practical Source, Source Conversion, Inductance, Induced EMF, Capacitance, Energy Stored in Inductor, Energy Stored in Capacitor, Electric Power. (2 Lectures)
2. *DC Networks* : Kirchoff's Laws (KCL and KVL), Node voltage and Mesh current Analysis, Delta-Star and Star-Delta conversions, Superposition principle, Thevanin's theorem, Nortion's theorem. Transients in R L and R-C circuits with D.C. excitation. (4 Lecturers)
3. *Magnetic Circuits* : Introduction to Electromagnetism, B-H curve for magnetic materials, Permeability. Reluctance. Solution of simple magnetic circuits, Hysteresis loss and Eddy current loss. (3 Lectures)
4. *D.C. Machines* : Construction and principle of operation as DC generator and DC Motor, EMF equation of a DC generator, Speed Equation of a DC Motor, Methods of excitation, Speed control of D.C Motor, Open circuit characteristics of DC shunt generator. (4 Lectures.)

MODULE - II

(16 Lectures)

5. *Single-Phase AC Circuits* : Single-phase EMF Generation, Sinusoidal waveform and Phasor representation of EMF. Average and Effective value of sinusoids, Peak factor, Form factor, Complex voltage and current in R, L, C and RLC circuits, Complex representaton of impedance and Power using j-operator, Power factor, Introduction to resonance in RLC series circuits. Band width and Quality factor. (6 Lectures)
6. *Three-Phase AC Circuits* : Comparison between single-phase and three-phase systems, Three-phase EMF Generation, Line and Phase quantities in star and delta connected networks, Solution of balanced three-phase networks, Power and its measurement in three-phase balanced circuits. (5 Lectures)
7. *Transformers* : Construction and principle of operation of single-phase transformers. EMF Equaton of single-phase transformers, Transformation ratio, Practical and Ideal transformers, Transformer losses, Transformer, Phasor Diagram at no-load and load condition, Transformer ratings. (5 Lectures)

MODULE - III

(13 Lectures)

8. *Induction Motors* : Introduction to three-phase Induction Motors, Principle of Operation of Three-phase induction Motors, Introduction to Single-phase Induction Motors, Concept of Slip, Slip-Torque characteristics (no derivations). (4 Lectures)
9. *Measuring Instruments* : Introduction, PMMC Ammeters and Voltmeters, Moving Iron Ammeters and voltmeters, Block diagrammatic study of Digital Voltmeters, Analog Multi-meters, Dynamometer type wattmeter, Energy meter. (6 Lectures)
10. *Power Systems* : Introduction, Types of generating plants, Thermal power plant, Hydro Electric (Hydel) Power Plant, Nuclear Power Plant, Transmission and Distribution Systems, Utilization of Electric Energy. (3 Lectures)

CONTENTS

1

Fundamentals of Current Electricity

Introduction

The study of nature of electricity has been attracting the attention of scientists for hundreds of years. Several theories about electricity were developed through experiments and by observation of its behaviour. The only theory that has survived over the years to explain the nature of electricity is the *Modern Electron Theory of Matter*. This theory has been the result of research work conducted by scientists like Sir William Crooks, J.J. Thomson, Robert A. Millikan, Sir Earnest Rutherford and Neils Bohr. In this chapter, we shall deal with some basic concepts concerning electricity.

1.1. Modern Electron Theory

The nature of electricity can be beautifully explained on the basis of *Modern Electron Theory of Matter. According to this theory, all matter whether solid, liquid or gas is composed of minute particles called *molecules*. A molecule is in turn made up of **atoms.** Those substances whose molecules consist of the same kind of atoms are called ***elements*** while those whose molecules consist of atoms of different kinds are called *compounds*. The number of known stable elements is 105 while the number of compounds is unlimited.

Atoms are the building blocks of all matter. An atom consists of a central part called *nucleus* and around the nucleus (called *extranucleus*), there are a number of electrons revolving in different paths or orbits. The size of the nucleus is very small as compared to the size of the atom. The diameter of an atom is of the order of 10^{-10} metre compared to the diameter of the nucleus of the order of 10^{-15} metre. Thus, there is vast empty space existing in an atom.

1. Nucleus. It is the central part of an atom and contains *protons* and *neutrons*. A proton is a positively charged particle having mass 1837 times that of an electron. A neutron has the same mass as proton but no charge. Obviously, the nucleus of an atom bears a positive charge. Since the mass of an electron is very small as compared to that of a proton or neutron, the nucleus of an atom constitutes the entire weight of atom *i.e.*,

Atomic Weight = No. of protons + No. of neutrons

---

* When this theory was stated, it was believed that all atoms were made up of electrons and protons. Hence, the name electron theory. Although it is well known to-day that an atom contains a number of other particles such as neutrons, mesons, positrons etc. yet the term electron theory has been retained.

** For example, oxygen is an element and its molecule has two atoms of the same kind. On the other hand, water is a compound and its molecule contains two atoms of hydrogen and one atom of oxygen.

2. Extranucleus. It is the outer part of an atom and contains electrons only. An electron is a negatively charged particle having negative charge equal to the positive charge on a proton. The electrons move round the nucleus in different paths or orbits obeying the following rules :

(*i*) The maximum number of electrons that can be accommodated in an orbit is given by $2n^2$ where n is the number of orbit. Thus, the first orbit can accommodate $2 \times 1^2 = 2$ electrons ; the second orbit $2 \times 2^2 = 8$ electrons ; third orbit $2 \times 3^2 = 18$ electrons and so on.

(*ii*) The last orbit cannot accommodate more than 8 electrons.

(*iii*) The last but one orbit cannot accommodate more than 18 electrons.

Under normal conditions, the number of electrons is equal to the number of protons in an atom. Therefore, an atom is neutral as a whole ; the negative charge on the electrons cancelling the positive charge on the protons. The number of protons or electrons in an atom is called atomic number *i.e.*

Atomic number = No. of protons or electrons in an atom

*Fig. 1.1 (*i*) shows the structure of a copper atom. The atomic weight and atomic number of copper are 64 and 29 respectively. Therefore, a copper atom has 29 protons, 29 electrons and 35 neutrons. The 29 electrons are distributed among the various orbits as : first orbit will have 2 electrons ; second 8 electrons, 3rd 18 electrons and fourth orbit 1 electron. Fig. 1.1 (*ii*) shows the structure of an aluminium atom. It has 13 electrons which are distributed in the first, second and third orbits as 2, 8 and 3 electrons respectively.

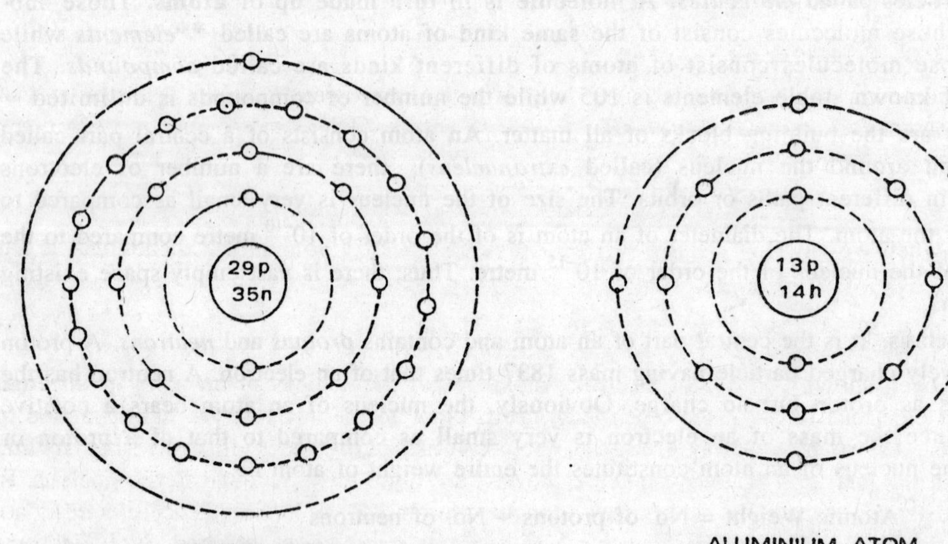

(*i*) (*ii*)

Fig. 1.1

* Copper and aluminium have been intentionally chosen because these two elements are widely used in electrical engineering.

Conclusion. The reader may note the following points carefully :

(*i*) Every matter is electrical in nature *i.e.*, it contains particles of electricity *viz.*, protons and electrons.

(*ii*) In the normal state, the number of electrons is equal to the number of protons in an atom. Therefore, an atom is neutral as a whole. This explains why a body does not exhibit any charge under ordinary conditions.

(*iii*) One substance differs from another in the number and arrangement of particles of atom *viz.* protons, neutrons and electrons. Thus the number of these particles for copper and aluminium atoms is different (see Fig. 1.1). Hence, the two substances are different.

1.2. Nature of Electricity

The above discussion shows that matter is electrical in nature *i.e.*, it contains particles of electricity *viz.* protons and electrons. Whether a given body exhibits electricity (*i.e.*, charge) or not depends upon the relative number of these particles of electricity.

(*i*) If the number of protons is equal to the number of electrons in a body, the resultant charge is zero and the body will be electrically neutral. Thus, the paper of this book is electrically neutral (*i.e.*, paper exhibits no charge) because it has the same number of protons and electrons.

(*ii*) If from a neutral body, some *electrons are removed, there occurs a deficit of electrons in the body. Consequently, the body attains a *positive charge*

(*iii*) If a neutral body is supplied with electrons, there occurs an excess of electrons. Consequently, the body attains a *negative charge.*

1.3. Unit of Charge

The charge on an electron is so small that it is not convenient to select it as the unit of charge. In practice, *Coulomb* is used as the unit of charge. One Coulomb of charge is equal to the charge on 628×10^{16} electrons *i.e.*,

$$1 \text{ Coulomb} = \text{Charge on } 628 \times 10^{16} \text{ electrons}$$

Thus, when we say that a body has a positive charge of 1 coulomb, it means that it has a deficit of 628×10^{16} electrons.

1.4. Free Electrons

It has been shown above that electrons move round the nucleus of an atom in different orbits. The electrons in the inner orbits (*i.e.*, orbits close to the nucleus) are tightly bound to the nucleus. As we move away from the nucleus, this binding goes on decreasing so that electrons in the last orbit (called valence electrons) are quite loosely bound to the nucleus. In certain substances, especially metals (*e.g.*, copper, aluminium *etc.*), the valence electrons are so weakly attached to their nuclei that they can be easily removed or detached. Such electrons are called free electrons.

Those valence electrons which are very loosely attached to the nucleus of an atom are called **free electrons.**

The free electrons move at random from one atom to another in the material. In fact, they are so loosely attached that they do not know the atom to which they belong. It may be noted

* Electrons have very small mass and, therefore, are much more mobile than protons. On the other hand, protons are powerfully held in the nucleus and cannot be removed or detached.

here that all valence electrons in a metal are not free electrons. It has been found that one atom of a metal can provide at the most one free electron. Since a small piece of metal has billions of atoms, one can except a very large number of free electrons in metals. For instance, one cubic centimetre of copper has about 8.5×10^{22} free electrons at room temperature.

1.5. Electric Current

The flow of free electrons is called *electric current.* The flow of electric current can be beautifully explained by referring to Fig. 1.2. The copper strip has a large number of free electrons. When electric pressure or voltage is applied, then free electrons, being negatively charged, will start moving towards the positive terminal round the circuit as shown in Fig. 1.2. This directed flow of free electrons is called *electric current.*

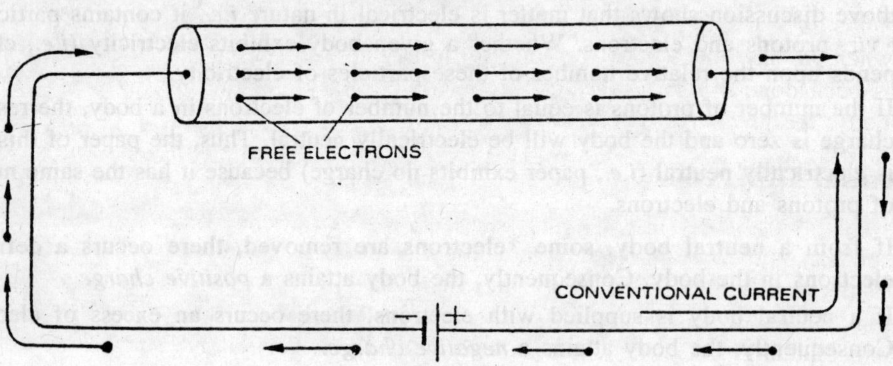

Fig. 1.2

The actual direction of current (*i.e.*, flow of electrons) is from negative terminal to the positive terminal through that part of the circuit external to the cell. However, prior to Electron Theory, it was assumed that current flowed from positive terminal to the negative terminal of the cell *via* the circuit. This* convention is so firmly established that it is still in use. This assumed direction is now called *conventional current.*

Unit of Current. The strength of electric current I is the rate of flow of electrons *i.e.*, charge flowing per second.

$$\therefore \qquad \text{Current, } I = \frac{Q}{t}$$

The charge Q is measured in coulombs and time t in seconds. Therefore, the unit of electric current will be *coulombs/sec* or *ampere.* If $Q = 1$ coulomb, $t = 1$ sec, then $I = 1/1 = 1$ ampere.

One ampere *of current is said to flow through a wire if at any section one coulomb of charge flows in one second.*

Thus, if 5 amperes current is flowing through a wire, it means that 5 coulombs per second flow past any section of the wire.

1.6. Electric Potential

When a body is charged, work is done in charging it. This work done is stored in the body in the form of potential energy. The charged body has the capacity to do work by mov-

* This convention was adopted because prior to electron theory, it was believed that electric current was the movement of positive electricity from the positive to the negative terminal. Now, it is well established that electric current is the movement of free electrons which bear negative charge.

ing other charges either by attraction or repulsion. The ability of the charged body to do work is called electric potential.

The capacity of a charged body to do work is called **electric potential.**

The greater the capacity of a charged body to do work, the greater is its electric potential. Obviously, the work done to charge a body to 1 coulomb will be a measure of its electric potential *i.e.*,

$$\text{Electric potential, } V = \frac{\text{Work done}}{\text{Charge}} = \frac{W}{Q}$$

The work done is measured in joules and charge in coulombs. Therefore, the unit of electric potential will be *joules/coulomb* or *volt.* If $W = 1$ joule, $Q = 1$ coulomb, then, $V = 1/1 = 1$ volt.

Hence a body is said to have an electric potential of **1 volt** *if 1 joule of work is done to give it a charge of 1 coulomb.*

Thus, when we say that a body has an electric potential of 5 volts, it means that 5 joules of work has been done to charge the body to 1 coulomb. The greater the joules/coulomb on a charged body, the greater is its electric potential.

1.7. Potential Difference

The difference in the potentials of two charged bodies is called **potential difference.**

If two bodies have different electric potentials, a potential difference exists between the bodies. Consider two bodies *A* and *B* having potentials of 5 volts and 3 volts respectively as shown in Fig. 1.3 (*i*). Each coulomb of charge on body *A* has an energy of 5 joules while each coulomb of charge on body *B* has an energy of 3 joules. Clearly, body *A* is at higher potential than the body *B*.

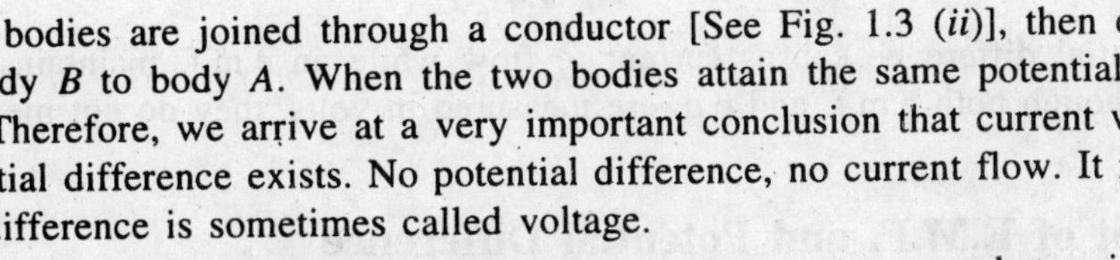

Fig. 1.3

If the two bodies are joined through a conductor [See Fig. 1.3 (*ii*)], then electrons will *flow from body *B* to body *A*. When the two bodies attain the same potential, the flow of current stops. Therefore, we arrive at a very important conclusion that current will flow in a circuit if potential difference exists. No potential difference, no current flow. It may be noted that potential difference is sometimes called voltage.

Unit. Since the unit of electric potential is volt, one can expect that unit of potential difference will also be *volt.* It is defined as under :

The potential difference between two points is **1 volt** *if one joule of work is* ***done in transferring 1 coulomb of charge from one point to the other.*

---

* The conventional current flow will be in the opposite direction *i.e.*, from body *A* to body *B*.

** 1 joule of work will be done in this case if 1 coulomb is transferred from point of lower potential to that of higher potential. However, 1 joule of work will be released (as heat) if 1 coulomb of charge moves from a point of higher potential to a point of lower potential.

1.8. Maintaining Potential Difference

A device that maintains potential difference between two points is said to develop electromotive force (e.m.f.). A simple example is that of a cell. Fig. 1.4 shows the familiar voltaic cell. It consists of a copper plate (called anode) and a zinc rod (called cathode) immersed in dilute H_2SO_4.

The chemical action taking place in the cell removes electrons from copper plate and transfers them to the zinc rod. This transference of electrons takes place through the agency of dil. H_2SO_4 (called electrolyte). Consequently, the copper plate attains a positive charge of $+Q$ coulombs and zinc rod a charge of $-Q$ coulombs. The chemical action of the cell has done a certain amount of work (say W joules) to do so. Clearly, the potential difference between the two plates will be W/Q volts. If the two plates are joined through a wire, some electrons from zinc rod will be attracted through the wire to copper plate. The chemical action of the cell now transfers an equal amount of electrons from copper plate to zinc rod internally through the cell to maintain original potential difference (*i.e.*, W/Q). This process continues so long as the circuit is complete or so long as there is chemical energy. The flow of electrons through the external wire from zinc rod to copper plate is the electric current.

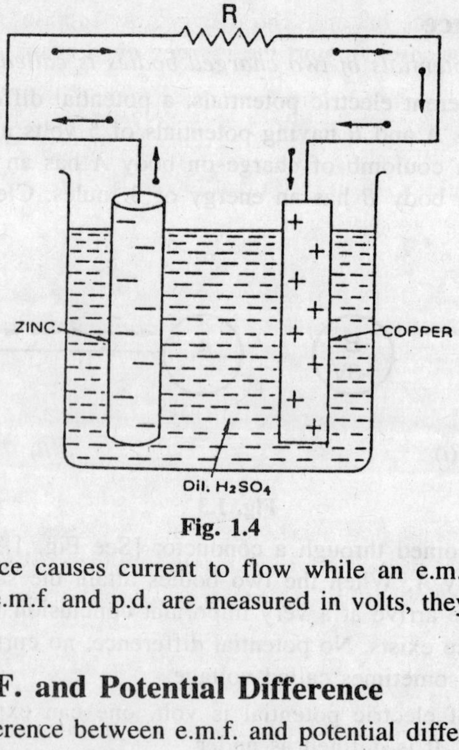

Fig. 1.4

Thus potential difference causes current to flow while an e.m.f. maintains the potential difference. Although both e.m.f. and p.d. are measured in volts, they do not mean exactly the same thing.

1.9. Concept of E.M.F. and Potential Difference

There is a distinct difference between e.m.f. and potential difference. The e.m.f. of a device, say a battery, is a measure of the energy the battery gives to each coulomb of charge. Thus if a battery supplies 4 joules of energy per coulomb, we say that it has an e.m.f. of 4 volts. The energy given to each coulomb in a battery is due to the chemical action.

The potential difference between two points, say A and B, is a measure of the energy used by one coulomb in moving from A to B. Thus if potential difference between points A and B is 2 volts, it means that each coulomb will give up an energy of 2 joules in moving from A to B.

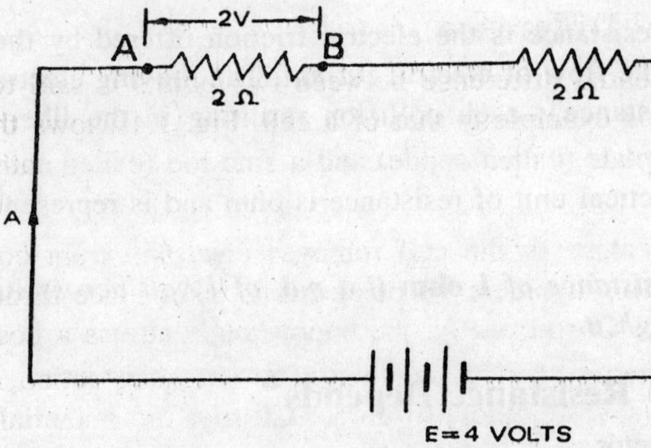

Fig. 1.5

Illustration. The difference between e.m.f. and p.d. can be made more illustrative by referring to Fig. 1.5. Here battery has an e.m.f. of 4 volts. It means that battery supplies 4 joules of energy to each coulomb continuously. As each coulomb travels from the positive terminal of the battery, it gives up its most of energy to resistances (2 Ω and 2 Ω in this case) and remaining to connecting wires. When it returns to the negative terminal, it has lost all its energy originally supplied by the battery. The battery now supplies fresh energy to each coulomb (4 joules in the present case) to start the journey once again.

The p.d. between any two points in the circuit is the energy used by one coulomb in moving from one point to another. Thus in Fig. 1.5, p.d. between *A* and *B* is 2 volts. It means that 1 coulomb will give up an energy of 2 joules in moving from *A* to *B*. This energy will be released as heat from the part *AB* of the circuit.

The following points may be noted carefully :

(*i*) The name e.m.f. at first sight implies that it is a force that causes current to flow. This is not correct because it is not a force but energy supplied to charge by some active device such as a battery.

(*ii*) Electromotive force (e.m.f.) maintains potential difference while p.d. causes current to flow.

Example 1.1. *A charge of 4 coulombs is flowing between points A and B of a circuit. If the potential difference between A and B is 2 volts, how many joules will be released by part AB of the circuit ?*

Solution. The p.d. of 2 volts between points *A* and *B* means that each coulomb of charge will give up an energy of 2 joules in moving from *A* to *B*. As the charge flowing is 4 coulombs, therefore, total energy released by part *AB* of the circuit is

$$= 4 \times 2 = \mathbf{8\ joules}$$

1.10. Resistance

The opposition offered by a substance to the flow of electric current is called **resistance.**

Since current is the flow of free electrons, resistance is the opposition offered by the substance to the flow of free electrons. This opposition occurs because atoms and molecules of the substance obstruct the flow of these electrons. Certain substances (*e.g.*, metals such as silver, copper, aluminium *etc.*) offer very little opposition to the flow of electric current and are called conductors. On the other hand, those substances which offer high opposition to the flow of electric current (*i.e.*, flow of free electrons) are called insulators *e.g.*, glass, rubber, mica, dry wood *etc.*

It may be noted here that resistance is the electric friction offered by the substance and causes production of heat with the flow of electric current. The moving electrons collide with atoms or molecules of the substance ; each collision resulting in the liberation of minute quantity of heat.

Unit of resistance. The practical unit of resistance is ohm and is represented by the symbol Ω. It is defined as under :

A wire is said to have a resistance of **1 ohm** *if a p.d. of 1 volt across its ends causes 1 ampere of current to flow through it.*

1.11. Factors upon which Resistance Depends

The resistance R of a conductor

(*i*) is directly proportional to its length (l).
(*ii*) is inversely proportional to its area of cross-section (a).
(*iii*) depends upon the nature of material.
(*iv*) changes with temperature.

From the first three points (leaving temperature for the time being), we have,

$$R \propto \frac{l}{a}$$

or $$R = \rho \frac{l}{a}$$

where ρ (Greek letter 'Rho') is a constant and is known as *resistivity* or *specific resistance* of the material. Its value depends upon the nature of material.

1.12. Specific Resistance or Resistivity

We have seen above that $R = \rho \frac{l}{a}$

If $l = 1$ m, $a = 1$ m^2, then $R = \rho$

Hence **specific resistance** *of a material is the resistance offered by 1 m length of wire of material having an area of cross-section of 1 m^2 [See Fig. 1.6 (i)].*

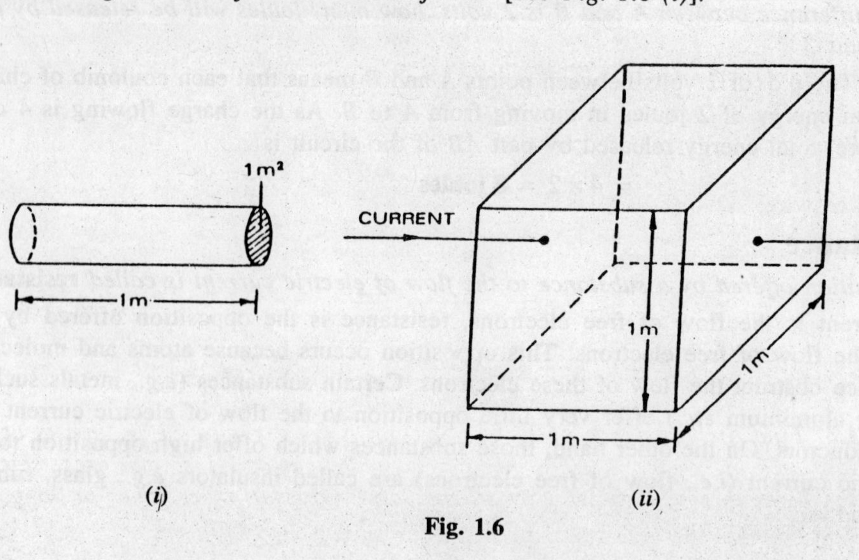

Fig. 1.6

Specific resistance can also be defined in another way. Take a cube of the material having each side 1 m. Considering any two opposite faces, the area of cross-section is 1 m^2 and length is 1 m [See Fig. 1.6(*ii*)] *i.e.* l = 1 m, a = 1 m^2.

Hence **specific resistance** *may be defined as the resistance between the opposite faces of a metre cube of the material.*

Unit of resistivity

We know $$R = \frac{\rho l}{a}$$

or $$\rho = \frac{R a}{l}$$

In SI units, length is measured in metres and area of x-section in m^2 so that the unit of resistivity will be ohm-m.

$$\text{Unit of } \rho = \frac{\text{ohm} \times \text{m}^2}{\text{m}} = \text{ohm–m}$$

1.13. Conductance

Conductance (G) is the reciprocal of resistance.

We know $$R = \rho \frac{l}{a}$$

∴ $$G = \frac{1}{R} = \frac{1}{\rho} \frac{a}{l}$$

or $$G = \sigma \frac{a}{l}$$

where σ (Greek letter 'sigma') is called the conductivity or specific conductance of the material. The unit of conductance is mho *i.e.*, ohm spelt backward. A little reflection shows that SI unit of conductivity will be mho/m.

Example 1.2. *If 10 metres of manganin-wire, 0.13 cm in diameter has a resistance of 3.4 ohms, find the specific resistance of the material.*

Solution.

Length of wire, $l = 10 \text{ m} = 1000 \text{ cm}$

Area of wire, $a = \frac{\pi}{4} d^2 = \frac{\pi}{4} (0.13)^2 = 132.8 \times 10^{-4} \text{ cm}^2$

Resistance of wire, $R = 3.4\ \Omega$

Resistivity of wire, $\rho = ?$

We know $$R = \rho \frac{l}{a}$$

or $$\rho = \frac{R a}{l} = \frac{3.4 \times 132.8 \times 10^{-4}}{1000} = \mathbf{45 \times 10^{-6}\ \Omega\ cm}$$

Example 1.3. *Find the resistance of 1000 metres of a copper wire 25 sq. mm is cross-section. The resistance of copper is 1/58 ohm per metre length and 1 sq-mm cross-section. What will be the resistance of another wire of the same material, three times as long and one-half the cross-sectional area ?*

Solution.

For the first case, $R_1 = ?\,;\quad a_1 = 25\text{ mm}^2\,;\quad l_1 = 1000\text{ m}$

For the second case, $R_2 = 1/58\,\Omega\,;\quad a_2 = 1\text{ mm}^2\,;\quad l_2 = 1\text{ m}$

$$R_1 = \rho\,(l_1/a_1)\,;\; R_2 = \rho\,(l_2/a_2)$$

$$\therefore \qquad \frac{R_1}{R_2} = \frac{l_1}{l_2}\times\frac{a_2}{a_1} = \left(\frac{1000}{1}\right)\times\left(\frac{1}{25}\right) = 40$$

or $$R_1 = 40\,R_2 = 40\times\frac{1}{58} = \mathbf{\frac{20}{29}\,\Omega}$$

For the third case, $R_3 = ?\,;\quad a_3 = a_1/2\,;\quad l_3 = 3l_1$

$$\therefore \qquad \frac{R_3}{R_1} = \left(\frac{l_3}{l_1}\right)\times\left(\frac{a_1}{a_3}\right) = (3)\times(2) = 6$$

or $$R_3 = 6R_1 = 6\times\frac{20}{29} = \mathbf{\frac{120}{29}\,\Omega}$$

Example 1.4. *A length of wire has a resistance of 4.5 ohms. Find the resistance of another wire of the same material three times as long and twice the cross-sectional area.*

Solution.

For the first case, $R_1 = 4.5\,\Omega\,;\; l_1 = l\,;\; a_1 = a$

For the second case, $R_2 = ?\,;\; l_2 = 3l\,;\; a_2 = 2a$

Now $$R_1 = \rho\frac{l_1}{a_1}\,;\; R_2 = \rho\frac{l_2}{a_2}$$

$$\therefore \qquad \frac{R_2}{R_1} = \left[\frac{l_2}{l_1}\right]\times\left[\frac{a_1}{a_2}\right] = \left[\frac{3l}{l}\right]\times\left[\frac{a}{2a}\right] = 1.5$$

or $$R_2 = 1.5\,R_1 = 1.5\times 4.5 = \mathbf{6.75\,\Omega}$$

Example 1.5. *10 c.c. of copper are (i) drawn into a wire 100 metres long (ii) rolled into a square sheet of 10 cm side. Find the resistance of wire and the resistance between opposite faces of the sheet if specific resistance of copper is $1.7\times 10^{-6}\ \Omega$–cm.*

Solution.

(*i*) As seen from Fig. 1.7 (*i*), in this case,

Length of wire, $l = 100\text{ m} = 10^4\text{ cm}$

Area of cross-section, $a = \dfrac{\text{volume}}{\text{length}} = \dfrac{10}{10^4} = 0.001\text{ cm}^2$

$\therefore$ Resistance of wire,

$$R = \frac{\rho l}{a} = \frac{1.7\times 10^{-6}\times 10^4}{0.001} = \mathbf{17\,\Omega}$$

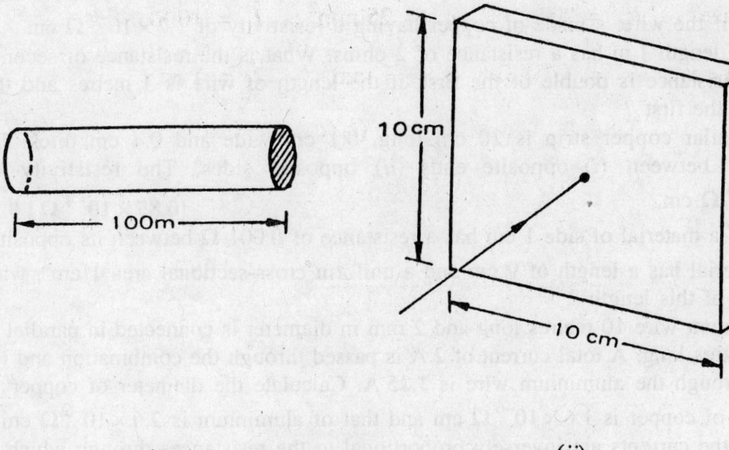

Fig. 1.7

(*ii*) As seen from Fig. 1.7 (*ii*), in this case,

Area of cross-section, $a = 10 \times 10 = 100 \text{ cm}^2$

Length (*i.e.*, thickness), l = volume/area = 10/100 = 0.1 cm

$\therefore$ Resistance between opposite faces of square sheet

$$R = \rho \frac{l}{a} = 1.7 \times 10^{-6} \times \frac{0.1}{100} = \mathbf{17 \times 10^{-10}\ \Omega}$$

Example 1.6. *A copper wire of diameter 1 cm had a resistance of 0.15 Ω. It was drawn under pressure so that its diameter was reduced to 50%. What is the new resistance of the wire ?*

Solution.

Area of wire before drawing, $a_1 = \frac{\pi}{4}(1)^2 = 0.785 \text{ cm}^2$

Area of wire after drawing, $a_2 = \frac{\pi}{4}(0.5)^2 = 0.196 \text{ cm}^2$

As the volume of wire remains the same before and after drawing,

$$\therefore \quad a_1 l_1 = a_2 l_2 \quad or \quad l_2/l_1 = a_1/a_2 = 0.785/0.196 = 4$$

For the first case, $R_1 = 0.15\ \Omega$; $a_1 = 0.785 \text{ cm}^2$; $l_1 = l$

For the second case, $R_2 = ?$; $a_2 = 0.196 \text{ cm}^2$; $l_2 = 4l$

Now $$R_1 = \rho \frac{l_1}{a_1}; \; R_2 = \rho \frac{l_2}{a_2}$$

$$\therefore \quad \frac{R_2}{R_1} = \left(\frac{l_2}{l_1}\right) \times \left(\frac{a_1}{a_2}\right) = (4) \times (4) = 16$$

or $$R_2 = 16 R_1 = 16 \times 0.15 = \mathbf{2.4\ \Omega}$$

Tutorial Problems

1. Calculate the resistance of 915 metres length of a wire having a uniform cross-sectional area of 0.77 cm^2 if the wire is made of copper having a resistivity of 1.7×10^{-6} Ω cm. **[0.08 Ω]**
2. A wire of length 1 m has a resistance of 2 ohms. What is the resistance of second wire, whose specific resistance is double of the first, if the length of wire is 3 metres and the diameter is double of the first ? **[3 Ω]**
3. A rectangular copper strip is 20 cm long, 0.1 cm wide and 0.4 cm thick. Determine the resistance between (*i*) opposite ends (*ii*) opposite sides. The resistivity of copper is 1.7×10^{-6} Ω cm. **[0.85×10^{-4} Ω ; 0.212×10^{-6} Ω]**
4. A cube of a material of side 1 cm has a resistance of 0.001 Ω between its opposite faces. If the same material has a length of 9 cm and a uniform cross-sectional area 1 cm^2, what will be the resistance of this length ? **[0.009 Ω]**
5. An aluminium wire 10 metres long and 2 mm in diameter is connected in parallel with a copper wire 6 metres long. A total current of 2 A is passed through the combination and it is found that current through the aluminium wire is 1.25 A. Calculate the diameter of copper wire. Specific resistance of copper is 1.6×10^{-6} Ω cm and that of aluminium is 2.6×10^{-6} Ω cm . **[0.94 mm]**
Hint. As the currents are inversely proportional to the resistances through which they flow,

$$\therefore \text{ Resistance of copper} = \frac{5}{3} \times \text{Resistance of aluminium}$$

1.14. Effect of Temperature on Resistance

In general, the resistance of a material changes with the change in temperature. The effect of temperature upon resistance varies according to the type of material as discussed below :

(*i*) The resistance of pure metals (*e.g.*, copper, aluminium) increases with the increase of temperature. The change in resistance is fairly regular for normal range of temperatures so that temperature/resistance graph is a straight line as shown in Fig. 1.8 (for *copper). Since the resistance of metals increases with the rise in temperature, they have ***positive temperature co-efficient of resistance.***

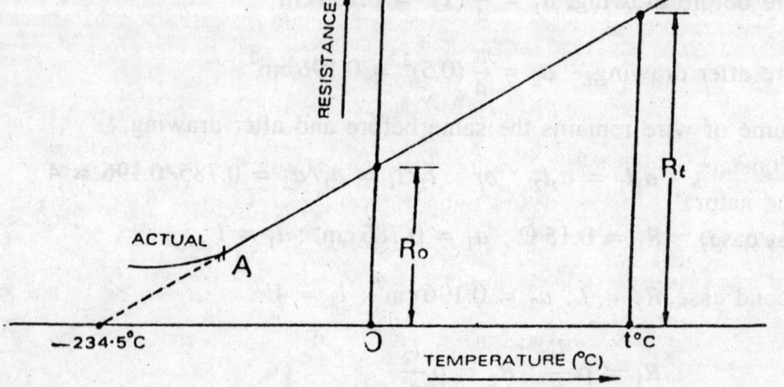

Fig. 1.8

---

* Fig. 1.8 shows temperature/resistance graph for copper which is a straight line. If this line is extended backward, it would cut the temperature axis at –234.5°C. It means that theoretically, the resistance of copper is zero at –234.5°C. However, in actual practice, the curve departs (point *A*) from the straight line path at very low temperatures.

(*ii*) The resistance of electrolytes, insulators (*e.g.*, glass, mica, rubber *etc.*) and semiconductors (*e.g.*, germanium, silicon etc.) decreases with the increase in temperature. Hence these materials have *negative temperature co-efficient of resistance.*

(*iii*) The resistance of alloys increases with the rise in temperature but this increase is very small and irregular. For some high resistance alloys (*e.g.*, Eureka, manganin, constantan *etc.*), the change in resistance is practically negligible over a wide range of temperatures.

The above behaviour of materials can be explained as follows. When the temperature of a substance increases, the molecules vibrate more rapidly, impeding the movement of free electrons through the substance. This causes the resistance of the substance to rise. With the rise in temperature of a conductor (*e.g.*, metals), there is no increase in the number of free electrons and the sole effect of temperature rise is to increase resistance due to the increased molecular vibration. On the other hand, a temperature rise in insulators and semiconductors creates many more free electrons than existed in the cooler state. Often this increase in the number of free electrons more than offsets the interference to the drift movement caused by the increased molecular activity. Hence, the resistance of such materials decreases with the increase in temperature. In case of alloys, these two effects almost cancel each other so that there is negligible increase in resistance due to rise in temperature.

1.15. Temperature Co-efficient of Resistance

Consider a conductor having resistance R_0 at 0°C and R_t at t°C. It has been found that in the normal range of temperatures, the increase in resistance (*i.e.*, $R_t - R_0$).

(*i*) is directly proportional to the initial resistance *i.e.*,

$$R_t - R_0 \propto R_0$$

(*ii*) is directly proportional to the rise in temperature *i.e.*,

$$R_t - R_0 \propto t$$

(*iii*) depends upon the nature of material.

Combining the first two, we get,

$$R_t - R_0 \propto R_0\, t$$

or
$$R_t - R_0 = {}^*\alpha_0 R_0 t \quad \text{...}(i)$$

where α_0 is a constant and is called temperature co-efficient of resistance at 0°C. Its value depends upon the nature of material and temperature.

Rearranging eq. (*i*), we get,
$$R_t = R_0\,(1 + \alpha_0 t) \quad \text{...}(ii)$$

Definition of α_0. From eq. (*i*), we get,

$$\alpha_0 = \frac{R_t - R_0}{R_0 \times t}$$

= Increase in resistance/ohm original resistance/°C rise in temperature

---

* It will be shown in Art. 1.17 that value of α depends upon temperature. Therefore, it is referred to the original temperature *i.e.*, 0°C in this case. Hence, the symbol α_0.

Hence **temperature co-efficient of resistance** *of a material is the increase in resistance per ohm original resistance per °C rise in temperature.*

A little reflection shows that unit of α will be ohms/ohms/°C *i.e.*, /°C. Thus, copper has a temperature co-efficient of resistance of 0.00426/°C. It means that if a copper wire has a resistance of 1 Ω at 0°C, then it will increase by 0.00426 Ω for 1°C rise in temperature *i.e.*, it will become 1.00426 Ω at 1°C. Similarly, if temperature is raised to 10°C, then resistance will become $1 + 10 \times 0.00426 = 1.0426$ ohms.

If a conductor has a resistance R_0, R_1 and R_2 at 0°C, t_1°C and t_2°C respectively, then

$$R_1 = R_0(1 + \alpha_0 t_1) \; ; \qquad R_2 = R_0(1 + \alpha_0 t_2)$$

$$\therefore \qquad \frac{R_2}{R_1} = \frac{1 + \alpha_0 t_2}{1 + \alpha_0 t_1}$$

1.16. Graphical Determination of α

The value of temperature co-efficient of resistance can also be determined graphically from temperature/resistance graph of the material. Fig. 1.9 shows the temperature/resistance graph for a conductor. The graph is a straight line AX as is the case with all conductors. The resistance of the conductor is R_0 (represented by OA) at 0°C and it becomes R_t at t°C. By definition,

$$\alpha_0 = \frac{R_t - R_0}{R_0 \times t}$$

But $\quad R_t - R_0 = BC$

and $\quad t = \text{rise in temp.} = AB$

$$\therefore \qquad \alpha_0 = \frac{BC}{R_0 \times AB}$$

But BC/AB is the slope of temperature/resistance graph.

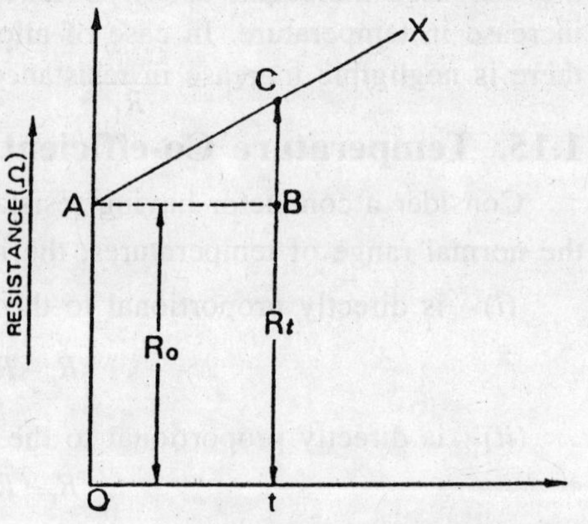

Fig. 1.9

$$\therefore \qquad \alpha_0 = \frac{\text{Slope of temp./resistance graph}}{\text{Original resistance}} \qquad ...(i)$$

Hence, **temperature co-efficient of resistance** *of a material at 0°C is the slope of temp./resistance graph divided by resistance at 0°C (i.e., R_0).*

The following points may be particularly noted :

(*i*) The value of α depends upon temperature. At any temperature, α can be calculated by using eq. (*i*).

Thus $\qquad \alpha_0 = \dfrac{\text{Slope}^* \text{ of temp./resistance graph}}{R_0}$

and $\qquad \alpha_t = \dfrac{\text{Slope of temperature/resistance graph}}{R_t}$

(*ii*) The value of α_0 is maximum and it decreases as the temperature is increased. This is clear from the fact that the slope of temperature/resistance graph is constant and R_0 has the minimum value.

---

* The slope of temp./resistance graph of a conductor is always constant (slope of a straight line.)

1.17. Temperature Co-efficient at Various Temperatures

Consider a conductor having resistance R_0 and R_1 at temperatures 0°C and t_1°C respectively. Let α_0 and α_1 be the temperature co-efficients of the conductor at 0°C and t_1°C respectively. It is desired to establish the relationship between α_1 and α_0. Fig. 1.10 shows the temperature/resistance graph of the conductor. As proved in Art. 1.16,

$$\alpha_0 = \frac{\text{Slope of graph}}{R_0}$$

$\therefore$ Slope of graph $= \alpha_0 R_0$

Similarly, $\alpha_1 = \dfrac{\text{Slope of graph}}{R_1}$

or Slope of graph $= \alpha_1 R_1$

Since the slope of temperature/ resistance graph is constant,

$$\therefore \quad \alpha_0 R_0 = \alpha_1 R_1$$

$$or \quad \alpha_1 = \frac{\alpha_0 R_0}{R_1}$$

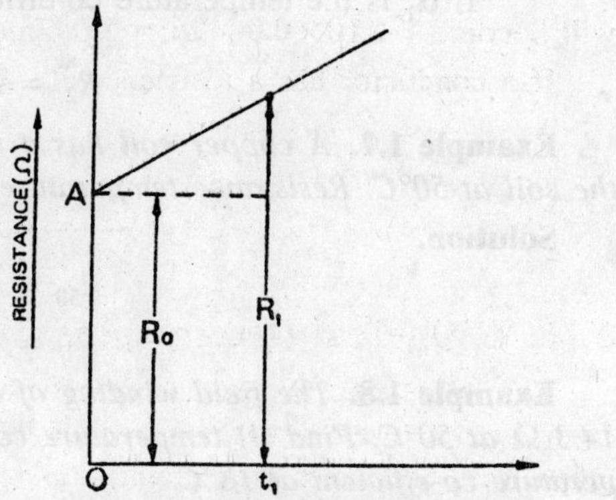

Fig. 1.10

$$= \frac{\alpha_0 R_0}{R_0(1+\alpha_0 t_1)} \qquad [\because R_1 = R_0(1+\alpha_0 t_1)]$$

$$\therefore \quad \alpha_1 = \frac{\alpha_0}{1+\alpha_0 t_1} \qquad ...(i)$$

$$\text{Similarly,*} \quad \alpha_2 = \frac{\alpha_0}{1+\alpha_0 t_2} \qquad ...(ii)$$

Subtracting the reciprocal of Eq. (*i*) from the reciprocal of Eq. (*ii*),

$$\frac{1}{\alpha_2} - \frac{1}{\alpha_1} = \frac{1+\alpha_0 t_2}{\alpha_0} - \frac{1+\alpha_0 t_1}{\alpha_0} = t_2 - t_1$$

$$\therefore \quad \alpha_2 = \frac{1}{\frac{1}{\alpha_1} + (t_2 - t_1)} \qquad ...(iii)$$

Eq. (*i*) gives the relation between α_1 and α_0 while Eq. (*iii*) gives the relation between α_2 and α_1.

1.18. Summary of Temperature Co-efficient Relations

(*i*) If R_0 and α_0 are the resistance and temperature co-efficient of a conductor at 0°C, then its resistance R_t at t°C is given by ;

$$R_t = R_0(1+\alpha_0 t)$$

---

* $\alpha_0 R_0 = \alpha_2 R_2$ where R_2 is the resistance at t_2 °C.

$$or \quad \alpha_2 = \frac{\alpha_0 R_0}{R_2} = \frac{\alpha_0 R_0}{R_0(1+\alpha_0 t_2)} = \frac{\alpha_0}{1+\alpha_0 t_2}$$

(*ii*) If α_0, α_1 and α_2 are the temperature co-efficients at 0°C, t_1°C and t_2°C respectively, then

$$\alpha_1 = \frac{\alpha_0}{1+\alpha_0 t_1} \qquad \alpha_2 = \frac{\alpha_0}{1+\alpha_0 t_2} \; ; \qquad \alpha_2 = \frac{1}{\frac{1}{\alpha_1}+(t_2-t_1)}$$

(*iii*) Suppose R_1 and R_2 are the resistances of a conductor at t_1°C and t_2°C respectively. If α_1 is the temperature co-efficient at t_1°C, then

$$R_2^* = R_1\,[1+\alpha_1\,(t_2-t_1)]$$

Example 1.7. *A copper coil has a resistance of 40 ohms at 0°C. Find the resistance of the coil at 50°C. Resistance temperature co-efficient of copper is 0.0043/°C at 0°C.*

Solution.

$$R_{50} = R_0\,(1+\alpha_0 t)$$
$$= 40\,(1+0.0043\times 50) = \mathbf{48.6\ \Omega}$$

Example 1.8. *The field winding of a generator has a resistance of 12.7 Ω at 18°C and 14.3 Ω at 50°C. Find (i) temperature co-efficient at 0°C (ii) resistance at 0°C and (iii) temperature co-efficient at 18°C.*

Solution .

(*i*)
$$R_{18} = R_0\,(1+\alpha_0\times 18)$$
$$R_{50} = R_0\,(1+\alpha_0\times 50)$$

or
$$\frac{R_{18}}{R_{50}} = \frac{1+18\,\alpha_0}{1+50\,\alpha_0}$$

or
$$\frac{12.7}{14.3} = \frac{1+18\,\alpha_0}{1+50\,\alpha_0}$$

On solving, $\alpha_0 = \dfrac{\mathbf{1}}{\mathbf{236}}\,\mathbf{/°C}$

(*ii*)
$$R_{18} = R_0\,(1+\alpha_0\times 18)$$

$\therefore$
$$R_0 = \frac{R_{18}}{1+18\,\alpha_0} = \frac{12.7}{1+18\times(1/236)} = \mathbf{11.8\ \Omega}$$

(*iii*)
$$\alpha_{18} = \frac{\alpha_0}{1+\alpha_0\times 18} = \frac{1}{(1/\alpha_0)+18} = \frac{1}{236+18} = \frac{\mathbf{1}}{\mathbf{254}}\,\mathbf{/°C}$$

Example 1.9. *A certain length of aluminium wire has a resistance of 28.3 Ω at 20°C. What is its resistance at 60°C ? The temperature co-efficient of resistance of aluminium is 0.00403/°C at 20°C.*

Solution.

$$R_{60} = R_{20}\,[1+\alpha_{20}\,(t_2-t_1)]$$
$$= 28.3\,[1+0.00403(60-20)] = \mathbf{32.86\ \Omega.}$$

---

* Slope of graph, $\tan\theta = R_0\alpha_0 = R_1\alpha_1 = R_2\alpha_2$

Increase in resistance as temperature is raised from t_1°C to t_2°C

$$= \tan\theta\,(t_2-t_1) = R_1\alpha_1\,(t_2-t_1)$$

$\therefore$ Resistance at t_2°C, $R_2 = R_1+R_1\alpha_1\,(t_2-t_1) = R_1\,[1+\alpha_1\,(t_2-t_1)]$

Example 1.10. *The shunt winding of a motor has a resistance of 80 Ω at 15°C. Find its resistance at 50°C. Resistance temperature co-efficient of copper is 0.004/°C at 0°C.*

Solution.

$$R_{15} = R_0 (1 + \alpha_0 \times 15)$$

$$\therefore \quad R_0 = \frac{R_{15}}{1 + 15\alpha_0} = \frac{80}{1 + 15 \times 0.004} = 75.47\ \Omega$$

Now,

$$R_{50} = R_0 (1 + \alpha_0 \times 50)$$

$$= 75.47 (1 + 0.004 \times 50) = \mathbf{90.56\ \Omega}$$

Example 1.11. *The resistance of the field coils of a dynamo is 173 Ω at 16°C. After working for 6 hours on full load, the resistance of the coil increases to 212 Ω. Calculate the mean temperature rise of field coils. Assume temperature co-efficient of resistance of copper to be 0.00426/°C at 0°C.*

Solution. Let t°C be the final temperature.

$$\therefore \quad \frac{R_{16}}{R_t} = \frac{R_0 (1 + \alpha_0 \times 16)}{R_0 (1 + \alpha_0 \times t)}$$

$$or \quad \frac{173}{212} = \frac{1 + 0.00426 \times 16}{1 + 0.00426 \times t}$$

$$or \quad 0.816 = \frac{1.068}{1 + 0.00426\, t}$$

$$\therefore \quad t = 72.5°C$$

$$\text{Rise in temperature} = t - 16 = 72.5 - 16 = \mathbf{56.5°C}$$

Example 1.12. *The field circuit of a 440V shunt motor took 2.3 A when first switched on, the ambient temperature being 17°C. Later the field current was found to remain steady at 1.9 A. Determine the temperature of the winding, assuming $\alpha_0 = 1/234.5$ per °C at 0°C.*

Solution. Let t°C be the final temperature of the winding.

$$R_{17} = 440/2.3 = 191.3\ \Omega\ ; \ R_t = 440/1.9 = 231.5\ \Omega$$

$$\therefore \quad \frac{R_t}{R_{17}} = \frac{R_0 (1 + \alpha_0 \times t)}{R_0 (1 + \alpha_0 \times 17)}$$

$$or \quad \frac{231.5}{191.3} = \frac{1 + (t/234.5)}{1 + (17/234.5)}$$

$$\therefore \quad \mathbf{t = 69.85°C}$$

Example 1.13. *The filament of a 60-watt, 230 V lamp has a normal working temperature of 2000°C. Find the current flowing in the filament at the instant of switching, when the lamp is cold. Assume the temperature of cold lamp to be 15°C and $\alpha_{15} = 0.005/°C$.*

Solution. Resistance of lamp at 2000°C, $R_{2000} = V^2/W = (230)^2/60 = 881.67\ \Omega$

$$R_{2000} = R_{15} [1 + \alpha_{15} (2000 - 15)]$$

$$\therefore \quad R_{15} = \frac{R_{2000}}{1 + 0.005\,(1985)} = \frac{881.67}{10.925} = 80.7\ \Omega$$

∴ Current taken by cold lamp (*i.e.*, at the time of switching)

$$I = V/R_{15} = 230/80.7 = \mathbf{2.85\ A}$$

Example 1.14. *Two coils connected in series have resistances of 600 Ω and 300 Ω and temperature co-efficients of 0.001 and 0.004 respectively at 20°C. Find the resistance of combination at 50°C. What is the effective temperature co-efficient of the combination at 20°C ?*

Solution.

Resistance of 600 Ω coil at 50°C $= 600\,[1 + \alpha_{20}\,(50 - 20)]$

$= 600\,[1 + 0.001 \times 30] = 618\ \Omega$

Resistance of 300 Ω coil at 50°C $= 300\,[1 + \alpha_{20}\,(50 - 20)]$

$= 300\,[1 + 0.004 \times 30] = 336\ \Omega$

∴ Resistance of the combination at 50°C $= 618 + 336 =$ **954 Ω**

Let β be the temperature co-efficient of the combination at 20°C. Since the equivalent resistance of the combination is 954 Ω and 900 Ω at 50°C and 20°C respectively,

$$954 = 900\,[1 + \beta\,(50 - 20)]$$

or $$954/900 = 1 + 30\beta$$

∴ $$\beta = \mathbf{0.002/°C}$$

Tutorial Problems

1. A platinum coil has a resistance of 3.146 Ω at 40°C and 3.767 Ω at 100°C. Find the resistance at 0°C and the temperature co-efficient of resistance at 40°C. **[2.732 Ω ; 1/280 per °C]**
2. A copper coil has a resistance of 0.4 Ω at 12°C. Find its resistance at 52°C. Resistance-temperature co-efficient of copper is given to be 0.004/°C at 0°C. **[0.4612 Ω]**
3. A copper coil has a resistance of 50 Ω at 15°C and 58 Ω at 55°C. Calculate the temperature co-efficient of resistance at 0°C. **[1/235 per °C]**
4. The field coils of a shunt motor have resistance of 45 Ω at 20°C. Find the average temperature of the winding at the end of the run when the resistance is increased to 48.5 Ω. Temperature co-efficient of resistance is 0.004/°C at 0°C. **[41°C]**
5. A coil which has an initial temperature of 20°C, is connected to a 180-V supply and the initial current is found to be 4A. Sometime later, the current is found to have decreased to 3.4A, the supply voltage being unchanged at 180 V. Determine the temperature rise of the coil. Assume the temperature co-efficient of copper to be 0.0043/°C at 0°C. **[44.6°C]**
6. A nichrome heater is operated at 1500°C. What is the percentage increase in its resistance over that at room temperature (20°C) ? Temperature co-efficient of resistance of nichrome is 0.00016/°C at 20°C. **[23.6%]**
7. A copper coil has a resistance of 200 Ω at 80°C. Find the resistance at 20°C. The temperature co-efficient of resistance of copper is 0.0039/°C at 20°C. **[162 Ω]**
8. A piece of resistance wire 15.6 m long and of cross-sectional area 12 mm^2 at a temperature of 0°C passes a current of 7.9 A when connected to a *d.c.* supply at 240 V. Calculate (*i*) resistivity of wire (*ii*) the current which will flow when the temperature rises to 55°C. Temperature co-efficient of resistance of wire is 0.00029/°C. **[23.37×10^{-6} Ω m ; 7.78 A]**
9. When working normally, the temperature of the filament in a 230 V. 150 W gas-filled tungsten lamp is 2750°C. Assuming a room temperature of 16°C, calculate (*i*) the normal current taken by the lamp (*ii*) the current taken at the moment of switching on. Temperature co-efficient of tungsten is 0.0047/°C. **[0.652 A, 8.45 A]**
10. The field coil of a motor has a resistance of 250 Ω at 15°C. By how much will the resistance increase if the motor attains an average temperature of 45°C when running. Take $\alpha_0 = 0.00428$/°C. **[30 Ω]**

1.19. Ohm's Law

The relationship between voltage (*V*) the current (*I*) and resistance (*R*) in a d.c. circuit was first discovered by German scientist George Simon Ohm. This relationship is called Ohm's law and may be stated as under :

The ratio of potential difference (V) between the ends of a conductor to the current (I) flowing between them is constant, provided the physical conditions (e.g., temperature etc.) do not change i.e.

$$\frac{V}{I} = \text{Constant} = R$$

where R is the resistance of the conductor between the two points considered.

For example, if in Fig. 1.11, the voltage between points A and B is V volts and current flowing is I amperes, then V/I will be constant and equal to R, the resistance between points A and B. If the voltage is doubled up, the current will also be doubled up so that the ratio V/I remains constant. It may be noted here that if voltage is measured in volts and current in amperes, then resistance will be in *ohms. Ohm's law can be expressed in three forms *viz.*

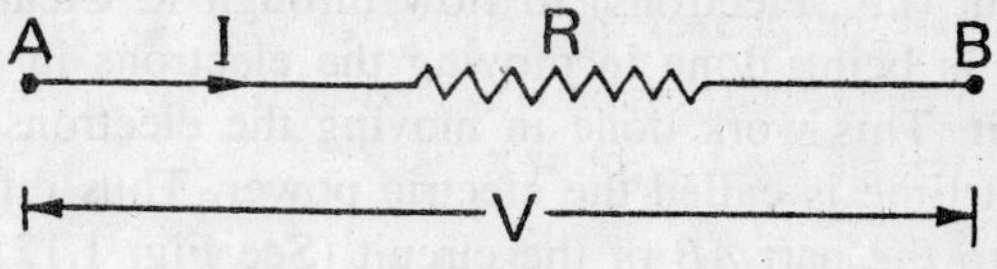

Fig. 1.11

$$I = V/R\,; \qquad V = IR\,; \qquad R = V/I$$

Example 1.15. *A battery has an e.m.f. of 12.8 volts and supplies a current of 3.24 A. What is the resistance of the circuit ? How many coulombs leave the battery in 5 minutes ?*

Solution.

Circuit resistance, $R = V/I = 12.8/3.2 = 4\ \Omega$

Charge flowing in 5 minutes = current × time in seconds

$= 3.2 \times 5 \times 60 = \mathbf{960\ C.}$

Example 1.16. *A metal filament lamp takes 0.3A at 230V. If the voltage is reduced to 115 V, will the current be halved ? Explain your answer.*

Solution. No. It is because Ohm's law is applicable only if the resistance of the circuit does not change. In the present case, when voltage is reduced from 230 V to 115 V, the temperature of the lamp will decrease too much, resulting in an enormous decrease of lamp resistance. Consequently, Ohm's law ($I = V/R$) cannot be applied. To give an idea to the reader, the hot resistance (*i.e.*, at normal operating temperature) of an incandescent lamp is more than 10 times its cold resistance.

Example 1.17. *An incandescent lamp has a resistance of 10 Ω when not burning. When burning with a voltage of 120 V, its resistance is 144 Ω. Find the current at the instant of switching and the normal current when burning. What is the wattage of the lamp ?*

Solution.

Current at the instant of switching $\quad I_1 = V/R_1 = 120/10 = \mathbf{12\ A}$

Normal current when burning $\quad i_2 = V/R_2 = 120/144 = \mathbf{0.833\ A}$

Wattage of lamp = $VI_2 = 120 \times 0.833 = \mathbf{100\ W}$

The reader may note that when lamp is first connected to supply, the current is very high (12 A in this case). This current raises the temperature of the filament rapidly and the resistance increases due to the increased temperature. By the time normal operating temperature of the lamp is reached, the resistance is increased many times (144 Ω in this case) the cold

* *The unit of resistance (i.e., ohm) was named in honour of George Simon Ohm.*

resistance. If the resistance of the filament did not increase, the power input would remain $P = VI_1 = 120 \times 12 = 1440$ W and the lamp would be destroyed by its own heat.

1.20. Electric Power

The rate at which work is done in an electric circuit is called **electric power** *i.e.,*

$$\text{Electric power} = \frac{\text{Work done in electric circuit}}{\text{Time}}$$

When voltage is applied to a circuit, it causes current (*i.e.,* electrons) to flow through it. Clearly, work is being done in moving the electrons in the circuit. This work done in moving the electrons in a unit time is called the electric power. Thus referring to the part *AB* of the circuit (See Fig. 1.12),

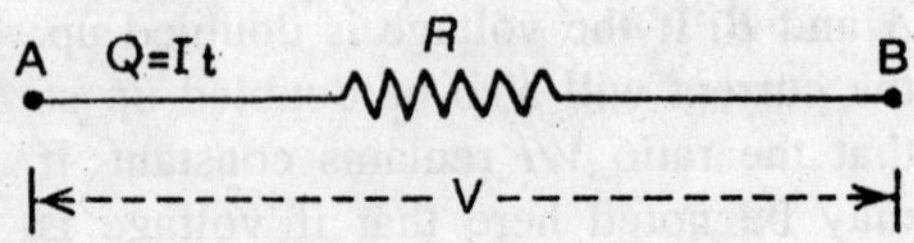

Fig. 1.12

$$V = P.D. \text{ across } AB \text{ in volts}$$

$$I = \text{Current in amperes}$$

$$R = \text{Resistance of } AB \text{ in } \Omega$$

$$t = \text{Time in sec. for which current flows}$$

The total charge that flows in t seconds is $Q = I \times t$ coulombs and by definition (See Art. 1.6),

$$V = \frac{\text{Work}}{Q}$$

or $$\text{Work} = VQ = VIt \qquad (\because Q = It)$$

$$\therefore \text{ Electric power, } \quad P = \frac{\text{Work}}{t} = \frac{VIt}{t} = VI \text{ Joule/sec } or \text{ watts}$$

$$\therefore \qquad P = VI = I^2R = \frac{V^2}{R}$$

$$(\because V = IR \text{ or } I = \frac{V}{R})$$

The above three formulae are equally valid for calculation of electric power in a d.c. circuit. Which one is to be used depends simply on which quantities are known or most easily determined.

Unit of electric power. The basic unit of electric power is *Joules/sec* or *watt*. The power consumed in a circuit is 1 watt if a p.d. of 1 *V* causes 1*A* current to flow through the circuit.

$$\text{Power in watts} = \text{Voltage in volts} \times \text{Current in amperes}$$

The bigger units of electric power are kilowatts (kW) and megawatts (MW)

$$1 \text{ kW} = 1000 \text{ watts}; \ 1 \text{ MW} = 10^6 \text{ watts or } 10^3 \text{ kW}.$$

1.21. Electrical Energy

The total work done in an electric circuit is called **electrical energy** *i.e.,*

$$\text{Electrical energy} = \text{Electrical power} \times \text{time}$$

$$= VIt = I^2Rt = \frac{V^2}{R}t$$

In practice, electrical energy is measured in kilowatt hour (kWh).

$$\text{Energy in kWh} = \text{Power in kW} \times \text{Time in hours}$$

One kilowatt-hour (kWh) *of electrical energy is expended in a circuit if 1 kW (1000 watts) of power is supplied for 1 hour.*

The electricity bills are made on the basis of total electrical energy consumed by the consumer. The unit for charge of electricity is 1 kWh. One kWh is also called Board of Trade (B.O.T.) unit or simply unit. Thus when we say that a consumer has consumed 100 units of electricity, it means that electrical energy consumption is 100 kWh.

Example 1.18. *A heating element supplies 300 kilojoules in 50 minutes. Find the p.d. across the element when current is 2 amperes.*

Solution.

Total charge, $Q = I \times t = 2 \times 50 \times 60 = 6000 \text{ C}$

P.D., $V = \dfrac{\text{Work}}{\text{charge}} = \dfrac{300 \times 10^3}{6000} = \mathbf{50 \text{ V}}$

Example 1.19. *The following are the details of load on a circuit connected through a supply meter :*

(*i*) *Six lamps of 40 watts each working for 4 hours per day*

(*ii*) *Two fluorescent tubes 125 watts each working for 2 hours per day*

(*iii*) *One 1000 watt heater working for 3 hours per day.*

If each unit of energy costs 70 P, what will be the electricity bill for the month of June ?

Solution.

Total wattage of lamps $= 40 \times 6 = 240$ watts

Total wattage of tubes $= 125 \times 2 = 250$ watts

Wattage of heater $= 1000$ watts

Energy consumed by the appliances per day

$$= (240 \times 4) + (250 \times 2) + (1000 \times 3)$$

$$= 4460 \text{ watt–hours} = 4.46 \text{ kWh}$$

Total energy consumed in the month of June (*i.e.*, in 30 days)

$$= 4.46 \times 30 = 133.8 \text{ kWh}$$

Bill for the month of June $= \text{Rs } 0.7 \times 133.8 =$ **Rs 93.66**

Tutorial Problems

1. A resistor of 50 Ω has a p.d. of 100 volts d.c. across it for 1 hour. Calculate (*i*) power and (*ii*) energy. **[(*i*) 200 watts (*ii*) 7.2×10^5 J]**
2. A current of 10 A flows through a resistor for 10 minutes and the power dissipated by the resistor is 100 watts. Find the p.d. across the resistor and the energy supplied to the circuit. **[10 V ; 6×10^4 J]**
3. A factory is supplied with power at 210 volts through a pair of feeders of total resistance 0.0225 Ω. The load consists of 250, 60-Watt lamps and 4 motors each taking 40 amperes. Find :
   (*i*) total current required
   (*ii*) voltage at the station end of feeders
   (*iii*) power wasted in feeders. **[231.4 A (*ii*) 215.78 V (*iii*) 1.4 kW]**
4. The electrical load in a small workshop consists of 14 lamps, each rated at 240 V, 60 W and 3 electric fires, each rated at 240 V, 1 kW. What is the effective resistance of the load ? **[15 Ω]**

OBJECTIVE QUESTIONS

1. The diameter of an atom is about :
(i) 10^{-10} m (ii) 10^{-8}m
(iii) 10^{-2}m (iv) 10^{-15}m
2. The number of compounds available in nature is
(i) 105 (ii) 300
(iii) 1000 (iv) unlimited
3. The mass of a proton is..... the mass of electron.
(i) equal to (ii) less than
(iii) about 1837 times (iv) 200 times
4. The electrons in the last orbit of an atom are called
(i) free electrons
(ii) bound electrohs
(iii) valence electrons
(iv) thermionic electrons
5. If the number of valence electrons of an atom is less than 4, the substance is usually
(i) a conductor (ii) an insulator
(iii) a semiconductor (iv) none of the above
6. If the number of valence electrons of an atom is more than 4, the substance is usually
(i) a semiconductor
(ii) a conductor
(iii) an insulator
(iv) none of the above.
7. If the number of valence electrons of an atom is 4, then the substance is usually
(i) a semiconductor (ii) an insulator
(iii) a conductor (iv) none of the above
8. One coulomb of charge is equal to the charge on
(i) 625 × 10^{16} electrons
(ii) 6.25 × 10^{16} electrons
(iii) 62.5 × 10^{16} electrons
(iv) 0.625 × 10^{16} electrons
9. 1 cm^3 of copper at room temperature has about
(i) 200 free electrons
(ii) 20 × 10^{10} free electrons
(iii) 8.5 × 10^{22} free electrons
(iv) 3 × 10^5 free electrons
10. The electric current is due to the flow of
(i) positive charges only
(ii) negative charges only
(iii) both positive and negative charges
(iv) neutral particles only
11. E.M.F. in a circuit
(i) increases circuit resistance
(ii) decreases circuit resistance
(iii) maintains potential difference
(iv) none of the above
12. The electric current in a metallic conductor is due to the flow of
(i) electrons only
(ii) positive ions only
(iii) electrons and positive ions only
(iv) neutral atoms only
13. Potential difference has the unit of
(i) charge (ii) power
(iii) energy (iv) none of the above
14. The electric current in a liquid is due to the flow of
(i) electrons only
(ii) positive ions only
(iii) negative and positive ions both
(iv) electrons and positive ions both
15. Electric current is a
(i) scalar quantity (ii) vector quantity
(iii) sometimes scalar and sometimes vector
(iv) number only
16. A billion electrons pass through a cross-section of a conductor in 10^{-3} s. The current is
(i) 10^{-7}A (ii) 1.6 × 10^{-7} A
(iii) 2× 10^{-4} A (iv) 2.6 × 10^{-3} A
17. The electric current in a discharge tube containing a gas is due to
(i) electrons only
(ii) positive ions only
(iii) negative and positive ions both
(iv) electrons and positive ions both
18. How long does it take 50 μC of charge to pass a point in a circuit if the current flow is 15 mA?
(i) 3.33× 10^{-3} s (ii) 6× 10^2
(iii) 2 s (iv) none of the above
19. The specific resistance of a wire depends upon

(*i*) its length
(*ii*) its cross- sectional area
(*iii*) its dimensions
(*iv*) its material

20. The quantity of charge that will be transferred by a current flow of 10 A over 1 hour period is
(*i*) 10 C (*ii*) 3.6×10^4 C
(*iii*) 2.4×10^3 C (*iv*) 1.6×10^2 C

21. The resistance of a straight conductor is independent of
(*i*) temperature (*ii*) material
(*iii*) cross-sectional area
(*iv*) shape of cross- section

22. A length of wire has a resistance of 6 ohms. The resistance of a wire of the same material three times as long and twice the cross-sectional area will be
(*i*) 36 ohms (*ii*) 9 ohms
(*iii*) 1 ohm (*iv*) 12ohms

23. The drift velocity of electrons of the order of
(*i*) $1 ms^{-1}$ (*ii*) 10^{-3} ms^{-1}
(*iii*) $10^6 ms^{-1}$ (*iv*) 3×10^8 ms^{-1}

24. Conductivity is the reciprocal of
(*i*) current density (*ii*) resistance
(*iii*) resistivity (*iv*) none of the above

25. The specific resistance of a metallic conductor...... with rise in temperature
(*i*) increases
(*ii*) decreases
(*iii*) remains unchanged
(*iv*) none of the above

26. The potential difference of an energy source that provides 50 mJ of energy for every micro coulomb of charge that flows is
(*i*) 5 V (*ii*) 50 V
(*iii*) 500 V (*iv*) 50 kV

27. Insulators have............. temperature coefficient of resistance.
(*i*) zero (*ii*) positive
(*iii*) negative (*iv*) none of above

28. Eureka has............ temperature co-efficient of resistance.
(*i*) almost zero (*ii*) negative
(*iii*) positive (*iv*) none of the above

29. What quantity of charge must be delivered by a battery with a potential difference of 100 V to do 500 J of work?
(*i*) 5 C (*ii*) 0.5 C
(*iii*) 50 C (*iv*) 500 C

30. A piece of aluminium wire is stretched to reduce its diameter to half of its original value. Its resistance will become.
(*i*) two times (*ii*) four times
(*iii*) eight times (*iv*) sixteen times

31. Semiconductors have .... temperature co-efficient of resistance
(*i*) zero (*ii*) negative
(*iii*) positive (*iv*) none of the above

32. The thermal speed of electrons is of the order of
(*i*) 1 ms^{-1} (*ii*) 10^{-3} ms^{-1}
(*iii*) 10^6 ms^{-1} (*iv*) 3×10^8 ms^{-1}

33. Constantan wire is used for making standard resistance because it has
(*i*) low specific resistance
(*ii*) high specific resistance
(*iii*) negligibly small temperature co- efficient of resistance
(*iv*) high melting point

34. The resistance of a material 2 m long and 2 m^2 in area of x- section is 1.6×10^{-8} Ω. Its specific resistance will be
(*i*) 1.6×10^{-8} Ωm (*ii*) 6.4×10^{-8} Ωm
(*iii*) 3.2×10^{-8} Ωm (*iv*) 0.16×10^{-8} Ωm

35. The SI unit of conductivity is
(*i*) ohm - m (*ii*) ohm/ m
(*iii*) mho-m (*iv*) mho/m

ANSWERS TO OBJECTIVE QUESTIONS

1. (*i*)	**2.** (*iv*)	**3.** (*iii*)	**4.** (*iii*)	**5.** (*i*)	**6.** (*iii*)	**7.** (*i*)	**8.** (*i*)
9. (*iii*)	**10.** (*iii*)	**11.** (*iii*)	**12.** (*i*)	**13.** (*iii*)	**14.** (*iii*)	**15.** (*i*)	**16.** (*ii*)
17. (*iv*)	**18.** (*i*)	**19.** (*iv*)	**20.** (*ii*)	**21.** (*iv*)	**22.** (*ii*)	**23.** (*ii*)	**24.** (*iii*)
25. (*i*)	**26.** (*iv*)	**27.** (*iii*)	**28.** (*i*)	**29.** (*i*)	**30.** (*iv*)	**31.** (*ii*)	**32.** (*iii*)
33. (*iii*)	**34.** (*i*)	**35.** (*iv*)					

HINTS TO SELECTED OBJECTIVE QUESTIONS

16. $I = \dfrac{ne}{t} = \dfrac{10^9 \times 1.6 \times 10^{-19}}{10^{-3}} = 1.6 \times 10^{-7}\,A$

18. $t = \dfrac{q}{I} = \dfrac{50 \times 10^{-6}}{15 \times 10^{-3}} = 3.33 \times 10^{-3}\,s$

20. $q = It = (10) \times (60 \times 60) = 3.6 \times 10^{4}\,\text{C}$

22. $R_1 = \rho\,\dfrac{l_1}{a_1}$; $R_2 = \rho\,\dfrac{l_2}{a_2}$

$$\therefore \quad \frac{R_2}{R_1} = \left(\frac{l_2}{l_1}\right) \times \left(\frac{a_1}{a_2}\right) = \left(\frac{3l}{l}\right) \times \left(\frac{a}{2a}\right) = 1.5$$

or $R_2 = 1.5\ R_1 = 1.5 \times 6 = 9$ ohms

26. $V = \dfrac{W}{Q} = \dfrac{50 \times 10^{-3}}{1 \times 10^{-6}} = 50 \times 10^{3}\,\text{V} = 50\text{kV}$

29. $Q = \dfrac{W}{V} = \dfrac{500}{100} = 5C$

30. $R_1 = \rho\,\dfrac{l_1}{a_1}$; $R_2 = \rho\,\dfrac{l_2}{a_2}$

$$\frac{R_2}{R_1} = \left(\frac{l_2}{l_1}\right) \times \left(\frac{a_1}{a_2}\right)$$

Now, $l_1 a_1 = l_2 a_2$ or $l_1 a_1 = l_2 \left(\dfrac{a_1}{4}\right)$ $\therefore\ l_2 = 4l_1$

$$\therefore \quad \frac{R_2}{R_1} = \left(\frac{4l_1}{l_1}\right) \times \frac{a_1}{(a_1/4)} = 16 \text{ or } \quad R_2 = 16\,R_1$$

2

D.C. Circuits

Introduction

The closed path followed by direct current (*d.c.*) is called a d.c. circuit. A d.c. circuit essentially consists of a source of d.c. power (*e.g.*, battery, d.c. generator *etc.*), the conductors used to carry current and the load. The source supplies electrical energy to the load which converts it into heat or other forms of energy. The load for a d.c. circuit is usually a *resistance. In a d.c. circuit, loads (*i.e.*, resistances) may be connected in series, parallel or series-parallel. In this chapter, we shall confine our discussion to d.c. circuits only.

2.1. Resistances in Series

The circuit in which resistances are connected end to end so that the same current flows through all the resistances is called a series circuit as shown in Fig. 2.1 (*i*). By ohm's law, voltages across various resistances are :

$$V_1 = IR_1 \; ; \; V_2 = IR_2 \; ; \; V_3 = IR_3$$

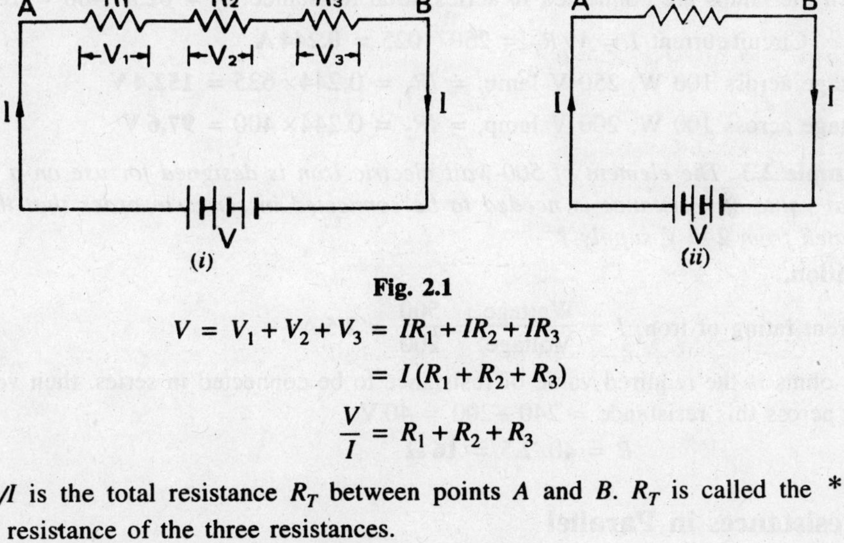

Fig. 2.1

Now, $$V = V_1 + V_2 + V_3 = IR_1 + IR_2 + IR_3$$
$$= I(R_1 + R_2 + R_3)$$

or $$\frac{V}{I} = R_1 + R_2 + R_3$$

But V/I is the total resistance R_T between points A and B. R_T is called the **total or equivalent resistance of the three resistances.

$\therefore$ $$R_T = R_1 + R_2 + R_3$$

---

* Other passive elements *viz.* inductance and capacitance are relevant only in a.c. circuits.

** Total or equivalent resistance is the single resistance, which, if substituted for the series resistances, would provide the same current in the circuit.

Hence when a number of resistances are connected in series, the total resistance is equal to the sum of individual resistances.

Example 2.1. *Two filament lamps A and B take 0.8 A and 0.9 A respectively when connected across 110 V supply. Calculate the value of current when they are connected in series across a 220-V supply, assuming the filament resistances to remain unaltered. Also find the voltage across each lamp.*

Solution.

For lamp A, $R_A = 110/0.8 = 137.5\,\Omega$; For lamp B, $R_B = 110/0.9 = 122.2\,\Omega$

When the lamps are connected in series, total resistance $R_T = 137.5 + 122.2 = 259.7\,\Omega$

$\therefore$ Circuit current, $I = V/R_T = 220/259.7 = \mathbf{0.827\ A}$

Voltage across lamp $A = IR_A = 0.847 \times 137.5 = \mathbf{116.5\ V}$

Voltage across lamp $B = IR_B = 0.847 \times 122.2 = \mathbf{103.5\ V}$

Example 2.2. *A 100-watt 250 V lamp is connected in series with a 100 watt, 200 V lamp across 250 V supply. Calculate (i) circuit current and (ii) voltage across each lamp. Assume the lamp resistances to remain unaltered.*

Solution.

Resistance $R = V^2/W$

Resistance of 100 watt, 250 V lamp, $R_1 = (250)^2/100 = 625\,\Omega$

Resistance of 100 watt, 200 V lamp, $R_2 = (200)^2/100 = 400\,\Omega$

When the lamps are connected in series, total resistance $R_T = 625 + 400 = 1025\,\Omega$

$\therefore$ Circuit current, $I = V/R_T = 250/1025 = \mathbf{0.244\ A}$

Voltage across 100 W, 250 V lamp, $= IR_1 = 0.244 \times 625 = \mathbf{152.4\ V}$

Voltage across 100 W, 200 V lamp, $= IR_2 = 0.244 \times 400 = \mathbf{97.6\ V}$

Example 2.3. *The element of 500-watt electric iron is designed for use on a 200 V supply. What value of resistance is needed to be connected in series in order that the iron can be operated from 240 V supply ?*

Solution.

$$\text{Current rating of iron, } I = \frac{\text{Wattage}}{\text{Voltage}} = \frac{500}{200} = 2.5\ \text{A}$$

If R ohms is the required value of resistance to be connected in series, then voltage to be dropped across this resistance = 240 − 200 = 40 V

$$\therefore \quad R = 40/2.5 = \mathbf{16\ \Omega}$$

2.2. Resistances in Parallel

When one end of each resistance is joined to a common point and the other end of each resistance is joined to another common point so that there are as many paths for current flow as the number of resistances, it is called a **parallel circuit** as shown in Fig. 2.2 (*i*). Note that voltage across each resistance is the same (*i.e.* V volts in this case) and there are as many current paths as the number of resistances. By ohm's law, current through each resistance is :

$$I_1 = V/R_1,\ I_2 = V/R_2\ ;\ I_3 = V/R_3$$

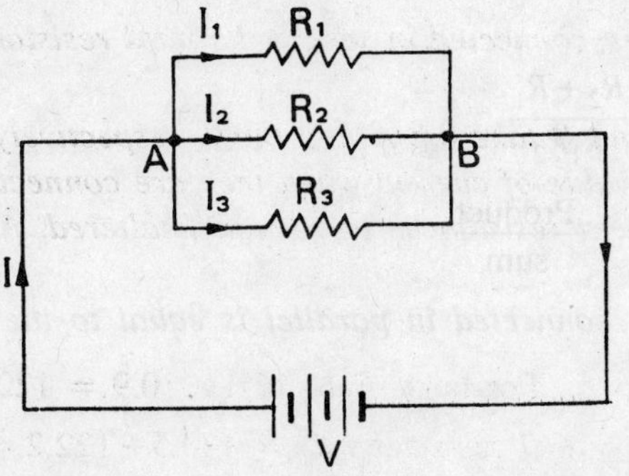

(i)

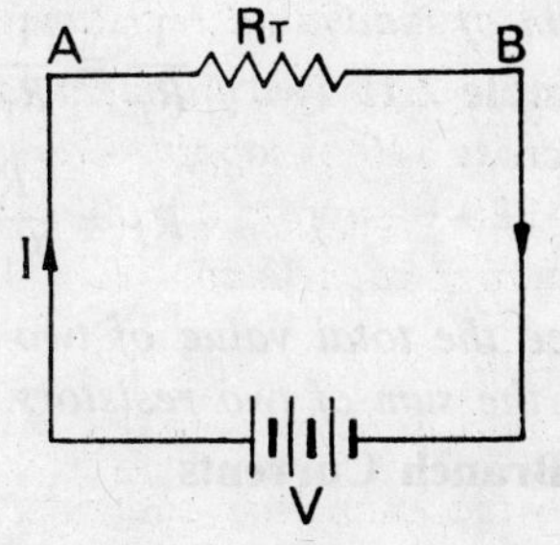

(ii)

Fig. 2.2

Now, $$I = I_1 + I_2 + I_3 = \frac{V}{R_1} + \frac{V}{R_2} + \frac{V}{R_3}$$

$$= V\left(\frac{1}{R_1} + \frac{1}{R_2} + \frac{1}{R_3}\right)$$

or $$\frac{I}{V} = \frac{1}{R_1} + \frac{1}{R_2} + \frac{1}{R_3}$$

But V/I is the total resistance R_T of the parallel resistances [See Fig. 2.2 (*ii*)] so that $I/V = 1/R_T$

$$\therefore \quad \frac{1}{R_T} = \frac{1}{R_1} + \frac{1}{R_2} + \frac{1}{R_3}$$

Hence when a number of resistances are connected in parallel, the reciprocal of total resistance is equal to the sum of reciprocals of individual resistances.

Note.

(*i*) The total resistance of a parallel circuit is always less than the smallest of the resistances.

(*ii*) If n resistors, each of $R\ \Omega$, are connected in parallel, then, $R_T = R/n$

2.3. Special Case of Parallel Circuit

A frequent special case of parallel resistors is a circuit that contains two resistances in parallel. Fig. 2.3 shows two resistances R_1 and R_2 connected in parallel across a battery of V volts. The total current I divides into two parts ; I_1 flowing through R_1 and I_2 flowing through R_2.

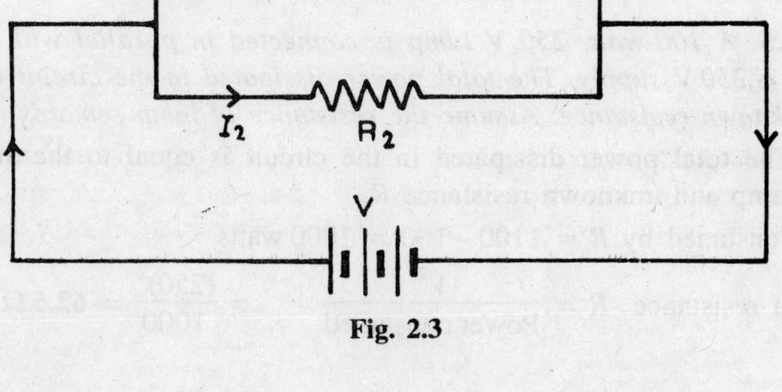

Fig. 2.3

(*i*) **Total resistance R_T**

$$\frac{1}{R_T} = \frac{1}{R_1} + \frac{1}{R_2} = \frac{R_2 + R_1}{R_1 R_2}$$

$$\therefore \qquad R_T = \frac{R_1 R_2}{R_1 + R_2} \quad \textit{i.e.,} \quad \frac{\text{Product}}{\text{sum}}$$

Hence the total value of two resistors connected in parallel is equal to the product divided by the sum of two resistors.

(*ii*) **Branch Currents**

$$R_T = \frac{R_1 R_2}{R_1 + R_2}$$

$$V = IR_T = I\frac{R_1 R_2}{R_1 + R_2}$$

Current through R_1, $$I_1 = \frac{V}{R_1} = I\frac{R_2}{R_1 + R_2} \qquad \left[\text{Putting } V = I\frac{R_1 R_2}{R_1 + R_2}\right]$$

Current through R_2, $$I_2 = \frac{V}{R_2} = I\frac{R_1}{R_1 + R_2}$$

Hence in a parallel circuit of two resistors, the current in one resistor is the line current (i.e., total current) times the opposite resistor divided by the sum of two resistors.

Example 2.4. *Two coils connected in series have a resistance of 18 Ω and when connected in parallel have a resistance of 4Ω. Find the value of resistances.*

Solution. Let R_1 and R_2 be the resistances of the coils. When resistances are connected in series, $R_T = 18\,\Omega$.

$$\therefore \qquad R_1 + R_2 = 18 \qquad ...(i)$$

When resistances are connected in parallel, $R_T = 4\,\Omega$

$$\therefore \qquad \frac{R_1 R_2}{R_1 + R_2} = 4 \qquad ...(ii)$$

Multiplying Eqns. (*i*) and (*ii*), we get, $R_1 R_2 = 18 \times 4 = 72$

Now $$R_1 - R_2 = \sqrt{(R_1 + R_2)^2 - 4R_1R_2} = \sqrt{(18)^2 - 4 \times 72}$$

$$\therefore \qquad R_1 - R_2 = \pm 6 \qquad ...(iii)$$

Solving Eqns. (*i*) and (*iii*), we get,

$$R_1 = \mathbf{12\,\Omega}\text{; or } \mathbf{6\,\Omega}\text{;} \quad R_2 = \mathbf{6\,\Omega} \text{ or } \mathbf{12\,\Omega}$$

Example 2.5. *A 100-watt, 250 V lamp is connected in parallel with an unknown resistance R across a 250 V supply. The total power dissipated in the circuit is 1100 watts. Find the value of unknown resistance. Assume the resistance of lamp remains unaltered.*

Solution. The total power dissipated in the circuit is equal to the sum of powers consumed by the lamp and unknown resistance *R*.

∴ Power consumed by $R = 1100 - 100 = 1000$ watts

Value of resistance $$R = \frac{V^2}{\text{Power consumed}} = \frac{(250)^2}{1000} = \mathbf{62.5\,\Omega}$$

Example 2.6. *A battery having an e.m.f. of 12 V is connected across terminals AB of the circuit shown in Fig. 2.4. Find (i) current flowing in each resistance and (ii) total power absorbed by the circuit.*

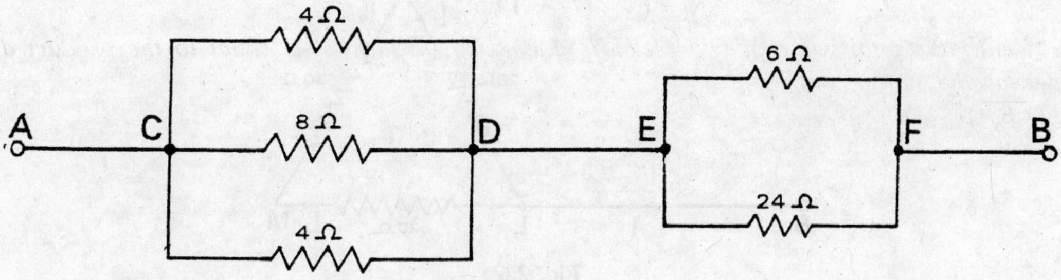

Fig. 2.4

Solution.

Resistance between C and D is given by

$$\frac{1}{R_{CD}} = \frac{1}{4}+\frac{1}{8}+\frac{1}{4} = \frac{5}{8}\,\Omega$$

$\therefore$ $R_{CD} = 8/5 = 1.6\,\Omega$

Resistance between E and F, $R_{EF} = \dfrac{6\times 24}{6+24} = 4.8\,\Omega$

Resistance between A and B, $R_{AB} = R_{CD}+R_{EF} = 1.6+4.8 = 6.4\,\Omega$

Total circuit current, $I = V/R_{AB} = 12/6.4 = 1.875$ A

Voltage between C and D, $V_{CD} = I\,R_{CD} = 1.875\times 1.6 = 3$ V

Voltage between E and F, $V_{EF} = I\,R_{EF} = 1.875\times 4.8 = 9$ V

(*i*) Current through 4 Ω resistor = 3/4 = **0.75 A**

Current through 8 Ω resistor = 3/8 = **0.375 A**

Current through 6 Ω resistor = 9/6 = **1.5 A**

Current through 24 Ω resistor= 9/24 = **0.375 A**

(*ii*) Power absorbed $= I^2 R_{AB} = (1.875)^2\times 6.4$ = **22.5 watts**

Example 2.7. *Three resistors 4Ω, 12 Ω and 6 Ω are connected in parallel. If the total current taken is 12 A, find the current through each resistor.*

Solution.

Fig. 2.5 shows the circuit arrangement.

$$\frac{1}{R_T} = \frac{1}{4}+\frac{1}{12}+\frac{1}{6} = \frac{6}{12}$$

$\therefore$ $R_T = 12/6 = 2\,\Omega$

P.D. across the parallel circuit $= I\,R_T = 12\times 2 = 24$ V

Current through 4 Ω = 24/4 = **6 A**

Current through 12 Ω = 24/12 = **2 A**

Current through 6 Ω = 24/6 = **4 A**

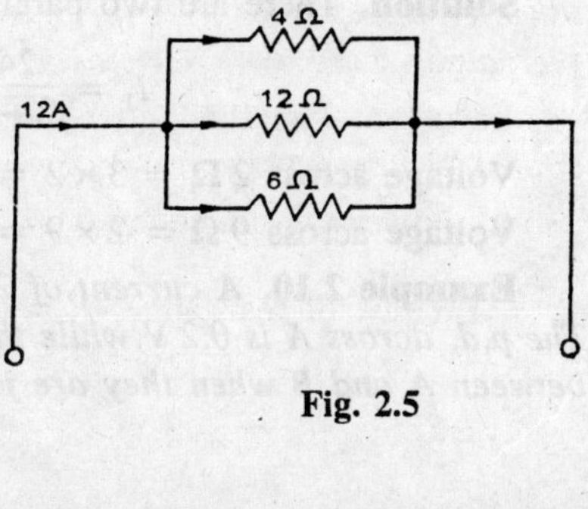

Fig. 2.5

Example 2.8. *A battery connected across AB in Fig. 2.6 is delivering 28 watts into the network of resistors. Calculate the voltage across AB. If the e.m.f. of the battery is 22 V, find the internal resistance of the battery.*

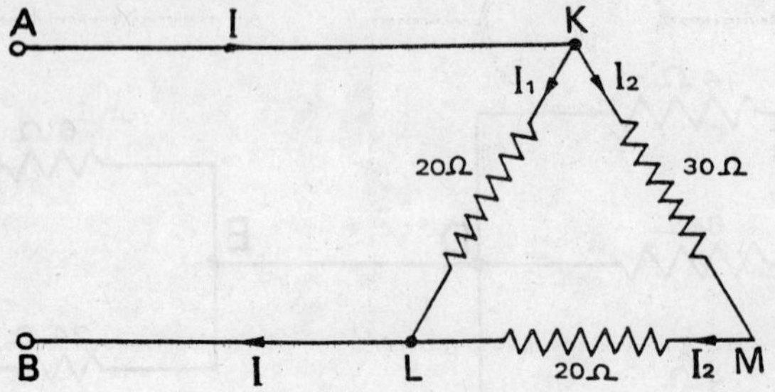

Fig. 2.6

Solution. The resistors in branch *KML* are in series and this branch is in parallel with branch *KL*.

$\therefore$ Total resistance between *A* and *B* is, $R_T = \dfrac{20 \times (30+20)}{20+(30+20)} = \dfrac{100}{7}\ \Omega$

If *V* is the voltage across *AB*, then power delivered to the network of resistors is V^2/R_T.

$\therefore$ $V^2/R_T = 28$ *or* $V^2 = 28 \times R_T = 28 \times 100/7 = 400$ volts

$\therefore$ $V = \sqrt{400} = \mathbf{20\ volts}$

Current in branch *KL*, $I_1 = 20/20 = 1$ A ; Current in branch *KML*, $I_2 = 20/50 = 0.4$ A

$\therefore$ Total current, $I = 1 + 0.4 = 1.4$ A

Now, $E = V + Ir$

$\therefore$ Internal resistance of supply, $r = \dfrac{E-V}{I} = \dfrac{22-20}{1.4} = \mathbf{1.429\ \Omega}$

Example 2.9. *In the circuit shown in Fig. 2.7, find the voltage and the current in each element.*

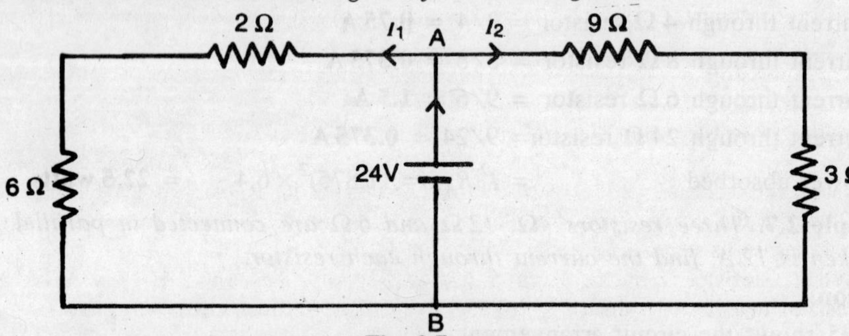

Fig. 2.7.

Solution. There are two parallel paths between points *A* and *B*.

$$I_1 = \frac{24}{2+6} = \frac{24}{8} = \mathbf{3\ A}\ ; \quad I_2 = \frac{24}{9+3} = \frac{24}{12} = \mathbf{2\ A}$$

Voltage across 2 Ω = 3 × 2 = **6 V** ; Voltage across 6 Ω = 3 × 6 = **18 V**

Voltage across 9 Ω = 2 × 9 = **18 V** ; Voltage across 3 Ω = 2 × 3 = **6 V**

Example 2.10. *A current of 20 A flows through two ammeters A and B joined in series. The p.d. across A is 0.2 V while that across B is 0.3 V. Find how the same current will divide between A and B when they are joined in parallel.*

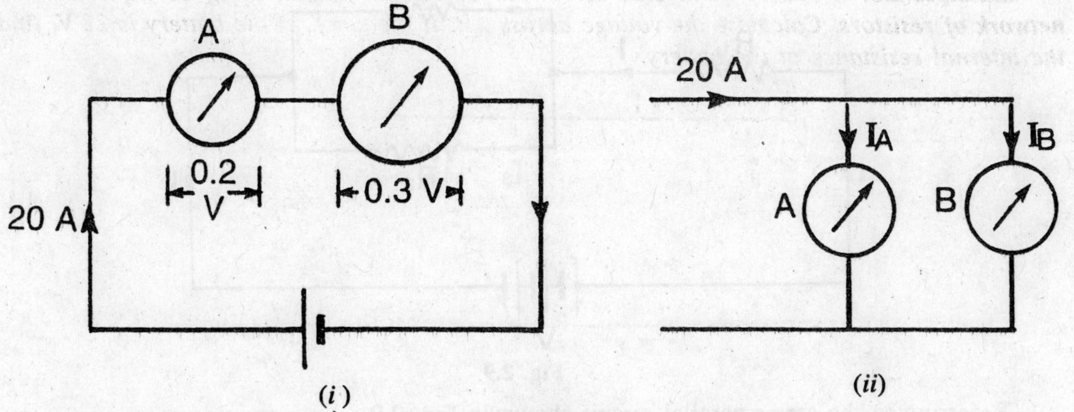

Fig. 2.8

Solution. We can find the resistances of the ammeters from Fig. 2.8 (*i*).

$$R_A = 0.2/20 = 0.01\ \Omega\ ;\ R_B = 0.3/20 = 0.015\Omega$$

When the two ammeters are joined in parallel [See Fig. 2.8 (*ii*)], 20 A current will be shared as under :

$$I_A = 20 \times \frac{0.015}{0.01+0.015} = \mathbf{12\ A}\ ;\quad I_B = 20 \times \frac{0.01}{0.01+0.015} = \mathbf{8\ A}$$

Tutorial Problems

1. Two resistors of 4 Ω and 6 Ω are connected in parallel. If the total current is 30 A, find the current through each resistor. **[18 A ; 12 A]**

2. Four resistors are in parallel. The currents in the first three resistors are 4 mA, 5 mA and 6 mA respectively. The voltage drop across the fourth resistor is 200 Volts. The total power dissipated is 5 watts. Determine the values of the resistances of the branches and the total resistance. **[50 KΩ, 40 KΩ, 33.33 KΩ, 8 KΩ ; 5 KΩ]**

3. Four resistors of 2Ω, 3 Ω, 4 Ω and 5 Ω respectively are connected in parallel. What potential difference must be applied to the group in order that total power of 100 watts may be absorbed ? **[8.826 Volts]**

4. Three resistors are in parallel. The current in the first resistor is 0.1 A. The power dissipated in the second is 3 watts. The voltage drop across the third is 100 volts. Determine the ohmic values of resistors and the total resistance if total current is 0.2 A. **[1000 Ω, 3333.3 Ω, 1428.5 Ω ; 500 Ω]**

5. Two coils each of 250 Ω resistance are connected in series across a constant voltage mains. Calculate the value of resistance to be connected in parallel with one of the coils to reduce the p.d. across its terminals by 1%. **[12,375 Ω]**

6. When a resistor is placed across a 230-volt supply, the current is 12 A. What is the value of resistor that must be placed in parallel to increase the load current to 16 A ? **[57.5 Ω]**

7. A 50-ohm resistor is in parallel with a 100-ohm resistor. The current in 50 Ω resistor is 7.2 A. What is the value of third resistance to be added in parallel to make the line current 12.1 A. **[276.9 Ω]**

2.4. Series-Parallel Circuit

As the name suggests, this circuit is a combination of series and parallel circuits. A simple example of such a circuit is illustrated in Fig. 2.9. Note that R_2 and R_3 are connected in parallel with each other and that both together are connected in series with R_1. One simple rule to solve such circuits is to first reduce the parallel branches to an equivalent series branch and then solve the circuit as a simple series circuit.

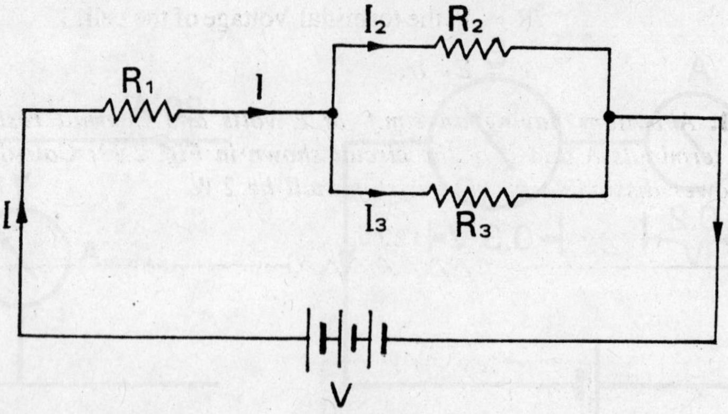

Fig. 2.9

Referring to the series-parallel circuit shown in Fig. 2.9,

R_T for parallel combination $= \dfrac{R_2 R_3}{R_2 + R_3}$

Total circuit resistance $= R_1 + \dfrac{R_2 R_3}{R_2 + R_3}$

Voltage across parallel combination $= I \times \dfrac{R_2 R_3}{R_2 + R_3}$

The reader can now readily find the values I, I_2 and I_3.

2.5. Internal Resistance of a Supply

All supplies (*e.g.*, a cell) must have some internal resistance, however small. This is shown as a series resistor external to the supply. Fig. 2.10 shows a cell of *e.m.f.* E volts and internal resistance r. When the cell is delivering no current (*i.e.*, on no load), the p.d. across the terminals will be equal to *e.m.f.* E of the cell as shown in Fig. 2.10 (*i*).

When some load resistance R is connected across the terminals of the cell, then current I starts flowing in the circuit. This current causes a voltage drop across internal resistance r of the cell so that terminal voltage V available will be less than E. The relationship between E and V can be easily established [See Fig. 2.10 (*ii*)].

$$I = \frac{E}{R + r}$$

or $\qquad IR = E - Ir$

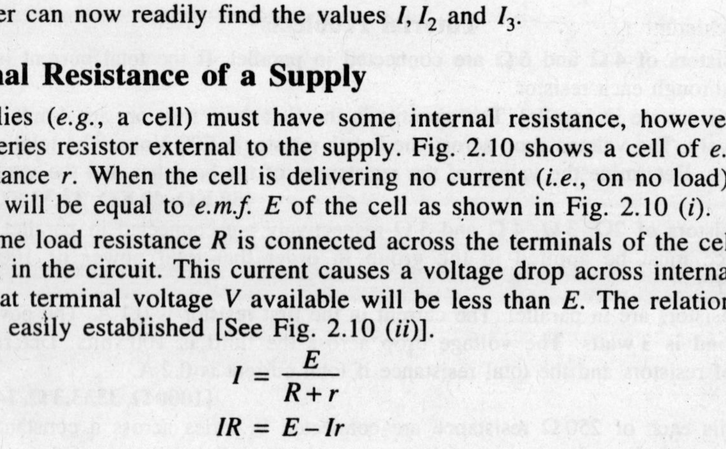

(*i*) (*ii*)

Fig. 2.10.

But $IR = V$, the terminal voltage of the cell.

$\therefore$ $V = E - Ir.$

Example 2.11. *A battery having an e.m.f. of E volts and internal resistance 0.1 Ω is connected across terminals A and B of the circuit shown in Fig. 2.11. Calculate the value of E in order that power dissipated in 2 Ω resistor shall be 2 W.*

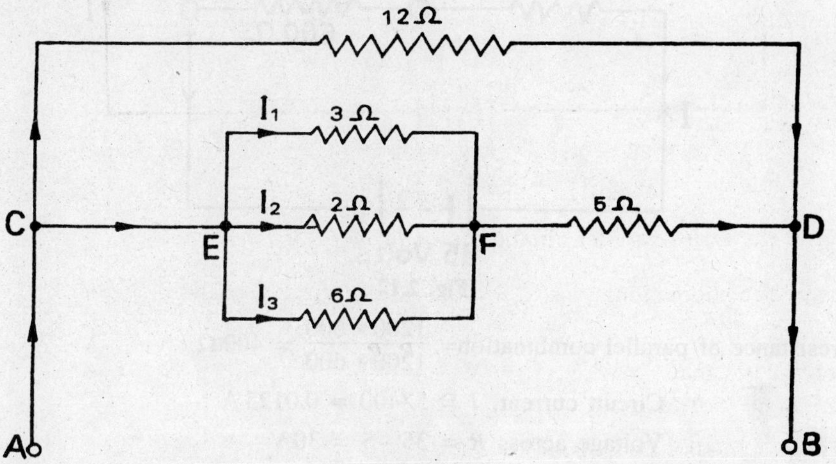

Fig. 2.11

Solution. Resistance between *E* and *F* is given by :

$$\frac{1}{R_{EF}} = \frac{1}{3} + \frac{1}{2} + \frac{1}{6} = \frac{6}{6}$$

$\therefore$ $R_{EF} = 6/6 = 1\ \Omega$

Resistance of branch $CEFD = 1 + 5 = 6\ \Omega$

Current through 2 Ω $= \sqrt{\dfrac{\text{Power loss}}{\text{Resistance}}} = \sqrt{\dfrac{2}{2}} = 1$ A

P.D. across *EF* $= 1 \times 2 = 2$ V

Current through 3 Ω $= 2/3 = 0.67$ A

Current through 6 Ω $= 2/6 = 0.33$ A

Current in branch *CED* $= 1 + 0.67 + 0.33 = 2$ A

P.D. across *CD* $= 6 \times 2 = 12$ V

Current through 12 Ω $= 12/12 = 1$ A

Current supplied by battery $= 2 + 1 = 3$ A

$\therefore$ E = P.D. across *AB* or *CD* + Drop in battery resistance

$= 12 + 0.1 \times 3 =$ **12.3 V**

Example 2.12. *A 35 V d.c. supply is connected across a resistance of 600 Ω in series with an unknown resistance R. A voltmeter having a resistance of 1200 Ω is connected across 600 Ω and shows a reading of 5 V. Calculate the value of resistance R.*

Solution. Fig. 2.12 shows the circuit arrangement.

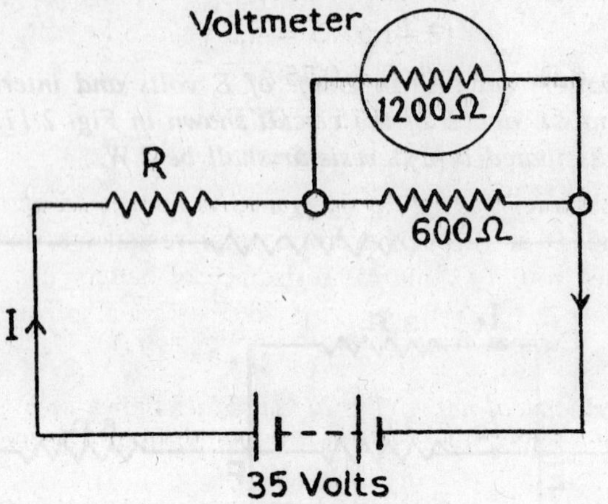

Fig. 2.12

Total resistance of parallel combination= $\dfrac{1200 \times 600}{1200 + 600} = 400\ \Omega$

$\therefore$ Circuit current, $I = 5/400 = 0.0125$ A

Voltage across $R = 35 - 5 = 30$ V

$\therefore$ Value of $R = 30/0.0125 =$ **2400Ω**

Example 2.13. *Six resistors are connected as shown in Fig. 2.13. If a battery having an e.m.f. of 24 volts and internal resistance of 1 Ω is connected to the terminals A and B, find (i) the current from the battery (ii) p.d. across 8 Ω and 4 Ω resistors (iii) the current taken from the battery if a conductor of negligible resistance is connected in parallel with 8 Ω resistor.*

Solution.

Resistance between E and F, $R_{EF} = \dfrac{(4+2) \times 6}{(4+2)+6} = 3\ \Omega$

Resistance between C and D, $R_{CD} = \dfrac{(5+3) \times 8}{(5+3)+8} = 4\ \Omega$

Resistance between A and B, $R_{AB} = 3 + 4 = 7\ \Omega$

Total circuit resistance, $R_T = R_{AB}$ + supply resistance $= 7 + 1 = 8\ \Omega$

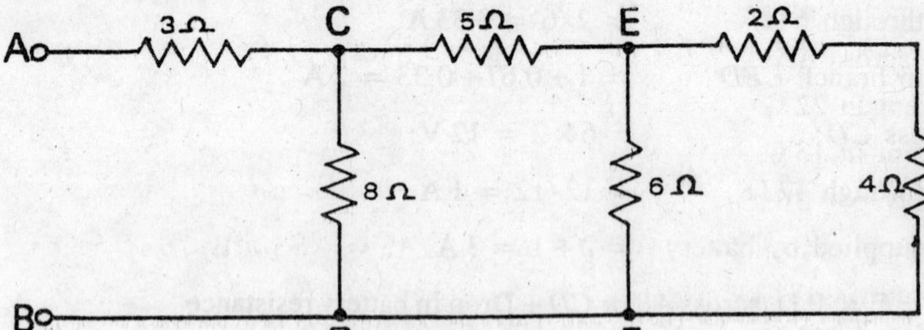

Fig. 2.13

(*i*) Current from battery, $I = E/R_T = 24/8 =$ **3 A**

(*ii*) P.D. across 8 Ω $= E - I(3+1) = 24 - 3\ (4) =$ **12 V**

Current through 8 Ω $= 12/8 = 1.5$ A

Current through 5 Ω	$= 3 - 1.5 = 1.5$ A
P.D. across *EF*	$= 12 - 1.5 \times 5 = 4.5$ V
Current through 6 Ω	$= 4.5/6 = 0.75$ A
∴ Current through 4 Ω	$= 1.5 - 0.75 = 0.75$ A
∴ Voltage across 4 Ω	$= 0.75 \times 4 =$ **3 V**

(*iii*) When a conductor of negligible resistance is connected across 8 Ω, then resistance between *C* and *D* is zero. Therefore, total resistance in the circuit is now 3 Ω resistor in series with 1 Ω internal resistance of battery.

$$\therefore \quad \text{Current from battery} = \frac{24}{3+1} = \mathbf{6\ A}$$

Example 2.14. *An electrical network is arranged as shown in Fig. 2.14. Find (i) the current in branch AF (ii) the power absorbed in branch BE and (iii) p.d. across branch CD.*

Solution.

Resistance between *E* and *C*, $R_{EC} = \dfrac{(5+9) \times 14}{(5+9)+14} = 7\ \Omega$

Resistance between *B* and *E*, $R_{BE} = \dfrac{(11+7) \times 18}{(11+7)+18} = 9\ \Omega$

Resistance between *A* and *E*, $R_{AE} = \dfrac{(13+9) \times 22}{(13+9)+22} = 11\ \Omega$

Total circuit resistance, $R_T = 11 + 1 = 12\ \Omega$

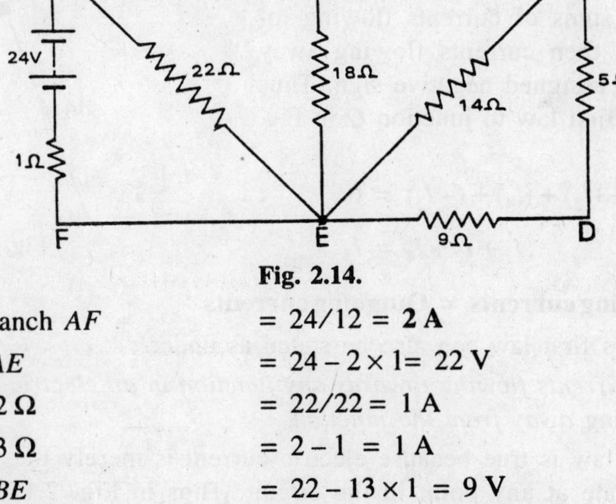

Fig. 2.14.

(*i*) Current in branch *AF*	$= 24/12 =$ **2 A**
(*ii*) P.D. across *AE*	$= 24 - 2 \times 1 = 22$ V
Current in 22 Ω	$= 22/22 = 1$ A
Current in 13 Ω	$= 2 - 1 = 1$ A
P.D. across *BE*	$= 22 - 13 \times 1 = 9$ V
Power in branch *BE*	$= (9)^2/18 =$ **4.5 watts**
(*iii*) Current in 18 Ω	$= 9/18 = 0.5$ A
Current in 11 Ω	$= 1 - 0.5 = 0.5$ A

At Point *C*, current finds two parallel paths *viz.* path *CE* of 14 Ω and path *CDE* of $9 + 5 = 14\ \Omega$.

∴ Current in branch *CDE* $= 0.5/2 = 0.25$ A

∴ P.D. across *CD* $= 0.25 \times 5 =$ **1.25 V**

Tutorial Problems

1. A resistor of 3.6 Ω is connected in series with another of 4.56 Ω. What resistance must be placed across 3.6 Ω so that the total resistance of the circuit shall be 6 Ω. **[2.4 Ω]**
2. A circuit consists of three resistors of 3 Ω, 4 Ω and 6 Ω in parallel and a fourth resistor of 4 Ω in series. A battery of e.m.f. 12 V and internal resistance 6 Ω is connected across the circuit. Find the total current in the circuit and terminal voltage across the battery. **[1.059 A, 5.65 V]**
3. A resistance R is connected in series with a parallel circuit comprising two resistors of 12 Ω and 8 Ω respectively. The total power dissipated in the circuit is 70 W when the applied voltage is 22 volts. Calculate the value of R. **[0.91 Ω]**
4. Two resistors R_1 and R_2 of 12 Ω and 6 Ω are connected in parallel and this combination is connected in series with a 6.25 Ω resistance R_3 and a battery which has an internal resistance of 0.25 Ω. Determine the e.m.f. of the battery. **[12.6 V]**

2.6. Kirchhoff's Laws

Gustav Kirchhoff (1824—1887), an eminent German Physicist did a considerable amount of work on the principles governing the behaviour of electric circuits. He gave his findings in a set of two laws—a current and a voltage law—which together are known as *Kirchhoff's laws.* It is upon these laws that electric theory is based.

First law. This law relates to the currents at the junction points of a circuit and is stated below :

The algebraic sum of currents flowing towards a junction in an electric circuit is zero.

An algebraic sum is one in which the sign of the quantity is taken into account. For example, consider four conductors carrying currents I_1, I_2, I_3 and I_4 and meeting at point O as shown in Fig. 2.15. If we take the signs of currents flowing towards O as positive, then currents flowing away from point O will be assigned negative sign. Thus applying Kirchhoff's first law to junction O in Fig. 2.15,

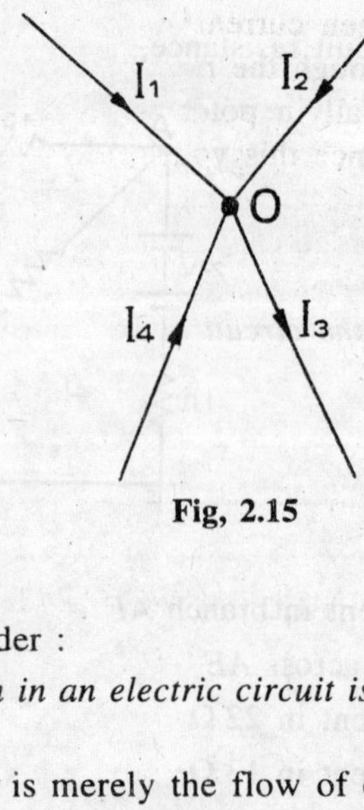

Fig, 2.15

$$(I_1) + (I_2) + (I_4) + (-I_3) = 0$$

or

$$I_1 + I_2 + I_4 = I_3$$

i.e., **Incoming currents = Outgoing currents**

Hence Kirchhoff's first law can also be stated as under :

The sum of the currents flowing towards any junction in an electric circuit is equal to the sum of currents flowing away from the junction.

Kirchhoff's first law is true because electric current is merely the flow of electrons and they cannot accumulate at any point in the circuit. Thus in Fig. 2.15, currents I_1, I_2 and I_4 are flowing towards point O and naturally $(I_1 + I_2 + I_4)$ must flow out of point O. Therefore $I_1 + I_2 + I_4 = I_3$.

Second Law. This law relates to e.m.f.s and voltage drops in a circuit and is sometimes called voltage law. It is stated as follows :

In any closed circuit or mesh, the algebraic sum of all the electromotive forces (e.m.fs) and the voltage drops is equal to zero i.e.

In any closed circuit or mesh,

Algebraic sum of e.m.fs + Algebraic sum of voltage drops = 0

The validity of Kirchhoff's second law can be readily established. If we start from any point in a closed circuit and go back to that point after going round the circuit, there is no increase or decrease in the potential. This means that the sum of e.m.fs of all the sources met on the way plus the voltage drops in the resistances must be zero.

2.7. Signs of e.m.f.s and Voltage Drops

While applying Kirchhoff's second law to a closed circuit, it is very important to assign proper signs to e.m.f.s and voltage drops in the closed circuit. The following sign convention may be followed :

A ***rise in potential** should be considered **positive** while a **fall in potential** should be considered **negative.**

(*i*) Thus if we go from the positive terminal of battery or source to the negative terminal [See Fig. 2.16 (*i*)], there is a fall in potential and the e.m.f. should be assigned negative sign. On the other hand, if we go from the negative terminal of the battery or source to the positive terminal [See Fig. 2.16 (*ii*)], there is a rise in potential and the e.m.f. should be given positive sign. *It may be noted that sign of e.m.f. is independent of the direction of current through that branch.*

(*ii*) When current flows through a resistor, there is a voltage drop across it. If we go through the resistor in the same direction as the current [See Fig. 2.17 (*i*)], there is a fall in potential because current flows from higher potential to lower potential. Hence this voltage drop should be given negative sign. On the other hand, if we go against the current flow [See Fig. 2.17 (*ii*)], there is a rise in potential and the voltage drop should be given positive sign. *It may be noted that sign of voltage drop depends upon the direction of current and is independent of the polarity of the e.m.f. in the circuit under consideration.*

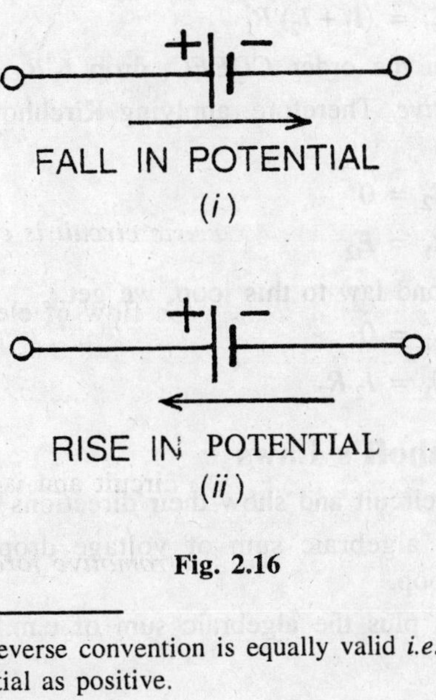

Fig. 2.16

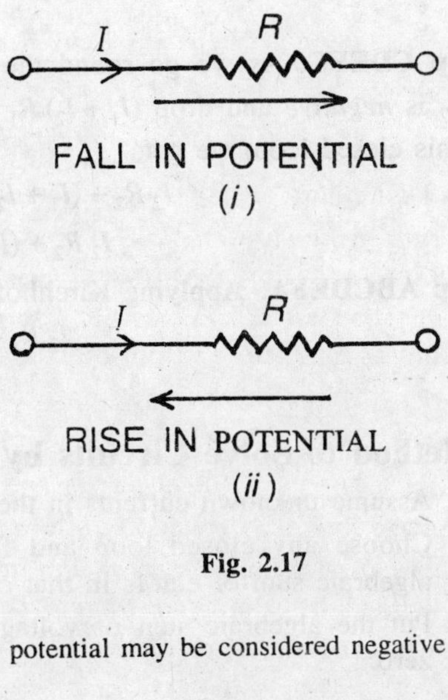

Fig. 2.17

* The reverse convention is equally valid *i.e.*, rise in potential may be considered negative and fall in potential as positive.

2.8. Illustration of Kirchhoff's Laws

Kirchoff's Laws can be beautifully explained by referring to Fig. 2.18. Mark the direction of currents as indicated. The direction in which currents are assumed to flow is unimportant, since if wrong direction is chosen, it will be indicated by a negative sign in the result.

(*i*) The magnitude of current in any branch of the circuit can be found by applying Kirchhoff's first law. Thus at junction *C* in Fig. 2.18, the incoming currents to the junction are I_1 and I_2. Obviously, the current in branch *CF* will be $I_1 + I_2$.

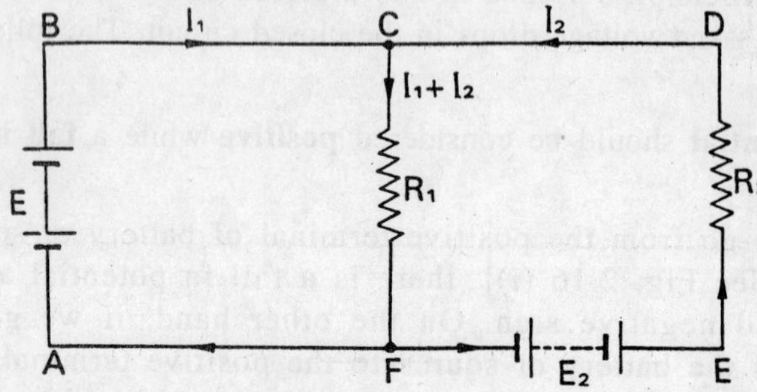

Fig. 2.18

(*ii*) There are three closed loops in Fig. 2.18 *viz.* *ABCFA*, *CDEFC* and *ABCDEFA*. Kirchhoff's second law can be applied to these closed loops to get the desired equations.

Loop ABCFA. In this loop, e.m.f. E_1 will be given *positive* sign. It is because as we consider the loop in the order *ABCFA*, we go from *–ve* terminal to the positive terminal of the battery in the branch *AB* and hence there is a rise in potential. The voltage drop in branch *CF* is $(I_1 + I_2)\, R_1$ and shall bear *negative* sign. It is because as we consider the loop in the order *ABCFA*, we go with current in branch *CF* and there is a fall in potential. Applying Kirchhoff's second law to the closed loop *ABCFA*,

$$-(I_1 + I_2)\, R_1 + E_1 = 0$$

or
$$E_1 = (I_1 + I_2)\, R_1$$

Loop CDEFC. As we go round the loop in the order *CDEFC*, drop $I_2 R_2$ is *positive*, e.m.f. E_2 is *negative* and drop $(I_1 + I_2)\, R_1$ is *positive*. Therefore, applying Kirchhoff's second law to this closed loop, we get,

$$I_2 R_2 + (I_1 + I_2)\, R_1 - E_2 = 0$$

or
$$I_2 R_2 + (I_1 + I_2)\, R_1 = E_2$$

Loop ABCDEFA. Applying Kirchhoff's second law to this loop, we get,

$$I_2 R_2 - E_2 + E_1 = 0$$

or
$$E_2 - E_1 = I_2 R_2$$

2.9. Method to Solve Circuits by Kirchhoff's Laws

(*i*) Assume unknown currents in the given circuit and show their directions by arrows.

(*ii*) Choose any closed loop and find the algebraic sum of voltage drops *plus* the algebraic sum of e.m.fs in that closed loop.

(*iii*) Put the algebraic sum of voltage drops plus the algebraic sum of e.m.fs equal to zero.

(*iv*) Write equations for as many closed loops as the number of unknown quantities.

(*v*) If the value of the assumed current comes out to be negative, it means that actual direction of current is opposite to that of assumed direction.

Example 2.15. *Calculate the current in 2 Ω resistor in Fig. 2.19.*

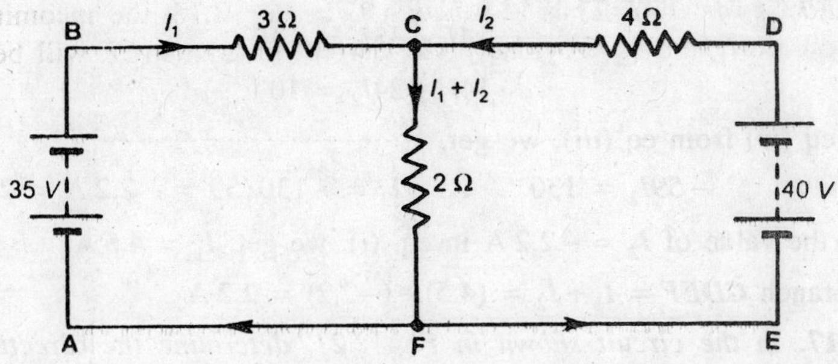

Fig. 2.19

Solution. Mark the currents in the various branches as shown in Fig. 2.19. Since there are two unknown quantities I_1 and I_2, two equations must be formed by considering two closed loops.

Loop ABCFA $$35 - 3I_1 - 2(I_1 + I_2) = 0$$

or $$5I_1 + 2I_2 = 35 \quad ...(i)$$

Loop ABCDEFA $$35 - 3I_1 + 4I_2 - 40 = 0$$

or $$3I_1 - 4I_2 = -5 \quad ...(ii)$$

Multiplying eq. (*i*) by 2 and then adding it to eq. (*ii*), we get,

$$13I_1 = 65 \qquad \therefore\ I_1 = 65/13 = 5\text{ A}$$

Substituting the value of $I_1 = 5$ A in eq. (*i*), we get, $I_2 = 5$ A

$\therefore$ Current in 2 Ω resistor $= I_1 + I_2 = 5 + 5 =$ **10 A**

Example 2.16. *For the circuit shown in Fig. 2.20, find the current flowing in all branches.*

Solution. Mark the currents in various branches as shown in Fig. 2.20. Since there are two unknown quantities I_1 and I_2, two closed loops will be considered.

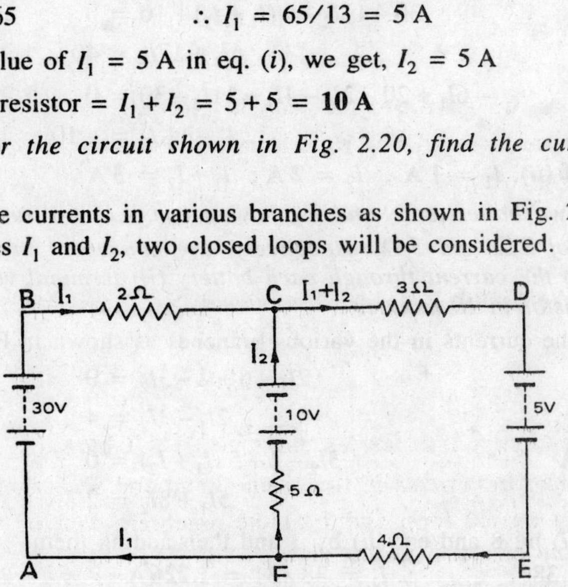

Fig. 2.20

Loop ABCFA $30 - 2I_1 - 10 + 5I_2 = 0$

or $2I_1 - 5I_2 = 20$ ...(*i*)

Loop FCDEF $-5I_2 + 10 - 3(I_1 + I_2) - 5 - 4(I_1 + I_2) = 0$

or $7I_1 + 12I_2 = 5$ ...(*ii*)

Multiplying eq (*i*) by 7 and eq (*ii*) by 2, we get,

$$14I_1 - 35I_2 = 140 \quad ...(iii)$$

$$14I_1 + 24I_2 = 10 \quad ...(iv)$$

Subtracting eq (*iv*) from eq (*iii*), we get,

$$-59I_2 = 130 \qquad \therefore \quad I_2 = -130/59 = -2.2 \text{ A} = \textbf{2.2 A} \text{ from } C \text{ to } F$$

Substituting the value of $I_2 = -2.2$ A in eq. (*i*), we get, $I_1 = \textbf{4.5 A}$

Current in branch $CDEF = I_1 + I_2 = (4.5) + (-2.2) = \textbf{2.3 A}$

Example 2.17. *In the circuit shown in Fig. 2.21, determine the direction and value of current in each of the batteries.*

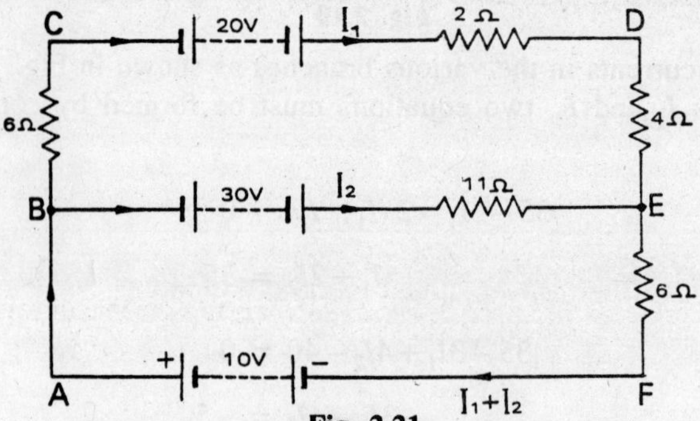

Fig. 2.21

Solution. Mark the currents in the various branches as shown in Fig. 2.21.

Loop ABEFA $30 - 11I_2 - 6(I_1 + I_2) + 10 = 0$

or $6I_1 + 17I_2 = 40$ ...(*i*)

Loop BCDEB $-6I_1 + 20 - 2I_1 - 4I_1 + 11I_2 - 30 = 0$

or $12I_1 - 11I_2 = -10$ ...(*ii*)

From eqs. (*i*) and (*ii*), $I_1 = \textbf{1 A}$; $I_2 = \textbf{2 A}$; $I_1 + I_2 = \textbf{3 A}$

Example 2.18. *Two batteries E_1 and E_2, having e.m.fs of 6 V and 2 V respectively and internal resistances of 2 Ω and 3 Ω respectively, are connected in parallel across a 5 Ω resistor. Calculate (i) the current through each battery (ii) terminal voltage and (iii) energy dissipated in 5 Ω resistor in 10 minutes.*

Solution. Mark the currents in the various branches as shown in Fig. 2.22.

Loop HCDFH $2I_1 - 6 + 2 - 3I_2 = 0$

or $2I_1 - 3I_2 = 4$ ...(*i*)

Loop ABHFEGA $3I_2 - 2 + 5(I_1 + I_2) = 0$

or $5I_1 + 8I_2 = 2$ ...(*ii*)

Multiplying eq. (*i*) by 8 and eq. (*ii*) by 3 and then adding them,

$$31I_1 = 38 \qquad \therefore I_1 = 38/31 = \textbf{1.226 A}$$

i.e., battery E_1 is being *discharged* at 1.226 A.

Substituting I_1 = 1.226 A in eq. (*i*), we get, I_2 = **– 0.516 A** *i.e.*, battery E_2 is being *charged* at 0.516 A.

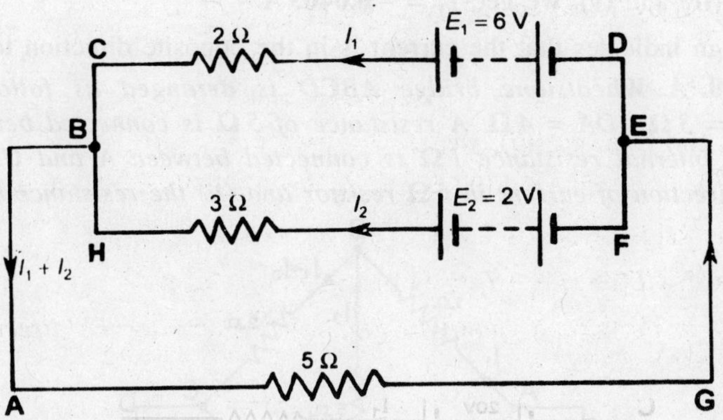

Fig. 2.22

Terminal voltage, $V = (I_1 + I_2) \times 5 = (1.226 - 0.516) \times 5 = $ **3.55 V**

Energy dissipated in 10 min. $= V \times (I_1 + I_2) \times t$

$= (3.55)(0.71)(10 \times 60)$ watt–sec or joules = **1512 joules**

Example 2.19. *Calculate the current in 5 Ω resistor shown in Fig. 2.23 (i).*

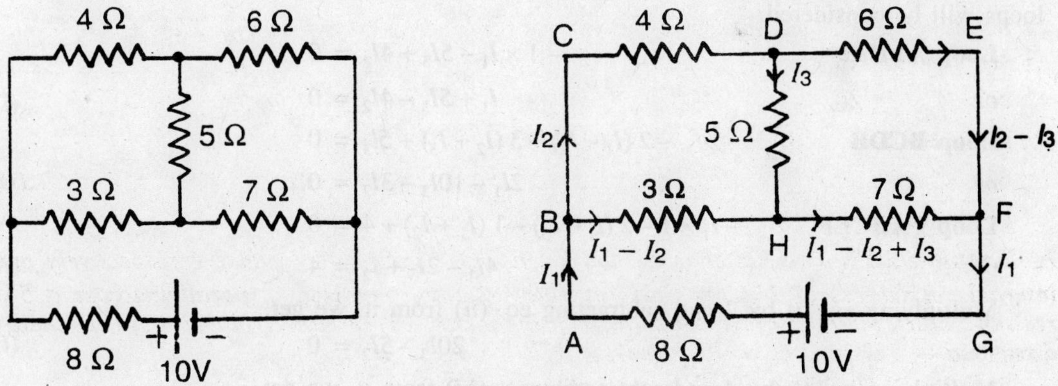

(*i*) (*ii*)

Fig. 2.23

Solution. Mark the direction of currents in the branches as shown in Fig. 2.23 (*ii*).

Loop ABHFGA $\quad -3(I_1 - I_2) - 7(I_1 - I_2 + I_3) + 10 - 8I_1 = 0$

or $\quad 18I_1 - 10I_2 + 7I_3 = 10 \quad$ ...(*i*)

Loop BCDHB $\quad -4I_2 - 5I_3 + 3(I_1 - I_2) = 0$

or $\quad -3I_1 + 7I_2 + 5I_3 = 0 \quad$ ...(*ii*)

Loop HDEFH $5I_3 - 6(I_2 - I_3) + 7(I_1 - I_2 + I_3) = 0$

or $-7I_1 + 13I_2 - 18I_3 = 0$ ...(*iii*)

Adding [(*ii*) × 6 + (*i*)], we get, $32I_2 + 37I_3 = 10$ ...(*iv*)

Subtracting [(*ii*) × 7 − (*iii*) × 3], we get, $10I_2 + 89I_3 = 0$ ...(*v*)

Solving eqs. (*iv*) and (*v*), we get, $I_3 = -$ **0.0403 A**

The minus sign indicates that the current is in the opposite direction to that shown.

Example 2.20. *A Wheatstone bridge ABCD is arranged as follows : AB = 1 Ω; BC = 2 Ω; CD = 3 Ω; DA = 4 Ω. A resistance of 5 Ω is connected between B and D. A 4-volt battery of internal resistance 1 Ω is connected between A and C. Calculate (i) the magnitude and direction of current in 5 Ω resistor and (ii) the resistance between A and C.*

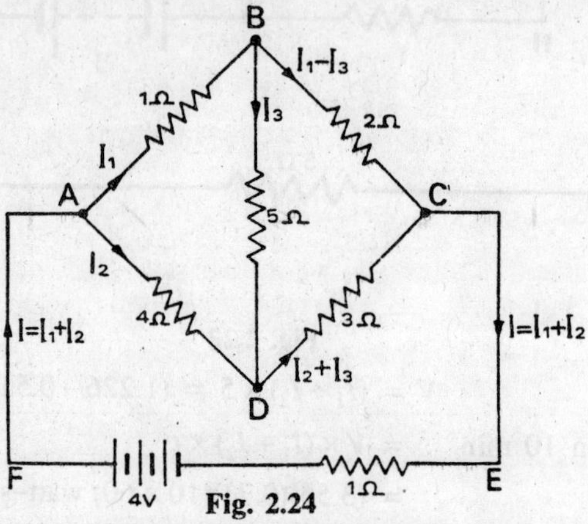

Fig. 2.24

Solution. Fig. 2.24 shows the Wheatstone bridge *ABCD*. Mark the currents in the various branches as shown. Since there are three unknown quantities (*viz.* I_1, I_2 and I_3), three closed loops will be considered.

Loop ABDA $-1 \times I_1 - 5I_3 + 4I_2 = 0$

or $I_1 + 5I_3 - 4I_2 = 0$ ...(*i*)

Loop BCDB $-2(I_1 - I_3) + 3(I_2 + I_3) + 5I_3 = 0$

or $2I_1 - 10I_3 - 3I_2 = 0$ ...(*ii*)

Loop FABCEF $-I_1 \times 1 - 2(I_1 - I_3) - 1(I_1 + I_2) + 4 = 0$

or $4I_1 - 2I_3 + I_2 = 4$ ...(*iii*)

Multiplying eq. (*i*) by 2 and subtracting eq. (*ii*) from it, we get,

$20I_3 - 5I_2 = 0$ ...(*iv*)

Multiplying eq. (*i*) by 4 and subtracting eq. (*iii*) from it, we get,

$22I_3 - 17I_2 = -4$ ...(*v*)

Multiplying eq. (*iv*) by 17 and eq. (*v*) by 5, we get,

$340I_3 - 85I_2 = 0$ ...(*vi*)

$110I_3 - 85I_2 = -20$ ...(*vii*)

Subtracting eq. (*vii*) from eq. (*vi*), we get $230I_3 = 20$ *or* $I_3 = 20/230 = 0.087$ A

i.e., Current in 5 Ω, I_3 = **0.087 A** from *B* to *D*

Substituting the value of $I_3 = 0.087$ A in eq. (*iv*), we get, $I_2 = 0.348$ A.

Substituting values of $I_3 = 0.087$ A and $I_2 = 0.348$ A in eq. (*ii*), $I_1 = 0.957$ A.

Current supplied by battery, $I = I_1 + I_2 = 0.957 + 0.348 = 1.305$ A.

P.D. between A and C = *E.M.F. of battery – Drop in battery*

$$= 4 - 1.305 \times 1 = 2.695 \text{ V}$$

$$\therefore \text{ Resistance between } A \text{ and } C = \frac{\text{P.D. Across } AC}{\text{Battery current}} = \frac{2.695}{1.305} = \mathbf{2.065\ \Omega}$$

Example 2.21. *In the Wheatstone bridge network shown in Fig. 2.25, find the current in each resistor.*

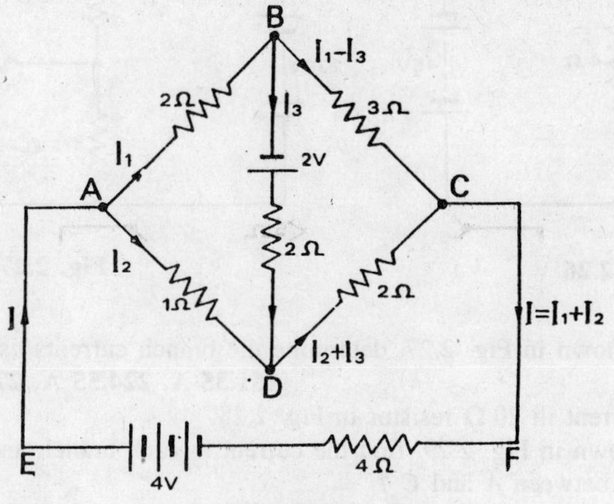

Fig. 2.25

Solution. Fig. 2.25 shows the Wheatstone bridge network. Mark the currents in various branches as shown. Since there are three unknown quantities (*viz.* I_1, I_2 and I_3), three closed loops will be considered.

Loop ABDA $\quad -2I_1 + 2 - 2I_3 + I_2 = 0$

or $\quad 2I_1 + 2I_3 - I_2 = 2 \quad$ ...(*i*)

Loop BCDB $\quad -3(I_1 - I_3) + 2(I_2 + I_3) + 2I_3 - 2 = 0$

or $\quad 3I_1 - 7I_3 - 2I_2 = -2 \quad$ ...(*ii*)

Loop EABCFE $\quad -2I_1 - 3(I_1 - I_3) - 4(I_1 + I_2) + 4 = 0$

or $\quad 9I_1 - 3I_3 + 4I_2 = 4 \quad$ ...(*iii*)

Multiplying eq. (*i*) by 2 and subtracting eq. (*ii*) from it, we get,

$$I_1 + 11I_3 = 6 \quad \text{...(iv)}$$

Multiplying eq. (*i*) by 4 and adding eq. (*iii*) to it, we get,

$$17I_1 + 5I_3 = 12 \quad \text{...(v)}$$

Multiplying eq. (*iv*) by 17 and subtracting eq. (*v*) from it, we get,

$$182I_3 = 90 \quad \text{or} \quad I_3 = 90/182 = \mathbf{0.494\ A}$$

Substituting $I_3 = 0.494$ A in eq. (*iv*), we get, $I_1 = \mathbf{0.566\ A}$

Substituting $I_1 = 0.566$ A and $I_3 = 0.494$ A in eq. (*i*) ; $I_2 = \mathbf{0.12\ A}$

Current in branch $BC = I_1 - I_3 = 0.566 - 0.494 = \mathbf{0.072\ A}$

Current in branch $DC = I_2 + I_3 = 0.12 + 0.494 =$ **0.614 A**

Total current, $I = I_1 + I_2 = 0.556 + 0.12 =$ **0.686 A**

Tutorial Problems

1. Using Kirchhoff's laws, find the current in various resistors in the circuit shown in Fig. 2.26.
[6.574 A, 3.611 A, 10.185 A]

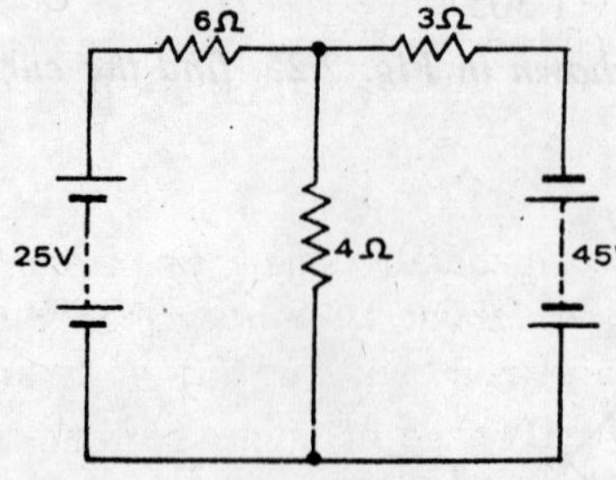

Fig. 2.26

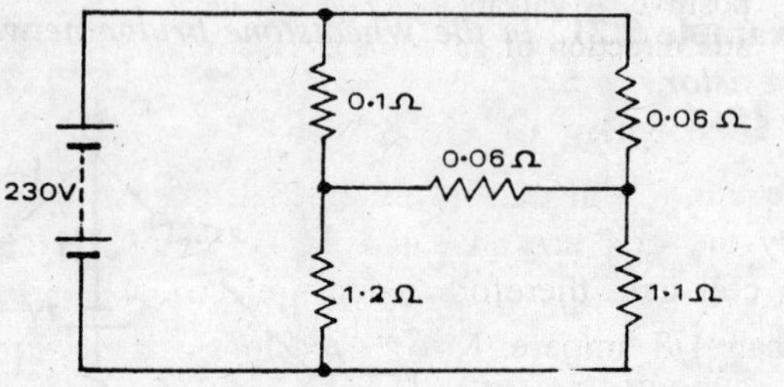

Fig. 2.27

2. For the circuit shown in Fig. 2.27, determine the branch currents using Kirchhoff's laws.
[151.35 A, 224.55 A, 27.7 A, 179.05 A, 196.84 A]

3. Calculate the current in 20 Ω resistor in Fig. 2.28. **[26.67 mA]**

4. In the circuit shown in Fig. 2.29, find the current in each branch and the current in the battery. What is the p.d. between *A* and *C* ?
[Branch *ABC* = **0.581 A** ; Branch *ADC* = **0.258 A** ; V_{AC} = **0.45 V**]

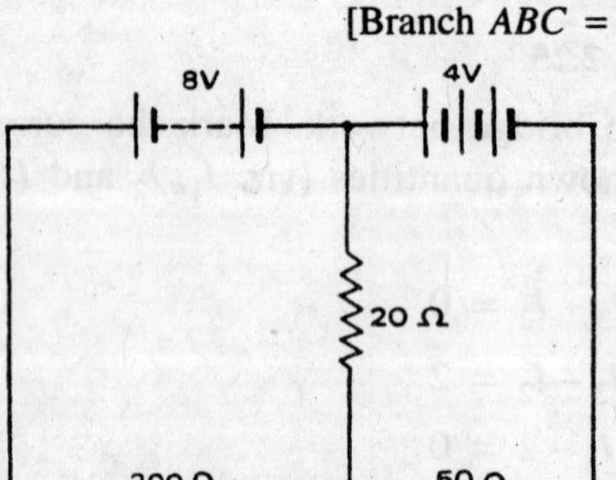

Fig. 2.28

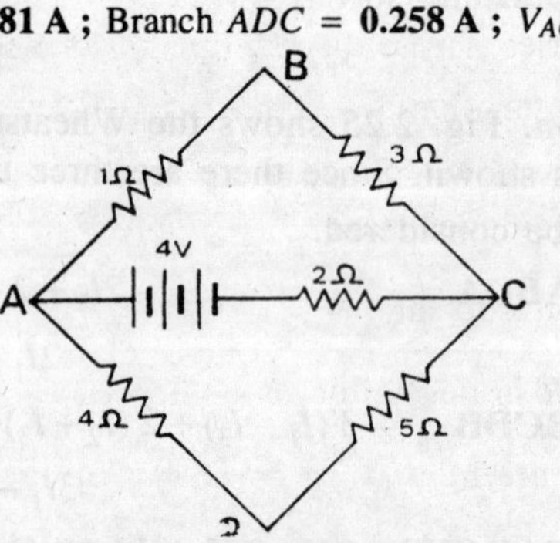

Fig. 2.29

5. Two batteries *A* and *B* having e.m.fs of 20 V and 21 V respectively and internal resistances of 0.8 Ω and 0.2 Ω respectively, are connected in parallel across 50 Ω resistor. Calculate (*i*) the current through each battery and (*ii*) the terminal voltage.
[(*i*) Battery *A* = **0.4725 A** ; Battery *B* = **0.0714 A** (*ii*) **20 V**]

6. A battery having an e.m.f. of 10 V and internal resistance 0.01 Ω is connected in parallel with a second battery of e.m.f. 10 V and internal resistance 0.008 Ω. The two batteries in parallel are properly connected for charging from a d.c. supply of 20 V through a 0.9 Ω resistor. Calculate the current taken by each battery and the current from the supply. **[4.91 A, 6.14 A, 11.05 V]**

7. Two batteries *A* and *B* having e.m.fs. 12 V and 8 V respectively and internal resistances of 2 Ω and 1 Ω respectively, are connected in parallel across 10 Ω resistor. Calculate (*i*) the current in each of the batteries and the external resistor and (*ii*) p.d. across external resistor.
[(*i*) I_A = **1.625 A** discharge ; I_B = **0.75 A** Charge, **0.875 A** (*ii*) **8.75 V**]

8. A Wheatstone bridge *ABCD* is arranged as follows : *AB* = 20 Ω, *BC* = 5 Ω, *CD* = 4 Ω and *DA* = 10 Ω. A galvanometer of resistance 6 Ω is connected between *B* and *D*. A 100-volts supply of negligible resistance is connected between A and C with A positive. Find the magnitude and direction of galvanometer current. **[0.667 A**, from D to B]

9. A network $ABCD$ consists of the following resistors : AB = 5 KΩ, BC = 10 KΩ, CD = 15 KΩ and DA = 20 KΩ. A fifth resistor of 10 KΩ is connected between A and C. A dry battery of e.m.f. 120 V and internal resistance 500 Ω is connected across the resistor AD. Calculate (*i*) the total current supplied by the battery (*ii*) the p.d. across points C and D and (*iii*) the magnitude and direction of current through branch AC.
[(*i*) **11.17 mA** (*ii*) **81.72 V** (*iii*) **3.27 mA** *from A to C*]

10. A Wheatstone bridge $ABCD$ is arranged as follows, AB = 10 Ω, BC = 30 Ω CD = 15 Ω and DA = 20 Ω. A 2 volt battery of internal resistance 2 Ω is connected between A and C with A positive. A galvanometer of resistance 40 Ω is connected between B and D. Find the magnitude and direction of galvanometer current. [**11.5 mA** *from B to D*]

2.10. Grouping of Cells

The e.m.f. and current obtained from a single cell are generally small. For instance, an ordinary dry cell has an e.m.f. of 1.5 volts and can deliver about 1/8 ampere continuously. Such a cell can, therefore, supply electrical energy to a circuit requiring 1.5 volts and not more than 1/8 ampere. Many occasions arise where higher voltage or higher current or both are required. To meet these needs, a number of cells are suitably connected or grouped. The combination of cells thus obtained is called a *battery*. The following are the three methods of grouping of cells to form a battery :

(*i*) Series grouping ; (*ii*) Parallel grouping ; (*iii*) Series-parallel grouping.

2.11. Series Grouping of Cells

In series grouping, the negative terminal of one cell is connected to the positive terminal of the next cell and so on. Consider n cells, each of e.m.f. E and internal resistance r, connected in series across an external resistance R as shown in Fig. 2.30.

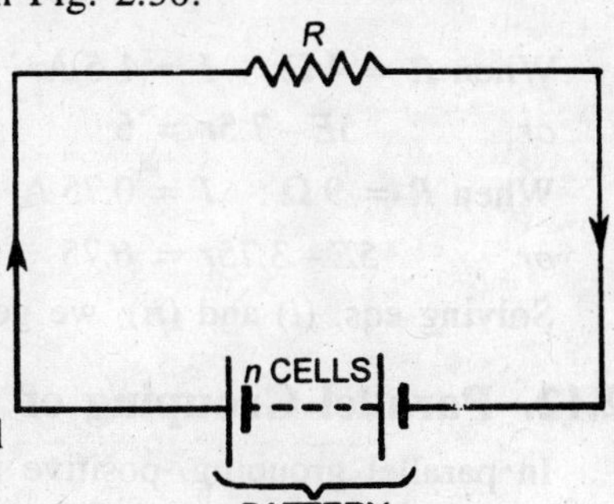

Fig. 2.30

Total battery e.m.f. $= nE$ volts

Internal resistance of battery $= nr\ \Omega$

Total circuit resistance $= R + nr$

$\therefore$ Circuit current, $I = \dfrac{nE}{R + nr}$

(*i*) If R is negligible as compared to nr, then,

Circuit current, $I = \dfrac{nE}{nr} = \dfrac{E}{r}$ = Current due to one cell

(*ii*) If nr (*i e.*, battery resistance) is negligible as compared to R, then,

Circuit current, $I = \dfrac{nE}{R} = n \times \dfrac{E}{R} = n \times$ current due to one cell

Hence, series grouping is useful when battery resistance is negligible as compared to external resistance.

Example 2.22. *Twelve cells, each of e.m.f. 2 volts and internal resistance 0.5 Ω, are connected in series across an external resistance of 4.5 Ω. Determine (i) the current supplied by the battery (ii) terminal voltage of the battery and (iii) fall in voltage per cell.*

Solution.

E.M.F. of each cell, E = 2 V; Internal resistance of each cell, r = 0.5 Ω ;

Number of cells, n = 12 ; External resistance, R = 4.5 Ω

Total e.m.f. of battery $= nE = 12 \times 2 = 24$ V ; Internal resistance of battery $= nr = 12 \times 0.5 \times 6\ \Omega$

Total circuit resistance $= R + nr = 4.5 + 6 = 10.5\ \Omega$

(*i*) Current supplied by battery, $I = 24/10.5 = \mathbf{2.286\ A}$

(*ii*) Terminal voltage of battery $= IR = 2.286 \times 4.5 = \mathbf{10.287\ V}$

(*iii*) Fall in voltage per cell $= Ir = 2.286 \times 0.5 = \mathbf{1.143\ V}$

Example 2.23. *A battery consisting of 10 cells in series has two of the cells reversed. E.M.F. of each cell is 1.5 V and internal resistance 0.1 Ω. The resistance of the external circuit is 4 Ω. Find the reduction in current due to the two cells being reverse connected.*

Solution.

When 10 cells are properly connected in series,

Total e.m.f. of battery $= nE = 10 \times 1.5 = 15$ V

Total circuit resistance $= R + nr = 4 + 10 \times 0.1 = 5\ \Omega$

∴ Circuit current $= 15/5 = 3$ A

When two of the cells are reverse connected,

Total e.m.f. of battery $= 8 \times 1.5 - 2 \times 1.5 = 9$ V

Total circuit resistance $= 5\ \Omega$ *i.e.*, same as before

∴ Circuit current $= 9/5 = 1.8$ A

∴ Reduction in current $= 3 - 1.8 = \mathbf{1.2\ A}$

Example 2.24. *A battery is formed by five cells in series. When the external resistance is 4 Ω, the current is 1.5 A and when it is 9 Ω, the current falls to 0.75 A. Find the e.m.f. and internal resistance of each cell.*

Solution. In series grouping of cells, $I = \dfrac{nE}{R + nr}$ or $nE = I(R + nr)$

When $R = 4\ \Omega$, $I = 1.5$ A ∴ $5E = 1.5(4 + 5 \times r)$

or $5E - 7.5r = 6$ ...(*i*)

When $R = 9\ \Omega$; $I = 0.75$ A ∴ $5E = 0.75(9 + 5 \times r)$

or $5E - 3.75r = 6.75$ ...(*ii*)

Solving eqs. (*i*) and (*ii*), we get $E = \mathbf{1.5\ V}$ and $r = \mathbf{0.2\ \Omega}$

2.12. Parallel Grouping of Cells

In parallel grouping, positive terminals of all the cells are joined together and in like manner, all the negative terminals are connected together. Consider *m* rows in parallel, each containing one cell. Let *E* and *r* be the e.m.f. and internal resistance respectively of each cell. Further, let this parallel group be connected across an external resistance *R* as shown in Fig. 2.31.

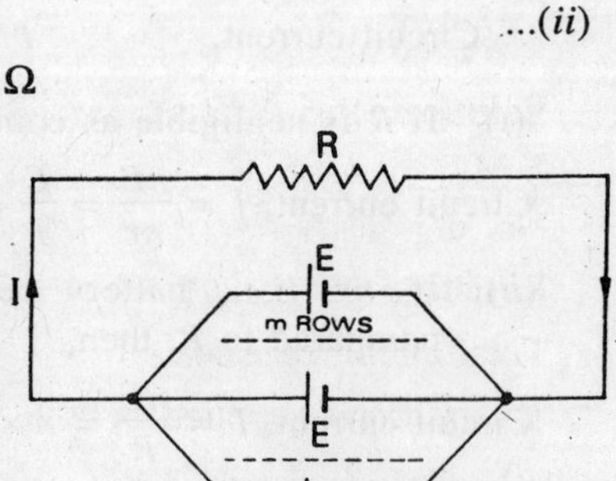

Fig. 2.31

E.M.F. of battery $= E$

Since the cells are connected in parallel, their internal resistances are also in parallel. If r_T is the total resistance of the battery, then,

$$\frac{1}{r_T} = \frac{1}{r} + \frac{1}{r} + \frac{1}{r} + \text{ ....}m \text{ terms}$$

$$= \frac{1+1+1+\ldots\ldots m \text{ terms}}{r} = \frac{m}{r}$$

$$\therefore \quad r_T = \frac{r}{m}\,\Omega$$

Total circuit resistance $= R + (r/m)$

$$\therefore \text{ Circuit current,} \quad I = \frac{E}{R+(r/m)}$$

(*i*) If r/m (*i.e.*, battery resistance) is negligible as compared to R, then,

$$\text{Circuit current, } I = \frac{E}{R} = \text{current due to one cell.}$$

(*ii*) If R is negligible as compared to r/m, (*i.e.*, battery resistance), then,

$$\text{Circuit current, } I = \frac{E}{r/m} = m \times \frac{E}{R} = m \times \text{current due to one cell}$$

Hence, parallel grouping is useful when external resistance is negligible as compared to battery resistance.

Example 2.25. *Four cells, each of e.m.f. 1.5 V and internal resistance 0.5 Ω, are connected in parallel across an external resistance of 2.5 Ω. Determine (i) the current supplied by the battery (ii) current supplied by each cell and (iii) the terminal voltage of the battery.*

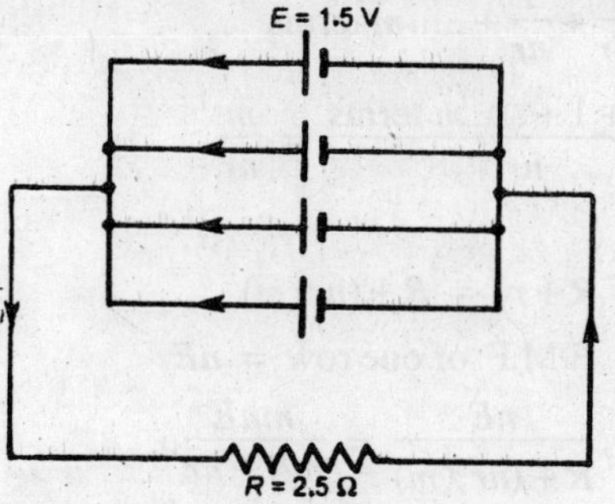

Fig. 2.32

Solution. Fig. 2.32 shows the arrangement of cells in parallel.

E.M.F. of each cell, $E = 1.5$ V ; Resistance of each cell, $r = 0.5\ \Omega$

No. of rows of cells, $m = 4$; Battery e.m.f. = 1.5 V

Resistance of battery $= r/m = 0.5/4 = 0.125\ \Omega$

Total circuit resistance $= R + r/m = 2.5 + 0.125 = 2.625\ \Omega$

(*i*) Current supplied by battery, $I = 1.5/2.625 =$ **0.571 A**

(*ii*) Current supplied by each cell $= 0.571/4 =$ **0.1427 A**

(*iii*) Terminal voltage of battery $= IR = 0.571 \times 2.5 =$ **1.4275 V**

Example 2.26. *Sixteen cells, each having an internal resistance of 4 Ω, when joined in parallel send a current of 2.72 A through an external resistance of 0.3 Ω. Find the e.m.f. of each cell.*

Solution.

$m = 16$; $r = 4\ \Omega$; $I = 2.72$ A ; $R = 0.3\ \Omega$; $E = ?$

Total circuit resistance $= R + (r/m) = 0.3 + (4/16) = 0.55\ \Omega$

$\therefore$ Circuit current, $I = E/0.55$

or $2.72 = E/0.55$

$\therefore$ Battery e.m.f. $E = 2.72 \times 0.55 = $ **1.496 V**

Since the cells are connected in parallel, battery e.m.f. is equal to e.m.f. of each cell.

2.13. Series-Parallel Grouping of Cells

In series-parallel grouping, a number of cells are connected in series and a number of sets of such series connected cells in parallel, as shown in Fig. 2.33. Let there be n cells connected in series and m such rows connected in parallel across an external resistance R. Further, let E and r be the e.m.f. and internal resistance respectively of each cell.

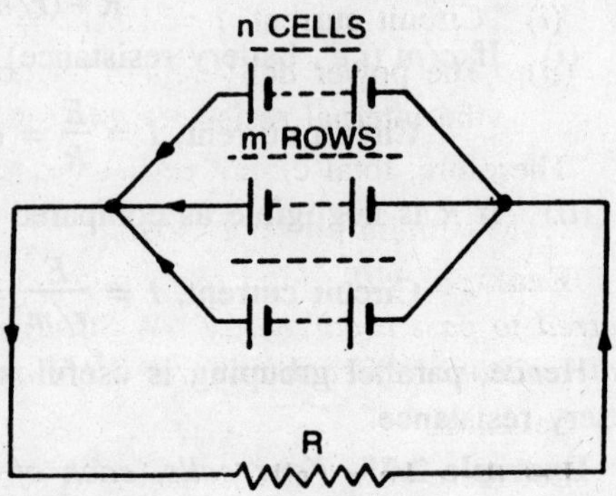

Fig. 2.33

Resistance of one row of cells

$$= r + r + r + \ldots\ldots n \text{ terms} = nr$$

Now, there are m rows in parallel, each having a resistance of nr. If r_T is the total resistance of the battery, then,

$$\frac{1}{r_T} = \frac{1}{nr} + \frac{1}{nr} + \frac{1}{nr} + \ldots\ldots m \text{ terms}$$

$$= \frac{1 + 1 + 1 + \ldots\ldots m \text{ terms}}{nr} = \frac{m}{nr}$$

$\therefore$ $r_T = nr/m$

Total circuit resistance $= R + r_T = R + (nr/m)$

Battery e.m.f. $=$ E.M.F. of one row $= nE$

$$\therefore \text{ Circuit current, } \quad I = \frac{nE}{R + (nr/m)} = \frac{mnE}{mR + nr}$$

Example 2.27. *Thirty two cells are arranged in rows of eight in series and four rows in parallel. The e.m.f. of each cell is 2 V and internal resistance of each cell is 2 Ω. Determine (i) the current supplied to an external resistance of 8 Ω and (ii) maximum power that can be delivered to the load.*

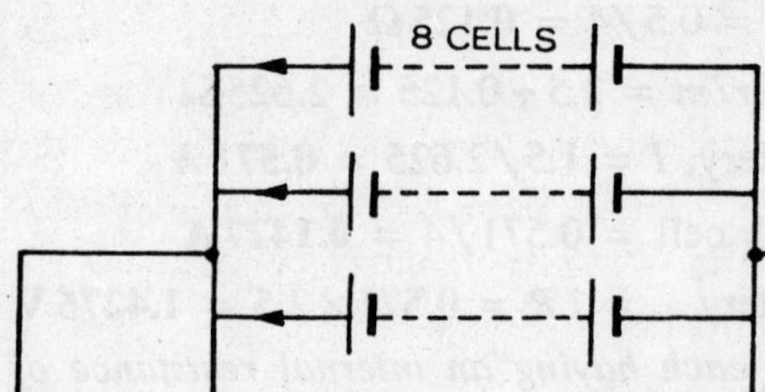

Fig. 2.34.

Solution. Fig. 2.34, shows the arrangement of cells.

No. of cells per row, $n = 8$; Number of rows, $m = 4$

E.m.f. of each cell, $E = 2$ V ; Internal resistance/cell, $r = 2\,\Omega$

External resistance, $R = 8\,\Omega$; Battery e.m.f. $= nE = 8 \times 2 = 16$ V

Battery resistance $= nr/m = 8 \times 2/4 = 4\,\Omega$

Total circuit resistance $= R + (nr/m) = 8 + 4 = 12\,\Omega$

(*i*) Circuit current, $I = 16/12 =$ **1.33 A**

(*ii*) The power delivered to the load (8 Ω) will be maximum when the load, $R = nr/m$ the internal resistance of battery $= 4\,\Omega$.

Therefore, total circuit resistance for maximum power delivered to load is $= 4 + 4 = 8\,\Omega$

∴ Maximum power to load $= (2)^2 \times 4 =$ **16 watts**

Example 2.28. *Find the minimum number of cells connected in two rows in parallel required to pass a current of 6 A through an external resistance of 0.7 Ω. The e.m.f. and internal resistance of each cell are 2.1 V and 0.5 Ω respectively.*

Solution.

Let there be n cells in series in each row.

Battery e.m.f. $= nE = 2.1\,n$; Total circuit resistance $= R + (nr/m) = 0.7 + 0.25\,n$

$$\text{Circuit current} = \frac{2.1\,n}{0.7 + 0.25\,n}$$

or
$$6 = \frac{2.1\,n}{0.7 + 0.25\,n} \qquad \therefore\ n = 7$$

Since there are two rows in parallel, the total number of cell is $= 7 \times 2 =$ **14**

Example 2.29. *A battery of twelve cells is required to send the largest possible current through a resistance of 10 ohms. Find the type of grouping which will give the largest current and estimate its value. The resistance of each cell is 6.3 Ω and its e.m.f. is 1.4 volt.*

Solution.

According to maximum power transfer theorem, maximum power (and hence current) is transferred to a load if load resistance is equal to the internal resistance of source (*i.e.*, battery in this case). Therefore, to solve such problems, find the internal resistance of the battery and see how much it deviates from external resistance (*i.e.*, load).

For series grouping, battery resistance $= nr = 12 \times 6.3 = 75.6\,\Omega$

For parallel grouping, battery resistance $= r/m = 6.3/12 = 0.525\,\Omega$

The reader may note that internal resistance of battery for series as well as parallel grouping is far different from external resistance of 10 Ω. Hence these types of groupings cannot provide strong current. This leaves series-parallel grouping to be used.

Let n = number of cells in series in each row

m = number of rows in parallel

Internal resistance of battery $= nr/m$

For maximum current in series-parallel grouping,

$$R = nr/m$$

or
$$10 = n \times 6.3/m$$

$$\therefore \quad \frac{m}{n} = \frac{6.3}{10} = 0.63 \qquad ...(i)$$

But total number of cells is $m \times n$, *i.e.*, $m \times n = 12$ ...(*ii*)

Multiplying Eq. (*i*) and Eq. (*ii*), we get $m^2 = 12 \times 0.63$ *or* $m = \sqrt{12 \times 0.63} \approx 3$

$\therefore$ No. of rows, $m = 3$; No. of cells per row, $n = 12/3 = 4$

Tutorial Problems

1. Ten cells, each of e.m.f. 1.5 V and internal resistance 0.2 Ω are connected in series. What current will they send through an external resistance of 4 Ω. **[2.5 A]**
2. Three cells of e.m.f. 1.1, 1.3 and 1.5 volts are connected in series to supply current to an external resistance of 1.6 Ω. The internal resistances of three cells are 0.25, 0.35 and 0.4 Ω respectively. Determine the current through the external resistance. **[1.5 A]**
3. A battery consists of 5 cells, each of e.m.f. 1.5 V and internal resistance 0.25 Ω, connected in series. Through what external resistance will it send a current of 2 A ? **[2.5 Ω]**
4. Six cells, each of e.m.f. 1.5 volts and internal resistance of 2 Ω are joined in parallel and connected to an external resistance of 5 Ω. What current will flow ? **[0.529 A]**
5. A battery consists of 4 cells, each of e.m.f. 1.46 volts. The cells are connected in parallel to an external resistance of 0.525 Ω. If the current supplied by the battery is 0.8 A, find the internal resistance of each cell. **[5.2 Ω]**
6. Two cells each having e.m.f. of 2 V and internal resistance of 1 Ω are given. Show how will you connect them to a resistance of 0.5 Ω so that current through external resistance is 2 *A*. **[In parallel]**
7. Four cells, each of e.m.f. 1.5 V and resistance 0.5 Ω, are connected in series-parallel across an external resistance of 2.5 Ω. Find the current in external resistance. **[1 A]**
8. 24 cells each having an e.m.f. of 2.1 V and internal resistance of 2 Ω are to supply the maximum current to a 12 Ω external resistance. Show how the cells are to be connected and what is the value of current to external resistance ?**[Series-parallel, m = 3 ; n = 8 ; 1.625 A]**
9. Four secondary cells each having an e.m.f. of 1.3 V and internal resistance 0.025 Ω are to supply the maximum current to a 12 Ω external resistance. Show how the cells are to be connected and what will be the value of maximum current ? **[Series group ; 0.43 A]**
10. Twenty-four cells are arranged in rows of six in series and the four rows in parallel. The battery sends a current of 2.53 A through an external resistance of 3.2 Ω. If the internal resistance of each cell is 0.08 Ω, find the e.m.f. of each cell. **[1.4 V]**

OBJECTIVE QUESTIONS

1. A d.c. circuit usually has .... as the load.
   (*i*) resistance (*ii*) inductance
   (*iii*) capacitance
   (*iv*) both inductance and capacitance
2. The purpose of load in an electric circuit is to
   (*i*) increase the circuit current
   (*ii*) utilise electrical energy
   (*iii*) decrease the circuit current
   (*iv*) none of the above
3. Electrical appliances are not connected in series because
   (*i*) series circuit is complicated
   (*ii*) power loss is more
   (*iii*) appliances have different current ratings
   (*iv*) none of the above
4. Electrical appliances are connected in parallel because it
   (*i*) is a simple circuit
   (*ii*) draws less current
   (*iii*) results in reduced power loss
   (*iv*) makes the operation of appliances independent of each other
5. Voltage drop across 14.5 Ω resistor in Fig. 2.35 is

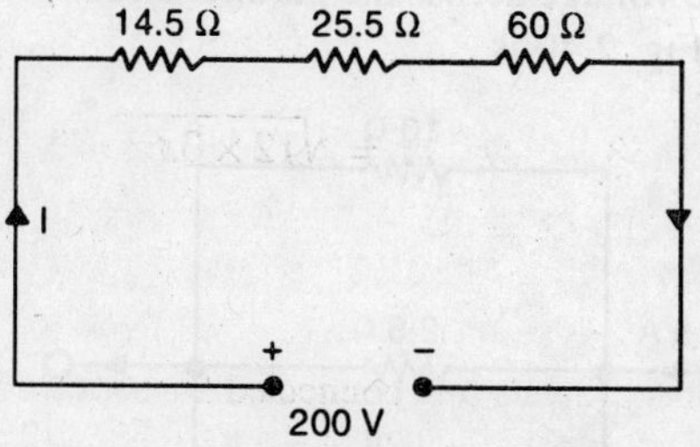

Fig. 2.35

(*i*) 29 V (*ii*) 30.5 V
(*iii*) 14 V (*iv*) 18 V

6. Referring to Fig. 2.36 the total circuit resistance will be

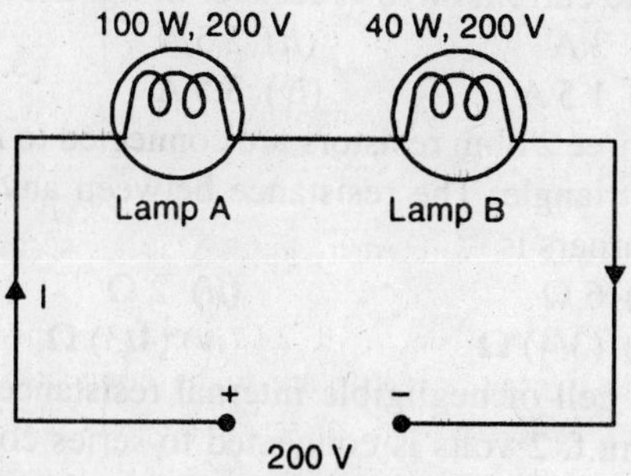

Fig. 2.36

(*i*) 1000 Ω (*ii*) 400 Ω
(*iii*) 1400 Ω (*iv*) 135 Ω

7. In Fig. 2.36,
(*i*) the lamp *A* will be brighter than lamp *B*
(*ii*) the lamp *B* will be brighter than lamp *A*
(*iii*) the two lamps will be equally bright
(*iv*) none of the above

8. When cells are arranged in parallel
(*i*) the current capacity increases
(*ii*) the current capacity decreases
(*iii*) the e.m.f. increases
(*iv*) the e.m.f. decreases

9. When a number of resistances are connected in parallel, the total resistance is
(*i*) greater than the smallest resistance
(*ii*) between the smallest and greatest resistance
(*iii*) less than the smallest resistance
(*iv*) none of the above

10. If a d.c. supply of 180 V is connected across terminals *AB* in Fig. 2.37, then current in 6 Ω resistor will be

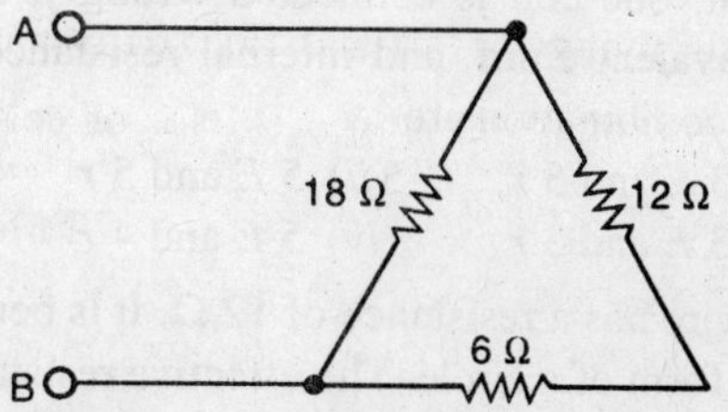

Fig. 2.37

(*i*) 10 A (*ii*) 5 A
(*iii*) 12 A (*iv*) 6 A

11. A battery of 24 V is applied across terminals *AB* of the circuit shown in Fig. 2.38. The current in 2 Ω resistor will be

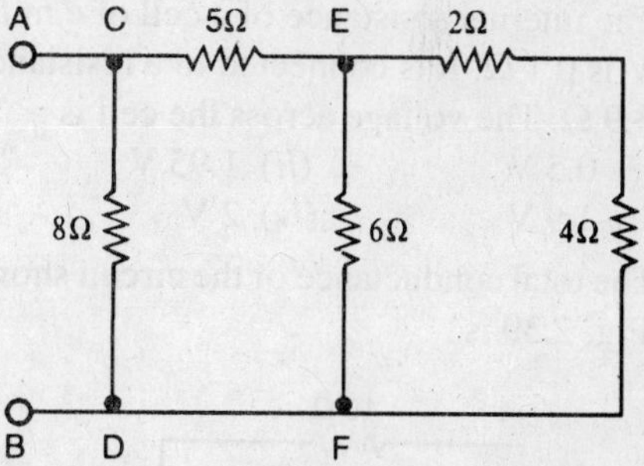

Fig. 2.38

(*i*) 3 A (*ii*) 6 A
(*iii*) 2.5 A (*iv*) 1.5 A

12. A cell of e.m.f. *E* is connected across a resistance *r*. The potential difference between the terminals of the cell is found to be *V*. The internal resistance of the cell is

(*i*) $\frac{2(E-V)V}{r}$ (*ii*) $\frac{2(E-V)r}{E}$

(*iii*) $(E-V)r$ (*iv*) $\left(\frac{E-V}{V}\right)r$

13. An external resistance *R* is connected to a cell of internal resistance *r*. The maximum current flows in the external resistance when
(*i*) $R < r$ (*ii*) $R > r$
(*iii*) $R = r$
(*iv*) any other value of *R*

14. Five cells each of e.m.f. E and internal resistance r are connected in series. If due to oversight, one cell is connected wrongly, then equivalent e.m.f. and internal resistance of the combination are

(*i*) 3 E and 5 r (*ii*) 5 E and 5 r
(*iii*) 3 E and 3 r (*iv*) 5 E and 4 r

15. A wire has a resistance of 12 Ω. It is bent in the form of a circle. The effective resistance between the two points on any diameter of the circle is

(*i*) 6 Ω (*ii*) 3 Ω
(*iii*) 12 Ω (*iv*) 24 Ω

16. The smallest resistance obtained by connecting 50 resistances of 1/4 ohm each is

(*i*) 50/4 Ω (*ii*) 4/50 Ω
(*iii*) 200 Ω (*iv*) 1/200 Ω

17. The internal resistance of a cell of e.m.f. 2.0 V is 0.1 Ω. It is connected to a resistance of 3.9 Ω. The voltage across the cell is

(*i*) 0.5 V (*ii*) 1.95 V
(*iii*) 1.9 V (*iv*) 2 V

18. The total conductance of the circuit shown in Fig. 2.39 is

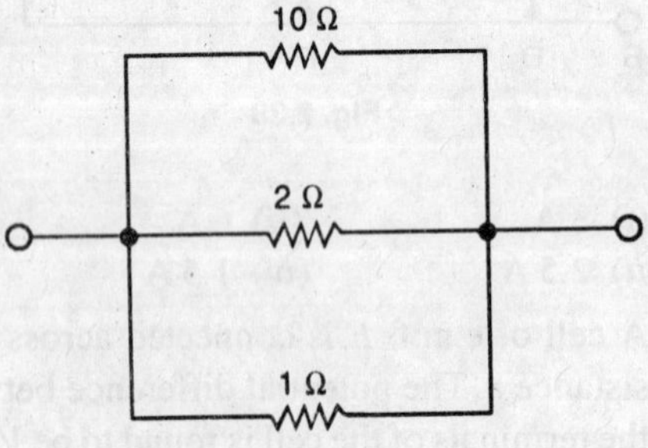

Fig. 2.39

(*i*) 13 S (*ii*) 1.6 S
(*iii*) 6 S (**iv**) 2.5 S

19. If 10 Ω resistance is removed in Fig. 2.39 then total conductance of the circuit will be

(*i*) 3 S (*ii*) 6S
(*iii*) 2 S (*iv*) 1.5 S

20. The voltage across the parallel circuit shown in Fig. 2.40 is

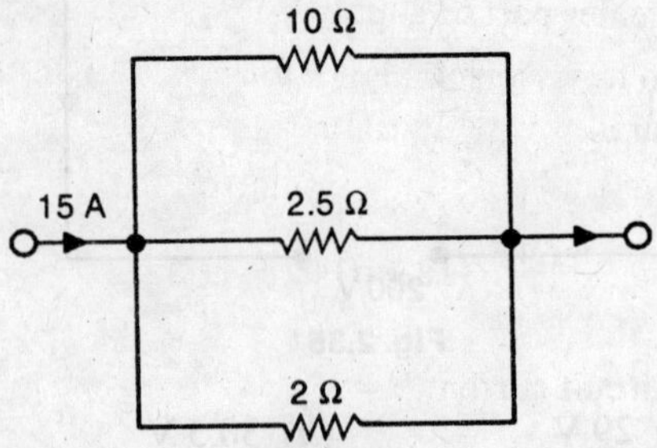

Fig. 2.40

(*i*) 15 V (*ii*) 10 V
(*iii*) 30 V (*iv*) 12.5 V

21. The current in 10 Ω resistor in fig. 2.40 is

(*i*) 3 A (*ii*) 2.5 A
(*iii*) 1.5 A (*iv*) 3.5 A

22. Three 2 ohm resistors are connected to form a triangle. The resistance between any two corners is

(*i*) 6 Ω (*ii*) 2 Ω
(*iii*) (3/4) Ω (*iv*) (4/3) Ω

23. A cell of negligible internal resistance and e.m.f. 2 volts is connected to series combination of 2, 3 and 5 ohms. The potential difference across the terminals of 3 Ω resistor will be

(*i*) (2/3) V (*ii*) 0.6 V
(*iii*) 3 V (*iv*) 6 V

24. A 200 W and 100 W bulb both meant for operation at 220 V are connected in series. When connected to a 220 V supply, the power consumed by them is

(*i*) 66 W (*ii*) 33 W
(*iii*) 100 W (*iv*) 300 W

25. An electric fan and a heater are marked as 100 W, 220 V and 1000 W, 220 V respectively. The resistance of heater is

(*i*) zero
(*ii*) greater than that of fan
(*iii*) less than that of fan
(*iv*) equal to that of fan

ANSWERS TO OBJECTIVE QUESTIONS

1. (*i*)	**2.** (*ii*)	**3.** (*iii*)	**4.** (*iv*)	**5.** (*i*)	**6.** (*iii*)	**7.** (*ii*)	**8.** (*i*)
9. (*iii*)	**10.** (*i*)	**11.** (*iv*)	**12.** (*iv*)	**13.** (*iii*)	**14.** (*i*)	**15.** (*ii*)	**16.** (*iv*)
17. (*ii*)	**18.** (*ii*)	**19.** (*iv*)	**20.** (*i*)	**21.** (*iii*)	**22.** (*iv*)	**23.** (*ii*)	**24.** (*i*)
25. (*iii*)							

HINTS TO SELECTED OBJECTIVE QUESTIONS

7. The resistance of 40 W lamp is much more (2.5 times) than that of 100 W lamp. Consequently, a greater part of supply voltage will appear across 40 W lamp.

10. So far as terminals *AB* are concerned, 12 Ω and 6 Ω are in series and this series combination is in parallel with 18 Ω resistor.

$$\therefore R_{AB} = \frac{(12+6)\times 18}{(12+6)+18} = \frac{18\times 18}{36} = 9\,\Omega$$

Circuit current, $I = \dfrac{V}{R_{AB}} = \dfrac{180}{9} = 20$ A

Since there are two parallel paths of equal resistance, current in each path is 10 A.

11. $R_{EF} = \dfrac{6\times 6}{6+6} = 3\,\Omega$; $\quad R_{CD} = R_{AB} = \dfrac{8\times 8}{8+8} = 4\,\Omega$

$\therefore$ Circuit current $= \dfrac{24}{R_{AB}} = \dfrac{24}{4} = 6$ A

Current in 5 Ω = 6/2 = 3 A

Current is 2 Ω = 3/2 = 1.5 A

15. Fig. 2.41 shows the conditions of the problem. Each parallel path has a resistance of 6 Ω.

$\therefore$ Effective resistance between two points on the diameter

$$= 6 \parallel 6 = \frac{6\times 6}{6+6} = \frac{36}{12} = 3\,\Omega$$

17. Circuit current, $\quad I = \dfrac{E}{R+r} = \dfrac{2.0}{3.9+0.1} = \dfrac{2.0}{4} = 0.5\,A$

$\therefore$ Voltage across cell, $V = E - Ir = 2.0 - 0.5 \times 0.1 = 1.95$ V

18. Conductance *G* is the reciprocal of resistance *R i.e.* $G = 1/R$

$$\frac{1}{R_T} = \frac{1}{R_1} + \frac{1}{R_2} + \frac{1}{R_3}$$

or $\quad G = G_1 + G_2 + G_3 = \dfrac{1}{10} + \dfrac{1}{2} + \dfrac{1}{1} = 1.6$ S

20. $G = G_1 + G_2 + G_3 = \dfrac{1}{10} + \dfrac{1}{2.5} + \dfrac{1}{2} = 1$ S

$\therefore \quad V = IR_T = \dfrac{I}{G} = \dfrac{15\text{ A}}{1\text{ S}} = 15$ V

24. $R_T = R_1 + R_2$

or $\quad \dfrac{R_T}{V^2} = \dfrac{R_1}{V^2} + \dfrac{R_2}{V^2}$

$$\frac{1}{P_T} = \frac{1}{P_1} + \frac{1}{P_2}$$

$$\therefore \quad P_T = \frac{P_1 P_2}{P_1 + P_2} = \frac{200\times 100}{200+100} = 66\text{ W}$$

3

Network Theorems (D.C.)

Introduction

Any arrangement of electrical energy sources, resistances and other circuit elements is called an electrical network. The terms *circuit* and *network* are used synonymously in electrical literature. Thus Fig. 3.1 may be referred to as an electrical circuit or network. In the text so far we employed two network laws *viz.* Ohm's law and Kirchhoff's laws to solve network problems. Occasions arise when these laws applied to certain networks do not yield quick and easy solution. To overcome this difficulty, other network theorems and techniques have been developed which are very useful in analysing both simple and complex electrical circuits. Through the use of these theorems and techniques, it is possible either to simplify the network itself or render the analytical solution easy. In this chapter, we shall focus our attention on important network theorems and techniques with special reference to their utility in solving network problems.

3.1. Network Terminology

While discussing network theorems and techniques, one often comes across the following terms :

(i) **Active element.** An active element is one which supplies electrical energy to the circuit. Thus in Fig. 3.1, E_1 and E_2 are the active elements because they supply energy to the circuit.

(ii) **Passive element**. A passive element is one which receives electrical energy and then either converts it into heat (resistance) or stores in an electric field (capacitance) or magnetic field (inductance). In Fig. 3.1, there are three passive elements, namely R_1, R_2 and R_3. These passive elements (*i.e.* resistances in this case) receive energy from the active elements (*i.e.* E_1 and E_2) and convert it into heat.

(iii) **Node**. A node of a network is an equipotential surface at which *two or more* circuit elements are joined. Thus in fig. 3.1, circuit elements R_1 and E_1 are joined at A and hence A is the node. Similarly B, C and D are nodes.

(iv) **Junction**. A junction is that point in a network where *three or more* circuit elements are joined. In Fig. 3.1, there are only two junction points *viz.* B and D. That B is a junction is clear from the fact that three circuit elements R_1, R_2 and R_3 are joined at it. Similarly, point D is a junction because it joins three circuit elements R_2, E_1 and E_2.

(v) **Branch**. A branch is that part of a network which lies between two junction points. Thus referring to Fig. 3.1, there are a total of three branches *viz.* BAD, BCD and BD. The branch BAD consists of R_1 and E_1; the branch BCD consists of R_3 and E_2 and branch BD merely consists of R_2.

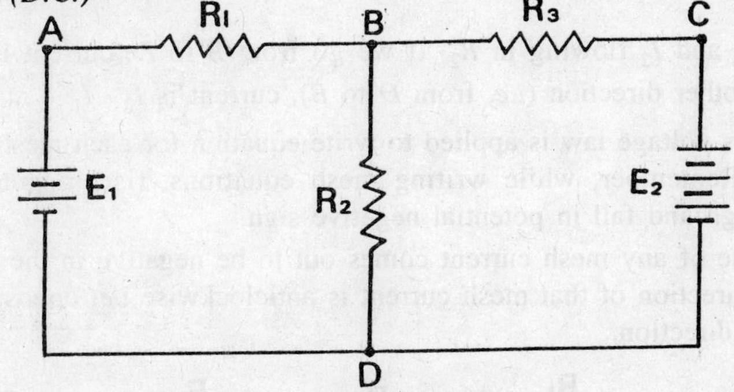

Fig. 3.1

(*vi*) **Loop**. A loop is any closed path of a network. Thus in Fig. 3.1, *ABDA*. *BCDB* and *ABCDA* are the loops.

(*vii*) **Mesh**. A mesh is the most elementary form of a loop and cannot be further divided into other loops. In Fig. 3.1, both loops *ABDA* and *BCDB* qualify as meshes because they cannot be further divided into other loops. However, the loop *ABCDA* cannot be called a mesh because it encloses two loops *ABDA* and *BCDB*.

3.2. Network Theorems And Techniques

Having acquainted himself with network terminology, the reader is set to study the various network theorems and techniques. In this chapter, we shall discuss the following network theorems and techniques :

(*i*) Maxwell's mesh current method (*ii*) Nodal analysis
(*iii*) Superposition theorem (*iv*) Thevenin's theorem
(*v*) Norton's theorem (*vi*) Max. power transfer theorem
(*vii*) Delta/star or star/delta transformation.

The network theorem or technique to be used will depend upon the network arrangement. The general rule is this. Use that theorem or technique which requires a smaller number of independent equations to obtain the solution or which can yield easy solution.

3.3. Maxwell's Mesh Current Method

In this method, Kirchhoff's voltage law is applied to a network to write mesh equations in terms of *mesh currents* instead of branch currents. Each mesh is assigned a separate mesh current. This mesh current is assumed to flow *clockwise* around the perimeter of the mesh without splitting at a junction into branch currents. Kirchhoff's voltage law is then applied to write equations in terms of unknown mesh currents. The branch currents are then found by taking the algebraic sum of the mesh currents which are common to that branch.

Explanation. Maxwell's mesh current method consists of following steps :

(*i*) Each mesh is assigned a separate mesh current. For convenience, all mesh currents are assumed to flow in *clockwise direction. For example, in Fig. 3.2, meshes *ABDA* and *BCDB* have been assigned mesh currents I_1 and I_2 respectively.

(*ii*) If two mesh currents are flowing through a circuit element, the actual current in the circuit element is the algebraic sum of the two. Thus in Fig. 3.2, there are two mesh

---

* It is convenient to consider all mesh currents in one direction (clockwise or anticlockwise). The same result will be obtained if mesh currents are given arbitrary directions.

currents I_1 and I_2 flowing in R_2. If we go from B to D, current is $I_1 - I_2$ and if we go in the other direction (*i.e.* from D to B), current is $I_2 - I_1$.

(*iii*) Kirchhoff's voltage law is applied to write equation for each mesh in terms of mesh currents. Remember, while writing mesh equations, rise in potential is assigned positive sign and fall in potential negative sign.

(*iv*) If the value of any mesh current comes out to be negative in the solution, it means that true direction of that mesh current is anticlockwise *i.e.* opposite to the assumed clockwise direction.

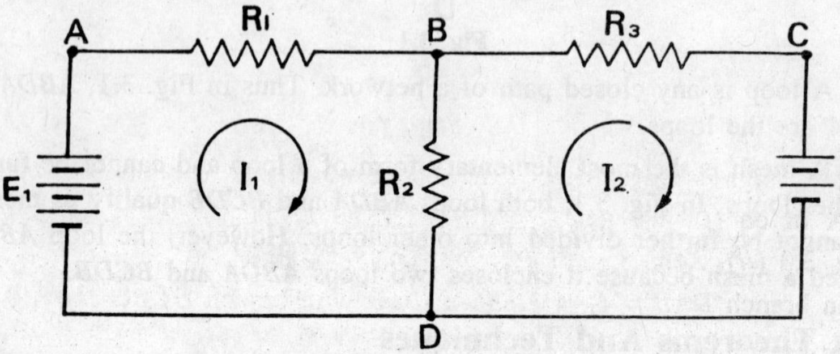

Fig. 3.2

Applying Kirchhoff's voltage law to Fig. 3.2, we have,

Mesh ABDA

$$-I_1R_1 - (I_1 - I_2)\,R_2 + E_1 = 0$$

or $$I_1\,(R_1 + R_2) - I_2R_2 = E_1 \qquad ...(i)$$

Mesh BCDB

$$-I_2R_3 - E_2 - (I_2 - I_1)\,R_2 = 0$$

or $$-I_1R_2 + (R_2 + R_3)\,I_2 = -E_2 \qquad ...(ii)$$

Solving eq (*i*) and eq (*ii*) simultaneously, mesh current I_1 and I_2 can be found out. Once the mesh currents are known, the branch currents can be readily obtained. The advantage of this method is that it usually reduces the number of equations to solve a network problem.

Example 3.1. *In the network shown in Fig. 3.3, find the magnitude and direction of each branch current by mesh current method.*

Solution. Assign mesh currents I_1 and I_2 to meshes $ABDA$ and $BCDB$ respectively as shown in Fig. 3.3 (*i*).

Mesh ABDA

$$-40\,I_1 - 20\,(I_1 - I_2) + 120 = 0$$

or $$60\,I_1 - 20\,I_2 = 120 \qquad ...(i)$$

Mesh BCDB

$$-60\,I_2 - 65 - 20\,(I_2 - I_1) = 0$$

or $$-20\,I_1 + 80\,I_2 = -65 \qquad ...(ii)$$

Multiplying eq (*ii*) by 3 and adding it to eq (*i*), we get,

$$220\,I_2 = -75 \qquad \therefore \quad I_2 = -75/220 = -0.341\text{A}$$

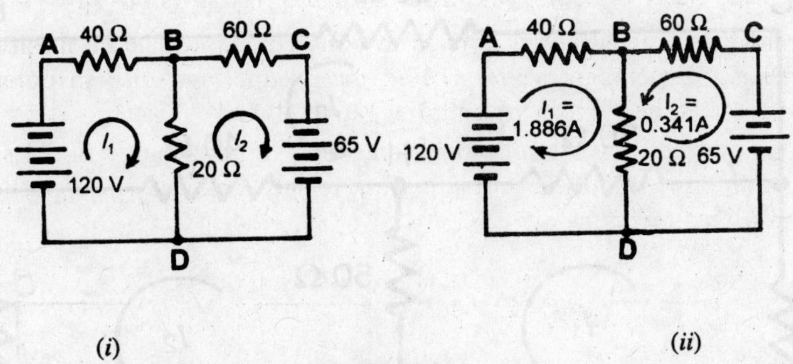

Fig. 3.3

The minus sign shows that true direction of I_2 is anticlockwise. Substituting $I_2 = -0.341$A in eq (*i*), we get $I_1 = 1.886$A. The actual direction of flow of currents is shown in Fig. 3.3 (*ii*). Referring to Fig. 3.3 (*ii*),

Current in branch $DAB = I_1 =$ **1.886A**; Current in branch $DCB = I_2 =$ **0.341A**

Current in branch $BD = I_1 + I_2 = 1.886 + 0.341 =$ **2.227A**

Example 3.2. *Calculate the current in each branch of the circuit shown in Fig. 3.4*

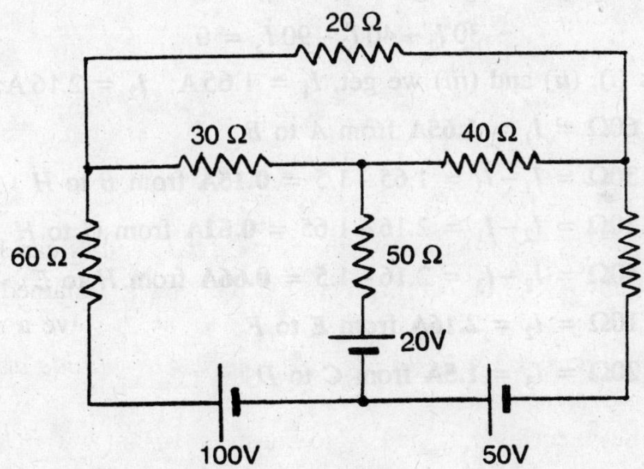

Fig. 3.4

Solution. Assign mesh current I_1, I_2 and I_3 to meshes *ABHGA*, *HEFGH* and *BCDEHB* respectively as shown in fig. 3.5.

Mesh ABHGA

$$-60\,I_1 - 30\,(I_1 - I_3) - 50\,(I_1 - I_2) - 20 + 100 = 80$$

or
$$140\,I_1 - 50\,I_2 - 30\,I_3 = 8 \quad \text{... } (i)$$

Mesh GHEFG

$$20 - 50\,(I_2 - I_1) - 40\,(I_2 - I_3) - 10\,I_2 + 50 = 0$$

or
$$-50\,I_1 + 100\,I_2 - 40\,I_3 = 70 \quad \text{...} (ii)$$

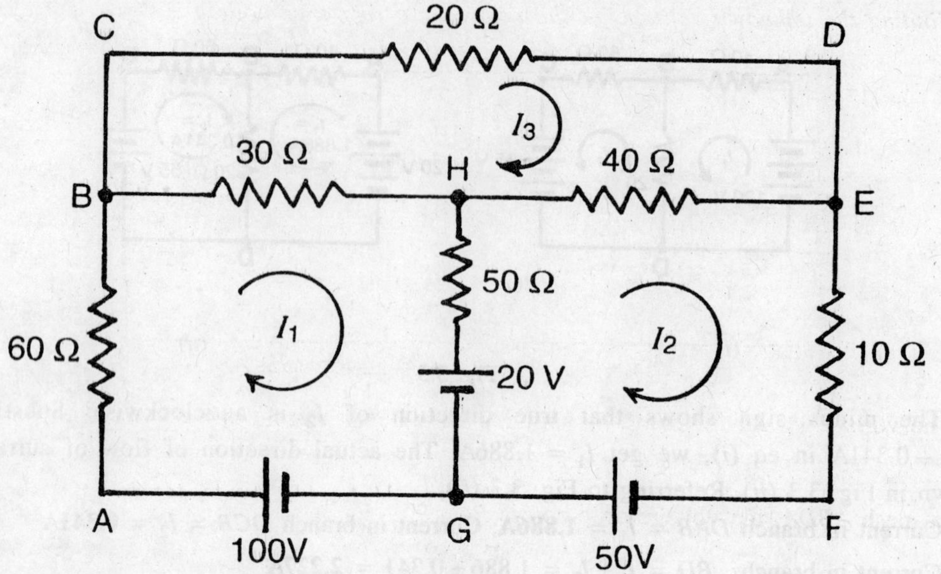

Fig. 3.5

Mesh BCDEHB

$$-20\,I_3 - 40\,(I_3 - I_2) - 30\,(I_3 - I_1) = 0$$

or

$$30\,I_1 + 40\,I_2 - 90\,I_3 = 0 \qquad \text{... (iii)}$$

Solving for equations (*i*), (*ii*) and (*iii*) we get, $I_1 = 1.65\,\text{A}$; $I_2 = 2.16\,\text{A}$; $I_3 = 1.5\text{A}$

∴ Current in 60Ω = I_1 = **1.65A** from *A* to *B*

Current in 30Ω = $I_1 - I_3 = 1.65 - 1.5$ = **0.15A** from *B* to *H*

Current in 50Ω = $I_2 - I_1 = 2.16 - 1.65$ = **0.51A** from *G* to *H*

Current in 40Ω = $I_2 - I_3 = 2.16 - 1.5$ = **0.66A** from *H* to *E*

Current in 10Ω = I_2 = **2.16A** from *E* to *F*

Current in 20Ω = I_3 = **1.5A** from *C* to *D*

3.4. Nodal Analysis

Consider the circuit shown in Fig. 3.6. The branch currents in the circuit can be found by Kirchhoff's laws or Maxwell's mesh current method. There is another method, called *nodal analysis,* for determining the branch currents in the circuit. In this method, one of the nodes (Remember a node is a point in a network where two or more circuit elements meet) is taken as the *reference node*. The potentials of all the points in the circuit are measured *w.r.t.* this reference node. In Fig. 3.6 *A, B, C* and *D* are four nodes and node *D* has been taken as the *reference node. A glance at the circuit shows that voltages at nodes *A* and *C* *w.r.t.* reference node *D* are known. These are E_1 = 120V and E_2 = 65V respectively. The only potential of node *B* *w.r.t D* (call it V_B) is unknown. If this potential V_B can be found, each branch current can be determined because the voltage across each resistor will then be known.

* An obvious choice would be ground or common, if such a point exists.

Hence **nodal analysis** *essentially aims at choosing a reference node in the network and then finding the unknown voltages at the nodes w.r.t. reference node.*

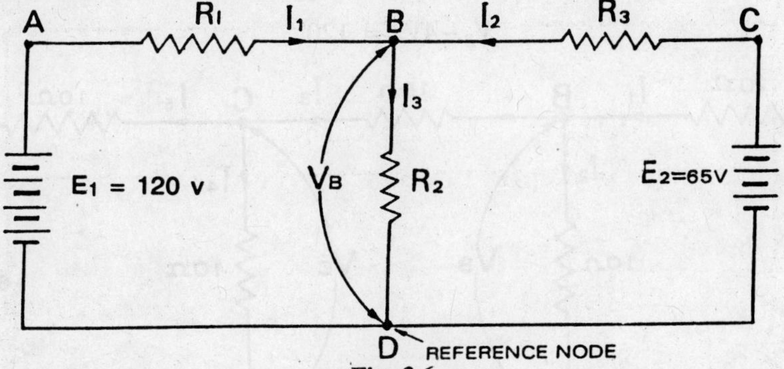

Fig. 3.6

The voltage V_B can be found by applying Kirchhoff's current law at point *B*.

$$I_1 + I_2 = I_3 \qquad \text{...(i)}$$

In mesh *ABDA*, the voltage drop across R_1 is $E_1 - V_B$.

$$\therefore \qquad I_1 = \frac{E_1 - V_B}{R_1}$$

In mesh *CBDC*, the voltage drop across R_3 is $E_2 - V_B$.

$$\therefore \qquad I_2 = \frac{E_2 - V_B}{R_3}$$

Also current $I_3 = V_B/R_2$

Putting the values of I_1, I_2 and I_3 in eq (*i*), we get,

$$\frac{E_1 - V_B}{R_1} + \frac{E_2 - V_B}{R_3} = \frac{V_B}{R_2} \qquad \text{...(ii)}$$

All quantities except V_B are known. Hence V_B can be found out. Once V_B is known, all branch currents can be calculated. It may be seen that nodal analysis requires only one equation [eq (*ii*)] for determining the branch currents in this circuit. However, Kirchhoff's or Maxwell's solution would have needed two equations.

Example 3.3. *Find the currents in the various branches of the circuit shown in Fig. 3.7 by nodal analysis.*

Solution. Mark the currents in the various branches as shown in Fig. 3.7. If the value of any current comes out to be negative in the solution, it means that actual direction of current is opposite to that of assumed. Take point *E* (or *F*) as the reference node. We shall find the voltage at nodes *B* and *C*.

At node B $\qquad I_2 + I_3 = I_1$

or $\qquad \frac{V_B}{10} + {}^*\frac{V_B - V_C}{15} = \frac{100 - V_B}{20}$

or $\qquad 13V_B - 4V_C = 300 \qquad \text{...(i)}$

At node C $\qquad I_4 + I_5 = I_3$

---

* Note that the current I_3 is assumed to flow from *B* to *C*. Therefore, with this assumption, $V_B > V_C$.

$$\text{or} \quad \frac{V_C}{10} + \frac{V_C + 80}{10} = \frac{V_B - V_C}{15}$$

$$\text{or} \quad V_B - 4V_C = 120 \qquad ...(ii)$$

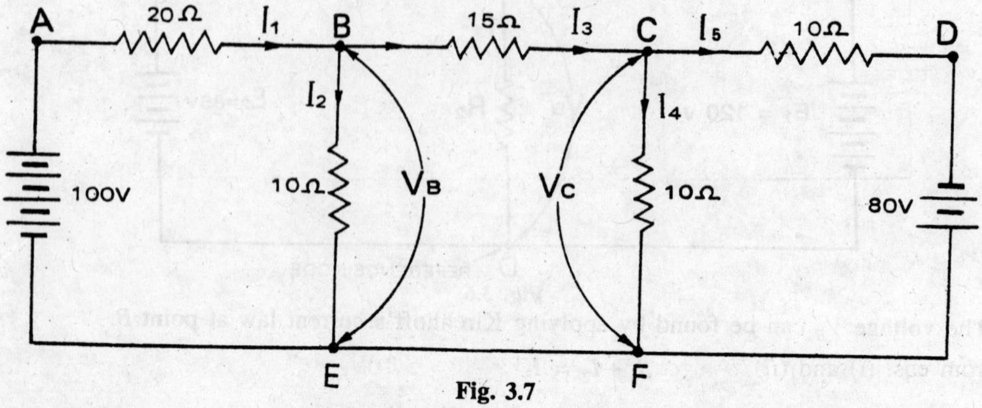

Fig. 3.7

Subtracting eq (*ii*) from eq (*i*), we get, $12V_B = 180 \quad \therefore \quad V_B = 180/12 = 15$ V

Putting $V_B = 15$ volts in eq (*i*), we get, $V_C = -26.25$ volts.

$$\therefore \quad \text{Current} \rightarrow I_1 = \frac{100 - V_B}{20} = \frac{100 - 15}{20} = \mathbf{4.25A}$$

$$\text{Current} \rightarrow I_2 = V_B/10 = 15/10 = \mathbf{1.5A}$$

$$\text{Current} \rightarrow I_3 = \frac{V_B - V_C}{15} = \frac{15 - (-26.25)}{15} = \mathbf{2.75A}$$

$$\text{Current} \rightarrow I_4 = V_C/10 = -26.25/10 = \mathbf{-2.625A}$$

$$\text{Current} \rightarrow I_5 = \frac{V_C + 80}{10} = \frac{-26.25 + 80}{10} = \mathbf{5.375A}$$

The negative sign for I_4 shows that actual current flow is opposite to that of assumed.

Example 3.4. *For the circuit shown in Fig. 3.8, find (i) node pair voltages V_B and V_C and (ii) currents in various branches.*

Solution. Mark the currents in the various branches as shown in Fig. 3.8. Take point *D* (or *E*) as the reference node.

At node B

$$I_1 + I_3 = I_2$$

$$\text{or} \quad \frac{12 - V_B}{2} + \frac{V_C + 4 - V_B}{4} = \frac{V_B}{10}$$

$$\text{or} \quad (6 - 0.5V_B) + (0.25V_C + 1 - 0.25V_B) = 0.1V_B$$

$$\text{or} \quad 0.85V_B - 0.25V_C = 7 \qquad ...(i)$$

At node C

$$I_3 + I_4 = I_5$$

$$\text{or} \quad \frac{V_C + 4 - V_B}{4} + \frac{V_C}{20} = \frac{8 - V_C}{16}$$

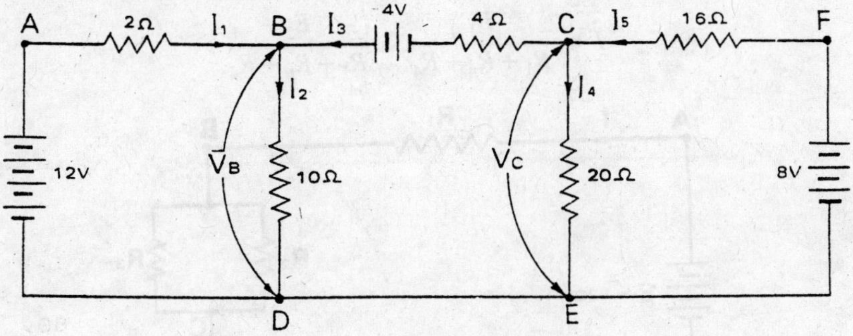

Fig. 3.8

or $$(0.25V_C + 1 - 0.25V_B) + 0.05V_C = 0.5 - 0.0625V_C$$

or $$-0.25V_B + 0.362V_C = -0.5 \qquad \text{.. } (ii)$$

From eqs. (*i*) and (*ii*), we get, V_B = **9.82V**; V_C = **5.40V**

$\therefore$ Current $I_1 = \dfrac{12 - V_B}{2} = \dfrac{12 - 9.82}{2} = \mathbf{1.09A}$

Current $I_2 = V_B/10 = 9.82/10 = \mathbf{0.982A}$

Current $I_3 = \dfrac{V_C + 4 - V_B}{4} = \dfrac{5.40 + 4 - 9.82}{4} = \mathbf{-\,0.105A}$

Current $I_4 = V_C/20 = 5.40/20 = \mathbf{0.27A}$

Current $I_5 = \dfrac{8 - V_C}{16} = \dfrac{8 - 5.40}{16} = \mathbf{0.162A}$

The negative sign for I_3 simply means that current is in opposite direction to that assumed.

3.5. Superposition Theorem

The essence of superposition theorem is that an e.m.f. acting in a *linear network produces the same effect whether it acts alone or in conjunction with other e.m.f.s. The theorem may be stated as under:

In a linear network containing more than one source of e.m.f., the resultant current in any branch is the algebraic sum of the currents that would be produced by each e.m.f., acting alone, all other sources of e.m.f.. being replaced meanwhile by their respective internal resistances.

Explanation. To show that this statement is true, consider the circuit shown in Fig. 3.9. The total circuit resistance R_T is given by :

$$R_T = R_1 + R_p + R_4 \text{ where } R_p = \frac{R_2 R_3}{R_2 + R_3}$$

Circuit current $$I = \frac{\text{Net e.m.f}}{R_T} = \frac{E_1 - E_2}{R_1 + R_p + R_4}$$

---

* A linear circuit is one whose parameters (*e.g.*, resistances) are constant *i.e., they do not change with current and voltage.*

or $$I = \frac{E_1}{R_1 + R_p + R_4} - \frac{E_2}{R_1 + R_p + R_4} \quad ...(i)$$

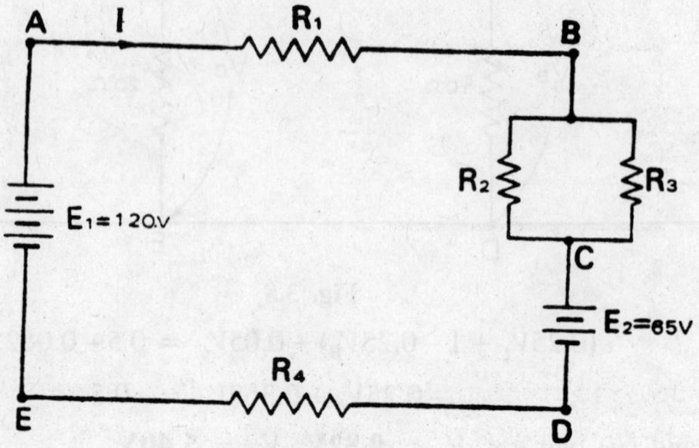

Fig. 3.9

Let us now find the circuit current by superposition theorem. Consider that e.m.f. E_1 is acting alone by replacing E_2 with a short circuit between C and D.

$\therefore$ Current due to E_1 alone, $I_1 = \dfrac{E_1}{R_1 + R_p + R_4}$

Now replace E_1 with a short circuit between A and E so that now E_2 is acting alone.

$\therefore$ Current due to E_2 alone, $I_2 = \dfrac{-E_2}{R_1 + R_p + R_4}$

The minus sign with I_2 indicates that current produced by E_2 alone is in a direction opposite to that produced by E_1 alone.

$\therefore$ Circuit current, I = Algebraic sum of I_1 and I_2

$$= \left(\frac{E_1}{R_1 + R_p + R_4}\right) + \left(\frac{-E_2}{R_1 + R_p + R_4}\right)$$

$$= \frac{E_1}{R_1 + R_p + R_4} - \frac{E_2}{R_1 + R_p + R_4}$$

which is the same as in eq. (*i*). This establishes the validity of superposition theorem.

Example 3.5. *In the network shown in Fig. 3.10 (i), find the different branch currents by superposition theorem.*

Solution. Since there are two sources of e.m.f., two circuits [3.10 (*ii*) and 3.11] are required for analysis by superposition theorem.

In Fig. 3.10 (*ii*), 40V source is replaced by a short so that e.m.f. of 35V is acting alone. The various currents can be found as under [Refer to Fig. 3.10 (*ii*)] :

Total resistance across source $= 3 + \dfrac{2 \times 4}{2 + 4} = 4.33\ \Omega$

Total circuit current, $I_1' = 35/4.33 = 8.08\text{A}$

Current in 4 Ω resistor, $^*I_2' = 8.08 \times 2/(2+4) = 2.69\text{A}$

---

* Remember (Refer back to Art. 2.3) current distribution in parallel circuit is as under :

Current in one of the two parallel resistors = Total current × $\dfrac{\text{Resistance of other resistor}}{\text{Sum of two resistances}}$

Current in 2 Ω resistor, $I_3' = 8.08 \times 4/(2+4) = 5.39\text{A}$

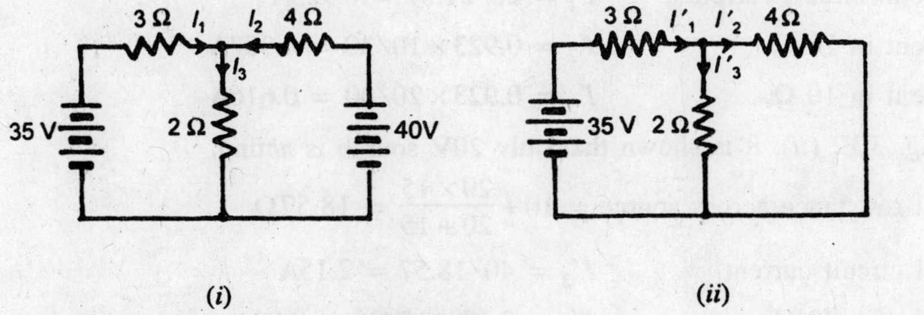

Fig. 3.10

In Fig. 3.11, 35 V source is replaced by a short so that now 40V source is acting alone.

Total resistance across source= $4 + \dfrac{3 \times 2}{3+2} = 5.2\ \Omega$

Total circuit current, $I''_2 = 40/5.2 = 7.69\text{A}$

Current in 3Ω, $I''_1 = 7.69 \times 2/(2+3) = 3.08\text{A}$

Current in 2Ω, $I''_3 = 7.69 \times 3/(2+3) = 4.61\text{A}$

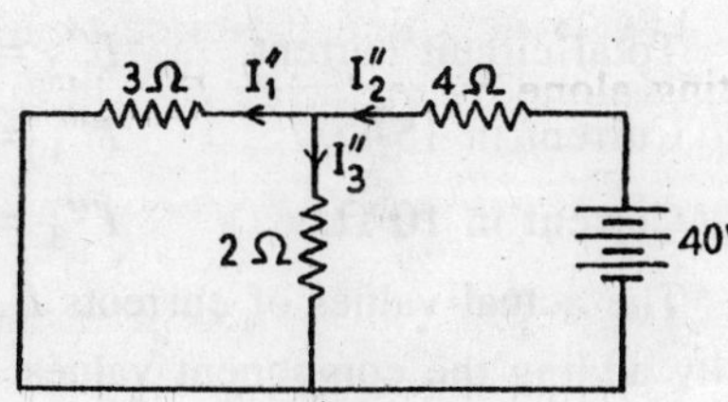

Fig. 3.11

The actual current values of I_1, I_2 and I_3 shown in fig. 3.10 (*i*) can be found by algebraically adding the component values.

$$\therefore \quad I_1 = I'_1 - I''_1 = 8.08 - 3.08 = \mathbf{5A}$$

$$I_2 = I''_2 - I'_2 = 7.69 - 2.69 = \mathbf{5A}$$

$$I_3 = I'_3 + I''_3 = 5.4 + 4.61 = \mathbf{10.01A}$$

Example 3.6. *Using superposition theorem, find the current in each branch of the network shown in Fig. 3.12 (i).*

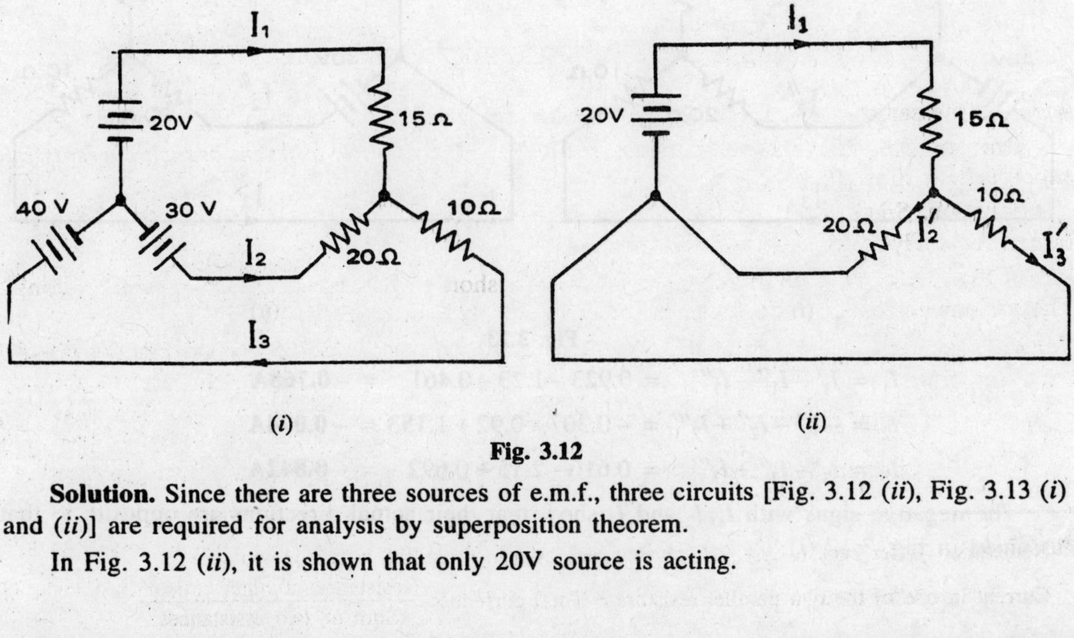

Fig. 3.12

Solution. Since there are three sources of e.m.f., three circuits [Fig. 3.12 (*ii*), Fig. 3.13 (*i*) and (*ii*)] are required for analysis by superposition theorem.

In Fig. 3.12 (*ii*), it is shown that only 20V source is acting.

Total resistance across source $= 15 + \frac{20 \times 10}{20 + 10} = 21.67\,\Omega$

$\therefore$ Total circuit current, $I'_1 = 20/21.67 = 0.923\text{A}$

Current in 20 Ω, $I'_2 = 0.923 \times 10/30 = 0.307\text{A}$

Current in 10 Ω, $I'_3 = 0.923 \times 20/30 = 0.616\text{A}$

In Fig. 3.12 (*ii*), it is shown that only 20V source is acting.

Total resistance across source $= 10 + \frac{20 \times 15}{20 + 15} = 18.57\Omega$

Total circuit current, $I''_3 = 40/18.57 = 2.15\text{A}$

Current in 20 Ω, $I''_2 = 2.15 \times 15/35 = 0.92\text{A}$

Current in 15 Ω, $I''_1 = 2.15 \times 20/35 = 1.23\text{A}$

In Fig. 3.13 (*ii*), only 30 V source is acting in the circuit.

Total resistance across source $= 20 + 10 \times 15/(10 + 15) = 26\,\Omega$

Total circuit current, $I'''_2 = 30/26 = 1.153\text{A}$

Current in 15 Ω, $I'''_1 = 1.153 \times 10/25 = 0.461\text{A}$

Current in 10 Ω, $I'''_3 = 1.153 \times 15/25 = 0.692\text{A}$

The actual values of currents I_1, I_2 and I_3 shown in fig. 3.12 (*i*) can be found by algebraically adding the component values.

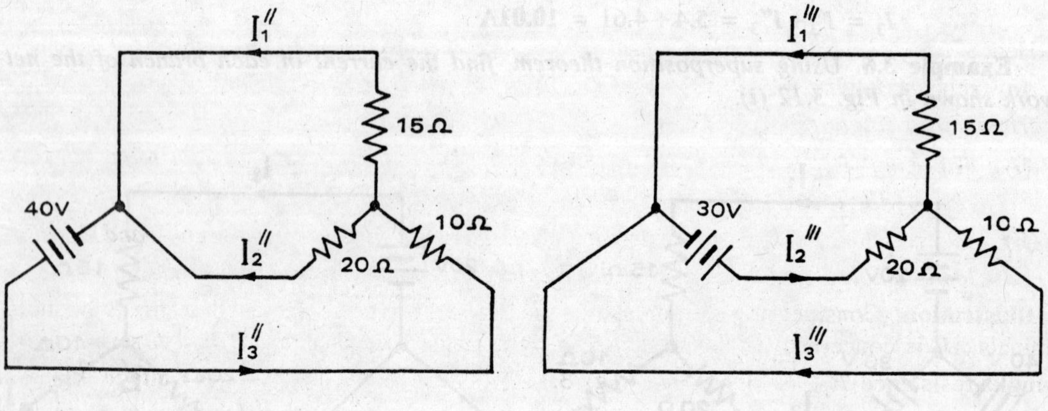

Fig. 3.13

$$I_1 = I_1' - I_1'' - I_1''' = 0.923 - 1.23 - 0.461 = \mathbf{-0.768A}$$

$$I_2 = -I_2' - I_2'' + I_2''' = -0.307 - 0.92 + 1.153 = \mathbf{-0.074A}$$

$$I_3 = I_3' - I_3'' + I_3''' = 0.616 - 2.15 + 0.692 = \mathbf{-0.842A}$$

The negative signs with I_1, I_2 and I_3 show that their actual directions are opposite to that assumed in Fig. 3.12 (*i*).

3.6. Thevenin's theorem

Fig. 3.14 (*i*) shows a network enclosed in a box with two terminals *A* and *B* brought out. The network in the box may consist of any number of resistors and e.m.f. sources connected in any manner. But according to Thevenin the entire circuit behind terminals *A* and *B* can be replaced by a single source of e.m.f. E_{Th} (called Thevenin voltage) in series with a single resistance R_{Th} (called Thevenin resistance) as shown in Fig. 3.14 (*ii*). The values of E_{Th} and R_{Th} are determined as mentioned in Thevenin's theorem. Once Thevenin's equivalent circuit is obtained [See Fig. 3.14 (*ii*)], then current through any load resistance R_L connected across *AB* is given by ;

$$I = \frac{E_{th}}{R_{th} + R_L}$$

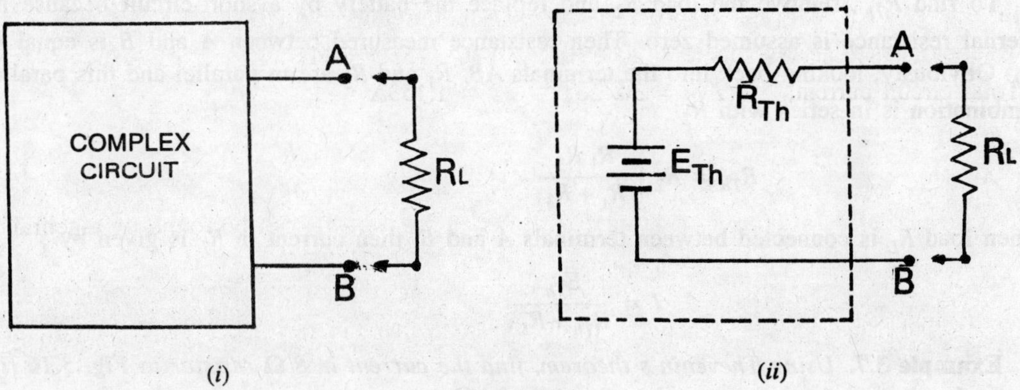

Fig. 3.14

Hence Thevenin's theorem as applied to d.c. circuits may be stated as under :

Any network having terminals A and B can be replaced by a single source of e.m.f. E_{Th} *in series with a single resistance* R_{Th}.

(*i*) *The e.m.f.* E_{Th} *is the voltage obtained across terminals A and B with load, if any, removed i.e., it is the open-circuited voltage between A and B.*

(*ii*) *The resistance* R_{Th} *is the resistance of the network measured between A and B with load removed and sources of e.m.f. replaced by their internal resistances.*

Illustration. Consider the circuit shown in Fig. 3.15 (*i*). As far as the circuit behind terminals *AB* is concerned, it can be replaced by a single source of e.m.f. E_{Th} in series with a single resistance R_{Th} as shown in Fig. 3.15 (*ii*). The e.m.f E_{Th} is the voltage across terminals *AB* with R_L removed. With R_L disconnected, there is no current in R_2 and E_{Th} will be the voltage appearing across R_3.

$$\therefore \quad E_{Th} = \text{Voltage across } R_3$$

$$= \text{Current through } R_3 \times \text{Resistance } R_3$$

$$= \frac{V}{R_1 + R_3} \times R_3$$

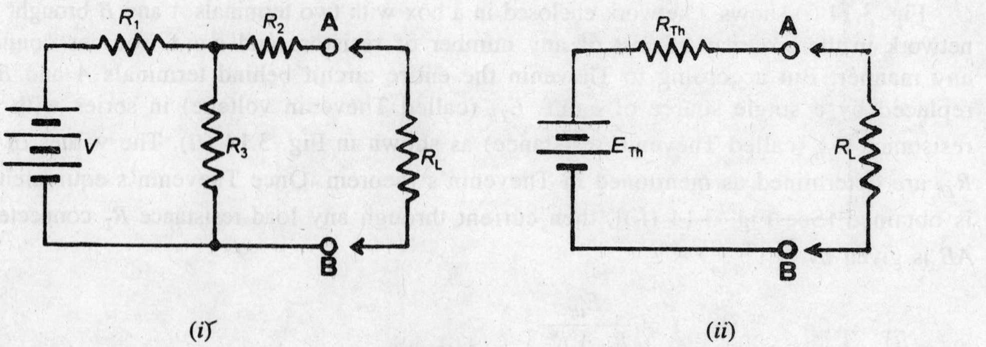

Fig. 3.15

To find R_{Th}, remove the load R_L and replace the battery by a short-circuit because its internal resistance is assumed zero. Then resistance measured between A and B is equal to R_{Th}. Obviously, looking back into the terminals AB, R_1 and R_3 are in parallel and this parallel combination is in series with R_2.

$$\therefore \qquad R_{Th} = R_2 + \frac{R_1 R_3}{R_1 + R_3}$$

When load R_L is connected between terminals A and B, then current in R_L is given by ;

$$I = \frac{E_{Th}}{R_{Th} + R_L}$$

Example 3.7. *Using Thevenin's theorem, find the current in 8 Ω resistor in Fig. 3.16 (i). Given that battery has an internal resistance of 1 Ω.*

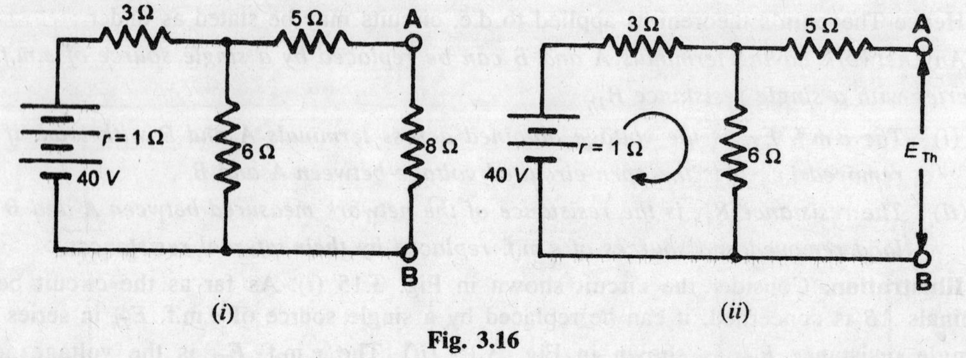

Fig. 3.16

Solution. It is required to find current in 8 Ω resistor by Thevenin's theorem. The solution involves two steps :

(*i*) The first step is to find the open-circuited voltage E_{Th} at terminals AB. It is the voltage appearing across AB when 8 Ω resistor is removed as shown in Fig. 3.16 (*ii*). With 8 Ω resistor removed, there is no drop in 5 Ω and E_{Th} will be the voltage appearing across 6 Ω resistor. Referring to Fig. 3.16 (*ii*),

$$\therefore \qquad \text{Circuit current} = \frac{40}{1+3+6} = 4\text{A}$$

$$\text{Voltage across } 6\,\Omega,\ E_{Th} = 4 \times 6 = 24\text{V}$$

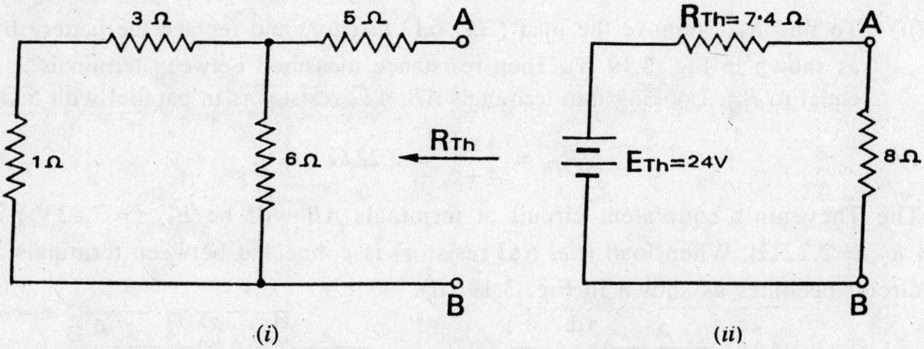

Fig. 3.17

(*ii*) The second step is to find Thevenin's resistance R_{Th}. For this purpose, remove the load (*i.e* 8 Ω) and replace the battery by its internal resistance of 1 Ω as shown in Fig. 3.17 (*i*). The resistance now measured between *A* and *B* is equal to R_{Th}. Looking into the terminals *AB*, 5 Ω resistor is in series with parallel combination of (3 + 1) Ω and 6 Ω.

$$\therefore \qquad R_{Th} = 5 + \frac{4 \times 6}{4 + 6} = 7.4\,\Omega$$

The Thevenin's equivalent circuit behind terminals *AB* is E_{Th} (= 24*V*) in series with R_{Th} (= 7.4 Ω). When load of 8 Ω is connected between terminals *A* and *B*, the circuit becomes as shown in Fig. 3.17 (*ii*).

$$\therefore \qquad \text{Current in } 8\,\Omega = \frac{E_{Th}}{R_{Th} + 8} = \frac{24}{7.4 + 8} = \mathbf{1.56A}$$

Example 3.8. *Using Thevenin's theorem, find the current in* 6 Ω *resistor in Fig. 3.18 (i).*

Solution. Since internal resistances of batteries are not given, it will be assumed that they are zero. The solution involves two steps *viz.* finding Thevenin voltage E_{Th} and Thevenin resistance R_{Th}.

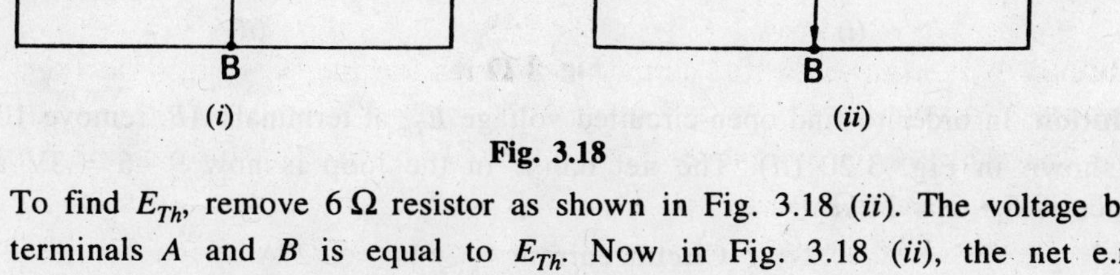

Fig. 3.18

(*i*) To find E_{Th}, remove 6 Ω resistor as shown in Fig. 3.18 (*ii*). The voltage between terminals *A* and *B* is equal to E_{Th}. Now in Fig. 3.18 (*ii*), the net e.m.f. is 4.5 − 3 = 1.5V and total circuit resistance is 9 Ω.

∴ Circuit current = 1.5/9 = 0.167A

The voltage across *AB* is equal to 4.5 V *less* drop in 4Ω resistor.

∴ Voltage across *AB*, E_{Th} = 4.5 − 0.167 × 4 = 3.83V

(*ii*) To find R_{Th}, remove the load (*i.e.* 6 Ω resistor) and replace the battery by a short as shown in Fig. 3.19 (*i*). Then resistance measured between terminals *A* and *B* is equal to R_{Th}. Looking into terminals *AB*, 4 Ω resistor is in parallel with 5 Ω resistor.

$$\therefore \qquad R_{Th} = \frac{4 \times 5}{4 + 5} = 2.22\ \Omega$$

The Thevenin's equivalent circuit at terminals *AB* will be E_{Th} (= 3.83 V) in series with R_{Th} (= 2.22 Ω). When load (*i.e.* 6 Ω resistor) is connected between terminals *A* and *B*, the circuit becomes as shown in Fig. 3.19 (*ii*).

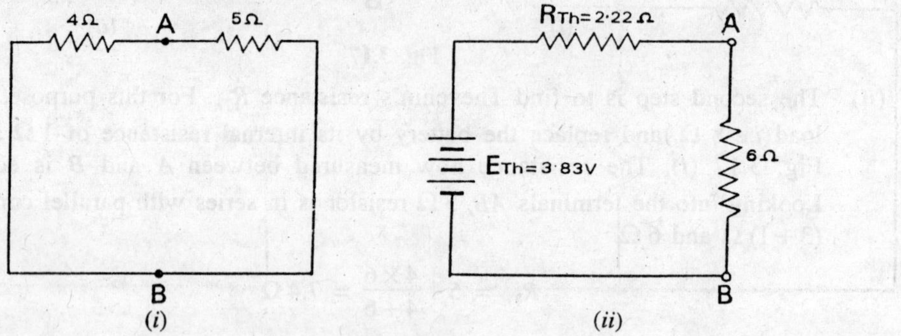

Fig. 3.19

$$\therefore \text{ Current in } 6\ \Omega = \frac{E_{Th}}{R_{Th} + 6} = \frac{3.83}{2.22 + 6} = \mathbf{0.466A}$$

Example 3.9. *Find the p.d. across AB in the circuit shown in Fig. 3.20 (i). Use Thevenin's theorem.*

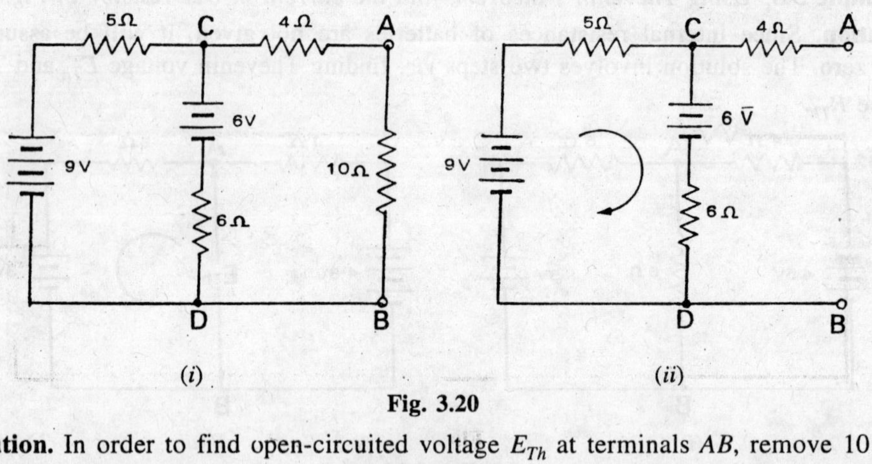

Fig. 3.20

Solution. In order to find open-circuited voltage E_{Th} at terminals *AB*, remove 10 Ω resistor as shown in Fig. 3.20 (*ii*). The net e.m.f. in the loop is now 9 − 6 = 3V and total resistance is 5 + 6 = 11 Ω.

$$\text{Circuit current} = 3/11 = 0.27\text{A}$$

Voltage across *AB* or *CD*, E_{Th} = 9 − 5 × 0.27 = 7.64 V

In order to find Thevenin resistance R_{Th}, remove the load (*i.e.* 10 Ω resistor) and replace the batteries by short as shown in Fig. 3.21 (*i*). The resistance measured between terminals *A* and B is equal to R_{Th}. Looking into terminals AB, 4 Ω resistor is in series with parallel combination of 5 Ω and 6 Ω resistors.

$$\therefore \qquad R_{Th} = 4 + \frac{5 \times 6}{5+6} = 6.72\ \Omega$$

The Thevenin's equivalent circuit at terminals AB is E_{Th} (= 7.64V) in series with R_{Th} (= 6.72 Ω). When load (*i.e.* 10 Ω resistor) is connected between terminals A and B, the circuit becomes as shown in Fig. 3.21 (*ii*).

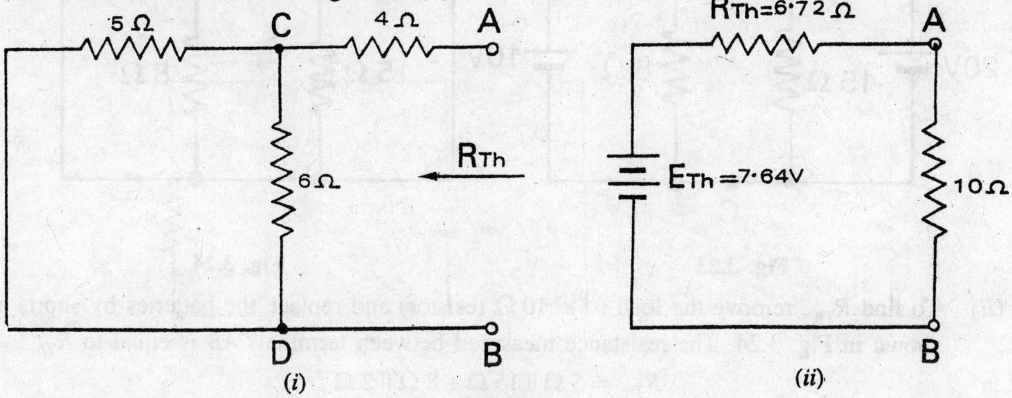

Fig. 3.21

$$\therefore \qquad \text{Current in } 10\ \Omega = \frac{E_{Th}}{R_{Th}+10} = \frac{7.64}{6.72+10} = 0.457\text{A}$$

$$\therefore \qquad \text{P.D. across } 10\ \Omega = 0.457 \times 10 = \mathbf{4.57V}$$

Example 3.10. *Find the current in the 10 Ω resistor in the network shown in Fig. 3.22. Use Thevenin's theorem.*

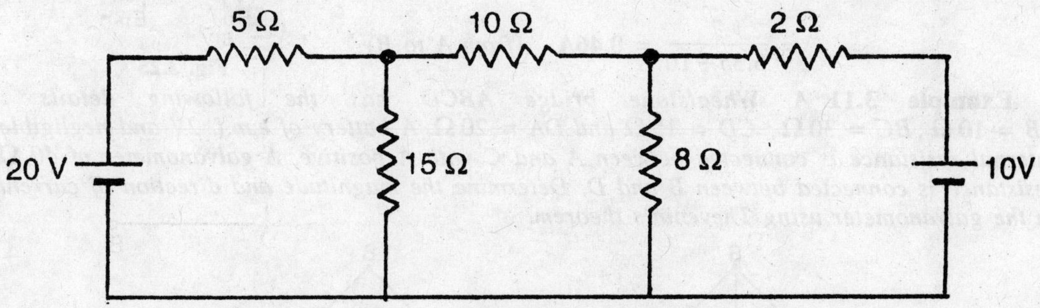

Fig. 3.22

Solution. The solution involves two steps *viz* finding Thevenin voltage E_{Th} and Thevenin resistance R_{Th}.

(*i*) To find E_{Th}, remove 10 Ω resistor as shown in Fig. 3.23. The voltage between terminals A and B is equal to E_{Th}. Referring to Fig. 3.23, we have,

$$I_1 = \frac{20}{5+15} = 1\text{A} \qquad \therefore \quad V_{AC} = 1 \times 15 = 15\text{V}$$

$$I_2 = \frac{10}{2+8} = 1\text{A} \qquad \therefore \quad V_{BC} = 1 \times 8 = 8\text{V}$$

$$\therefore \quad V_{AB} = E_{Th} = V_{AC} - V_{BC} = 15 - 8 = 7\text{V}$$

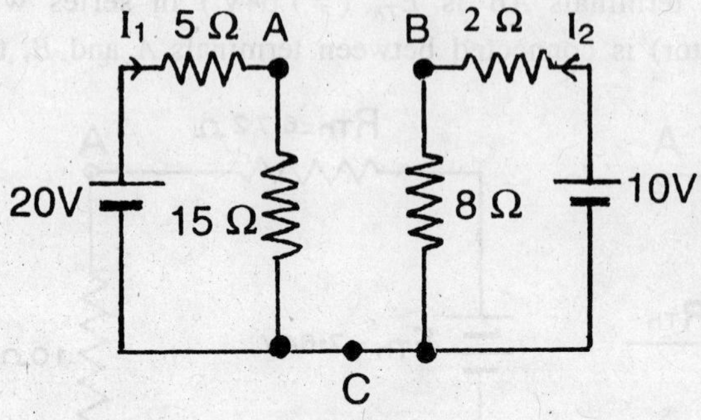

Fig. 3.23

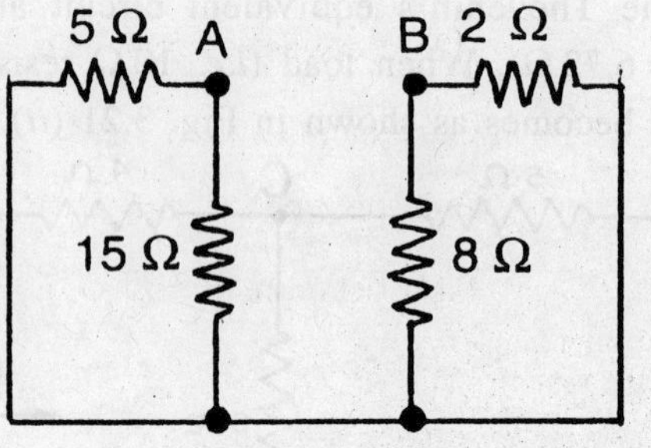

Fig. 3.24

(*ii*) To find R_{Th}, remove the load (*i.e.* 10 Ω resistor) and replace the batteries by shorts as shown in Fig. 3.24. The resistance measured between terminals *AB* is equal to R_{Th}.

$$\therefore \quad R_{Th} = 5\,\Omega \,||\, 15\,\Omega + 8\,\Omega \,||\, 2\,\Omega$$

$$= \frac{5 \times 15}{5+15} + \frac{8 \times 2}{8+2} = 5.35\,\Omega$$

The Thevenin's equivalent circuit at terminals *AB* is $E_{Th}\,(=7V)$ in series with $R_{Th}\,(=5.35\,\Omega)$. When load (*i.e.* 10 Ω resistor) is connected between terminals *A* and *B*, the circuit becomes as shown in Fig. 3.25.

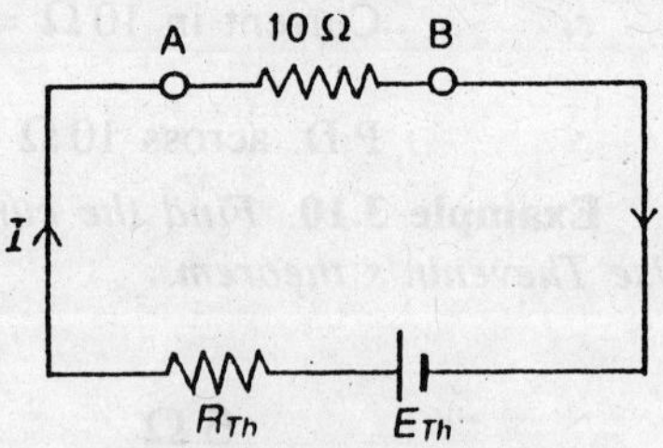

Fig. 3.25

$$\text{Current in} \quad 10\,\Omega = \frac{E_{Th}}{R_{Th} + 10}$$

$$= \frac{7}{5.35 + 10} = \mathbf{0.46A} \quad (\textit{from A to B})$$

Example 3.11. *A Wheatstone bridge ABCD has the following details : AB = 10 Ω ; BC = 30 Ω ; CD = 15 Ω and DA = 20 Ω. A battery of e.m.f. 2V and negligible internal resistance is connected between A and C with A positive. A galvanometer of 40 Ω resistance is connected between B and D. Determine the magnitude and direction of current in the galvanometer using Thevenin's theorem.*

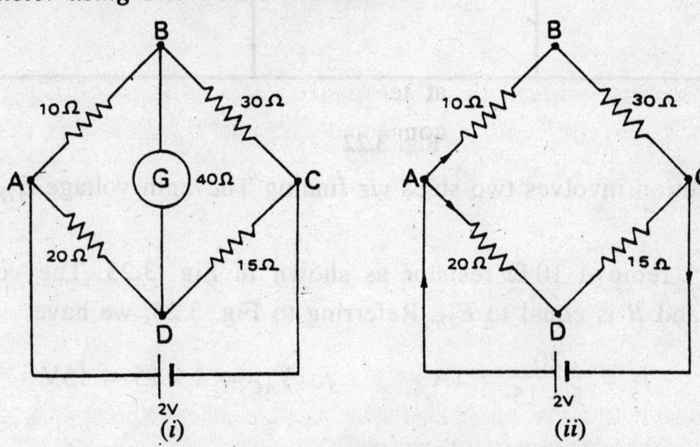

Fig. 3.26

Solution. Fig. 3.26 (*i*) shows the Wheatstone bridge. To find the open-circuited voltage E_{Th} at terminals *BD*, remove 40 Ω resistor as shown in Fig. 3.26 (*ii*). Then voltage between terminals *B* and *D* is equal to E_{Th}.

$$\text{Current in branch } ABC = 2/(10+30) = 0.05\text{A}$$

$$\text{P.D. between } A \text{ and } B, V_{AB} = 10 \times 0.05 = 0.5\text{ V}$$

$$\text{Current in branch } ADC = 2/(20+15) = 0.0571\text{A}$$

$$\text{P.D. between } A \text{ and } D, V_{AD} = 0.0571 \times 20 = 1.142\text{V}$$

$$\therefore \quad \text{P.D. between } B \text{ and } D, E_{Th} = V_{AD} - V_{AB}$$

$$= 1.142 - 0.5 = 0.642\text{V}$$

Obviously, point *B** *is positive w.r.t. D i.e.* current in galvanometer, when connected between *B* and *D*, will flow from *B* to *D*.

In order to find. Thevenin resistance R_{Th}, remove 40 Ω resistor and replace the battery by a short (as internal resistance of battery is assumed zero) as shown in Fig. 3.27 (*i*). Then resistance measured between terminals *B* and *D* is equal to R_{Th}. Looking into terminals *BD*, parallel combination of 10 Ω and 30 Ω resistors is in series with parallel combination of 20 Ω and 15 Ω resistors.

$$\therefore \quad R_{Th} = \frac{10 \times 30}{10+30} + \frac{20 \times 15}{20+15} = 7.5 + 8.57 = 16.07\,\Omega$$

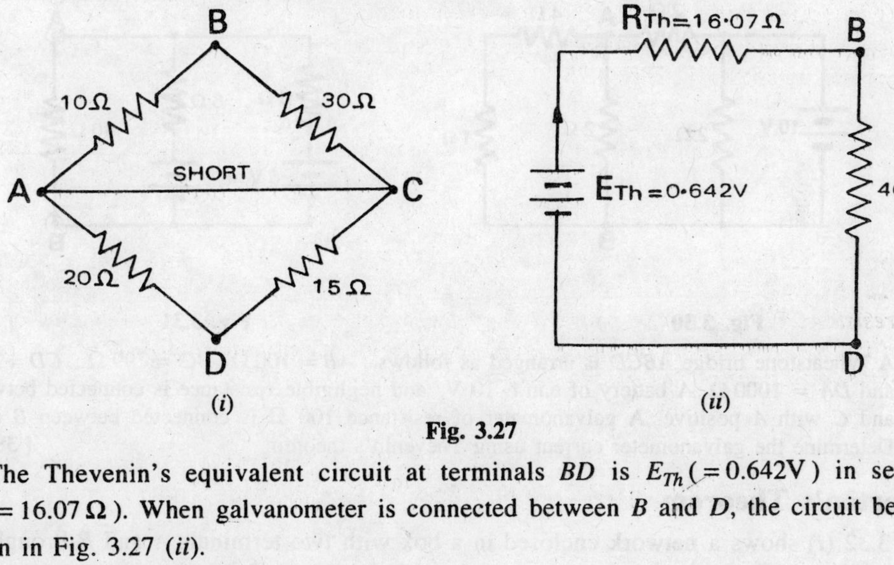

Fig. 3.27

The Thevenin's equivalent circuit at terminals *BD* is E_{Th} (= 0.642V) in series with R_{Th} (= 16.07 Ω). When galvanometer is connected between *B* and *D*, the circuit becomes as shown in Fig. 3.27 (*ii*).

∴ Galvanometer Current

$$= \frac{E_{Th}}{R_{Th}+40} = \frac{0.642}{16.07+40} = 11.5 \times 10^{-3}\text{ A} = \mathbf{11.5\text{ mA}} \quad \textit{from B to D}$$

---

* The potential at point *D* is 1.42V lower than at *A*. Also potential of point *B* is 0.5 V lower than at *A*. Hence, point *B* is at higher potential than point *D*.

Tutorial Problems

1. Using Thevenin's theorem, find the current in 10 Ω resistor in the circuit shown in Fig. 3.28. **[0.481A]**

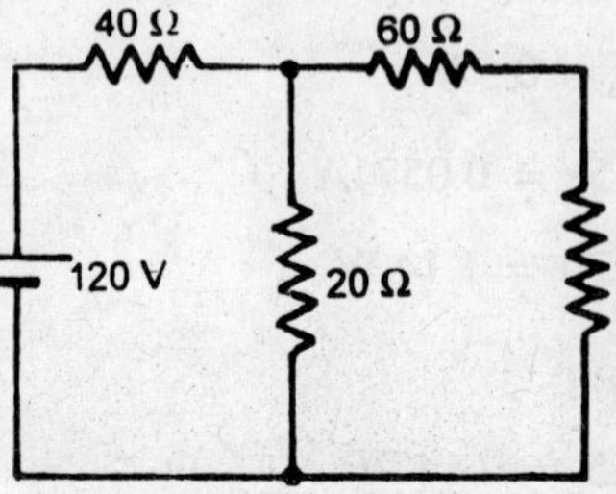

Fig. 3.28

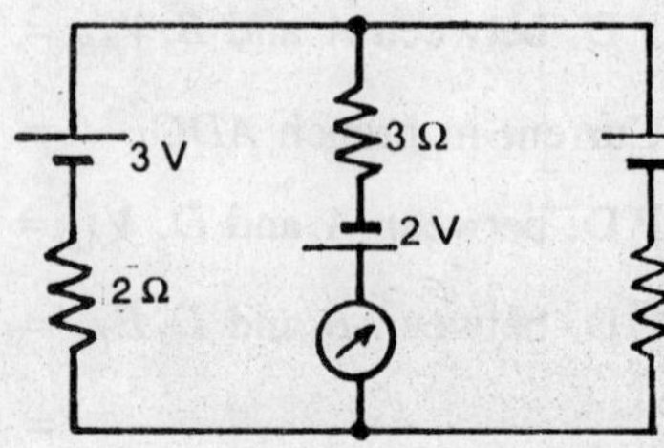

Fig. 3.29

2. Using Thevenin's theorem, find the current in the ammeter shown in Fig. 3.29. **[1A]**

3. Use Thevenin's theorem to find the p.d. across *AB* branch of the network shown in Fig. 3.30. **[4.16V]**

4. Determine Thevenin's equivalent circuit to the left of *AB* in Fig. 3.31. **[A 6V source in series with 3 Ω]**

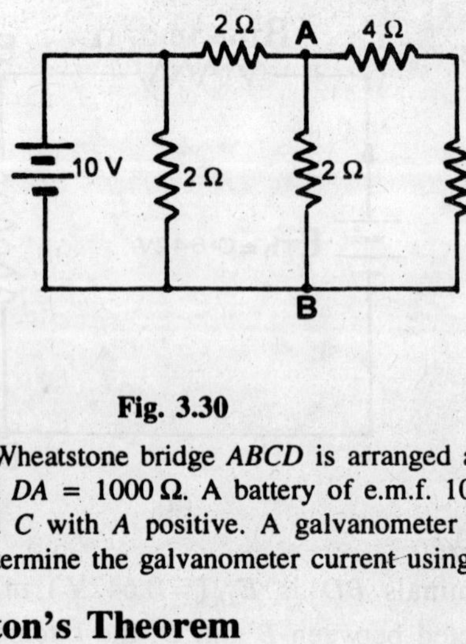

Fig. 3.30

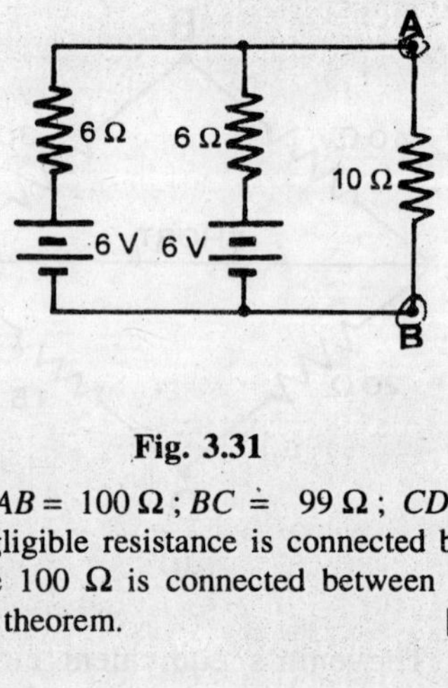

Fig. 3.31

5. A Wheatstone bridge *ABCD* is arranged as follows : *AB* = 100 Ω ; *BC* = 99 Ω ; *CD* = 1000 Ω and *DA* = 1000 Ω. A battery of e.m.f. 10 V and negligible resistance is connected between *A* and *C* with *A* positive. A galvanometer of resistance 100 Ω is connected between *B* and *D*. Determine the galvanometer current using Thevenin's theorem. **[38.6 μA]**

3.7. Norton's Theorem

Fig. 3.32 (*i*) shows a network enclosed in a box with two terminals *A* and *B* brought out. The network in the box may contain any number of resistors and e.m.f. sources connected in any manner. But according to Norton, the entire circuit behind *AB* can be replaced by a current source I_N in parallel with a resistance R_N as shown in Fig. 3.32 (*ii*). The resistance R_N is the same as Thevenin resistance R_{Th}. The value of I_N is determined as mentioned in Norton's theorem. Once Norton's equivalent circuit is determined [See Fig. 3.32 (*ii*)], then current through any load R_L connected across *AB* can be readily obtained.

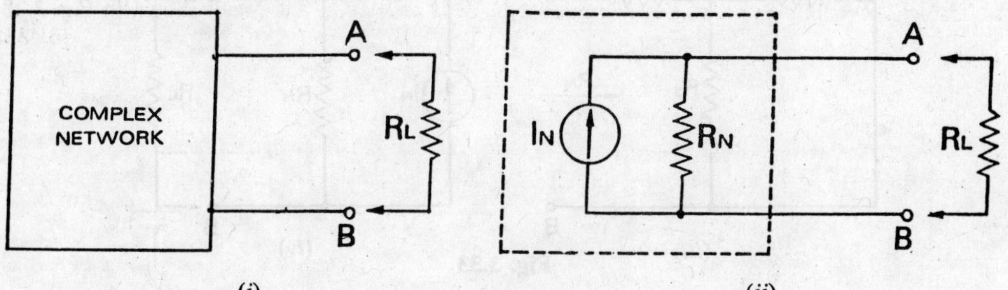

(i) (ii)

Fig. 3.32

Hence Norton's theorem as applied to d.c. circuits may be stated as under :

Any network having two terminals A and B can be replaced by a current source of current output I_N in parallel with a resistance R_N.

(i) *The output I_N of the current source is equal to the current that would flow through AB when A and B are short-circuited.*

(ii) *The resistance R_N is the resistance of the network measured between A and B with load removed and the sources of e.m.f., replaced by their internal resistances.*

Norton's theorem is *converse* of Thevenin's theorem in that Norton equivalent circuit uses a current generator instead of voltage generator and the resistance R_N (which is the same as R_{Th}) in parallel with the generator instead of being in series with it.

Illustration. Fig. 3.33 illustrates the application of Norton's theorem. As far as the circuit behind terminals *AB* is concerned [see Fig. 3.33 (*i*)], it can be replaced by a current source I_N in parallel with a resistance R_N as shown in Fig. 3.33 (*iv*). The output I_N of the current generator is equal to the current that would flow when terminals *A* and *B* are short-circuited as shown in Fig. 3.33 (*ii*). The load on the source when terminals *AB* are short-circuited is given by ;

$$R' = R_1 + \frac{R_2 R_3}{R_2 + R_3} = \frac{R_1R_2 + R_1R_3 + R_2R_3}{R_2 + R_3}$$

Source current, $$I' = \frac{V}{R'} = \frac{V(R_2 + R_3)}{R_1R_2 + R_1R_3 + R_2R_3}$$

Short-circuit current, I_N = Current in R_2 in Fig. 3.33 (*ii*)

$$= I' \times \frac{R_3}{R_2 + R_3} = \frac{VR_3}{R_1R_2 + R_1R_3 + R_2R_3}$$

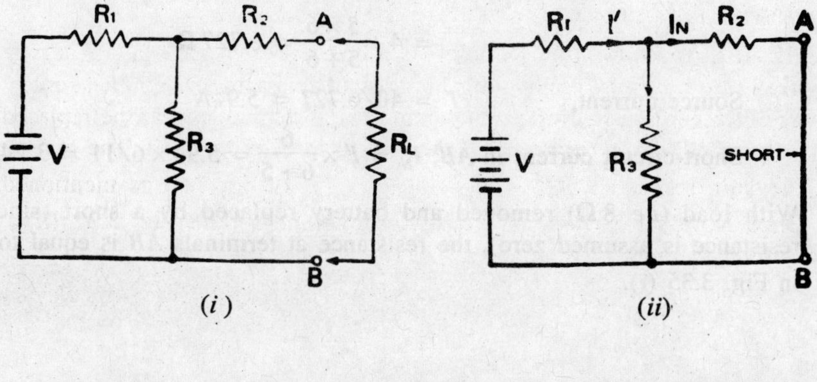

(i) (ii)

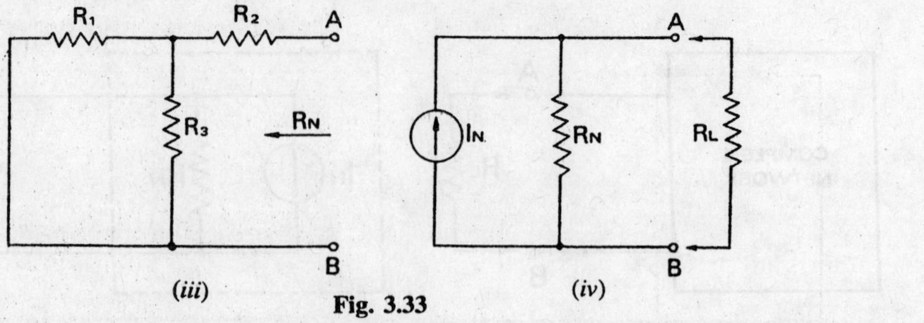

Fig. 3.33

To find R_N, remove the load R_L and replace battery by a short because its internal resistance is assumed zero [see Fig. 3.33 (*iii*)].

$$\therefore \quad R_N = \text{ Resistance at terminals } AB \text{ in Fig 3.33 } (iii)$$

$$= R_2 + \frac{R_1 R_3}{R_1 + R_3}$$

Thus the values of I_N and R_N are known. The Norton equivalent circuit will be as shown in Fig. 3.33 (*iv*).

Example 3.12. *Using Norton's theorem, find the current in* 8 Ω *resistor of the network shown in Fig. 3.34 (i).*

Solution. We shall reduce the network to the left of *AB* in Fig. 3.34 (*i*) to Norton equivalent circuit. For this purpose, we are required to find I_N and R_N.

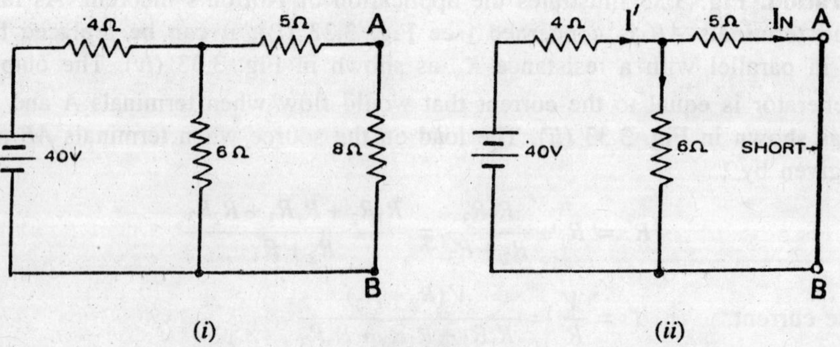

Fig. 3.34

(*i*) With load removed (*i.e.* 8 Ω) and terminals *AB* short-circuited [see Fig. 3.34 (*ii*)], the current that flows through *AB* is equal to I_N. Referring to Fig. 3.34 (*ii*),

$$\text{Load on source} = 4\,\Omega + 5\,\Omega \parallel 6\,\Omega$$

$$= 4 + \frac{5 \times 6}{5+6} = 6.727\,\Omega$$

$$\text{Source current,} \quad I' = 40/6.727 = 5.94\text{A}$$

$$\therefore \quad \text{Short-circuit current in } AB, \; I_N = I' \times \frac{6}{6+5} = 5.94 \times 6/11 = 3.24\text{A}$$

(*ii*) With load (*i.e* 8 Ω) removed and battery replaced by a short (since its internal resistance is assumed zero), the resistance at terminals *AB* is equal to R_N as shown in Fig. 3.35 (*i*).

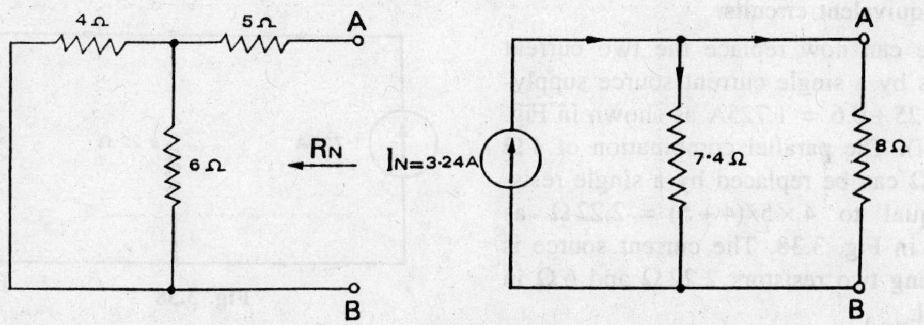

(i) (ii)

Fig. 3.35

$$\therefore \qquad R_N = 5\,\Omega + 4\,\Omega \parallel 6\,\Omega = 5 + \frac{4\times 6}{4+6} = 7.4\,\Omega$$

The Norton equivalent circuit behind terminals *AB* is $I_N\,(=3.24\text{A})$ in parallel with $R_N\,(=7.4\,\Omega)$. When load 8 Ω is connected across terminals *AB*, the circuit becomes as shown in Fig. 3.35 (*ii*). The current source is supplying two resistors 7.4 Ω and 8 Ω in parallel.

$$\therefore \qquad \text{Current in } 8\,\Omega = 3.24 \times \frac{7.4}{8+7.4} = \mathbf{1.55A}$$

Example 3.13. *Using Norton's theorem, find the current in the branch AB containing 6 Ω resistor of the network shown in Fig. 3.36 (i).*

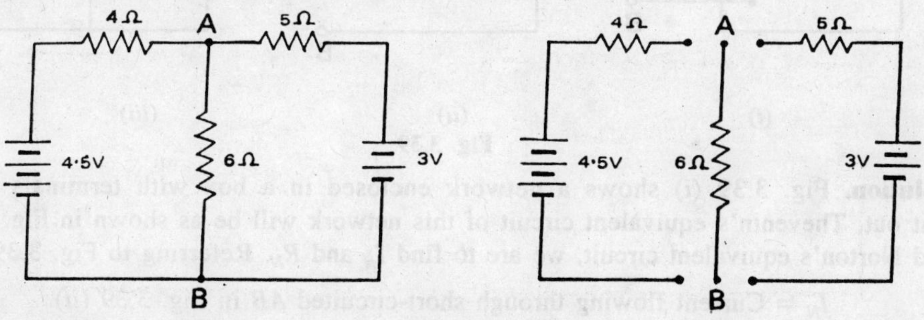

(i) **Fig. 3.36** (ii)

Solution. The load of 6 Ω is supplied by two sources. We shall find Norton's equivalent circuit for each source. Applying Norton's theorem to the left of *AB* in Fig. 3.36 (*ii*),

$$I_N = 4.5/4 = 1.125\text{A}\ ; \qquad R_N = 4\,\Omega$$

Applying Norton's theorem to the right of *AB* in Fig. 3.36 (*ii*),

$$I_N = 3/5 = 0.6\text{A}\ ; \qquad R_N = 5\,\Omega$$

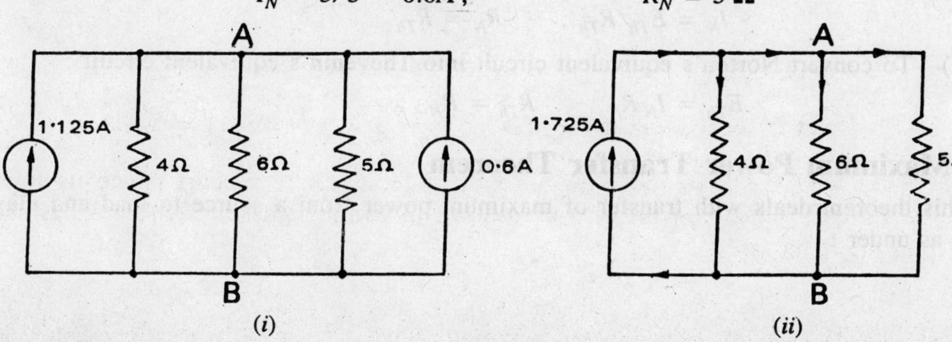

(i) (ii)

Fig. 3.37

The circuit shown in Fig. 3.37 (*i*) has been redrawn, replacing voltage sources by Norton's equivalent circuits.

We can now replace the two current sources by a single current source supplying 1.125 + 0.6 = 1.725A as shown in Fig. 3.37 (*ii*). The parallel combination of 4 Ω and 5 Ω can be replaced by a single resistor equal to $4 \times 5/(4+5) = 2.22\,\Omega$ as shown in Fig. 3.38. The current source is supplying two resistors 2.22 Ω and 6 Ω in parallel.

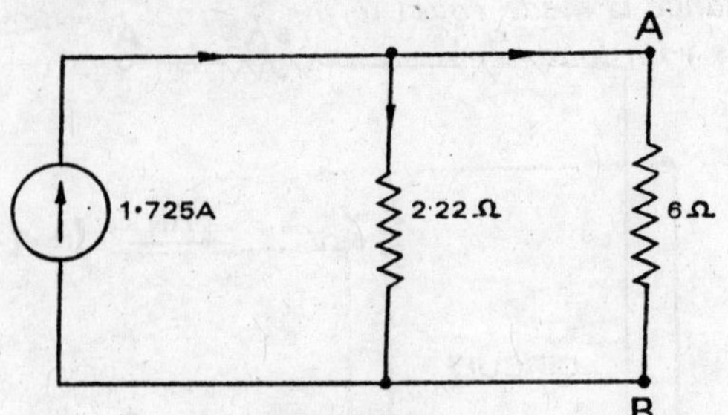

Fig. 3.38

$$\therefore \qquad \text{Current in } 6\,\Omega = 1.725 \times \frac{2.22}{6+2.22} = \mathbf{0.466A}$$

Example 3.14. *Show that when Thevenin equivalent circuit of a network is converted into Norton equivalent circuit,* $I_N = E_{Th}/R_{Th}$ *and* $R_N = R_{Th}$.

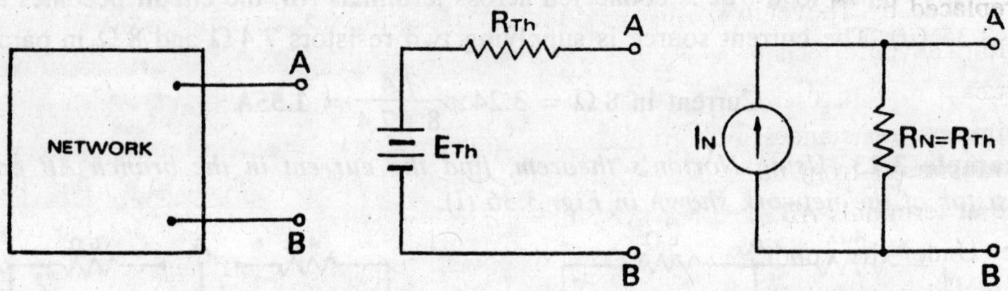

(*i*) (*ii*) (*iii*)

Fig. 3.39

Solution. Fig. 3.39 (*i*) shows a network enclosed in a box with terminals *A* and *B* brought out. Thevenin's equivalent circuit of this network will be as shown in Fig. 3.39 (*ii*). To find Norton's equivalent circuit, we are to find I_N and R_N. Referring to Fig. 3.39 (*iii*),

I_N = Current flowing through short-circuited *AB* in Fig. 3.39 (*ii*).

$= E_{Th}/R_{Th}$

R_N = Resistance at terminals *AB* in Fig. 3.39 (*ii*).

$= R_{Th}$

Hence we arrive at the following important conclusions :

(*i*) To convert Thevenin's equivalent circuit into Norton's equivalent circuit :

$$I_N = E_{Th}/R_{Th} \,; \qquad R_N = R_{Th}$$

(*ii*) To convert Norton's equivalent circuit into Thevenin's equivalent circuit :

$$E_{Th} = I_N R_N \,; \qquad R_{Th} = R_N$$

3.8. Maximum Power Transfer Theorem

This theorem deals with transfer of maximum power from a source to load and may be stated as under :

In d.c. circuits, maximum power is transferred from a source to load when the load resistance is made equal to the internal resistance of the source as viewed from the load terminals with load removed and all e.m.f. sources replaced by their internal resistances.

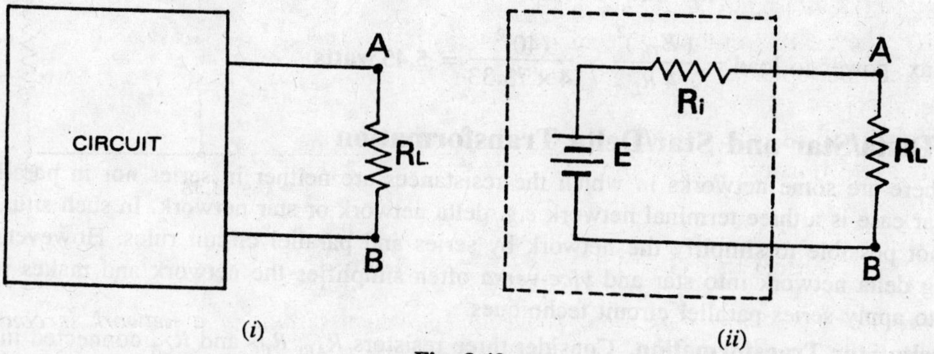

Fig. 3.40

Fig. 3.40 (*i*) shows a circuit supplying power to a load R_L. The circuit enclosed in the box can be replaced by Thevenin's equivalent circuit consisting of Thevenin voltage $E\,(=E_{Th})$ in series with Thevenin resistance $R_i\,(=R_{Th})$ as shown in Fig. 3.40 (*ii*). Clearly resistance R_i is the resistance measured between terminals *AB* with R_L removed and e.m.f. sources replaced by their internal resistances. According to maximum power transfer theorem, maximum power will be transferred from the circuit to the load when R_L is made equal to R_i, the Thevenin resistance at terminals *AB*. The proof of this theorem is left as an exercise for the reader.

Note. Under the conditions of maximum power transfer, the efficiency is only 50% as one-half of the total power generated is dissipated in the internal resistance R_i of the source.

Example 3.15. *Calculate the value of load resistance R_L to which maximum power may be transferred from the circuit shown in Fig. 3.41 (i). Determine also the value of maximum power.*

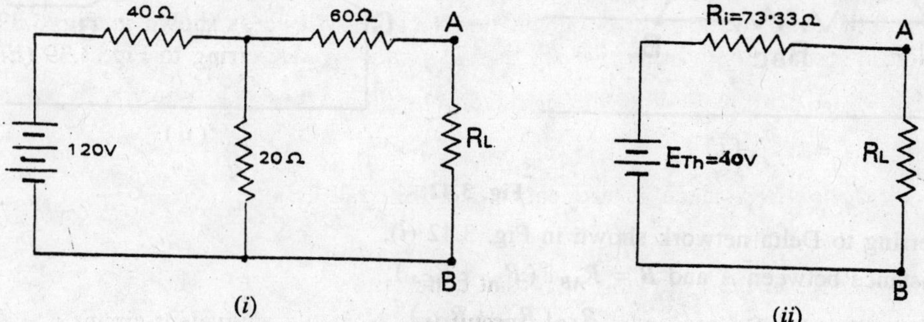

Fig. 3.41

Solution. We shall first find Thevenin's equivalent circuit to the left of terminals *AB*.

E_{Th} = Voltage at *AB* with R_L removed

$$= \frac{120}{40+20} \times 20 = 40 \text{ volts}$$

R_i = Resistance at *AB* with R_L removed and 120 V replaced by a short

$$= 60\,\Omega + 40\,\Omega \parallel 20\,\Omega$$

$$= 60 + 40 \times 20/60 = 73.33\Omega$$

The Thevenin's equivalent circuit to the left of AB in Fig. 3.41 (i) will be E_{Th} ($=40$V) in series with R_i ($=73.33\,\Omega$). When R_L is connected between A and B, the circuit becomes as shown in Fig, 3.41 (ii). It is clear that maximum power will be transferred when :

$$R_L = R_i = \mathbf{73.33\,\Omega}$$

$$\text{Max. power to load} = \frac{(E_{Th})^2}{4\,R_L} = \frac{(40)^2}{4 \times 73.33} = \mathbf{5.45\ watts}$$

3.9. Delta/Star and Star/Delta Transformation

There are some networks in which the resistances are neither in series nor in parallel. A familiar case is a three terminal network *e.g.* delta network or star network. In such situations, it is not possible to simplify the network by series and parallel circuit rules. However, converting delta network into star and *vice-versa* often simplifies the network and makes it possible to apply series-parallel circuit techniques.

Delta/Star Transformation. Consider three resistors R_{AB}, R_{BC} and R_{CA} connected in delta to three terminals A, B and C as shown in Fig 3.42 (i). It is desired to replace these three delta-connected resistors by three resistors R_A, R_B and R_C connected in star [See Fig. 3.42 (ii)] so that the two networks are electrically equivalent. The two arrangements will be electrically equivalent if resistance between any two terminals of one network is equal to the resistance between the corresponding terminals of the other network.

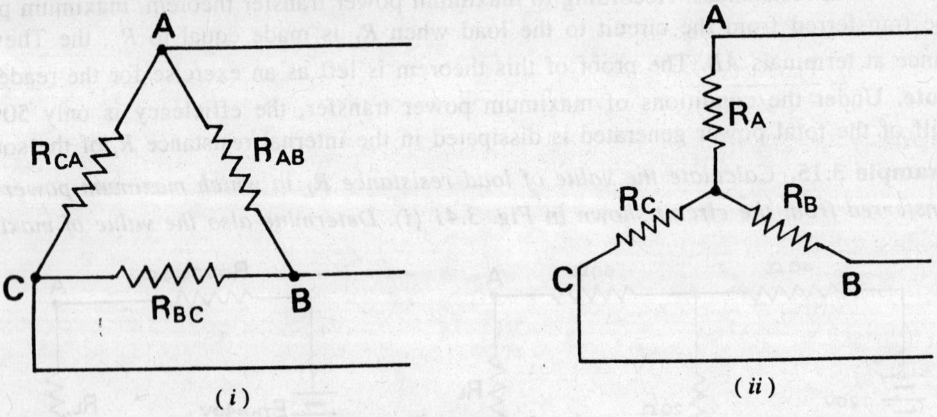

Fig. 3.42

Referring to Delta network shown in Fig. 3.42 (i),

Resistance between A and $B = R_{AB} \parallel (R_{BC} + R_{CA})$

$$= \frac{R_{AB}(R_{BC}+R_{CA})}{R_{AB}+R_{BC}+R_{CA}} \qquad ...(i)$$

Referring to star network shown in Fig. 3.42 (ii),

Resistance between A and $B = R_A + R_B$ $\qquad ...(ii)$

Since the two arrangements are electrically equivalent,

$$\therefore \quad R_A + R_B = \frac{R_{AB}(R_{BC}+R_{CA})}{R_{AB}+R_{BC}+R_{CA}} \qquad ...(iii)$$

Similarly it can be shown that between terminals B and C and terminals C and A,

$$R_B + R_C = \frac{R_{BC}(R_{CA}+R_{AB})}{R_{AB}+R_{BC}+R_{CA}} \quad \text{...(iv)}$$

and

$$R_C + R_A = \frac{R_{CA}(R_{AB}+R_{BC})}{R_{AB}+R_{BC}+R_{CA}} \quad \text{... (v)}$$

Subtracting eq (*iv*) from eq (*iii*) and adding the result to eq (*v*),

$$R_A = \frac{R_{AB}R_{CA}}{R_{AB}+R_{BC}+R_{CA}} \quad \text{...(vi)}$$

Similarly,

$$R_B = \frac{R_{BC}R_{AB}}{R_{AB}+R_{BC}+R_{CA}} \quad \text{... (vii)}$$

and

$$R_C = \frac{R_{CA}R_{BC}}{R_{AB}+R_{BC}+R_{CA}} \quad \text{... (viii)}$$

How to remember ? There is an easy way to remember it. Referring to Fig. 3.43, star-connected resistances R_A, R_B and R_C are electrically equivalent to delta connected resistances R_{AB}, R_{BC} and R_{CA}. We have seen above that :

$$R_A = \frac{R_{AB}R_{CA}}{R_{AB}+R_{BC}+R_{CA}}$$

i.e. Any arm of star connection

$$= \frac{\text{Product of two adjacentarms of }\Delta}{\text{Sum of arms of }\Delta}$$

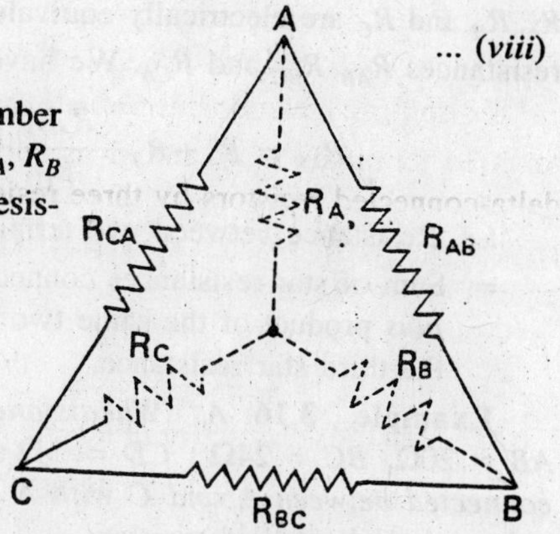

Fig. 3.43

Star/delta Transformation. Now let us consider how to replace star-connected network of Fig. 3.42 (*ii*) by the equivalent delta connected network shown in Fig. 3.42 (*i*). This can be easily done by considering eqs. (*vi*), (*vii*) and (*viii*) above.

Dividing eq. (*vi*) by eq. (*vii*), we have,

$$R_A/R_B = R_{CA}/R_{BC} \qquad \therefore \qquad R_{CA} = \frac{R_A R_{BC}}{R_B}$$

Dividing eq. (*vi*) by eq. (*viii*), we have,

$$R_A/R_C = R_{AB}/R_{BC} \qquad \therefore \qquad R_{AB} = \frac{R_A R_{BC}}{R_C}$$

Substituting the values of R_{CA} and R_{AB} in eq. (*vi*) we have,

$$R_A = \frac{\left(\frac{R_A}{R_C}R_{BC}\right)\left(\frac{R_A}{R_B}R_{BC}\right)}{\frac{R_A}{R_C}R_{BC}+R_{BC}+\frac{R_A}{R_B}R_{BC}}$$

$$= \frac{R_A^2/R_C R_B}{(R_A/R_C)+1+(R_A/R_B)} \times R_{BC}$$

$$= \frac{R_A^2}{R_A R_B + R_C R_B + R_A R_C} \times R_{BC}$$

or $$R_{BC} = \frac{R_A R_B + R_C R_B + R_A R_C}{R_A}$$

$\therefore$ $$R_{BC} = R_B + R_C + \frac{R_B R_C}{R_A}$$

Similarly, $$R_{CA} = R_C + R_A + \frac{R_C R_A}{R_B}$$

and $$R_{AB} = R_A + R_B + \frac{R_A R_B}{R_C}$$

How to remember ? There is an easy way to remember it. Referring to Fig. 3.44, star-connected resistances R_A, R_B and R_C are electrically equivalent to delta connected resistances R_{AB}, R_{BC} and R_{CA}. We have seen above that :

$$R_{AB} = R_A + R_B + \frac{R_A R_B}{R_C}$$

i.e. Resistance between two terminals of delta

= Sum of star resistances connected to those terminals *plus* product of the same two resistances divided by the third star resistance.

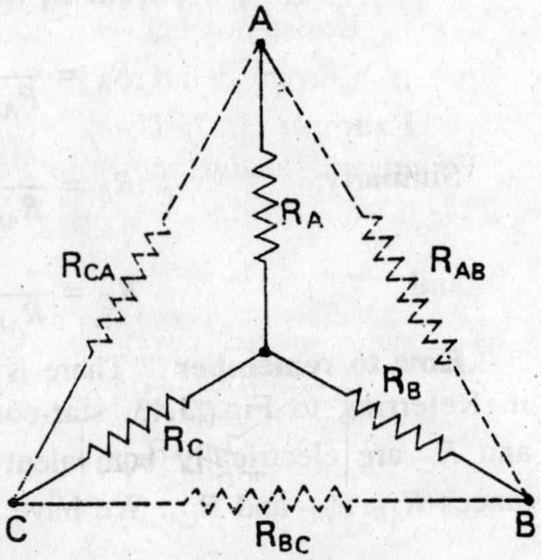

Fig. 3.44

Example 3.16. *A Wheatstone bridge ABCD has the following details : AB* = 20Ω; *BC* = 24Ω ; *CD* = 5 Ω ; *DA* = 50 Ω *and BD* = 30Ω. *A d.c. source of 220V is connected between A and C with A positive. Determine the current delivered by the source using star/delta transformation.*

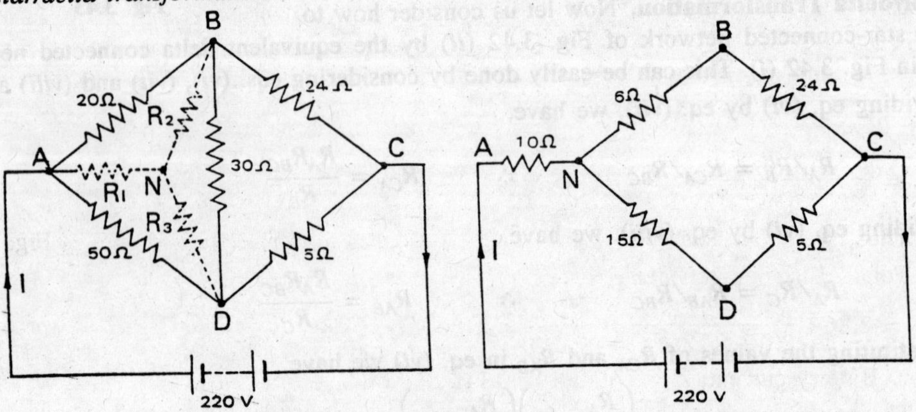

Fig. 3.45

Solution. The network *ABDA* consisting of resistances 20 Ω, 30 Ω and 50 Ω forms a delta with corners at *A, B* and *D*. These delta connected resistances can be replaced by equivalent star-connected resistances R_1, R_2 and R_3 as shown in Fig. 3.45 (*i*).

$$R_1 = \frac{R_{AB}\,R_{DA}}{R_{AB} + R_{BD} + R_{DA}} = \frac{20 \times 50}{20 + 30 + 50} = 10\,\Omega$$

$$R_2 = \frac{R_{AB}\,R_{BD}}{R_{AB} + R_{BD} + R_{DA}} = \frac{20 \times 30}{20 + 30 + 50} = 6\,\Omega$$

$$R_3 = \frac{R_{DA}\,R_{BD}}{R_{AB}+R_{BD}+R_{DA}} = \frac{50\times 30}{20+30+50} = 15\,\Omega$$

Thus the network shown in Fig. 3.45 (*i*) reduces to the network shown in Fig. 3.45 (*ii*). As seen from Fig. 3.45 (*ii*), there are two parallel paths between *N* and *C*.

$\therefore$ Resistance between *N* and $C = (6+24)\,\Omega \parallel (15+5)\,\Omega = \dfrac{30\times 20}{30+20} = 12\,\Omega$

Resistance between *A* and $C = 10+12 = 22\,\Omega$

$\therefore$ Current delivered by source, $I = 220/22 =$ **10A**

Example 3.17. *Using delta/star transformation, find the galvanometer current in the Wheatstone bridge shown in Fig. 3.46 (i).*

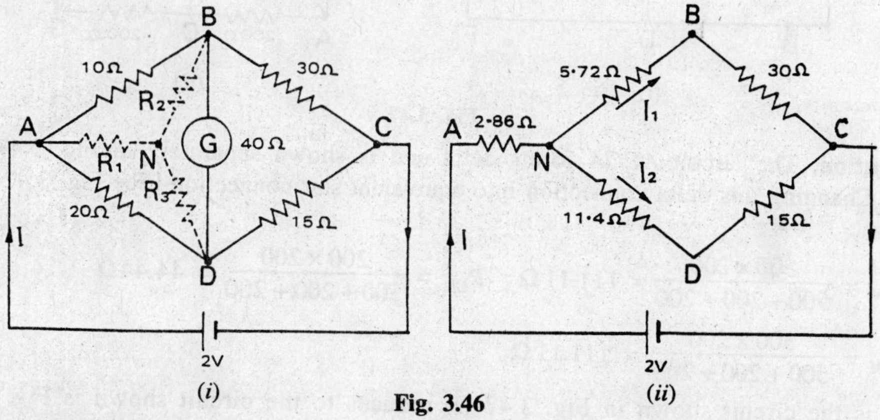

Fig. 3.46

Solution. The network *ABDA* in Fig. 3.46 (*i*) forms a delta. These delta connected resistances can be replaced by equivalent star-connected resistances R_1, R_2 and R_3 as shown in Fig. 3.46 (*i*).

$$R_1 = \frac{R_{AB}\,R_{AD}}{R_{AB}+R_{BD}+R_{DA}} = \frac{10\times 20}{10+40+20} = 2.86\,\Omega\ ;\ R_2 = \frac{R_{AB}\,R_{BD}}{R_{AB}+R_{BD}+R_{DA}} = \frac{10\times 40}{10+40+20} = 5.72\,\Omega$$

$$R_3 = \frac{R_{DA}\,R_{BD}}{R_{AB}+R_{BD}+R_{DA}} = \frac{20\times 40}{10+40+20} = 11.4\,\Omega$$

Thus, the network shown in Fig. 3.46 (*i*) reduces to the network shown in Fig. 3.46 (*ii*).

Total resistance between *A* and *C*

$$= 2.86 + \frac{(30+5.72)\,(15+11.4)}{(30+5.72)+(15+11.4)} = 18.04\,\Omega$$

$\therefore$ Battery current, $I = 2/18.04 = 0.11$ A

The battery current at *N* divides into two parallel paths [See Fig. 3.46(*ii*)].

$\therefore$ Current in branch *NBC*, $\quad I_1 = 0.11 \times \dfrac{26.4}{26.4+35.72} = 0.047\text{A}$

Current in branch *NDC*, $\quad I_2 = 0.11 \times \dfrac{35.72}{26.4+35.72} = 0.063\text{A}$

Potential of *B w.r.t. C* $= 30 \times 0.047 = 1.41$V

Potential of *D w.r.t. C* $= 15 \times 0.063 = 0.945$V

Clearly, point *B* is at a higher potential than point *D* by $1.41 - 0.945 = 0.465$V.

$\therefore$ Galvanometer current $= \dfrac{\text{P.D. between } B \text{ and } D}{\text{Galvanometer resistance}} = 0.465/40 = 11.6 \times 10^{-3}$ A

$=$ **11.6 mA** *from B to D*

Example 3.18. *Determine the load current in branch EF in the circuit shown in Fig. 3.47 (i).*

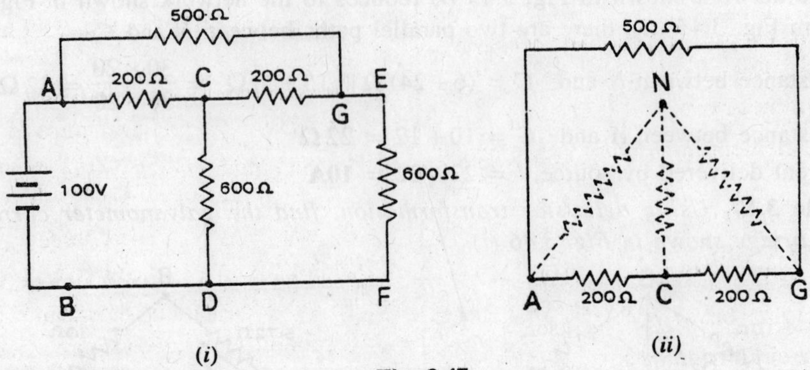

Fig. 3.47

Solution. The circuit *ACGA* forms delta and is shown separately in Fig. 3.47 (*ii*) for clarity. Changing this delta connection into equivalent star connection [See Fig. 3.47 (*ii*)], we have,

$$R_{AN} = \frac{500 \times 200}{500+200+200} = 111.11\,\Omega\,; \quad R_{CN} = \frac{200 \times 200}{500+200+200} = 44.44\Omega$$

$$R_{GN} = \frac{500 \times 200}{500+200+200} = 111.11\,\Omega$$

Thus the circuit shown in Fig. 3.47 (*i*) reduces to the circuit shown is Fig. 3.48 (*i*). The branch $NEF (= 111.11 + 600 = 711.11\Omega)$ is in parallel with branch *NCD* $(= 44.44 + 600 = 644.44\,\Omega)$ and the equivalent resistance of this parallel combination is

$$= \frac{711.11 \times 644.44}{711.11 + 644.44} = 338\,\Omega$$

The circuit shown in Fig. 3.48 (*i*) reduces to the circuit shown in Fig. 3.48 (*ii*).

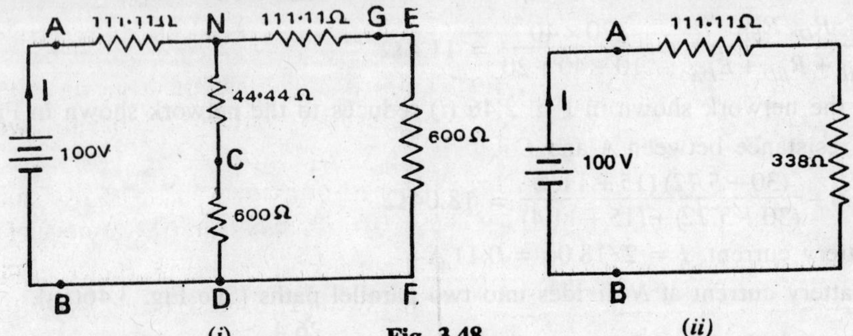

Fig. 3.48

$$\therefore \text{ Battery current, } I = \frac{100}{338 + 111.11} = 0.222\text{A}$$

This battery current divides into two parallel paths [See Fig. 3.48 (*i*)] *viz.* branch *NEF* and branch *NCD*.

∴ Current in branch *NEF i.e.* in branch *EF*

$$= 0.222 \times \frac{644.44}{711.11 + 644.44} = \mathbf{0.1055A}$$

OBJECTIVE QUESTIONS

1. An active element in a circuit is one which ....
   (*i*) receives energy (*ii*) supplies energy
   (*iii*) both receives and supplies energy
   (*iv*) none of the above.
2. A passive element in a circuit is one which ....
   (i) supplies energy
   (*ii*) receives energy
   (*iii*) both supplies and receives energy
   (*iv*) none of the above
3. An electric circuit contains.............
   (*i*) active elements only
   (*ii*) passive elements only
   (*iii*) both active and passive elements
   (*iv*) none of the above
4. A linear circuit is one whose parameters (e.g. resistances etc.).......................
   (*i*) change with change in current
   (*ii*) change with change in voltage
   (*iii*) do not change with voltage and current
   (*iv*) none of the above
5. In the circuit shown in Fig. 3.49 the number of nodes is ....
   (*i*) one (*ii*) two
   (*iii*) three (*iv*) four

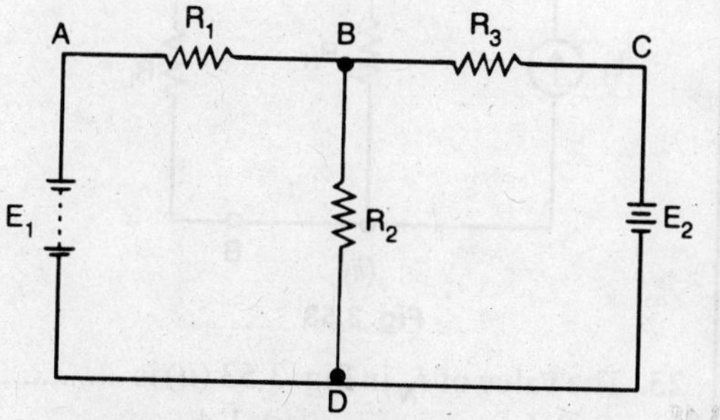

Fig. 3.49

6. In the circuit shown in Fig. 3.49 there are ...... junctions.
   (*i*) three (*ii*) four
   (*iii*) two (*iv*) none of the above
7. The circuit shown in Fig 3.49 has .... branches.
   (*i*) two (*ii*) four
   (*iii*) three (*iv*) none of these
8. The circuit shown in Fig. 3.49 has .... loops.
   (*i*) two (*ii*) four
   (*iii*) three (*iv*) none of the above
9. In the circuit shown in Fig. 3.49, there are .... meshes.
   (*i*) two (*ii*) three
   (*iii*) four (*iv*) five
10. To solve the circuit shown in Fig. 3.50 by Kirchhoff's laws, we require ....
    (*i*) one equation (*ii*) two equations
    (*iii*) three equations (*iv*) none of the above
11. To solve the circuit shown in Fig. 3.50 by nodal analysis, we require ....
    (*i*) one equation(*ii*) two equations
    (*iii*) three equations (*iv*) none of the above

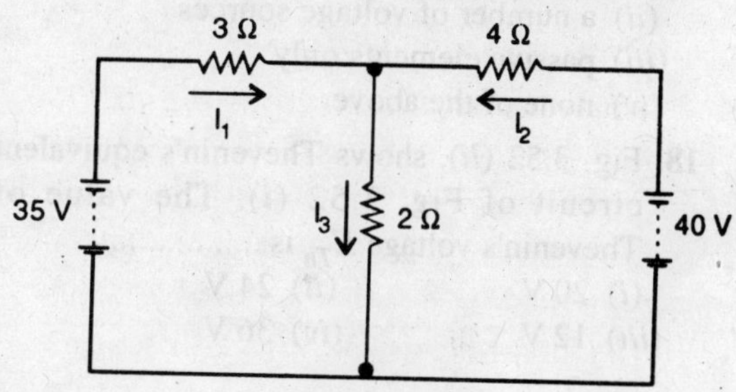

Fig. 3.50

12. To solve the circuit shown in Fig. 3.50 by superposition theorem, we require ....
    (*i*) one circuit (*ii*) two circuits
    (*ii*) three circuits (*iv*) none of the above
13. To solve the circuit shown in Fig. 3.50 by Maxwell's mesh current method, we require ....
    (*i*) one equation (*ii*) three equations
    (*iii*) two equations (*iv*) none of the above
14. In the circuit shown in Fig. 3.51, the voltage at node B *w.r.t. D* is calculated to be 15 V. The current in 3 Ω resistor will be ....
    (*i*) 2 *A* (*ii*) 5 *A*
    (*iii*) 2.5 *A* (*iv*) none of the above
15. The current in 2 Ω horizontal resistor in Fig.

3.51 is ....

(*i*) 10 *A* (*ii*) 5 *A*

(*iii*) 2 *A* (*iv*) 2.5 *A*

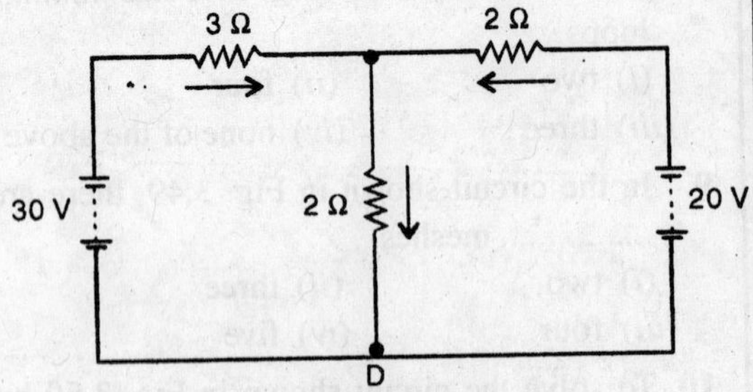

Fig. 3.51

16. In order to solve the circuit shown in Fig. 3.51 by nodal analysis, we require ....

(*i*) one equation (*ii*) two equations

(*iii*) three equations (*iv*) none of the above

17. The superposition theorem is used when the circuit contains ....

(*i*) a single voltage source

(*ii*) a number of voltage sources

(*iii*) passive elements only

(*iv*) none of the above

18. Fig. 3.52 (*ii*), shows Thevenin's equivalent circuit of Fig. 3.52 (*i*). The value of Thevenin's voltage E_{Th} is ....

(*i*) 20 V (*ii*) 24 V

(*iii*) 12 V (*iv*) 36 V

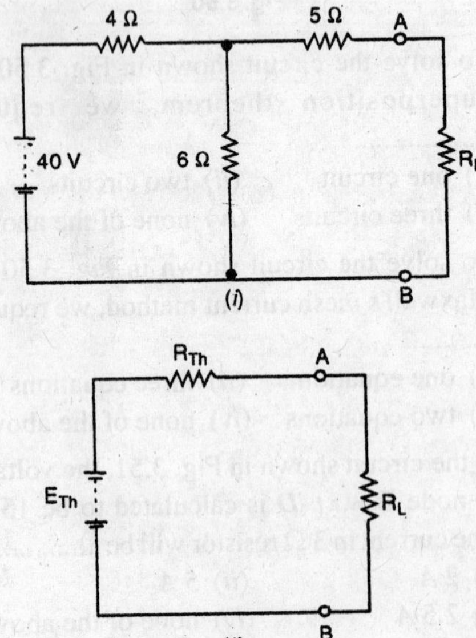

Fig. 3.52

19. The value of R_{TH} in Fig. 3.52 (*ii*) is ....

(*i*) 15 Ω (*ii*) 3.5 Ω

(*iii*) 6.4 Ω (*iv*) 7.4 Ω

20. The open-circuited voltage at terminals *AB* in Fig. 3.52 (i) is ....

(*i*) 12 V (*ii*) 20 V

(*iii*) 24 V (*iv*) 40 V

21. For transfer of maximum power in the circuit shown in fig. 3.52 (i), the value of R_L should be ....

(*i*) 3.5 Ω (*ii*) 6.4 Ω

(*iii*) 7.4 Ω (*iv*) 15 Ω

22. Fig. 3.53 (*ii*) shows Norton's equivalent circuit of Fig. 3.53 (*i*). The value of R_N is ....

(*i*) 5 Ω (*ii*) 4.5 Ω

(*iii*) 10.5 Ω (*iv*) none of the above

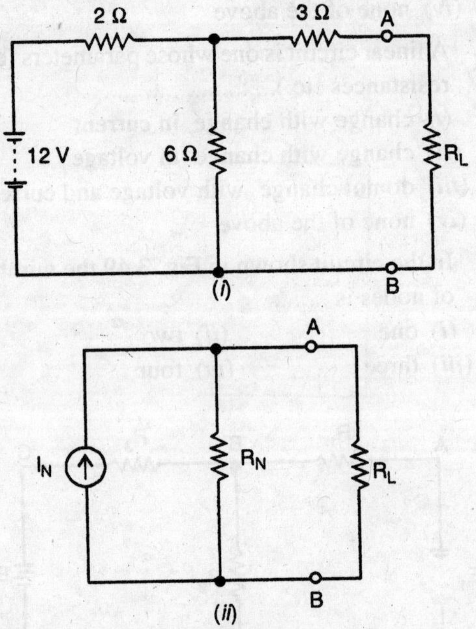

Fig. 3.53

23. The value of I_N in Fig. 3.53 (*ii*) is ....

(*i*) 3 *A* (*ii*) 1 *A*

(*iii*) 2 *A* (*iv*) none of the above

24. Thevenin's theorem is .... form of an equivalent circuit.

(*i*) voltage (*ii*) current

(*iii*) both voltage and current

(*iv*) none of the above

25. Norton's theorem is .... Thevenin's theorem.

(i) the same as (ii) converse of
(iii) none of the above (iv) cannot say

26. In the analysis of a vacuum tube circuit, we generally use .... theorem.
(i) superposition (ii) Norton's
(iii) Thevenin's (iv) reciprocity

27. Norton's theorem is .... form of an equivalent circuit.
(i) both voltage and current
(ii) Current
(iii) voltage (iv) none of the above

28. In the analysis of a transistor circuit, we usually use .... theorem.
(i) Norton's (ii) Thevenin's
(iii) reciprocity (iv) superposition

29. Fig. 3.54 (i) shows Norton's equivalent circuit of a network whereas Fig. 3.54 (ii) shows its Thevenin's equivalent circuit. The value of E_{Th} is ....
(i) 1.5 V (ii) 0.866 V
(iii) 3 V (iv) 6 V

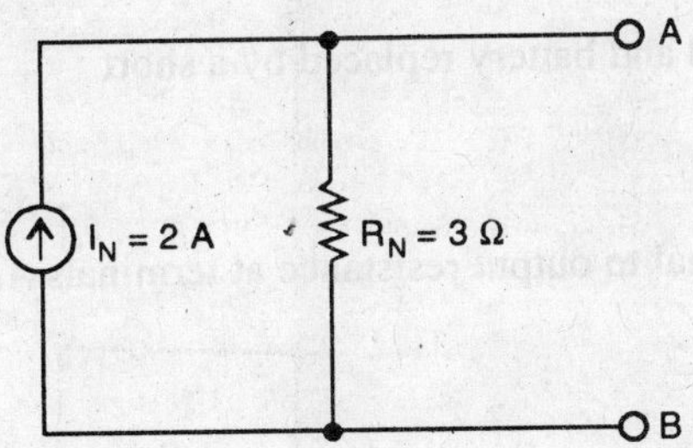

(i)

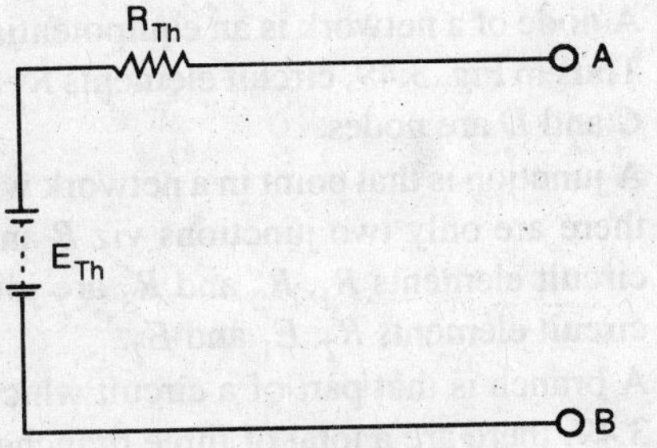

(ii)

Fig. 3.54

30. The value of R_{Th} in Fig. 3.54 (ii) is ....
(i) 3 Ω (ii) 2 Ω
(iii) 1.5 Ω (iv) 6 Ω

31. If in Fig. 3.54 (i), the value of I_N is 3 A, then value of E_{TH} in Fig. 3.54 (ii) will be ....
(i) 1 V (ii) 9 V
(iii) 5 V (iv) none of the above

32. For transfer of maximum power, the relation between load resistance R_L and internal resistance R_i of the voltage source is ....
(i) $R_L = 2 R_i$ (ii) $R_L = 0.5 R_i$
(iii) $R_L = 1.5 R_i$ (iv) $R_L = R_i$

33. Under the conditions of maximum power transfer, the efficiency is ....
(i) 75 % (ii) 100 %
(iii) 50 % (iv) 25 %

34. The open-circuited voltage at the terminals of load R_L is 30 V. Under the conditions maximum power transfer, the load voltage will be ....
(i) 30 V (ii) 10 V
(iii) 5 V (iv) 15 V

35. The maximum power transfer theorem is used in ....
(i) electronic circuits
(ii) power system
(iii) home lighting circuits
(iv) none of the above

ANSWERS TO OBJECTIVE QUESTIONS

1. (ii)	**2.** (ii)	**3.** (iii)	**4.** (iii)	**5.** (iv)	**6.** (iii)	**7.** (iii)	**8.** (iii)
9. (i)	**10.** (ii)	**11.** (i)	**12.** (ii)	**13.** (iii)	**14.** (ii)	**15.** (iv)	**16.** (i)
17. (ii)	**18.** (ii)	**19.** (iv)	**20.** (iii)	**21.** (iii)	**22.** (ii)	**23.** (iii)	**24.** (i)
25. (ii)	**26.** (iii)	**27.** (ii)	**28.** (i)	**29.** (iv)	**30.** (i)	**31.** (ii)	**32.** (iv)
33. (iii)	**34.** (iv)	**35.** (i)					

HINTS TO SELECTED OBJECTIVE QUESTIONS

5. A node of a network is an equipotential surface at which *two or more* circuit elements are joined. Thus in Fig. 3.49, circuit elements R_1 and E_1 are joined at A and hence A is the node. Similarly B, C and D are nodes.

6. A junction is that point in a network where three or more circuit elements are joined. In Fig. 3.49, there are only two junctions viz B and D. That B is a junction is clear from the fact that three circuit elements R_1, R_2 and R_3 are joined at it. Similarly D is a junction because it joins three circuit elements R_2, E_1 and E_2.

7. A branch is that part of a circuit which lies between two junction points. Thus referring to Fig. 3.49, there are a total of three branches viz *BAD*, *BCD* and *BD*.

8. A loop is any closed path of a network. Thus in Fig. 3.49, there are only three loops viz. *ABDA*, *BCDB* and *ABCDA*.

9. A mesh is the most elementary form of a loop and cannot be further divided into other loops. In Fig. 3.49, both loops *ABDA* and *BCDB* qualify as meshes because they cannot be further divided into other loops. However, the loop *ABCDA* cannot be called a mesh because it encloses two loops *ABDA* and *BCDB*.

10. Since there are two unknown quantities (*viz* I_1 and I_2), we require two equations.

12. Since there are two sources of voltage, two circuits are required for analysis by superposition theorem.

14. Current in 3 Ω resistor $= \dfrac{30-15}{3} = 5\ A$

18. E_{Th} = Open-circuited voltage at terminals AB

$= \text{Current in } 6\ \Omega \times 6\ \Omega = 4\ A \times 6\ \Omega = 24\ \text{V}$

19. R_{Th} = Resistance at terminals AB with R_L removed and battery replaced by a short.

$$= 4\ \Omega \parallel 6\ \Omega + 5\ \Omega = \frac{4\times 6}{4+6} + 5 = 7.4\ \Omega$$

21. For transfer of maximum power, R_L should be equal to output resistance at terminals AB. Now output resistance at terminals AB

$= R_{Th} = 7.4\ \Omega.$

22. $R_N = R_{Th} = 2\ \Omega \parallel 6\ \Omega + 3\ \Omega = \dfrac{2\times 6}{2+6} + 3 = 4.5\ \Omega$

23. I_n = Current through terminals AB with load removed and terminals AB short-circuited.

$$\text{Load on source} = 2\ \Omega + 3\ \Omega \parallel 6\ \Omega = 2\ \Omega + \frac{3\times 6}{3+6} = 4\ \Omega$$

Source current, $I = 12/4 = 3$ A; ∴ $I_N = 3\times \dfrac{6}{3+6} = 2$ A

29. $E_{TH} = I_N R_N = 2\ \text{A} \times 3\ \Omega = 6\ V.$

30. $R_{Th} = R_N = 3\ \Omega$

33. Suppose a source of voltage V and internal resistance R_i is supplying a current of I to load R_L under maximum power transfer conditions.

$$\text{Efficiency} = \frac{I^2 R_L}{I^2 (R_L + R_i)} = \frac{I^2 R_L}{I^2 \times 2R_L}$$

$$[\because R_L = R_i] = \frac{1}{2} = 50\ \%$$

4

Units—Work, Power and Energy

Introduction

Engineering is an applied science dealing with a very large number of *physical quantities like distance, time, speed, temperature, force, voltage, resistance *etc.* Although it is possible to assign a standard unit for each quantity, it is rarely necessary to do so because many of the quantities are functionally related through experiment, derivation or definition. In the study of mechanics, for example, the units of only three quantities (*viz. mass, length* and *time*) need to be selected. All other quantities (*e.g.* area, volume, velocity, force *etc.*) can be expressed in terms of the units of these three quantities by means of experimental, derived and defined **relationship between the physical quantities. The units selected for these three quantities are called *fundamental units*. In order to cover the entire subject of engineering, three more fundamental quantities have been selected *viz electric current, temperature* and *luminous intensity*. Thus there are in all six fundamental quantities (*viz.* mass, length, time, current, temperature and luminous intensity) which need to be assigned proper and standard units. The units of all other physical quantities can be derived from the units of these six fundamental quantities. In this chapter, we shall focus our attention on the mechanical, electrical and thermal units of work, power and energy.

4.1. International System of Units

Although several systems were evolved to assign units to the above mentioned six fundamental quantities, only international system of units (abbreviated as SI) has been universally accepted. The units assigned to these six fundamental quantities in this system are given below in the tabular form :

Quantity	Symbol	Unit name	Unit symbol
Length	l, L	metre	m
Mass	m	kilogram	kg
Time	t	second	s
Electric Current	I	ampere	A
Temperature	T	degree kelvin	K
Luminous Intensity	I	candela	Cd

The units of all other physical quantities in science and engineering (*i.e.* other than six fundamental quantities above) can be derived from the above six units. Thus the unit of velocity (= 1m/s) results when unit of length (= 1m) is divided by the unit of time (= 1s).

---

* A physical quantity is one which can be measured.

** For example, by definition, speed is the distance travelled per second. Therefore, speed is related to distance (*i.e* length) and time.

Similarly, the unit of force (= 1 Newton) results when the unit of mass (= 1 *kg*) is multiplied by the unit of acceleration (= $1 m/s^2$).

4.2. Important Physical Quantities

It is profitable to give a brief description of the following physical quantities much used in science and engineering :

(*i*) **Mass.** It is the quantity of matter possessed by a body. The SI unit of mass is kilogram (kg). The mass of a body is a constant quantity and is independent of place and position of the body. Thus the mass of a body is the same whether it is on Earth's surface, the Moon's surface, on the top of a mountain or down a deep well.

(*ii*) **Force.** It is the product of mass (kg) and acceleration (m/s^2). The unit of force is newton (N) ; being the force required to accelerate a mass of 1kg through an acceleration of $1 m/s^2$.

$$\therefore \quad F = ma \text{ newtons}$$

where m = mass of the body in kg

a = acceleration in m/s^2

(*iii*) **Weight.** The force with which a body is attracted towards the centre of Earth is called the weight of the body. Now, force = mass × acceleration. If m is the mass of a body in kg and g is the acceleration due to gravity in m/s^2, then

$$\text{Weight, } W = mg \text{ newtons}$$

As the value of g* varies from place to place on earth's surface, therefore, the weight of the body varies accordingly. However, for practical purposes, we take $g = 9.81 \text{ m/s}^2$ so that weight of body = $9.81\,m$ newtons. Thus if a mass of 1 kg rests on a table, the downward force on the table *i.e.* weight of the body is $W = 9.81 \times 1 = 9.81$ newtons.

The following points may be noted carefully :

(*a*) The mass of a body is a constant quantity whereas its weight depends upon the place or position of the body. However, it is reasonably accurate to express weight $W = 9.81\,m$ newtons where m is the mass of the body in kg.

(*b*) Sometimes weight is given in kg. wt. units. One kg. wt means weight of mass of 1 kg *i.e.* $9.81 \times 1 = 9.81$ newtons. Therefore, 1 kg. wt. = 9.81 newtons

Thus when we say that a body has a weight of 100 kg, it means that it has a mass of 100 kg and that it exerts a downward force of $100 \times 9.81 = 981$ newtons

4.3. Unit of Work or Energy

Work is said to be done on a body when a force acts on it and the body moves through some distance. This work done is stored in the body in the form of energy. Therefore, work and energy are measured in the same units. The SI unit of work or energy is *joule* or *newton-metre* and is defined as under :

The work done on a body is **one joule** *if a force of one newton moves the body through 1*m *in the direction of force.*

---

* The value of g is about 9.81 m/s^2 at sea-level whereas at equator, it is about 9.78 m/s^2 and at each pole it is about 9.832 m/s^2

It may be noted that work done or energy possessed in an electrical circuit or mechanical system or thermal system is measured in the same units *viz.* joules. This is expected because mechanical, electrical and thermal energies are interchangeable. For example, when mechanical work is transferred into heat or heat into work, the quantity of work in joules is equal to the quantity of *heat in joules.

4.4. Some Cases of Mechanical Work or Energy

It may be helpful to give a few important cases of work done or energy possessed in a mechanical system :

(*i*) When a force of F newtons is exerted on a body through a distance 'd' metres in the direction of force, then

$$\text{Work done} = F \times d \text{ joules or Nm}$$

(*ii*) Suppose a force of F newtons is maintained tangentially at a radius r metres from O as shown in Fig. 4.1. In one revolution, the point of application of force travels through a distance of $2\pi r$ metres.

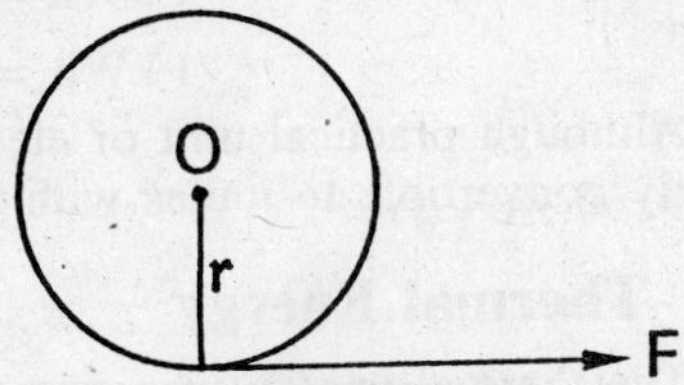

Fig. 4.1

∴ Work done in one revolution

$$= \text{Force} \times \text{Distance moved in 1 revolution}$$
$$= F \times 2\pi r$$
$$= 2\pi \times T \text{ joules or Nm}$$

where $T = Fr$ is the torque. Clearly, the SI unit of torque will be joules or Nm. If the body makes N revolutions per minute, then

$$\text{Work done/minute} = 2\pi \text{ NT joules}$$

(*iii*) If a body of mass m kg is moving linearly with a speed of v m/s, then kinetic energy possessed by the body is given by ;

$$\text{K.E. of the body} = \frac{1}{2} mv^2 \text{ joules}$$

(*iv*) If a body having a mass of m kg is lifted vertically through a height of h metres and g is acceleration due to gravity in m/s^2, then

$$\text{Potential energy of body} = \text{Work done in lifting the body}$$
$$= \text{Force required} \times \text{height}$$
$$= \text{Weight of body} \times \text{height}$$
$$= mg \times h$$
$$= mgh \text{ joules}$$

4.5. Electrical Energy

The SI unit of electrical work done or electrical energy expended in a circuit is also joule—exactly the same as for mechanical energy. It is defined as under :

One joule *of energy is expended electrically when one coulomb is moved through a p.d. of 1 volt.*

---

* Although heat energy was assigned a separate unit *viz.* calorie but the reader remembers that 1 calorie = 4.186 joules. In fact, the thermal unit calorie is obsolete and now-a-days heat is expressed in joules.

Suppose a charge of Q coulomb moves through a p.d. of V volts as shown in Fig. 4.2. Then electrical energy expended is given by ;

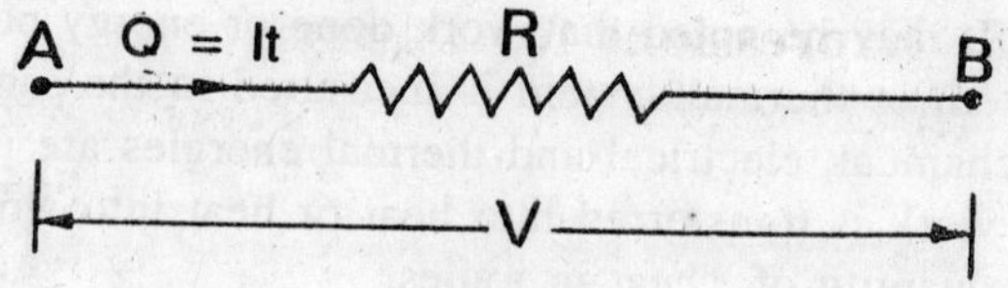

Fig. 4.2

Electrical energy expended

$$= VQ = VIt = I^2Rt = \frac{V^2t}{R} \text{ joules}$$

$$(\because Q = It \; ; \; V = IR \; ; \; I = V/R)$$

It may be mentioned here that joule is also known as watt-second *i.e.* 1 joule = 1 watt-sec. When we are dealing with large amount of electrical energy, it is often convenient to express it in *kilowatt hours (kWh).*

$$1 \text{ kWh} = 1000 \text{ watt–hours} = 1000 \times 3600 \text{ watt-sec or joules}$$

$$\therefore \quad 1 \text{ kWh} = 36 \times 10^5 \text{ joules or watt-sec}$$

Although practical unit of electrical energy is kWh, yet it is easy to see that this unit is readily convertible to joules with the help of above relation.

4.6. Thermal Energy

The thermal energy was originally assigned the unit 'calorie'. One calorie is the amount of heat required to raise the temperature of 1 gm of water through 1°C. If S is the specific heat of a body, then amount of heat required to raise the temperature of m gm of body through θ°C is given by ;

$$\text{Heat gained} = mS\theta \text{ calories}$$

It has been found experimentally that 1 calorie = 4.186 joules so that heat energy in calories can be expressed in joules as under :

$$\text{Heat gained} = (mS\theta) \times 4.186 \text{ Joules}$$

The reader may note that SI unit of heat is also Joule. In fact, the thermal unit calorie is obsolete and unit Joule is preferred these days.

4.7. Units of Power

The rate of doing work is called power. In other words, power is the work done per unit time. Since work is measured in joules and time in seconds, the unit of power will be *joule/sec* or *watt.*

(*i*) In practice, watt is often found to be inconveniently small, consequently kilowatt (KW) is used. One kW is equal to 1000 watts *i.e.*

$$1 \text{ kW} = 1000 \text{ watts}$$

For larger powers, the unit megawatt (*MW*) is used. One megawatt is equal to 1000 kW *i.e.*

$$1 \text{ MW} = 1000 \text{ kW} = 1000 \times 1000 \text{ watts}$$

$$\therefore \quad 1 \text{ MW} = 10^6 \text{ watts}$$

(*ii*) Sometimes power is measured in *horse power (H.P.)

$$1 \text{ H.P.} = 746 \text{ watts}$$

---

* This unit for power was conceived by James Watt, a Scottish scientist who invented the steam engine. In his experiments, he compared the output of his engine with the power a horse could put out. He found that an average horse could do work at the rate of 746 joules/sec. Although power can be expressed in watts or kW, the unit H.P. is still used.

4.8. Expressions for Power

(*i*)
$$\text{Power} = \text{Work done/sec.}$$
$$= \text{Force} \times \text{distance/sec}$$
$$= \text{Force} \times \text{velocity}$$

(*ii*) If a body makes *N r.p.m.* and the torque acting is *T* newton-metre, then

$$\text{Work done/minute} = 2\pi NT \text{ joules [See Art 4.4]}$$

$$\text{Work done/sec} = \frac{2\pi NT}{60} \text{ joules/sec or watts}$$

i.e.
$$\text{Power} = \frac{2\pi NT}{60} \text{ watts}$$

$$\text{Since 746 watts} = 1.\, H.P., \text{we have,}$$

$$\text{Power} = \frac{2\pi NT}{60 \times 746} \text{ H.P.}$$

where *T* is in newton-m and *N* is in *r.p.m.*

Example 4.1. *A mass of 10 kg is raised by means of rope passing round a pulley wheel of 0.5 m diameter. Calculate the work done when the pulley makes 50 revolutions.*

Solution.

$$\text{Force required, } F = mg = 10 \times 9.81 = 98.1 \text{ N}$$

$$\text{Torque exerted, } T = F \times r = 98.1 \times 0.5/2 = 24.525 \text{joules}$$

Distance covered in N revolutions is $= 2\pi\, r \times N$

$$\therefore \quad \text{Total work done} = F \times 2\pi\, rN$$

$$= T \times 2\pi N = (24.525) \times (2\pi \times 50) = \mathbf{7704.75\ joules}$$

Example 4.2. *An electric motor is developing 10 H.P. at a speed of 1200 r.p.m. Calculate the torque in newton-metre.*

Solution.
$$H.P. = \frac{2\pi \times NT}{60 \times 746} \quad \text{(See Art. 4.8)}$$

or
$$10 = \frac{2\pi \times 1200 \times T}{60 \times 746}$$

$$\therefore \quad \text{Torque,} \quad T = \frac{10 \times 60 \times 746}{2\pi \times 1200} = \mathbf{59.4\ Nm}$$

Example 4.3. *A lift of 250 kg mass is raised with a velocity of 5 m/s. If the driving motor has an efficiency of 85%, calculate the input power to the motor.*

Solution.

$$\text{Weight of lift, } F = mg = 250 \times 9.81 \text{ N}$$

$$\text{Output power of motor} = \text{Force} \times \text{Velocity} = (250 \times 9.81) \times 5 = 12260 \text{ W}$$

$$\therefore \quad \text{Input power to motor} = \frac{\text{output power}}{\text{efficiency}} = 12260/0.85 = \mathbf{14420\ W}$$

Example 4.4. *An electrically driven pump lifts 80 m^3 of water per minute through a height of 12 m. Allowing an overall efficiency of 70% for the motor and pump, calculate the input power to motor. If the pump is in operation for an average of 2 hours per day for 30 days, calculate the energy consumption in kWh and the cost of energy at the rate of 50 P per kWh. Assume 1 m^3 of water has a mass of 1000 kg and $g = 9.81\, m/s^2$.*

Solution.

Mass of 80 m^3 of water, $m = 80 \times 1000 = 8 \times 10^4$ kg

Weight of water lifted, $W = mg = 8 \times 10^4 \times 9.81$ N

Height through which water lifted, $h = 12$ m

(*i*) W.D. by motor/minute $= mgh = 8 \times 10^4 \times 9.81 \times 12$ joules

W.D. by motor/second $= \dfrac{8 \times 10^4 \times 9.81 \times 12}{60} = 156960$ watts

$\therefore$ Output power of motor $= 156960$ watts

Input power to motor $= \dfrac{\text{Motor output}}{\text{Efficiency}} = \dfrac{156960}{0.7}$

$= 2{,}24{,}228$ W $=$ **224.228 kW**

(*ii*) Total energy consumption = Input power × Time of operation

$= (224.228) \times (2 \times 30)$ kWh $=$ **13453 kWh**

(*iii*) Total cost of energy = Rs 0.5×13453 = **Rs 6726.5**

Example 4.5. *An electric-motor driven pump lifts 10 m^3 of water per minute to a height of 20 metres. The efficiencies of the pump and motor are 80% and 90% respectively. Calculate the current taken by the motor at 440 V. Mass of 1 m^3 of water is 1000 kg.*

Solution. Mass of 10 m^3 of water, $m = 10 \times 1000 = 10^4$ kg

Wt. of water lifted $= mg = 10^4 \times 9.81$ N

Work done/minute $= mgh = 10^4 \times 9.81 \times 20 = 196.2 \times 10^4$ joules

Work done/second $= 196.2 \times 10^4/60 = 32.7 \times 10^3$ watts

$\therefore$ Power output $= 32.7$ kW

Input to pump $= 32.7/0.8 = 40.87$ kW

Motor input $= 40.87/0.9 = 45.41$ kW

Current taken by motor $= \dfrac{45.41 \times 1000}{440} =$ **103.2A**

Example 4.6. *A 100 MW hydro-electric station is supplying full-load for 10 hours a aay. Calculate the volume of water which has been used. Assume effective head of station as 200m and overall efficiency of the station as 80%.*

Solution.

Energy supplied by the station in 10 hours

$= (100 \times 10^3) \times 10 = 10^6$ kWh

$= 36 \times 10^5 \times 10^6 = 36 \times 10^{11}$ joules

Energy input of station $= 36 \times 10^{11}/0.8 = 45 \times 10^{11}$ joules

Suppose m kg is the mass of water used in 10 hours.

Then, $mgh = 45 \times 10^{11}$

or $m = \dfrac{45 \times 10^{11}}{9.81 \times 200} = 22.93 \times 10^8$ kg

Since 1 m^3 of water has a mass of 1000 kg,

Volume of water used $= 22.93 \times 10^8/10^3 =$ $\mathbf{22.93 \times 10^5\ m^3}$

Example 4.7. *A hydro-electric power station has a reservoir of area 2.4 square kilometres and capacity $5 \times 10^6\ m^3$. The effective head of water is 100 m. The penstock, turbine and generator efficiencies are 95%, 90%, and 85% respectively.*

(*i*) *Calculate the total energy in kWh which can be generated from the power station.*

(*ii*) *If a load of 15,000 kW has been supplied for 3 hours, find the fall in reservoir level.*

Solution.

(*i*) Wt. of water available, $W = \text{Volume of reservoir} \times 1000 \times 9.81\ \text{N}$

$$= (5 \times 10^6) \times (1000) \times (9.81) = 49.05 \times 10^9\ \text{N}$$

Overall efficiency, $\eta_{overall} = 0.95 \times 0.90 \times 0.85 = 0.726$

Electrical energy that can be generated from the station

$$= W \times \text{Effective head} \times \eta_{overall}$$

$$= (49.05 \times 10^9) \times (100) \times (0.726) = 35.61 \times 10^{11}\ \text{watt-sec}$$

$$= \frac{35.61 \times 10^{11}}{1000 \times 3600} = \mathbf{9,89,166\ kWH}$$

(*ii*) Level of reservoir $= \dfrac{\text{Volume of reservoir}}{\text{Area of reservoir}} = \dfrac{5 \times 10^6}{2.4 \times 10^6} = 2.083\ \text{m}$

kWh generated in 3 hrs $= 15000 \times 3 = 45,000\ \text{kWh}$

Using unity method, we get,

Fall in reservoir level $= \dfrac{2.083}{9,89,166} \times 45,000 = 0.0947\ \text{m} = \mathbf{9.47\ cm}$

Example 4.8. *Calculate the current required by a 500 V d.c. locomotive when drawing 100 tonne load at 25 km/hr with a tractive resistance of 7 kg/tonne along (i) level road and (ii) a gradient 1 in 100. Given that the efficiency of motor and gearing is 70%.*

Solution.

Weight of locomotive, $W = 100\ \text{tonne} = 100,000\ \text{kg}$

Tractive resistance, $F = 7 \times 100 = 700\ \text{kg-wt} = 700 \times 9.81 = 6867\ \text{N}$

(*i*) **Level Track.** In this case, the force required is equal to the tractive resistance *F* [See Fig. 4.3 (*i*)].

$$\text{Distance travelled/sec} = \frac{25 \times 1000}{3600} = 6.94\ \text{m}$$

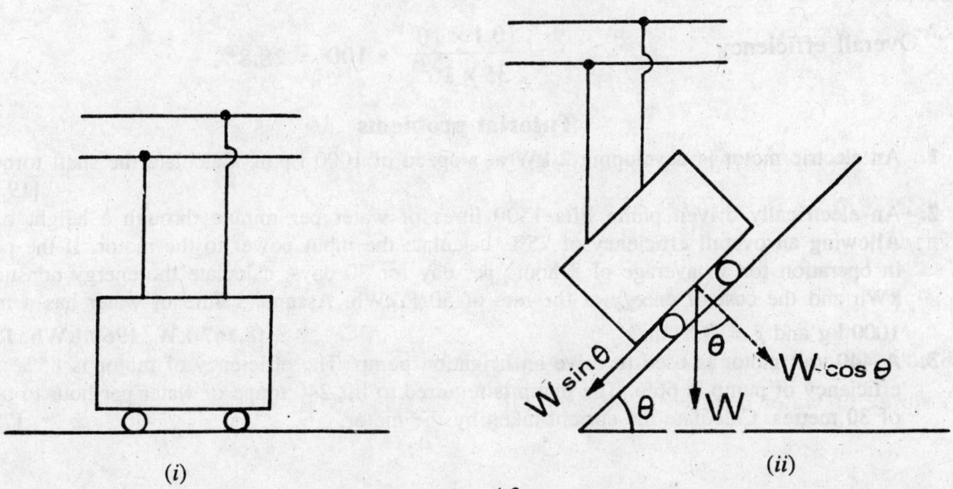

(*i*) (*ii*)

Work done/see = Force × Distance/sec

or Motor output = 6867 × 6.94 = 47,657 watts

Motor input = 47,657/0.7 = 68,081 watts

∴ Current drawn = 68,081/500 = **136.16A**

(*ii*) **Inclined plane.** In this case, the total force required is the sum of tractive resistance F and component $W \sin\theta$ of locomotive weight. Clearly $\sin\theta = 1/100 = 0.01$.

∴ Force required = $W \sin\theta + F$

$= (100{,}000 \times 0.01 + 700)\ 9.81\ \text{N} = 16{,}677\ \text{N}$

Work done/sec = Force × distance travelled/sec

= 16,677 × 6.94 = 1,15,738 watts

∴ Motor output = 1,15,738 watts

Motor input = 1,15,738/0.7 = 1,65,340 watts

∴ Current drawn = 1,65,340/500 = **330.68A**

Example 4.9. *A steam power station has an overall efficiency of 20% and 0.6 kg of coal is burnt per kWh of electrical energy generated. Calculate the calorific value of fuel.*

Solution.

Let x kcal/kg be the calorific value of the fuel.

Heat produced by 0.6 kg of coal = 0.6 x kcal

Heat equivalent of 1 kWh = 860 kcal

$$\therefore \quad \text{Overall efficiency} = \frac{\text{Electrical output in heat units}}{\text{Heat of combustion}}$$

$$or \quad 0.2 = \frac{860}{0.6x}$$

$$\therefore \quad x = \mathbf{7166.67\ kcal/kg}$$

Example 4.10. *A generating station has a daily output of 280 MWh and uses 500 tonne of coal in the process.The coal releases 7×10^6 J/kg when burnt. Calculate the overall efficiency.*

Solution.

Input energy/day, $W_i = (7 \times 10^6) \times (500 \times 1000) = 35 \times 10^{11}$ J

Output energy/day, $W_0 = 280\ \text{MWh} = (280 \times 10^3) \times (36 \times 10^5)\ \text{J} = 10.1 \times 10^{11}$ J

$$\therefore \text{Overall efficiency} = \frac{10.1 \times 10^{11}}{35 \times 10^{11}} \times 100 = \mathbf{28.8\%}$$

Tutorial problems

1. An electric motor is developing 2 kW at a speed of 1000 r.p.m. Calculate the shaft torque. **[19.1 Nm]**
2. An electrically driven pump lifts 1500 litres of water per minute through a height of 25m. Allowing an overall efficiency of 75%, calculate the input power to the motor. If the pump is in operation for an average of 8 hours per day for 30 days, calculate the energy consumed in kWh and the cost of energy at the rate of 50 P/kWh. Assume 1 litre of water has a mass of 1000 kg and $g = 9.81\ \text{m/s}^2$. **[8.167 kW, 1960 kWh, Rs 980]**
3. A 440-volt motor is used to drive an irrigation pump. The efficiency of motor is 85% and the efficiency of pump is 66%. The pump is required to lift 240 tonne of water per hour to a height of 30 metres. Calculate the current taken by the motor. **[79.48A]**

4. A 250 V d.c. electric motor operates a lift and raises a mass of 1 tonne through a vertical distance of 17 m in 8 seconds. If the overall efficiency of motor and lift mechanism is 64%, calculate the current taken by the motor. **[130.3 A]**
5. The effective water head for a 100 MW station is 200 metres. The station supplies full-load for 12 hours a day. If the overall efficiency of the station is 86.4%, find the volume of water used. **[25.48×10^5 m^3]**

4.9. Heating Effect of Electric Current

When electric current (*i.e.* flow of free electrons) passes through a conductor, there is a considerable 'friction' between the moving electrons and the molecules of the conductor. The electrical energy supplied to the conductor to overcome this 'electrical friction' (which we refer to as electrical resistance) is converted into heat. This is known as heating effect of electric current. For example, if I amperes is flowing through a conductor of resistance R ohms for t seconds, then electrical energy supplied is I^2Rt joules. This energy is not destroyed but the whole of I^2Rt joules is changed into heat.

The heating effect of electric current is utilised in the manufacture of many heating appliances such as electric heater, electric toaster, electric kettle, soldering iron *etc*. The basic principle of all these appliances is the same. Electric current is passed through a high resistance (called heating element), thus producing the required heat. There are a number of substances used for making a heating element. One that is commonly used is an alloy of nickel and chromium, called **nichrome**. This alloy has a resistance more than 50 times that of copper. The heating element may be either nichrome wire or ribbon wound on some insulating material that is able to withstand heat. While dealing with problems on heating appliances, the following points may be kept in mind :

(*i*) The electrical energy in kWh can be converted into joules by the following relation :

$$1\text{kWh} = 36 \times 10^5 \text{ joules}$$

(*ii*) The heat energy in calories can be converted into joules by the following relation :

$$1 \text{ calories} = 4.186 \text{ joules} \; ; \; 1 \text{ kcal} = 4186 \text{ joules}$$

(*iii*) The electrical energy in kWh can be converted into calories (or kilocalories) by the following relation :

$$1 \text{ kWh} = 36 \times 10^5 \text{ joules} = \frac{36 \times 10^5}{4.186} \text{ calories} = 860 \times 10^3 \text{ calories}$$

$$\therefore \quad 1 \text{ kWh} = 860 \text{ kcal}$$

(*iv*) The electrical energy supplied to the heating appliance forms the *input energy*. The heat obtained from the device is the *output energy*. The difference between the two, if any, represents the loss of energy during conversion from electrical into heat energy.

Example 4.11. *A soldering iron is rated at 50 watts when connected to a 250 V supply. If the soldering iron takes 5 minutes to heat to a working temperature of 190°C from 20 °C, find its mass, assuming it to be made of copper. Given specific heat capacity of copper is 390 J/Kg °C.*

Solution. Let m kg be the mass of soldering iron.

Heat gained by the soldering iron= $ms\theta = m \times 390 \times (190 - 20) = 66{,}300\, m$ joules

Heat released by the heating element = power × time = $(50) \times (5 \times 60) = 15{,}000$ joules

Assuming all the heat released by the element is absorbed by the copper *i.e.* soldering iron is 100% efficient,

$$15{,}000 = 66{,}300\,m$$

$$\therefore \quad m = 15{,}000/66{,}300 = \mathbf{0.226\ kg}$$

Example 4.12. *An electric kettle marked 1kW, 230 V, takes 7.5 minutes to bring 1 kg of water at 15°C to boiling point (100°C). Find the efficiency of the kettle.*

Solution. Heat received by water (*i.e.* output energy)

$$= \text{mass} \times \text{specific heat} \times \text{rise in temp.}$$

$$= 1 \times 1 \times (100 - 15) = 85 \text{ kcal}$$

Electrical energy supplied to the kettle (*i.e* input energy)

$$= \text{wattage} \times \text{time} = 1 \text{ kW} \times 7.5/60 \text{ hr} = 0.125 \text{ kWh}$$

$$= 0.125 \times 860 \text{ kcal} = 107.5 \text{ kcal} \quad (\because 1 \text{ kWh} = 860 \text{ kcal})$$

$$\therefore \quad \text{Efficiency of kettle} = \frac{85}{107.5} \times 100 = \mathbf{79.07\%}$$

Example 4.13. *The cost of boiling 2 kg of water in an electric kettle is 12 paise. The kettle takes 6 minutes to boil water from an ambient temperature of 20°C. Calculate (i) the efficiency of kettle and (ii) the wattage of kettle if cost of 1 kWh is 40 paise.*

Solution.

(*i*) Heat received by water (*i.e.* output energy) $= 2 \times 1 \times 80 = 160$ kcal

Electrical energy supplied (*i.e.* input energy) $= 12/40$ kWh $= 860 \times 12/40 = 258$ kcal

$$\therefore \quad \text{Kettle efficiency} = \frac{160}{258} \times 100 = \mathbf{62\%}$$

(*ii*) Let *W* kilowatt be the power rating of the kettle.

$$\text{Input energy} = W \times \text{time in hours}$$

or

$$12/40 = W \times 6/60$$

$$\therefore \text{ Wattage of kettle,} \quad W = \frac{12}{40} \times \frac{60}{6} = \mathbf{3\ kW}$$

Example 4.14. *How long will it take to raise the temperature of 880 gm of water from 16°C to boiling point ? The heater takes 2 amperes at 220 V and its efficiency is 90% ?*

Solution. Heat received by water (*i.e.* output energy)

$$= 0.88 \times 1 \times (100 - 16) = 73.92 \text{ kcal} = 73.92/860 = 0.086 \text{ kWh}$$

Electrical energy supplied to the heater (*i.e.* input energy)

$$= 0.086/0.9 = 0.096 \text{ kWh}$$

The heater is supplying a power of $220 \times 2 = 440$ watts $= 0.44$ kW. Let *t* hours be the required time.

$$\text{Input energy} = \text{wattage} \times \text{time}$$

$$0.096 = 0.44 \times t$$

$$\therefore \quad t = 0.096/0.44 = 0.218 \text{ hours} = \mathbf{13.08\ minutes}$$

Example 4.15. *An electric heater contains 40 litres of water initially at a mean temperature of 15°C. The heater supplies an energy of 2.5 kWh to water. Assuming no heat losses, what is the final mean temperature of water ?*

Solution. Let θ°C be the final mean temperature of water.

Electrical energy supplied to water (*i.e.* input energy)

$$= 2.5 \text{ kWh} = 0.25 \times 36 \times 10^5 = 9 \times 10^6 \text{ J}$$

Heat received by water (*i.e.* output energy)

$$= ms\,(\theta - 15) = 40 \times 4186\,(\theta - 15)\text{ J}$$

As there are no heat losses, input energy is equal to the output energy.

$$\therefore \quad 40 \times 4186\,(\theta - 15) = 9 \times 10^6$$

$$\text{or} \quad \theta - 15 = \frac{9 \times 10^6}{40 \times 4186} = 53.8$$

$$\therefore \quad \theta = 53.8 + 15 = \mathbf{68.8°C}$$

Example 4.16. *An electric kettle is required to raise the temperature of 2 kg of water from 20°C to 100°C in 15 minutes. Calculate the resistance of the heating element if the kettle is to be used on a 240 volts supply. Assume the efficiency of the kettle to be 80%.*

Solution.

Heat received by water (*i.e.* output energy)

$$= 2 \times 1 \times (100 - 20) = 160\text{ kcal} = 160/860 = 0.186\text{ kWh}$$

Electrical energy supplied to the kettle

$$= 0.186/0.8 = 0.232\text{ kWh}$$

The electrical energy of 0.232 kWh is supplied in 15/60 = 0.25 hours.

$$\therefore \text{ Power rating of kettle} = 0.232/0.25 = 0.928\text{ kW} = 928\text{ watts}$$

Let R ohms be the resistance of the heating element.

$$\therefore \quad V^2/R = 928 \quad \text{or} \quad R = \frac{240 \times 240}{928} = \mathbf{62\ \Omega}$$

Example 4.17. *An electric kettle contains 1.5 kg of water at 15°C. It takes 2.5 hours to raise the temperature to 90°C. Assuming the heat losses due to radiation and heating the kettle to be 15 kcal, find (i) wattage of kettle and (ii) current taken if supply voltage is 230 V.*

Solution. Heat received by water (*i.e.* output energy)

$$= 1.5 \times 1 \times (90 - 15) = 112.5\text{ kcal}$$

Electrical energy supplied to the kettle

$$= 112.5 + 15 = 127.5\text{ kcal} = 127.5/860 = 0.1482\text{ kWh}$$

(*i*) Input energy $=$ Wattage in kW $\times$ time in hours

$0.1482 =$ Wattage $\times$ 2.5

$\therefore$ Wattage of kettle $= 0.1482/2.5 = 0.0592$ kW $=$ **59.2 watts**

(*ii*) Current taken $= 59.2/230 =$ **0.257A**

Example 4.18. *Two heaters A and B are in parallel across the supply voltage V. Heater A produces 500 kcal in 20 minutes and B produces 1000 kcal in 10 minutes. The resistance of heater A is 10 Ω. What is the resistance of heater B ? If the same heaters are connected in series, how much heat will be produced in 5 minutes ?*

Solution.

$$\text{Heat produced} = \frac{V^2 t}{R \times J}\text{ kcal}$$

$$\text{For heater } A,\ 500 = \frac{V^2 \times (20 \times 60)}{10 \times J} \qquad ...(i)$$

$$\text{For heater } B,\ 1000 = \frac{V^2 \times (10 \times 60)}{R \times J} \qquad ...(ii)$$

Dividing Eq (*i*) by (*ii*), we get,

$$\frac{500}{1000} = \frac{20 \times 60}{10 \times 60} \times \frac{R}{10}$$

$$\therefore \qquad R = 2.5\ \Omega$$

When the heaters are connected in series, the total resistance becomes $10 + 2.5 = 12.5\ \Omega$

$\therefore$ Heat produced in 5 minutes

$$= \frac{V^2 t}{R \times J} = \frac{V^2}{J} \times \frac{t}{R}$$

$$= \frac{5{,}000}{20 \times 60} \times \frac{5 \times 60}{12.5} \qquad \left[\text{From Eq. } (i) \quad \frac{V^2}{J} = \frac{5000}{20 \times 60}\right]$$

$$= \mathbf{100\ kcal}$$

Tutorial Problems

1. Calculate the time taken for a 3 kW electrical immersion heater to raise the temperature of 136 litres of water from 10°C to 70°C. Assume that overall efficiency of the equipment is 90%. Given that 1 litre of water has a mass of 1kg. **[3 hours, 30 minutes, 54 seconds]**
2. An electrical heater takes 5 minutes to convert 1 kg of water at room temperature of 18°C into steam at 100°C. Calculate the power rating of the heater. Assume the heating equipment to have an efficiency of 90%. Take the latent heat of evaporation of water as 2.257 MJ/kg. **[9.63 kW]**
3. An electric heater rated at 2 kW contains 68 kg of water at 20°C. Assuming the heat losses to be 15% of the input, find the temperature of water after the heater has been switched for 2.5 hours. **[73.75°C]**
4. Find the amount of electrical energy expended in raising the temperature of 45 litres of water by 75°C. To what height could a weight of 5 tonnes be raised with the expenditure of the same energy. Assume efficiencies of heating and lifting equipment to be 90% and 70% respectively. **[4.36 kWh, 224m]**
5. Calculate the time taken for a 25 kW furnace, having an overall efficiency of 80% to melt 20kg of aluminium. Take the specific heat capacity, melting point and latent heat of fusion of aluminium as 896 J/kg °C, 657°C and 402 kJ/kg respectively. **[16 min, 13 sec]**

OBJECTIVE QUESTIONS

1. A kilowatt hour is the unit of
(*i*) energy (*ii*) power
(*iii*) electric charge (*iv*) electric current

2. Heat produced in a conductor depends
(*i*) inversely on time
(*ii*) directly on time
(*iii*) both (*i*) and (*ii*)
(*iv*) none of the above

3. If potential difference across a conductor of a material of resistivity ρ remains constant, then heat produced in the conductor is proportional to
(*i*) ρ (*ii*) ρ^2
(*iii*) $\frac{1}{\sqrt{\rho}}$ (*iv*) $\frac{1}{\rho}$

4. One kilowatt hour is equal to
(*i*) 36×10^3 J (*ii*) 10^3 J
(*iii*) 36×10^5 J (*iv*) 10^5 J

5. The heater element in an electric iron is made of
(*i*) nichrome (*ii*) iron
(*iii*) tungsten (*iv*) constantan

6. The heat produced (in kcals) in a resistance *R* ohms when a current *I* amperes flows through it for *t* seconds is given by
(*i*) $I^2 R t$ (*ii*) $\frac{I^2 Rt}{4.2 \times 10^3}$
(*iii*) $I^2 R t \times 4.2$ (*iv*) $\frac{I R t}{J}$

7. Three electric bulbs 40 W, 60 W and 100 W are designed to work on a 220 V mains. Which bulb will burn most brightly if they are connectd in series across 220 V mains?

(*i*) 60 W (*ii*) 40 W
(*iii*) 100 W
(*iv*) all bulbs will burn equally brightly

8. Two electric bulbs have tungsten filament of the same length. If one of them gives 60 W and the other 100 W, then,
(*i*) 100 W bulb has thicker filament
(*ii*) 60 W bulb has thicker filament
(*iii*) both filaments are of the same thickness
(*iv*) information insufficient

9. Two electric lamps of 40 W each are connected in parallel. Take power consumed by the combination is
(*i*) 20 W (*ii*) 60 W
(*iii*) 80 W (*iv*) 100 W

10. The time required for 1 kW heater to raise the temperature of 10 litres of water through 10°C is
(*i*) 210 sec (*ii*) 840 sec
(*iii*) 42 sec (*iv*) 420 sec

11. Two electric lamps 40W, 220 V each are connected in series across 220 V. The power consumed by the combination is
(*i*) 80 W
(*ii*) 20 W
(*iii*) 160 W
(*iv*) 40 W

12. Two bulbs, 100 W, 250 V and 200 W, 250 V, are connected in parallel across a 500 V line. Then,
(*i*) 100 W bulbs will be fused
(*ii*) 200 W bulb will be fused
(*iii*) Both bulbs will be fused
(*iv*) No bulb will be fused

13. Two bulbs, 100 W, 250 V and 200 W, 250 V are connecated in series to a 500 V line. Then,
(*i*) 100 W bulb will be fused
(*ii*) 200 W bulb will be fused
(*iii*) both the bulbs will be fused
(*iv*) no bulb will be fused

14. Three bulbs B_1, B_2 and B_3 are connected to the mains as shown in Fig. 4.4. If B_3 is disconnected from the circuit by opening switch *S*, then incandescence of bulb B_1 will

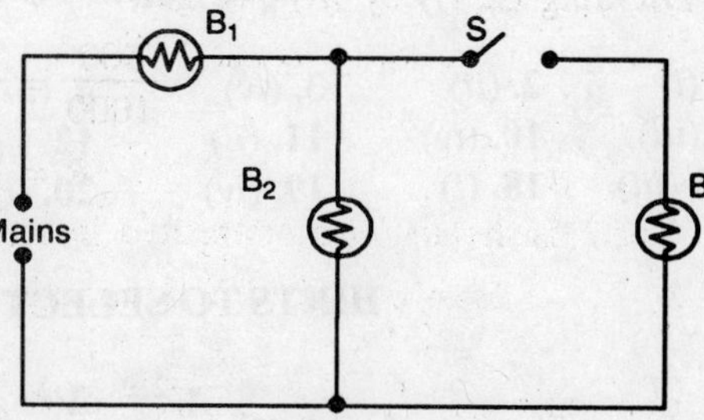

Fig. 4.4

(*i*) decrease (*ii*) increase
(*iii*) No change (*iv*) cannot say

15. You are given the following electrical appliances
(*a*) 1 kW, 250 V electric heater
(*b*) 1 kW, 250 V, electric kettle
(*c*) 1 kW, 250 V, electric bulb
Which of these has the highest resistance?
(*i*) heater (*ii*) kettle
(*iii*) bulb
(*iv*) all have equal resistances

16. If current in an electric bulb drops by 2%, then power decreases by
(*i*) 2 % (*ii*) 4 %
(*iii*) 1 % (*iv*) 16 %

17. Two bulbs of 500 W and 300 W are manufactured to operate on 220 V line. The ratio of resistance of 500 W bulb to that of 300 W bulb is
(*i*) 5 : 3 (*ii*) 9 : 25
(*iii*) 3 : 5 (*iv*) 25 : 9

18. The same mass of copper is drawn into two wires 1 mm thick and 3mm thick. Two wires are connectaed in series and current is passed. Heat produced in the wires is in the ratio.
(*i*) 81 : 1 (*ii*) 1 : 81
(*iii*) 3 : 1 (*iv*) 9 : 1

19. The power of a heater is 500 W at 800° C. What will be its power at 200° C ?
Temperature coefficient of resistance is 4×10^{-4}/°C.
(*i*) 484 W (*ii*) 672 W
(*iii*) 526 W (*iv*) 620 W

20. A constant voltage is applied between two ends of a uniform metallic wire. Some heat is developed. If both length and radius of wire are halved, the heat developed in the same duration is
(*i*) half (*ii*) twice
(*iii*) one-fourth (*iv*) same

ANSWERS TO OBJECTIVE QUESTIONS

1. (*i*)	**2.** (*ii*)	**3.** (*iv*)	**4.** (*iii*)	**5.** (*i*)	**6.** (*ii*)	**7.** (*ii*)	**8.** (*i*)
9. (*iii*)	**10.** (*iv*)	**11.** (*ii*)	**12.** (*iii*)	**13.** (*i*)	**14.** (*i*)	**15.** (*iv*)	**16.** (*ii*)
17. (*iii*)	**18.** (*i*)	**19.** (*iv*)	**20.** (*i*)				

HINTS TO SELECTED OBJECTIVE QUESTIONS

3. Rate of heat produced, $H = \dfrac{V^2}{R} = \dfrac{V^2}{\rho l/a} = \dfrac{V^2 a}{\rho l}$

$\therefore \quad H \propto 1/\rho$ ($\because$ V, a and l are constant)

8. $R = \rho\, l/a$. The filament of that bulb is thicker which has low resistance ($\because$ l is constant in the two cases). Now 100 W bulb has less resistance than that of 60 W bulb. Hence 100 W bulb has thicker filament.

9. $\dfrac{1}{R_T} = \dfrac{1}{R_1} + \dfrac{1}{R_2}$

or $\dfrac{V^2}{R_T} = \dfrac{V^2}{R_1} + \dfrac{V^2}{R_2}$

or $P_T = P_1 + P_2 = 40 + 40 = 80$ W

11. $R_T = R_1 + R_2$ or $\dfrac{R_T}{V^2} = \dfrac{R_1}{V^2} + \dfrac{R_2}{V^2}$

or $\dfrac{1}{P_T} = \dfrac{1}{P_1} + \dfrac{1}{P_2} \quad \therefore \quad P_T = \dfrac{P_1 P_2}{P_1 + P_2} = \dfrac{40 \times 40}{40 + 40} = 20$ W

14. When bulb B_3 is disconnected from the circuit, the total circuit resistance is increased and total circuit current is decreased. As power dissipation = $I^2 R$, the incandescence of bulb B_1 wire decrease.

19. Power, $P \propto 1/R$ so that

$P_{200} = P_{800}\,[1 + \alpha\,(\theta_2 - \theta_1)] = 500\,[1 + 4 \times 10^{-4} \times 600] = 620$ W

5

Electrostatics

Introduction

A charged body has either deficiency or excess of electrons and is said to possess *electricity*. If two oppositely charged bodies are connected through a conductor, electrons will flow from the negative charge (excess of electrons) to the positive charge (deficit of electrons). This flow of electrons is called electric current. The electric current, will continue to flow so long as 'excess' and deficiency of electrons exist in the bodies. The branch of electrical engineering which deals with the flow of electrons (*i.e.*, current) is called *current electricity* and is important in many ways. For example, it is the electric current by means of which electrical energy can be transferred from one point to another for utilisation.

There can be another situation where charges (*i.e.*, electrons) do not move but remain static or stationary on the bodies. Such a situation will arise when the charged bodies are separated by some insulating medium, disallowing the movement of electrons. This is called *static electricity* and the branch of engineering which deals with static electricity is called *electrostatics.* Although current electricity is of greater practical use, yet the importance of static electricity cannot be ignored. Many of the advancements made in the field of electricity owe their developments to the knowledge scientists obtained from electrostatics. The most useful outcomes of static electricity are the development of lightning rod and the capacitor. In this chapter, we shall confine our attention to the behaviour and applications of static electricity.

5.1. Coulomb's Laws of Electrostatics

First law. This law relates to the nature of force between two charged bodies and may be stated as under :

Like charges repel each other while unlike charges attract each other.

Second law. This law tells about the magnitude of force between two charged bodies and may be stated as under :

*The force between two *point charges is directly proportional to the product of their magnitudes and inversely proportional to the square of distance between them.*

Mathematically,

$$F \propto \frac{Q_1 Q_2}{d^2}$$

or

$$F = k\frac{Q_1 Q_2}{d^2} \quad ...(i)$$

* Charged bodies approximate point charges if they are small compared to the distance between them.

where k is a constant whose value depends upon the medium in which the charges are placed and the system of units employed. In SI units, force is measured in newtons, charge in coulombs, distance in metres and the value of k is given by :

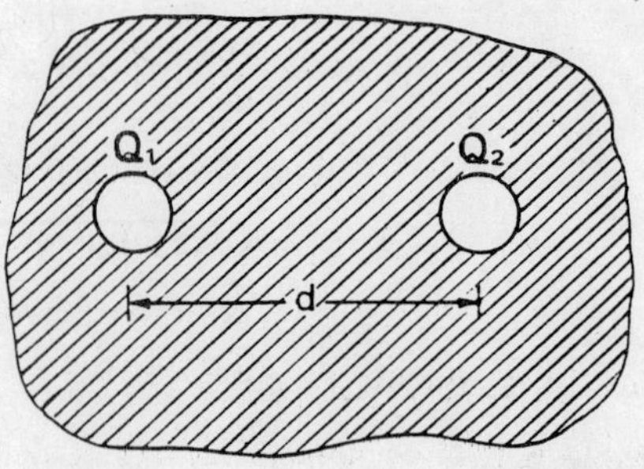

Fig. 5.1

$$k = \frac{1}{4\pi\,\varepsilon_0\,\varepsilon_r}$$

where ε_0 = Absolute permittivity of vacuum or air

ε_r = Relative permittivity of the medium in which the charges are placed. For vacuum or air, its value is 1.

The value of $\varepsilon_0 = 8.854 \times 10^{-12}$ *F/m* and the value of ε_r is different for different media.

$$\therefore \quad F = \frac{Q_1\,Q_2}{4\pi\,\varepsilon_0\,\varepsilon_r\,d^2} \quad \text{...(ii)}$$

$$\text{Now} \quad \frac{1}{4\pi\,\varepsilon_0} = \frac{1}{4\pi \times 8.854 \times 10^{-12}} = 9 \times 10^9$$

$$\therefore \quad F = 9 \times 10^9 \frac{Q_1 Q_2}{\varepsilon_r\,d^2} \quad \text{...in a medium}$$

$$= 9 \times 10^9 \frac{Q_1 Q_2}{d^2} \quad \text{...in air}$$

Unit charge. The unit of charge (*i.e.*, 1 coulomb) can also be defined from Coulomb's second law of electrostatics. Suppose two equal charges placed 1 m apart in *air* exert a force of 9×10^9 newtons *i.e.*,

$$Q_1 = Q_2 = Q\,;\, d = 1\text{ m}\,;\; F = 9 \times 10^9\text{ N}$$

$$\therefore \quad F = 9 \times 10^9 \frac{Q_1 Q_2}{d^2}$$

$$\text{or} \quad 9 \times 10^9 = 9 \times 10^9 \frac{Q^2}{(1)^2}$$

$$\text{or} \quad Q^2 = 1$$

$$\text{or} \quad Q = \pm 1 = 1\text{ coulomb}$$

Hence, **one coulomb** *is that charge which when placed in air at a distance of one metre from an equal and similar charge repels it with a force of* 9×10^9 *N.*

5.2. Absolute and Relative Permittivity

Permittivity is the property of a medium and affects the magnitude of force between two point charges. The greater the permittivity of a medium, the lesser the force between the charged bodies placed in it. Air or vacuum has a minimum value of permittivity. The absolute (or actual) permittivity ε_0 (Greek letter 'epsilon') of air or vacuum is 8.854×10^{-12} F/m. The absolute (or actual) permittivity ε of all other insulating materials is greater than ε_0. The ratio $\varepsilon/\varepsilon_0$ is called the **relative permittivity* of the material and is denoted by ε_r *i.e.*,

---

* Thus when we say that relative permittivity of a material is 10, it means that its absolute or actual permittivity $\varepsilon = \varepsilon_0\,\varepsilon_r = 8.854 \times 10^{-12} \times 10 = 8.854 \times 10^{-11}$ F/m.

$$\varepsilon_r = \frac{\varepsilon}{\varepsilon_0}$$

ε = Absolute (or actual) permittivity of material

ε_0 = Absolute (actual) permittivity of air or vacuum (8.854×10^{-12})

ε_r = Relative permittivity of material.

Obviously, ε_r for air would be $\varepsilon_0 / \varepsilon_0 = 1$

Example 5.1. *Two spheres charged with equal but opposite charges experience a force of 25×10^5 N when they are placed 2 cm apart in a medium of relative permittivity 5. Determine the charge on each sphere.*

Solution. Let Q and $-Q$ coulomb be the charges on the spheres. Since the force is attractive, it will be assigned a minus sign.

$$F = 9 \times 10^9 \frac{Q_1 Q_2}{\varepsilon_r d^2}$$

or
$$-25 \times 10^5 = 9 \times 10^9 \frac{Q(-Q)}{5 \times (0.02)^2}$$

or
$$Q = \sqrt{\frac{25 \times 10^5 \times 5 \times (0.02)^2}{9 \times 10^9}} = 0.0745 \times 10^{-2} \text{ C} = \mathbf{745\ \mu C}$$

Example 5.2. *A small sphere is given a charge of + 20 μC and a second sphere of equal diameter is given a charge of – 5 μC. The two spheres are allowed to touch each other and are then spaced 10 cm apart. What force exists between them ? Assume air as the medium.*

Solution. When the two spheres touch each other, the resultant charge = (20) + (– 5) = 15 μC. When the spheres are separated, charge on each sphere, $Q_1 = Q_2 = 15/2 = 7.5$ μC.

$\therefore$ Force,
$$F = 9 \times 10^9 \times \frac{Q_1 Q_2}{d^2}$$

$$= 9 \times 10^9 \times \frac{(7.5 \times 10^{-6})(7.5 \times 10^{-6})}{(0.1)^2} = \mathbf{50.62\ N} \textit{ repulsive}$$

Example 5.3. *Find the force between two charges 10 cm apart in vacuum. The charges are 4×10^{-5} C and 6×10^{-8} C. If the same charges are separated by the same distance in kerosene ($\varepsilon_r = 2$), what is the force between them ?*

Solution.

$$F = 9 \times 10^9 \frac{Q_1 Q_2}{d^2} = 9 \times 10^9 \frac{(4 \times 10^{-5})(6 \times 10^{-8})}{(0.1)^2} = \mathbf{2.16\ N} \quad \textit{...in vacuum}$$

$$= 9 \times 10^9 \frac{Q_1 Q_2}{\varepsilon_r d^2} = \frac{1}{2}(2.16) = \mathbf{1.08\ N} \quad \textit{...in a medium}$$

Example 5.4. *Calculate the force on a unit positive charge at P on X-axis whose co-ordinates are (x = 2 m, y = 0) due to the following two charges :*

(i) a positive charge of 10^{-9} C situated in air at the origin (x = 0, y = 0).

(ii) a negative charge of -2×10^{-9} C situated on X-axis (x = 1 m, y = 0).

Solution. The conditions of the problem are represented in Fig. 5.2.

Repulsive force on unit positive charge at P due to $+10^{-9}$ C Charge

$$= 9 \times 10^9 \frac{1 \times 10^{-9}}{(2)^2}$$

$$= 2.25 \text{ N}$$

Attractive force on unit positive charge at P due to -2×10^{-9} C charge

$$= 9 \times 10^9 \times \frac{2 \times 10^{-9}}{(1)^2} = 18 \text{ N}$$

$\therefore$ Resultant attractive force on unit positive charge at P = $18 - 2.25$ = **15.75 N**

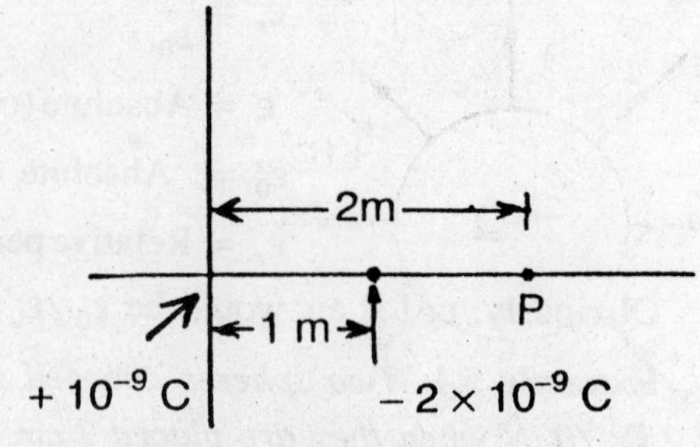

Fig. 5.2

Example 5.5. *Three point charges of* $+5\,\mu C$, $+5\,\mu C$ *and* $+5\mu C$ *are placed at the vertices of an equilateral triangle which has sides 10 cm long. Find the force on each charge.*

Solution. Fig. 5.3 shows the equilateral triangle with the three charges placed at the vertices. Consider the charge $+5\mu$ C placed at corner C. It is being repelled by charges placed at A and B along ACD and BCE respectively. These two forces are equal, each being given by ;

$$F = 9 \times 10^9 \times \frac{(5 \times 10^{-6}) \times (5 \times 10^{-6})}{(0.1)^2}$$

$$= 22.5 \text{ N}$$

Since the angle between these two equal forces is 60°, their resultant is

$$= 2\,F \cos 30° = 2 \times 22.5 \times \sqrt{3}/2 = \mathbf{38.97\ N}$$

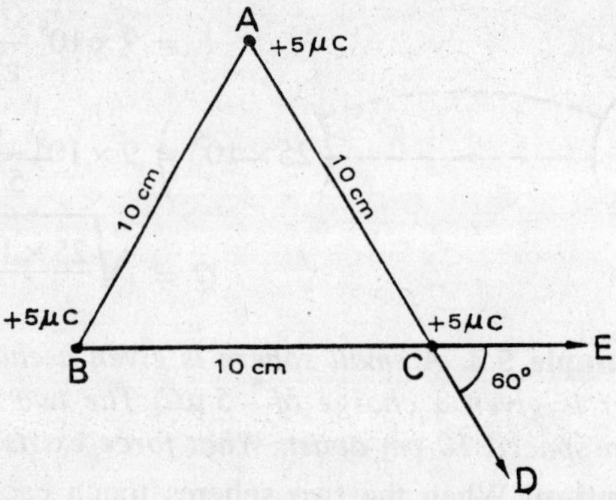

Fig. 5.3

The force acting on the charges placed at corners A and B will also be the same.

5.3. Electric Field

The region surrounding a charged body is always under stress and strain because of the electrostatic charge. If a small charge is placed in this region, it will experience a force according to Coulomb's laws. This stressed region around a charged body is called electric field. Theoretically, electric field due to a charge extends upto infinity but its effect practically dies away very quickly as the distance from the charge increases.

The space (or field) in which a charge experiences a force is called an **electric field** or **electrostatic field.**

The electric field around a charged body is represented by imaginary lines, called *electric lines of *force*. By convention, the direction of these lines of force at any point is the direction along which a unit positive charge (*i.e.*, positive charge of 1C) placed at that point would move or tend to move. The unit positive charge is sometimes called a *test charge* because it is used as an indicator to find the direction of electric field. Following this convention, it is clear that electric lines of force would always originate from a positive charge and end on a negative charge. The electric lines of force leave or enter the charged surface **normally.

* So called because forces are experienced by charges in this region.

** If a line of force is at an angle other that 90°, it will have a tangential component. This tangential component would cause redistribution (*i.e.*, movement) of charge. By definition, electrostatic charge is static and hence tangential component cannot exist.

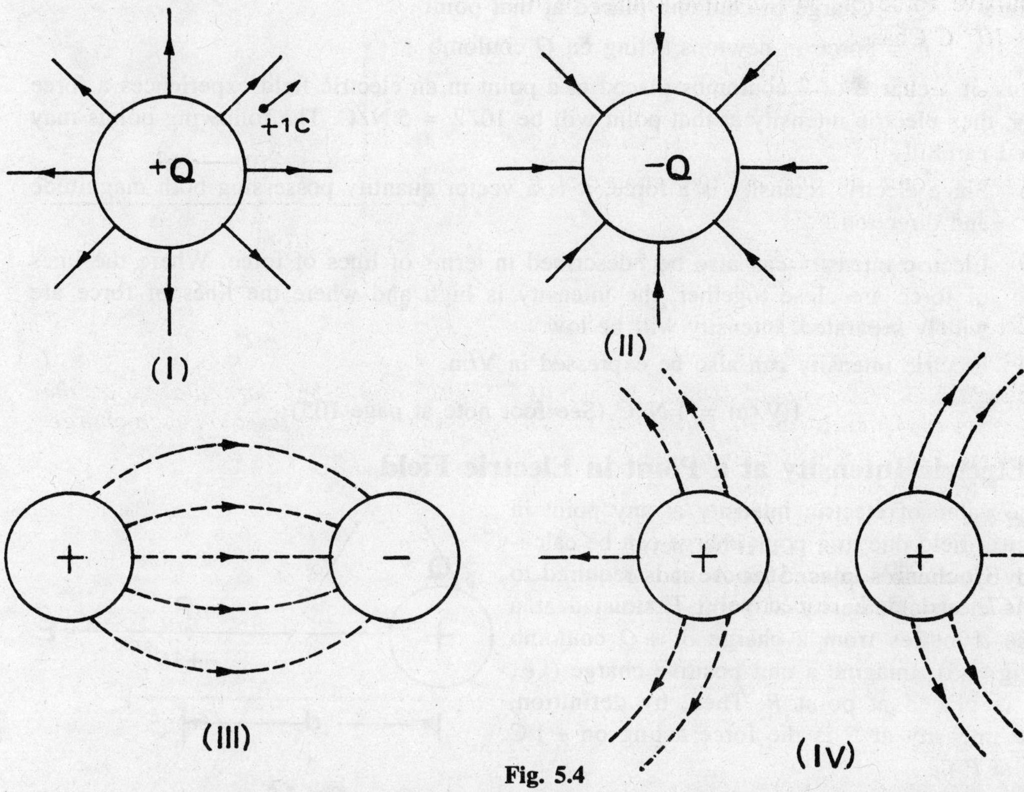

Fig. 5.4

Fig. 5.4 shows typical field distribution. Fig. 5.4 (*i*) shows electric field due to an isolated positively charged sphere. A unit positive charge placed near it will experience a force directed radially away from the sphere. Therefore the direction of electric field will be radially outward as shown in Fig. 5.4 (*i*). For the negatively charged sphere [See Fig. 5.4 (*ii*)], the force acting on the unit positive charge would be directed radially towards the sphere. Fig. 5.4 (*iii*) shows the electric field between a positive charge and a negative charge while Fig. 5.4 (*iv*) shows electric field between two similarly charged (*i.e.*, +vely charged) bodies.

5.4. Electric Flux

The total electric lines of force which flow outward from a positive charge is called **electric flux.**

Electric flux is a measure of the overall size of electric field and is measured in coulomb (C). The symbol for electric flux is Greek letter ψ (psi).

5.5. Electric Intensity or Field Strength (E)

To describe an electric field, we must specify its intensity or strength. The intensity of electric field at any point is determined by the force acting on a unit positive charge placed at that point.

Electric intensity (*or* **field strength**) *at a point in an electric field is the force acting on a unit positive charge placed at that point. Its direction is the direction along which the force acts i.e.,*

$$\text{Electric intensity at a point, } E = \frac{F}{+Q} \; N/C$$

where Q = Charge in coulomb placed at that point

F = Force in newtons acting on Q coulomb

Thus, if a charge of 2 coulombs placed at a point in an electric field experiences a force of 10 N, then electric intensity at that point will be 10/2 = 5 N/C. The following points may be noted carefully :

(*i*) Since electric intensity is a force, it is a vector quantity possessing both magnitude and direction.

(*ii*) Electric intensity can also be *described in terms of lines of force. Where the lines of force are close together, the intensity is high and where the lines of force are widely separated, intensity will be low.

(*iii*) Electric intensity can also be expressed in V/m.

1 V/m = 1 N/C (See foot note at page 105)

5.6. Electric Intensity at a Point in Electric Field

The value of electric intensity at any point in an electric field due to a point charge can be calculated by Coulomb's laws. Suppose it is required to find the electric intensity at point P situated at a distance d metres from a charge of $+Q$ coulomb (See Fig. 5.5). Imagine a unit positive charge (*i.e.*, $+1$ C) is placed at point P. Then, by definition, electric intensity at P is the force acting on $+1$ C placed at P *i.e.*,

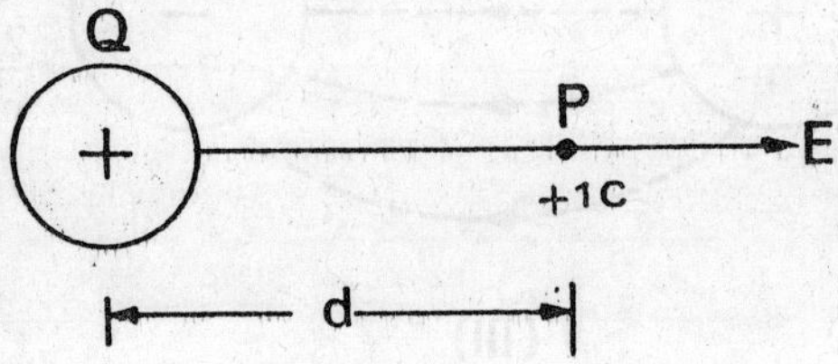

Fig. 5.5

Electric intensity at P,

$$E = \text{Force on } +1\text{ C placed at } P$$

$$= 9 \times 10^9 \frac{Q \times 1}{\varepsilon_r d^2}$$

$$\therefore \quad E = 9 \times 10^9 \frac{Q}{\varepsilon_r d^2} \quad \text{...in a medium}$$

$$= 9 \times 10^9 \frac{Q}{d^2} \quad \text{...in air}$$

Note the direction of electric intensity. It is acting radially away from $+Q$. For a negative charge (*i.e.*, $-Q$), its direction would have been radially towards the charge.

Example 5.6. *(i) A charge of 2 μC placed in an electric field experience a force of 0.08 N. What is the magnitude of electric intensity ?*

(ii) A charge of 0.52 μC is placed in an electric field where field intensity is 4.5×10^5 N/C. What is the magnitude of force acting on the charge ?

Solution. (*i*) Charge, $Q = 2\mu C = 2 \times 10^{-6}$ C

---

* It may be noted that electric lines of force do not actually exist. It is only a way of representing an electric field. However, it is a useful method of representation. It is a usual practice to indicate high field strength by drawing lines of force close together and low field strength by widely spaced lines.

Force acting, $F = 0.08$ N

$\therefore$ Electric intensity, $E = \frac{F}{Q} = \frac{0.08}{2\times10^{-6}} = \mathbf{4\times10^4\ N/C}$

(*ii*) If electric intensity at a point in an electric field is *E*, it means that a force of *E* newtons acts on a charge of 1 C placed at that point. If a charge of *Q* coulomb is placed at that point, then force (*F*) acting on the charge is

$$F = EQ = (4.5\times10^5)\times(0.52\times10^{-6}) = \mathbf{0.23\ N}$$

Example 5.7. *Two equal and opposite charges of magnitude 2×10^{-7} C are placed 15 cm apart. What is the magnitude and direction of electric intensity (E) at a point mid-way between the charges. What force would act on a proton (charge = $+1.6\times10^{-19}$ C) placed there ?*

Solution. Fig. 5.6 shows two equal and opposite charges separated by a distance of 15 cm *i.e.*, 0.15 m. Let *M* be the mid point *i.e.*, $AM = MB = 0.075$ m.

(*i*) Imagine a charge of + 1 C placed at *M*.

Electric intensity at *M* due to charge $+2\times10^{-7}$ C

$$E_1 = 9\times10^9\times\frac{2\times10^{-7}}{(0.075)^2} = 0.32\times10^6 \text{ N/C } \textit{along AM}$$

Electric intensity at *M* due to charge -2×10^{-7} C

$$E_2 = 9\times10^9\times\frac{2\times10^{-7}}{(0.075)^2} = 0.32\times10^6 \text{ N/C } \textit{along MB.}$$

Since electric intensities are acting in the same direction, the resultant intensity *E* is the sum of E_1 and E_2.

$\therefore$ Resultant intensity at point *M*

$$E = 0.32\times10^6 + 0.32\times10^6 = \mathbf{0.64\times10^6\ N/C}\ \textit{along AB}$$

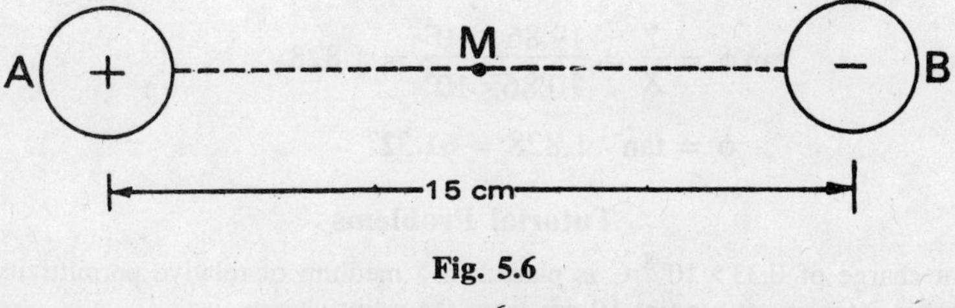

Fig. 5.6

(*ii*) Electric intensity *E* at *M* is 0.64×10^6 N/C. Therefore, force *F* acting on a proton (charge, $Q = +1.6\times10^{-19}$ C) placed at *M* is

$$F = EQ = (0.64\times10^6)\times(1.6\times10^{-19}) = \mathbf{1.024\times10^{-13}\ N}\ \textit{along AB}$$

Example 5.8. *Three point charges of $+8\times10^{-9}$ C, $+32\times10^{-9}$ C and $+24\times10^{-9}$ C are placed at the corners A, B and C of a square ABCD having each side 4 cm. Find the electric intensity at the corner D. Assume the medium is air.*

Solution. The various charges are shown in Fig. 5.7. In order to find electric intensity at the corner *D*, imagine a unit positive charge (*i.e.*, + 1 C) is placed at *D*. It is clear from Fig. 5.7 that $BD = \sqrt{2}\times0.04$ m.

Electric intensity at D due to charge $+8 \times 10^{-9}$ C

$$= 9 \times 10^9 \times \frac{8 \times 10^{-9} \times 1}{(0.04)^2} \text{ N/C} = 4.5 \times 10^4 \text{ N/C } \textit{along } AX$$

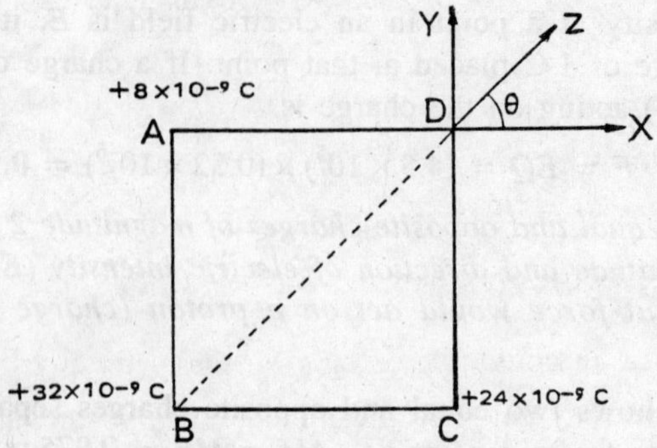

Fig. 5.7

Electric intensity at D due to charge $+32 \times 10^{-9}$ C

$$= 9 \times 10^9 \times \frac{32 \times 10^{-9} \times 1}{(\sqrt{2} \times 0.04)^2} \text{ N/C} = 9 \times 10^4 \text{ N/C } \textit{along } DZ$$

Electric intensity at D due to charge $+24 \times 10^{-9}$ C

$$= 9 \times 10^9 \times \frac{24 \times 10^{-9} \times 1}{(0.04)^2} \text{ N/C} = 13.5 \times 10^4 \text{ N/C } \textit{along } DY$$

The three forces are acting at D ; θ being 45°. In order to find the resultant, resolve the forces along X-axis and Y-axis.

Total X-component, $X = 4.5 \times 10^4 + 9 \times 10^4 \cos 45° + 0 = 10.86 \times 10^4$ N/C

Total Y-component, $Y = 13.5 \times 10^4 + 9 \times 10^4 \cos 45° + 0 = 19.86 \times 10^4$ N/C

Resultant intensity at $D = \sqrt{(10.86 \times 10^4)^2 + (19.86 \times 10^4)^2} = \mathbf{22.63 \times 10^4}$ **N/C**

Let the resultant intensity make an angle ϕ with DX.

$$\therefore \quad \tan\phi = \frac{Y}{X} = \frac{19.86 \times 10^4}{10.86 \times 10^4} = 1.828$$

or $\phi = \tan^{-1} 1.828 = \mathbf{61.32°}$

Tutorial Problems

1. A point charge of 0.33×10^{-8} C is placed in a medium of relative permittivity of 5. Calculate the electric intensity at a point 10 cm from the point charge. **[0.525×10^3 N/C]**
2. What is the magnitude of a point charge chosen so that electric field 20 cm away has a magnitude of 18×10^6 N/C ? **[80 μC]**
3. Three point charges of $+0.33 \times 10^{-8}$ C, $+0.33 \times 10^{-8}$ C and -0.165×10^{-8} C are placed at the corners of a square of 5 cm side. Calculate the electric intensity at the fourth corner. **[1.635×10^4 N/C]**
4. A vacuum tube contains two plane parallel plate electrodes 7.5 mm apart. If a p.d. of 150 V (d.c.) is maintained between them, what is the electric field strength in the gap ? What is the force acting on an electron in the gap ? Charge on an electron = 1.6×10^{-19} C. **[2×10^4 N/C ; 3.2×10^{-15} N]**

5. A tiny charged droplet carries one surplus electron (1.6×10^{-19} C). If this droplet is balanced between two parallel plates in a field of 10^5 N/C, find the mass of droplet. **[1.63×10^{-15} kg]**
6. Three equal charges of $+4 \times 10^{-7}$ C are located at the corners of a right triangle *ABC* whose sides are $AB = 6$ cm, $BC = 8$ cm and $CA = 10$ cm. Find the force exerted on the charge located at the 90° angle. **[0.459 N repulsion at 29.4°]**

5.7. Electric Flux Density

The **electric flux density** *at any section in an electric field is the electric flux crossing normally per unit area of that section i.e.*

$$\text{Electric flux density, } D = \frac{\psi}{A} \text{ C/m}^2$$

where ψ is the electric flux in coulombs passing normally through an area A m^2.

For example, when we say that electric flux density in an electric field is 4 C/m^2, it means that 4 C of electric flux passes normally through an area of 1 m^2. Electric flux density is a vector quantity ; possessing both magnitude and direction. Its direction is the same as the direction of electric intensity.

Relation between D and E. Consider a charge of $+Q$ coulombs placed in a medium of relative permittivity ε_r as shown in Fig. 5.8. The electric flux density at P at a distance d metres from the charge can be found as follows. With centre at the charge and radius d metres, an imaginary sphere can be considered. The electric flux of Q coulombs will pass normally through this imaginary sphere.

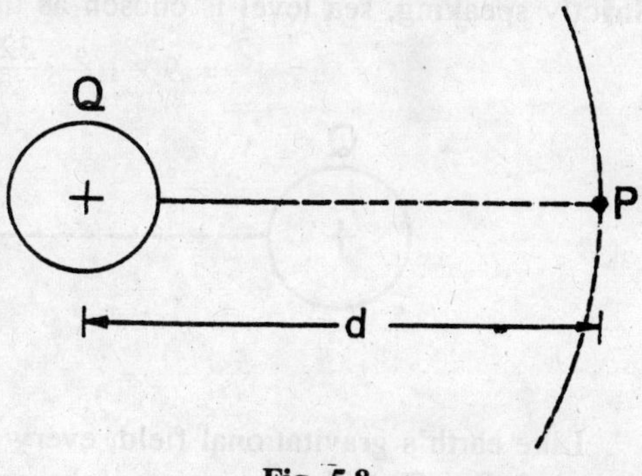

Fig. 5.8

Now area of sphere is $= 4\pi d^2$.

$\therefore$ Flux density at P, $D = \dfrac{\text{Flux}}{\text{area}} = \dfrac{Q}{4\pi d^2}$

Also Electric intensity at P, $E = \dfrac{Q}{4\pi\varepsilon_0\varepsilon_r d^2} = \dfrac{Q}{4\pi d^2} \times \dfrac{1}{\varepsilon_0\varepsilon_r} = \dfrac{D}{\varepsilon_0\varepsilon_r}$ $\left[\because D = \dfrac{Q}{4\pi d^2}\right]$

$$\therefore \qquad D = \varepsilon_0\varepsilon_r E$$

Hence flux density at any point in an electric field is $\varepsilon_0\varepsilon_r$ times the electric intensity at that point.

It may be noted that in a uniform field (*e.g.*, between the plates of a capacitor), the flux density D (= $\varepsilon_0\varepsilon_r E$) remains the same and is independent of relative permittivity of the medium.

Example 5.9. *In the electric field of a parallel plate air capacitor, the values of D and E are 0.885 μC/m^2 and 10^5 N/C respectively. If area of each plate is 1 m^2, find (i) electric flux (ii) value of absolute permittivity of free space.*

Solution.

(*i*) Electric flux, $\psi = D \times A = (0.885 \times 10^{-6}) \times 1 = \mathbf{0.885 \times 10^{-6}\ C}$

(*ii*) Now, $D = \varepsilon_0\varepsilon_r E$ [See Art. 5.7]

$= \varepsilon_0 E$ (For air, $\varepsilon_r = 1$)

$$\therefore \qquad \varepsilon_0 = \frac{D}{E} = \frac{0.885 \times 10^{-6}}{10^5} = \mathbf{8.85 \times 10^{-12}\ F/m.}$$

Example 5.10. *Two parallel plates are charged to have a p.d. of 100 V. Each has an area of 0.05 m² and they are separated 0.5 mm apart. Calculate the electric charge on each plate.*

Solution.

Electric intensity, $E = \frac{V}{d} = \frac{100}{0.5 \times 10^{-3}} = 2 \times 10^5$ V/m

Flux density, $D = \varepsilon_0 \varepsilon_r E = (8.854 \times 10^{-12}) \times 1 \times 2 \times 10^5 = 17.7 \times 10^{-7}$ C/m²

Charge on each plate, $Q = D \times A = (17.7 \times 10^{-7}) \times (0.05) = \mathbf{0.885 \times 10^{-7}}$ **C**

5.8. Electric Potential

We know that earth has gravitational field which attracts the bodies towards earth. When a body is raised above the ground level, it possesses mechanical potential energy which is equal to the amount of work done in raising the body to that point. The greater the height to which the body is raised, the greater will be its potential energy. Thus the potential energy of the body depends upon its position in the gravitational field ; being zero on earth's surface. Strictly speaking, sea level is chosen as the place of zero potential.

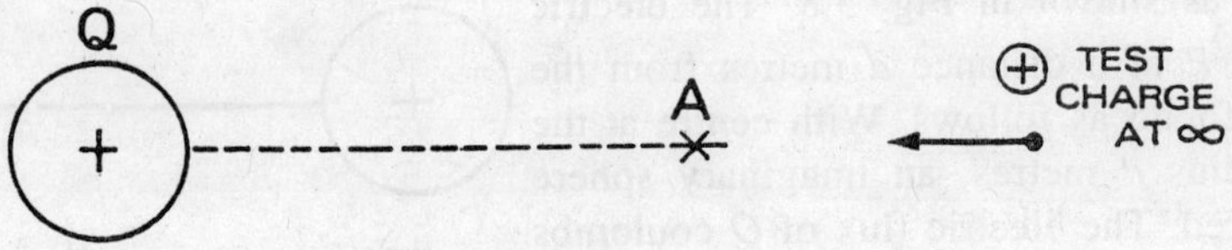

Fig. 5.9

Like earth's gravitational field, every charge has electric field which theoretically extends upto infinity. Consider an isolated charge $+Q$ fixed in space as shown in Fig. 5.9. If a test charge (*i.e.*, + 1 C) is placed at infinity, the force on it due to charge $+Q$ is *zero. If the test charge at infinity is moved towards $+Q$, a force of repulsion acts on it (like charges repel) and hence work is required to be done to bring it to a point like *A*. Hence when the test charge is at *A*, it has some amount of electric potential energy which we commonly call electric potential. The closer the point to the charge, the higher will be the electric potential at that point. Therefore, electric potential at a point due to a charge depends upon the position of the point ; being zero if the point is situated at infinity. Obviously, in electric field, infinity is chosen as the point of **zero potential.

Hence **electric potential** *at a point in an electric field is the amount of work done in bringing a unit positive charge (i.e., +1 C) from infinity to that point i.e.*

$$\text{Electric potential} = \frac{\text{Work}}{\text{Charge}} = \frac{W}{Q}$$

where *W* is the work done to bring a charge of *Q* coulombs from infinity to the point under consideration.

---

* $F = 9 \times 10^9 \times \frac{Q \times 1}{d^2}$ · As $d \to \infty$, $F \to 0$

** In practice, earth is chosen as zero potential. It is because earth is such a huge conductor that its potential practically remains constant.

Unit. The *SI* unit of electric potential is *volt and may be defined as under :

The potential at a point in an electric field is **1 volt** *if 1 Joule of work is done in bringing a unit positive charge (i.e., +1 C) from infinity to that point* **against the electric field.

Thus when we say that potential at a point in an electric field is 5V, it simply means that 5 joules of work has been done in bringing a unit positive charge from infinity to that point.

5.9. Electric Potential Difference

In practice, we are more concerned with potential difference between two points rather than their ***absolute potentials. The potential difference (p.d.) between two points may be defined as under :

The potential difference between two points is the amount of work done in moving a unit positive charge (i.e., +1 C) from the point of lower potential to the point of higher potential.

Consider two points *A* and *B* in the electric field of a charge $+Q$ as shown in Fig. 5.10. Let V_2 and V_1 be the absolute potentials at *A* and *B* respectively. Clearly $V_2 > V_1$. The potential V_1 at *B* means that V_1 Joules of work has been done in bringing a unit positive charge from infinity to point *B*. Let the extra work done to bring the unit positive charge from *B* to *A* be *W* Joules.

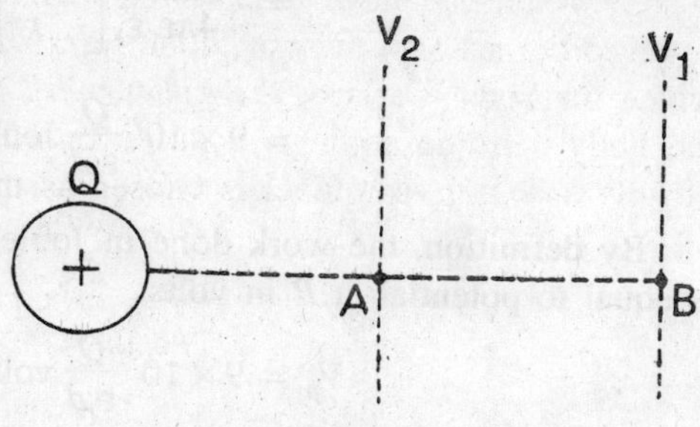

Fig. 5.10.

$\therefore$ Potential at $A = V_1 + W$

$\therefore$ P.D. between *A* and $B = (V_1 + W) - V_1$

or $V_2 - V_1 = W = W.D.$ to move + 1 C from *B* to *A*

The SI unit of potential difference is volt and may be defined as under :

The p.d. between two points is **1V** *if 1 Joule of work is done in bringing a unit positive charge (i.e., +1 C) from the point of lower potential to the point of higher potential.*

Thus when we say that p.d. between two points is 5 volts, it simply means that 5 Joules of work will have to be done to bring +1 C of charge from the point of lower potential to the point of higher potential. Conversely, 5 Joules of work or energy will be released if +1 C charge moves from the point of higher potential to the point of lower potential.

5.10. Potential at a Point

Consider an isolated positive charge of *Q* coulombs placed in a medium of relative permittivity ε_r. It is desired to find the electric potential at point *P* due to this charge. Let *P* be at a distance *d* metres from the charge. Imagine a unit positive charge (*i.e.*, +1 C) placed at *A* and situated *x* metres from the charge. Then the force acting on this unit charge (*i.e.*, electric intensity) is given by ;

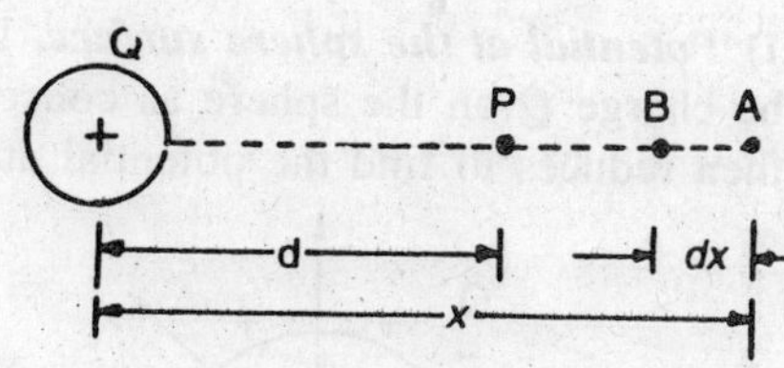

Fig. 5.11

---

* Electric potential $= W/Q$ = Joules/Coulomb. Now Joule/Coulomb has been given a special name *viz.* volt.

** Note if the field is due to a positive charge (as is in this case), work will be done against the electric field. However, if the field is due to a negative charge, work is done by the electric field.

***The potential at a point with infinity as reference is termed as absolute potential.

$$F = E = \frac{Q}{4\pi\varepsilon_0\varepsilon_r x^2}$$

If this unit positive charge at A is moved through a small distance dx towards the charge $+Q$, then work done is given by :

$$dW = \frac{Q}{4\pi\varepsilon_0\varepsilon_r x^2} \times (- {}^* dx) = - \frac{Q}{4\pi\varepsilon_0\varepsilon_r x^2} dx$$

Total work done in bringing a unit positive charge from infinity to point P is

$$\text{Total work done, } W = \int_\infty^d - \frac{Q}{4\pi\varepsilon_0\varepsilon_r x^2} dx = - \frac{Q}{4\pi\varepsilon_0\varepsilon_r} \int_\infty^d \frac{1}{x^2} dx$$

$$= - \frac{Q}{4\pi\varepsilon_0\varepsilon_r} \left[- \frac{1}{x} \right]_\infty^d = \frac{-Q}{4\pi\varepsilon_0\varepsilon_r} \left[- \frac{1}{d} - \left(- \frac{1}{\infty} \right) \right] = \frac{Q}{4\pi\varepsilon_0\varepsilon_r d}$$

$$= 9 \times 10^9 \frac{Q}{\varepsilon_r d} \text{ Joules} \qquad \left[\because \frac{1}{4\pi\varepsilon_0} = 9 \times 10^9 \right]$$

By definition, the work done in Joules to bring a unit positive charge from infinity to P is equal to potential at P in volts.

$$\therefore \quad V_p = 9 \times 10^9 \frac{Q}{\varepsilon_r d} \text{ volts} \qquad \text{...in a medium}$$

$$= 9 \times 10^9 \frac{Q}{d} \text{ volts} \qquad \text{...in air}$$

Electric potential is a scalar quantity. Therefore, electric potential at a point due to a number of charges is equal to the algebraic sum of potentials due to each charge.

5.11. Potential of a Charged Sphere

Consider an isolated sphere of radius r metres placed in air and charged uniformly with Q coulombs. The field has spherical symmetry *i.e.*, lines of force spread out normally from the surface and meet at the centre of the sphere if produced backward. Outside the sphere, then field is exactly the same as though the charge Q on sphere were concentrated at its centre.

(*i*) ***Potential at the sphere surface.*** Due to spherical symmetry of the field, we can imagine the charge Q on the sphere as concentrated at its centre O [See Fig. 5.12 (*i*)]. The problem then reduces to find the potential at a point r metres from a charge Q.

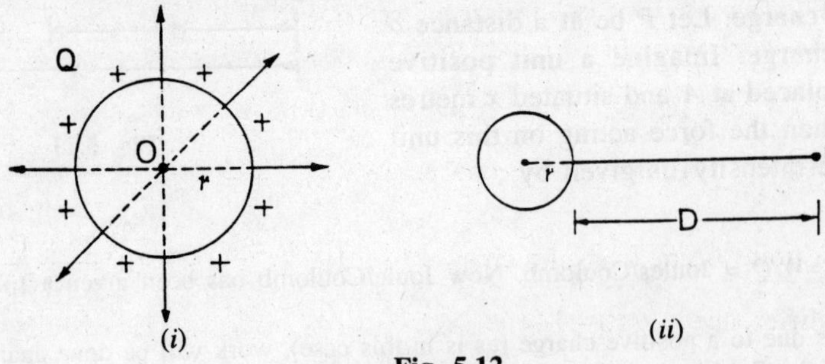

Fig. 5.12

* The negative sign is taken because dx is considered in the negative direction of distance (x).

$\therefore$ Potential at the surface of sphere $= \dfrac{Q}{4\pi\varepsilon_0 r}$ volts [See Art 5.10]

$$= 9 \times 10^9 \frac{Q}{r}^* \text{ volts}$$

(ii) ***Potential outside the sphere.*** Consider a point *P* outside the sphere as shown in Fig. 5.12 *(ii)*. Let this point be at a distance of *D* metres from the surface of sphere.

$$\text{Then potential at} \quad P = 9 \times 10^9 \frac{Q}{(D+r)} \text{ volts}$$

(iii) **Potential inside the sphere.** Since there is no electric flux inside the sphere, electric intensity inside the sphere is zero.

Now, $$\text{electric intensity} = \frac{\text{Change in potential}}{r}$$

or $$0 = \text{Change in potential}$$

Hence all the points inside the sphere are at the same potential as the points on the surface.

Example 5.11. *A point charge of 1 μC is placed between points A and B 3 m apart. Point A is 2 m from the charge and B is 1 m from the charge. What is the potential difference* $V_A - V_B$*?*

Solution.

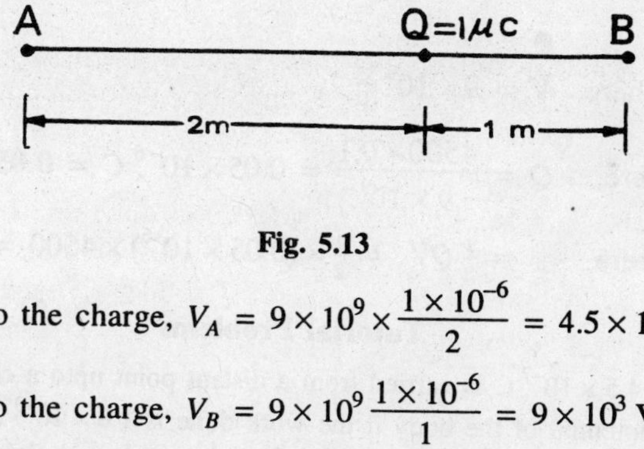

Fig. 5.13

Potential at *A* due to the charge, $V_A = 9 \times 10^9 \times \dfrac{1 \times 10^{-6}}{2} = 4.5 \times 10^3\ V$

Potential at *B* due to the charge, $V_B = 9 \times 10^9 \dfrac{1 \times 10^{-6}}{1} = 9 \times 10^3$ V

$$\therefore \quad V_A - V_B = 4.5 \times 10^3 - 9 \times 10^3 = \mathbf{-4.5 \times 10^3\ V}$$

Example 5.12. *Two positive point charges of* 16×10^{-10} *C and* 12×10^{-10} *C are placed 10 cm apart. Find the work done in bringing the two charges 4 cm closer.*

Solution. Suppose the charge 16×10^{-10} C to be fixed.

Potential of a point 10 cm from the charge 16×10^{-10} C$= 9 \times 10^9 \dfrac{16 \times 10^{-10}}{0.1} = 144$ V

Potential of a point 6 cm from the charge 16×10^{-10} C $= 9 \times 10^9 \dfrac{16 \times 10^{-10}}{0.06} = 240$ V

$\therefore$ Potential difference $= 240 - 144 = 96$ V

---

* If sphere is placed in a medium (ε_r), then potential is $= 9 \times 10^9 \dfrac{Q}{\varepsilon_r r}$

Work done = Charge × p.d. = $12\times10^{-10}\times96$ = **11.52×10^{-8} Joules**

Example 5.13. *A square ABCD has each side of 1 m. Four point charges of +0.01 μC, – 0.02 μC, + 0.03 μC and +0.02 μC are placed at A, B, C and D respectively. Find the potential at the centre of the square.*

Solution. Fig. 5.14 shows the square *ABCD* with charges placed at its corners. The diagonals of the square intersect at point *P*. Clearly point *P* is the centre of the square. The distance of each charge from point *P* is

$$= \frac{1}{2}\sqrt{1^2+1^2} = \frac{1}{2}\times\sqrt{2} = 0.707 \text{ m}$$

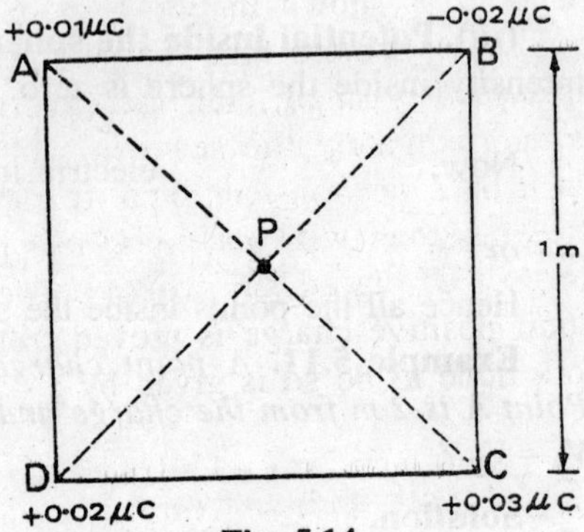

Fig. 5.14

The potential at point *P* due to all charges is equal to the algebraic sum of potentials due to each charge.

∴ Potential at *P* due to all charges

$$= 9\times10^9\left[\frac{Q_1}{0.707}+\frac{Q_2}{0.707}+\frac{Q_3}{0.707}+\frac{Q_4}{0.707}\right]$$

$$= \frac{9\times10^9}{0.707}[(0.01-0.02+0.03+0.02)\,10^{-6}]$$

$$= \frac{9\times10^9}{0.707}\times0.04\times10^{-6} = \mathbf{509.2\ V}$$

Example 5.14. *A sphere of radius 10 cm is charged to a potential of 4500 volts. Calculate (i) the charge and (ii) energy of sphere.*

Solution.

(*i*) Potential of sphere, $V = 9\times10^9\dfrac{Q}{r}$

∴ Charge on sphere, $Q = \dfrac{4500\times0.1}{9\times10^9} = 0.05\times10^{-6}\ C = \mathbf{0.05\ \mu C}$

(*ii*) Energy of sphere $= \frac{1}{2}QV = \frac{1}{2}\times(0.05\times10^{-6})\times4500 = \mathbf{112.5\times10^{-6}}$ **Joules**

Tutorial Problems

1. A charge of -4.5×10^{-7} C is carried from a distant point upto a charged metal sphere. What is the electrical potential of the body if the work done is 1.8×10^{-3} Joule ? **[4×10^3 V]**
2. The difference of potentials between two points in an electric field is 6 volts. How much work is required to move a charge of 300 μC between these points. **[1.8×10^{-3} Joule]**
3. A force of 0.032 N is required to move a charge of 42 μC in an electric field between two points 25 cm apart. What potential difference exists between the two points ? **[1.9×10^2 V]**
4. What is the magnitude of an isolated positive charge to give an electric potential of 100 V at 10 cm from the charge ? **[1.11×10^{-9} C]**
5. A square *ABCD* has each side of 1 m. Four charges of + 0.02 μC, + 0.04 μC, + 0.06 μC and + 0.02 μC are placed at *A, B, C* and *D* respectively, Find the potential at the centre of square. **[1000 V]**
6. A sphere of radius 0.1 m has a charge of 5×10^{-8} C. Determine the potential (*i*) at the surface of sphere, (*ii*) inside the sphere and (*iii*) at a distance of 1 m from the surface of the sphere. Assume air as the medium. **[(*i*) 4500 V (*ii*) 4500 V (*iii*) 409 V]**

5.12. Potential Gradient

The change of potential per unit distance is called **potential gradient** *i.e.,*

$$\text{Potential gradient} = \frac{V_2 - V_1}{S}$$

where $V_2 - V_1$ is the change in potential (or p.d.) between two points S metres apart. Obviously, the unit of potential gradient will be volts/m.

Consider a charge $+Q$ and let there be two points A and B situated S metres apart in its electric field as shown in Fig. 5.15. Clearly, potential at point A is more than the potential at point B. If distance S is small, then the electric intensity will be approximately the same in this small distance. Let it be E newtons/coulomb. It means that a force of E newtons will act on a unit positive charge (*i.e.*, + 1 C) placed any where between A and B. If a unit positive charge is moved from B to A, then work done to do so is given by ;

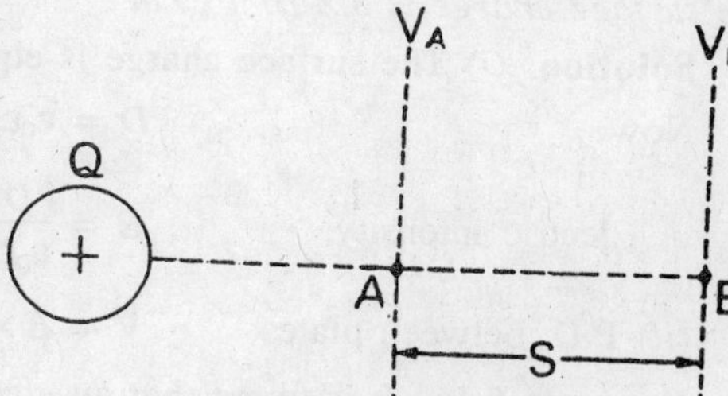

Fig. 5.15

$$\text{Work done} = E \times S \text{ Joules}$$

But work done in bringing a unit positive charge from B to A is the potential difference $(V_A - V_B)$ between A and B

$$\therefore \quad E \times S = V_A - V_B$$

or $$E = \frac{V_A - V_B}{S} = \text{Potential gradient}$$

In differential form, $$E = -\overset{*}{\frac{dV}{dS}}$$

Hence electric intensity at a point is numerically equal to the potential gradient at that point.

Since electric intensity is numerically equal to potential gradient at any point, both must be measured in the same units. Clearly, electric intensity can also be measured in V/m. For example, when we say that potential gradient at a point is 1000 V/m, it means that electric intensity at that point is also **1000 V/m or 1000 N/C.

5.13. Breakdown Voltage or Dielectric Strength

In an insulator or dielectric, the valence electrons are tightly bound so that no free electrons are available for current conduction. However, when voltage applied to a dielectric is gradually increased, a point is reached when these electrons are torn away, a large current (much large than the usual leakage current) flows through the dielectric and the material loses its insulating properties. Usually, a spark or arc occurs which burns up the material. The minimum voltage required to breakdown a dielectric is called breakdown voltage or dielectric strength.

---

* Since work done in moving +1 C from B to A is against electric field, a negative sign must be used to make the equation technically correct.

** It can be shown that 1 V/m = 1 N/C as under :

$$1\text{ V/m} = \frac{\text{joule/coulomb}}{\text{metre}} = \frac{\text{newton} \times \text{metre}}{\text{metre} \times \text{coulomb}} = 1\text{ N/C}$$

The maximum voltage which a unit thickness of a dielectric can withstand without being punctured by a spark discharge is called **dielectric strength** *of the material.*

The dielectric strength (or breakdown voltage) is generally measured in kV/cm or kV/mm. For example, air has a dielectric strength of 30 kV/cm. It means that maximum p.d. which 1 cm thickness of air can withstand across it without breaking down is 30 kV.

Example 5.15. *A parallel plate capacitor has plates 1 mm apart and a dielectric with relative permittivity of 3.39. Find (i) electric intensity and (ii) the voltage between plates if the surface charge is 3×10^{-4} C/m^2.*

Solution. (*i*) The surface charge is equal to electric flux density D.

Now $$D = \varepsilon_0 \varepsilon_r E$$

$\therefore$ Electric intensity, $$E = \frac{D}{\varepsilon_0 \varepsilon_r} = \frac{3 \times 10^{-4}}{8.854 \times 10^{-12} \times 3.39} = \mathbf{10^7\ V/m}$$

(*ii*) P.D. between plates, $$V = E \times dx = 10^7 \times (1 \times 10^{-3}) = \mathbf{10^4\ V}$$

Example 5.16. *It is found that an electric field of 3×10^6 V/m in air will cause electrical breakdown of air. What is the greatest charge that can be placed on the sphere of 1m diameter ? What is the potential of the sphere for that charge ?*

Solution. (*i*) Electric intensity E is equal to the breakdown voltage (= 3×10^6 V/m).

Let Q coulomb be the maximum charge that can be placed on the sphere.

Electric intensity at the surface of sphere, $$E = 9 \times 10^9 \frac{Q}{r^2} \quad \textit{(medium is air)}$$

or $$Q = \frac{E \times r^2}{9 \times 10^9} = \frac{(3 \times 10^6) \times (0.5)^2}{9 \times 10^9} = 83.3 \times 10^{-6}\ \text{C} = \mathbf{83.3\ \mu C}$$

(*ii*) Potential of Sphere, $$V = 9 \times 10^9 \frac{Q}{r} = 9 \times 10^9 \times \frac{83.3 \times 10^{-6}}{0.5} = \mathbf{15 \times 10^5\ V}$$

Example 5.17. *A sheet of glass 1.5 cm thick and of relative permittivity 7 is introduced between two parallel brass plates 2 cm apart. The remainder of the space between the plates is occupied by air. If a p.d. of 10,000 V is applied between the plates, calculate (i) electric intensity of air film between glass and plate and (ii) in the glass sheet.*

Solution. Fig. 5.16 shows the arrangement. Let V_1 and V_2 be the p.d. across air and glass respectively and E_1 and E_2 the corresponding electric intensities.

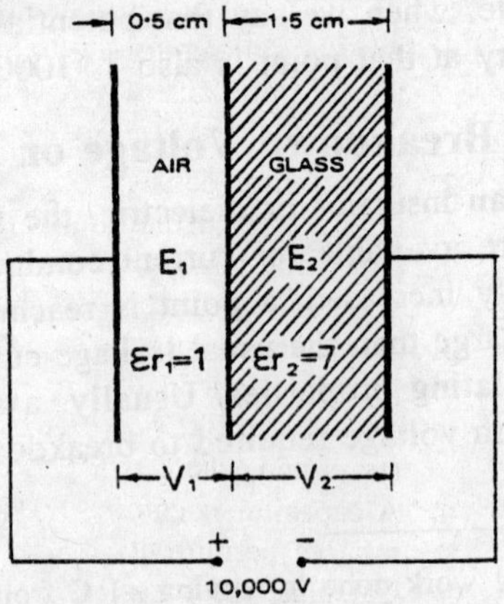

Fig. 5.16

Now, $$V_1 = E_1 x_1 = E_1 \times (0.5 \times 10^{-2})$$

and $$V_2 = E_2 x_2 = E_2 \times (1.5 \times 10^{-2})$$

Now $$V = V_1 + V_2$$

or $$10{,}000 = (0.5\, E_1 + 1.5\, E_2)\, 10^{-2}$$

or $$E_1 + 3\, E_2 = 2 \times 10^6 \quad \ldots(i)$$

Now electric flux density D (= $\varepsilon_0 \varepsilon_r E$) is the same in the two media because it is independent of the surrounding medium.

$$\therefore \qquad \varepsilon_0\, \varepsilon_{r1}\, E_1 = \varepsilon_0\, \varepsilon_{r2}\, E_2$$

$$or \qquad E_1 = 7\, E_2 \qquad ...(ii)$$

From exps (*i*) and (*ii*), we get,

Electric intensity in air $= \mathbf{1.4 \times 10^6\ V/m}$; Electric intensity in glass $= \mathbf{0.2 \times 10^6\ V/m}$

Example 5.18. *A capacitor has two dielectrics 1 mm and 2 mm thick. The relative permittivities of these dielectrics are 3 and 6 respectively. Calculate the potential gradient along the dielectrics if a p.d. of 1000 V is applied between the plates.*

Solution. Fig. 5.17 shows the arrangement. Finding the potential gradient means to find the electric intensity (or electric stress).

$$V_1 = E_1\, x_1$$

$$= E_1 \times (1 \times 10^{-3})$$

$$V_2 = E_2\, x_2$$

$$= E_2 \times (2 \times 10^{-3})$$

Now $\qquad V = V_1 + V_2$

or $\qquad 1000 = (E_1 + 2E_2)\, 10^{-3}$

or $\qquad E_1 + 2E_2 = 10^6 \qquad ...(i)$

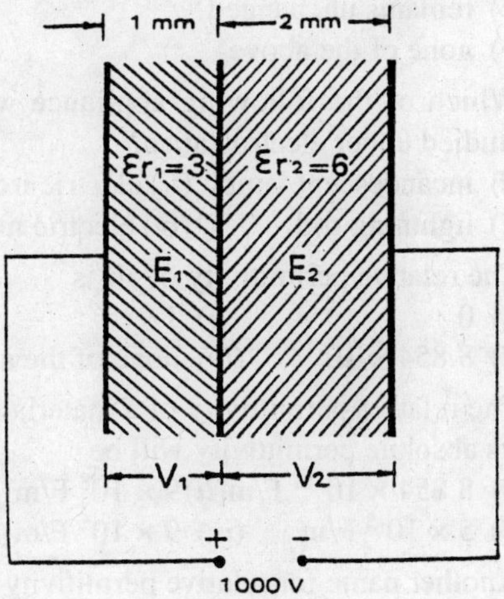

Fig. 5.17

Since flux density $D\ (= \varepsilon_0 \varepsilon_r E)$ is the same in the two media,

$$\therefore \qquad \varepsilon_0 \varepsilon_{r1} E_1 = \varepsilon_0 \varepsilon_{r2} E_2$$

$$or \qquad 3\, E_1 = 6\, E_2 \qquad ...(ii)$$

From exps. (*i*) and (*ii*), we get,

$$E_1 = \mathbf{0.5 \times 10^6\ V/m}\ ; \quad E_2 = \mathbf{0.25 \times 10_6\ V/m}$$

Tutorial Problems

1. A parallel plate capacitor is built up from two metal plates, each of area 0.05 m^2, separated 0.02 mm by a dielectric. If the p.d. between plates is 200 V, and charge on each plate is 10 μC, find (*i*) the flux density between the plates and (*ii*) absolute permittivity of dielectric. **[(*i*) 2×10^{-4} C/m^2 (*ii*) 20 pF/m]**

2. A capacitor consists of two parallel plates 5 mm apart. The space between the plates is filled with a layer of paper 2 mm thick and a sheet of glass 3 mm thick. The relative permittivities of paper and glass are 2 and 8 respectively. A p.d. of 5 kV is applied between the plates. Find the potential gradient in each dielectric. **[1.82 kV/mm** in paper ; **0.453 kV/mm** in glass**]**

3. Two parallel metal plates of large area are spaced at a distance of 1 cm from each other in air and a p.d. of 5 kV is maintained between them. If a sheet of glass, 0.5 cm thick and having a relative permittivity of 6, is introduced between the plates, what will be the maximum electric stress and where will it occur ? **[8.58 kV/cm** in air gap**]**

4. A capacitor is composed of two plates separated by three dielectrics each of thickness 1 mm and relative permittivity 2, 4, and 5 respectively. A p.d. of 1000 V is applied between the plates. Calculate the potential gradient in each dielectric.

[5.26×10^5 V/m ; 2.63×10^5 V/m ; 2.11×10^5 V/m]

OBJECTIVE QUESTIONS

1. The force between two electrons separated by a distance r varies as
(*i*) r^2 (*ii*) r
(*iii*) r^{-1} (*iv*) r^{-2}

2. Two charges are placed at a certain distance apart. A brass sheet is placed between them. The force between them will
(*i*) increase (*ii*) decrease
(*iii*) remains unchanged
(*iv*) none of the above

3. Which of the following appliance will be studied under electrostatics?
(*i*) incandescent lamp (*ii*) electric iron
(*iii*) lightning rod (*iv*) electric motor

4. The relative permittivity of air is
(*i*) 0 (*ii*) 1
(*iii*) 8.854×10^{-12} (*iv*) none of the above

5. The relative permittivity of a material is 10. Its absolute permittivity will be
(*i*) 8.854×10^{-11} F/m (*ii*) 9×10^8 F/m
(*iii*) 5×10^{-5} F/m (*iv*) 9×10^5 F/m

6. Another name for relative permitivity is
(*i*) dielectric constant
(*ii*) dielectric strength
(*iii*) potential gradient
(*iv*) none of the above

7. The relative permittivity of most materials lies between
(*i*) 20 and 100 (*ii*) 10 and 20
(*iii*) 100 and 200 (*iv*) 1 and 10

8. When the relative permittivity of the medium is increased, the force between two charges placed at a given distance apart
(*i*) increases (*ii*) decreases
(*iii*) remains the same
(*iv*) none of the above

9. Two charges are placed at a distance apart. If a glass slab is placed between them, the force between the charges will
(*i*) be zero (*ii*) increase
(*iii*) decrease (*iv*) remain the same

10. There are two charges of + 1 μC and + 5 μC. The ratio of the forces acting on them will be
(*i*) 1 : 1 (*ii*) 1 : 5
(*iii*) 5 : 1 (*iv*) 1 : 25

11. A soap bubble is given a negative charge. Its radius
(*i*) decreases (*ii*) increases
(*iii*) remains unchanged
(*iv*) information is incomplete to say anything

12. The ratio of force between two small spheres with constant charge in air and in a medium of relative permittivity K is
(*i*) $K^2 : 1$ (*ii*) $1 : K$
(*iii*) $1 : K^2$ (*iv*) $K : 1$

13. An electric field can deflect
(*i*) x-rays (*ii*) neutrons
(*iii*) α-particles (*iv*) γ-rays

14. Electric lines of force enter or leave a charged surface at an angle
(*i*) of 90° (*ii*) of 30°
(*iii*) of 60°
(*iv*) depending upon surface conditions

15. The relation between absolute permittivity of vacuum (ε_0), absolute permeability of vacuum (μ_0) and velocity of light (c) in vacuum is
(*i*) $\mu_0\varepsilon_0 = c^2$ (*ii*) $\mu_0/\varepsilon_0 = c$
(*iii*) $\varepsilon_0/\mu_0 = c$ (*iv*) $\dfrac{1}{\mu_0/\varepsilon_0} = c^2$

16. As one penetrates a uniformly charged sphere, the electric field strength E
(*i*) increases (*ii*) decreases
(*iii*) is zero at all points
(*iv*) remains the same as at the surface

17. If the relative permittivity of the medium increases, the electric intensity at a point due to a given charge
(*i*) decreases (*ii*) increases
(*iii*) remains the same
(*iv*) none of the above

18. Electric lines of force about a negative point charge are
(*i*) circular, anticlockwise
(*ii*) circular, clockwise
(*iii*) radial, inward
(*iv*) radial, outward

19. A hollow sphere of charge does not produce an electric field at any
(*i*) outer point (*ii*) interior point

(*iii*) beyond 2 m (*iv*) beyond 10 m

20. Two charged spheres of radii 10 cm and 15 cm are connected by a thin wire. No current will flow if they have
(*i*) the same charge
(*ii*) the same energy
(*iii*) the same field on their surface
(*iv*) the same potential

21. Electric potential is a
(*i*) scalar quantity
(*ii*) a vector quantity
(*iii*) dimensionless
(*iv*) nothing can be said

22. A charge Q_1 exerts some force on a second charge Q_2. A third charge Q_3 is brought near. The force of Q_1 exerted on Q_2
(*i*) decreases
(*ii*) increases
(*iii*) remains unchanged
(*iv*) increases if Q_3 is of the same sign as Q_1 and decreases if Q_3 is of opposite sign

23. The potential at a point due to a charge is 9 V. If the distance is increased three times, the potential at that point will be
(*i*) 27 V (*ii*) 3 V
(*iii*) 12 V (*iv*) 18 V

24. A hollow metal sphere of radius 5 cm is charged such that the potential on its surface is 10 V. The potential at the centre of the sphere is
(*i*) 10 V
(*ii*) 0 V
(*iii*) same as at point 5 cm away from the surface
(*iv*) same as at point 25 cm away from the surface

25. If a unit charge is taken from one point another over an equipotential surface, the
(*i*) work is done on the charge
(*ii*) work is done by the charge
(*iii*) work on the charge is constant
(*iv*) no work is done

26. The electric field intensity on the surface of a charged conductor is
(*i*) zero
(*ii*) directed normally to the surface
(*iii*) directed tangentially to the surface
(*iv*) directed along 45° to the surface

27. The force between two charges separated by a distance *d* in air is 10 N. When the charges are placed same distance apart in a medium of dielectric constant ε_r the force between them is 2 N. What is the value of ε_r?
(*i*) 0.5 (*ii*) 20
(*iii*) 5 (*iv*) none of above

28. Two small spheres, each carrying a charge *Q*, placed *d* metres apart repel each other with a force *F*. If one of the spheres is taken around the other one in a circular path of radius *r*, the work done will be
(*i*) zero (*ii*) $F \times 2\pi r$
(*iii*) $F \times r$ (*iv*) $F/2\pi$

29. The electric potential across part *AB* of a circuit is 2 V; point *A* being at higher potential. If a charge 2 *C* moves from *A* to *B*, then energy released is
(*i*) 2 Joules (*ii*) 12 Joules
(*iii*) 8 Joules (*iv*) 4 Joules

30. Out of the materials given below, which has the highest dielectric strength?
(*i*) glass (*ii*) mica
(*iii*) oiled paper (*iv*) air

31. Two equal and similar charges are placed a finite distance apart. A third equal and dissimilar charge is placed mid-way between them. The third charge will be
(*i*) in stable equilibrium
(*ii*) in unstable equilibrium
(*iii*) neutralised (*iv*) oscillating

32. The positive terminal of a 12 V battery is earthed. The negative terminal will be at
(*i*) zero potential (*ii*) + 12 V
(*iii*) – 12 V (*iv*) + 24 V

33. A tiny particle carrying a charge of 0.3 C is accelerated through a potential difference of 1000 V. The kinetic energy acquired by the particle is
(*i*) 900 J (*ii*) 400 J
(*iii*) 100 J (*iv*) 300 J

34. Two charges + 3 μC and – 12 μC are separated by a distance of 0.4 m. Where should a third charge of + 3 μC be placed form + 3 μC so that it experiences zero force?
(*i*) 0.4 m (*ii*) 0.2 m
(*iii*) 0.3 m (*iv*) 0.1 m

35. For a dipole electric field varies as
(*i*) r^{-2} (*ii*) r^{-3}
(*iii*) r^{-1} (*iv*) r^{-4}

ANSWERS TO OBJECTIVE QUESTIONS

1. (*iv*)	**2.** (*ii*)	**3.** (*iii*)	**4.** (*ii*)	**5.** (*i*)	**6.** (*i*)	**7.** (*iv*)	**8.** (*ii*)
9. (*iii*)	**10.** (*i*)	**11.** (*ii*)	**12.** (*iv*)	**13.** (*iii*)	**14.** (*i*)	**15.** (*iv*)	**16.** (*iii*)
17. (*i*)	**18.** (*iii*)	**19.** (*ii*)	**20.** (*iv*)	**21.** (*i*)	**22.** (*iii*)	**23.** (*ii*)	**24.** (*i*)
25. (*iv*)	**26.** (*ii*)	**27.** (*iii*)	**28.** (*i*)	**29.** (*iv*)	**30.** (*i*)	**31.** (*ii*)	**32.** (*iii*)
33. (*iv*)	**34.** (*i*)	**35.** (*ii*)					

HINTS TO SELECTED OBJECTIVE QUESTIONS

5. $\varepsilon_r = \varepsilon/\varepsilon_0$ $\therefore$ Absolute permittivity, $\varepsilon = \varepsilon_0\,\varepsilon_r = 8.854 \times 10^{-12} \times 10 = 8.854 \times 10^{-11}$ F/m

8. Force between two charges in a medium is given by :

$$F = 9 \times 10^9 \frac{Q_1\,Q_2}{\varepsilon_r\,d^2}$$

It is clear that force is inversely proportional to the relative permittivity of the medium in which the charges are placed.

12. $F_{air} = \frac{1}{4\pi\varepsilon_0}\frac{Q_1\,Q_2}{d^2};\ F_m = \frac{1}{4\pi\varepsilon_0 K}\frac{Q_1\,Q_2}{d^2}$ $\therefore$ $\frac{F_{air}}{F_m} = K$

17. Electric intensity (E) at a point in a medium due to charge Q is

$$E = 9 \times 10^9 \frac{Q}{\varepsilon_r\,d^2}$$

Therefore, as the relative permittivity (ε_r) of the medium increases, the electric field intensity decreases.

22. Superposition principle.

23. Electric potential at a point due to charge Q is

$$V = 9 \times 10^9 \frac{Q}{\varepsilon_r\,d} \text{ volts}$$

It is clear that electric potential is inversely proportional to the distance of the point from the given charge.

27. $\varepsilon_r = \frac{F_{air}}{F_m} = \frac{10}{2} = 5$

31. If the third charge is moved slightly towards one of the charges, it would experience greater attraction due to that charge compared to the other one. Therefore, the third charge is in unstable equilibrium.

33. K.E. acquired = charge × p.d. = 0.3 × 1000 = 300 Joules

6

Capacitance

Introduction

It is well known that different bodies hold different charge when given the same potential. This charge holding property of a body is called *capacitance* or *capacity* of the body. In order to store sufficient charge, a device called capacitor is purposely constructed. A capacitor essentially consists of two conducting surfaces (say metal plates) separated by an insulating material (*e.g.*, air, mica, paper *etc.*). It has the property to store electrical energy in the form of electrostatic charge. The capacitor can be connected in a circuit so that this stored energy can be made to flow in a desired circuit to perform a useful function. Capacitance plays an important role in d.c. as well as a.c. circuits. In many circuits (*e.g.*, radio and television circuits), capacitors are intentionally inserted to introduce the desired capacitance. In this chapter, we shall confine our attention to the role of capacitance in d.c. circuits only.

6.1. Capacitor

Any two conducting surfaces separated by an insulating material is called a ***capacitor*** *or* **condenser.**

The conducting surfaces are called the *plates* of the capacitor and the insulating material is called the **dielectric*. The most commonly used dielectrics are air, mica, waxed paper, ceramics *etc.* The following points may be noted carefully :

(*i*) The ability of a capacitor to store charge (*i.e.*, its capacitance) depends upon the area of plates, distance between plates and the nature of insulating material (or dielectric).

(*ii*) A capacitor is generally named after the dielectric used *e.g.* air capacitor, paper capacitor, mica capacitor *etc.*

(*iii*) The capacitor may be in the form of parallel plates, concentric cylinder or other arrangement.

6.2. How Does a Capacitor Store Charge ?

Fig. 6.1 shows how a capacitor stores charge when connected to a d.c. supply. The parallel plate capacitor having plates *A* and *B* is connected across a battery of *V* volts as shown in Fig. 6.1 (*i*). With the switch *S* open as shown in Fig. 6.1 (*i*), the capacitor plates are neutral *i.e.*, there is no charge on the plates. When the switch is closed as shown in Fig. 6.1 (*ii*), the electrons from plate *A* will be attracted by the +ve terminal of the battery and these electrons start **accumulating on plate *B*. The result is that plate *A* attains more and more

* A steady current cannot pass through an insulator but an electric field can. For this reason, an insulator is often referred to as a dielectric.

** The electrons cannot flow from plate *B* to *A* as there is insulating material between the plates. Hence electrons detached from plate *A* start piling up on plate *B*.

positive charge and plate B gets more and more negative charge. This action is referred to as charging a capacitor because the capacitor plates are becoming charged. This process of electron flow or charging (*i.e.*, detaching electrons from plate A and accumulating on B) continues till p.d. across capacitor plates becomes equal to battery voltage V. When the capacitor is charged to battery voltage V, the current flow ceases as shown in Fig. 6.1 (*iii*). If now the switch is opened as shown in Fig. 6.1 (*iv*), the capacitor plates will retain the charges. Thus the capacitor plates which were neutral to start with now have charges on them. This shows that a capacitor stores charge.

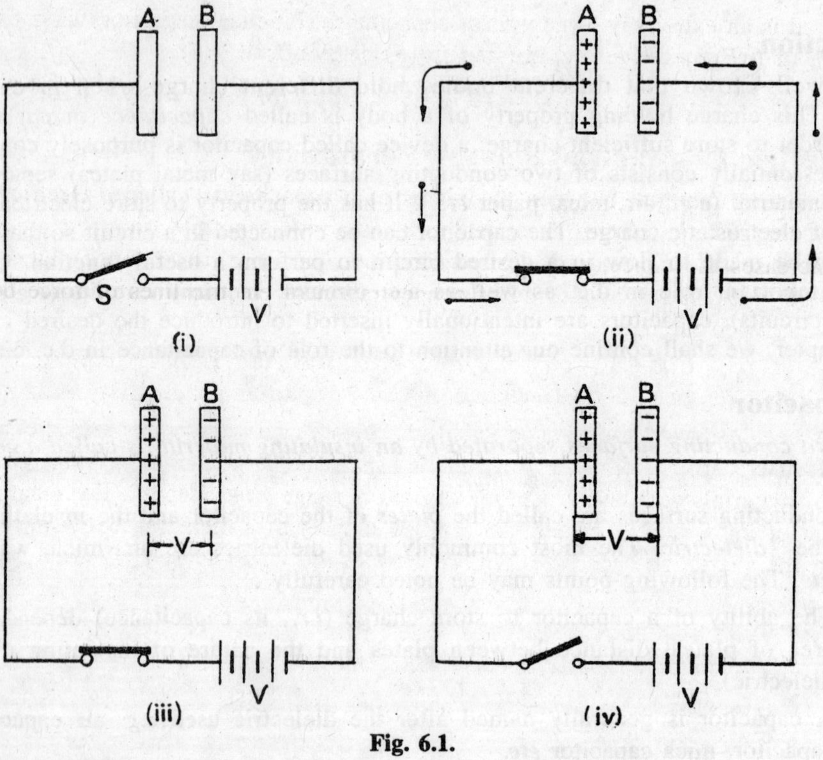

Fig. 6.1.

Note. When a capacitor is charged, the two plates carry equal and opposite charges (say $+Q$ and $-Q$). This is expected because one plate loses as many electrons as the other plate gains. Thus charge on a capacitor means charge on *either* plate.

6.3. Capacitance

The ability of a capacitor to store charge is known as its capacitance. It has been found experimentally that charge Q stored in a capacitor is directly proportional to the p.d. across it *i.e.*

$$Q \propto V$$

or $$\frac{Q}{V} = \text{Constant} = C$$

The constant C is called the capacitance of the capacitor. Hence, capacitance of a capacitor can be defined as under :

The ratio of charge on capacitor plates to the p.d. across the plates is called **capacitance** *of the capacitor.*

Unit of capacitance

We know $C = Q/V$

The SI unit of charge is 1 coulomb and that of voltage is 1 volt. Therefore, the SI unit of capacitance is 1 coulomb/volt which is also called *farad* (Symbol *F*) in honour of Michael Faraday.

$$1 \text{ farad} = 1 \text{ coulomb/volt}$$

A capacitor is said to have a capacitance of **1 farad** *if a charge of 1 coulomb accumulates on each plate when a p.d. of 1 volt is applied across the plates.*

The farad is an extremely large unit of capacitance. Practical capacitors have capacitances of the order of (μF) and micro-microfarad ($\mu\mu F$) or picofarad (pF).

$$1\ \mu F = 10^{-6} \text{ F} \ ; \quad 1\ \mu\mu F \text{ (or 1 pF)} = 10^{-12}\ F$$

6.4. Dielectric Constant or Relative Permittivity

The insulating material between the plates of a capacitor is called dielectric. When the capacitor is charged, the electrostatic field extends across the dielectric. The presence of dielectric* increases the concentration of electric lines of force between the plates and hence the charge on each plate. The degree of concentration of electric lines of force between the plates depends upon the nature of dielectric.

The ability of a dielectric material to concentrate electric lines of force between the plates of a capacitor is called **dielectric constant** *or* **relative permittivity** *of that material.*

Air has been assigned a reference value of dielectric constant (or relative permittivity) as 1. The dielectric constant of all other insulating materials is greater than unity. The dielectric constants of materials commonly used in capacitors range from 1 to 10. For example, dielectric constant of mica is 6. It means that if mica is used as a dielectric between the plates of a capacitor, the charge on each plate will be 6 times the value when air is used ; other things remaining equal. In other words, with mica as dielectric, the capacitance of the capacitor becomes 6 times as great as when air is used.

Let V = Potential difference between capacitor plates

Q = Charge on capacitor when air is dielectric

Then $C_{air} = Q/V$

When mica is used as a dielectric in the same capacitor and the same p.d. is applied, the capacitor will now hold a charge of $6\,Q$.

$$\therefore \quad C_{\text{mica}} = \frac{6\,Q}{V} = 6\frac{Q}{V} = 6\,C_{\text{air}}$$

$$\text{or} \quad \frac{C_{mica}}{C_{air}} = 6 = \text{Dielectric Constant of mica}$$

Hence **dielectric constant** (*or* **relative permittivity**) of a dielectric material is the ratio of capacitances of a capacitor with that material as a dielectric to the capacitance of the same capacitor with air as dielectric.

---

* Normally the electrons of the atoms of the dielectric revolve round their nuclei in their regular orbits. When the capacitor is charged, the electro-static field causes distortion of the orbits of the electrons of the dielectric. This distortion of orbits causes more electrons to be transferred from one plate to the other. Hence, the presence of dielectric increases the charge on the capacitor plates and hence the capacitance.

6.5. Parallel Plate Capacitor with Uniform Medium

Consider a parallel plate capacitor consisting of two plates, each of area A square metres and separated by a *uniform* dielectric of thickness d metres and relative permittivity ε_r as shown in Fig. 6.2. Let a p.d. of V volts applied between the plates place a charge of $+Q$ and $-Q$ on the plates. Suppose E and D respectively are the electric intensity and flux density between the plates.

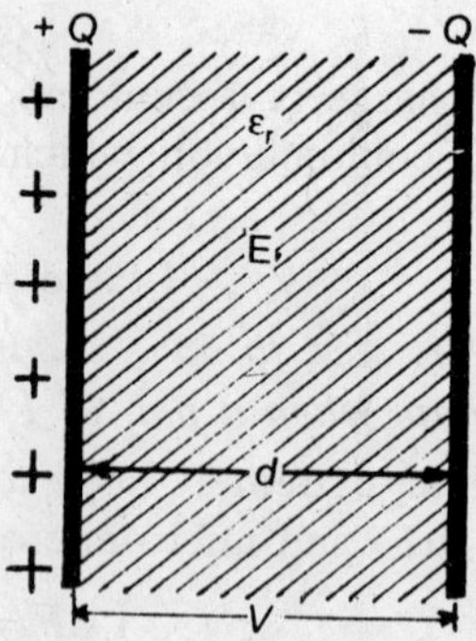

Fig. 6.2

Now, $D = Q/A$ coulomb/m^2

and $E = V/d$

But $D = \varepsilon_0 \varepsilon_r E$

$$\therefore \quad \frac{Q}{A} = \varepsilon_0 \varepsilon_r \frac{V}{d}$$

$$\text{or} \quad \frac{Q}{V} = \frac{\varepsilon_0 \varepsilon_r A}{d}$$

The ratio Q/V is the capacitance C of the capacitor.

$$\therefore \quad C = \frac{\varepsilon_0 \varepsilon_r A}{d} \quad \text{...in a medium}$$

$$= \frac{\varepsilon_0 A}{d} \quad \text{...in air}$$

It may be seen that capacitance is directly proportional to relative permittivity of the medium (ε_r) and area (A) of capacitor plate. However, it is inversely proportional to the distance (d) between the plates.

6.6. Parallel-Plate Capacitor with Composite Medium

Suppose the space between the plates is occupied by three dielectrics of thickness d_1, d_2 and d_3 metres and relative permittivities ε_{r1}, ε_{r2} and ε_{r3} respectively as shown in Fig. 6.3. The electric flux density D in the dielectrics remains the *same and is equal to Q/A. However, the electric intensities in the three dielectrics will be different and are given by ;

$$E_1 = \frac{D}{\varepsilon_0 \varepsilon_{r1}}; \quad E_2 = \frac{D}{\varepsilon_0 \varepsilon_{r2}}; \quad E_3 = \frac{D}{\varepsilon_0 \varepsilon_{r3}}$$

If V is the total p.d. across the capacitor and V_1, V_2 and V_3 the p.d.s across the three dielectrics respectively, then

$$V = V_1 + V_2 + V_3 = E_1 d_1 + E_2 d_2 + E_3 d_3$$

$$= \frac{D}{\varepsilon_0 \varepsilon_{r1}} d_1 + \frac{D}{\varepsilon_0 \varepsilon_{r2}} d_2 + \frac{D}{\varepsilon_0 \varepsilon_{r3}} d_3 = \frac{D}{\varepsilon_0}\left[\frac{d_1}{\varepsilon_{r1}} + \frac{d_2}{\varepsilon_{r2}} + \frac{d_3}{\varepsilon_{r3}}\right]$$

$$= \frac{Q}{\varepsilon_0 A}\left[\frac{d_1}{\varepsilon_{r1}} + \frac{d_2}{\varepsilon_{r2}} + \frac{d_3}{\varepsilon_{r3}}\right] \quad \left(\because D = \frac{Q}{A}\right)$$

---

* The total charge on each plate is Q. Hence Q coulombs is also the total electric flux through each dielectric.

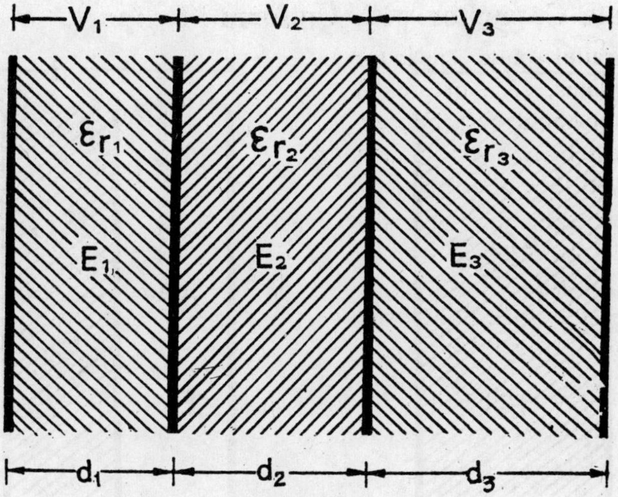

Fig. 6.3

or
$$\frac{Q}{V} = \frac{\varepsilon_0 A}{\left(\frac{d_1}{\varepsilon_{r1}} + \frac{d_2}{\varepsilon_{r2}} + \frac{d_3}{\varepsilon_{r3}}\right)}$$

But Q/V is the capacitance C of the capacitor.

$$\therefore \quad C = \frac{\varepsilon_0 A}{\left(\frac{d_1}{\varepsilon_{r1}} + \frac{d_2}{\varepsilon_{r2}} + \frac{d_3}{\varepsilon_{r3}}\right)} \text{ farad}$$

In general,
$$C = \frac{\varepsilon_0 A}{\sum \frac{d}{\varepsilon_r}} \text{ farad} \qquad \text{...}(i)$$

Different Cases

(*i*) **Medium partly air.** Fig. 6.4 shows a parallel plate capacitor having plates *d* metres apart. Suppose the medium between the plates consists partly of air and partly of dielectric of thickness *t* metres and relative permittivity ε_{r2}. Then thickness of air is $d - t$. Using the relation (*i*), above, we have,

$$C = \frac{\varepsilon_0 A}{\frac{d-t}{1} + \frac{t}{\varepsilon_{r2}}} = \frac{\varepsilon_0 A}{d - \left(t - \frac{t}{\varepsilon_{r2}}\right)} \text{ farad}$$

(*ii*) **When dielectric slab introduced.** Fig. 6.5 shows a parallel-plate air capacitor having plates *d* metres apart. Suppose a dielectric slab of thickness *t* metres and relative permittivity ε_{r2} is introduced between the plates of the capacitor.

Using the relation (*i*) above, we have,

$$C = \frac{\varepsilon_0 A}{\frac{d-t}{1} + \frac{t}{\varepsilon_{r2}}} = \frac{\varepsilon_0 A}{d - \left(t - \frac{t}{\varepsilon_{r2}}\right)} \text{ farad}$$

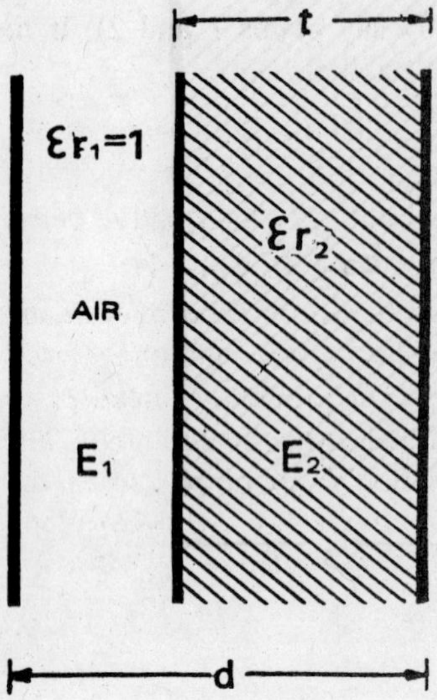

Fig. 6.4

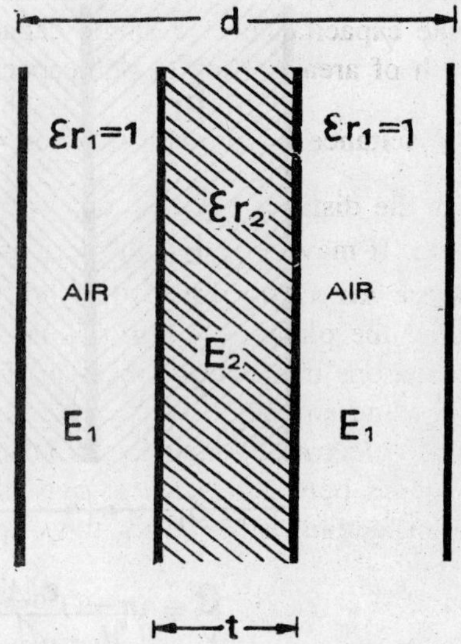

Fig. 6.5

6.7. Multiplate Capacitor

The most *convenient way of achieving large capacitance is by using large plate area. Increasing the plate area may increase the physical size of the capacitor enormously. In order to obtain a large area of plate surface without using too bulky a capacitor, multiplate construction is employed. In this construction, the capacitor is built up of alternate sheets of metal foil (*i.e.*, plates) and thin sheets of dielectric. The odd-numbered metal sheets are connected together to form one terminal T_1 and even-numbered metal sheets are connected together to form the second terminal T_2.

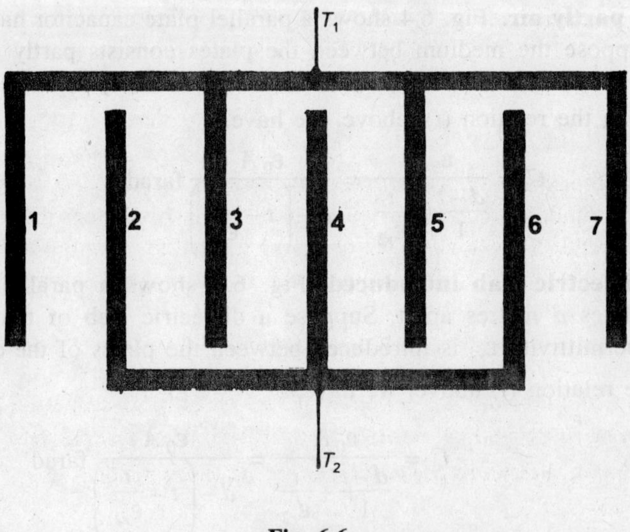

Fig. 6.6

* The capacitance of a capacitor can also be increased by (*i*) using a dielectric of high ε_r and (*ii*) decreasing the distance between plates. High cost limits the choice of dielectric and dielectric strength of the insulating material limits the reduction in spacing between the plates.

Fig. 6.6 shows a multiplate capacitor with seven plates. A little reflection shows that this arrangement is equivalent to 6 capacitors in parallel. The total capacitance will, therefore, be 6 times the capacitance of a single capacitor (formed by say plates 1 and 2). If there are n plates, each of area A, then $(n-1)$ capacitors will be in parallel.

$$\therefore \text{ Capacitance of } n \text{ plate capacitor} = (n-1)\frac{\varepsilon_0 \varepsilon_r A}{d}$$

where d is the distance between any two adjacent plates and ε_r is the relative permittivity of the medium. It may be seen that plate area is increased from A to $A(n-1)$.

Variable Air Capacitor. It is a multiplate air capacitor whose capacitance can be varied by changing the plate area. Fig. 6.7 shows a variable air capacitor commonly used to "tune in" radio stations in the radio receiver. It consists of a set of stationary metal plates Y fixed to the frame and another set of movable metal plates X fixed to the central shaft. The two sets of plates are electrically insulated from each other. Rotation of the shaft moves the plates X into the spaces between plates Y, thus changing the *common (or effective) plate area and hence the capacitance. The capacitance of such a capacitor is given by ;

$$C = (n-1)\frac{\varepsilon_0 A}{d} \qquad (\because \varepsilon_r = 1)$$

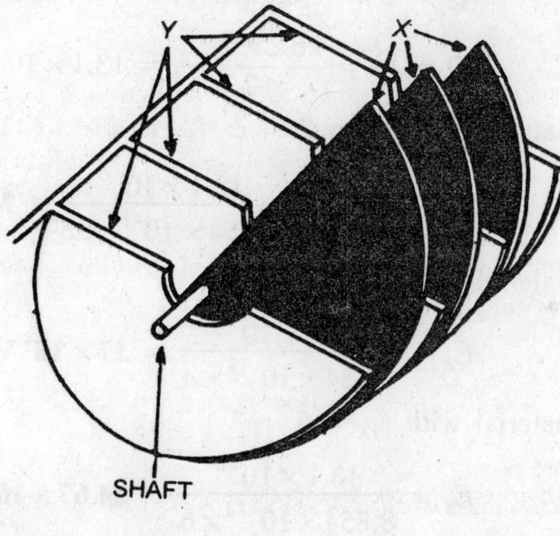

Fig. 6.7

When the movable plates X are completely rotated in (*i.e.*, the two sets of plates completely overlap each other), the common plate area 'A' is maximum and so is the capacitance of the capacitor. Minimum capacitance is obtained when the movable plates X are completely rotated out of stationary plates Y. The capacitance of such variable capacitors is from zero to about 4000 pF.

Note. In all the formulas derived for capacitance, capacitance will be in farad if area is in m^2 and the distance between plates is in *m*.

Example 6.1. *Calculate the capacitance of two metal plates of area 30 m^2 and separated by a dielectric 2 mm thick and of relative permittivity 6. If the electric field strength in the dielectric is 500 V/mm, calculate the total charge on each plate.*

---

* Remember in the formula for capacitance, A is the common plate area *i.e.*, plate area facing the opposite polarity plate area.

Solution.

Capacitance, $C = \dfrac{\varepsilon_0 \varepsilon_r A}{d} = \dfrac{8.854 \times 10^{-12} \times 6 \times 30}{2 \times 10^{-3}} = 0.797 \times 10^{-6}\ \text{F} = \mathbf{0.797\ \mu F}$

Voltage across plates, $V = E \times d = 500 \times 2 = 1000\ \text{V}$

Charge on each plate, $Q = CV = (0.797 \times 10^{-6}) \times (1000) = 0.797 \times 10^{-3}\ \text{C} = \mathbf{0.797\ mC}$

Example 6.2. *A parallel plate capacitor has plates of area 2 m² spaced by three layers of different dielectric materials. The relative permittivities are 2, 4, 6 and thicknesses are 0.5, 1.5 and 0.3 mm respectively. Calculate the combined capacitance and the electric stress in each material when applied voltage in 1000 V.*

Solution.

Capacitance, $$C = \frac{\varepsilon_0 A}{\dfrac{d_1}{\varepsilon_{r1}} + \dfrac{d_2}{\varepsilon_{r2}} + \dfrac{d_3}{\varepsilon_{r3}}}$$

$$= \frac{8.854 \times 10^{-12} \times 2}{\dfrac{0.5 \times 10^{-3}}{2} + \dfrac{1.5 \times 10^{-3}}{4} + \dfrac{0.3 \times 10^{-3}}{6}} = \mathbf{0.0262 \times 10^{-6}\ F}$$

Charge on each plate, $Q = CV = (0.0262 \times 10^{-6}) \times 1000 = 26.2 \times 10^{-6}\ \text{C}$

Electric flux density, $D = \dfrac{Q}{A} = \dfrac{26.2 \times 10^{-6}}{2} = 13.1 \times 10^{-6}\ \text{C/m}^2$

Electric stress in the material with $\varepsilon_{r1} = 2$

$$E_1 = \frac{D}{\varepsilon_0 \varepsilon_{r1}} = \frac{13.1 \times 10^{-6}}{8.854 \times 10^{-12} \times 2} = \mathbf{74 \times 10^4\ V/m}$$

Electric stress in the material with $\varepsilon_{r2} = 4$

$$E_2 = \frac{13.1 \times 10^{-6}}{8.854 \times 10^{-12} \times 4} = \mathbf{37 \times 10^4\ V/m}$$

Electric stress in the material with $\varepsilon_{r3} = 6$

$$E_3 = \frac{13.1 \times 10^{-6}}{8.854 \times 10^{-12} \times 6} = \mathbf{24.67 \times 10^4\ V/m}$$

It is clear from the above example that electric stress is greatest in the material having the least relative permittivity. Since air has the lowest relative permittivity, efforts should be made to avoid air pockets in the dielectric materials.

Example 6.3. *A capacitor is composed of two plates separated by 3 mm of dielectric of permittivity 4. An additional piece of insulation 5 mm thick is now inserted between the plates. If the capacitor now has capacitance one-third of its original capacitance, find the relative permittivity of the additional dielectric.*

Solution. Fig. 6.8 (*i*) and Fig. 6.8 (*ii*) respectively show the two cases.

For the first case, $$C = \frac{\varepsilon_0 \varepsilon_r A}{d} = \frac{\varepsilon_0 \times 4 \times A}{3 \times 10^{-3}} \quad ...(i)$$

For the second case, $$\frac{C}{3} = \frac{\varepsilon_0 A}{\dfrac{d_1}{\varepsilon_{r1}} + \dfrac{d_2}{\varepsilon_{r2}}} = \frac{\varepsilon_0 A}{\dfrac{3 \times 10^{-3}}{4} + \dfrac{5 \times 10^{-3}}{\varepsilon_{r2}}} \quad ...(ii)$$

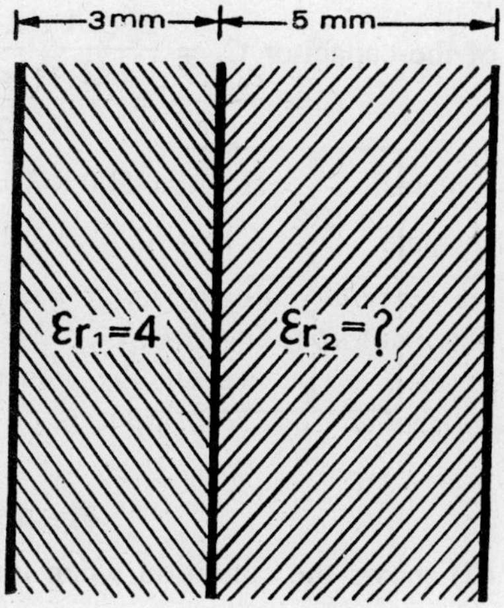

Fig. 6.8

Dividing eq (*i*) by eq (ii), we get,

$$3 = \frac{4}{3}\left(\frac{3}{4} + \frac{5}{\varepsilon_{r2}}\right)$$

or
$$9 = 3 + 20/\varepsilon_{r2} \qquad \therefore \varepsilon_{r2} = 20/6 = \mathbf{3.33}$$

Example 6.4. *An air capacitor has two parallel plates of 1500 cm² in area and 5 mm apart. If a dielectric slab of area 1500 cm², thickness 2 mm and relative permittivity 3 is now introduced between the plates, what must be the new separation between the plates to bring the capacitance to the original value.*

Solution. This is a case of introduction of dielectric slab into an air capacitor. As proved in Art 6.6, the capacitance under this condition becomes :

$$C = \frac{\varepsilon_0 A}{d - (t - t/\varepsilon_r)} \qquad ...(i)$$

If the medium were totally air, capacitance would have been

$$C_{air} = \frac{\varepsilon_0 A}{d} \qquad ...(ii)$$

Inspection of eqs (*i*) and (*ii*) shows that with the introduction of dielectric slab between the plates of air capacitor, its capacitance increases. The distance between the plates is effectively reduced by $t - (t/\varepsilon_r)$. In order to bring the capacitance to the original value, the plates must be separated by this much distance in air.

∴ New separation between the plates $= d + (t - t/\varepsilon_r) = 5 + (2 - 2/3) = \mathbf{6.33\ mm}$

Example 6.5. *A parallel plate capacitor has three similar parallel plates. Find the ratio of capacitance when the inner plate is mid-way between the outers to the capacitance when inner plate is three times as near one plate as the other.*

Solution. (*i*) Fig. 6.9 (*i*) shows the condition when the inner plate is mid-way between the outer plates. This arrangement is equivalent to two capacitors in parallel.

$$\text{Capacitance of the capacitor } C_1 = \frac{\varepsilon_0 \varepsilon_r A}{d/2} + \frac{\varepsilon_0 \varepsilon_r A}{d/2} = \frac{4\varepsilon_0 \varepsilon_r A}{d}$$

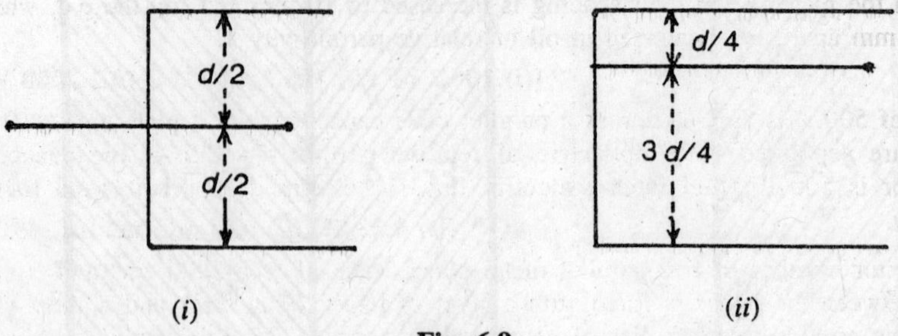

Fig. 6.9

(*ii*) Fig. 6.9 (*ii*) shows the condition when inner plate is three times as near as one plate as the other.

$$\text{Capacitance of the capacitor, } C_2 = \frac{\varepsilon_0 \varepsilon_r A}{d/4} + \frac{\varepsilon_0 \varepsilon_r A}{3d/4} = \frac{16\varepsilon_0 \varepsilon_r A}{3d}$$

$$\therefore \quad \frac{C_1}{C_2} = \mathbf{0.75}$$

Example 6.6. *A mica dielectric parallel plate capacitor has 21 plates each having an effective area of 5 cm² and each separated by a gap of 0.005 mm. Find the capacitance. Take the relative permittivity of mica as 6.*

Solution. For a multiplate capacitor, the capacitance is given by (See Art 6.7) ;

$$C = (n-1)\frac{\varepsilon_0 \varepsilon_r A}{d} = (21-1)\frac{(8.854\times 10^{-12})\times 6\times (5\times 10^{-4})}{0.005\times 10^{-3}}$$

$$= 0.1062\times 10^{-6}\text{ F} = \mathbf{0.1062\ \mu F}$$

Example 6.7. *A variable air capacitor has 11 movable plates and 12 stationary plates. The area of each plate is 0.0015 m² and separation between opposite plates is 0.001 m. Determine the maximum capacitance of this variable capacitor.*

Solution. The capacitance will be maximum when the moveable plates are completely rotated in *i.e.*, when the two sets of plates completely overlap each other. Under this condition, the common (or effective) area is equal to the physical area of each plate.

$$C = (n-1)\frac{\varepsilon_0 \varepsilon_r A}{d}$$

Here $n = 11 + 12 = 23$; $\varepsilon_r = 1$; $A = 0.0015\text{ m}^2$; $d = 0.001\text{ m}$

$$\therefore \quad C = (23-1)\times\frac{8.854\times 10^{-12}\times 1\times 0.0015}{0.001} = 292\times 10^{-12}\text{ F} = \mathbf{292\ pF}$$

Tutorial Problems

1. A capacitor consisting of two parallel plates 0.5 mm apart in air and each of effective area 500 cm² is connected to a 100 V battery. Calculate (*i*) the capacitance and (*ii*) the charge. [(*i*) **885 pF** (*ii*) **0.0885 μC**]
2. A capacitor consisting of two parallel plates in air, each of effective area 50 cm² and 1 mm apart, carries a charge of 1770×10^{-12} C. Calculate the p.d. between the plates. If the distance between the plates is increased to 5 mm, what will be the electrical effect ? [(*i*) **40 V** (*ii*) p.d.across plates is increased to **200 V**]

3. Two insulated parallel plates each of 600 cm^2 effective area and 5 mm apart in air are charged to a p.d. of 1000 V. Calculate (*i*) the capacitance and (*ii*) the charge on each plate.
The source of supply is now disconnected, the plates remaining insulated. Calculate (*iii*) the p.d. between the plates when their spacing is increased to 10 mm and (*iv*) the p.d. when the plates, still 10 mm apart, are immersed in oil of relative permittivity 5.
[(*i*) **106.2 pF** (*ii*) **106.2×10^{-12} C** (*iii*) **2000 V** (*iv*) **400 V**]
4. A p.d. of 500 V is applied across a parallel plate capacitor with a plate area of 0.025 m^2. The plates are separated by a dielectric of relative permittivity 2.5. If the capacitance of the capacitor is 500 μF, find (*i*) the electric flux (*ii*) electric flux density and (*iii*) the electric intensity. [(*i*) **0.25 μC** (*ii*) **0.01 mC/m^2** (*iii*) **45.3×10^6 V/m**]
5. A capacitor consists of two parallel metal plates, each of area 2000 cm^2 and 5 mm apart. The space between the plates is filled with a layer of paper 2 mm thick and a sheet of glass 3 mm thick. The relative permittivities of paper and glass are 2 and 8 respectively. A p.d. of 5 kV is applied across the plates. Calculate (*i*) the capacitance of the capacitor and (*ii*) the potential gradient in each dielectric. [(*i*) **1290 pF** (*ii*) **1820 V/mm** (paper) ; **453 V/mm** (glass)]
6. A parallel plate capacitor has a plate area of 20 cm^2 and the plates are separated by three dielectric layers each 1 mm thick and of relative permittivity 2, 4 and 5 respectively. Find the capacitance of the capacitor and the electric stress in each dielectric if applied voltage is 1000 V. [**18.6 pF ; 5.26×10^5 V/m ; 2.63×10^5 V/m ; 2.11×10^5 V/m**]
7. A 1 μF parallel plate capacitor that can just withstand a p.d. of 6000 V uses a dielectric having a relative permittivity 5, which breaks down if the electric intensity exceeds 30×10^6 V/m. Find (*i*) the thickness of dielectric required and (*ii*) the effective area of each plate.
[(*i*) **0.2 mm** (*ii*) **4.5 m^2**]
8. An air capacitor has two parallel plates 10 cm^2 in area and 5 mm apart. When a dielectric slab of area 10 cm^2 and thickness 5 mm was inserted between the plates ; one of the plates has to be moved by 0.4 cm to restore the capacitance. What is the dielectric constant of the slab ? [**5**]
9. A multiplate parallel capacitor has 6 fixed plates connected in parallel, interleaved with 5 similar plates ; each plate has an effective area of 120 cm^2. The gap between the adjacent plates is 1 mm. The capacitor is immersed in oil of relative permittivity 5. Calculate the capacitance.
[**5.31 pF**]
10. Calculate the number of sheets of tin foil and mica for a capacitor of 0.33 μF capacitance if area of each sheet of tin foil is 82 cm^2, the mica sheets are 0.2 mm thick and have relative permittivity 5. [**182** sheets of mica ; **183** sheets of tin foil]

6.8. Cylindrical Capacitor

A cylindrical capacitor consists of two co-axial cylinders separated by an insulating medium. This is an important practical case since a single core cable is in effect a capacitor of this kind. The conductor (or core) of the cable is the inner cylinder while the outer cylinder is represented by lead sheath which is at earth potential. The two co-axial cylinders have insulation between them.

Consider a single core cable with conductor diameter d metres and inner sheath diameter D metres (See Fig. 6.10). Let the charge per metre axial length of the cable be Q coulombs and ε_r be the relative permittivity of the insulating material. Consider a cylinder of radius x metres. According to Gauss's theorem, electric flux passing through this cylinder is Q coulombs. The surface area of this cylinder is

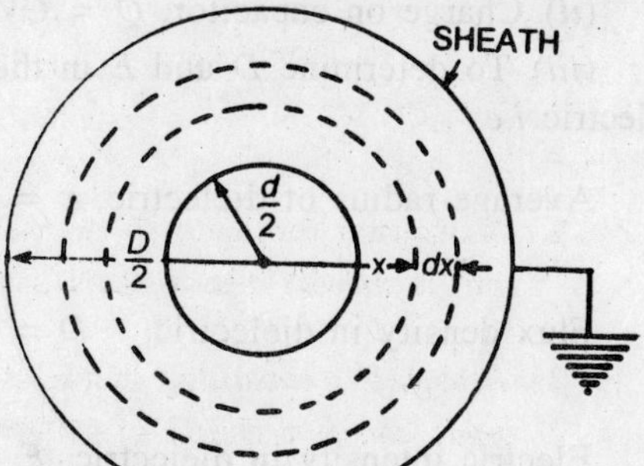

Fig. 6.10

$$= 2\pi x \times 1 = 2\pi x \text{ m}^2$$

$\therefore$ Electric flux density at any point P on the considered cylinder

$$D_x = \frac{Q}{2\pi x} \text{ C/m}^2$$

Electric intensity at point P is

$$E_x = \frac{D_x}{\varepsilon_0 \varepsilon_r} = \frac{Q}{2\pi x \varepsilon_0 \varepsilon_r} \text{ V/m}$$

The work done in moving a unit positive charge from point P through a distance dx in the direction of electric field is $E_x\, dx$. Hence the work done in moving a unit positive charge from conductor to sheath, which is the p.d. V between the conductor and sheath, is given by :

$$V = \int_{d/2}^{D/2} E_x\, dx = \int_{d/2}^{D/2} \frac{Q}{2\pi x \varepsilon_0 \varepsilon_r} dx = \frac{Q}{2\pi \varepsilon_0 \varepsilon_r} \log_e \frac{D}{d}$$

Capacitance of cable,

$$C = \frac{Q}{V} = \frac{Q}{\frac{Q}{2\pi\varepsilon_0\varepsilon_r}\log_e \frac{D}{d}} \text{ F/m} = \frac{2\pi\varepsilon_0\varepsilon_r}{\log_e (D/d)} \text{ F/m}$$

$$= \frac{2\pi \times 8.854 \times 10^{-12} \times \varepsilon_r}{2.303 \log_{10} (D/d)} \text{ F/m} = \frac{\varepsilon_r}{41.4 \log_{10} (D/d)} \times 10^{-9} \text{ F/m}$$

If the cable has a length of l metres, then capacitance of the cable is

$$= \frac{\varepsilon_r l}{41.4 \log_{10} (D/d)} \times 10^{-9} \text{ F} = \frac{24\, \varepsilon_r l}{\log_{10} (D/d)} \text{ pF}$$

Example 6.8. *In a concentric cable 20 cm long, the diameters of inner and outer cylinders are 15 cm and 15.4 cm respectively. The relative permittivity of the insulation is 5. If a p.d. of 5000 V is maintained between the two cylinders, calculate :*

(i) *capacitance of cylindrical capacitor*

(ii) *the charge*

(iii) *the electric flux density and electric intensity in the dielectric.*

Solution.

(i) Capacitance of the cylindrical capacitor

$$C = \frac{\varepsilon_r l}{41.4 \log_{10} (D/d)} \times 10^{-9} = \frac{5 \times 0.2}{41.4 \log_{10} (15.4/15)} \times 10^{-9} \text{ F} = \mathbf{2.11 \times 10^{-9} \text{ F}}$$

(ii) Charge on capacitor, $Q = CV = (2.11 \times 10^{-9}) \times 5000 = 10.55 \times 10^{-6} \text{ C} = \mathbf{10.55\ \mu C}$

(iii) To determine D and E in the dielectric, we shall consider the average radius of dielectric *i.e.*

$$\text{Average radius of dielectric, } x = \frac{1}{2}\left[\frac{15}{2} + \frac{15.4}{2}\right] = 7.6 \text{ cm} = 0.076 \text{ m}$$

$$\text{Flux density in dielectric, } D = \frac{Q}{2\pi x l} \text{ C/m}^2 = \frac{10.55 \times 10^{-6}}{2\pi \times 0.076 \times 0.2} = \mathbf{110.47 \times 10^{-6} \text{ C/m}^2}$$

$$\text{Electric intensity in dielectric, } E = \frac{D}{\varepsilon_0 \varepsilon_r} = \frac{110.47 \times 10^{-6}}{8.854 \times 10^{-12} \times 5} = \mathbf{2.5 \times 10^6 \text{ V/m}}$$

6.9. Capacitors in Series

Consider three capacitors, having capacitances C_1, C_2 and C_3 farad respectively, connected in series across a p.d. of V volts [see Fig. 6.11 (*i*)]. In series connection, charge on each capacitor is the *same (*i.e.*, $+Q$ on one plate and $-Q$ on the other) but p.d. across each is different.

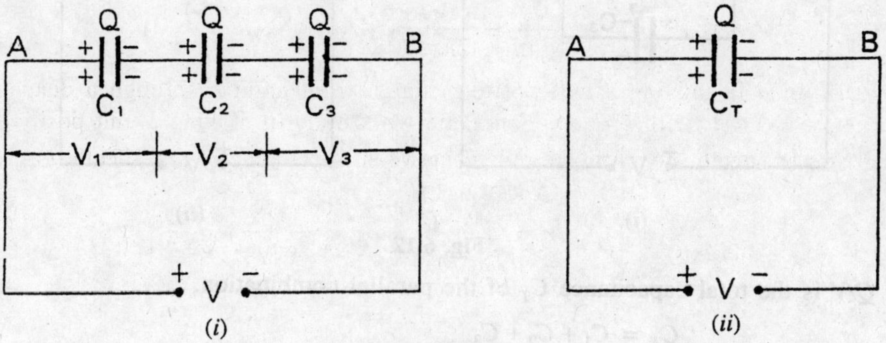

Fig. 6.11

Now, $$V = V_1 + V_2 + V_3 = \frac{Q}{C_1} + \frac{Q}{C_2} + \frac{Q}{C_3} = Q\left(\frac{1}{C_1} + \frac{1}{C_2} + \frac{1}{C_3}\right)$$

or $$\frac{V}{Q} = \frac{1}{C_1} + \frac{1}{C_2} + \frac{1}{C_3}$$

But Q/V is the total capacitance C_T between points A and B so that $V/Q = 1/C_T$ [See 6.11 (*ii*)].

$$\therefore \quad \frac{1}{C_T} = \frac{1}{C_1} + \frac{1}{C_2} + \frac{1}{C_3}$$

Thus capacitors in series are treated in the same manner as are resistors in parallel.

Special Case. Frequently we come across two capacitors in series. The total capacitance in such a case is given by ;

$$\frac{1}{C_T} = \frac{1}{C_1} + \frac{1}{C_2} = \frac{C_1 + C_2}{C_1 C_2}$$

or $$C_T = \frac{C_1 C_2}{C_1 + C_2} \quad \textit{i.e.,} \quad \frac{\text{Product}}{\text{sum}}$$

6.10. Capacitors in Parallel

Consider three capacitors, having capacitances C_1, C_2 and C_3 farad respectively, connected in parallel across a p.d. of V volts [See Fig. 6.12 (*i*)]. In parallel connection, p.d. across each capacitor is the same but charge on each is different.

Now, $Q = Q_1 + Q_2 + Q_3 = C_1V + C_2V + C_3V = V(C_1 + C_2 + C_3)$

or $$\frac{Q}{V} = C_1 + C_2 + C_3$$

---

* When voltage V is applied, a similar electron movement occurs on each plate. Hence the same charge is stored by each capacitor. Alternatively, current (charging) in a series circuit is the same. Since $Q = It$ and both I and t are the same for each capacitor, the charge on each capacitor is the same.

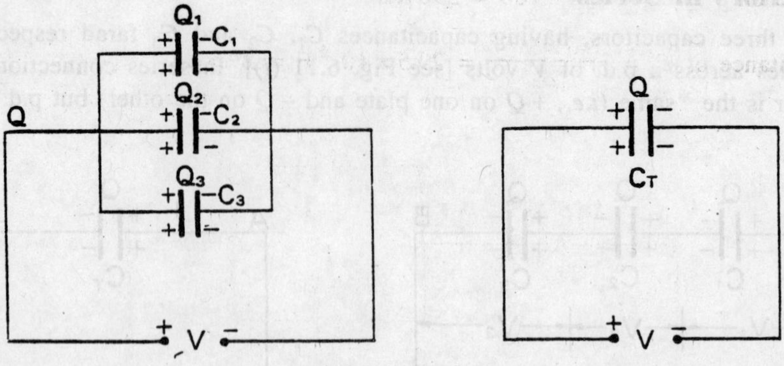

Fig. 6.12

But Q/V is the total capacitance C_T of the parallel combination.

$$\therefore \quad C_T = C_1 + C_2 + C_3$$

Thus capacitors in parallel are treated in the same manner as are resistors in series.

Example 6.9. *Three capacitors of capacitance 2 μF, 4μF and 6 μF respectively are connected in series to a 220 V d.c. supply. Find (i) the total capacitance (ii) charge on each capacitor and (iii) p.d. across capacitor.*

Solution.

(*i*)
$$\frac{1}{C_T} = \frac{1}{2} + \frac{1}{4} + \frac{1}{6} = \frac{6+3+2}{12} = \frac{11}{12}$$

$$\therefore \quad C_T = 12/11 = \mathbf{1.091\ \mu F}$$

(*ii*) In series connection, charge on each capacitor is the same.

$\therefore$ Charge on each capacitor, $Q = C_T V = (1.091 \times 10^{-6}) \times 220 = \mathbf{240 \times 10^{-6}\ C}$

(*iii*) P.D. across 2μF $= \dfrac{Q}{C_1} = \dfrac{240 \times 10^{-6}}{2 \times 10^{-6}} = \mathbf{120\ V}$

P.D. across 4μF $= \dfrac{Q}{C_2} = \dfrac{240 \times 10^{-6}}{4 \times 10^{-6}} = \mathbf{60\ V}$

P.D. across 6μF $= \dfrac{Q}{C_3} = \dfrac{240 \times 10^{-6}}{6 \times 10^{-6}} = \mathbf{40\ V}$

It is clear that in series connection, greatest voltage will appear across the capacitor having the least capacitance.

Example 6.10. *In the circuit shown in Fig. 6.13, the total charge is 750 μC. Determine the value of V_1, V and C_2.*

Solution.

$$V_1 = \frac{Q}{C_1} = \frac{750 \times 10^{-6}}{15 \times 10^{-6}} = \mathbf{50\ V}$$

$$V = V_1 + V_2 = 50 + 20 = \mathbf{70\ V}$$

$$\text{Charge on } C_3 = C_3 \times V_2 = (8 \times 10^{-6}) \times 20$$

$$= 160 \times 10^{-6} = 160\ \mu\text{C}$$

$\therefore$ Charge on $\quad C_2 = 750 - 160 = 590\,\mu C$

$\therefore$ Capacitance of $C_2 = \dfrac{590 \times 10^{-6}}{20} = 29.5 \times 10^{-6}$ F $=$ **29.5 μF**

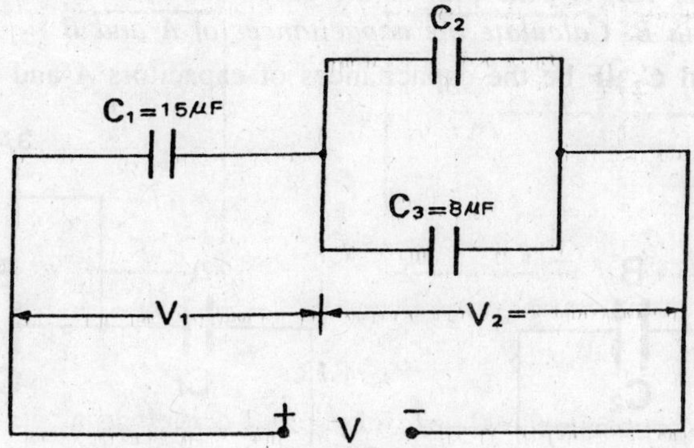

Fig. 6.13

Example 6.11. *A 5μ F capacitor is charged to a p.d. of 100 V and then connected in parallel with an uncharged 3μF capacitor. Calculate the p.d. across the parallel capacitors.*

Solution.

Charge on 5 μF capacitor, $Q = CV = (5 \times 10^{-6}) \times 100 = 0.0005$ C

When the capacitors are connected in parallel, the total capacitance is $C_T = 5 + 3 = 8$ μF. The charge 0.0005 C is distributed between the two capacitors to have a common p.d.

$\therefore$ P.D. across parallel capacitors $= \dfrac{Q}{C_T} = \dfrac{0.0005}{8 \times 10^{-6}} =$ **62.5 V**

Example 6.12. *Two capacitors of capacitance 4 μF and 6 μF respectively are connected in series across a p.d. of 250 V. The capacitors are disconnected from the supply and are reconnected in parallel with each other. Calculate the new p.d. and charge on each capacitor.*

Solution.

Series connection. Let V_1 and V_2 be the voltages across the capacitors.

Charge on each capacitor $= C_1 V_1 = C_2 V_2$

$\therefore \qquad 4\,V_1 = 6\,V_2$

or $\qquad V_1 = 1.5\,V_2 \qquad$ ...(*i*)

Also $\qquad V_1 + V_2 = 250 \qquad$ ...(*ii*)

From eqs. (*i*) and (*ii*), we have $V_1 = 150$ volts and $V_2 = 100$ volts

Charge on each capacitor $= C_1 V_1 = (4 \times 10^{-6}) \times 150 = 0.0006$C

Charge on both capacitors $= 2 \times 0.0006 = 0.0012$C

Parallel connection. When the capacitors are connected in parallel, the total capacitance $C_T = 4 + 6 = 10$ μF. The total charge 0.0012 C is distributed between the two capacitors to have a common p.d.

P.D. across capacitors $= \dfrac{0.0012}{10 \times 10^{-6}} =$ **120 V**

Charge on 4 μF capacitor $= (4 \times 10^{-6}) \times 120 = 480 \times 10^{-6}$ C = **480 μC**

Charge on 6 μF capacitor $= (6 \times 10^{-6}) \times 120 = 720 \times 10^{-6}$ C = **720 μC**

Example 6.13.. *Two capacitors A and B are connected in series across a 200 V d.c. supply. The p.d. across A is 120 V. This p.d. is increased to 140 V when a 3 μF capacitor is connected in parallel with B. Calculate the capacitances of A and B.*

Solution. Let C_1 and C_2 μF be the capacitances of capacitors A and B respectively.

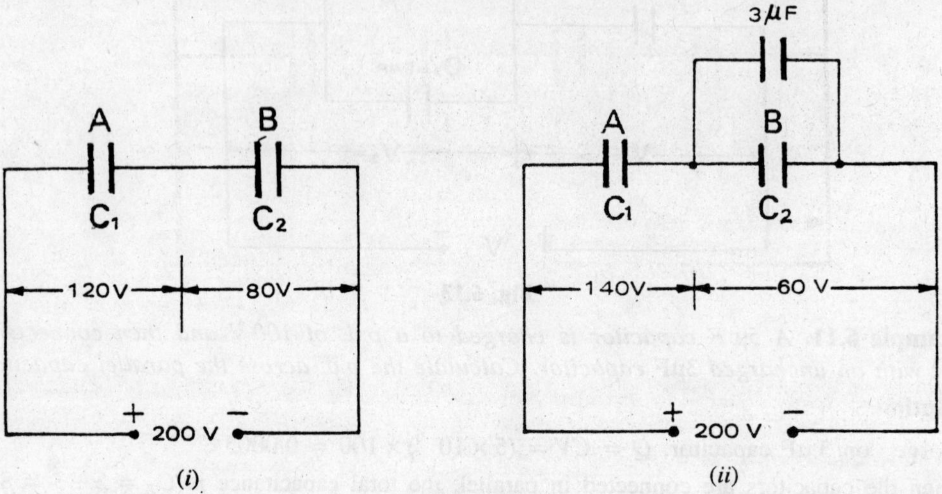

Fig. 6.14

When the capacitors are connected in series [see Fig. 6.14 (*i*)], charge on each capacitor is the same.

$$\therefore \quad C_1 \times 120 = C_2 \times 80$$

or $$C_2 = 1.5\, C_1 \qquad ...(i)$$

When a 3 μF capacitor is connected in parallel with B [see Fig. 6.14 (*ii*)], the combined capacitance of this parallel branch is $(C_2 + 3)$. Thus the circuit shown in Fig. 6.13 (*ii*) can be thought as a series circuit consisting of capacitances C_1 and $(C_2 + 3)$ connected in series.

$$\therefore \quad C_1 \times 140 = (C_2 + 3)\, 60$$

or $$7\, C_1 - 3\, C_2 = 9 \qquad ...(ii)$$

Solving eqs (*i*) and (*ii*), we get, C_1 = **3.6 μF** ; C_2 = **5.4 μF**.

Example 6.14. *Given some capacitors of 0.1μ F capable of withstanding upto 15 V. Calculate the number of capacitors needed if it is desired to obtain a capacitance of 0.1 μF for use in a circuit involving 60 V.*

Solution.

Capacitance of each capacitor, $C = 0.1$ μF

Voltage rating of each capacitor, $V_C = 15$ V

Supply voltage, $V = 60$ V

Since each capacitor can withstand 15 V only, the number of capacitors to be connected in series = 60/15 = 4.

Capacitance of 4 series-connected capacitors, $C_S = C/4 = 0.1/4 = 0.025$ μF

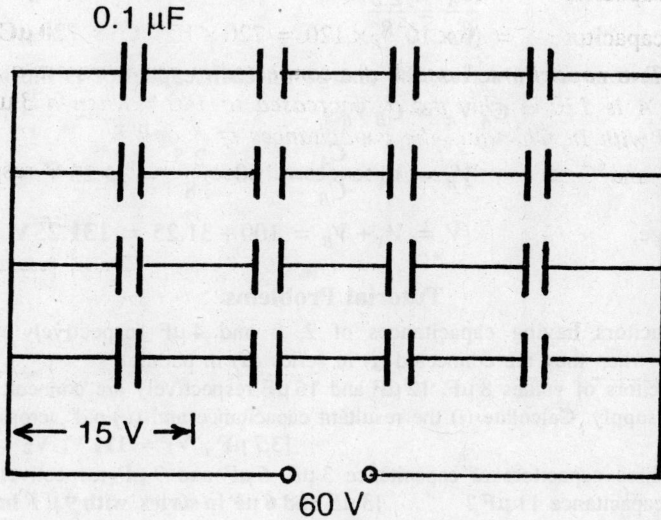

Fig. 6.15

Since it is desired to have total capacitance of 0.1 μF, number of such rows $= C/C_s = 0.1/0.025 = 4$.

$\therefore$ Total number of capacitors $= 4 \times 4 =$ **16**

Fig. 6.15 shows the arrangement of capacitors.

Example 6.15. *Two parallel-plate capacitors A and B are connected in series. Capacitor A has a plate area of 5000 mm², an air dielectric and the distance between the plates is 1 mm. Capacitor B has a plate area of 2000 mm², a solid dielectric of relative permittivity 4 and of thickness 0.5 mm. Find the voltage across the combination if potential gradient associated with capacitor A is 100 kV/m.*

Solution. Fig. 6.16 shows the conditions of the problem.

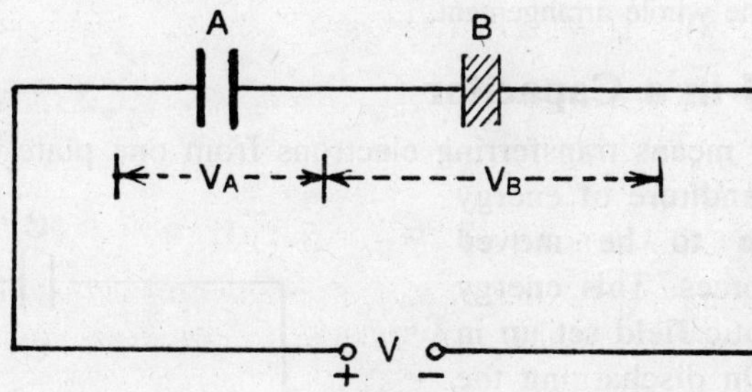

Fig. 6.16

$$V_A = E_A \times d_A = (100 \times 10^3) \times (1 \times 10^{-3}) = 100 \text{ V}$$

$$C_A = \frac{\varepsilon_0 A}{d_A} = \frac{\varepsilon_0 \times 5000 \times 10^{-6}}{1 \times 10^{-3}}$$

$$C_B = \frac{\varepsilon_0 \varepsilon_r A}{d_B} = \frac{\varepsilon_0 \times 4 \times 2000 \times 10^{-6}}{0.5 \times 10^{-3}}$$

$$\therefore \qquad \frac{C_A}{C_B} = \frac{2.5}{8}$$

Since the two capacitors are in series, the charge on each capacitor is the same *i.e.*,

$$C_A V_A = C_B V_B$$

or $$V_B = V_A \times \frac{C_A}{C_B} = 100 \times \frac{2.5}{8} = 31.25 \text{ V}$$

$\therefore$ Total voltage, $$V = V_A + V_B = 100 + 31.25 = \mathbf{131.25 V}$$

Tutorial Problems

1. Three capacitors having capacitances of 2, 3 and 4 μF respectively. Calculate the total capacitance when they are connected (*i*) in series (*ii*) in parallel. **[0.923 μF ; 9μF]**
2. Three capacitors of values 8 μF, 12 μF and 16 μF respectively are connected in series across a 240 V d.c. supply. Calculate (*i*) the resultant capacitance and (*ii*) p.d. across each capacitor. **[3.7 μF ; V_1 = 111 V, V_2 = 74 V, V_3 = 55 V]**
3. How can three capacitors of capacitance 3 μF, 6 μF and 9 μF respectively be arranged to a system of capacitance 11 μF ? **[3 μF and 6 μF in series, with 9 μ F in parallel with both]**
4. Two capacitors of capacitance 0.5 μF and 0.3 μF are joined in series. What value of capacitance joined in parallel with this combination would give a capacitance of 0.5 μF ? **[0.31 μF]**
5. Three capacitors *A, B* and *C* are connected in series across a 200 V d.c. supply. The p.d.s. across the capacitors are 40 V, 70 V and 90 V respectively. If the capacitance of *A* is 8 μF, what are the capacitances of *B* and *C* ? **[4.57 μF, 3.56 μF]**
6. A capacitor of 4 μF capacitance is charged to a p.d. of 400 V and then connected in parallel with an uncharged capacitor of 2 μF capacitance. Calculate the p.d. across the parallel capacitors. **[267 V]**
7. Circuit *ABC* is made up as follows. *AB* consists of a 3 μF capacitor. *BC* consists of a 3 μF capacitor in parallel with 5 μF capacitor. If a d.c. supply of 100 V is connected between *A* and *C*, determine the charge on each capacitor. **[160 μC (*AB*) ; 60 μC (3μF in *BC*) ; 100 μC]**
8. Two capacitors, *A* and *B*, having capacitances of 20 μF and 30 μF respectively, are connected in series to a 600 V d.c. supply. If a third capacitor *C* is connected in parallel with *A*, it is found that p.d., across *B* is 400 V. Determine the capacitance of capacitor *C*. **[40 μF]**
9. Three series connected capacitors of equal dimensions have dielectrics of relative permittivities 1, 2 and 3 respectively. Determine the voltage across each capacitor given that a p.d. of 1.5 kV is applied across the whole arrangement. **[820 V ; 410 V ; 270 V]**

6.11. Energy Stored in a Capacitor

Charging a capacitor means transferring electrons from one plate of the capacitor to the other. This involves expenditure of energy because electrons have to be moved against the *opposing forces. This energy is stored in the electrostatic field set up in the dielectric medium. On discharging the capacitor, the field collapses and the stored energy is released.

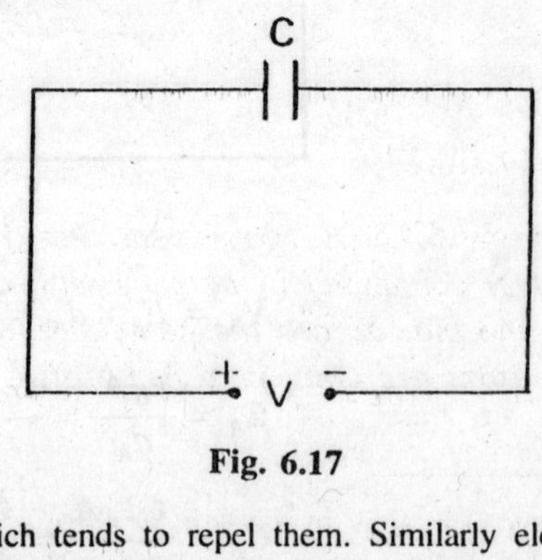

Fig. 6.17

Consider a capacitor of *C* farad being charged from a d.c. source of V volts as

* Electrons are being pushed to the negative plate which tends to repel them. Similarly electrons are removed from the positive plate which tends to attract them. In either case, forces oppose the transfer of electrons from one plate to the other. This opposition increases as the charge on the plates increases.

shown in Fig. 6.17. Suppose at any stage of charging, the charge on the capacitor is q coulomb and p.d. across the plates is V volts.

Then $$C = \frac{q}{V}$$

At this instant, V Joules (by definition of V) of work will be done in transferring 1 C of charge from one plate to the other. If further small charge dq is transferred, then work done is

$$dW = V\,dq = CV\,dV \qquad \left[\begin{array}{l} \because\ q = CV \\ \therefore\ dq = C\,dV \end{array}\right]$$

$\therefore$ Total work done in raising the potential of uncharged capacitor to V volts is

$$W = \int_0^V CV\,dV = C\left[\frac{V^2}{2}\right]_0^V$$

or $$W = \frac{1}{2}CV^2 \text{ Joules}$$

This work done is stored in the electrostatic field set up in the dielectric.

$\therefore$ Energy stored in the capacitor

$$E = \frac{1}{2}CV^2 = \frac{1}{2}{}^{*}QV = {}^{**}\frac{Q^2}{2C} \text{ Joules}$$

Example 6.16. *A capacitor consists of two metal plates, each 40 cm × 40 cm, spaced 6 mm apart. The space between the metal plates is filled with a glass plate 5 mm thick and a layer of paper 1 mm thick. The relative permittivities of glass and paper are 8 and 2 respectively. Find the capacitance of the system. If a p.d. of 10 kV is applied to the capacitor, determine the energy stored in it.*

Solution.

$$C = \frac{\varepsilon_0 A}{\dfrac{d_1}{\varepsilon_{r1}} + \dfrac{d_2}{\varepsilon_{r2}}}$$

Here $A = 0.4 \times 0.4 = 0.16\text{ m}^2$; $d_1 = 5\text{ mm} = 5 \times 10^{-3}\text{ m}$

$d_2 = 1\text{ mm} = 10^{-3}\text{ m}$; $\varepsilon_{r1} = 8$; $\varepsilon_{r2} = 2$

$$\therefore \qquad C = \frac{8.854 \times 10^{-12} \times 0.16}{10^{-3}\left(\dfrac{5}{8} + \dfrac{1}{2}\right)} = \mathbf{0.001259 \times 10^{-6}\ F}$$

Energy stored, $$E = \tfrac{1}{2}CV^2 = \tfrac{1}{2}(0.001259 \times 10^{-6}) \times (10^4)^2 = \mathbf{62.95 \times 10^{-3}\ J}$$

Example 6.17. *A 16 μF capacitor is charged to 100 V. After being disconnected, it is immediately connected in parallel with an uncharged capacitor of capacitance 4 μ F. Determine (i) the p.d. across the combination and (ii) the electrostatic energies before and after the capacitors are connected in parallel.*

---

* Putting $C = Q/V$ in the exp. $E = \frac{1}{2}CV^2$

** Putting $V = Q/C$ in the exp. $E = \frac{1}{2}CV^2$

Solution.

$$C_1 = 16\ \mu F\ ;\ C_2 = 4\ \mu F$$

Before joining

Charge on 16 μF capacitor, $Q = C_1V_1 = (16 \times 10^{-6}) \times 100 = 1.6 \times 10^{-3}$ C

Energy stored, $E_1 = \frac{1}{2} C_1 V_1^2 = \frac{1}{2}(16 \times 10^{-6}) \times 100^2 = \mathbf{0.08\ J}$

After joining

When the capacitors are connected in parallel, the total capacitance $C_T = C_1 + C_2 = 16 + 4 = 20\ \mu F$. The charge 1.6×10^{-3} C distributes between the two capacitors to have a common p.d. of V volts.

P.D. across parallel combination

$$V = \frac{Q}{C_T} = \frac{1.6 \times 10^{-3}}{20 \times 10^{-6}} = \mathbf{80\ V}$$

Energy stored, $E_2 = \frac{1}{2} C_T V^2 = \frac{1}{2}(20 \times 10^{-6}) \times (80)^2 = \mathbf{0.064\ J}$

It may be noted that there is a loss of energy. This is due to the heat dissipated in the conductor connecting the capacitors.

Example 6.18. *A metal sphere 4 m in diameter is charged to a potential of 3 MV. Calculate the heat generated when the sphere is earthed through a long resistance wire.*

Solution.

Potential at the surface of sphere $V = 9 \times 10^9 \frac{Q}{r}$

∴ Charge on sphere, $Q = \frac{V \times r}{9 \times 10^9} = \frac{(3 \times 10^6) \times 2}{9 \times 10^9} = 0.67 \times 10^{-3}$ C

Energy stored in sphere $= \frac{1}{2} QV = \frac{1}{2}(0.67 \times 10^{-3}) \times (3 \times 10^6) = \mathbf{1005\ J}$

Energy stored in the sphere will be dissipated as heat in the resistance wire.

Example 6.19. *Two capacitors A and B are placed (i) in series (ii) in parallel. Capacitor $C_A = 100\ \mu F$ and $C_B = 50\ \mu F$. Find the maximum energy stored in the circuit when 240 V, 50Hz supply is applied.*

Solution. $V_{max} = 240\sqrt{2}$ Volts

(*i*)
$$C_s = \frac{C_A C_B}{C_A + C_B} = \frac{100 \times 50}{100 + 50} = \frac{100}{3}\ \mu F$$

$$E_{max} = \frac{1}{2} C_S V_{max}^2 = \frac{1}{2}\left(\frac{100}{3} \times 10^{-6}\right)(240\sqrt{2})^2 = \mathbf{1.92\ J}$$

(*ii*)
$$C_P = C_A + C_B = 100 + 50 = 150\ \mu F$$

$$E_{max} = \frac{1}{2} C_P V_{max}^2 = \frac{1}{2}(150/10^6)(240\sqrt{2})^2 = \mathbf{8.64\ J}$$

6.12. Force on Charged Plates

When two plates carry opposite charges, a force of attraction exists between the plates. Thus the plates of a capacitor are attracted towards one another. Consider two parallel conducting plates x metres apart and carrying constant charges of $+Q$ and $-Q$ coulombs respectively as shown in Fig. 6.18. Let the force of attraction between the two plates be F newtons.

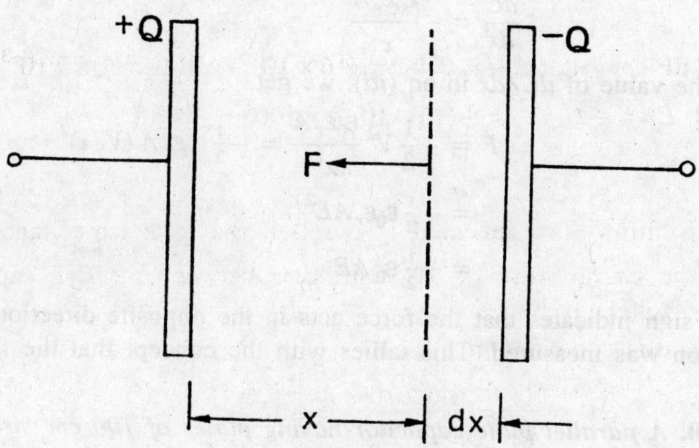

Fig. 6.18

If one of the plates is moved away from the other by a small distance dx, then work done is

$$= F \times dx \text{ joules} \quad \ldots(i)$$

Since the charges on plates remain constant, no electrical energy can enter or leave the system during the movement dx.

$$\therefore \quad \text{Work done} = \text{Change in stored energy}$$

$$\text{Initial stored energy} = \frac{1}{2}\frac{Q^2}{C} \text{ joules}$$

Since the separation of the plates has increased, the capacitance will decrease by dC. The final capacitance is, therefore, $(C - dC)$.

$$\text{Final stored energy} = \frac{1}{2}\frac{Q^2}{(C-dC)} = \frac{Q^2\,(C+dC)}{2\,[C^2 - (dC)^2]}^{*}$$

Since dC is small as compared to C, $(dC)^2$ can be neglected compared to C^2.

$$\therefore \text{ Final stored energy} = \frac{Q^2\,(C+dC)}{2C^2} = \frac{Q^2}{2C} + \frac{Q^2}{2C^2}\,dC$$

$$\therefore \text{ Change in stored energy} = \left(\frac{Q^2}{2C} + \frac{Q^2}{2C^2}\,dC\right) - \frac{Q^2}{2C} = \frac{Q^2}{2C^2}\,dC \quad \ldots(ii)$$

Equating eqs (*i*) and (*ii*), we get,

$$F \times dx = \frac{Q^2}{2C^2}\,dC$$

or

$$F = \frac{Q^2}{2C^2}\frac{dC}{dx} = \frac{1}{2}V^2\frac{dC}{dx} \quad \ldots(iii)$$

$$(\because V = Q/C)$$

---

* Note this step. Multiply the numerator and denominator by $(C + dC)$

Now $$C = \frac{\varepsilon_0 \varepsilon_r A}{x}$$

$$\therefore \quad \frac{dC}{dx} = -\frac{\varepsilon_0 \varepsilon_r A}{x^2}$$

Substituting the value of dC/dx in eq (*iii*), we get,

$$F = -\frac{1}{2} V^2 \frac{\varepsilon_0 \varepsilon_r A}{x^2} = -\tfrac{1}{2} \varepsilon_0 \varepsilon_r A \, (V/x)^2$$

$$= -\tfrac{1}{2} \varepsilon_0 \varepsilon_r A E^2 \qquad \text{...in a medium}$$

$$= -\tfrac{1}{2} \varepsilon_0 A E^2 \qquad \text{...in air}$$

The negative sign indicates that the force acts in the opposite direction to that in which the plate separation was measured. This tallies with the concept that the force is one of attraction.

Example 6.20. *A parallel plate capacitor having plates of 100 cm² area is immersed in oil of relative permittivity 10. The plates are charged to a p.d. of 5 kV and the distance between them is 2 cm. Permittivity of free space = 8.854×10^{-12} F/m. Determine the force between the plates.*

Solution.

Force of attraction between the plates of capacitor

$$F = \frac{1}{2} \varepsilon_0 \varepsilon_r A E^2 \text{ newtons}$$

Here $\varepsilon_r = 10$; $A = 100\,\text{cm}^2 = 0.01\,\text{m}^2$;

Now $$E = \frac{V}{d} = \frac{5 \times 10^3}{2 \times 10^{-2}} = 2.5 \times 10^5 \text{ V/m}$$

$$\therefore \quad F = \frac{1}{2}(8.854 \times 10^{-12}) \times 10 \times 0.01 \times (2.5 \times 10^5)^2 = \mathbf{2.77 \times 10^{-2}\ N}$$

Example 6.21. *A parallel plate capacitor has its plates separated by 0.5 mm of air. The area of plates is 2 m² and they are charged to a p.d. of 100 V. The plates are pulled apart until they are separated by 1 mm of air. Assuming the p.d. to remain unchanged, what is the mechanical force experienced in separating the plates ?*

Solution.

Before plate movement

Capacitance, $$C_1 = \frac{\varepsilon_0 A}{d} = \frac{8.854 \times 10^{-12} \times 2}{0.5 \times 10^{-3}} = 35.4 \times 10^{-9} \text{ F}$$

Stored energy, $$E_1 = \frac{1}{2} C_1 V^2 = \frac{1}{2}(35.4 \times 10^{-9}) \times (100)^2 = 17.7 \times 10^{-5} \text{ J}$$

After plate movement

Capacitance, $$C_2 = \frac{1}{2} C_1 = \frac{1}{2}(35.4 \times 10^{-9}) = 17.7 \times 10^{-9} \text{ F}$$

Stored energy, $$E_2 = \frac{1}{2} C_2 V^2 = \frac{1}{2}(17.7 \times 10^{-9}) \times (100)^2 = 8.85 \times 10^{-5} \text{ J}$$

Change in stored energy $= (17.7 - 8.85) \times 10^{-5} = 8.85 \times 10^{-5}$ J

Suppose F newtons is the average mechanical force between the plates. The plates are separated by a distance $dx = 1 - 0.5 = 0.5$ mm.

$$\therefore \quad F \times dx = \text{Change in stored energy}$$

$$\text{or} \quad F = \frac{8.85 \times 10^{-5}}{0.5 \times 10^{-3}} = \mathbf{17.7 \times 10^{-2}\ N}$$

6.13. Charging of a Capacitor

Consider an uncharged capacitor of capacitance C connected in series with a resistor R to a d.c. supply of V volts as shown in Fig. 6.19. When the switch is closed, the capacitor starts charging up and charging current flows in the circuit. The charging current in maximum at the instant of switching and decreases gradually as the voltage across the capacitor increases. When the capacitor is charged to applied voltage V, the charging current becomes zero.

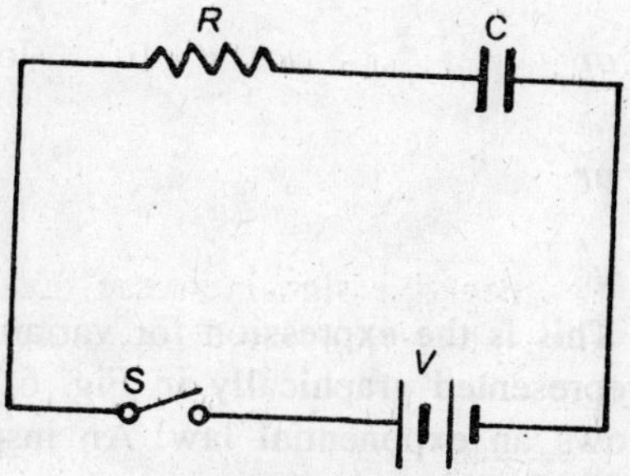

Fig. 6.19

1. At switching instant. At the instant the switch is closed, the voltage across capacitor is zero since we started with an uncharged capacitor. The entire voltage V is dropped across resistance R and charging current is maximum (Call it I_m).

$\therefore$ Initial charging current, $I_m = V/R$

Voltage across capacitor = 0

Charge on capacitor = 0

2. At any instant. After having closed the switch, the charging current starts decreasing and the voltage across capacitor gradually increases. Let at any instant during charging :

i = Charging current

v = P.D. across C

q = Charge on capacitor = $C v$

(i) Voltage across capacitor

According to Kirchhoff's voltage law, the applied voltage V is equal to the sum of voltage drops across resistor and capacitor.

$$\therefore \quad V = v + iR \qquad ...(i)$$

$$\text{or} \quad V = v + CR^{*}\frac{dv}{dt}$$

$$\text{or} \quad -\frac{dv}{V - v} = -\frac{dt}{RC}$$

Integrating both sides, we get,

$$\int -\frac{dv}{V - v} = \int -\frac{dt}{RC}$$

$$\log_e (V - v) = -\frac{t}{RC} + K \qquad ...(ii)$$

---

* $i = \dfrac{dq}{dt} = \dfrac{d}{dt}(q) = \dfrac{d}{dt}(Cv) = C\dfrac{dv}{dt}$

where K is a constant whose value can be determined from the initial conditions. At the instant of closing the switch S, $t = 0$ and $v = 0$.

Substituting these values in eq. (*ii*), we get, $\log_e V = K$

Putting the value of $K = \log_e V$ in eq (*ii*), we get,

$$\log_e (V - v) = -\frac{t}{RC} + \log_e V$$

or $$\log_e \frac{V - v}{V} = -\frac{t}{RC}$$

or $$\frac{V - v}{V} = e^{-t/RC}$$

$\therefore$ $$v = V[1 - e^{-t/RC}] \qquad ...(iii)$$

This is the expression for variation of voltage across the capacitor (v) *w.r.t.* time (t) and is represented graphically in Fig. 6.20 (*i*). Note that growth of voltage across the capacitor follows an exponential law. An inspection of eq (*iii*) reveals that as t increases, the term $e^{-t/RC}$ gets smaller and voltage v across capacitor gets larger.

(*ii*) *Charge on Capacitor*

q = Charge at any time t

Q = Final charge.

Since $v = q/C$ and $V = Q/C$, the exp (*iii*) becomes :

$$\frac{q}{C} = \frac{Q}{C}[1 - e^{-t/RC}]$$

or $$q = Q(1 - e^{-t/RC}) \qquad ...(iv)$$

Again the increase of charge on capacitor plates follows exponential law.

(*iii*) *Charging current*

From exp (*i*), $V - v = iR$

From exp (*iii*), $V - v = Ve^{-t/RC}$

$\therefore$ $$iR = Ve^{-t/RC}$$

or $$i = \frac{V}{R}e^{-t/RC}$$

$\therefore$ $$i = I_m e^{-t/RC}$$

where $I_m (= V/R)$ is the initial charging current. Again the charging current decreases following exponential law. This is also represented graphically in Fig. 6.20 (*ii*).

(*iv*) *Rate of rise of voltage across capacitor*

We have seen above that

$$V = v + CR\frac{dv}{dt}$$

At the instant the switch is closed, $v = 0$

$\therefore$ $$V = CR\frac{dv}{dt}$$

or Initial rate of rise of voltage across capacitor,

$$\frac{dv}{dt} = \frac{V}{CR} \text{ volts/sec} \qquad ...(iv)$$

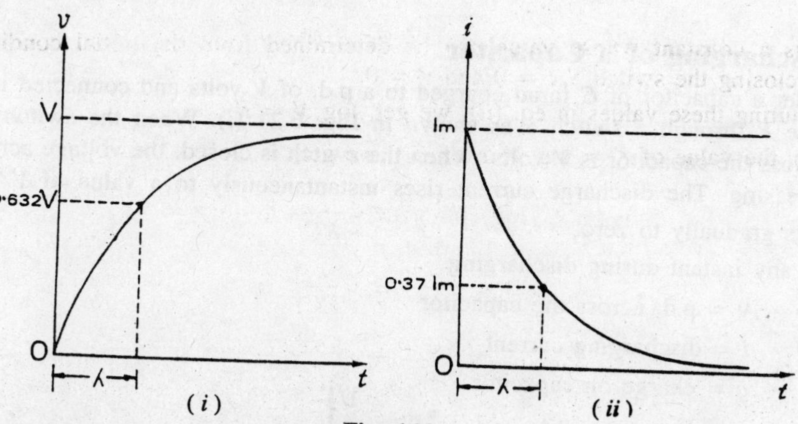

Fig. 6.20

Note. The capacitor is almost fully charged in a time equal to 5 *RC i.e.*, 5 time constants.

6.14. Time Constant

Consider the eq. (*iii*) above showing the rise of voltage across the capacitor :

$$v = V(1 - e^{-t/RC})$$

The exponent of *e* is *t/RC*. The quantity *RC* has the *dimensions of time so that exponent of *e* is a number. The quantity *RC* is called the *time constant* of the circuit and affects the charging (or discharging) time. It is represented by λ (or *T* or τ).

∴ Time constant, $\lambda = RC$ seconds

Time constant may be defined in one of the following ways :

(*i*) At the instant of closing the switch, p.d. across capacitor is zero. Therefore, putting $v = 0$ in the expression $V = v + CR\dfrac{dv}{dt}$,

$$V = CR\frac{dv}{dt}$$

or $$\frac{dv}{dt} = \frac{V}{CR}$$

If this rate of rise of voltage could continue, the capacitor voltage will reach the final value V in time = $V \div V/CR = RC$ seconds = time constant λ.

Hence ***time constant*** *may be defined as the time required for the capacitor voltage to rise to its final steady value V if it continued rising at its initial rate (i.e., V/CR).*

(*ii*) If the time interval $t = \lambda$ (or, *RC*), then

$$v = V(1 - e^{-t/t}) = V(1 - e^{-1}) = 0.632\,\text{V}$$

Hence ***time constant*** *can also be defined as the time required for the capacitor voltage to reach 0.632 of its final steady value V.*

(*iii*) If the time interval $t = \lambda$ (or, *RC*), then

$$i = I_m e^{-t/t} = I_m e^{-1} = 0.37\,I_m$$

Hence ***time constant*** *can also be defined as the time required for the charging current to fall to 0.37 of its initial maximum value* I_m.

---

* $RC = \left(\dfrac{\text{Volt}}{\text{Ampere}}\right) \times \left(\dfrac{\text{Coulomb}}{\text{Volt}}\right) = \dfrac{\text{Volt}}{(\text{Coulomb/sec})} \times \left(\dfrac{\text{Coulomb}}{\text{Volt}}\right) = \text{seconds}$

6.15. Discharging of a Capacitor

Consider a capacitor of C farad charged to a p.d. of V volts and connected in series with a resistance R through a switch S as shown in Fig. 6.21 (*i*). When the switch is open, the voltage across the capacitor is V volts. When the switch is closed, the voltage across capacitor starts decreasing. The discharge current rises instantaneously to a value of V/R (= I_m) and then decays gradually to zero.

Let at any instant during discharging

v = p.d. across the capacitor

i = discharging current

q = charge on capacitor

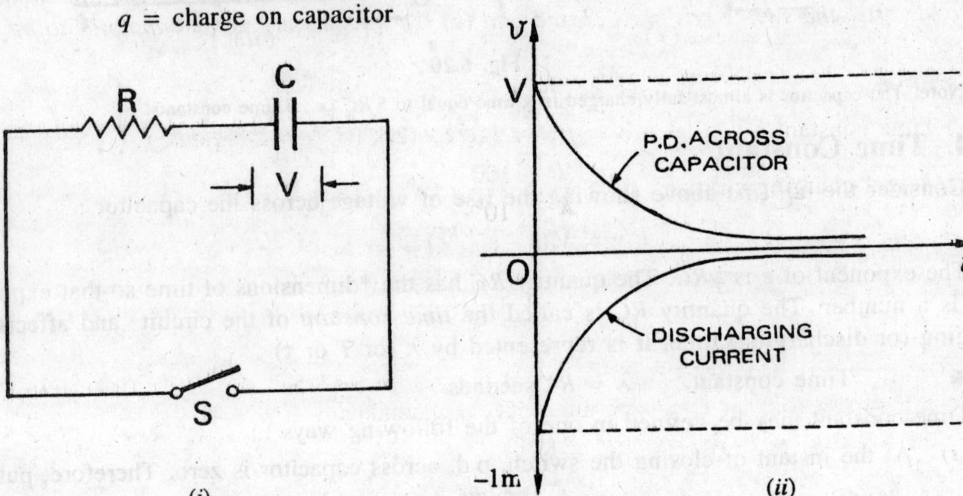

(*i*) (*ii*)

Fig. 6.21

By Kirchhoff's voltage law, we have,

$$0 = v + RC\frac{dv}{dt}$$

or
$$\frac{dv}{v} = -\frac{dt}{RC}$$

Integrating both sides, we get,

$$\int\frac{dv}{v} = -\frac{1}{RC}\int dt$$

$\therefore$
$$\log_e v = -\frac{t}{RC} + K \qquad \text{...}(i)$$

At the instant of closing the switch, $t = 0$ and $v = V$. Putting these values in Eq (*i*), we get,

$$\log_e V = K$$

$\therefore$ Equation (*i*) becomes : $\log_e v = (-t/RC) + \log_e V$

or
$$\log_e \frac{v}{V} = -\frac{t}{RC}$$

or
$$\frac{v}{V} = e^{-t/RC}$$

$$\therefore \qquad v = V e^{-t/\lambda} \qquad ...(ii)$$

Again RC (= λ) is the time constant and has the dimensions of time.

Similarly $$q = Q e^{-t/RC}$$

and $$i = -I_m e^{-t/RC}$$

Note that negative sign is attached to I_m. This is because the discharging current flows in the opposite direction to that in which the charging current flows.

Example 6.22. *A 2 μF capacitor is connected, by closing a switch, to a supply of 100 volts through a 1 MΩ series resistance. Calculate (i) the time constant (ii) initial charging current (iii) the initial rate of rise of p.d. across capacitor (iv) voltage across the capacitor 6 seconds after the switch has been closed and (v) the time taken for the capacitor to be fully charged.*

Solution.

(*i*) Time constant, $\lambda = RC = (10^6) \times (2 \times 10^{-6}) = \mathbf{2\ seconds}$

(*ii*) Initial charging current, $I_m = \dfrac{V}{R} = \dfrac{100}{10^6} \times 10^6 = \mathbf{100\ \mu A}$

(*iii*) Initial rate of rise of voltage across capacitor

$$\frac{dv}{dt} = \frac{V}{CR} = \frac{100}{(2 \times 10^{-6}) \times 10^6} = \mathbf{50\ V/s}$$

(*iv*) $$v = V(1 - e^{-t/RC})$$

Here V = 100 volts ; t = 6 seconds; RC = 2 seconds

$$\therefore \qquad v = 100(1 - e^{-6/2})$$

$$= 100(1 - e^{-3}) = \mathbf{95.1\ V}$$

(*v*) Time taken for the capacitor to be fully charged

$$= 5\,RC = 5 \times 2 = \mathbf{10\ seconds.}$$

Example 6.23. *A capacitor of 8 μ F capacitance is connected to a d.c. source through a resistance of 1 megaohm. Calculate the time taken by the capacitor to receive 95% of its final charge. How long will it take the capacitor to be fully charged ?*

Solution.

$$q = Q(1 - e^{-t/RC})$$

Here $RC = (10)^6 \times 8 \times 10^{-6} = 8$ seconds ; $q/Q = 0.95$

$$\therefore \qquad 0.95 = 1 - e^{-t/8}$$

or $$e^{-t/8} = 0.05$$

or $$e^{t/8} = 1/0.05 = 20$$

or $$(t/8) \log_e e = \log_e 20$$

$$\therefore \qquad t = 8 \log_e 20 = \mathbf{23.96\ seconds}$$

Time taken for the capacitor to be fully charged

$$= 5\,RC = 5 \times 8 = \mathbf{40\ seconds}$$

Example 6.24. *A resistance R and a 4 μF capacitor are connected in series across a 200 V d.c. supply. Across the capacitor is connected a neon lamp that strikes at 120 V. Calculate the value of R to make the lamp strike after 5 seconds.*

Solution. The voltage across the neon lamp has to rise to 120 V in 5 seconds

$$v = V(1 - e^{-t/\lambda})$$

or $$120 = 200(1 - e^{-5/\lambda})$$

or $$e^{-5/\lambda} = 1 - (120/200) = 0.4$$

or $$e^{5/\lambda} = 1/0.4 = 2.5$$

or $$(5/\lambda)\log_e e = \log_e 2.5$$

$\therefore$ $$\lambda = \frac{5}{\log_e 2.5} = 5.457 \text{ seconds}$$

or $$RC = 5.457$$

$\therefore$ $$R = \frac{5.457}{4 \times 10^{-6}} = 1.364 \times 10^6\ \Omega = \mathbf{1.364\ M\Omega}$$

Example 6.25. *A capacitor of 1 μF and resistance 82 KΩ are connected in series with an e.m.f. of 100 V. Calculate the magnitude of energy and the time in which energy stored in the capacitor will reach half of its equilibrium value.*

Solution.

(*i*) Equilibrium value of energy $= \frac{1}{2} CV^2$

$\therefore$ Energy stored $\propto V^2$

Half energy of the equilibrium value will be stored when voltage across capacitor is $v = 100/\sqrt{2} = 70.7$ volts.

$\therefore$ Energy stored $= \frac{1}{2} Cv^2 = \frac{1}{2}(1 \times 10^{-6}) \times (70.7)^2 = \mathbf{0.0025 J}$

(*ii*) $$v = V(1 - e^{-t/RC})$$

Here, $RC = (82 \times 10^3) \times (1 \times 10^{-6}) = 0.082\ s$; $v = 70.7$; $V = 100$

$\therefore$ $$70.7 = 100(1 - e^{-t/0.082})$$

or $$e^{-t/0.082} = 1 - (70.7/100) = 0.293$$

or $$e^{t/0.082} = 1/0.293 = 3.413$$

or $$(t/0.082)\log_e e = \log_e 3.413$$

$\therefore$ $$t = 0.082 \times \log_e 3.413 = \mathbf{0.1\ second}$$

Tutorial Problems

1. A capacitor is being discharged from a d.c. source through a resistance of 2 MΩ. If it takes 0.2 second for the charge to reach 75% of its final value, what is the capacitance of the capacitor ? **[18×10^{-4} F]**
2. An 8 μF capacitor is connected in series with 0.5 MΩ resistance across 200-V supply. Calculate (*i*) initial charging current (*ii*) the current and p.d. across capacitor 4 seconds after it is connected to the supply. **[(*i*) 400 μA (*ii*) 147 μA, 126.4 V]**
3. What resistance connected in series with a capacitance of 4 μF will give the circuit a time constant of 2 seconds ? **[500 KΩ]**
4. A series *RC* circuit is to have an initial charging current of 4 mA and a time constant of 3.6 seconds when connected to 120 V d.c. supply. Calculate the values of *R* and *C*. What will be the energy stored in the capacitor ? **[30 KΩ ; 120 μF ; 0.864 J]**

5. A 20 μF capacitor initially charged to a p.d. of 500 V is discharged through an unknown resistance. After one minute, the p.d. at the terminals of the capacitor is 200 V. What is the value of the resistance ? **[3.274 MΩ]**

OBJECTIVE QUESTIONS

1. The capacitance of a capacitor is .... relative permittivity.
   (*i*) directly proportional to
   (*ii*) inversely proportional to
   (*iii*) independent of
   (*iv*) directly proportional to square of
2. An air capacitor has the same dimensions as that of a mica capacitor. If the capacitance of mica capacitor is 6 times that of air capacitor, then relative permittivity of mica is
   (*i*) 36 (*ii*) 12
   (*iii*) 3 (*iv*) 6
3. The most convenient way of achieving large capacitance is by using
   (*i*) multiplate construction
   (*ii*) decreased distance between plates
   (*iii*) air as dielectric
   (*iv*) dielectric of low permittivity
4. Another name for relative permittivity is
   (*i*) dielectric strength
   (*ii*) breakdown voltage
   (*iii*) specific inductive capacity
   (*iv*) potential gradient
5. A capacitor opposes
   (*i*) change in current
   (*ii*) change in voltage
   (*iii*) both change in current and voltage
   (*iv*) none of the above
6. If a multiplate capacitor has 7 plates each of area 6 cm^2, then
   (*i*) 6 capacitors will be in parallel
   (*ii*) 7 capacitors will be in parallel
   (*iii*) 7 capacitors will be in series
   (*iv*) 6 capacitors will be in series
7. The capacitance of three-plate capacitor [See Fig. 6.22 (*ii*)] is .... that of 2-plate capacitor.

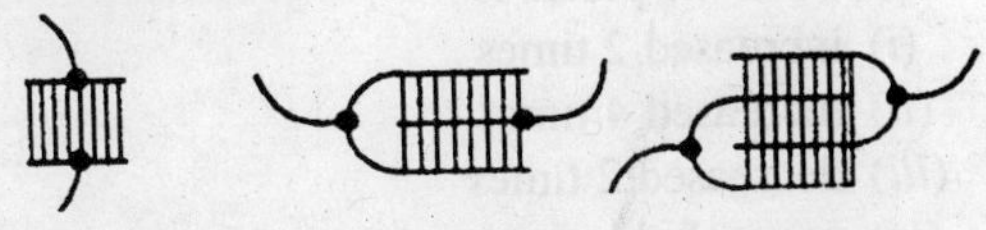

2-plate capacitor (*i*) 3-plate capacitor (*ii*) 4-plate capacitor (*iii*)

Fig. 6.22

   (*i*) 3 times (*ii*) 6 times
   (*iii*) 4 times (*iv*) 2 times
8. The capacitance of a 4-plate capacitor [See Fig. 6.22 (*iii*)] is .... that of 2-plate capacitor.
   (*i*) 2 times (*ii*) 4 times
   (*iii*) 3 times (*iv*) 6 times
9. Two capacitors of capacitances 3 μF and 6 μF in series will have a total capacitance of
   (*i*) 9 μF (*ii*) 2 μF
   (*iii*) 18 μF (*iv*) 24 μF
10. The capacitance of a parallel-plate capacitor does not depend upon
   (*i*) area of plates
   (*ii*) medium between plates
   (*iii*) separation between plates
   (*iv*) metal of plates
11. In order to increase the capacitance of a parallel-plate capacitor, one should introduce between the plates a sheet of
   (*i*) mica (*ii*) tin
   (*iii*) copper (*iv*) stainless steel
12. The capacitance of a parallel-plate capacitor depends upon :
   (*i*) the type of metals used
   (*ii*) separation between plates
   (*iii*) thickness of plates
   (*iv*) potential difference between plates
13. The force between the plates of a parallel plate capacitor of capacitance *C* and distance of separation of plates *d* with a potential difference *V* between the plates is
   (*i*) $\frac{CV^2}{2d}$ (*ii*) $\frac{C^2V^2}{2d^2}$

(*iii*) $\frac{C^2V^2}{d^2}$ (*iv*) $\frac{V^2d}{C}$

14. A parallel-plate air capacitor is immersed in oil of dielectric constant 2. The electric field between the plates is
(*i*) increased 2 times
(*ii*) increased 4 times
(*iii*) decreased 2 times
(*iv*) none of above

15. Two capactiors of capacitance C_1 and C_2 are connected in parallel. A charge Q given to them is shared. The ratio of charges Q_1/Q_2 is
(*i*) C_2/C_1 (*ii*) C_1/C_2
(*iii*) $C_1C_2/1$ (*iv*) $1/C_1$

16. The dimensional formula of capacitance is
(*i*) $M^{-1}L^{-2}T^{-4}A^2$ (*ii*) $M^{-1}L^2T^4A^2$
(*iii*) $ML^2T^{-4}A$ (*iv*) $M^{-1}L^{-2}T^4A^2$

17. Four capacitors are connected as shown in Fig. 6.23. What is the equivalent capacitance between *A* and *B* ?

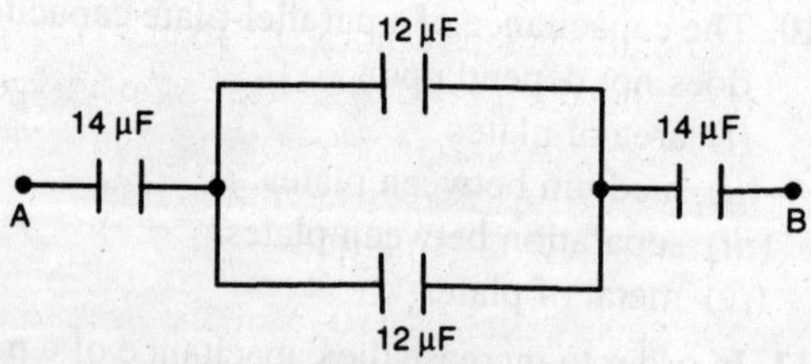

Fig. 6.23

(*i*) 36 μF (*ii*) 5.4 μF
(*iii*) 52 μF (*iv*) 11.5 μF

18. The empty space between the plates of a capacitor is filled with a liquid of dielectric constant *K*. The capacitance of capacitor
(*i*) increases by a factor *K*
(*ii*) decreases by a factor *K*
(*iii*) increases by a factor K^2
(*iv*) decreases by a factor K^2

19. A parallel plate capacitor is made by stacking *n* equally spaced plates connected alternately. If the capacitance between any two plates is C, then the resulting capacitance is
(*i*) *C* (*ii*) *nC*
(*iii*) $(n - 1)\,C$ (*iv*) $(n + 1)\,C$

20. 64 drops of radius *r* combine to form a bigger drop of radius *R*. The ratio of capacitances of bigger to smaller drop is
(*i*) 1 : 4 (*ii*) 2 : 1
(*iii*) 1 : 2 (*iv*) 4 : 1

21. Two capacitors have capacitance 25 μF when in parallel and 6 μF when in series. Their individual capacitances are
(*i*) 12 μF and 13 μF (*ii*) 15 μF and 10 μF
(*iii*) 10 μF and 8 μF (*iv*) none of the above

22. A capacitor of 20 μF charged to 500 V is connected in parallel with another capacitor of 10 μF capacitance and charged to 200 V. The common potential is
(*i*) 200 V (*ii*) 250 V
(*iii*) 400 V (*iv*) 300 V

23. Which of the following does not change when a glass slab is introduced between the plates of a charged parallel plate capacitor?
(*i*) electric charge (*ii*) electric energy
(*iii*) capacitance
(*iv*) electric field intensity

24. A capacitor of 1 μF is charged to a potential of 50 V. It is now connected to an uncharged capacitor of capacitance 4 μF. The common potential is
(*i*) 50 V (*ii*) 20 V
(*iii*) 15 V (*iv*) 10 V

25. Three parallel plates each of area *A* with separation d_1 between first and second and d_2 between second and third are arranged to form a capacitor. If the dielectric constants are K_1 and K_2, the capacitance of this capacitor is

(*i*) $\frac{\varepsilon_0 K_1 K_2}{A(d_1+d_2)}$ (*ii*) $\frac{\varepsilon_0}{A\left(\frac{d_1}{K_1}+\frac{d_2}{K_2}\right)}$

(*iii*) $\frac{\varepsilon_0 A K_1 K_2}{d_1+d_2}$ (*iv*) $\frac{\varepsilon_0 A}{\frac{d_1}{K_1}+\frac{d_2}{K_2}}$

26. The plates of a charged parallel-plate capacitor are pulled apart. Now
(*i*) p.d. will remain unchanged
(*ii*) p.d. will decrease
(*iii*) p.d. will increase
(*iv*) the capacitance will increase

27. A dielectric is introduced between the plates of a capacitor kept at a constant potential difference. The charge on the capacitor
(*i*) increases (*ii*) decreases
(*iii*) remains the same (*iv*) none of above

28. Two capacitors of capacitances 2 pF and 6 pF are connected in parallel across a 120 V d.c. source. The sum total of the charges on the capactiors is
(*i*) 15 pC (*ii*) 1980 pC
(*iii*) 180 pC (*iv*) 960 pC

29. Three capactiors of capacitances 3 μF, 9μF and 18 μF are connected once in series and another time in parallel. The ratio of equivalent capacitances in the two cases (C_s/C_p) will be
(*i*) 1 : 15 (*ii*) 1 : 3
(*iii*) 1 : 9 (*iv*) 1 : 12

30. A dielectric is placed in between two parallel plates of a capacitor as shown in Fig 6.24. The dielectric constant of the dielectric is *K*. If the initial capacitance is *C*, then the new capacitance will be

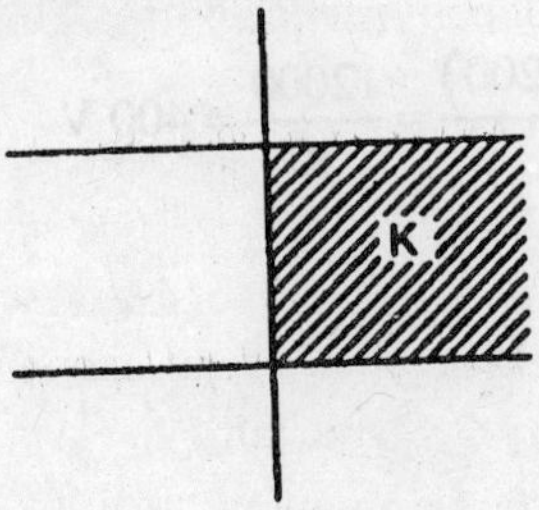

Fig. 6.24

(*i*) *KC* (*ii*) $\frac{C(K+1)}{2}$
(*iii*) $(K-1)C$ (*iv*) $(K+1)C$

31. A capacitor is charged through a p.d. of 200 V and possesses charge of 0.1 C. When discharged, it would release an energy of
(*i*) 20 J (*ii*) 1 J
(*iii*) 10 J (*iv*) 12 J

32. A parallel-plate capacitor with air as medium between plates has a capacitance of 10 μF. The area of the capacitor is divided into equal halves and filled with two media having dielectric constant $K_1 = 2$, $K_2 = 4$. The capacitance of the system will now be
(*i*) 10 μF (*ii*) 20 μF
(*iii*) 40 μF (*iv*) 30 μF

33. Force acting on a charged particle kept between the plates of a charged capacitor is *F*. If one of the plates of the capacitor is removed, the force acting on the same particle will become
(*i*) *F*/2 (*ii*) 0
(*iii*) *F* (*iv*) 2*F*

34. Two insulated charged spheres of radii 20 cm and 25 cm respectively and having an equal charge *Q* are connected by a copper wire and are then separated. Then,
(*i*) both spheres will have the same charge
(*ii*) charge on 20 cm sphere is greater
(*iii*) charge on 25 cm sphere is greater
(*iv*) charge on each will be 2 *Q*.

35. The capacitance of a parallel plate capacitor is 5 μF. When a glass plate is inserted between its two plates, its potential reduces to 1/8 of the original value. The value of dielectric constant of glass is
(*i*) 4 (*ii*) 8
(*iii*) 40 (*iv*) 1.6

ANSWERS TO OBJECTIVE QUESTIONS

1. (*i*)	2. (*iv*)	3. (*i*)	4. (*iii*)	5. (*ii*)	6. (*i*)	7. (*iv*)	8. (*iii*)
9. (*ii*)	10. (*iv*)	11. (*i*)	12. (*ii*)	13. (*i*)	14. (*iii*)	15. (*ii*)	16. (*iv*)
17. (*ii*)	18. (*i*)	19. (*iii*)	20. (*iv*)	21. (*ii*)	22. (*iii*)	23. (*i*)	24. (*iv*)
25. (*iv*)	26. (*iii*)	27. (*iii*)	28. (*iv*)	29. (*i*)	30. (*ii*)	31. (*iii*)	32. (*iv*)
33. (*i*)	34. (*iii*)	35. (*ii*)					

HINTS TO SELECTED OBJECTIVE QUESTIONS

2. $\frac{C_{mica}}{C_{air}}$ = dielectric constant of mica = 6

7. The 3-plate capacitor forms two similar capacitors in parallel. Hence the capacitance of a 3-plate capacitor is twice that of a 2-plate capacitor.

9. When two capacitors are connected in series, the total capacitance C_T is given by :

$$C_T = \frac{\text{Product}}{\text{sum}} = \frac{3 \times 6}{3+6} = 2\mu\text{F}$$

13. Due to opposite charges on the two plates of a capacitor, there is a force of attraction between them

$\therefore$ Work done = Energy stored

or $F \times d = \frac{1}{2} CV^2$

or $F = \frac{CV^2}{2d}$

17. $C_p = 12 + 12 = 24$ μF. Now $\frac{1}{C_s} = \frac{1}{14} + \frac{1}{24} + \frac{1}{14} = \frac{31}{168}$ $\quad \therefore C_s = 168/31 = 5.4$ μF

20. $\frac{4}{3}\pi R^3 \rho = 64 \times \left(\frac{4}{3}\pi r^3\right)\rho$ or $R = (64)^{1/3} \times r = 4$ $\therefore$ $R/r = 4r$

If C_b and C_s are the capacitances of bigger and smaller drops respectively, then $C_b = 4\pi\varepsilon_0 R$ and $C_s = 4\pi\varepsilon_0 r$.

$$\therefore \frac{C_b}{C_s} = \frac{4\pi\varepsilon_0 R}{4\pi\varepsilon_0 r} = \frac{R}{r} = \frac{4}{1}$$

22. Common potential, $V = \frac{C_1V_1 + C_2V_2}{C_1 + C_2} = \frac{10^{-6}(20 \times 500 + 10 \times 200)}{10^{-6}(20+10)} = \frac{12000}{30} = 400$ V

25. The system constitutes two capacitors C_1 and C_2 in series.

$$\therefore \frac{1}{C_s} = \frac{1}{C_1} + \frac{1}{C_2} = \frac{d}{\varepsilon_0 K_1 A} + \frac{d_1}{\varepsilon_0 K_2 A} = \frac{1}{\varepsilon_0 A}\left(\frac{d_1}{K_1} + \frac{d_2}{K_2}\right)$$

$$\therefore C_s = \frac{\varepsilon_0 A}{\frac{d_1}{K_1} + \frac{d_2}{K_2}}$$

26. In order to pull the plates apart, work will have to be done. This work done is stored in the electric field of the capacitor ($U = CV^2/2$). Therefore p.d. across plates increases.

30. In the absence of dielectrics, capacitance of the two capacitors are C/2 and C/2. With dielectric present as shown, the capacitances of the two capacitors are $C/2$ and $KC/2$. Since the two capacitors are in parallel,

$$\text{Resultant capacitance} = \frac{C}{2} + \frac{KC}{2} = \frac{C(K+1)}{2}$$

33. The electric field will be halved.

35. $C = q/V$. When potential difference is reduced by 8 times, then for the same charge, capacitance increases by 8 times.

$$\therefore K = \frac{8 \times 5 \times 10^{-6}}{5 \times 10^{-6}} = 8$$

7

Magnetism and Electromagnetism

Introduction

In the ancient times people believed that the invisible force of magnetism was purely a magical quality and hence they showed little practical interest. However, with steadily increasing scientific knowledge over the passing centuries, magnetism assumed a larger and larger role. To-day magnetism has attained a place of pride in electrical engineering. Without the aid of magnetism, it is impossible to operate such devices as electric generators, electric motors, transformers, electrical instruments *etc*. Without the use of magnetism, we should be deprived of such valuable assets as the radio, television, telephone, telegraph and the ignition systems of our cars, airplanes, trucks *etc*. In fact, electrical engineering is so much dependent on magnetism that without it a very few of our modern devices would be possible. The purpose of this chapter is to present the salient features of magnetism so that a reader may be able to understand the function of magnetism in the electrical equipment.

7.1. Poles of a Magnet

If we take a bar magnet and dip it into iron filings, it will be observed that the iron filings cluster about the ends of the bar magnet. The ends of the bar magnet are apparently points of maximum magnetic effect and for convenience we call them the **poles* of the magnet. A magnet has two poles *viz* north pole and south pole. In order to determine the polarity of a magnet, suspend or pivot it at the centre. The magnet will then come to rest north-south direction. The end of the magnet pointing north is called *north pole* of the magnet while the end pointing south is called the *south pole*. The following points may be noted about the poles of a magnet :

(*i*) The poles of a magnet cannot be separated. If a bar magnet is broken into two parts, each part will be a complete magnet with poles at its ends. No matter how many times a magnet is broken, each piece will contain *n*-pole at one end and *s*-pole at the other.

(*ii*) The two poles of a magnet are of equal strength. The pole strength is represented by *m*.

7.2. Laws of Magnetic Force

(*i*) *Like poles repel each other while unlike poles attract each other.*

(*ii*) *The force between two magnetic poles is directly proportional to the product of their pole strengths and inversely proportional to the square of distance between them.*

* Magnetic poles have no physical reality, but the concept enables us to appreciate magnetic effects more easily.

Consider two poles of magnetic strength m_1 and m_2 placed at a distance d apart in a medium as shown in Fig 7.1. According to Coulomb's laws, the force between the two poles is given by;

$$F \propto \frac{m_1 m_2}{d^2}$$

$$= K \frac{m_1 m_2}{d^2}$$

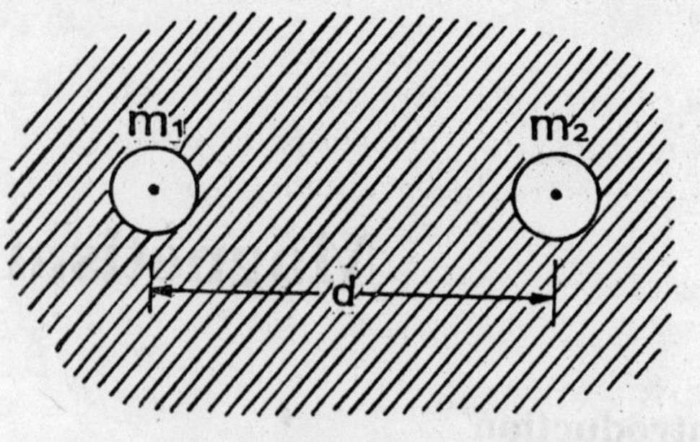

Fig. 7.1

where K is a constant whose value depends upon the surrounding medium and the system of units employed. In *SI* units, force is measured in newtons, pole strength in weber, distance in metres and the value of K is given by;

$$K = \frac{1}{4\pi \mu_0 \mu_r}$$

where μ_0 = Absolute permeability of vacuum or air

μ_r = Relative permeability of the surrounding medium

For vacuum or air, its value is 1.

The value of $\mu_0 = 4\pi \times 10^{-7}$ H/m and the value of μ_r is different for different media.

$$\therefore \quad F = \frac{m_1 m_2}{4\pi \mu_0 \mu_r d^2} \text{ newtons} \qquad \textit{... in a medium}$$

$$= \frac{m_1 m_2}{4\pi \mu_0 d^2} \text{ newtons} \qquad \textit{... in air}$$

Unit of pole strength. By unit pole strength we mean 1 weber. It can be defined from Coulomb's laws of magnetic force. Suppose two equal point poles placed 1m apart in *air* exert a force of 62800 newtons *i.e.*

$$m_1 = m_2 = m \,;\; d = 1 \text{ m};\; F = 62800 \text{ N}$$

$$\therefore \quad F = \frac{m_1 m_2}{4\pi \mu_0 d^2} \qquad (\because \text{ For air, } \mu_r = 1)$$

$$\text{or} \quad 62800 = \frac{m^2}{4\pi \times 4\pi \times 10^{-7} \times (1)^2}$$

$$\text{or} \quad m^2 = (62800) \times (4\pi \times 4\pi \times 10^{-7} \times 1) = 1$$

$$\therefore \quad m = \pm 1 \text{ wb}$$

Hence a **pole of unit strength** (*i.e. 1wb*) *is that pole which when placed in air 1 m from an identical pole, repels it with a force of 62800 newtons.*

7.3. Magnetic Field

Just as electric field exists near a charged object, similarly magnetic field exists around a magnet. If an isolated magnetic pole is brought near a magnet, it experiences a force according to Coulomb's laws. This region near the magnet where forces act on magnetic poles is called a *magnetic field*. The magnetic field is strongest near the pole and goes on decreasing in strength as we move away from the magnet.

The space (or field) in which a magnetic pole experiences a force is called a **magnetic field**

The magnetic field around a magnet is represented by imaginary lines called *magnetic lines of force*. By convention, the direction of these lines of force at any point is the direction along which an *isolated unit *n*-pole (*i.e.* *n*-pole of 1wb) placed at that point would move or tends to move. Following this convention, it is clear that magnetic lines of force would emerge from *n*-pole of the magnet, pass through the surrounding medium and re-enter the *s*-pole. Inside the magnet, each line of force passes from *s*-pole to *n*-pole (see Fig 7.2), thus forming a closed loop or magnetic circuit.

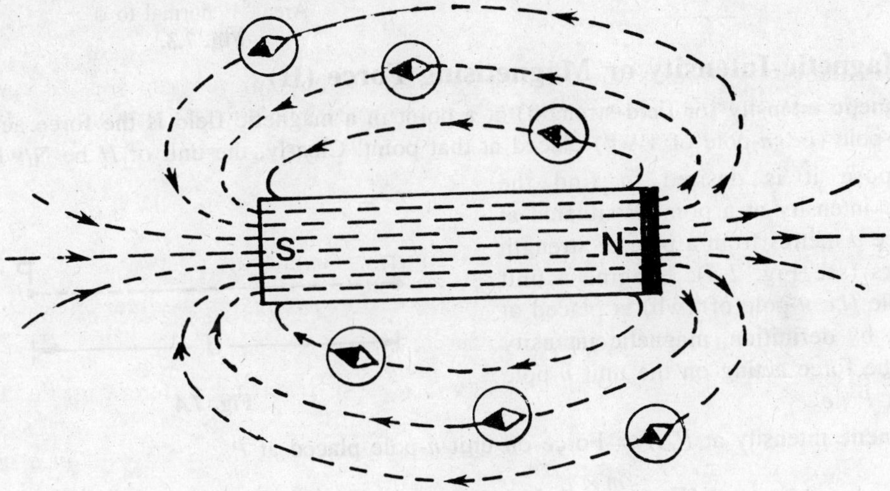

Fig. 7.2

Although magnetic lines of force have no real existence and are purely imaginary, yet they are a useful concept to describe the various magnetic effects.

7.4. Magnetic Flux

The magnetic field cannot be detected by any of our personal senses but its effects can be observed in many ways. To identify the magnetic field quantitatively (*i.e.* with numbers), we generally use the term magnetic flux.

The amount of magnetic field produced by a magnetic source is called **magnetic flux.**

Magnetic flux is denoted by Greek letter ϕ. If 10 magnetic lines come out of the north pole or enter the south pole of a magnet, then magnetic flux ϕ = 10 lines or maxwells. The *SI* unit of magnetic flux is *weber*.

$$1\text{wb} = 10^8 \text{ lines}$$

7.5. Magnetic Flux Density

The magnetic flux density is the flux per unit area at right angles to the flux (see Fig 7.3) *i.e.*

$$\text{Flux density, } B = \frac{\phi}{A} \; wb/m^2$$

---

* Theoretically, it is not possible to get an isolated *n*-pole. However, a small compass needle well approximates to an isolated *n*-pole. The marked end (*n*-pole) of the compass needle indicates the direction of magnetic lines of force as shown in Fig. 7.2.

where ϕ = flux in wb

A = area in m^2 normal to flux.

The *SI* unit of magnetic flux density is wb/m^2 or *tesla. Flux density is a measure of field concentration *i.e.* amount of flux in each square metre of the field. In practice, it is much more important than the total amount of flux.

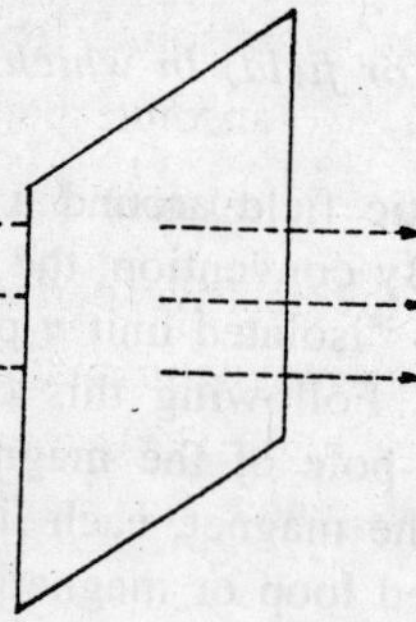

Area '*A*' normal to ϕ

Fig. 7.3.

7.6. Magnetic Intensity or Magnetising Force (H)

Magnetic intensity (or field strength) at a point in a magnetic field is the force acting on a unit *n*-pole (*i.e. n*-pole of 1 wb) placed at that point. Clearly, the unit of *H* be N/wb.

Suppose it is desired to find the magnetic intensity at a point *P* situated at a distance *d* metres from a pole of strength *m* webers (see Fig. 7.4). Imagine a unit north pole (*i.e n*-pole of 1wb) is placed at *P*. Then, by definition, magnetic intensity at *P* is the force acting on the unit *n*-pole placed at *P i.e.*

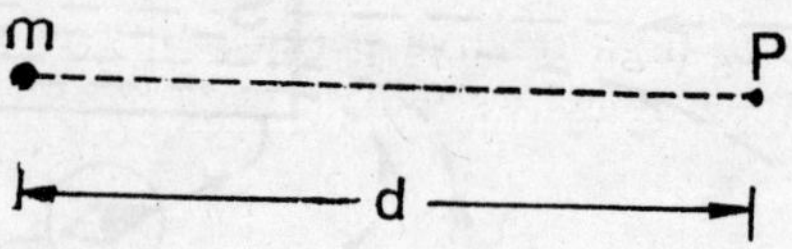

Fig. 7.4

Magnetic intensity at *P*, *H* = Force on unit *n*-pole placed at *P*

or $$H = \frac{m \times 1}{4\pi\,\mu_0\,d^2}\ \text{N/wb} \qquad [\because\ \mu_r = 1 \text{ for air}]$$

or $$H = \frac{m}{4\pi\,\mu_0\,d^2}\ \text{N/wb}$$

The reader may note the following points carefully :

(*i*) Magnetic intensity is a vector quantity, possessing both magnitude and direction.

(*ii*) If a pole of *m* wb is placed in a uniform magnetic field of strength *H* newtons/wb, then force acting on the pole, $F = m\,H$ newtons.

7.7. Absolute and Relative Permeability

Permeability of a material means its conductivity for magnetic flux. The greater the permeability of a material, the greater is its conductivity for magnetic flux and *vice-versa*. Air or vacuum is the poorest conductor of magnetic flux. The absolute (or actual) permeability **μ_0 (Greek letter "mu") of air or vacuum is $4\pi \times 10^{-7}$ H/m. The absolute (or actual) permeability μ of magnetic materials is much greater than μ_0. The ratio μ/μ_0 is called the relative permeability of the material and is denoted by μ_r *i.e.*

$$\mu_r = \frac{\mu}{\mu_0}$$

* Named in honour of Nikola Tesla (1857-1943), an American electrician and inventor.

** The absolute (or actual) permeability of all non-magnetic materials is also $4\,\pi \times 10^{-7}$ H/m .

where μ = absolute (or actual) permeability of the material

μ_0 = absolute permeability of air or vacuum

μ_r = relative permeability of the material

Obviously, the relative permeability for air or vacuum would be $\mu_0/\mu_0 = 1$. The value of μ_r for all non-magnetic materials is also 1. However, relative permeability of magnetic materials is very high. For example, soft iron (*i.e.* pure iron) has a relative permeability of 8000 whereas its value for permalloy (an alloy containing 22% iron and 78% nickel) is as high as 50,000.

Concept of relative permeability. The relative permeability of a material is a measure of the relative ease with which that material conducts magnetic flux compared with the conduction of flux in air. Fig 7.5 illustrates the concept of relative permeability. In Fig. 7.5 (*i*), the magnetic flux passes between the poles of a magnet in air. Consider a soft iron ring ($\mu_r = 8000$) placed between the same poles as shown in Fig. 7.5 (*ii*). Since soft iron is a very good conductor of magnetic flux, the flux follows a path entirely within the soft iron itself. The flux density in the soft iron is much greater than it is in air. In fact, flux density in soft iron will be 8000 times (*i.e.* μ_r times) the flux density in air.

Due to high relative permeability of magnetic materials (*e.g.* iron, steel and other magnetic alloys), they are widely used for the cores of all electromagnetic equipment.

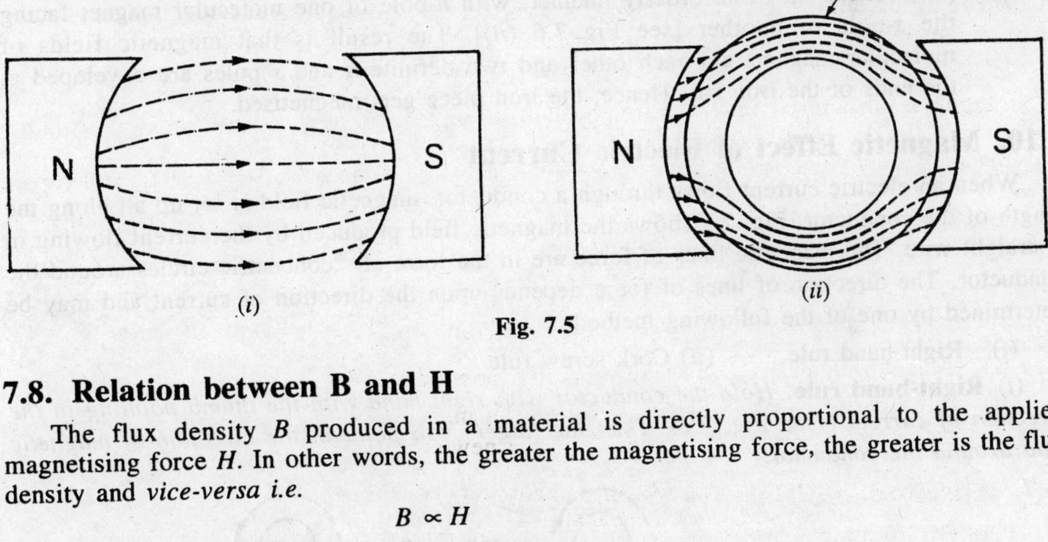

Fig. 7.5

7.8. Relation between B and H

The flux density *B* produced in a material is directly proportional to the applied magnetising force *H*. In other words, the greater the magnetising force, the greater is the flux density and *vice-versa i.e.*

$$B \propto H$$

or
$$\frac{B}{H} = \text{Constant} = \mu$$

The ratio *B/H* in a material is always constant and is equal to the absolute permeability $\mu (= \mu_0 \mu_r)$ of the material. This relation gives yet another definition of absolute permeability of a material.

Obviously,
$$B = \mu_0 \mu_r H \quad \text{... in a medium}$$
$$= \mu_0 H \quad \text{... in air}$$

7.9. Molecular Theory of Magnetism

There have been various theories developed from time to time for the explanation of magnetism. The theory proposed by *Weber* in 1852 and modified by Ewing in 1890, is the most popular explanation. According to this theory, *molecules of all substances are basically magnets in themselves, each having a n and s pole.* In other words, every substance consists of a very large number of tiny magnets called *molecular magnets.*

(*i*) Before a piece of iron has been magnetised, these molecular magnets lie in disorderly positions [see Fig 7.6 (*i*)] so that poles of molecular magnets neutralise each other. Hence, the iron piece does not show any magnetism *i.e.* no poles are developed at the ends.

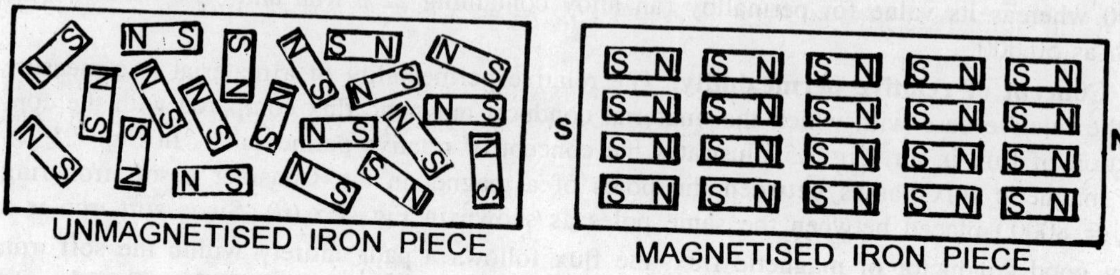

(*i*) (*ii*)

Fig. 7.6

(*ii*) When a magnetising force is applied to the iron bar (by rubbing a magnet or passing electric current through a wire wound over it), the molecular magnets are turned and tend to line up in an orderly manner, with *n*-pole of one molecular magnet facing the *s*-pole of another [see Fig 7.6 (*ii*)]. The result is that magnetic fields of molecular magnets aid each other and two definite *n* and *s* poles are developed at the ends of the iron bar. Hence, the iron piece gets magnetised.

7.10. Magnetic Effect of Electric Current

When an electric current flows through a conductor, magnetic field is set up all along the length of the conductor. Fig. 7.7 shows the magnetic field produced by the current flowing in a straight wire. The magnetic lines of force are in the form of *concentric circles around the conductor. The direction of lines of force depends upon the direction of current and may be determined by one of the following methods :

(*i*) Right-hand rule (*ii*) Cork screw rule

(*i*) **Right-hand rule**. *Hold the conductor with right hand with the thumb pointing in the direction of current (see Fig. 7.7). Then the fingers will point in the direction of magnetic field around the conductor.*

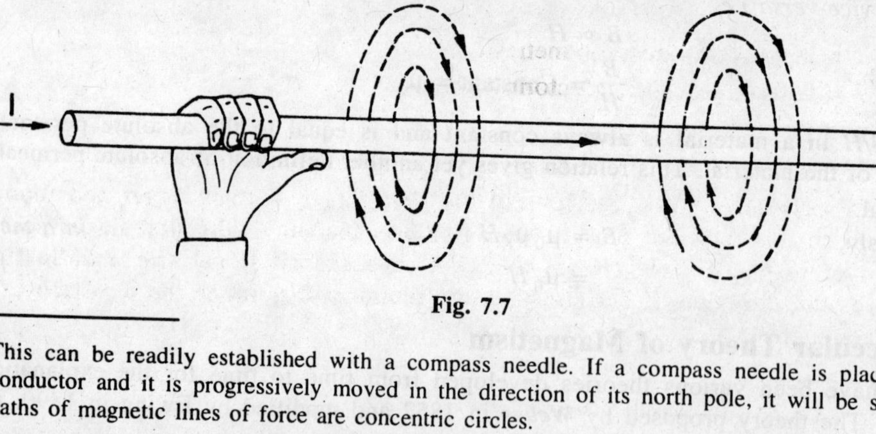

Fig. 7.7

* This can be readily established with a compass needle. If a compass needle is placed near the conductor and it is progressively moved in the direction of its north pole, it will be seen that the paths of magnetic lines of force are concentric circles.

Applying this rule to Fig 7.7, it is clear that when viewed from left hand side, the direction of magnetic lines of force will be clockwise.

(*ii*) **Cork screw rule.** Hold the cork screw in your right hand and rotate it in such a way that it advances in the direction of current. Then the direction in which the hand rotates will be the direction of magnetic lines of force.

Referring to Fig 7.7 and applying cork screw rule, the direction of magnetic lines of force will be clockwise.

Right-hand rule for coil. The magnetic polarity of a coil of several turns or a *solenoid can be determined by right-hand rule stated below :

Grasp the whole coil with right hand so that the fingers are curled in the direction of current. The thumb stretched parallel to the axis of the coil will point towards the n-pole end of the coil (See Fig. 7.8)

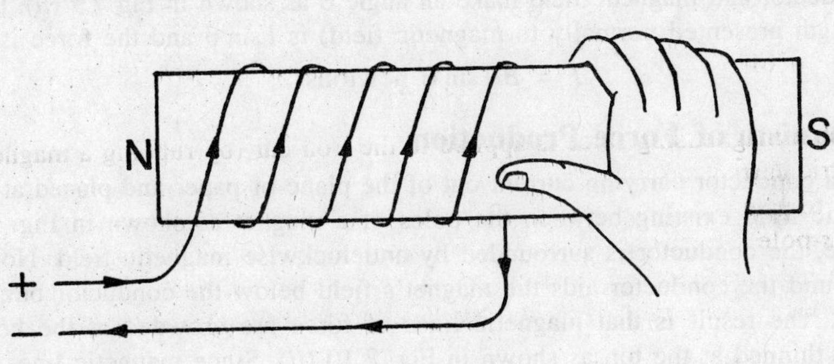

Fig. 7.8

It may be noted that both right-hand rules (for a conductor and for a coil) discussed so far can be applied in reverse. If we know the direction of magnetic field encircling a conductor or the magnetic polarity of a coil, we can determine the direction of current by applying appropriate right-hand rule.

7.11. Current Carrying Conductor in Magnetic Field

When a current-carrying conductor is placed at right angles to a magnetic field, it experiences a mechanical force F given by;

$$F = BIl \text{ newtons}$$

where l = length of conductorin metres

I = current through conductorin amperes

B = flux density in wb/m^2

The direction of this force can be found by Fleming's left-hand rule.

Fleming's Left-hand Rule. Stretch out the first finger, second finger and thumb of your left hand so that they are at right angles to one another. If the first finger points in the direction of magnetic field (North to South) and second finger (*i.e.* middle finger) points towards the direction of current, then the thumb will point in the direction of motion of the conductor.

---

* Solenoid is a Greek word meaning "tube-like".

Applying Fleming's left-hand rule to Fig. 7.9 (*i*), it is clear that the force F on the conductor will act vertically upwards. Note that if direction of any two of F, B and I is known, the direction of the third can be found by this rule.

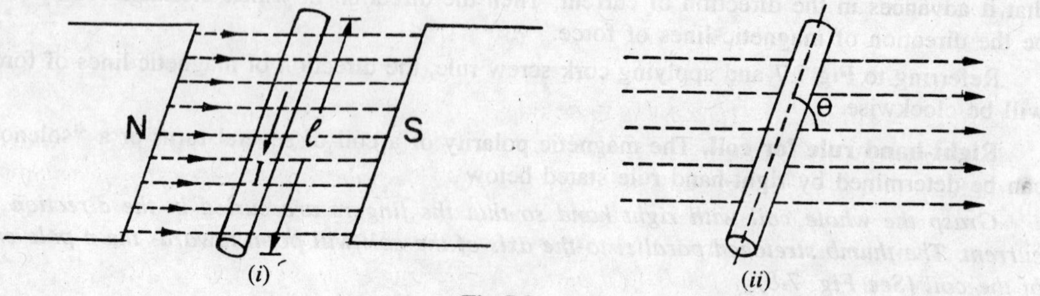

Fig 7.9

If the conductor and magnetic field make an angle θ as shown in Fig 7.9 (*ii*), then effective length (*i.e.* length presented normally to magnetic field) is $l \sin \theta$ and the force is given by;

$$F = BIl \sin \theta \text{ newtons}$$

7.12. Mechanism of Force Production

Consider a conductor carrying current out of the plane of paper and placed at right angles to the magnetic field existing between the poles of a magnet as shown in Fig. 7.10 (*i*). By right-hand rule, the conductor is surrounded by anticlockwise magnetic field. Note that magnetic field around the conductor aids the magnet's field below the conductor but opposes the field above it. The result is that magnetic lines of force are crowded at the bottom of the conductor and thinned at the top as shown in Fig. 7.10 (*ii*). Since magnetic lines of force act somewhat like stretched rubber bands, a force F acts on the conductor from bottom towards top *i.e.* the force moves the conductor upwards.

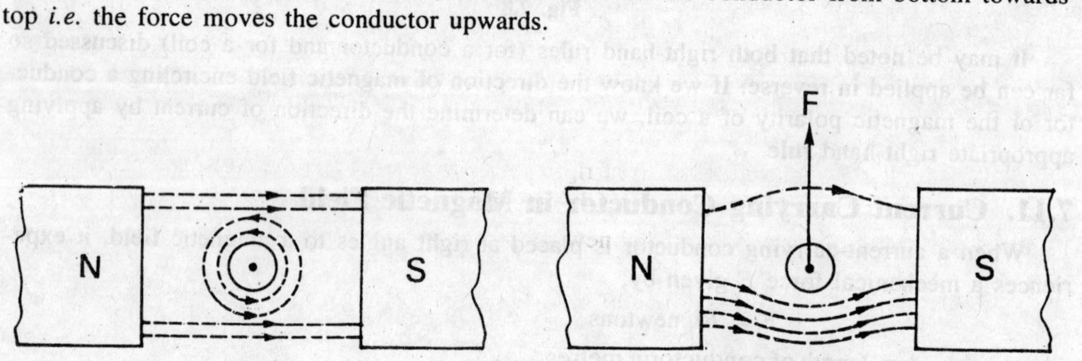

Fig. 7.10

If the direction of current in the conductor or direction of field is reversed, the direction of force is also reversed. It may be noted that direction of force can also be determined by applying Fleming's left-hand rule.

Example 7.1. *In a certain electric motor, wires that carry a current of 5A are perpendicular to a magnetic field of 0.8T. What is the force on each cm of these wires ?*

Solution.

Force on each cm of wires, $F = BIl = (0.8) \times (5) \times (0.01) = \mathbf{0.04N}$

Example 7.2. *A straight wire 0.5m long carries a current of 100A and lies at right angles to a uniform field of 1.5T. Find the mechanical force on the conductor when (i) it lies in the given position (ii) it lies in a position such that it is inclined at an angle of 30° to the direction of field.*

Solution.

(*i*) $F = BIl = 1.5 \times 100 \times 0.5 = \mathbf{75N}$

(*ii*) $F = BIl \sin\theta = 1.5 \times 100 \times 0.5 \times \sin 30° = \mathbf{37.5N}$

Example 7.3. *A conductor of length 100 cm and carrying 100A is situated in and at right angles to a uniform magnetic field produced by the pole core of an electrical machine. If the pole core has a circular cross-section of 120 mm diameter and the total flux in the core is 16 mwb, find (i) the mechanical force on the conductor and (ii) power required to move the conductor at a speed of 10m/s in a plane at right angles to the magnetic field.*

Solution.

X-sectional area of pole core $= (\pi/4) \times (0.12)^2 = 0.0113 \text{ m}^2$

Flux density of field, $B = \dfrac{\text{Flux}}{\text{Pole core area}} = \dfrac{16 \times 10^{-3}}{0.0113} = 1.416 \text{ wb/m}^2$

(*i*) Force on the conductor, $F = BIl = 1.416 \times 100 \times 1 = \mathbf{141.6N}$

(*ii*) Power required = Force × distance/second $= 141.6 \times 10 = \mathbf{1416\ watts}$

Example 7.4. *The plane of a rectangular coil makes an angle of 60° with the direction of a uniform field of flux density 4×10^{-2} wb/m². The coil is of 20 turns, measuring 20 cm by 10 cm, and carries a current of 0.5A. Calculate the torque acting on the coil.*

Solution. Consider a rectangular coil, measuring *b* by *l*, of N turns carrying a current of *I* amperes and placed in a uniform magnetic field of *B*wb/ m². The coil is pivoted about the mid points of the sides *b* and is free to rotate about an axis in its own plane; this axis being at right angles to the field density *B* [See Fig. 7.11 (*i*)]. When current is passed through the coil, forces act on the coil sides.

(*i*) The forces developed on each half of coil sides *b* are equal and produce torques of opposing sense. They, therefore, cancel each other.

(*ii*) The coil sides *l* always remain at right angles to the field as the coil rotates. The force *F* acting on each of the coil sides *l* gives rise to a torque as shown in Fig. 7.11 (*ii*).

Force on each coil side l, $F = BIlN$ newtons

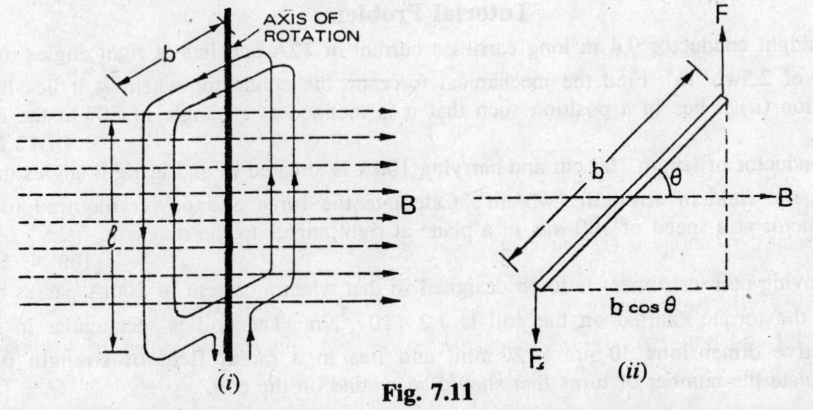

Fig. 7.11

The perpendicular distance between the lines of action of two forces is $b \cos\theta$.

$\therefore$ Torque, $T = F \times b \cos\theta = (BIlN)\, b \cos\theta$

$\therefore$ $T = {}^{*}BINA \cos\theta$ newton–metre

where $A\,(= l \times b)$ is the area of the coil. By an extension of this reasoning, the expression may be proved quite generally for a coil of area A and of any shape.

In the given problem :

$B = 4 \times 10^{-2}$ wb/m^2; $A = 20 \times 10 = 200$ cm^2 $= 2 \times 10^{-2}$ m^2; $I = 0.5$ A ; $\theta = 60°$; $N = 20$

$\therefore$ Torque, $T = (4 \times 10^{-2}) \times (0.5) \times (20) \times (2 \times 10^{-2}) \times \cos 60° = \mathbf{4 \times 10^{-3}\ Nm}$

Example 7.5. *A d.c. motor consists of an armature winding of 400 turns (equivalent to 800 conductors). The effective length of conductor in the field is 160mm and the conductors are situated at a radius of 100mm from the centre of the motor shaft. The magnetic flux density is 0.6 wb/ m^2 and a current of 25A flows through the winding. Calculate the torque available at the motor shaft.*

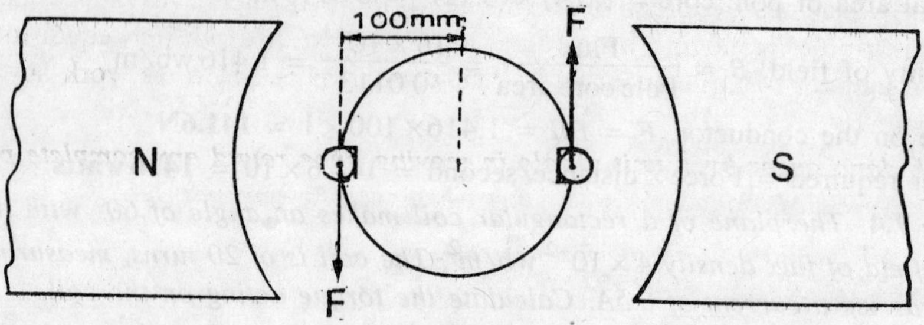

Fig. 7.12

Solution. In a d.c. motor, conductors carrying direct current are wound on an iron core (called armature) placed in a magnetic field. The mechanical forces acting on the conductors rotate the armature. Fig. 7.12 shows a single turn on the armature of the d.c. motor.

Force on each conductor, $F = BIl = (0.6) \times (25) \times (0.16) = 2.4$N

The perpendicular distance between the two conductors of a turn is twice the radius of armature *i.e* $2 \times 100 = 200$ mm $= 0.2$ m.

$\therefore$ Torque provided by 1 turn $= 2.4 \times 0.2 = 0.48$ Nm

Total torque $= 400 \times 0.48 =$ **192 Nm**

Tutorial Problems

1. A straight conductor 0.4 m long carries a current of 12A and lies at right angles to a uniform field of 2.5wb/ m^2. Find the mechanical force on the conductor when (*i*) it lies in the given position (*ii*) it lies in a position such that it is inclined at an angle of 30° to the direction of field. **[(*i*)12 N (*ii*) 6 N]**

2. A conductor of length 100 cm and carrying 100A is situated in and at right angles to a uniform magnetic field of strength 1wb/ m^2. Calculate the force and power required to move the conductor at a speed of 100 m/s in a plane at right angles to the magnetic field. **[100 N; 1000 watts]**

3. A moving coil instrument is to be designed so that when a current of 20mA passes through the coil, the torque exerted on the coil is 1.2×10^{-3}Nm. The coil is rectangular in shape with effective dimensions 30 mm × 20 mm, and lies in a radial field of strength 0.8 wb/ m^2. Calculate the number of turns that should be wound on the coil. **[125 turns]**

---

* How to remember ? Torque is equal to BINA (name of *girl*) multiplied by cos θ.

4. The control springs of a moving coil instrument provide a resisting torque of 12×10^{-6} Nm per degree. Calculate the torque exerted when the coil deflects through 75°.
If the coil has effective dimensions 40 mm × 15 mm, is wound with 400 turns, and the deflection of 45° occurs when a current of 5mA passes through the coil, calculate the strength of radial field in which the coil lies. **[900×10^{-6} Nm; 0.75 wb/m^2]**

5. A d.c. motor is to provide a torque of 540 Nm. The armature winding consists of 600 turns (equivalent to 1200 conductors). The effective length of a conductor in the field is 250 mm and the conductors are situated at a radius of 150 mm from the centre of the motor shaft. Each conductor carries a current of 10A. Calculate the flux density which must be provided by the radial field in which the conductors lie. **[1.2 wb/m^2]**

7.13. Work Law

This law relates to work done in a magnetic circuit *i.e.* closed magnetic flux path. Consider N straight conductors, each carrying a current of I amperes as shown in Fig. 7.13. This current carrying conductor arrangement will set up magnetic field. If a unit n-pole is moved in opposition to the lines of force, work will be done ****against*** the magnetic field. The work done on a unit n-pole in moving once round any complete path will be equal to the product of current and number of turns enclosed by that path. This is known as work law and may be stated as under :

The work done on or by a unit n-pole in moving once round any complete path is equal to the product of current and number of turns enclosed by that path i.e.

$$** \oint H_r d_r = NI$$

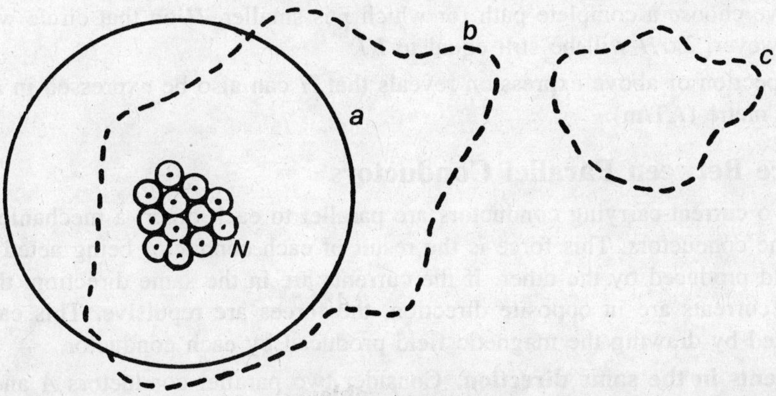

Fig. 7.13

where H_r is the magnetising force at a distance r. The circle round the integral sign indicates that the integral is round a complete path.

The work law is applicable regardless of the shape of complete path. Thus in Fig. 7.13, paths a and b completely enclose N conductors. If a unit n-pole is moved once round any of these complete paths, the work done in each case will be equal to NI. Although path c is a complete path, it fails to enclose any current carrying conductor. Hence no work is done in moving a unit n-pole round such a path.

---

* If the unit n-pole is moved in the direction of magnetic field, work will be done by the magnetic force on whatever force is restraining the movement of the pole.

** This law can also be stated as *the closed line integral of magnetic field intensity (H) is equal to the enclosed ampere turns that produce the magnetic field.*

Applications. The work law can be profitably used to find the field intensity (*H*) near simple current carrying conductor arrangements. Let us consider the case of a long straight conductor carrying a current of *I* amperes as shown in Fig 7.14. The conductor will set up magnetic lines of force which encircle it. Consider a circular path of radius *r* metres. By symmetry, the field intensity *H* on all the points of this circular path will be the same. If a unit *n*-pole is moved once round this circular path, then work done is $= 2\pi\, rH$. By work law, this must be equal to the ampere turns enclosed bythis circular path.

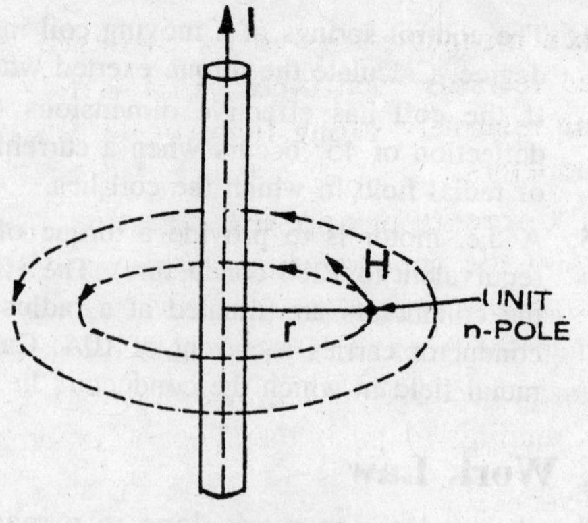

Fig. 7.14

$$\therefore \qquad 2\pi rH = I \qquad (\because \text{ Number of turns, } N = 1)$$

or
$$H = \frac{I}{2\pi r}$$

If there had been *N* conductors enclosed by the path then,

$$H = N\frac{I}{2\pi r}$$

The following points may be noted carefully :

(*i*) If we choose a complete path for which *r* is smaller, *H* on that circle will be larger. However, $2\pi rH$ will be still equal to *NI*.

(*ii*) Inspection of above expression reveals that *H* can also be expressed in ampere turns per metre (AT/m).

7.14. Force Between Parallel Conductors

When two current-carrying conductors are parallel to each other, a mechanical force acts on each of the conductors. This force is the result of each conductor being acted upon by the magnetic field produced by the other. If the currents are in the same direction, the forces are attractive; if currents are in opposite direction, the forces are repulsive. This can be beautifully illustrated by drawing the magnetic field produced by each conductor.

(*i*) **Currents in the same direction.** Consider two parallel conductors *A* and *B* carrying current in the same direction (*i.e* into the plane of paper) as shown in Fig. 7.15 (*i*). Each conductor will set up its own magnetic field as shown. It is clear that in the space between *A* and *B*, the two fields are in opposition and hence they tend to cancel each other. However, in the space outside *A* and *B*, the two fields assist each other. Hence the resultant field distribution will be as shown in Fig. 7.15 (*ii*).

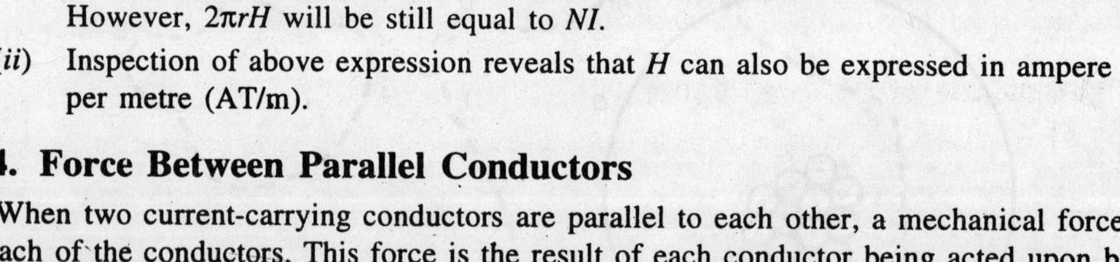

(*i*) Fig. 7.15 (*ii*)

Since magnetic lines of force behave as stretched elastic cords, the two conductors are attracted towards each other. Alternatively, the conductors can be viewed as moving away from the relatively strong field (in the space outside *A* and *B*) into the weaker field between the conductors.

(*ii*) **Currents in opposite direction.** Consider two parallel conductors *A* and *B* carrying currents in the opposite direction as shown in Fig. 7.16. Each conductor will set up its own field as shown. It is clear that in the space outside *A* and *B*, the two fields are in opposition and hence they tend to cancel each other. However, in the space between *A* and *B*, the two fields assist each other. The lateral pressure between lines of force exerts a force on the conductors tending to push them apart. In other words, the conductors experience a repulsive force.

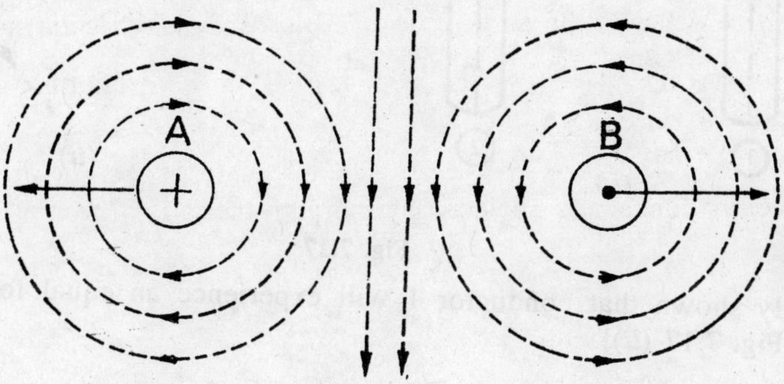

Fig. 7.16

7.15. Magnitude of Mutual Force

Fig 7.17 (*i*) shows two parallel conductors placed in air and carrying current in the same direction.

I_1 = current in conductor 1 in amperes

I_2 = current in conductor 2 in amperes

l = length of each conductor in *m*

d = distance between conductors in *m*

It is clear that each of the two parallel conductors lies in the magnetic field of the other conductor. In order to determine the magnitude of force, we can consider conductor 2 placed in the magnetic field produced by conductor 1 as shown in Fig. 7.17 (*ii*). The flux density *B* at the centre of conductor 2 due to current I_1 in conductor 1 is

$$B = \frac{\mu_0 I_1}{2\pi d}$$ ... *See Art. 7.13*

Force acting on conductor 2 is given by :

$$F = BI_2 l = \left(\frac{\mu_0 I_1}{2\pi d}\right) I_2 l = \frac{4\pi \times 10^{-7} I_1 I_2 l}{2\pi d}$$

$$= \frac{2 I_1 I_2 l}{d} \times 10^{-7} \text{ newtons}$$

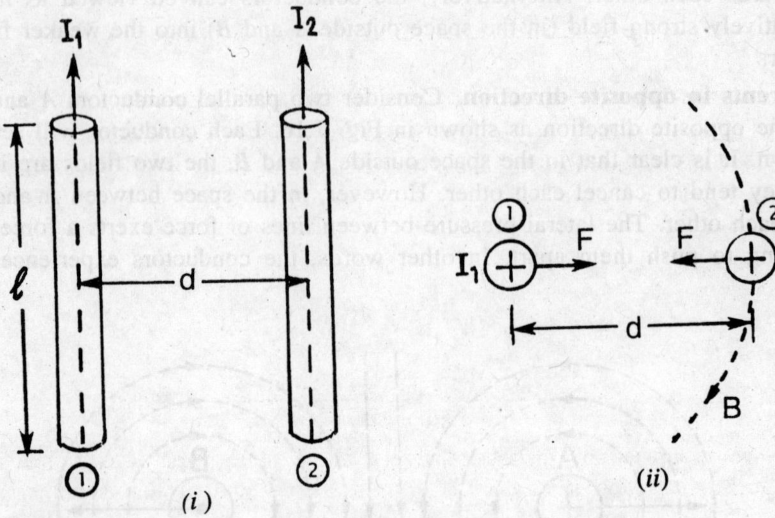

Fig. 7.17

It can be easily shown that conductor 1 will experience an equal force in the opposite direction [See Fig. 7.17 (*ii*)].

7.16. Definition of Ampere

The force acting between two parallel conductors has led to the modern definition of an ampere. We have seen above that force between two parallel current-carrying conductors is

$$F = \frac{2 I_1 I_2 l}{d} \times 10^{-7} \text{ newtons}$$

If $I_1 = I_2 = 1A$; $l = 1\ m$; $d = 1\ m$, then

$$F = \frac{2 \times 1 \times 1 \times 1}{1} \times 10^{-7} = 2 \times 10^{-7}\text{N}$$

Hence **one ampere** *is that current which, if maintained in two long parallel conductors, and placed 1m apart in vacuum, would produce between these conductors a force equal to* 2×10^{-7} *newton per metre of length (See Fig. 7.18).*

Historically, the ampere was fixed originally in a very different way. The constant 2×10^{-7} that appears in the modern definition was chosen so as to keep the magnitude of ampere the same as formerly.

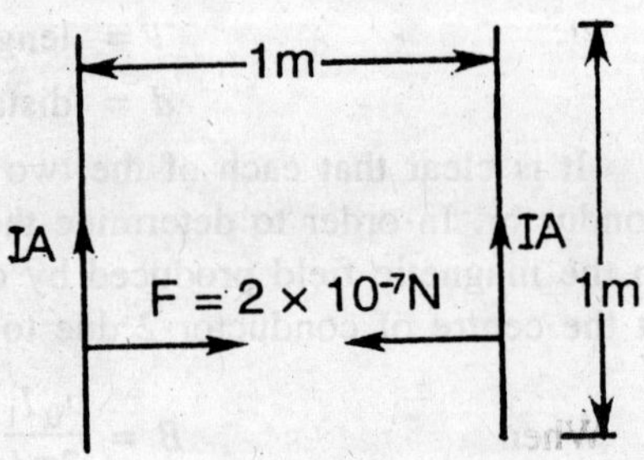

Fig. 7.18

Example 7.6 *The starting motor of a certain car is connected to a battery by a pair of cables 8 mm apart for a distance of 50 cm. Find the force between the cables when the current in them is 100A.*

Solution. Force between the cables (see Art. 7.15) is

$$F = \frac{2 I_1 I_2 l}{d} \times 10^{-7} \text{ N}$$

Here $I_1 = I_2 = 100\text{A}$; $l = 0.5$ m ; $d = 8 \times 10^{-3}$ m

$$\therefore \quad F = \frac{2 \times 100 \times 100 \times 0.5}{8 \times 10^{-3}} \times 10^{-7} = \mathbf{0.125\,N}$$

Since the currents are in opposite directions, the force is repulsive.

Example 7.7 *Two bus-bars 80 mm apart are supported by insulators every metre along their length. The busbars each carry a current of 20 kA. What is the force acting on each insulator ?*

Solution.

Force on each metre of busbar

$$F = \frac{2\,I_1\,I_2}{d} \times 10^{-7}\ \text{N} \qquad (\because l = 1\ \text{m})$$

$$= \frac{2 \times (20 \times 10^3) \times (20 \times 10^3) \times 10^{-7}}{80 \times 10^{-3}} = \mathbf{1000\,N}$$

Since busbars are supported by insulators, every metre along their length, this is the force (*i.e.* 1000 N) acting on each insulator. It should be noted that the force will be that of repulsion.

Example 7.8 *Force between two parallel conductors carrying current in opposite direction is 3.2 N/m when they are placed 50 mm apart in air. If current flowing in one conductor is 1000A, find the current in the other conductor. Mention whether it is a force of attraction or repulsion.*

Solution.

Force between the conductors per metre length

$$F = \frac{2\,I_1\,I_2}{d} \times 10^{-7}\ \text{N}$$

Here $F = 3.2\,\text{N}$; $I_1 = 1000\text{A}$; $I_2 = ?$; $d = 50 \times 10^{-3}$ m

$$\therefore \quad 3.2 = \frac{2 \times 1000 \times I_2}{50 \times 10^{-3}} \times 10^{-7}$$

or $$3.2 = (40 \times 10^{-4})\,I_2$$

or $$I_2 = \frac{3.2}{40 \times 10^{-4}} = \mathbf{800\,A}$$

Since the two conductors carry currents in the opposite direction, the force is repulsive.

OBJECTIVE QUESTIONS

1. When a magnet is heated
   (*i*) it gains magnetism
   (*ii*) it loses magnetism
   (*iii*) it neither loses nor gains magnetism
   (*iv*) none of the above
2. The magnetic material used in permanent magnets is
   (*i*) iron (*ii*) soft steel
   (*iii*) nickel (*iv*) hardened steel
3. The magnetic material used in temporary magnets is
   (*i*) hardened steel (*ii*) cobalt steel
   (*iii*) soft iron (*iv*) tungsten steel
4. Magnetic flux density is a
   (*i*) vector quantity (*ii*) scalar quantity
   (*iii*) phasor (*iv*) none of the above
5. The relative permeability of a ferromagnetic material is 1000. Its absolute permeability will be
   (*i*) 10^6H/m (*ii*) $4\pi \times 10^{-3}$H/m

(*iii*) $4\pi \times 10^{-11}$ H/m (*iv*) none of the above

6. The main advantage of temporary magnets is that we can
(*i*) change the magnetic flux
(*ii*) use any magnetic material
(*iii*) decrease the hysteresis loss
(*iv*) none of the above

7. One weber is equal to
(*i*) 10^6 lines (*ii*) $4\pi \times 10^{-7}$ lines
(*iii*) 10^{12} lines (*iv*) 10^8 lines

8. Magnetic field intensity is a
(*i*) scalar quantity (*ii*) vector quantity
(*iii*) phasor (*iv*) none of the above

9. The absolute permeability of a material having a flux density of 1 Wb/m^2 is 10^{-3} H/m. The value of magnetising force is
(*i*) 10^{-3} AT/m (*ii*) $4\pi \times 10^{-3}$ AT/m
(*iii*) 1000 AT/m (*iv*) $4\pi \times 10^3$ AT/m

10. When the relative permeability of a material is slightly less than 1, it is called a
(*i*) diamagnetic material
(*ii*) paramagnetic material
(*iii*) ferromagnetic material
(*iv*) none of the above

11. The greater percentage of substances are
(*i*) diamagnetic (*ii*) paramagnetic
(*iii*) ferromagnetic (*iv*) none of the above

12. When the relative permeability of material is much greater than 1, it is called
(*i*) diamagnetic material
(*ii*) paramagnetic material
(*iii*) ferromagnetic material
(*iv*) none of the above

13. The magnetic flux density in an air-cood coil is 10^{-2} Wb/m^2. With a cast iron core of relative permeability 100 inserted, the flux density will become
(*i*) 10^{-4} Wb/m^2 (*ii*) 10^4 Wb/m^2
(*iii*) 10^{-2} Wb/m^2 (*iv*) 1 Wb/m^2

14. Which of the following is more suitable for the core of an electromagnet
(*i*) soft iron (*ii*) air
(*iii*) steel (*iv*) tungsten steel

15. The source of a magnetic field is
(*i*) an isolated magnetic pole
(*ii*) static electric charge
(*iii*) magnetic substances
(*iv*) current loop

16. A magnetic needle is kept in a uniform magnetic field. It experiences
(*i*) a force and a torque
(*ii*) a force but not a torque
(*iii*) a torque but not a force
(*iv*) neither a torque nor a force

17. The unit of pole strength is
(*i*) A/m^2 (*ii*) Am
(*iii*) Am^2 (*iv*) Wb/m^2

18. When the relative permeability of a material is slightly more than 1, it is called a
(*i*) diamagnetic material
(*ii*) paramagnetic material
(*iii*) ferromagnetic material
(*iv*) none of the above

19. AT/m is the unit of
(*i*) m.m.f. (*ii*) reluctance
(*iii*) magnetising force
(*iv*) magnetic flux density

20. A magnetic needle is kept in a non-uniform magnetic field. It experiences
(*i*) a force and a torque
(*ii*) a force but not a torque
(*iii*) a torque but not a force
(*iv*) neither a force nor a torque

21. Magnetic flux passes more readily through
(*i*) air (*ii*) wood
(*iii*) vacuum (*iv*) iron

22. Iron is ferromagnetic
(*i*) above 770°C (*ii*) below 770° C
(*iii*) at all temperatures
(*iv*) none of the above

23. The relative permeability of a material is 0.9998. It is
(*i*) diamagnetic (*ii*) paramagnetic
(*iii*) ferromagnetic (*iv*) none of the above

24. Magnetic lines of force
(*i*) intersect at infinity
(*ii*) intersect within the magnet
(*iii*) cannot intersect at all
(*iv*) none of the above

25. Demagnetising of magnets can be done by
(*i*) rough handling (*ii*) heating
(*iii*) magnetising in opposite direction

(*iv*) all of the above

26. The relative permeability of iron is of the order of

(*i*) zero (*ii*) 10^4
(*iii*) 1 (*iv*) 10^{-4}

27. Which of the following has the highest permeability?

(*i*) paramagnetic (*ii*) diamagnetic
(*iii*) ferromagnetic (*iv*) vacuum

28. The unit of relative permeability is

(*i*) A/m (*ii*) H/m
(*iii*) Wb/m^2 (*iv*) none of the above

29. A magnet is kept in air surrounded by an iron ring. The magnetic lines of force from the magnet will be

(*i*) crowded in the ring
(*ii*) crowded in air
(*iii*) evenly distributed
(*iv*) none of the above

30. The distance between two magnetic poles is doubled and their pole strengths are also doubled. The force between them

(*i*) increases four times
(*ii*) decreases four times
(*iii*) remains unchanged
(*iv*) none of the above

31. Fig. 7.19 shows the magnetic field around a conductor. In which direction is the current flowing?

(*i*) from right to left
(*ii*) from left to right
(*iii*) can be either of above
(*iv*) data incomplete

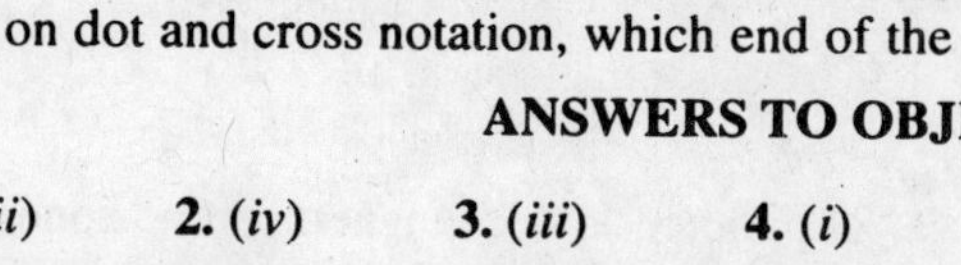

Fig. 7.19

32. Fig 7.20 shows a cross-section of a coil. Based on dot and cross notation, which end of the coil is the north pole?

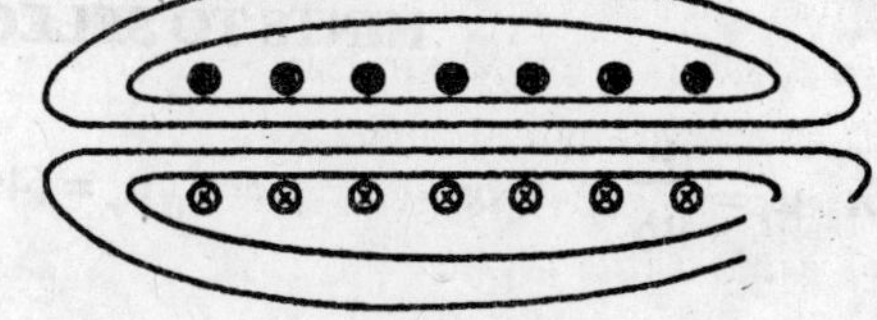

Fig. 7.20

(*i*) left (*ii*) right
(*iii*) left or right depending on magnitude of current
(*iv*) data incomplete

33. Will the two conductors in Fig. 7.21 repel or attract?

Fig. 7.21

(*i*) repel (*ii*) attract
(*iii*) neither repel nor attract
(*iv*) data insufficient

34. Will the two conductors in Fig. 7.22 will attract or repel?

Fig. 7.22

(*i*) repel (*ii*) attract
(*iii*) neither repel nor attract
(*iv*) data insufficient

35. Which end of the coil shown in Fig. 7.23 is the north pole?

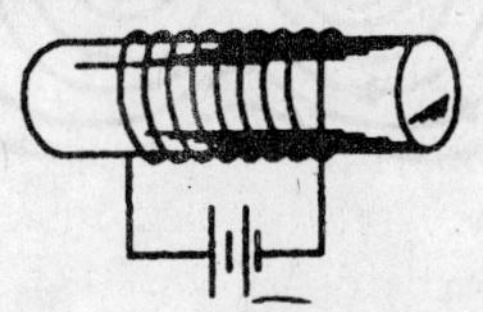

Fig. 7.23

(*i*) left (*ii*) right
(*iii*) neither left nor right
(*iv*) data incomplete

ANSWERS TO OBJECTIVE QUESTIONS

1. (*ii*)	**2.** (*iv*)	**3.** (*iii*)	**4.** (*i*)	**5.** (*ii*)	**6.** (*i*)	**7.** (*iv*)	**8.** (*ii*)
9. (*iii*)	**10.** (*i*)	**11.** (*ii*)	**12.** (*iii*)	**13.** (*iv*)	**14.** (*i*)	**15.** (*iv*)	**16.** (*iii*)
17. (*ii*)	**18.** (*ii*)	**19.** (*iii*)	**20.** (*i*)	**21.** (*iv*)	**22.** (*ii*)	**23.** (*i*)	**24.** (*iii*)

25. *(iv)** **26.** *(ii)* **27.** *(iii)* **28.** *(iv)* **29.** *(i)* **30.** *(iii)* **31.** *(ii)* **32.** *(ii)*
33. *(ii)* **34.** *(i)* **35.** *(ii)*

HINTS TO SELECTED OBJECTIVE QUESTIONS

5. $\mu_r = \dfrac{\mu}{\mu_0}$ or $\mu = \mu_0 \mu_r = (4\pi \times 10^{-7}) \times 10000 = 4\pi \times 10^{-3}$ H/m

9. $\mu = \dfrac{B}{H}$ or $H = \dfrac{B}{\mu} = \dfrac{1}{10^{-3}} = 1000$ AT / m

13. $\mu_r = \dfrac{B_{iron}}{B_{air}}$ or $B_{iron} = \mu_r \times B_{air} = 100 \times 1$[illegible]

29. Magnetic fields are distorted by ferromagnetic materials in their vicinity. If a piece of iron is placed near a magnet, most of the flux lines will be crowded in the iron piece. It is because, iron is a good conductor of magnetic flux.

30. $F \propto \dfrac{m_1 m_2}{d^2}$; $F' \propto \dfrac{(2m_1)(2m_2)}{(2d)^2}$ $\therefore$ $\dfrac{F'}{F} = 1$

32. Apply right-hand rule for coils. The coil is grasped in the right hand with the fingers pointing in the direction of current. Then thumb points in the direction of north pole.

33. Both the conductors are marked with dots. The currents are both flowing toward the observer. Use your right hand to verify that the field around each conductor travels in a counterclockwise direction (see Fig. 7.24). The fields join in this case, becoming one field. This results in the force of attraction between the conductors.

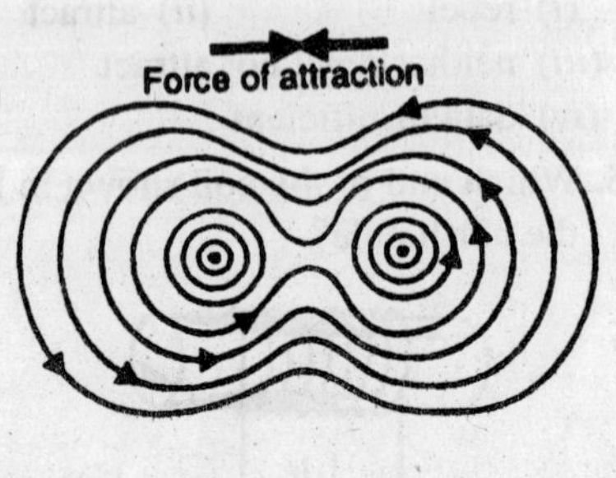

Currents flowing in the same directions

Fig. 7.24

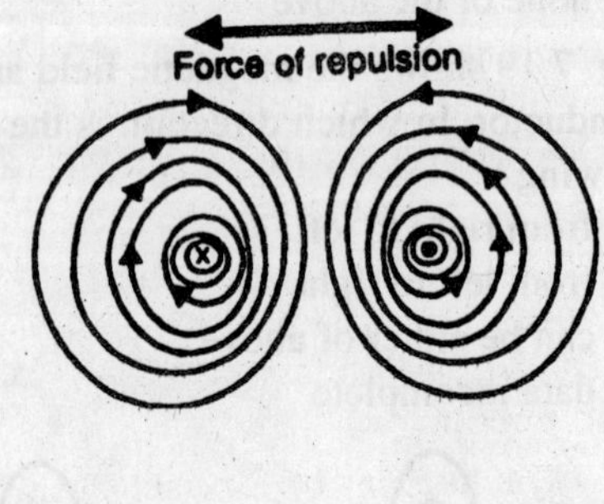

Currents flowing in opposite directions

Fig. 7.25

34. The currents are flowing in opposite directions. The fields cannot join in this case and the lines repel each other (see Fig. 7.25). There is a force of repulsion between the conductors.

8

Magnetic Circuits

Introduction

We have seen that magnetic lines of force form closed loops around and through the magnetic material. The closed path followed by magnetic flux is called a magnetic circuit just as the closed path followed by current is called an electric circuit. Many electrical devices (*e.g.* generator, motor, transformer *etc.*) depend upon magnetism for their operation. Therefore, such devices have magnetic circuits *i.e.* closed flux paths. In order that these devices function efficiently, their magnetic circuits must be properly designed to obtain the required magnetic conditions. In this chapter, we shall focus our attention on the basic principles of magnetic circuits and methods to obtain their solution.

8.1. Magnetic Circuit

The closed path followed by magnetic flux is called a **magnetic circuit.**

In a magnetic circuit, the magnetic flux leaves the *n*-pole, passes through the entire circuit and returns to the starting point. A magnetic circuit usually consists of materials having high permeability *e.g.* iron, soft steel *etc.*. It is because these materials offer very small opposition to the 'flow' of magnetic flux. Consider a coil of N turns wound on an iron core as shown in Fig. 8.1. When current I is passed through the coil, magnetic flux ϕ is set up in the core. The flux follows the closed path $ABCDA$ and hence $ABCDA$ is the magnetic circuit.

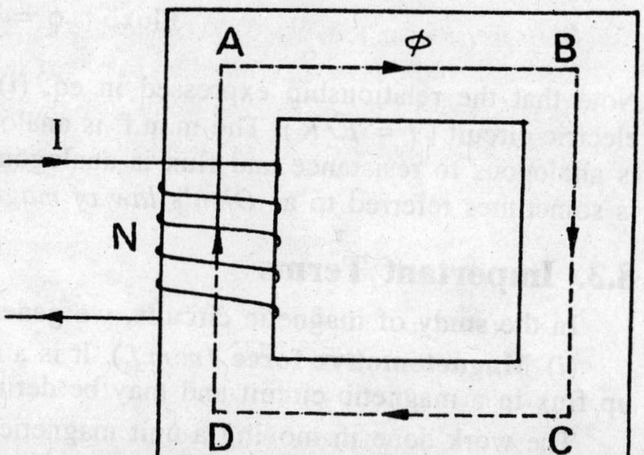

Fig. 8.1

(*i*) The amount of flux set up in the core depends upon current (I) and number of turns (N). The product NI is called *magneto-motive force* (m.m.f.) and determines the amount of flux set up in the magnetic circuit.

$$m.m.f. = NI \text{ ampere turns}$$

It can just be compared to electromotive force (e.m.f.) which sends current in an electric circuit.

(*ii*) The opposition that the magnetic circuit offers to the flux is called *reluctance*. It depends upon length of magnetic circuit (*i.e.* length $ABCDA$ in this case), area of *x*-section of the circuit and the nature of material that makes up the magnetic circuit.

8.2. Analysis of Magnetic Circuit

Consider the magnetic circuit shown in Fig. 8.1. Suppose the mean length of the magnetic circuit (*i.e.* length *ABCDA*) is *l* metres, cross-sectional area of the core is '*a*' m^2 and relative permeability of core material is μ_r. When current *I* is passed through the coil, it will set up flux ϕ in the material.

Flux density in the material, $B = \dfrac{\phi}{a}$ wb/m^2

Magnetising force in the material, $H = \dfrac{B}{\mu_0\mu_r} = \dfrac{\phi}{a\mu_0\,\mu_r}$ AT/m

According to work law (See Art. 7.13) the work done in moving a unit magnetic pole once round the magnetic circuit (*i.e.* path *ABCDA* in this case) is equal to the ampere-turns enclosed by the magnetic circuit.

$$\therefore \qquad {}^{*}H \times l = NI$$

or
$$\frac{\phi}{a\mu_0\,\mu_r} \times l = NI$$

or
$$\phi = \frac{NI}{(l/a\,\mu_0\mu_r)}$$

The quantity *NI* which produces the magnetic flux is called the magnetomotive force (m.m.f.) and is measured in ampere-turns. The quantity $l/a\,\mu_0\mu_r$ is called the reluctance of the magnetic circuit. Reluctance is the opposition that the magnetic circuit offers to flux.

$$\therefore \qquad \text{Flux,} \quad \phi = \frac{\text{m.m.f.}}{\text{reluctance}} \qquad \text{... } (i)$$

Note that the relationship expressed in eq. (*i*) has a strong resemblance to Ohm's law for electric circuit ($I = E/R$). The m.m.f. is analogous to e.m.f. in the electric circuit, reluctance is analogous to resistance and flux is analogous to current. Because of this similarity, eq. (*i*) is sometimes referred to as *Ohm's law of magnetic circuit.*

8.3. Important Terms

In the study of magnetic circuits, we generally come across the following terms :

(*i*) **Magnetomotive force** (*m.m.f.*). It is a magnetic pressure which sets up or tends to set up flux in a magnetic circuit and may be defined as under :

The work done in moving a unit magnetic pole once round the magnetic circuit is called the magnetomotive force (*m.m.f.*). It is equal to the product of current and number of turns of the coil *i.e.*.

$$\text{m.m.f.} = NI \text{ ampere–turns (or AT)}$$

Magnetomotive force in a magnetic circuit corresponds to e.m.f. in an electric circuit.

(*ii*) **Reluctance.** The opposition that the magnetic circuit offers to flux is called reluctance. Its unit is **AT/wb.

$$\text{Reluctance, } S = \frac{l}{a\,\mu_0\,\mu_r}$$

Reluctance in a magnetic material corresponds to resistance ($R = \rho\, l/a$) in an electric circuit. Both of them vary as length ÷ area and are dependent upon the nature of material of

---

* You may recall that *H* means force acting on a unit magnetic pole. If the unit pole is moved once round the magnetic circuit (*i.e.* distance covered is *l*), then work done $= H \times l$

** Reluctance $= \dfrac{\text{m.m.f.}}{\text{flux}} = \dfrac{AT}{wb} = AT/wb$

the circuit. Magnetic materials (*e.g.* iron, steel *etc.*) have a low reluctance because the value of μ_r is very large in their case. On the other hand, non-magnetic materials (*e.g.* air, wood, copper, brass *etc.*) have a high reluctance because they possess least value of μ_r; being 1 in case of all non-magnetic materials.

(*iii*) **Permeance**. It is the reciprocal of reluctance and is a measure of the ease with which flux can pass through the material. Its unit is wb/AT.

$$\text{Permeance} = \frac{1}{\text{reluctance}} = \frac{a\,\mu_0\mu_r}{l}$$

Permeance of a magnetic circuit corresponds to conductance (reciprocal of resistance) in an electric circuit.

8.4. Comparison between Magnetic and Electric Circuits

Magnetic Circuit

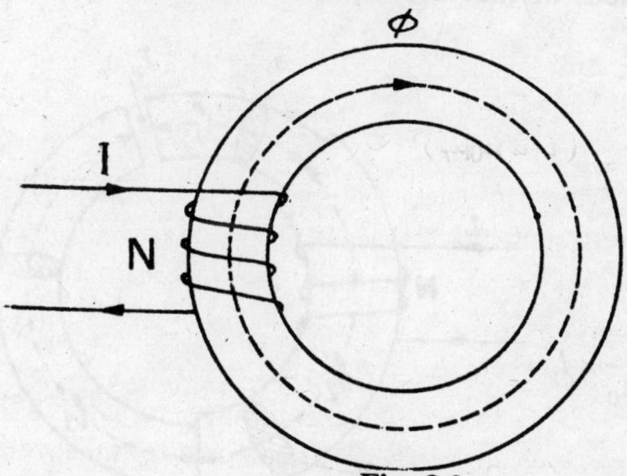

Fig. 8.2

Electric circuit

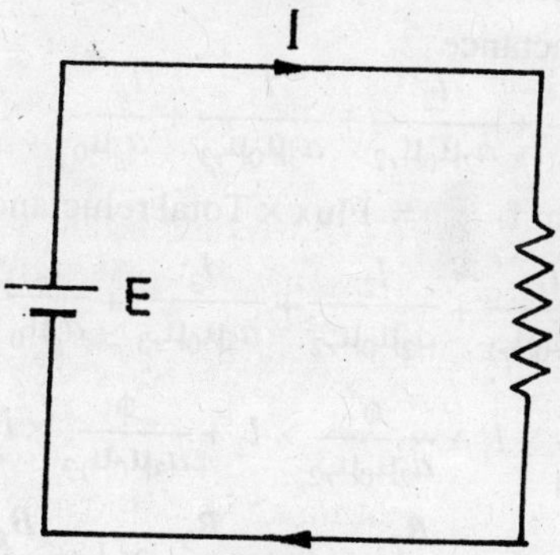

Fig. 8.3

Similarities

Magnetic Circuit	Electric Circuit
1. The closed path for magnetic flux is called a magnetic circuit.	1. The closed path for electric current is called an electric circuit.
2. Flux, $\phi = \frac{\text{m.m.f.}}{\text{reluctance}}$	2. Current, $I = \frac{\text{e.m.f.}}{\text{resistance}}$
3. m.m.f. (ampere-turns)	3. e.m.f. (volts)
4. Reluctance $S = \frac{l}{a\,\mu_0\,\mu_r}$	4. Resistance $R = \rho\frac{l}{a}$
5. Flux density $B = \frac{\phi}{a}$ wb/m^2	5. Current density $J = \frac{I}{a}$ A/m^2
6. m.m.f drop = ϕS	6. Voltage drop = IR
7. Magnetic intensity, $H = NI/l$	7. Electric intensity, $E = V/d$

Dissimilarities

Magnetic Circuit	Electric Circuit
1. Truly speaking magnetic flux does not flow.	1. The electric current actually flows in an electric circuit.
2. There is no magnetic insulator. For example, flux can be set up even in air (the best known magnetic insulator) with reasonable m.m.f.	2. There are a number of electric insulators. For instance, air is a very good insulator and current cannot pass through it.

3. The value of μ_r is not constant for a given magnetic material. It varies considerably with flux density (B) in the material. This implies that reluctance of a magnetic circuit is not constant rather it depends upon B.	3. The value of resistivity (ρ) varies very slightly with temperature. Therefore, the resistance of an electric circuit is practically constant. This salient feature calls for different approach to the solution of magnetic and electric circuits.
4. No energy is expended in a magnetic circuit. In other words, energy is required in creating the flux, and not in maintaining it.	4. When current flows through an electric circuit ; energy is expended so long as the current flows. The expended energy is dissipated in the form of heat.

8.5. Composite Magnetic Circuit

A series magnetic circuit that has parts of different dimensions and materials is called a composite magnetic circuit as shown in Fig 8.4. Each part will have its own reluctance. The total reluctance is equal to the sum of reluctances of individual parts *i.e.*

Total reluctance

$$= \frac{l_1}{a_1\mu_0\mu_{r1}} + \frac{l_2}{a_2\mu_0\mu_{r2}} + \frac{l_3}{a_3\mu_0\mu_{r3}} + \frac{l_g}{a_g\mu_0}\,^*$$

Total m.m.f. = Flux × Total reluctance

$$= \phi\left[\frac{l_1}{a_1\mu_0\mu_{r1}} + \frac{l_2}{a_2\mu_0\mu_{r2}} + \frac{l_3}{a_3\mu_0\mu_{r3}} + \frac{l_g}{a_g\mu_0}\right]$$

$$= \frac{\phi}{a_1\mu_0\mu_{r1}} \times l_1 + \frac{\phi}{a_2\mu_0\mu_{r2}} \times l_2 + \frac{\phi}{a_3\mu_0\mu_{r3}} \times l_3 + \frac{\phi}{a_g\mu_0} \times l_g$$

$$= \frac{B_1}{\mu_0\mu_{r1}} \times l_1 + \frac{B_2}{\mu_0\mu_{r2}} \times l_2 + \frac{B_3}{\mu_0\mu_{r3}} \times l_3 + \frac{B_g}{\mu_0} \times l_g$$

$$= H_1 l_1 + H_2 l_2 + H_3 l_3 + H_g l_g \qquad (\because H = B/\mu_0\mu_r)$$

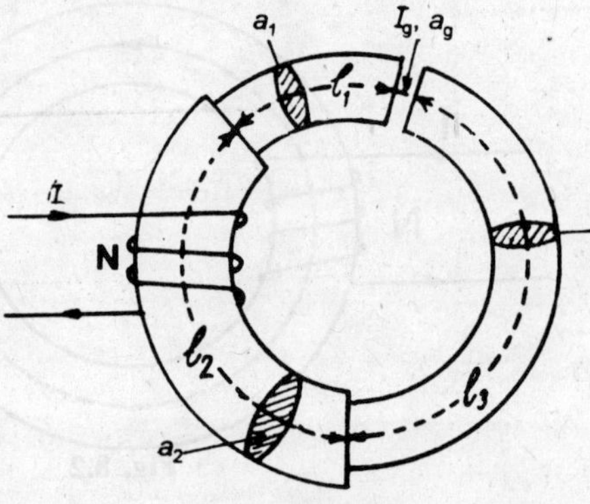

Fig. 8.4

How to find ampere-turns?

Hence the total ampere-turns required for a series magnetic circuit can be found as under :

(*i*) Find H for each part of the series magnetic circuit. For air, $H = B/\mu_0$ whereas for a magnetic material, $H = B/\mu_0\mu_r$.

(*ii*) Find the *mean* length (l) of magnetic path for each part of the circuit.

(*iii*) Find AT required for each part of the magnetic circuit using the relation, $AT = H \times l$

(*iv*) The total AT required for the entire series circuit is equal to the sum of AT for various parts.

8.6. Parallel Magnetic Circuits

A magnetic circuit which has more than one path for flux is called a parallel magnetic circuit. It can just be compared to a parallel electric circuit which has more than one path for electric current.

The concept of parallel magnetic circuit is illustrated in Fig. 8.5. Here a coil of N turns wounded on limb AF carries a current of I amperes. The flux ϕ_1 set up by the coil divides at B into two paths, namely ;

* For air, $\mu_r = 1$

(*i*) flux ϕ_2 passes along the path *BE*

(*ii*) flux ϕ_3 follows the path *BCDE*

Clearly, $\phi_1 = \phi_2 + \phi_3$

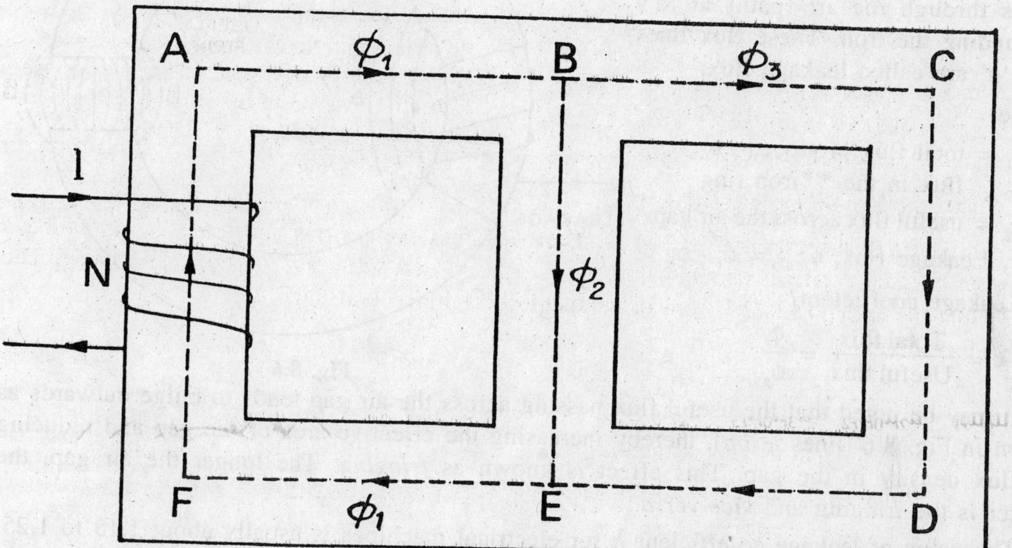

Fig. 8.5

The magnetic paths *BE* and *BCDE* are in parallel and form a parallel magnetic circuit. The *AT* required for this parallel circuit is equal to *AT* required for any *one of the paths.

Let S_1 = reluctance of path *EFAB*

S_2 = reluctance of path *BE*

S_3 = reluctance of path *BCDE*

$\therefore$ Total m.m.f. required = m.m.f. for path *EFAB* + m.m.f. for path *BE* or path *BCDE*

or

$$NI = \phi_1 S_1 + \phi_2 S_2$$

$$= \phi_1 S_1 + \phi_3 S_3$$

The reluctances S_1, S_2 and S_3 must be determined from a calculation of $l/a\,\mu_0\mu_r$ for those paths of the magnetic circuit in which ϕ_1, ϕ_2 and ϕ_3 exist respectively.

8.7. Leakage Flux

The flux that does not follow the desired path in a magnetic circuit is called a **leakage flux.**

In most of practical magnetic circuits, a large part of flux path is through a magnetic material and the remainder part of flux path is through air. The flux in the air gap is known as *useful flux* because it can be utilised for various useful purposes. Fig. 8.6 shows an iron

* This means that we may consider either path, say path *BE*, and calculate *AT* required for it. The same *AT* will also send the flux (ϕ_3 in this case) through the other parallel path *BCDE*. The situation is similar to that of two resistors R_1 and R_2 in parallel in an electric circuit. The voltage *V* required to send currents (say I_1 and I_2) in the resistors is equal to that appearing across either resistor *i.e.* $V = I_1 R_1 = I_2 R_2$.

ring wound with a coil and having a narrow air gap. The total flux produced by the coil does not pass through the air gap as some of it *leaks through the air (paths at 'a') surrounding the iron. These flux lines as at '*a*' are called leakage flux.

Let

ϕ_i = total flux produced *i.e.* flux in the **iron ring

ϕ_g = useful flux across the air gap.

∴ Leakage flux, $\phi_{leak} = \phi_i - \phi_g$

Leakage coefficient,

$$\lambda = \frac{\text{Total flux}}{\text{Useful flux}} = \frac{\phi_i}{\phi_g}$$

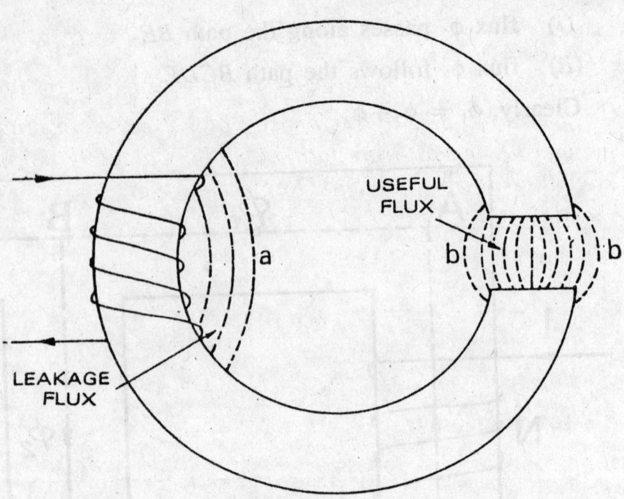

Fig. 8.6

It may be noted that the useful flux passing across the air gap tends to bulge outwards as shown in Fig. 8.6 (lines at *bb*), thereby increasing the effective area of the gap and reducing the flux density in the gap. This effect is known as *fringing*. The longer the air gap, the greater is the fringing and *vice-versa*.

The value of leakage co-efficient λ for electrical machines is usually about 1.15 to 1.25.

Example 8.1. *An iron ring has a cross-sectional area of 400 mm² and a mean diameter of 25 cm. It is wound with 500 turns. If the value of relative permeability is 250, find the total flux set up in the ring. The coil resistance is 474 Ω and the supply voltage is 20V.*

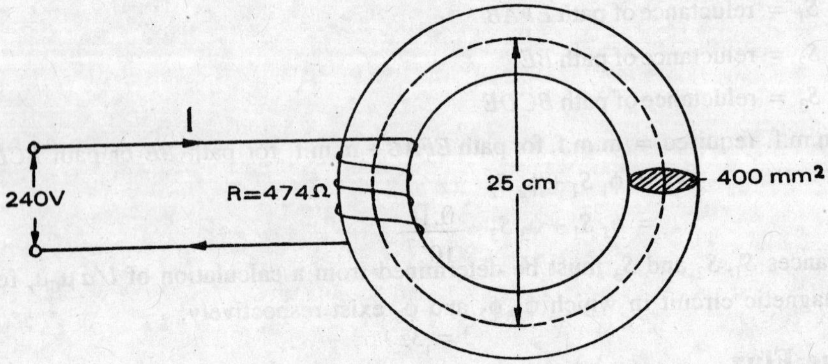

Fig. 8.7

Solution. The conditions of the problem are represented in Fig. 8.7.

$$I = V/R = 240/474 = 0.506\text{A}$$

$$l = \pi D = \pi \times (25 \times 10^{-2}) = 0.7854\ \text{m}$$

$$H = \frac{NI}{l} = \frac{500 \times 0.506}{0.7854} = 322.13\ \text{AT/m}$$

* Air is not a good magnetic insulator. Therefore, leakage of flux from iron to air takes place easily.

** The flux ϕ_i is not constant all round the ring. However, for reasonable accuracy, it is assumed that the iron carries the whole of the flux produced by the coil.

$$B = \mu_0\mu_r H = 4\pi \times 10^{-7} \times 250 \times 322.13 = 0.1012 \text{wb/m}^2$$

$$\therefore \quad \phi = B \times a = 0.1012 \times (400 \times 10^{-6}) = \mathbf{40.48 \times 10^{-6} wb}$$

Example 8.2. *A flux density of 1.2 wb/m² is required in the 2 mm air gap of an electromagnet having an iron path 1m long. Calculate the m.m.f. required, assuming a relative permeability of iron as 1500. Neglect leakage.*

Solution

AT for air gap

$$H_g = \frac{B}{\mu_0} = \frac{1.2}{4\pi \times 10^{-7}} = 955 \times 10^3 \text{ AT/m}$$

AT required $= H_g \times l = (955 \times 10^3)(2 \times 10^{-3}) = 1910$

AT for iron part

$$H_i = \frac{B}{\mu_0 \mu_r} = \frac{1.2}{4\pi \times 10^{-7} \times 1500} = 637 \text{ AT/m}$$

AT required $= H_i \times l_i = 637 \times 1 = 637$

$\therefore$ Total AT required $= 1910 + 637 = \mathbf{2547}$

Example 8.3. *An iron ring of cross-sectional area 6 cm² is wound with a wire of 100 turns and has a saw cut of 2 mm. Calculate the magnetising current required to produce a flux of 0.1 mwb if mean length of magnetic path is 30 cm and relative permeability of iron is 470.*

Solution.

$$B = \frac{\phi}{a} = \frac{0.1 \times 10^{-3}}{6 \times 10^{-4}} = 0.167 \text{ wb/m}^2$$

AT for air gap

$$H_g = \frac{B}{\mu_0} = \frac{0.167}{4\pi \times 10^{-7}} = 132.9 \times 10^3 \text{ AT/m}$$

AT required $= H_g \times l_g = 132.9 \times 10^3 \times (2 \times 10^{-3}) = 265.8$

AT for iron part

$$H_i = \frac{B}{\mu_0 \mu_r} = \frac{0.167}{4\pi \times 10^{-7} \times 470} = 282.75 \text{ AT/m}$$

AT required $= H_i \times l_i = 282.75 \times 0.3 = 84.83$

Total AT $= 265.8 + 84.83 = 350.63$

Magnetising current, $I = 350.63/100 = \mathbf{3.51A}$

It may be seen that many more ampere-turns are required to produce the magnetic flux through 2mm of air gap than through the iron part. This is expected because reluctance of air is much more than that of iron.

Example 8.4. *A circular iron ring has a mean circumference of 1.5 m and a cross-sectional area of 0.01 m². A saw-cut of 4 mm wide is made in the ring. Calculate the magnetising current required to produce a flux of 0.8 mwb in the air gap if the ring is wound with a coil of 175 turns. Assume relative permeability of iron as 400 and leakage factor 1.25.*

Solution.

$$\phi_g = 0.8 \times 10^{-3} \text{ wb}; \quad a = 0.01 \text{ m}^2; \quad l_i = 1.5 \text{ m}; \quad l_g = 4 \times 10^{-3} \text{ m}$$

AT for air gap

$$B_g = \frac{\phi_g}{a} = \frac{0.8 \times 10^{-3}}{0.01} = 0.08 \text{ wb/m}^2$$

$$H_g = \frac{B_g}{\mu_0} = \frac{0.08}{4\pi \times 10^{-7}} = 63662 \text{ AT/m}$$

$$AT_g = H_g \times l_g = 63662 \times (4 \times 10^{-3}) = 254.6 \text{ AT}$$

AT for iron path

$$\phi_i = \phi_g \times \lambda = 0.8 \times 10^{-3} \times 1.25 = 10^{-3} \text{ wb}$$

$$B_i = \phi_i / a = 10^{-3}/0.01 = 0.1 \text{ wb/m}^2$$

$$H_i = \frac{B_i}{\mu_0 \mu_r} = \frac{0.1}{4\pi \times 10^{-7} \times 400} = 199 \text{ AT/m}$$

$\therefore$ $AT_i = H_i \times l_i = 199 \times 1.5 = 298.5$

$\therefore$ Total $AT = 254.6 + 298.5 = 553.1$

$\therefore$ Magnetising current, $I = 553.1/N = 553.1/175 =$ **3.16**A

Example 8.5. *An iron ring has a mean diameter of 15 cm, a cross-section of 20 cm² and a radial gap of 0.5 mm cut in it. It is uniformly wound with 1500 turns of insulated wire and a magnetising current of 1A produces a flux of 1mwb. Neglecting the effect of magnetic leakage and fringing, calculate (i) reluctance of the magnetic circuit and (ii) relative permeability of iron.*

Solution.

$$B = \frac{\phi}{a} = \frac{1 \times 10^{-3}}{20 \times 10^{-4}} = 0.5 \text{ wb/m}^2$$

AT for Air-gap

$$H_g = B/\mu_0 = 0.5/4\pi \times 10^{-7} = 398 \times 10^3 \text{ AT/m}$$

AT required $= H_g \times l_g = (398 \times 10^3) \times 0.5 \times 10^{-3} = 199$

AT for iron part

Total *AT* $= NI = 1500 \times 1 = 1500$

AT for iron part, $AT_i = 1500 - 199 = 1301$

$$H_i = \frac{AT_i}{l_i} = \frac{1301}{0.471} = 2762 \text{ AT/m}$$

$$\therefore \quad \mu_r = \frac{B}{\mu_0 H_i} = \frac{0.5}{4\pi \times 10^{-7} \times 2762} = \mathbf{144}$$

$$\text{Reluctance} = \frac{\text{Total m.m.f}}{\text{flux}} = \frac{1500}{1 \times 10^{-3}} = \mathbf{15 \times 10^5 \text{ AT/wb}}$$

Example 8.6. *A steel ring 30 cm mean diameter and of circular section 2 cm in diameter has an air gap 1 mm long. It is wound uniformly with 600 turns of wire carrying current of 2.5A. Find (i) total m.m.f. (ii) total reluctance and (iii) flux. Neglect magnetic leakage. The iron path takes 40% of the total m.m.f.*

Solution.

(*i*) Total m.m.f. $= NI = 600 \times 2.5 =$ **1500 AT**

(*ii*) Let M_1 and M_2 be the m.m.f for iron part and air-gap respectively and S_1 and S_2 their corresponding reluctances.

$$M_1 = 40\% \text{ of } 1500 = (40/100) \times 1500 = 600 \text{ AT}$$

$$M_2 = 1500 - 600 = 900 \text{ AT}$$

Now, $M_1 = \phi S_1$ and $M_2 = \phi S_2$

$$\therefore \quad \frac{S_1}{S_2} = \frac{M_1}{M_2} = \frac{600}{900} = 0.67$$

$$S_2 = \frac{l_g}{a\mu_0} = \frac{1 \times 10^{-3}}{\pi (1 \times 10^{-2})^2 \times 4\pi \times 10^{-7}} = 2.5 \times 10^6 \text{ AT/wb}$$

$$\therefore \quad S_1 = 0.67 S_2 = 0.67 \times (2.5 \times 10^6) = 1.675 \times 10^6 \text{ AT/wb}$$

Total reluctance $= S_1 + S_2$

$= (1.675 + 2.5)\, 10^6 = \mathbf{4.175 \times 10^6}$ **AT/wb**

(*iii*) Flux $= \dfrac{\text{Total m.m.f.}}{\text{Total reluctance}} = \dfrac{1500}{4.175 \times 10^6}$

$= 0.36 \times 10^{-3}$ wb $=$ **0.36 mwb**

Example 8.7. *The magnetic circuit shown in Fig. 8.8 is built up of iron of square cross-section 3 cm side. Each air gap is 2 mm wide. Each coil is wound with 1000 turns and exciting current is 1A. The relative permeability of part A and part B may be taken as 1000 and 1200 respectively. Find (i) reluctance of part A (ii) reluctance of part B (iii) reluctance of two air gaps (iv) total reluctance and (v) total m.m.f.*

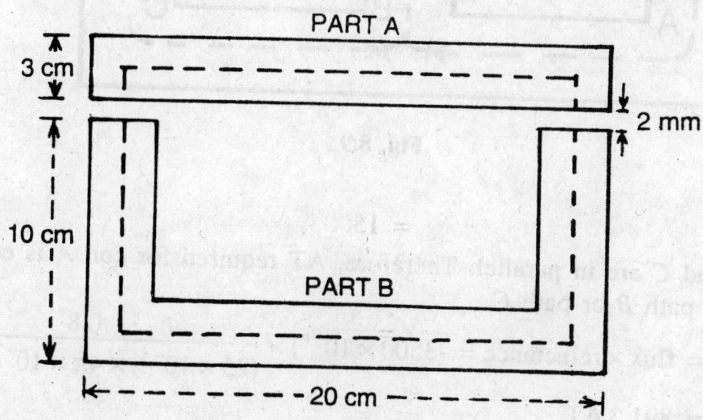

Fig. 8.8

Solution. The dotted line shows the mean path of flux

(*i*) **Part A**

Mean length, $l_A = 20 - 1.5 - 1.5 + 1.5 + 1.5 = 20 \text{ cm} = 0.2 \text{ m}$

Area of *X*-section, $a = 3 \times 3 = 9 \text{ cm}^2 = 9 \times 10^{-4} \text{ m}^2$

Reluctance, $S_A = \dfrac{l_A}{a\mu_0\mu_r} = \dfrac{0.2}{(9 \times 10^{-4}) \times 4\pi \times 10^{-7} \times (1000)} =$ **176838 AT/wb**

(ii) **Part B**

Mean length, $l_B = (20-1.5-1.5)+2(10-1.5) = 34 \text{ cm} = 0.34 \text{ m}$

Reluctance, $S_2 = \dfrac{l_B}{a\mu_0\mu_r} = \dfrac{0.34}{(9\times10^{-4})\times(4\pi\times10^{-7})\times1200} = \mathbf{250521\ AT/wb}$

(iii) **Air gaps**

Air gap length, $l_g = 2+2 = 4 \text{ mm} = 0.004 \text{m}$

Reluctance, $S_g = \dfrac{l_g}{a\mu_0} = \dfrac{0.004}{(9\times10^{-4})\times4\pi\times10^{-7}} = \mathbf{3536776\ AT/wb}$

(iv) Total reluctance $= 176838+250521+3536776 = \mathbf{3964135\ AT/wb}$

(v) Total m.m.f. $= NI = (2\times1000)\times1 = \mathbf{2000\ AT}$

Example 8.8. *A magnetic core made of annealed sheet steel has the dimensions as shown in Fig. 8.9. The X-section is 25 cm² everywhere. The flux in branches A and B is 3500 μwb but that in the branch C is zero. Find the required ampere-turns for coil A and for coil C. Relative permeability of sheet steel is 1000.*

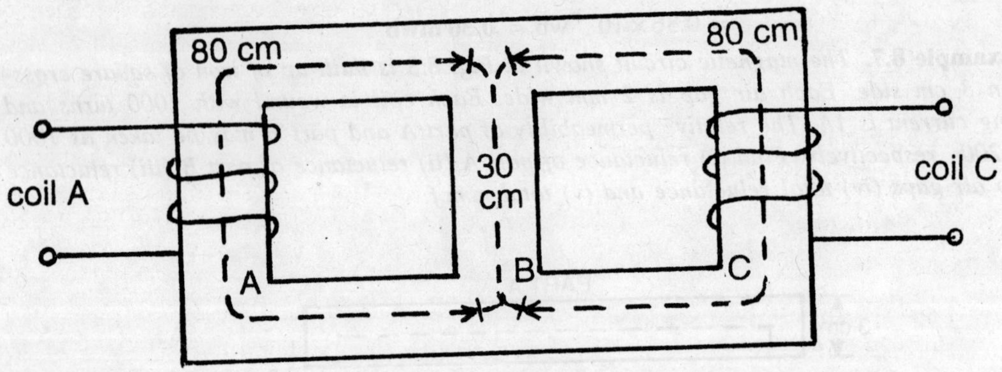

Fig. 8.9

Solution

AT for coil A

Flux paths *B* and *C* are in parallel. Therefore, AT required for coil *A* is equal to AT for path *A* plus AT for path *B* or path *C*.

$$AT \text{ for path } A = \text{flux}\times\text{reluctance} = (3500\times10^{-6})\times\frac{0.8}{(25\times10^{-4})\times4\pi\times10^{-7}\times1000}$$

$$= 891.3 \text{ AT}$$

$$AT \text{ for path } B = \text{Flux}\times\text{reluctance} = (3500\times10^{-6})\times\frac{0.3}{(25\times10^{-4})\times4\pi\times10^{-7}\times1000}$$

$$= 334.2 \text{ AT}$$

Total AT for coil $A = 891.3+334.2 = \mathbf{1225.5\ AT}$

AT for coil C

The coil *C* produces flux ϕ_C μwb in the opposite direction to that produced by coil A.

$$\text{m.m.f. of path } B = \text{m.m.f. of path } C$$

$$\phi_B S_B = \phi_C S_C$$

$$(3500 \times 10^{-6} \times \frac{l_B}{a\mu_0\mu_r} = \phi_c \times \frac{l_c}{a\mu_0\mu_r}$$

$$\therefore \quad \phi_C = (3500\times 10^{-6}) \times l_B / l_C = (3500 \times 10^{-6}) \times 0.3/0.8 = 1312.5\,\mu\text{ wb}$$

Total AT for coil $C = \phi_C \times$ reluctance

$$= (1312.5 \times 10^{-6}) \frac{0.8}{(25 \times 10^{-4}) \times 4\pi \times 10^{-7} \times 1000} = \mathbf{334.22\ AT}$$

Tutorial Problems

1. A core forms a closed magnetic loop of path length 32 cm. Half of this path has a cross-sectional area of 2 cm^2 and relative permeability 800. The other half has a cross-sectional area of 4 cm^2 and relative permeability 400. Find the current needed to produce a flux of 0.4 wb in the core if it is wound with 1000 turns of insulated wire. Ignore leakage and fringing effects. **[636.8A]**
2. An iron ring has a cross-sectional area of 400 mm^2 and a mean diameter of 250 mm. An air gap of 1 mm has been made by a saw-cut across the section of the ring. If a magnetic flux of 0.3 mwb is required in the air gap, find the current necessary to produce this flux when a coil of 400 turns is wound on the ring. The iron has a relative permeability of 500. **[3.84 A]**
3. An iron ring has a mean circumferential length of 60 cm and a uniform winding of 300 turns. An air gap has been made by a saw cut across the section of the ring. When a current of 1 A flows through the coil, the flux density in the air gap is found to be 0.126 wb/m^2. How long is the air-gap. Assume iron has a relative permeability of 300. **[1 mm]**
4. An iron magnetic circuit has a uniform cross-sectional area of 5 cm^2 and a length of 25 cm. A coil of 120 turns is wound uniformly over the magnetic circuit. When the current in the coil is 1.5 A, the total flux is 0.3 wb. Find the relative permeability of iron. **[663]**
5. A steel ring of 10 cm mean radius and of circular x-section 1 cm in radius has an air-gap of 1 mm length. It is wound uniformly with 500 turns of wire carrying a current of 3A. Neglect magnetic leakage. The air gap takes 60% of the total m.m.f. Find total m.m.f. and total reluctance.

8.8. B-H Curve

The *B/H* curve (or magnetisation curve) indicates the manner in which the flux density (*B*) varies with magnetising force (*H*). Fig 8.10 (*i*) shows the general shape of *B-H* curve of a magnetic material. The non-linearity of the curve indicates that the relative permeability $\mu_r (= B/\mu_0 H)$ of a magnetic material is not *constant but depends very largely upon the flux density. Fig. 8.10 (*ii*) shows how relative permeability μ_r of a magnetic material (cast steel) changes with flux density.

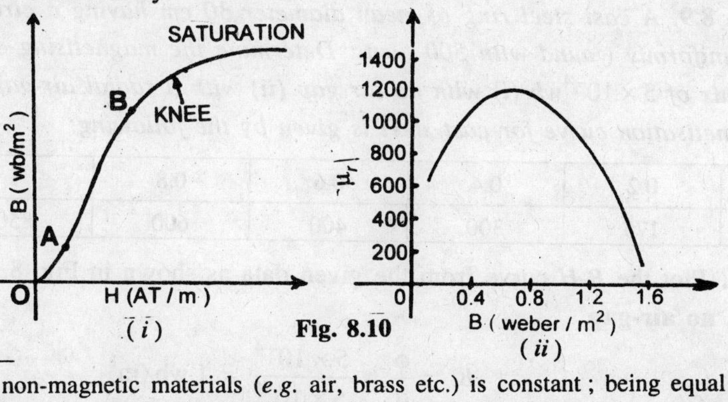

Fig. 8.10

---

* The μ_r for all non-magnetic materials (*e.g.* air, brass etc.) is constant ; being equal to 1. Therefore, for a non-magnetic material $B = \mu_0 H$. Since μ_0 is constant, $B \propto H$ or *B–H* curve for a non-magnetic material is a straight line passing through the origin.

While carrying out magnetic calculations, it should be ensured that the values of μ_r and H are taken at the working flux density. For this purpose, the *B-H* curve of the material in question may be very helpful. In fact, the use of *B-H* curves permits the calculations of magnetic circuits with a fair degree of ease.

8.9. Magnetic Calculations from B-H Curves

The solution of magnetic circuits can be easily obtained by the use of *B-H* curves. The procedure is as under :

(*i*) Corresponding to the flux density B in the material, find the magnetising force H from the *B-H* curve of the material.

(*ii*) Compute the magnetic length l.

(*iii*) m.m.f. required = $H \times l$

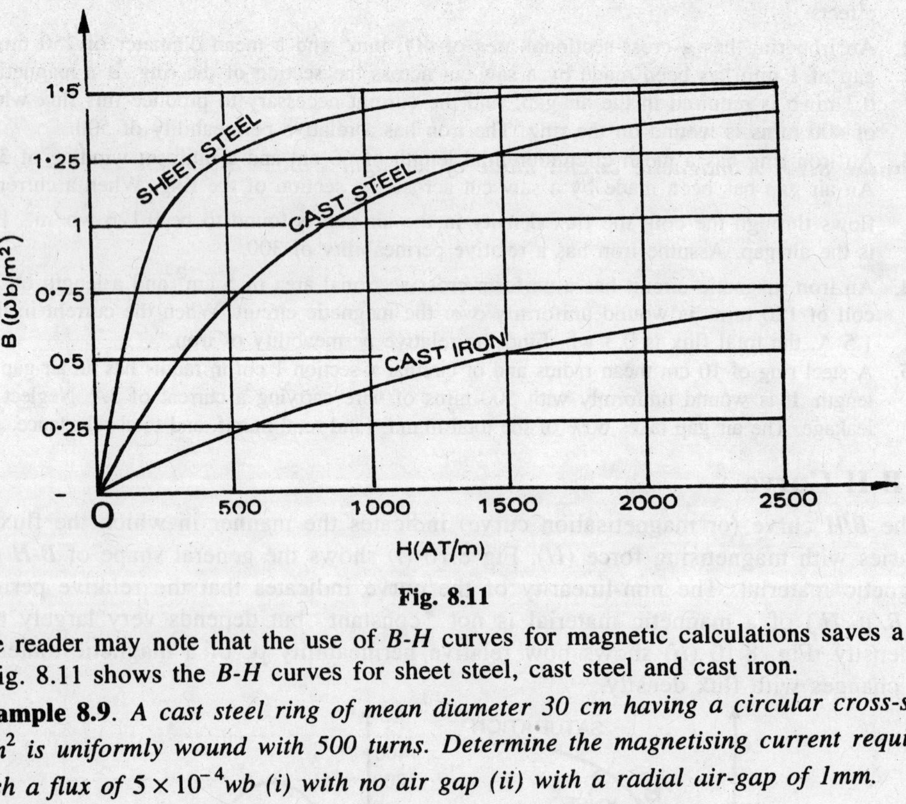

Fig. 8.11

The reader may note that the use of *B-H* curves for magnetic calculations saves a lot of time. Fig. 8.11 shows the *B-H* curves for sheet steel, cast steel and cast iron.

Example 8.9. *A cast steel ring of mean diameter 30 cm having a circular cross-section of 5 cm² is uniformly wound with 500 turns. Determine the magnetising current required to establish a flux of* 5×10^{-4} *wb (i) with no air gap (ii) with a radial air-gap of 1mm.*

The magnetisation curve for cast steel is given by the following:

B (Wb/m²)	0.2	0.4	0.6	0.8	1	1.2
H (AT/m)	175	300	400	600	850	1250

Solution. Plot the *B-H* curve from the given data as shown in Fig. 8.12

(*i*) **With no air-gap**

$$B_i = \frac{\phi}{a} = \frac{5 \times 10^{-4}}{5 \times 10^{-4}} = 1 \text{ wb/m}^2$$

From the *B-H* curve, we find that for a flux density of 1wb/ m^2, the value of

$$H_i = 850 \text{ AT/m}$$

$$l_i = \pi D = \pi \times 30 \times 10^{-2} = 0.942 \text{ m}$$

$\therefore$ Total AT required $= H_i \times l_i$

$= 850 \times 0.942 = 800.7$

$\therefore$ Magnetising current,

$I = 800.7/500 =$ **1.6A**

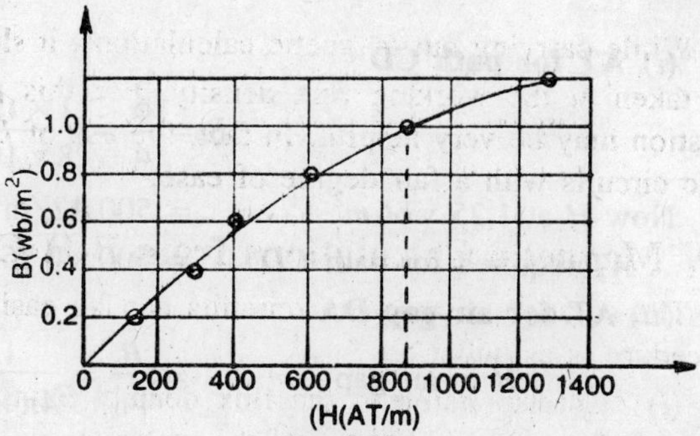

Fig. 8.12

(*ii*) **With air-gap of 1mm**

Flux density in air-gap, $B_g = 1 \text{wb/m}^2$ (Same as in steel)

Magnetising force required, $^*H_g = \dfrac{B}{\mu_0} = \dfrac{l}{4\pi \times 10^{-7}} = 7.96 \times 10^5$ AT/m

AT required for air gap $= H_g \times l_g = (7.96 \times 10^5) \times (1 \times 10^{-3}) = 796$

Total AT required $= 800.7 + 796 = 1596.7$

$\therefore$ Magnetising current, $I = 1596.7/500 =$ **3.19A**

Example 8.10. *A magnetic circuit made of wrought iron is arranged as shown in Fig. 8.13. The central limb has a cross-sectional area of 8 cm^2 and each of the side limbs has a cross-sectional area of 5 cm^2. Calculate the ampere-turns required to produce a flux of 1 mwb in the central limb, assuming the magnetic leakage is negligible. Given that for wrought iron (from B-H curve), H = 500AT/m at B = 1.25wb/ m^2 and H = 200 AT / m at B = 1wb/ m^2.*

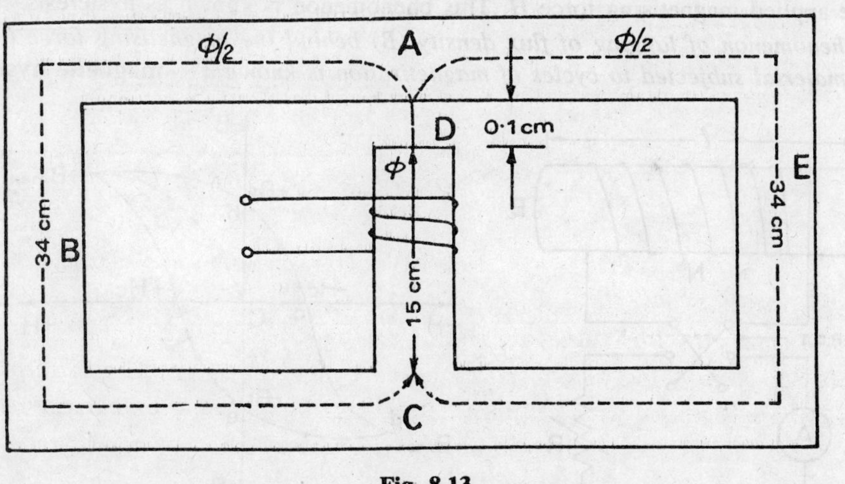

Fig. 8.13

Solution. The flux φ set up in the central limb divides equally into two identical parallel paths *viz.* path *ABC* and path *AEC*. The total m.m.f. required for the entire circuit is the sum of the following three m.m.fs :

(*i*) that required for path *CD*

(*ii*) that required for air gap *DA*

(*iii*) that required for either of parallel paths (*i.e.* path *ABC* or path *AEC*).

* We do not use *B-H* curve to find *AT* for air gap. It is because μ_r for air (in fact for all non-magnetic materials) is constant, being equal to 1, and *AT* can be calculated directly.

(*i*) **AT for path CD**

$$B = \frac{\phi}{a} = \frac{1 \times 10^{-3}}{8 \times 10^{-4}} = 1.25 \text{ wb/m}^2$$

Now H at 1.25 wb/ m^2 $= 500\, AT/\text{m}$ (*given*)

$\therefore$ *AT* required for path *CD* $= 500 \times 0.15 = 75$

(*ii*) **AT for air-gap DA**

$$H \text{ in air gap} = \frac{B}{\mu_0} = \frac{1.25}{4\pi \times 10^{-7}} = 994.7 \times 10^3 \text{ AT/m}$$

$\therefore$ *AT* required for air-gap $= (994.7 \times 10^3) \times (0.1 \times 10^{-2}) = 994.7$

(*iii*) **AT for path ABC**

Flux in path *ABC* $= \phi/2 = 1/2 = 0.5$ mwb

$$\text{Flux density in path } ABC = \frac{0.5 \times 10^{-3}}{5 \times 10^{-4}} = 1 \text{ wb/m}^2$$

Now H at 1 wb/ m^2 $= 200$ AT/m (*given*)

$\therefore$ *AT* required for path *ABC* $= 200 \times 0.34 = 68$

$\therefore$ Total *AT* required $= 75 + 994.7 + 68 =$ **1137.7**

The reader may note that air gap "grabs" 87 per cent of the applied ampere-turns.

8.10. Magnetic Hysteresis

When a magnetic material is subjected to a cycle of magnetisation (*i.e.* it is magnetised first in one direction and then in the other), it is found that flux density B in the material lags behind the applied magnetising force H. This phenomenon is known as hysteresis.

The phenomenon of lagging of flux density (B) behind the magnetising force (H) in a magnetic material subjected to cycles of magnetisation is known as ***magnetic hysteresis**.

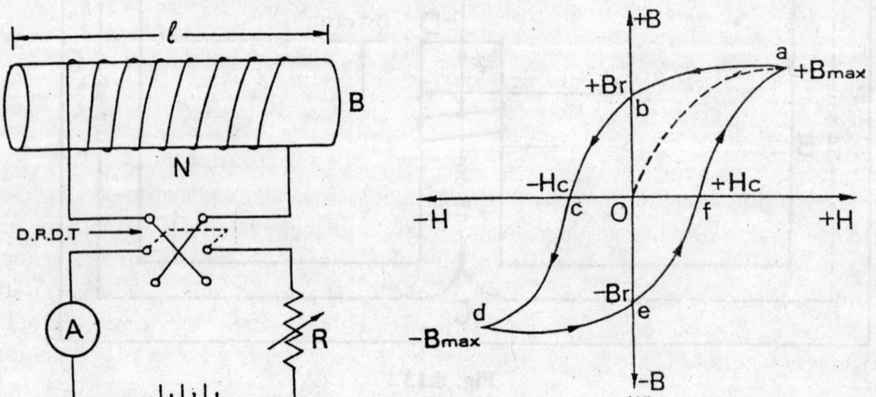

Fig. 8.14

Hysteresis Loop. Consider an unmagnetised iron bar *AB* wound with *N* turns as shown in Fig. 8.14 (*i*). The magnetising force $H\,(= NI/l)$ produced by this solenoid can be changed by varying the current through the coil. We shall see that when the iron piece is subjected to**one cycle

---

* Hysteresis is derived from the Greek work *hysterein* meaning to lag behind.

** If we start with unmagnetised iron piece, then magnetise it in one direction and then in the other direction and finally demagentise it (*i.e.* obtain the original condition we started with), the piece is said to go through one cycle of magnetisation. Compare with one cycle of alternating current or voltage.

of magnetisation, the resultant *B-H* curve traces a loop *abcdefa* called *hysteresis loop.* [See Fig. 8.14 (*ii*)].

(*i*) When current in the solenoid is zero, $H = 0$ and hence B in the iron is zero. As H is increased (by increasing solenoid current), the flux density (B) also increases until the point of maximum flux density ($+ B_{max}$) is reached. The material is saturated and beyond this point, the flux density will not increase regardless of any increase in current or magnetising force. Note that *B-H* curve of iron follows the path *oa*.

(*ii*) If now H is gradually reduced (by reducing solenoid current), it is found that flux density B does not decrease along *oa* but follows the path *ab*. At point *b*, the magnetising force H is zero but flux density in the material has a finite value $+ B_r$ ($= ob$) called *residual flux density*. In other words, B lags behind H. The greater the lag, the greater is the residual magnetism (*i.e.* ordinate *ob*) retained by the iron piece. The power of retaining residual magnetism is called *retentivity* of the material.

(*iii*) To demagnetise the iron piece (*i.e.* to remove the residual magnetism *ob*), the magnetising force H is reversed by reversing the current through the coil. When H is gradually increased in the reverse direction, the *B-H* curve follows the path *bc* so that when $H = oc$, the residual magnetism is zero. The value of H ($= oc$) required to wipe out residual magnetism is known as *coercive force* (H_c).

(*iv*) If H is further increased in the negative direction, the material again saturates (point *d*) in the negative direction. Reducing H to zero and then increasing it in the positive direction completes the curve *defa*. Thus when an iron piece is subjected to one cycle of magnetisation, the *B-H* curve traces a closed loop *abcdefa* called *hysteresis loop*.

It is clear from the hysteresis loop (*i.e.* *B-H* curve of iron for one cycle of magnetisation) that B lags behind H. The two never attain zero value simultaneously.

Note. For one cycle of magnetisation, one hysteresis loop is traced. If a magnetic material is located within a coil through which alternating current (50 Hz) flows, 50 loops will be formed every second.

8.11. Hysteresis Loss

When a magnetic material is subjected to a cycle of magnetisation (*i.e.* it is magnetised first in one direction and then in the other), an energy loss takes place due to the *molecular friction in the material. That is, the domains (or molecular magnets) of the material resist being turned first in one direction and then the other. Energy is thus expended in the material in overcoming this opposition. This loss is in the form of heat and is called *hysteresis loss.* Hysteresis loss is present in all those electrical machines whose iron parts are subjected to cycles of magnetisation. The obvious effect of hysteresis loss is the rise of temperature of the machine.

(*i*) Transformers and most electric motors operate on alternating current. In such devices, the flux is the iron changes continuously, both in value and direction. Hence hysteresis loss occurs in such machines.

(*ii*) Hysteresis loss also occurs when an iron part rotates in a constant magnetic field *e.g.* d.c. machines.

---

* The opposition offered by the magnetic domains (or molecular magnets) to the turning effect of magnetising force is sometimes referred to as the molecular friction.

8.12. Calculation of Hysteresis Loss

We will now show that area of hysteresis loop represents the *energy loss /m³/cycle.

Let l = length of the iron bar

A = area of x–section of bar

N = No. of turns of wire of solenoid

Suppose at any instant the current in the solenoid is i. Then,

$$H = \frac{Ni}{l} \quad or \quad i = \frac{Hl}{N}$$

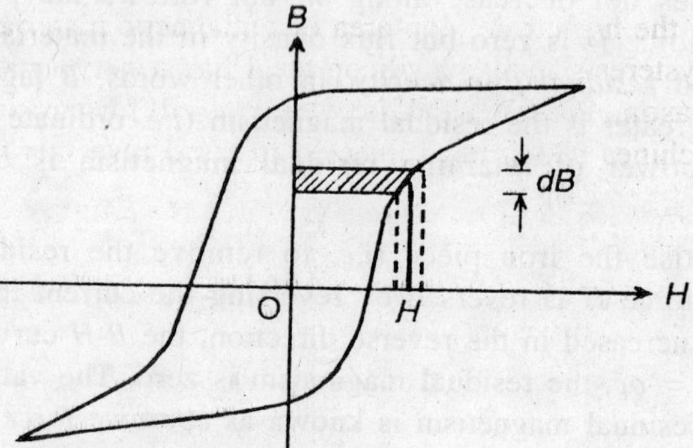

Fig. 8.15

Suppose the current increases by di in a small time dt. This will cause the flux density to increase by dB and hence an increase in flux $d\phi\,(= AdB)$. This causes an e.m.f. e to be induced in the solenoid.

$$e = N\frac{d\phi}{dt} = NA\frac{dB}{dt}$$

By Lenz's law, this e.m.f. opposes the current i so that energy dW is spent in overcoming this opposing e.m.f.

$$\therefore \quad dW = ei\,dt \text{ joules}$$

$$= \left(NA\frac{dB}{dt}\right) \times \left(\frac{Hl}{N}\right) \times dt \text{ joules}$$

$$= Al \times H \times dB \text{ Joules}$$

$$= V \times (H \times dB) \text{ Joules}$$

where $Al = V$ = volume of iron bar

Now $H \times dB$ is the area of the shaded strip (See Fig. 8.15). For one cycle of magnetisation, the area $H \times dB$ will be equal to the area of hysteresis loop.

$\therefore$ Hysteresis energy loss/cycle, $W_h = V \times$ (area of loop) Joules

If f is the frequency of reversal of magnetisation, then,

Hysteresis power loss, $P_h = W_h \times f = V \times$ (area of loop) $\times f$

---

* In order to set up magnetic field, certain amount of energy has to supplied which is stored in the field. If the field is in free space, the stored energy is returned to the circuit when the field collapses. If the field is in a magnetic material, not all the energy supplied can be returned ; part of it having been converted into heat due to hysteresis effect.

Note While calculating the area of hysteresis loop, proper scale factors of B and H must be considered, For example if the scales are : 1cm = $x\,AT/m$ ...for H

1cm = $y\,wb/m^2$ ... for B

Then, $W_h = xy \times$ (area of loop in cm^2) $\times V$ Joules

where x and y are the scale factors.

8.13. Importance of Hysteresis Loop

The shape and size of the hysteresis loop *largely depends upon the nature of the material. The choice of a magnetic material for a particular application often depends upon the shape and size of the hysteresis loop. A few cases are discussed below by way of illustration:

(*i*) The smaller the hysteresis loop area of a magnetic material, the less is the hysteresis loss. The hysteresis loop for silicon steel has a very small area [See Fig. 8.16 (*i*)]. For this reason, silicon steel is widely used for making transformer cores and rotating machines which are subjected to rapid reversals of magnetisation.

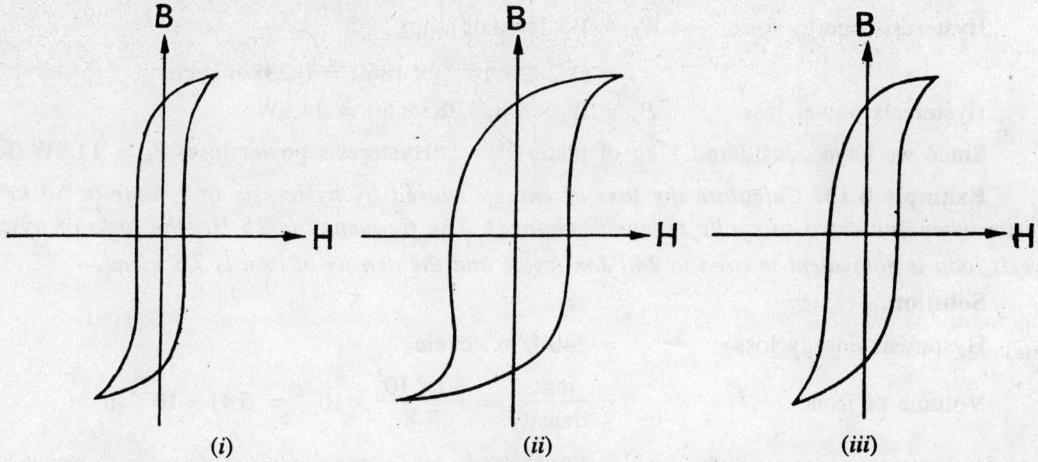

Fig. 8.16

(*ii*) The hysteresis loop for hard steel [see Fig 8.16 (*ii*)] indicates that this material has high retentivity and coercivity. Therefore, hard steel is quite suitable for making permanent magnets. But due to the large area of the loop, there is greater hysteresis loss. For this reason, hard steel is not suitable for the construction of electrical machines

(*iii*) The hysteresis loop for wrought iron [see Fig. 8.16 (*iii*)] shows that this material has fairly good residual magnetism and coercivity. Hence, it is suitable for making cores of electromagnets.

Example 8.11. *A magnetic circuit is made of silicon steel and has a volume of* $2 \times 10^{-3}\,m^3$. *The area of hysteresis loop of silicon steel is found to be 7.25* cm^2; *the scales being 1 cm* = $10\,AT/m$ *and* 1 cm = $4\,wb/m^2$. *Calculate the hysteresis power loss when the flux is alternating at 50 Hz.*

Solution.

Area of hysteresis loop = (Area in cm^2) $\times$ (Scale factors) = $(7.25) \times (xy)$

---

* It also depends upon (*i*) the maximum value of flux density established and (*ii*) the initial magnetic state of the material.

$= (7.25) \times (10 \times 4) = 290\ J/m^3/cycle$

Hysteresis power loss, P_h = Volume × area of loop × frequency

$= (2 \times 10^{-3}) \times (290) \times (50)\ W = \mathbf{29W}$

Example 8.12. *The area of hysteresis loop obtained with a certain magnetic material was 9.3 cm². The co-ordinates were such that 1 cm = 1000AT/m and 1 cm = 0.2 wb/m². If the density of the given material is 7.8 g/cm³, calculate the hysteresis loss in watts/kg at 50Hz.*

Solution.

Volume of 1 kg of material, $V = \frac{10^3}{7.8} \times 10^{-6} = 1.282 \times 10^{-4}\ m^3$

Area of hysteresis loop = Area in cm² × scale factors

$= (9.3) \times (1000 \times 0.2) = 1860\ J/m^3/cycle$

Hysteresis energy loss, $W_h = V \times$ (area of loop)

$= (1.282 \times 10^{-4}) \times 1860 = 0.238\ J/cycle$

Hysteresis power loss, $P_h = W_h \times f = 0.283 \times 50 = 11.9W$

Since we have considered 1 kg of material, ∴ Hysteresis power loss, P_h = **11.9W/kg**

Example 8.13. *Calculate the loss of energy caused by hysteresis in 1 hour in 50 kg of iron when subjected to cyclic magnetic changes. The frequency is 25 Hz, the area of hysteresis loop is equivalent in area to 240 J/m³/cycle and the density of iron is 7.8 g/cm³.*

Solution.

Hysteresis energy loss $= 240\ J/m^3/cycle$

Volume of iron $= \frac{\text{mass}}{\text{density}} = \frac{50 \times 10^3}{7.8} \times 10^{-6} = 6.41 \times 10^{-3}\ m^3$

No. of cycles/hour $= 25 \times 60 \times 60 = 9 \times 10^4$

∴ Energy loss/hour = volume × (area of loop) × cycles/hour

$= (6.41 \times 10^{-3}) \times (240) \times (9 \times 10^4) = \mathbf{138456\ J}$

Example 8.14. *The armature of a 4-pole d.c. generator has a volume of $12 \times 10^{-3}\ m^3$. During rotation, the armature is taken through a hysteresis loop whose area is 20 cm² when plotted to a scale of 1 cm = 100AT/m, 1cm = 0.1 wb/m². Determine the hysteresis loss in watts when the armature rotates at a speed of 900 r.p.m.*

Solution. Since it is a 4-pole machine, two hysteresis loops will be formed in one revolution of the armature.

∴ No. of loops generated/second, $f = 2 \times 900/60 = 30$

Hysteresis energy loss/cycle = Area of loop in cm² × scale factors

$= 20 \times (100 \times 0.1) = 200\ J/m^3$

Total hysteresis energy loss/second = volume × (area of loop in J/m³) × f

$= (12 \times 10^{-3}) \times 200 \times 30 = 72\ W$

i.e. Hysteresis power loss = **72W**

Tutorial Problems

1. A hysteresis loop is plotted against a horizontal axis which scales 1 cm = 1000AT/m and a vertical axis which scales 1cm = 0.2wb/m^2. If the area of the loop is 9 cm^2 and the overall height is 14 cm, calculate :

(*i*) the hysteresis loss in J/m^3/cycle;

(*ii*) the maximum flux density ;

(*iii*) the hysteresis loss in watt/kg assuming the density of the material to be 7.8g/cm^3.

[(*i*) **1800** (*ii*) **1.4 wb/m^2** (*iii*) **11.55 W**]

2. The hysteresis loop for a specimen of mass 12 kg is equivalent to 30W/mm^3. Find the loss of energy in kwh in one hour at 50 Hz. The density of the specimen is 7.8 g/cm^3. **[0.024kWh]**

3. A transformer is made of 200 kg of steel plate with a specific gravity of 7.5. It may be assumed that the maximum operating flux density is 1.1wb/m^2 for all parts of the steel. When a specimen of the steel was tested, it was found to have a hysteresis loop of area 100cm^2 for a maximum flux density of 1.1 wb/m^2. If the scales of the hysteresis loop graph were 1cm = 50AT/m and 1cm = 0.1wb/m^2, calculate the hysteresis power loss when the transformer is operated on 50 Hz mains. **[667W]**

4. A magnetic core is made from sheet steel, the hysteresis loop of which has an area of 2.1 cm^2; the scales being 1cm = 400 AT/m and 1cm = 0.4 wb/m^2. The core measures 100 cm long and has an average cross-sectional area of 10 cm^2. The hysteresis loss is 16.8 W. Calculate the frequency of alternating flux. **[50Hz]**

8.14. Steinmetz Hysteresis Law

To eliminate the need of finding the area of hysteresis loop for computing the hysteresis loss, Steinmetz devised an empirical law for finding the hysteresis loss. He found that the area of hysteresis loop of a magnetic material is directly proportional to 1.6the power of the maximum flux density established *i.e.*.

$$\text{Area of hysteresis loop} \propto {}^{*}B_{max}^{1.6}$$

or $$\text{Hysteresis energy loss} \propto B_{max}^{1.6} \text{ Joules/m}^3\text{/cycle}$$

or $$\text{Hysteresis energy loss} = \eta\, B_{max}^{1.6} \text{ Joules/m}^3\text{/cycle}$$

where η is a constant called **hysteresis coefficient**. Its value depends upon the nature of material. The smaller the value of η of a magnetic material, the lesser is the hysteresis loss. The armatures of electrical machines and transformer cores are made of magnetic materials having low hysteresis co-efficient in order to reduce the hysteresis loss. The best transformer steels have η values around 130, for cast steel they are around 2500 and for cast iron about 3750.

If V is the volume of the material in m^3 and f is the frequency of magnetisation, then

Hysteresis power loss, $P_h = \eta f B_{max}^{1.6} V$ J/s or watts

Example 8.15. *A power transformer has a core material for which hysteresis coefficient is 130 J/m^3. Its volume is 8,000 cm^3 and the maximum flux density is 1.25 wb/m^2. What is the hysteresis loss in watts if the frequency of the alternating current is 50Hz.*

---

* The index 1.6 is called **Steinmetz index**. In fact, the value of this index depends upon the nature of material and may vary from 1.6 to 2.5. However, reasonable accuracy is obtained if it is taken as 1.6.

Solution.

Hysteresis power loss, $P_h = \eta f B_{max}^{1.6} V = (130) \times (50) \times (1.25)^{1.6} \times (8000 \times 10^{-6}) = \mathbf{74.31\,W}$

Example 8.16. *The volume of a transformer core built up of sheet steel laminations is 5000 cm³ and the gross cross-sectional area is 240 cm². Because of the insulation between the plates, the net cross-sectional area is 90% of the gross. The maximum value of flux is 22mwb and the frequency is 50Hz. Find (i) the hysteresis loss/m³/cycle and (ii) power loss in watts. Take hysteresis coefficient as 250.*

Solution.

$$a = 0.9 \times 240 = 216\,\text{cm}^2; \; B_{max} = \frac{22 \times 10^{-3}}{216 \times 10^{-4}} = 1.019\ \text{wb/m}^2$$

(*i*) Hysteresis energy loss $= \eta B_{max}^{1.6} = 250 \times (1.019)^{1.6} = \mathbf{257.6\ J/m^3/cycle}$

(*ii*) Hysteresis power loss $P_h = \eta f B_{max}^{1.6} \times V = (257.6) \times (50) \times (5000 \times 10^{-6}) = \mathbf{64.4W}$

Example 8.17. *The area of hysteresis loop obtained with a certain specimen of iron was 9.3 cm². The co-ordinates were such that 1 cm = 1000AT/m and 1 cm = 0.2wb/m². Calculate (i) the hysteresis loss in J/m³/cycle (ii) hysteresis loss in W/m³ at a frequency of 50Hz (iii). If the maximum flux density was 1.5 wb/m², calculate the hysteresis loss/ m³ for a maximum flux density of 1.2wb/m², and a frequency of 30Hz, assuming the loss to be proportional to* $B_{max}^{1.8}$.

Solution

(*i*) Hysteresis energy loss $= (xy) \times (\text{area of loop})\ \text{J/m}^3/\text{cycle}$

$= (1000 \times 0.2) \times 9.3 = \mathbf{1860\ J/m^3/cycle}$

(*ii*) Hysteresis power loss $= 1860 \times 50 = \mathbf{93{,}000\ W/m^3}$

(*iii*) Hysteresis loss/ $m^3 = \eta f (B_{max})^{1.8}$

or $93000 = \eta \times 50 \times (1.5)^{1.8}$

$$\therefore \quad \eta = \frac{93{,}000}{50 \times (1.5)^{1.8}} = 896.5$$

For $B_{max} = 1.2\ \text{wb/m}^2$ and $f = 30\text{Hz}$

Hysteresis loss / $m^3 = \eta f (B_{max})^{1.8}\ \text{W} = 896.5 \times 30 \times (1.2)^{1.8} = \mathbf{37342W}$

Example 8.18. *In a certain transformer, the hysteresis loss was found to be 160 watts when the maximum flux density was 1.1wb / m² and the frequency 60Hz. What will be the loss when the maximum flux density is reduced to 0.9wb/m² and frequency to 50 Hz ?*

Solution. According to Steinmetz hysteresis law,

Hysteresis loss, $P_h \propto f (B_{max})^{1.6}$

For the first case, $P_1 \propto 60 \times (1.1)^{1.6}$

For the second case, $P_2 \propto 50 \times (0.9)^{1.6}$

$$\therefore \quad \frac{P_2}{P_1} = \frac{50 \times (0.9)^{1.6}}{60 \times (1.1)^{1.6}} = 0.604$$

$\therefore \quad P_2 = 0.604\,P_1$

$= 0.604 \times 160 = \mathbf{96.64\ W}$

OBJECTIVE QUESTIONS

1. In Fig. 8.17, the magnetic circuit is the path
(*i*) *DAB* (*ii*) *ABCDA*
(*iii*) *ABC* (*iv*) *ABCD*

2. If *l* is the magnetic path in Fig. 8.17, then magnetising force is
(*i*) *NI* (*ii*) *NI* × *l*
(*iii*) *l/NI* (*iv*) *NI/l*

3. The reluctance of the magnetic circuit shown in Fig. 8.17 is

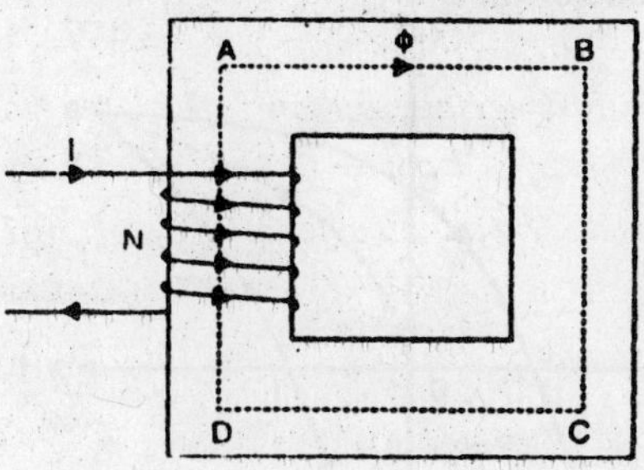

Fig. 8.17

(*i*) *NI/l* (*ii*) ϕ/*N*1
(*iii*) *NI*/1ϕ (*iv*) ϕ1/*l*

4. The SI unit of reluctance is
(*i*) AT/Wb (*ii*) AT/m
(*iii*) AT (*iv*) N/Wb

5. A magnetic circuit has a m.m.f. of 400 AT and a reluctance of 2×10^5 AT/Wb. The magnetic flux in the magnetic circuit is
(*i*) 3×10^{-5} Wb (*ii*) 2×10^{-3} Wb
(iii) 1.5×10^{-2} Wb (iv) 2.5×10^{-4} Wb

6. A 2 cm long coil has 10 turns and carries a current of 750 mA. The magnetising force of the coil is
(*i*) 225 AT/m (*ii*) 675 AT/m
(*iii*) 450 AT/m (*iv*) 375 AT/m

7. A magnetic device has a core with cross- section of 1 inch². If the flux in the core is 1 mWb, then flux density (1 inch = 2.54 cm) is
(*i*) 2.5 T (*ii*) 1.3 T
(*iii*) 1.55 T (*iv*) 0.25 T

8. The reluctance of a magnetic circuit varies as.......
(*i*) length × area (*ii*) length ÷ area
(*iii*) area ÷ length (*iv*) $(\text{length})^2$ + area

9. The reluctance of a magnetic circuit is........ relataive permeability of the material comprising the circuit.
(*i*) directly proportional to
(*ii*) inversely proportional to
(*iii*) independent of
(*iv*) none of the above

10. M.M.F. in a magnetic circuit corresponds to .... in an electric circuit.
(*i*) voltage drop
(*ii*) potential difference
(*iii*) electric intensity
(*iv*) e.m.f.

11. Permeance of a magnetic circuit is .... area of x-section of the circuit.
(*i*) inversely proportional to
(*ii*) directly proportional to
(*iii*) independent of
(*iv*) none of the above.

12. The magnitude of AT required for air gap is much greater than that required for iron part of a magnetic circuit because.............
(*i*) air is a gas
(*ii*) air has the lowest relative permeability
(*iii*) air is a conductor of magnetic flux
(*iv*) none of the above

13. In electro-mechanical conversion devices (*e.g.* motors and generators), a small air gap is left between the rotor and stator in order to ....
(*i*) complete the magnetic path
(*ii*) decrease the reluctance of magnetic path
(*iii*) permit mechanical clearance
(*iv*) increase flux density in air gap

14. A magnetic circuit carries a flux ϕ_i in the iron part and a flux ϕ_g in the air gap. Then leakage co-efficient is..........

(i) ϕ_i / ϕ_g (ii) ϕ_g / ϕ_i
(iii) $\phi_g \times \phi_i$ (iv) none of the above

15. The value of leakage co-efficient for electrical machines is usually about ....

(i) 0.5 to 1 (ii) 4 to 10
(iii) above 10 (iv) 1.15 to 1.25

16. The reluctance of a magnetic circuit depends upon ....

(i) current in the coil
(ii) no. of turns of coil
(iii) flux density in the circuit
(iv) none of the above

17. The B-H curve for ....will be a straight line passing through the origin.

(i) air (ii) soft iron
(iii) hardened steel (iv) silicon steel

18. Whatever may be the flux density in...........the material will never saturate.

(i) soft iron (ii) cobalt steel
(iii) air (iv) silicon steel

19. The B-H curve of ....will not be a straight line.

(i) air (ii) copper
(iii) wood (iv) soft iron

20. The B-H curve is used to find the m.m.f. of .... in a magnetic circuit.

(i) air gap (ii) iron part
(iii) both air gap and iron part
(iv) none of the above

21. A magnetising force of 800 AT/m will produce a flux density of ....in air.

(i) 1mWb/m^2 (ii) 1 Wb/m^2
(iii) 10 mWb/m^2 (iv) 0.5 Wb/m^2

22. The saturation flux density for most magnetic materials is about ....

(i) 0.5 Wb/m^2 (ii) 10 Wb/m^2
(iii) 2 Wb/m^2 (iv) 1 Wb/m^2

23. Hysteresis is the phenomenon of .... in a magnetic circuit.

(i) lagging of *B* behind *H*
(ii) lagging of *H* behind *B*
(iii) setting up constant flux
(iv) none of the above

24. In Fig. 8.18, the point .... represents the saturation condition.

(i) b (ii) c
(iii) a (iv) e

25. In Fig. 8.18, ... represents the residual magnetism.

(i) of (ii) oc
(iii) ob (iv) none of the above

26. In Fig. 8.18 oc represents the ....

(i) residual magnetism
(ii) coercive force (iii) retentivity
(iv) none of the above

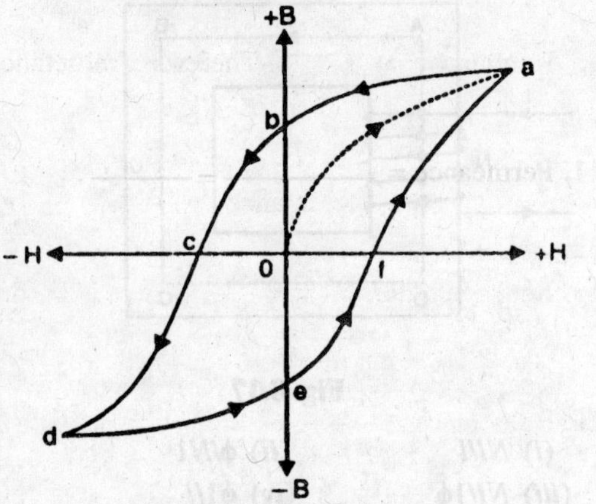

Fig. 8.18

27. If a magnetic material is located within a coil through which alternating current (50 Hz frequency) flows, then .... hysteresis loops will be formed every second.

(i) 50 (ii) 25
(iii) 100 (iv) 150

28. Out of the following materials, the area of hysteresis loop will be least for ....

(i) wrought iron (iii) hard steel
(iii) silicon steel (iv) soft iron

29. The materials used for the core of a good relay should have .... hysteresis loop.

(i) large (ii) very large
(iii) narrow (iv) none of the above

30. The magnetic material used for .... should have a large hysteresis loop

(i) transformers (ii) d.c. generators
(iii) a.c. motors
(iv) permanent magnets

ANSWERS TO OBJECTIVE QUESTIONS

1. (*ii*)	**2.** (*iv*)	**3.** (*iii*)	**4.** (*i*)	**5.** (*ii*)	**6.** (*iv*)	**7.** (*iii*)	**8.** (*ii*)
9. (*ii*)	**10.** (*iv*)	**11.** (*ii*)	**12.** (*ii*)	**13.** (*iii*)	**14.** (*i*)	**15.** (*iv*)	**16.** (*iii*)
17. (*i*)	**18.** (*iii*)	**19.** (*iv*)	**20.** (*ii*)	**21.** (*i*)	**22.** (*iii*)	**23.** (*i*)	**24.** (*iii*)
25. (*iii*)	**26.** (*ii*)	**27.** (*i*)	**28.** (*iii*)	**29.** (*iii*)	**30.** (*iv*)		

HINTS TO SELECTED OBJECTIVE QUESTIONS

4. Unit of reluctance $= \dfrac{\text{Unit of m. m. f.}}{\text{Unit of flux}} = \text{AT / Wb}$.

6. Magnetising force, $H = \dfrac{NI}{l} = \dfrac{10 \times (750 \times 10^{-3})}{2 \times 10^{-2}} = 375 \text{ AT / m}$.

8. Reluctance $= \dfrac{l}{a\,\mu_0\mu_r}$. Therefore, reluctance varies as length ÷ area

11. Permeance $= \dfrac{l}{\text{Reluctance}} = \dfrac{a\,\mu_0\mu_r}{l}$.

14. Leakage co-efficient $= \dfrac{\phi_i}{\phi_g}$; where ϕ_i = flux in iron part, ϕ_g = flux in air gap.

16. Reluctance $= \dfrac{l}{a\,\mu_0\,\mu_r}$. Clearly reluctance depends upon μ_r. Now the value of μ_r is not constant for a magnetic substance ; it varies with flux density.

17. All non-magnetic materials have *B-H* curve a straight line passing through the origin. For non-magnetic materials,

$$B = \mu_0 H$$

Since μ_o (= $4\pi \times 10^{-7}$ H/m) is constant, $B \propto H$

Or *B-H* curve is a straight line passing through the origin.

19. For a magnetic material,

$$B = \mu_0\mu_r H$$

Since μ_r is not a constant quantity, the relation between *B* and *H* is non- linear.

21. For air (or any other non-magnetic material),

$B = \mu_0 H = (4\pi \times 10^{-7}) H$

Now $\quad 4\pi \times 10^{-7} = \dfrac{1}{800000}$

$\therefore \quad H = 800000\,B$

or $\quad B = \dfrac{800}{800000} = 10^{-3} \text{ Wb/m}^2 = 1 \text{ mWb/m}^2$.

30. A large hysteresis loop means high residual flux density and large coercive force. Both these features mean that demagnetizing energy required is very large.

9

Electromagnetic Induction

Introduction

In the beginning of nineteenth century, Oersted discovered that a magnetic field exists around a current carrying conductor. In other words, magnetism can be created by means of an electric current. Can a magnetic field create an electric current in a conductor ? In 1831, Michael Faraday, the famous English scientist discovered that this could be done. He demonstrated that when the magnetic flux linking a conductor changes, an e.m.f. is induced in the conductor. This phenomenon is known as *electromagnetic induction*. The great discovery of electromagnetic induction by Faraday through a series of brilliant experiments has brought a revolution in the engineering world. Most of the electrical devices (*e.g.* electric generator, transformer, telephones *etc.*) are based on this principle. In this chapter, we shall confine our attention to the various aspects of electromagnetic induction.

9.1. Electomagnetic Induction

When the magnetic flux *linking a conductor changes, an e.m.f. is induced in the conductor. If the conductor forms a complete loop or circuit, a current will flow in it. This phenomenon is known as **electromagnetic induction. Two things are worth noting. First, the basic requirement for electromagnetic induction is the change in flux linking the conductor (or coil). Secondly, the e.m.f. and hence the current in this conductor (or coil) will persist so long as this change is taking place.

To demonstrate the phenomenon of electromagnetic induction, consider a coil C of several turns connected to a centre zero galvanometer G as shown in Fig. 9.1. If a permanent magnet is moved towards the coil, it will be observed that the galvanometer shows deflection in one direction. If the magnet is moved away from the coil, the galvanometer again shows deflection but in the opposite direction. In either case, *the deflection will persist so long as the magnet is in motion.* The production of e.m.f. and hence current in coil C is due to the fact that when the magnet is in motion (towards or away from the coil), the amount of flux linking the coil changes—the basic requirement for inducing e.m.f. in the coil. If the movement of the magnet is stopped, though the flux is linking the coil, there is *no change in flux* and hence no e.m.f. is induced in the coil. Consequently, the deflection of the galvanometer reduces to zero.

It is emphasised here that the basic requirement for inducing e.m.f. in a coil is not the flux linking the coil *but* the change in flux linking the coil. No change in flux, no e.m.f. induced in the coil.

---

* Magnetic lines of force form closed loops. Flux linking the conductor means that the flux embraces it *i.e.* it encircles the conductor.

** So called because electricity is produced from magnetism (*i.e. electromagnetic*) and that there is no physical connection (*induction*) between the magnetic field and the conductor.

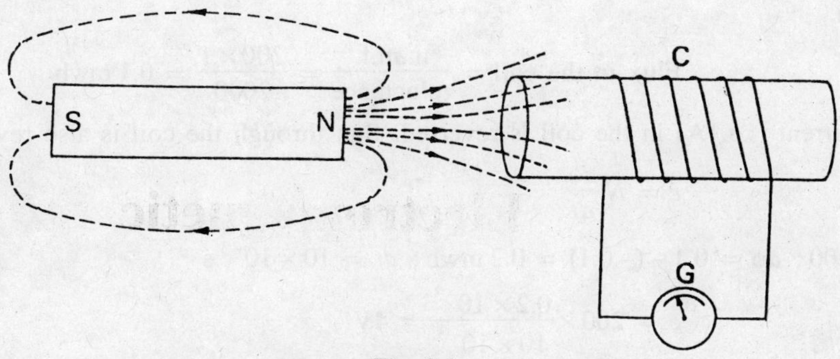

Fig. 9.1

9.2. Faraday's Laws of Electromagnetic Induction

Faraday performed a series of experiments to demonstrate the phenomenon of electromagnetic induction. He summed up his conclusions into two laws, known as Faraday's laws of electromagnetic induction.

First law. It states :

When the flux linking a conductor or coil changes, an e.m.f. is induced in it.

Second law. It states :

The magnitude of induced e.m.f. in a coil is equal to the rate of change of flux linkages.

Suppose a coil has N turns and flux linking the coil increases from ϕ_1 wb to ϕ_2 wb in t second. Now flux linkages means the product of flux and number of turns.

$$\text{Initial flux linkages} = N\phi_1\text{ ; Final flux linkages} = N\phi_2$$

$$\therefore \quad e = \text{Rate of change of flux linkages} = \frac{N\phi_2 - N\phi_1}{t}$$

or $$e = N\frac{(\phi_2 - \phi_1)}{t}$$

In differential form, we have, $e = N\dfrac{d\phi}{dt}$ volts

It is a usual practice to give a minus sign to the right-hand side expression. The minus sign comes from Lenz's law (See Art 9.3) and indicates that the voltage is induced in a direction to oppose the change in flux that produced it.

$$\therefore \quad e = -N\frac{d\phi}{dt} \text{ volts}$$

Example 9.1. *A coil of 100 turns is linked by a flux of 20 mwb. If this flux is reversed in a time of 2 ms, calculate the average e.m.f. induced in the coil.*

Solution.

Change in flux, $\quad d\phi = 20 - (-20) = 40 \text{ mwb} = 40 \times 10^{-3}$ wb

Time taken for the change, $\quad dt = 2\text{ms} = 2 \times 10^{-3}$ s

$$\therefore \quad e = N\frac{d\phi}{dt} = 100 \times \frac{40 \times 10^{-3}}{2 \times 10^{-3}} = \mathbf{2000V}$$

Example 9.2. *A coil of 200 turns of wire is wound on a magnetic circuit of reluctance 2000 AT/mwb. If a current of 1A flowing in the coil is reversed in 10 ms, find the average e.m.f induced in the coil.*

Solution.

$$\text{Flux in the coil} = \frac{\text{m.m.f.}}{\text{reluctance}} = \frac{200 \times 1}{2000} = 0.1 \text{ mwb}$$

When current (*i.e.* 1A) in the coil is reversed, flux through the coil is also reversed.

$$e = N\frac{d\phi}{dt}$$

Here, $N = 200$; $d\phi = 0.1 - (-0.1) = 0.2$ mwb ; $dt = 10 \times 10^{-3}$ s

$$\therefore \quad e = 200 \times \frac{0.2 \times 10^{-3}}{10 \times 10^{-3}} = \mathbf{4V}$$

Example 9.3 *The field winding of 4-pole d.c. generator consists of 4 coils connected in series, each coil being wound with 1200 turns. When the field is excited, there is a magnetic flux of 0.04wb/pole. If the field switch is opened at such a speed that the flux falls to the residual value of 0.004wb/pole in 0.1 second, calculate the average value of e.m.f. induced across the field winding terminals.*

Solution.

Total no. of turns, $N = 1200 \times 4 = 4800$

Total initial flux $= 4 \times 0.04 = 0.16$ wb

Total residual flux $= 4 \times 0.004 = 0.016$ wb

Change in flux, $d\phi = 0.16 - 0.016 = 0.144$ wb

Time taken, $dt = 0.1$ second

$$\therefore \text{ Induced e.m.f.,} \quad e = N\frac{d\phi}{dt} = 4800 \times \frac{0.144}{0.1} = \mathbf{6912V}$$

9.3. Direction of Induced E.M.F. and Current

The direction of induced e.m.f. and hence current in a conductor or coil can be determined by one of the following two methods :

1. Lenz's law
2. Fleming's Right-hand rule

1. Lenz's Law. Emil Lenz, a German scientist gave the following simple rule (known as Lenz's law) to find the direction of induced current :

An induced current will flow in such a direction so as to oppose the cause that produces it.

The cause that produces the current is the change of flux linking the coil. Therefore, the direction of induced current will be such that its own magnetic field opposes the change in flux that produced the induced current. Let us apply Lenz's law to Fig. 9.2. Here *N*-pole of the magnet is approaching the coil. According to Lenz's law, the direction of induced current will be such that flux set up by it will oppose the change in original flux. This is possible only if the left-hand face of the coil becomes *N*-pole. Once we know the magnetic polarity of the coil face, the direction of induced current can be easily determined by applying Right hand rule for the coil (See Art 7.10).

It may be noted here that Lenz's law directly follows from the law of conservation of energy *i.e.* in order to set up induced current,some energy must be expended. In the above case, for example, when the *N*-pole of the magnet is approaching the coil, the induced current will flow in the coil in such a direction that the left-hand face of the coil becomes *N*-pole. The result is that motion of the magnet is opposed. The mechanical energy spent in overcoming this opposition is converted into electrical energy which appears in the coil.

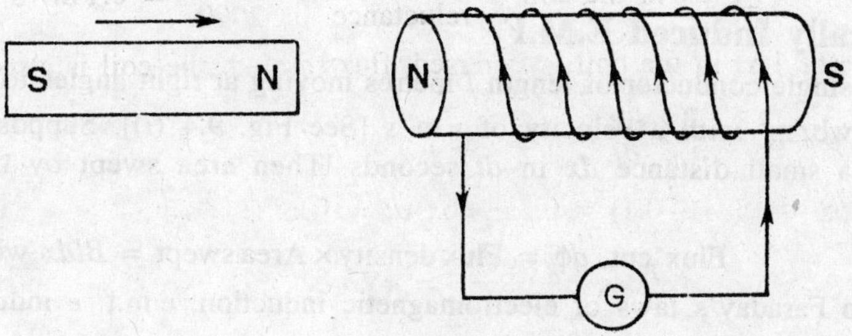

Fig. 9.2

2. Fleming's Right-hand Rule. This rule is particularly suitable to find the direction of induced e.m.f. and hence current when the conductor moves at right angles to a stationary magnetic field. It may be stated as under :

Stretch out the **forefinger, middle finger** *and* **thumb** *of your right hand so that they are at right angles to one another. If the forefinger points in the direction of magnetic field, thumb in the direction of motion of the conductor, then the middle finger will point in the direction of induced current.*

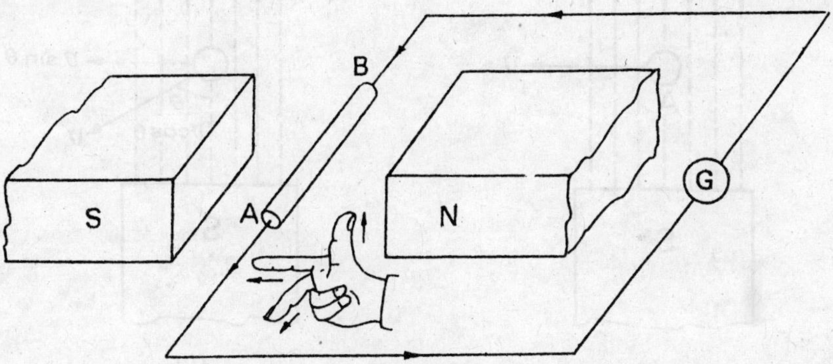

Fig. 9.3

Consider a conductor *AB* moving upwards at right angles to a uniform magnetic field as shown in Fig. 9.3. Applying Fleming's right hand rule, it is clear that the direction of induced current is from *B* to *A*. If the motion of the conductor is downward, keeping the direction of magnetic field unchanged, then the direction of induced current will be from *A* to *B*.

9.4. Induced E.M.F.

When the flux linking a conductor (or coil) changes, an e.m.f. is induced in it. This change in flux linkages can be brought about in the following two ways :

(*i*) The conductor is moved in a stationary magnetic field in such a way that the flux linking it changes in magnitude. The e.m.f. induced in this way is called *dynamically induced e.m.f.* (as in an a.c. or d.c. generator). It is so called because e.m.f. is induced in the conductor which is in motion.

(*ii*) The conductor is stationary and the magnetic field is moving or changing. The e.m.f. induced in this way is called *statically induced e.m.f.* (as in a transformer). It is so called because the e.m.f. is induced in a conductor which is stationary.

It may be noted that in either case the magnitude of induced e.m.f. is given by $N d\phi/dt$ or derivable from this relation.

9.5. Dynamically Induced E.M.F.

Consinder a single conductor of length l metres moving at right angles to a uniform magnetic field of B wb/m^2 with a velocity of v m/s [See Fig. 9.4 (*i*)]. Suppose the conductor moves through a small distance dx in dt seconds. Then area swept by the conductor is $= l \times dx$

$$\therefore \quad \text{Flux cut, } d\phi = \text{Flux density} \times \text{Area swept} = Bldx \text{ wb}$$

According to Faraday's laws of electromagnetic induction, e.m.f. e induced in the conductor is given by;

$$e = N\frac{d\phi}{dt} = \frac{B\,l\,dx}{dt} \qquad (\because N = 1)$$

$$\therefore \quad e = B\,l\,v \text{ volts} \qquad (\because dx/dt = v)$$

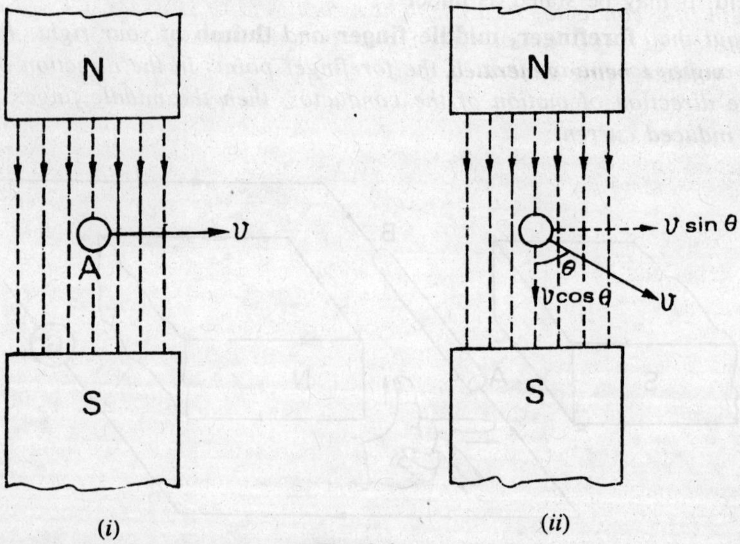

Fig. 9.4

If the conductor moves at angle θ to the magnetic field [See Fig. 9.4 (*ii*)], then the velocity at which the conductor moves across the field is *$v \sin\theta$.

$$\therefore \quad e = Bl\,v \sin\theta$$

The direction of induced e.m.f. can be determined by Fleming's Right hand rule.

Example 9.3. *A conductor of length 0.5 m situated in and at right angles to a uniform magnetic field of flux density 1 wb/m^2 moves with a velocity of 40 m/s. Calculate the e.m.f. induced in the conductor. What will be the e.m.f. induced if the conductor moves at an angle 60° to the field ?*

Solution.

(*i*) $e = Bl\nu = 1 \times 0.5 \times 40 = \mathbf{20V}$

(*ii*) $e = Bl\nu \sin\theta = 1 \times 0.5 \times 40 \times \sin 60° = \mathbf{17.32\ V}$

---

* If the conductor is moved parallel to the magnetic field, there would be no change in flux and hence no e.m.f. would be induced.

Example 9.4. *An aircraft has a wing span of 56 m. It is flying horizontally at a speed of 810 km/hr and the vertical component of earth's magnetic field is 4×10^{-4} wb/m^2. Calculate the potential difference between the wing tips of the aircraft.*

Solution. Induced e.m.f. = Blv

$$B = 4\times10^{-4}\ \text{wb/m}^2;\ l = 56\ \text{m};\ v = 810\times10^3/3600 = 225\ \text{m/s}$$

$\therefore$ Induced e.m.f. $= (4\times10^{-4})\times56\times(225) = 5.04$V

or Potential difference = **5.04V**

Example 9.5. *Each field pole of a 4-pole d.c. generator produces a magnetic flux of 20 mwb. The armature conductors rotate at 1200 r.p.m. and may be assumed to cut the flux at right angles. What is the average value of induced e.m.f. in each armature conductor ?*

Solution. In one revolution, each conductor cuts a total flux of $20\times4 = 80$ mwb. Since the conductor is rotating at $1200/60 = 20$ r.p.s., time taken for one revolution is $= 1/20 = 0.05$ second.

$\therefore$ Average e.m.f. induced in each conductor $= N\dfrac{d\phi}{dt} = 1\times\dfrac{80\times10^{-3}}{0.05} = \mathbf{1.6\ V}$

Example 9.6. *A d.c. generator consists of conductors lying on a radius of 10 cm and the effective length of a conductor in a constant radial field of strength 0.9 wb/ m^2 is 12 cm. The armature rotates at 1400 r.p.m. Given that the generator has 152 conductors in series, calculate the voltage being generated.*

Solution. Since the magnetic field is radial, the conductors cut the magnetic lines of force at right angles.

$$\text{Velocity, } v = \omega\times r = \frac{2\pi N}{60}\times r = \frac{2\pi\times1400}{60}\times0.1 = 14.66\text{m/s}$$

Voltage generated in each conductor= $Blv = 0.9\times0.12\times14.66 = 1.583$V

Voltage generated in 152 conductors in series $= 1.583\times152 =$ **240.6 V**

Tutorial Problems

1. A wire 10 cm long is moved at a uniform speed of 4 m/s at right angles to its length and to a uniform field. Calculate the density of the field if the e.m.f. generated in the wire is 0.15V. **[0.375 wb/ m^2]**

2. Calculate the e.m.f generated in the axle of a car travelling at 108 km/hr, assuming the length of the axle to be 1.5 m and the vertical component of earth's field to be 320μ wb/m^2. **[14.4 mV]**

3. A d.c. generator has 136 conductors in series. Each conductor has an effective length of 15cm, is constrained to move in a circular path of radius 12 cm and lies in a uniform field of flux density 1.2wb/m^2. Calculate the e.m.f. generated when the armature rotates at 1400 r.p.m. **[430 V]**

4. In a d.c. generator, the conductors are constrained to move in a circular path of radius 12 cm. Each conductor has an effective length of 20 cm and the magnetic field has a uniform flux density of 0.4 wb/m^2. Calculate the speed of rotation in r.p.m. so that an e.m.f. of 0.8 V is induced in each conductor. **[796 r.p.m]**

5. An aeroplane with a wing span of 25 m is flying horizontally at a speed of 900 km/hr. Calculate the p.d. between the wingtips if the vertical component of earth's field is 4×10^{-5} wb/m^2. **[0.25 V]**

9.6. Statically Induced E.M.F.

When the conductor is stationary and the field is moving or changing, the e.m.f. induced in the conductor is called statically induced e.m.f. A statically induced e.m.f. can be further sub-divided into :

1. Self-induced e.m.f.   2. Mutually induced e.m.f.

1. Self-induced e.m.f. The e.m.f. induced in a coil due to the change of its own flux linked with it is called *self-induced e.m.f.*

When a coil is carrying current (See Fig. 9.5), a magnetic field is established through the coil. If the current in the coil changes, then flux linking the coil also changes. Hence by Faraday's laws, an e.m.f. will be induced in the coil. This is known as self-induced e.m.f. The magnitude of this self-induced e.m.f = $N\ d\phi/dt$. The direction of induced e.m.f. (by Lenz's law) is always such so as to oppose the cause responsible for inducing the e.m.f; namely change of current (and hence field) in the coil.

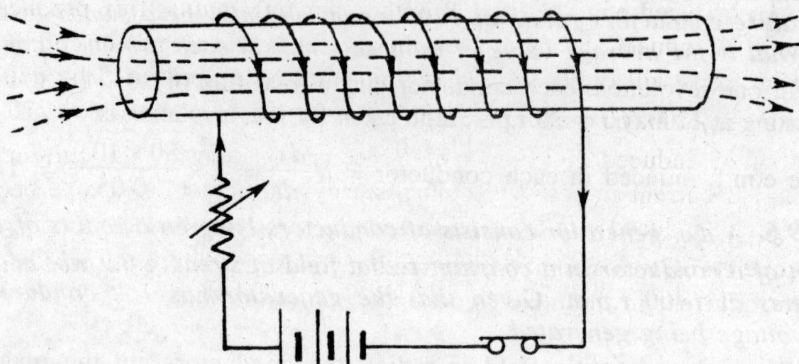

Fig. 9.5

Thus when current in a coil changes, an e.m.f. (self-induced e.m.f.) is induced in it which opposes the change of current in the coil. This property of the coil is called its *self-inductance* or *inductance*. It may be noted that self-induced e.m.f (and hence inductance) *does not prevent the current from changing; it serves only to delay the change.* Thus, after the switch is closed (See Fig. 9.5), the current will rise from zero to final steady value in sometime. This delay is due to self-induced e.m.f. or inductance of the coil.

Note. Self-induced e.m.f. (and hence inductance) comes into picture only when current in the coil changes (decreases or increases). When current in the coil is steady or constant, the magnetic field also becomes constant and self-induced e.m.f. drops to zero. Under such condition (*i.e.* constant current), the circuit does not exhibit inductance *i.e.* no self-induced e.m.f.

2. Mutually Induced e.m.f. The e.m.f. induced in a coil due to the changing current in the neighbouring coil is called *mutually induced e.m.f.*

Consider two coils *A* and *B* placed adjacent to each other as shown in Fig. 9.6. A part of the magnetic flux produced by coil *A* passes through or links with coil *B*. This flux which is

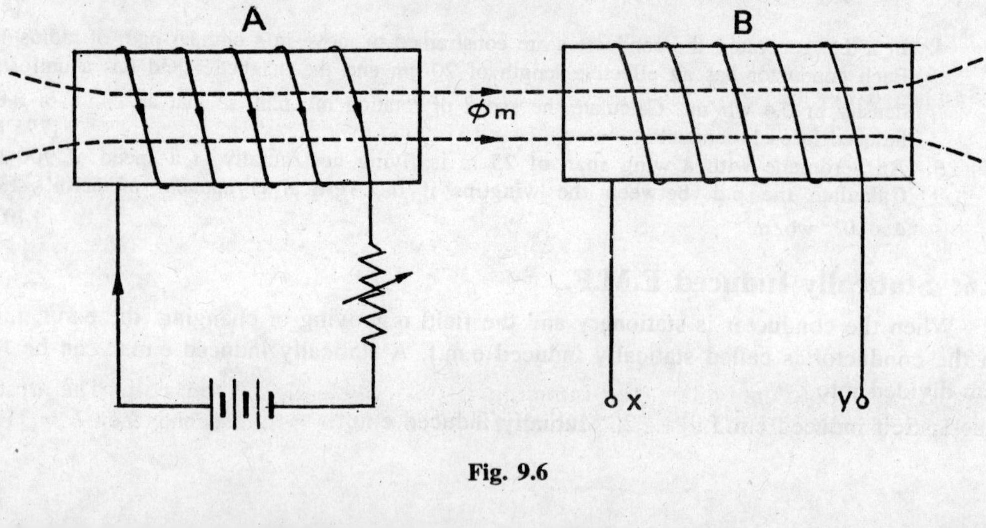

Fig. 9.6

common to both the coils *A* and *B* is called *mutual flux* (ϕ_m). If current in coil *A* is varied, the mutual flux also varies and hence e.m.f. is induced in both the coils. The e.m.f. induced in coil *A* is called self-induced e.m.f. as already discussed. The e.m.f. induced in coil *B* is known as *mutually induced e.m.f.*

The magnitude of mutually induced e.m.f. is given by Faraday's laws *i.e.* $e_M = N_B \, d\phi m/dt$ where N_B is the number of turns of coil *B* and $d\phi_m/dt$ is the rate of change of mutual flux *i.e.* flux common to both the coils. The direction of mutually induced e.m.f. (by Lenz's law) is always such so as to oppose the very cause producing it. The cause producing the mutually induced e.m.f. in coil *B* is the changing mutual flux produced by coil *A*. Hence the direction of induced current (when the circuit is completed) in coil *B* will be such that the flux set up by it will oppose the changing mutual flux produced by coil *A*.

The following points may be noted carefully :

(*i*) The mutually induced e.m.f. in coil *B* persists so long as the current in coil *A* is changing. If current in coil *A* becomes steady, the mutual flux also becomes steady and mutually induced e.m.f. drops to zero.

(*ii*) The property of two neighbouring coils to induce voltage in one coil due to the change of current in the other is called *mutual inductance.*

9.7. Self-inductance (L)

The property of a coil that opposes any change in the amount of current flowing through it is called its **self-inductance** *or* **inductance.**

This property (*i.e.* inductance) is due to the self-induced e.m.f. in the coil itself by the changing current. If the current in the coil is increasing, the self-induced e.m.f. is set up in such a direction so as to oppose the rise of current *i.e.* direction of self-induced e.m.f. is opposite to that of the applied voltage. Similarly, if the current in the coil is decreasing, self-induced voltage will be such so as to oppose the decrease in current *i.e.* self-induced e.m.f. will be in the same direction as the applied voltage. It may be noted that self-inductance does not prevent the current from changing; it serves only to delay the change.

The greater the self-induced e.m.f. ($= Nd\,\phi/dt$), the greater the self-inductance of the coil' and hence larger is the opposition to the changing current. Hence inductance of a coil depends upon the following factors :

(*i*) Shape and number of turns.

(*ii*) μ_r of the material surrounding the coil

(*iii*) The speed with which magnetic field changes

Expression for L. Consider a coil of *N* turns carrying a current of *I* amperes. If current in the coil changes, the flux linkages of the coil will also change. This will set up a self-induced e.m.f. *e* in the coil given by;

$$e = N\frac{d\phi}{dt} = \frac{d}{dt}(N\phi)$$

Since flux is due to current in the coil, it follows that flux linkages ($= N\phi$) will be proportional to *I*.

$$\therefore \qquad e \propto \frac{dI}{dt} = \text{constant} \times \frac{dI}{dt}$$

$$\therefore \qquad e = L\frac{dI}{dt} \qquad \text{... (i)}$$

where *L* is a constant called **self-inductance** or **inductance** of the coil. The unit of inductance is henry (H). If in Eq. (*i*) above, e = 1 volt, dI/dt = 1A/second, then L = 1H.

Hence a coil (or circuit) has an inductance of **1 henry** *if an e.m.f. of 1 volt is induced in it when current through it changes at the rate of 1 ampere per second.*

9.8 Other Expressions for L

Apart from exp (*i*) above, the value of L can be determined by one of the following two ways :

(*i*) **First method** If the flux linkages of the coil and current are known, then inductance can be determined as under :

$$e = L\frac{dI}{dt} = \frac{d}{dt}(LI)$$

Also
$$e = N\frac{d\phi}{dt} = \frac{d}{dt}(N\phi)$$

From the two expressions, we have, $LI = N\phi$

$$\therefore \quad L = \frac{N\phi}{I} \qquad \text{... (i)}$$

Thus, inductance is the flux linkages of the coil per ampere. If $N\phi$ = 1wb-turn and I = 1A, then L = 1H.

Hence a coil has an inductance of **1 henry** *if a current of 1 A in the coil sets up flux linkages of 1 wb-turn.*

(*ii*) **Second method.** The inductance of a magnetic circuit can be found in terms of its physical dimensions. Consider an iron-cored solenoid of dimensions as shown in Fig. 9.7. Inductance of the solenoid is given by [from exp. (*i*) above] :

$$L = N\frac{d\phi}{dI}$$

Now
$$\phi = \frac{\text{m.m.f.}}{\text{reluctance}} = \frac{NI}{l/a\,\mu_0\mu_r}$$

Differentiating ϕ w.r.t. I, we get,
$$\frac{d\phi}{dI} = \frac{N\,a\,\mu_0\mu_r}{l}$$

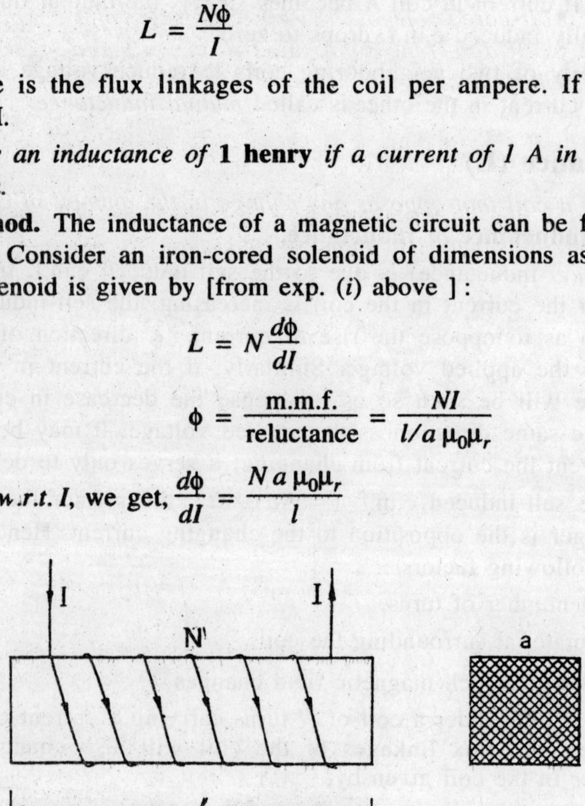

Fig. 9.7

$$\therefore \quad L = N\frac{(N\,a\,\mu_0\,\mu_r)}{l} \text{ or } L = \frac{N^2\,a\mu_0\,\mu_r}{l} \qquad \text{...(ii)}$$

$$= \frac{N^2}{l/a\mu_0\mu_r} = \frac{N^2}{\text{Reluctance (S)}} \qquad \text{...(iii)}$$

It may be noted that inductance is directly proportional to turns squared and inversely proportional to reluctance of the magnetic path.

Example 9.7. *A wooden toroid of mean diameter 400 mm and cross-sectional area 400 mm^2 is uniformly wound with a coil of 1000 turns which carries a current of 2A. Determine (i) self-inductance of the coil and (ii) the e.m.f. induced in the coil when the current is uniformly reduced to zero in 10 ms.*

Solution.

$$l = 0.4\pi \text{ m} ; \quad a = 400 \times 10^{-6} \text{ m}^2$$

(i) $$L = \frac{N^2 a \mu_0 \mu_r}{l} = \frac{(1000)^2 \times 400 \times 10^{-6} \times 4\pi \times 10^{-7} \times 1}{0.4\pi} = \mathbf{0.4 \times 10^{-3} \text{ H}}$$

(ii) $$e = L\frac{dI}{dt} = 0.4 \times 10^{-3} \frac{2-0}{10 \times 10^{-3}} = \mathbf{0.08V}$$

Example 9.8. *An air-cored solenoid having a diameter of 4 cm and a length of 60 cm is wound with 4000 turns. If a current of 5A flows in the solenoid, calculate (i) the inductance and (ii) energy stored in joules.*

Solution.

$$a = (\pi/4)\, d^2 = (\pi/4)\,(16 \times 10^{-4}) = 4\pi \times 10^{-4} \text{ m}^2$$

(i) $$L = \frac{N^2 a \mu_0 \mu_r}{l} = \frac{(4000)^2 \times (4\pi \times 10^{-4}) \times 4\pi \times 10^{-7} \times 1}{0.6} = \mathbf{0.0421H}$$

(ii) $$\text{Energy stored} = \frac{1}{2} LI^2 = \frac{1}{2}(0.0421) \times (5)^2 = \mathbf{0.526\ J}$$

Example 9.9. *A coil wound on an iron core of permeability 400 has 150 turns and a cross-sectional area of 5 cm^2. Calculate the inductance of the coil. Given that a steady current of 3mA produces a magnetic field of 10 lines/cm^2 when air is present as the medium.*

Solution.

$$\mu_i = \frac{\text{Flux density in iron}}{\text{Flux density in air}} = \frac{B_i}{10}$$

$$\therefore \quad B_i = 10 \times \mu_i = 10 \times 400 = 4000 \text{ lines/cm}^2$$

Flux produced by 3mA current in the iron core,

$$\phi = B_i \times a = 4000 \times 5 = 20{,}000 \text{ lines} = 2 \times 10^{-4} \text{ wb}$$

$$\therefore \quad L = \frac{N\phi}{I} = \frac{150 \times 2 \times 10^{-4}}{3 \times 10^{-3}} = \mathbf{10H}$$

Example 9.10. *A 300-turn coil has a resistance of 6 Ω and an inductance of 0.5H. Determine the new resistance and new inductance if one-third of the turns are removed. Assume all the turns have the same circumference.*

Solution. Since the resistance of a coil is directly proportional to its length,

$$R_1/R_2 = N_1/N_2 \quad or \quad 6/R_2 = 300/200$$

$$\therefore \quad R_2 = \frac{200}{300} \times 6 = \mathbf{4\Omega}$$

Now $$\frac{L_1}{L_2} = \frac{N_1^2/S}{N_2^2/S}$$

Assuming the reluctance of magnetic path to be constant,

$$\frac{0.5}{L_2} = \frac{(300)^2}{(200)^2} \qquad \therefore \; L_2 = \mathbf{0.22H}$$

Tutorial Problems

1. A coil has self-inductance of 10 *H*. If a current of 200 mA is reduced to zero in a time of 1 ms, find the average value of induced e.m.f. across the terminals of the coil. **[2000 V]**
2. A coil consists of 750 turns and a current of 10A in the coil gives rise to a magnetic flux of 1200 μwb. Calculate the inductance of the coil and determine the average e.m.f. induced in the coil when this current is reversed in 0.01 second. **[0.09 H; 180V]**
3. Calculate the inductance of a solenoid of 2000 turns wound uniformly over a length of 50 cm on a cylindrical paper tube 4 cm in diameter. The medium is air. **[12.62 mH]**
4. An air-cored solenoid 1m long and 10 cm in diameter has 5000 turns. Calculate (*i*) the inductance and (*ii*) energy stored in it when a current of 2A passes through it. **[0.2468 H; 0.4936 J]**
5. A circular iron ring of mean diameter 100 mm and cross-sectional area 500 mm^2 has 200 turns of wire uniformly wound around the circumference. If the relative permeability of iron is assumed to be 1200, find the self-inductance of the coil. **[96 mH]**

9.9. Mutual Inductance (M)

The two coil so arranged that a change of current in one coil causes an e.m.f. to be induced in the other are said to have **mutual inductance.**

Consider two coils *A* and *B* placed adjacent to each other as shown in Fig. 9.8. If current I_1 flows in the coil *A*, a flux is set up and a part ϕ_{12} (*mutual flux*) of this flux links the coil *B*. If current in coil *A* changes, the mutual flux also changes and hence e.m.f. is induced in coil *B*. This e.m.f. induced in coil *B* is termed as mutually induced e.m.f. The converse of this action is also true *i.e.*, a change of current in coil *B* will produce a mutually induced e.m.f. in coil *A*.

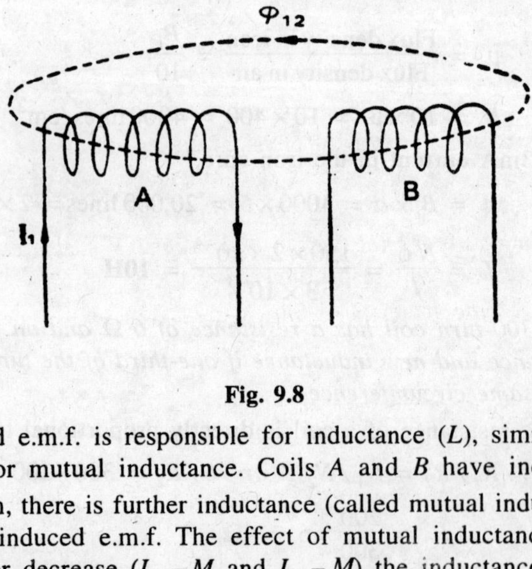

Fig. 9.8

Just as self-induced e.m.f. is responsible for inductance (*L*), similarly mutually induced e.m.f. is responsible for mutual inductance. Coils *A* and *B* have inductance of L_A and L_B respectively. In addition, there is further inductance (called mutual inductance *M*) between the coils due to mutually induced e.m.f. The effect of mutual inductance is to either increase ($L_A + M$ and $L_B + M$) or decrease ($L_A - M$ and $L_B - M$) the inductance of the two coils; this being decided by the arrangement of the coils.

Expression for M

Mutually induced e.m.f. in coil *B* ∝ Rate of change of current in coil *A*

or
$$e_M \propto \frac{dI_1}{dt}$$

or
$$e_M = M\frac{dI_1}{dt} \quad \text{... (i)}$$

where M is a constant called **mutual inductance** between the two coils. The unit of mutual inductance is henry (H). If in exp. (i), e_M = 1volt, dI_1/dt = 1A/sec, then M = 1H.

Hence mutual inductance between two coils is **1 henry** *if current changing at the rate of 1A/sec in one coil induces an e.m.f. of 1V in the other coil.*

9.10. Other Expressions for M

Apart from exp (i) above for M, mutual inductance can also be determined by onc of the following two methods :

(i) **First method**. Fig 9.9 shows two magnetically coupled coils A and B.

$$e_M = M\frac{dI_1}{dt} = \frac{d}{dt}(MI_1)$$

Also
$$e_M = N_2\frac{d\phi_{12}}{dt} = \frac{d}{dt}(N_2\phi_{12})$$

From these two expressions, we have,

$$MI_1 = N_2\phi_{12}$$

or
$$M = \frac{N_2\phi_{12}}{I_1} \quad \text{... (i)}$$

(i)

If $N_2\phi_{12}$ = 1wb-turn and I_1 = 1A, then M = 1H

Hence mutual inductance between two coils is **1 henry** *if a current of 1A flowing in one coil produces flux linkages of 1wb-turn in the other.*

(ii) **Second method.** The mutual inductance between the two coils can be determined in terms of physical dimensions of the magnetic circuit. Fig. 9.9 shows two magnetically coupled coils A and B having N_1 and N_2 turns respectively. Suppose l and 'a' are the length and area of cross-section of the magnetic circuit respectively. Let μ_r be the relative permeability of the material of which the magnetic circuit is composed.

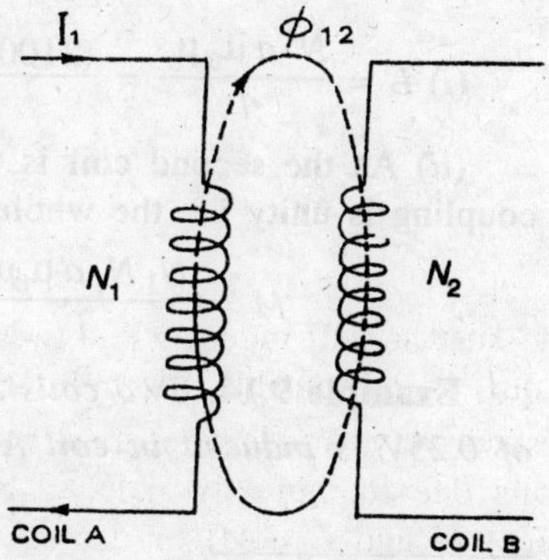

Fig. 9.9

Mutual flux,
$$\phi_{12} = \frac{\text{m.m.f.}}{\text{reluctance}} = \frac{N_1 I_1}{l/a\mu_0\mu_r}$$

or
$$\frac{\phi_{12}}{I_1} = \frac{N_1\, a\, \mu_0\mu_r}{l}$$

Now,
$$M = \frac{N_2\,\phi_{12}}{I_1}$$

$$\therefore \quad M = \frac{N_1 N_2 a \mu_0 \mu_r}{l} \qquad \text{...(ii)}$$

$$= \frac{N_1 N_2}{l/a \mu_0 \mu_r} = \frac{N_1 N_2}{\text{Reluctance (S)}} \qquad \text{... (iii)}$$

It may be noted that mutual inductance is inversely proportional to the reluctance of the magnetic circuit.

Example 9.11. *Two identical coils A and B of 1000 turns each lie in parallel planes such that 80% of flux produced by one coil links with the other. A current of 5A flowing in coil A produces a flux of 0.05 mwb in it. If the current in coil A changes from + 12A to –12A in 0.02 second, calculate (i) the mutual inductance and (ii) the e.m.f. induced in coil B.*

Solution.

$$*\phi_{12} = 0.8 \times 0.05 \times 10^{-3} = 0.4 \times 10^{-4} \text{ wb}$$

(*i*) $$M = \frac{N_2 \phi_{12}}{I_1} = \frac{1000 \times 0.4 \times 10^{-4}}{5} = \mathbf{0.008\ H}$$

(*ii*) $$e_B = M \frac{dI_1}{dt} = 0.008 \times \frac{12 - (-12)}{0.02} = \mathbf{9.6V}$$

Example 9.12. *Two coils A and B having turns 100 and 1000 respectively are wound side by side on a closed iron circuit of cross-sectional area 8 cm² and mean length 80 cm. The relative permeability of iron is 900. Calculate the mutual inductance between the coils. What will be the induced e.m.f. in coil B if current in coil A is increased uniformly from zero to 10A in 0.02 second ?*

Solution.

(*i*) $$M = \frac{N_1 N_2 a \mu_0 \mu_r}{l} = \frac{100 \times 1000 \times 8 \times 10^{-4} \times 4\pi \times 10^{-7} \times 900}{0.8} = \mathbf{0.113\ H}$$

(*ii*) $$e_B = M \frac{dI_1}{dt} = 0.113 \times \frac{10 - 0}{0.02} = \mathbf{56.5V}$$

Example 9.13. *A solenoid 70 cm in length and of 2100 turns has a radius of 4.5 cm. A second coil of 750 turns is wound upon the middle part of the solenoid. Find (i) self-inductance of solenoid (ii) mutual inductance between the two coils.*

Solution. Since μ_r is not given, it will be assumed to be 1.

$$a = \pi r^2 = \pi (4.5 \times 10^{-2})^2 \ m^2$$

(*i*) $$L = \frac{N^2 a \mu_0 \mu_r}{l} = \frac{(2100)^2 \times \pi (4.5 \times 10^{-2})^2 \times 4\pi \times 10^{-7} \times 1}{0.7} = \mathbf{51 \times 10^{-3}\ H}$$

(*ii*) As the second coil is wound on the middle part of the solenoid, the co-efficient of coupling is unity *i.e.* the whole of the flux produced by solenoid links the second coil.

$$M = \frac{N_1 N_2 a \mu_0 \mu_r}{l} = \frac{2100 \times 750 \times \pi (4.5 \times 10^{-2})^2 \times 4\pi \times 10^{-7} \times 1}{0.7} = \mathbf{18.2 \times 10^{-3}\ H}$$

Example 9.14. *Two coils A and B wound side by side on a paper tube former. An e.m.f. of 0.25V is induced in coil A when the flux linking it changes at the rate of 10^{-3} wb/s. A*

---

* Note that 80% of flux produced in coil *A* links with coil *B*. Therefore, mutual flux (ϕ_{12}) is 80% of 0.05 mwb.

current of 2A in coil B causes a flux of 10^{-5}wb to link coil A. What is the mutual inductance between the coils ?

Solution.

Induced e.m.f. in coil $A = N_1 \dfrac{d\phi}{dt}$

or $0.25 = N_1 \times 10^{-3} \quad \therefore \quad N_1 = 0.25/10^{-3} = 250 \text{ turns}$

Flux linkages in coil A due to 2 A in coil B is $= 250 \times 10^{-5}$ *wb–turns*

$$\therefore \quad M = \frac{\text{Flux linkages in coil } A}{\text{Current in coil } B} = 250 \times 10^{-5}/2 = \mathbf{1.25 \times 10^{-3}\ H}$$

9.11. Co-efficient of Coupling

The **co-efficient of coupling** *(k) between two coils is defined as the fraction of magnetic flux produced by the current in one coil that links the other coil.*

The co-efficient coupling has a maximum value of 1 (or 100%) when the entire flux of one coil links the other. If one-half of the flux set up in one coil links the other, then co-efficient of coupling is 0.5 or 50%. Suppose the two coils have inductances of L_1 and L_2 and M is the mutual inductance between them. Then it can be shown that :

$$M = k\sqrt{L_1 L_2}$$

Obviously, M between the two coils will be maximum when $k = 1$

$\therefore$ Maximum mutual inductance between the two coils $= \sqrt{L_1 L_2}$

Example 9.15. *Two magnetically coupled coils have self-inductances, $L_1 = 100\,mH$ and $L_2 = 400\,mH$. If the co-efficient of coupling is 0.8, find the value of mutual inductance between the coils. What would be the maximum possible mutual inductance ?*

Solution

(*i*) $M = k\sqrt{L_1 L_2} = 0.8\sqrt{100 \times 10^{-3} \times 400 \times 10^{-3}} = \mathbf{160 \times 10^{-3}\ H}$

(*ii*) The value of M will be maximum when $k = 1$.

$\therefore$ Maximum $M = \sqrt{L_1 L_2} = \sqrt{100 \times 10^{-3} \times 400 \times 10^{-3}} = \mathbf{200 \times 10^{-3}\ H}$

Example 9.16. *An iron ring of cross-sectional area 800 mm^2 and of mean radius 170 mm has two windings connected in series, one of 500 turns and the other of 700 turns. If the relative permeability of iron is 1200, find (i) the self-inductance of each coil, and (ii) mutual inductance, assuming that there is no leakage.*

Solution.

(*i*) Reluctance, $S = \dfrac{l}{a\mu_0\mu_r} = \dfrac{2\pi \times 170 \times 10^{-3}}{(800 \times 10^{-6}) \times 4\pi \times 10^{-7} \times (1200)}$

$= 8.85 \times 10^5$ AT/ Wb

$L_1 = \dfrac{N_1^2}{S} = \dfrac{(500)^2}{8.85 \times 10^5} = \mathbf{0.282\ H}$ ...See Art. 9.8

$L_2 = \dfrac{N_2^2}{S} = \dfrac{(700)^2}{8.85 \times 10^5} = \mathbf{0.554\ H}$

(*ii*) Since there is no leakage flux, it means that the flux produced by one coil entirely links the other *i.e.* $k = 1$.

$$\therefore \quad M = k\sqrt{L_1 L_2} = 1\sqrt{0.282 \times 0.554} = \mathbf{0.395\ H}$$

Example 9.17. *A battery of 24 V is connected to the primary (coil 1) of a two-winding transformer as shown in Fig. 9.10 and the secondary (coil 2) is open-circuited. The coil parameters are :*

$$R_1 = 10\,\Omega \qquad R_2 = 30\,\Omega$$
$$N_1 = 100 \textit{ turns} \qquad N_2 = 160 \textit{ turns}$$
$$\phi_1 = 0.01 \textit{ wb} \qquad \phi_2 = 0.008 \textit{ wb}$$

Calculate (i) the self-inductance of coil 1 (ii) the mutual inductance (iii) the co-efficient of coupling and (iv) the self-inductance of coil 2.

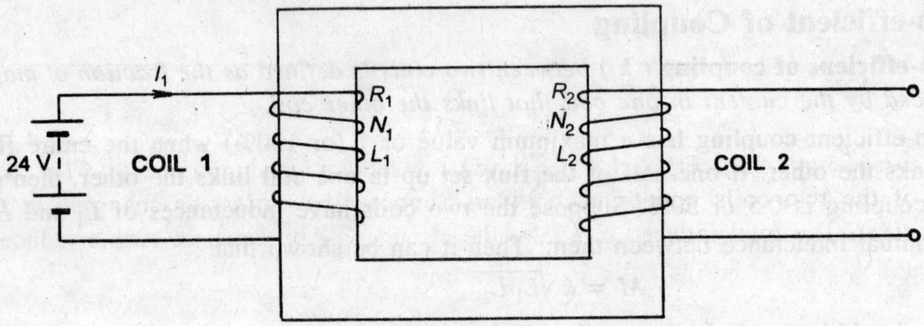

Fig. 9.10

Solution.

(i) $I_1 = V/R_1 = 24/10 = 2.4\text{A}$

$$\therefore \quad L_1 = \frac{N_1 \phi_1}{I_1} = \frac{100 \times 0.01}{2.4} = \mathbf{0.417\ H}$$

(ii) $$M = \frac{N_2 \phi_2}{I_1} = \frac{160 \times 0.008}{2.4} = \mathbf{0.533\ H}$$

(iii) $k = 0.008/0.01 = \mathbf{0.8}$

(iv) $M = k\sqrt{L_1 L_2}$

or $0.533 = 0.8\sqrt{0.417 \times L_2} \quad \therefore \quad L_2 = \mathbf{1.064\ H}$

9.12. Inductances in Series

Consider two coils connected in series as shown in Fig. 9.11.

Let L_1 = inductance of first coil

L_2 = inductance of second coil

M = mutual inductance between the coils

(i) **Series-aiding.** This is the case when the coils are so arranged that their fluxes aid each other *i.e.* in the same direction as shown in Fig. 9.11 *(i)*. Suppose the current is changing at the rate *di/dt*. The total induced e.m.f. in the circuit will be equal to the e.m.fs' induced in L_1 and L_2 *plus* the mutually induced e.m.fs, *i.e.*

$$e = L_1\frac{di}{dt} + L_2\frac{di}{dt} + M\frac{di}{dt} + M\frac{di}{dt} \qquad \textit{... in magnitude}$$
$$= (L_1 + L_2 + 2M)\,di/dt$$

If L_T is the total inductance of the circuit, then,

$$e = L_T \frac{di}{dt}$$

$$\therefore \qquad L_T = L_1 + L_2 + 2M \qquad \text{... fluxes additive}$$

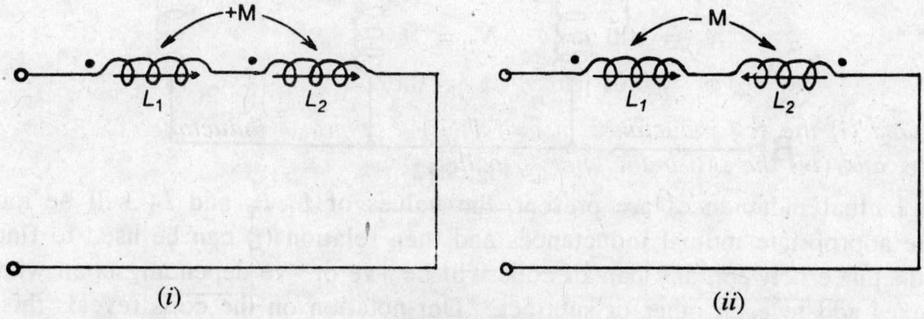

Fig. 9.11

(*ii*) **Series-opposing**. Fig. 9.11. (*ii*) shows the series-opposing connection *i.e.* the fluxes of the two coils oppose each other. Suppose the current is changing at the rate *di/dt*. The total induced e.m.f. in the circuit will be equal to e.m.fs induced in L_1 and L_2 *minus* the mutually induced e.m.fs.

$$e = L_1 \frac{di}{dt} + L_2 \frac{di}{dt} - M\frac{di}{dt} - M\frac{di}{dt} = (L_1 + L_2 - 2M)\frac{di}{dt}$$

If L_T is the total inductance of the circuit, then,

$$e = L_T \frac{di}{dt}$$

$$\therefore \qquad L_T = L_1 + L_2 - 2M \qquad \text{... fluxes subtractive}$$

Note. If the coils are so positioned that $^*M = 0$, then $L_T = L_1 + L_2$

9.13. Inductances in Parallel

Consider three inductances L_1, L_2 and L_3 in parallel as shown in Fig. 9.12. Assume for a moment that mutual inductance between the coils is zero. Referring to Fig. 9.12, we have,

$$i_T = i_1 + i_2 + i_3$$

or $$\frac{di_T}{dt} = \frac{di_1}{dt} + \frac{di_2}{dt} + \frac{di_3}{dt}$$

But $$e = L\frac{di}{dt} \quad \text{or} \quad \frac{di}{dt} = \frac{e}{L}$$

$$\therefore \qquad \frac{e}{L_T} = \frac{e}{L_1} + \frac{e}{L_2} + \frac{e}{L_3}$$

or $$\frac{1}{L_T} = \frac{1}{L_1} + \frac{1}{L_2} + \frac{1}{L_3} \qquad \text{... (i)}$$

---

* If the coils are so placed that fluxes produced by them are at right angles to each other, then mutual flux will be zero and hence $M = 0$

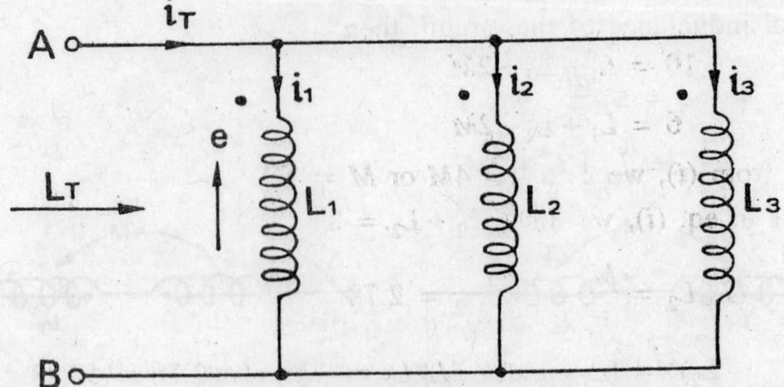

Fig. 9.12

When mutual inductances are present, the values of L_1, L_2 and L_3 will be modified to include the appropriate mutual inductances and then relation (*i*) can be used to find L_T. The mutual inductance between any pair of coils will be +ve or –ve depending upon whether their mutual fluxes add to each other or subtract. *Dot notation on the coils reveals this information. Thus mutual inductance M_{12} between L_1 and L_2 in Fig. 9.12 is positive because their dotted ends are connected to the same terminal *A* of the supply. Similarly mutual inductance M_{13} between L_1 and L_3 is positive. The result is that value of L_1 is modified and becomes $L_1 + M_{12} + M_{13}$. The same procedure can be followed to find the modified values of L_2 and L_3. By inserting the modified values of L_1, L_2 and L_3 in relation (*i*), we can find L_T.

Example 9.18. *Two identical coils, when connected in series, have total inductance of 12H and 4H depending upon their method of connection. Find (i) self-inductance of the coils and (ii) mutual inductance between the coils.*

Solution.

$$L_T = L_1 + L_2 \pm 2M$$

For series-aiding,
$$12 = L + L + 2M \quad \text{... (i)}$$

For series-opposing,
$$4 = L + L - 2M \quad \text{... (ii)}$$

Solving eqs. (*i*) and (*ii*), we get, L = **4H** and M = **2H**

Example 9.19. *Two coils of self-inductances 3H and 2H respectively are connected in series. If the co-efficient of coupling between the coils is 0.5, find the inductance of the circuit when the coils are connected in (i) series-aiding (ii) series-opposing.*

Solution.

$$L_T = L_1 + L_2 \pm 2M$$

$$= L_1 + L_2 \pm 2k\sqrt{L_1 L_2} \qquad (\because M = k\sqrt{L_1 L_2})$$

For series-aiding,
$$L_T = 3 + 2 + 2 \times 0.5\sqrt{3 \times 2} = \mathbf{7.45\ H}$$

For series-opposing,
$$L_T = 3 + 2 - 2 \times 0.5\sqrt{3 \times 2} = \mathbf{2.55\ H}$$

Example 9.20. *When two coils are connected in series, their effective inductance is found to be 10H. When the connections of one coil are reversed, the effective inductance is 6 H. If the co-efficient of coupling is 0.6, calculate the self-inductance of each coil and the mutual inductance.*

---

* **Dot notation:** It is generally not possible to state from the figure whether the fluxes of the two coils are additive or in opposition. Dot notation removes this confusion. The end of the coil through which the current enters is indicated by placing a dot behind it. If the current after leaving the dotted end of coil L_1 enters the dotted end of coil L_2, it means the fluxes of the two coils are additive otherwise in opposition.

Solution.

$$10 = L_1 + L_2 + 2M \quad \text{... }(i)$$

$$6 = L_1 + L_2 - 2M \quad \text{... }(ii)$$

Subtracting (*ii*) from (*i*), we get, $4 = 4M$ or $M =$ **1H**

Putting $M = 1H$ in eq. (*i*), we have, $L_1 + L_2 = 8$ ... (*iii*)

Also $$*L_1 L_2 = \frac{M^2}{k^2} = \frac{1}{(0.6)^2} = 2.78 \quad \text{... }(iv)$$

Now $$(L_1 - L_2)^2 = (L_1 + L_2)^2 - 4L_1 L_2 = (8)^2 - 4 \times 2.78 = 52.88$$

$$\therefore \quad L_1 - L_2 = \sqrt{52.88} = 7.27 \quad \text{... }(v)$$

Solving eq. (iii) and (v), $L_1 =$ **7.635 H** and $L_2 =$ **0.365 H**

Example 9.21. *Two coils of self-inductances 15H and 20H are connected in parallel. If the mutual inductance between the coils is 8H, find the total inductance of the circuit when (i) the mutual fluxes aid each other (ii) the mutual fluxes oppose each other.*

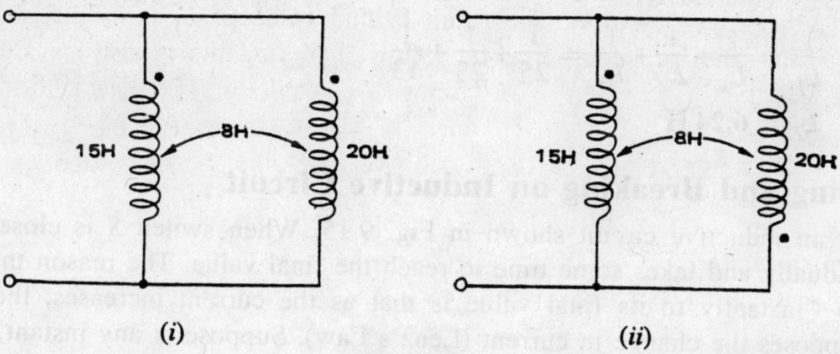

Fig. 9.13

Solution.

(*i*) When the coils are so arranged that their mutual fluxes aid each other [See Fig. 9.13 (*i*)], the mutual inductance between the coils is positive. The net inductance of each branch becomes :

$$L_1 = 15 + 8 = 23 \text{ H}; \quad L_2 = 20 + 8 = 28 \text{ H}$$

$$\therefore \quad \frac{1}{L_T} = \frac{1}{L_1} + \frac{1}{L_2} = \frac{L_1 + L_2}{L_1 L_2}$$

or $$L_T = \frac{L_1 L_2}{L_1 + L_2} = \frac{23 \times 28}{23 + 28} = \textbf{12.63 H}$$

(*ii*) When the coils are so arranged that their mutual fluxes oppose each other [See Fig. 9.13 (*ii*)], the mutual inductance is negative. The net inductance of each branch becomes :

$$L_1 = 15 - 8 = 7 \text{ H}; \quad L_2 = 20 - 8 = 12 \text{ H}$$

$$\therefore \quad L_T = \frac{L_1 L_2}{L_1 + L_2} = \frac{7 \times 12}{7 + 12} = \textbf{4.42 H}$$

Example 9.22. *Three coils of inductances 30H, 25H and 40H are connected in parallel. The mutual inductances among the coils are as shown in Fig. 9.14. Find the total inductance of the circuit.*

---

* Remember $M = k\sqrt{L_1 L_2}$ $\quad \therefore M^2 = k^2 (L_1 L_2)$

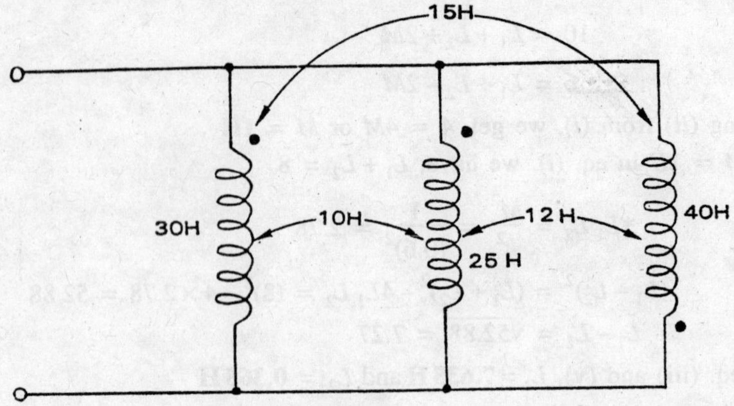

Fig. 9.14

Solution. The net branch inductances are :

$$L_1 = 30 + 10 - 15 = 25 \text{ H}; \quad L_2 = 25 + 10 - 12 = 23 \text{ H}; \quad L_3 = 40 - 12 - 15 = 13 \text{ H}$$

Now $\dfrac{1}{L_T} = \dfrac{1}{L_1} + \dfrac{1}{L_2} + \dfrac{1}{L_3} = \dfrac{1}{25} + \dfrac{1}{23} + \dfrac{1}{13}$

or $L_T = \mathbf{6.24\ H}$

9.14. Closing and Breaking an Inductive Circuit

Consider an inductive circuit shown in Fig. 9.15. When switch *S* is closed, the current increases gradually and takes some time to reach the final value. The reason the current does not build up *instantly to its final value is that as the current increases, the self induced e.m.f. in *L* opposes the change in current (Lenz's Law). Suppose at any instant, the current is *i* and is increasing at the rate of *di/dt*.

Then,

$$V = v_R + v_L$$

$$= iR + L\frac{di}{dt}$$

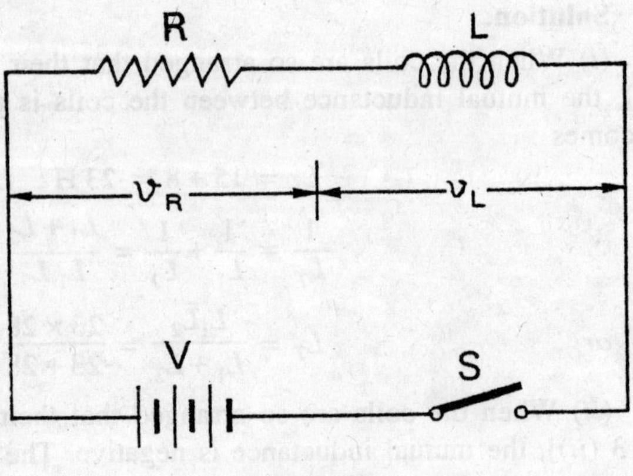

Fig. 9.15

As the current increases, v_R ($= iR$) increases and v_L decreases since *V* is constant. The decrease in v_L ($= L\,di/dt$) means that *di/dt* decreases because *L* is constant. The result is that after some time *di/dt* becomes zero and so does the self-induced e.m.f. v_L ($= L\,di/dt$). At this stage, the current attains the final fixed value *I* given by;

* The current is zero at the instant the switch is closed because it must start from zero.

Now, self-induced e.m.f., $v_L = L\dfrac{di}{dt}$

If current change (*i.e.*, *di*) is instant, it means $di/dt = 0$. This means that *L* is infinite which is impossible. So it is not possible for current in inductance to change from one value to the other in zero time.

$$V = IR + 0$$

or $$I = V/R$$

Thus, when a d.c. circuit containing inductance is switched on, the current takes some time to reach the final value $I (= V/R)$. Note that the role of inductance is to delay the change; it cannot prevent the current from attaining the final value. Similarly, when an inductive circuit is opened, the current does not jump to zero but falls gradually. In either case, the delay in change depends upon the values of L and R as explained in the next article.

9.15. Rise of Current in an Inductive Circuit

Consider an inductive circuit shown in Fig. 9.15. When switch S is closed, the current rises from zero to the final value $I (= V/R)$ in a small time t. Suppose at any instant, the current is i and is increasing at the rate of di/dt. Then,

$$V = iR + L\frac{di}{dt}$$

or $$V - iR = L\frac{di}{dt}$$

or $$\frac{di}{V - iR} = \frac{dt}{L}$$

*Multiplying both sides by $-R$, we get,

$$\frac{-R\,di}{V - iR} = \frac{-R}{L}dt$$

Integrating both sides, we get,

$$\int -\frac{R\,di}{V - iR} = -\frac{R}{L}\int dt$$

or $$\log_e (V - iR) = -\frac{R}{L}t + K \qquad \text{... (i)}$$

where K is a constant whose value can be determined from the initial conditions. At $t = 0$, $i = 0$. Putting these values in exp (*i*), we have, $\log_e V = K$

$\therefore$ Equation (*i*) becomes : $$\log_e (V - iR) = -\frac{R}{L}t + \log_e V$$

or $$\log_e \frac{V - iR}{V} = -\frac{R}{L}t$$

or $$\frac{V - iR}{V} = e^{-Rt/L}$$

or $$V - iR = Ve^{-Rt/L}$$

or $$i = \frac{V}{R}(1 - e^{-Rt/L})$$

But $V/R = I$, the final value of current attained by the circuit.

$\therefore$ $$i = I(1 - e^{-Rt/L}) \qquad \text{... (ii)}$$

---

* This step makes the numerator on the L.H.S. a differential of the denominator.

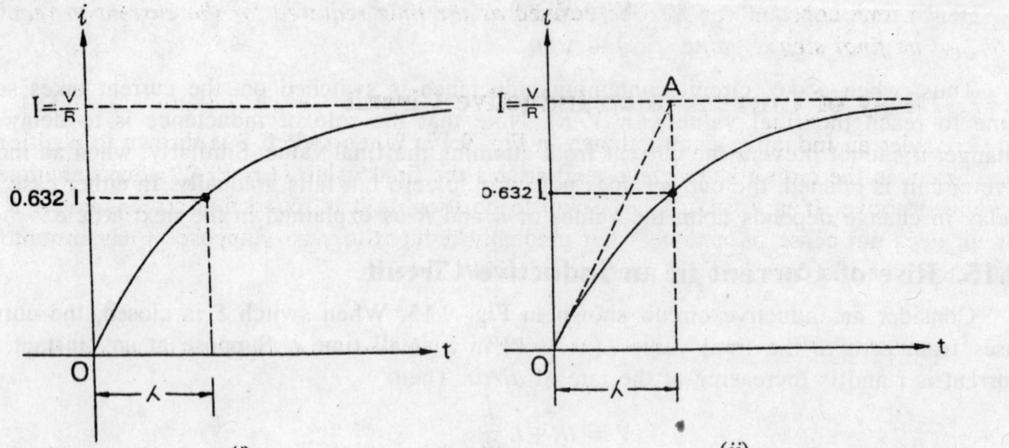

Fig. 9.16

Eq. (*ii*) shows that rise of current follows an exponential law (See Fig. 9.16). As t increases, the term $e^{-Rt/L}$ gets smaller and current i in the circuit gets larger. Theoretically, the current will reach its final value $I\,(= V/R)$ in an infinite time. However, practically it reaches this value in a short time.

Note. $V = iR + L\,di/dt$

At the instant the switch is closed, $i = 0$ $\quad \therefore \quad V = L\,di/dt$

Initial rate of rise of current, $\dfrac{di}{dt} = \dfrac{V}{L}$ A/sec.

9.16. Time Constant

Consider the eq (*ii*) above showing the rise of current *w.r.t* time *t*.

$$i = I\,(1 - e^{-Rt/L})$$

The exponent of e is Rt/L. The quantity L/R has the dimensions of time so that exponent of e (*i.e.* Rt/L) is a number. The quantity L/R is called the *time constant* of the circuit and affects the rise of current in the circuit. It is represented by λ.

$\therefore$ Time constant, $\lambda = L/R$ seconds

Time constant of an inductive circuit can be defined in the following ways :

(*i*) Consider the graph showing the rise of current *w.r.t.* time t [See Fig. 9.16 (*ii*)]. The initial rate of rise of current (*i.e.* at $t = 0$) in the circuit is

$$\frac{di}{dt} = \frac{V}{L}$$

If this rate of rise of current were maintained, the graph would be linear [*i.e.* OA in Fig. 9.16 (*ii*)] instead of exponential. If this rate of rise could continue, the circuit current will reach the final value $I\,(= V/R)$ in time

$$= \frac{V}{R} \div \frac{V}{L}$$

$$= L/R = \text{Time constant } \lambda$$

Hence time constant may be defined *as the time required for the current to rise to is final steady value if it continued rising at its initial rate (i.e. V/L).*

(*ii*) If time interval, $t = \lambda$ (or L/R), then

$$i = I\,(1 - e^{-Rt/L}) = I\,(1 - e^{-1}) = 0.632I$$

Hence time constant can also be defined *as the time required for the current to reach 0.632 of its final steady value.*

9.17. Decay of Current in an Inductive Circuit

Consider an inductive circuit shown in Fig. 9.17. When switch S is thrown to position 2, the current in the circuit starts rising and attains the final value $I\,(= V/R)$ after sometime as explained above. If now switch is thrown to position 1, it is found that current in the $R-L$ circuit does not cease immediately but gradually reduces to zero. Suppose at any instant, the current is i and is decreasing at the rate of di/dt. Then,

$$0 = iR + L\frac{di}{dt}$$

or $$\frac{di}{i} = -\frac{R}{L}\,dt$$

Integrating both sides, we get, $\log_e i = -\frac{R}{L}t + K$ ...(*i*)

where K is a constant whose value can be determined from the initial conditions. When $t = 0$, then $i = I\,(= V/R)$.

Putting these values in eq (*i*), we have,

$$\log_e I = 0 + K \quad or \quad K = \log_e I$$

$\therefore$ Equation (*i*) becomes : $\log_e i = -\frac{R}{L}t + \log_e I$

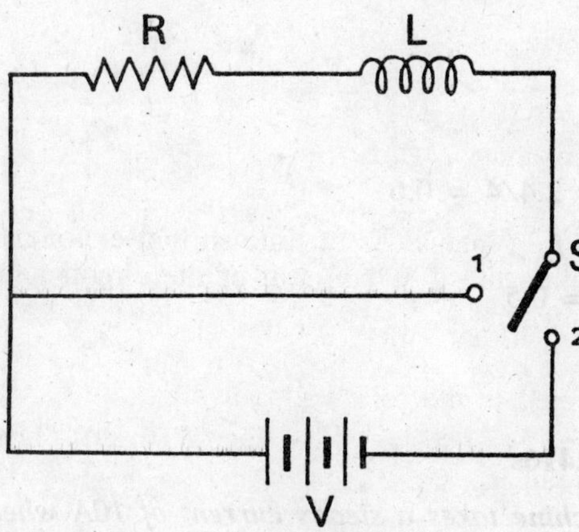

Fig. 9.17

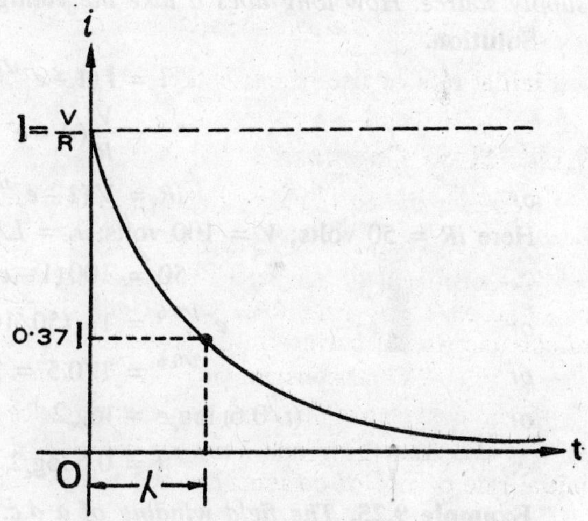

Fig. 9.18

or $$\log_e \frac{i}{I} = -\frac{R}{L}t$$

or $$\frac{i}{I} = e^{-Rt/L}$$

or $$i = Ie^{-Rt/L}$$ ... (*ii*)

Eq. (*ii*) gives the decay of current in an $R-L$ series circuit with time t and is represented graphically in Fig. 9.18. Note that decay of current follows the exponential law.

Time constant. The quantity L/R in eq. (*ii*) is known as time constant of the circuit. When $t = \lambda\,(= L/R)$,

$$i = I\,e^{-1} = 0.37\,I$$

Hence, time constant may also be defined *as the time taken by the current to fall to 0.37 of its final steady value ($I = V/R$) while decaying.*

Example 9.23. *A coil having a resistance of 15 Ω and inductance of 10H is connected to a constant 75V supply. Find (i) the rate of change of current at the instant of closing the switch (ii) final steady value (iii) time constant of the circuit (iv) the time taken for the current to reach a value of 4A.*

Solution.

(*i*) Initial rate of rise of current

$$\frac{di}{dt} = \frac{V}{L} = \frac{75}{10} = \mathbf{7.5A/s}$$

(*ii*) Final steady value, $I = V/R = 75/15 = \mathbf{5A}$

(*iii*) Time constant, $\lambda = L/R = 10/15 = \mathbf{0.67s}$

(*iv*) $i = I(1 - e^{-t/\lambda})$

$$4 = 5(1 - e^{-t/0.67})$$

or $e^{-t/0.67} = 1 - (4/5) = 0.2$

or $e^{t/0.67} = 1/0.2 = 5$

or $(t/0.67) \log_e e = \log_e 5$

$\therefore$ $t = 0.67 \log_e 5 = \mathbf{1.078s}$

Example 9.24. *A coil having L = 2.4H and R = 4Ω is connected to a constant 100V supply source. How long does it take the voltage across the resistance to reach 50V ?*

Solution.

$$i = I(1 - e^{-t/\lambda})$$

$$= \frac{V}{R}(1 - e^{-t/\lambda})$$

or $iR = V(1 - e^{-t/\lambda})$

Here iR = 50 volts; V = 100 volts; $\lambda = L/R = 2.4/4 = 0.6$

$\therefore$ $50 = 100(1 - e^{-t/0.6})$

or $e^{-t/0.6} = 1 - (50/100) = 0.5$

or $e^{t/0.6} = 1/0.5 = 2$

or $(t/0.6) \log_e e = \log_e 2$

$\therefore$ $t = 0.6 \log_e 2 = \mathbf{0.416s}$

Example 9.25. *The field winding of a d.c. machine takes a steady current of 10A when connected to a d.c. source of 230 volts. On switching the voltage to the field, it is observed that it takes 0.3 second to reach 5A.*

(*i*) *Find the value of inductance of the field winding.*

(*ii*) *Find also the time for the current to fall to 30% of the final steady value if the field terminals are short-circuited.*

Solution.

(*i*) **Growth of current**

$$i = I(1 - e^{-t/\lambda})$$

Here i = 5 A; I = 10 A; $R = 230/10 = 23\Omega$; t = 0.3 s

$\therefore$ $5 = 10(1 - e^{-0.3/\lambda})$

or $e^{-0.3/\lambda} = 1 - (5/10) = 0.5$

or $$e^{0.3/\lambda} = 1/0.5 = 2$$

or $$(0.3/\lambda)\log_e e = \log_e 2$$

$\therefore$ $$\lambda = \frac{0.3 \log_e e}{\log_e 2} = 0.433 \text{ s}$$

Now, $$\lambda = L/R \text{ or } L = \lambda R = 0.433 \times 23 = \mathbf{9.96\ H}$$

(*ii*) **Decay of current**

$$i = Ie^{-t/\lambda}$$

Here $i = 30\%$ of $10\text{ A} = 3\text{ A}$; $\lambda = 0.433\text{ s}$

$\therefore$ $$3 = 10e^{-t/0.433}$$

or $$e^{t/0.433} = 10/3 = 3.33$$

or $$(t/0.433)\log_e e = \log_e 3.33$$

$\therefore$ $$t = \frac{0.433 \log_e 3.33}{\log_e e} = \mathbf{0.521s}$$

Example 9.26. *A coil has a time constant of 0.1s and an inductance of 10 H. Find (i) the value of current 0.1s after switching to a steady p.d. of 100 v (ii) the time taken for the current to reach half its steady-state value.*

Solution.

$$\lambda = L/R = 1 \quad \therefore \; R = L = 10\,\Omega$$

(*i*) $$i = I(1 - e^{-t/\lambda})$$

Here $$I = V/R = 100/10 = 10\text{ A};\; t = 0.1\text{ s};\; \lambda = 1\text{ s}$$

$\therefore$ $$i = 10(1 - e^{-0.1/1}) = \mathbf{0.952A}$$

(*ii*) $$i = I(1 - e^{-t/\lambda})$$

Here $$i = 5\text{ A}, I = 10\text{ A}, \lambda = 1\text{ s}$$

$\therefore$ $$5 = 10(1 - e^{-t/1})$$

or $$e^{-t} = 1 - (5/10) = 0.5$$

or $$e^{t} = 1/0.5 = 2$$

or $$t \log_e e = \log_e 2$$

$\therefore$ $$t = \log_e 2/\log_e e = \mathbf{0.693s}$$

Example 9.27. *A coil of resistance 200 Ω and inductance 10H is connected to a battery of 100V. Find (i) the current in the coil 40 ms after closing the switch (ii) the rate at which the current will then be increasing.*

Solution.

$$I = 100/200 = 0.5\text{ A};\; \lambda = L/R = 10/200 = 0.05\text{ s};\; t = 0.04\text{ s}$$

(*i*) $$i = I(1 - e^{-t/\lambda}) = 0.5(1 - e^{-0.04/0.05}) = 0.5(1 - e^{-0.8}) = \mathbf{0.275A}$$

(*ii*) At the instant the current is 0.275A,

$$V = iR + L\frac{di}{dt}$$

$\therefore$ $$\frac{di}{dt} = \frac{V - iR}{L} = \frac{100 - 0.275 \times 200}{10} = \mathbf{4.5A/s}$$

Note. Back e.m.f. at this instant $= L\,di/dt = 10 \times 4.5 = 45V$

Example 9.28. *The time constant of a certain inductive coil was found to be 2.5 ms. With a resistance of 80 Ω added in series, a new time constant of 0.5 ms was obtained. Find the inductance and resistance of the coil.*

Solution. Time constant, $\lambda = L/R$

For the first case, $L/R = 2.5$

For the second case, $L/(R+80) = 0.5$

$$\therefore \quad \frac{R+80}{R} = \frac{2.5}{0.5} = 5$$

or $R+80 = 5R \quad \therefore \quad R = \mathbf{20\Omega}$

Now $L/R = 2.5 \quad \therefore \quad L = 2.5\,R = 2.5 \times 20 = \mathbf{50\,H}$

Tutorial Problems

1. An inductive coil of inductance 1H and resistance 100 Ω is supplied from a 50 V d.c. source of internal resistance 5Ω. Calculate :
   (*i*) the time constant of the circuit
   (*ii*) the time elapsed from closing the circuit at which the circuit current has reached 35% of the steady state value. **[(*i*) 9.52 ms (*ii*) 4.1 ms]**
2. The current in a d.c. circuit composed of a 10 H coil and 100 Ω resistor in series reached 40% of the maximum value 20 ms after closing the circuit. Calculate the coil resistance. **[255.4 Ω]**
3. A coil has a time constant of 1s and an inductance of 10H. What is the value of the current 0.1s after switching on to a d.c. potential of 100V ? Find also the time taken for the current to reach half its steady state value. **[(*i*) 0.952 A (*ii*) 0.697s]**
4. A coil of inductance 5H and resistance 100Ω carries a steady current of 2A. Calculate the initial rate of fall of current in the coil after a short circuiting switch connected across its terminals has been suddenly closed. What was the energy stored in the coil and in what form was it dissipated ? **[(*i*) -40A/s (*ii*) 10 J, heat]**
5. A constant voltage is applied to *R-L* series circuit at $t = 0$ by closing a switch. The voltage across *L* is 25 V at $t = 0$ and drops to 5*V* at $t = 0.025$ second. If $L = 2$ H, find (*i*) the applied voltage and (*ii*) the value of *R*. **[(*i*) 25V (*ii*) 128.56 Ω]**
6. Determine the value of resistor which when connected in series with a 10H, 200Ω coil will allow the current to change by 50% of the maximum value in 20 ms from the instant of applying the supply voltage. **[146.6 Ω]**

9.18. Energy Stored in a Magnetic Field

In order to establish a magnetic field around a coil, energy is *required, though no energy is needed to ***maintain* it. This energy is stored in the magnetic field and is not used up. When the current is decreased, the flux surrounding the coil is decreased, causing the stored energy to be returned to the circuit. Consider an inductor connected to a d.c. source as shown in Fig. 9.19 (*i*). The inductor is equivalent to inductance *L* in series with a small resistance *R* as shown in Fig. 9.19 (*ii*). The energy supplied to the circuit is spent in two ways :

(*i*) A part of supplied energy is spent to meet $I^2 R$ losses and cannot be recovered.

---

* When the coil is connected to supply, current increases from zero gradually and reaches the final value $I\,(= V/R)$ after some time. During this change of current, an e.m.f. is induced in *L* due to the change in flux linkages. This induced e.m.f. opposes the rise of current. Electrical energy must be supplied to meet this opposition. This supplied energy is stored in the magnetic field.

** To impart a kinetic energy of $\frac{1}{2}mv^2$ to a body, energy is required but no energy is required to maintain it at that energy level.

(*ii*) The remaining part is spent to create flux around the coil (or inductor) and is stored in the magnetic field. When the field collapses, the stored energy is returned to the circuit.

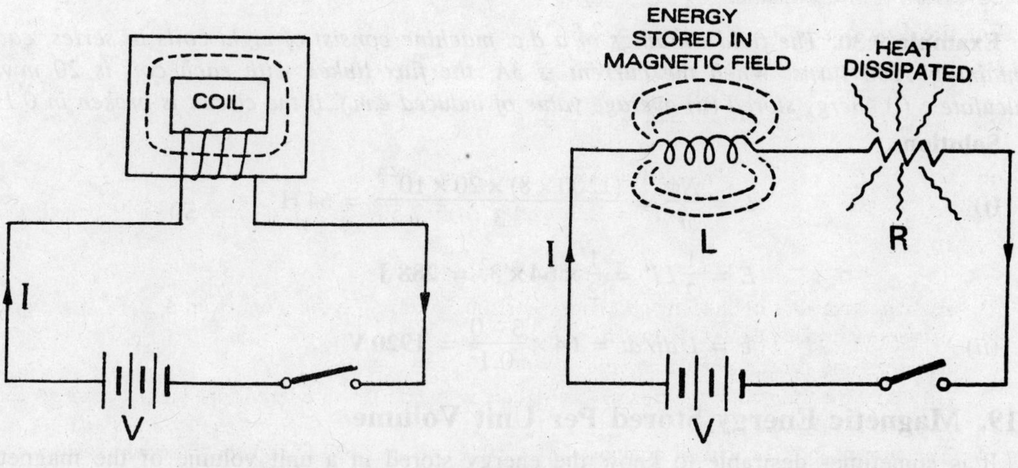

(*i*) (*ii*)

Fig. 9.19

Mathematical Expression. Suppose at any instant the current in the coil is i and is increasing at the rate of di/dt. Then e.m.f. e across L is given by ;

$$e = L\frac{di}{dt}$$

$\therefore$ Instantaneous power, $$p = ei = Li\frac{di}{dt}$$

During a short interval of time dt, the energy dw put into the magnetic field is equal to power multiplied by time *i.e.*

$$dw = p.dt = \left(Li\frac{di}{dt}\right)dt = Li\,di$$

The total energy put into the magnetic field from the time current is zero until it has attained the final steady value I is :

$$W = \int_0^I Li\,di = \frac{1}{2}LI^2$$

$\therefore$ Energy stored in magnetic field $= \frac{1}{2}LI^2$ joules

Note that energy stored will be in joules if inductance (L) and current (I) are in henry and amperes respectively.

Note. If current in an inductor varies, the stored energy rises and falls in step with the current. Thus, whenever current increases, the coil absorbs energy and whenever current falls, energy is returned to the circuit.

Example 9.29. *Calculate the inductance and energy stored in the magnetic field of an air-cored solenoid 50cm long, 5cm in diameter and wound with 1000 turns, if carrying a current of 5A.*

Solution.

$$a = (\pi/4)\,(5 \times 10^{-2})^2 = 19.63 \times 10^{-4}\ m^2$$

$$L = \frac{N^2 a \mu_0 \mu_r}{l} = \frac{(1000)^2 \times (19.63 \times 10^{-4}) \times 4\pi \times 10^{-7} \times 1}{0.5} = \mathbf{0.005\ H}$$

$$E = \frac{1}{2} LI^2 = \frac{1}{2} \times 0.005 \times 5^2 = \mathbf{0.0625\ J}$$

Example 9.30. *The field wnidings of a d.c. machine consist of eight coils in series, each containing 1200 turns. When the current is 3A, the flux linked with each coil is 20 mwb. Calculate : (i) energy stored (ii) average value of induced e.m.f. if the circuit is broken in 0.1s*

Solution.

(*i*)
$$L = \frac{N\phi}{I} = \frac{(1200 \times 8) \times 20 \times 10^{-3}}{3} = 64\ \text{H}$$

$$E = \frac{1}{2} LI^2 = \frac{1}{2} \times 64 \times 3^2 = \mathbf{288\ J}$$

(*ii*)
$$e = L\, di/dt = 64 \times \frac{3-0}{0.1} = \mathbf{1920\ V}$$

9.19. Magnetic Energy Stored Per Unit Volume

It is sometimes desirable to know the energy stored in a unit volume of the magnetic field. Consider a coil of N turns wound over a magnetic circuit of length l metres and uniform X-sectional area of $a\ m^2$.

$$\text{Energy stored} = \frac{1}{2} LI^2 = \frac{1}{2} \left(\frac{N^2 a \mu_0 \mu_r}{l} \right) I^2$$

$$= \frac{1}{2} (a \mu_0 \mu_r)\, l \left(\frac{NI}{l} \right)^2$$

$$= \frac{1}{2} (\mu_0 \mu_r)(al) H^2 \qquad (\because H = NI/l)$$

Now al = Volume of magnetic field in m^3

$$\therefore \quad \text{Energy stored}/m^3 = \frac{1}{2} \mu_0 \mu_r H^2$$

$$= \frac{1}{2} \mu_0 \mu_r \left(\frac{B}{\mu_0 \mu_r} \right)^2 \qquad \left[\because H = \frac{B}{\mu_0 \mu_r} \right]$$

or
$$\text{Energy stored}/m^3 = \frac{B^2}{2 \mu_0 \mu_r} \text{ Joules} \qquad \text{.... in a medium}$$

$$= \frac{B^2}{2\mu_0} \text{ Joules} \qquad \text{..... in air}$$

Note that energy stored will be in joules if flux density B is in wb/m^2.

9.20. Lifting Power of a Magnet

Consider two poles, north and south, each of area 'a' square metres as shown in Fig. 9.20. Let P newtons be the force of attraction between the two poles. If one of the poles, say S-pole, is pulled apart through a small distance dx, then work will have to be done against the force of attraction.

$$\text{Work done} = P \times dx \text{ Joules}$$

This work done is stored in the additional volume of the magnetic field created. Additional volume of magnetic field created

$$= a \times dx \text{ m}^3$$

$$\therefore \text{ Increase in stored energy} = \frac{B^2}{2\mu_0} \times a\,dx$$

Now, increase in stored energy = Work done

$$\therefore \quad \frac{B^2}{2\mu_0} \times a\,dx = P \times dx$$

$$or \quad P = \frac{B^2 a}{2\mu_0} \text{ newtons}$$

It may be noted that P will be in newtons if B is in wb/m^2 and 'a' in m^2.

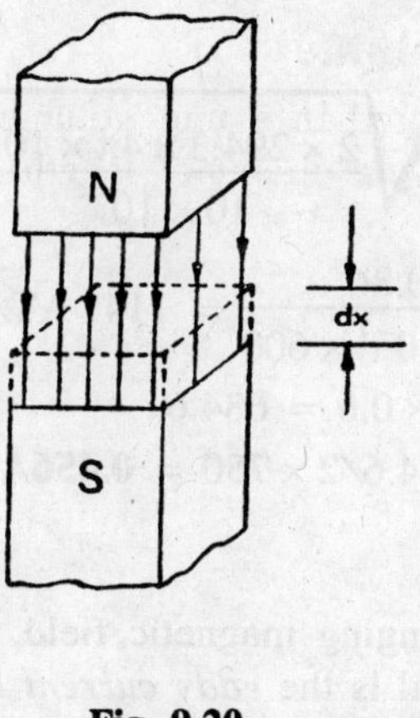

Fig. 9.20

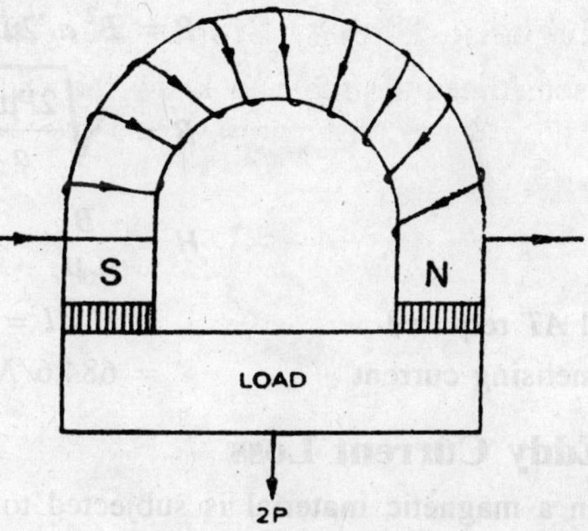

Fig. 9.21

Note that P is the force of attraction at each pole. In a practical magnet, there are two poles [See Fig. 9.21] so that the total force of attraction is 2 P. Electromagnets are widely used for commercial lifting jobs such as loading scrap iron into a truck or raising an armature to a higher position.

Example 9.31. *In a telephone receiver, the size of each pole of the electromagnet is 1.2 cm × 0.2 cm and the flux between each pole and the diaphragm is 4×10^{-6} wb. With what force is the diaphragm attracted towards the poles ?*

Solution.

$$a = 1.2 \times 0.2 = 0.24 \text{ cm}^2 = 0.24 \times 10^{-4} \text{ m}^2$$

$$B = \frac{\phi}{a} = \frac{4 \times 10^{-6}}{0.24 \times 10^{-4}} = 0.167 \text{ wb/m}^2$$

$$P = \frac{B^2 a}{2\mu_0} = \frac{(0.167)^2 \times 0.24 \times 10^{-4}}{2 \times 4\pi \times 10^{-7}} = 0.266 \text{ newtons}$$

$$\text{Total force} = 2P = 2 \times 0.266 = \mathbf{0.532\ newtons}$$

Example 9.32. *A lifting magnet of inverted U-shape is formed out of an iron bar 60 cm long and 10 cm^2 in cross-sectional area. Exciting coils of 750 turns each are wound on the two side limbs and are connected in series. Calculate the exciting current necessary for the* *magnet to lift a load of 60 kg, assuming that the load has negligible reluctance and makes close contact with the magnet. Relative permeability of the material of magnet is 600.*

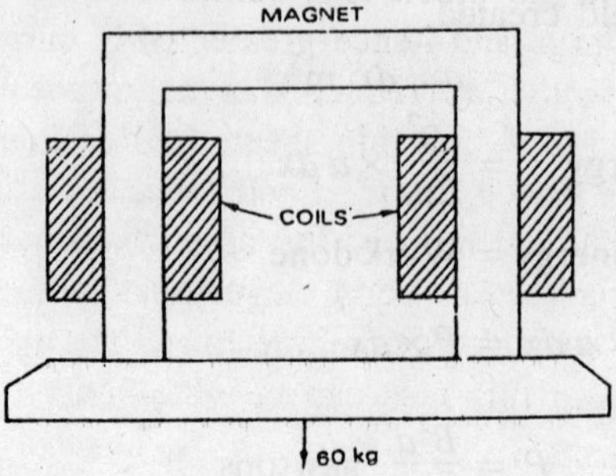

Fig. 9.22

Solution.

$$P = 60 \times 9.81/2 = 294.3$$

Now $P = B^2 a/2\mu_0$

$$\therefore \quad B = \sqrt{\frac{2P\mu_0}{a}} = \sqrt{\frac{2 \times 294.3 \times 4\pi \times 10^{-7}}{10 \times 10^{-4}}} = 0.86\ wb/m^2$$

$$H = \frac{B}{\mu_0 \mu_r} = \frac{0.86}{4\pi \times 10^{-7} \times 600} = 1141\ \text{AT/m}$$

Total AT required $= H \times l = 1141 \times 0.6 = 684.6$

Magnetising current $= 684.6/N = 684.6/2 \times 750 = \mathbf{0.456A}$

9.21. Eddy Current Loss

When a magnetic material is subjected to a changing magnetic field, in addition to the hysteresis loss, another loss that occurs in the material is the *eddy current loss*. The changing flux induces voltages in the material according to Faraday's laws of electromagnetic induction. Since the material is conducting, these induced voltages circulate currents within the body of the material. These induced currents do no useful work and are known as eddy currents. These eddy currents develop i^2R loss in the material. Like hysteresis loss, the eddy current loss also results in the rise of temperature of the material. The hysteresis and eddy current losses in a magnetic material are sometimes called core losses or iron losses.

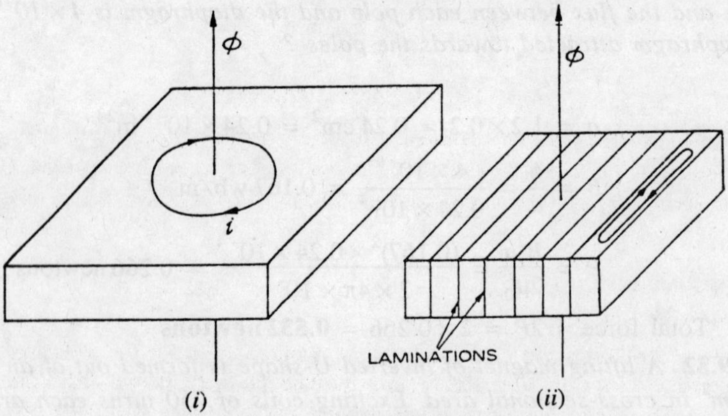

Fig. 9.23

Fig. 9.23 (*i*) shows a solid block of iron subjected to a changing magnetic field. The eddy current power loss in the block will be i^2R where i is the eddy current and R is the resistance of the eddy current path. Since the block is a continuous iron piece of large *x*-section, the magnitude of i will be very *large and hence greater eddy current loss will result. The obvious method of reducing this loss is to reduce the magnitude of eddy current. This can be achieved by splitting the solid block into thin sheets (called *laminations*) in planes parallel to the magnetic flux as shown in Fig. 9.23 (*ii*). Each lamination is insulated from the other by a layer of varnish. This arrangement reduces the area of each section and hence the induced e.m.f. It also increases the resistance of eddy current paths since the area through which the currents can pass is smaller. Both effects combine to reduce the eddy current and hence eddy current loss. Further reduction in this loss can be obtained by using a magnetic material of high resistivity (*e.g.* **silicon steel).

9.22. Formula for Eddy Current Loss

It is difficult to determine the eddy current power loss because the current and resistance values cannot be determined directly. Experiments have shown that eddy current power loss in a magnetic material can be expressed as :

Eddy current power loss, $P_e = k_e B_m^2 t^2 f^2 v$ watts

where k_e = eddy-current co-efficient and its value depends upon the nature of the material

B_m = maximum flux density in wb/m²

t = thickness of lamination in m

f = frequency of flux in Hz

v = volume of material in m³

Example 9.33. *The flux in a magnetic core is alternating sinusoidally at 50Hz. The maximum flux density is 1.5wb/m². The eddy current loss then amounts to 140 watts. Find the eddy current loss in the core when the frequency is 75Hz and the flux density 1.2 wb/m².*

Solution.

Eddy current power loss, $P_e \propto B_m^2 f^2$

For the first case, $P_{e1} \propto (1.5)^2 \times (50)^2$

For the second case, $P_{e2} \propto (1.2)^2 \times (75)^2$

$$\therefore \quad \frac{P_{e2}}{P_{e1}} = \left(\frac{1.2}{1.5}\right)^2 \times \left(\frac{75}{50}\right)^2 = 1.44$$

or $P_{e2} = 140 \times 1.44 = \mathbf{201.6\ W}$

Example 9.34. *A transformer connected to 25Hz supply has a core loss of 1500 watts of which 1000 watts are due to hysteresis and 500 watts due to eddy currents. If the flux density is kept constant and frequency is increased to 50 Hz, find the new value of the core loss.*

Solution.

Hysteresis power loss, $P_h \propto B_m^{1.6} f$

---

* The large area of the block will have greater e.m.f. induced in it. Larger *x*-section also means smaller resistance to eddy current path. Both these effects increase the magnitude of eddy current to a great extent.

** Alloying iron with silicon increases resistivity and has a further beneficial effect in reducing the hysteresis loss.

Eddy current power loss, $P_e \propto B_m^2 f^2$

Hysteresis loss

$$\frac{P_{h2}}{P_{h1}} = \left(\frac{B_{m2}}{B_{m1}}\right)^{1.6} \times \frac{f_2}{f_1} = (1)^{1.6} \times 50/25 = 2$$

$$\therefore \quad P_{h2} = 2 \times 1000 = 2000 \text{ W}$$

Eddy current loss

$$\frac{P_{e2}}{P_{e1}} = \left(\frac{B_{m2}}{B_{m1}}\right)^2 \times \left(\frac{f_2}{f_1}\right)^2 = (1)^2 \times \left(\frac{50}{25}\right)^2 = 4$$

$$\therefore \quad P_{e2} = 4\, P_{e1} = 4 \times 500 = 2000 \text{ W}$$

$\therefore$ New core loss $= P_{h2} + P_{e2} = 2000 + 2000 =$ **4000W**

OBJECTIVE QUESTIONS

1. The basic requirement for inducing *e.m.f.* in a coil is that ....
   (*i*) flux should link the coil
   (*ii*) there should be change in flux linking the coil.
   (*iii*) coil should form a closed loop
   (*iv*) none of the above
2. The e.m.f. induced in a coil is ....the rate of change in flux linkages.
   (*i*) directly proportional to
   (*ii*) inversely proportional to
   (*iii*) independent of
   (*iv*) none of the above
3. The e.m.f. induced in a coil of N turns is given by ....
   (*i*) $d\phi/dt$ (*ii*) $N d\phi/dt$
   (*iii*) $-N d\phi/dt$ (*iv*) $N\, dt/d\phi$
4. The direction of induced *e.m.f.* in a conductor (or coil) can be determined by ....
   (*i*) work law (*ii*) Ampere's law
   (*iii*) Fleming's right hand rule
   (*iv*) Fleming's left hand rule
5. In Fig. 9.24, the conductor is moving upward. The direction of induced *e.m.f.* is ....

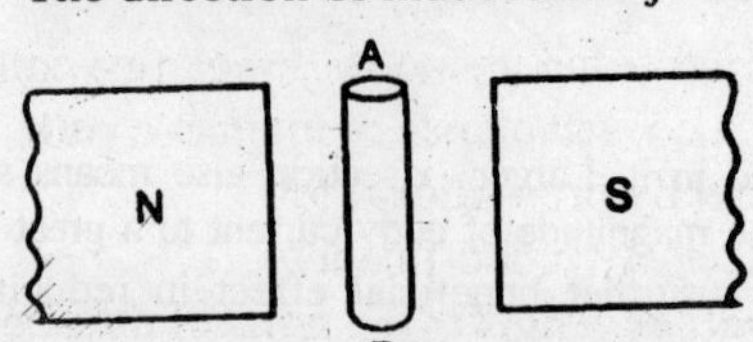

Fig. 9.24

   (*i*) from *A* to *B* (*ii*) from *B* to *A*
   (*iii*) none of the above
6. In Fig. 9.25, the direction of induced *e.m.f.* in the conductor A is ....

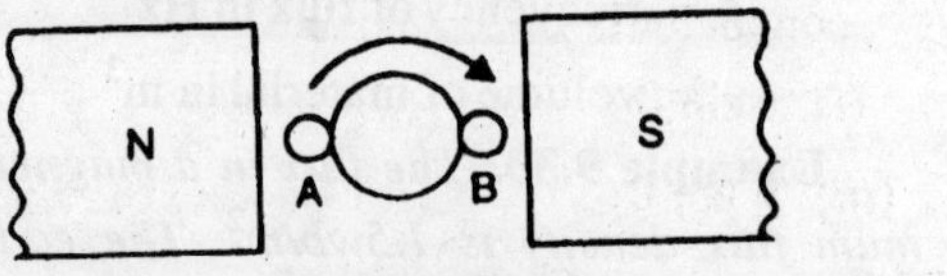

Fig. 9.25

   (*i*) into the plane of paper
   (*ii*) out of plane of paper
   (*iii*) none of the above
7. In Fig. 9.25, the rate of change of flux linkages of conductors *A* and *B* is ....
   (*i*) minimum (*ii*) maximum
   (*iii*) mid-way between *(a)* and *(b)*
   (*iv*) none of the above
8. The *e.m.f.* induced in a ....is the statically induced *e.m.f.*
   (*i*) d.c. generator
   (*ii*) transformer
   (*iii*) d.c. motor
   (*iv*) none of the above
9. The *e.m.f.* induced in a .... is dynamically induced *e.m.f.*
   (*i*) alternator
   (*ii*) transformer
   (*iii*) d.c. generator
   (*iv*) none of the above

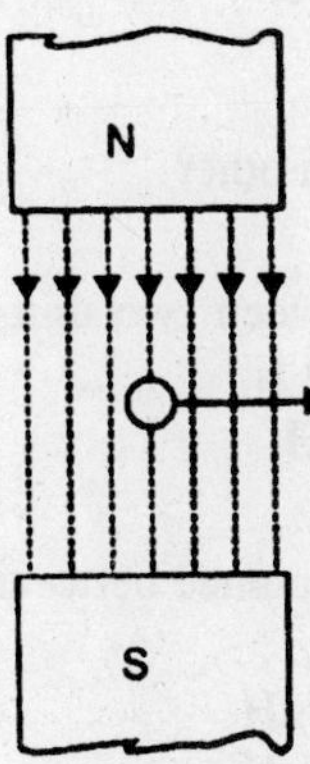

Fig. 9.26

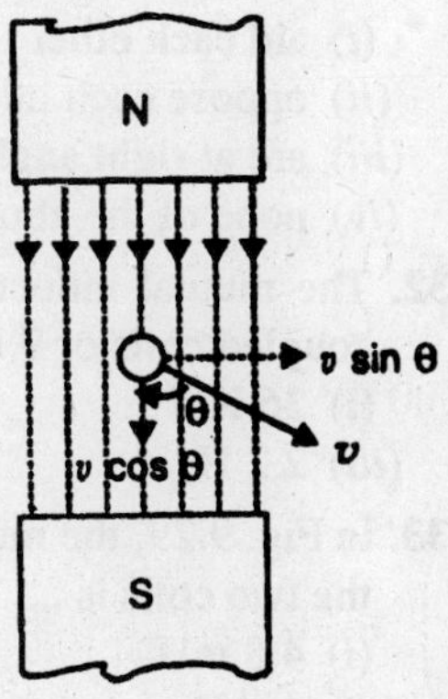

Fig. 9.27

10. In Fig. 9.26, 1 single conductor of length *l* metres moves at right angles to a uniform field of *B* wb/m^2 with a velocity fo *v* m/s. The *e.m.f.* induced is ....
(*i*) *B l/v* (*ii*) *B v/l*
(*iii*) *B/v* (*iv*) *lv/B*

11. In Fig. 9.27, the component of velocity that does not induce any *e.m.f.* in the conductor is ....
(*i*) *v* sin θ (*ii*) *v* cos θ
(*iii*) *v* tan θ (*iv*) none of the above

12. Inductance opposes .... in current in a circuit.
(*i*) only increase (*ii*) only decrease
(*iii*) change (*iv*) none of the above

13. If the number of turns of a coil is increased, its inductance ....
(*i*) remains the same (*ii*) is increased
(*iii*) is decreased (*iv*) none of theabove

14. If the relative permeability of the material surrounding the coil is increased, the inductance of the coil ....
(*i*) is increased (*ii*) is decreased
(*iii*) remains unchanged
(*iv*) none of the above

15. Inductance in a circuit ....
(*i*) prevents the currents from changing
(*ii*) delays the change in current
(*iii*) causes power loss
(*iv*) causes the current to lead the voltage

16. The inductance of a coil is ....the reluctance of magnetic path.
(*i*) independent of
(*ii*) directly proportional to
(*iii*) inversely proportional to
(*iv*) none of the above

17. If the number of turns of a coil is increased two times, its inductance is ....
(*i*) increased two times
(*ii*) decreased two times
(*iii*) decreased four times
(*iv*) increased four times

18. A circuit has inductance of 2H. If the circuit current changes at the rate of 10 A/second, then self-induced *e.m.f.* is ....
(*i*) 5 V (*ii*) 0.2 V
(*iii*) 20 V (*iv*) 10 V

19. A current of 2 A through a coil sets up flux linkages of 4 Wb-turn. The inductance of the coil is ....
(*i*) 8 H (*ii*) 0.5 H
(*iii*) 2 H (*iv*) 1 H

20. An air-cored choke is used for .... applications.
(*i*) radio frequency (*ii*) audio frequency
(*iii*) power frequency (*iv*) none of the above

21. If a 10-turn coil has a second layer of 10 turns wound over the first, then total inductance will be about .... the original inductance.
(*i*) two times (*ii*) four times
(*iii*) six times (*iv*) three times

22. An iron-cored coil of 10 turns has reluctance of 100 AT/Wb. The inductance of the coil is ....
(*i*) 1H (*ii*) 10 H
(*iii*) 0.1 H (*iv*) 5 H

23. An iron-cored coil has an inductance of 2 H. If the reluctance of the magnetic path is 200 AT/Wb, the number of turns on the coils is ....
(*i*) 100 (*ii*) 400
(*iii*) 50 (*iv*) 20

24. The mutual inductance between two coils is ....reluctance of magnetic path.
(*i*) directly proportional to
(*ii*) inversely proportional to
(*iii*) independent of (*iv*) none of the above

25. Mutual inductance between two coils can be decreased by ....
(i) increasing the number of turns of either coil
(ii) by moving the coils closer
(iii) by moving the coils apart
(iv) none of the above

26. Mutual inductance between two coils is 4H. If current in one coil changes at the rate of 2 A/second, then *e.m.f.* induced in the order coil is....................
(i) 8 V (ii) 2 V
(ii) 0.5 V (iv) none of the above

27. If in Fig. 9.28, $\phi_{12} = 2$ Wb, $N_2 = 20$ and $I_2 =$ 20 A, then mutual inductance between the coils is ....

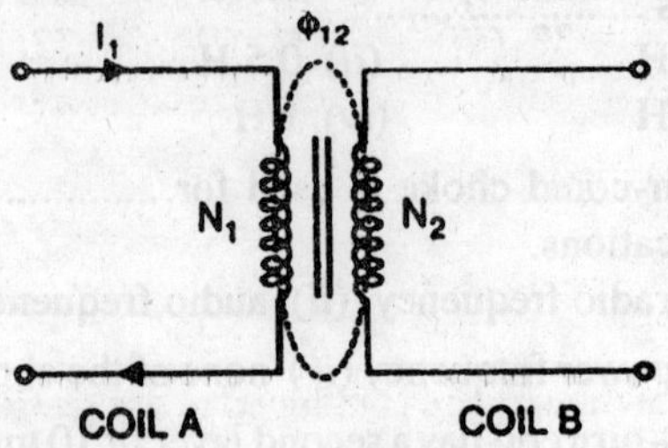

Fig. 9.28

(i) 200 H (ii) 20 H
(iii) 4 H (iv) 2 H

28. If in Fig. 9.28, $N_1 = 100$, $N_2 = 1000$ and mutual inductance between the coils is 2 H, the reluctance of magnetic circuit is ....
(i) 5×10^4 AT/Wb (ii) 10^5 AT/Wb
(iii) 20 AT/Wb (iv) 5 AT/Wb

29. If the co-efficient of coupling between two coils is increased, mutual inductance between the coils ....
(i) is increased (ii) is decreased
(iii) remains unchanged
(iv) none of the above

30. The maximum mutual inductance between the coils shown in Fig. 9.29 is given by ....
(i) $L_A L_B$ (ii) L_A / L_B
(iii) $\sqrt{L_A L_B}$ (iv) $(L_A L_B)^2$

31. The mutual inductance between two coils is zero when fluxes produced by them ....
(i) aid each other
(ii) oppose each other
(iii) are at right angles to each other
(iv) none of the above

32. The mutual inductance between two unity coupled coils of 9 H and 4 H is ....
(i) 36 H (ii) 13 H
(iii) 2.2 H (iv) 6 H

33. In Fig. 9.29, the mutual inductance between the two coils is ....
(i) 4.8 mH (ii) 15 mH
(iii) 4 mH (iv) 24 mH

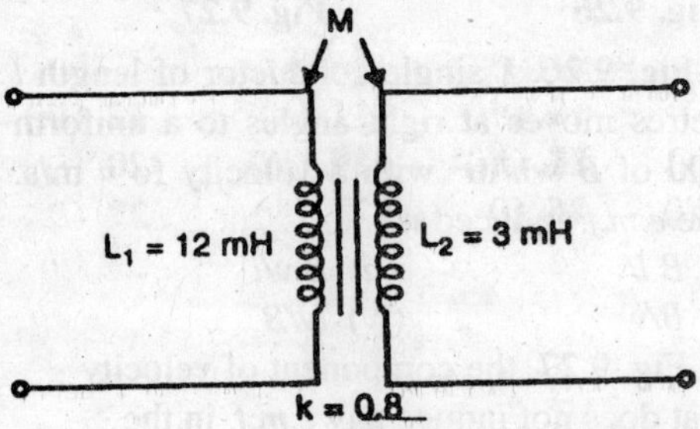

Fig. 9.29

34. In Fig. 9.29, the maximum inductance between the coils can be ....
(i) 15 mH (ii) 24 mH
(iii) 6 mH (iv) 4.8 mH

35. In Fig. 9.30, the inductance of coil 1 is ....
(i) 2 H (ii) 5 H
(iii) 50 H (iv) none of the above

36. In Fig. 9.30, the mutual inductance between the coils is ....

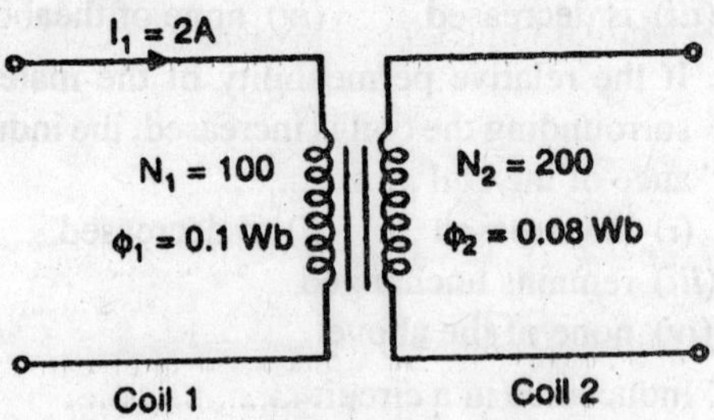

Fig. 9.30

(i) 4 H (ii) 100 H
(iii) 8 H (iv) none of the above

37. In Fig. 9.30, the co-efficient of coupling between the coils is ....

(*i*) 1.25 (*ii*) 0.8
(*iii*) 0.008 (*iv*) none of the above

38. The co-efficient of coupling between the primary and secondary of a practical iron-cored transformer is about..........................
(*i*) 0.9 (*ii*) 0.1
(*iii*) 1.5 (*iv*) 0.2

39. In Fig. 9.31, the inductance of coil L_1 will become ....
(*i*) 9 H (*ii*) 6 H
(*iii*) 12 H (*iv*) 3 H

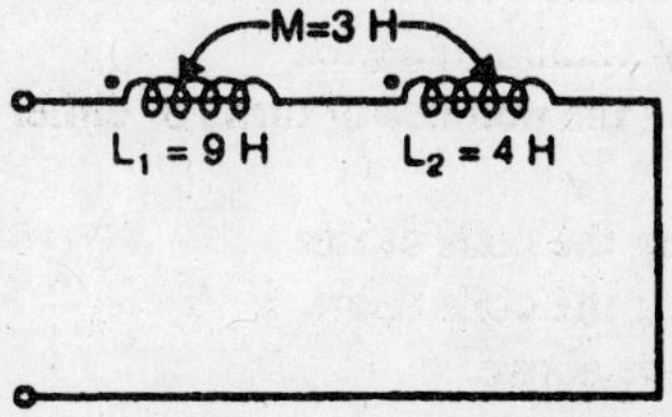

Fig. 9.31

40. In Fig. 9.31, the inductance of coil L_2 will become ....
(*i*) 7 H (*ii*) 1 H
(*iii*) 12 H (*iv*) 1.33 H

ANSWERS TO OBJECTIVE QUESTIONS

1. (*ii*)	**2.** (*i*)	**3.** (*iii*)	**4.** (*iv*)	**5.** (*ii*)	**6.** (*ii*)	**7.** (*ii*)	**8.** (*ii*)
9. (*iii*)	**10.** (*iii*)	**11.** (*ii*)	**12.** (*iii*)	**13.** (*ii*)	**14.** (*i*)	**15.** (*ii*)	**16.** (*iii*)
17. (*iv*)	**18.** (*iii*)	**19.** (*iii*)	**20.** (*i*)	**21.** (*ii*)	**22.** (*i*)	**23.** (*iv*)	**24.** (*ii*)
25. (*iii*)	**26.** (*i*)	**27.** (*iv*)	**28.** (*i*)	**29.** (*i*)	**30.** (*iii*)	**31.** (*iii*)	**32.** (*iv*)
33. (*i*)	**34.** (*iii*)	**35.** (*ii*)	**36.** (*iii*)	**37.** (*ii*)	**38.** (*i*)	**39.** (*iii*)	**40.** (*i*)

HINTS TO SELECTED OBJECTIVE QUESTIONS

3. $e = -N\dfrac{d\phi}{dt}$

The minus sign comes from Lenz's Law and indicates that the voltage is induced in a direction to oppose the change in flux that produced it.

11. The component $v \cos \theta$ is parallel to the magnetic field and hence no e.m.f. is induced in the conductor due to this component.

13. Inductance of coil = $\dfrac{N^2}{\text{Reluctance}}$

18. $e = L\dfrac{di}{dt} = 2 \times 10 = 20\text{ V}$ **19.** $L = \dfrac{N\phi}{I} = \dfrac{4}{2} = 2H$

24. Mutual inductance between two coils = $\dfrac{N_1 N_2}{\text{reluctance}}$

26. $e_M = M\dfrac{dI}{dt} = 4 \times 2 = 8\text{V}$ **27.** $M = \dfrac{N_2\phi_{12}}{I_1} = \dfrac{20 \times 2}{20} = 2H$

29. $M = k\sqrt{L_1 L_2}$ where k = co-efficient of coupling between two coils of inductance L_1 and L_2.

32. $M = k\sqrt{L_1 L_2} = 1\sqrt{9 \times 4} = 6H$

35. Inductance of coil 1 = $\dfrac{N_1\phi_1}{I_1} = \dfrac{100 \times 0.1}{2} = 5H$

37. Co-efficient of coupling, $k = \dfrac{\phi_2}{\phi_1} = \dfrac{0.08}{0.1} = 0.8$

39. Since the coils are connected series-aiding (See dot notation), the inductance of each coil will increase by M (*i.e.*, 3*H*).

10

D.C. Generators

Introduction

Although a far greater percentage of the electrical machines in service are a.c. machines, the d.c. machines are of considerable industrial importance. The principal advantage of the d.c. machine, particularly the d.c. motor, is that it provides a fine control of speed. Such an advantage is not claimed by any a.c. motor. However, d.c. generators are not as common as they used to be, because direct current, when required, is mainly obtained from an a.c. supply by the use of rectifiers. Nevertheless, an understanding of d.c. generator is important because it represents a logical introduction to the behaviour of d.c. motors. Indeed many d.c. motors in industry actually operate as d.c. generators for a brief period. In this chapter, we shall deal with various aspects of d.c. generators.

10.1. Generator Principle

An electric generator is a machine that converts mechanical energy into electrical energy.

An electric generator is based on the principle that whenever flux is cut by a conductor, an e.m.f. is induced which will cause a current to flow if the conductor circuit is closed. The direction of induced e.m.f. (and hence current) is given by *Fleming's right hand rule. Therefore, the essential components of a generator are :

(*i*) a magnetic field

(*ii*) conductor or a group of conductors

(*iii*) motion of conductor *w.r.t.* magnetic field

10.2. Simple Loop Generator

Consider a single turn loop *ABCD* rotating clockwise in a uniform magnetic field with a constant speed as shown in Fig. 10.1 As the loop rotates, the flux linking the coil sides *AB* and *CD* changes continuously. Hence the e.m.f. induced in these coil sides also changes but the e.m.f. induced in one coil side **adds to that induced in the other.

(*i*) When the loop is in position no. 1 [See Fig. 10.1], the generated e.m.f. is zero because the coil sides (*AB* and *CD*) are cutting no flux but are moving parallel to it.

(*ii*) When the loop is in position no. 2, the coil sides are moving at an angle to the flux and, therefore, a low e.m.f. is generated as indicated by point 2 in Fig. 10.2.

---

* *Fleming's Right hand rule.* Stretch the thumb, fore-finger and middle finger of your right hand so that they are at right angles to each other. If the fore-finger points in the direction of field, thumb in the direction of motion of the conductor, then middle finger will point in the direction of induced e.m.f.

** It is because the coil sides always remain under the influence of opposite poles *i.e.* if one coil side is under the influence of the *N*-pole, then the other coil side will be under the influence of *S*-pole and *vice-versa.*

(*iii*) When the loop is in position no. 3, the coil sides (*AB* and *CD*) are at right angle to the flux and are, therefore, cutting the flux at a maximum rate. Hence at this instant, the generated e.m.f. is maximum as indicated by point 3 in Fig. 10.2.

(*iv*) At position 4, the generated e.m.f. is less because the coil sides are cutting the flux at an angle.

(*v*) At position 5, no magnetic lines are cut and hence, induced e.m.f. is zero as indicated by point 5 in Fig. 10.2.

(*vi*) At position 6, the coil sides move under a pole of opposite polarity and hence the direction of generated e.m.f. is reversed. The maximum e.m.f. in this direction (*i.e.* reverse direction, See Fig. 10.2) will be when the loop is at position 7 and zero when at position 1. This cycle repeats with each revolution of the coil.

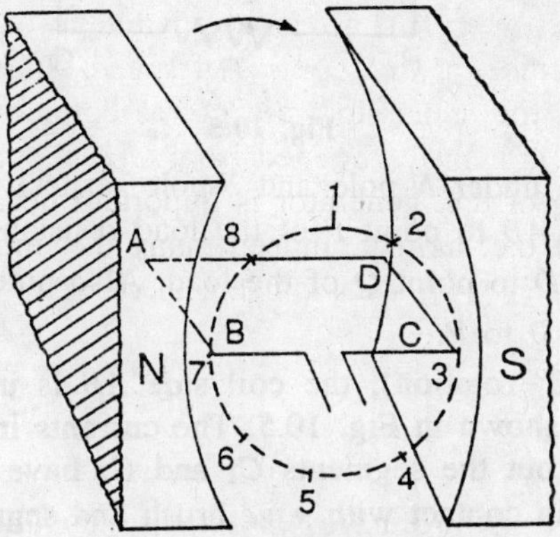

Fig. 10.1

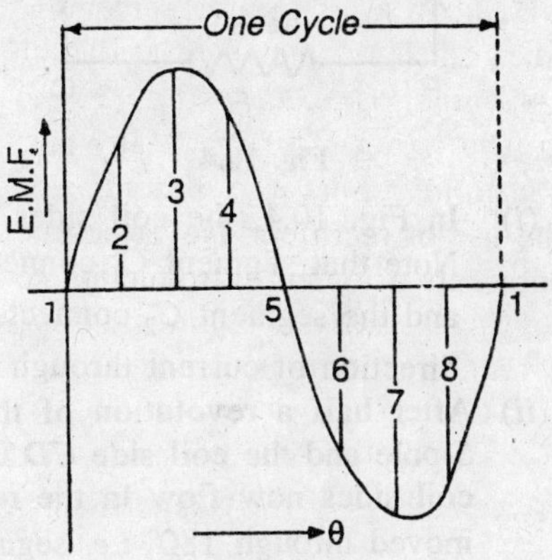

Fig. 10.2

Note that e.m.f. generated in the loop is alternating one. It is because any coil side, say *AB*, has e.m.f. in one direction when under the influence of *N*-pole and in the other direction when under the influence of *S*-pole. If a load is connected across the ends of the loop, then alternating current will flow through the load. The alternating voltage generated in the loop can be converted into direct voltage by a device called *commutator*. We then have the d.c. generator. In fact, a commutator is a mechanical rectifier.

10.3. Action of Commutator

If, somehow, connection of the coil side to the external load is reversed at the same instant the current in the coil side reverses, the current through the load will be direct current. This is what a commutator does. Fig. 10.3 shows a commutator having two segments C_1 and C_2. It consists of a cylindrical metal ring cut into two halves or segments C_1 and C_2 respectively separated by a thin sheet of mica. The commutator is mounted on, but insulated from, the rotor shaft. The ends of coil sides *AB* and CD are connected to the segments C_1 and C_2 respectively as shown in Fig. 10.4. Two stationary carbon brushes rest on the commutator and lead current to the external load. With this arrangement, the commutator at all times connects the coil side under *S*-pole to the *+ve* brush and that under *N*-pole to the *-ve* brush.

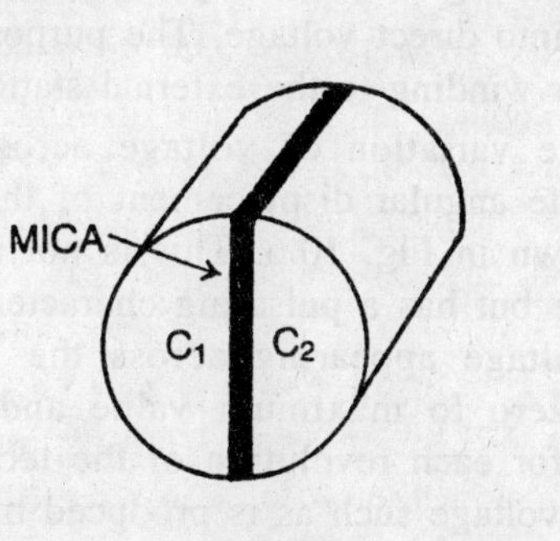

Fig. 10.3

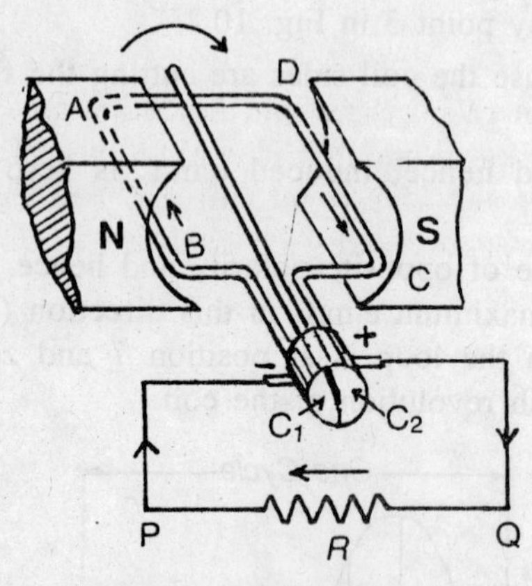

Fig. 10.4

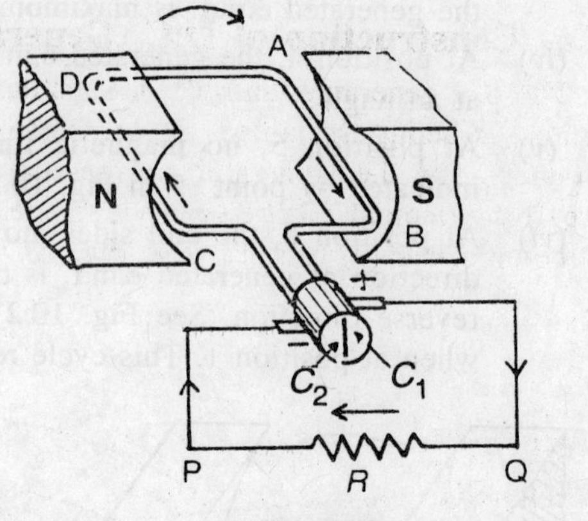

Fig. 10.5

(*i*) In Fig. 10.4, the coil sides *AB* and *CD* are under *N*-pole and *S*-pole respectively. Note that segment C_1 connects the coil side *AB* to point *P* of the load resistance *R* and the segment C_2 connects the coil side *CD* to point *Q* of the load. Also note the direction of current through load. It is from *Q* to *P*.

(*ii*) After half a revolution of the loop (*i.e.* 180° rotation), the coil side *AB* is unde *S*-pole and the coil side *CD* under *N*-pole as shown in Fig. 10.5. The currents in the coil sides now flow in the reverse direction but the segments C_1 and C_2 have also moved through 180° *i.e.* segment C_1 is now in contact with + *ve* brush and segment C_2 in contact with –*ve* brush. Note that commutator has reversed the coil connections to the load *i.e.* coil side *AB* is now connected to point *Q* of the load and coil side *CD* to the point *P* of the load. Also note the direction of current through the load. It is *again from *Q* to *P*.

Thus the alternating voltage generated in the loop will appear as direct voltage across the brushes. The reader may note that e.m.f. generated in the armature winding of a d.c. generator is alternating one. It is by the use of commutator that we convert the generated alternating e.m.f. into direct voltage. The purpose of brushes is simply to lead current from the rotating loop or winding to the external stationary load.

The variation of voltage across the brushes with the angular displacement of the loop will be as shown in Fig. 10.6. This is not a steady direct voltage but has a pulsating character. It is because the voltage appearing across the brushes varies from zero to maximum value and back to zero twice for each revolution of the loop. A pulsating direct voltage such as is produced by a single loop is not suitable for many commercial uses. What we require is the steady direct voltage. This can be

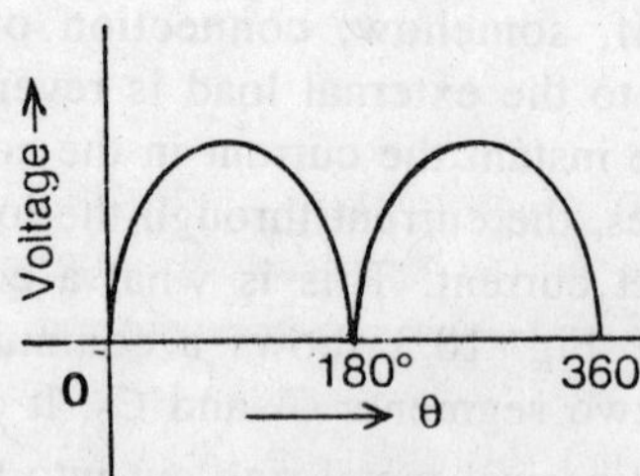

Fig. 10.6

---

* The situation is that currents in the coil sides are reversed and at the same time the connections of the coil sides to the external load are reversed. This means that current will flow in the same direction through the load.

achieved by using a large number of coils connected in series. The resulting arrangement is known as *armature winding.*

10.4. Construction of D.C. Generator

The d.c. generators and d.c. motors have the same general construction. In fact, when the machine is being assembled, the workmen usually do not know whether it is a d.c. generator or motor. Any d.c. generator can be run as a d.c. motor and *vice-versa.* All d.c. machines have five principal components *viz* (*i*) field system (*ii*) armature core (*iii*) armature winding (*iv*) commutator (*v*) brushes [See Fig. 10.7]

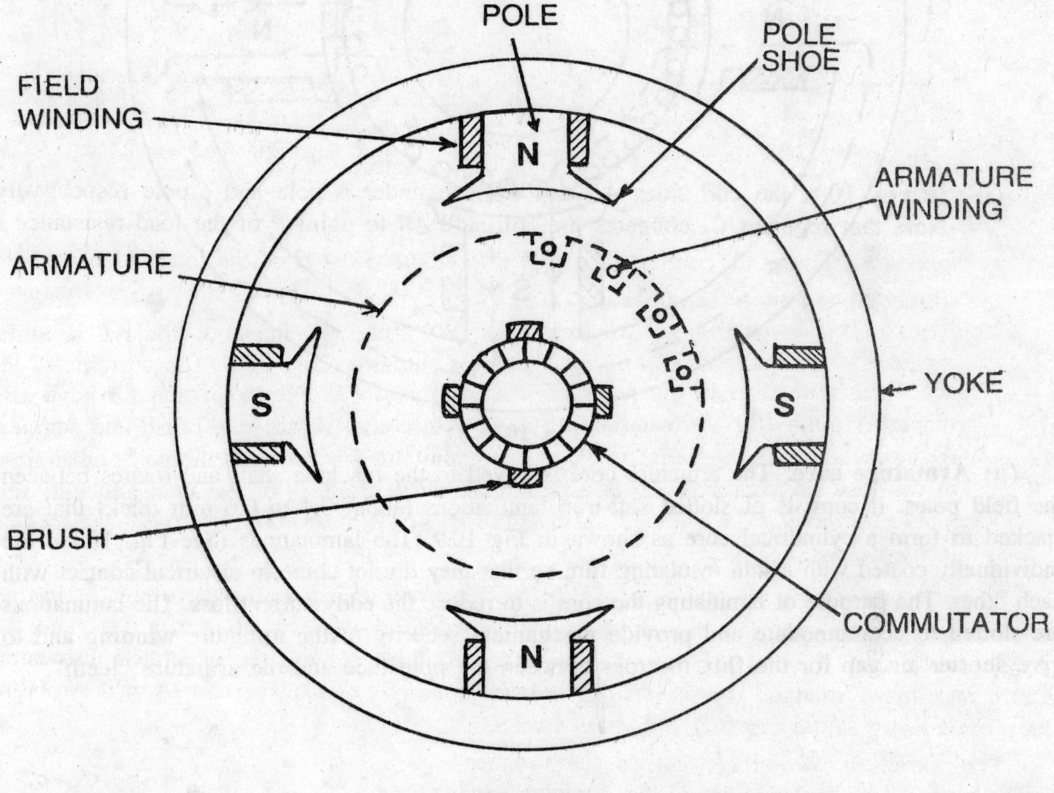

Fig. 10.7

(*i*) **Field system.** The function of the field system is to produce uniform magnetic field within which the armature rotates. It consists of a number of salient poles (of course, even number) bolted to the inside of circular frame (generally called *yoke*). The yoke is usually made of solid cast steel whereas the pole pieces are composed of stacked laminations. Field coils are mounted on the poles and carry the d.c. exciting current. The field coils are connected in such a way that adjacent poles have opposite polarity.

The m.m.f. developed by the field coils produces a magnetic flux that passes through the pole pieces, the air gap, the armature and the frame (See Fig. 10.8). Practical d.c. machines have air gaps ranging from 0.5 mm to 1.5 mm. Since armature and field systems are composed of materials that have high permeability, most of the m.m.f. of field coils is required to set up flux in the air gap. By reducing the length of air gap, we can reduce the size of field coils (*i.e.* number of turns).

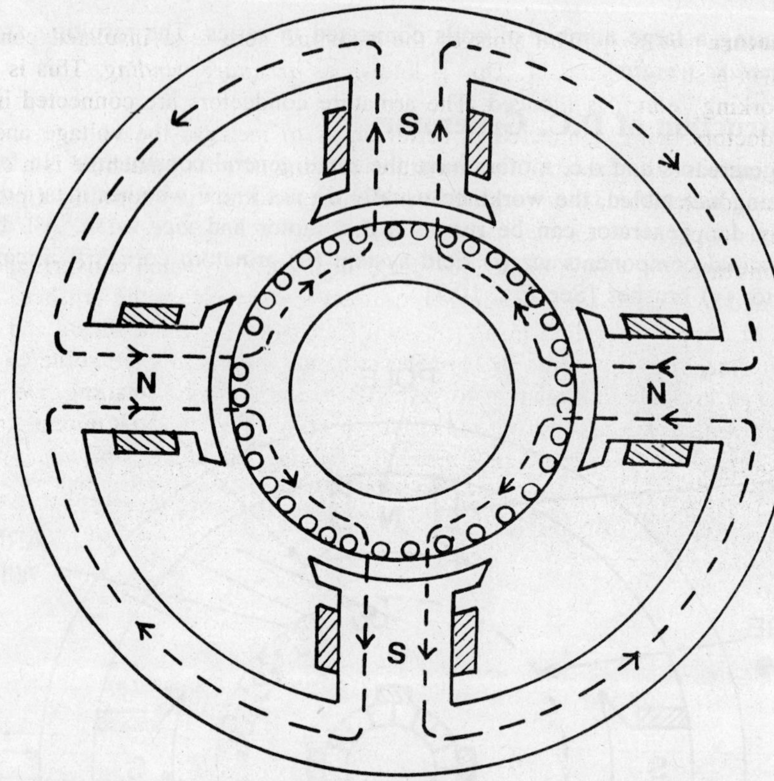

Fig. 10.8

(*ii*) **Armature core.** The armature core is keyed to the machine shaft and rotates between the field poles. It consists of slotted soft-iron laminations (about 0.4 to 0.6 mm thick) that are stacked to form a cylindrical core as shown in Fig 10.9. The laminations (See Fig. 10.10) are individually coated with a thin insulating film so that they do not come in electrical contact with each other. The purpose of laminating the core is to reduce the eddy current loss. The laminations are slotted to accommodate and provide mechanical security to the armature winding and to give shorter air gap for the flux to cross between the pole face and the armature "teeth".

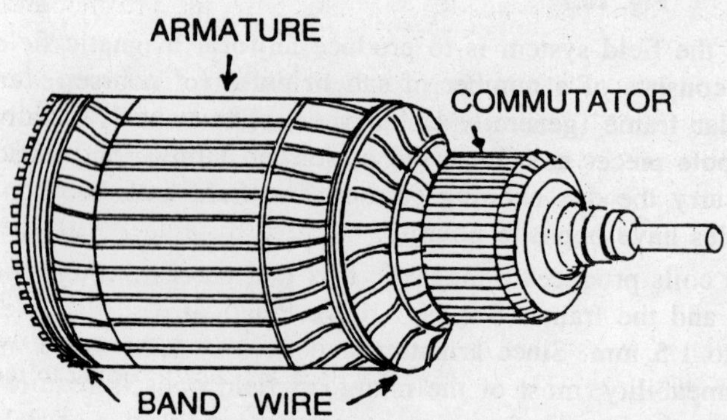

Fig. 10.9

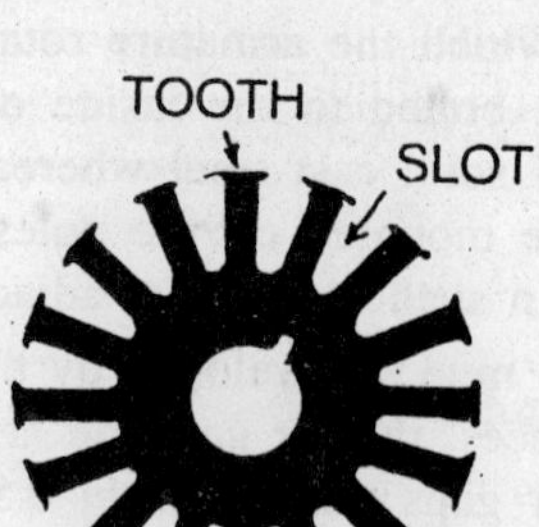

Fig. 10.10

(*iii*) **Armature winding.** The slots of the armature core hold insulated conductors that are connected in a suitable manner. This is known as *armature winding*. This is the winding in which " working" e.m.f. is induced. The armature conductors are connected in series-parallel; the conductors being connected in series so as to increase the voltage and in parallel paths so as to increase the current. The armature winding of a d.c. machine is a closed-circuit winding; the conductors being connected in a symmetrical manner forming a closed loop or series of closed loops.

(*iv*) **Commutator.** A commutator is a mechanical rectifier which converts the alternating voltage generated in the armature winding into direct voltage across the brushes. The commutator is made of copper segments insulated from each other by mica sheets and mounted on the shaft of the machine (See Fig. 10.11). The armature conductors are soldered to the commutator segments in a suitable manner to give rise to the armature winding. Depending upon the manner in which the armature conductors are connected to the commutator segments, there are two types of armature winding in a d.c. machine *viz* (*a*) lap winding (*b*) wave winding.

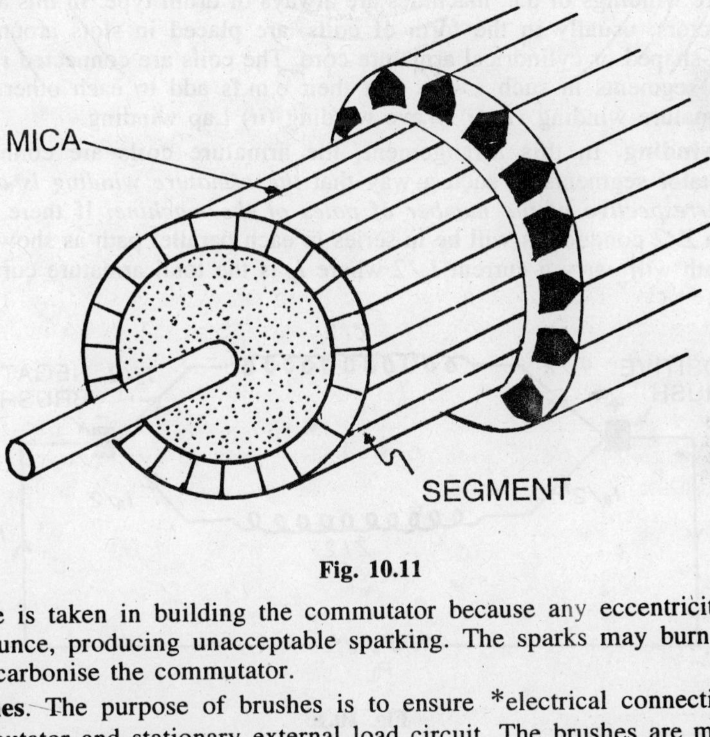

Fig. 10.11

Great care is taken in building the commutator because any eccentricity will cause the brushes to bounce, producing unacceptable sparking. The sparks may burn the brushes and overheat and carbonise the commutator.

(*v*) **Brushes**. The purpose of brushes is to ensure *electrical connections between the rotating commutator and stationary external load circuit. The brushes are made of **carbon and rest on the commutator. The brush pressure is adjusted by means of adjustable springs (See Fig. 10.12). If the brush pressure is very large, the friction produces heating of the commutator and the brushes. On the other hand, if it is too weak, the imperfect contact with the commutator may produce sparking.

---

* When the machine is acting as a generator, the brushes carry current from the commutator to the external stationary load. In case the machine is acting as a motor, they feed the supply current to the commutator.

** Because carbon has good electrical conductivity and is soft enough not to scratch the commutator. To improve the conductivity, sometimes we add a small amount of copper.

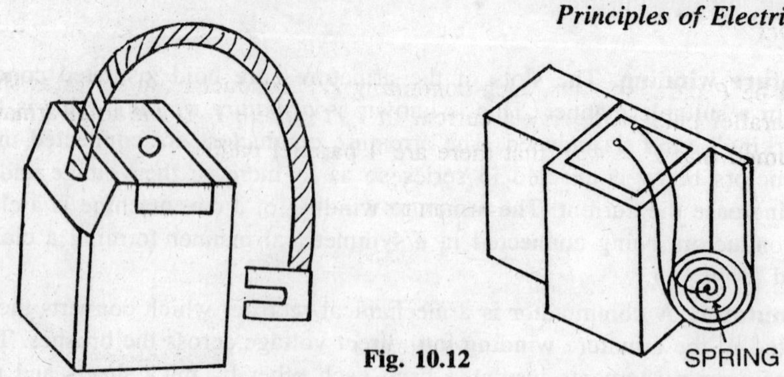

Fig. 10.12

Multipole machines have as many brushes as they have poles. For example, a 4-pole machine has 4 brushes. As we go round the commutator, the successive brushes have positive and negative polarities. Brushes having the same polarity are connected together so that we have two terminals *viz* the *+ve* terminal and the *–ve* terminal.

10.5. Types of Armature Winding

The armature windings of d.c. machines are always of drum type. In this arrangement, the armature conductors, usually in the form of coils, are placed in slots around the complete surface of drum-shaped or cylindrical armature core. The coils are connected in series through the commutator segments in such a way that their e.m.fs add to each other. There are two types of d.c. armature winding *viz* (*i*) Wave winding (*ii*) Lap winding.

(*i*) **Wave winding.** In this arrangement, the armature coils are connected in series through commutator segments in such a way that *the armature winding is divided into two parallel paths irrespective of the number of poles of the machine.* If there are Z armature conductors, then $Z/2$ conductors will be in series in each parallel path as shown in Fig. 10.13. Each parallel path will carry a current $I_a/2$ where I_a is the total armature current.

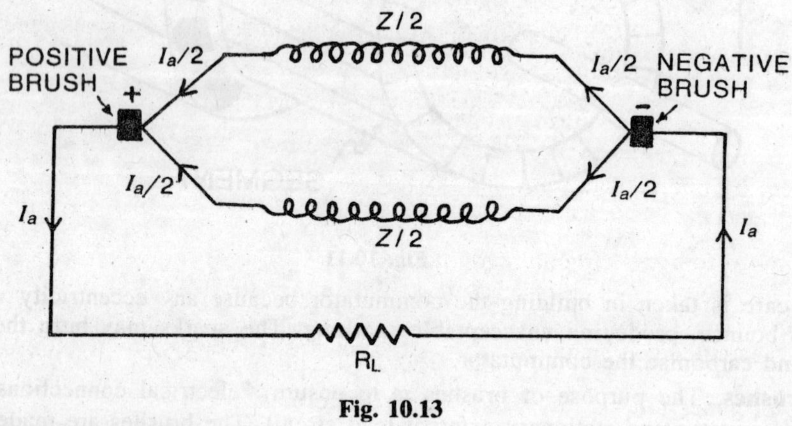

Fig. 10.13

To sum up, in a wave winding;

(*a*) there are two parallel paths irrespective of number of poles of the machine.

(*b*) each parallel path has $Z/2$ conductors in series; Z being total number of armature conductors.

(*c*) e.m.f. generated = e.m.f./parallel path

(*d*) total armature current, $I_a = 2 \times$ current/parallel path

(*ii*) **Lap winding.** In this arrangement, the armature coils are connected in series through commutator segments in such a way that *the armature winding is divided into as many parallel paths as the number of poles of the machine.* If there are Z conductors and P poles, then there will be P parallel paths, each containing Z/P conductors in series as shown in Fig. 10.14. Each parallel path will carry a current of I_a/P where I_a is the total armature current. Here it is assumed that $P = 4$ so that there are 4 parallel paths.

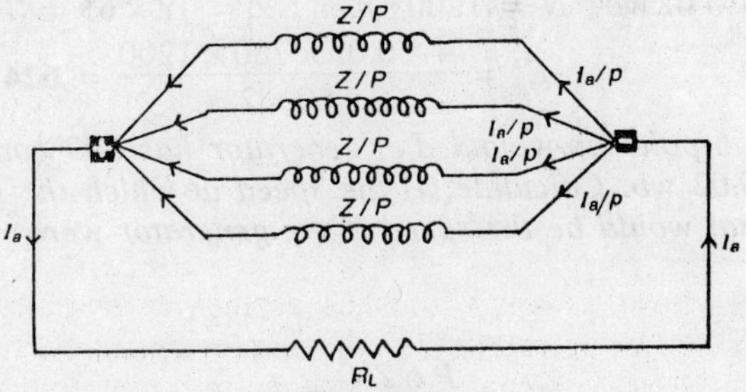

Fig. 10.14

To sum up, in a lap winding;

(*a*) there are as many parallel paths as the number of poles (P) of the machine.

(*b*) each parallel path has Z/P conductors in series where Z and P are the total number of armature conductors and poles respectively.

(*c*) e.m.f. generated = e.m.f./parallel path.

(*d*) total armature current, $I_a = P \times$ current/parallel path.

10.6. E.M.F. Equation of a D.C. Generator

We shall now derive an expression for the e.m.f. generated in a d.c. generator.

Let
ϕ = flux/pole in Wb
Z = total number of armature conductors
P = number of poles
A = number of parallel paths = 2.... for wave winding
= P .... for lap winding
N = speed of armature in r.p.m.
E_g = e.m.f. of the generator = e.m.f./parallel path

Flux cut by one conductor in one revolution of the armature,

$$d\phi = P\,\phi \text{ webers}$$

Time taken to complete one revolution.

$$dt = 60/N \text{ second}$$

e.m.f. generated/conductor $= \dfrac{d\phi}{dt} = \dfrac{P\phi}{60/N} = \dfrac{P\,\phi\,N}{60}$ volts

e.m.f. of generator, E_g = e.m.f. per parallel path

= (e.m.f./conductor) × no. of conductors in series per parallel path

$$= \frac{P\phi N}{60} \times \frac{Z}{A}$$

$$\therefore \quad E_g = \frac{P\,\phi Z N}{60\,A}$$

where $A = 2$ *... for wave winding*

$= P$ *... for lap winding*

Example 10.1. *Calculate the e.m.f. generated by 4-pole wave-wound generator having 65 slots with 12 conductors per slot when driven at 1200 r.p.m. The flux per pole is 0.02 wb.*

Solution.

$$E_g = \frac{P\phi ZN}{60A}$$

Here $P = 4$; $\phi = 0.02\ wb$; $N = 1200\ r.p.m.$; $Z = 12 \times 65 = 780$; $A = 2$

$$\therefore \quad E_g = \frac{4 \times 0.02 \times 780 \times 1200}{60 \times 2} = \textbf{624 volts}$$

Example 10.2. *A 6-pole lap-wound d.c. generator has 600 conductors on its armature. The flux per pole is 0.02 wb. Calculate (i) the speed at which the generator must be run to generate 300V (ii) What would be the speed if the generator were wave-wound ?*

Solution.

(*i*) **Lap wound**

$$E_g = \frac{P\phi ZN}{60A}$$

$$\therefore \quad N = \frac{E_g \times 60A}{P\phi Z} = \frac{300 \times 60 \times 6}{6 \times 0.02 \times 600} = \textbf{1500 r.p.m.}$$

(*ii*) **Wave wound**

$$N = \frac{E_g \times 60A}{P\phi Z} = \frac{300 \times 60 \times 2}{6 \times 0.02 \times 600} = \textbf{500 r.p.m.}$$

Example 10.3. *An 8-pole, lap-wound armature rotated at 350 r.p.m. is required to generate 260 V. The useful flux per pole is 0.05 wb. If the armature has 120 slots, calculate the number of conductors per slot.*

Solution.

$$E_g = \frac{P\phi ZN}{60A}$$

$$\therefore \quad Z = \frac{E_g \times 60A}{P\phi N} = \frac{260 \times 60 \times 8}{8 \times 0.05 \times 350} = \textbf{890}$$

$\therefore$ No. of conductors/slot = 890/120 = **7.14**

This value must be an even number.

Hence, conductors/slot = **8**

Example 10.4. *The armature of a 6-pole, 600 r.p.m. lap-wound generator has 90 slots. If each coil has 4 turns, calculate the flux per pole required to generate an e.m.f. of 288 volts.*

Solution. Each turn has two active conductors and 90 coils are required to fill 90 slots.

$$\therefore \quad Z = 90 \times 4 \times 2 = 720$$

$$E_g = \frac{P\phi ZN}{60A}$$

or

$$\phi = \frac{E_g \times 60A}{PZN} = \frac{288 \times 60 \times 6}{6 \times 720 \times 600} = \textbf{0.04 wb}$$

Example 10.5. *The armature of a d.c. generator has 81 slots and the commutator has 243 segments. It is wound to give lap winding having 1 turn per coil. If the flux per pole is 30 mwb, calculate the generated e.m.f. at a speed of 1200r.p.m. Number of poles = 6.*

Solution. The number of coils is equal to the number of commutator segments. Each turn has 2 active conductors.

$$\therefore \quad Z = 243 \times 2 = 486; \quad \phi = 30 \times 10^{-3} = 0.03\ \text{wb}$$

$$E_g = \frac{P\phi ZN}{60A} = \frac{6 \times 0.03 \times 486 \times 1200}{60 \times 6} = \textbf{291.6 volts}$$

10.7. Armature Resistance (R_a)

The resistance offered by the armature circuit is known as armature resistance (R_a) and includes :

(*i*) resistance of armature winding

(*ii*) resistance of brushes.

The armature resistance depends upon the construction of machine. Except for small machines, its value is generally less than 1Ω.

10.8. Types of D.C. Generators

The magnetic field in a d.c. generator is normally produced by electromagnets rather than permanent magnets. Generators are generally classified according to their methods of field excitation. On this basis, d.c. generators are divided into the following two classes:

(*i*) Separately excited d.c. generators

(*ii*) Self-excited d.c. generators

The behaviour of a d.c. generator on load depends upon the method of field excitation adopted.

10.9. Separately Excited D.C. Generators

A d.c. generator whose field magnet winding is supplied from an independent external d.c. source (*e.g.* a battery *etc.*) is called *a separately excited generator.* Fig. 10.15 shows the connections of a separately excited generator. The voltage output depends upon the speed of rotation of armature and the field current ($E_g = P\phi ZN/60A$). The greater the speed and field current, greater is the generated e.m.f. It may be noted that separately excited d.c. generators are rarely used in practice. The d.c. generators are normally of self-excited type.

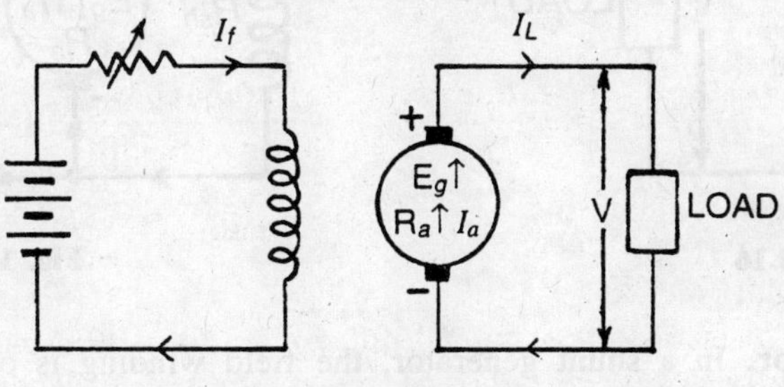

Fig. 10.15

Armature current, $I_a = I_L$

Terminal voltage, $V = E_g - I_a R_a$

Electric power developed $= E_g I_a$

Power delivered to load $= E_g I_a - I_a^2 R_a = I_a (E_g - I_a R_a) = V I_a$

10.10. Self-excited D.C. Generators

A d.c. generator whose field magnet winding is supplied current from the output of the generator itself is called a **self-excited generator.* There are three types of self-excited gen-

* How is self-excitation achieved ? When the armature is rotated, a small voltage is induced in the armature winding due to residual flux in the poles. This voltage produces a small field current in the field winding and causes the flux per pole to increase. The increased flux increases the induced voltage which further increases the field current. These events take place rapidly and the generator builds up to the rated generated voltage.

erators depending upon the manner in which the field winding is connected to the armature, namely;

(*i*) Series generator

(*ii*) Shunt generator

(*iii*) Compound generator

(*i*) **Series generator.** In a series-wound generator, the field winding is connected in series with armature winding so that whole armature current flows through the field winding as well as the load. Fig. 10.16 shows the connections of a series-wound generator. Since the field winding carries the whole of load current, it has a few turns of thick wire having low resistance. Series generators are rarely used except for special purposes *e.g.* as boosters.

Armature current, $I_a = I_{se} = I_L = I$ (say)

Terminal voltage, $V = E_g - I(R_a + R_{se})$

Power developed in armature $= E_g I_a$

Power delivered to load $= E_g I_a - I_a^2 (R_a + R_{se}) = I_a [E_g - I_a (R_a + R_{se})] = VI_a$ or VI_L

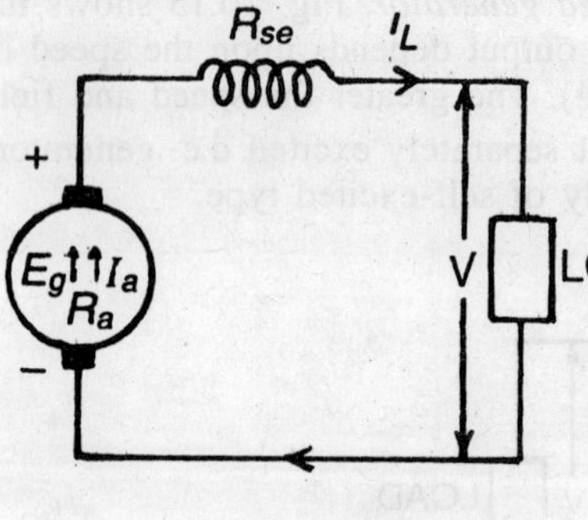

Fig. 10.16

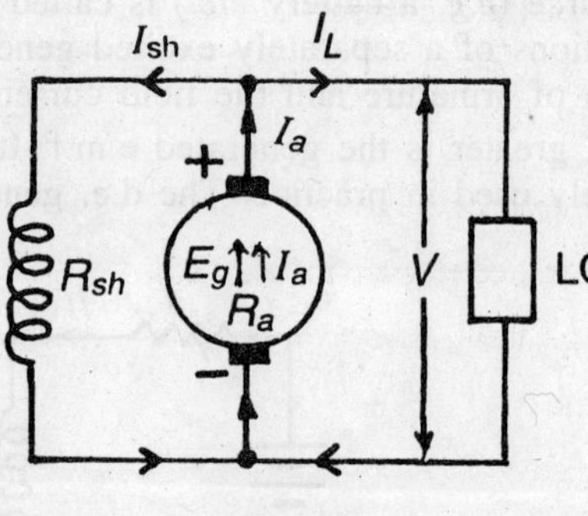

Fig. 10.17

(*ii*) **Shunt generator.** In a shunt generator, the field winding is connected in parallel with the armature winding so that terminal voltage of the generator is applied across it. The shunt field winding has many turns of fine wire having high resistance. Therefore, only a part of armature current flows through shunt field winding and the rest flows through the load. Fig 10.17 shows the connections of a shunt-wound generator.

Shunt field current, $I_{sh} = V/R_{sh}$

Armature current, $I_a = I_L + I_{sh}$

Terminal voltage, $V = E_g - I_a R_a$

Power developed in armature $= E_g I_a$

Power delivered to load $= VI_L$

(*iii*) **Compound generator.** In a compound-wound generator, there are *two sets of field windings on each pole—one is in series and the other in parallel with the armature. A compound wound generator may be :

---

* Normally, the majority of m.m.f. is provided by the shunt field. The two windings may be connected to aid each other (*cumulative compounding*) or they may oppose each other (*differential compounding*).

(a) **Short Shunt** in which only shunt field winding is in parallel with the armature winding [See Fig 10.18 (*i*)].

(b) **Long shunt** in which shunt field winding is in parallel with both series field and armature winding [See Fig. 10.18 (*ii*)].

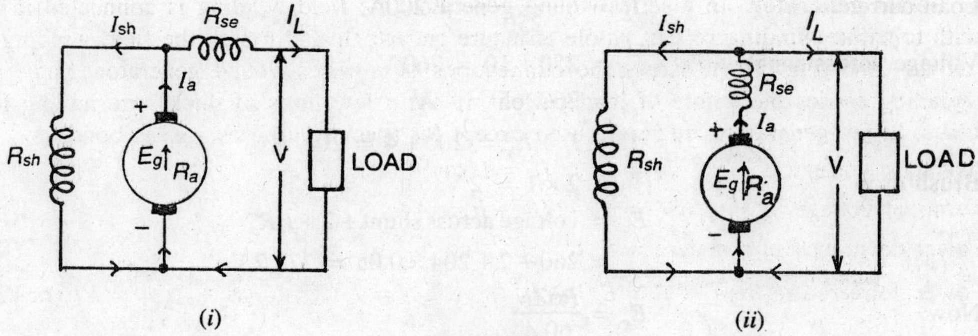

Fig. 10.18

Short shunt

Series field current, $I_{se} = I_L$

Shunt field current $I_{sh} = \dfrac{V + I_{se}R_{se}}{R_{sh}}$

Terminal voltage, $V = E_g - I_aR_a - I_{se}R_{se}$

Power developed in armature $= E_gI_a$

Power delivered to load $= VI_L$

Long shunt

Series field current, $I_{se} = I_a = I_L + I_{sh}$

Shunt field current, $I_{sh} = V/R_{sh}$

Terminal voltage, $V = E_g - I_a(R_a + R_{se})$

Power developed in armature $= E_gI_a$

Power delivered to load $= VI_L$

Example 10.6. *A 100 kW, 240V shunt generator has a field resistance of 55 Ω and armature resistance of 0.067 Ω. Find the full-load generated voltage.*

Solution. Fig. 10.19 shows the shunt generator circuit.

$$I_L = \frac{100 \times 10^3}{240} = 416.7\text{A}$$

$$I_{sh} = 240/55 = 4.36\text{ A}$$

$$I_a = I_L + I_{sh}$$

$$= 416.7 + 4.36 = 421.1\text{A}$$

$$E_g = V + I_aR_a$$

$$= 240 + 421.1 \times 0.067 = \mathbf{268.2\ V}$$

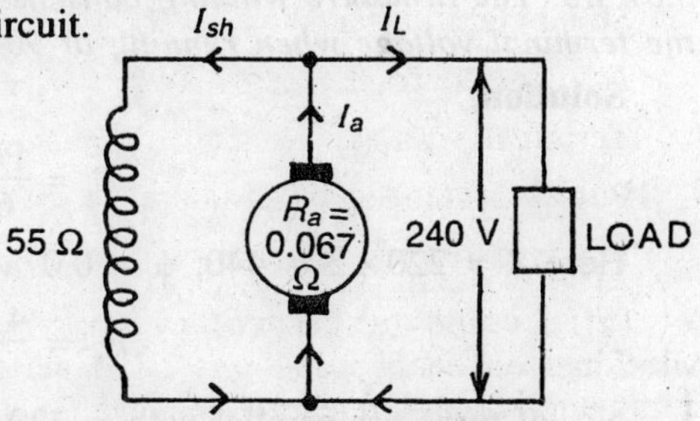

Fig. 10.19

Example 10.7. *A 4-pole d.c. shunt generator with a wave-wound armature has to supply a load of 500 lamps each of 100W at 250V. Allowing 10V for the voltage drop in the con-*

necting leads between the generator and the load and drop of 1V per brush, calculate the speed at which the generator should be driven. The flux per pole is 30 mwb and the armature and shunt field resistances are respectively 0.05 Ω and 65 Ω. The number of armature conductors is 390.

Solution.

Load current, $I_L = \dfrac{500 \times 100}{250} = 200\text{A}$

Voltage across shunt $= 250 + 10 = 260 \text{ V}$

$I_{sh} = 260/65 = 4 \text{ A}$

$I_a = I_L + I_{sh} = 200 + 4 = 204 \text{ A}$

Brush drop $= 2 \times 1 = 2 \text{ V}$

$E_g = \text{voltage across shunt} + 2 + I_a R_a$

$= 260 + 2 + 204 \times 0.05 = 272.2 \, V$

Now, $E_g = \dfrac{P\phi ZN}{60\,A}$

$\therefore \quad N = \dfrac{E_g \times 60A}{P\phi Z} = \dfrac{272.2 \times 60 \times 2}{4 \times 30 \times 10^{-3} \times 390} = \mathbf{698 \text{ r.p.m}}$

[For wave winding, A = 2]

Example 10.8. *A 30 kW, 300V d.c. shunt generator has armature and field resistances of 0.05 Ω and 100 Ω respectively. Calculate the total power developed by the armature when it delivers full load output.*

Solution. Fig. 10.20 shows the shunt generator circuit.

$I_L = \dfrac{30 \times 10^3}{300} = 100 \text{ A}$

$I_{sh} = 300/100 = 3 \text{ A}$

$I_a = I_L + I_{sh} = 100 + 3 = 103 \text{ A}$

$E_g = V + I_a R_a$

$= 300 + 103 \times 0.05 = 305.15 \text{ V}$

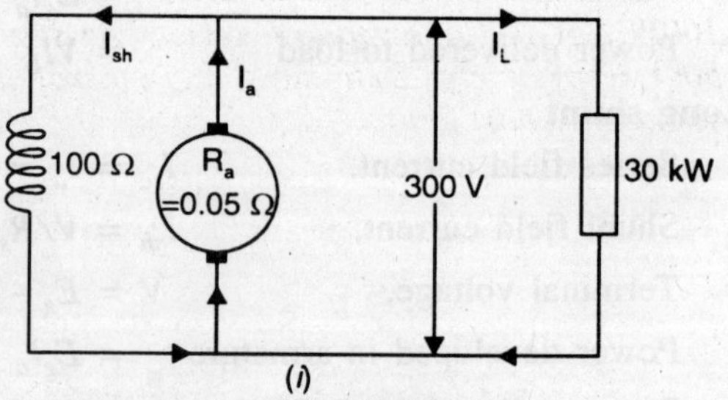

(*i*)

Fig. 10.20

Power developed by armature $= E_g I_a$

$= 305.15 \times 103 = 31.43 \times 10^3 \text{ W} = \mathbf{31.43 \text{ kW}}$

Example 10.9. *A 4-pole lap-wound d.c. shunt generator has a useful flux per pole of 0.07 wb. The armature winding consists of 220 turns, each of 0.004 Ω resistance. Calculate the terminal voltage when running at 900 r.p.m. if the armature current is 50A.*

Solution.

$E_g = \dfrac{P\phi ZN}{60\text{A}}$

Here $Z = 220 \times 2 = 440$; $\phi = 0.07\,wb$; $N = 900 \text{ r.m.p}$; $A = P = 4$

$\therefore \quad E_g = \dfrac{4 \times 0.07 \times 440 \times 900}{60 \times 4} = 462 \text{ V}$

No. of turns per parallel path $= 220/4 = 55$

Resistance per parallel path $= 0.004 \times 55 = 0.22\,\Omega$

Now $E_g = V + I_a R_a$

or $462 = V + 50 \times 0.055$

$\therefore$ $V = \mathbf{459.25\ volts}$

Example 10.10. *A d.c. shunt generator runs at 400 r.p.m. and delivers 500 kW to bus-bars having a constant voltage of 400 V. Assuming the field excitation to be constant at 5A, calculate the speed at which the generator must run if the load on it is to be reduced to 300 kW. The armature resistance is 0.015Ω.*

Solution.

$$^{*}E_g \propto \phi N$$

In the present problem, ϕ is given to be constant. Therefore, $E_g \propto N$

Initial conditions

Load current, $I_1 = 500 \times 10^3/400 = 1250\text{ A}$

Armature current $= 1250 + 5 = 1255\text{ A}$

$\therefore$ Generated e.m.f., $E_1 = 400 + 1255 \times 0.015 = 418.8\text{ V}$

Final conditions

Load current, $I_2 = 300 \times 10^3/400 = 750\text{ A}$

Armature current $= 750 + 5 = 755\text{ A}$

$\therefore$ Generated e.m.f., $E_2 = 400 + 755 \times 0.015 = 411.3\text{ V}$

$$E_2/E_1 = N_2/N_1$$

$\therefore$ $N_2 = (E_2/E_1) \times N_1 = (411.3/418.8) \times 400 = \mathbf{393\ r.p.m.}$

Example 10.11. *A compound generator is to supply a load of 250 lamps, each rated at 100W, 250V. The armature, series and shunt windings have resistances of 0.06 Ω, 0.04 Ω and 50Ω respectively. Determine the generated e.m.f. when the machine is connected in (i) long shunt (ii) short shunt. Take drop per brush as 1V.*

Solution.

(*i*) Fig. 10.21 shows the connections of the machine when connected in long shunt.

$$I_L = \frac{250 \times 100}{250} = 100\text{ A}$$

$$I_{sh} = 250/50 = 5\text{ A}$$

$$I_a = I_L + I_{sh} = 100 + 5 = 105\text{ A}$$

$$E_g = V + I_a(R_a + R_{se}) + \text{Total brush drop}$$

$$= 250 + 105\,(0.06 + 0.04) + 2 \times 1 = \mathbf{262.5\ V}$$

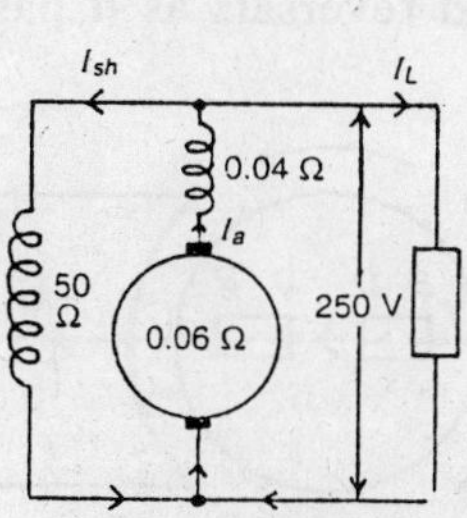

Fig. 10.21

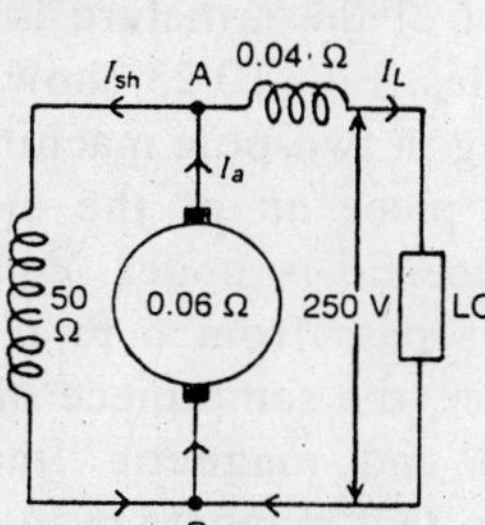

Fig. 10.22

* $E_g = P\phi ZN/60A$. For a given machine, P, Z and A are constant so that $E_g \propto \phi N$

(*ii*) Fig 10.22 shows the connections when the machine is connected in short shunt.

$$I_L = 100 \text{ A} \qquad \text{... same as before}$$

Voltage across shunt, $V_{AB} = 250 + 100 \times 0.04 = 254 \text{ V}$

$$I_{sh} = 254/50 = 5.08 \text{ A}$$

$$I_a = I_L + I_{sh} = 100 + 5.08 = 105.08 \text{ A}$$

$$E_g = V_{AB} + I_a R_a + \text{Total brush drop}$$

$$= 254 + 105.08 \times 0.06 + 2 \times 1 = \mathbf{262.3 \text{ V}}$$

10.11. Losses in a D.C. Machine

The losses in a d.c. machine (generator or motor) may be divided into three classes *viz* (*i*) copper losses (*ii*) iron or core losses and (*iii*) mechanical losses. All these losses appear as heat and thus raise the temperature of the machine. They also lower the efficiency of the machine.

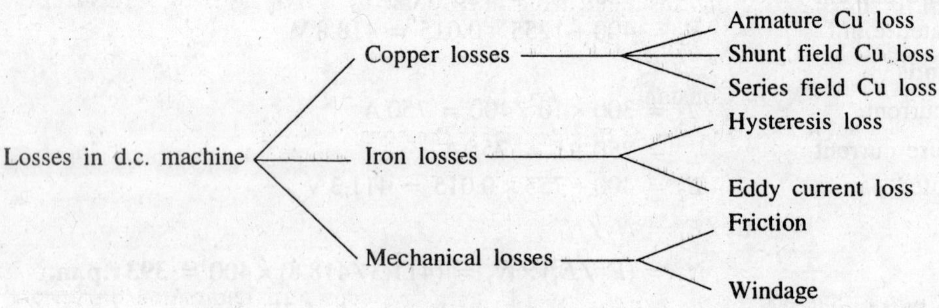

1. Copper losses. These losses occur due to currents in the various windings of the machine.

(*i*) Armature copper loss $= I_a^2 R_a$

(*ii*) Shunt field copper loss $= I_{sh}^2 R_{sh}$

(*iii*) Series field copper loss $= I_{se}^2 R_{se}$

Note. There is also brush contact loss due to brush contact resistance (*i.e.* resistance between the surface of brush and surface of commutator). This loss is generally included in armature copper loss.

2. Iron or Core losses. These losses occur in the armature of a d.c. machine and are due to the rotation of armature in the magnetic field of the poles. They are of two types *viz* (*i*) hysteresis loss (*ii*) eddy current loss.

(*i*) *Hysteresis loss*. Hysteresis loss occurs in the armature of the d.c. machine since any given part of the armature is subjected to magnetic field reversals as it passes under successive poles. Fig. 10.23 shows an armature rotating in two-pole machine. Consider a small piece *ab* of the armature. When the piece *ab* is under *N*-pole, the magnetic lines pass from *a* to *b*. Half a revolution later, the same piece of iron is under *S*-pole and magnetic lines pass from *b* to *a* so that magnetism in the iron is reversed. In order to reverse continuously

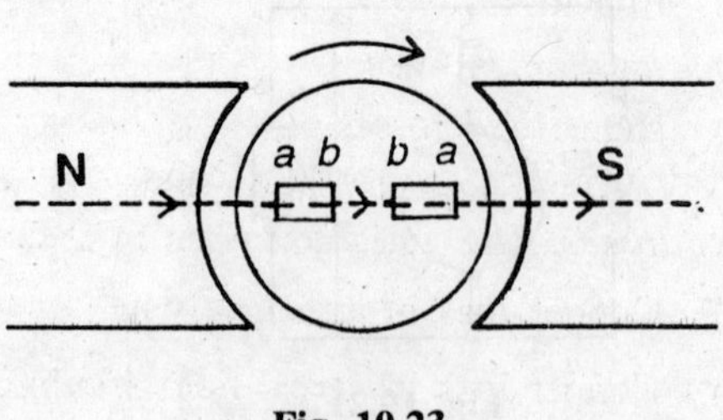

Fig. 10.23

the molecular magnets in the armature core, some amount of power has to be spent which is called hysteresis loss. It is given by :

Hysteresis loss, $P_h = \eta B_{max}^{1.6} f V$ watts

where
$$B_{max} = \text{Maximum flux density in armature}$$
$$f = \text{Frequency of magnetic reversals}$$
$$= NP/120 \text{ where } N \text{ is in r.p.m.}$$
$$V = \text{Volume of armature in m}^3$$
$$\eta = \text{Steinmetz hysteresis co-efficient}$$

In order to reduce this loss in a d.c. machine, armature core is made of such materials which have a low value of steinmetz hysteresis co-efficient *e.g.* silicon steel.

(*ii*) *Eddy current loss*. When armature rotates in the magnetic field of the poles, an e.m.f. is induced in it which circulates eddy currents in the armature core. The power loss due to these eddy currents is called eddy current loss. In order to reduce this loss, the armature core is built up of thin laminations insulated from each other by a thin layer of varnish.

Eddy current loss, $P_e = K_e B_{max}^2 f^2 t^2 V$ watts

where
$$K_e = \text{Constant}$$
$$B_{max} = \text{Maximum flux density in the core}$$
$$f = \text{Frequency of magnetic reversals}$$
$$t = \text{Thickness of lamination}$$
$$V = \text{Volume of core in } m^3$$

It may be noted that eddy current loss depends upon the square of lamination thickness. For this reason, lamination thickness should be kept as small as possible.

3. Mechanical losses. These losses are due to fricition and windage.

(*i*) friction loss *e.g.* bearing friction, brush friction etc.

(*ii*) windage loss *i.e.* air friction of rotating armature.

These losses depend upon the speed of the machine. But for a given speed, they are practically constant.

Note. Iron losses and mechanical losses together are called **stray losses.**

10.12. Constant and Variable Losses

The losses in a d.c. generator (or d.c. motor) may be sub-divided into (*i*) constant losses (*ii*) variable losses.

(*i*) **Constant losses**. Those losses in a d.c. generator which remain constant at all loads are known as *constant losses*. The constant losses in a d.c. generator are :

(*a*) iron losses

(*b*) mechanical losses

(*c*) shunt field losses

(*ii*) **Variable losses.** Those losses in a d.c. generator which vary with load are called *variable losses.* The variable losses in a d.c. generator are :

(*a*) Copper loss in armature winding ($I_a^2 R_a$)

(*b*) Copper loss in series field winding ($I_{se}^2 R_{se}$)

$$\text{Total losses} = \text{Constant losses} + \text{Variable losses}$$

10.13. Power Stages

The various power stages in a d.c. generator are represented diagrammatically in Fig. 10.24.

$$A - B = \text{Iron and fricition losses}$$

$$B - C = \text{Copper losses}$$

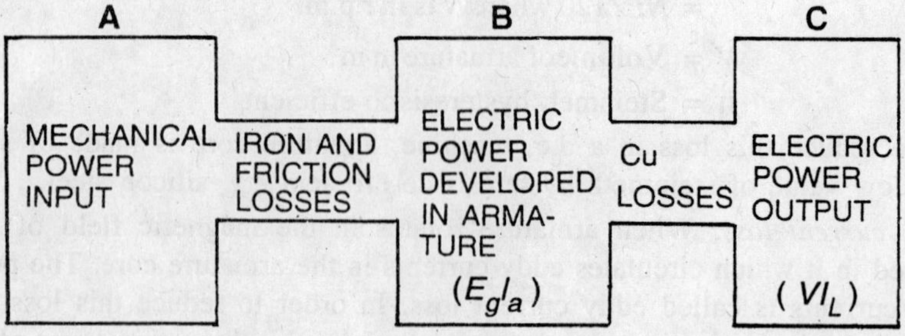

Fig. 10.24

(*i*) **Mechanical efficiency**

$$\eta_m = \frac{B}{A} = \frac{E_g I_a}{\text{Mechanical power input}}$$

(*ii*) **Electrical efficiency**

$$\eta_e = \frac{C}{B} = \frac{VI_L}{E_g I_a}$$

(*iii*) **Commercial or overall efficiency**

$$\eta_c = \frac{C}{A} = \frac{VI_L}{\text{Mechanical power input}}$$

Clearly, $\eta_c = \eta_m \times \eta_e$

Unless otherwise stated, commercial efficiency is always understood.

Now, commercial efficiency, $\eta_c = \dfrac{C}{A} = \dfrac{\text{output}}{\text{input}} = \dfrac{\text{input} - \text{losses}}{\text{input}}$

10.14. Condition for Maximum Efficiency

The efficiency of a d.c. generator is not constant but varies with load. Consider a shunt generator delivering a load current I_L at a terminal voltage V.

$$\text{Generator output} = VI_L$$

$$\begin{aligned}\text{Generator input} &= \text{Output} + \text{Losses} \\ &= VI_L + \text{Variable losses} + \text{Constant losses} \\ &= VI_L + I_a^2 R_a + W_C \\ &= VI_L + (I_L + I_{sh})^2 R_a + W_C \quad [\because I_a = I_L + I_{sh}]\end{aligned}$$

The shunt field current I_{sh} is generally small as compared to I_L and, therefore, can be neglected.

$$\therefore \quad \text{Generator input} = VI_L + I_L^2 R_a + W_C$$

$$\therefore \quad \eta = \frac{\text{output}}{\text{input}} = \frac{VI_L}{VI_L + I_L^2 R_a + W_C}$$

$$= \frac{1}{1+\left(\frac{I_L R_a}{V}+\frac{W_C}{VI_L}\right)} \quad \text{... (i)}$$

The efficiency will be maximum when the denominator of eq (*i*) is minimum *i.e.*

$$\frac{d}{dI_L}\left(\frac{I_L R_a}{V}+\frac{W_C}{VI_L}\right) = 0$$

or $$\frac{R_a}{V}-\frac{W_C}{VI_L^2} = 0$$

or $$\frac{R_a}{V} = \frac{W_C}{VI_L^2}$$

or $$I_L^2 R_a = W_C$$

i.e. **Variable loss = Constant loss** $(\because I_L \simeq I_a)$

The load current corresponding to maximum efficiency is given by ;

$$I_L = \sqrt{\frac{W_C}{R_a}}$$

Hence, the efficiency of a d.c. generator will be maximum when the load current is such that variable loss is equal to the constant loss.

Example 10.12. *A shunt generator supplies 96A at a terminal voltage of 200 volts. The armature and shunt field resistances are 0.1 Ω and 50 Ω respectively. The iron and frictional losses are 2500W. Find (i) e.m.f. generated (ii) copper losses (iii) commercial efficiency.*

Solution. Fig. 10.25 shows the connections of shunt generator.

(*i*) $I_{sh} = 200/50 = 4\text{ A}$

$I_a = I_L + I_{sh} = 96 + 4 = 100\text{ A}$

$E_g = V + I_a R_a = 200 + 100 \times 0.1 = \textbf{210 V}$

(*ii*) Armature Cu loss

$= I_a^2 R_a = (100)^2 \times 0.1 = 1000\text{ W}$

Shunt Cu loss $= I_{sh}^2 R_{sh} = (4)^2 \times 50 = 800\text{ W}$

Total Cu loss $= 1000 + 800 = \textbf{1800 W}$

(*iii*) Total losses = Stray losses + Cu losses

$= 2500 + 1800 = 4300\text{ W}$

Output power $= 96 \times 200 = \textbf{19200 W}$

Input power $= 19200 + 4300 = 23500\text{ W}$

$\therefore \quad \eta = \frac{19200}{23500} \times 100 = \textbf{81.7\%}$

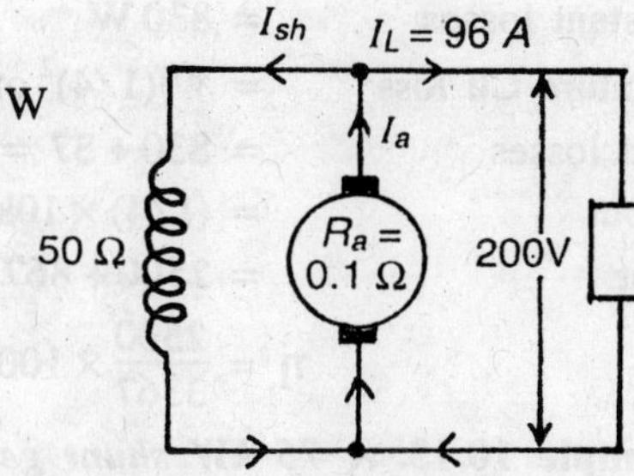

Fig. 10.25

Example 10.13. *The shunt generator delivers full load current of 200A at 240V. The shunt field resistance is 60Ω and full-load efficiency is 90%. The stray losses are 800W. Find (i) armature resistance (ii) current at which maximum efficiency occurs.*

Solution

(*i*) Generator output $= 240 \times 200 = 48000\text{ W}$

Generator input $= 48000/0.9 = 53333$ W

Total losses $= 53333 - 48000 = 5333$ W

Shunt field current, $I_{sh} = 240/60 = 4$ A

Armature current, $I_a = 200 + 4 = 204$ A

Shunt Cu loss $= I_{sh}^2 R_{sh} = (4)^2 \times 60 = 960$ W

Constant losses = Stray losses + Shunt Cu loss $= 800 + 960 = 1760$ W

∴ Armature Cu loss = Total losses – Constant losses $= 5333 - 1760 = 3573$ W

i.e. $I_a^2 R_a = 3573$

or $R_a = 3573/I_a^2 = 3573/(204)^2 = \mathbf{0.0858\ \Omega}$

(*ii*) For maximum efficiency, the condition is :

Variable losses = Constant losses

or $* I_L^2 R_a = 1760$

$$\therefore \quad I_L = \sqrt{\frac{1760}{0.0858}} = \mathbf{143.22\ A}$$

Example 10.14. *A 10kW d.c. shunt generator has the following losses at full load :*

Mechanical losses = 290W; *Iron losses* = 420 W

Shunt Cu loss = 120 W; *Armature Cu loss* = 595 W

Calculate the efficiency at (i) no load (ii) 25% of full-load.

Solution.

(*i*) **At no-load**

The stray losses remain constant at no load or at any other load.

Stray losses = Mechanical losses + Iron losses $= 290 + 420 = 710$ W

Total losses at no load = 710 + Shunt Cu loss $= 710 + 120 = 830$ W

The efficiency at no load is **zero** because there is no output.

(*ii*) **At 25% of Full-load**

Constant losses = 830 W ... *same as before*

Armature Cu loss $= **(1/4)^2$ of F.L. value $= (1/4)^2 \times 595 = 37$ W

Total losses $= 830 + 37 = 867$ W

Output $= (1/4) \times 10$kW $= 2.5$ kW $= 2500$ W

Input $= 2500 + 867 = 3367$ W

$$\therefore \quad \eta = \frac{2500}{3367} \times 100 = \mathbf{74\%}$$

Example 10.15. *A 75 kW shunt generator is operated at 230V. The stray losses are 1810W and shunt field circuit draws 5.35A. The armature circuit has a resistance of 0.035Ω and brush drop is 2.2V. Calculate (i) total losses (ii) input of prime mover (iii) efficiency at rated load.*

---

* Variable losses in armature are $I_a^2 R_a$. In the derivation of expression for maximum efficiency, I_{sh} was neglected so that $I_a = I_L$

** When the generator is loaded to 25% of its normal rating, armature current is 25% (or 1/4) of its full load value. Since Cu loss varies as square of current, we have $(1/4)^2$ term here.

Solution.

(*i*) Load current, $I_L = 75 \times 10^3/230 = 326.1$ A

Armature current, $I_a = I_L + I_{sh} = 326.1 + 5.35 = 331.5$ A

Armature Cu loss $= I_a^2 R_a = (331.5)^2 \times 0.035 = 3846$ W

Shunt Cu loss $= VI_{sh} = 230 \times 5.35 = 1230$ W

Brush power loss $=$ Brush drop $\times I_a = 2.2 \times 331.5 = 729$ W

$\therefore$ Total losses $= 1810 + 3846 + 1230 + 729 =$ **7615 W**

(*ii*) Output power $= 75$ kW $= 75000$ W

Input of prime mover $= 75000 + 7615 =$ **82615 W**

(*iii*) $\eta = \dfrac{75000}{82615} \times 100 = \mathbf{90.8\%}$

Tutorial Problems

1. A 4-pole shunt generator with wave wound armature has 41 slots, each having 12 conductors. The armature resistance is 0.05 Ω and shunt field resistance is 200 Ω. The flux per pole is 25 mwb. If a load resistance of 10 Ω is connected across the armature terminals, calculate the voltage across the load when generator is driven at 1000 r.p.m. **[389.5V]**
2. A long shunt compound generator has full load output of 100 kW at 250 volts. The armature, series and shunt windings have resistances of 0.05 Ω, 0.03 Ω and 55Ω respectively. Find the armature current and generated e.m.f. **[404.5A, 282.3V]**
3. A shunt generator supplies 75A at 200V through feeders of resistance 0.08 Ω. The armature and shunt field windings have resistance of 0.04 Ω and 80Ω respectively. Find the terminal voltage and generated e.m.f. **[206V, 209.1V]**
4. A 4-pole short shunt compound generator has armature, shunt field and series field resistances of 0.4 Ω, 160 Ω and 0.2 Ω respectively. The armature is lap connected with 440 conductors and is driven at 600 r.p.m. Calculate the flux per pole when the machine is delivering 120A at 400V. **[0.108 wb]**
5. The resistance of field circuit of a shunt-excited d.c. generator is 200 Ω. When the output of the generator is 100 kW, the terminal voltage is 500V and the generated e.m.f. 525V. Calculate /(*i*) the armature resistance (*ii*) the value of generated e.m.f. when the output is 60kW, if the terminal voltage is then 520V. **[(*i*) 0.123 Ω (*ii*) 534.5V]**

10.15. Armature Reaction

So far we have assumed that the only flux acting in a d.c. machine is that due to the main poles called *main flux*. However, current flowing through armature conductors also creates a magnetic flux (called *armature flux*) that distorts and weakens the flux coming from the poles. This distortion and field weakening takes place in both generators and motors. The action of armature flux on the main flux is known as *armature reaction.*

The phenomenon of armature reaction in a d.c. generator is shown in Fig. 10.26. Only one pole is shown for clarity. When the generator is on no-load, a small current flowing in the armature does not appreciably affect the main flux ϕ_1 coming from the pole [See Fig 10.26 (*i*)]. When the generator is loaded, the current flowing through armature conductors sets up flux ϕ_2. Fig 10.26. (*ii*) shows flux due to armature current alone. By superimposing ϕ_1 and ϕ_2, we obtain the resulting flux ϕ_3 as shown in Fig 10.26 (*iii*).

Referring to Fig 10.26 (*iii*), it is clear that flux density at the trailing pole tip (point *B*) is increased while at the leading pole tip (point *A*), it is decreased. This unequal field distribution produces the following two effects :

(*i*) The main flux is distorted.

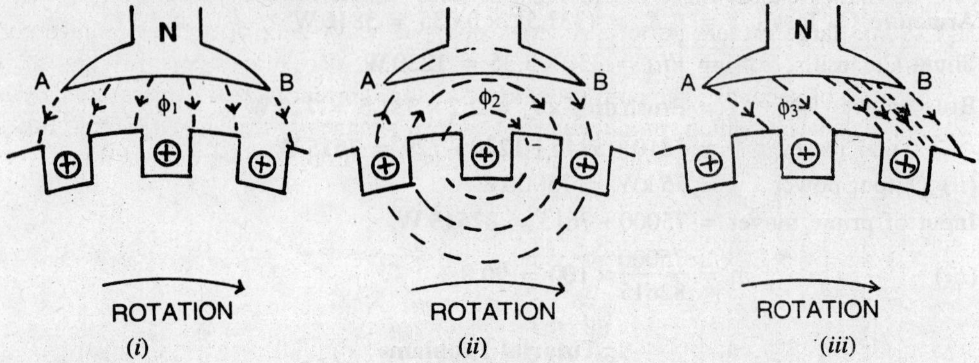

(*i*) (*ii*) (*iii*)

Fig. 10.26

(*ii*) Due to higher flux density at pole tip *B*, saturation sets in. Consequently, the increase in flux at pole tip *B* is less than the decrease in flux under pole tip *A*. Flux ϕ_3 at full load is, therefore, less than flux ϕ_1 at no load.

10.16 Problems of Armature Reaction

Armature reaction in a d.c. machine distorts and weakens the main flux. This invites the following two problems :

(*i*) The armature reaction may decrease the main flux by 10%. This causes the generated e.m.f. in a generator to fall with the increase in load current.

(*ii*) To ensure good commutation, the brushes must be placed in the neutral zone. However, due to distortion of flux, neutral zone is shifted. If the load varies frequently, the shifting of brush position is not practicable.

10.17. Correcting Armature Reaction Effect

There are four main ways of overcoming the armature reaction problem:

(*i*) Rotate the brush hanger mechanism to find the correct neutral zone position. Such a method can be applied only for fixed load current.

(*ii*) Shape or otherwise modify the ends of the field pole shoes so that high flux cannot exist on the ends because of the high reluctance path.

(*iii*) To counter the effect of armature reaction, we usually place a set of *interpoles* or *commutating poles* mid-way between the main poles of a d.c. machine as shown in Fig. 10.27. The polarity of the commutating pole must be that of the main pole just ahead of it (*i.e.* in the direction of rotation) for a generator (See Fig 10.27) and just behind for a motor. The interpoles set up flux equal and opposite to the armature flux. The commutating-pole windings are connected

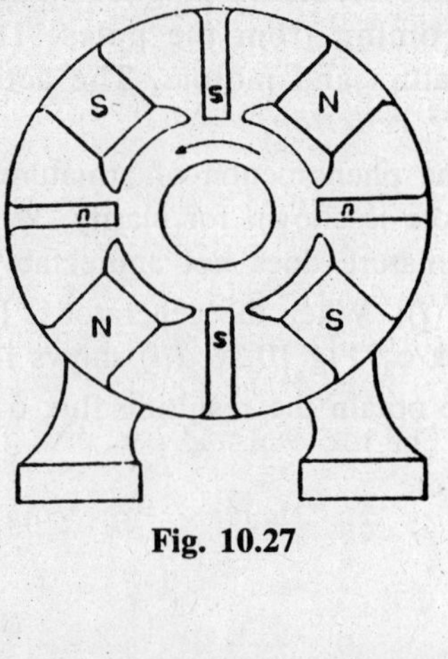

Fig. 10.27

in series with the armature so that respective fluxes rise and fall together as the load current changes. By cancelling the armature effect in this way, we no longer have to shift the brushes. Commutating poles are now invariably fitted to all d.c. machines except those of the smallest sizes or the lowest speeds.

(*iv*) Some large motors perform a series of rapid heavy-duty operations (*e.g.* motors in steel mills, rolling mills *etc.*). They accelerate, decelerate, stop, reverse, all in a matter of seconds. The corresponding armature current increases, decreases, reverses in step wise fashion, producing very sudden changes in armature reaction. For such motors, commutating poles do not adequately neutralise the armature flux.

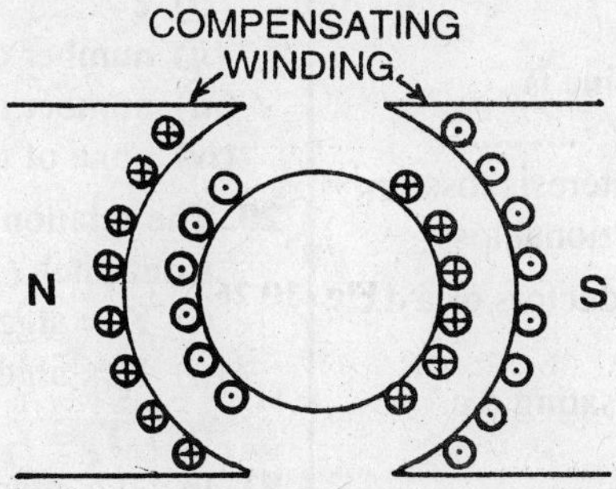

Fig. 10.28

To eliminate this problem, additional *compensating windings* are used. A compensating winding is an auxiliary winding embedded in slots in the pole faces of the main poles. The compensating winding is connected in series with the armature in a manner so that direction of current through the compensating conductors in any one pole face will be opposite to the direction of current through the adjacent armature conductors as shown in Fig. 10.28. The compensating windings produce a flux equal and opposite to the armature flux and thus completely neutralise the armature reaction. The addition of a compensating winding considerably increases the cost of the machine and is justified only for machines intended for unusually severe service.

OBJECTIVE QUESTIONS

1. The yoke of a d.c. machine is made of........
   (*i*) silicon steel (*ii*) soft iron
   (*iii*) aluminium (*iv*) cast steel
2. The armature of a d.c. machine is made of ....
   (*i*) silicon steel (*ii*) wrought iron
   (*iii*) cast steel (*iv*) soft iron
3. The coupling field between electrical and mechanical systems of a d.c. machine is ......
   (*i*) electric field (*ii*) magnetic field
   (*iii*) both electric and magnetic fields
   (*iv*) none of the above
4. The real working part of a d.c. machine is the ....
   (*i*) commutator (*ii*) field winding
   (*iii*) armature winding (*iv*) none of the above
5. ....d.c. machines are most common.
   (*i*) 2-pole (*ii*) 4-pole
   (*iii*) 6-pole (*iv*) 8-pole
6. The armature winding of a d.c. machine is place on the rotor to ....
   (*i*) save iron (*ii*) reduce losses
   (*iii*) facilitate commutation
   (*iv*) reduce armature reaction
7. The yoke of a d.c. machine carries ....pole flux.
   (*i*) the (*ii*) one-half of
   (*iii*) two times the (*iv*) none of the above
8. A 4-pole d.c. machine has ....magnetic circuits.

(*i*) 2 (*ii*) 8
(*iii*) 4 (*iv*) none of the above

9. The field structure of a d.c. machine uses .....
(*i*) salient-pole arrangement
(*ii*) non-salient pole arrangement
(*iii*) both (*i*) and (*ii*)
(*iv*) none of the above

10. Small d.c. machines generally have ....poles.
(*i*) 4 (*ii*) 6
(*iii*) 2 (*iv*) 8

11. The armature of a d.c. machine is laminated in order to reduce...............
(*i*) eddy current loss (*ii*) hysteresis loss
(*iii*) copper loss (*iv*) frictional loss

12. The current in armature conductors of a d.c. machine is ....
(*i*) pure d.c. (*ii*) pulsating d.c.
(*iii*) a.c.
(*iv*) pure d.c. plus pulsating d.c.

13. The greatest eddy current loss occurs in the ....of a d.c. machine.
(*i*) field poles (*ii*) yoke
(*ii*) commutating poles
(*iv*) armature

14. The brush voltage drop in a d.c. machine is about................
(*i*) 0.1 V (*ii*) 10 V
(*iii*) 2 V (*iv*) 20 V

15. Carbon brushes are used in a d.c. machine because ....
(*i*) carbon lubricates and polishes the commutator
(*ii*) contact resistance is decreased
(*iii*) carbon is cheap
(*iv*) none of the above

16. The armature winding of a d.c. machine is ....winding.
(*i*) an open-circuit (*ii*) a closed-circuit
(*iii*) partly open-circuit and partly closed circuit
(*iv*) none of the above

17. The commutator pitch for a simplex lap winding is equal to ....
(*i*) number of poles of the machine
(*ii*) pole pairs
(*iii*) 1
(*iv*) none of the above

18. In a simplex wave winding, the number of parallel paths is equal to ....
(*i*) number of poles in the machine
(*ii*) 2
(*iii*) number of pairs of poles
(*iv*) none of the above

19. The number of parallel paths in a simplex lap winding is equal to ....
(*i*) 2
(*ii*) number of pairs of poles
(*iii*) number of poles
(*iv*) none of the above

20. The relation among commutator pitch (Y_c), front pitch (Y_f) and back pitch (Y_b) is ....
(*i*) Y_c = algebraic sum of Y_f and Y_b
(*ii*) Y_c = algebraic difference of Y_f and Y_b
(*iii*) $Y_c = Y_b/Y_f$ (*iv*) none of the above

21. In a d.c. machine, the number of commutator segments is equal to ....
(*i*) number of conductors
(*ii*) twice the number of poles
(*iii*) number of coils
(*iv*) none of the above

22. The nature of armature winding of a d.c. machine is decided by ....
(*i*) front pitch (*ii*) commutator pitch
(*iii*) back pitch (*iv*) none of the above

23. High-voltage d.c. machines use ....winding.
(*i*) lap (*ii*) wave
(*iii*) either lap or wave
(*iv*) none of the above

24. In a lap winding, the number of the brushes required is equal to ....
(*i*) number of poles
(*ii*) number of pairs of poles
(*iii*) commutator pitch
(*iv*) none of the above

25. In a wave winding, the commutator pitch is approximately equal to ....
(*i*) pole pitch
(*ii*) twice the pole pitch
(*iii*) thrice the pole pitch
(*iv*) none of the above

26. For a given number of poles (> 2) and armature conductors, a lap winding will carry

.... a wave winding.
(*i*) more current than
(*ii*) less current than
(*iii*) same current as
(*iv*) none of the above.

27. An 8-pole, 250 V, wave-wound generator has 400 conductors. If the generator is to be lap-wound, the number of conductors required is ....
(*i*) 800 (*ii*) 100
(*iii*) 3200 (*iv*) 1600

28. An 8-pole duplex lap winding will have ....parallel paths.
(*i*) 8 (*ii*) 4
(*iii*) 32 (*iv*) 16

29. A triplex wave winding will have ....parallel paths.
(*i*) 6 (*ii*) 2
(*iii*) 4 (*iv*) none of the above

30. For a given d.c. generator, the generated voltage depends upon.........
(*i*) flux only (*ii*) speed only
(*iii*) both speed and flux
(*iv*) none of the above

31. A 6-pole lap-wound generator has 300 conductors; the e.m.f. induced per conductor being 5 V. The generated voltage of the generator is ....
(*i*) 60 V (*ii*) 1500 V
(*iii*) 360 V (*iv*) 250 V

32. For the same rating, a d.c. machine has ....an a.c. machine.
(*i*) the same weight as
(*ii*) more weight than
(*iii*) less weight than
(*iv*) none of the above.

33. The field winding of a d.c. shunt machine usually carries .... of the rated current of the machine.
(*i*) 2% to 5%
(*ii*) 15% to 20%
(*iii*) more than 20%
(*iv*) less than 0.5%

34. A separately excited d.c. generator is not used because ....
(*i*) it is costly
(*ii*) separate d.c. source is required for field circuit
(*iii*) voltage drops considerably with load
(*iv*) none of the above.

35. The greatest percentage of heat loss in a d.c. machine is due to ....
(*i*) eddy current loss (*ii*) hysteresis loss
(*iii*) copper loss (*iv*) frictional loss

36. The size of a d.c. generator can be reduced by using ....
(*i*) lap winding
(*ii*) high-resistance winding material
(*iii*) iron commutator
(*iv*) magnetic material of high permeability

37. Hysteresis loss in a d.c. machine is directly proportional to ....
(*i*) speed (*ii*) $(\text{speed})^2$
(*iii*) $(\text{speed})^{1.6}$ (*iv*) none of the above

38. The physical size of 100 kW, 2000 r.p.m. generator is .... that of 10 kW, 200 r.p.m. generator.
(*i*) more than (*ii*) less than
(*iii*) $(\text{speed})^{1.6}$
(*iv*) none of the above

39. ....together are called stray losses.
(*i*) copper and iron losses
(*ii*) iron and mechanical losses
(*iii*) field copper loss and iron loss
(*iv*) frictional loss and copper loss

40. The armature conductors of a 6-pole, lap wound d.c. generator are divided into
(*i*) 2 parallel paths (*ii*) 4 parallel paths
(*iii*) 3 parallel paths (*iv*) 6 parallel paths

ANSWERS TO OBJECTIVE QUESTIONS

1. (*iv*)	**2.** (*i*)	**3.** (*ii*)	**4.** (*iii*)	**5.** (*ii*)	**6.** (*iii*)	**7.** (*ii*)	**8.** (*iii*)
9. (*i*)	**10.** (*iii*)	**11.** (*i*)	**12.** (*iii*)	**13.** (*iv*)	**14.** (*iii*)	**15.** (*i*)	**16.** (*ii*)
17. (*iii*)	**18.** (*ii*)	**19.** (*iii*)	**20.** (*ii*)	**21.** (*iii*)	**22.** (*ii*)	**23.** (*ii*)	**24.** (*i*)
25. (*ii*)	**26.** (*i*)	**27.** (*iv*)	**28.** (*iv*)	**29.** (*i*)	**30.** (*iii*)	**31.** (*iv*)	**32.** (*ii*)
33. (*i*)	**34.** (*ii*)	**35.** (*iii*)	**36.** (*iv*)	**37.** (*i*)	**38.** (*iii*)	**39.** (*ii*)	**40.** (*iv*)

HINTS TO SELECTED OBJECTIVE QUESTIONS

2. It is desired that armature should provide a path of low reluctance for the flux. For this reason, armature is made of material having high permeability *e.g.* silicon steel.

8. In a d.c. machine, there are as many magnetic circuits as there are poles. Thus, a 4-pole d.c. machine will have 4 magnetic circuits. However, m.m.f. acting in each circuit is due to the ampere-turns of a pair of poles.

13. Since every part of the armature sees cyclic reversals in magnetic flux direction, it is potentially subjected to more severe eddy current-losses than the field pole shoes. Hence armature is invariably laminated to reduce the eddy current loss.

17. The commutator pitch for a simplex lap winding is given by ;

$$Y_c = \pm 1$$

The plus sign applies if, as is usual, the winding is *progressive* ; the minus sign, if it is *retrogressive*.

21. In a d.c. armature winding, the two ends of a coil are connected to two different commutator segments and two coils are connected to each segment so that number of commutator segments is equal to the number of coils.

22. It is the commutator pitch that determines the grouping of the coils in series and parallel. In other words, it is the commutator pitch that determines as to how many parallel paths, each consisting of equal number of coils in series, will be formed. Hence, the nature of winding is determined by the commutator pitch.

28. With a *n*-plex lap winding, the number of parallel paths is $P\,n$ where

P = number of poles

n = 2, 3 etc and stand for duplex, triplex etc.

Hence 8-pole duplex lap winding will have $8 \times 2 = 16$ parallel paths.

29. With an *n*-plex wave winding, the number of parallel paths is $2n$ where n = 2, 3 etc. and stands for duplex, triplex etc. Hence a triplex wave winding will have $2 \times 3 = 6$ parallel paths.

31. Number of conductors per parallel path = 300/6 = 50. Hence the generator voltage = 50 × 5 = 250 V.

35. The armature current of a d.c. machine is very large. Since copper loss is proportional to the square of current, hence a greater part of power loss occurs in the armature.

11

D.C. Generator Characteristics

Introduction

The speed of a d.c. machine operated as a generator is fixed by the prime mover. For general-purpose operation, the prime mover is equipped with a speed governor so that the speed of the generator is practically constant. Under such condition, the generator performance deals primarily with the relation between excitation, terminal voltage and load. These relations can be best exhibited graphically by means of curves known as generator characteristics. These characteristics show at a glance the behaviour of the generator under different load conditions.

11.1. D.C. Generator Characteristics

The following are the three most important characteristics of a d.c. generator :

1. Open Circuit Characteristic (O.C.C.)

This curve shows the relation between the generated *e.m.f.* at no-load (E_o) and the field current (I_f) at constant speed. It is also known as magnetic characteristic or no-load saturation curve. Its shape is practically the same for all generators whether separately or self-excited.

2. Internal or Total Characteristic (E/I_a)

This curve shows the relation between the generated e.m.f. on load (E) and the armature current (I_a). The e.m.f. E will be less than E_o due to the effects of armature reaction. Therefore, this curve will lie below the open circuit characteristic.

3. External Characteristic (V/I_L)

This curve shows the relation between the terminal voltage (V) and load current (I_L). The terminal voltage V will be less than E due to voltage drop in the armature circuit. Therefore, this curve will lie below the internal characteristic.

11.2. Open Circuit Characteristic

The O.C.C. for a d.c. generator is determined as follows. The field winding of the d.c. generator (series or shunt) is disconnected from the machine and is separately excited from an external d.c. source as shown in Fig. 11.1 (*ii*). The generator is run at fixed speed (*i.e.*, normal speed). The field current (I_f) is increased from zero in steps and the corresponding values of generated e.m.f. (E_o) read off on a voltmeter connected across the armature terminals. On plotting the relation between E_o and I_f, we get the open circuit characteristic as shown in fig. 11.1 (*i*).

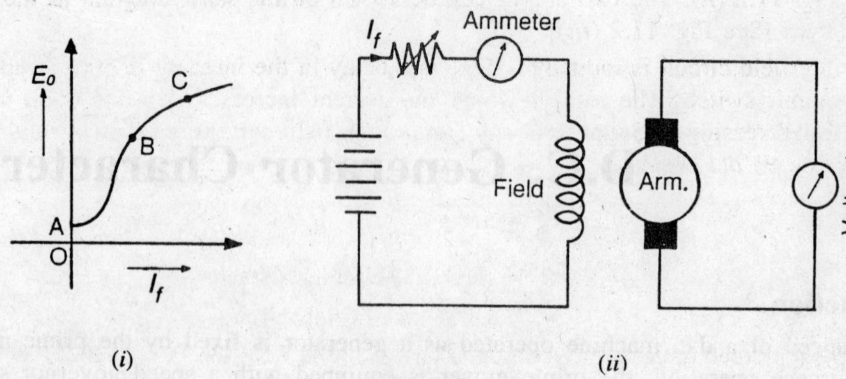

Fig. 11.1

The following points may be noted from O.C.C. :

(*i*) When the field current is zero, there is some generated e.m.f. *OA*. This is due to the residual magnetism in the field poles.

(*ii*) Over a fairly wide range of field current (upto point *B* in the curve), the curve is linear. It is because in this range, *reluctance of iron is negligible as compared with that of air gap. The air gap reluctance is constant and hence linear relationship.

(*iii*) After point *B* on the curve, the reluctance of iron also comes into picture. It is because at higher flux densities, μ_r for iron decreases and reluctance of iron is no longer negligible. Consequently, the curve deviates from linear relationship.

(*iv*) After point *C* on the curve, the magnetic saturation of poles begins and E_0 tends to level off.

11.3. Generator Build-up

Let us see how voltage builds up in a self excited d.c. generator. Consider a shunt generator. If the generator is run at a constant speed, some e.m.f. will be generated due to residual magnetism in the main poles. This small e.m.f. circulates a field current which in turn produces additional flux to reinforce the original residual flux (provided field winding connections are correct). This process continues and the generator builds up the normal generated voltage following the O.C.C. shown in Fig. 11.2 (*i*).

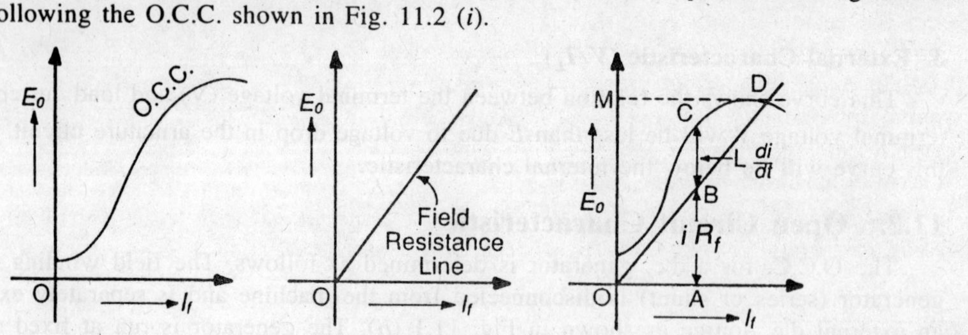

Fig. 11.2

* At low flux density, μ_r for iron is high and hence iron path offers negligible reluctance. Remember μ_r for iron depends upon flux density. As flux density increases, μ_r decreases.

The field resistance R_f can be represented by a straight line passing through the origin as shown in Fig. 11.2 (*ii*). The two curves can be shown on the same diagram as they have the same ordinates [See Fig. 11.2 (*iii*)].

Since the field circuit is inductive, there is a delay in the increase in current upon closing the field circuit switch. The rate at which the current increases depends upon the voltage available for increasing it. Suppose at any instant, the field current is i (= OA) and is increasing at the rate di/dt. Then.

$$E_o = i R_f + L\frac{di}{dt}$$

where R_f = total field circuit resistance

L = inductance of field circuit.

At the considered instant, the total e.m.f. available is AC [See Fig. 11.2 (*iii*)]. An amount AB of the e.m.f. AC is absorbed by the voltage drop $i R_f$ and the remainder part BC is available to overcome $L\, di/dt$. Since this surplus voltage is available, it is possible for the field current to increase above the value OA. However, at point D, the available voltage is OM and is all absorbed by $i R_f$ drop. Consequently, the field current cannot increase further and the generator build up stops.

We arrive at a very important conclusion that the voltage build up of the generator is given by the point of intersection of O.C.C. and field resistance line. Thus in Fig. 11.2 (*iii*), D is point of intersection of the two curves. Hence, the generator will build up a voltage OM.

Note : The voltage build up in a series generator is exactly like that of a shunt generator. Here, voltage build up will occur if the whole series circuit is closed.

11.4. Critical Field Resistance for a Shunt Generator

We have seen above that voltage build up in a shunt generator depends upon field circuit resistance. If the field circuit resistance is R_1 (line OA), then generator will build up a voltage OM as shown in *Fig. 11.3. If the field circuit resistance is **increased to R_2 (line OB), the generator will build up a voltage OL, slightly less than OM. As the field circuit resistance is increased, the slope of resistance line also increases. When the field resistance line becomes tangent (line OC) to O.C.C., the generator would *just* excite. If the field circuit resistance is increased beyond this point (say line OD), the generator will †fail to excite. The field circuit resistance represented by line OC (tangent to O.C.C.) is called *critical field resistance* R_C for the shunt generator. It may be defined as under :

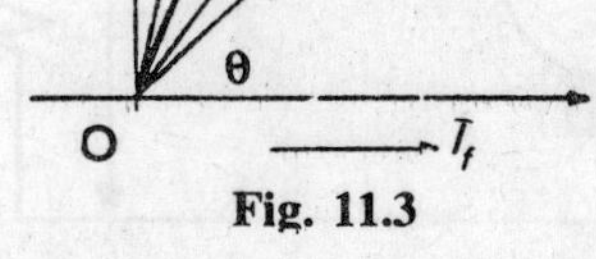

Fig. 11.3

The maximum field circuit resistance (for a given speed) with which the shunt generator would just excite is known as its **critical field resistance.**

* The effect of residual magnetism has been neglected in drawing O.C.C. so that it starts from the origin.

** Resistance = slope of line = $\tan\theta$

Here θ is the angle which the lines makes with x-axis. As θ increases, $\tan\theta$ and hence resistance increases.

† Since now all the available e.m.f. is immediately absorbed by $i R_f$ drop.

It should be noted that shunt generator will build up voltage only if field circuit resistance is less than critical field resistance.

11.5. Critical Resistance for a Series Generator

Fig. 11.4 shows the voltage build up in a series generator. Here R_1, R_2 *etc.* represent the total circuit resistance (load resistance and field winding resistance). If the total circuit resistance is R_1, then series generator will build up a voltage *OL*. The line *OC* is tangent to O.C.C. and represents the critical resistance R_C for a series generator. If the total resistance of the circuit is more than R_C, (say line *OD*), the generator will fail to build up voltage. Note that Fig. 11.4 is similar to Fig. 11.3 with the following differences :

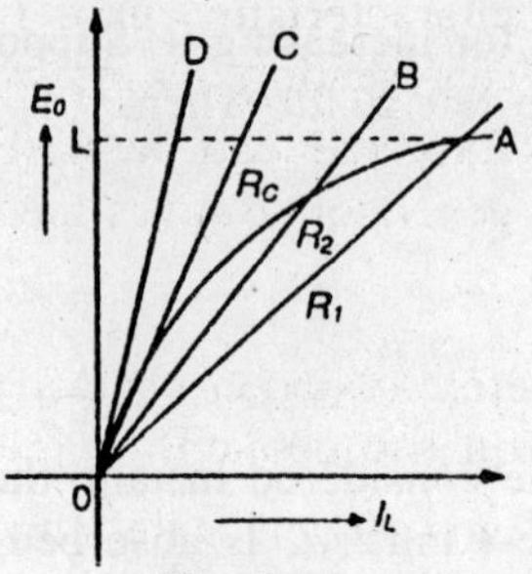

Fig. 11.4

(*i*) In Fig. 11.3, R_1, R_2 *etc.* represent the total field circuit resistance. However, R_1, R_2 *etc.* in Fig. 11.4 represent the total circuit resistance (load resistance and series field winding resistance *etc.*).

(*ii*) In Fig. 11.3, field current alone is represented along *X*-axis. However, in Fig. 11.4, load current I_L is represented along *X*-axis. Note that in a series generator, field current = load current I_L

11.6. Characteristics of Series Generator

Fig. 11.5(*i*) shows the connections of a series wound generator. Since there is only one current (that which flows through the whole machine), the load current is the same as the exciting current.

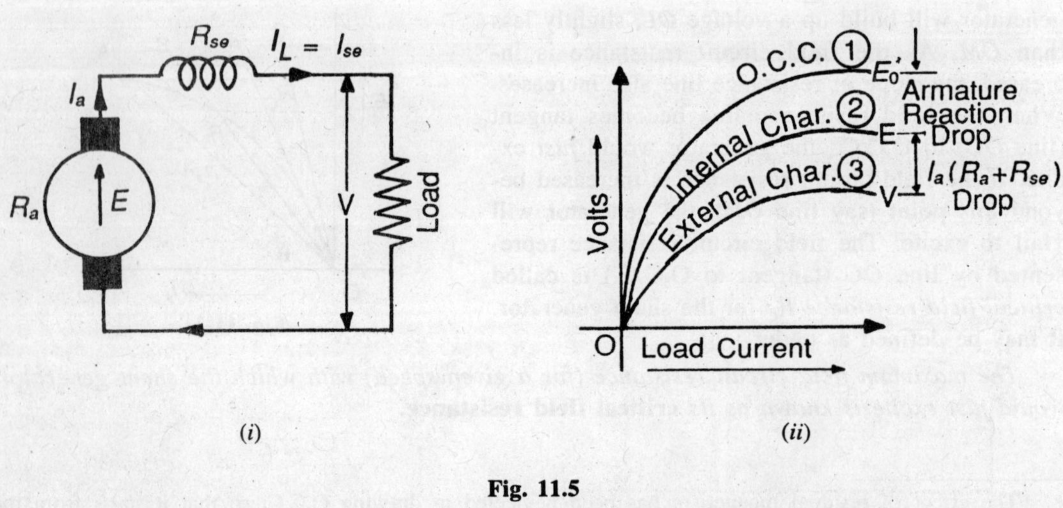

Fig. 11.5

(*i*) **O.C.C.** Curve 1 shows the open circuit characteristic (O.C.C.) of a series generator. It can be obtained experimentally by disconnecting the field winding from the machine and exciting it from a separate d.c. source as discussed in Art 11.2.

(*ii*) **Internal characteristic** Curve 2 shows the total or internal characteristic of a series generator. It gives the relation between the generated e.m.f. E on load and *armature current. Due to armature reaction, the flux in the machine will be less than the flux at no load. Hence, e.m.f. E generated under load conditions will be less than the e.m.f. E_0 generated under no load conditions. Consequently, internal characteristic curve lies below the O.C.C. curve; the difference between them representing the effect of armature reaction [See Fig. 11.5 (*ii*)].

(*iii*) **External characteristic** Curve 3 shows the external characteristic of a series generator. It gives the relation between terminal voltage V and load current I_L.

$$V = E - I_a (R_a + R_{se})$$

Therefore, external characteristic curve will lie below internal characteristic curve by an amount equal to ohmic drop [*i.e.* $I_a (R_a + R_{se})$] in the machine as shown in Fig. 11.5 (*ii*).

The internal and external characteristics of a d.c. series generator can be plotted from one another as shown in Fig. 11.6. Suppose we are given the internal characteristic of the generator. Let the line OC represent the resistance of the whole machine *i.e.* $R_a + R_{se}$. If the load current is OB, drop in the machine is AB *i.e.*

$$AB = \text{ohmic drop in the machine}$$
$$= OB\,(R_a + R_{se})$$

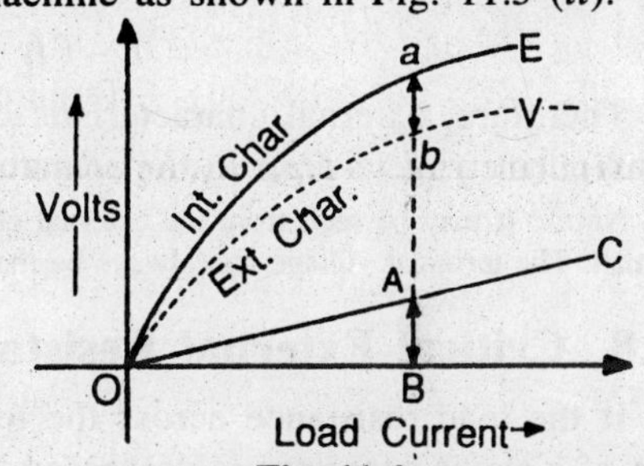

Fig. 11.6

Now raise a perpendicular from point B and mark a point b on this line such that $ab = AB$. Then point b will lie on the external characteristic of the generator. Following similar procedure, other points of external characteristic can be located. It is easy to see that we can also plot internal characteristic from the external characteristic.

11.7. Characteristics of a Shunt Generator

Fig. 11.7. (*i*) shows the connections of a shunt wound generator. The armature current I_a splits up into two parts; a small fraction I_{sh} flowing through shunt field winding while the major part I_L goes to the external load.

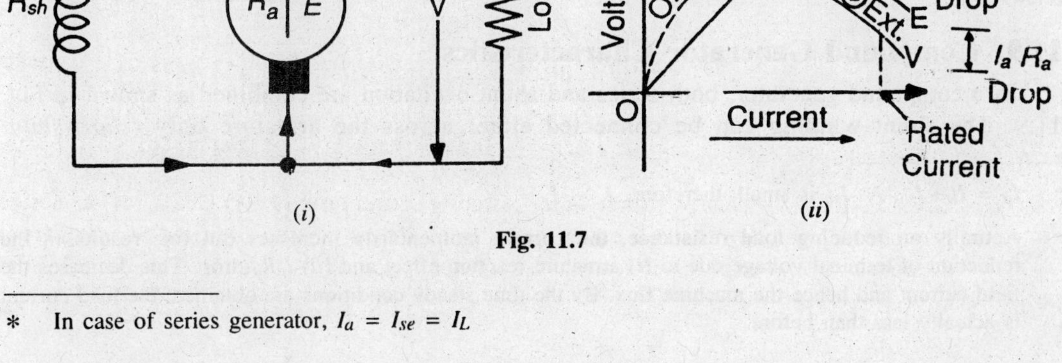

Fig. 11.7

---

* In case of series generator, $I_a = I_{se} = I_L$

(*i*) **O.C.C.** The O.C.C. of a shunt generator is similar in shape to that of a series generator as shown in Fig. 11.7 (*ii*). The line *OA* represents the shunt field circuit resistance. When the generator is run at normal speed, it will build up a voltage *OM*. At no-load, the terminal voltage of the generator will be constant (= *OM*) represented by the horizontal dotted line *MC*.

(*ii*) **Internal characteristic**. When the generator is loaded, flux per pole is reduced due to armature reaction. Therefore, e.m.f. *E* generated on load is less than the e.m.f. generated at no load. As a result, the internal characteristic *(E/I_a) drops down slightly as shown in Fig. 11.7 (*ii*).

(*iii*) **External characteristic**. Curve 2 shows the external characteristic of a shunt generator. It gives the relation between terminal voltage *V* and load current I_L.

$$V = E - I_a R_a$$

$$= E - (I_L + I_{sh})\, R_a$$

Therefore, external characteristic curve will lie below the internal characteristic curve by an amount equal to drop in the armature circuit [*i.e.*, $(I_L + I_{sh})\, R_a$] as shown in Fig. 11.7 (*ii*).

Note : It may be seen from the external characteristic that change in terminal voltage from no-load to full load is small. The terminal voltage can always be maintained constant by adjusting the field rheostat R.

11.8. Critical External Resistance for Shunt Generator

If the load resistance across the terminals of a shunt generator is decreased, then load current increases. However, there is a limit to the increase in load current with the decrease of load resistance. Any decrease of load resistance beyond this point, instead of increasing the current, ultimately results in **reduced current. Consequently, the external characteristic turns back (dotted curve) as shown in Fig. 11.8.

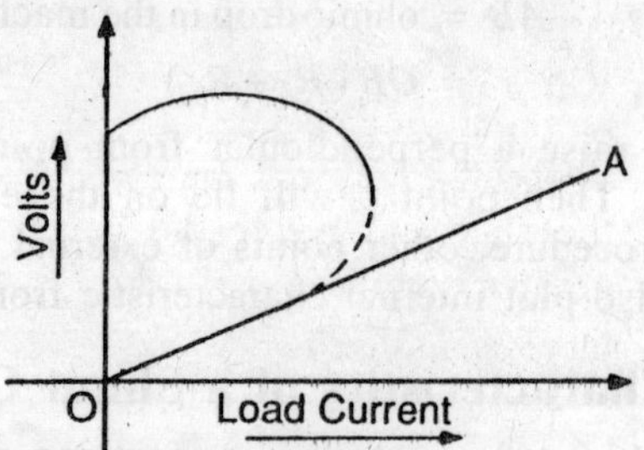

Fig. 11.8

The tangent *OA* to the curve represents the minimum external resistance required to excite the shunt generator on load and is called *critical external resistance*. If the resistance of the external circuit is less than the critical external resistance (represented by tangent *OA* in Fig. 11.8), the machine will refuse to excite or will de-excite if already running. This means that external resistance is so low as virtually to short circuit the machine and so doing away with its excitation.

Note : There are two critical resistances for a shunt generator *viz.* (*i*) critical field resistance (*ii*) critical external resistance.

For the shunt generator to build up voltage, the former should not be exceeded and the latter must not be gone below.

11.9. Compound Generator Characteristics

In a compound generator, both series and shunt excitation are combined as shown in Fig. 11.9. The shunt winding can be connected either across the armature only (*short-shunt*

---

* $I_a = I_L + I_{sh}$. As I_{sh} is small, therefore, $I_a \simeq I_L$

** Actually on reducing load resistance, the current momentarily increases but this results in the reduction of terminal voltage due to (*i*) armature reaction effect and (*ii*) $I_a R_a$ drop. This decreases the field current and hence the machine flux. By the time steady conditions are obtained, the load current is actually less than before.

connection S) or across armature plus series field (*long-shunt connection* G). The compound generator can be *cumulatively compounded or differentially compounded generator. The latter is rarely used in practice. Therefore, we shall discuss the characteristics of cumulatively-compounded generator. It may be noted that external characteristics of long and short shunt compound generators are almost identical.

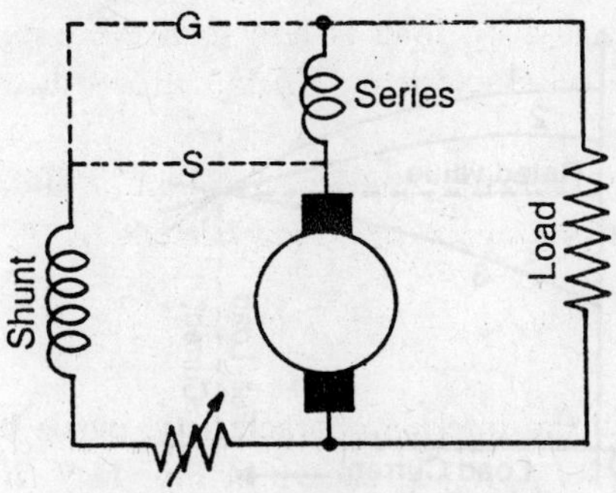

Fig. 11.9

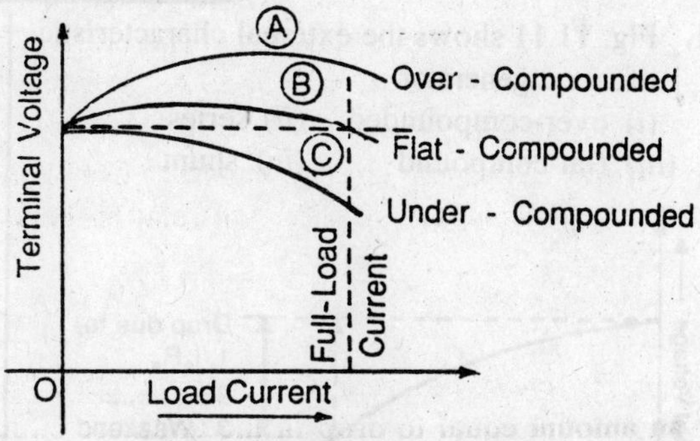

Fig. 11.10

External characteristic. Fig. 11.10 shows the external characteristics of a cumulatively compounded generator. The series excitation aids the shunt excitation. The degree of compounding depends upon the increase in series excitation with the increase in load current.

(*i*) If series winding turns are so adjusted that with the increase in load current the terminal voltage increases, it is called *over-compounded generator*. In such a case, as the load current increases, the series field m.m.f. increases and tends to increase the flux and hence the generated voltage. The increase in generated voltage is greater than the $I_a R_a$ drop so that instead of decreasing, the terminal voltage increases as shown by curve *A* in Fig. 11.10.

(*ii*) If series winding turns are so adjusted that with the increase in load current, the terminal voltage substantially remains constant, it is called *flat-compounded generator*. The series winding of such a machine has lesser number of turns than the one in over-compounded machine and, therefore, does not increase the flux as much for a given load current. Consequently, the full-load voltage is nearly equal to the no-load voltage as indicated by curve *B* in Fig. 11.10.

(*iii*) If series field winding has lesser number of turns than for a flat-compounded machine, the terminal voltage falls with increase in load current as indicated by curve *C* in Fig. 11.10. Such a machine is called *under-compounded generator*.

11.10. Applications of Compound Generators

In some applications, we can tolerate a reasonable drop in terminal voltage as the load current increases but this is unacceptable in lighting circuits. For example, the distribution system of a ship supplies power to both d.c. machinery and incandescent lamps. The current delivered by the generator changes continuously in response to the varying loads. These current variations produce the corresponding changes in the generator terminal voltage, causing the lights to flicker. To eliminate such voltage fluctuations, we use *over-compounded generators*. In an over-compounded generator, as the load current increases, the generator terminal

---

* (*i*) If the series and shunt excitation help each other, it is called cumulatively-compounded generator.
(*ii*) If shunt and series excitation oppose each other, it is called differentially-compounded generator

voltage rises above the no-load value. The rise in voltage compensates for (*i*) $I_a R_a$ drop in the generator and (*ii*) the *IR* drop in the line between the generator and load.

Note : If compounding is too strong, a low resistance can be placed in parallel with the series field winding. This reduces the field current and has the same effect as reducing the number of turns of series winding.

OBJECTIVE QUESTION

1. Fig. 11.11 shows the external characteristic of ....generator.
   (*i*) over-compounded (*ii*) series
   (*iii*) flat-compound (*iv*) shunt

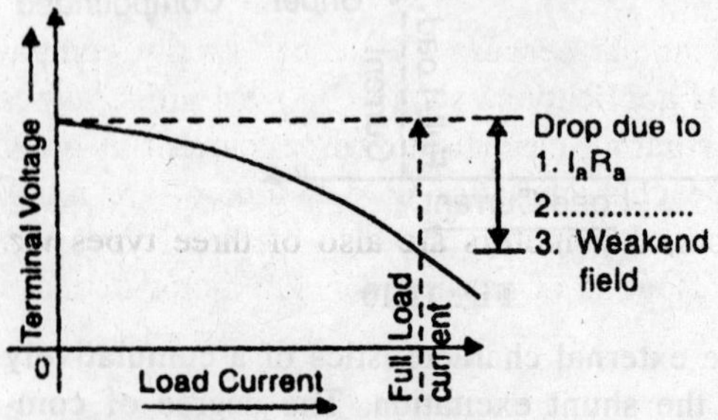

Fig. 11.11

2. Referring to Fig. 11.11, the voltage drop at no. 2 is due to.........
   (*i*) friction (*ii*) armature reaction
   (*iii*) field circuit (*iv*) none of the above
3. Fig. 11.12, shows the external characteistics of a .... generator.

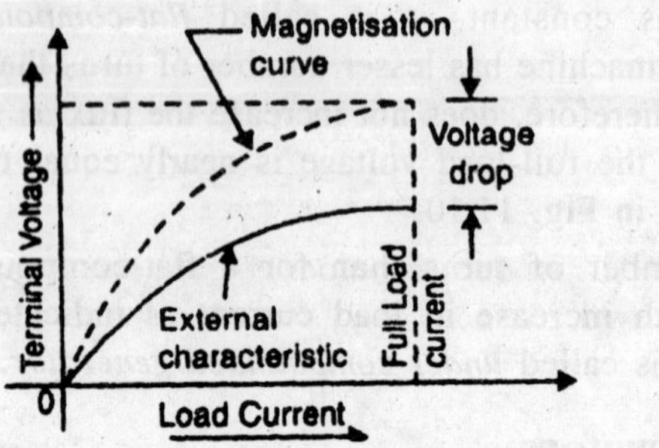

Fig. 11.12

   (*i*) shunt (*ii*) flat-compounded
   (*iii*) series (*iv*) over-compounded
4. Fig. 11.13, shows the external characteristics of three types of generators having the same rating. Curve 2 represents for .... generator.
   (*i*) shunt (*ii*) separately-excited
   (*iii*) compound (*iv*) none of the above

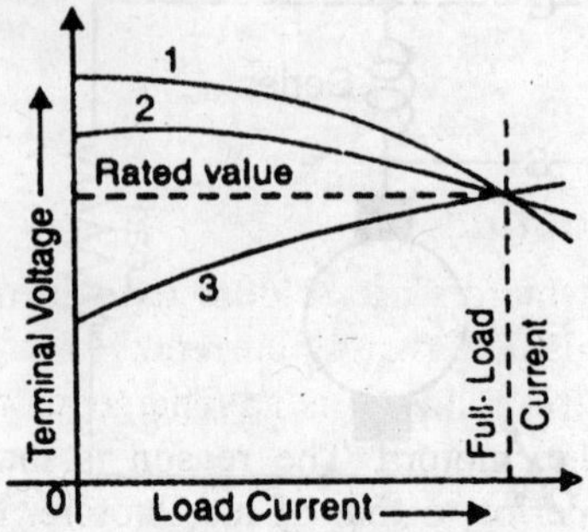

Fig. 11.13

5. Curve 3 in Fig. 11.13 represents the external characteristic of a.........generator.
   (*i*) compound (*ii*) separately-excited
   (*iii*) shunt (*iv*) none of the above
6. Curve 1 in Fig. 11.13 represents the external characteristic of a ....generator.
   (*i*) shunt (*ii*) series
   (*iii*) flat-compounded (*iv*) none of the above.
7. The curve B in Fig. 11.14 is the external characteristic of a ....generator.
   (*i*) shunt (*ii*) series
   (*iii*) flat-compounded
   (*iv*) under-compounded

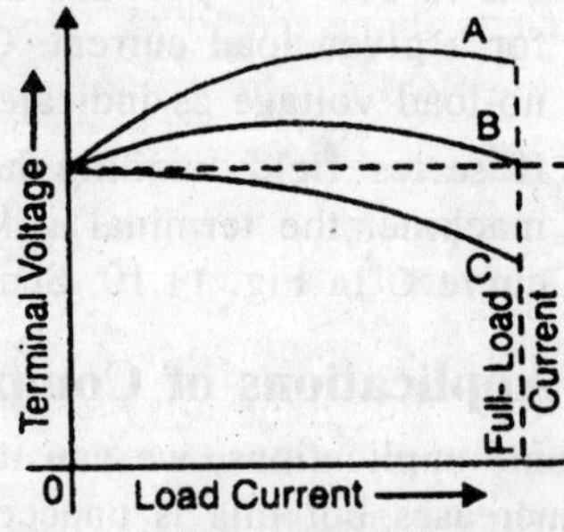

Fig. 11.14

8. The curve A in Fig. 11.14 represents the external characteristic of ....generator.
   (*i*) shunt (*ii*) flat-compounded
   (*iii*) under-compounded
   (*iv*) an over-compounded

ANSWERS TO OBJECTIVE QUESTIONS

1. (*iv*) **2.** (*ii*) **3.** (*iii*) **4.** (*ii*) **5.** (*i*) **6.** (*i*) **7.** (*iii*) **8.** (*iv*)

12

D.C. Motors

Introduction

D.C. motors are seldom used in ordinary applications because all electric supply companies furnish alternating current. However, for special applications such as in steel mills, mines and electric trains, it is advantageous to convert alternating current into direct current in order to use d.c. motors. The reason is that speed/torque characteristics of d.c. motors are much more superior to that of a.c. motors. Like generators, d.c. motors are also of three types *viz.* series-wound, shunt-wound and compound-wound. The use of a particular d.c. motor depends upon the type of mechanical load it has to drive.

12.1. D.C. Motor Principle

A machine that converts d.c. power into mechanical power is known as a d.c. motor. Its operation is based on the principle that when a current carrying conductor is placed in a magnetic field, the conductor experiences a mechanical force. The direction of this force is given by Fleming's left hand rule and magnitude is given by ;

$$F = BIl \text{ newtons}$$

Basically, there is no constructional difference between a d.c. motor and a d.c. generator. The same d.c. machine can be run as a generator or motor.

12.2. Working of D.C. Motor

Consider a part of a multipolar d.c. motor as shown in Fig. 12.1. When the terminals of the motor are connected to an external source of d.c. supply;

(*i*) the field magnets are excited developing alternate *N* and *S* poles.

(*ii*) the armature conductors carry *currents. All conductors under *N*-pole carry currents in one direction while all the conductors under *S*-pole carry currents in the opposite direction.

Suppose the conductors under *N*-pole carry currents into the plane of the paper and those under *S*-pole carry currents out of the plane of the paper as shown in Fig. 12.1. Since each armature conductor is carrying current and is placed in the magnetic field, mechanical force acts on it. Referring to Fig. 12.1 and applying Fleming's left hand rule, it is clear that force on each conductor

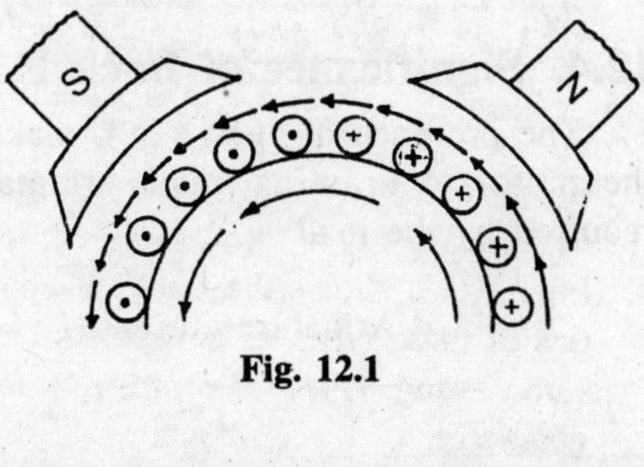

Fig. 12.1

* Since the armature of a motor is the same as that of generator, the current from the supply line must divide and pass through several paths of the armature—2 for wave wound armature and P for lap wound armature. As in a generator, currents through all conductors under a pole will be in the same direction.

is tending to rotate the armature in anticlockwise direction. All these forces add together to produce a driving torque which sets the armature rotating. When the conductor moves from one side of a brush to the other, the current in that conductor is reversed and at the same time it comes under the influence of next pole which is of opposite polarity. Consequently, the direction of force on the conductor remains the same.

12.3. Back or Counter E.M.F.

When the armature of a d.c. motor rotates under the influence of the driving torque, the armature conductors move through the magnetic field and hence e.m.f. is induced in them as in a generator. The induced e.m.f. acts in opposite direction to the applied voltage V (*Lenz's law) and is known as *back or counter e.m.f.* E_b. The back e.m.f. E_b ($= P\phi ZN/60A$) is *always* less than the applied voltage V, although this difference is small when the motor is running under normal conditions.

Consider a shunt wound motor shown in Fig. 12.2. When d.c. voltage V is applied across the motor terminals, the field magnets are excited and armature conductors are supplied with current. Therefore, driving torque acts on the armature which begins to rotate. As the armature rotates, back e.m.f. E_b is induced which opposes the applied voltage V. The applied V has to force current through the armature against the back e.m.f. E_b. The electric work done in overcoming and causing the current to flow against E_b is converted into mechanical energy developed in the armature. It follows, therefore, that energy conversion in a d.c. motor is only possible due to the production of back e.m.f. E_b.

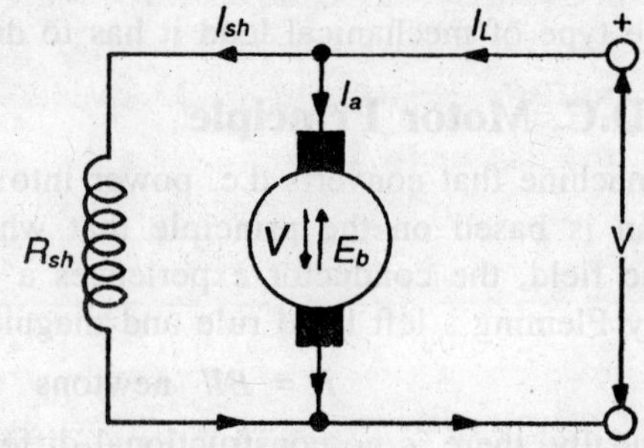

Fig. 12.2

Net voltage across armature circuit $= V - E_b$.

If R_a is the armature circuit resistance, then $I_a = \dfrac{V - E_b}{R_a}$

Since V and R_a are usually fixed, the value of E_b will determine the current drawn by the motor. If the speed of the motor is high, then back e.m.f. E_b ($= P\phi ZN/60A$) is large and hence the motor will draw less armature current and *vice-versa*.

12.4. Significance of Back E.M.F.

The presence of back e.m.f. makes the d.c. motor a *self-regulating* machine *i.e.*, it makes the motor to draw as much armature current as is just sufficient to develop the torque required by the load.

$$\text{Armature current, } I_a = \frac{V - E_b}{R_a}$$

---

* According to Lenz's law, the direction of induced e.m.f. is such that it opposes the cause producing it. The cause producing the back e.m.f. E_b is the applied voltage V. Hence E_b opposes the applied voltage V.

(*i*) When the motor is running on no load, small torque is required to overcome the friction and windage losses. Therefore, the armature current I_a is small and the back e.m.f. is nearly equal to the applied voltage.

(*ii*) If the motor is suddenly loaded, the first effect is to cause the armature to slow down. Therefore, the speed at which the armature conductors move through the field is reduced and hence the back e.m.f. E_b falls. The decreased back e.m.f. allows a larger current to flow through the armature and larger current means increased driving torque. Thus, the driving torque increases as the motor slows down. The motor will stop slowing down when the armature current is just sufficient to produce the increased torque required by the load.

(*iii*) If the load on the motor is decreased, the driving torque is momentarily in excess of the requirement so that armature is accelerated. As the armature speed increases, the back e.m.f. E_b also increases and causes the armature current I_a to decrease. The motor will stop accelerating when the armature current is just sufficient to produce the reduced torque required by the load.

It follows, therefore, that back e.m.f. in a d.c. motor regulates the flow of armature current i.e. it automatically changes the armature current to meet the load requirement.

12.5. Voltage Equation of D.C. Motor

Let in a d.c. motor (See Fig. 12.3),

$$V = \text{applied voltage}$$
$$E_b = \text{back e.m.f.}$$
$$R_a = \text{armature resistance}$$
$$I_a = \text{armature current}$$

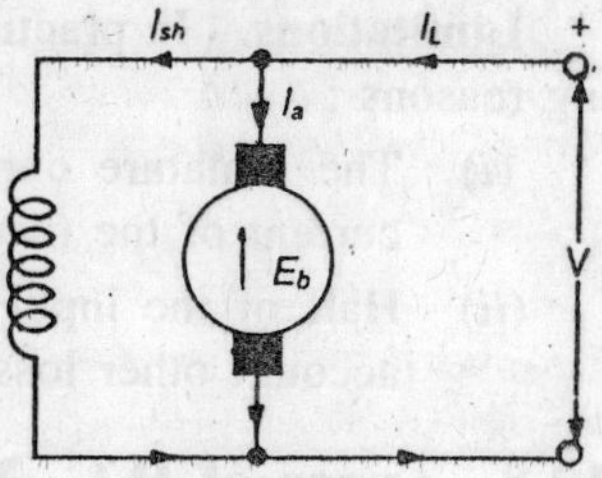

Fig. 12.3

Since back e.m.f. E_b acts in opposition to the applied voltage V, the net voltage across the armature circuit is $V - E_b$. The armature current I_a is given by ;

$$I_a = \frac{V - E_b}{R_a}$$

or $$V = E_b + I_a R_a \quad ...(i)$$

This is known as voltage equation of the d.c. motor.

12.6. Power Equation

If eq. (*i*) above is multiplied by I_a throughout, we get,

$$V I_a = E_b I_a + I_a^2 R_a$$

This is known as power equation of the d.c. motor.

$V I_a$ = electric power supplied to armature (*armature input*)

$E_b I_a$ = *power developed by armature (*armature output*)

$I_a^2 R_a$ = electric power wasted in armature (*armature Cu loss*)

Thus out of the armature input, a small portion (about 5%) is wasted as $I_a^2 R_a$ and the remaining portion $E_b I_a$ is converted into mechanical power within the armature.

---

* This will be the mechanical power output in watts.

12.7. Condition for Maximum Power

The mechanical power developed by the motor is $P_m = E_b I_a$

Now, $$P_m = V I_a - I_a^2 R_a$$

Since V and R_a are fixed, power developed by the motor depends upon armature current. For maximum power, dP_m/dI_a should be zero.

$$\therefore \quad \frac{dp_m}{dI_a} = V - 2 I_a R_a = 0$$

or $$I_a R_a = V/2$$

Now, $$V = E_b + I_a R_a$$

$$= E_b + V/2 \qquad [\because I_a R_a = V/2]$$

$$\therefore \quad E_b = V/2$$

Hence mechanical power developed by the motor is maximum when back e.m.f. is equal to half the applied voltage.

Limitations. In practice, we never aim at achieving maximum power due to the following reasons :

(*i*) The armature current under this condition is very large — much excess of rated current of the machine.

(*ii*) Half of the input power is wasted in the armature circuit. In fact, if we take into account other losses (iron and mechanical), the efficiency will be well below 50%.

12.8. Types of D.C. Motors

Like generators, there are three types of d.c. motors characterised by the connections of field winding in relation to the armature *viz.* :

(*i*) *shunt-wound motor* in which the field winding is connected in parallel with the armature.

(*ii*) *series-wound motor* in which the field winding is connected in series with the armature.

(*iii*) *compound-wound motor* which has two field windings ; one connected in parallel with the armature and the other in series with it.

Example 12.1. *A 250V shunt motor takes a total current of 20A. The shunt field and armature resistances are 200Ω and 0.3Ω respectively. Determine (i) value of back e.m.f. (ii) gross mechanical power in the armature.*

Solution.

(*i*) Shunt current, $I_{sh} = 250/200 = 1.25\text{ A}$

Armature current, $I_a = 20 - 1.25 = 18.75\text{ A}$

Back e.m.f., $E_b = V - I_a R_a$

$= 250 - 18.75 \times 0.3 = \mathbf{244.4V}$

(*ii*) Mechanical power developed $= E_b I_a = 244.4 \times 18.75 = \mathbf{4582.5W}$

Example 12.2. *A 230V motor has an armature circuit resistance of 0.6Ω. If the full-load armature current is 30A and no load armature current is 4A, find the change in back e.m.f. from no load to full-load.*

Solution.

$$E_b = V - I_a R_a$$

At no load, $E_b = 230 - 4 \times 0.6 = 227.6\text{V}$

At full-load, $E_b = 230 - 30 \times 0.6 = 212\text{V}$

Change in back e.m.f. $= 227.6 - 212 = \mathbf{15.6V}$

Example 12.3. *A 4-pole motor is fed at 440V and takes an armature current of 50A. The resistance of the armature circuit is 0.28Ω. The armature winding is wave-connected with 888 conductors and useful flux per pole is 0.023Wb. Calculate the speed of the motor.*

Solution.

$$V = E_b + I_a R_a$$

or $440 = E_b + 50 \times 0.28$

∴ $E_b = 440 - 50 \times 0.28 = 426\text{V}$

Now $E_b = \dfrac{P\phi ZN}{60A}$

∴ $N = \dfrac{E_b \times 60A}{P\phi Z} = \dfrac{426 \times 60 \times 2}{4 \times 0.023 \times 888} = \mathbf{626\ r.p.m.}$

Example 12.4. *The counter e.m.f. of a shunt motor is 227V, the field resistance is 160Ω and field current is 1.5A. If the line current is 39.5 A, find the armature resistance. Also find the armature current when the motor is stationary.*

Solution.

Applied voltage, $V = I_{sh} R_{sh} = 1.5 \times 160 = 240\text{ V}$

Armature current, $I_a = I_L - I_{sh} = 39.5 - 1.5 = 38\text{ A}$

$$V = E_b + I_a R_a$$

∴ $R_a = \dfrac{V - E_b}{I_a} = \dfrac{240 - 227}{38} = \mathbf{0.342\ \Omega}$

At the moment of start-up, the armature is stationary so that $E_b = 0$

∴ $I_a = V/R_a = 240/0.342 = \mathbf{701.5A}$

Example 12.5. *A 20KW, 250V d.c. shunt generator has armature and field resistances of 0.1Ω and 125Ω respectively. Calculate the total armature power developed when running (i) as a generator delivering 20kW output (ii) as a motor taking 20kW input.*

Solution.

(*i*) **As a generator.** Fig. 12.4 (*i*) shows the connections of shunt generator.

$$I_L = 20 \times 10^3/250 = 80\,A$$

$$I_{sh} = 250/125 = 2\,A$$

$$I_a = I_L + I_{sh} = 80 + 2 = 82\,A$$

$$E_g = V + I_a R_a = 250 + 82 \times 0.1 = 258.2\,V$$

∴ Power developed in the armature

$$= E_g I_a = 258.2 \times 82 = 21.17 \times 10^3\,W = \mathbf{21.17\,kW}$$

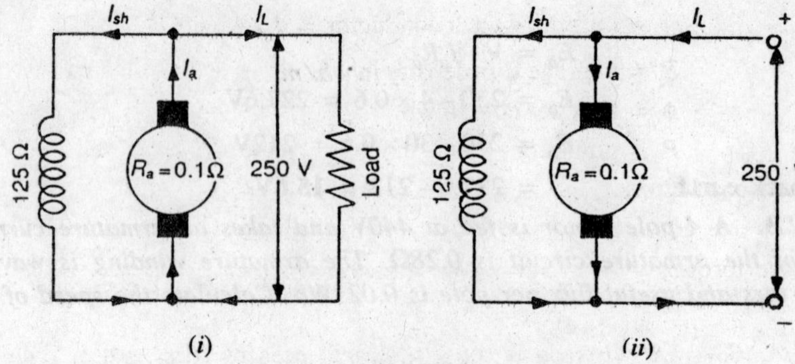

Fig. 12.4.

(*ii*) **As a motor** Fig. 12.4(*ii*) shows the connections of shunt motor.

$$I_L = 20 \times 10^3/250 = 80 \text{ A}$$

$$I_{sh} = 250/125 = 2 \text{ A}$$

$$I_a = I_L - I_{sh} = 80 - 2 = 78 \text{ A}$$

$$E_b = V - I_a R_a = 250 - 78 \times 0.1 = 242.2 \text{ V}$$

Power developed in the armature

$$= E_b I_a = 242.2 \times 78 = 18.9 \times 10^3 \ W = \mathbf{18.9\ kW}$$

Tutorial Problems

1. A 220V shunt motor takes a total current of 20A. The shunt field resistance is 250Ω while armature resistance is 0.3Ω. Calculate the back e.m.f. **[214.26V]**
2. The armature of a d.c. machine has a resistance of 0.1Ω and is connected to a 230V supply. Calculate the back e.m.f. when it is running (*i*) as a generator giving 80A (*ii*) as a motor taking 80A. **[(*i*) 238V (*ii*) 222V]**
3. A 4-pole d.c. motor is connected to a 500V d.c. supply and takes an armature current of 80A. The resistance of the armature circuit is 0.4Ω. The armature is wave wound with 522 conductors and useful flux per pole is 0.025wb. Determine the speed of the motor. **[1075 r.p.m.]**

12.9. Armature Torque of D.C. Motor

Torque is the turning moment of a force about an axis and is measured by the product of force (F) and radius (r) at right angle to which the force acts *i.e.*

$$T = F \times r$$

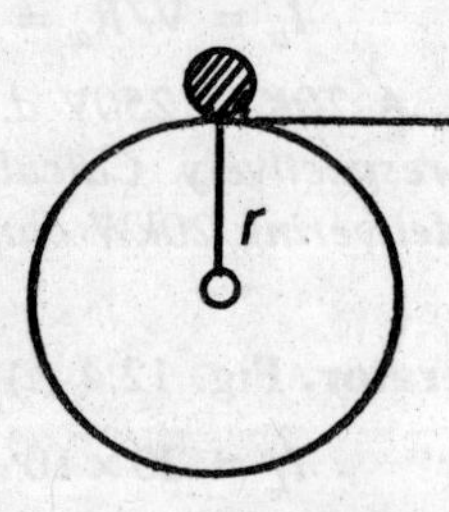

Fig. 12.5

In a d.c. motor, each conductor is acted upon by a circumferential force F at a distance r, the radius of the armature (See Fig. 12.5). Therefore, each conductor exerts a torque, tending to rotate the armature. The sum of the torques due to all armature conductors is known as *gross or armature torque* (T_a).

Let in a d.c. motor,

r = average radius of armature in m

l = effective length of each conductor in m

Z = total number of armature conductors

A = number of parallel paths

i = current in each conductor = I_a/A

B = average flux density in wb/m^2

ϕ = flux per pole in wb

P = number of poles

Force on each conductor, $F = B\,i\,l$ newtons

Torque due to one conductor $= F \times r$ newton - metre

Total armature torque, $T_a = Z\,F\,r$ newton - metre

$= Z\,B\,i\,l\,r$

Now $i = I_a/A$, $B = \phi/a$ where a is the x-sectional area of flux path at radius r. Clearly, $a = 2\pi rl/P$.

$$\therefore \quad T_a = Z \times (\phi/a) \times (I_a/A) \times l \times r$$

$$= Z \times \frac{\phi}{2\pi rl/P} \times \frac{I_a}{A} \times l \times r$$

$$= \frac{Z\,\phi\,I_a\,P}{2\pi\,A} Nm$$

or $$T_a = 0.159 Z\,\phi\,I_a\,(P/A) Nm \qquad ...(i)$$

Since Z, P and A are fixed for a given machine,

$$\therefore \quad T_a \propto \phi\,I_a$$

Hence *torque in a d.c. motor is directly proportional to flux per pole and armature current.*

(*i*) For a *shunt motor*, flux ϕ is practically constant.

$$\therefore \quad T_a \propto I_a$$

(*ii*) For a *series motor*, flux ϕ is directly proportional to armature current I_a provided magnetic saturation does not take place.

$$\therefore \quad T_a \propto I_a^2 \qquad \text{...upto magnetic saturation}$$

Alternative expression for T_a

$$E_b = \frac{P\,\phi\,Z\,N}{60A}$$

$$\therefore \quad \frac{P\,\phi\,Z}{A} = \frac{60 \times E_b}{N}$$

From eq. (*i*), we get the expression of T_a as :

$$T_a = 0.159 \times \left(\frac{60 \times E_b}{N}\right) \times I_a$$

or $$T_a = 9.55 \times \frac{E_b\,I_a}{N} Nm$$

12.10. Shaft Torque (T_{sh})

The torque which is available at the motor shaft for doing useful work is known as *shaft torque*. It is represented by T_{sh}.

Fig. 12.6 illustrates the concept of shaft torque. The total torque T_a developed in armature of a motor is not available at the shaft as a part of it is lost in overcoming the iron and frictional losses in the motor. Therefore, shaft torque T_{sh} is somewhat less than the total armature torque T_a. The difference $T_a - T_{sh}$ is known as lost torque.

$$T_{sh} = 9.55 \times \frac{\text{Output}}{\text{N}} \; Nm$$

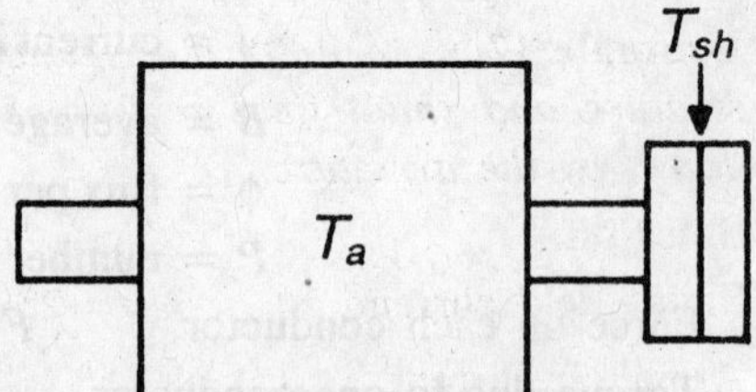

Fig. 12.6

12.11. Brake Horse Power (B.H.P.)

The horse power developed by the shaft torque is known as *brake horse power* (B.H.P.). If the motor is running at *N r.p.m.* and the shaft torque is T_{sh} newton metres, then,

$$\text{W.D./revolution} = \text{force} \times \text{distance moved in 1 revolution}$$

$$= F \times 2\pi r = 2\pi \times T_{sh} \text{ J}$$

W.D./minute $= 2\pi N T_{sh}$ J

W.D./sec. $= \dfrac{2\pi N T_{sh}}{60}$ *watts*

$$= \frac{2\pi N T_{sh}}{60 \times 746} \; H.P.$$

∴ Useful output power $= \dfrac{2\pi N \, T_{sh}}{60 \times 746}$ *H.P.*

or $B.H.P. = \dfrac{2\pi N \, T_{sh}}{60 \times 746}$

Example 12.6. *Calculate the value of torque established by the armature of a 4-pole motor having 774 conductors, two paths in parallel, 24mwb flux per pole, when the total armature current is 50A.*

Solution.

Armature torque, $T_a = 0.159 \, Z \, \phi \, I_a \, (P/A)$

Here, $Z = 774$; $\phi = 24 \times 10^{-3}$ *Wb* ; $I_a = 50A$; $P = 4$; $A = 2$

∴ $T_a = 0.159 \times 774 \times 24 \times 10^{-3} \times 50 \times (4/2) =$ **295.35Nm**

Example 12.7. *An armature of a 6-pole machine 75cm in diameter has 664 conductors each having an effective length of 30cm and carrying a current of 100A. If 70% of total conductors lie simultaneously in the field of average flux density 0.85wb/m², calculate (i) armature torque (ii) horse power output at 250 r.p.m.*

Solution.

(*i*) Force on one conductor, $F = B \, i \, l = 0.85 \times 100 \times 0.3 = 25.5 \, N$

Torque due to one conductor $= F \times r = 25.5 \times 0.375 = 9.56 \, Nm$

No. of effective conductors, $Z = 664 \times 70/100 = 465$

∴ Total armature torque, $T_a = 9.56 \times 465 =$ **4445.4 *Nm***

(*ii*) Power output $= \dfrac{2\pi N T_a^*}{60 \times 746}$ H.P.

$$= \frac{2 \, \pi \times 250 \times 4445.4}{60 \times 746} = \textbf{156 H.P.}$$

* Iron and frictional losses neglected so that $T_a = T_{sh}$

Example 12.8. *A 230V d.c. shunt motor takes a current of 40A and runs at 1100 r.p.m. If armature and shunt field resistances are 0.25Ω and 230Ω respectively, find the torque developed by the armature.*

Solution.

Shunt field current, $I_{sh} = 230/230 = 1A$

Armature current, $I_a = 40 - 1 = 39$ A

Back e.m.f. $E_b = V - I_a R_a = 230 - 39 \times 0.25 = 220.25$ V

Now, $T_a = 9.55 \times E_b I_a / N$ ...See Art. 12.9

$= 9.55 \times (220.25 \times 39/1100)\ Nm =$ **74.66 Nm**

12.12. Speed of a D.C. Motor

$$E_b = V - I_a R_a$$

But $$E_b = P\phi Z N/60A$$

$$\therefore \quad \frac{P\phi Z N}{60A} = V - I_a R_a$$

or $$N = \frac{(V - I_a R_a)}{\phi} \frac{60A}{PZ}$$

or $$N = K \frac{(V - I_a R_a)}{\phi} \quad \text{where } K = 60A/PZ$$

But $$V - I_a R_a = E_b$$

$$\therefore \quad N = K \frac{E_b}{\phi}$$

or $$N \propto \frac{E_b}{\phi}$$

Therefore, in a d.c. motor, speed is directly proportional to back e.m.f. E_b and inversely proportional to flux per pole ϕ.

12.13. Speed Relations

If a d.c. motor has initial values of speed, flux per pole and back e.m.f. as N_1, ϕ_1 and E_{b_1} respectively and the corresponding final values are N_2, ϕ_2 and E_{b_2}, then,

$$N_1 \propto \frac{E_{b_1}}{\phi_1} \quad \text{and} \quad N_2 \propto \frac{E_{b_2}}{\phi_2}$$

$$\therefore \quad \frac{N_1}{N_2} = \frac{E_{b_1}}{E_{b_2}} \times \frac{\phi_2}{\phi_1}$$

(*i*) For a **shunt motor**, flux practically remains constant so that $\phi_1 = \phi_2$.

$$\therefore \quad \frac{N_1}{N_2} = \frac{E_{b_1}}{E_{b_2}}$$

(*ii*) For a **series motor**, $\phi \propto I_a$ prior to saturation.

$$\therefore \quad \frac{N_1}{N_2} = \frac{E_{b_1}}{E_{b_2}} \times \frac{I_{a_2}}{I_{a_1}}$$

where I_{a_1} = initial armature current

I_{a_2} = final armature current

12.14. Speed Regulation

The **speed regulation** *of a motor is the change in speed from full-load to no load and is expressed as a percentage of the speed at full-load i.e.*

$$\% \text{ Speed regulation} = \frac{N.L.\text{speed} - F.L.\text{Speed}}{F.L.\text{speed}} \times 100$$

$$= \frac{N_0 - N}{N} \times 100$$

where N_0 = NO – load speed

N = Full-load speed

Example 12.9. *A 230V d.c. shunt motor takes 5A at no load and runs at 1000 r.p.m.. Calculate the speed when loaded and taking a current of 30A. The armature and field resistances are 0.2Ω and 230Ω respectively.*

Solution.

Let N_2 r.p.m. be the speed under loaded conditions.

No load conditions

$$I_{sh} = 230/230 = 1 \text{ A}$$

$$I_{a_1} = I_L - I_{sh} = 5 - 1 = 4 \text{ A}$$

$$E_{b_1} = V - I_{a_1} R_a = 230 - 4 \times 0.2 = 229.2\text{V}$$

$$N_1 = 1000 \; r.p.m.$$

Loaded conditions

$$I_{a_2} = I_L - I_{sh} = 30 - 1 = 29 \text{ A}$$

$$\mathbf{E_{b_2} = V - I_{a_2} R_a = 230 - 29 \times 0.2 = 224.2 \text{ V}}$$

Since it is a shunt motor,

$$\therefore \quad \frac{N_2}{N_1} = \frac{E_{b_2}}{E_{b_1}}$$ ...See Art 12.13

$$\therefore \quad N_2 = \left(\frac{E_{b_2}}{E_{b_1}}\right) N_1 = \frac{224.2}{229.2} \times 1000 = \mathbf{978 \text{ r.p.m.}}$$

Example 12.10. *A 6-pole lap wound shunt motor has 500 conductors. The armature and shunt field resistances are 0.05Ω and 25Ω respectively. Find the speed of the motor if it takes 120A from a d.c. supply of 100V. Flux per pole is 20mwb.*

Solution.

$$I_{sh} = 100/25 = 4A$$

$$I_a = I_L - I_{sh} = 120 - 4 = 116A$$

$$E_b = V - I_a R_a = 100 - 116 \times 0.05 = 94.2 \text{ V}$$

Now,
$$E_b = \frac{P \phi ZN}{60A}$$

$$\therefore \quad N = \frac{E_b \times 60A}{P\phi Z} = \frac{94.2 \times 60 \times 6}{6 \times 20 \times 10^{-3} \times 500} = \mathbf{565 \ r.p.m.}$$

Example 12.11. *A d.c. shunt generator delivers an output of 100kW at 500V when running at 800 r.p.m. The armature and field resistances are 0.1Ω and 100Ω respectively. Calculate the speed of the same machine when running as a shunt motor and taking 100kW input at 500V. Allow 1 volt per brush for contact drop.*

Solution.

As a generator. Fig. 12.7 (*i*) shows the connections.

$$I_L = 100 \times 10^3/500 = 200 \text{ A}$$

$$I_{sh} = 500/100 = 5 \text{ A}$$

$$I_{a_1} = I_L + I_{sh} = 200 + 5 = 205 \text{ A}$$

Generated e.m.f.,
$$E_1 = V_1 + I_{a_1} R_a + \text{Brush drop}$$

$$= 500 + 205 \times 0.1 + 2 \times 1 = 522.5 \text{ V}$$

$$N_1 = 800 \text{ r.p.m.}$$

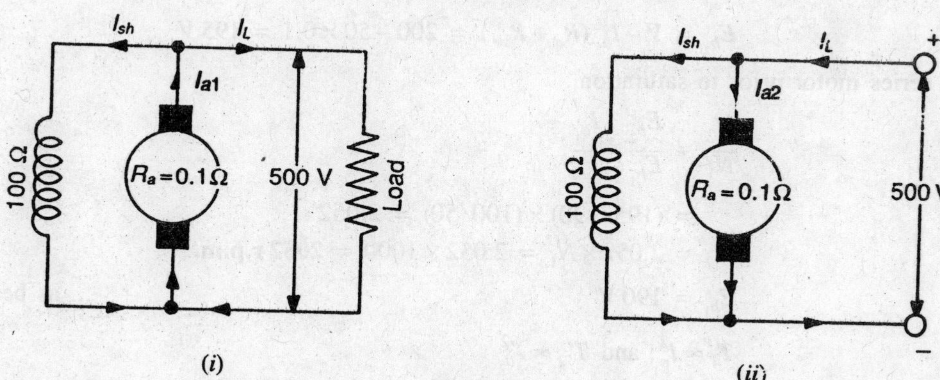

Fig. 12.7

As a motor. Fig. 12.7 (*ii*) shows the connections.

$$I_{a_2} = I_L - I_{sh} = 200 - 5 = 195 \text{ A}$$

Back e.m.f.,
$$E_2 = V - I_{a_2} R_a - \text{Brush drop}$$

$$= 500 - 195 \times 0.1 - 2 \times 1 = 478.5 \text{ V}$$

Let N_2 be the speed of the machine in r.p.m. when working as a motor.

For a shunt machine,
$$\frac{N_2}{N_1} = \frac{E_2}{E_1}$$

...See Art 12.13

$$\therefore \quad N_2 = \left(\frac{E_2}{E_1}\right) N_1 = \frac{478.5}{522.5} \times 800 = \mathbf{732.6 \ r.p.m.}$$

Example 12.12. *A 200V series motor takes a current of 100A and runs at 1000 r.p.m. The total resistance of the motor is 0.1Ω and the field is unsaturated. Calculate :*

(*i*) *the percentage change in torque and speed if the load is so changed that motor current is 50A.*

(*ii*) *the motor current and speed if the torque is halved.*

Solution.

(*i*) $T \propto \phi I_a$

Since the field is unsaturated, $\phi \propto I_a$

Initial armature current, $I_{a_1} = 100$ A

New armature current, $I_{a_2} = 50$ A

Now $T_1 \propto I_{a_1}^2$, and $T_2 \propto I_{a_2}^2$

$$\therefore \quad T_2 = T_1\left(\frac{I_{a_2}}{I_{a_1}}\right)^2 = T_1\left(\frac{50}{100}\right)^2 = 0.25\,T_1$$

$$\therefore \text{ \% change in torque} = \frac{T_1 - T_2}{T_1} \times 100 = \frac{T_1 - 0.25\,T_1}{T_1} \times 100 = \mathbf{75\%}$$

$$E_{b_1} = V - I_{a_1}(^*R_a + R_{se}) = 200 - 100 \times 0.1 = 190\ V$$

$$E_{b_2} = V - I_{a_2}(R_a + R_{se}) = 200 - 50 \times 0.1 = 195\ V$$

For series motor prior to saturation,

$$\frac{N_2}{N_1} = \frac{E_{b_2}}{E_{b_1}} \times \frac{I_{a_1}}{I_{a_2}}$$

$$= (195/190) \times (100/50) = 2.052$$

$$\therefore \quad N_2 = 2.052 \times N_1 = 2.052 \times 1000 = \mathbf{2052\ r.p.m.}$$

(*ii*) $E_{b_1} = 190\ V$ ... as before

$$T_1 \propto I_{a_1}^2 \text{ and } T'_2 \propto I'^2_{a_2}$$

$$\therefore \quad I'_{a_2} = I_{a_1}\sqrt{T'_2/T_1} = 100\sqrt{1/2} = \mathbf{70.7A}$$

$$E'_{b_2} = 200 - 70.7 \times 0.1 = 192.93\ V$$

$$\frac{N'_2}{N_1} = (E'_{b_2}/E_{b_1}) \times (I_{a_1}/I'_{a_2})$$

$$= (192.93/190) \times (100/70.7) = 1.436$$

$$\therefore \quad N'_2 = 1.436 \times N_1 = 1.436 \times 1000 = \mathbf{1436\ r.p.m.}$$

Tutorial Problems

1. A 4.pole 200V d.c. shunt motor has 360 wave connected conductors on its armature. The full load armature current is 30A and flux per pole is 0.03Wb. The armature resistance is 0.2Ω and the contact drop is 1V per brush. Determine the full load speed of the motor. **[672 r.p.m.]**
2. A 4-pole, 460V shunt motor has its armature wave-connected with 880 conductors. The useful flux per pole is 0.02Wb and armature resistance is 0.7Ω. If the armature current is 40A, calculate (*i*) speed and (*ii*) torque. **[(*i*) 730 r.p.m. (*ii*) 226 Nm]**
3. A 230V d.c. shunt motor tales a no load current of 2A and runs at 1200 r.p.m. If the full-load current is 40A, find (*i*) speed on full-load (*ii*) percentage speed regulation. Resistance of armature is 0.25Ω. **[(*i*) 1150 r.p.m. (*ii*) 4.35%]**

* Given that $R_a + R_{se} = 0.1\ \Omega$

4. A 230V shunt motor runs at 1000 r.p.m. and takes 5A. The armature resistance of the motor is 0.25Ω and shunt field resistance is 230Ω. Calculate the drop in speed when the motor takes a line current of 41A. **[40 r.p.m.]**
5. A shunt generator has an output of 10kW at 500V; the speed being 1000 r.p.m. The armature circuit resistance is 0.5Ω and the field resistance is 250Ω. Calculate the speed when running as a shunt motor taking 50kW at 500V. **[961 r.p.m.]**

12.15. Losses in a D.C. Motor

The losses occurring in a d.c. motor are the same as in a d.c. generator (see Art 10.9). These are :

(*i*) copper losses

(*ii*) mechanical losses

(*iii*) iron losses

As in a generator, these losses cause (*a*) an increase of temperature and (*b*) reduction in efficiency of the d.c. motor.

12.16. Efficiency of a D.C. Motor

Like a d.c. generator, the efficiency of a d.c. motor is the ratio of output power to the input power *i.e.*

$$\text{Efficiency, } \eta = \frac{\text{output}}{\text{input}} \times 100 = \frac{\text{output}}{\text{output} + \text{losses}} \times 100$$

As for a generator (See Art 10.13), the efficiency of a d.c. motor will be maximum when :

$$\text{Variable losses} = \text{Constant losses}$$

12.17. Power Stages

The power stages in a d.c. motor are represented diagrammatically in Fig. 12.8.

$$A - B = \text{Copperlosses}$$

$$B - C = \text{Iron and friction losses}$$

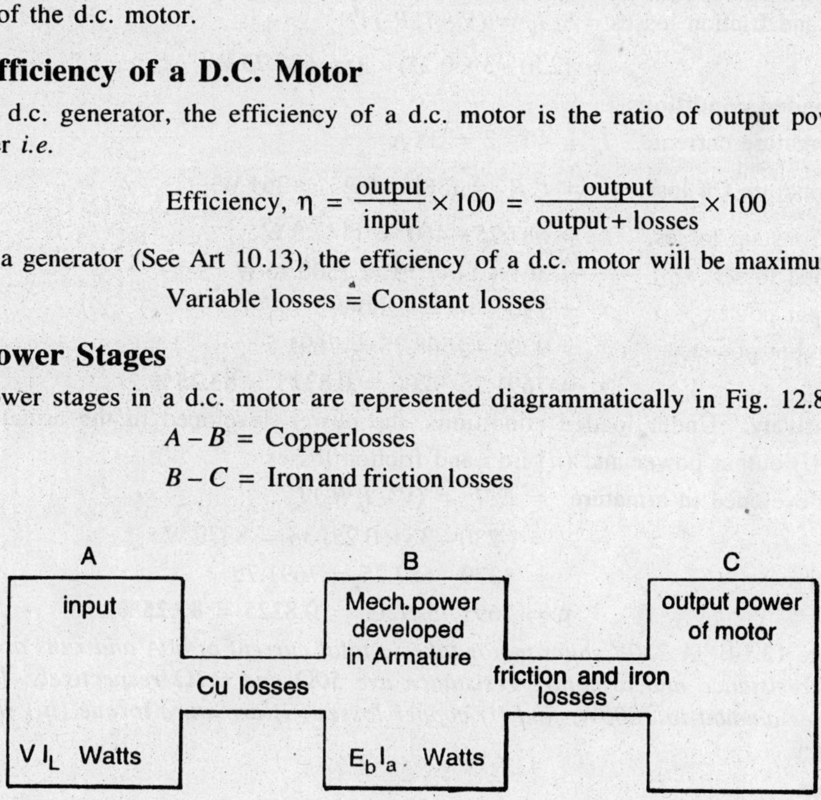

Fig. 12.8

Overall efficiency, $\eta_c = C/A$

Electrical efficiency, $\eta_e = B/A$

Mechanical efficiency, $\eta_m = C/B$

Example 12.13. *A 230V shunt motor takes 5A at no load. The resistances of the armature and field circuit are 0.25Ω and 115Ω respectively. If the motor is loaded so as to carry 40A, determine (i) iron and friction losses and (ii) efficiency.*

Solution.

(*i*) **No load conditions**

$$\text{Input power} = V I_L = 230 \times 5 = 1150 \text{ W}$$

At no load, the total input power supplied is used up in overcoming (*a*) small armature Cu loss (*b*) shunt field loss and (*c*) iron and friction losses.

$$I_{sh} = 230/115 = 2\,A; \quad I_a = 5 - 2 = 3 \text{ A}$$

$$\text{Armature Cu loss} = I_a^2 R_a = (3)^2 \times 0.25 = 2.25 \text{ W}$$

$$\text{Field Cu loss} = I_{sh}^2 R_{sh} = (2)^2 \times 115 = 460 \text{ W}$$

$$\therefore \text{ Iron and friction losses} = 1150 - (2.25 + 460) = \mathbf{687.25\ W}$$

Alternatively. At no load, the power developed in the armature ($E_b I_a$) is entirely used up in overcoming iron and friction losses.

$$\therefore \text{ Iron and friction losses} = E_b I_a = (V - I_a R_a) I_a$$

$$= (230 - 3 \times 0.25) \times 3 = \mathbf{687.75\ W}$$

(*ii*) **Loaded conditions**

Armature current, $I_a = 40 - 2 = 38$ A

Armature Cu loss, $= I_a^2 R_a = (38)^2 \times 0.25 = 361$ W

*Constant losses, $= 687.75 + 460 = 1147.5$ W

Total losses, $= 361 + 1147.75 = 1508.75$ W

Input power, $= 230 \times 40 = 9200$ W

Output power, $= 9200 - 1508.75 = 7691.5$

$$\therefore \quad \eta = 7691.25/9200 = 0.8325 = \mathbf{83.25\%}$$

Alternatively. Under loaded conditions, the power developed in the armature ($E_b I_a$) consists of (*i*) output power and (*ii*) iron and friction losses.

Power developed in armature $= E_b I_a = (V - I_a R_a) I_a$

$= (230 - 38 \times 0.25)\, 38 = 8379$ W

Output power $= 8379 - 687.25 = 7691.75$

$$\therefore \quad \eta = 7691.75/9200 = 0.8325 = \mathbf{83.25\%}$$

Example 12.14. *A 220V shunt motor takes a total current of 80A and runs at 800 r.p.m. Shunt field resistance and armature resistance are 50Ω and 0.1Ω respectively. If iron and friction losses amount to 1600W, find (i) copper losses (ii) armature torque (iii) shaft torque (iv) efficiency.*

Solution.

Shunt field current, $I_{sh} = 220/50 = 4.4$ A

Armature current, $I_a = 80 - 4.4 = 75.6$ A

Back e.m.f. $E_b = V - I_a R_a = 220 - 75.6 \times 0.1 = 212.44$ V

(*i*) Input power $= V I_L = 220 \times 80 = 17600$ W

Power developed in armature $= E_b I_a = 212.44 \times 75.6 = 16050$ W

---

* Constant losses = Iron and friction losses + shunt field losses

$\therefore$ Copper losses $= 17600 - 16050 = \mathbf{1550\,W}$

(*ii*) Armature torque, $T_a = 9.55 \times E_b I_a / N = \dfrac{9.55 \times 212.44 \times 75.6}{800} = \mathbf{192\,Nm}$

(*iii*) Output power $= 16050 - 1600 = 14450\,W$

Shaft torque, $T_{sh} = 9.55 \times \text{output}/N = 9.55 \times 14450/800 = \mathbf{172.5\,Nm}$

(*iv*) Efficiency $= \dfrac{14450}{17600} \times 100 = \mathbf{82.1\%}$

Tutorial Problems

1. A 220V shunt motor is taking a current of 30A. Armature and shunt field resistances are 0.2Ω and 100Ω respectively. Iron and friction losses amount to 500W. Find the efficiency of the motor. **[82.4%]**
2. A shunt motor running at 600 r.p.m. takes 80A at 250V. The armature and shunt field resistances are 0.1Ω and 50Ω respectively. Iron and friction losses amount to 2188W. Find (*i*) armature torque (*ii*) shaft torque (*iii*) copper losses (*iv*) efficiency. **[(*i*) 288 Nm (*ii*) 254 Nm (*iii*) 1812 W (*iv*) 80%]**
3. The output of certain shunt motor is 3.73 kW when taking 50A at 100 V and running at 750 r.p.m. Armature resistance 0.06Ω and shunt field resistance 50Ω. Find (*i*) iron and friction losses (*ii*) armature torque (*iii*) shaft torque. **[(*i*) 932 W (*ii*) 59.2 Nm (*iii*) 47.4 Nm]**

12.18. D.C. Motor Characteristics

The performance of a d.c. motor can be judged from its characteristic curves known as motor characteristics. Following are the three important characteristics of a d.c. motor :

(*i*) **Torque and Armature current characteristic (T_a/I_a)**

It is the curve between armature torque T_a and armature current I_a of a d.c. motor. It is also known as *electrical characteristic* of the motor.

(*ii*) **Speed and armature current characteristic (N/I_a)**

It is the curve between speed N and armature current I_a of a d.c. motor. It is very important characteristic as it is often the deciding factor in the selection of the motor for a particular application.

(*iii*) **Speed and torque characteristic (N/T_a)**

It is the curve between speed N and armature torque T_a of a d.c. motor. It is also known as *mechanical characteristic*.

12.19. Characteristics of Shunt Motors

Fig. 12.9 shows the connections of a d.c. shunt motor. The field current I_{sh} is constant since the field winding is directly connected to the supply voltage V which is assumed to be constant. Hence, the flux in a shunt motor is *approximately constant.

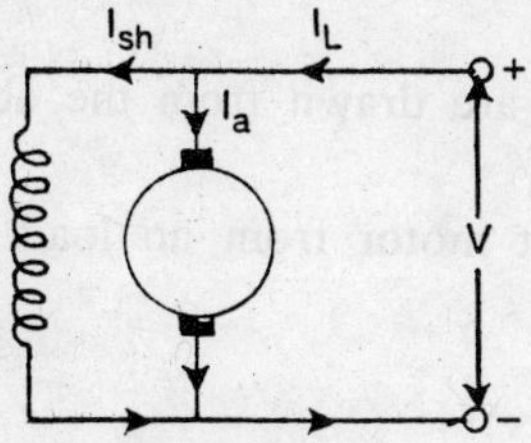

Fig. 12.9

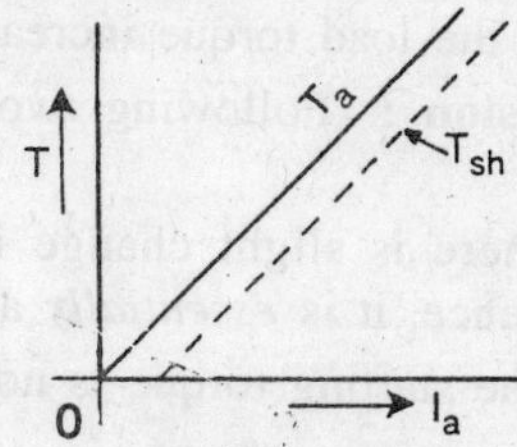

Fig. 12.10

* Due to armature reaction, flux decreases a little but the decrease in flux is usually negligible under normal conditions.

(*i*) **T_a/I_a Characteristic**. We know that in a d.c. motor,

$$T_a \propto \phi I_a$$

Since the motor is operating from a constant supply voltage, flux ϕ is constant (neglecting armature reaction).

$$\therefore \qquad T_a \propto I_a$$

Hence T_a/I_a characteristic is a straight line passing through the origin as shown in Fig. 12.10. The shaft torque (T_{sh}) is less than T_a and is shown by a dotted line. It is clear from the curve that a very large current is required to start a heavy load. Therefore, a shunt motor should not be started on heavy load.

(*ii*) **N/I$_a$ Characteristic.** The speed N of a motor is given by ;

$$N \propto \frac{E_b}{\phi}$$

The flux ϕ and back e.m.f. E_b in a shunt motor are almost constant under normal conditions. Therefore, speed of a shunt motor will remain constant as the armature current varies (dotted line *AB* in Fig. 12.11). Strictly speaking, when load is increased, E_b ($= V - I_a R_a$) and ϕ decrease due to the armature resistance drop and armature reaction respectively. However, E_b decreases slightly more than ϕ so that the *speed of the motor decreases slightly with load (line *AC*).

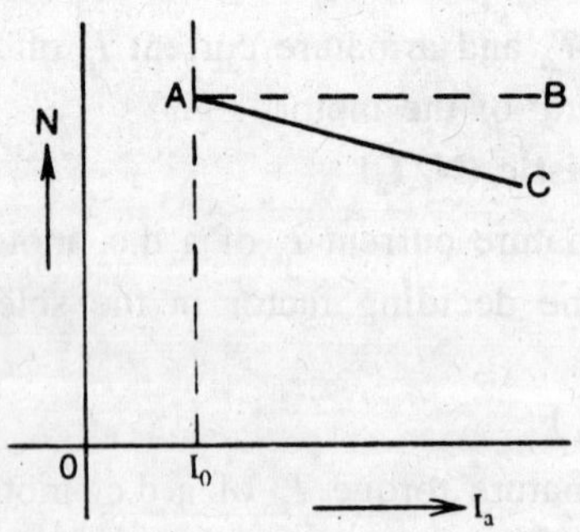

Fig. 12.11

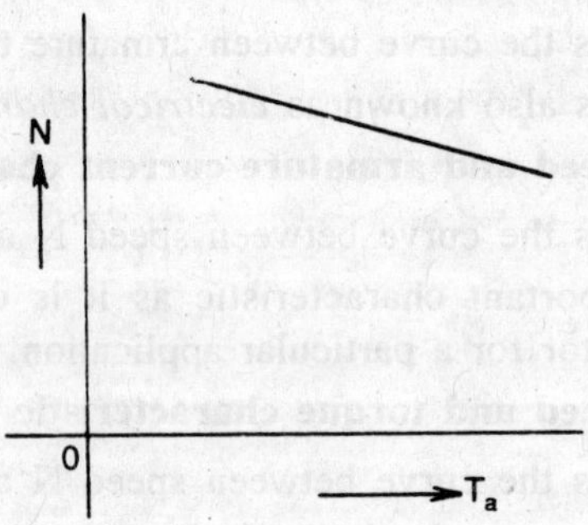

Fig. 12.12

(*iii*) **N/T$_a$ characteristic** This curve is obtained by plotting the values of N and T_a for various armature currents (See Fig. 12.12). It may be seen that speed falls somewhat as the load torque increases.

Conclusion : Following two important conclusions are drawn from the above characteristics :

(*i*) There is slight change in the speed of a shunt motor from no load to full-load. Hence, it is *essentially* a constant-speed motor.

(*ii*) The starting torque is not high because $T_a \propto I_a$.

* It may be noted that characteristic does not have a point of zero armature current because a small current (no load current I_0) is necessary to maintain rotation of the motor at no load.

12.20. Characteristics of Series Motors

Fig. 12.13 shows the connections of a series motor. Note that current passing through the field winding is the same as that in the armature. If the mechanical load on the motor increases, the armature current also increases. Hence, the flux in a series motor increases with the increase in armature current and *vice-versa.*

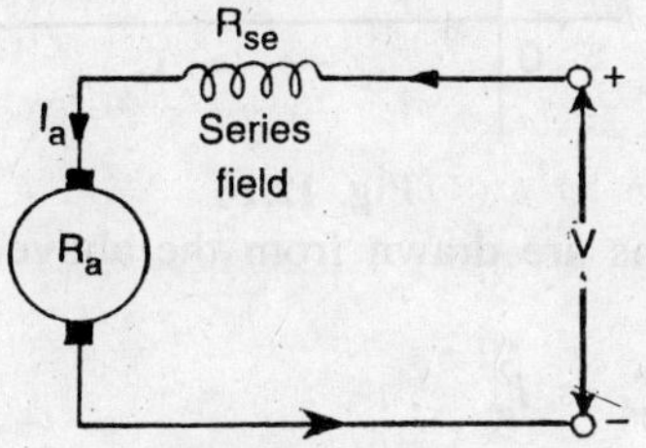

Fig. 12.13

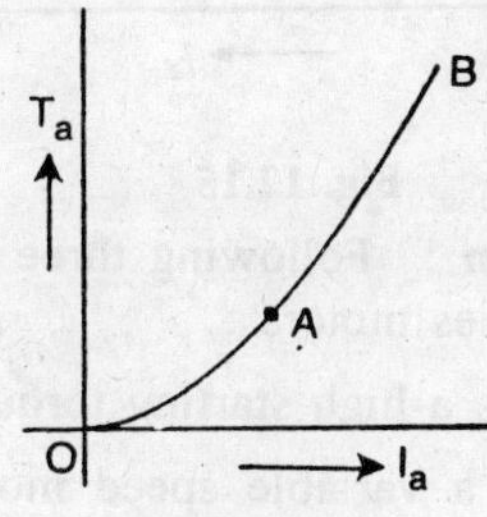

Fig. 12.14

(*i*) **T_a/I_a Characteristic** We know that :

$$T_a \propto \phi I_a$$

Upto magnetic saturation, $\phi \propto I_a$ so that $T_a \propto I_a^2$

After magnetic saturation, ϕ is constant so that $T_a \propto I_a$

Thus upto magnetic saturation, the armature torque is directly proportional to the square of armature current. If I_a is doubled, T_a is almost quadrupled. Therefore, T_a/I_a curve upto magnetic saturation is a parabola (portion *OA* of the curve in Fig. 12.14). However, after magnetic saturation, torque is directly proportional to the armature current. Therefore, T_a/I_a curve after magnetic saturation is a straight line (portion *AB* of the curve).

It may be seen that in the initial portion of the curve (*i.e.* upto magnetic saturation), $T_a \propto I_a^2$. This means that starting torque of a d.c. series motor will be very high as compared to a shunt motor (where $T_a \propto I_a$).

(*ii*) **N/I_a Characteristic** The speed *N* of a series motor is given by ;

$$N \propto \frac{E_b}{\phi} \quad \text{where } E_b = V - I_a(R_a + R_{se})$$

When the armature current increases, the back e.m.f. E_b decreases due to $I_a(R_a + R_{se})$ drop while the flux ϕ increases. However, $I_a(R_a + R_{se})$ drop is quite small under normal conditions and may be neglected.

$$\therefore \quad N \propto \frac{1}{\phi}$$

$$\propto \frac{1}{I_a} \quad \text{up to magnetic saturation}$$

Thus, upto magnetic saturation, the N/I_a curve follows the hyperbolic path as shown in Fig. 12.15. After saturation, the flux becomes constant and so does the speed.

(*iii*) **N/T_a Characteristic.** The N/T_a characteristic of a series motor is shown in Fig. 12.16. It is clear that series motor develops high torque at low speed and *vice-versa.*

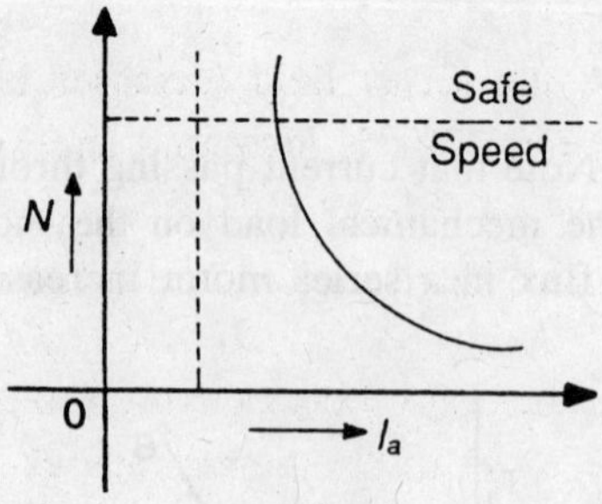

Fig. 12.15

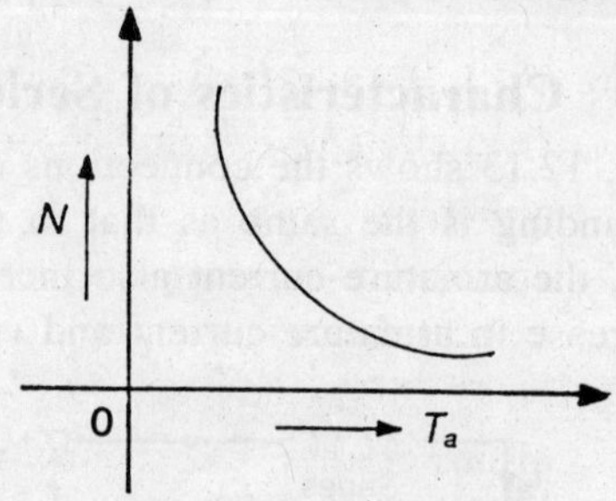

Fig. 12.16

Conclusion : Following three important conclusions are drawn from the above characteristics of series motors :

(*i*) It has a high starting torque because initially $T_a \propto I_a^2$

(*ii*) It is a variable speed motor (See N/I_a curve in Fig. 12.15). *i.e.* it automatically adjusts the speed as the load changes. Thus if the load decreases, its speed is automatically raised and *vice-versa.*

(*iii*) At no load, the armature current is very small and so is the flux. Hence, the speed rises to an excessive high value ($\because N \propto 1/\phi$). This is dangerous for the machine which may be destroyed due to centrifugal forces set up in the rotating parts. *Therefore, a series motor should never be started on no load.* However, to start a series motor, mechanical load is first put and then the motor is started.

Note. The minimum load on a d.c. series motor should be great enough to keep the speed within limits. If the speed becomes dangerously high, then motor must be disconnected from the supply.

12.21. Compound Motors

A compound motor has both series field and shunt field. The shunt field is always stronger than the series field. Compound motors are of two types :

(*i*) *Cumulative-compound motors* in which series field aids the shunt field.

(*ii*) *Differential compound motors* in which series field opposes the shunt field.

Differential-compound motors are rarely used due to their poor torque characteristics at heavy loads.

12.22. Characteristics of Cumulative Compound Motors

Fig. 12.17 shows the connections of a cumulative-compound motor. Each pole carries a series as well as shunt field winding ; the series field aiding the shunt field.

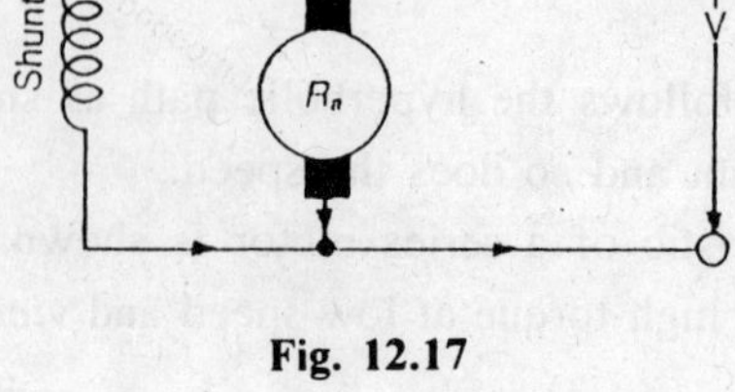

Fig. 12.17

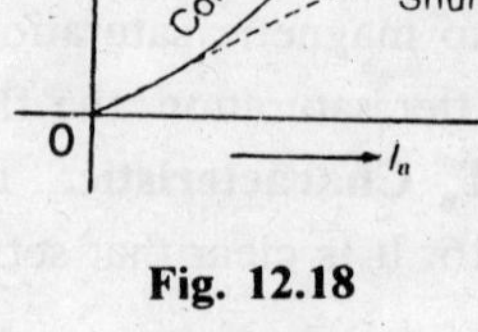

Fig. 12.18

(*i*) **T_a/I_a Characteristic.** As the load increases, the series field increases but shunt field strength remains constant. Consequently, total flux is increased and hence the armature torque ($\because T_a \propto \phi I_a$). It may be noted that torque of a cumulative-compound motor is greater than that of shunt motor for a given armature current due to series field [see Fig. 12.18].

(*ii*) **N/I_a Characteristic.** As explained above, as the load increases, the flux per pole also increases. Consequently, the speed ($N \propto 1/\phi$) of the motor falls as the load increases (See Fig. 12.19). It may be noted that as the load is added, the increased amount of flux causes the speed to decrease more than does the speed of a shunt motor. Thus the speed regulation of a cumulative compound motor is poorer than that of a shunt motor.

Note : Due to shunt field, the motor has a definite no load speed and can be operated safely at no load.

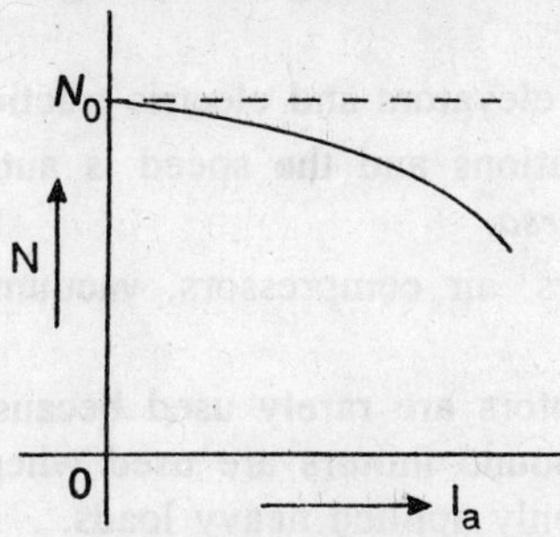

Fig. 12.19

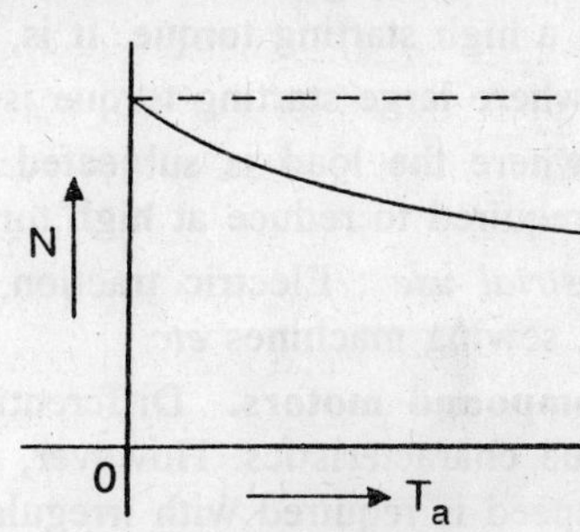

Fig. 12.20

(*iii*) **N/T_a Characteristic.** Fig. 12.20 shows N/T_a characteristic of a cumulative compound motor. For a given armature current, the torque of a cumulative compound motor is more than that of a shunt motor but less than that of a series motor.

Conclusion : A cumulative compound motor has characteristics intermediate between series and shunt motors.

(*i*) Due to the presence of shunt field, the motor is prevented from running away at no load.

(*ii*) Due to the presence of series field, the starting torque is increased.

12.23. Comparison of Three Types of Motors

(*i*) The speed regulation of a shunt motor is better than that of a series motor. However, speed regulation of a cumulative compound motor lies between shunt and series motors (See Fig. 12.21).

(*ii*) For a given armature current, the starting torque of a series motor is more than that of a shunt motor. However, the starting torque of a cumulative compound motor lies between series and shunt motors (See Fig. 12.22).

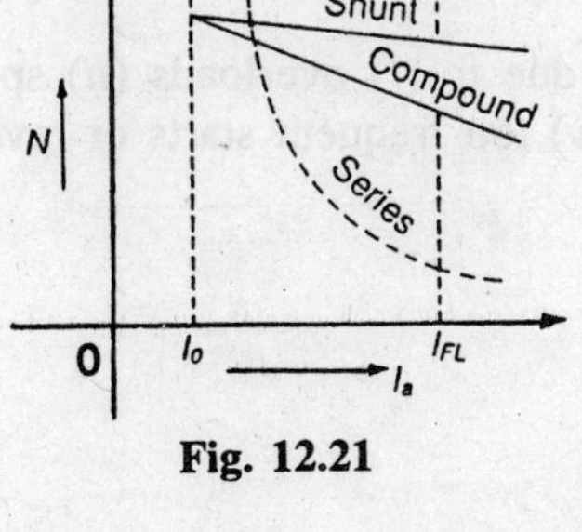

Fig. 12.21

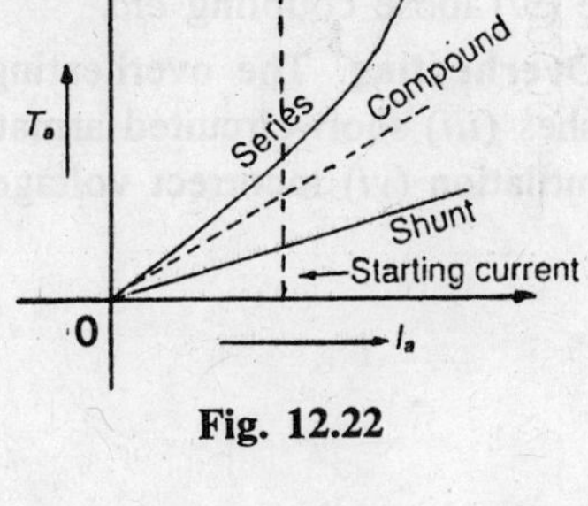

Fig. 12.22

(*iii*) Both shunt and cumulative cumpound motors have definite no load speed. However, a series motor has dangerously high speed at no load.

12.24. Applications of D.C. Motors

1. Shunt motors. The characteristics of a shunt motor reveal that it is an approximately constant speed motor. It is, therefore, used

(*i*) where the speed is required to remain almost constant from no load to full load.

(*ii*) where the load has to be driven at a number of speeds and any one of which is required to remain nearly constant.

Industrial use : Lathes, drills, boring mills, shapers, spinning and weaving machines *etc.*

2. Series motors. It is a variable speed motor *i.e.* speed is low at high torque and *vice-versa.* However, at light or no load, the motor tends to attain dangerously high speed. The motor has a high starting torque. It is, therefore, used

(*i*) where large starting torque is required *e.g.* in elevators and electric traction.

(*ii*) where the load is subjected to heavy fluctuations and the speed is automatically required to reduce at high torques and *vice-versa.*

Industrial use : Electric traction, cranes, elevators, air compressors, vacuum cleaners, hair drier, sewing machines *etc.*

3. Compound motors. Differential-compound motors are rarely used because of their poor torque characteristics. However, cumulative-compound motors are used where a fairly constant speed is required with irregular loads or suddenly applied heavy loads.

Industrial use : Presses, shears, reciprocating machines *etc.*

12.25. Troubles in D.C. Motors

Several troubles may arise in a d.c. motor and a few of them are discussed below :

1. Failure to start. This may be due to (*i*) ground fault (*ii*) open or short-circuit fault (*iii*) wrong connections (*iv*) too low supply voltage (*v*) frozen bearing or (*vi*) excessive load.

2. Sparking at brushes. This may be due to (*i*) troubles in brushes (*ii*) troubles in commutator (*iii*) troubles in armature or (*iv*) excessive load.

(*i*) Brush troubles may arise due to insufficient contact surface, too short a brush, too little spring tension or wrong brush setting.

(*ii*) Commutator troubles may be due to dirt on the commutator, high mica, rough surface or eccentricity.

(*iii*) Armature troubles may be due to an open armature coil. An open armature coil will cause sparking each time the open coil passes the brush. The location of this open coil is noticeable by a burnt line between segments connecting the coil.

3. Vibrations and pounding noises. These may be due to (*i*) worn bearings (*ii*) loose parts (*iii*) rotating parts hitting stationary parts, (*iv*) armature unbalanced (*v*) misalignment of machine (*vi*) loose coupling *etc.*

4. Overheating. The overheating of motor may be due to (*i*) overloads (*ii*) sparking at the brushes (*iii*) short-circuited armature or field coils (*iv*) too frequent starts or reversals (*v*) poor ventilation (*vi*) incorrect voltage.

OBJECTIVE QUESTIONS

1. A d.c. motor is used to ....
   (i) generate power
   (ii) change mechanical energy to electrical energy
   (iii) change electrical energy to mechanical energy
   (iv) increase energy put into it.
2. Carbon brushes are preferable to copper brushes because...............
   (i) they have longer life
   (ii) they reduce armature reaction
   (iii) they have lower resistance
   (iv) they reduce sparking
3. The field poles and armature of a d.c. machine are laminated to ....
   (i) reduce the weight of the machine
   (ii) decrease the speed
   (iii) reduce eddy currents
   (iv) reduce armature reaction
4. The back *e.m.f.* in a d.c. motor ....
   (i) opposes the applied voltage
   (ii) aids the applied voltage
   (iii) aids the armature current
   (iv) none of the above
5. The value of back *e.m.f.* (E_b) in a d.c. motor is maximum at ....
   (i) no load (ii) full load
   (iii) half full-load (iv) none of the above
6. The motor equation is given by ....
   (i) $V = E_b - I_a R_a$ (ii) $V = E_b + I_a R_a$
   (iii) $E_b = I_a R_a - V$ (iv) none of the above.
7. Fig. 12.23, shows the d.c. shunt motor on no load. The quntity $E_b I_a$ represents ....
   (i) input power to armature
   (ii) copper losses in armature
   (iii) core losses
   (iv) total friction, windage and core losses

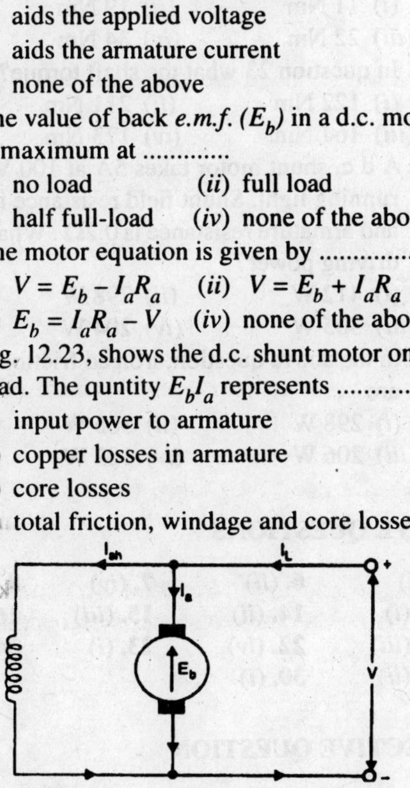

Fig. 12.23. Shunt motor on no load

8. In Fig. 12.23, the quantity $V I_a$ represents ....
   (i) mechanical output of motor
   (ii) core losses
   (iii) all losses
   (iv) friction and windage losses
9. The mechanical power developed in a d.c. motor is maximum when back e.m.f. (E_b) is equal to ....the applied voltage (V).
   (i) twice (ii) half
   (iii) one-third
   (iv) none of the above
10. The amount of back e.m.f. of a shunt motor will increase when ....
    (i) the load is increased
    (ii) the field is weakened
    (iii) the field is strengthened
    (iv) none of the above
11. The torque developed by a d.c. motor is directly proportional to ....
    (i) flux per pole × armature current
    (ii) armature resistance × applied voltage
    (iii) armature resistance × armature current
    (iv) none of the above
12. The shaft torque (T_{sh}) in a d.c. motor is less than total armature torque (T_a) because of .... in the motor.
    (i) Cu losses (ii) field losses
    (iii) iron and friction losses
    (iv) none of the above
13. Armature reaction in a d.c. motor is increased ....
    (i) when the armature current increases
    (ii) when the armature current decreases
    (iii) when the field current increases
    (iv) by interpoles
14. With respect to the direction of rotation, interpoles on a d.c. motor must have the same polarity as the main poles ....
    (i) ahead of them
    (ii) behind them
    (iii) none of the above
15. In a d.c. motor, the brushes are shifted from the mechanical neutral plane in a direction opposite to the rotation to ....
    (i) decrease speed (ii) increase speed

(*iii*) reduce sparking
(*iv*) produce flat characteristics

16. In very large d.c. motors with severe heavy duty, armature reaction effects are corrected by ....
(*i*) using interpoles only
(*ii*) using compensatory windings in addition to interpoles
(*iii*) shifting the brush position
(*iv*) none of the above

17. The deciding factor in the selection of a d.c. motor for a particular application is its ....characteristic.
(*i*) speed-torque
(*ii*) speed-armature current
(*iii*) torque-armature current
(*iv*) none of the above

18. The demand for a large increase in torque of a d.c. shunt motor is met by a ....
(*i*) large decrease in speed
(*ii*) large increase in speed
(*iii*) large increase in current
(*iv*) small increase in current.

19. The demand for a large increase in torque of a d.c. series motor is met by a ....
(*i*) large decrease in current
(*ii*) large decrease in speed
(*iii*) large increase in speed
(iv) small decrease in speed

20. A 440 V shunt motor has an armature resistance of 0.8Ω and a field resistance of 200Ω. Find the back e.m.f. when giving an output of 7.46 kW at 85% efficiency.
(*i*) 222.4 V (*ii*) 425.8 V
(*iii*) 312.6 V (*iv*) 392.7 V

21. A 250 V d.c. shunt motor takes a total current of 20 A. Resistance of shunt field winding is 200Ω and that of armature is 0.3Ω. What is the current in the armature ?
(*i*) 9.65 A (*ii*) 11.25 A
(*iii*) 18.75 A (*iv*) 16.62 A

22. In the above question, what is the approximate value of back e.m.f.?
(*i*) 212 V (*ii*) 232 V
(*iii*) 227 V (*iv*) 244 V

23. A 220 V d.c. shunt motor takes a total current of 80 A and runs at 800 r.p.m. Resistance of shunt field is 50Ω and that of armature 0.1Ω. The iron and friction losses amount to 1600 W. What is the driving power of the motor?
(*i*) 16050 W (*ii*) 14500 W
(*iii*) 12600 W (*iv*) 18500 W

24. In question 23, what is B.H. P. ?
(*i*) 15.75 (*ii*) 11.78
(*iii*) 14.28 (*iv*) 19.37

25. In question 23, find the copper losses?
(*i*) 1120 W (*ii*) 1550 W
(*iii*) 1775 W (*iv*) 1140 W

26. In question 23, what is the armature torque?
(*i*) 212 Nm (*ii*) 176 Nm
(*iii*) 192 Nm (*iv*) 242 Nm

27. In question 23, what is the lost torque?
(*i*) 11 Nm (*ii*) 19 Nm
(*iii*) 22 Nm (*iv*) 34 Nm

28. In question 23 what the shaft torque?
(*i*) 122 Nm (*ii*) 211 Nm
(*iii*) 164 Nm (*iv*) 173 Nm

29. A d.c. shunt motor takes 5A at 100 V when running light. Shunt field resistance is 50 Ω and armature resistance is 0.2Ω . What is the driving power?
(*i*) 412 W (*ii*) 298 W
(*iii*) 305 W (*iv*) 204 W

30. In the above question, iron ad friction losses are
(*i*) 298 W (*ii*) 305 W
(*iii*) 206 W (*iv*) 102 W

ANSWERS TO OBJECTIVE QUESTIONS

1. (*iii*)	**2.** (*iv*)	**3.** (*iii*)	**4.** (*i*)	**5.** (*i*)	**6.** (*ii*)	**7.** (*iv*)	**8.** (*iii*)
9. (*ii*)	**10.** (*iii*)	**11.** (*i*)	**12.** (*iii*)	**13.** (*i*)	**14.** (*ii*)	**15.** (*iii*)	**16.** (*ii*)
17. (*i*)	**18.** (*iii*)	**19.** (*ii*)	**20.** (*ii*)	**21.** (*iii*)	**22.** (*iv*)	**23.** (*i*)	**24.** (*iv*)
25. (*ii*)	**26.** (*iii*)	**27.** (*ii*)	**28.** (*iv*)	**29.** (*ii*)	**30.** (*i*)		

HINTS TO SELECTED OBJECTIVE QUESTION

2. Since the contact resistance of carbon brushes is very high, sparking at the brushes can be prevented.

7. At no load, there is no output. Consequently, the entire power developed in the armature is used up in overcoming friction, windage and core losses.

8. At no load, there is no output. Consequently, the motor input (VI_a) is all used up to overcome the losses.

12. The total torque developed in the armature is not available at the shaft as some percentage of it is used up in overcoming mechanical losses (friction) and iron losses in the motor.

16. Some large d.c. motors employed in steel mills perform a series of heavy-duty operations. They accelerate, decelerate, stop, reverse all in a mater of seconds. The corresponding armature current increases, decreases, reverses in stepwise fashion, producing very sudden changes in armature reaction. For such motors, interpoles do not adequately neutralize the armature m.m.f. To eliminate this problem, additional *compensating windings* are connected in series with the armature. They are distributed in slots, cut into the pole faces of the main field poles. Like interpoles, these windings produce an m.m.f. equal and opposite to the m.m.f. of the armature.

18. For a d.c. motor, $T_a \propto \phi I_a$. But in a shunt motor, ϕ is practically constant so that $T_a \propto I_a$. If torque is doubled, armature current is also doubled.

19. *D.C.* series motors exert largest torque at low speeds *i.e.*, when starting. As the load decreases, they automatically raise the speed.

20. Motor input power $= \dfrac{7.46 \times 1000}{0.85} W$

Motor input current $= \dfrac{7.46 \times 1000}{0.85 \times 440} = 19.95$ A

$I_{sh} = 440/200 = 2.2$ A ; $I_a = 19.95 - 2.2 = 17.75$ A

$\therefore$ Back e.m.f., $E_b = V - I_a R_a = 440 - 17.75 \times 0.8 = 425.8$ V

21. $I_{sh} = 250/200 = 1.25$ A ; $I_a = 20 - 1.25 = 18.75$ A

22. $E_b = V - I_a R_a = 250 - 18.75 \times 0.3 \simeq 244$ V

23. $I_{sh} = 220/50 = 4.4$ A ; $I_a = 80 - 4.4 = 75.6$ A

$E_b = V - I_a R_a = 220 - 75.6 \times 0.1 = 212.44$ A

$\therefore$ Driving power $= E_b I_a = 212.44 \times 75.6 \simeq 16050$ W

24. Output power = Driving power – Iron and friction losses

= 16050 – 1600 = 14450 W

$\therefore$ $B.H.P. = \dfrac{14450}{746} = 19.37$

25. Input power = 220 × 80 = 17600 W

Copper losses = Input power – driving power = 17600 – 16050 = 1550 W

26. Armature torque, $T_a = \dfrac{9.55 E_b I_a}{N} = \dfrac{9.55 \times 16050}{800} = 192$ Nm.

27. The whole of armature torque (T_a) is not available for doing useful work ; a certain part being required to overcome the iron and friction losses.

Lost Torque $= 9.55 \times \dfrac{\text{Iron and friction losses}}{N} = 9.55 \times \dfrac{1600}{800} = 19$ Nm

28. Shaft torque, $T_{sh} = T_a$ – Lost torque = 192 – 19 = 173 Nm

29. $I_{sh} = 100/50 = 2$ A ; $I_a = 5 - 2 = 3$ A ; $E_b = V - I_a R_a = 100 - 3 \times 0.2 = 99.4$ V

Driving power $= E_b I_a = 99.4 \times 3 = 298$ W

30. Output is zero. Therefore, iron and friction losses = driving power = 298 W.

13

Speed Control of D.C. Motors

Introduction

Although a far greater percentage of electric motors in service are a.c. motors, the d.c. motor is of considerable industrial importance. The principal advantage of a d.c. motor is that its speed can be changed over a wide range by a variety of simple methods. Such a fine speed control is generally not possible with a.c. motors. In fact, fine speed control is one of the reasons for the strong competitive position of d.c. motors in modern industrial applications. In this chapter, we shall discuss the various methods of speed control of d.c. motors.

13.1. Speed Control of D.C. Motors

The speed of a d.c. motor is given by ;

$$N \propto \frac{E_b}{\phi}$$

or $$N = K\frac{(V - I_a R)}{\phi} \text{ r.p.m.} \quad ...(i)$$

where $R = R_a$ ...*for shunt motor*

$= R_a + R_{se}$ ...*for series motor*

From exp. (*i*), it is clear that there are two main methods of controlling the speed of a d.c. motor, namely ;

(*i*) By varying the flux per pole (ϕ). This is known as *flux control method.*

(*ii*) By varying the resistance in the armature circuit. This is known as *armature control method.*

13.2. Speed Control of Shunt Motors

The speed of a shunt motor can be changed by (*i*) flux control method (*ii*) armature control method. The former method is frequently used because it is simple and inexpensive.

1. Flux control method. It is based on the fact that by varying the flux ϕ, the motor speed, ($N \propto 1/\phi$) can be changed and hence the name flux control method. In this method, a variable resistance (known as *shunt field rheostat*) is placed in series with shunt field winding as shown in Fig. 13.1.

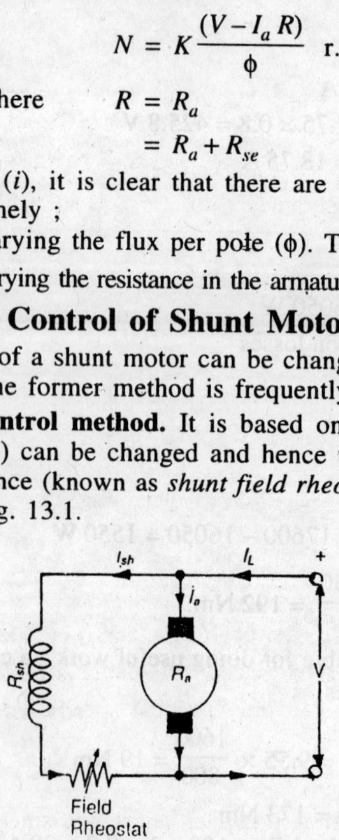

Fig. 13.1

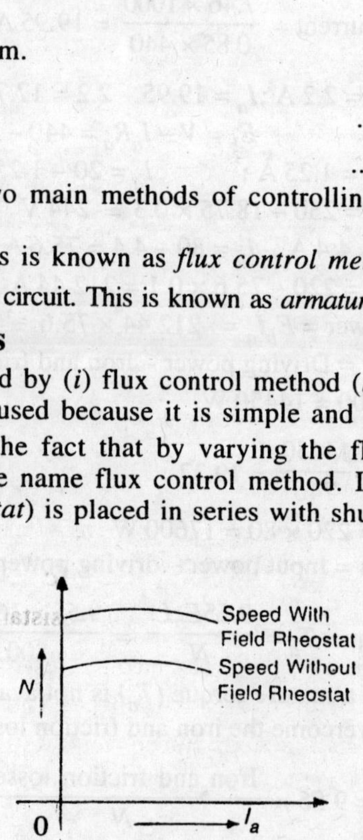

Fig. 13.2

The shunt field rheostat reduces the shunt field current I_{sh} and hence the flux ϕ. Therefore, we can only raise the speed of the motor above the normal speed (See Fig. 13.2). Generally, this method permits to increase the speed in the ratio 3 : 1. Wider speed ranges tend to produce instability and poor commutation.

Advantages

(*i*) This is an easy and convenient method.

(*ii*) It is an inexpensive method since very little power is wasted in the shunt field rheostat due to relatively small value of I_{sh}.

(*iii*) The speed control exercised by this method is independent of load on the machine.

Disadvantages

(*i*) Only speeds higher than the normal speed can be obtained since the total field circuit resistance cannot be reduced below R_{sh}—the shunt field winding resistance.

(*ii*) There is a limit to the maximum speed obtainable by this method. It is because if the flux is too much weakened, commutation becomes poorer.

2. Armature control method. This method is based on the fact that by varying the voltage available across the armature, the back e.m.f. and hence the speed of the motor can be changed. This is done by inserting a variable resistance R_C (known as *controller resistance*) in series with the armature as shown in Fig. 13.3.

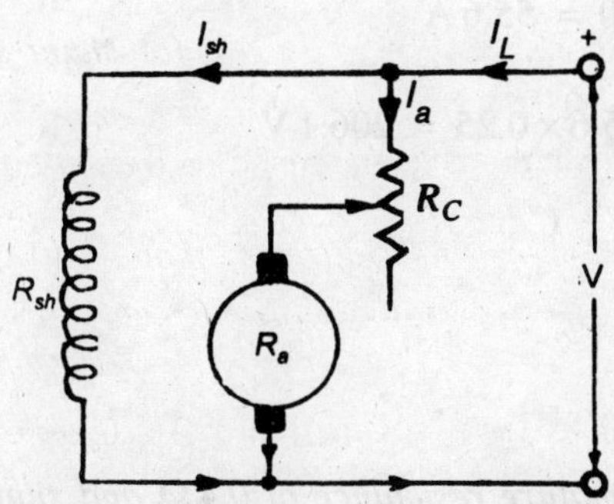

Fig. 13.3

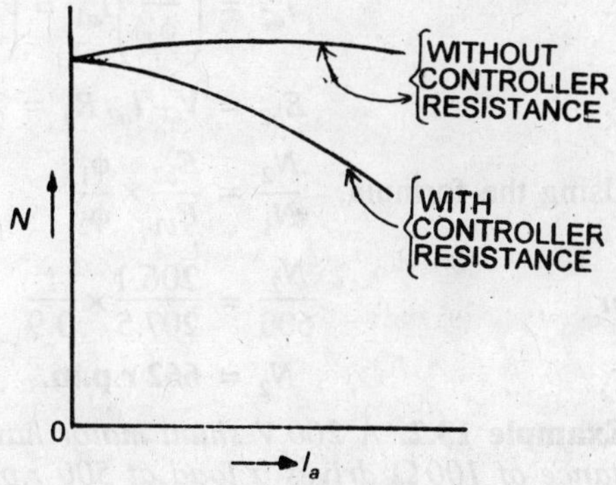

Fig. 13.4

$$N \propto V - I_a (R_n + R_C)$$

where R_C = controller resistance.

Due to voltage drop in the controller resistance, the back e.m.f. (E_b) is *decreased. Since $N \propto E_b$, the speed of the motor is reduced. The highest speed obtainable is that corresponding to $R_C = 0$ *i.e.*, normal speed. Hence, this method can only provide speeds below the normal speed (See Fig. 13.4).

Disadvantages

(*i*) A large amount of power is wasted in the controller resistance since it carries full armature current I_a.

---

* $E_b = V - I_a (R_a + R_C)$

Since V is constant, E_b will decrease.

(*ii*) The speed varies widely with load since the speed depends upon the voltage drop in the controller resistance and hence on the armature current demanded by the load.

(*iii*) The output and efficiency of the motor are reduced.

(*iv*) This method results in poor speed regulation.

Due to above disadvantages, this method is seldom used to control the speed of shunt motors.

Note. The armature control method is a very common method for the speed control of d.c. series motors. The disadvantage of poor speed regulation is not important in a series motor which is used only where varying speed service is required.

Example 13.1. *A 220V d.c. shunt motor having an armature resistance of 0.25 Ω carries an armature current of 50A and runs at 600 r.p.m. If the flux is reduced by 10% by field regulator, find the speed assuming load torque remains the same.*

Solution.

Initial conditions

$$N_1 = 600 \text{ r.p.m.}, \; I_{a1} = 50 \text{ A}$$

$$E_{b1} = V - I_{a1} R_a = 220 - 50 \times 0.25 = 207.5 \text{ V}$$

Final conditions

$$\phi_2 = 0.9\, \phi_1 \text{ so that } \phi_2/\phi_1 = 0.9$$

As the load torque remains unchanged, $\phi_1 I_{a1} = \phi_2 I_{a2}$

$$\therefore \qquad I_{a2} = \left(\frac{\phi_1}{\phi_2}\right) I_{a1} = \left(\frac{1}{0.9}\right) 50 = 55.6 \text{ A}$$

$$\therefore \qquad E_{b2} = V - I_{a2} R_a = 220 - 55.6 \times 0.25 = 206.1 \text{ V}$$

Using the formula, $$\frac{N_2}{N_1} = \frac{E_{b2}}{E_{b1}} \times \frac{\phi_1}{\phi_2}$$

or $$\frac{N_2}{600} = \frac{206.1}{207.5} \times \frac{1}{0.9}$$

$$\therefore \qquad N_2 = \mathbf{662 \text{ r.p.m.}}$$

Example 13.2. *A 200 V shunt motor having armature resistance of 0.4 Ω and shunt field resistance of 100 Ω drives a load at 500 r.p.m. taking 27 A. It is desired to run the motor at 700 r.p.m. Assuming the load torque to be constant, find the value of resistance to be used as field regulator. Neglect saturation effect.*

Solution.

Initial conditions

$$N_1 = 500 \text{ r.p.m.}, \; I_L = 27 \text{ A} \; ; \; I_{sh1} = 200/100 = 2 \text{ A}$$

$$I_{a1} = I_L - I_{sh1} = 27 - 2 = 25 \text{ A}$$

$$E_{b1} = 200 - 25 \times 0.4 = 190 \text{ V}$$

Final conditions

$$E_{b2} = V - I_{a2} R_a = 200 - 0.4\, I_{a2}$$

As the load torque is constant, $\phi_1 I_{a1} = \phi_2 I_{a2}$. But $\phi \propto I_{sh}$ so that we have,

$$I_{sh1}\, I_{a1} = I_{sh2}\, I_{a2}$$

$$\therefore \qquad I_{sh2} = \frac{I_{sh1}\, I_{a1}}{I_{a2}} = \frac{2 \times 25}{I_{a2}} = \frac{50}{I_{a2}}$$

Now, $$\frac{N_1}{N_2} = \frac{E_{b1}}{E_{b2}} \times \frac{I_{sh2}}{I_{sh1}}$$

or $$\frac{500}{700} = \frac{190}{200 - 0.4\,I_{a2}} \times \frac{(50/I_{a2})}{2}$$

or $$I_{a2}^2 - 500\,I_{a2} + 16625 = 0$$

Solving this equation, we get, $I_{a2} = 35\text{ A}$

$\therefore$ $$I_{sh2} = 50/I_{a2} = 50/35 = 1.43\text{ A}$$

$\therefore$ $$R_{sh2} = V/I_{sh2} = 200/1.43 = 139.86\,\Omega$$

Field rheostat resistance $= 139.86 - 100 = \mathbf{39.86\,\Omega}$

Example 13.3. *A shunt motor supplied at 230 V runs at 900 r.p.m. while taking armature current of 30 A ; the resistance of armature circuit being 0.4 Ω. Calculate the resistance required in series with the armature circuit to reduce the speed to 500 r.p.m., assuming that the armature current is 25 A.*

Solution.

Initial conditions

$$N_1 = 900\text{ r.p.m.},\ I_{a1} = 30\text{ A}$$
$$E_{b1} = V - I_{a1} R_a = 230 - 30 \times 0.4 = 218\text{ V}$$

Final conditions

$$N_2 = 500\text{ r.p.m.},\ I_{a2} = 25\text{ A},\ R_t = R_a + R_C$$
$$E_{b2} = V - I_{a2} R_t = 230 - 25R_t$$

Since excitation is unchanged, $\phi_1 = \phi_2$

$\therefore$ $$\frac{N_2}{N_1} = \frac{E_{b2}}{E_{b1}}$$

or $$\frac{500}{900} = \frac{230 - 25\,R_t}{218}$$

$\therefore$ $$R_t = 4.356\,\Omega$$

$\therefore$ Additional series resistance required in the armature

$$R_C = R_t - R_a = 4.356 - 0.4 = \mathbf{3.956\,\Omega}$$

Tutorial Problems

1. A 200 V shunt motor having an armature resistance of 0.2 Ω carries an armature current of 50A and runs at 960 r.p.m. If the flux is reduced by 10%, find the speed, assuming load torque remains the same. **[1059 r.p.m.]**
2. A shunt motor supplied at 200 V runs at 500 r.p.m. while taking armature current of 30 A. The armature resistance is 0.5 Ω. Calculate the resistance required in series with the armature to reduce the speed to 300 r.p.m. ; assuming the current in the armature remains the same. **[2.47 Ω]**
3. A shunt motor connected across 440 V supply takes an armature current of 20 A and runs at 500 r.p.m. The armature resistance is 0.6 Ω. If the magnitude of flux is reduced by 30% and the torque developed by the armature increases by 40%, what is the speed of the motor ? **[695 r.p.m.]**

13.3. Speed Control of D.C. Series Motors

The speed control of d.c. series motors can be obtained by (*i*) flux control method (*ii*) armature-resistance control method. The latter method is mostly used.

1. Flux control method. In this method, the flux produced by the series motor is varied and hence the speed. The variation of flux can be achieved in the following ways :

(*i*) **Field diverters.** In this method, a variable resistance (called *field diverter*) is connected in parallel with series field winding as shown in Fig. 13.5. Its effect is to shunt some portion of the line current from the series field winding, thus weakening the field and increasing the speed ($\because N \propto 1/\phi$). The lowest speed obtainable is that corresponding to zero current in the diverter (*i.e.*, diverter is open). Obviously, the lowest speed obtainable is the normal speed of the motor. Consequently, this method can only provide speeds above the normal speed. The series field diverter method is often employed in traction work.

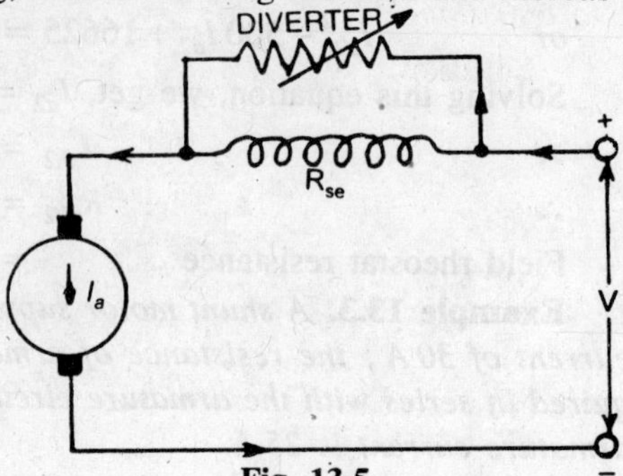

Fig. 13.5

(*ii*) **Armature diverter.** In order to obtain speeds below the normal speed, a variable resistance (called *armature diverter*) is connected in parallel with the armature as shown in Fig. 13.6. The diverter shunts some of the line current, thus reducing the armature current. Now for a given load, if I_a is decreased, the flux ϕ must increase ($\because T \propto \phi I_a$). Since $N \propto 1/\phi$, the motor speed is decreased. By adjusting the armature diverter, any speed lower than the normal speed can be obtained.

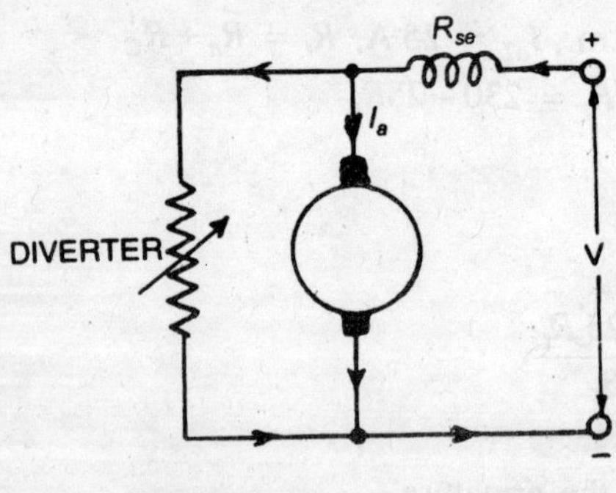

Fig. 13.6

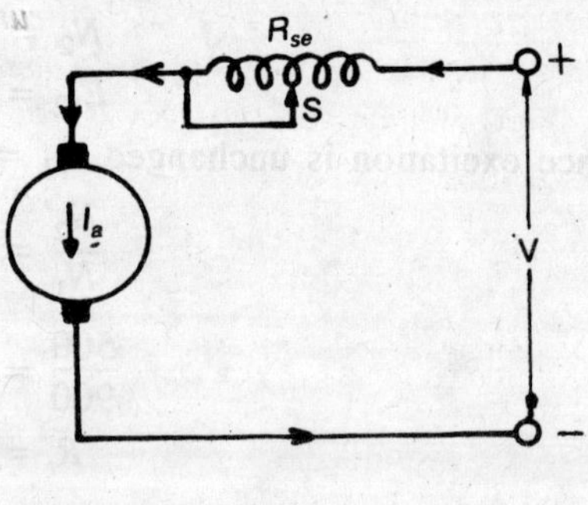

Fig. 13.7

(*iii*) **Tapped field control. In this method, the flux is reduced (and hence speed is increased) by decreasing the number of turns of the series field winding as shown in Fig. 13.7. The switchs can short circuit any part of the field winding, thus decreasing the flux and raising the speed. With full turns of the field winding, the motor runs at normal speed and as the field turns are cut out, speeds higher than normal speed are achieved.**

2. Armature-resistance control. In this method, a variable resistance is directly connected in series with the supply to the complete motor as shown in Fig. 13.8. This reduces the voltage available across the armature and hence the speed falls. By changing the value of variable resistance, any speed below the normal speed can be obtained. This is the most common method employed to control the speed of d.c. series motors. Although this method has poor speed regulation, this has no significance for series motors because they are used in varying speed applications. The loss of power in the series resistance for many applications of series motors is not too serious since in these applications, the control is util-

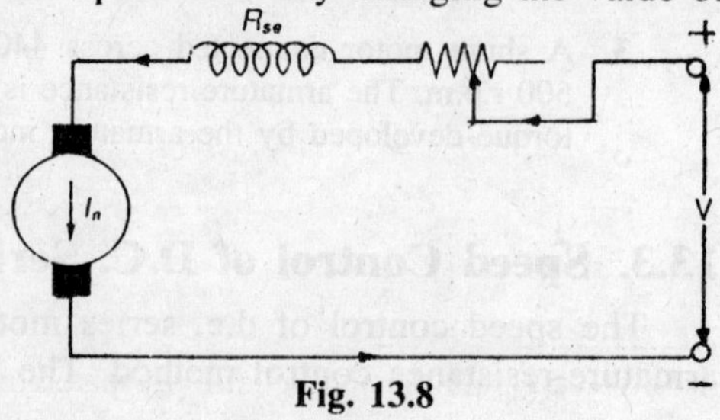

Fig. 13.8

ised for a large portion of the time for reducing the speed under light-load conditions and is only used intermittently when the motor is carrying full load.

Example 13.4. *A 220V d.c. series motor runs at 900 r.p.m. when taking a line current of 40A. The armature resistance and series field resistance are 0.06Ω and 0.04Ω respectively. If current taken remains the same, calculate the series resistance required to reduce the speed to 600 r.p.m.*

Solution. $R_a + R_{se} = 0.04 + 0.06 = 0.1\ \Omega$

$$N_1 = 900\text{ r.p.m.}\ ;\ N_2 = 600\text{ r.p.m.}\ ;\ I_{a1} = I_{a2} = 40\text{ A}$$

$$E_{b1} = 200 - 40 \times 0.1 = 196\text{ V}$$

$$E_{b2} = (200 - 40R)\text{ V}$$

where R = Series resistance + $R_a + R_{se}$

Now
$$\frac{N_2}{N_1} = \frac{E_{b2}}{E_{b1}} \qquad (\because \phi_2 = \phi_1)$$

or
$$\frac{600}{900} = \frac{200 - 40R}{196}$$

$$\therefore \quad R = 1.73\ \Omega$$

∴ Series resistance = $1.73 - 0.1 = \mathbf{1.63\ \Omega}$

Example 13.5. *A 440 V d.c. series motor of negligible resistance and with unsaturated magnetic circuit takes 50 A when running at a certain speed on a given load. If the load torque varies as the cube of the speed, find the resistance necessary to reduce the speed by 50%. What is then the current ?*

Solution.

$$*T \propto I_a^2$$

Also
$$T \propto N^3 \qquad \text{...given}$$

For the first case, $I_{a1}^2 \propto N_1^3$

For the second case, $I_{a2}^2 \propto N_2^3$

$$\therefore \quad \left(\frac{I_{a2}}{I_{a1}}\right)^2 = \left(\frac{N_2}{N_1}\right)^3 = \left(\frac{1}{2}\right)^3 \qquad \left(\because \frac{N_2}{N_1} = \frac{1}{2}\right)$$

or
$$I_{a2} = I_{a1}\sqrt{\frac{1}{8}} = 50\sqrt{\frac{1}{8}} = \mathbf{17.68\ A}$$

Now
$$E_{b1} = 440 - I_{a1} \times 0 = 440\text{ V}$$

$$E_{b2} = (440 - 17.68R)\text{ V}$$

where R = series resistance inserted

$$\frac{N_2}{N_1} = \frac{E_{b2}}{E_{b1}} \times \frac{I_{a1}}{I_{a2}} \qquad (\because \phi \propto I_a)$$

or
$$\frac{1}{2} = \frac{440 - 17.68R}{440} \times \frac{50}{17.68}$$

$$\therefore \quad R = \mathbf{20.5\ \Omega}$$

---

* $T \propto \phi I_a$. As the series field is unsaturated, $\phi \propto I_a$

∴ $T \propto I_a^2$.

Tutorial Problems

1. A 220 V series motor takes 10 A and runs at 600 r.p.m., its total resistance is 0.8 Ω. At what speed will it run when a 5-ohm resistance is connected in series, the motor is taking the same current at the same supply voltage ? **[458 r.p.m.]**
2. A series motor with unsaturated field and negligible resistance, when running at a certain speed on a given load takes 60 A at 460 V. If the load torque varies as the cube of the speed, calculate the series resistance required to reduce the speed by 25%. **[6 Ω]**
3. A d.c. series motor is connected to a 220 V and runs at 500 r.p.m. taking a current of 50 A. Calculate the value of series resistance which will reduce the speed to 300 r.p.m. ; the load torque then being one-half of its previous value. Resistance of motor is 0.1 Ω. Assume the flux to be proportional to field current. **[3.54 Ω]**

13.4. Necessity of Motor Starters

At starting, when the motor is stationary, there is no back e.m.f. in the armature. Consequently, if the motor is directly switched on to the mains, the armature will draw a heavy current ($I_a = V/R_a$) because of small armature resistance. As an example, 5 H.P., 220 V shunt motor has a full-load current of 20 A and an armature resistance of about 0.5 Ω. If this motor is directly switched on to supply, it would take an armature current of 220/0.5 = 440 A which is 22 times the full-load current. This high starting current may result in :

(*i*) burning of armature due to excessive heating effect.

(*ii*) damaging the commutator and brushes due to heavy sparking.

(*iii*) excessive voltage drop in the line to which the motor is connected. The result is that the operation of other appliances connected to the line may be impaired and in particular cases, they may refuse to work.

In order to avoid excessive current at starting, a variable resistance (known as *starting resistance*) is inserted in series with the armature circuit. This resistance is gradually reduced as the motor gains speed (and hence E_b increases) and eventually it is cut out completely when the motor has attained full speed.

Note. The value of starting resistance is generally such that starting current is limited to 1.25 to 2 times the full-load current.

13.5. Shunt Motor Starter

Fig. 13.9 shows the schematic diagram of a shunt motor starter with protective devices. It consists of starting resistance divided into several sections and connected in series with the armature. The tapping points of the starting resistance are brought out to a number of studs. One end of shunt field winding is connected to the first stud and the other end to the far side of the supply. Included in this circuit is the *no-volt release coil*. One end of the starting handle *A* is connected to one side of supply through *overload release coil*. The other end of starting handle moves against spring and makes contact with each stud during starting operation, cutting out more and more starting resistance as it passes over each stud in clockwise direction.

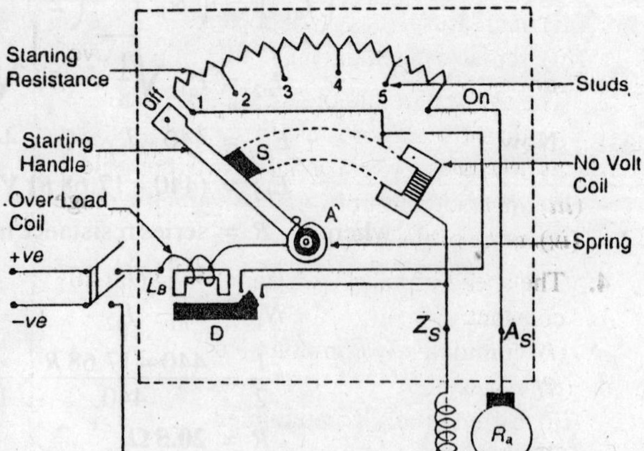

Fig. 13.9

Operation

(*i*) To start with, the d.c. supply is switched on with starting handle in the OFF position.

(*ii*) **The handle is now moved clockwise to the first stud. As soon as it comes in contact with the first stud, the shunt field winding is directly connected across the supply, while the whole starting resistance is inserted in series with the armature circuit.**

(*iii*) **As the handle is gradually moved over to the final stud, the starting resistance is cut out of the armature circuit in steps. The handle is now held magnetically by the no-volt release coil which is energised by shunt field current.**

(*iv*) **If the supply voltage is suddenly interrupted or if the field excitation is accidentally cut, the no-volt release coil is demagnetised and the handle goes back to the OFF position under the pull of the spring. If no-volt release coil were not used, then in case of failure of supply, the handle would remain on the final stud. If then supply is restored, the motor will be directly connected across the supply, resulting in an excessive armature current.**

(*v*) **If the motor is over loaded (or a fault occurs), it will draw excessive current from the supply. This current will increase the ampere-turns of the over-load release coil and pull the armature *D*, thus short-circuiting the no-volt release coil. The no-volt coil is demagnetised and the starting handle is pulled to the OFF position by the spring. Thus, the motor is automatically disconnected from the supply.**

OBJECTIVE QUESTIONS

1. A d.c. motor is still used in industrial applications because it.............
(*i*) is cheap
(*ii*) is simple in construction
(*iii*) provides fine speed control
(*iv*) none of the above

2. When the speed of a d.c. motor increases, its armature current ....
(*i*) increases (*ii*) decreses
(*iii*) remains constant (*iv*) none of the above

3. The speed of a d.c. motor is ....
(*i*) directly proportional to flux per pole
(*ii*) inversely proportional to flux per pole
(*iii*) inversely proportional to applied voltage
(*iv*) none of the above

4. The speed of a..............motor is practically constant.
(*i*) cumulatively compounded
(*ii*) series
(*iii*) differentially compounded
(*iv*) shunt

5. ....motor is variable speed motor.
(*i*) series (*ii*) shunt
(*iii*) cumulatively compounded
(*iv*) differentially compounded.

6. The most commonly used method of speed control of a d.c. motor is by varying...........
(*i*) voltage applied to the motor
(*ii*) field strength
(*iii*) effective number of conductors in series
(*iv*) armature circuit resistance

7. The running speed of a d.c. series motor is basically determined by ....
(*i*) field excitation (*ii*) load
(*iii*) armature resistance
(*iv*) none of the above

8. After a shunt motor is up to speed, the speed may be increased considerably by ....
(*i*) increasing field circuit resistance
(*ii*) decreasing field circuit resistance
(*iii*) increasing armature circuit resistance
(*iv*) reducing the load.

9. .... motor has the best speed regulation.
(*i*) series
(*ii*) cumulatively compounded
(*iii*) shunt
(*iv*) differentially compounded.

10. A series motor will over speed when ....
(*i*) the load is increased
(*ii*) the field is opened
(*iii*) the armature circuit is opened
(*iv*) load is removed

11. D.C. shunt motor are used in those applications where ....is required.
(*i*) high starting torque
(*ii*) practically constant speed
(*iii*) high no-load speed
(*iv*) variable speed

12. When load is removed, the motor that will run at the highest speed is the .... motor.
(*i*) shunt
(*ii*) series
(*iii*) cumulatively compounded
(*iv*) differentially compounded.

13. A series motor of resistance 1Ω between the terminals runs at 800 r.p.m. at 200 V with a current of 15 A. Find the speed at which it will run when connected in series with a 5Ω resistance and taking the same current at the same supply voltage.
(*i*) 476 r.p.m. (*ii*) 384 r.p.m.
(*ii*) 572 r.p.m. (*iv*) 602 r.p.m.

14. What happens to speed when the flux is reduced by 10% in a 200 V. d.c. shunt motor having an armature resistance of 0.2 Ω, carrying a current of 50 A and running at 960 r.p.m. prior to weakening of field? The total torque may be assumed to be constant and iron and friction losses may be neglected.
(*i*) 1250 r.p.m. (*ii*) 1058 r.p.m.
(*iii*) 920 r.p.m. (*iv*) 576 r.p.m.

15. By flux control method of speed control of a d.c. shunt motor, we can obtain speeds
(*i*) above the normal speed only
(*ii*) below the normal speed only
(*iii*) above as well as below the normal speed
(*iv*) none of the above.

16. By putting controller resistance in series with the armature of a d.c. motor, we can obtain speeds
(*i*) above the normal speed only
(*ii*) below the normal speed only
(*iii*) above as well as below the normal speed
(*iv*) none of the above.

17. The speed of a d.c. motor can be controlled by changing
(*i*) its flux
(*ii*) armature circuit resistance
(*iii*) applied voltage
(*iv*) all of the above

18. Motor starters are essential for
(*i*) accelerating the motor
(*ii*) starting the motor
(*iii*) avoiding excessive starting current
(*iv*) preventing fuse blowing

19. The only disadvantage of field control method for controlling the speed of a d.c. shunt motor is that it
(*i*) gives speeds lower than the normal speed
(*ii*) is wasteful
(*iii*) needs a large rheostat
(*iv*) adversely affects commutation

20. The rheostatic speed control method is very
(*i*) economical (*ii*) efficient
(*iii*) unsuitable for rapidly changing loads
(*iv*) suitable for getting speeds above the normal

ANSWERS TO OBJECTIVE QUESTIONS

1. (*iii*)	**2.** (*ii*)	**3.** (*ii*)	**4.** (*iv*)	**5.** (*i*)	**6.** (*ii*)	**7.** (*ii*)	**8.** (*ii*)
9. (*iii*)	**10.** (*iv*)	**11.** (*ii*)	**12.** (*ii*)	**13.** (*i*)	**14.** (*ii*)	**15.** (*i*)	**16.** (*ii*)
17. (*iv*)	**18.** (*iii*)	**19.** (*iv*)	**20.** (*iii*)				

HINTS TO SELECTED OBJECTIVE QUESTIONS

1. One potent reason for the widespread use of d.c. motors in industry is that their speed can be changed over a wide range by simple methods. Such a fine speed control is not possible with a.c. motors.

2. $I_a = \dfrac{V - E_b}{R_a}$

If the speed of the d.c. motor increases, then the back e.m.f. E_b (= $P\phi ZN/60$ A) also increses. It is clear from the above equation that armature current (I_a) will decrease.

3. $N \propto \dfrac{V - I_a R_a}{\phi}$

It is clear that speed (*N*) of a d.c. motor is inversely proportional to flux per pole (ϕ).

6. The fact that the speed of a d.c. motor varies with field excitation provides a convenient means for controlling the speed of shunt and compound motors.

13. $E_b \propto \phi N$ *or* $E_b \propto N$ **($\because$ Armature current and hence ϕ is the same)**

$E_{b_1} = 200 - 15 \times 1 = 185$ V ; $E_{b_2} = 200 - 15(5 + 1) = 110$ V.

Now $\dfrac{E_{b_2}}{E_{b_1}} = \dfrac{N_2}{N_1}$ or $\dfrac{110}{185} = \dfrac{N_2}{800}$ $\therefore N_2 =$ 476 r.p.m.

14. When the flux is decreased, speed $(N \propto E_b / \phi)$ of the motor increases.

Now $\dfrac{N_2}{N_1} = \dfrac{E_{b_2}}{E_{b_1}} \times \dfrac{\phi_1}{\phi_2}$

Also $T_a \propto \phi\, I_a$. As torque is the same in the two cases,

$\therefore \phi_1 I_{a_1} = \phi_2 I_{a_2}$ or $\phi_1 \times 50 = 0.9\,\phi_1 \times I_{a_2}$ $\therefore I_{a_2} = 50/0.9 = 55.55$ A

$E_{b_1} = 200 - 50 \times 0.2 = 190$ V ;

$E_{b_2} = 200 - 55.55 \times 0.2 = 188.9$ V

$\therefore \dfrac{N_2}{960} = \dfrac{188.9}{190} \times \dfrac{\phi_1}{0.9\,\phi_1}$ $\therefore N_2 = 1058$ r.p.m.

16. This method gives speeds below the normal speed, not above it because armature voltage can be decreased (not increased) by the controller resistance.

19. The only disadvantage of this method is that commutation becomes unsatisfactory because the effect of armature reaction is greater on a weakened field.

14

Chemical Effects of Electric Current

Introduction

There are two types of electric conductors *viz.* (*i*) electronic or metallic conductors and (*ii*) electrolytic or ionic conductors. In the former case, current conduction is due to the movement of free electrons and there is no physical or chemical change except the rise in temperature *e.g.*, metals, alloys *etc.* In electrolytic or ionic conductors, current conduction is due to the movement of ions together with the resulting chemical changes *e.g.*, electrolytes. In this chapter, we shall confine our attention to this type of conductors *i.e.*, ionic conductors.

14.1. Electrolytes

A conducting liquid is called an electrolyte *e.g.*, acids (H_2SO_4, HCl *etc.*), solutions of inorganic compounds (*e.g.*, $CuSO_4$, NaCl *etc.*), hydroxides of metals (*e.g.*, KOH, NaOH *etc.*). In an electrolyte, there are no free electrons and current conduction takes place through ions (*i.e.*, charged atoms).

Let us see how an electrolyte provides ions. For example, take the case of $CuSO_4$ (electrolyte). The reader may recall that copper sulphate is an ionic compound *i.e.*, each molecule of $CuSO_4$ is formed due to the attraction between oppositely charged atoms *viz.* Cu^{++} and SO_4^{--}. When dissolved in water, the force of attraction between them is tremendously reduced due to *high permittivity of water. The result is that Cu^{++} and SO_4^{--} get separated. These charged atoms are called ions and the process is called ionisation.

14.2. Electrolysis

*The conduction of electric current through the solution of an electrolyte together with the resulting chemical changes is called **electrolysis**.*

Fig. 14.1 shows the phenomenon of electrolysis. When copper sulphate is dissolved in water, it splits up into its components *viz.* the positive copper ions (Cu^{++}) and negative sulphate ions (SO_4^{--}). This process is called *ionisation*.

If two metal plates, say copper plates, (called *electrodes*) are set at the opposite ends and a **d.c. potential difference is applied to these plates [See Fig. 14.1], an electric field is created between these electrodes. The negative sulphate ions SO_4^{--} move towards the anode

* $$F = \frac{Q_1 Q_2}{4\pi \varepsilon_0 \varepsilon_r d^2}$$

Relative permittivity of water is 80. It means that when $CuSO_4$ is dissolved in water, attraction force is reduced 80 times than in air.

** Note that d.c. supply must be used to attract ions of only one kind to each electrode.

(*i.e.*, +ve electrode) and positive copper ions (Cu^{++}) travel towards the cathode (*i.e.*, –ve electrode).

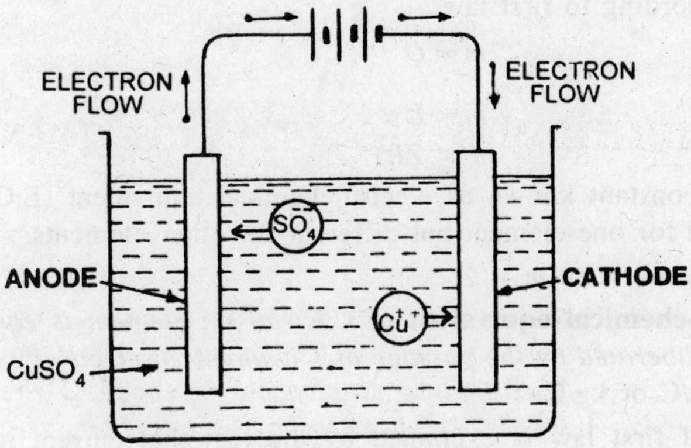

Fig. 14.1

(*i*) A sulphate ion (SO_4^{--}) on reaching the anode gives its two extra electrons to it and becomes sulphate radical. These given up electrons continue their journey towards the cathode *via* the external circuit. Now the sulphate radical cannot exist and therefore, it acts chemically on the anode material to form copper sulphate according to the following reaction :

$$Cu + SO_4 \longrightarrow CuSO_4$$

Thus, copper from anode continuously dissolves into the solution so long as this action takes place.

(*ii*) At the same time, a copper ion (Cu^{++}) on reaching the cathode takes two electrons from it (these are the same electrons given by the sulphate ion at the anode and have come to cathode *via* the external circuit). The copper ion (Cu^{++}) combines with these two electrons to become copper atom and gets deposited on the cathode.

$$Cu^{++} + 2e \longrightarrow \text{Cu atom}$$

Thus, copper from the solution ($CuSO_4$) gets deposited on the cathode.

Two things are worth nothing about the conduction of current in electrolytes. First, the electric current in an electrolyte is due to the movement of *ions (Cu^{++} and SO_4^{--} in the present case) unlike metals where current is due to the movement of free electrons. Secondly, the passage of electric current through an electrolyte causes chemical changes. Thus in the present case, the anode material (*i.e.*, copper) continuously dissolves into the solution and gets deposited on the cathode.

14.3. Faraday's Laws of Electrolysis

Faraday performed a series of experiments to determine the factors which govern the mass of an element deposited or liberated during electrolysis. He summed up his conclusions into two laws, known as Faraday's laws of electrolysis.

---

* Note that the flow of current in the external circuit (being metallic wires) is electronic *i.e.*, by electrons. It is only within the electrolyte that conduction is by ions.

First law. *The mass of an element deposited or liberated at an electrode is directly proportional to the quantity of electricity that passes through the electrolyte.*

If m is the mass of an element deposited or liberated due to the passage of I amperes for t seconds, then according to first law,

$$m \propto Q$$

or $$m \propto It \qquad (\because Q = It)$$

or $$m = ZIt$$

where Z is a constant known as electro-chemical equivalent (E.C.E.) of the element. It has the same value for one element but different for other elements.

If $Q = 1$ Coulomb, then $m = Z$

Hence **electro-chemical equivalent** *(E.C.E.) of an element is equal to the mass of element deposited or liberated by the passage of 1 coulomb of electricity through the electrolyte.* Its unit will be gm/C or kg/C.

The validity of first law is explained by the fact that current inside the electrolyte is carried by the ions themselves. Hence the masses of the chemical substances reaching the anode and cathode are proportional to the quantity of electricity carried by the ions *i.e.*, mass of an ion liberated at any electrode is proportional to the quantity of electricity passed through the electrolyte.

Second law. *The mass of an element deposited or liberated during electrolysis is directly proportional to the chemical equivalent weight of that element i.e.*

$$m \propto \text{Chemical equivalent weight of the element}$$

Faraday's second law is illustrated in Fig. 14.2 where silver and copper voltameters are connected in series. When the same current is passed for the same time through the two voltameters, it will be seen that the masses of silver (Ag) and copper (Cu) deposited on the respective cathodes are in the ratio of 108 : 32. These values of 108 and 32 are respectively the equivalent weights of silver and copper.

$$\frac{\text{Mass of silver deposited}}{\text{Mass of copper deposited}} = \frac{\text{Eq. wt. of Ag}}{\text{Eq. wt. of Cu}} = \frac{108}{32}$$

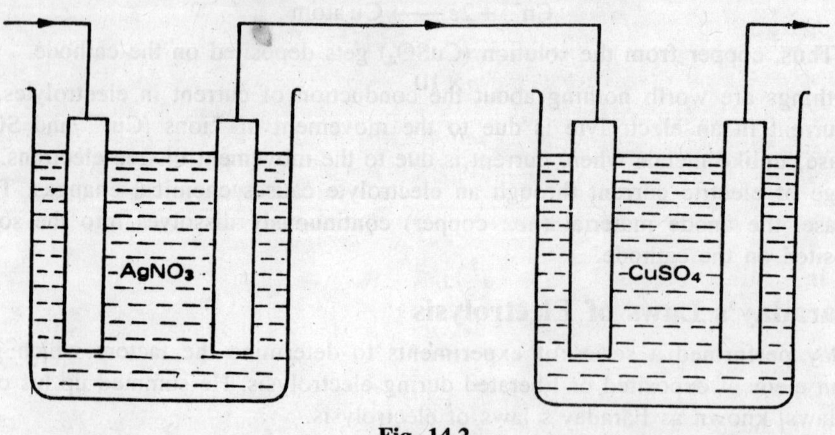

Fig. 14.2

Faraday's second law can be explained as follows. The negative ions (*i.e.*, NO_3^- and SO_4^{--}) from the solutions give up their respective extra electrons to the anodes. These electrons come to cathode *via* the external circuit and are taken up by the positive ions (Ag^+ and Cu^{++}) to become metallic atoms and get deposited on the respective cathodes. Suppose 10 electrons are flowing in the external circuit. Since silver is monovalent (*i.e.*, its valency is 1), 10 silver ions must be liberated at the cathode of silver voltameter. Again copper is bivalent (*i.e.*, its valency is 2) and hence 5 copper ions must be liberated at the cathode of copper voltameter. This means that mass of an element (silver or copper) liberated is directly proportional to the atomic weight and inversely proportional to the valency of that element *i.e.*,

$$\text{Mass liberated,} \quad m \propto \frac{\text{Atomic weight}}{\text{Valency}}$$

$$\propto \text{Chemical eq. wt. of the element}$$

Example 14.1. *Find the value of electro-chemical equivalent of chromium if 2.16 gm of chromium are liberated by the passage of a current of 5A for 1 hour and 20 minutes.*

Solution.

Mass of chromium liberated, $m = 2.16 \text{ gm}$

Current, $I = 5 \text{ A}$

Time for which current flows, $t = 1 \text{ hour, } 20 \text{ min} = 4800 \text{ s}$

Now $m = ZIt$ ...See Art 14.3

$$\therefore \quad Z = \frac{m}{It} = \frac{2.16}{5 \times 4800} = \mathbf{9 \times 10^{-5} \ gm/C}$$

Example 14.2. *A current passes through two voltameters in series, one having silver plates and a solution of $AgNO_3$, and the other copper plates and a solution of $CuSO_4$. After the current has ceased to flow, 3.6 gm of silver have been deposited. How much copper will have deposited in the other voltameter ? Take E.C E. of silver as 0.001118 gm/C and that of copper as 328.86×10^{-6} gm/C.*

Solution.

For silver voltameter, $m_1 = Z_1 It$

$$or \quad It = \frac{m_1}{Z_1} = \frac{3.6}{0.001118}$$

For copper voltameter, $m_2 = Z_2 It$

$$= (328.86 \times 10^{-6}) \times \frac{3.6}{0.001118} = \mathbf{1.06 \ gm}$$

Example 14.3. *If 16 amperes deposit 12 gm of silver in 9 minutes, how much copper would 10 amperes deposit in 15 minutes ? At. wt. of silver = 108 and At. wt. of copper = 63.5.*

Solution.

16 A in 9 minutes deposit silver = 12 gm

10 A in 15 minutes deposit silver = $12 \times (10/16) \times (15/9) = 12.5 \text{ gm}$

Eq. wt. of silver = At wt/Valency = 108/1 = 108

Eq. wt. of copper = 63.5/2 = 31.75

Let m gm be the mass of copper deposited by 10 A in 15 minutes. Then by Faraday's second law of electrolysis,

$$\frac{\text{mass of Cu deposited}}{\text{mass of Ag deposited}} = \frac{\text{Eq. wt. of Cu}}{\text{Eq. wt. of Ag}}$$

or $$\frac{m}{12.5} = \frac{31.75}{108}$$

$\therefore$ $$m = \frac{31.75}{108} \times 12.5 = \mathbf{3.67\ gm}$$

Example 14.4. *A coating of nickel 1 mm thick is to be deposited on a cylinder 2 cm in diameter and 30 cm in length. Calculate the time taken if the current used is 100A. The following data may be taken. Specific gravity of nickel = 8.9, At. wt. of nickel = 58.7 (divalent), E.C.E. of silver = 1.12 mg/C, At. wt. of silver = 108.*

Solution.

Area of curved surface of cylinder $= \pi D \times l = \pi \times 2 \times 30 = 188.5\ \text{cm}^2$

Volume of Ni to be deposited $= 188.5 \times 0.1 = 18.85\ \text{cm}^3$

Mass of Ni to be deposited, $m = 18.85 \times 8.9 = 167.7\ \text{gm}$

Eq. wt. of Ni $= 58.7/2 = 29.35$

Eq. wt. of Ag $= 108/1 = 108$

$$\frac{\text{E.C.E. of Ni}}{\text{E.C.E. of Ag}} = \frac{\text{Eq. wt. of Ni}}{\text{Eq. wt. of Ag}}$$

$\therefore$ $$\text{E.C.E. of Ni} = \frac{29.35}{108} \times 1.12 = 0.304\ \text{mg/C}$$

Now $$m = ZIt$$

$\therefore$ $$t = \frac{m}{ZI} = \frac{167.7}{0.304 \times 10^{-3} \times 100} = 5516\ \text{seconds} = \mathbf{91.93\ minutes}$$

Example 14.5. *A square plate has a side of 10 cm. It is connected to the cathode in a copper electroplating bath and a current of 4 A is passed through the bath. Neglecting the thickness of plate, how long does it take to cover both sides of the plate with copper to a thickness of 0.1 mm. Take density of copper as 9000 kg/m³ and its E.C.E. as 330 × 10⁻⁹ kg/C.*

Solution.

Area of both sides of plates $= 2 \times (10 \times 10^{-2}) \times (10 \times 10^{-2}) = 2 \times 10^{-2}\ \text{m}^2$

Thickness of copper layer $= 0.1\ \text{mm} = 0.1 \times 10^{-3}\ \text{m}$

Volume of copper to be deposited $= (2 \times 10^{-2}) \times (0.1 \times 10^{-3}) = 2 \times 10^{-6}\ \text{m}^3$

Mass of copper to be deposited, $m = 2 \times 10^{-6} \times 9000 = 18 \times 10^{-3}\ \text{kg}$

Now, $$m = ZIt$$

$\therefore$ $$t = \frac{m}{ZI} = \frac{18 \times 10^{-3}}{330 \times 10^{-9} \times 4} = 13636\ \text{seconds} = \mathbf{227\ min,\ 16\ s}$$

Tutorial Problems

1. A current of 5A flows for 40 minutes through an electrolyte which is a solution of a salt of chromium in water. Calculate the mass of chromium liberated. The electro-chemical equivalent of chromium is 90×10^{-9} kg/C. **[1.08 gm]**
2. How long will it take to deposit, from a copper sulphate solution, a coating of copper 0.05 mm thick on an area of 118 cm² if the supply p.d. is 4.5 volts and the total resistance of the circuit is 2.3 Ω. Specific gravity of copper is 8.93 and E.C.E. of copper = 0.329 mg/C. **[2.269 hr]**

3. A metal plate having a surface area of 115 cm^2 is to be silver plated. If a current of 1.5A is passed for 1 hour and 30 minutes, what thickness of copper will be deposited ? Specific gravity of silver = 10.5 and E.C.E. of silver = 1.118 mg/C. **[0.075 mm]**
4. A worn shaft is to be reconditioned by depositing chromium on its curved surface to a thickness of 0.1 mm. The shaft has a diameter of 3.5 cm and a length of 80 cm. If a current of 4.4A is passed, calculate how long the process will take. Density of chromium = 6600 kg/m^3 and E.C.E. of chromium = 90×10^{-9} kg/C. **[41 hours, 44 minutes]**
5. Due to an error, a car battery is overcharged with a current of 5A for 10 hours. Given that the electro-chemical equivalents of hydrogen and oxygen are 10.4×10^{-9} kg/C and 83.2×10^{-9} kg/C respectively, calculate the volume of distilled water which must be added to compensate for the loss. **[16.8 c.c.]**

14.4. Cell

We have seen that when a current passes through the solution of an electrolyte, chemical changes take place *i.e.*, electrical energy is converted into chemical energy. The converse of this is also true *i.e.*, we can convert chemical energy into electrical energy. The device which accomplishes this job is called a cell.

A **cell** *is a source of e.m.f. in which chemical energy is converted into electrical energy.*

A cell essentially consists of two metal plates of different materials immersed in a suitable solution. The plates are called electrodes and the solution is called electrolyte. The magnitude of e.m.f. of a cell depends upon (*i*) the nature of material of electrodes used and (*ii*) the nature of electrolyte. For example, if plates of zinc and lead are placed in a solution of sulphuric acid, the e.m.f. developed is about 0.5 V ; zinc and copper in sulphuric acid about 1.1 V and zinc and silver in sulphuric acid about 1.2 V. It may be noted that the e.m.f. of a cell does not depend upon the size and spacing of the plates. However, with the increase in the size of plates, the capacity of the cell increases *i.e.*, the cell will deliver current for a longer period.

14.5. Types of Cells

Using various metals and methods of construction, a large variety of cells has been developed. However, the cells can be divided into two main classes *viz.*

1. Primary cells 2. Secondary cells

1. Primary cells. A cell in which chemical action is not reversible is called a **primary cell** *e.g.*, voltaic cell, Daniel cell, Lachlanche cell, dry cell *etc.* As a primary cell delivers current, the active materials are used up. When the active materials are nearly consumed, the cell stops delivering current. In order to renew the cell, fresh active materials are provided. Another drawback of a primary cell is that it cannot provide large and steady current for a longer period. This fact makes the primary cell rather an expensive source of electrical energy. Due to these drawbacks, the use of primary cells is limited to torch batteries and for experimental purposes in the laboratories.

2. Secondary cells. A cell in which chemical action is reversible is called a **secondary cell** or **storage cell.** A secondary cell operates on the same principle as a primary cell but differs in the method in which it may be renewed. In a secondary cell, there is no actual consumption of any plate and that the chemical process is reversible. When the cell is delivering current (*i.e., discharging*), the chemical action changes the composition of plates. When the cell is exhausted, the chemical action can be reversed (*i.e.*, plates can be restored to the original condition) by passing current through the cell in the reverse direction to that in which the cell provided current. This process is called charging. In other words, charging process reverses the chemical action and enables the plates to acquire original composition. There are several types of secondary or storage cells ; the more common ones being :

(*i*) Lead-acid cell

(*ii*) Nickel-iron-alkaline cell (or Edison cell)

(*iii*) Nickel-cadmium-alkaline cell.

As in a primary cell, the e.m.f. of a secondary cell also depends upon the materials used for electrodes and the nature of electrolyte. The e.m.f. of a lead-acid cell is about 2V and that of Nickel-cadmium-alkaline and Edison cells is about 1.2V.

Note. When a secondary cell is charged, electrical energy is converted into chemical energy which is stored in the cell. When the cell discharges, the stored chemical energy starts converting into electrical energy. For this reason, a secondary cell is sometimes called a **storage cell.**

14.6. Lead-Acid Cell

The most inexpensive secondary cell is the lead acid cell and is widely used for commercial purposes. A lead-acid cell when ready for use contains two plates immersed in a dilute sulphuric acid (H_2SO_4) of specific gravity about 1.28. The positive plate (*anode*) is of **lead-peroxide** (PbO_2) which has chocolate brown colour and the negative plate (*cathode*) is **lead** (Pb) which is of grey colour.

(*i*) When the cell supplies current to a load (*i.e.*, discharging), the chemical action that takes place forms lead sulphate ($PbSO_4$) on both the plates with water being formed in the electrolyte. After a certain amount of energy has been withdrawn from the cell, both plates are transformed into the same material (*i.e.*, $PbSO_4$) and the specific gravity of the electrolyte (H_2SO_4) is lowered. The cell is then said to be discharged. There are several methods (See Art. 14.11) to ascertain whether the cell is discharged or not.

(*ii*) To charge the cell, direct current is passed through the cell in the reverse direction to that in which the cell provided current. This reverses the chemical process and again forms a lead peroxide (PbO_2) positive plate and a pure lead (Pb) negative plate. At the same time, H_2SO_4 is formed at the expense of water, restoring the electrolyte (H_2SO_4) to its original condition.

The chemical changes that occur during discharging and recharging of a lead-acid cell are discussed in the articles that follow.

14.7. Chemical Changes During Discharging

By discharging of a cell we mean that it is delivering current to the external circuit. Consider a charged lead acid cell with anode of PbO_2 and cathode of Pb ; the electrolyte being dilute H_2SO_4 [See Fig. 14.3]. Sulphuric acid splits up into hydrogen ions (H^+H^+) and sulphate ions (SO_4^{--}). The sulphate ions move towards the cathode and hydrogen ions move towards the anode causing the following chemical actions :

At cathode. On reaching the cathode, a sulphate ion (SO_4^{--}) gives up its two extra electrons to become sulphate radical. These electrons given up at the cathode move through the external circuit to the anode where they are available to neutralise the positive ions (H^+H^+) arriving there. Since sulphate radical cannot exist, it enters into chemical action with cathode material (Pb) to form lead sulphate ($PbSO_4$).

$$SO_4^{--} - 2e \longrightarrow SO_4 \text{ (radical)}$$

$$\underset{\text{(Grey colour)}}{Pb + SO_4} \longrightarrow \underset{\text{(Whitishin colour)}}{PbSO_4}$$

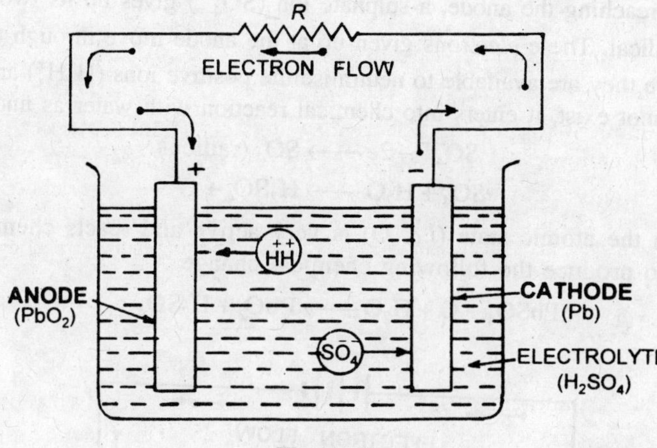

Fig. 14.3

At anode. On reaching the anode, each hydrogen ion takes one electron from it to become hydrogen gas. This electron is given by the sulphate ion at the cathode and has come to the anode *via* the external circuit.

$$H^+H^+ + 2e \longrightarrow 2H$$

The hydrogen gas liberated at the anode acts chemically on the anode material (PbO_2) and reduces it to lead oxide (PbO).

$$PbO_2 + 2H \longrightarrow PbO + H_2O$$

Sulphuric acid reacts PbO to form $PbSO_4$

$$PbO + H_2SO_4 \longrightarrow PbSO_4 + H_2O$$

The chemical changes that take place during discharging can be summed up as under :

(*i*) Both the plates are converted into lead sulphate ($PbSO_4$) which is whitish in colour.

(*ii*) Water is formed which lowers the specific gravity of the electrolyte (H_2SO_4). When the cell is fully discharged, the specific gravity of H_2SO_4 falls to about 1.18.

(*iii*) The e.m.f. of the cell falls. The lead-acid cell should not be discharged beyond the point where its e.m.f. falls to about 1.8 volts.

(*iv*) The chemical energy stored in the cell is converted into electrical energy.

It is important to note that e.m.f. of the cell provides little indication to the state of discharge of the cell since it remains close to 2V for 90% of the discharge period. In practice, specific gravity of the electrolyte (H_2SO_4) is used to know the state of discharge. The cell should be recharged when specific gravity of H_2SO_4 falls to 1.18.

14.8. Chemical Changes During Recharging

Consider a discharged lead-acid cell having both the plates converted to lead sulphate ($PbSO_4$). In order to recharge the cell, direct current is passed through the cell in the reverse direction to that in which the cell provided current. To do so, the anode is connected to the positive terminal of d.c. source and cathode to the negative terminal of the source as shown in Fig. 14.4. The electrolyte (H_2SO_4) breaks up into hydrogen ions (H^+H^+) and sulphate ions

(SO_4^{--}). Hydrogen ions move towards cathode and sulphate ions move towards anode causing the following chemical reactions :

At anode. On reaching the anode, a sulphate ion (SO_4^{--}) gives up its two extra electrons to become sulphate radical. These electrons given up at the anode move through the external circuit to the cathode where they are available to neutralise the positive ions (H^+H^+) arriving there. Since sulphate radical cannot exist, it enters into chemical reaction with water as under :

$$SO_4^{--} - 2e \longrightarrow SO_4 \text{ (radical)}$$

$$SO_4 + H_2O \longrightarrow H_2SO_4 + O$$

The oxygen in the atomic state (*i.e.* O) is very active and reacts chemically with anode material ($PbSO_4$) to produce the following chemical change :

$$PbSO_4 + O + H_2O \longrightarrow PbO_2 + H_2SO_4$$

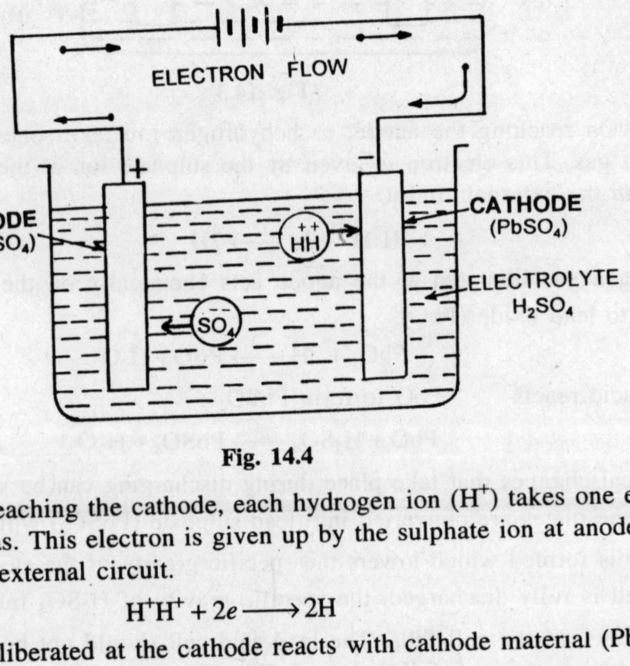

Fig. 14.4

At Cathode. On reaching the cathode, each hydrogen ion (H^+) takes one electron from it to become hydrogen gas. This electron is given up by the sulphate ion at anode and has come to the cathode *via* the external circuit.

$$H^+H^+ + 2e \longrightarrow 2H$$

The hydrogen gas liberated at the cathode reacts with cathode material ($PbSO_4$) to reduce it to lead (Pb) as under :

$$PbSO_4 + 2H \longrightarrow Pb + H_2SO_4$$

As the charging process goes on, the anode is converted into PbO_2 and cathode into Pb. The H_2SO_4 produced in the chemical reactions above increases the specific gravity of the electrolyte. The chemical changes that occur during recharging can be summed up as under :

(*i*) The positive plate (anode) is converted into PbO_2 and the negative plate (cathode) into Pb.

(*ii*) H_2SO_4 is formed in the reactions. Therefore, the specific gravity of the electrolyte (H_2SO_4) is raised. When the cell is fully charged, the specific gravity of H_2SO_4 rises to about 1.28.

(*iii*) The e.m.f. of the cell rises. The e.m.f. of a fully charged lead-acid cell is about 2 volts.

(*iv*) Electrical energy supplied is converted into chemical energy which is stored in the cell.

14.9. Construction of a Lead-acid Battery

In a 6-volt lead-acid battery, three cells are connected in series whereas for 12-volt type, six cells are series connected. Fig. 14.5 shows the cut-away view of a lead-acid battery.

(*i*) **Container.** The container houses the plates and the electrolyte. It is made of glass or transparent synthetic material or hard rubber depending upon service requirements. The container is sealed off at the top to prevent the spilling of electrolyte.

(*ii*) **Plates.** The capacity of a lead-acid cell *depends upon the plate area. To increase the effective area of plates without increasing the size of the cell, we use a large number of thin plates in the cell instead of two. The alternate positive and negative plates are sandwiched together with insulators called *separators*. The negative plates are connected together as the positive plates. A commercial cell always has odd number of plates such as 11, 13, 15 or 17. The number of negative plates is always one more than the number of positive plates ; the outside plates being negative. A separate compartment is provided for each cell and each compartment has large space at the bottom so that any sediment from the plate may collect there.

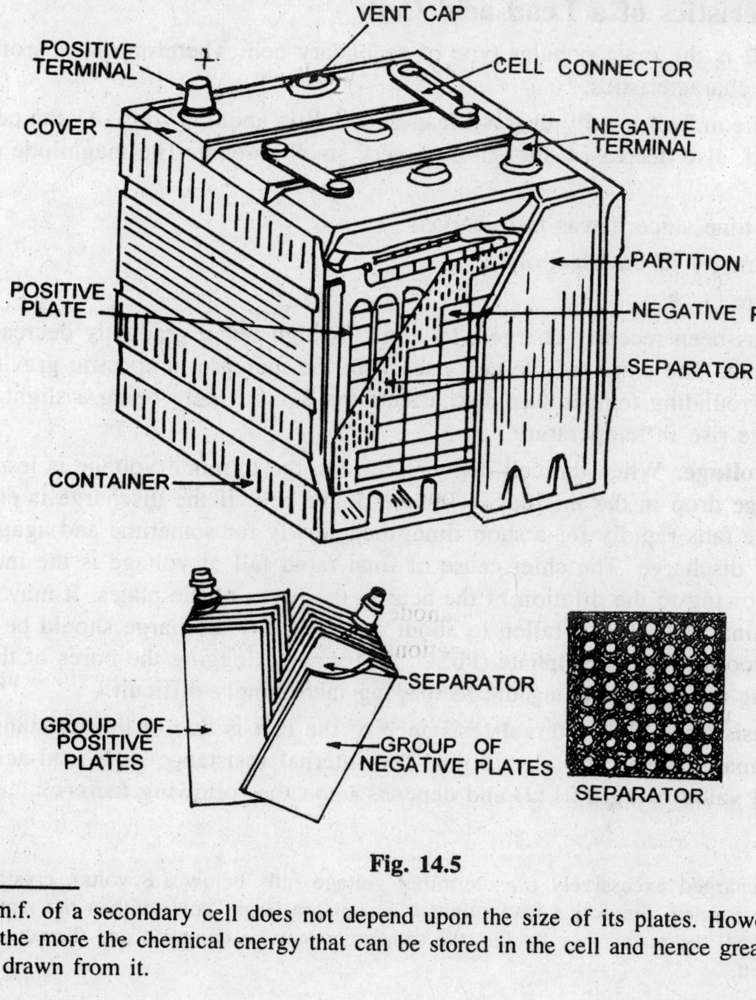

Fig. 14.5

* The e.m.f. of a secondary cell does not depend upon the size of its plates. However, the larger the plates, the more the chemical energy that can be stored in the cell and hence greater the current that can be drawn from it.

(iii) **Separators.** In order to save space and to reduce the internal resistance of the cell, the plates are placed close together. To prevent the plates touching each other if they wrap or buckle, they are separated by non-conducting materials (*e.g.*, wood, rubber *etc.*) called separators.

(iv) **Electrolyte.** The electrolyte is dilute sulphuric acid (H_2SO_4) solution mixed in such a proportion so that with a fully charged battery, its specific gravity is about 1.28. Sometimes, we replace water that has evaporated. To ensure normal battery life, only pure water should be used.

(v) **Cell cover.** Each cell compartment has a cover ; usually made of moulded hard rubber. Openings are provided in these covers for two terminal posts and vent cap.

(vi) **Vent caps.** Each cell has a hole into which is fitted the vent cap. This cap has a vent hole to allow the free exit of the gas formed in the cell. The vent caps can be easily removed for adding water or taking hydrometer reading.

(vii) **Inter-cell connector.** It is a lead alloy link that joins the cells in series. The positive terminal of the cell is marked by a large + sign or with a red colour.

(viii) **Cell terminals.** Each cell has two terminals. The terminals are generally made of lead as it does not erode due to the acid electrolyte.

14.10. Characteristics of a Lead-acid Cell

A lead-acid cell is the most popular type of secondary cell. Therefore, it is profitable to study its important characteristics.

1. E.M.F. The e.m.f. of a fully charged lead-acid cell is about 2 volts. As the cell delivers current, its e.m.f. also decreases, though by a very small amount. The magnitude of e.m.f. depends upon :

(i) length of time since it was last charged

(ii) specific gravity of the electrolyte

(iii) temperature

When a cell has been recently charged, its e.m.f. is high but it gradually decreases even if left on open-circuit. The e.m.f. of the cell rises with the increase in specific gravity of the electrolyte. The surrounding temperature also affects the e.m.f. ; there being a slight increase in its value with the rise in temperature.

2. Terminal voltage. When the cell delivers current, the terminal voltage is less than its e.m.f. due to voltage drop in the internal resistance of the cell. If the discharge is continued, the terminal voltage falls rapidly for a short time, then slowly for sometime and again rapidly towards the end of discharge. The chief cause of final rapid fall of voltage is the increase of internal resistance owing to the dilution of the acid in the pores of the plates. It may be noted that when the terminal voltage has fallen to about 1.8 volts, the discharge should be stopped. If it is continued, too much lead sulphate ($PbSO_4$) is formed, clogging the pores of the plates, possibly *damaging them and making the recharging increasingly difficult.

3. Internal resistance. The internal resistance of the cell is due to the resistance of the plates, the active material and the electrolyte. The internal resistance of a lead-acid cell is very small (typical value being 0.01 Ω) and depends upon the following factors :

---

* If a cell is discharged excessively (*i.e.*, terminal voltage falls below 1.8 volts), crystalline lead sulphate will be formed. Since this lead sulphate occupies a larger volume than the material from which it is formed, the plates may buckle, the separators may be damaged and the active material may be dislodged.

(*i*) Area of plates—decreases with the increase in plate area

(*ii*) Spacing between plates— decreases with the decrease in spacing

(*iii*) Sp. Gravity of H_2SO_4— decreases with the increase in sp. gravity.

The internal resistance of a lead-acid cell should be minimum in order to reduce the internal drop. This is achieved by using multiplate construction in a cell. As explained in Art 14.9, the negative plates of a cell are connected together as are the positive plates. The effect of this arrangement is as if we have connected a number of cells in parallel. At the same time, the length of electrolyte between the plates is reduced. The result is that internal resistance of the cell is lowered.

4. Capacity. The capacity of a cell is the quantity of electricity which it can give out during single discharge until its terminal voltage falls to 1.8 volts. It is measured by the product of current in amperes and the time in hours *i.e.*

Capacity of cell = $I_d \times T_d$ ampere-hours (or *Ah*)

where I_d = Steady discharging current in amperes

T_d = Time in hours for which the cell can supply current until its p.d. falls to 1.8 volts.

The capacity of a cell depends upon the following factors :

(*i*) Rate of discharge—Higher the rate of discharge, less the capacity

(*ii*) Temperature—Increases with temperature

(*iii*) Area of plates—Increases with plate area

(*iv*) Sp. gravity of electrolyte—Increases with Sp. gravity

5. Efficiency. There are two ways of expressing the efficiency of a secondary cell *viz.* (*i*) ampere-hour efficiency (*ii*) watt-hour efficiency.

(*i*) Ampere-hour efficiency,

$$\eta_{Ah} = \frac{\text{Ampere-hours provided on discharge}}{\text{Ampere-hours of charge}} \times 100$$

$$= \frac{I_d \times T_d}{I_c \times T_c} \times 100$$

Since ampere-hours is the quantity of electricity, this efficiency is sometimes called *quantity efficiency*. The ampere-hour efficiency of a lead-acid cell is about 90%.

(*ii*) Watt-hour efficiency,

$$\eta_{wh} = \frac{\text{Energy given on discharge}}{\text{Energy input of charge}} \times 100$$

$$= \frac{V_d \times I_d \times T_d}{V_c \times I_c \times T_c} \times 100$$

$$= \frac{I_d \times T_d}{I_c \times T_c} \times 100 \times \frac{V_d}{V_c}$$

$$= \eta_{Ah} \times \frac{V_d}{V_c}$$

Since average p.d. during discharge (*i.e.*, V_d) is lower than the average p.d. during charge (*i.e.*, V_c) due to the effects of internal resistance and polarisation, η_{wh} is always less than η_{Ah}. The watt-hour (or energy efficiency) of a lead-acid cell is about 75%.

Example 14.6. *A lead-acid cell maintains a constant current of 1.5A for 20 hours before its terminal voltage falls to 1.8 volts. What is the capacity of the cell ?*

Solution. The capacity of a lead-acid cell is the ampere-hours provided by it until its terminal voltage falls to 1.8 volts.

$\therefore$ Capacity of cell = $I_d \times T_d = 1.5 \times 20 =$ **30 Ah**

Example 14.7. *A lead-acid cell has 13 plates, each 25 cm × 20 cm. The clearance between the neighbouring plates is 1.2 mm. If the resistivity of the acid is 1.6 Ω-cm, find the internal resistance of the cell.*

Solution. Since there are 13 plates, the arrangement constitutes 12 tiny cells in parallel.

Resistance of each tiny cell, $R = \rho \dfrac{l}{a}$

Here $\rho = 1.6\,\Omega\text{-cm}$; $l = 0.12$ cm ; $a = 25 \times 20 = 500\,\text{cm}^2$

$$\therefore \quad R = 1.6 \times (0.12/500) = 384 \times 10^{-6}\,\Omega$$

Since the cells are connected in parallel, total internal resistance R_T of the cell is :

$$R_T = R/12 = 384 \times 10^{-6}/12 = \mathbf{32 \times 10^{-6}\,\Omega}$$

Example 14.8. *A lead-acid cell is charged at the rate of 18 A for 10 hours at an average voltage of 2.26 volts. It is discharged in the same time at the rate of 17.2A ; the average voltage during discharge being 1.98 volts. Calculate (i) ampere-hour efficiency (ii) watt-hour efficiency.*

Solution.

(*i*) Ampere-hour efficiency,

$$\eta_{Ah} = \frac{I_d \times T_d}{I_c \times T_c} \times 100 = \frac{17.2 \times 10}{18 \times 10} \times 100 = \mathbf{95.55\%}$$

(*ii*) Watt-hour efficiency,

$$\eta_{wh} = \eta_{Ah} \times \frac{V_d}{V_c} = 95.55 \times \frac{1.98}{2.26} = \mathbf{83.71\%}$$

14.11. Indications of a Fully Charged Cell

During the charging process, it is very essential that the battery is taken out from the charging circuit as soon as it is fully charged. Overcharging as well as undercharging are undesirable and should always be avoided. The indications of a fully charged cell (or battery) are :

(*i*) Voltage (*ii*) Specific gravity of electrolyte

(*iii*) Gassing (*iv*) Colour of plates

(*i*) **Voltage.** During charging, the terminal potential of a cell increases and provides an indication to its state of charge. A fully charged lead-acid cell has a terminal voltage of about 2 volts.

(*ii*) **Specific gravity.** During the charging process, the specific gravity of the electrolyte (H_2SO_4) increases and provides an important indication to the state of charge of the cell. The specific gravity of the electrolyte of a fully charged lead-acid cell is about 1.28. This can be measured by means of a *hydrometer.* It may be noted that the state of charge of a lead-acid cell (or battery) is determined from the specific gravity of the electrolyte (H_2SO_4). The following table may be useful in this regard :

Sp. Gr. of H_2SO_4	*State of Charge*
1.13	Discharged
1.19	25% charged
1.22	50% charged
1.25	75% charged
1.28	100% charged

(*iii*) **Gassing.** When the cell is fully charged, the charging current starts electrolysis of water. The result is that hydrogen is given off at the cathode and oxygen at the anode ; the process being known as **gassing*. Gassing at both the plates indicates that the charging current is doing no useful work and hence should be stopped.

(*iv*) **Colour of plates.** The visual examination of colour of the plates of a lead-acid cell provides yet another important indication to its state of charge. When the cell is fully charged, the positive plate gets converted into PbO_2 (chocolate brown) and the negative plate to spongy lead (grey).

14.12. Charging of Lead-Acid Batteries

In order to ensure normal life for a battery, it should be maintained in the charged condition. While charging a battery, the following points may be kept in view :

(*i*) The charging source must be d.c. one. If a.c. supply is available, it must be converted to d.c. before being applied to the battery.

(*ii*) During charging, it should be ensured that polarity is correct. The polarity is correct when the positive terminal of d.c. charging source is connected to the positive terminal of the battery.

(*iii*) The charging voltage must be more than the e.m.f. of the battery to be charged. Approximately 2.5 volts per cell should be applied. For example, if a battery to be charged has 6 cells,then applied direct voltage should be about 15 volts.

(*iv*) The charging current should be set at proper value.

Charging rate. What should be the value of charging current ? The answer is that it may be as high as the battery can take without excessive "gassing" or heat. Too great a charging current may produce excessive heat. This will cause plates to buckle and short circuit may occur. Also the water of electrolyte will be lost due to electrolysis and evaporation, raising the concentration of sulphuric acid left behind. The strong acid then may char the separators, causing permanent damage to the battery. The temperature of the electrolyte should not be permitted to rise above 40°C. Efforts should be made to use the charging current value recommended by the manufacturer. In case it is not available, one of the following thumb rules may be applied :

(*i*) The charging current should be 1A for every positive plate of a single cell. Thus, if the cell contains 13 plates, six of them will be positive. The charging current for the battery should, therefore, be 6A.

(*ii*) The charging rate should be such that full charge can be obtained in 8 hours. Thus 100 Ah battery should be charged at the rate of 100/8 = 12.5A. This will ensure the maximum life of the battery.

---

* Gassing is harmless so long as it is not violent enough to dislodge the active material from the plates. However, it leads to waste of energy and the usual practice is to gradually reduce the charging current towards the finish of charge in order to reduce gassing and to avoid overheating of battery.

14.13. Battery Charging Circuit

Fig. 14.6 shows the battery charging circuit. A d.c. source of suitable magnitude is connected in series with a rheostat R, ammeter and the battery to be charged. Ensure that polarity is correct *i.e.* positive terminal of d.c. source should be connected to the positive terminal of the battery. The charging current is adjusted to the required value with the help of rheostat. As the charging process proceeds, the terminal voltage of the battery rises but the charging current is kept constant by adjusting the value of rheostat R. The terminal voltage of the battery and specific gravity of electrolyte are checked at regular intervals of time. When the terminal voltage ceases to rise, the specific gravity of electrolyte reaches the value 1.28 and there is enough gassing at the plates, the battery is fully charged. It is then taken out of the charging circuit. The entire charging process may take several hours.

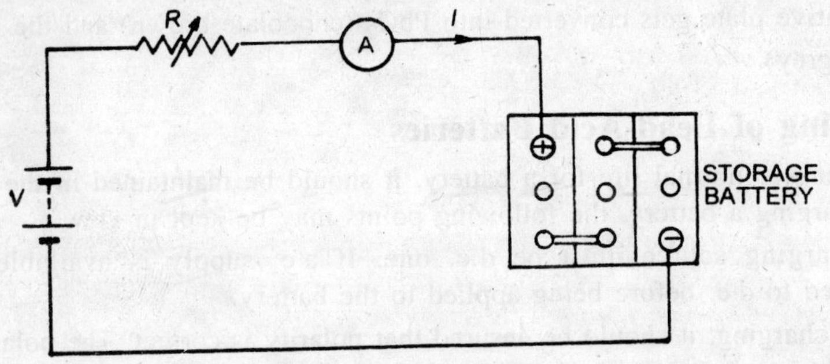

Fig. 14.6

Calculations. When the battery is being charged, its e.m.f. acts in opposition to the applied voltage. The applied voltage V sends a charging current I against the back e.m.f. E_b of the battery. The input power is VI but the power being supplied to the battery is $E_b I$. The power $E_b I$ is converted into chemical energy which is stored in the battery.

Charging current, $$I = \frac{V - E_b}{R + r}$$

where R = resistance of rheostat in the circuit

r = internal resistance of the battery

The charging current is kept constant throughout (by adjusting R) except towards the finish of charge.

The following points may be kept in mind during charging :

(*i*) When the battery is being charged, the vents must be open so that the gases (H_2 and O_2) may be able to escape. Otherwise the case may be cracked.

(*ii*) The mixture of hydrogen and oxygen is explosive. Therefore, care must be taken not to carry an open flame or lighted cigarette near a battery being charged.

(*iii*) The charging current should be such that battery temperature does not exceed 40°C and that violent gassing does not take place. Instead of a constant charging current, the usual practice is to charge the battery at a tapered rate *i.e.* at high rate at first but at a gradually reduced rate as the battery becomes fully charged.

(*iv*) After charging, water should be added to compensate for the loss of water by gassing and evaporation. The level of the electrolyte should be 1 cm above the tops of the plates. If water is not added, the excessive concentration of H_2SO_4 may char the separators, causing permanent damage to the battery.

Example 14.9. *A battery of 40 cells is to be charged from a 180V supply. The internal resistance of each cell is 0.05Ω and the charging current is to be 4A. If the average e.m.f. of each cell during charge is 2.5V, what should be the value of series resistor ?*

Solution.

Applied voltage, $V = 180$ volts

Charging current, $I = 4A$

Back e.m.f. of battery, $E_b = 40 \times 2.5 = 100$ volts

Internal resistance of battery, $r = 40 \times 0.05 = 2\Omega$

Let R ohms be the required value of series resistor.

Charging current, $I = \dfrac{V - E_b}{R + r}$

or $4 = \dfrac{180 - 100}{R + 2}$

$\therefore$ $R = \mathbf{18\Omega}$

Example 14.10. *It is desired to charge a 12-V car battery at 6A from a 230-V d.c. source. The d.c. source and battery are connected in series with a group of 60-watt, 220-V bulbs in parallel. How many lamps are required for the purpose ?*

Solution.

Charging current, $I = \dfrac{V - E_b}{R}$ Neglecting internal resistance

or $6 = \dfrac{230 - 12}{R}$

$\therefore$ Required circuit resistance, $R = 218/6 = 36.33\,\Omega$

Resistance of each bulb $= (220)^2/60 = 806.67\,\Omega$

Let N be the required number of bulbs in parallel. It means that N bulbs in parallel should give an equivalent resistance of 36.33Ω.

$\therefore$ $36.33 = 806.67/N$ *or* $N = 806.67/36.33 =$ **22 bulbs**

Example 14.11. *A 6-cell, 12-V battery is to be charged at a constant rate of 10A from a 24-V supply. If the e.m.f. of each cell at the beginning of charge and at the end of charge is 1.9V and 2.4V, what should be the maximum and minimum resistance to be connected in series with the battery ? Resistance of battery is negligible.*

Solution.

Beginning of charge

Back e.m.f. of battery, $E_b = 6 \times 1.9 = 11.4$ volts

Charging current, $I = (V - E_b)/R$

or $R = \dfrac{V - E_b}{I} = \dfrac{24 - 11.4}{10} = \mathbf{1.26\Omega}$

End of charging

Back e.m.f of battery, $E_b = 6 \times 2.4 = 14.4$ volts

$\therefore$ $R = \dfrac{V - E_b}{I} = \dfrac{24 - 14.4}{10} = \mathbf{0.96\Omega}$

Example 14.12. *Calculate the supply voltage necessary for charging a battery of 110 cells at 30A (i) at the beginning of charge and (ii) at the end of the charge, if each cell possesses p.d. of 2.1 volts at the beginning and 2.7 volts at the end of charge. Allow 0.06Ω for the resistance of the connecting leads.*

Solution. The internal resistance of battery is assumed to be zero.

Beginning of charge

Back e.m.f. of battery, $E_b = 110 \times 2.1 = 231$ volts

Charging current, $I = \dfrac{V - E_b}{0.06}$

or $30 = \dfrac{V - 231}{0.06}$

∴ Supply voltage $V = 231 + 30 \times 0.06 = \mathbf{232.8\ volts}$

End of charging

Back e.m.f. of battery, $E_b = 110 \times 2.7 = 297$ volts

∴ Supply voltage, $V = 297 + 30 \times 0.06 = \mathbf{298.9\ volts}$

Example 14.13. *Thirty-five lead-acid cells, each of discharging capacity 100Ah at the 10-hour rate are to be fully charged at constant current for 8 hours. The d.c. supply is 120 V, the ampere-hour efficiency is 80% and the e.m.f. of each cell at the beginning and end of charge is 1.9V and 2.6V respectively. Calculate the maximum and minimum value of the necessary resistance. Ignore internal resistance of the cells.* [C & G]

Solution. The output of each cell is 100Ah. Since Ah efficiency is 80%, the input ampere hours will be = 100/0.8 = 125. As the number of hours of charge is 8,

∴ Charging current, $I = 125/8 = 15.62\,A$

Beginning of charge

Back e.m.f. of battery, $E_b = 35 \times 1.9 = 66.5$ volts

Charging current, $I = (V - E_b)/R$

∴ External resistance, $R = \dfrac{V - E_b}{I} = \dfrac{120 - 66.5}{15.62} = \mathbf{3.42\ \Omega}$

The value of external resistance will be maximum at the beginning of charge because then back e.m.f. of the battery is minimum.

End of charging

Back e.m.f. of battery, $E_b = 35 \times 2.6 = 91$ volts

∴ External resistance $R = \dfrac{V - E_b}{I} = \dfrac{120 - 91}{15.62} = \mathbf{1.86\ \Omega}$

Example 14.14. *A shunt generator is used to charge a battery of 100 cells in series. If each cell has a terminal p.d. of 2.5V at the completion of charge and is of internal resistance 0.001Ω, calculate the e.m.f. to be generated to give a charging current of 15A at the end of charge.*

Assume that the armature and field resistances are 0.2Ω and 200Ω respectively and the cables connecting the generator to the battery have a resistance of 0.6Ω.

Solution. Fig. 14.7 shows the whole arrangement. The charging current I = 15A.

Back e.m.f. of battery,	$E_b = 100 \times 2.5 = 250$ volts
Battery terminal voltage,	$V = 250 + \text{Drop in battery}$
	$= 250 + (15 \times 0.001 \times 100) = 251.5$ volts
Voltage at generator terminals	$= 251.5 + \text{Drop in connecting cables}$
	$= 251.5 + 15 \times 0.6 = 260.5$ volts

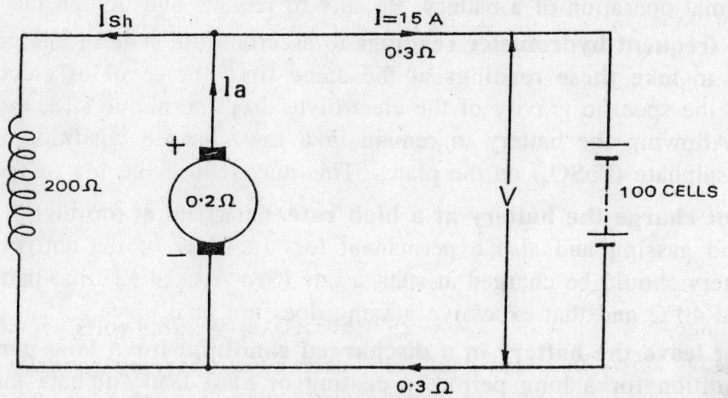

Fig. 14.7

Shunt current,	$I_{sh} = 260.5/200 = 1.3\,A$
Armature current,	$I_a = 15 + 1.3 = 16.3\,A$
∴ E.M.F. of generator	$= 260.5 + 16.3 \times 0.2 =$ **263.76volts**

Tutorial Problems

1. A battery of accumulators of e.m.f. 50 volts and internal resistance 2Ω is charged on 100-V direct mains. What series resistance will be required to give a charging current of 2A? If the cost of energy is 70 paise per kWh, what will it cost to charge the battery for 8 hours and what percentage of energy supplied will be wasted in the form of heat ? [**23Ω; Rs. 1.12; 50%**]
2. A battery of 4 lead-acid cells in series is to be charged and the only available source is 210V d.c. supply. The desired charging current is 0.5 A and each cell on charge has an e.m.f. of 2.5 volts. If an electric lamp is used as the controlling resistance for the circuit, what voltage and size of the lamp will be necessary ? [**200V; 100W**]
3. A lead-acid battery of 50 cells is to be charged at 25A from a 200V d.c. supply using a series resistor. If the e.m.f. of each cell is 1.95 volts at the commencement of charge and 2.55 volts at the end of the charge, find the range of series resistance value required. The internal resistance of each cell may be assumed to remain constant at 0.002Ω and the resistance of the connecting leads may be neglected. [**4Ω to 2.8 Ω**]
4. Determine the variation in the applied voltage required to charge a battery of 100 lead-acid cells if charged from an e.m.f. of 1.8 volts to 2.6 volts per cell with a constant charging current of 10A. The internal resistance of each cell is 0.01Ω.
[Initial voltage = **190V**; Final voltage = **270 V**; Variation = **80V**]

14.14. Care of Lead-acid Batteries

The average life of a lead-acid battery is two to four years, depending upon its quality and the kind of care exercised in its use. To get the maximum life out of a lead-acid battery, the following procedures are recommended :

(*i*) **Keep the top of battery clean and dry** at all times to prevent corrosion and leakage of current. If corrosion has started, the battery may be cleaned by the use of a stiff brush and then wiped with a rag moistened in a solution of household ammonia.

(*ii*) **Keep the electrolyte at the proper level.** In the normal operation of a lead-acid battery, a certain amount of water is lost from the electrolyte (H_2SO_4) due to evaporation and gassing. The level of the electrolyte should be checked at regular intervals of one or two weeks and when found to be too low, it should be brought to the proper level by adding water free from impurities. It should always be 1 cm above the tops of plates and separators as this prevents decay of separators. As only water evaporates, it is never necessary to add acid during normal operation of a battery. Be sure to replace and tighten the vent caps.

(*iii*) **Take frequent hydrometer readings** to ascertain the state of charge of the battery. It is advisable to take these readings at the same time the level of electrolyte is being checked. When the specific gravity of the electrolyte drops to about 1.18, the battery should be recharged. Allowing the battery to remain in a low charged condition will produce an excess of lead sulphate ($PbSO_4$) on the plates. This may reduce the life of the battery.

(*iv*) **Do not charge the battery at a high rate.** Charging at too high a rate causes excessive heat and gassing and also a permanent loss in some of the active material of the plates. The battery should be charged at such a rate (See Art. 14.12) that battery temperature does not exceed 40°C and that excessive gassing does not take place.

(*v*) **Do not leave the battery in a discharged condition for a long period.** If left in a discharged condition for a long period, a coating of hard lead sulphate may form on the plates. The sulphate coating cannot be converted to active materials and the battery is permanently damaged. The action is commonly called **sulphation** and may also be caused due to excessive heat and too much concentration of sulphuric acid in the electrolyte.

(*vi*) **If a lead-acid battery is to be stored** for several months, the level of the electrolyte should be periodically checked. After intervals of 4 to 6 weeks, the battery should be given a freshening charge so that it will be kept fully charged at all times.

(*vii*) **Do not short circuit the battery.** Since the internal resistance of a lead-acid battery is very small (typical value being 0.01Ω), a short-circuit will give a damaging current of several hundred amperes. This may cause the plates to buckle, thus ruining the battery.

14.15. Applications of Lead-acid Batteries

The chief application of lead-acid batteries is at places where a d.c. generator cannot be conveniently installed. Such batteries are employed whenever low-voltage, high-current d.c. source is required. Some of the important applications of lead-acid batteries are given below :

(*i*) Lead-acid batteries are extensively used in automobiles. A d.c. generator, battery and load are connected in parallel. Since the internal resistance of a lead-acid battery is very low, it provides a large output current required for *starting the engine of the car. In addition, the battery furnishes the power of lights, radio *etc.* when the engine is stopped. When the engine is running, it operates the d.c. generator which takes over the duties of the battery. The generator also charges up the battery so that it will be ready when needed.

(*ii*) Railway train lighting system is similar to the above mentioned system. When the train is in motion, axle-driven d.c. generators furnish power for lighting, fans *etc.* When the train runs at slow speeds or is stopped, lead-acid batteries take over the duties of the generators.

---

* The starting motor of an automobile may require more than 300 A at start. Because of its low internal resistance, a lead-acid battery can furnish currents of such large magnitude but only for a few seconds at a time.

(*iii*) In a.c. generating plants, lead-acid batteries are used to energise the contr apparatus *e.g.* switching. During shut down of generators, these batteries are used supply emergency lights.

(*iv*) Hospitals and other places (*e.g.* theatres, banks *etc.*)where continuous supply is absolutely essential often use batteries as an emergency supply.

(*v*) These are used for lighting purpose in remote rural areas where there are no power lines. The batteries are kept in charge by means of a d.c. generator. When the generator fails, the batteries supply the load.

14.16. Nickel-Iron Cell or Edison Cell

Nickel-iron-alkaline cell was developed by the American scientist Thomas A. Edison in 1909. When the nickel-iron cell is in the charged condition, the active material on the positive plate is nickel oxide (Ni_2O_3) and that on the negative plate is iron (Fe). The electrolyte is 21% solution of potassium hydroxide (KOH) with a small amount of lithium added. When the cell is discharged, the positive plate is reduced to lower oxide of nickel (*i.e.* NiO) and the negative plate is oxidised to iron oxide (FeO). When the cell is recharged, the chemical process is reversed *i.e.* positive plate is converted to Ni_2O_3 and the negative plate to Fe. The equation of the chemical action is as under :

$$\underset{\text{positive plate}}{Ni_2O_3} + 2KOH + \underset{\text{negative plate}}{Fe} \underset{\text{charge}}{\overset{\text{discharge}}{\rightleftharpoons}} \underset{\text{positive plate}}{2NiO} + 2KOH + \underset{\text{negative plate}}{FeO}$$

The equation is read from left to right for discharging and from right to left for charging. The following points may be noted in these electro-chemical reactions :

(*i*) There is merely transfer of oxygen from one plate to the other. Thus during discharging, oxygen (O_2^-) from the positive plate is transferred to the negative plate to form FeO. The reverse happens during charging.

(*ii*) The specific gravity of the electrolyte (KOH) remains unchanged during charge or discharge.

The e.m.f. of a nickel-iron cell is about 1.2V when charged and falls to 0.9V when discharged. Since the electrolyte does not undergo any change in specific gravity during charge or discharge, the state of charge of this cell cannot be determined by the specific gravity of the electrolyte. Instead, a voltmeter is employed to ascertain whether the cell is charged upto its rated voltage.

Advantages

(*i*) Since the specific gravity of the electrolyte (KOH) does not change, a nickel-iron cell is not damaged by being left in a fully discharged condition for considerable period of time.

(*ii*) It is more rugged and can stand more mechanical and electrical abuse than a lead-acid cell.

(*iii*) It weighs only about half as much as a lead-acid cell of equivalent capacity.

(*iv*) It does not produce acid fumes that corrode nearby objects.

(*v*) Its life is much longer than that of a lead-acid cell.

(*vi*) it can be discharged at a high rate for long periods without danger to the battery and can be recharged safely at a rapid rate.

(*vii*) It can withstand high temperatures better than a lead-acid cell.

Disadvantages

(*i*) It is costlier than a lead-acid cell.

(*ii*) The e.m.f. of a nickel cell is about 1.2 V against 2V of the lead-acid cell. A supply of 12V needs a battery of 10 nickel-iron cells in series but only 6 lead-acid cells in series.

(*iii*) It has higher internal resistance than the lead-acid cell and cannot provide a large current. For this reason, it is unsuitable for automobile starting service.

(*iv*) The efficiencies of a nickel-iron cell are lower than that of a lead-acid cell. The ampere-hour and watt-hour efficiencies of such a cell are about 80% and 65% respectively.

Applications. Due to their high mechanical strength, lightness, ability to withstand arduous conditions and free from acid fumes, nickel-iron batteries have been found to be more satisfactory for traction purposes, mine locomotives, *submarines *etc.* They are particularly suitable in cases where exceptional vibrations are encountered as on motor cycles *etc.*

14.17. Construction of Nickel-Iron Cell

Like a lead-cell, a nickel-iron cell also has a multiplate construction. The positive plates consist of a number of perforated steel tubes mounted on a nickel-plated steel grid. Each tube is filled with nickel oxide (Ni_2O_3). The negative plates are constructed of many small perforated steel pockets, each filled with powdered iron (See Fig. 14.8). The perforations in the tubes and pockets of the plates permit the electrolyte to get at the active materials. The alternate positive and negative plates of the cell are sandwiched together with separators. The negative plates are connected together as are the positive plates. The number of negative plates is always one more than the number of positive plates; the outside plates being negative.

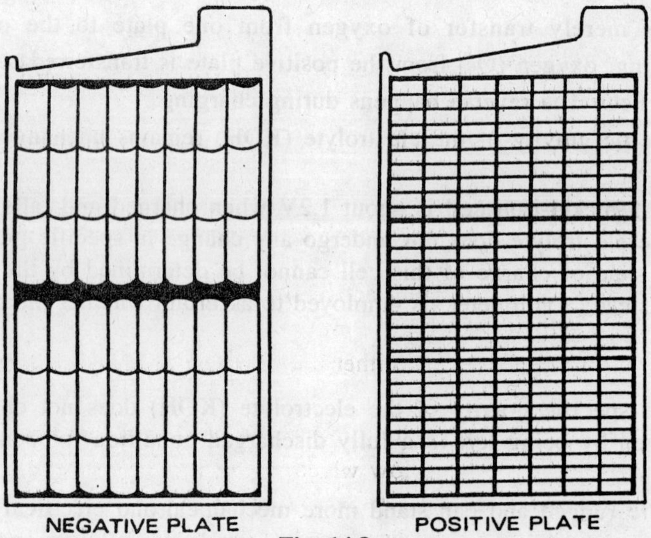

Fig. 14.8

A nickel-iron cell embraces the following additional arrangements for better performance :

(*i*) The plates are made thinner than that of a lead-acid cell. For this reason, a nickel-iron cell occupies less volume than the equivalent lead-acid cell.

---

* When a submarine is submerged, it cannot use those engines which consume air *e.g.* petrol engine or diesel engine. Hence it runs on electric motors operated by nickel-iron batteries. The lead-acid battery is not suitable because the acid fumes may corrode the equipment. On the surface, the submarine is run by diesel engines. These engines also operate a d.c. generator which recharges the batteries.

(*ii*) The assembled plates are housed in a nickel plated steel container. A steel container is used because there is no acid to attack it.

(*iii*) In order to decrease the internal resistance of the cell, nickel oxide is intimately mixed with flakes of pure nickel and the iron with mercuric oxide.

14.18. Nickel-Cadmium Cell

Nickel-Cadmium cell was developed by Waldermar Junger, a Swede, in 1899. When the nickel-cadmium cell is in the charged condition, the active material on the positive plate is $Ni(OH)_3$ and that on the negative plate is cadmium (Cd). The electrolyte is the same as in the nickel-iron cell *i.e.* potassium hydroxide (KOH). When the cell is discharged, the positive plate is converted to $Ni(OH)_2$ and the negative plate to $Cd(OH)_2$. When the cell is re-charged, the chemical process is reversed *i.e.* positive plate is converted to $Ni(OH)_3$ and negative plate to Cd. The equation of chemical action is :

$$\underset{\text{positive plate}}{2Ni(OH)_3} + \underset{\text{negative plate}}{Cd} \underset{\text{charge}}{\overset{\text{discharge}}{\rightleftharpoons}} \underset{\text{positive plate}}{2Ni(OH)_2} + \underset{\text{negative plate}}{Cd(OH)_2}$$

The equation is read from left to right for discharging and from right to left when charging. The following points may be noted in these electro-chemical reactions :

(*i*) There is merely transfer of hydroxyl ions ($2OH^-$) from one plate to the other. Thus during discharging, hydroxyl ions from the positive plate are transferred to the negative plate to form $Cd(OH)_2$. The reverse happens during charging.

(*ii*) The specific gravity of the electrolyte (KOH) remains unchanged during charge or discharge.

The e.m.f. of a nickel-cadmium cell is about 1.2V when charged and falls to 1.1V when discharged. Since the specific gravity of the electrolyte (KOH) remains unchanged during charge or discharge, it cannot be used as an indication to the state of charge of the cell. Instead, a voltmeter is employed to determine whether the cell is charged upto its rated voltage.

Advantages. The nickel-cadmium cell combines the best features of both the lead-acid and Edison types.

(*i*) It is rugged and has a very long active life. Periods upto 20 years of service can be expected from nickel-cadmium battery.

(*ii*) It can be stored indefinitely in either a discharged or charged state without suffering any ill effects.

(*ii*) A nickel-cadmium (Ni-Cd) battery can be charged in a shot time (say 1 hour).

(*iv*) Its internal resistance is very low which means that it can deliver very large current at constant terminal voltage.

(*v*) It can stand peak rates of discharge and charge upto 20 times the normal operating rate.

The chief disadvantage of a nickel-cadmium cell is that it is expensive.

OBJECTIVE QUESTIONS

1. The current conduction through the solution of an electrolyte is by
(*i*) free electrons (*ii*) ions
(*iii*) atoms (*iv*) valence electrons

2. For the proces of electrolysis is, we require
(*i*) d. c. supply (*ii*) a.c. supply

(*iii*) varying voltage
(*iv*) both d.c. and a.c. supply

3. A positively charged atom is sometimes called ....
(*i*) donor atom (*ii*) accéptor atom
(*iii*) cation (*iv*) anion

4. The mass of an element deposited or liberated at an electrode during electrolysis is .... the quantity of electricity passed through the electrolyte.
(*i*) directly proportional to
(*ii*) inversely proportional to
(*iii*) independent of (iv) none of the above

5. The E.C.E. of the copper is 304×10^{-6}gm/C. If during electrolysis, 2C of electricty is passed through the solution of $CuSO_4$, then, mass or copper deposited on cathode is ....
(*i*) 2 gm (*ii*) 152×10^{-6} gm
(*iii*) 608×10^{-6} gm (*iv*) 304×10^{-6} gm

6. Out of the following, E.C.E. of .... is largest.
(*i*) copper (*ii*) iron
(*iii*) nickel (*iv*) silver

7. During electrolysis, mass of an element liberated at the electrode is .... of the element.
(*i*) directly proportional to the valency
(*ii*) inversely proportional to atomic weight
(*iii*) directly proportional to chemical eq. wt.
(*iv*) none of the above

8. The relation between E.C.E. (z) and chemical equivalent weight (E) is that ....
(*i*) E/z is constant for all elements
(*ii*) E/z is different for different elements
(*iii*) E^2/z is constant for all elements.
(*iv*) E/z^2 is constant for all elements

9. When the same quantity of electricity is passed through silver and copper voltameters in series, the masses of silver (Ag) and copper (Cu) deposited on the respective cathodes will be in the ratio of their ....
(*i*) atomic weights (*ii*) valency
(*iii*) chemical equivalent weights
(*iv*) none of the above

10. The e.m.f. of a cell does not depend upon....
(*i*) nature of electrolyte
(*ii*) nature of material of electrodes
(*iii*) concentration of electrolyte
(*iv*) size and spacing of electrodes

11. If two different elements in the electrochemical series are used as the electrodes in a cell, the one higher in the series............
(*i*) will be positive plate
(*ii*) will be negative plate
(*iii*) can be positive or negative plate
(*iv*) none of the above .

12. The major drawback of a primary cell is that ....
(*i*) chemical action is not reversible
(*ii*) chemical action is reversible
(*iii*) the electrolyte used is very costly
(*iv*) it is not portable

13. The efficiency of a primary cell is about ......
(*i*) 25 % (*ii*) 15 %
(*iii*) 70 % (*iv*) 35 %

14. The most commonly used cell is ....
(*i*) lead-acid cell (*ii*) nickel-iron cell
(*iii*) nickel-cadmium cell
(*iv*) fuel cell

15. In practice, the state of discharge of a lead-acid cell is determined by ....
(*i*) e.m.f. of the cell
(*ii*) specific gravity of the electrolyte
(*iii*) colour of plates of the cell
(*iv*) none of the above

16. When a lead-acid cell is fully charged ......
(*i*) anode is converted into Pb
(*ii*) cathode is converted into PbO_2
(*iii*) specific gravity of H_2SO_4 rises to about 1.28
(*iv*) both plates become red

17. A multiplate construction is used in a lead-acid cell to ....
(*i*) increase the capacity of the cell
(*ii*) increase the e.m.f. of the cell

(*iii*) increase the internal resistance of the cell
(*iv*) facilitate cell connections

18. The commercial lead-acid cell has 15 plates. The number of negative plates will be ......
(*i*) 7 (*ii*) 9
(*iii*) 10 (*iv*) 8

19. The life of the positive plates of a lead-acid cell is roughly .... that of negative plates.
(*i*) the same as (*ii*) half
(*iii*) twice (*iv*) thrice

20. If the specific gravity of electrolyte (H_2SO_4) in a lead-acid cell increases, the internal resistance of the cell ....
(*i*) remains unchanged
(*ii*) is increased
(*iii*) is decreased
(*iv*) none of the above

21. The material of positive plates of a lead-acid cell is ....than those of negative plates.
(*i*) mechanically stronger
(*ii*) poorer in conductivity
(*iii*) better in conductivity
(*iv*) less chemically active

22. The specific gravity of the electrolyte (H_2SO_4) of a lead-acid cell is about 1.25. The cell is about ....
(*i*) 25 % charged (*ii*) 50 % charged
(*iii*) 75% charged (*iv*) 100% charged

23. The internal resistance of a lead-acid cell is mainly due to ....
(*i*) positive plates (*ii*) negative plates
(*iii*) both positive and negative plates
(*iv*) electrolyte

24. A lead-acid cell has 13 plates. In the absence of manufacturer's data, the charging current should be
(*i*) 13 A (*ii*) 2 A
(*iii*) 6 A (*iv*) 3 A

25. The ampere hour efficiency of a lead-acid cell is about ....
(*i*) 25 % (*ii*) 50 %
(*iii*) 75 % (*iv*) 90 %

26. The state of charge of a nickel-iron cell is determined by ....
(*i*) the specific gravity of electrolyte (KOH)
(*ii*) e.m.f. of the cell
(*iii*) colour of plates
(*iv*) none of the above

27. Nickel-iron batteries are used in ....
(*i*) starting of car engine
(*ii*) hospitals as an emergency supply
(*iii*) submarines
(*iv*) railway train lighting system

28. The state of charge of nickel-cadmium cell is determined by ....
(*i*) e.m.f. of the cell
(*ii*) specific gravity of the electrolyte (KOH)
(*iii*) colour of plates
(*iv*) none of the above

29. The chief disadvantage of a nickel-cadmium cell is that it ....
(*i*) takes a long time for charging
(*ii*) has high internal resistance
(*iii*) has a short life
(iv) is very expensive

30. The Faraday constant F is equal to (N is Avogadro number and e = 1.6×10^{-19} C)
(i) $F = Ne$ (ii) $F = N/e$
(*iii*) $F = N^2e$ (iv) $F = Ne^2$

ANSWERS TO OBJECTIVE QUESTIONS

1. (*ii*) **2.** (*i*) **3.** (*iii*) **4.** (*i*) **5.** (*iii*) **6.** (*iv*) **7.** (*iii*) **8.** (*i*) **9.** (*iii*) **10.** (*iv*) **11.** (*ii*)
12. (*i*) **13.** (*iii*) **14.** (*i*) **15.** (*ii*) **16.** (*iii*) **17.** (*i*) **18.** (*iv*) **19.** (*ii*) **20.** (*iii*) **21.** (*ii*) **22.** (*iii*)
23. (*iv*) **24.** (*iii*) **25.** (*iv*) **26.** (*ii*) **27.** (*iii*) **28.** (*i*) **29.** (*iv*) **30.** (*i*)

HINTS TO SELECTED OBJECTIVE QUESTIONS

4. $m = zQ$

where z = electrochemical equivalent

Q = quantity of electricity passed through the electrolyte

7. Mass liberated $\propto \dfrac{\text{Atomic weight}}{\text{Valency}}$

$\propto$ Chemical eq. wt of the element

14. Due to relatively low cost,lead-acid cell is most widely used.

15. The *e.m.f.* of a lead-acid cell provides little indication to the state of discharge of the cell since it remains close to 2 V for 90% of the discharge period. In practice, specific gravity of electrolyte is uscd to know the state of discharge.

18. The number of negative plates in a lead-acid cell is one more than the number of positive plates; the outside plates being negative.

19. The positive plate active material (*i.e.*, PbO_2) undergoes a large change in volume during charge and discharge as compared with negative plate active material (*i.e.*, Pb). Therefore, the tendency of PbO_2 to disintegrate and fall to the bottom of the cell is much greater than for Pb. The life of a positive plate is roughly half of the negative plate.

24. The charging current should be lA for every positive plate of a single cell.

26. The specific gravity of the electrolyte (KOH) remains unchanged during charge or discharge.

15

A.C. Fundamentals

Introduction

We have dealt so far with cases in which the currents are steady and in one direction; this is called direct current (d.c.). The use of direct currents is limited to a few applications *e.g.* charging of batteries, electroplating, electric traction *etc.* For large-scale power distribution there are, however, many advantages in using alternating current (a.c.). In an a.c. system, the voltage acting in the circuit changes polarity at regular intervals of time and the resulting current (called alternating current) changes direction accordingly. The a.c. system has offered so many advantages that at present, electrical energy is universally generated, transmitted and used in the form of alternating current. Even when d.c. energy is necessary, it is a common practice to convert a.c. into d.c. by means of rotary converters or rectifiers.

Three principal advantages are claimed for a.c. system over the d.c. system. First, alternating voltages can be stepped up or stepped down efficiently by means of a transformer. This permits the transmission of electric power at high voltages to achieve economy and distribute the power at utilization voltages. Secondly, a.c. motors (induction motors) are cheaper and simpler in construction than d.c. motors. Thirdly, the switchgear (*e.g.* switches, circuit breakers *etc.*) for a.c. system is simpler than the d.c. system. In this chapter, we shall confine our attention to the fundamentals of alternating currents.

15.1. Sinusoidal Alternating Voltage and Current

Commercial alternators produce sinusoidal alternating voltage *i.e.* alternating voltage is a sine wave. A sinusoidal alternating voltage can be produced (For proof, refer to Art 15.3) by rotating a coil with a constant angular velocity (say ω rad/sec) in a uniform magnetic field. The sinusoidal alternating voltage can be expressed by the equation :

$$e = E_m \sin \omega t$$

where e = Instantaneous value of alternating voltage

E_m = Maximum value of alternating voltage

ω = Angular velocity of the coil

Sinusoidal voltages always produce sinusoidal currents, unless the circuit is non-linear. Therefore, a sinusoidal current can be expressed in the same way as voltage *i.e.*, $i = I_m \sin \omega t$. Fig. 15.1 (*i*) shows the waveform of sinusoidal voltage whereas Fig. 15.1 (*ii*) shows the waveform of sinusoidal current. Note that sinusoidal voltage or current not only changes direction at regular intervals but the magnitude is also changing continuously.

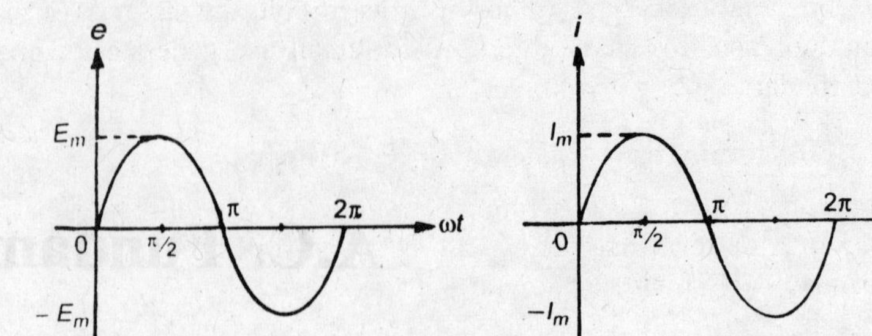

(i) (ii)

Fig. 15.1

Why Sine Wave form ? Why do we choose sinusoidal wave rather than a simple curve such as a square or triangular wave ? The following are the reasons for doing so :

(*i*) In a.c. machines and transformers, sinusoidal voltages and currents respectively produce the least iron and copper losses for a given output. The efficiency is, therefore, better.

(*ii*) Sinusoidal voltages and currents produce less interference (noise) on telephone lines.

(*iii*) The sine waveform produces the least disturbance in the electrical circuit and is smoothest and efficient waveform.

Due to above advantages, electric supply companies all over the world generate sinusoidal alternating voltages and currents. *It may be noted that alternating voltage and current mean sinusoidal voltage and current unless stated otherwise.*

15.2. Generation of Alternating Voltages and Currents

An alternating voltage may be generated

(*i*) by rotating a coil at constant angular velocity in a uniform magnetic field as shown in Fig. 15.2.

Or

(*ii*) by rotating a magnetic field at a constant angular velocity within a stationary coil as shown in Fig. 15.3.

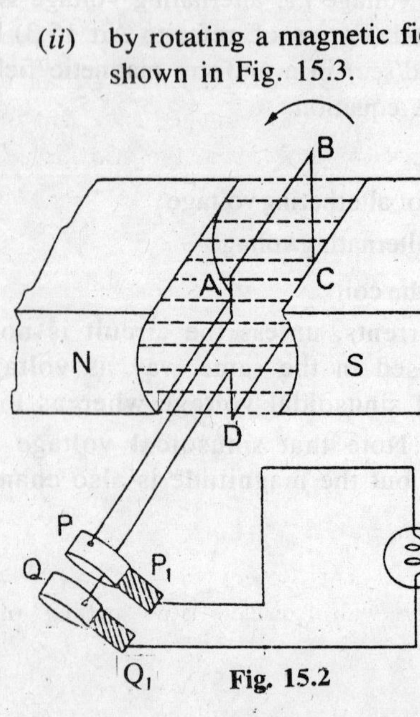

Fig. 15.2

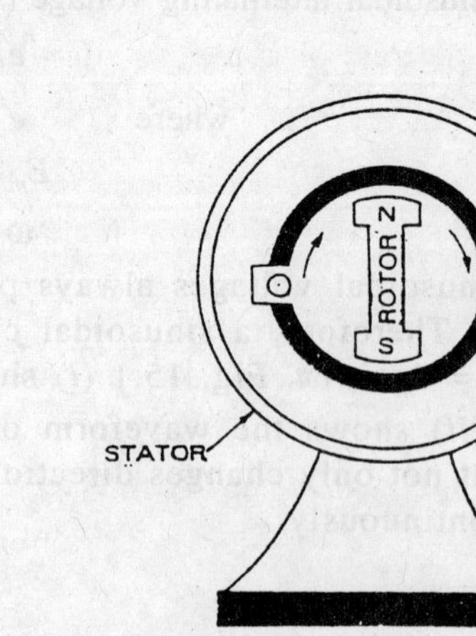

Fig. 15.3

In either case, the generated voltage will be of sinusoidal waveform. The magnitude of generated voltage will depend upon the number of turns of coil, the strength of magnetic field and the speed of rotation. The first method is used for small a.c. generators while the second method is employed for large a.c. generators.

15.3. Equation of Alternating Voltage and Current

Consider a rectangular coil of n turns rotating in anticlockwise direction with an angular velocity of ω rad/sec in a uniform magnetic field as shown in Fig. 15.4. The e.m.f. induced in the coil will be sinusoidal. This can be readily established.

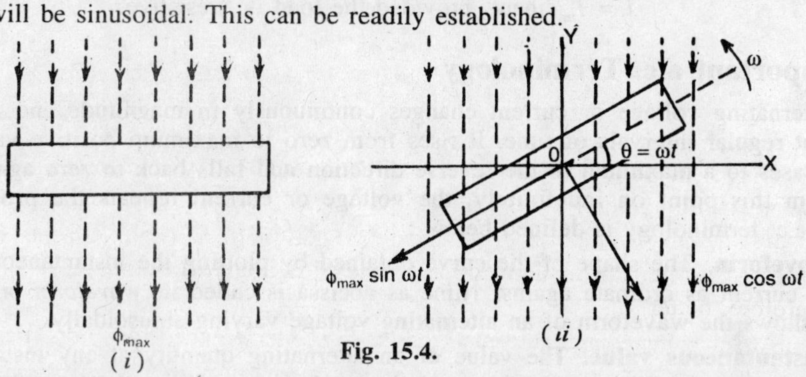

Fig. 15.4.

Let the time be measured from the instant the plane of the coil coincides with *OX*-axis. In this position of the coil [See Fig. 15.4 (*i*)], the flux linking with the coil has its maximum value ϕ_{max}. Let the coil turn through an angle $\theta\ (= \omega t)$ in anticlockwise direction in t seconds and assume the position shown in Fig. 15.4 (*ii*). In this position, the maximum flux ϕ_{max} acting vertically downward can be resolved into two perpendicular components *viz*.

(*i*) component $\phi_{max} \sin \omega t$ parallel to the plane of the coil. This component induces *no e.m.f. in the coil.

(*ii*) component $\phi_{max} \cos \omega t$ perpendicular to the plane of the coil. This component induces e.m.f. in the coil.

$\therefore$ Flux linkages of the coil at the considered instant (*i.e.* at $\theta°$)

$$= \text{No. of turns} \times \text{Flux linking}$$

$$= n\,\phi_{max} \cos \omega t$$

According to Faraday's laws of electromagnetic induction, the e.m.f. induced in a coil is equal to the rate of change of flux linkages of the coil. Hence the e.m.f. e at the considered instant is given by;

$$e = -\frac{d}{dt}(n\,\phi_{max} \cos \omega t)$$

$$= -n\,\phi_{max}\,\omega\,(-\sin \omega\,t)$$

$$= n\,\phi_{max}\,\omega \sin \omega\,t$$

The value of e will be maximum (call it E_m) when $\sin \omega t = 1$ *i.e.* when the coil has turned through 90° in anticlockwise direction from the reference axis (*i.e.* *OX*-axis).

$$\therefore \qquad E_m = n\,\phi_{max}\,\omega$$

---

* A coil moving parallel to the flux has no change of flux linkage with it or there is no "cutting" of flux.

$$\therefore \quad e = E_m \sin \omega t \quad \text{where } E_m = n\phi_{max}\,\omega$$
$$= E_m \sin \theta$$

It is clear that e.m.f. induced in the coil is sinusoidal *i.e.* instantaneous value of e.m.f. varies as the sine function of time angle (θ or ωt). If this alternating voltage ($e = E_m \sin \omega t$) is applied across a load, alternating current flows through the circuit which would also vary sinusoidally *i.e.*, following a sine law. The equation of the alternating current is given by;

$$i = I_m \sin \omega t \quad \text{provided the load is *resistive.}$$

15.4. Important a.c. Terminology

An alternating voltage or current changes continuously in magnitude and alternates in direction at regular intervals of time. It rises from zero to maximum positive value, falls to zero, increases to a maximum in the reverse direction and falls back to zero again (See Fig. 15.5). From this point on indefinitely, the voltage or current repeats the procedure. The important a.c. terminology is defined below :

(*i*) **Waveform**. The shape of the curve obtained by plotting the instantaneous values of voltage or current as ordinate against †time as abcissa is called its *waveform* or *waveshape*. Fig. 15.5 shows the waveform of an alternating voltage varying sinusoidally.

(*ii*) **Instantaneous value.** The value of an alternating quantity at any instant is called instantaneous value. The instantaneous values of alternating voltage and current are represented by *e* and *i* respectively. As an example, the instantaneous values of voltage (See Fig. 15.5) at 0°, 90° and 270° are 0, $+E_m$, $-E_m$ respectively.

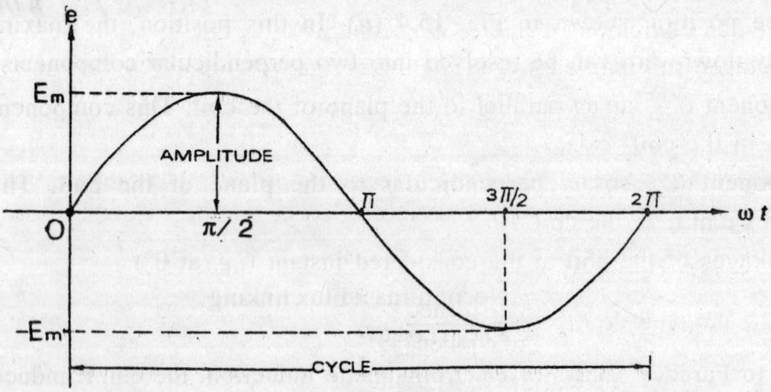

Fig. 15.5

(*iii*) **Cycle.** One complete set of positive and negative values of an alternating quantity is known as a cycle. Fig. 15.5 shows one cycle of an alternating voltage.

A cycle can also be defined in terms of angular measure. One cycle corresponds to 360° **electrical or 2π radians. The voltage or current generated in a conductor will span 360° electrical (or complete one cycle) when the conductor moves past a north and south pole.

---

* It will be shown later that if the load is inductive or capacitive, the current equation is changed in time angle.

† We know $\theta = \omega t$. Since ω is constant, $\theta \propto t$. Hence we may take time t or θ or ωt along X-axis.

** The time required to generate one cycle is divided into 360 divisions called electrical degrees. They differ from mechanical degrees. In one revolution, the coil will always traverse 360° mechanical. However, the electrical degrees spanned will depend upon the number of poles. For a 2-pole generator, electrical degrees spanned in one revolution of the coil are 360° *i.e.*, one cycle is generated. For a 4-pole generator, 2 cycles (*i.e.* 720°) will be generated for one revolution of the coil.

(*iv*) **Alternation.** One-half cycle of an alternating quantity is called an alternation. An alternation spans 180° electrical. Thus in Fig. 15.5, the positive or negative half of alternating voltage is the alternation.

(*v*) **Time period.** The time taken in seconds to complete one cycle of an alternating quantity is called its time period. It is generally represented by T.

(*vi*) **Frequency.** The number of cycles that occur in one second is called the frequency (f) of the alternating quantity. It is measured in cycles/sec (*c/s*) or Hertz (*Hz*). One Hertz is equal to 1 c/s.

The frequency of power system is low; the most common being 50 *c/s* or 50Hz. It means that alternating voltage or current completes 50 cycles in one second. The 50 Hz frequency is the most popular because it gives the best results when used for operating both lights and machinery.

(*vii*) **Amplitude.** The maximum value (positive or negative) attained by an alternating quantity is called its amplitude or peak value. The amplitude of an alternating voltage or current is designated by E_m (or V_m) or I_m.

15.5. Important Relations

Having become familiar with a.c. terminology, we shall now establish some important relations.

(*i*) **Time period and frequency.** Consider an alternating quantity having a frequency of f *c/s* and time period T second.

Time taken to complete f cycles = 1 second (*By definition*)

Time taken to complete 1 cycle = $1/f$ second

But time taken to complete one cycle is the time period T (by definition).

$$\therefore \quad T = \frac{1}{f} \quad or \quad f = \frac{1}{T}$$

(*ii*) **Angular velocity and frequency.** Referring to Fig. 15.4, the coil is rotating with an angular velocity of ω rad/sec. In one revolution of the coil, the angle turned is 2π radians and the voltage wave completes 1 cycle. The time taken to complete one cycle is the time period T of the alternating voltage.

$$\therefore \text{ Angular velocity, } \omega = \frac{\text{Angle turned}}{\text{Time taken}} = \frac{2\pi}{T}$$

$$or \quad \omega = 2\pi f \qquad (\because f = 1/T)$$

(*iii*) **Frequency and speed.** Consider a coil rotating at a speed of N r.p.m. in the field of P poles. As the coil moves past a north and south pole, one complete cycle is generated. Obviously, in one revolution of the coil, $P/2$ cycles will be generated.

$$\text{Now, Frequency, } f = \text{No. of cycles/sec}$$

$$= (\text{No. of cycles/revolution}) \times (\text{No. of revolutions/sec})$$

$$= \left(\frac{P}{2}\right) \times \left(\frac{N}{60}\right) = \frac{PN}{120}$$

$$\therefore \quad f = \frac{PN}{120}$$

For example, an a.c. generator having 10 poles and running at 600 r.p.m. will generate alternating voltage and current whose frequency is :

$$f = \frac{PN}{120} = \frac{10 \times 600}{120} = 50 \text{ Hz}$$

15.6. Different Forms of Alternating Voltage

The standard form of an alternating voltage is given by;

$$e = E_m \sin \theta = E_m \sin \omega t$$

$$= E_m \sin 2\pi ft = E_m \sin \frac{2\pi}{T} t$$

Which of the above form of equations is to be used will depend upon the given data.

The following points may be noted carefully:

(i) *The maximum value of alternating voltage is given by the co-efficient of sine of the time angle i.e..*

Maximum value of voltage, E_m = Co-efficient of sine of time angle

(ii) *The frequency f of alternating voltage is given by dividing the co-efficient of time in the angle by* 2π *i.e.*

$$\text{Frequency, } f = \frac{\text{Co-efficient of time in the angle}}{2\pi}$$

For example, suppose the equation of an alternating voltage is given by $e = 100 \sin 314t$. Then the maximum value of voltage, $E_m = 100$; frequency $f = 314/2\pi = 50$Hz and time period $T = 1/f = 1/50 = 0.02$ second.

Following similar procedure, maximum value, frequency, time period *etc.* can be found out from the various forms of current equation.

Example 15.1. *A rectangular coil measuring 30 cm by 20 cm and having 40 turns is rotated about an axis coinciding with one of its longer sides at a speed of 1500 r.p.m. in a uniform magnetic field of flux density 0.075 wb/m² Calculate (i) the frequency of the e.m.f. (ii) maximum value of e.m.f. and (iii) the value of e.m.f. when the coil plane makes an angle of 45° with the field direction.*

Solution. The statement of the problem suggests that the number of poles is 2.

(i) Frequency, $f = \frac{NP}{120} = \frac{1500 \times 2}{120} = \mathbf{25\ Hz}$

(ii) As proved in Art 15.3,

Maximum e.m.f., $E_m = n\,\phi_{max}\,\omega$

Here $\phi_{max} = B \times A = 0.075 \times (30 \times 20 \times 10^{-4}) = 0.0045$ wb

$$\omega = 2\pi f = 2\pi \times 25 = 157$$

$$\therefore \quad E_m = 40 \times (0.0045) \times 157 = \mathbf{28.26\ V}$$

(iii) When the coil plane makes an angle of 45° with the direction of field,

$$e = E_m \sin\theta = 28.26 \sin 45° = \mathbf{19.98\ V}$$

Example 15.2. *An alternating current i is given by;*

$$i = 141.4 \sin 314\,t$$

Find (i) the maximum value (ii) frequency (iii) time period and (iv) the instantaneous value when t is 3ms.

Solution. Comparing the given equation of alternating current with the standard form $i = I_m \sin \omega t$, we have, (*i*) Maximum value, I_m = **141.4 A**

(*ii*) Frequency, $f = \omega/2\pi = 314/2\pi =$ **50 Hz**

(*iii*) Time period, $T = 1/f = 1/50 =$ **0.02 s**

(*iv*) $i = 141.4 \sin 314\, t$

When $t = 3 \text{ ms} = 3 \times 10^{-3}$ s

$i = 141.4 \sin 314 \times 3 \times 10^{-3} =$ **114.35 V**

Example 15.3. *An alternating current of frequency 60Hz has a maximum value of 120 A.*

(*i*) *Write down the equation for the instantaneous value.*

(*ii*) *Reckoning time from the instant the current is zero and becoming positive, find the instantaneous value after 1/360 second.*

(*iii*) *Time taken to reach 96A for the first time.*

Solution.

(*i*) The instantaneous value of current is given by ;

$$i = I_m \sin \omega t$$

$$= I_m \sin 2\pi ft = 120 \sin 2\pi \times 60 \times t$$

$$\therefore \quad \mathbf{i = 120 \sin 120\, \pi t}$$

(*ii*) Fig. 15.6 shows the waveform of the given alternating current. Since point *O* has been taken as the reference, the current equation is :

$$i = 120 \sin 120\, \pi t$$

When $t = 1/360$ second

$$i = 120 \sin 120\, \pi \times 1/360$$

$$= 120 \sin (\pi/3) = \mathbf{103.92\ A}$$

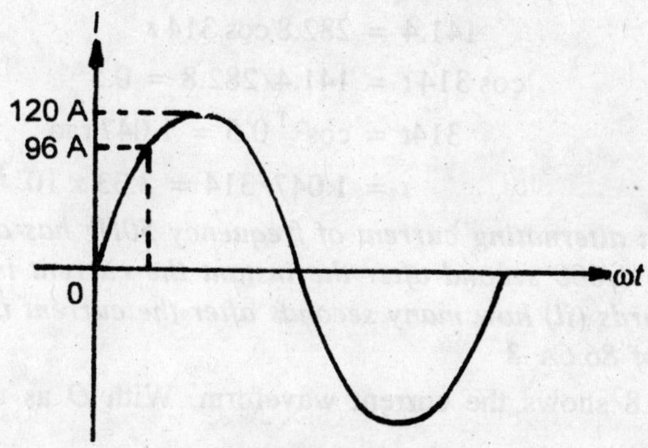

Fig. 15.6

(*iii*) Suppose the current becomes 96A for the first time after *t* second as shown in Fig. 15.6.

Then $96 = 120 \sin 120\, \pi t$

or $\sin 120\, \pi t = 96/120 = 0.8$

or $120\, \pi t = \sin^{-1} 0.8 = 0.927$ rad.

$$\therefore \quad t = \frac{0.927}{120 \times \pi} = \mathbf{0.00246\ s}$$

Example 15.4. *A sinusoidal voltage of 50Hz has a maximum value of $200\sqrt{2}$ volts. At what time measured from a positive maximum value will the instantaneous voltage be 141.4 volts ?*

Solution. Fig. 15.7 shows the voltage wave-form. The equation of this wave with point O as the reference is given by;

$$e = E_m \sin \omega t = E_m \sin 2\pi ft = 200\sqrt{2} \sin 2\pi \times 50 \times t$$

$$\therefore \quad e = 282.8 \sin 314\, t$$

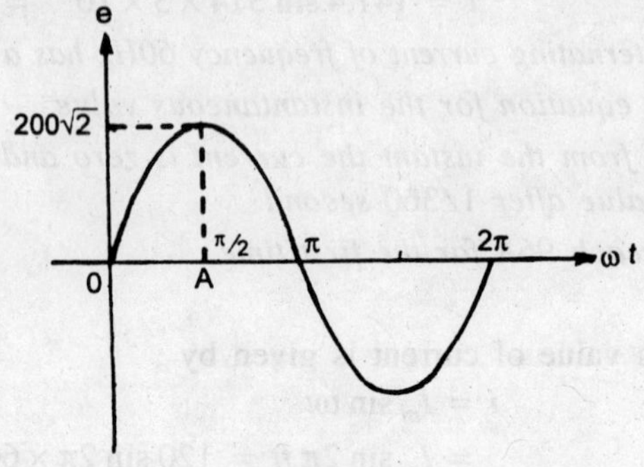

Fig. 15.7

This equation is valid when time is measured from the instant the voltage is zero *i.e.* point O. Since the time is measured from the positive maximum value (point A in Fig. 15.7), the above equation is modified to :

$$e = 282.8 \sin (314\, t + \pi/2) = 282.8 \cos 314\, t$$

Let the value of voltage become 141.4 volts t second after passing through the maximum positive value. Then,

$$141.4 = 282.8 \cos 314\, t$$

or $$\cos 314 t = 141.4/282.8 = 0.5$$

or $$314t = \cos^{-1} 0.5 = 1.047 \text{ rad}$$

$$\therefore \quad t = 1.047/314 = 3.33 \times 10^{-3} \text{ s} = \mathbf{3.33 \text{ ms}}$$

Example 15.5. *An alternating current of frequency 50Hz has a maximum value of 100A. Calculate (i) its value 1/600 second after the instant the current is zero and its value is decreasing there afterwards (ii) how many seconds after the current is zero and then increasing will attain the value of 86.6A ?*

Solution. Fig. 15.8 shows the current waveform. With O as the origin, the equation of the current is :

$$i = I_m \sin \omega t = 100 \sin 2\pi \times 50 \times t$$

$$\therefore \quad i = 100 \sin 100\pi\, t$$

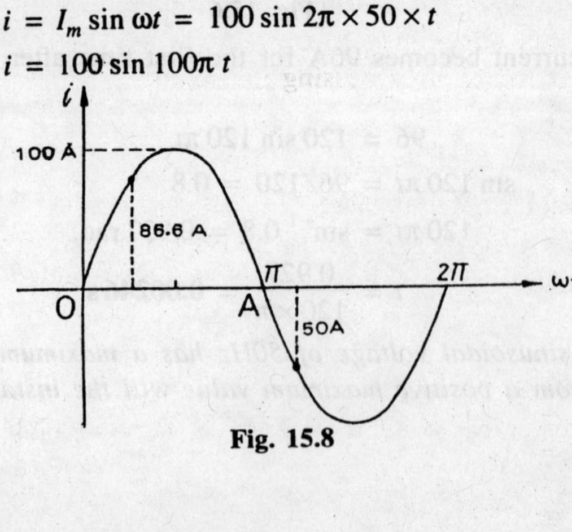

Fig. 15.8

(*i*) Since current is measured from the instant the current is zero and is decreasing there afterwards (point *A* in Fig. 15.8), the equation of the current with point *A* as the origin becomes:

$$i = 100 \sin (100\pi t + \pi) = -100 \sin 100\pi t$$

When $t = 1/600$ second, then,

$$i = -100 \sin 100 \times 180° \times \frac{1}{600} = \mathbf{-50A}$$

(*ii*) When current is measured from the instant the current is zero and is increasing thereafter (*i.e.* point *O* in Fig. 15.8), the current equation is given by ;

$$i = 100 \sin 100\pi t$$

or $$86.6 = 100 \sin 100\pi t$$

or $$\sin 100\pi t = 86.6/100 = 0.866$$

or $$100\pi t = \sin^{-1} 0.866 = 1.047 \text{ rad}$$

$\therefore$ $$t = 1.047/100\pi = 3.33 \times 10^{-3} \text{ s} = \mathbf{3.33 \text{ ms}}$$

Tutorial Problems

1. An alternating voltage is represented by ;

$$e = 141.4 \sin 377 t$$

Find (*i*) the maximum value (*ii*) frequency (*iii*) time period and (*iv*) the instantaneous value of voltage when *t* is 3 ms. **[(*i*) 141.4 V (*ii*) 60 Hz (*iii*) 16.67 ms (*iv*) 127.8 V]**

2. An alternating current of frequency 50Hz has a maximum value of $200\sqrt{2}A$. Reckoning the time from the instant the current is zero and becoming positive, find the time taken by the current to reach a value of 141.4A for a first and second time. **[1.67 ms; 21.67 ms]**
3. An alternating current takes 3.375ms to reach 15A for the first time after becoming instantaneously zero. The frequency of current is 40 Hz. Find the maximum value of alternating current. **[20A]**
4. A 50Hz sinusoidal voltage has a maximum value of 56.56V. Find the value of voltage 0.0025 second after passing through maximum positive value. At what time measured from a positive maximum value will instantaneous voltage be 14.14V ? **[40V; 4.2 ms]**
5. An alternating current of frequency 50Hz has a maximum value of 100A. Calculate its value 1/300 second after the instant the current is zero and its value is decreasing thereafter. **[–86.6 A]**

15.7. Values of Alternating Voltage and Current

In a d.c. system, the voltage and current are constant so that there is no problem of specifying their magnitudes. However, an alternating voltage or current varies from instant to instant. A natural question arises how to express the magnitude of an alternating voltage or current. There are three ways of expressing it, namely;

(*i*) Peak Value

(*ii*) Average value or Mean value

(*iii*) R.M.S value or Effective value.

Although peak and average values may be important in some engineering applications, it is the r.m.s. or effective value which is used to express the magnitude of an alternating voltage or current.

15.8. Average Value

The arithmetical average of all the values of an alternating quantity over one cycle is called its average value *i.e.*

$$\text{Average value} = \frac{\text{Area under the curve}}{\text{Base}}$$

(*i*) In case of *symmetrical waves (*e.g.* sinusoidal voltage or current), the average value over one cycle is zero. It is because positive half is exactly equal to the negative half so that net area is zero. However, the average value of positive or negative half is not zero. Hence in case of symmetrical waves, average value means the average value of half-cycle or one alternation.

$$\text{Average value of a symmetrical wave} = \frac{\text{Area of one alternation}}{\text{Base length of one alternation}}$$

$$= \frac{\text{Sum of **mid–ordinates over one alternation}}{\text{No. of mid–ordinates}}$$

(*ii*) In case of unsymmetrical waves (*e.g.* half-wave rectified voltage *etc.*), the average value is taken over the full cycle.

$$\text{Average value of an unsymmetrical wave} = \frac{\text{Area over one cycle}}{\text{Base length of 1 cycle}}$$

15.9. Average Value of Sinusoidal Current

The equation of an alternating current varying sinusoidally is given by :

$$i = I_m \sin\theta$$

Consider an elementary strip of width $d\theta$ in the first half-cycle of current wave as shown in Fig. 15.9. Let i be the mid-ordinate of this strip.

Area of strip $= i d\theta$

Area of half-cycle

$$= \int_0^{\pi} i\, d\theta = \int_0^{\pi} I_m \sin\theta\, d\theta = I_m\,[-\cos\theta]_0^{\pi} = 2I_m$$

$$\therefore \text{ Average value, } I_{av} = \frac{\text{Area of half–cycle}}{\text{Base length of half–cycle}} = \frac{2I_m}{\pi}$$

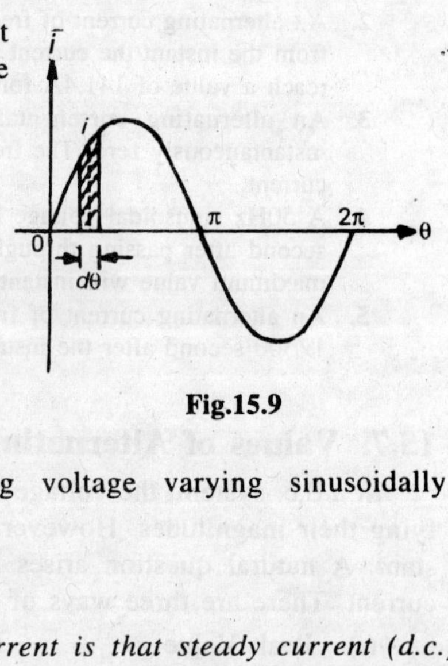

Fig.15.9

or $\mathbf{I_{av} = 0.637\, I_m}$

Similarly, it can be proved that for alternating voltage varying sinusoidally, $E_{av} = 0.637\, E_m$.

15.10. R.M.S. or Effective Value

The **effective** *or* **r.m.s. value** *of an alternating current is that steady current (d.c.) which when flowing through a given resistance for a given time produces the same amount of heat as produced by the alternating current when flowing through the same resistance for the same time.*

---

* A symmetrical wave is one which has positive half-cycle exactly equal to the negative half-cycle.

** Suppose positive half-cycle is divided into *n* equal parts. The middle value of each part is the mid-ordinate. The average value will be the sum of mid-ordinates divided by the number of mid-ordinates (*i.e. n* in this case).

For example, when we say that the r.m.s. or effective value of an alternating current is 5A, it means that the alternating current will do work (or produce heat) at the same rate as 5A direct current under similar conditions.

Illustration. The r.m.s. or effective value of an alternating current (or voltage) can be determined as follows. Consider the half-cycle of a non-sinusoidal alternating current i [See Fig. 15.10 (*i*)] flowing through a resistance $R\Omega$ for t seconds. Divide the time t in n equal intervals of time, each of duration t/n second. Let the mid-ordinates be $i_1, i_2, i_3, ...i_n$. Each current $i_1, i_2, i_3, ...i_n$ will produce heating effect when passed through the resistance R as shown in Fig. 15.10 (*ii*). Suppose the heating effect produced by current i in R is the same as produced by some direct current I flowing through the resistance R for the same time t seconds. Then direct current I is the r.m.s. or effective value of alternating current i.

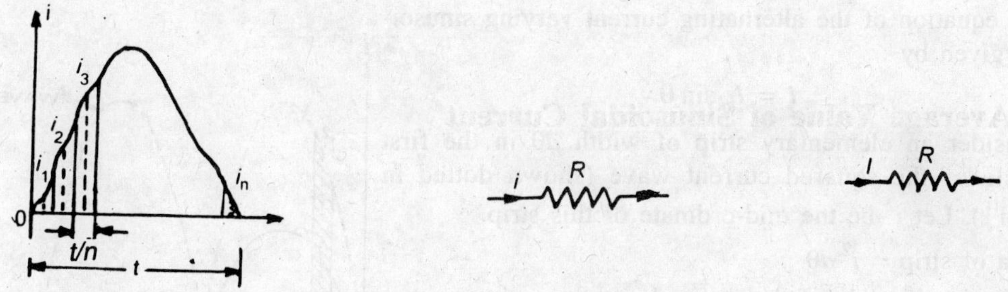

(*i*) (*ii*) (*iii*)

Fig. 15.10

The heating effect of various components of alternating current will be $i_1^2\,Rt/n, i_2^2\,Rt/n ... i_n^2\,Rt/n$ joules. Since the alternating current is varying, the heating effect will also vary.

Total heat produced by alternating current i is

$$= (\,i_1^2\,R + i_2^2\,R + i_3^2\,R + ...\, i_n^2\,R\,)\,t/n$$

$$= \left(\frac{i_1^2\,R + i_2^2\,R + i_3^2\,R + ...\, i_n^2\,R}{n}\right) t \text{ joules}$$

Heat produced by equivalent direct current I is $= I^2\,Rt$ joules

Since the heat produced in both cases is the same,

$$\therefore \qquad I^2Rt = \left(\frac{i_1^2\,R + i_2^2\,R + i_3^2\,R + ... + i_n^2\,R}{n}\right) t$$

$$\text{or} \qquad I = \sqrt{\frac{i_1^2 + i_2^2 + i_3^2 + ... + i_n^2}{n}} \qquad ...(i)$$

$= \sqrt{\text{mean value of } i^2}$

= Square **root** of the **mean** of the **squares** of the current

= root–mean–square (r.m.s.) value

The r.m.s. or effective value of alternating voltage can similarly be expressed as :

$$E = \sqrt{\frac{e_1^2 + e_2^2 + e_3^2 + ... + e_n^2}{n}}$$

(*i*) For symmetrical waves, the r.m.s. or effective value can be found by considering half-cycle or full-cycle. However, for unsymmetrical waves, full-cycle should be considered.

(*ii*) The r.m.s. value of a wave can also be expressed as :

$$\text{*R.M.S. value} = \sqrt{\frac{\text{Area of half-cycle wave squared}}{\text{Half-cycle base}}}$$

15.11. R.M.S. Value of Sinusoidal Current

The equation of the alternating current varying sinusoidally is given by :

$$i = I_m \sin \theta$$

Consider an elementary strip of width $d\theta$ in the first half-cycle of the squared current wave (shown dotted in Fig. 15.11). Let i^2 be the mid-ordinate of this strip.

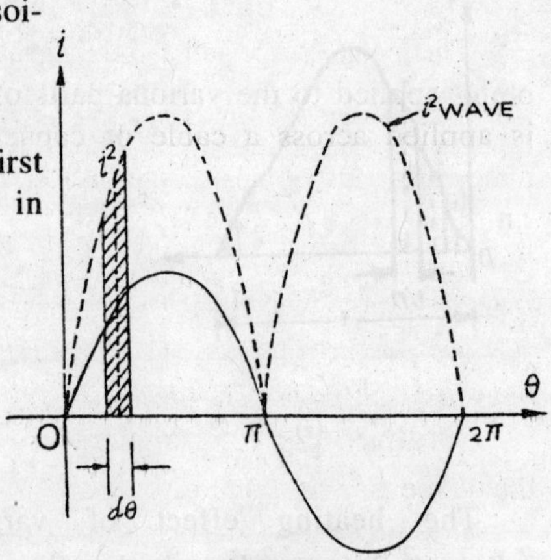

Fig. 15.11

Area of strip = $i^2 \, d\theta$

Area of half-cycle of the squared wave

$$= \int_0^{\pi} i^2 \, d\theta = \int_0^{\pi} I_m^2 \sin^2\theta \, d\theta$$

$$= I_m^2 \int_0^{\pi} \sin^2\theta d\theta = \frac{^{**}\pi I_m^2}{2}$$

$$\therefore \quad I_{r.m.s.} = \sqrt{\frac{\text{Area of half-cycle squared wave}}{\text{Half-cycle base}}} = \sqrt{\frac{\pi I_m^2/2}{\pi}} = \frac{I_m}{\sqrt{2}} = 0.707\, I_m$$

$$\therefore \quad \mathbf{I_{r.m.s.} = 0.707\, I_m}$$

Similarly, it can be proved that for alternating voltage varying sinusoidally, $E_{r.m.s.} = 0.707\, E_m$

In a.c. circuits, voltages and currents are normally given in r.m.s. values unless stated otherwise. A.c. ammeters and voltmeters are callibrated to record r.m.s. values.

15.12. Form Factor and Peak Factor

There exists a definite relation among the peak value, average value and r.m.s. value of an alternating quantity. The relationship is expressed by two factors, namely; form factor and peak factor.

---

* This result is readily obtained if the numerator and denominator of exp (*i*) on page 301 under the root is multiplied by *t/n*.

$$** \int_0^{\pi} \sin^2\theta\, d\theta = \int_0^{\pi} \frac{1 - \cos 2\theta}{2}\, d\theta = \frac{1}{2}\left[\theta - \frac{\sin 2\theta}{2}\right]_0^{\pi} = \frac{\pi}{2}$$

(*i*) **Form factor.** The ratio of r.m.s. value to the average value of an alternating quantity is known as form factor *i.e.*

$$\text{Form factor} = \frac{\text{R.M.S. value}}{\text{Average value}}$$

For a sinusoidal voltage or current,

$$\text{Form Factor} = \frac{0.707 \times \text{Max. value}}{0.637 \times \text{Max. value}} = 1.11$$

The form factor is useful in rectifier service because it enables us to find the r.m.s. value from average value and *vice-versa*

(*ii*) **Peak factor.** The ratio of maximum value to the r.m.s. value of an alternating quantity is known as peak factor *i.e.* $\text{Peak factor} = \frac{\text{Max. value}}{\text{R.M.S. value}}$

For a sinusoidal voltage or current : $\text{Peak factor} = \frac{\text{Max. value}}{0.707 \times \text{Max. value}} = 1.414$

The peak factor is of much greater importance because it indicates the maximum voltage being applied to the various parts of the apparatus. For instance, when an alternating voltage is applied across a cable or capacitor, the breakdown of insulation will depend upon the maximum voltage. The insulation must be able to withstand the maximum rather than the r.m.s. value of voltage.

Example 15.6. *An alternating current, when passed through a resistor immersed in water for 5 minutes, just raised the temperature of water to boiling point. When a direct current of 4A was passed through the same resistor under indentical conditions, it took 8 minutes to boil the water. Find the r.m.s. value of the alternating current.*

Solution. Let $I_{r.m.s.}$ amperes be the r.m.s. value of the alternating current and R ohms be the value of the resistor.

Heat produced by the alternating current

$$= I_{r.m.s.}^2 \times R \times t$$

$$= I_{r.m.s.}^2 \times R \times 5 \times 60 \text{ Joules} \qquad ...(i)$$

Heat produced by the direct current

$$= 4^2 \times R \times 8 \times 60 \text{ Joules} \qquad ...(ii)$$

As the heat produced in the two cases is the same,

$$\therefore \qquad I_{r.m.s.}^2 \times R \times 5 \times 60 = 4^2 \times R \times 8 \times 60$$

$$\text{or} \qquad I_{r.m.s.} = \sqrt{\frac{16 \times 8}{5}} = \mathbf{5.06\ A}$$

Example 15.7. *Find the average value, r.m.s. value, form factor and peak factor for (i) half-wave rectified alternating current and (ii) full-wave rectified alternating current.*

Solution.

(*i*) **Half-wave rectified a.c.** Fig. 15.12 shows half-wave rectified a.c. in which one half-cycle is suppressed *i.e.*, current flows for half the time during complete cycle.

$$I_{av} = \frac{\text{Area of one cycle}}{\text{Base}} = \frac{{}^{*}2I_m + 0}{2\pi} = \frac{I_m}{\pi}$$

* Since the wave is not symmetrical, complete cycle is to be considered. It has been shown by integral method (See Art. 15.9) that the area of a sinusoidal wave over one alternation is = $2 \times$ Max. value.

$$I_{r.m.s.} = \left[\frac{\text{Area of squared wave over one cycle}}{\text{Base}}\right]^{1/2}$$

$$= \left[\frac{{}^{*}(\pi I_m^2)/2 + 0}{2\pi}\right]^{1/2} = \frac{\mathbf{I_m}}{\mathbf{2}}$$

$$\text{Form factor} = \frac{I_{r.m.s.}}{I_{av}} = \frac{I_m/2}{I_m/\pi} = \mathbf{1.57}$$

$$\text{Peak factor} = \frac{I_{max}}{I_{r.m.s.}} = \frac{I_m}{I_m/2} = \mathbf{2}$$

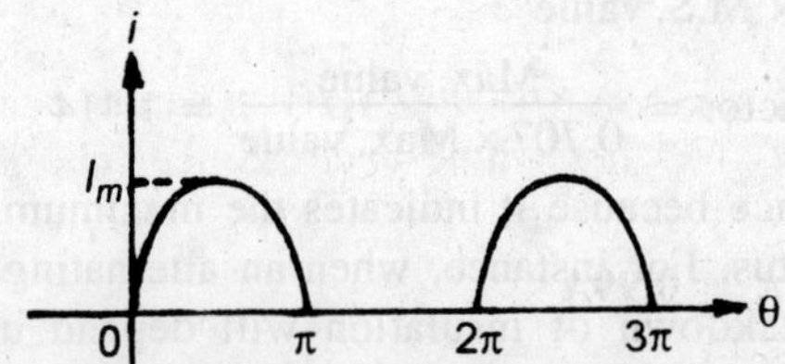

Fig. 15.12

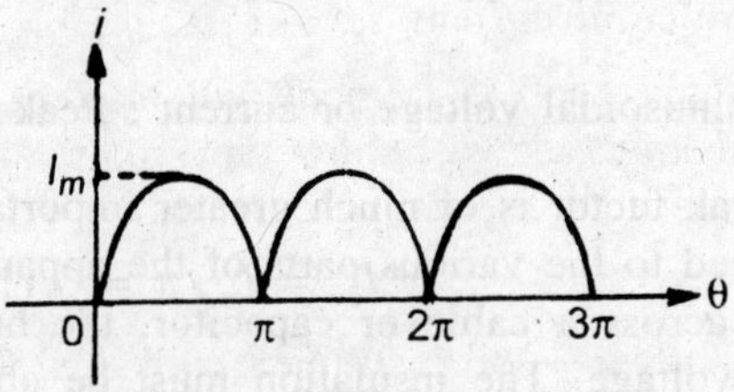

Fig. 15.13

(*ii*) **Full-wave rectified a.c.** Fig. 15.13 shows full-wave rectified a.c. in which both half-cycles appear in the output *i.e.*, current flows in the same direction for both half-cycles. Since the wave is symmetrical, half-cycle may be considered for various computations.

$$I_{av} = \frac{\text{Area of half–cycle}}{\text{Base}} = \frac{\mathbf{2I_m}}{\pi}$$

$$I_{r.m.s.} = \left[\frac{\text{Area of squared wave over half–cycle}}{\text{Base}}\right]^{1/2}$$

$$= \left[\frac{\pi I_m^2/2}{\pi}\right]^{1/2} = \frac{\mathbf{I_m}}{\sqrt{\mathbf{2}}}$$

$$\text{Form factor} = \frac{I_{r.m.s.}}{I_{av}} = \frac{I_m/\sqrt{2}}{(2/\pi)\,I_m} = \mathbf{1.11}$$

$$\text{Peak factor} = \frac{I_m}{I_{r.m.s.}} = \frac{I_m}{I_m/\sqrt{2}} = \mathbf{1.414}$$

Example 15.8. *An alternating voltage $v = 200 \sin 314t$ is applied to a device which offers an ohmic resistance of 20 Ω to the flow of current in one direction while entirely preventing the flow of current in the opposite direction. Calculate the r.m.s. value, average value and form factor.*

Solution. It is clear that the device is doing half-wave rectification. The maximum value of the rectified current is

$$I_m = E_m/R = 200/20 = 10\text{ A}$$

For a half-wave rectified a.c. (See example 15.7),

* Area of sinusoidal squared wave over one alternation is $= \pi I_m^2/2$ (See Art 15.11).

$$I_{r.m.s.} = I_m/2 = 10/2 = \mathbf{5\ A}$$

$$I_{av} = I_m/\pi = 10/\pi = \mathbf{3.18\ A}$$

$$\text{Form factor} = I_{r.m.s.}/I_{av} = 5/3.18 = \mathbf{1.57}$$

Example 15.9. *A moving coil ammeter, a thermal ammeter and a half-wave rectifier are connected in series with a resistor across 110 V a.c. supply. The circuit offers a resistance of 50Ω in one direction and an infinite resistance in the reverse direction. Calculate (i) the readings on the ammeters and (ii) the form factor and peak factor of current wave. Assume the supply voltage to be sinusoidal.*

Solution.

Max. value of voltage, E_m = 110/0.707 = 155.59V

Max. value of current, I_m = 155.59/50 = 3.11 A

The moving coil instrument will indicate the average value over the whole cycle while the thermal ammeter will show the r.m.s. value over the whole cycle. Since it is a case of half-wave rectified alternating current, the various values are (See example 15.7) :

(*i*) $I_{av} = I_m/\pi = 3.11/\pi = \mathbf{0.99A}$

Hence, moving coil instrument will read 0.99 A.

$$I_{r.m.s.} = I_m/2 = 3.11/2 = \mathbf{1.555\ A}$$

Hence, thermal ammeter will read 1.555A

(*ii*) Form factor = 1.555/0.99 = **1.57**

Peak factor = 3.11/1.555 = **2**

Example 15.10. *A current has the following steady values in amperes for equal intervals of time changing instantaneously from one value to the next :*

0, 10, 20, 30, 20, 10, 0, –10, –20, –30, –20, –10, 0 etc. Calculate (i) average value (ii) r.m.s. value (iii) form factor and (iv) peak factor.

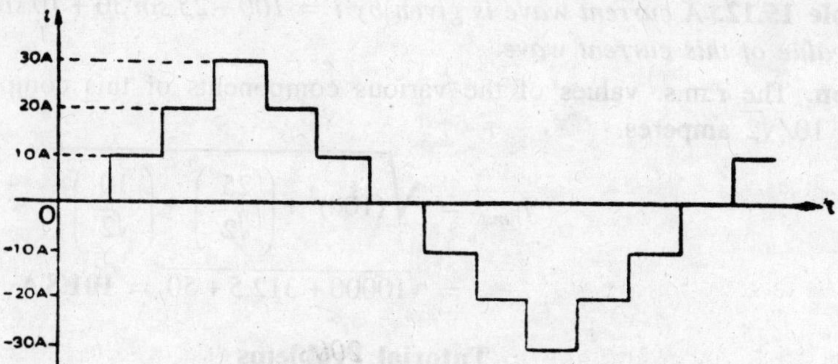

Fig. 15.14

Solution.

(*i*) $$I_{av} = \frac{i_1 + i_2 + i_3 + i_4 + i_5 + i_6}{6}$$

$$= \frac{0 + 10 + 20 + 30 + 20 + 10}{6} = \frac{90}{6} = \mathbf{15A}$$

(*ii*) $$I^2_{r.m.s.} = \frac{i_1^2 + i_2^2 + i_3^2 + i_4^2 + i_5^2 + i_6^2}{6} = \frac{1900}{6} = 316.67$$

or $$I_{r.m.s.} = \sqrt{316.67} = \mathbf{17.8\ A}$$

(*iii*) $$\text{Form factor} = \frac{I_{r.m.s.}}{I_{av}} = \frac{17.8}{15} = \mathbf{1.19}$$

(*iv*) $$\text{Peak factor} = \frac{I_m}{I_{r.m.s.}} = \frac{30}{17.8} = \mathbf{1.68}$$

Example 15.11. *A voltage wave is given by equation* $v = 50 + 30 \sin\theta$ *volts. Find the r.m.s. value of the wave.*

Solution.

$$E^2_{r.m.s.} = \frac{1}{2\pi}\int_0^{2\pi} (50 + 30\sin\theta)^2\, d\theta$$

$$= \frac{1}{2\pi}\int_0^{2\pi} (2500 + 3000\sin\theta + 900\sin^2\theta)\, d\theta$$

$$= \frac{1}{2\pi}\int_0^{2\pi} 2500\, d\theta + \frac{1}{2\pi}\int_0^{2\pi} 3000\sin\theta\, d\theta + \frac{1}{2\pi}\int_0^{2\pi} 900\sin^2\theta\, d\theta$$

$$= 2500 + {}^*0 + \frac{900}{2} = 2950$$

$$\therefore \quad E_{r.m.s.} = \sqrt{2950} = \mathbf{54.3\ V}$$

Another method

$$E_{r.m.s.} = \sqrt{50^2 + (30/\sqrt{2})^2} = \sqrt{2950} = 54.3\ \text{V}$$

Note. The r.m.s. value of a complex wave is equal to the square root of the sum of the squares of the r.m.s. values of its individual components.

Example 15.12. *A current wave is given by* $i = 100 + 25\sin 3\theta + 10\sin 5\theta$ *amperes. Find the r.m.s. value of this current wave.*

Solution. The r.m.s. values of the various components of this complex wave are 100, $25/\sqrt{2}$ and $10/\sqrt{2}$ amperes.

$$\therefore \quad I_{r.m.s.} = \sqrt{(100)^2 + \left(\frac{25}{\sqrt{2}}\right)^2 + \left(\frac{10}{\sqrt{2}}\right)^2}$$

$$= \sqrt{10000 + 312.5 + 50} = \mathbf{101.8\ A}$$

Tutorial Problems

1. An alternating voltage $e = 100 \sin 314t$ is applied to a device which offers an ohmic resistance of 50 Ω in one direction while entirely preventing the flow of current in the opposite direction. Calculate the average and r.m.s. values of current. Also find the form factor. **[0.637A ; 1A ; 1.57]**

2. A hot-wire ammeter, a moving coil ammeter and a rectifier are connected in series. The combination is connected across a sinusoidal source of 50V r.m.s. The forward resistance of rectifier is 20 Ω and the reverse resistance is infinite. Calculate the reading on each ammeter. **[1.767A ; 1.124A]**

---

* Remember average value of $\sin\theta$ over whole cycle is zero. Also the average value of $\sin^2\theta$ over the whole cycle is 1/2. These facts can be ascertained by carrying out the actual integration.

3. The equation of an alternating current is given by $i = 141.4 \sin 314t$. Find (*i*) r.m.s. current (*ii*) frequency and (*iii*) instantaneous value of current when t is 3.6*ms*. **[100A ; 50Hz ; 128A]**

4. A current has the following steady values (in amperes) for equal intervals of time, changing instantaneously from one value to the next.
0, 10, 20, 30, 20, 10, 0, −10, −20, −30, −20, −10 etc.
Calculate the r.m.s. value of current and form factor. **[17.8A; 1.18]**

15.13. Phase

Waves of alternating voltage and current are continuous. They do not stop after one cycle is completed but continue to repeat as long as the generator is operating. Consider an alternating voltage wave of time period T second as shown in Fig. 15.15. Note that the time is counted from the instant the voltage is zero and becoming positive. The maximum positive value $(+V_m)$ occurs at $T/4$ second or $\pi/2$ radians. We say that *phase* of maximum positive value is $T/4$ second or $\pi/2$ radians. It means that as the fresh cycle starts, $+V_m$ will occur at $T/4$ second or $\pi/2$ radians. Similarly, the phase of negative peak $(-V_m)$ is $3T/4$ second or $3\pi/2$ radians.

Hence **phase** *of a particular value of an alternating quantity is the fractional part of time period or cycle through which the quantity has advanced from the selected zero position of reference.*

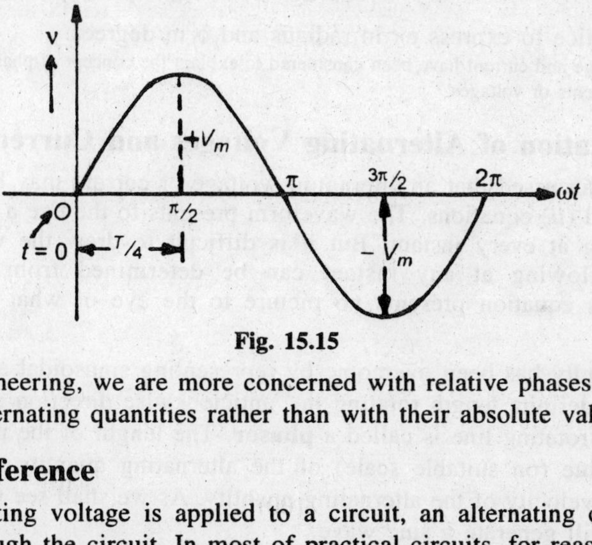

Fig. 15.15

In electrical engineering, we are more concerned with relative phases or phase difference between different alternating quantities rather than with their absolute values.

15.14 Phase Difference

When an alternating voltage is applied to a circuit, an alternating current of the same frequency flows through the circuit. In most of practical circuits, for reasons we will discuss later, voltage and current have different phases. In other words, they do not pass through a particular point, say *zero point in the same direction at the same instant. Thus voltage may be passing through its zero point while the current has passed or it is yet to pass through its zero point in the same direction. We say that voltage and current have a phase difference.

Hence when two alternating quantities of the same frequency have different zero points, they are said to have a **phase difference.**

The angle between zero points is the angle of phase difference ϕ. It is generally measured in degrees or radians. The quantity which passes through its zero point earlier is said to be leading while the other is said to be lagging. It should be noted that those zero points of

* We may select any point but it is more convenient to determine the phase difference by considering zero points or positive maximum values or negative maximum values of the two alternating quantities.

alternating quantities are to be considered where they pass in the same direction. Thus if voltage has passed through its zero point and is rising in the positive direction, then zero point considered for the current should have similar situation. Since both alternating quantities have the same frequency, the phase difference between them remains the same.

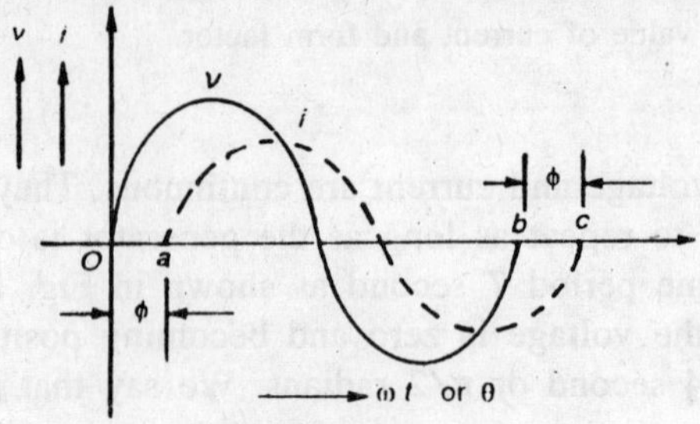

Fig. 15.16

Consider an a.c. circuit in which current i lags behind the voltage v by $\phi°$. This phase relationship is shown by waves in Fig. 15.16. The equations of voltage and current are :

$$v = V_m \sin \omega t$$

$$i = I_m \sin (\omega t - \phi)$$

It is a usual practice to express ωt in radians and ϕ in degrees.

Note. Although voltage and current have been considered to explain the concept of phase difference, it is equally valid for two or more currents or voltages.

15.15. Representation of Alternating Voltages and Currents

So far we have discussed that an alternating voltage or current may be represented in the form of (*i*) waves and (*ii*) equations. The waveform presents to the eye a very definite picture of what is happening at every instant. But it is difficult to draw the wave accurately. No doubt the current flowing at any instant can be determined from the equation form $i = I_m \sin \omega t$ but this equation presents no picture to the eye of what is happening in the circuit.

The above difficulty has been overcome by representing sinusoidal alternating voltage or current by a line of definite length rotating in *anticlockwise direction at a constant angular velocity (ω). Such a rotating line is called a **phasor**. The length of the phasor is taken equal to the maximum value (on suitable scale) of the alternating quantity and angular velocity equal to the angular velocity of the alternating quantity. As we shall see presently, this phasor (*i.e.*, rotating line) will generate a sine wave.

15.16. Phasor Representation of Sinusoidal Quantities

Consider an alternating current represented by the equation $i = I_m \sin \omega t$. Take a line *OP* to represent to scale the maximum value I_m. Imagine the line *OP* (or ***phasor*, as it is called) to be rotating in anticlockwise direction at an angular velocity ω rad/sec about the point *O*. Measuring the time from the instant when *OP* is horizontal, let *OP* rotate through an angle θ (= ωt) in the anticlockwise direction. The projection of *OP* on *Y*-axis is *OM*.

$$OM = OP \sin\theta$$

* It is a standard convention that the phasor is rotated anticlockwise—a convention that is in harmony with the general use of polar co-ordinates.

** An arrowhead is drawn at the outer end of phasor *OP* partly to indicate which end is assumed to move and partly to indicate precise length of the phasor when two or more phasors happen to coincide.

$= I_m \sin \omega t$

$= i$, the value of current at that instant.

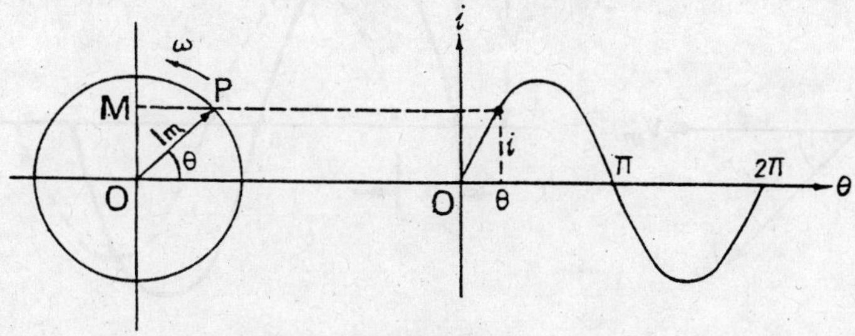

Fig. 15.17

Hence the projection of the phasor OP on Y-axis at any instant gives the value of current at that instant. Thus when $\theta = 90°$, the projection on Y-axis is OP ($= I_m$) itself. That the value of current at this instant (*i.e.*, at θ or $\omega t = 90°$) is I_m can be readily established if we put $\theta = 90°$ in the current equation. If we plot the projections of the phasor on Y-axis *versus* its angular position point-by-point, a sinusoidal alternating current wave is generated as shown in Fig. 15.17. Thus the phasor represents the sine wave for every instant of time.

The following points are worth noting :

(*i*) The length of the phasor represents the maximum value and the angle with axis of reference (*i.e.*, X-axis) indicates the phase of the alternating quantity *i.e.*, current in this case.

(*ii*) The phasor representation enables us to quickly obtain the numerical values and, at the same time, have a picture before the eye of the events taking place in the circuit. Thus in the position of the phasor OP shown in Fig.15.17, the instantaneous value is OM, the phase is θ and frequency is $\omega/2\pi$.

(*iii*) A phasor diagram permits addition and subtraction of alternating voltages or currents with a fair degree of ease.

Note. Alternating voltages and currents are not vector quantities. Voltage is simply energy or work per coulomb and cannot be classified as a vector. Current is also not a vector quantity because it is merely the flow of electrons through a wire. When we insert an ammeter in a circuit to measure current or connect a voltmeter between two points to measure the potential difference (*i.e.*, voltage), direction with reference to any set of axes is of no consequence. Therefore, neither alternating voltage nor current is a vector quantity. Instead, they are *phasors.

15.17. Phasor Diagram of Sine Waves of Same Frequency

Consider a sinusoidal voltage wave v and sinusoidal current i of the same frequency. Suppose the current lags behind the voltage by $\phi°$. The two alternating quantities can be represented on the same phasor diagram because the phasors V_m and I_m [See Fig. 15.18 (*i*)] rotate at the same angular velocity ω** and hence phase difference ϕ between them remains the same at all times. When each phasor completes one revolution, it generates the corresponding cycle [See Fig. 15.18 (*ii*)]. The equations of the two waves can be represented as :

---

* Remember that a vector has space co-ordinates while a phasor is derived from the time varying sinusoid.

** Remember $\omega = 2\pi f$

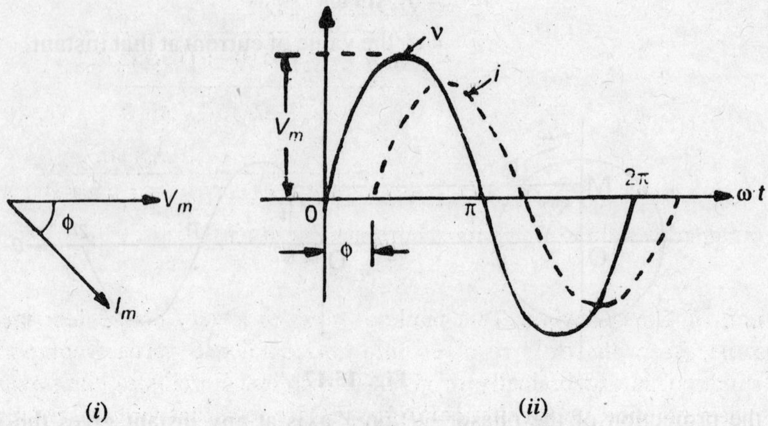

Fig. 15.18

$$v = V_m \sin \omega t$$

$$i = I_m \sin (\omega t - \phi)$$

15.18. Addition of Alternating Quantities

Alternating voltages and currents are phasors. They are added in the same manner as forces are added. Only phasors of the same kind may be added. Common sense tells us that we should not try to add volts to amperes. Addition of alternating currents or voltages can be accomplished by one of the following two methods :

1. Parallelogram method 2. Method of components.

1. Parallelogram method. This method is used for the addition of two phasors at a time. The two phasors are represented in magnitude and direction by the adjacent sides of a parallelogram. Then the diagonal of the parallelogram represents the maximum value of the resultant.

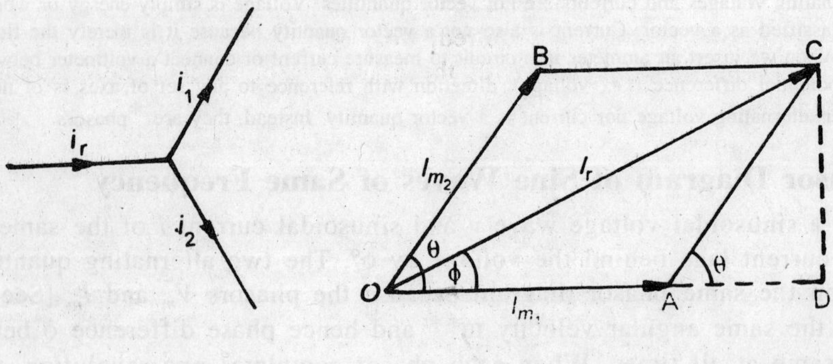

Fig. 15.19

Consider two alternating currents i_1 and i_2 flowing in the two branches of a circuit [See Fig. 15.19 (*i*)]. Let they be represented by :

$$i_1 = I_{m1} \sin \omega t$$

$$i_2 = I_{m2} \sin (\omega t + \theta)$$

$$OC = \sqrt{(OD)^2 + (CD)^2}$$

$$= \sqrt{(I_{m1} + I_{m2} \cos\theta)^2 + (I_{m2} \sin\theta)^2}$$

$$\therefore \quad I_r = \sqrt{I_{m1}^2 + I_{m2}^2 + 2I_{m1} I_{m2} \cos\theta}$$

Also $$\tan\phi = \frac{CD}{OD} = \frac{CD}{OA + AD} = \frac{I_{m2} \sin\theta}{I_{m1} + I_{m2} \cos\theta}$$

The instantaneous value of resultant current i_r is given by :

$$i_r = I_r \sin(\omega t + \phi)$$

2. Method of Components. This method provides a very convenient means to add two or more phasors. Each phasor is resolved into horizontal and vertical components. The horizontals are summed up algebraically to give the resultant horizontal component X. The verticals are likewise summed up algebraically to give the resultant vertical component Y.

Then, Resultant $= \sqrt{X^2 + Y^2}$

Phase angle of resultant, $\tan\phi = Y/X$

Thus referring to Fig. 15.19 (*ii*), we have,

$$X = I_{m1} + I_{m2} \cos\theta$$

$$Y = 0 + I_{m2} \sin\theta$$

$$\therefore \quad \text{Resultant,} \quad I_r = \sqrt{(I_{m1} + I_{m2} \cos\theta)^2 + (I_{m2} \sin\theta)^2}$$

$$= \sqrt{I_{m1}^2 + I_{m2}^2 + 2I_{m1} Im_2 \cos\theta}$$

which is the same as derived by parallelogram method.

$$\tan\phi = \frac{Y}{X} = \frac{I_{m2} \sin\theta}{I_{m1} + I_{m2} \cos\theta}$$

$$\therefore \quad i_r = I_r \sin(\omega t + \phi)$$

15.19. Subtraction of Alternating Quantities

If difference of two phasors is required, then one of the phasors is reversed and this reversed phasor is then compounded with the other phasor using parallelogram method or method of components.

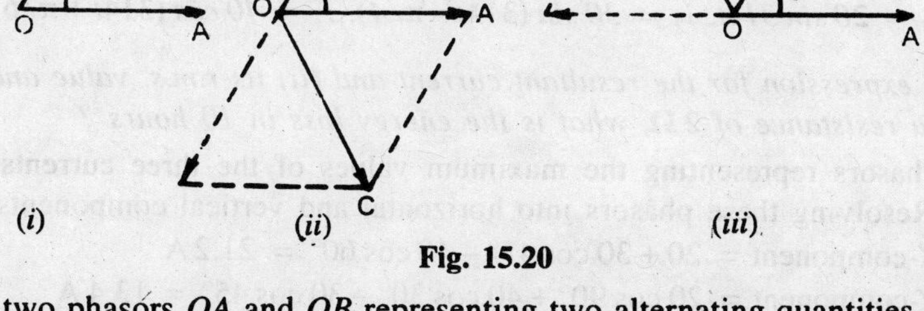

Fig. 15.20

Consider two phasors OA and OB representing two alternating quantities of the same kind [See Fig. 15.20(*i*)]. The phasor OB leads the phasor OA by θ. If it is required to subtract the phasor OB from OA, then OB is reversed and is compounded with phasor OA as shown in

Fig. 15.20 (*ii*). The phasor difference *OA*–*OB* is given by the phasor *OC*. In Fig. 15.20 (*iii*), phasor *OC* represents the phasor difference *OB*–*OA*.

Example 15.13. *The following expressions represent the instantaneous values of e.m.f. in three coils connected in series :* $e_1 = 50 \sin \omega t$; $e_2 = 40 \sin (\omega t + 60°)$; $e_3 = 60 \sin (\omega t - 30°)$. *Find an expression for the resultant e.m.f. when the coils are connected to give the sum of three e.m.fs.*

Solution. Phasors representing the maximum values of the three e.m.fs. are shown in Fig. 15.21 (*i*). Resolving these phasors into horizontal and vertical components, we get,

$$X\text{-component} = \text{Algebraic sum of components along } X\text{-axis}$$
$$= 50 + 40 \cos 60° + 60 \cos 30°$$
$$= 50 + 20 + 51.9 = 121.9 \text{ V}$$
$$Y\text{-component} = \text{Algebraic sum of components along } Y\text{-axis}$$
$$= {}^*50 \cos 90° + 40 \cos 30° - 60 \cos 60°$$
$$= 0 + 34.6 - 30 = 4.6 \text{ V}$$

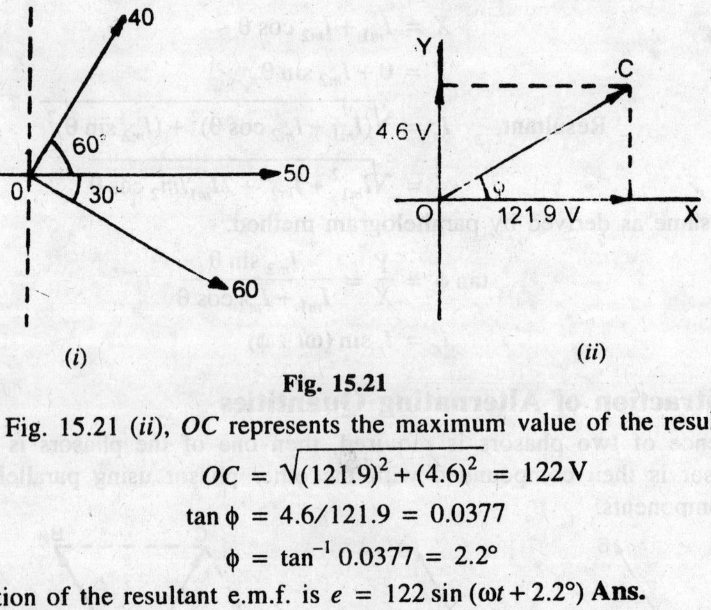

Fig. 15.21

As shown in Fig. 15.21 (*ii*), *OC* represents the maximum value of the resultant e.m.f.

$$OC = \sqrt{(121.9)^2 + (4.6)^2} = 122 \text{ V}$$
$$\tan \phi = 4.6/121.9 = 0.0377$$
$$\therefore \quad \phi = \tan^{-1} 0.0377 = 2.2°$$

$\therefore$ The equation of the resultant e.m.f. is $e = 122 \sin (\omega t + 2.2°)$ **Ans.**

Example 15.14. *Three circuits in parallel take the following currents :*

$$i_1 = 20 \sin 314t, \; i_2 = 30 \sin (314t - \pi/4), \; i_3 = 40 \cos (314t + \pi/6)$$

Find (i) the expression for the resultant current and (ii) its r.m.s. value and frequency. If the circuit has a resistance of 2 Ω, what is the energy loss in 10 hours ?

Solution. Phasors representing the maximum values of the three currents are shown in Fig. 15.22 (*i*). Resolving these phasors into horizontal and vertical components, we get,

$$X\text{-component} = 20 + 30 \cos 45° - 40 \cos 60° = 21.2 \text{ A}$$
$$Y\text{-component} = 20 \cos 90° + 40 \cos 30° - 30 \cos 45° = 13.4 \text{ A}$$

---

* Remember that rectangular component of a phasor *OA* (or a vector) along a direction

= *OA* × cosine of angle between *OA* and that direction

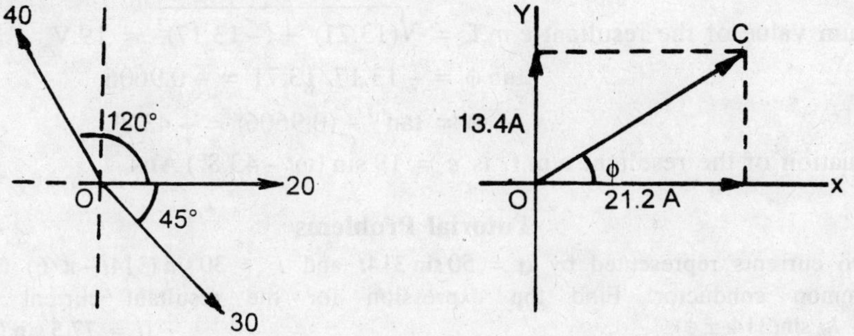

(i) (ii)

Fig. 15.22

(i) As shown in Fig. 15.22 (ii), OC represents the maximum value of the resultant current.

$$OC = \sqrt{(21.2)^2 + (13.4)^2} = 25.1\text{A}$$

$$\tan\phi = 13.4/21.2 = 0.632$$

$$\therefore \quad \phi = \tan^{-1} 0.632 = 32.3°$$

The equation of the resultant current is $i = 25.1 \sin(314t + 32.3°)$ **Ans.**

(ii) R.M.S. value of current, $I_{r.m.s.} = 25.1/\sqrt{2} = \mathbf{17.75\,A}$

Frequency $= \omega/2\pi = 314/2\pi = \mathbf{50\,Hz}$

The frequency of the resultant current is the same as that of the three currents.

Energy loss in 10 hours $= I^2_{r.m.s.} \times R \times t$

$$= (17.75)^2 \times 2 \times 10 = \mathbf{6301.25\,Wh.}$$

Example 15.15. *The following four e.m.fs. act together in a circuit :*

$$e_1 = 10 \sin \omega t\,; \qquad e_2 = 8 \sin(\omega t + \pi/3)$$

$$e_3 = 4 \sin(\omega t - \pi/6)\,; \qquad e_4 = 6 \sin(\omega t + 3\pi/4)$$

Calculate the e.m.f. represented by $e_1 - e_2 + e_3 - e_4$.

Solution. Phasors representing the maximum values of the four e.m.fs. are shown in Fig. 15.23 (*i*). In order to determine $e_1 - e_2 + e_3 - e_4$, reverse the phasors representing e_2 and e_4 as shown in Fig. 15.23 (*ii*). Referring to Fig. 15.23 (*ii*) and resolving the phasors into horizontal and vertical components, we get,

$$X\text{-component} = 10 \cos 0° + 4 \cos 30° - 8 \cos 60° + 6 \cos 45°$$

$$= 10 + 3.464 - 4 + 4.243 = 13.71\,\text{V}$$

$$Y\text{-component} = 10 \cos 90° - 4 \cos 60° - 6 \cos 45° - 8 \cos 30°$$

$$= 0 - 2 - 4.243 - 6.928 = -13.17\,\text{V}$$

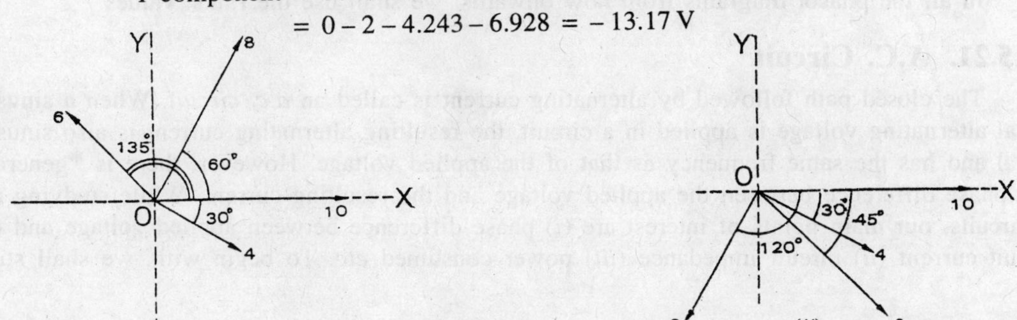

(i) (ii)

Fig. 15.23

Maximum value of the resultant e.m.f. $= \sqrt{(13.71)^2 + (-13.17)^2} = 19$ V

$$\tan\phi = -13.17/13.71 = -0.9606$$

$$\therefore \qquad \phi = \tan^{-1} - (0.9606) = -43.8°$$

The equation of the resultant e.m.f. is $e = 19 \sin(\omega t - 43.8°)$ **Ans.**

Tutorial Problems

1. Two currents represented by $i_1 = 50 \sin 314t$ and $i_2 = 30 \sin(314t - \pi/6)$ are fed into a common conductor. Find the expression for the resultant current in the form $i = I_m \sin(314t \pm \phi)$. **[$i$ = 77.5 sin (314t – 11°10′)]**
2. Find the sum of the following two e.m.fs. and express the answer in the similar form :
$$e_1 = 100 \sin \omega t \; ; \; e_2 = 100 \cos \omega t$$ **[e = 141.4 sin (ωt + 45°)]**
3. The following four currents are fed into a common conductor :
$$i_1 = 50 \sin \omega t \quad ; \quad i_3 = 40 \cos \omega t$$
$$i_2 = 25 \sin(\omega t + \pi/3) \quad ; \quad i_4 = 30 \sin(\omega t - \pi/2)$$
Find the expression for the resultant in the form $i = I_m \sin(\omega t \pm \phi)$. **[$i$ = 94 sin (ωt + 25°)]**
4. The instantaneous voltages across each of four series connected coils are given by :
$v_1 = 100 \sin \omega t$; $v_2 = 250 \cos \omega t$; $v_3 = 150 \sin(\omega t + \pi/6)$; $v_4 = 200 \sin(\omega t - \pi\sqrt{4})$.
Find the total p.d. and express the answer in the similar form. What will be the resultant p.d. if the polarity of v_2 is reversed ? **[v = 414 sin (ωt + 26.5°) ; v = 486 sin (ωt – 40°)]**

15.20. Phasor Diagrams Using r.m.s. Values

Instead of drawing the phasor diagram using maximum values, it is a common practice to draw it using r.m.s. values. This does not alter the phase difference (ϕ) between the phasors because only the lengths of the phasors are changed. Fig. 15.24(*i*) shows the phasor diagram using maximum values while Fig. 15.24 (*ii*) shows the same phasor diagram in terms of r.m.s. equivalent.

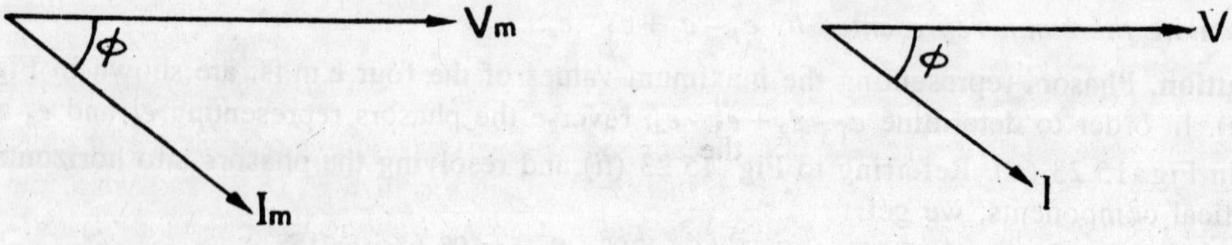

(*i*) (*ii*)

Fig. 15.24

In all the phasor diagrams from now onwards, we shall use the r.m.s. values.

15.21. A.C. Circuit

The closed path followed by alternating current is called an *a.c. circuit*. When a sinusoidal alternating voltage is applied in a circuit, the resulting alternating current is also sinusoidal and has the same frequency as that of the applied voltage. However, there is *generally a phase difference between the applied voltage and the resulting current. While studying a.c. circuits, our main points of interest are (*i*) phase difference between applied voltage and circuit current (*ii*) circuit impedance (*iii*) power consumed *etc*. To begin with, we shall study

---

* Except the case when the circuit contains resistance only.

these characteristies for simple a.c. circuits containing one circuit element only (R or L or C) and extend our discussion to the combination of these circuit elements in the later chapters.

15.22. A.C. Circuit Containing Resistance Only

Consider a circuit containing a pure resistance of $R\,\Omega$ connected across an alternating voltage source. Let the alternating voltage be given by the equation :

$$v = V_m \sin \omega t \quad \text{...(i)}$$

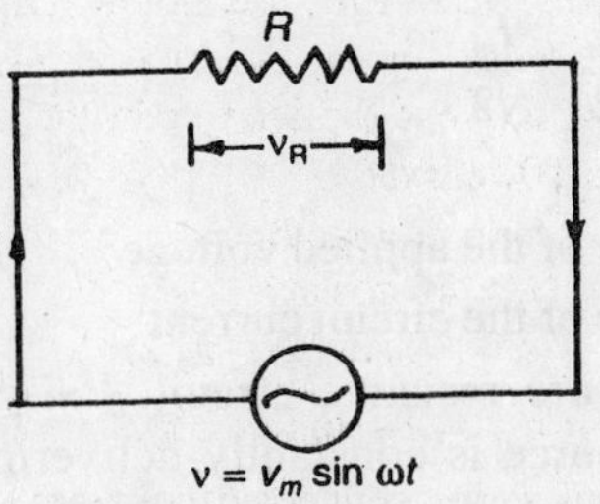

Fig. 15.25

Fig. 15.26

As a result of this voltage, an alternating current i will flow in the circuit. The applied voltage has to overcome the drop in the resistance only *i.e.*,

$$v = iR$$

or
$$i = \frac{v}{R}$$

Substituting the value of v, we get,

$$i = \frac{V_m}{R} \sin \omega t \quad \text{...(ii)}$$

The value of i will be maximum (*i.e.*, I_m) when $\sin \omega t = 1$

$$\therefore \quad I_m = V_m / R$$

$\therefore$ Eq. (*ii*) becomes,
$$i = I_m \sin \omega t \quad \text{...(iii)}$$

It is clear from eqs. (*i*) and (*iii*) that the applied voltage and circuit current are in phase with each other. This fact is also shown by the phasor diagram in Fig. 15.26 and wave diagram in Fig. 15.27. Note that r.m.s. values have been used in drawing the phasor diagram.

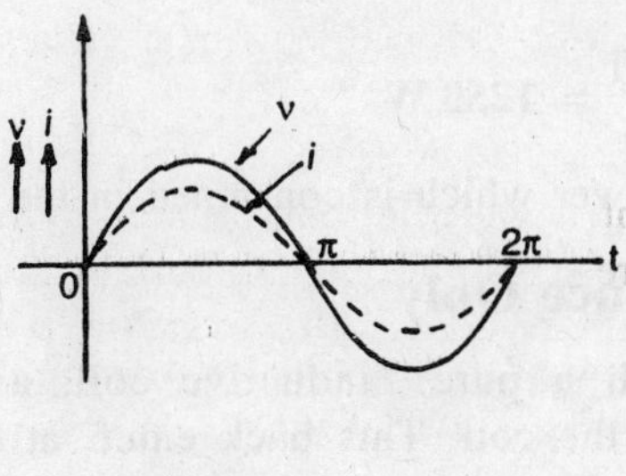

Fig. 15.27

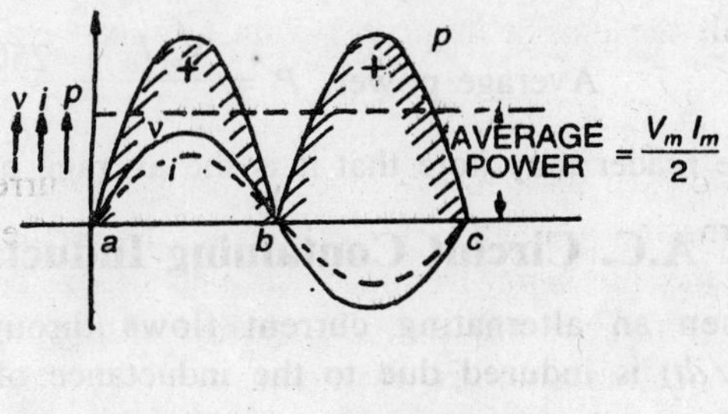

Fig. 15.28

Power

Instantaneous power, $p = vi = (V_m \sin \omega t)(I_m \sin \omega t) = V_m I_m \sin^2 \omega t$

$$= V_m I_m \frac{(1 - \cos 2\omega t)}{2}$$

$$= \frac{V_m I_m}{2} - \frac{V_m I_m}{2} \cos 2\omega t$$

Thus power consists of two parts *viz.* a constant part $(V_m I_m / 2)$ and a fluctuating part $(V_m I_m / 2) \cos 2\omega t$. Since power is a scalar quantity, average power over a complete cycle is to be considered. For a complete cycle, the average value of $(V_m I_m / 2) \cos 2\omega t$ is zero.

$\therefore$ Power consumed, $\quad P = \dfrac{V_m I_m}{2} = \dfrac{V_m}{\sqrt{2}} \times \dfrac{I_m}{\sqrt{2}}$

or $\quad P = VI$

where V = r.m.s. value of the applied voltage

I = r.m.s. value of the circuit current

Fig. 15.28 shows the power curve for a pure resistive circuit. It is clear that power is always positive. This means that the voltage source is constantly delivering power to the circuit which is consumed by the circuit.

Example 15.16. *An a.c. circuit consists of a pure resistance of 10 Ω and is connected across an a.c. supply of 230 V, 50 Hz. Calculate (i) current (ii) power consumed and (iii) equations for voltage and current.*

Solution.

(*i*) Current, $\quad I = V/R = 230/10 = \mathbf{23\ A}$

(*ii*) Power, $\quad P = VI = 230 \times 23 = \mathbf{5290\ W}$

(*iii*) Now, $\quad V_m = \sqrt{2}\, V = \sqrt{2} \times 230 = 325.27$ volts

$I_m = \sqrt{2}\, I = \sqrt{2} \times 23 = 32.52$ A

$\omega = 2\pi f = 2\pi \times 50 = 314$

$\therefore$ Equations of voltage and current are :

$$e = 325.27 \sin 314t\,; \quad i = 32.52 \sin 314t$$

Example 15.17. *In a pure resistive circuit, the instantaneous voltage and current are given by :*

$$v = 250 \sin 314t\,; \quad i = 10 \sin 314t.$$

Determine (i) the peak power and (ii) average power.

Solution.

In a pure resistive a.c. circuit,

(*i*) Peak power $= V_m I_m = 250 \times 10 = \mathbf{2500\ W}$

(*ii*) Average power, $P = \dfrac{V_m I_m}{2} = \dfrac{2500}{2} = \mathbf{1250\ W}$

The reader may note that it is the average power which is consumed in the circuit.

15.23. A.C. Circuit Containing Inductance Only

When an alternating current flows through a pure *inductive coil, a back e.m.f. (= $L\, di/dt$) is induced due to the inductance of the coil. This back e.m.f. at every instant

---

* Any circuit that is capable of producing flux has inductance as was pointed out in chapter 9. When alternating current flows through such a circuit, there is change in flux linking it and hence back e.m.f. (= $L\, di/dt$) is induced in the circuit. This e.m.f. opposes the applied voltage at every instant.

opposes the change in current through the coil. Since there is no ohmic drop, the applied voltage has to overcome the back e.m.f. only.

$\therefore$ Applied alternating voltage = Back e.m.f.

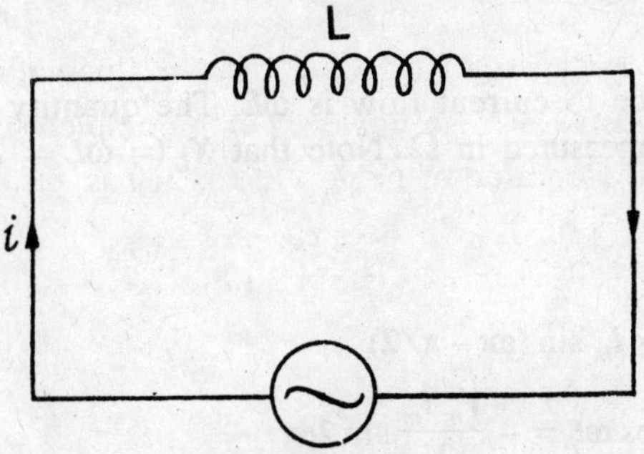

Fig. 15.29

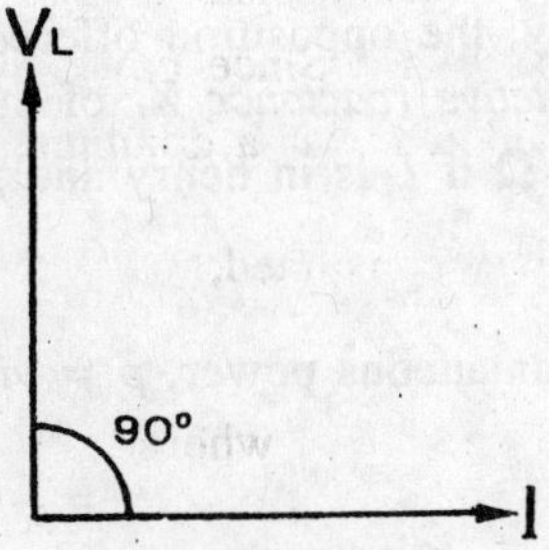

Fig. 15.30

Consider an alternating voltage applied to a pure inductance of L henry as shown in Fig. 15.29. Let the equation of the applied alternating voltage be :

$$v = V_m \sin \omega t \qquad ...(i)$$

Clearly, $V_m \sin \omega t = L \dfrac{di}{dt}$

or $di = \dfrac{V_m}{L} \sin \omega t \, dt$

Integrating both sides, we get,

$$i = \frac{V_m}{L} \int \sin \omega t \, dt = \frac{V_m}{\omega L} (-\cos \omega t)$$

or $$i = \frac{V_m}{\omega L} \sin (\omega t - \pi/2) \qquad ...(ii)$$

The value of i will be maximum (*i.e.*, I_m) when $\sin (\omega t - \pi/2)$ is unity.

$\therefore$ $I_m = V_m / \omega L$

$\therefore$ Eq. (*ii*) becomes, $i = I_m \sin (\omega t - \pi/2)$ where $I_m = V_m/\omega_L$ ...(*iii*)

It is clear from eqs. (*i*) and (*iii*) that current lags behind the voltage by $\pi/2$ radians or 90°. *Hence in a pure inductance, current lags behind the voltage by 90°.* This fact is also shown in the phasor diagram in Fig. 15.30 and wave diagram in Fig. 15.31.

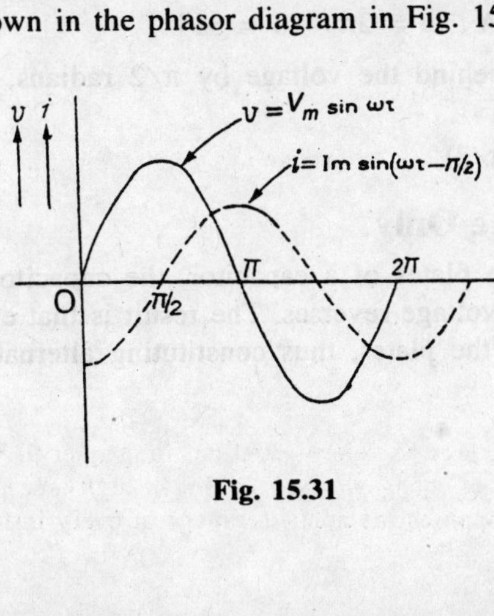

Fig. 15.31

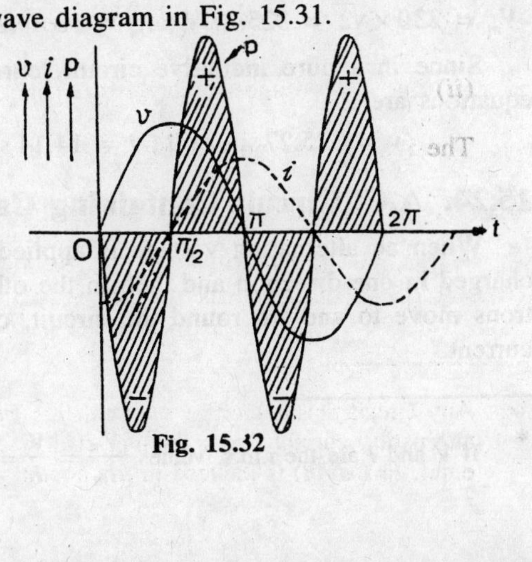

Fig. 15.32

Inductive reactance. We have seen above that :

$$I_m = V_m / \omega L$$

or

$$\frac{{}^{*}V_m}{I_m} = \omega L$$

Clearly, the opposition offered by inductance to current flow is ωL. The quantity ωL is called *inductive reactance* X_L of the coil and is measured in Ω. Note that X_L (= $\omega L = 2\pi f L$) will be in Ω if L is in henry and f in Hz.

Power

Instantaneous power, $p = vi = V_m \sin \omega t \times I_m \sin(\omega t - \pi/2)$

$$= -V_m I_m \sin \omega t \cos \omega t = -\frac{V_m I_m}{2} \sin 2\omega t$$

$\therefore$ Average power, P = Average of p over one cycle

$$= \frac{1}{2\pi} \int_0^{2\pi} -\frac{V_m I_m}{2} \sin 2\omega t \, d(\omega t) = 0$$

Hence, power absorbed in pure inductance is zero.

Fig. 15.32 shows the power curve for a pure inductive circuit. An examination of power curve over one cycle shows that positive power is equal to the negative power. Hence the resultant power over one cycle is zero *i.e.*, pure inductance consumes no power. The electric power merely flows from the source to the coil and back again.

Example 15.18. *A pure inductive coil allows a current of 10 A to flow from a 230 V, 50 Hz supply. Find (i) inductive reactance (ii) inductance of the coil (iii) power absorbed. Write down the equations for voltage and current.*

Solution.

(*i*) Circuit current, $I = V/X_L$

$\therefore$ Inductive reactance, $X_L = V/I = 230/10 = \mathbf{23\ \Omega}$

(*ii*) Now, $X_L = 2\pi f L$

$$\therefore \quad L = \frac{X_L}{2\pi f} = \frac{23}{2\pi \times 50} = \mathbf{0.073\ H}$$

(*iii*) Power absorbed = **Zero**

$V_m = 230 \times \sqrt{2} = 325.27$ V ; $I_m = 10 \times \sqrt{2} = 14.14$ A ; $\omega = 2\pi \times 50 = 314$

Since in a pure inductive circuit, current lags behind the voltage by $\pi/2$ radians, the equations are :

$$v = 325.27 \sin 314t \; ; \; i = 14.14 \sin(314t - \pi/2)$$

15.24. A.C. Circuit Containing Capacitance Only

When an alternating voltage is applied across the plates of a capacitor, the capacitor is charged in one direction and then in the other as the voltage reverses. The result is that electrons move to and fro round the circuit, connecting the plates, thus constituting alternating current.

---

* If V and I are the r.m.s. values, $\frac{V_m}{I_m} = \frac{V}{I} = \omega L$

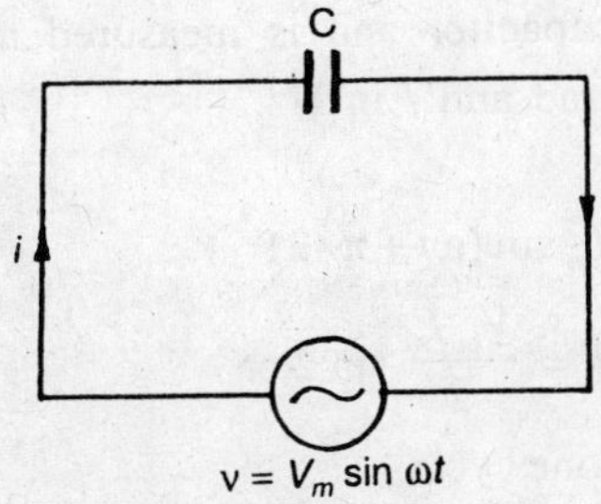

Fig. 15.33

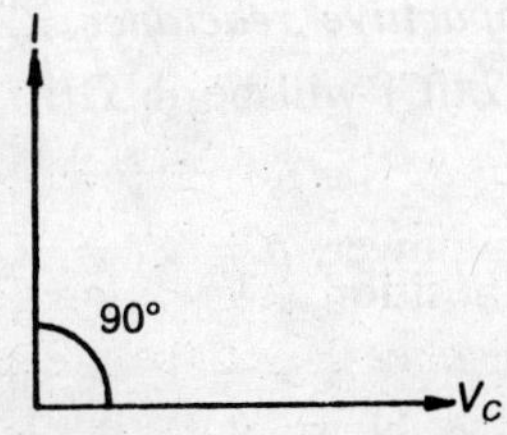

Fig. 15.34

Consider an alternating voltage applied to a capacitor of capacitance C farad as shown in Fig. 15.33. Let the equation of the applied alternating voltage be :

$$v = V_m \sin \omega t \qquad ...(i)$$

As a result of this alternating voltage, alternating current will flow through the circuit. Let at any instant i be the current and q be the charge on the plates.

Charge on capacitor, $q = C\,v = C\,V_m \sin \omega t$

$\therefore$ Circuit current, $i = \frac{d}{dt}(q) = \frac{d}{dt}(C\,V_m \sin \omega t) = \omega C\,V_m \cos \omega t$

or $$i = \omega C\,V_m \sin(\omega t + \pi/2) \qquad ...(ii)$$

The value of i will be maximum (*i.e.*, I_m) when $\sin(\omega t + \pi/2)$ is unity.

$\therefore$ $$I_m = \omega C\,V_m$$

Substituting the value $\omega C\,V_m = I_m$ in eq. (*ii*), we get,

$$i = I_m \sin(\omega t + \pi/2) \qquad ...(iii)$$

It is clear from eqs. (*i*) and (*iii*) that current leads the voltage by $\pi/2$ radians or 90°. *Hence in a pure capacitance, current leads the voltage by* 90°. This fact is also shown in the phasor diagram in Fig. 15.34 and wave diagram in Fig. 15.35.

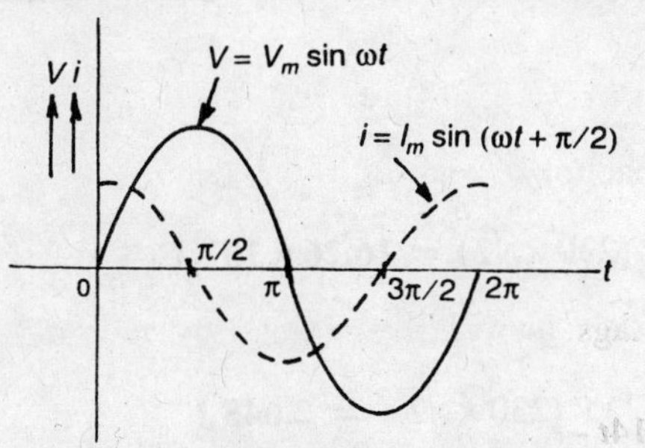

Fig. 15.35

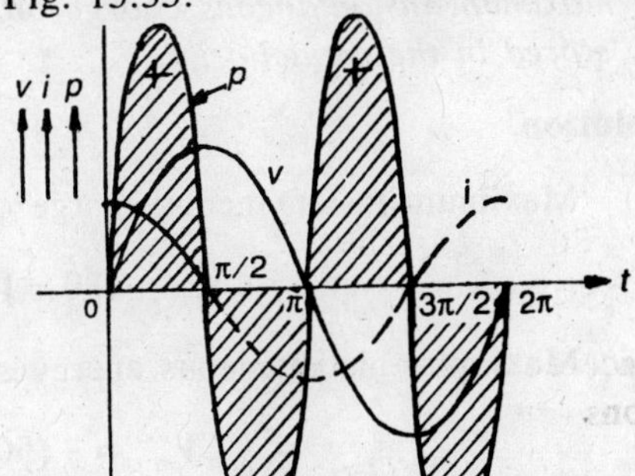

Fig. 15.36

Capacitive reactance. We have seen above that :

$$I_m = \omega C\,V_m$$

or $$\frac{{}^{*}V_m}{I_m} = \frac{1}{\omega C}$$

---

* If V and I are the r.m.s. values, $\frac{V_m}{I_m} = \frac{V}{I} = \frac{1}{\omega C}$

Clearly, the opposition offered by capacitance to current flow is $1/\omega C$. The quantity $1/\omega C$ is called *capacitive reactance* X_C of the capacitor and is measured in Ω. Note that $X_C (= 1/\omega C = 1/2\pi fC)$ will be in Ω if C is in farad and f in Hz.

Power

Instantaneous power, $p = v\,i = V_m \sin \omega t \times I_m \sin(\omega t + \pi/2)$

$$= V_m I_m \sin \omega t \cos \omega t = \frac{V_m I_m}{2} \sin 2\omega t$$

∴ Average power, P = Average of p over one cycle

$$= \frac{1}{2\pi} \int_0^{2\pi} \frac{V_m I_m}{2} \sin 2\omega t \; d(\omega t) = 0$$

Hence, power absorbed in a pure capacitance is zero.

Fig 15.36 shows the power curve for a pure capacitive circuit. The power curve is similar to that for a pure inductor because now current leads the voltage by 90°. It is clear that positive power is equal to the negative power over one cycle. Hence net power absorbed in a pure capacitor is zero.

Example 15.19. *A 318μF capacitor is connected across a 230V, 50Hz system. Determine (i) the capacitive reactance (ii) r.m.s. value of current and (iii) equations for voltage and current.*

Solution.

(*i*) Capacitive reactance, $X_C = \frac{1}{2\pi fC} = \frac{10^6}{2\pi \times 50 \times 318} = \mathbf{10\ \Omega}$

(*ii*) R.M.S. value of current, $I = V/X_C = 230/10 = \mathbf{23\ A}$

(*iii*) $V_m = 230 \times \sqrt{2} = 325.27$ volts ; $I_m = \sqrt{2} \times 23 = 32.53$ A ; $\omega = 2\pi \times 50 = 314$

∴ Equations for voltage and current are : $v = 325.27 \sin 314t$; $i = 32.53 \sin(314t + \pi/2)$

Example 15.20. *A 50μF capacitor is connected across a 230V, 50Hz supply. Determine (i) the maximum instantaneous charge on the capacitor and (ii) the maximum instantaneous energy stored in the capacitor.*

Solution.

(*i*) Maximum instantaneous charge on capacitor

$$= CV_m = (50 \times 10^{-6}) \times (230 \times \sqrt{2}) = \mathbf{16.26 \times 10^{-3}\ C}$$

(*ii*) Maximum instantaneous energy stored

$$= \tfrac{1}{2} CV_m^2 = \tfrac{1}{2}(50 \times 10^{-6}) \times (230 \times \sqrt{2})^2 = \mathbf{2.645\ J}$$

Example 15.21. *The instantaneous current in a pure inductance of 5H is expressed as* $i = 10 \sin(314t - \pi/2)$. *A capacitor is connected in parallel with the inductor. What should be the capacitance of the capacitor to receive the same amount of energy as inductance at the same terminal voltage ?*

Solution. The current through pure inductance is

$$i = 10 \sin(314t - \pi/2)$$

Here $\qquad I_m = 10$ A ; $\omega = 314$

Maximum energy stored in inductance

$$= \frac{1}{2} L I_m^2 = \frac{1}{2} \times 5 \times 10^2 = 250 \text{ J}$$

Now, $V_m = \omega L I_m = 314 \times 5 \times 10 = 15700$ volts

Maximum energy stored in the capacitor

$$= \frac{1}{2} C V_m^2 = \frac{1}{2} \times C \times (15700)^2$$

$\therefore$ $\frac{1}{2} \times C \times (15700)^2 = 250$

or $C = \frac{250 \times 2}{(15700)^2} = 2.028 \times 10^{-6} \text{ F} = \mathbf{2.028\ \mu F}$

OBJECTIVE QUESTIONS

1. The a.c. system is preferred to d.c. system because ....
   (*i*) a.c. voltages can be easily changed in magnitude
   (*ii*) d.c. motors do not have fine speed control
   (*iii*) high-voltage a.c. transmission is less efficient
   (*iv*) d.c. voltage cannot be used far domestic appliances
2. In a.c. system, we generate sinewave form because ....
   (*i*) it can be easily drawn
   (*ii*) it produces least disturbance in electrical circuits
   (*iii*) it is nature's standard
   (*iv*) other waves cannot be produced easily
3. .... will work only on d.c. supply.
   (*i*) Electric lamp (*ii*) Refrigerator
   (*iii*) Heater (*iv*) Electroplating
4. .... will produce a.c. voltage.
   (*i*) Friction
   (*ii*) Photoelectric effect
   (*iii*) Thermal energy (*iv*) Crystal
5. A coil is rotating in the uniform field of an 8-pole generator. In one revolution of the coil, the number af cycles generated by the voltage is ....
   (*i*) one (*ii*) two
   (*iii*) four (*iv*) eight
6. An alternating voltage is given by $v = 20$ sin 157 *t*. The frequency of the alternating voltage is ....
   (*i*) 50 Hz (*ii*) 25 Hz
   (*iii*) 100 Hz (*iv*) 75 Hz
7. An alternating current is given by $i = 10$ sin 314 *t*. The time taken to generate two cycles of current is ....
   (*i*) 0.02 second (*ii*) 0.01 second
   (*iii*) 0.04 second (*iv*) 0.05 second
8. An alternating voltage is given by $v = 30$ sin 314 *t*. The time taken by the voltage to reach – 30V for the first time is ....
   (*i*) 0.02 second (*ii*) 0.1 second
   (*iii*) 0.03 second (*iv*) 0.015 second
9. A sine wave has a maximum value of 20 V. Its value at 135° is ....
   (*i*) 10 V (*ii*) 14.14 V
   (*iii*) 15 V (*iv*) 5 V
10. A sinusoidal current has a magnitude of 3A at 120°. Its maximum value will be ....
    (*i*) $\sqrt{3}$ A (*ii*) $\sqrt{3}/2$ A
    (*iii*) $2\sqrt{3}$ A (*iv*) 6 A
11. An alternating current is given by $i = 10$ sin 314 *t*. Measuring time from $t = 0$, the time taken by the current to reach + 10 V for the second time is ....
    (*i*) 0.05 second (*ii*) 0.1 second
    (*iii*) 0.025 second (*iv*) 0.02 second
12. An. a.c. generator having 10 poles and running at 600 r.p.m. will generate an alternating voltage of frequency ....
    (*i*) 25 Hz (*ii*) 100 Hz
    (*iii*) 50 Hz (*iv*) 200 Hz
13. We have assigned a frequency of 50 Hz to power system because it ....
    (*i*) can be easily obtained
    (*ii*) gives best result when used for operating both lights and machinery
    (*iii*) leads to easy calculations
    (*iv*) none of the above
14. An alternating voltage is given by $v = 100$ sin 314 *t* volts. Its average value will be............
    (*i*) 70.7 V (*ii*) 50 V
    (*iii*) 63.7 V (*iv*) 100V

15. An alternating current whose average value is 1 A will produce............1 *A* d.c. under similar conditions.
(*i*) less heat than (*ii*) more heat than
(*iii*) the same heat as (*iv*) none of the above

16. A sinusoidal alternating current has a maximum value of I_m. Its average value will be ....
(*i*) I_m/π (*ii*) $I_m/2\pi$
(*iii*) $2I_m/\pi$ (*iv*) none of the above

17. The area of a sinusoidal wave over a half cycle is ....
(*i*) max. value ÷ 2 (*ii*) 2 × max. value
(*iii*) max. value ÷ π (*iv*) max. value ÷ 2π

18. An alternating voltage is given by $v = 200 \sin 314\, t$. Its r.m.s value will be ....
(i) 100 V (*ii*) 282.8 V
(*iii*) 141.4 V (*iv*) 121.4 V

19. The r.m.s. value of sinusoidally varying current is .... that of its average value.
(*i*) more than (*ii*) less than
(*iii*) same as (*iv*) none of the above

20. Alternating voltages and currents are expressed in r.m.s. values because ....
(*i*) they can be easily determined
(*ii*) calculations become very simple
(*iii*) they give comparison with d.c.
(*iv*) none of the above

21 The average value of $\sin^2\theta$ over a complete cycle is ....
(*i*) + 1 (*ii*) –1
(*iii*) 1/2 (*iv*) zero

22. The average value of sin θ over a complete cycle is ....
(*i*) zero (*ii*) + 1
(*iii*) – 1 (*iv*) 1/2

23. An alternating current is given by $i = I_m \sin \theta$. The average value of squared wave of this current over a complete cycle is.
(*i*) $I^2_m/2$ (*ii*) I_m/π
(*iii*) $2I_m/\pi$ (*iv*) $2I_m$

24. The form factor of a sinusoidal wave is ....
(*i*) 1.414 (*ii*) 1.11
(*iii*) 2 (*iv*) 1.5

25. The filament of a vacuum tube requires 0.4 A d.c. to heat it. The r.m.s. value of a.c. required is ....
(*i*) $0.4 \times \sqrt{2}$ A (*ii*) 0.4 ÷ 2 A
(*iii*) $0.8 \div \sqrt{2}$ A (*iv*) 0.4 A.

26. A 100 V peak a.c. is as effective as .... d.c.
(*i*) 100 V (*ii*) 50 V
(*iii*) 70.7 V (*iv*) none of the above

27. The form factor .... of a wave is 1.
(*i*) sinusoidal (*ii*) square
(*iii*) triangular (*iv*) tooth

28. Out of the following .... wave is the peakiest.
(*i*) sinusoidal (*ii*) square
(*iii*) rectangular (*iv*) triangular

29. The peak factor of a sinewave form is ...-..........
(*i*) 1.11 (*ii*) 1.414
(*iii*) 2 (*iv*) 1.5

30. When a 15.V square wave is connected across a 50.V a.c. voltmeter, it will read
(*i*) 15 V (*ii*) $15 \times \sqrt{2}$ V
(*iii*) $15/\sqrt{2}$ V (*iv*) none of the above

31. The breakdown voltage of an insulation depends upon .... value of alternating voltage.
(*i*) average (*ii*) r.m.s.
(*iii*) peak (*iv*) twice the r.m.s.

32. The peak factor of a half-wave rectified a.c. is ....
(*i*) 1.57 (*ii*) 2
(*iii*) 1.11 (*iv*) 1.4142

33. The form factor of a half-wave rectified a.c. is...........
(*i*) 2 (*ii*) 1.11
(*iii*) 1.414 (*iv*) 1.57

34. When 200 V sinusoidal peak-to-peak is connected across an a.c. voltmeter, it will read ....
(*i*) 141.4 V (*ii*) 50 V
(*iii*) 70.7 V (*iv*) none of the above

35. In Fig. 15.37, the wave that will produce maximum heat under similar conditions is

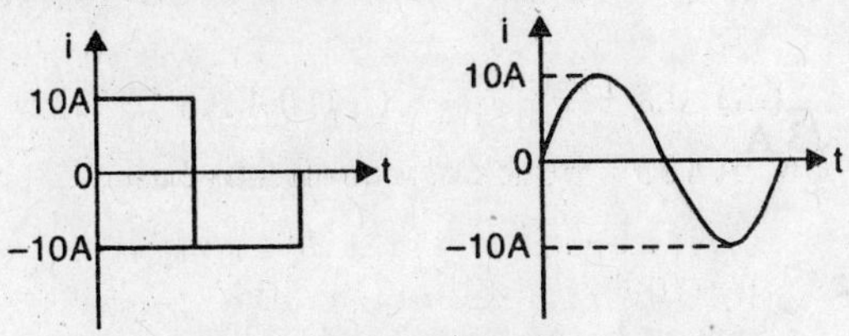

Fig. 15. 37

(*i*) square wave (*ii*) sinusoidal wave
(*iii*) triangular wave (*iv*) saw tooth wave

36. In Fig. 15.38 .... wave will have the highest half-cycle average value.
(*i*) saw tooth (*ii*) square
(*iii*) triangular (*iv*) sinusoidal

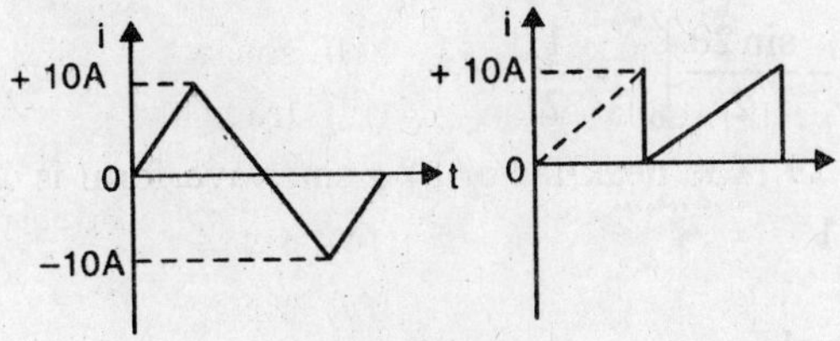

Fig. 15.38

37. The average value of a sinusoidal current is 100 A. Its r.m.s. value is ....
(*i*) 63.7A (*ii*) 70.7 A
(*iii*) 141.4 A (*iv*) 111 A

38. A current wave is given by $i = 4 + 2\sqrt{2}$ sin $3\theta + 4\sqrt{2}$ sin 5θ. The r.m.s. value of current wave is ....
(*i*) 10A. (*ii*) 6 A
(*iii*) $\sqrt{56}$ A (*iv*) 5 A

39. In Fig.15.39, current is given by $i = I_m \sin\theta$. The voltage equation will be ....
(*i*) $V_m \sin\theta$ (*ii*) $V_m \sin(\theta + \phi)$
(*iii*) $V_m \sin(\theta - \phi)$ (*iv*) $V_m \sin(\theta - 2\phi)$

40. The waveforms of voltage and current shown in Fig. 15.39 would exist in .... circuit.

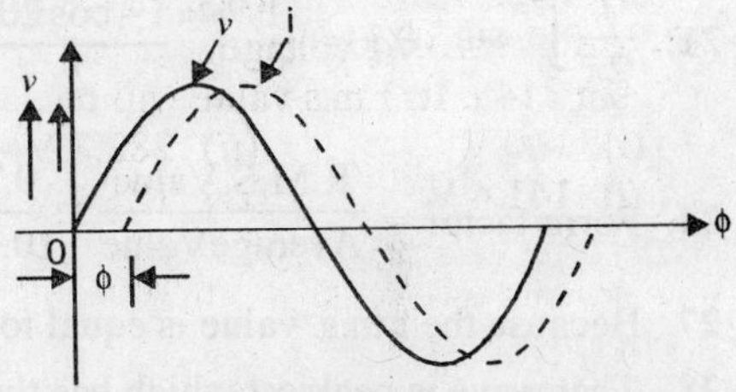

Fig. 15. 39

(*i*) a resistive circuit (*ii*) a capacitive circuit
(*iii*) an inductive circuit (*iv*) none of the above

ANSWERS TO OBJECTIVE QUESTIONS

1. (*i*)	**2.** (*ii*)	**3.** (*iv*)	**4.** (*iv*)	**5.** (*iii*)
6. (*ii*)	**7.** (*iii*)	**8.** (*iv*)	**9.** (*ii*)	**10.** (*iii*)
11. (*iii*)	**12.** (*iii*)	**13.** (*ii*)	**14.** (*iii*)	**15.** (*i*)
16. (*iii*)	**17.** (*ii*)	**18.** (*iii*)	**19.** (*i*)	**20.** (*iii*)
21. (*iii*)	**22.** (*i*)	**23.** (*i*)	**24.** (*ii*)	**25.** (*iv*)
26. (*iii*)	**27.** (*ii*)	**28.** (*iv*)	**29.** (*ii*)	**30.** (*i*)
31. (*iii*)	**32.** (*ii*)	**33.** (*iv*)	**34.** (*iii*)	**35.** (*i*)
36. (*ii*)	**37.** (*iv*)	**38.** (*ii*)	**39.** (*ii*)	**40.** (*iii*)

HINTS TO SELECTED OBJECTIVE QUESTIONS

5. For a 2–pole generator, electrical degrees spanned in one revolution of the coil are 360° *i.e.* one cycle is generated. For an 8-pole generator, 4 cycles (*i.e.* 1440° electrical) will be generated for one revolution of the coil.

6. Comparing with standard equation, $v = V_m \sin \omega t$, $\omega = 157$

or $f = 157/2\pi = 25$ Hz

7. $f = 314/2\pi = 50$ Hz ; T = 1/50 = 0.02 second

∴ $2T = 2 \times 0.02 = 0.04$ second

8. Time taken $= \dfrac{T}{2} + \dfrac{T}{4} = \dfrac{3T}{4} = \dfrac{3}{4} \times 0.02 = 0.015$ second

9. $v = V_m \sin\theta = 20 \sin 135° = 14.14$ V

10. $i = I_m \sin\theta$ or $3 = I_m \sin 120°$

$\therefore \qquad I_m = \dfrac{3}{\sin 120°} = \dfrac{3}{\sin 60°} = 2\sqrt{3}$ A

12. $f = \dfrac{NP}{120} = \dfrac{600 \times 10}{120} = 50 Hz$

14. $V_{av} = 0.637\ V_m = 0.637 \times 100 = 63.7$ V

17. Area of half-cycle $= \int_0^{\pi} i d\theta = \int_0^{\pi} I_m \sin\theta d\theta = I_m[-\cos\theta]_0^{\pi} = 2I_m$

18. $V_{r.m.s} = \dfrac{V_m}{\sqrt{2}} = 0.707\ V_m = 0.707 \times 200 = 141.4$ V

21. $\dfrac{1}{2\pi}\int_0^{2\pi} \sin^2\theta d\theta = \dfrac{1}{2\pi}\int_0^{2\pi}\dfrac{1-\cos 2\theta}{2} d\theta = \dfrac{1}{4\pi}\left[\theta - \dfrac{\sin 2\theta}{2}\right]_0^{2\pi} = \dfrac{1}{2}$

24. Form factor $= \dfrac{\text{R.M.S.Value}}{\text{Average Value}} = \dfrac{0.707\, I_m}{0.637 I_m} = 1.11$

27. Because the r.m.s. value is equal to the average value.

28. That wave is peakiest which has the greatest form factor.

29. Peak factor $= \dfrac{\text{Max. Value}}{\text{R.M.S. value}} = \dfrac{I_m}{0.707 I_m} = 1.414$

30. A voltmeter records the r.m.s. value and the r.m.s. value of 15-V square wave is 15 V.

34. $V_m = 200/2 = 100$ V $\therefore V_{r.m.s.} = 0.707\ V_m = 0.707 \times 100 = 70.7$ V

35. Since maximum value is the same in each case, square wave will have the greatest r.m.s. value (*i.e.* 10 A in this case) and hence will produce the largest heat.

38. $I_{r.m.s.} = \sqrt{(4)^2 + (2)^2 + (4)^2} = 6$ A

16

Series A.C. Circuits

Introduction

A circuit in which the same alternating current flows through all the circuit elements (*i.e.* *R, L, C*) is called a series a.c. circuit. Again in the study of these circuits, our points of interest will be (*i*) phase angle ϕ between the applied voltage and circuit current (*ii*) circuit impedance and current and (*iii*) power consumed. Since current is common in a series circuit, it shall be taken as the reference phasor in drawing the phasor diagrams.

16.1. R-L Series Circuit

Fig. 16.1 (*i*) shows a pure resistance R ohms connected in series with a coil of pure inductance of *L* henry.

Let V = r.m.s. value of the applied voltage

I = r.m.s. value of the circuit current

$\therefore$ $V_R = IR$ .... where V_R is in phase with I

$V_L = IX_L$ .... where V_L leads I by 90°

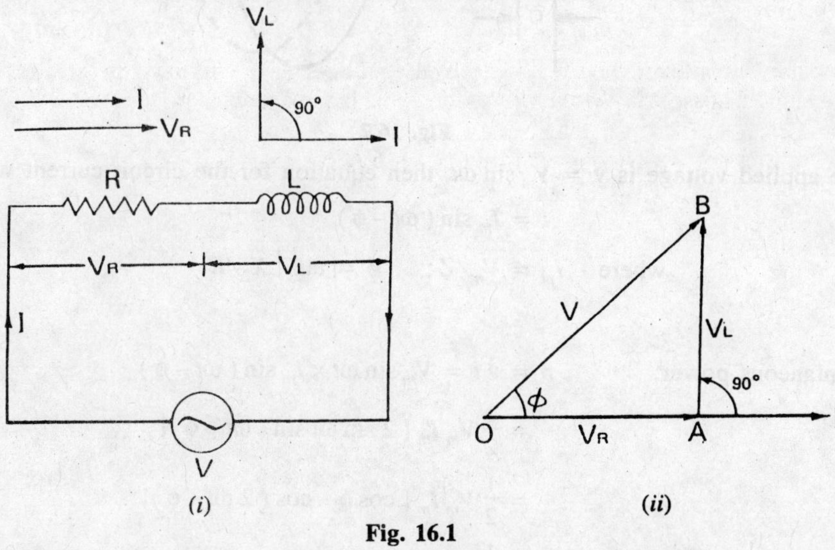

Fig. 16.1

Taking current as the reference phasor, the phasor diagram of the circuit can be drawn as shown in Fig. 16.1 (*ii*). The voltage drop V_R (= IR) is in phase with current and is represented in magnitude and direction by the phasor *OA*. The voltage drop V_L (= IX_L) leads current by 90° and is represented in magnitude and direction by the phasor *AB*. The applied voltage *V* is the phasor sum of these two drops *i.e.*

$$V = \sqrt{V_R^2 + V_L^2} = \sqrt{(IR)^2 + (IX_L)^2} = I\sqrt{R^2 + X_L^2}$$

$$\therefore \quad I = \frac{V}{\sqrt{R^2 + X_L^2}}$$

The quantity $\sqrt{R^2 + X_L^2}$ offers opposition to current flow and is called *impedance* of the circuit. It is represented by Z and is measured in ohms (Ω).

$$\therefore \quad I = V/Z \text{ where } Z = \sqrt{R^2 + X_L^2}$$

It is clear from the phasor diagram that *circuit current I lags the applied voltage V by* $\phi°$. This fact is also illustrated in the wave diagram shown in Fig. 16.2. The value of phase angle ϕ can be determined from the phasor diagram as :

$$\tan\phi = \frac{V_L}{V_R} = \frac{IX_L}{IR} = \frac{X_L}{R}$$

Since X_L and R are known, the value of ϕ can be calculated.

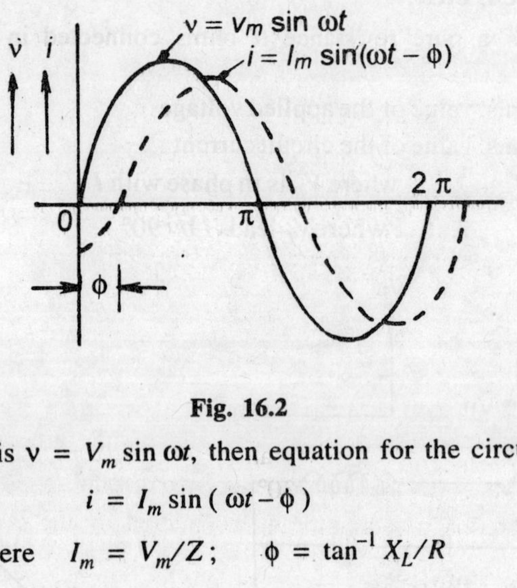

Fig. 16.2

If the applied voltage is $v = V_m \sin\omega t$, then equation for the circuit current will be :

$$i = I_m \sin(\omega t - \phi)$$

$$\text{where} \quad I_m = V_m/Z; \quad \phi = \tan^{-1} X_L/R$$

Power

Instantaneous power, $\quad p = v\,i = V_m \sin\omega t \times I_m \sin(\omega t - \phi)$

$$= \frac{1}{2} V_m I_m [2\sin\omega t \sin(\omega t - \phi)]$$

$$= \frac{1}{2} V_m I_m [\cos\phi - \cos(2\omega t - \phi)]$$

$$= \frac{1}{2} V_m I_m \cos\phi - \frac{1}{2} V_m I_m \cos(2\omega t - \phi)$$

Thus instantaneous power consists of two parts :

(*i*) Constant part $\frac{1}{2} V_m I_m \cos\phi$ whose average value is the same.

(*ii*) A pulsating component $\frac{1}{2} V_m I_m \cos(2\omega t - \phi)$ whose average value over one complete cycle is zero.

$\therefore$ Average power, $$P = \frac{V_m I_m}{2} \cos\phi = \frac{V_m}{\sqrt{2}} \times \frac{I_m}{\sqrt{2}} \times \cos\phi$$

or $$P = VI \cos\phi$$

where *V* and I are the r.m.s. values of voltage and current. The term $\cos\phi$ is called **power factor** of the circuit and its value is given by (from phasor diagram);

Power factor, $$\cos\phi = \frac{IR}{IZ} = \frac{R}{Z}$$

Alternatively $$P = VI\cos\phi$$

$$= (IZ)\,I\,(R/Z) \qquad [\because \cos\phi = R/Z \text{ and } V = IZ]$$

$$= I^2R$$

This is expected because power is consumed in resistance only; inductance does not consume any power.

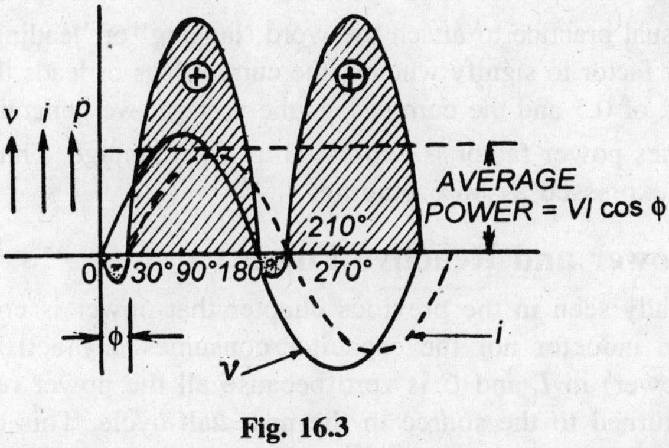

Fig. 16.3

Fig. 16.3 shows the power curve for an R-L series circuit. It is clear that power is negative between 0° and 30° and between 180° and 210°. During the rest of the cycle, the power is positive. This means that power is being consumed by the circuit.

16.2. Impedance Triangle

The phasor diagram of a *R-L* series circuit is shown in Fig. 16.4. Dividing each side of the phasor diagram by the same factor *I*, we get a triangle whose sides represent *R*, X_L and Z. Such a triangle is known as *impedance triangle* (See Fig. 16.5). Just as in Fig. 16.4, the impedance triangle is also a right-angled triangle.

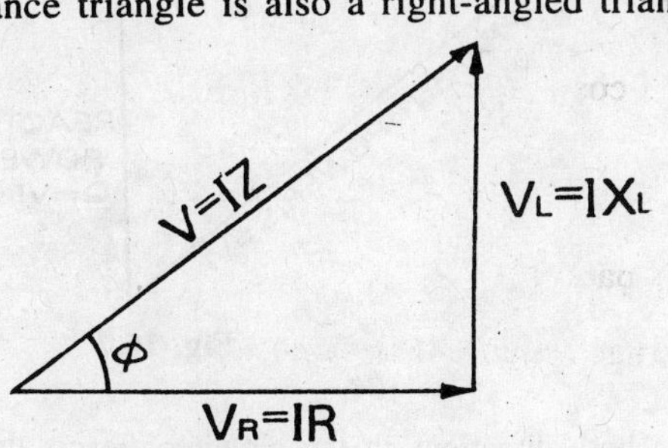

Fig. 16.4.

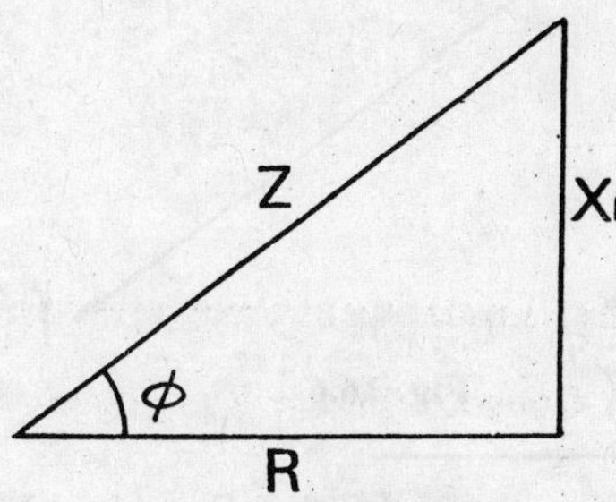

Fig. 16.5

Impedance triangle is a useful concept in a.c. circuits as it enables us to calculate,

(*i*) the impedance of the circuit *i.e.* $Z = \sqrt{R^2 + X_L^2}$

(*ii*) power factor of the circuit *i.e.* $\cos\phi = R/Z$

(*iii*) phase angle ϕ *i.e.* $\tan\phi = X_L/R$

(*iv*) whether current leads or lags the voltage.

Therefore, it is always profitable to draw the impedance triangle while analysing an a.c. circuit.

16.3 Power Factor

The power factor (*i.e.* $\cos\phi$) of a circuit can be defined in one of the following ways :

(*i*) Power factor = $\cos\phi$ = cosine of angle between V and I

(*ii*) Power factor = $\dfrac{R}{Z} = \dfrac{\text{Resistance}}{\text{Impedance}}$ [See Fig. 16.5]

(*iii*) Power factor = $\dfrac{VI\cos\phi}{VI} = \dfrac{\text{True power}}{\text{Apparent power}}$

It may be noted that power factor can never have a value greater than 1.

(*a*) It is a usual practice to attach the word 'lagging' or 'leading' with the numerical value of power factor to signify whether the current lags or leads the voltage. Thus if a circuit has a p.f. of 0.5 and the current lags the voltage, we generally write p.f. as 0.5 lagging.

(*b*) Sometimes power factor is expressed as a percentage. Thus 0.8 lagging power factor may be expressed as 80% lagging.

16.4. True Power and Reactive Power

We have already seen in the previous chapter that power is consumed only in resistance since neither pure inductor nor the capacitor consumes any active power. The power consumed (or true power) in L and C is zero because all the power received from the source in a half-cycle is returned to the source in the next half-cycle. This circulating power is called the ***reactive power** and does no useful work in the circuit. The reader may recall that current and voltage are in phase in a resistance whereas they are 90° out of phase in L or C. Therefore, we come to the conclusion that current in phase with voltage produces true or active power whereas current 90° out of phase with voltage contributes to reactive power *i.e.*

True power = Voltage × Current in phase with voltage

Reactive power = Voltage × Current 90° out of phase with voltage

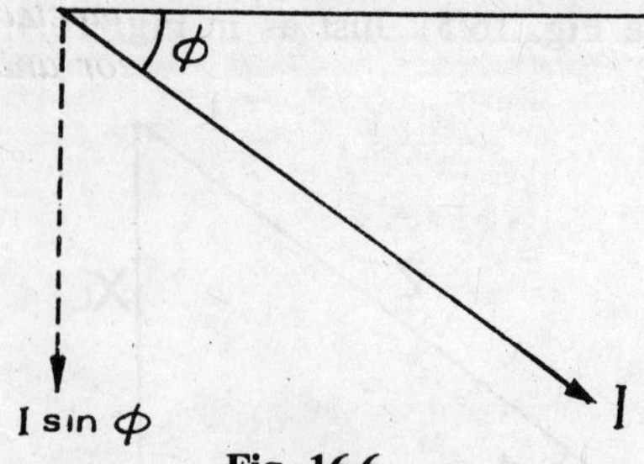

Fig. 16.6

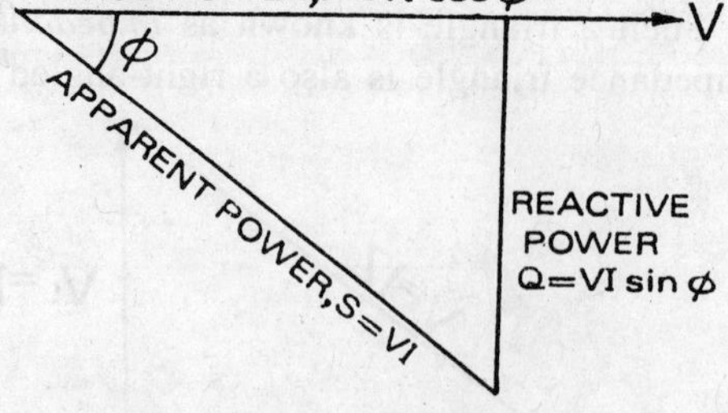

Fig. 16.7

---

* This is the power that flows back and forth in both directions in the circuit or reacts upon itself. Hence the name.

Consider an inductive circuit in which current I lags behind the applied voltage V by $\phi°$. The phasor diagram of the circuit is shown in Fig. 16.6. The current I can be resolved into two rectangular components *viz.* (*i*) $I\cos\phi$ in phase with V and (*ii*) $I\sin\phi$; 90° out of phase with V.

$\therefore$ True power, $P = V \times I\cos\phi = VI\cos\phi$ watts or kW

Reactive power, $Q = V \times I\sin\phi = VI\sin\phi$ VAR or kVAR

Apparent power, $S = V \times I = VI$ VA or kVA

(*i*) The component $I\cos\phi$ is called the *in phase component* or *wattful component*. It is this component of total current which contributes to true power (*i.e.* $VI\cos\phi$).

(*ii*) The component $I\sin\phi$ is called the *reactive component* and contributes to reactive power (*i.e.* $VI\sin\phi$). The reactive power is neither consumed in the circuit nor it does any useful work. It merely flows back and forth in both directions in the circuit. A wattmeter does not measure the reactive power.

(*iii*) The product of voltage (V) and actual current (I) in the circuit is called the apparent power (*i.e.* VI). To avoid confusion, it is measured in volt-amperes (VA).

16.5. Power Triangle

If we multiply each of the current phasors in Fig. 16.6 by V, we get the power triangle shown in Fig. 16.7. This is a right-angled triangle and indicates the relation among apparent power, true power and reactive power. It reveals the following facts about the circuit :

(*i*) Power factor, $\cos\phi = \dfrac{\text{True power}}{\text{Apparent power}} = \dfrac{VI\cos\phi}{VI}$

(*ii*) $(\text{Apparent power})^2 = (\text{True power})^2 + (\text{Reactive power})^2$

or $S^2 = P^2 + Q^2$

(*iii*) True power, $P = \text{Apparent power} \times \cos\phi = \text{VI}\cos\phi$

Reactive power, $\theta = \text{Apparent power} \times \sin\phi = \text{VI}\sin\phi$

Let us illustrate the power relations in an a.c. circuit with an example. Suppose a circuit draws a current of 10A at a voltage of 200V and its p.f. is 0.8 lagging. Then,

Apparent power, $S = VI = 200 \times 10 = 2000$ VA

True power, $P = VI\cos\phi = 200 \times 10 \times 0.8 = 1600$ W

Reactive power, $Q = VI\sin\phi = 200 \times 10 \times 0.6 = 1200$ VAR

The circuit receives an apparent power of 2000VA and is able to convert only 1600 watts into true power. The reactive power of 1200 VAR does no useful work, it merely flows into and out of the circuit periodically. In fact, reactive power is a liability on the source because the source has to supply the additional current (*i.e.*, $I\sin\phi$).

Example 16.1. *A coil having a resistance of 7Ω and an inductance of 31.8 mH is connected to 230V, 50Hz supply. Calculate (i) the circuit current (ii) phase angle (iii) power factor and (iv) power consumed.*

Solution.

(*i*) $X_L = 2\pi f L = 2\pi \times 50 \times 31.8 \times 10^{-3} = 10\Omega$

$Z = \sqrt{R^2 + X_L^2} = \sqrt{7^2 + 10^2} = 12.2\Omega$

$I = V/Z = 230/12.2 = \mathbf{18.85A}$

(*ii*) $\phi = \tan^{-1} X_L/R = \tan^{-1}(10/7) = \mathbf{55°}$ *lag*

(*iii*) $p.f. = \cos\phi = \cos 55° = \mathbf{0.573\ lag}$

(*iv*) $P = VI\cos\phi = 230 \times 18.85 \times 0.573 = \mathbf{2484.24W}$

Example 16.2. *A choke coil takes a current of 2.5A when connected across 250V, 50Hz mains and consumes 400 watts. Find (i) the power factor (ii) resistance of the coil and (iii) inductance of the coil.*

Solution.

(*i*) $P = VI\cos\phi$

$$\therefore \quad \cos\phi = \frac{P}{VI} = \frac{400}{250 \times 2.5} = \mathbf{0.64}\ lag$$

(*ii*) $Z = V/I = 250/2.5 = 100\,\Omega$

$\therefore \quad R = Z\cos\phi = 100 \times 0.64 = \mathbf{64\,\Omega}$

(*iii*) $X_L = \sqrt{Z^2 - R^2} = \sqrt{100^2 - 64^2} = 76.84\,\Omega$

$$\therefore \quad L = \frac{X_L}{2\pi f} = \frac{76.84}{2\pi \times 50} = \mathbf{0.245\,H}$$

Example 16.3. *An e.m.f. given by 326 sin 314t is applied to a certain circuit and current is* 2 sin (314*t* – 1.3736). *Find (i) frequency of voltage (ii) phase angle between voltage and current in degrees and (iii) resistance of the circuit.*

Solution.

(*i*) $$f = \frac{\omega}{2\pi} = \frac{314}{2\pi} = \mathbf{50\,Hz}$$

(*ii*) $\phi = 1.3736 \text{ radians} = 1.3736 \times 180°/\pi = \mathbf{78.7°}\ lag$

(*iii*) $Z = V_m/I_m = 326/2 = 163\,\Omega$

$\therefore \quad R = Z\cos\phi = 163\cos 78.7° = \mathbf{32.2\,\Omega}$

Example 16.4. *A 200V, 50Hz inductive circuit takes a current of 10A, lagging 30°. Find (i) the resistance (ii) reactance and (iii) inductance of the circuit.*

Solution.

(*i*) $Z = V/I = 200/10 = 20\,\Omega$

$R = Z\cos\phi = 20\cos 30° = \mathbf{17.32\,\Omega}$

(*ii*) $X_L = Z\sin\phi = 20\sin 30° = \mathbf{10\,\Omega}$

(*iii*) $$L = \frac{X_L}{2\pi f} = \frac{10}{2\pi \times 50} = \mathbf{0.0318H}$$

Example 16.5. *In an R-L series circuit, a voltage of 100V at 25 Hz produces one ampere while the same voltage at 75Hz produces half ampere. Find the values of R and L.*

Solution.

$$Z_1 = 100/1 = 100\,\Omega\ ;\ Z_2 = 100/0.5 = 200\,\Omega$$

$$\therefore \quad R^2 + (2\pi \times 25L)^2 = 100^2 \qquad ...(i)$$

and $$R^2 + (2\pi \times 75L)^2 = 200^2 \qquad ...(ii)$$

From eqs. (*i*) and (*ii*), $R = \mathbf{79\,\Omega}$, $L = \mathbf{0.39\,H}$

Example 16.6. *Find the inductance of a coil of negligible resistance which when connected in series with a non-inductive resistor of 100 Ω reduces the current to one half of its original value on a supply of frequency 50Hz.*

Solution. The impedance Z of the series combination should be twice that of resistance $R\ (= 100\Omega)$ *i.e.* $Z = 2R$

$$Z^2 = R^2 + X_L^2$$

or $\quad 4R^2 = R^2 + X_L^2$

or $\quad X_L = \sqrt{3}\,R = \sqrt{3} \times 100 = 173.2\,\Omega$

$\therefore \quad L = \dfrac{X_L}{2\pi f} = \dfrac{173.2}{2\pi \times 50} = \mathbf{0.551\,H}$

Example 16.7. *A pure inductance of 318mH is connected in series with a pure resistance of 75 Ω. The circuit is supplied from 50Hz source and the voltage across 75 Ω resistor is found to be 150V. Calculate the supply voltage and the phase angle.*

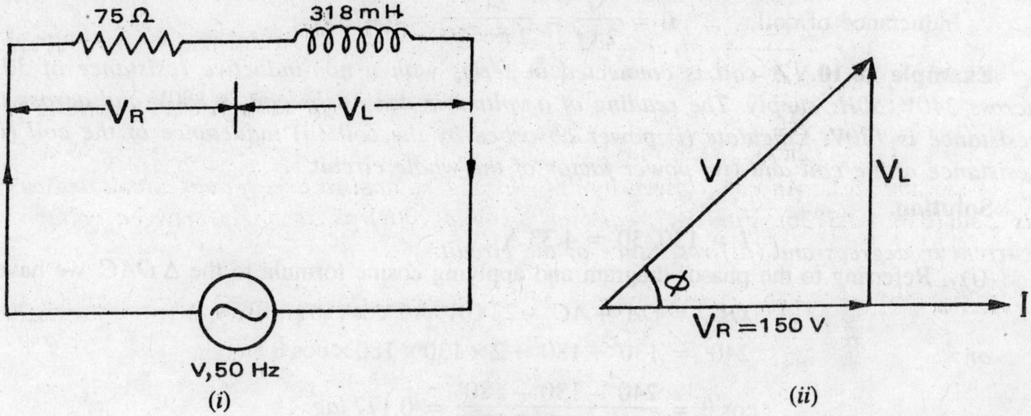

Fig. 16.8

Solution. The circuit diagram and the phasor diagram are shown in Fig. 16.8.

$$I = V_R/R = 150/75 = 2\text{A}$$

$$X_L = 2\pi f L = 2\pi \times 50 \times 318 \times 10^{-3} = 100\,\Omega$$

$$V_L = IX_L = 2 \times 100 = 200\text{V}$$

Referring to the phasor diagram in Fig. 16.8(*ii*),

Supply voltage, $\quad V = \sqrt{V_R^2 + V_L^2} = \sqrt{150^2 + 200^2} = \mathbf{250\,V}$

Phase angle, $\quad \phi = \tan^{-1}(X_L/R) = \tan^{-1}(100/75) = \mathbf{53.06°}\ \textit{lag}$

Example 16.8. *A voltage is given by the expression* v = 340 sin ωt *and is applied to a circuit, the current being given by i = 14.14 sin (ωt – π/6). Calculate (i) impedance (ii) resistance (iii) reactance and (iv) power consumed.*

Solution.

(*i*) $Z = V_m/I_m = 340/14.14 = \mathbf{24\,\Omega}$

(*ii*) $R = Z\cos\phi = 24\cos\pi/6 = \mathbf{20.8\,\Omega}$

(*iii*) $X_L = Z\sin\phi = 24\sin\pi/6 = \mathbf{12\,\Omega}$

(*iv*) $P = I^2R = (14.14/\sqrt{2})^2 \times 20.8 = \mathbf{2080\,W}$

Example 16.9. *A coil when connected across a 100V d.c. supply dissipates 500 W of power. When connected across a 100 V a.c. supply of frequency 50 Hz, it dissipates 200 W. Calculate the values of resistance and inductance of the coil.*

Solution.

D.C. supply

Resistance of coil, $\quad R = V^2/P = 100^2/500 = \mathbf{20\,\Omega}$

A.C. supply

Power consumed, $P = VI\cos\phi = V\times\frac{V}{Z}\times\frac{R}{Z} = \frac{V^2R}{Z^2}$

Impedance of coil, $Z = \sqrt{\frac{V^2R}{P}} = \sqrt{\frac{100^2\times 20}{200}} = 31.62\,\Omega$

Reactance of coil, $X_L = \sqrt{Z^2 - R^2} = \sqrt{(31.62)^2 - 20^2} = 24.5\,\Omega$

$\therefore$ Inductance of coil, $L = \frac{X_L}{2\pi f} = \frac{24.5}{2\pi\times 50} = \mathbf{0.078\,H}$

Example 16.10. *A coil is connected in series with a non-inductive resistance of 30 Ω across 240V, 50Hz supply. The reading of a voltmeter across the coil is 180V and across the resistance is 130V. Calculate (i) power absorbed by the coil (ii) inductance of the coil (iii) resistance of the coil and (iv) power factor of the whole circuit.*

Solution.

$$I = 130/30 = 4.33\text{ A}$$

(*i*) Referring to the phasor diagram and applying cosine formula to the Δ *OAC*, we have,

$$OC^2 = OA^2 + AC^2 - 2\,.\,OA\,.\,AC\cos(180-\theta)$$

or $$240^2 = 130^2 + 180^2 + 2\times 130\times 180\times\cos\theta$$

$\therefore$ $$\cos\theta = \frac{240^2 - 130^2 - 180^2}{2\times 130\times 180} = 0.177\ lag$$

$$P_{coil} = V_L I\cos\theta = 180\times 4.33\times 0.177 = \mathbf{137.95\ W}$$

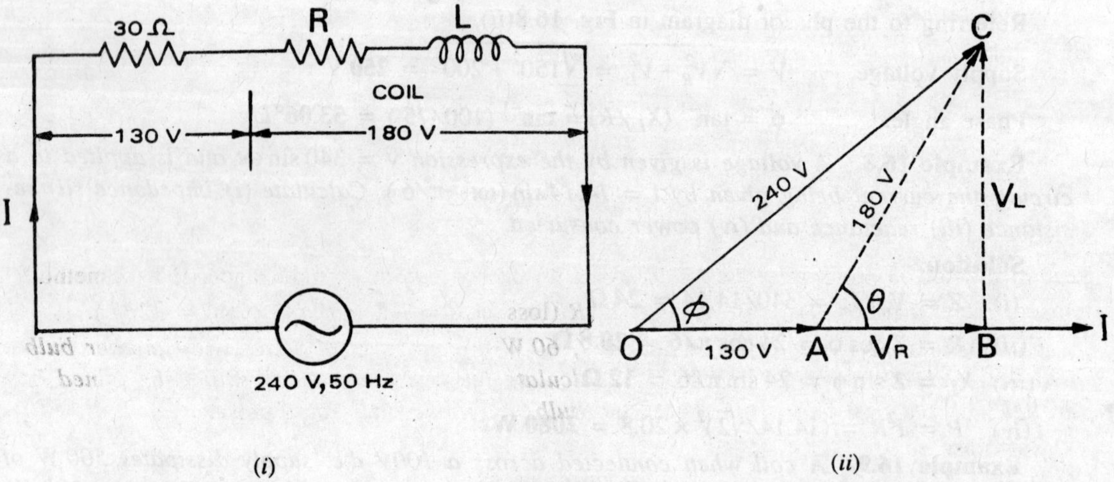

Fig. 16.9

(*ii*) $V_L = AC\sin\theta = 180\sqrt{1-(0.177)^2} = 177.16\ V$

$X_L = V_L/I = 177.16/4.33 = 40.9\,\Omega$

$\therefore$ $L = X_L/2\pi f = 40.9/2\pi\times 50 = \mathbf{0.13\ H}$

(*iii*) $V_R = AC\cos\theta = 180 \times 0.177 = 31.86\text{ V}$

$\therefore \quad R = V_R/I = 31.86/4.33 = \mathbf{7.36\,\Omega}$

(*iv*) Circuit *p.f.* $= \cos\phi = \dfrac{OB}{OC} = \dfrac{130 + 31.86}{240} = \mathbf{0.674}\,lag$

Example 16.11. *A 100 V, 60 W lamp is to be operated on 220 V, 50 Hz supply. Find the value of (i) non-inductive resistance (ii) pure inductance in series with the lamp so that the lamp is not overrun. Which would be preferable ?*

Solution.

Rated current of lamp, $\quad I = 60/100 = 0.6\text{ A}$

(*i*) ***Non-inductive resistance.*** Fig. 16.10 shows a resistance R connected in series with the lamp.

$$V_R = 220 - 100 = 120\text{V}$$

Both V_R and lamp voltage are in phase with I.

$$\therefore \quad R = V_R/I = 120/0.6 = \mathbf{200\,\Omega}$$

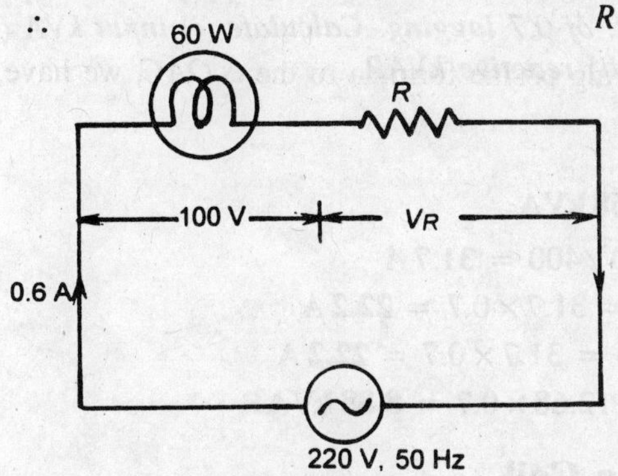

Fig. 16.10

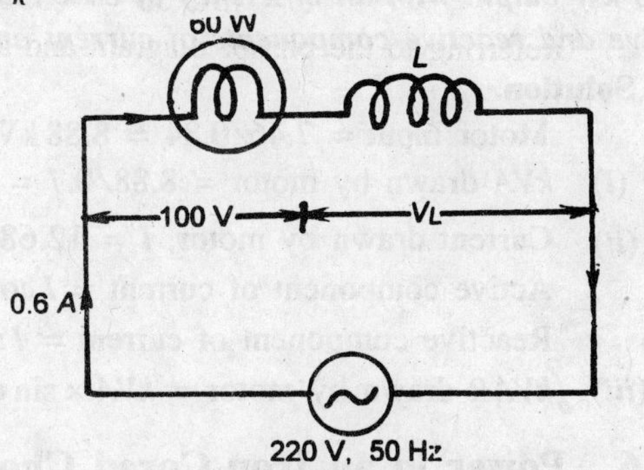

Fig. 16.11

(*ii*) ***Pure inductance***. Fig. 16.11 shows pure inductance L in series with the lamp. The lamp voltage (*i.e.* 100V) is in phase with I while V_L leads I by 90°

$$\therefore \quad V_L = \sqrt{220^2 - 100^2} = 196\,V$$

$$\therefore \quad X_L = V_L/I = 196/0.6 = 326.67\,\Omega$$

$$L = \frac{X_L}{2\pi f} = \frac{326.67}{2\pi \times 50} = \mathbf{1.04\,H}$$

Second method is preferable because there is no power loss in an inductance. If first method is used, there will be a large power loss in R (loss in $R = I^2R = (0.6)^2 \times 200 = 72\,W$).

Example 16.12. *A bulb rated at 110 V, 60 W is connected in series with another bulb rated 110 V, 100 W across 220 V mains. Calculate the resistance which should be joined in parallel with the first bulb so that both the bulbs may take their rated power.*

Solution.

Rated current of bulb L_1, $\quad I_1 = 60/110 = 0.545\text{ A}$

Rated current of bulb L_2, $\quad I_2 = 100/110 = 0.91\text{ A}$

In order that the bulbs take their rated power, they must carry their rated currents. This can be done by connecting a resistance R in parallel with bulb L_1 as shown in Fig. 16.12.

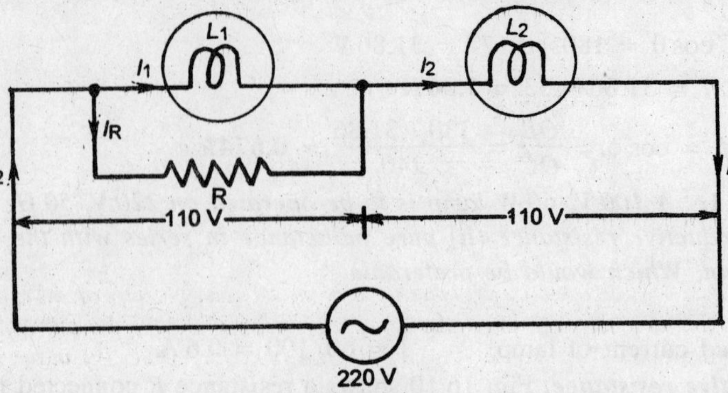

Fig. 16.12

$$I_R = I_2 - I_1 = 0.91 - 0.545 = 0.365\text{A}$$

$$\therefore \quad R = 110/0.365 = \mathbf{301.4\Omega}$$

Example 16.13. *A single phase motor operating from 400 V, 50 Hz supply is developing 7.46 kW output with an efficiency of 84% and p.f. of 0.7 lagging. Calculate (i) input kVA (ii) active and reactive components of current and (iii) reactive kVAR.*

Solution.

Motor input = 7.46/0.84 = 8.88 kW

(*i*) *kVA* drawn by motor = 8.88/0.7 = **12.68 kVA**

(*ii*) Current drawn by motor, $I = 12.68\times 10^3/400 = 31.7\,A$

Active component of current = $I\cos\phi = 31.7\times 0.7 = \mathbf{22.2\ A}$

Reactive component of current = $I\sin\phi = 31.7\times 0.7 = \mathbf{22.2\ A}$

(*iii*) *kVAR* drawn by motor = $kVA\times\sin\phi = 12.68\times 0.7 = \mathbf{8.88\ kVAR}$

16.6. Power in an Iron-Cored Choking Coil

The power *P* taken by an iron-cored coil has to supply :

(*i*) power loss in ohmic resistance *i.e.* I^2R where *R* is the d.c. resistance or true resistance of the coil.

(*ii*) the iron loss P_i (*i.e.* hysteresis and eddy current loss).

$$\therefore \quad P = I^2R + P_i \quad \text{...(i)}$$

The iron loss can be thought as additional resistance in series with the resistance of the coil. Thus the effect of iron core is to increase the effective resistance of the coil. We can say that input to an iron-cored coil is $I^2 R_{eff}$ *i.e.*

$$P = I^2 R_{eff} \quad \text{...(ii)}$$

From eqs. (*i*) and (*ii*), we have,

$$I^2 R_{eff} = I^2 R + P_i$$

or

$$R_{eff} = R + \frac{P_i}{I^2}$$

i.e. Effective resistance = True (or d.c.) resistance + $\frac{P_i}{I^2}$

Example 16.14. *An alternating current flows through an iron-cored coil. The d.c. resistance of the coil is 25Ω and the coil takes 10mA and 3.5 mW of power. What is the effective resistance of the coil ?*

Solution.

$$I^2 R_{eff} = P$$

or $$R_{eff} = \frac{P}{I^2} = \frac{3.5 \times 10^{-3}}{(10 \times 10^{-3})^2} = \mathbf{35\ \Omega}$$

The reader may note that with an iron-core, the resistance of the coil is increased from 25 Ω to 35 Ω.

Example 16.15. *An iron-cored choking coil has a resistance of 4 Ω when measured by a d.c. supply. On a 240 V, 50 Hz mains supply, it dissipated 500 W, the current taken being 10 A. Calculate (i) impedance (ii) the power factor (iii) the iron loss and (iv) inductance of the coil.*

Solution.

(*i*) Impedance of coil, $Z = V/I = 240/10 = \mathbf{24\ \Omega}$

(*ii*) Power factor, $\cos\phi = \frac{P}{VI} = \frac{500}{240 \times 10} = \mathbf{0.208\ lag}$

(*iii*) Total loss = Loss in resistance + Iron loss

$$500 = 10^2 \times 4 + P_i$$

$$\therefore \quad P_i = 500 - 400 = \mathbf{100W}$$

(*iv*) Effective resistance, $R_{eff} = R + P_i/I^2 = 4 + 100/10^2 = 5\ \Omega$

Reactance of coil, $X_L = \sqrt{Z^2 - R_{eff}^2} = \sqrt{24^2 - 5^2} = 23.47\ \Omega$

∴ Inductance of coil, $L = \frac{X_L}{2\pi f} = \frac{23.47}{2\pi \times 50} = \mathbf{74.71 \times 10^{-3}\ H}$

16.7. Circle Diagram of R-L Series Circuit

Consider an *R-L* series circuit supplied from a constant voltage *V* and fixed frequency. The only variable is *R* as shown in Fig. 16.13 (*i*). At any particular value of *R*, the phasor diagram will be as shown in Fig. 16.13 (*ii*). As *R* is varied, the circuit current *I* also changes. It can be shown that locus of current *I* is a semi-circle as shown in Fig. 16.13 (*iii*).

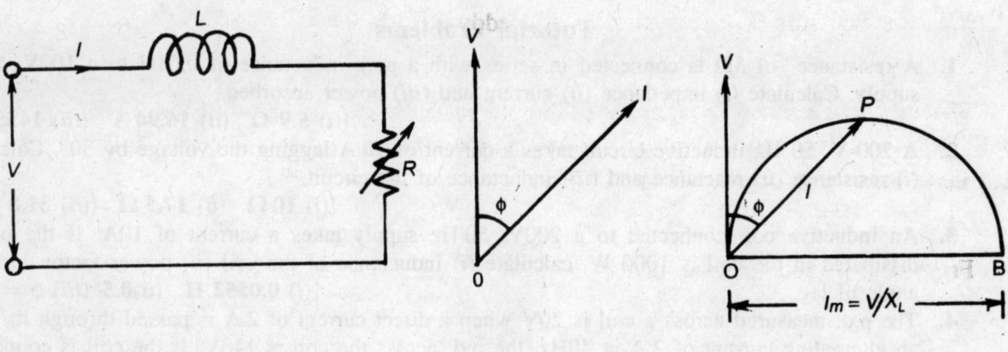

Fig. 16.13

(*i*) At a particular value of *R*, the current in the circuit is *OP*; the point *P* lying on a semicircle.

(*ii*) As *R* is decreased, the phase angle ϕ ($\tan\phi = X_L/R$) increases and point *P* shifts towards *B*. When *R* becomes zero, the phase angle is 90° lagging and *I* has the maximum value $I_m = V/X_L$.

(*iii*) As R is increased, ϕ decreases and point P moves towards O. When R becomes infinite, the phase angle is zero and point P approaches O. Consequently, circuit current becomes zero.

(*iv*) It can be shown that maximum power in the circuit is given by ;

$$\text{Max. Power} = \frac{V^2}{2X_L}$$

It will occur when point P is the mid-point of the arc of semicircle *i.e.* $\phi = 45°$.

Example 16.16. *Give vector diagram for 100 V, 50 Hz R-L series circuit having $\omega L = 1$ ohm and variable resistance. Find the maximum power. Draw the locus of current vector.*

Solution.

Fig. 16.14 (*i*) shows the circuit diagram whereas Fig. 16.14 (*ii*) shows the locus of circuit current.

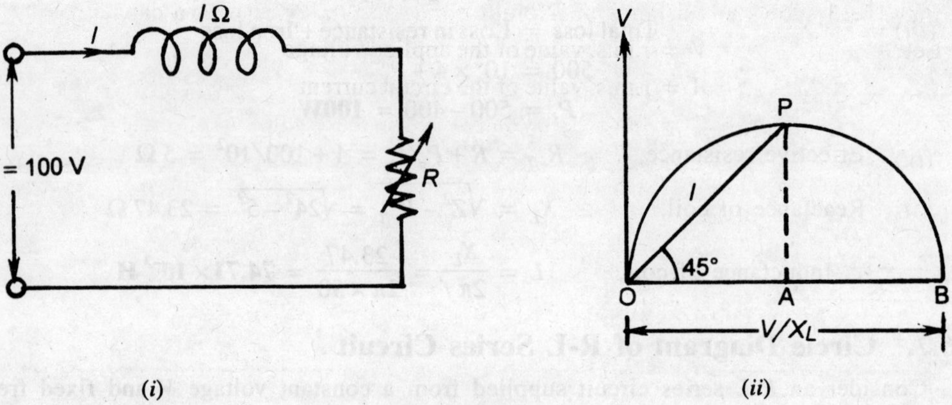

Fig. 16.14.

$$\text{Max. Power} = \frac{V^2}{2\,X_L} = \frac{100^2}{2 \times 1} = \mathbf{5000\ W}$$

Tutorial Problems

1. A resistance of 5Ω is connected in series with a pure inductance of 0.01H to a 100V, 50Hz supply. Calculate (*i*) impedance (*ii*) current and (*iii*) power absorbed. **[(*i*) 5.9 Ω (*ii*) 16.94 A (*iii*) 1435 W]**
2. A 200-V, 50 Hz inductive circuit takes a current of 10 A lagging the voltage by 30°. Calculate (*i*) resistance (*ii*) reactance and (*iii*) inductance of the circuit. **[(*i*) 10 Ω (*ii*) 17.3 Ω (*iii*) 31.8 mH]**
3. An inductive coil connected to a 200V, 50 Hz supply takes a current of 10A. If the power dissipated in the coil is 1000 W, calculate (*i*) inductance of the coil (*ii*) power factor and (*iii*) angle of lag. **[(*i*) 0.0552 H (*ii*)0.5 (*iii*) ϕ = 60°]**
4. The p.d. measured across a coil is 20V when a direct current of 2 A is passed through it. With an alternating current of 2 A at 40Hz, the p.d. across the coil is 140V. If the coil is connected to a 230V, 50Hz supply, calculate (*i*) current (*ii*) power and (*iii*) the power factor. **[(*i*) 2.64A (*ii*) 69.7 W (*iii*) 0.1147 lag]**
5. An a.c. circuit takes a power of 4.2 kW at a power factor of 0.6 lagging. Find (*i*) apparent power and (*ii*) reactive power. **[(*i*) 7 kVA (*ii*) 5.6 kVAR]**
6. A coil is joined in series with a pure resistor of resistance 800 Ω across a 100V, 50Hz supply The reading of a voltmeter across the coil is 45V and across the pure resistor is 80V. Find (*i*) inductance and (*ii*) resistance of the coil. **[(*i*) L = 1.4 H (*ii*) R = 98.5 Ω]**

7. When a certain inductive coil is supplied at 240V, 50Hz, the current is 6.45A. When the frequency is changed to 40Hz at 240V, the current taken is 7.48A. Calculate the inductance and resistance of the coil. **[(*i*) L = 0.1 H (*ii*) R = 20 Ω]**

8. A coil has a resistance of 75 Ω and an inductance of 1.4 H. When the applied voltage is 240V a.c., at what frequency is the current 0.3 A ? What is the power factor at this frequency ? **[(*i*) 90.55 Hz (*ii*) 0.0938 lag]**

9. When voltage of 240V, 50Hz is applied to a coil P, the current drawn is 10 A and power dissipated is 450W. When the same supply is connected to a coil Q, the current is 15 A and power dissipated is 950W. Calculate (*i*) inductive reactance of each coil and (*ii*) current and power taken when the two coils are connected in series across the above voltage. **[(*i*) X_P = 23.6 Ω ; X_Q = 15.4 Ω (*ii*) 6A ; 314 W]**

10. A metal filament lamp is constructed to take a current of 0.2A at 100V. Calculate the value of inductance which when placed in series with the lamp across 200V, 50Hz supply causes the lamp to work under normal conditions of current and voltage. **[2.76 H]**

16.8. R-C Series Circuit

Fig. 16.15 shows a resistance of R ohms connected in series with a capacitor of C farad.

Let V = r.m.s. value of the applied voltage

I = r.m.s. value of the circuit current

V_R = IR ....where V_R is in phase with I

V_C = IX_C ....where V_C lags I by 90°

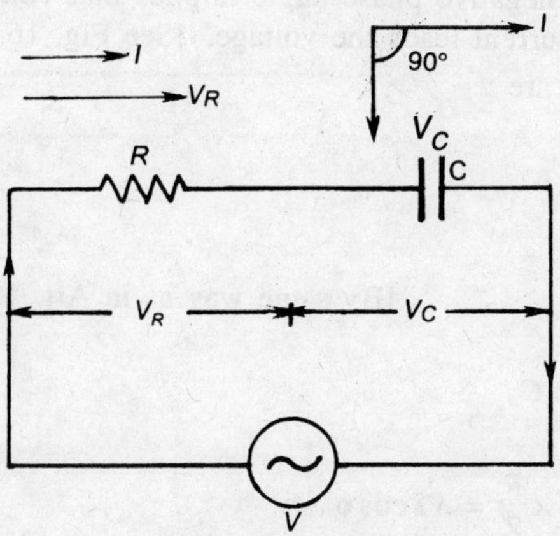

Fig. 16.15

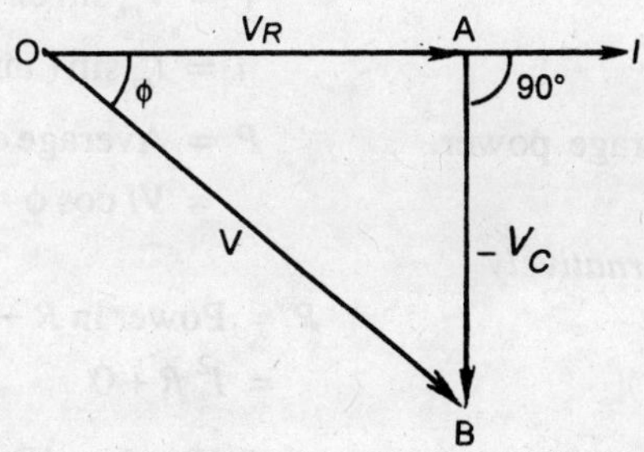

Fig. 16.16

The phasor diagram of the circuit is shown in Fig. 16.16. The supply voltage V is the phasor sum of V_R (= IR) and V_C (= IX_C) drops *i.e.*

$$V = \sqrt{V_R^2 + V_C^2} = \sqrt{(IR)^2 + (-IX_C)^2} = I\sqrt{R^2 + X_C^2}$$

$$\therefore \quad I = \frac{V}{\sqrt{R^2 + X_C^2}}$$

The quantity $\sqrt{R^2 + X_C^2}$ offers opposition to current flow and is called impedance of the circuit.

$$\therefore \quad I = V/Z \quad \text{where } Z = \sqrt{R^2 + X_C^2}$$

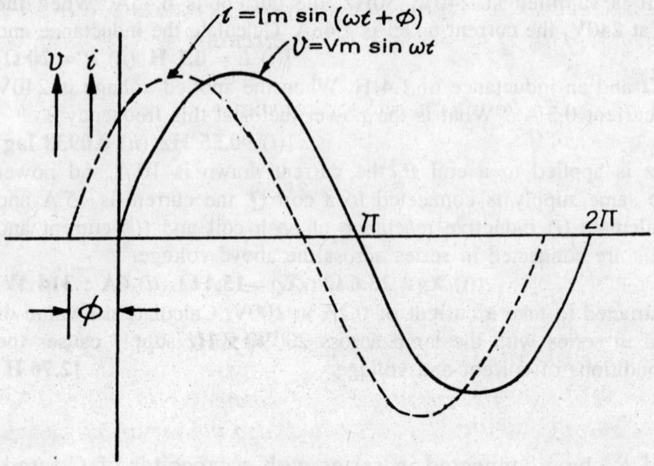

Fig. 16.17

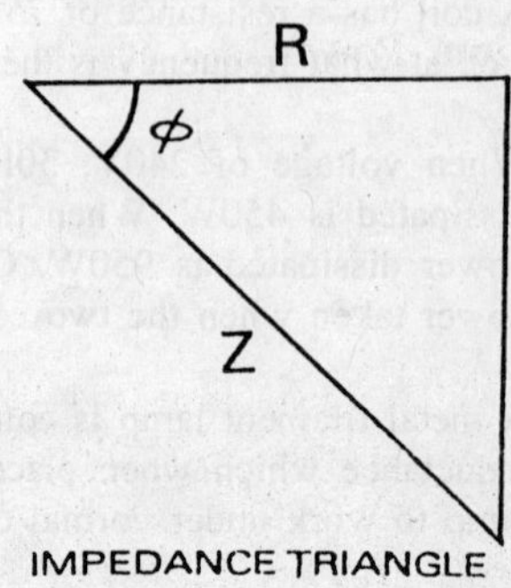

Fig. 16.18

It is clear from the phasor diagram that *circuit current I leads the applied voltage V by* $\phi°$ where :

$$\tan\phi = -\frac{V_C}{V_R} = -\frac{IX_C}{IR} = -\frac{X_C}{R}$$

Since current is taken as the reference phasor, negative phase angle implies that voltage lags behind the current. This is the same thing as current leads the voltage. [See Fig. 16.17]

Power. The equations for voltage and current are :

$$v = V_m \sin\omega t$$

$$i = I_m \sin(\omega t + \phi)$$

Average power, $P = \text{Average of } vi$

$= VI\cos\phi$ [By same way as in Art. 16.1]

Alternatively

$$P = \text{Power in } R + \text{Power in } C$$

$$= I^2 R + 0$$

$$= IR \times I = IR \times \frac{V}{Z} = VI \times \frac{R}{Z} = VI\cos\phi$$

Example 16.17. *A capacitor of capacitance 79.5μF is connected in series with a non-inductive resistance of 30 Ω across 100 V, 50 Hz supply. Find (i) impedance (ii) current (iii) phase angle and (iv) equation for the instantaneous value of current.*

Solution.

(*i*) $X_C = \frac{1}{2\pi f C} = \frac{10^6}{2\pi \times 50 \times 79.5} = 40\,\Omega$

$Z = \sqrt{R^2 + X_C^2} = \sqrt{30^2 + 40^2} = \mathbf{50\,\Omega}$

(*ii*) $I = V/Z = 100/50 = \mathbf{2\,A}$

(*iii*) $\phi = \tan^{-1} X_C/R = \tan^{-1} 40/30 = \mathbf{53°}$ *lead*

(*iv*) $I_m = 2 \times \sqrt{2} = 2.828$ A ; $\omega = 2\pi f = 2\pi \times 50 = 314$

$\therefore\ i = 2.828 \sin(314t + 53°)$

Example 16.18. *A capacitor and resistor are connected in series across a 120V, 50Hz supply. The circuit draws a current of 1.144A. If power loss in the circuit is 130.8W. find the values of resistance and capacitance.*

Solution. Power loss occurs in resistance only as capacitor consumes no power.

$$P = I^2R$$

$$\therefore \quad R = P/I^2 = 130.8/(1.144)^2 = \mathbf{100\Omega}$$

$$Z = V/I = 120/1.144 = 104.9\Omega$$

$$X_C = \sqrt{Z^2 - R^2} = \sqrt{(104.9)^2 - (100)^2} = 31.7\,\Omega$$

$$\therefore \quad C = \frac{1}{2\pi f X_C} = \frac{1}{2\pi \times 50 \times 31.7} = \mathbf{100 \times 10^{-6}\ F}$$

Example 16.19. *A resistor R in series with a capacitor C is connected to 50Hz, 240V supply. Find the value of C so that R absorbs 300 W at 100 V. Find also the maximum charge and maximum energy stored in C.*

Solution.

R absorbs 300W when connected to 100V so that $R = V^2/P = 100^2/300 = 100/3\ \Omega$

Now, $\quad I^2 R = 300W \quad \therefore I = \sqrt{300 \times 3/100} = 3A$

$$Z = V/I = 240/3 = 80\,\Omega$$

$$X_C = \sqrt{Z^2 - R^2} = \sqrt{80^2 - (100/3)^2} = 72.7\,\Omega$$

$$C = \frac{1}{2\pi f X_C} = \frac{1}{2\pi \times 50 \times 72.7} = \mathbf{43.7 \times 10^{-6}\ F}$$

$$Q_{max} = C\,V_{max} = (43.7 \times 10^{-6}) \times 240 \times \sqrt{2} = \mathbf{0.0148C}$$

$$E = \frac{1}{2}CV_{max}^2 = \frac{1}{2} \times 43.7 \times 10^{-6} \times (240 \times \sqrt{2})^2 = \mathbf{2.5\ J}$$

Example 16.20. *A 10 Ω resistor and 400 μF capacitor are connected in series to a 60 V sinusoidal supply. The circuit current is 5A. Calculate the supply frequency and phase angle between the current and voltage.*

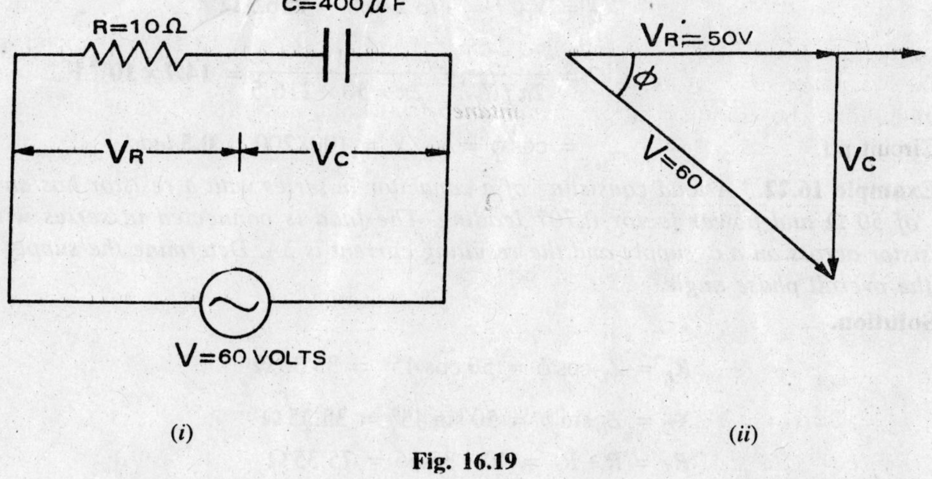

Fig. 16.19

Solution.

$$V_R = IR = 5 \times 10 = 50\ V$$

$$V_C = \sqrt{V^2 - V_R^2} = \sqrt{60^2 - 50^2} = 33.17\ \text{V}$$

$$X_C = V_C/I = 33.17/5 = 6.634\Omega$$

$$f = \frac{1}{2\pi C X_C} = \frac{10^6}{2\pi \times 400 \times 6.634} = \mathbf{60\ Hz}$$

$$\phi = \tan^{-1} V_C/V_R = \tan^{-1} 33.17/50 = \mathbf{33.6°}\ \textit{lead}$$

Example 16.21. *Calculate the capacitance of a condenser to be connected in series with a 100V, 80W lamp to enable it to be used on a 200V, 50Hz supply. What is p.f. of the circuit ?*

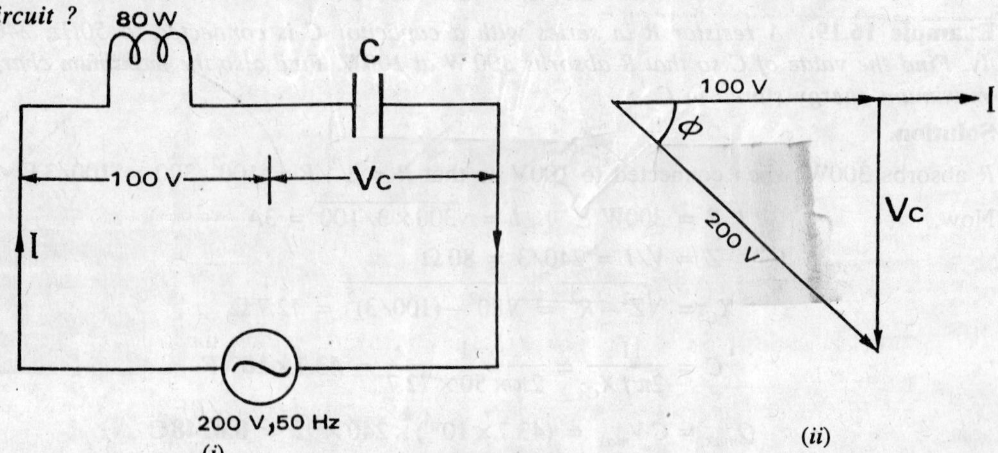

Fig. 16.20

Solution. The voltage across the bulb (pure resistance) should remain 100V for its proper operation. Fig. 16.20 (*i*) shows the required condenser of *C* farad in series with the lamp. The phasor diagram of the circuit is shown in Fig. 16.20 (*ii*).

Rated current of bulb, $I = 80/100 = 0.8\text{A}$

$$V_C = \sqrt{200^2 - 100^2} = 173.2\ \text{V}$$

$$X_C = V_C/I = 173.2/0.8 = 216.5\ \Omega$$

$$\therefore \quad C = \frac{1}{2\pi f X_C} = \frac{1}{2\pi \times 50 \times 216.5} = \mathbf{14.7 \times 10^{-6}\ F}$$

Circuit p.f. $= \cos\phi = V_R/V = 100/200 = \mathbf{0.5}\ \textit{lead}$

Example 16.22. *A load consisting of a capacitor in series with a resistor has an impedance of 50 Ω and power factor 0.707 leading. The load is connected in series with a 40 Ω resistor across an a.c. supply and the resulting current is 3A. Determine the supply voltage and the overall phase angle.*

Solution.

$$R_L = Z_L \cos\phi = 50 \cos 45° = 35.35\ \Omega$$

$$X_C = Z_L \sin\phi = 50 \sin 45° = 35.35\ \Omega$$

$$R_T = R + R_L = 40 + 35.35 = 75.35\ \Omega$$

$$Z = \sqrt{(75.35)^2 + (35.35)^2} = 83.23\,\Omega$$

$$V = IZ = 3 \times 83.23 = \mathbf{250\,V}$$

$$\phi = \tan^{-1} X_C/R_T = \tan^{-1} 35.35/75.35 = \mathbf{25.17°}\ \textit{lead}$$

16.9. Circle Diagram of R-C Series Circuit

Consider an *R-C* series circuit supplied from a constant voltage *V* and fixed frequency. The only variable is *R* as shown in Fig. 16.21 (*i*). At any particular value of *R*, the phasor diagram will be as shown in Fig. 16.21 (*ii*). As *R* is varied, the circuit current *I* also changes. It can be shown that locus of current *I* is a semicircle as shown in Fig. 16.21 (*iii*).

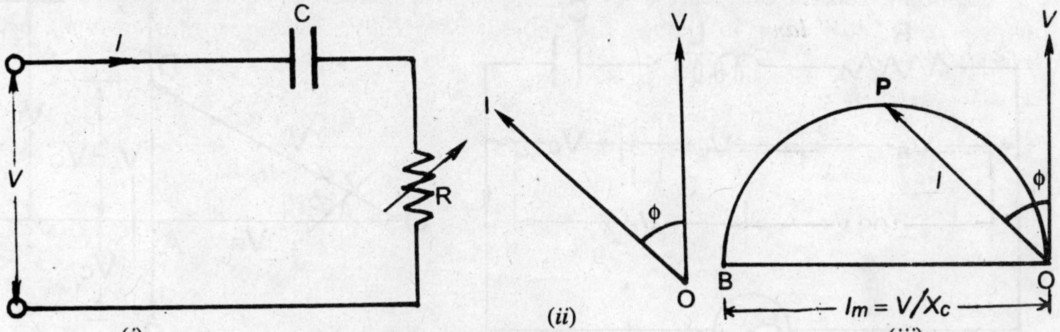

Fig. 16.21

(*i*) At a particular value of *R*, the current in the circuit is *OP*; the point *P* lying on a semicircle.

(*ii*) As *R* is decreased, the phase angle ϕ increases ($\tan\phi = X_C/R$) and point *P* shifts towards *B*. When *R* becomes zero, the phase angle is 90° leading and *I* has the maximum value, $I_m = V/X_C$

(*iii*) As *R* is increased, ϕ decreases and point *P* shifts towards *O*. When *R* becomes infinite, the phase angle is zero and point *P* approaches *O*. Consequently, circuit current becomes zero.

(*iv*) It can be shown that maximum power in the circuit is given by;

$$\text{Maximum power} = \frac{V^2}{2X_C}$$

It will occur when *P* is the mid-point of the arc of semicircle *i.e.* $\phi = 45°$

Example 16.23. *Give vector diagram for 100V, 50 Hz R-C series circuit having* $1/\omega C = 1$ *ohm and variable resistance. Find the maximum power. Draw the locus of current vector.*

Solution. Fig. 16.22 (*i*) shows the circuit diagram whereas Fig. 16.22 (*ii*) shows the locus of circuit current.

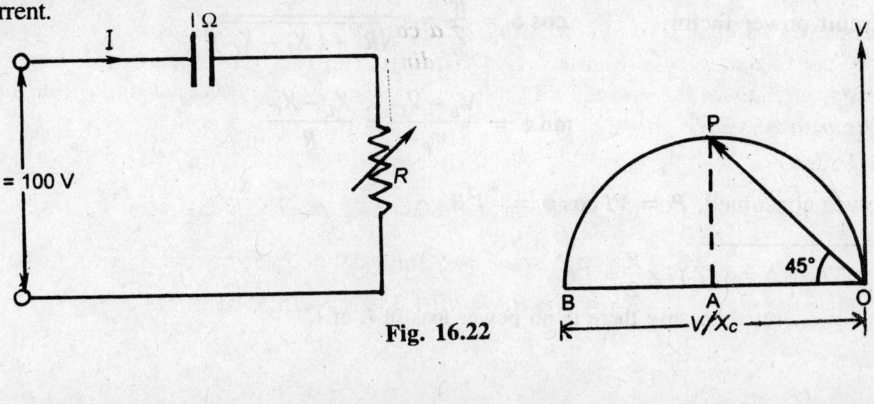

Fig. 16.22

$$\text{Maximum power} = \frac{V^2}{2X_C} = \frac{100^2}{2 \times 1} = \mathbf{5000\ W}$$

16.10. R-L-C Series Circuit

This is a general series a.c. circuit. Fig. 16.23 shows *R, L* and *C* connected in series across a supply voltage *V* (*r.m.s.*). The resulting circuit current is *I* (*r.m.s.*).

∴ Voltage across *R*, $V_R = IR$ ... V_R is in phase with I

Voltage across *L*, $V_L = IX_L$ ... where V_L leads L by 90°

Voltage across *C*, $V_C = IX_C$ ... where V_C lags I by 90°

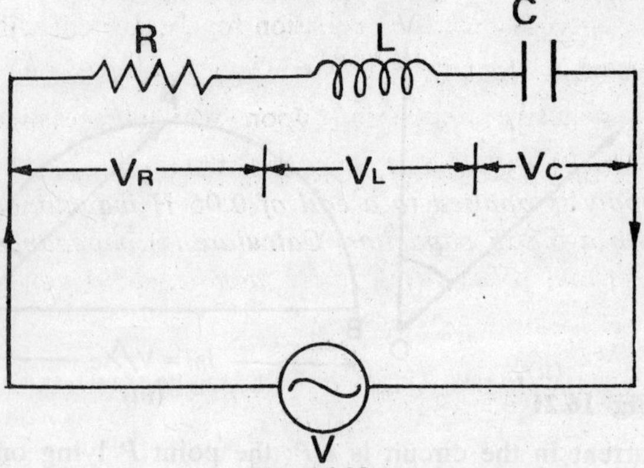

Fig. 16.23

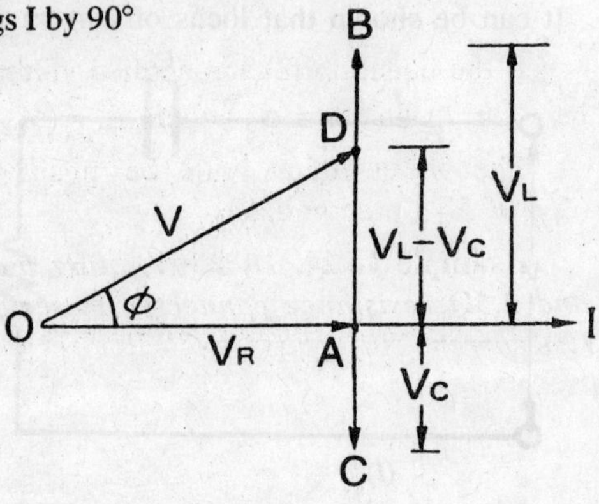

Fig. 16.24

As before, the phasor diagram is drawn taking current as the reference phasor. In the phasor diagram (See Fig. 16.24), *OA* represents V_R, *AB* represents V_L and *AC* represents V_C. It follows that the circuit can either be effectively inductive or capacitive depending upon which voltage drop (V_L or V_C) is predominant. For the case considered, $V_L > V_C$ so that net voltage drop across *L-C* combination is $V_L - V_C$ and is represented by *AD*. Therefore, the applied voltage *V* is the phasor sum of V_R and $V_L - V_C$ and is represented by *OD*.

$$V = \sqrt{V_R^2 + (V_L - V_C)^2}$$

$$= \sqrt{(IR)^2 + (IX_L - IX_C)^2} = I\sqrt{R^2 + (X_L - X_C)^2}$$

$$\therefore \quad I = \frac{V}{\sqrt{R^2 + (X_L - X_C)^2}}$$

The quantity $\sqrt{R^2 + (X_L - X_C)^2}$ offers opposition to current flow and is called impedance of the circuit.

Circuit power factor, $$\cos\phi = \frac{R}{Z} = \frac{R}{\sqrt{R^2 + (X_L - X_C)^2}} \quad ...(i)$$

$$\tan\phi = \frac{V_L - V_C}{V_R} = \frac{X_L - X_C}{R} \quad ...(ii)$$

Power consumed, $P = VI\cos\phi = {}^*I^2R$

---

* $P = VI\cos\phi = (IZ)\,I \times \frac{R}{Z} = I^2R$

This is expected because there is no power loss in *L* or *C*.

Three cases of R-L-C series circuit. We have seen that the impedance of a *R-L-C* series circuit is given by :

$$Z = \sqrt{R^2 + (X_L - X_C)^2}$$

(*i*) When $X_L - X_C$ is positive (*i.e.* $X_L > X_C$), the phase angle ϕ is positive and the circuit is inductive.

(*ii*) When $X_L - X_C$ is negative (*i.e.* $X_C > X_L$), the phase angle ϕ is negative and the circuit is capacitive.

(*iii*) When $X_L - X_C$ is zero (*i.e.* $X_L = X_C$), the phase angle ϕ is zero and the circuit is purely resistive.

If the equation for the applied voltage is $v = V_m \sin \omega t$, then equation for the current will be : $i = I_m \sin(\omega t \pm \phi)$ where $I_m = V_m/Z$

The value of ϕ will be positive or negative depending upon which reactance (X_L or X_C) predominates.

Example 16.24. *A 230V, 50Hz a.c. supply is applied to a coil of 0.06 H inductance and 2.5Ω resistance connected in series with a 6.8μF capacitor. Calculate (i) impedance (ii) current (iii) phase angle between current and voltage (iv) power factor and (v) power consumed.*

Solution.

$$X_L = 2\pi f L = 2\pi \times 50 \times 0.06 = 18.84\,\Omega$$

$$X_C = \frac{1}{2\pi f C} = \frac{10^6}{2\pi \times 50 \times 6.8} = 468\,\Omega$$

$$X = X_L - X_C = 18.84 - 468 = -449.16\Omega$$

(*i*)
$$Z = \sqrt{R^2 + X^2} = \sqrt{(2.5)^2 + (-449.16)^2} = \mathbf{449.2\,\Omega}$$

(*ii*)
$$I = V/Z = 230/449.2 = \mathbf{0.512\,A}$$

(*iii*)
$$\phi = \tan^{-1}\frac{X}{R} = \tan^{-1}\frac{-449.16}{2.5} = \mathbf{-89.7°}$$

The negative sign with ϕ shows that current is leading the voltage.

(*iv*)
$$p.f. = \cos\phi = R/Z = 2.5/449.2 = \mathbf{0.0056}\ \textit{lead}$$

(*v*)
$$P = VI\cos\phi = 230 \times 0.512 \times 0.0056 = \mathbf{0.66\,W}$$

Example 16.25. *A resistance R, an inductance L = 0.01H and a capacitance C are connected in series. When an alternating voltage v = 400 sin (3000t – 20°) is applied to the series combination, the current flowing is $10\sqrt{2}$ sin (3000t – 65°). Find the values of R and C.*

Solution.

$$\phi = 65° - 20° = 45°\ \textit{lag}\ \text{i.e. circuit is inductive.}$$

$$X_L = \omega L = 3000 \times 0.01 = 30\,\Omega$$

$$\tan 45° = X/R \quad \therefore X = R$$

$$Z = V_m/I_m = 400/10\sqrt{2} = 28.3\,\Omega$$

$$Z^2 = R^2 + X^2 = R^2 + R^2 = 2R^2$$

$\therefore$
$$R = Z/\sqrt{2} = 28.3/\sqrt{2} = \mathbf{20\,\Omega}$$

Now
$$X = X_L - X_C$$

$\therefore$
$$X_C = X_L - X = 30 - 20 = 10\,\Omega$$

$$\therefore \quad C = \frac{1}{\omega X_C} = \frac{1}{3000 \times 10} = \mathbf{33.3 \times 10^{-6} \ F}$$

Example 16.26. *A coil of resistance 8 Ω and inductance 0.03H is connected to an a.c. supply at 240V, 50Hz. Calculate the value of capacitance which when connected in series with the above coil, causes no change in the value of current and power taken from the supply.*

Solution.

$$X_L = 2\pi fL = 2\pi \times 50 \times 0.03 = 9.424 \ \Omega$$

To maintain the same current and power, the impedance of the circuit should not change. This is possible if the series capacitor has a capacitive reactance equal to twice the inductive reactance *i.e.*

$$X_C = 2X_L = 2 \times 9.424 = 18.848 \ \Omega$$

$$\therefore \quad C = \frac{1}{2\pi f X_C} = \frac{1}{2\pi \times 50 \times 18.848} = \mathbf{168.9 \times 10^{-6} \ F}$$

Example 16.27. *A coil of p.f. 0.8 is connected in series with a 110μF capacitor. The supply frequency is 50 Hz. The p.d. across the coil is found to be equal to the p.d. across the capacitor. Calculate the resistance and inductance of the coil.*

Solution.

$$X_C = \frac{1}{2\pi fC} = \frac{10^6}{2\pi \times 50 \times 110} = 29 \ \Omega$$

Now, $I Z_{coil} = IX_C$

$\therefore \quad Z_{coil} = X_C = 29 \ \Omega$

For the coil, $\cos \phi = R/Z_{coil}$

$\therefore \quad R = Z_{coil} \cos \phi = 29 \times 0.8 = \mathbf{23.2 \ \Omega}$

Reactance of coil, $X_L = Z_{coil} \sin \phi = 29 \times 0.6 = 17.4 \ \Omega$

$$\therefore \quad L = \frac{X_L}{2\pi f} = \frac{17.4}{2\pi \times 50} = \mathbf{0.055 \ H}$$

Example 16.28. *A resistance of 1 ohm, an inductance of 1 henry and a capacitance of 1 farad are connected in series across a voltage and the line current is 1A. Find the energy consumed in one hour.*

Solution. Energy is only consumed in R since L or C does not consume any energy.

Energy consumed $= I^2Rt = (1)^2 \times 1 \times 3600 = \mathbf{3600 \ J}$

Tutorial Problems

1. A circuit is made up of 5 Ω resistance, 0.2 *H* inductance and 60 μ*F* capacitance in series. If the circuit current is 20A at a supply frequency of 50*Hz*, find the applied voltage. **[219 V]**
2. A series circuit consists of a non-inductive resistor of 10Ω, an inductor having a reactance of 50 Ω and a capacitor having a reactance of 30Ω. It is connected to a 230V a.c. supply. Calculate (*i*) current (*ii*) the power consumed and (*iii*) the power factor. [(*i*) **10.3A** (*ii*) **1060W** (*iii*) **0.447** *lag*]
3. A coil of resistance 20Ω is in series with an inductance of 0.04H. A supply of 230V, 50Hz is applied to the combination. Determine the capacitance which when connected in series with the coil causes no change in the magnitude and power taken from the supply. **[67.5 μF]**

4. When a certain inductive coil is supplied at 240V, 50Hz, the current is 6.45A. When the frequency is changed to 40Hz at 240V, the current taken is 7.48A. Calculate the inductance and resistance of the coil. [**L = 0.1H ; R = 20 Ω**]

5. Two impedances Z_1 and Z_2 when connected separately across a 200V, 50Hz supply consume powers of 100 W, 60 W at power factors of 0.5 lagging and 0.6 leading respectively. If the two impedances are now connected together in series across the same supply, calculate (*i*) circuit current (*ii*) power absorbed (*iii*) circuit power factor. [(*i*) **0.54 A** (*ii*) **99 W** (*iii*) **0.787** ***lead***]

6. A circuit consists of a resistance of 12Ω, capacitance of 320 μF and an inductance of 0.08H, all in series. A supply of 240V, 50Hz is applied to the ends of the circuit. Calculate the frequency at which the circuit power factor would be unity. [**32 Hz**]

7. A high impedance voltmeter is used to measure the voltage drop across each of the ideal circuit elements of a R-L-C series circuit. If the r.m.s. readings are 40V, 25V and 60V for V_L, V_R and V_C respectively, determine the equation for the applied voltage. The frequency of the applied voltage is 50Hz. [**v = 45.28 sin (314t – 38.66°)**]

16.11. Resonance in R-L-C Circuit

A series circuit is said to be in *resonance when circuit power factor is unity *i.e.* $X_L = X_C$. The frequency f_r at which it occurs is called the *resonant frequency*. The resonance (*i.e.* $X_L = X_C$) in an *R-L-C* circuit can be achieved by changing the supply frequency because X_L and X_C are frequency dependent. At a certain frequency, called the resonant frequency f_r, X_L becomes equal to X_C and series resonance occurs.

At series resonance, $$X_L = X_C$$

or $$2\pi f_r L = \frac{1}{2\pi f_r C}$$

∴ Resonant frequency $$f_r = \frac{1}{2\pi\sqrt{LC}}$$

If L and C are measured in henry and farad respectively, then f will be in Hz.

16.12. Effects of Series Resonance

When series resonance occurs, the effect on the circuit is the same as though neither inductance nor capacitance is present. The current under this condition is dependent solely on the resistance of the circuit and voltage across it.

(*i*) The impedance of the circuit is minimum and is equal to the resistance of the circuit *i.e.*

$$Z_r = R$$

(*ii*) The current in the circuit is maximum as it is limited by the resistance of the circuit alone *i.e.*,

$$I_r = V/Z_r = V/R$$

As the current is at its maximum value, the power of the circuit will also be at its maximum value.

(*iii*) Since at series resonance the current flowing in the circuit is very large, the voltage drops across L and C are also very large. In fact, these drops are much greater than the applied voltage. However, voltage drop across *L-C* combination as a whole will be zero because these drops are equal in magnitude but 180° out of phase with each other.

---

* Resonance means to be in step with. When in *R-L-C* series circuit, current is in phase with applied voltage (*i.e.* circuit *p.f.* = 1), electrical resonance is said to occur. Clearly, series resonance in *R-L-C* circuit will occur when $X_L = X_C$.

16.13. Resonance Curve

The curve between current and frequency is known as resonance curve. Fig. 16.25 shows the resonance curve of a typical *R-L-C* series circuit. Note that current reaches its maximum value at the resonant frequency (f_r), falling off rapidly on either side at that point. It is because if the frequency is below f_r, $X_C > X_L$ and the net reactance is no longer zero. If the frequency is above f_r, then $X_L > X_C$ and the net reactance is again not zero. In both cases, the circuit impedance will be more than, the impedance Z_r ($= R$) at resonance. The result is that the magnitude of circuit current decreases rapidly as the frequency changes from the resonant frequency.

Note also the effect of resistance in the circuit. The smaller the resistance, the greater the current at resonance and sharper is the curve. On the other hand, the greater the resistance, the lower the resonant peak and flatter is the curve (See Fig. 16.25).

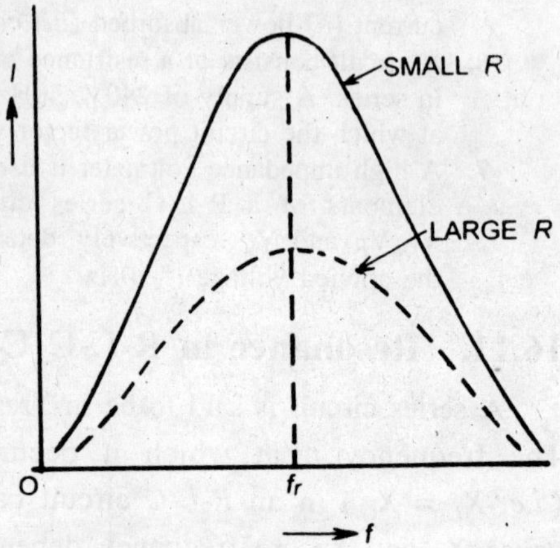

Fig. 16.25

16.14. Q-Factor of Series Resonant Circuit

At series resonance, the p.d. across *L* or *C* (the two drops being equal and opposite) builds up to a value many times greater than the applied voltage *V*. This voltage magnification produced by resonance is termed as *Q*-factor of the series resonant circuit (*Q* stands for quality) *i.e.*

$$Q\text{-factor} = \frac{\text{Voltage across } L \text{ or } C}{\text{Applied voltage}} = \frac{I_r X_L}{{}^*I_r R} = \frac{X_L}{R}$$

$$\therefore \quad Q\text{-factor} = \frac{\omega_r L}{R} \qquad \text{...(i) where } \omega_r = 2\pi f_r$$

The *Q*-factor of a series resonant circuit can also be expressed in terms of *L* and *C*.

We know, $$f_r = \frac{1}{2\pi\sqrt{LC}} \quad or \quad 2\pi f_r = \frac{1}{\sqrt{LC}}$$

or $$\omega_r = \frac{1}{\sqrt{LC}}$$

Substituting the value of ω_r in eq., (*i*), we get,

$$Q\text{-factor} = \frac{1}{R}\sqrt{\frac{L}{C}} \qquad \text{...(ii)}$$

The value of *Q*-factor depends entirely upon the design of coil (*i.e.* *R-L* part of the *R-L-C* circuit) because resistance arises in this rather than in a capacitor. With a well designed coil, the quality factor can be 200 or more.

Physical meaning of Q-factor. Let us now turn to the physical meaning of *Q*-factor. The *Q*-factor of a series circuit indicates how many times the p.d. across *L* or *C* is greater than the applied voltage at resonance. For example, consider a *R-L-C* series circuit connected to a 240*V* source. If *Q*-factor of the coil is 20, then voltage across the coil or a capacitor will be $20 \times 240 = 4800$ volts at resonance *i.e.*,

* At series resonance, applied voltage $V = I_r R =$ voltage across *R*

$$V_C = V_L = QV_R = 20 \times 240 = 4800 \text{ volts}$$

16.15. Bandwidth of a Series Resonant Circuit

The bandwidth of a series circuit is defined *as the range of frequency over which circuit current is equal to or greater than 70.7% of maximum current (i.e., I_r, current at resonance).*

The current in a series *R-L-C* circuit changes with frequency. Referring to the resonance curve (*i.e.* current/frequency curve) in fig. 16.26, it is clear that for any frequency lying between f_1 and f_2, the circuit current is equal to or greater than 70.7% of maximum current (*i.e.* $I_r = V/R$). Therefore $f_1 - f_2$ is the bandwidth of the circuit *i.e.*,

Band width, $BW = f_1 - f_2$

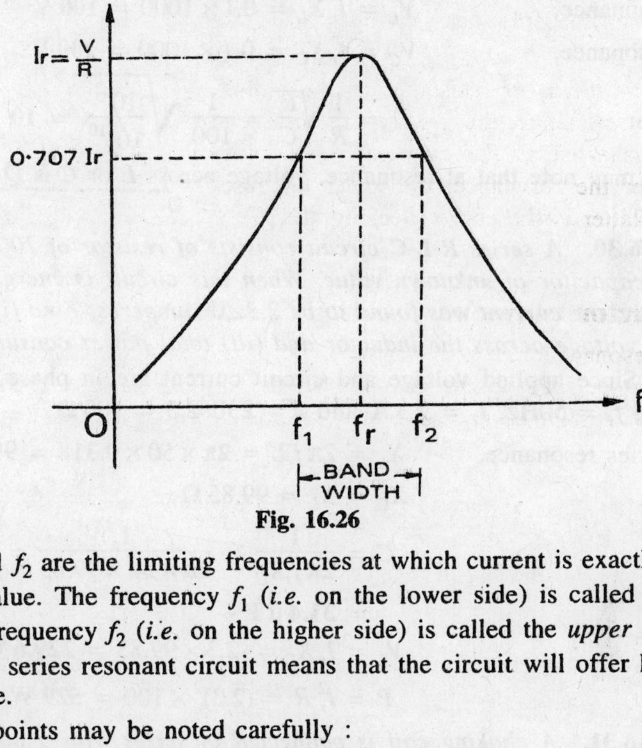

Fig. 16.26

Note that f_1 and f_2 are the limiting frequencies at which current is exactly equal to 70.7% of the maximum value. The frequency f_1 (*i.e.* on the lower side) is called the *lower cut off frequency* and the frequency f_2 (*i.e.* on the higher side) is called the *upper cut off frequency.* The bandwidth of a series resonant circuit means that the circuit will offer low impedance to this frequency range.

The following points may be noted carefully :

(*i*) If the resonant frequency is not centrally located *w.r.t.* lower and upper cut off frequencies, then f_r is the geometric mean of the two *i.e.*,

$$f_r = \sqrt{f_1 f_2}$$

(*ii*) If Q of a circuit is ≥ 10 and the resonant frequency is sufficiently centered *w.r.t.* the two cut off frequencies (f_1 and f_2), then,

$$f_2 = f_r + \frac{BW}{2} \qquad \text{For } Q \geq 10$$

$$f_1 = f_r - \frac{BW}{2}$$

(*iii*) It can be proved mathematically that the following relation exists between Q-factor and bandwidth of a series resonant circuit :

$$Q = \frac{f_r}{BW}$$

Example 16.29. *A coil of resistance 100Ω and inductance 100μH is connected in series with a 100 pF capacitor. The circuit is connected to a 10V variable frequency supply. calculate (i) the resonant frequency (ii) current at resonance (iii) voltage across L and C at resonance and (iv) Q-factor of the circuit.*

Solution.

(*i*) Resonant frequency, $f_r = \dfrac{1}{2\pi\sqrt{LC}} = \dfrac{1}{2\pi\sqrt{10^{-4}\times 10^{-10}}} = \mathbf{1.59\times 10^6\ Hz}$

(*ii*) Current at resonance, $I_r = V/R = 10/100 = \mathbf{0.1\ A}$

(*iii*) At resonance, $X_L = 2\pi f_r L = 2\pi \times 1.59\times 10^6 \times 10^{-4} = 1000\ \Omega$

At resonance, $V_L = I_r X_L = 0.1\times 1000 = \mathbf{100\ V}$

At resonance, $V_C = I_r X_C = 0.1\times 1000 = \mathbf{100\ V}$

(*iv*) *Q*-factor $= \dfrac{1}{R}\sqrt{\dfrac{L}{C}} = \dfrac{1}{100}\sqrt{\dfrac{10^{-4}}{10^{-10}}} = \mathbf{10}$

The reader may note that at resonance, voltage across *L* or *C* is *Q* times the applied voltage.

Example 16.30. *A series R-L-C circuit consists of resistor of 100 ohms, an inductor of 0.318H and a capacitor of unknown value. When this circuit is energised by 230∠0°, 50Hz sinusoidal supply, the current was found to be 2.3∠0° amperes. Find (i) value of capacitor in microfarad (ii) voltage across the inductor and (iii) total power consumed.*

Solution. Since applied voltage and circuit current are in phase, it is a case of series resonance. Here f_r = 50Hz, I_r = 2.3 A and R = 230/2.3 = 100 Ω

(*i*) At series resonance, $X_L = 2\pi f_r L = 2\pi\times 50\times 0.318 = 99.85\Omega$

Now $X_C = X_L = 99.85\ \Omega$

∴ $C = \dfrac{1}{2\pi f_r X_C} = \dfrac{1}{2\pi\times 50\times 99.85} = 31.4\times 10^{-6}\ F$

$= \mathbf{31.4\ \mu F}$

(*ii*) $V_L = I_r X_L = 2.3\times 99.85 = \mathbf{229.6\ V}$

(*iii*) $P = I_r^2 R = (2.3)^2\times 100 = \mathbf{529\ W}$

Example 16.31. *A choking coil is connected in series with a 20 μF capacitor. With a constant supply voltage of 200 V, it is found that the circuit takes its maximum current of 50A when the supply frequency is 100 Hz. Calculate (i) resistance and inductance of the choking coil and (ii) voltage across the capacitor. What is the Q-factor of the circuit ?*

Solution. Since current is maximum, the circuit is in resonance *i.e.*, I_r = 50A ; f_r = 100Hz

(*i*) $R = V/I_r = 200/50 = \mathbf{4\ \Omega}$

$X_C = 1/2\pi f_r C = 10^6/2\pi\times 100\times 20 = 79.62\ \Omega$

Now $X_C = X_L = 79.62\ \Omega$

∴ $L = X_L/2\pi f_r = 79.62/2\pi\times 100 = \mathbf{0.127\ H}$

(*ii*) $V_C = I_r X_C = 50\times 79.62 = \mathbf{3981\ V}$

Q-factor $= V_C/V = 3981/200 = \mathbf{19.9}$

Tutorial Problems

1. A resistance of 15 Ω and an inductance of 4H and a capacitance 25 μF are connected in series across 230V a.c. supply. Calculate (*i*) the frequency at which the current shall be maximum (*ii*) current at this frequency and (*iii*) p.d. across inductance. **[(*i*) 15.9Hz (*ii*) 15.33A (*iii*) 6123V]**
2. A 1200 Ω resistor, a 0.7 H coil and a 0.001μF capacitor are in series across a 120V source. Determine (*i*) the resonant frequency (*ii*) the voltage across the capacitor at resonance and (*iii*) *Q*-factor of the circuit at resonance. **[(*i*) 6015Hz (*ii*) 2646V (*iii*) 22]**
3. A capacitor C is in series with a 75Ω resistor and a 12H coil across a 220V, 60Hz supply. Determine the value of C that resonates the circuit. **[0.587 μ F]**
4. A series circuit consists of a 325Ω resistor, a 0.6 pF capacitor and a coil of 2.93H. Determine the bandwidth if the circuit resonates at 120×10^6 Hz. **[17.643×10^6 Hz]**

OBJECTIVE QUESTIONS

1. The impedance of an *R-L* series circuit is given by ....
   (*i*) $R + X_L$ (*ii*) $R^2 + X_L^2$
   (*iii*) $\sqrt{R^2 + X_L^2}$ (*iv*) none of the above
2. In an *R-L* series circuit, line current ....
   (*i*) leads the applied voltage
   (*ii*) lags behind the applied voltage
   (*iii*) is in phase with applied voltage
   (*iv*) none of the above
3. In an *R-L* series circuit, the phase difference ϕ between applied voltage and circuit current will increase if................
   (*i*) X_L is increased
   (*ii*) *R* is increased
   (*iii*) X_L is decreased
   (*iv*) supply frequency is decreased
4. The power consumed in an a.c. circuit is given by ....
   (*i*) $V_m I_m$ (*ii*) $\frac{V_m I_m}{2}\cos\phi$
   (*iii*) $IZ\cos\phi$ (*iv*) $V_m I_m \cos\phi$
5. The power factor of an a.c. circuit is given by ....
   (*i*) X_L / R (*ii*) Z/R
   (*iii*) R / X_L (*iv*) R/Z
6. In an R-L series circuit, $R = 10\Omega$ and $X_L = 10\Omega$. The phase angle between applied voltage and circuit current is ....
   (*i*) 45° (*ii*) 30°
   (*iii*) 60° (*iv*) 36.8°
7. For greater accuracy, the value of ϕ (*i.e.*, phase angle) should be determined from ....
   (*i*) $\cos\phi$ (*ii*) $\sin\phi$
   (*iii*) $\tan\phi$ (*iv*) $\sec\phi$
8. The power factor of a d.c. circuit is ....
   (*i*) 0 (*ii*) 1
   (*iii*) –1 (*iv*) none of the above
9. An inductive circuit draws a line current of 10 A. If the reactive component of line current is 6 A, then power factor of the circuit is ....
   (*i*) 0.6 lagging (*ii*) 0.8 lagging
   (*iii*) 0.5 lagging (*iv*) none of the above
10. As the power factor of a circuit is increased, ...
    (*i*) reactive power is decreased
    (*ii*) active power is decreased
    (*iii*) reactive power is increased
    (*iv*) both active and reactive powers are increased.
11. An R-L series a.c. circuit has 15 V across resistor and 20 V across the inductor. The supply voltage is ....
    (*i*) 35 V (*ii*) 5 V
    (*iii*) 25 V (*iv*) $\sqrt{175}$ V
12. The power factor of the circuit shown in Fig. 16.27 is ....
    (*i*) 0.8 lagging (*ii*) 0.4 lagging
    (*iii*) 0.75 lagging (*iv*) 0.6 lagging

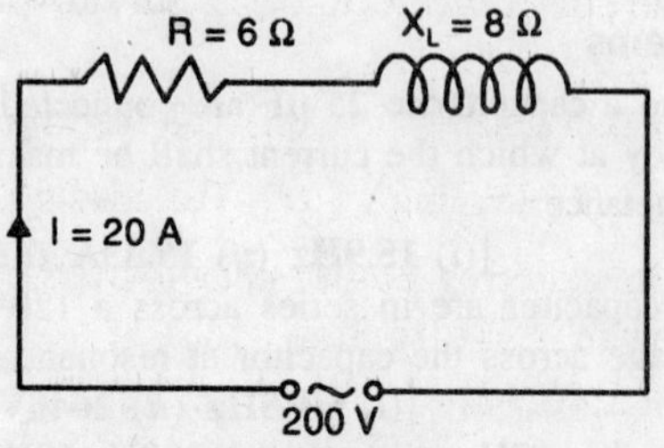

Fig. 16.27

13. The power consumed in the circuit shown in Fig. 16.27 is ....
(i) 4000 W (ii) 2400 W
(ii) 2000 W (iv) 1200 W

14. The reactive power drawn by the circuit shown in Fig. 16.27 is ....
(i) 2400 VAR (ii) 4000 VAR
(iii) 3200 VAR (iv) 2000 VAR

15. The active component of line current in Fig. 16.27 is ....
(i) 12 A (ii) 16 A
(iii) 20 A (iv) 4 A

16. The value of R in the circuit shown in Fig. 16.28 is ....
(i) 16Ω (ii) 10Ω
(iii) 3Ω (iv) 4Ω

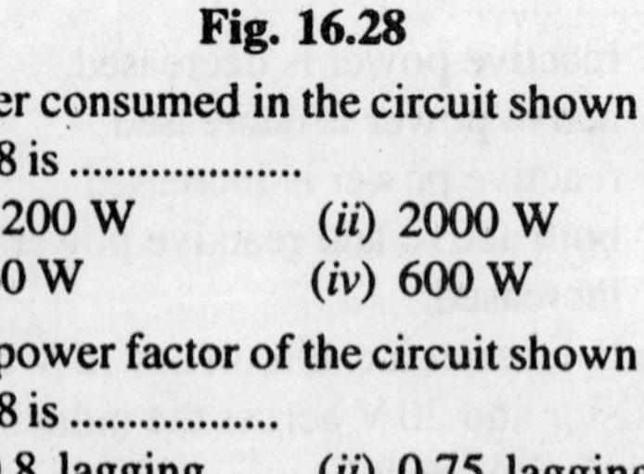

Fig. 16.28

17. Power consumed in the circuit shown in Fig. 16.28 is ....
(i) 1200 W (ii) 2000 W
(iii) 60 W (iv) 600 W

18. The power factor of the circuit shown in Fig. 16.28 is ....
(i) 0.8 lagging (ii) 0.75 lagging
(iii) 0.5 lagging (iv) 0.6 lagging

19. The wattful component of current in the circuit shown in Fig. 16.28 is ....
(i) 10 A (ii) 15 A
(iii) 12 A (iv) 16 A

20. The active and reactive components of line current of an inductive circuit are equal. The power factor of the circuit is ....
(i) 1 (ii) 0.707 lagging
(iii) 0.5 lagging (iv) 0.866 lagging

21. The active and reactive powers of an inductive circuit are 60 W and 80 VAR respectively. The power factor of the circuit is ....
(i) 0.8 lagging (ii) 0.75 lagging
(iii) 0.6 lagging (iv) 0.5 lagging

22. The active and apparent powers of an a.c. circuit are equal in magnitude. The circuit power factor is ....
(i) 0.707 (ii) 0.5
(iii) 0.8 (iv) 1

23. A wattmeter indicates .... power.
(i) active (ii) reactive
(iii) apparent (iv) none of the above

24. If the lagging reactive power of an a.c. circuit increases, the power factor of the circuit ....
(i) is increased (ii) is decreased
(iii) remains unchanged
(iv) none of the above

25. Reactive power in an a.c. circuit is ....
(i) measured by a wattmeter
(ii) the useful power
(iii) consumed in the circuit
(iv) a liability on the circuit

26. A 200 V, 50 Hz inductive circuit takes a current of 10 A, lagging 30°. The inductive reactance of the circuit is ....
(i) 20Ω (ii) 10Ω
(iii) 17.32Ω (iv) 16Ω

27. In an R-L series circuit, the two sides of the impedance triangle that form the phase angle are ....
(i) R and X_L (ii) R and Z
(iii) Z and X_L (iv) none of the above

28. Impedance of an a.c. circuit is a ....
(i) phasor (ii) vector quantity
(iii) scalar quantity (iv) none of the above

29. The power factor of the circuit shown in Fig. 16.29 is ....
(i) 0.6 lagging (ii) 0.3 lagging
(iii) 0.8 lagging (iv) 0.75 lagging

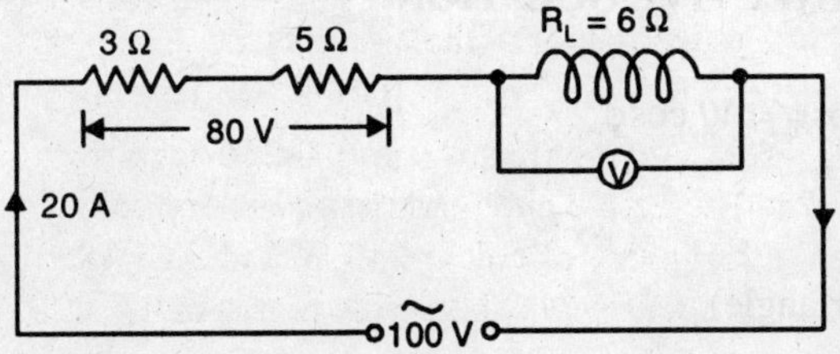

Fig. 16.29

30. In Fig. 16.29, the voltmeter will read ....

(i) 20 V (ii) 40 V
(iii) 120 V (iv) 60 V

31. The reactive component of line current in Fig. 16.29 is ....

(i) 8 A (ii) 6 A
(iii) 10 A (iv) 12 A

32. The power consumed in the circuit shown in Fig. 16.29 is ....

(i) 1000 W (ii) 800 W
(iii) 1400 W (iv) 600 W

33. A circuit when connected to 200 V mains takes a current of 20 A, leading the voltage by one-twelfth of time period. The circuit resistance is ....

(i) 10Ω (ii) 8.66Ω
(iii) 20Ω (iv) 17.32Ω

34. In Fig. 16.30, the voltmeter will read..........

(i) 100 V (ii) 50 V
(iii) 200 V (iv) 175 V

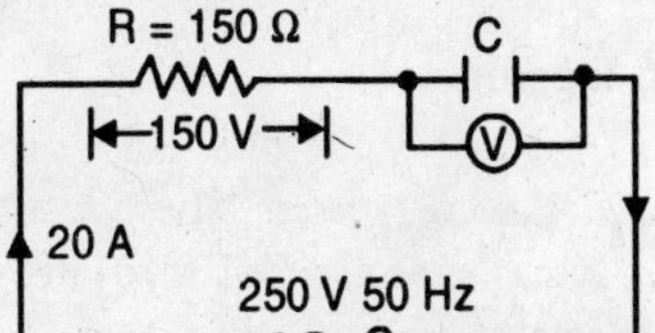

Fig. 16.30

35. The power factor of the circuit shown in Fig. 16.30. is ....

(i) 0.6 leading (ii) 0.8 leading
(iii) 0.75 leading (iv) 0.5 leading.

36. The reactive power drawn by the circuit shown in Fig. 16.30 is ....

(i) 250 VAR (ii) 150 VAR
(iii) 200 VAR (iv) 100 VAR

37. An a.c. system supplies an apparent power of 10 kVA to a circuit, the reactive power being 6 k VAR. The power factor of the circuit is ......

(i) 0.8 (ii) 0.6
(iii) 0.75 (iv) none of the above

38. A low power factor of the circuit means that it will ....

(i) draw more active power
(ii) draw less line current
(iii) draw more reactive power
(iv) cause less voltage drop in the line.

39. The impedance of an *R-C* series circuit is 100Ω. If circuit current leads the applied voltage by 45°, then capacitive reactance is ....

(i) 70.7Ω (ii) 141.42Ω
(iii) 100Ω (iv) 50Ω

40. In an R-C series circuit, $X_C = R$. The phase angle between applied voltage and circuit current is ....

(i) 30° (ii) 45°
(iii) 60° (iv) 90°

ANSWERS TO OBJECTIVE QUESTIONS

1. (iii)	**2.** (ii)	**3.** (i)	**4.** (ii)	**5.** (iv)	**6.** (i)	**7.** (iii)	**8.** (ii)
9. (ii)	**10.** (i)	**11.** (iii)	**12.** (iv)	**13.** (ii)	**14.** (iii)	**15.** (i)	**16.** (iii)
17. (i)	**18.** (iv)	**19.** (iii)	**20.** (ii)	**21.** (iii)	**22.** (iv)	**23.** (i)	**24.** (ii)
25. (iv)	**26.** (ii)	**27.** (ii)	**28.** (iii)	**29.** (iii)	**30.** (iv)	**31.** (ii)	**32.** (ii)
33. (ii)	**34.** (iii)	**35.** (i)	**36.** (iii)	**37.** (i)	**38.** (iii)	**39.** (i)	**40.** (ii)

HINTS TO SELECTED OBJECTIVE QUESTION

4. Power Consumed $= \dfrac{V_m I_m}{2}\cos\phi = \dfrac{V_m}{\sqrt{2}} \times \dfrac{I_m}{\sqrt{2}}\cos\phi = VI\cos\phi$

where V and I are the r.m.s. values.

5. Power factor $= \cos\phi = R/Z$ (From impedance triangle)

6. $\tan\phi = X_L / R = 10/10 = 1 \therefore \phi = 45°$

Remember, in an R-L series circuit, when $R = X_L$, phase angle will be 45°.

9. Active component $= \sqrt{10^2 - 6^2} = 8$

$$\text{Power factor} = \frac{\text{Active Component}}{\text{Total current}} = \frac{I\cos\phi}{I} = \frac{8}{10} = 0.8$$

11. Supply voltage

$$= \sqrt{V_R^2 + V_L^2} = \sqrt{(15)^2 + (20)^2} = 25 \text{ V}$$

13. Power consumed, $P = VI\cos\varphi$

$$= VI(R/Z) = 200 \times 20(6/10) = 2400 \text{ W}$$

14. Reactive power $= VI\sin\phi$

$$= VI(X/Z) = 200 \times 20(8/10) = 3200 \text{ VAR}$$

15. Active component $= I\cos\phi = 20 \times 0.6 = 12$ A

21. Apparent power

$$= \sqrt{(60)^2 + (80)^2} = 100$$

$$\text{Power factor} = \frac{\text{Active power}}{\text{Apparent power}} = \frac{VI\cos\phi}{VI} = \frac{60}{100} = 0.6 \text{ lagging}$$

26. $Z = 200/10 = 20\Omega$; $X_L = Z\sin\phi = 20 \times \sin 30° = 10\Omega$

30. $V_L = \sqrt{V^2 - V_R^2} = \sqrt{100^2 - 80^2} = 60$ V

31. Circuit current, $I = 80/3 + 5 = 10$ A

Reactive component $= I\sin\phi = I(X_L/Z) = 10(6/10) = 6A$

33. $Z = 200/20 = 10\Omega$; $\phi = 360°/12 = 30°$

$R = Z\cos\phi = 10\cos 30° = 8.66\Omega$

35. Power factor $= V_R/V = 150/250 = 0.6$ leading

39. $X_C = Z\sin\phi = 100\sin 45° = 70.7\Omega$

17

Phasor Algebra

Introduction

So far we used phasor diagrams to solve problems on a.c. circuits. A phasor diagram is a graphical representation of the phasors (*i.e.*, voltages and currents) of an a.c. circuit and may not yield quick results in case of complex circuits. Engineers have developed techniques to represent a phasor in an algebraic (*i.e.*, mathematical) form. Such a technique is known as *phasor algebra* or *complex algebra.* Phasor algebra has provided a relatively simple but powerful tool for obtaining quick solution of a.c. circuits. It simplifies the mathematical manipulation of phasors to a great extent. In this chapter, we shall discuss the various methods of representing phasors in a mathematical form and their applications to a.c. circuits.

17.1. Notation of Phasors on Rectangular Co-ordinate axes

Consider a phasor **V** lying along *OX*-axis as shown in Fig. 17.1. If we multiply this phasor by –1, the phasor is reversed *i.e.*, it is rotated through 180° in the counterclockwise (*CCW*) direction. Let us see with what factor the phasor be multiplied so that it rotates through 90° in *CCW* direction. Suppose this factor is *j*. As shown in Fig. 17.1, multiplying the phasor by j^2 rotates the phasor through 180° in *CCW* direction. This means that multiplying the phasor by j^2 is the same as multiplying by –1. It follows, therefore, that

$$j^2 = -1$$

or $$j = \sqrt{-1}$$

We arrive at a very important conclusion that *when a phasor is multiplied by j (= $\sqrt{-1}$), the phasor is rotated through 90° in CCW direction.* Each successive multiplication by *j* rotates the phasor through an additional 90° in the *CCW* direction. It is easy to see that multiplying a phasor by

$j = \sqrt{-1}$ ...90° *CCW* rotation from *OX*-axis

$j^2 = -1$ ...180° *CCW* rotation from *OX*-axis

$j^3 = j^2 \,.\, j = -j$ ...270° *CCW* rotation from *OX*-axis

$j^4 = j^2\, j^2 = 1$ ...360° *CCW* rotation from *OX*-axis

Fig. 17.1 shows the effect of multiplying a phasor by *j*.

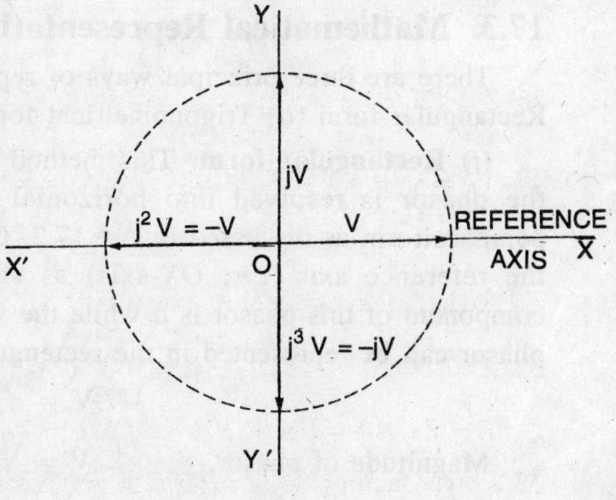

Fig. 17.1

17.2. Significance of Operator *j*

Just as the symbol + indicates the operation of adding two numbers, similarly *j* indicates an operation of rotating the phasor through 90° in *CCW* direction. The operator *j* does not change the magnitude of the phasor. Consider a phasor **V** displaced θ° counter-clockwise from *OX*-axis as shown in Fig. 17.2. This phasor can be resolved into two rectangular components *viz.* the horizontal component '*a*' along *X*-axis and the vertical component '*b*' along *Y*-axis. It can be seen that vertical component is displaced 90° *CCW* from *OX*-axis. Therefore, mathematically, we can express this component as *jb*, meaning that component *b* is displaced 90° *CCW* from component *a* (*i.e.*, *OX*-axis).

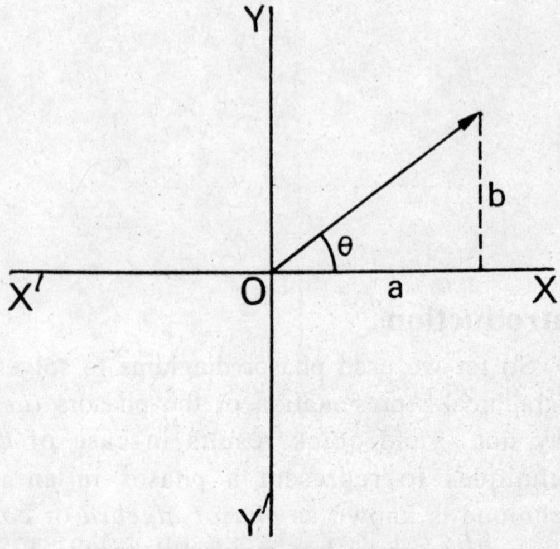

Fig. 17.2

$$\therefore \qquad \mathbf{V} = a + jb$$

Magnitude of **V**, $\quad V = \sqrt{a^2 + b^2}$

Its angle with *OX*-axis, $\quad \theta = \tan^{-1}(b/a)$

The reader may note that $a + jb$ is the mathematical form of the phasor **V**.

(*i*) The quantity $a + jb$ is called a complex number or complex quantity.

(*ii*) The horizontal component (*i.e.*, *a*) is called the *in-phase* (or active) component while the vertical component is called *quadrature* (or reactive) component.

Note. In this book, bold-faced letters (*e.g.*, **V**, **I** *etc.*) will be used to represent the phasor completely, including both magnitude and direction. However, only the magnitude of the phasor will be represented by the same letter in the ordinary type (*e.g.*, *V*, *I etc.*).

17.3. Mathematical Representation of Phasors

There are three principal ways of representing a phasor in the mathematical form *viz.* (*i*) Rectangular form (*ii*) Trigonometrical form and (*iii*) Polar form.

(*i*) **Rectangular form.** This method is also known as *symbolic notation.* In this method, the phasor is resolved into horizontal and vertical components and is expressed in the complex form as discussed in Art 17.2. Consider a voltage phasor **V** displaced θ° *CCW* from the reference axis (*i.e.*, *OX*-axis) as shown in Fig. 17.3 (*i*). The horizontal or in-phase component of this phasor is *a* while the vertical or quadrature component is *b*. Therefore, the phasor can be represented in the rectangular form as :

$$\mathbf{V} = a + jb$$

Magnitude of phasor, $\quad V = \sqrt{a^2 + b^2}$

Its angle *w.r.t.* *OX*-axis, $\quad \theta = \tan^{-1}(b/a)$

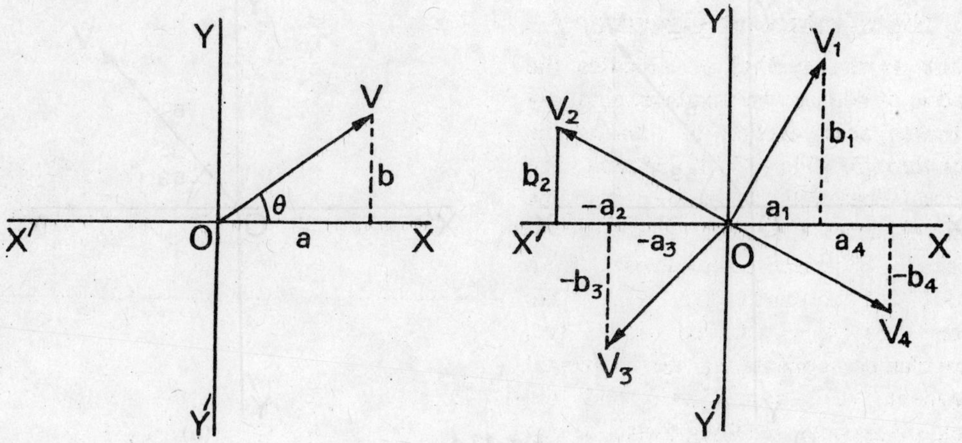

(*i*) (*ii*)

Fig. 17.3

Fig. 17.3 (*ii*) shows the phasors in the various quadrants. The phasors have been resolved into horizontal and vertical components. The various phasors can be represented in the rectangular form as :

$$\mathbf{V_1} = a_1 + jb_1 \qquad \mathbf{V_2} = -a_2 + jb_2$$
$$\mathbf{V_3} = -a_3 - jb_3 \qquad \mathbf{V_4} = a_4 - jb_4$$

The magnitude and phase angles of these phasors can be found out as explained above. Remember phase angles are to be measured from the reference axis *i.e., OX*-axis. If the angle is measured *CCW* from this reference axis, it is assigned a +ve sign. If the angle is measured clockwise from the reference axis, it is assigned a –ve sign.

(*ii*) **Trigonometrical form.** It is similar to the rectangular form except that in-phase and quadrature components of the phasor are expressed in the trigonometrical form. Thus referring to Fig. 17.3 (*i*) above, $a = V\cos\theta$ and $b = V\sin\theta$ where V is the magnitude of the phasor **V**. Hence phasor **V** can be expressed in the trigonometrical form as :

$$\mathbf{V} = V(\cos\theta + j\sin\theta)$$

It may be noted that in this form, we express the phasor in terms of its magnitude and phase angle θ.

(*iii*) **Polar form.** It is a usual practice to write the trigonometrical form $\mathbf{V} = V(\cos\theta + j\sin\theta)$ in what is called polar form as :

$$^*\mathbf{V} = V\angle\theta$$

where V is the magnitude of the phasor and θ is its angle measured *CCW* from the reference axis *i.e., OX*-axis. A negative angle in the polar form indicates clockwise measurement of the angle from the reference axis. Hence, polar form can be written in general as :

$$\mathbf{V} = V\angle \pm\theta$$

17.4. Conversion from One Form to the Other

The reader may see that the above three mathematical forms of representing a phasor convey the same information *i.e.*, magnitude of the phasor and its phase angle. Therefore, it is possible to convert one form to the other. Consider a phasor **V** having in-phase and quadrature components as 3 and 4 respectively as shown in Fig. 17.4 (*i*).

* meaning **V** equals V at angle θ (in *CCW* direction). There is mathematical explanation for this form. It is a short hand form of trigonometrical form.

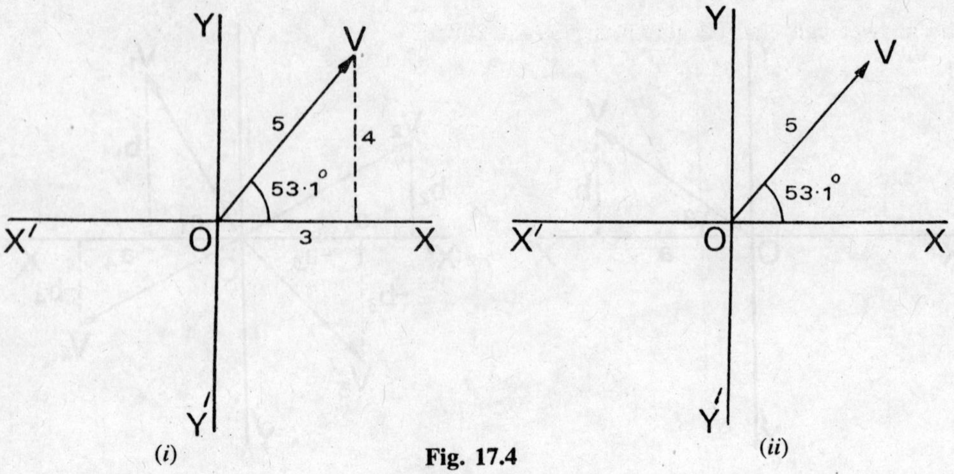

Fig. 17.4

Magnitude of the phasor, $V = \sqrt{3^2 + 4^2} = 5$

Its angle *w.r.t.* OX-axis, $\theta = \tan^{-1} 4/3 = 53.1°$

In rectangular form, $\mathbf{V} = 3 + j4$

In trigonometrical form, $\mathbf{V} = 5\,(\cos 53.1° + j \sin 53.1°)$

In polar form, $\mathbf{V} = 5 \angle 53.1°$ [See Fig. 17.4 (*ii*)]

This numerical illustrates how one form can be converted to the other.

Example 17.1. *Express the polar form of voltage* $\mathbf{V} = 50 \angle 36.87°$ *in trigonometrical and rectangular forms.*

Solution.

Trigonometrical form

$$\mathbf{V} = V(\cos\theta + j\sin\theta)$$
$$= 50\,(\cos 36.87° + j \sin 36.87°) \textbf{ Ans.}$$

Rectangular form

In-phase component $= V\cos\theta = 50 \times \cos 36.87° = 40$

Quadrature component $= V\sin\theta = 50 \times \sin 36.87° = 30$

$\therefore$ $\mathbf{V} = 40 + j\,30$ **Ans.**

Example 17.2. *Express the following in polar form (i) $3 + j7$ (ii) $-2 + j5$ (iii) $-50 - j75$ and (iv) $6 - j8$.*

Solution.

(*i*) Magnitude $= \sqrt{3^2 + 7^2} = 7.62$

Phase angle, $\theta = \tan^{-1}(7/3) = 66.8°$

$\therefore$ $3 + j7 = \mathbf{7.62 \angle 66.8°}$

Note that $3 + j7$ lies in the first quadrant. As seen in Fig. 17.5 (*i*), θ measured in *CCW* direction from OX-axis is 66.8° and is, therefore, positive. If θ is measured in clockwise direction from OX-axis, its value is $= 360° - 66.8° = 293.2°$ and this angle is negative. Hence the above answer can also be given as $7.62 \angle -293.2°$.

(*ii*) Magnitude $= \sqrt{(-2)^2 + (5)^2} = 5.39$

Phase angle, $\theta = \tan^{-1} -5/2 = 180° - 68.2° = 111.8°$

$\therefore$ $-2 + j5 = \mathbf{5.39 \angle 111.8°}$

The answer can also be given as $5.39 \angle -248.2°$.

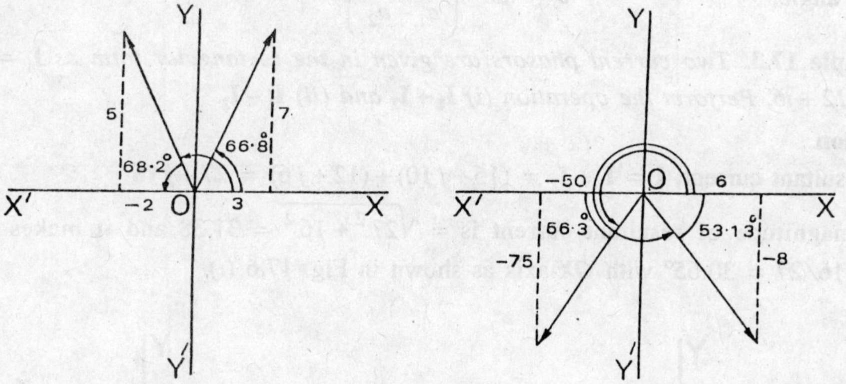

(*i*) (*ii*)

Fig. 17.5

(*iii*) Magnitude $= \sqrt{(-50)^2 + (-75)^2} = 90.1$

Phase angle, $\theta = \tan^{-1} -75/-50 = 180° + 56.3° = 236.3°$

$\therefore \quad -50 - j\,75 = \mathbf{90.1 \angle 236.3°}$

The answer can also be given as $90.1 \angle -123.7°$.

(*iv*) Magnitude $= \sqrt{6^2 + (-8)^2} = 10$

Phase angle, $\theta = \tan^{-1} -8/6 = 360° - 53.13° = 306.87°$

$\therefore \quad 6 - j\,8 = \mathbf{10 \angle 306.87°}$

The answer can also be given as $10 \angle -53.13°$.

17.5. Addition and Subtraction of Phasors

The rectangular form is best suited for addition or subtraction of phasors. If the phasors are given in polar form, they should be first converted to rectangular form and then addition or subtraction be carried out.

(*i*) **Addition.** For the addition of phasors in the rectangular form, the in-phase components are added together and the quadrature components are added together. Consider two voltage phasors represented as :

$$\mathbf{V}_1 = a_1 + jb_1\,; \qquad \mathbf{V}_2 = a_2 + jb_2$$

Then resultant, $\mathbf{V} = \mathbf{V}_1 + \mathbf{V}_2 = (a_1 + jb_1) + (a_2 + jb_2)$

$$= (a_1 + a_2) + j\,(b_1 + b_2)$$

Magnitude of resultant, $V = \sqrt{(a_1 + a_2)^2 + (b_1 + b_2)^2}$

Its angle from *OX*-axis, $\theta = \tan^{-1} \dfrac{(b_1 + b_2)}{a_1 + a_2}$

(*ii*) **Subtraction.** Like addition, the subtraction of phasors is done by using ordinary rules of phasor algebra.

$$\mathbf{V} = \mathbf{V}_1 - \mathbf{V}_2 = (a_1 + jb_1) - (a_2 + jb_2)$$

$$= (a_1 - a_2) + j\,(b_1 - b_2)$$

$\therefore \quad V = \sqrt{(a_1 - a_2)^2 + (b_1 - b_2)^2}$

Phase angle, $\theta = \tan^{-1}\left(\dfrac{b_1 - b_2}{a_1 - a_2}\right)$

Example 17.3. *Two current phasors are given in the rectangular form as* $\mathbf{I}_1 = 15 + j10$ *and* $\mathbf{I}_2 = 12 + j6$. *Perform the operation (i)* $\mathbf{I}_1 + \mathbf{I}_2$ *and (ii)* $\mathbf{I}_1 - \mathbf{I}_2$.

Solution.

(*i*) Resultant current, $\mathbf{I} = \mathbf{I}_1 + \mathbf{I}_2 = (15 + j\,10) + (12 + j\,6) = \mathbf{27 + j\,16}$

The magnitude of resultant current is $= \sqrt{27^2 + 16^2} = 31.38$ and it makes an angle $\theta = \tan^{-1} 16/27 = 30.65°$ with *OX*-axis as shown in Fig. 17.6 (*i*).

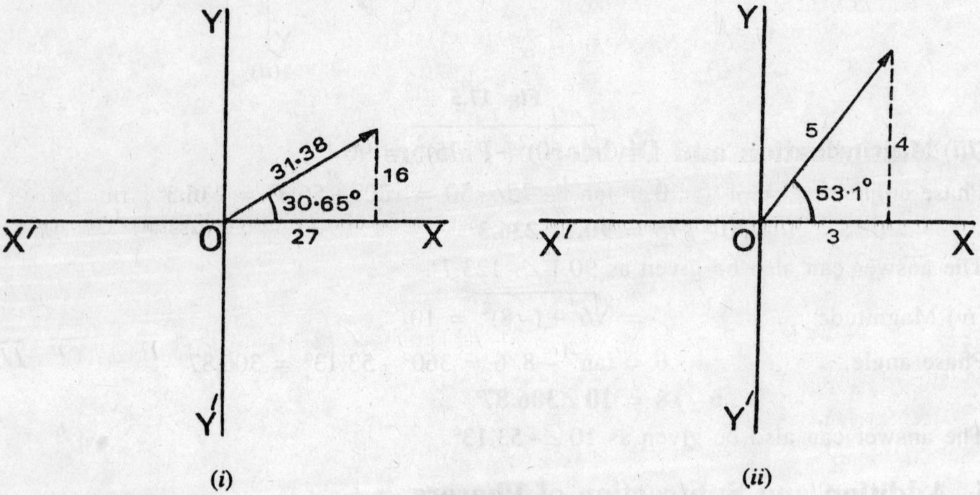

Fig. 17.6

(*ii*) Resultant current, $\mathbf{I} = \mathbf{I}_1 - \mathbf{I}_2 = (15 + j\,10) - (12 + j\,6) = \mathbf{3 + j\,4}$

The magnitude of resultant is $= \sqrt{3^2 + 4^2} = 5$ and it makes an angle $\theta = \tan^{-1} 4/3 = 53.1°$ with *OX*-axis as shown in Fig. 17.6 (*ii*).

Example 17.4. *Determine the resultant voltage of two sinusoidal generators in series whose voltages are* $\mathbf{V}_1 = 25\,\angle 15°$ *and* $\mathbf{V}_2 = 15\,\angle 60°$.

Solution. We shall first convert polar form to rectangular form and then carry out the addition.

$$\mathbf{V}_1 = 25\,(\cos 15° + j \sin 15°) = 24.15 + j\,6.47$$

$$\mathbf{V}_2 = 15\,(\cos 60° + j \sin 60°) = 7.5 + j\,12.99$$

$$\therefore \quad \mathbf{V} = \mathbf{V}_1 + \mathbf{V}_2 = (24.15 + j\,6.47) + (7.5 + j\,12.99) = 31.65 + j\,19.46$$

$$\therefore \quad V = \sqrt{(31.65)^2 + (19.46)^2} = 37.15$$

Phase angle, $\theta = \tan^{-1}(19.46/31.65) = 31.6°$

$$\therefore \quad \mathbf{V} = \mathbf{37.15\,\angle 31.6°}.$$

17.6. Conjugate of a Complex Number

Two complex numbers (or phasors) are said to be conjugate if they differ only in the algebraic sign of their quadrature components. Thus, the conjugate of $2 + j\,3$ is $2 - j\,3$. The conjugate of $-4+j3$ is $-4-j3$. In the polar form, conjugate of $5\angle 30°$ is $5\angle -30°$ and conjugate of $10\angle -40°$ is $10\angle 40°$. It is a usual practice to use an asterik (*)to indicate the conjugate.

Consider a complex number $\mathbf{A} = a+jb = A\angle\theta$. Then its conjugate will be $\overset{*}{\mathbf{A}} = a-jb = A\angle -\theta$. The conjugate numbers have the following properties :

(*i*) The sum of two conjugate numbers results in in-phase component only (*i.e.*, no *j* part).

$$\mathbf{A}+\overset{*}{\mathbf{A}} = (a+jb)+(a-jb) = 2a$$

(*ii*) The difference of two conjugate numbers results in quadrature component only.

$$\mathbf{A}-\overset{*}{\mathbf{A}} = (a+jb)-(a-jb) = j\,2b$$

(*iii*) When two conjugates are multiplied, the result has no *j*-part (*i.e.*, no quadrature component).

$$\mathbf{A}\times\overset{*}{\mathbf{A}} = (a+jb)(a-jb)$$

$$= a^2-j^2b^2 = a^2+b^2 \qquad (\because j^2 = -1)$$

17.7. Multiplication and Division of Phasors

It is easier to multiply and divide the phasors when they are in polar form than in the rectangular form. This will become apparent from the following discussion. Consider two phasors given by ;

$$\mathbf{V}_1 = a_1+jb_1 = V_1\angle\theta_1$$

$$\mathbf{V}_2 = a_2+jb_2 = V_2\angle\theta_2$$

1. Multiplication

(*i*) **Rectangular form.**

$$\mathbf{V}_1\times\mathbf{V}_2 = (a_1+jb_1)(a_2+jb_2)$$

$$= a_1a_2+ja_1b_2+ja_2b_1+j^2b_1b_2$$

$$= (a_1a_2-b_1b_2)+j\,(a_1b_2+a_2b_1) \qquad (\because j^2 = -1)$$

Magnitude of resultant $= \sqrt{(a_1a_2-b_1b_2)^2+(a_1b_2+a_2b_1)^2}$

Its angle *w.r.t.* *OX*-axis, $\theta = \tan^{-1}\left(\dfrac{a_1b_2+a_2b_1}{a_1a_2-b_1b_2}\right)$

(*ii*) **Polar form.** To multiply the phasors that are in polar form, multiply their magnitudes and add the angles (algebraically).

$$\mathbf{V}_1\times\mathbf{V}_2 = V_1\angle\theta_1\times V_2\angle\theta_2 = V_1V_2\angle\theta_1+\theta_2$$

The reader may see that multiplication of phasors becomes easier when they are expressed in polar form.

2. Division

(*i*) **Rectangular form**

$$\frac{\mathbf{V}_1}{\mathbf{V}_2} = \frac{a_1+jb_1}{a_2+jb_2}$$

$$= \frac{a_1+jb_1}{a_2+jb_2}\times\frac{a_2-jb_2}{a_2-jb_2} \qquad \left[\begin{array}{l}\text{Rationalising}\\ \text{the denominator}\end{array}\right]$$

$$= \frac{(a_1a_2 + b_1b_2) + j\,(b_1a_2 - a_1b_2)}{a_2^{\,2} + b_2^{\,2}}$$

$$= \frac{a_1a_2 + b_1b_2}{a_2^{\,2} + b_2^{\,2}} + j\frac{(b_1a_2 - a_1b_2)}{a_2^{\,2} + b_2^{\,2}}$$

(*ii*) **Polar form.** To divide the phasors that are in polar form, divide the magnitude of phasors and subtract the denominator angle from the numerator angle.

$$\therefore \qquad \frac{\mathbf{V}_1}{\mathbf{V}_2} = \frac{V_1\angle\theta_1}{V_2\angle\theta_2} = \frac{V_1}{V_2}\angle\theta_1 - \theta_2$$

Practically without exception, division is the easiest in the polar form.

Example 17.5. *Two phasors are given in the following form :*

$$\mathbf{V}_1 = 4 + j\,3\,; \qquad \mathbf{V}_2 = 5 + j\,6$$

Evaluate $\mathbf{V}_1 \times \mathbf{V}_2$ *and* $\mathbf{V}_1/\mathbf{V}_2$ *in (i) rectangular form (ii) polar form.*

Solution.

Rectangular form

$$\mathbf{V}_1 \times \mathbf{V}_2 = (4 + j\,3)\,(5 + j\,6) = 20 + j\,24 + j\,15 + j^2\,18 = \mathbf{2 + j\,39}$$

$$\frac{\mathbf{V}_1}{\mathbf{V}_2} = \frac{4 + j\,3}{5 + j\,6} = \frac{4 + j\,3}{5 + j\,6} \times \frac{5 - j\,6}{5 - j\,6}$$

$$= \frac{20 - j\,24 + j\,15 - j^2\,18}{5^2 - j^2\,6^2} = \frac{38 - j\,9}{61}$$

$$= \frac{38}{61} - j\frac{9}{61} = \mathbf{0.623 - j\,0.147}$$

Polar form. Convert the phasors to polar form.

$$\mathbf{V}_1 = 4 + j\,3 = 5\angle 36.87°$$

$$\mathbf{V}_2 = 5 + j\,6 = 7.81\angle 50.19°$$

$$\therefore \qquad \mathbf{V}_1 \times \mathbf{V}_2 = 5\angle 36.87° \times 7.81\angle 50.19°$$

$$= 5 \times 7.81\angle 36.87° + 50.19° = \mathbf{39.05\angle 87.06°}$$

$$\frac{\mathbf{V}_1}{\mathbf{V}_2} = \frac{5\angle 36.87°}{7.81\angle 50.19°} = \frac{5}{7.81}\angle 36.87° - 50.19° = \mathbf{0.64\ \angle -13.32°}$$

The negative angle means that the phasor is below *OX*-axis.

Example 17.6. *The following three phasors are given :*

$$\mathbf{A} = 5 + j\,5\,;\ \mathbf{B} = 50\angle 40°\,;\ \mathbf{C} = 4 + j\,0$$

Perform the following indicated operations :

(*i*) $\dfrac{\mathbf{AB}}{\mathbf{C}}$ (*ii*) $\dfrac{\mathbf{BC}}{\mathbf{A}}$

Solution. Expressing the three phasors in polar form, we have,

$$\mathbf{A} = 5 + j\,5 = 7.07\angle 45°\,;\ \mathbf{B} = 50\angle 40°\,;\ \mathbf{C} = 4 + j\,0 = 4\angle 0°$$

(*i*)
$$\frac{\mathbf{AB}}{\mathbf{C}} = \frac{7.07\angle 45° \times 50\angle 40°}{4\angle 0°} = \mathbf{88.37\angle 85°}$$

(*ii*)
$$\frac{\mathbf{BC}}{\mathbf{A}} = \frac{50\angle 40° \times 4\angle 0°}{7.07\angle 45°} = \mathbf{28.29\angle -5°}$$

Example 17.7. *The applied voltage in a series circuit is given by* $\mathbf{V} = 200 \angle 0°$ *volts. The impedance offered by the circuit is* $\mathbf{Z} = 83.3 \angle 40°\ \Omega$. *Find the magnitude of current. Is the circuit inductive or capacitive ?*

Solution.

$$\mathbf{I} = \frac{\mathbf{V}}{\mathbf{Z}} = \frac{200 \angle 0°}{83.3 \angle 40°} = \mathbf{2.4 \angle -40°\ A}$$

The voltage phasor lies along *OX*-axis. The negative angle of circuit current means that current phasor is below *OX*-axis. Hence the circuit is inductive.

17.8. Powers and Roots of Phasors

Powers and roots of complex numbers can be found very conveniently in polar form. If the complex number is not in polar form, it is always advisable to convert the number to this form and then carry out these algebraic operations.

(*i*) **Powers.** Suppose it is required to find the cube of the phasor $2 \angle 10°$. For this purpose, the phasor has to be multiplied by itself three times

$$\therefore \quad (2 \angle 10°)^3 = 2 \times 2 \times 2 \angle 10° + 10° + 10° = 8 \angle 30°$$

In general, $$\mathbf{A}^n = A^n \angle n\theta$$

(*ii*) **Roots.** Let us proceed backward with the above given example. Suppose we are to find the cube root of $8 \angle 30°$. It is clear that :

$$\sqrt[3]{8 \angle 30°} = 2 \angle 10°$$

In general, $$\sqrt[n]{\mathbf{A}} = \sqrt[n]{A} \angle \frac{\theta}{n}$$

Thus the square root of $64 \angle 40°$ is = $\sqrt[2]{64} \angle \frac{40°}{2} = 8 \angle 20°$

17.9. Phasor Algebra Applied to Series Circuit

Fig. 17.7 shows an *R-L* series circuit and its phasor diagram. Since the circuit current is taken as the reference phasor (*i.e.*, along *OX*-axis), we have,

$$\mathbf{I} = I + j\,0$$

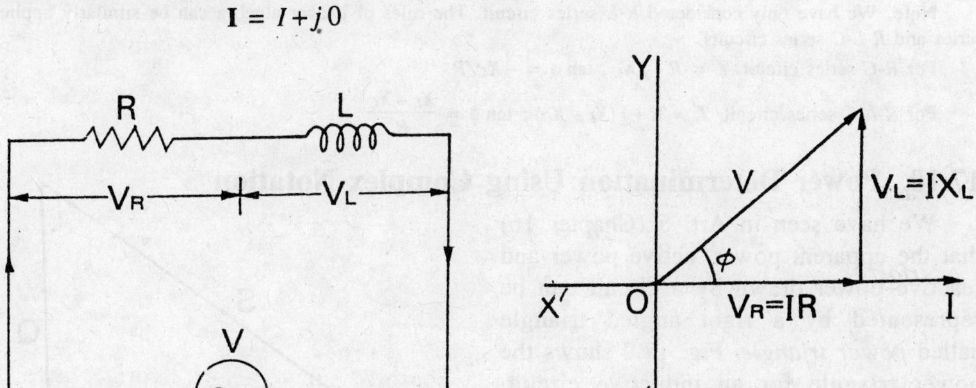

Fig. 17.7

Now, $$\mathbf{V} = \mathbf{V}_R + \mathbf{V}_L = (\mathbf{I}R + j\,0) + (0 + j\,\mathbf{I}X_L)$$

$$= \mathbf{I}(R + j\,X_L) = \mathbf{IZ}$$

$$\therefore \quad \mathbf{Z} = R + j\,X_L$$

Magnitude of impedance, $\quad Z = \sqrt{R^2 + X_L^{\,2}}$

Phase angle, $\quad \phi = \tan^{-1}(X_L/R)$

It is clear that current lags the voltage by $\phi°$.

Polar form. In the polar form, we have,

$$\mathbf{I} = I + j\,0 = I\angle 0°$$

$$\mathbf{Z} = R + j\,X_L = Z\angle\phi°$$

$$\therefore \quad \mathbf{V} = \mathbf{IZ} = I\angle 0° \times Z\angle\phi° = IZ\angle\phi°$$

This shows that applied voltage leads the current by $\phi°$. This is the same thing as the current lags the voltage by $\phi°$.

Voltage phasor as reference. In practice, we are given the applied voltage and circuit impedance and it is desired to find the circuit current. It is then profitable to take voltage as the reference phasor (*i.e.*, along *OX*-axis).

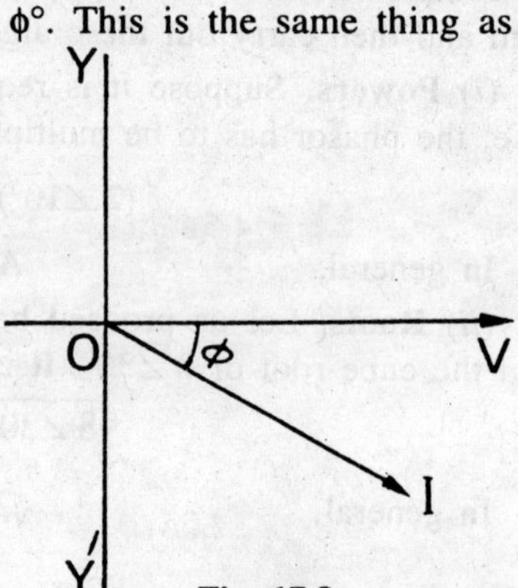

Fig. 17.8

$$\mathbf{V} = V + j\,0 = V\angle 0°$$

$$\mathbf{Z} = R + j\,X_L = Z\angle\phi°$$

$$\therefore \quad \mathbf{I} = \frac{\mathbf{V}}{\mathbf{Z}}$$

$$= \frac{V + j\,0}{R + j\,X_L} \quad \text{...Rectangular form}$$

$$= \frac{V}{Z}\angle -\phi° \quad \text{...Polar form}$$

Fig. 17.8 shows the phasor diagram of the circuit with voltage as the reference phasor.

Note. We have only considered *R-L* series circuit. The rules of phasor algebra can be similarly applied to *R-C* series and *R-L-C* series circuits.

For *R-C* series circuit, $\mathbf{Z} = R - j\,X_C$; $\tan\phi = -X_C/R$

For *R-L-C* series circuit, $\mathbf{Z} = R + j\,(X_L - X_C)$; $\tan\phi = \dfrac{X_L - X_C}{R}$

17.10. Power Determination Using Complex Notation

We have seen in Art. 5 (Chapter 16) that the apparent power, active power and reactive power drawn by a circuit can be represented by a right angled triangle called *power triangle.* Fig. 17.9 shows the power triangle for an inductive circuit. Here S (= VI) is the apparent power drawn by the circuit, P (= $VI\cos\phi$) is the active power (*i.e.*, power consumed by the circuit) and Q (= $VI\sin\phi$) is the reactive power. Using complex algebra, we can express the apparent power in complex form as :

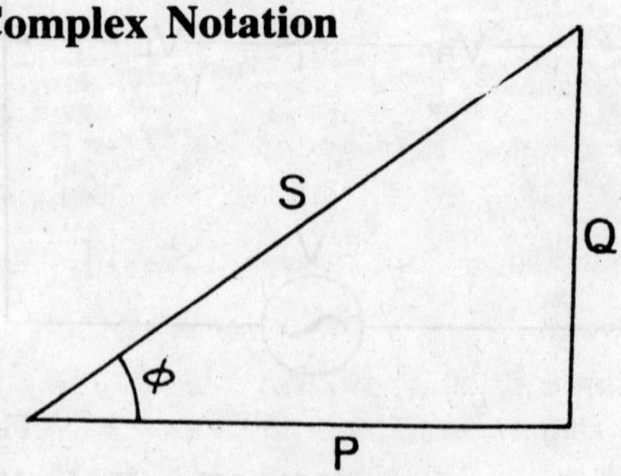

Fig. 17.9

$$S = P + jQ$$

It can be shown mathematically that *apparent power in complex form is equal to the product of phasor voltage and conjugate of phasor current i.e.*

$$S = \text{Phasor voltage} \times \text{Conjugate of phasor current}$$

or $$P + jQ = \mathbf{V} \times \overset{*}{\mathbf{I}}$$

where **V** is phasor voltage in complex form and $\overset{*}{\mathbf{I}}$ is the conjugate of phasor current **I** in complex form.

Active power, P = In-phase component

Reactive power, Q = Quadrature component

Example 17.8. *The current in a circuit is given by (4.5 + j12) A when the applied voltage is (100 + j150) V. Determine (i) the magnitude of impedance and (ii) phase angle.*

Solution.

$$\mathbf{V} = 100 + j\,150 = 180.28\angle 56.31°$$

$$\mathbf{I} = 4.5 + j\,12 = 12.82\angle 69.44°$$

(*i*) ∴ $$\mathbf{Z} = \mathbf{V}/\mathbf{I}$$

$$= \frac{180.28\angle 56.31°}{12.82\angle 69.44°}$$

$$= 14.06\angle -13.13°$$

∴ $$Z = \mathbf{14.06\,\Omega}$$

(*ii*) Phase angle, ϕ = **13.13°** *lead*

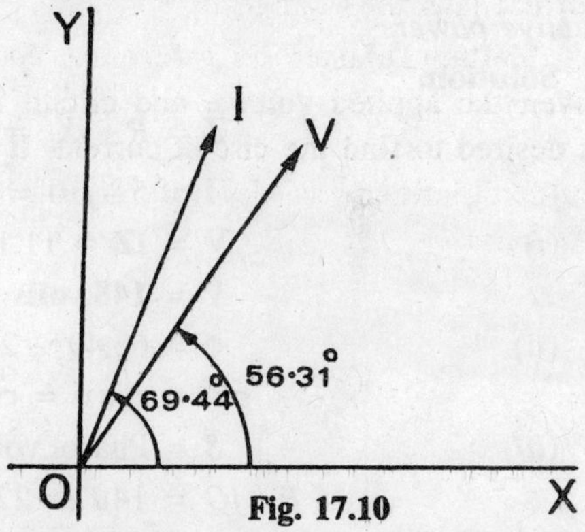

Fig. 17.10

Fig. 17.10 shows the phasor diagram. It is clear that current is leading the voltage by 13.13°. Therefore, the circuit is capacitive.

Note. The problem can also be solved using rectangular form. Since division is involved, it is easier to work out in polar form.

Example 17.9. *A resistance and an inductance are connected in series across a voltage of v = 566 sin 314t. The current expression is found to be i = 4 sin (314t − π/4). Find the value of resistance, inductance and power factor. What is the power drawn by the circuit ?*

Solution. From the equations of voltage and current, it is clear that :

$$\phi = 45° \text{ lag} \; ; \; V_m = 566 \text{ volts} \; ; \; I_m = 4 \text{ A} \; ; \; \omega = 314$$

∴ Power factor $= \cos\phi = \cos 45° = \mathbf{0.707}$ *lag*

Impedance, $Z = V_m/I_m = 566/4 = 141.5\,\Omega$

Resistance, $R = Z\cos\phi = 141.5 \times 0.707 = \mathbf{100\,\Omega}$

Reactance, $X_L = Z\sin\phi = 141.5 \times 0.707 = 100\,\Omega$

Inductance, $L = X_L/\omega = 100/314 = \mathbf{0.318\,H}$

Power, $P = I^2R = (4/\sqrt{2})^2 \times 100 = \mathbf{800\,W}$

Example 17.10. *A current of 120 − j 50 flows through a circuit when the applied voltage is 8 + j2. Determine (i) impedance (ii) power factor (iii) power consumed and reactive power.*

Solution. $$\mathbf{V} = 8 + j\,2 = 8.25\angle 14°$$

$$\mathbf{I} = 120 - j\,50 = 130\angle -22.62°$$

(i) $\mathbf{Z} = \dfrac{\mathbf{V}}{\mathbf{I}} = \dfrac{8.25\angle 14^\circ}{130\angle -22.62^\circ} = 0.0635\angle 36.62^\circ$

$\therefore$ $Z = \mathbf{0.0635\,\Omega}$

(ii) $\phi = 36.62^\circ\ lag$

$\therefore$ $p.f. = \cos\phi = \cos 36.62^\circ = \mathbf{0.803}\,lag$

(iii) As shown in Art. 17.10,

Complex VA, S = Phasor voltage × Conjugate of phasor current

or $P + jQ = 8.25\angle 14^\circ \times 130\angle 22.62^\circ = 1072.5\angle 36.62^\circ$

$= 1072.5\,(\cos 36.62^\circ + j\sin 36.62^\circ) = 860.8 + j\,639.75$

$\therefore$ Power consumed, $P = \mathbf{860.8\ W}$; Reactive power, $Q = \mathbf{639.75\ VAR}$

Example 17.11. *In an R-L series circuit, $R = 10\,\Omega$ and $X_L = 8.66\,\Omega$. If current in the circuit is $(5 - j\,10)$ A, find (i) the applied voltage (ii) power factor and (iii) active power and reactive power.*

Solution.

$$\mathbf{Z} = R + jX_L = 10 + j\,8.66 = 13.23\angle 40.9^\circ$$

$$\mathbf{I} = 5 - j10 = 11.18\angle -63.43^\circ$$

(i) $\mathbf{V} = \mathbf{IZ} = 11.18\angle -63.43^\circ \times 13.23\angle 40.9^\circ = 148\angle -22.53^\circ$

$\therefore$ $V = \mathbf{148\ volts}$

(ii) $\phi = 63.43^\circ - 22.53^\circ = 40.9^\circ$

$p.f. = \cos\phi = \cos 40.9^\circ = \mathbf{0.756}\,lag$

(iii) S = Phasor voltage × conjugate of phasor current

$P + jQ = 148\angle -22.53^\circ \times 11.18\angle 63.43^\circ = 1654.64\angle 40.9^\circ$

$= 1250.66 + j\,1083.36$

$\therefore$ Active power, $P = \mathbf{1250.66\ W}$; Reactive power, $Q = \mathbf{1083.36\ VAR}$

Example 17.12. *The complex volt amperes in a series circuit are $(4330 - j\,2500)$ and the current is $(25 + j\,43.3)$ A. Find the applied voltage.*

Solution. Let the applied voltage be $\mathbf{V} = (a + jb)$

Complex VA = Phasor voltage × conjugate of phasor current

or $4330 - j\,2500 = (a + jb) \times (25 - j\,43.3)$

$= (25a + 43.3b) + j\,(-43.3\,a + 25\,b)$

$\therefore$ $25\,a + 43.3\,b = 4330$ ...(i)

and $-43.3\,a + 25\,b = -2500$ ...(ii)

Solving eqs. (i) and (ii), we have, $a = 86.6$ and $b = 50$

$\therefore$ $\mathbf{V} = \mathbf{(86.6 + j\,50)\ volts}$

Example 17.13. *A coil of resistance $12\,\Omega$ and inductive reactance of $25\,\Omega$ is connected in series with a capacitive reactance of $41\,\Omega$. The combination is connected to a supply of 230V. 50 Hz. Using phasor algebra, find (i) circuit impedance (ii) current and (iii) power consumed.*

Solution.

(i) $\mathbf{Z} = R + j\,(X_L - X_C)$

$= 12 + j\,(25 - 41) = 12 - j16 = 20\angle -53.13^\circ$

$\therefore$ $Z = \mathbf{20\,\Omega}$

(*ii*) Taking voltage as the reference phasor,

$$\mathbf{V} = 230 + j\,0 = 230\angle 0°$$

$$\therefore \quad \mathbf{I} = \frac{\mathbf{V}}{\mathbf{Z}} = \frac{230\angle 0°}{20\angle -53.13°} = 11.5\angle 53.13°$$

$$\therefore \quad I = \mathbf{11.5\ A}$$

(*iii*) It is clear that current leads the voltage by 53.13° *i.e.*, ϕ = 53.13°

$\therefore$ Power factor = $\cos\phi = \cos 53.13° = 0.6$ *lead*

$\therefore$ Power consumed, $P = VI\cos\phi = 230 \times 11.5 \times 0.6 =$ **1587 W**

Tutorial Problems

1. Write the polar form for the following :
   (*i*) $-4 + j\,6.928$ (*ii*) $-50 - j\,75$ **[(*i*) 8 ∠120° (*ii*) 90.1 ∠–123.7°]**
2. Obtain the sum of $\mathbf{E}_1 = 100\angle 45°$ and $\mathbf{E}_2 = 80\angle 120°$ in (*i*) rectangular form and (*ii*) polar form. **[(*i*) 30.7 + j140 (*ii*) 143 ∠77.6°]**
3. Perform the indicated operations and express the results in polar form :
   (*i*) $(5 + j\,12)^2$
   (*ii*) Add $10\angle 40°$ to $15\angle 60°$ and divide the result by $(20\angle 45° - 12\angle 180°)$
   (*iii*) $3\angle 30° + 5\angle -50° - 7\angle 20° + (6 + j\,5)$
   [(*i*) 169 ∠134.8° (*ii*) 1.715 ∠– 29.35° (*iii*) 5.24 ∠3.06°]
4. Determine the resistance and inductance of series connected elements that will draw a current of $20\angle -30°$ A from a 60 Hz sinusoidal generator whose voltage is $100\angle 0°$. **[4.35 Ω ; 6.6 mH]**
5. A coil, a capacitor and a resistor are connected in series and supplied by 1.5 KHz generator. The capacitive reactance is 4 Ω, the resistor is 4 Ω and the coil has a resistance of 3 Ω and an inductive reactance of 4 Ω. If the current in the circuit is $10\angle 20°$, determine (*i*) complex impedance (*ii*) voltage drop across coil in polar form. **[(*i*) 7 ∠0° Ω ; (*ii*) 50 ∠73.13° V]**
6. The p.d. across and the current in a circuit are represented by $(100 + j\,200)$ V and $(10 + j\,5)$ A respectively. Calculate the power consumed and the reactive power. **[2000W ; 1500VAR]**
7. In an *R-L-C* series circuit, *L* = 25 mH, *C* = 59 μF. Find the value of *R* so that current leads the voltage by 63.4° at a supply frequency of $400/2\pi$. **[20 Ω]**
8. The current in a circuit is given by $(4.5 + j12)$ A when the applied voltage is $100 + j\,150$. Determine (*i*) the complex expression for the impedance (*ii*) power (*iii*) the phase angle between voltage and current. **[(*i*) 13.7 – j3.2 Ω (*ii*) 2250 W (*iii*) 13.9°]**

OBJECTIVE QUESTIONS

1. When a phasor is multiplied by $-j$, it is rotated through .... in the counter-clockwise direction.
   (*i*) 90° (*ii*) 180°
   (*iii*) 270° (*iv*) none of he above
2. The value of j^5 is equal to ....
   (*i*) 1
   (*ii*) $\sqrt{-1}$
   (*iii*) – 1 (*iv*) $-\sqrt{-1}$
3. When a phasor is multiplied by j^6, it is rotated through .... in counter-clockwise direction.
   (*i*) 540° (*ii*) 360°
   (*iii*) 90° (*iv*) 270°
4. If a phasor is multiplied by *j*, then ....
   (*i*) only its magnitude changes
   (*ii*) only its direction changes
   (*iii*) both magnitude and direction change
   (*iv*) none of the above
5. In the complex number $4 + j\,7$, 7 is called the .... component.
   (*i*) real (*ii*) imaginary
   (*iii*) in-phase (*iv*) quadrature
6. The reciprocal of a complex number results in a ....
   (*i*) complex number
   (*ii*) real component only
   (*iii*) quadrature component only
   (*iv*) none of the above
7. Rationalising the denominator of a complex number eliminates ....

(*i*) *j* component in the numerator
(*ii*) *j* component in the denominator
(*iii*) *j* component in both numeater and denominator
(*iv*) none of above

8. If two complex numbers are equal, then ....
(*i*) only their magnitudes will be equal
(*ii*) only their angles will be equal
(*iii*) their in-phase and quadrature components will be separately equal
(*iv*) None of the above

9. A phasor 2 ∠180° can be expressed as ....
(*i*) j 2 (*ii*) $-j$ 2
(*iii*) – 2 (*iv*) 2

10. A current of 3 + j 4 amperes through a circuit. The magnitude of current is ....
(*i*) 7 A (*ii*) 5 A
(*iii*) 1 A (*iv*) 1.33 A

11. A complex number is given by $a + j\,b$. The magntiude of this complex number will be ....
(*i*) less than either *a* or *b*
(*ii*) more than the sum of *a* and *b*
(*iii*) less than 1.414 × whichever is the greater *a* or *b*
(*iv*) none of the above

12. The voltage applied in a circuit is given by 100 ∠60° volts. It can be written as ....
(*i*) 100 ∠–60° volts
(*ii*) 100 ∠240° volts
(*iii*) 100 ∠–300° volts
(*iv*) none of the above

13. The phasor 8 ∠60° is identical to ....
(*i*) 8 ∠240° (*ii*) 8 ∠780°
(*iii*) 8 ∠310° (*iv*) none of above

14. For addition or subtraction of phasors, we use .... from.
(*i*) Polar (*ii*) rectangular
(*iii*) trigonometrical (*iv*) none of the above

15. The quantity 2 + j2 can be expressed in the polar form as ....
(*i*) $2\sqrt{2}$ ∠45° (*ii*) 4 ∠ 90°
(*iii*) $2\sqrt{2}$ ∠–45° (*iv*) 1 ∠90°

16. For multiplication or division of phasors, we use .... form.
(*i*) rectangular (*ii*) trigonometrical
(*iii*) polar (*iv*) none of the above

17. If the impedance of an a.c. circuit is 10 ∠60° Ω, then resistance in the circuit is ....
(*i*) 5 Ω (*ii*) 8.66 Ω
(*iii*) 10 Ω (*iv*) none of the above

18. The conjugate of – 4 + j 3 is ....
(*i*) 4 – j 3 (*ii*) – 4 – j 3
(*iii*) 4 + j 3 (*iv*) none of the above

19. The conjugate of 10 ∠–40° is ....
(*i*) 0.1 ∠40° (*ii*) 0.1 ∠–40°
(*iii*) 10 ∠40° (*iv*) 5 ∠–20°

20. The sum of two conjugate numbers results in ....
(*i*) a complex number
(*ii*) in-phase component only
(*iii*) quadrature component only
(*iv*) none of the above

21. The difference of two conjugate numbers results in ....
(*i*) a complex number
(*ii*) in-phase component only
(*iii*) quadrature component only
(*iv*) none of the above

22. When two conjugate numbers are multiplied, the result is ....
(*i*) a complex number
(*ii*) no in-phase component
(*iii*) no quadrature component
(*iv*) none of the above

23. The value of 5 ∠40° × 3 ∠20° is ....
(*i*) 15 ∠60° (*ii*) 15 ∠20°
(*iii*) 1.6 ∠2° (*iv*) 15 ∠800°

24. The value of 9 ∠30° ÷ 3 ∠10° is
(*i*) 27 ∠40° (*ii*) 3 ∠20°
(*iii*) 3 ∠40° (*iv*) 3 ∠3°

25. The square root of 64 ∠36° is ....
(*i*) 8 ∠6° (*ii*) 8 ∠18°
(*iii*) 8 ∠36°
(*iv*) none of the abaove

26. The cube root of 8 ∠30° is ....
(*i*) 2 ∠10° (*ii*) 2 ∠30°
(*iii*) 8 ∠10° (*iv*) none of the above

27. The kVA drawn by an a.c. circuit is given by (3 + j 4) kVA. The active power drawn by the circuit is ....
(*i*) 3 kW (*ii*) 4 kW
(*iii*) 5 kW (*iv*) none of the above

28. The impedance of an a.c. circuit is 45 $\angle -30° \Omega$. The circuit will be ....
(*i*) resistive (*ii*) inductive
(*iii*) capacitive (*iv*) none of the above

29. In an a.c. circuit, the apparent power inh complex form is equal to ....
(*i*) Phasor voltage × conjugate of phasor current
(*ii*) Phasor voltage ÷ conjugate of phasor current
(*iii*) phasor voltage × phasor current
(*iv*) none of the above

30. A series a.c. circuit has $R = 4\ \Omega$ and $X_L = 3\ \Omega$. It will be expressed in the rectangular form as ....
(*i*) $-4 - j3$ (*ii*) $-4 + j3$
(*iii*) $4 + j3$ (*iv*) $4 - j3$

31. The two complex numbers are given as : 5 (cos 30° + *j* sin 30°) ; 10 (cos 15° + *j* sin 15°). The value of product of these two numbers is equal to ....
(*i*) 50 (cos 45° + *j* sin 45°)
(*ii*) 50 (cos 15° + *j* sin 15°)
(*iii*) 50 (cos 450° + *j* sin 450°)
(*iv*) none of the above

32. The voltage and current in an a.c. series circuit are 230 $\angle 0°$ volts and 100 $\angle 30°$ A respectively. The circuit will be ....
(*i*) resistive (*ii*) inductive
(*iii*) capacitive (*iv*) in resonance

33. The value of $\dfrac{10 \angle 60° \times 2 \angle 20°}{5 \angle 40°}$ will be ....
(*i*) 4 $\angle 40°$ (*ii*) 4 $\angle 120°$
(*iii*) 4 $\angle 300°$ (*iv*) none of the above

34. The current through a 4 H inductor is $i = 0.6 \sin (100\, t - 10°)$ A. Using phasor algebra, voltage across the inductor is
(*i*) 24 $\angle 100°$ V (*ii*) 340 $\angle 70°$ V
(*iii*) 48 $\angle -90°$ V (*iv*) 240 $\angle 80°$ V

35. The polar form of the following wave form is
$v_1 = 16 \sin (\omega t + 35°)$ V
$v_2 = 32 \sin (\omega t - 55°)$ V
(*i*) $v_1 = 16 \angle 70°$ V; $v_2 = 32 \angle 110°$ V
(*ii*) $v_1 = 16 \angle 35°$ V; $v_2 = 32 \angle -55°$ V
(*iii*) $v_1 = 8 \angle 35°$ V; $v_2 = 32 \angle 55°$ V
(*iv*) none of the above

ANSWERS TO OBJECTIVE QUESTIONS

1. (*iii*)	**2.** (*ii*)	**3.** (*i*)	**4.** (*ii*)	**5.** (*iv*)	**6.** (*i*)	**7.** (*ii*)	**8.** (*iii*)
9. (*iii*)	**10.** (*ii*)	**11.** (*iii*)	**12.** (*iii*)	**13.** (*ii*)	**14.** (*ii*)	**15.** (*i*)	**16.** (*iii*)
17. (*i*)	**18.** (*ii*)	**19.** (*iii*)	**20.** (*ii*)	**21.** (*iii*)	**22.** (*iii*)	**23.** (*i*)	**24.** (*ii*)
25. (*ii*)	**26.** (*i*)	**27.** (*i*)	**28.** (*iii*)	**29.** (*i*)	**30.** (*iii*)	**31.** (*i*)	**32.** (*iii*)
33. (*i*)	**34.** (*iv*)	**35.** (*ii*)					

HINTS TO SELECTED OBJECTIVE QUESTIONS

2. $j^5 = j^4 \times j = 1 \times j = \sqrt{-1}$

8. Suppose two equal complex numbers are $a + jb$ and $c + jd$.

$$\therefore \quad a + jb = c + jd$$
$$\therefore \quad a = c \text{ and } b = d$$

10. Magnitude = $\sqrt{3^2 + 4^2} = 5\ A$

12. The angle measured in CCW direction from OX axis is 60°. If the angle is measured in clockwise direction from OX axis, its value is = 360° – 60° = 300° and this angle is negative. Hence 100 $\angle 60°$ can be written as 100 $\angle -300°$.

17. $R = Z \cos \phi = 10 \cos 60° = 5\ \Omega$

18. Two complex numbers (or phasors) are said to be conjugate if they differ only in the algebraic sign of their quadrature components. Thus the conjugate of $-4 + j3$ is $-4 - j3$.

20. Suppose a complex number is $2 + j3$. Its conjugate will be 2–j3.

Sum = $(2 + j\,3) + (2 - j\,3) = 4$

23. $5\angle 40° \times 3\angle 20° = 5 \times 3\angle 40° + 20° = 15\angle 60°$

24. $9\angle 30° \div 3\angle 10° = \frac{9}{3}\angle 30° - 10° = 3\angle 20°$

25. Square root of $64\angle 36° = \sqrt{64}\angle\frac{36°}{2} = 8\angle 18°$

34. $i = 0.6 \sin(100\,t - 10°)$. $\therefore$ $i = 0.6\angle -10°$ A

$X_L = \omega L = 100 \times 4 = 400\ \Omega$; $\mathbf{X_L} = 400\angle 90°\ \Omega$

$\mathbf{V_L} = \mathbf{i} \times \mathbf{X_L} = 0.6\angle -10° \times 400\angle 90° = 240\angle 80°$ V

35. Fig 17.11 shows the solution of the problem.

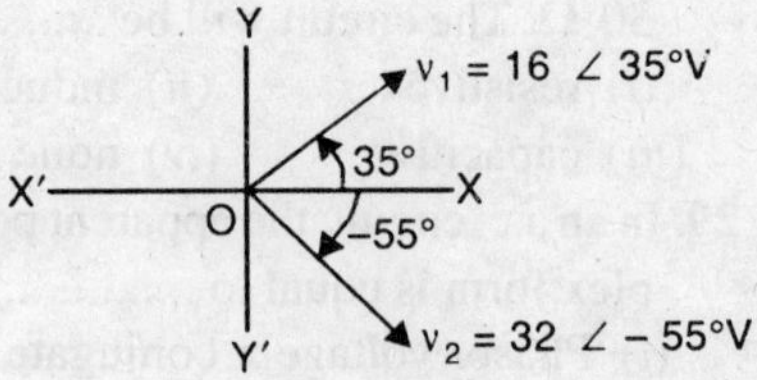

Fig. 17.11

18

Parallel A.C. Circuits

Introduction

As in parallel d.c. circuits, the voltage across all branches is the same in a parallel a.c. circuit. But current in any branch depends upon the impedance of that branch. The total line current supplied to the circuit is the *phasor sum* of the branch currents. Parallel circuits are used more frequently in electrical systems than are the series circuits. For example, electrical devices and equipment are connected in parallel across a.c. mains. In this chapter, we shall discuss the various methods of solving parallel a.c. circuits.

18.1. Solving Parallel A.C. Circuits

There are three main methods of solving parallel a.c. circuits, namely ;

(*i*) By phasor diagram

(*ii*) By pnasor algebra

(*iii*) By admittance method

The use of a particular method will depend upon the conditions of the problem. However, in general, that method should be used which yields quick results.

18.2. By Phasor Diagram

In this method, we draw the phasor diagram of the parallel circuit taking voltage as the reference phasor. The circuit or line current is determined by the method of components. Consider a parallel circuit consisting of two branches and connected to an alternating voltage of V volts (*r.m.s.*) as shown in Fig. 18.1

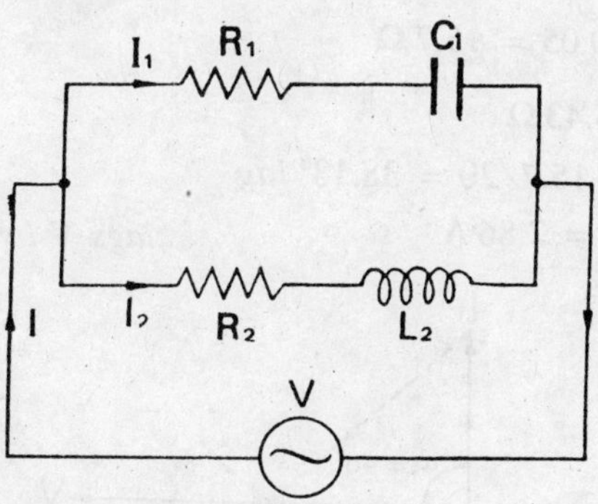

Fig. 18.1

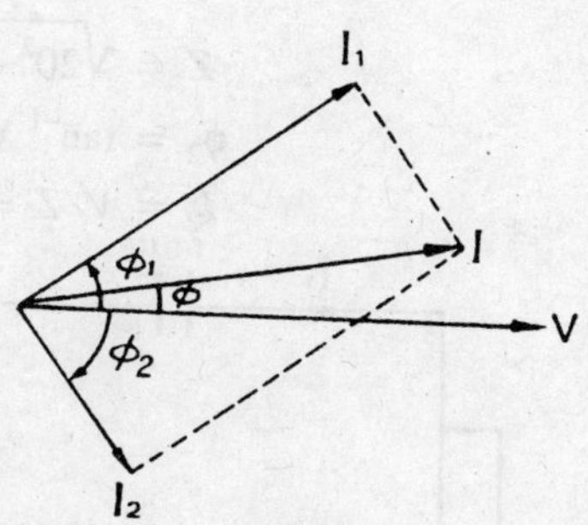

Fig. 18.2

Branch 1. $Z_1 = \sqrt{R_1^{\,2} + X_{C_1}^{\,2}}$; $\quad I_1 = \dfrac{V}{Z_1}$; $\quad \phi_1 = \tan^{-1}\dfrac{X_{C_1}}{R_1}$

The current I_1 leads the applied voltage V by $\phi_1°$ [See Fig. 18.2].

Branch 2. $Z_2 = \sqrt{R_2^2 + X_{L_2}^2}$; $I_2 = \dfrac{V}{Z_2}$; $\phi_2 = \tan^{-1}\dfrac{X_{L_2}}{R_2}$

The current I_2 lags behind V by $\phi_2°$ [See Fig. 18.2].

The line current I is the phasor sum of I_1 and I_2. Suppose its phase angle is ϕ as shown in Fig. 18.2. The values of I and ϕ can be determined by resolving the currents into rectangular components.

$$I\cos\phi = \text{Algebraic sum of components of } I_1 \text{ and } I_2 \text{ along } X\text{-axis}$$
$$= I_1\cos\phi_1 + I_2\cos\phi_2$$
$$I\sin\phi = \text{Algebraic sum of components of } I_1 \text{ and } I_2 \text{ along } Y\text{-axis}$$
$$= I_1\sin\phi_1 - I_2\sin\phi_2 -$$

But $$I^2 = (I\cos\phi)^2 + (I\sin\phi)^2$$

$\therefore$ $$I^2 = (I_1\cos\phi_1 + I_2\cos\phi_2)^2 + (I_1\sin\phi_1 - I_2\sin\phi_2)^2$$

or $$I = \sqrt{(I_1\cos\phi_1 + I_2\cos\phi_2)^2 + (I_1\sin\phi_1 - I_2\sin\phi_2)^2}$$

$$\tan\phi = \frac{I\sin\phi}{I\cos\phi} = \frac{I_1\sin\phi_1 - I_2\sin\phi_2}{I_1\cos\phi_1 + I_2\cos\phi_2} = \frac{Y\text{-Comp.}}{X\text{-Comp.}}$$

If ϕ is positive, line current I leads the voltage and if ϕ is negative, I lags behind the voltage.

$$\text{Circuit } p.f. = \frac{I\cos\phi}{I} = \frac{I_1\cos\phi_1 + I_2\cos\phi_2}{I} = \frac{X\text{-Comp.}}{I}$$

The phasor diagram method is suitable only when the parallel circuit is simple and contains two branches.

Example 18.1. *A capacitor of 50 μF is connected in parallel with a coil that has a resistance of 20 Ω and inductance of 0.05 H. If this parallel combination is connected across 200 V, 50 Hz supply, calculate (i) the line current (ii) power factor and (iii) power consumed.*

Solution.

$$X_C = \frac{1}{2\pi f C} = \frac{10^6}{2\pi \times 50 \times 50} = 63.7\ \Omega$$

$$I_1 = V/X_C = 200/63.7 = 3.14\ \text{A} \quad \text{...leads } V \text{ by } 90°$$

$$X_L = 2\pi f L = 2\pi \times 50 \times 0.05 = 15.7\ \Omega$$

$$Z = \sqrt{20^2 + 15.7^2} = 25.43\ \Omega$$

$$\phi_2 = \tan^{-1} X_L/R = \tan^{-1} 15.7/20 = 38.13° \text{ lag}$$

$$I_2 = V/Z = 200/25.43 = 7.86\ \text{A} \quad \text{...lags } V \text{ by } 38.13°$$

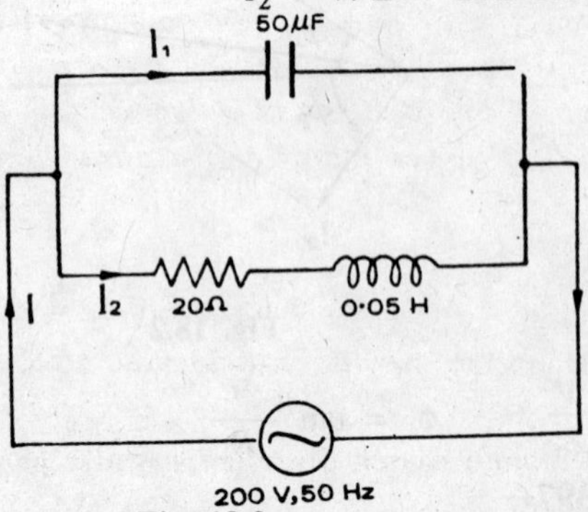

Fig. 18.3

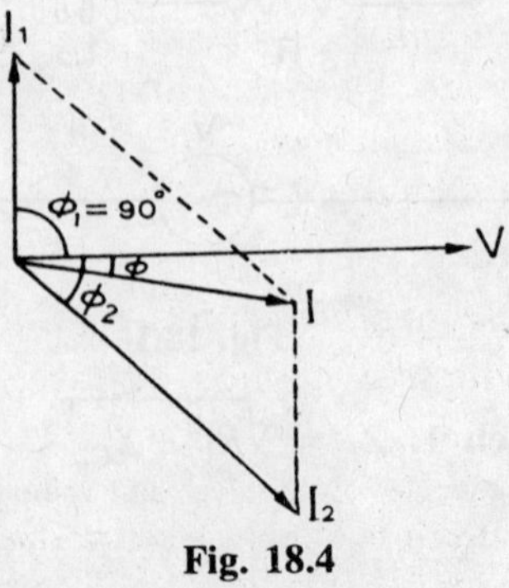

Fig. 18.4

Referring to the phasor diagram in Fig. 18.4, we have,

$$\text{Total } X\text{-component} = I_1 \cos\phi_1 + I_2 \cos\phi_2 = 3.14\cos 90° + 7.86\cos 38.13° = 0 + 6.18 = 6.18 \text{ A}$$

$$\text{Total } Y\text{-component} = I_1 \sin\phi_1 - I_2 \sin\phi_2 = 3.14\sin 90° - 7.86\sin 38.13° = 3.14 - 4.85 = -1.71\text{A}$$

(*i*) Line current, $I = \sqrt{(6.18)^2 + (-1.71)^2} = \mathbf{6.41A}$

(*ii*) Phase angle, $\phi = \tan^{-1}(-1.71/6.18) = -{}^*15.47°$

(*iii*) Power factor $= \cos\phi = \cos(-15.47°) = \mathbf{0.964}$ *lag*

(*iv*) Power consumed, $P = VI\cos\phi = 200 \times 6.41 \times 0.964 = \mathbf{1235.85\ W}$

Example 18.2. *A 10 Ω resistor, a 15.9 mH inductor and 159 μF capacitor are connected in parallel to a 200 V, 50 Hz source. Calculate the supply current and power factor.*

Solution. Fig. 18.5 shows the circuit diagram.

$$X_L = 2\pi fL = 2\pi \times 50 \times 15.9 \times 10^{-3} = 5\ \Omega$$

$$X_C = \frac{1}{2\pi fC} = \frac{10^6}{2\pi \times 50 \times 159} = 20\ \Omega$$

$I_R = V/R = 200/10 = 20$ A ...*in phase with V*

$I_L = V/X_L = 200/5 = 40$ A ...*lags V by 90°*

$I_C = V/X_C = 200/20 = 10$ A ...*leads V by 90°*

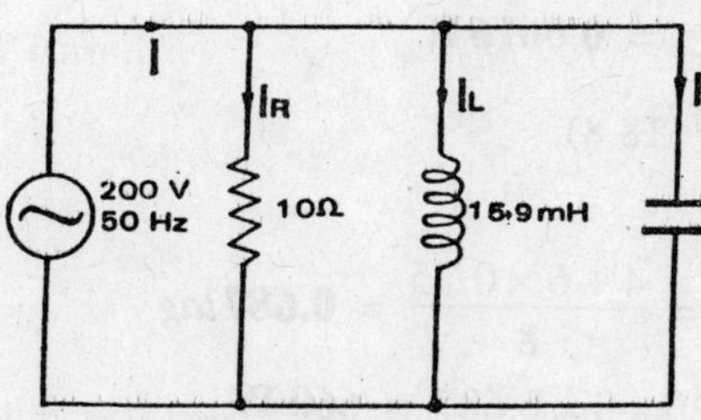

Fig. 18.5

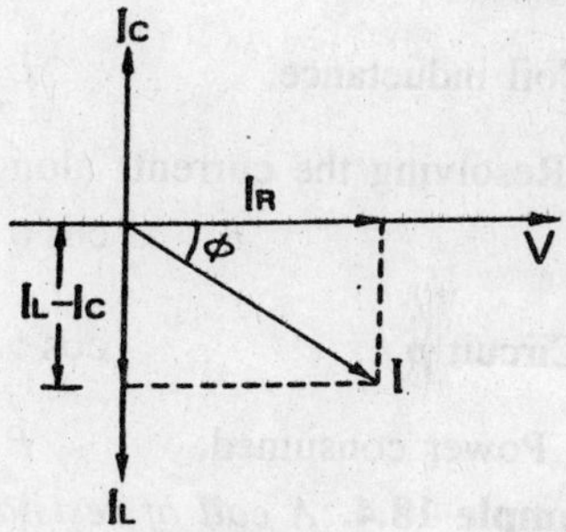

Fig. 18.6

Supply current, $I = \sqrt{I_R^2 + (I_L - I_C)^2} = \sqrt{20^2 + (40-10)^2} = \mathbf{36A}$

Circuit p.f. $= \cos\phi = I_R/I = 20/36 = \mathbf{0.56}$ *lag*

Example 18.3. *An inductive coil is connected in parallel with a pure resistor of 30 Ω and this parallel circuit is connected to a 50 Hz supply. The total current taken from the circuit is 8 A while the current in the resistor is 4 A and that in inductive coil is 6 A. Calculate (i) resistance and inductance of the coil (ii) power factor of the circuit and (iii) power taken by the circuit.*

Solution.

The second branch has a pure resistance (Z_2 = 30 Ω) so that current I_2 (= 4 A) will be in phase with the applied voltage. The first branch has an impedance of Z_1 and current

* In a parallel circuit, voltage is taken as the reference phasor. Therefore, negative phase angle means that circuit current lags behind the applied voltage.

I_1 (= 6 A) through it will lag behind the applied voltage by ϕ_1°. The line current I (= 8 A) is the phasor sum of I_1 and I_2 as shown in the phasor diagram in Fig. 18.8.

(*i*) Supply voltage, $V = I_2Z_2 = 4 \times 30 = 120$ V

Coil impedance, $Z_1 = V/I_1 = 120/6 = 20\,\Omega$

Referring to the phasor diagram in Fig. 18.8, we have,

$$I^2 = I_1^{\,2} + I_2^{\,2} + 2I_1I_2 \cos\phi_1$$

or

$$8^2 = 6^2 + 4^2 + 2 \times 6 \times 4 \times \cos\phi_1$$

$$\therefore \quad \cos\phi_1 = \frac{8^2 - 6^2 - 4^2}{2 \times 6 \times 4} = 0.25 \text{ and } \sin\phi_1 = 0.968$$

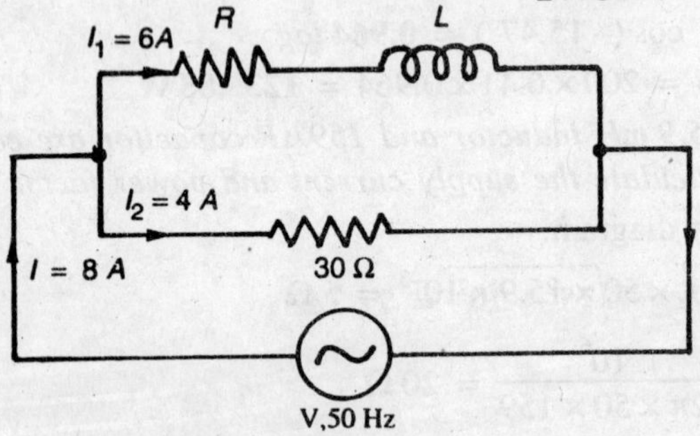

Fig. 18.7

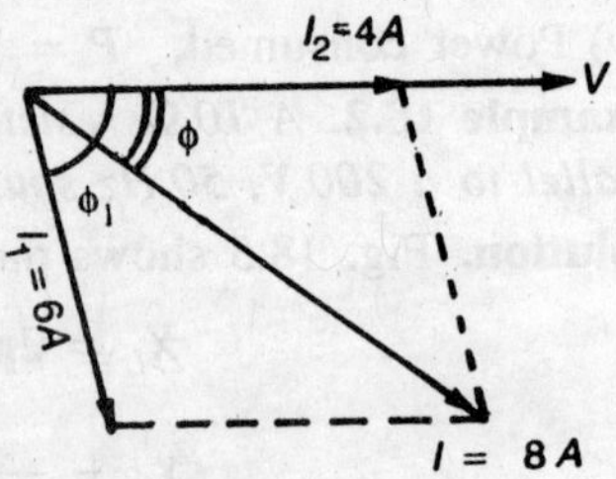

Fig. 18.8

∴ Coil resistance $R = Z_1 \cos\phi_1 = 20 \times 0.25 = \mathbf{5\,\Omega}$

Coil reactance, $X_L = Z_1 \sin\phi_1 = 20 \times 0.968 = 19.36\,\Omega$

Coil inductance, $L = \dfrac{X_L}{2\pi f} = \dfrac{19.36}{2\pi \times 50} = \mathbf{0.0616\,H}$

(*ii*) Resolving the currents along *X*-axis (see Fig. 18.8),

$$I\cos\phi = I_2 + I_1 \cos\phi_1$$

∴ Circuit *p.f.*, $\cos\phi = \dfrac{I_2 + I_1\cos\phi_1}{I} = \dfrac{4 + 6 \times 0.25}{8} = \mathbf{0.687}\ lag$

(*iii*) Power consumed, $P = VI\cos\phi = 120 \times 8 \times 0.687 = \mathbf{660\ W}$

Example 18.4. *A coil of resistance 50 Ω and inductance 0.318 H is connected in parallel with a circuit comprising a 75 Ω resistor in series with a 159 μF capacitor. The circuit is connected to 240 V, 50 Hz supply. Calculate :*

(*i*) *supply current*

(*ii*) *phase angle between supply current and applied voltage*

Find also the resistance and reactance of series circuit which will take the same current at the same p.f. as the parallel circuit.

Solution.

$$X_L = 2\pi fL = 2\pi \times 50 \times 0.318 = 100\,\Omega$$

$$Z_1 = \sqrt{R_1^{\,2} + X_L^{\,2}} = \sqrt{50^2 + 100^2} = 112\,\Omega$$

$$\phi_1 = \cos^{-1} R_1/Z_1 = \cos^{-1} 50/112 = 63.5° \ lag$$

$$I_1 = V/Z_1 = 240/112 = 2.15 \text{ A}$$ *...lags V by 63.5°*

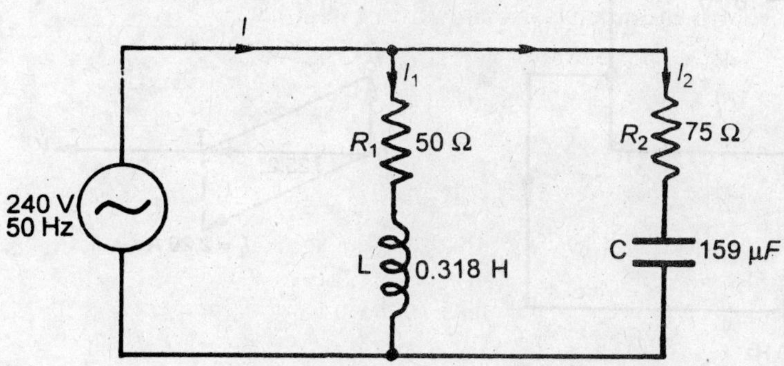

Fig. 18.9

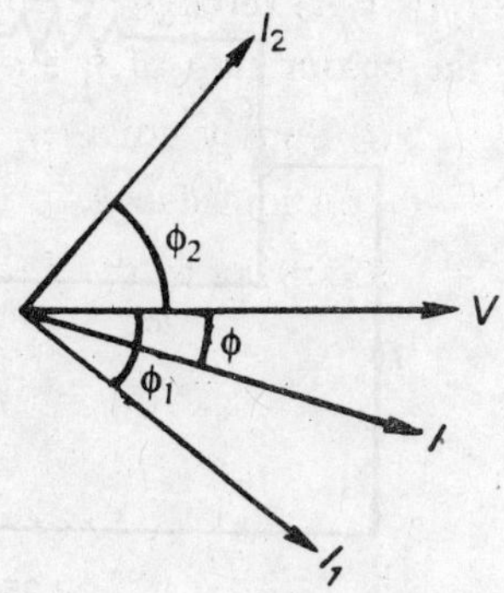

Fig. 18.10

$$X_C = \frac{1}{2\pi fC} = \frac{10^6}{2\pi \times 50 \times 159} = 20\ \Omega$$

$$Z_2 = \sqrt{R_2^{\ 2} + X_C^{\ 2}} = \sqrt{75^2 + 20^2} = 77.7\ \Omega$$

$$\phi_2 = \cos^{-1} R_2/Z_2 = \cos^{-1} 75/77.7 = 15°\ lead$$

$$I_2 = V/Z_2 = 240/77.7 = 3.09\text{ A} \quad \text{...leads } V \text{ by } 15°$$

Referring to the phasor diagram shown in Fig. 18.10,

$$\text{Total } X\text{-comp.} = I_1 \cos \phi_1 + I_2 \cos \phi_2$$
$$= 2.15 \cos 63.5° + 3.09 \cos 15° = 3.94\text{ A}$$
$$\text{Total } Y\text{-comp.} = -I_1 \sin \phi_1 + I_2 \sin \phi_2$$
$$= -2.15 \sin 63.5° + 3.09 \sin 15° = -1.13\text{ A}$$

(*i*) Supply current, $I = \sqrt{(3.94)^2 + (-1.13)^2} = \mathbf{4.1\ A}$

(*ii*) Phase angle, $\phi = \tan^{-1} - 1.13/3.94 = \mathbf{-16°}$

Since voltage is the reference phasor, negative angle means that supply current lags the applied voltage, *i.e.*, circuit is inductive.

Circuit impedance, $Z = V/I = 240/4.1 = 58.5\ \Omega$

Circuit resistance, $R = Z \cos \phi = 58.5 \cos (-16°) = \mathbf{56.2\ \Omega}$

Circuit reactance, $X_L = Z \sin \phi = 58.5 \sin (-16°) = \mathbf{16.12\ \Omega}$

Thus the parallel circuit is equivalent to 56 Ω resistor in series with 16.12 Ω inductive reactance.

Example 18.5. *A coil of resistance 100 Ω and inductance 0.15 H is connected to a 250 V, 50 Hz supply. Calculate the capacitance of the capacitor to be connected in parallel with the coil so that overall p.f. is unity. Also find the supply current then.*

Solution.

$$X_L = 2\pi fL = 2\pi \times 50 \times 0.15 = 47.1\ \Omega$$

$$Z = \sqrt{R^2 + X_L^{\ 2}} = \sqrt{(100)^2 + (47.1)^2} = 110.5\ \Omega$$

$$I_L = V/Z_L = 250/110.5 = 2.26\text{ A}$$

$$\phi_L = \tan^{-1} X_L/R = \tan^{-1} 47.1/100 = 25.2°\ lag$$

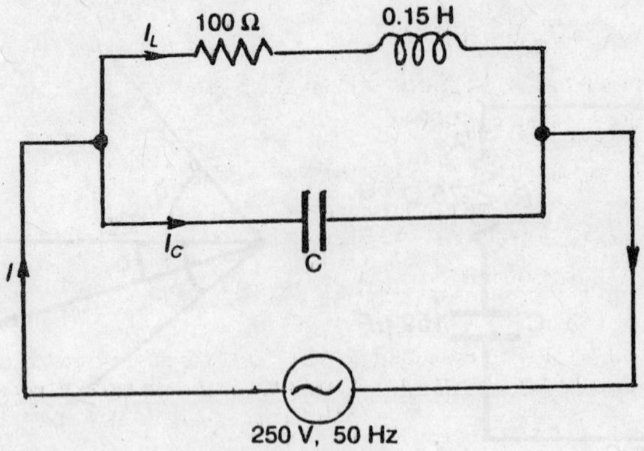

Fig. 18.11

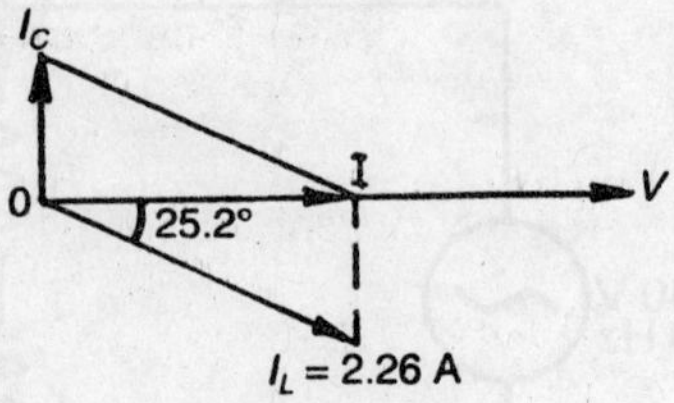

Fig. 18.12

Since the overall p.f. has to be unity, the magnitude of I_C should be such that line current I (*i.e.*, phasor sum of I_C and I_L) is in phase with V as shown in Fig. 18.12. In other words, the reactive component of line current should be zero. *i.e.*

$$I_C = I_L \sin \phi_L = 2.26 \sin 25.2° = 0.96 \text{ A}$$

Now $$I_C = V/X_C = V \times 2\pi f C$$

or $$C = \frac{I_C}{V \times 2\pi f} = \frac{0.96}{250 \times 2\pi \times 50} = \mathbf{12.2 \times 10^{-6}\ F}$$

When the capacitor is connected to the coil,

Line current, $I = I_L \cos \phi_L = 2.26 \cos 25.2° = \mathbf{2.04\ A}$

Example 18.6. *A 230 V, 50 Hz single phase motor takes 70 A and operates at a p.f. 0.75 lagging. If a capacitor of 159 μF is connected in parallel with the motor, find (i) the new line current (ii) circuit p.f.*

Solution.

$$I_m = 70 \text{ A}\ ;\ \phi_m = \cos^{-1} 0.75 = 41.4°$$

$$X_C = \frac{1}{2\pi f C} = \frac{10^6}{2\pi \times 50 \times 159} = 20\ \Omega$$

$$I_C = V/X_C = 230/20 = 11.5 \text{ A}$$ .....*leads V by 90°*

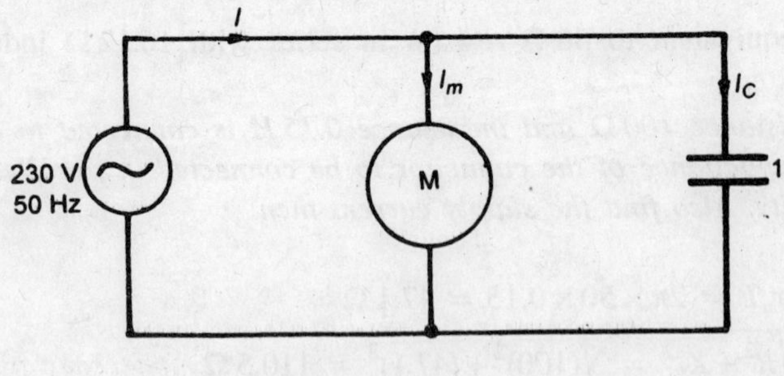

Fig. 18.13

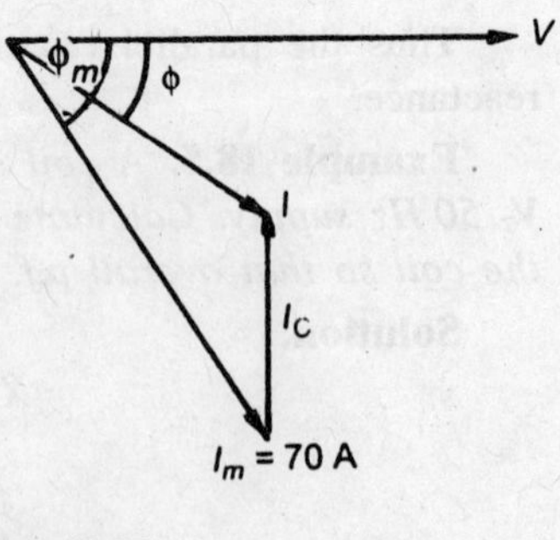

Fig. 18.14

As seen from the phasor diagram, the line current I lags the applied voltage by ϕ (less than ϕ_m).

$$\text{Total } X\text{-component} = I_m \cos\phi_m + I_C \cos 90^\circ$$
$$= 70 \times \cos 41.4^\circ + 11.5 \cos 90^\circ = 52.5 \text{ A}$$
$$\text{Total } Y\text{-component} = -I_m \sin\phi_m + I_C \sin 90^\circ$$
$$= -70 \sin 41.4^\circ + 11.5 \sin 90^\circ = -34.8 \text{ A}$$

(*i*) Line current, $I = \sqrt{(52.5)^2 + (-34.8)^2} = \mathbf{63\ A}$

(*ii*) Circuit *p.f.* $= \dfrac{\text{Total } X\text{-component}}{I} = \dfrac{52.5}{63} = \mathbf{0.833}\ \textit{lag}$

Note. Before connecting the capacitor, the line current was 70 A and circuit *p.f.* = 0.75 lagging. However, when capacitor is connected across the motor, the line current is reduced to 63 A and circuit *p.f.* improved to 0.833 lagging.

18.3. By Phasor Algebra

In this method, voltages, currents and impedances are expressed in the complex form *i.e.*, either in the rectangular or polar form. Since complex form includes both magnitude and phase angle, the solution of parallel-circuit problems can be obtained mathematically by using the rules of phasor algebra. This eliminates the need of phasor diagram. Referring to the parallel circuit shown in Fig. 18.15, we have,

$$\mathbf{V} = V + j\,0 = V \qquad \text{...Reference phasor}$$
$$\mathbf{Z}_1 = R_1 + j\,X_L\,; \qquad \mathbf{Z}_2 = R_2 - j\,X_C$$

(*i*) **Rectangular form**

$$\mathbf{I}_1 = \frac{\mathbf{V}}{\mathbf{Z}_1} = \frac{V}{R_1 + j\,X_L}$$
$$\mathbf{I}_2 = \frac{\mathbf{V}}{\mathbf{Z}_2} = \frac{V}{R_2 - jX_C}$$

Line current, $\mathbf{I} = \mathbf{I}_1 + \mathbf{I}_2 = \dfrac{V}{R_1 + j\,X_L} + \dfrac{V}{R_2 - j\,X_C}$

The solution of **I** can be obtained in the standard form $a \pm jb$ by using the rules of phasor algebra. Then it is an easy task to find the magnitude and phase angle of **I**.

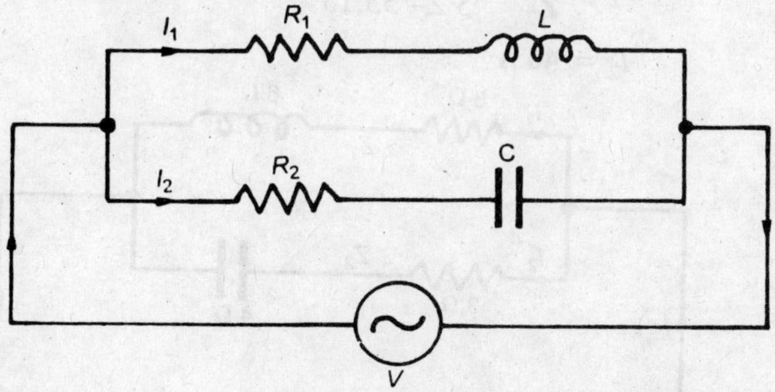

Fig. 18.15

(*ii*) **Polar form**

$$\mathbf{V} = V\,\angle 0^\circ \qquad \text{...Reference phasor}$$
$$\mathbf{Z}_1 = Z_1\,\angle\phi_1^\circ \text{ where } Z_1 = \sqrt{R_1^{\,2} + X_L^{\,2}}\,;\ \phi_1 = \tan^{-1} X_L/R$$
$$\mathbf{Z}_2 = Z_2\,\angle -\phi_2^\circ \text{ where } Z_2 = \sqrt{R_2^{\,2} + X_C^{\,2}}\,;\ \phi_2 = \tan^{-1} X_C/R$$

$$\therefore \quad \mathbf{I}_1 = \frac{\mathbf{V}}{\mathbf{Z}_1} = \frac{V\angle 0^\circ}{Z_1\angle\phi_1{}^\circ} = \frac{V}{Z_1}\angle -\phi_1{}^\circ$$

$$\mathbf{I}_2 = \frac{\mathbf{V}}{\mathbf{Z}_2} = \frac{V\angle 0^\circ}{Z_2\angle -\phi_2{}^\circ} = \frac{V}{Z_2}\angle\phi_2{}^\circ$$

$$\therefore \quad \mathbf{I} = \mathbf{I}_1 + \mathbf{I}_2 = \frac{V}{Z_1}\angle -\phi_1{}^\circ + \frac{V}{Z_2}\angle\phi_2{}^\circ$$

Example 18.7. *The voltage applied to a parallel circuit is 68 + j154 volts. The current in one branch is 10 + j14 amperes and the current in the other branch is 2 + j 8 amperes. What is the circuit power factor ?*

Solution.

Applied voltage, $\mathbf{V} = 68 + j\,154 = 168.34\angle 66.17^\circ$ volts

Line current, $\mathbf{I} = \mathbf{I}_1 + \mathbf{I}_2 = (10 + j\,14) + (2 + j\,8)$

$= 12 + j\,22 = 25.06\angle 61.39^\circ$

$\therefore$ Phase angle, $\phi = 66.17^\circ - 61.39^\circ = 4.78^\circ$

Circuit p.f. $= \cos\phi = \cos 4.78^\circ =$ **0.996** *lag*

Example 18.8. *Two impedances $Z_1 = 8 + j\,6$ and $Z_2 = 3 - j\,4$ are connected in parallel across 230 V, 50 Hz supply. Calculate (i) current in each branch (ii) line current (iii) circuit p.f. and (iv) power taken by the circuit.*

Solution. Fig. 18.16 shows the conditions of the problem. Taking voltage as the reference phasor, $\mathbf{V} = 230\angle 0^\circ$.

(*i*) $\mathbf{Z}_1 = 8 + j\,6 = 10\angle 36.87^\circ\ \Omega$

$$\mathbf{I}_1 = \frac{\mathbf{V}}{\mathbf{Z}_1} = \frac{230\angle 0^\circ}{10\angle 36.87^\circ} = 23\angle -36.87^\circ \text{ A}$$

$\therefore$ $I_1 =$ **23 A**

$\mathbf{Z}_2 = 3 - j\,4 = 5\angle -53.13^\circ\ \Omega$

$$\mathbf{I}_2 = \frac{\mathbf{V}}{\mathbf{Z}_2} = \frac{230\angle 0^\circ}{5\angle -53.13^\circ} = 46\angle 53.13^\circ \text{ A}$$

$\therefore$ $I_2 =$ **46 A**

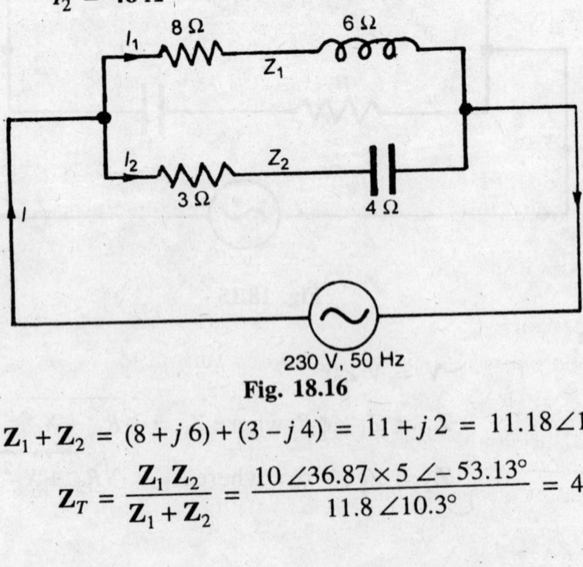

Fig. 18.16

$$\mathbf{Z}_1 + \mathbf{Z}_2 = (8 + j\,6) + (3 - j\,4) = 11 + j\,2 = 11.18\angle 10.3^\circ$$

$$\mathbf{Z}_T = \frac{\mathbf{Z}_1\,\mathbf{Z}_2}{\mathbf{Z}_1 + \mathbf{Z}_2} = \frac{10\angle 36.87 \times 5\angle -53.13^\circ}{11.8\angle 10.3^\circ} = 4.23\angle -26.56^\circ\ \Omega$$

$$\therefore \quad \mathbf{I} = \frac{\mathbf{V}}{\mathbf{Z}_T} = \frac{230\angle 0^\circ}{4.23\angle -26.56^\circ} = 54.4\angle 26.56^\circ \text{ A}$$

$$I = \mathbf{54.4\ A}$$

(*iii*) Circuit p.f. $= \cos 26.56 = \mathbf{0.894}$ *lead*

(*iv*) Power taken $= I_1^2R_1 + I_2^2R_2 = (23)^2 \times 8 + (46)^2 \times 3 = \mathbf{10580\ W}$

Example 18.9. *A voltage of 230 ∠30° V is applied to two circuits connected in parallel. The current in the branches are 20 ∠60° A and 40 ∠– 30° A. Find (i) the total impedance of the circuit (ii) power taken.*

Solution.

$$\mathbf{I}_1 = 20\angle 60^\circ = 20(\cos 60^\circ + j\sin 60^\circ) = (10 + j17.3) \text{ A}$$

$$\mathbf{I}_2 = 40\angle -30^\circ = 40(\cos 30^\circ - j\sin 30^\circ) = (34.6 - j\,20) \text{ A}$$

$$\mathbf{I} = \mathbf{I}_1 + \mathbf{I}_2 = (10 + j\,17.3) + (34.6 - j\,20)$$

$$= 44.6 - j2.7 = 44.7\angle -3.46^\circ \text{ A}$$

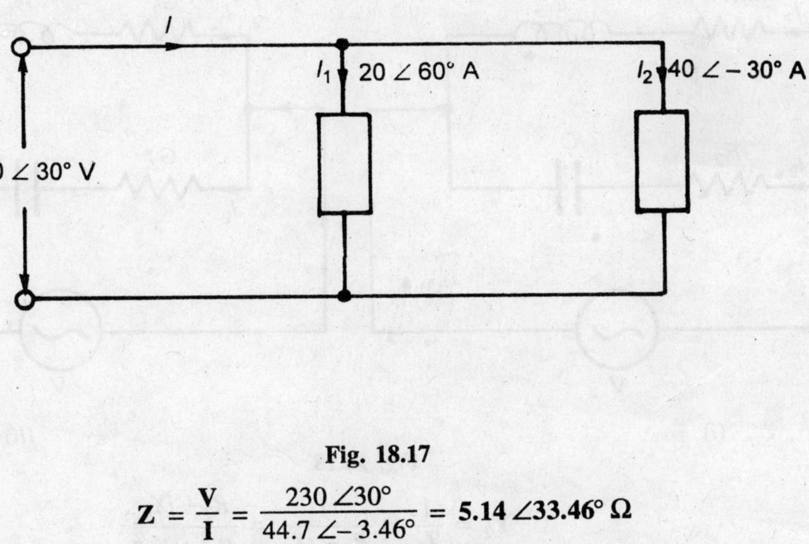

Fig. 18.17

$$\therefore \quad \mathbf{Z} = \frac{\mathbf{V}}{\mathbf{I}} = \frac{230\angle 30^\circ}{44.7\angle -3.46^\circ} = \mathbf{5.14\angle 33.46^\circ\ \Omega}$$

$$P = VI\cos\phi = 230 \times 44.7\cos 33.46^\circ = \mathbf{8577\ W}$$

18.4. Admittance

Whereas impedance is opposition to alternating current flow, admittance is the inducement to current flow. It is, therefore, defined as the reciprocal of impedance *i.e.*,

$$\text{Admittance, } Y = \frac{1}{Z} = \frac{I}{V} = \frac{\text{r.m.s.amperes}}{\text{r.m.s. volts}}$$

Just as impedance ($\mathbf{Z} = R \pm jX$) can be expressed in the complex form, similarly admittance can also be expressed in the *complex form as :

$$Y = G \pm jB$$

where G = conductance of the circuit

---

* The reciprocal of a complex number, after rationalising, results in a complex number.

B = susceptance of the circuit

Inductive susceptance is taken negative while capacitive susceptance is considered positive. The unit of Y, G and B is siemens (S).

18.5. Admittance Method

Consider the parallel circuit shown in Fig. 18.18 (*i*).

$$\mathbf{Y}_1 = \frac{1}{\mathbf{Z}_1} = \frac{1}{R_1 + jX_L} = \frac{{}^*R_1 - jX_L}{R_1^2 + X_L^2}$$

$$= \frac{R_1}{R_1^2 + X_L^2} - j\frac{X_L}{R_1^2 + X_L^2}$$

$$= G_1 - jB_1 \qquad \text{...(i)}$$

Thus for inductive branch, $G_1 = R_1/Z_1^2$ and $B_1 = X_L/Z_1^2$ as shown in Fig. 18.18(*ii*). Note from eq (*i*) that *inductive susceptance is negative.*

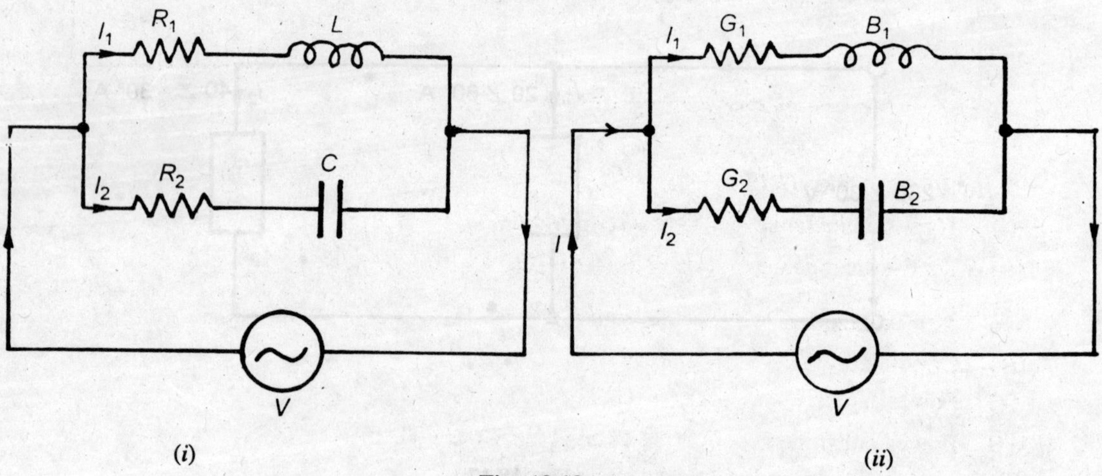

Fig. 18.18

$$\mathbf{Y}_2 = \frac{1}{\mathbf{Z}_2} = \frac{1}{R_2 - jX_C} = \frac{R_2 + jX_C}{R_2^2 + X_C^2}$$

$$= \frac{R_2}{R_2^2 + X_C^2} + j\frac{X_C}{R_2^2 + X_C^2}$$

$$= G_2 + jB_2 \qquad \text{...(ii)}$$

Thus for capacitive branch, $G_2 = R_2/Z_2^2$ and $B_2 = X_C/Z_2^2$ as shown in Fig. 18.18 (*ii*). Note from eq. (*ii*) that *capacitive susceptance is positive.*

$$\mathbf{Y} = \mathbf{Y}_1 + \mathbf{Y}_2 = (G_1 - jB_1) + (G_2 + jB_2)$$

$$= (G_1 + G_2) + j(-B_1 + B_2)$$

$$\therefore \qquad Y = \sqrt{(G_1 + G_2)^2 + (-B_1 + B_2)^2} \qquad \text{...in magnitude}$$

* After rationalising the denominator.

Line current, $I = VY$

Phase angle, $\phi = \tan^{-1} \dfrac{-B_1 + B_2}{G_1 + G_2}$

Power consumed, $P = VI \cos \phi$

18.6. Application of Admittance Method

Consider the parallel circuit shown in Fig. 18.19. The total conductance of the circuit is the sum of conductances of the branches. However, total susceptance is the *algebraic* sum of susceptances of the branches.

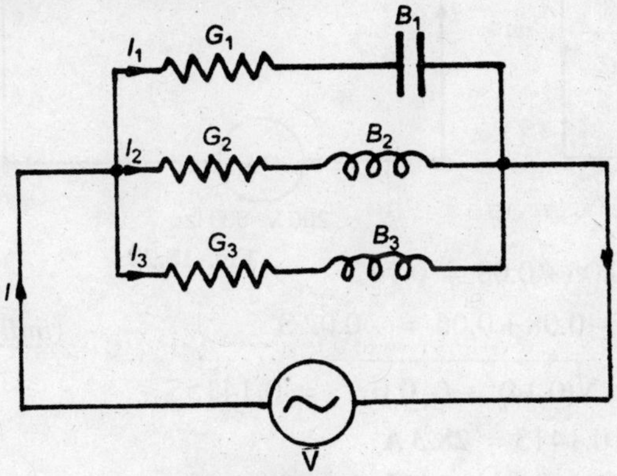

Fig. 18.19

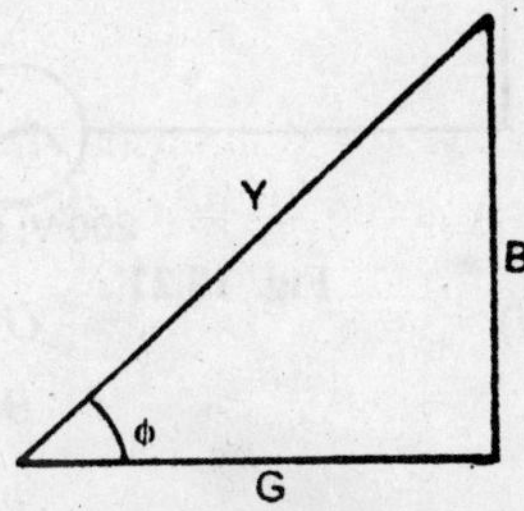

Fig. 18.20

(*i*) Total conductance, $G = G_1 + G_2 + G_3$

(*ii*) Total susceptance, $B = B_1 - B_2 - B_3$

(*iii*) Admittance, $Y = \sqrt{G^2 + B^2}$

(*iv*) Circuit current, $I = VY$

(*v*) Power factor, $\cos \phi = G/Y$

(*vi*) Power consumed, $P = VI \cos \phi = V (VY) (G/Y) = V^2 G$

The currents in the individual branches are :

$$I_1 = VY_1 \text{ where } Y_1 = \sqrt{G_1^2 + B_1^2}$$

$$I_2 = VY_2 \text{ where } Y_2 = \sqrt{G_2^2 + (-B_2)^2}$$

$$I_3 = VY_3 \text{ where } Y_3 = \sqrt{G_3^2 + (-B_3)^2}$$

Example 18.10. *A circuit having a resistance of 6 Ω and inductive reactance of 8 Ω is connected in parallel with another circuit having a resistance of 8 Ω and a capacitive reactance of 6 Ω. The parallel circuit is connected across 200 V, 50 Hz supply. Calculate :*

(*i*) *supply current*

(*ii*) *power factor of the whole circuit*

(*iii*) *power consumed*

(*iv*) *the resistance and reactance of a series circuit which will take the same current at the same p.f. as the parallel circuit.*

Solution.

$$Z_1^2 = 6^2 + 8^2 = 100 ; \qquad Z_2^2 = 8^2 + 6^2 = 100$$

$$G_1 = R_1/Z_1^2 = 6/100 = 0.06 \text{ S}$$

$$B_1 = X_1/Z_1^2 = 8/100 = -0.08 \text{ S}$$

$$G_2 = R_2/Z_2^2 = 8/100 = 0.08 \text{ S}$$

$$B_2 = X_2/Z_2^2 = 6/100 = 0.06 \text{ S}$$

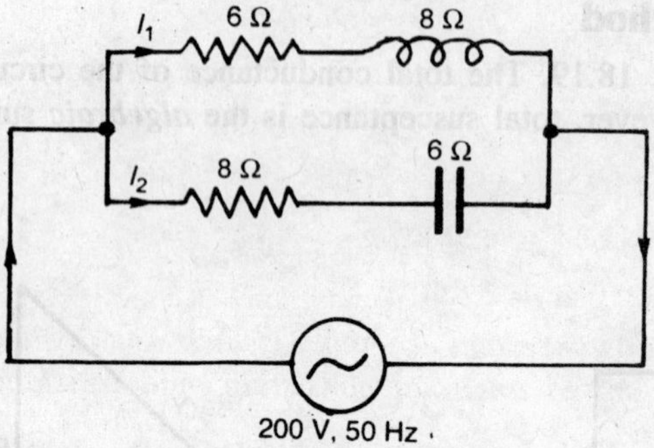

Fig. 18.21

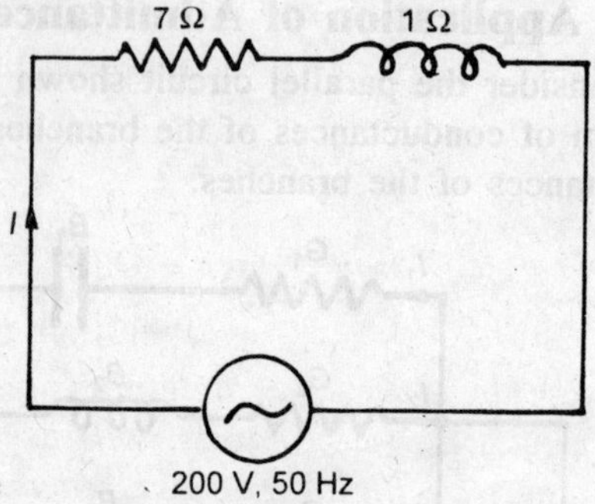

Fig. 18.22

(*i*)
$$G = G_1 + G_2 = 0.06 + 0.08 = 0.14 \text{ S}$$

$$B = -B_1 + B_2 = -0.08 + 0.06 = -0.02 \text{ S} \qquad (\textit{inductive})$$

$$Y = \sqrt{G^2 + B^2} = \sqrt{(0.14)^2 + (-0.02)^2} = 0.1415 \text{S}$$

$$I = VY = 200 \times 0.1415 = \mathbf{28.3\ A}$$

(*ii*)
$$p.f. = \cos\phi = G/Y = 0.14/0.1415 = \mathbf{0.99}\ \textit{lag}$$

(*iii*)
$$P = V^2G = (200)^2 \times 0.14 = \mathbf{5600\ W}$$

Also
$$P = VI\cos\phi = 200 \times 28.3 \times 0.99 = 5600 \text{W} \qquad \textit{...same as above}$$

(*iv*) Let the equivalent series resistance and reactance be R and X respectively

$$R = G/Y^2 = 0.14/(0.1415)^2 = 7\ \Omega$$

$$X = B/Y^2 = 0.02/(0.1415)^2 = 1\ \Omega$$

Since susceptance is negative, the reactance is inductive in nature. Fig. 18.22 shows the equivalent series circuit.

Example 18.11. *In the parallel circuit in Fig. 18.23 find (i) total admittance (ii) line current (iii) circuit p.f. and (iv) power consumed.*

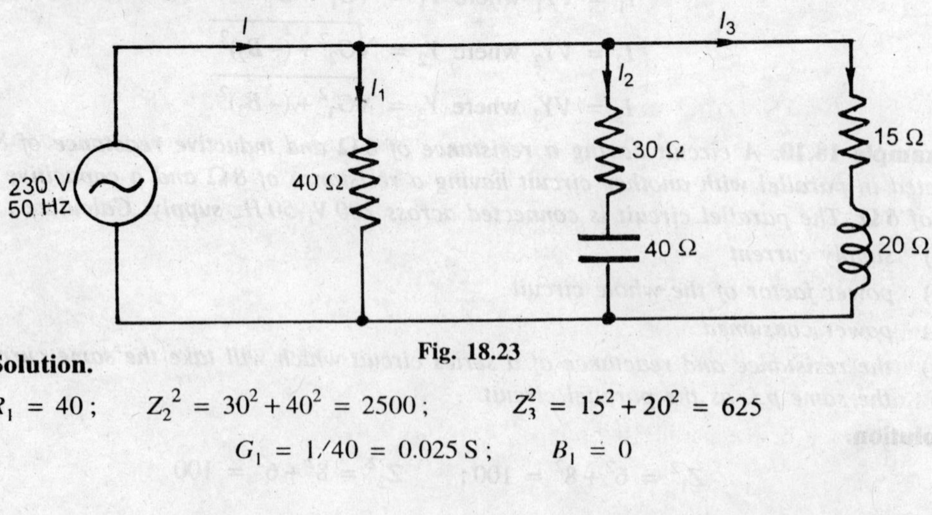

Fig. 18.23

Solution.

$$R_1 = 40\,; \qquad Z_2^2 = 30^2 + 40^2 = 2500\,; \qquad Z_3^2 = 15^2 + 20^2 = 625$$

$$G_1 = 1/40 = 0.025 \text{ S}\,; \qquad B_1 = 0$$

$$G_2 = 30/2500 = 0.012\text{ S}\;; \quad B_2 = 40/2500 = 0.016\text{ S}$$

$$G_3 = 15/625 = 0.024\text{ S}\;; \quad B_3 = 20/625 = -0.032\text{ S}$$

$$\therefore \quad G = G_1 + G_2 + G_3$$

$$= 0.025 + 0.012 + 0.024 = 0.061\text{ S}$$

$$B = B_1 + B_2 - B_3$$

$$= 0 + 0.016 - 0.032 = -0.016\text{ S}$$

(i) $$Y = \sqrt{G^2 + B^2} = \sqrt{(-0.061)^2 + (0.016)^2} = \mathbf{0.0631\text{ S}}$$

(ii) $$I = VY = 230 \times 0.0631 = \mathbf{14.5\text{ A}}$$

(iii) $$p.f. = G/Y = 0.061/0.0631 = \mathbf{0.966}\; lag$$

(iv) $$P = V^2G = (230)^2 \times 0.061 = \mathbf{3227\text{ W}}$$

Example 18.12. *Calculate the values of resistance and reactance, which when connected in parallel with one another, will be equivalent to a circuit consisting of a resistance of 10 Ω in series with an inductive reactance of 5 Ω.*

Solution.

Conductance of series circuit is

$$G = R/Z^2 = 10/10^2 + 5^2 = 0.08\text{ S}$$

Susceptance of series circuit is

$$B = X/Z^2 = 5/10^2 + 5^2 = -0.04\text{ S}$$

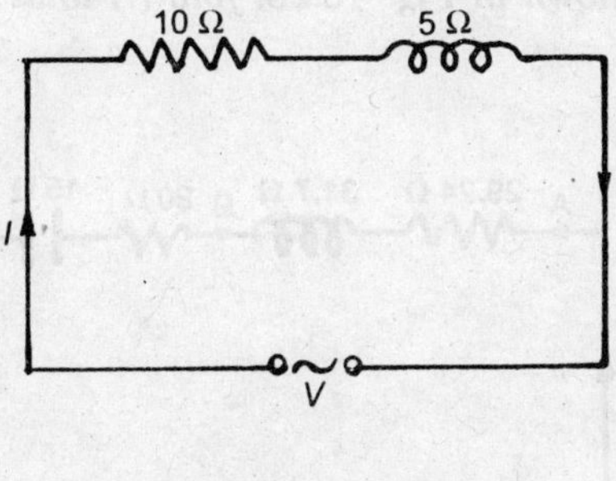

Fig. 18.24

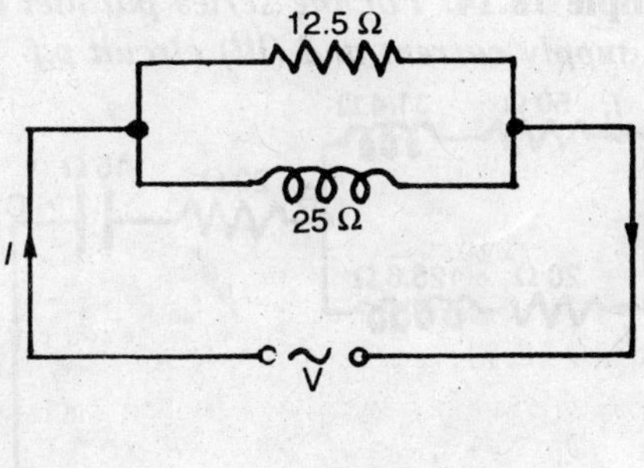

Fig. 18.25

Since the parallel circuit is to have the same conductance and susceptance as the series circuit, the values of parameters of parallel circuit are :

$$R = 1/G = 1/0.08 = \mathbf{12.5\ \Omega}\;; \quad X_L = 1/B = 1/0.04 = \mathbf{25\ \Omega}$$

Fig. 18.25 shows the equivalent parallel circuit.

Example 18.13. *A parallel circuit consisting of a 15 Ω resistance in parallel with an inductance of 0.06 H is connected to 230 V, 50 Hz source. Find the equivalent series circuit.*

Solution.

Conductance of the parallel circuit is

$$G = 1/R = 1/15 = 0.0667\text{ S}$$

Susceptance of the parallel circuit is

$$B = 1/X_L = 1/2\pi \times 50 \times 0.06 = -0.0531 \text{ S}$$

$$Y = \sqrt{G^2 + B^2} = \sqrt{(0.0667)^2 + (-0.0531)^2} = 0.0852 \text{ S}$$

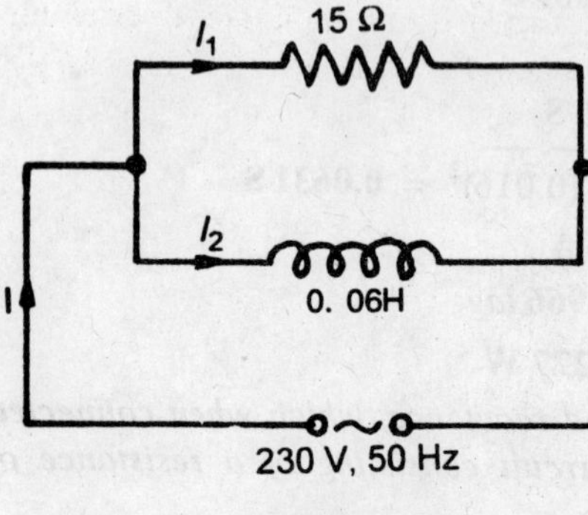

Fig. 18.26

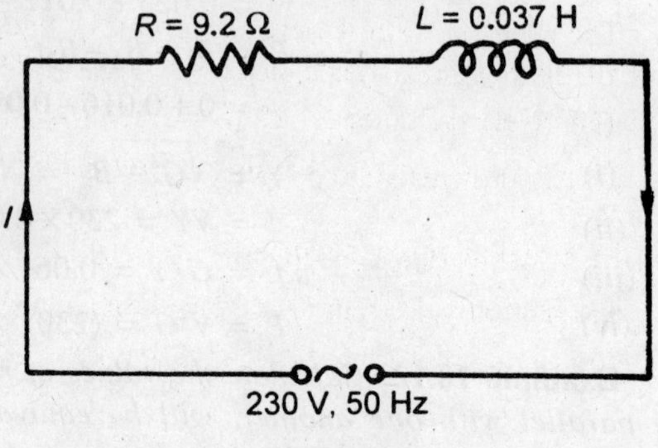

Fig. 18.27

Equivalent resistance of the series circuit

$$R' = G/Y^2 = 0.0667/(0.0852)^2 = \mathbf{9.2\ \Omega}$$

Equivalent inductive reactance of the series circuit

$$X_L' = B/Y^2 = 0.053/(0.0852)^2 = 11.75\Omega$$

$$\therefore \quad L' = X_L'/2\pi f = 11.75/2\pi \times 50 = \mathbf{0.037\ H}$$

The equivalent series circuit is shown in Fig. 18.27.

Example 18.14. *For the series parallel circuit shown in Fig. 18.28, find (i) total impedance (ii) supply current and (iii) circuit p.f.*

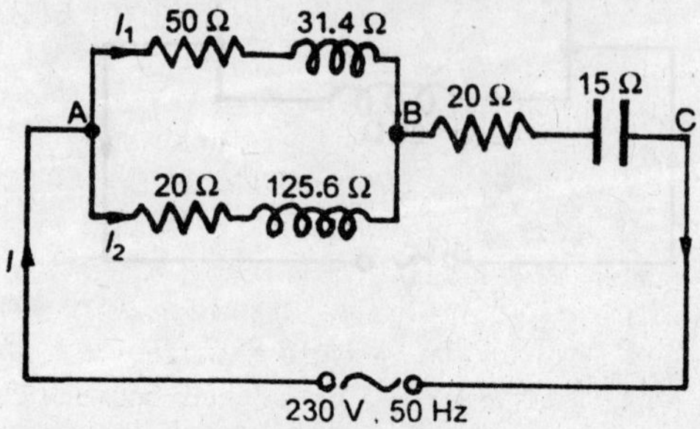

Fig. 18.28

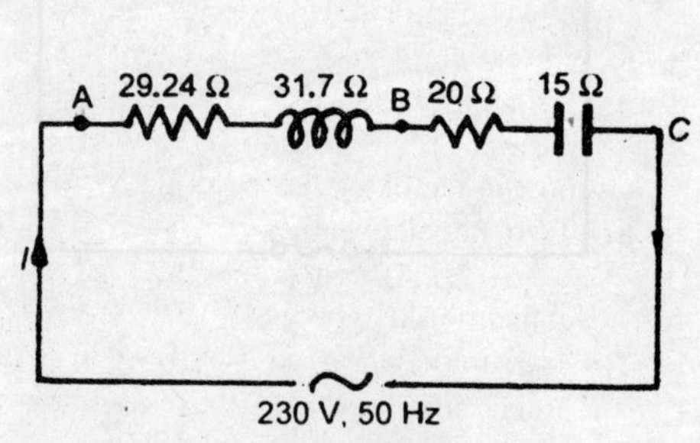

Fig. 18.29

Solution.

$$Z_1^2 = (50)^2 + (31.4)^2 = 3486 \; ; \; Z_2^2 = (20)^2 + (125.6)^2 = 16170$$

$$G_1 = R/Z_1^2 = 50/3486 = 0.0143\text{S}$$

$$B_1 = X_L/Z_1^2 = 31.4/3486 = -0.009 \text{ S}$$

$$G_2 = R/Z_2^2 = 20/16170 = 0.0012\text{S}$$

$$B_2 = X_L/Z_2^2 = 125.6/16170 = -0.0078 \text{ S}$$

$$G_{AB} = G_1 + G_2 = 0.0143 + 0.0012 = 0.0155 \text{ S}$$

$$B_{AB} = -B_1 - B_2 = -0.009 - 0.0078 = -0.0168 \text{ S}$$

$$Y_{AB} = \sqrt{G_{AB}^2 + B_{AB}^2} = \sqrt{(0.0155)^2 + (-0.0168)^2} = 0.023 \text{ S}$$

$$Z_{AB} = 1/Y_{AB} = 1/0.023 = 43.48\ \Omega$$

$$R_{AB} = G_{AB}/(Y_{AB})^2 = 0.0155/(0.023)^2 = 29.24\ \Omega$$

$$X_{AB} = B_{AB}/(Y_{AB})^2 = 0.0168/(0.023)^2 = 31.7\ \Omega \text{ (inductive)}$$

The parallel circuit between points A and B can be replaced by an equivalent series circuit of 29.24 Ω resistance in series with 31.7 Ω inductive reactance as shown in Fig. 18.29.

(*i*) Net reactance, $X_{AC} = 31.7 - 15 = 16.7\ \Omega$

Total resistance, $R_{AC} = 29.24 + 20 = 49.24\ \Omega$

Total impedance, $Z_{AC} = \sqrt{(16.7)^2 + (49.24)^2} = \mathbf{51.9\ \Omega}$

(*ii*) Supply current, $I = V/Z_{AC} = 230/51.9 = \mathbf{4.43\ A}$

(*iii*) Circuit p.f. $= R_{AC}/Z_{AC} = 49.24/51.9 = \mathbf{0.95}$ ***lag***

Tutorial Problems

1. A resistor of 10 Ω is connected in parallel with an inductor of 31.8 mH. A 200 V, 50 Hz supply is connected to the circuit. Determine (*i*) the line current (*ii*) power factor and (*iii*) power consumed by the circuit. **[(*i*) 28.28A (*ii*) 0.71*lag* (*iii*) 4015.76W]**
2. A coil of resistance 15 Ω and inductance 0.05 H is connected in parallel with a non-inductive resistance of 20 Ω. If the circuit is connected to 200V, 50Hz, find (*i*) current in each branch (*ii*) supply current (*iii*) power factor. **[(*i*) 9.22A ; 10A (*ii*) 17.7A (*iii*) 0.926 *lag*]**
3. A 10 Ω resistor, a 31.8 mH inductor and a 318 μF capacitor are connected in parallel and supplied from 200 V, 50 Hz supply. Determine the supply current and power factor. **[20 A ; 1]**
4. A parallel circuit consists of two branches A and B. Branch A has a resistance of 10 Ω and an inductance of 0.1 H in series. Branch B has a resistance of 20 Ω and a capacitor of 100 μF in series. The circuit is connected to 250 V, 50 Hz supply. Determine (*i*) the supply current and power factor (*ii*) power consumed by the circuit. **[(*i*) 6.04 A ; 0.965 *lag* (*ii*) 1457 W]**
5. An inductive circuit, in parallel with a non-inductive circuit of 20 Ω is connected across a 50 Hz supply. The inductive current is 4.3 A and the non-inductive current is 2.7 A. The total current is 5.8 A. Find (*i*) the power absorbed by the inductive branch (*ii*) its inductance and (*iii*) power factor of the combined circuit. **[(*i*) 78.6 W (*ii*) 0.0376 H (*iii*) 0.719 *lag*]**
6. Two impedances $Z_1 = (10 + j\,5)\ \Omega$ and $Z_2 = (8 + j\,6)\ \Omega$ are in parallel and connected to a 200 V, 50 Hz supply. Calculate (*i*) the supply current (*ii*) circuit power factor and (*iii*) power consumed by the circuit. **[(*i*) 37.74 A (*ii*) 0.848 *lag* (*iii*) 6400 W]**
7. Two impedances $Z_1 = (10 + j\,5)\ \Omega$ and $Z_2 = (25 - j\,10)\ \Omega$ are connected in parallel to a 100 V, 50 Hz supply. Find (*i*) circuit admittance (*ii*) circuit current (*iii*) the phase angle between circuit current and the applied voltage. **[(*i*) 0.1174 ℧ (*ii*) 11.74 A (*iii*) 12.9° *lagging*]**
8. An impedance of $(2 + j\,3)\ \Omega$ is connected in series with two impedances of $(4 + j\,2)\ \Omega$ and $(1 - j\,5)\ \Omega$, which are in parallel. Calculate the line current and the circuit power factor when the combined circuit is supplied at 10 V. **[1.71 A ; 0.963 *lag*]**
9. In a series-parallel circuit, the two parallel branches A and B are in series with C. The impedances are $Z_A = (10 + j\,8)\ \Omega$; $Z_B = (9 - j\,6)\ \Omega$ and $Z_C = (3 + j\,2)\ \Omega$. If the voltage across C is $100\ \angle 0°$ V, determine the values of I_A and I_B. **[I_A = 15.7 A ; I_B = 18.6 A]**

18.7. Resonance in Parallel Circuits

A parallel circuit containing reactive elements (L and C) is said to be in resonance when the circuit p.f. is unity *i.e.*, the reactive component of line current is zero. The frequency at which it occurs is called the *resonant frequency* f_r.

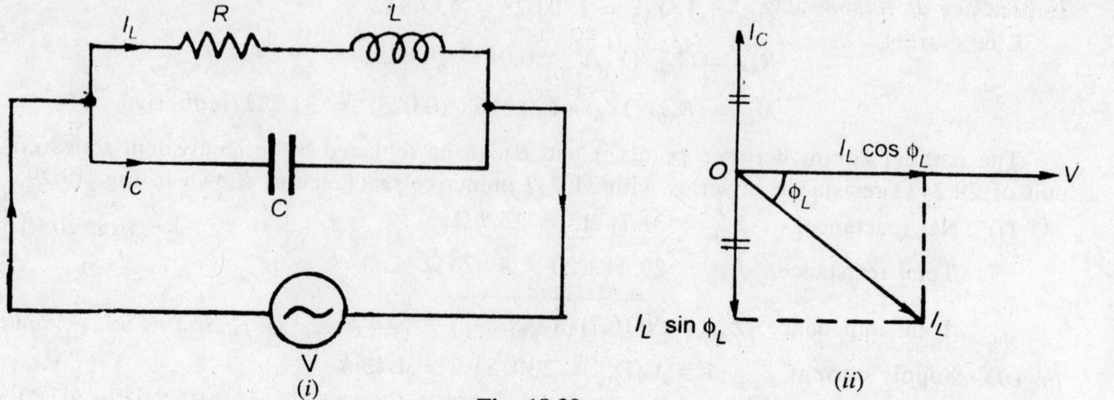

Fig. 18.30

Consider the most common parallel circuit consisting of a coil shunted by a capacitor as shown in Fig. 18.30 (*i*). The phasor diagram of this circuit is shown in Fig. 18.30 (*ii*). The circuit will be in resonance when the reactive component of line current is zero *i.e.*, $I_C - I_L \sin \phi_L = 0$. This can be achieved by changing the supply frequency because both I_C and $I_L \sin \phi_L$ are frequency *dependent. At some frequency called resonant frequency f_r, the reactive component of line current will be zero and resonance takes place.

At parallel resonance,

$$I_C - I_L \sin \phi_L = 0$$

or $$I_C = I_L \sin \phi_L$$

or $$\frac{V}{X_C} = \frac{V}{Z_L} \times \frac{X_L}{Z_L}$$

or $$X_L X_C = Z_L^2$$

or $$\omega L / \omega C = Z_L^2 \quad \text{...(i)}$$

or $$L/C = R^2 + (2\pi f_r L)^2$$

or $$(2\pi f_r L)^2 = \frac{L}{C} - R^2$$

or $$2\pi f_r L = \sqrt{\frac{L}{C} - R^2}$$

or $$f_r = \frac{1}{2\pi L}\sqrt{\frac{L}{C} - R^2}$$

or $$f_r = \frac{1}{2\pi}\sqrt{\frac{1}{LC} - \frac{R^2}{L^2}}$$

If the coil resistance is small (as is generally the case), then

$$f_r = \frac{1}{2\pi\sqrt{LC}} \quad \text{...as for series resonance}$$

The resonant frequency will be in *Hz* if *R*, *L* and *C* are measured in ohms, henry and farad respectively.

---

* $I_C = V/X_C = V \times 2\pi f C$

$I_L \sin \phi_L = \frac{V}{Z_L} \times \sin \phi_L$

where $Z_L = \sqrt{R^2 + (2\pi f L)^2}$; $\phi_L = \tan^{-1} X_L/R$

Impedance at Resonance

Line current, $I_r = I_L \cos\phi_L$

or $\dfrac{V}{Z_r} = \dfrac{V}{Z_L} \times \dfrac{R}{Z_L}$

or $\dfrac{1}{Z_r} = \dfrac{R}{Z_L^2}$

or $\dfrac{1}{Z_r} = \dfrac{R}{L/C}$ $\left[\because Z_L^2 = \dfrac{L}{C} \text{ from eq } (i)\right]$

$\therefore$ $Z_r = L/CR$

The impedance $Z_r (= L/CR)$ at parallel resonance is known as *equivalent* or *dynamic impedance*.

Note that $Z_r (= L/CR)$ is a pure resistance because there is no frequency term present. Further, the value of Z_r is very high because the ratio L/C is very large at parallel resonance.

Line current at resonance

At parallel resonance, the line current I_r is minimum and is given by ;

$$I_r = V/Z_r \quad \text{where } Z_r = L/CR$$

Because Z_r is very high, I_r will be very small. The small current I_r is only the amount needed to supply the resistance losses in the circuit.

18.8. Graphical Representation of Parallel Resonance

The action of a resonant circuit is best explained by referring to the curves illustrating the variation in the circuit conditions at or near resonance. We shall discuss two such curves *viz* impedance-frequency curve and current-frequency curve.

(*i*) **Impedance-frequency curve.** If we plot impedance-frequency graph for a parallel circuit shown in Fig. 18.30 (*i*), the shape of the curve will be as shown in Fig. 18.31. Note that the impedance of the circuit is maximum at resonance. As the frequency changes from resonance, the circuit impedance decreases very rapidly. This behaviour can be explained as follows. For frequencies below resonance, the capacitive reactance will be higher and thus *more current will flow through the coil. This causes the line current to lag behind the applied voltage and the circuit appears inductive. For frequencies above resonance, inductive reactance is higher and more current will flow through the capacitor. Consequently, line current leads the applied voltage and circuit appears capacitive. In either case, the circuit impedance is far less than its value at resonance.

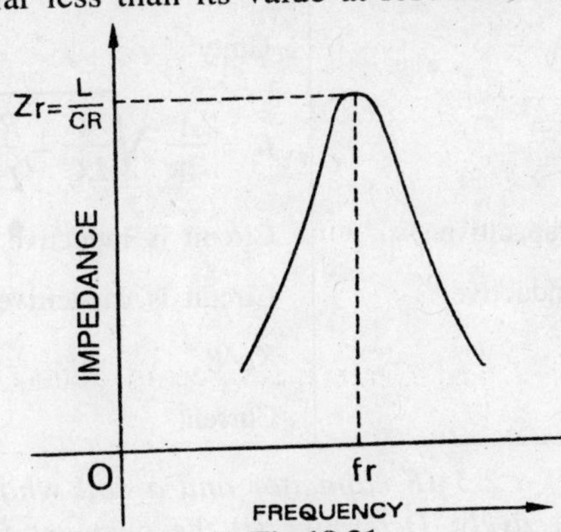

Fig.18.31

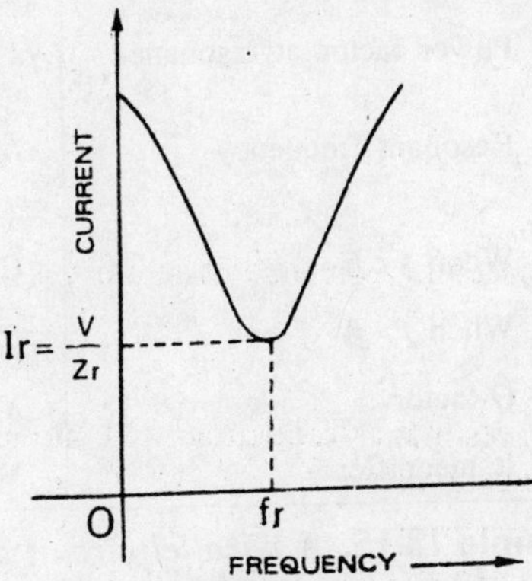

Fig. 18.32

* In the two parallel branches, more current will flow through that branch which has less impedance.

(*ii*) **Current-frequency curve.** Fig. 18.32 shows the current-frequency curve of the parallel circuit shown in Fig. 18.30(*i*). Note that value of line current ($I_r = V/Z_r$) is minimum at resonance. As the frequency changes from resonance, the line current increases rapidly. This action can be explained as follows. For frequencies other than the resonance, the reactive currents in the two branches of the circuit are not equal. The resultant reactive current must be supplied by the line. As the difference of the reactive currents in the two branches increases with the amount of deviation from the resonant frequency, the line current will increase in the same manner.

18.9. Q-factor of a Parallel Resonant Circuit

At parallel resonance, the current circulating between the two branches is many times greater than the line current. This current amplification produced by the resonance is termed as *Q*-factor of the parallel resonant circuit *i.e.*

$$Q\text{-factor} = \frac{\text{Circulating current between } L \text{ and } C}{I_r}$$

Now $$I_C = V/X_C = \omega_r CV \text{ and } I_r = V/(L/CR)$$

$$\therefore \quad Q\text{-factor} = \omega_r CV \div \frac{V}{L/CR} = \frac{\omega_r L}{R}$$

$$= \frac{2\pi f_r L}{R}$$

...same as for series circuit

The *Q*-factor of a parallel resonant circuit can also be expressed in terms of *L* and *C*. Neglecting resistance *R*, the resonant frequency is given by ;

$$f_r = \frac{1}{2\pi\sqrt{LC}}$$

$$\therefore \quad Q\text{-factor} = \frac{2\pi f_r L}{R} = \frac{2\pi L}{R} \times \frac{1}{2\pi\sqrt{LC}} = \frac{1}{R}\sqrt{\frac{L}{C}}$$

18.10. Comparison of Series and Parallel Resonant Circuits

S. No.	Particular	Series Circuit	Parallel Circuit
1.	Impedance at resonance	Minimum ($Z_r = R$)	Maximum ($Z_r = L/CR$)
2.	Current at resonance	Maximum ($I_r = V/R$)	Minimum ($I_r = V/Z_r$)
3.	Power factor at resonance	Unity	Unity
4.	Resonant frequency	$f_r = \frac{1}{2\pi\sqrt{LC}}$	$f_r = \frac{1}{2\pi}\sqrt{\frac{1}{LC} - \frac{R^2}{L^2}}$
5.	When $f < f_r$	Circuit is capacitive	Circuit is inductive
6.	When $f > f_r$	Circuit is inductive	Circuit is capacitive
7.	*Q*-factor	X_L/R	X_L/R
8.	It magnifies	Voltage	Current

Example 18.15. *A parallel circuit consists of a 2.5 μF capacitor and a coil whose resistance and inductance are 15 Ω and 260 mH respectively. Determine (i) the resonant frequency (ii) Q-factor of the circuit at resonance (iii) dynamic impedance of the circuit.*

Solution.

(*i*) Resonant frequency,

$$f_r = \frac{1}{2\pi}\sqrt{\frac{1}{LC} - \frac{R^2}{L^2}}$$

$$= \frac{1}{2\pi}\sqrt{\frac{10^6}{0.260 \times 2.5} - \frac{(15)^2}{(0.260)^2}} = \mathbf{197\ Hz}$$

(*ii*) $$Q\text{–factor} = \frac{2\pi f_r L}{R} = \frac{2\pi \times 197 \times 0.260}{15} = \mathbf{21.45}$$

(*iii*) $$Z_r = \frac{L}{CR} = \frac{0.260}{2.5 \times 10^{-6} \times 15} = \mathbf{6933\Omega}$$

Example 18.16. *A tuned circuit consisting of a coil having an inductance of 200 μH and a resistance of 20 Ω is in parallel with a variable capacitor. This combination is in series with a resistor of 8000 Ω. The entire circuit is connected to a 230V, 1MHz supply. Calculate (i) the value of C to give resonance (ii) the dynamic impedance and Q-factor of the tuned circuit and (iii) the current in each branch.*

Solution.

$$X_L = 2\pi fL = 2\pi \times 10^6 \times 200 \times 10^{-6} = 1256\ \Omega$$

$$Z_L = \sqrt{R^2 + X_L{}^2} = \sqrt{20^2 + (1256)^2} = 1256.16\Omega$$

(*i*) $$Z_L{}^2 = L/C \qquad \text{...at parallel resonance}$$

$$\therefore \quad C = \frac{L}{Z_L{}^2} = \frac{200 \times 10^{-6}}{(1256.16)^2} = 126.6 \times 10^{-12}\ \text{F} = \mathbf{126.6pF}$$

(*ii*) $$Z_r = \frac{L}{CR} = \frac{200 \times 10^{-6}}{126.6 \times 10^{-12} \times 20} = \mathbf{78957\Omega}$$

$$Q\text{-factor} = \frac{2\pi f_r L}{R} = \frac{2\pi \times 10^6 \times 200 \times 10^{-6}}{20} = \mathbf{62.8}$$

(*iii*) Total impedance $= 8000 + Z_r = 8000 + 78957 = 86957\Omega$

Line current, $I = 230/86957 = 2.64 \times 10^{-3}$ A

P.D. across tuned ckt. $= IZ_r = 2.64 \times 10^{-3} \times 78957 = 208.4$ V

Coil current, $I_L = 200/Z_L = 200/1256.16 = \mathbf{0.166\ A}$

Capacitor current, $I_C = 200/X_C = 200 \times 2\pi f_r C$

$= 200 \times 2\pi \times 10^6 \times 126.6 \times 10^{-12} = \mathbf{0.166\ A}$

Note that current in each branch is 62.8 times (*i.e.*, *Q*-factor times) the line current.

Tutorial Problems

1. A coil of 1000 Ω resistance and 0.15 H inductance is connected in parallel with a variable capacitor across a 2 V, 10 KHz a.c. supply. Calculate (*i*) the capacitance of the capacitor when the supply current is minimum (*ii*) the dynamic impedance of the network and (*iii*) the supply current at resonance. **[(*i*) 1690 μF (*ii*) 890 KΩ (*iii*) 2.25 μA]**
2. A 0.8 μF capacitor is in parallel with a coil that has a resistance of 4 Ω and an inductance of 0.2 H. Determine (*i*) the resonant frequency (*ii*) *Q*-factor at resonance (*iii*) input impedance at resonance. **[(*i*) 398 Hz (*ii*) 125 (*iii*) 62.5 KΩ]**

3. A coil that has an inductance L and a resistance of $8\ \Omega$ is in parallel with a variable capacitor. At $\omega = 1.5 \times 10^6$ rad/s, resonance is achieved when $C = 0.0037\ \mu F$. What is the inductance of the coil ? **[L = 120 μH]**
4. An inductor for which $L = 40$ mH is to be used in parallel with a 1 μF capacitor in a circuit that must have Q-factor at least as large as 25 when the frequency is 1 KHz. What is the minimum amount of resistance the coil may have ? What will the input impedance be when the minimum R is used ? **[R = 12.56 Ω ; Z = 3185 Ω]**
5. A coil has a resistance of 400 Ω and inductance of 318 μH. Find the capacitance of a capacitor which, when connected in parallel with the coil, will produce resonance with a supply frequency of 1 MHz. If a second capacitor of capacitance 23.5 pF is connected in parallel with the first capacitor, find the frequency at which resonance will occur. **[76.5 pF ; 870 KHz]**

OBJECTIVE QUESTIONS

1. Domestic appliances are connnected in parallel across a.c. mains because............
   (*i*) It is a simple arrangement
   (*ii*) operation of each appliance becomes independent of the other
   (*iii*) appliances have same currents ratings
   (*iv*) this arrangement occupies less space
2. When a parallel a.c. circuit contains a number of branches, then it is convenient to solve the circuit by ....
   (*i*) phasor diagram
   (*ii*) phasor algebra
   (*iii*) equivalent impedance method
   (*iv*) none of the above
3. The power taken by the circuit shown in Fig. 18.33 is ....
   (*i*) 480 W (*ii*) 1920 W
   (*iii*) 1200 W (*iv*) none of the above

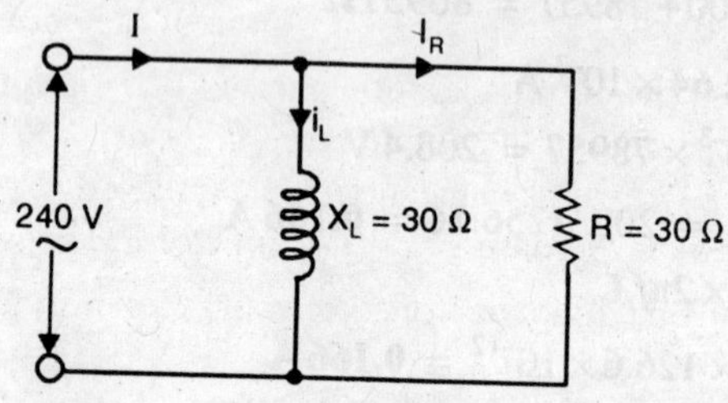

Fig. 18.33

4. The active component of line current in Fig. 18.33 is ....
   (*i*) 8 A (*ii*) 4 A
   (*iii*) 5.3 A (*iv*) none of the above
5. The power factor of the circuit shown in Fig. 18.33 is ....
   (*i*) 0.707 lagging (*ii*) 0.5 lagging
   (*iii*) 0.866 lagging (*iv*) none of the above
6. The total line current drawn by the circuit shown in Fig. 18.33 is ....
   (*i*) $8/\sqrt{2}$ A (*ii*) 16 A
   (*iii*) $8\sqrt{2}$ A (*iv*) none of the above
7. The power consumed in the circuit shown in Fig. 18.34 is ....
   (*i*) 480 W (*ii*) 960 W
   (*iii*) 1200 W (*iv*) none of the above

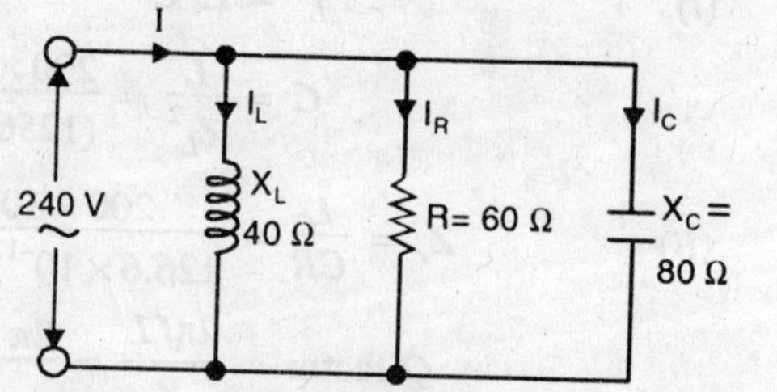

Fig. 18.34

8. The active component of line current in Fig. 18.34 is ....
   (*i*) 6A (*ii*) 3A
   (*iii*) 13 A (*iv*) 4 A
9. The line current drawn by the circuit shown in Fig. 18.34. is ....
   (*i*) 13 A (*ii*) 6 A
   (*iii*) 5A (*iv*) none of the above
10. The power factor of the circuit shown in Fig. 18.34 is ....
    (*i*) 0.8 (*ii*) 0.5
    (*iii*) 0.707 (*iv*) none of the above
11. The impedance of the circuit shown in Fig. 18.34 is ....
    (*i*) 180 ohms (*ii*) 24 ohms
    (*iii*) 48 ohms (*iv*) none of the above
12. The circuit shown in Fig. 18.34 is ....

(*i*) resistive (*ii*) capacitive
(*iii*) inductive (*iv*) in resonance

13. If in Fig. 18.34, X_L is made equal to X_C then line current will be ....

(*i*) 10 A (*ii*) 6 A
(*iii*) 4 A (*iv*) none of the above

14. The power consumed in the circuit shown in Fig. 18.35 is ....

(*i*) 8400 W (*ii*) 3600 W
(*iii*) 4000 W (*iv*) none of the above

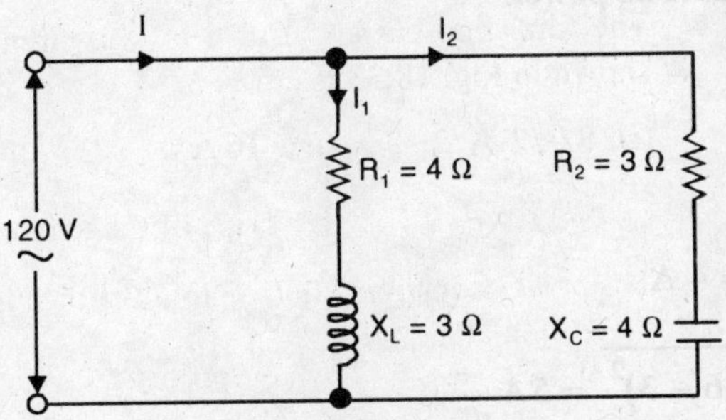

Fig. 18.35

15. If the circuit shown in Fig. 18.35 is connected to 120 V d.c., the current drawn by the circuit is ....

(*i*) 24 A (*ii*) 70 A
(*iii*) 48 A (*iv*) 30 A

16. The circuit shown in Fig. 18.35 is ....

(*i*) capacitive (*ii*) inductive
(*iii*) resistive (*iv*) in resonance

17. If the source frequency in Fig. 18.36. is low, then, ...

(*i*) coil takes a high lagging current
(*ii*) coil takes a low lagging current
(*iii*) capacitor takes a high leading current
(*iv*) circuit offers high impedance

18. If the source frequency in Fig. 18.36 is high then,.............

(*i*) coil takes a high lagging current
(*ii*) capacitor takes a high leading current
(*iii*) capacitor takes a low leading current
(*iv*) circuit offers high impedance

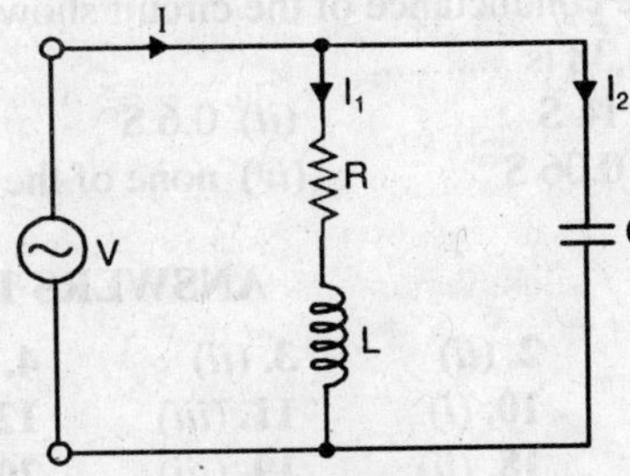

Fig. 18.36

19. The circuit shown in Fig. 18.36 will be in resonance when ....

(*i*) $X_L = X_C$ (*ii*) $I_1 = I_2$
(*iii*) *V* and *I* are in phase
(*iv*) none of the above

20. The circuit shown in Fig. 18.37 is ....

(*i*) in resonance (*ii*) resistive
(*iii*) inductive (*iv*) capacitive

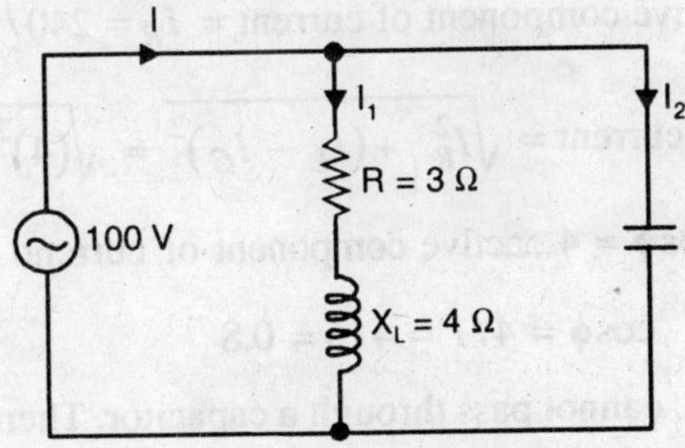

Fig. 18.37

21. The circuit shown in Fig. 18.37 will consume a power of ....

(*i*) 1200 W (*ii*) 2400 W
(*iii*) 500 W (*iv*) none of the above

22. If the admittance of a parallel a.c. circuit is increased, the circuit current ....

(*i*) remains constant (*ii*) is decreased
(*iii*) is increased (*iv*) none of the above

23. The admittance of the circuit shown in Fig. 18.38 is ....

(*i*) 10 S (*ii*) 14 S
(*iii*) 0.1 S (*iv*) none of the above

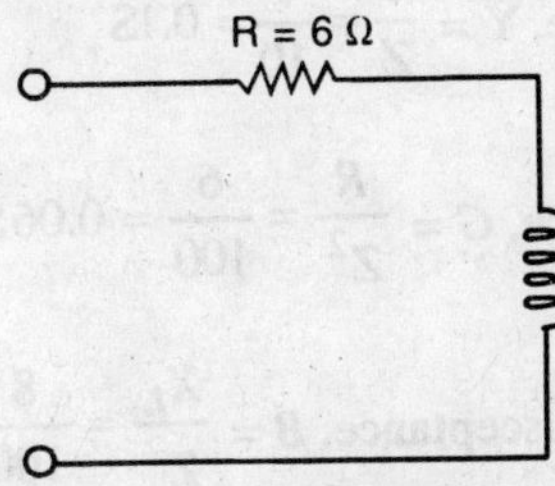

Fig. 18.38

24. The conductance of the circuit shown in Fig. 18.38 is ....
(*i*) 14 S (*ii*) 0.6 S
(*iii*) 0.06 S (*iv*) none of the above

25. The inductive susceptance of the circuit shown in Fig. 18.38 is ....
(*i*) 8 S (*ii*) 0.8 S
(*iii*) 0.08 S (*iv*) none of the above

ANSWERS TO OBJECTIVE QUESTIONS

1. (*ii*)	**2.** (*ii*)	**3.** (*ii*)	**4.** (*i*)	**5.** (*i*)	**6.** (*iii*)	**7.** (*ii*)	**8.** (*iv*)
9. (*iii*)	**10.** (*i*)	**11.** (*iii*)	**12.** (*iii*)	**13.** (*iii*)	**14.** (*iv*)	**15.** (*iv*)	**16.** (*i*)
17. (*i*)	**18.** (*ii*)	**19.** (*iii*)	**20.** (*iv*)	**21.** (*i*)	**22.** (*iii*)	**23.** (*iii*)	**24.** (*iii*)
25. (*iii*)							

HINTS TO SELECTED OBJECTIVE QUESTIONS

3. Power is consumed only in *R* as *L* and *C* consume no power.
Current in R = 240/30 = 8 A

$\therefore$ Power consumed = $(8)^2 \times 30 = 1920$ W

6. Total line current = $\sqrt{8^2 + 8^2} = 8\sqrt{2}\ A$

8. Active component of current = $I_R = 240/60 = 4$ A

9. Line current = $\sqrt{I_R^2 + (I_L - I_C)^2} = \sqrt{(4)^2 + (6-3)^2} = 5\,A$

10. $I\cos\phi = 4$...active component of current

$\therefore \quad \cos\phi = 4/I = 4/5 = 0.8$

15. D.C. cannot pass through a capacitor. Therefore, $R_2 - X_C$ branch will act as an open for d.c. The direct current will pass through the branch $R_l - X_L$. In this branch, only $R_l (= 4\Omega)$ will offer opposition as $X_L = 0$ for d.c.

$\therefore$ Circuit current = 120/4 = 30 A.

17. If the frequency is low, the reactance $(= 2\pi f L)$ of the coil will be low and that of capacitor $(= 1/2\pi f C)$ will be high. Hence, the coil will take a high lagging current.

19. In general, the condition for resonance in parallel circuit is not that inductive reactance (X_L) be equal to the capacitive reactance (X_C) but that the circuit should have unity power factor. In the special case, equal reactances result in unity power factor, but this may not always be true.

23. Circuit impedance, $Z = \sqrt{6^2 + 8^2} = 10\Omega$

$\therefore$ Admittance, $Y = \dfrac{1}{Z} = \dfrac{1}{10} = 0.1S$

24. Conductance, $G = \dfrac{R}{Z^2} = \dfrac{6}{100} = 0.06S$

25. Inductive susceptance, $B = \dfrac{X_L}{Z^2} = \dfrac{8}{100} = 0.08S$

19

Three-Phase Circuits

Introduction

The a.c. circuits discussed so far in the book are termed as single phase circuits because they contain a single alternating current and voltage wave. The generator producing a single phase supply (called single-phase generator) has only one armature winding. But if the generator is arranged to have two or more separate windings displaced from each other by equal electrical angles, it is called a *polyphase generator and will produce as many independent voltages as the number of windings or phases. The electrical displacement between the windings depends upon the number of windings or phases. For example, a 2-phase generator has two separate but identical windings that are 90° electrical apart and rotate in a common magnetic field. Obviously, such a generator will produce two alternating voltages of the same magnitude and frequency having a phase difference of 90°. Similarly, a 3-phase generator has three separate but identical windings that are 120° electrical apart and rotate in a common magnetic field. A 3-phase generator will, therefore, produce three voltages of the same magnitude and frequency but displaced 120° electrical from one another. Although several polyphase systems are possible, the 3-phase system is by far the most popular. In this chapter, we shall confine our attention to 3-phase system.

19.1. Polyphase Systems

Fig 19.1 shows the generation of single-phase, two-phase and 3-phase voltages.

(*i*) Fig. 19.1 (*i*) shows an elementary single-phase alternator. It has one winding or coil *A* rotating in anticlockwise direction with an angular velocity ω in the 2-pole field. The equation of the e.m.f. induced in the coil is given by ;

$$e_{a_1a_2} = E_m \sin \omega t$$

(*ii*) Fig 19.1 (*ii*) shows an elementary two-phase alternator. It has two identical coils *A* and *B* displaced 90° electrical from each other and rotating in anticlockwise direction with an angular velocity ω in the 2-pole field. The wave diagram is shown in Fig. 19.1 (*ii*). Note that the e.m.f. in coil *A* leads that in coil *B* by 90°. The equations of the two e.m.f.s are :

$$e_{a_1a_2} = E_m \sin \omega t$$

$$e_{b_1b_2} = E_m \sin (\omega t - 90°)$$

---

* The electrical displacement between the windings is determined by the number of phases or windings. For a 2-phase alternator, the two windings are displaced by 90 electrical degrees. For other polyphase systems (*e.g*,. 3-phase, 6-phase), the electrical displacement between different phases or windings is equal to 360/N where N is the number of phases. Thus for a 3-phase alternator, the three windings are 120 electrical degrees apart.

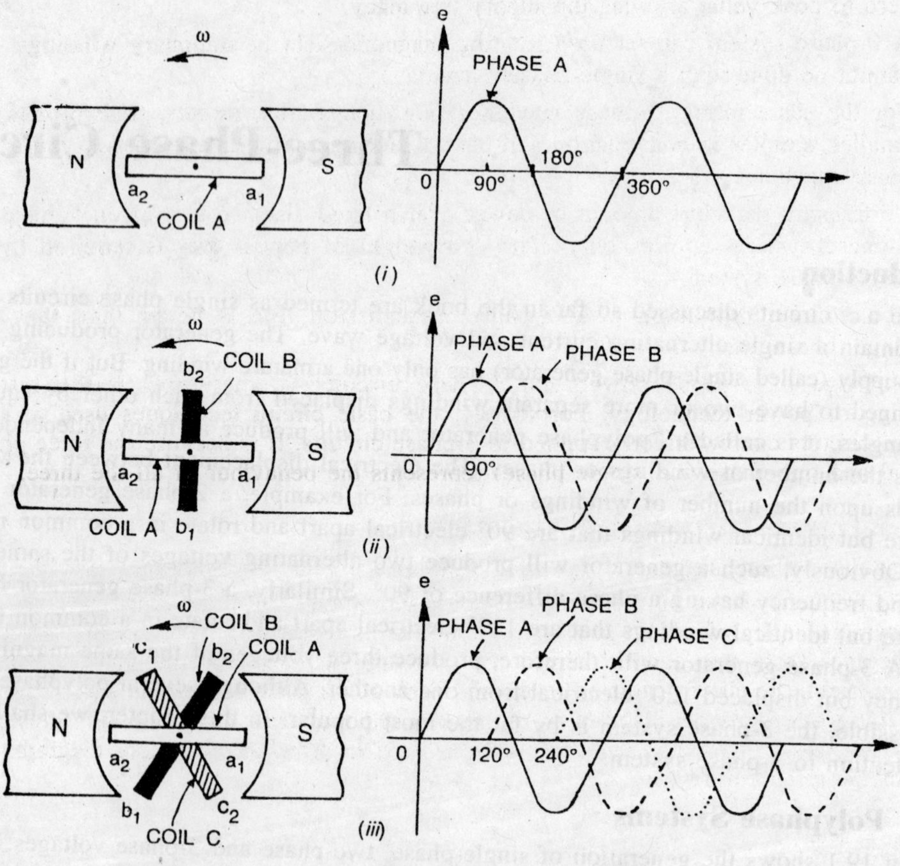

Fig. 19.1

(*iii*) Fig. 19.1 (*iii*) shows an elementary 3-phase alternator. It has three identical windings or coils *A*, *B* and *C* displaced *120 electrical degrees from each other and rotating in anticlockwise direction with angular velocity ω in the 2-pole field. Fig. 19.1 (*iii*) shows the wave diagram. Note that e.m.f. in coil *B* will be 120° behind that of coil *A* and e.m.f. in coil *C* will be 240° behind that of coil *A*. The equations of the three e.m.f.s can be represented as :

$$e_{a_1a_2} = E_m \sin \omega t$$

$$e_{b_1b_2} = E_m \sin (\omega t - 120°)$$

$$e_{c_1c_2} = E_m \sin (\omega t - 240°)$$

19.2. Reasons for the Use of 3-phase System

Electric power is generated, transmitted and distributed in the form of 3-phase power. Homes and small establishments are wired for single phase power but this merely represents a tap-off from the basic 3-phase system. Three-phase power is preferred over single-phase power for the following reasons :

---

* This can be accomplished by placing their start ends (denoted by a_1, b_1 and c_1) 120 electrical degrees apart.

(i) 3-phase power has a constant magnitude whereas single-phase power pulsates from zero to peak value at twice the supply frequency.

(ii) A 3-phase system can set up a rotating magnetic field in stationary windings. This cannot be done with a single-phase current.

(iii) For the same rating, 3-phase machines (*e.g.,* generators, motors, transformers) are smaller, simpler in construction and have better operating characteristics than single phase machines.

(iv) To transmit the same amount of power over a fixed distance at a given voltage, the 3-phase system requires only 3/4th the weight of copper that is required by the single-phase system.

(v) The voltage regulation of a 3-phase transmission line is better than that of a single-phase line.

A knowledge of 3-phase power and 3-phase circuits is, therefore, essential to an understanding of power technology. Fortunately, the basic circuit techniques used to solve single-phase circuits can be directly applied to 3-phase circuits. It is because the three phases are identical and one phase (*i.e.* single phase) represents the behaviour of all the three.

19.3. Elementary Three-Phase Alternator

In an actual 3-phase alternator, the three windings or coils are stationary and the field *rotates. Fig. 19.2 (*i*) shows an elementary 3-phase alternator. The three identical coils *A, B* and *C* are symmetrically placed in such a way that e.m.f.s induced in them are displaced 120 electrical degrees from one another. Since the coils are identical and are subjected to the same rotating field, the e.m.f.s induced in them will be of the same magnitude and frequency. Fig. 19.2 (*ii*) shows the wave diagram of the three e.m.f.s whereas Fig. 19.2 (*iii*) shows the phasor diagram. Note that r.m.s. values have been used in drawing the phasor diagram. The equations of the three e.m.f.s are :

$$e_A = E_m \sin\omega t$$

$$e_B = E_m \sin(\omega t - 120°)$$

$$e_C = E_m \sin(\omega t - 240°)$$

It can be proved in many ways that the sum of the three e.m.f.s at every instant is zero.

(i) Resultant $= e_A + e_B + e_C$

$$= E_m [\sin\omega t + \sin(\omega t - 120°) + \sin(\omega t - 240°)]$$

$$= E_m [\sin\omega t + 2\sin(\omega t - 180°)\cos 60°]$$

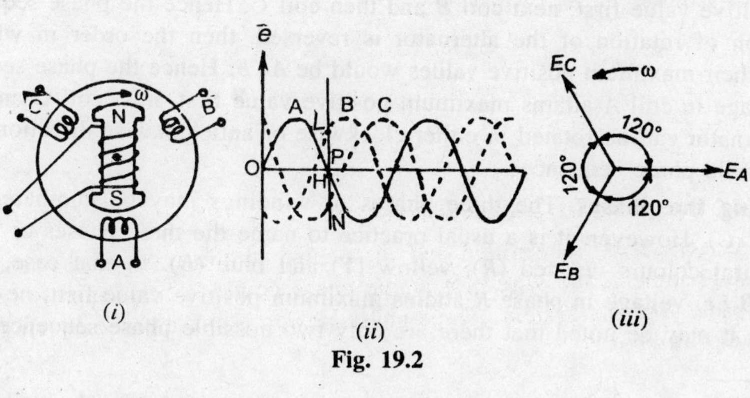

Fig. 19.2

* This arrangement has many technical and economical advantages.

$$= E_m [\sin \omega t - 2 \sin \omega t \cos 60°]$$
$$= 0$$

(ii) Referring to the wave diagram in Fig. 19.2 (ii), the sum of the three e.m.f.s at any instant is zero. For example, at the instant P, ordinate PL is positive while the ordinates PN and PH are negative. If you make actual measurements, it will be seen that :

$$PL + (-PN) + (-PH) = 0$$

(iii) Since the three windings or coils are identical, $E_A = E_B = E_C = E$ (in magnitude). As shown in Fig. 19.3, the resultant of E_A and E_B is E_r and its magnitude is $= 2E \cos 60° = E$. This resultant is equal and opposite to E_C. Hence the resultant of the three e.m.f.s is zero.

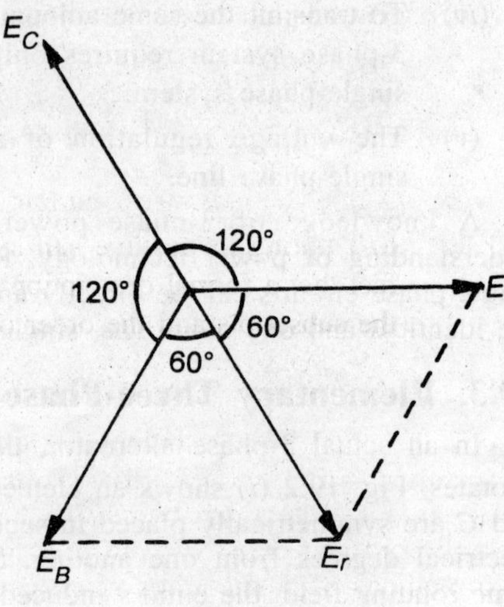

Fig. 19.3

(iv) Using complex algebra, we can again prove that the sum of the three e.m.f.s is zero. Thus taking E_A as the reference phasor, we have,

$$\mathbf{E}_A = E\angle 0° = E + j0$$
$$\mathbf{E}_B = E\angle -120° = E(-0.5 - j0.866)$$
$$\mathbf{E}_C = E\angle -240° = E(-0.5 + j0.866)$$
$$\therefore \mathbf{E}_A + \mathbf{E}_B + \mathbf{E}_C = (E + j0) + E(-0.5 - j0.866) + E(-0.5 + j0.866) = 0$$

19.4. Some Concepts

In the analysis of 3-phase system, we often come across the following terms :

(i) **Phase sequence.** The order in which the voltages in the three phases (or coils) reach their *maximum positive values is called the *phase sequence* or *phase order*. This is determined by the direction of rotation of the alternator. Thus, in Fig. 19.2 (i), the three coils A, B and C are producing voltages that are displaced 120 electrical degrees from one another. Referring to the wave diagram in Fig. 19.2 (ii), it is easy to see that voltage in coil A attains maximum positive value first, next coil B and then coil C. Hence the phase sequence is ABC. If the direction of rotation of the alternator is reversed, then the order in which the three phases attain their maximum positive values would be ACB. Hence the phase sequence is now ACB *i.e.*, voltage in coil A attains maximum positive value first, next coil C and then coil B. Since the alternator can be rotated in either clockwise or anticlockwise direction, there can be only two possible phase sequences.

(ii) **Naming the phases.** The three phases or windings may be numbered (1, 2, 3) or lettered (A, B, C). However, it is a usual practice to name the three phases or windings after the three natural colours *viz.* red (R), yellow (Y) and blue (B). In that case, the phase sequence is RYB *i.e.* voltage in phase R attains maximum positive value first, next phase Y and then phase B. It may be noted that there are only two possible phase sequences *viz* RYB and

* Instead of the positive maximum value, any other instantaneous value can be used to determine the phase sequence.

RBY. By convention, sequence *RYB* is taken as positive and *RBY* as negative. Throughout this book, the phase sequence considered is *RYB* unless stated otherwise.

(*iii*) **Double-subscript notation**. The double-subscript notation is a very useful concept and may be found advantageous in the analysis of 3-phase system. In this notation, two letters are placed at the foot of the symbol for voltage or current. The two letters indicate the two points between which voltage (or current) exists and the order of the letters indicates the relative polarity of voltage (or current)during *positive half-cycle.

(*a*) Thus V_{RY} indicates a voltage *V* between points *R* and *Y* with point *R* being positive *w.r.t.* point *Y* during its positive half-cycle. On the other hand, V_{YR} means that point *Y* is positive *w.r.t.* point *R* during its positive half-cycle.

Obviously, $$V_{RY} = -V_{YR}$$

(*b*) Again I_{RY} indicates a current *I* between points *R* and *Y* and that its direction is from *R* to *Y* during its positive half-cycle. The advantage of double-subscript notation lies in the fact that a formal description of voltage or current under consideration is not necessary; the subscripts and the order of the subscripts describe the quantity completely.

19.5. Interconnection of Three Phases

In a 3-phase alternator, there are three windings or phases. Each phase has two terminals *viz.* start and finish. If a separate load is connected across each winding as shown in Fig. 19.4, 6 conductors are required to transmit power. This will make the whole system complicated and expensive. In practice, the three windings are interconnected to give rise to two methods of connections *viz.*

(*i*) Star or Wye (*Y*) connection (*ii*) Mesh or Delta (Δ) connection

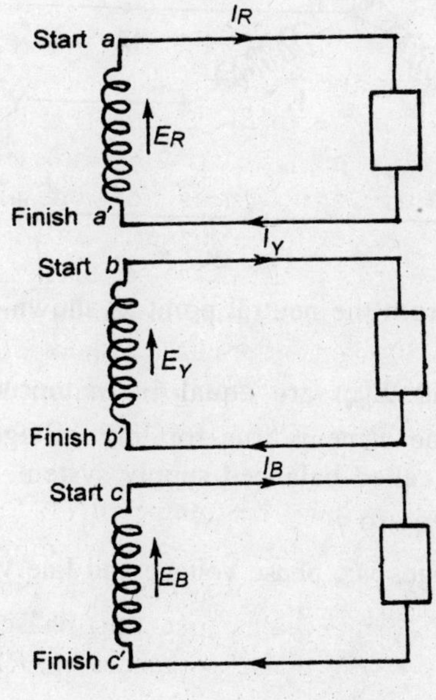

Fig. 19.4

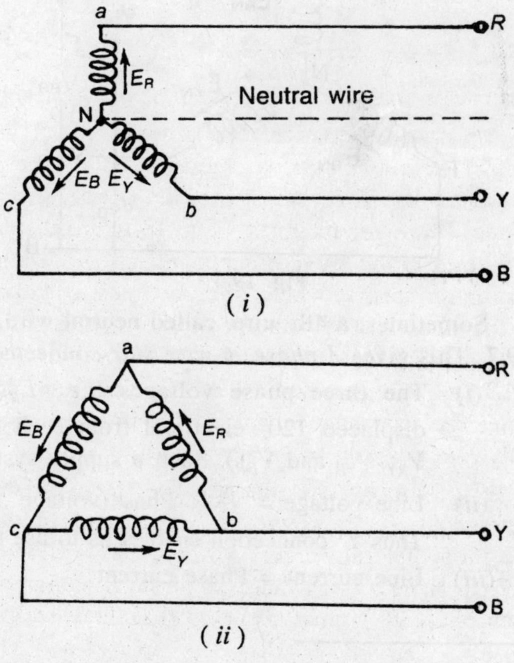

Fig. 19.5

---

* One may consider negative half-cycle. This criterion must be applied to all the three phases.

(*i*) In *Y*-connection, similar ends (start or finish) of the three phases are joined together within the alternator and three lines are run from the other free ends as shown in Fig. 19.5 (*i*). The common point *N* is called neutral point. In *Y*-connection, neutral conductor (shown dotted) may or may not be brought out. If a neutral conductor exists, the system is called *3-phase, 4-wire system*. If there is no neutral conductor, it is called *3-phase, 3-wire system*.

(*ii*) In Δ-connection, dissimilar ends (start to finish) of the phases are joined to form a closed mesh and the three lines are run from the junction points as shown in Fig. 19.5 (*ii*). In a Δ-connection, no neutral point exists and only *3-phase, 3-wire system* can be formed.

Note. The *Y* or delta connection serves substantially all the functions of three separate single phase circuits but with one important advantage that the number of conductors required is reduced. This results in the saving of conductor material and hence leads to economy.

19.6. Star or Wye Connection

In this method, *similar ends* (start or finish) of the three phases are joined together to form a common junction *N* as shown in Fig. 19.6. The common junction *N* is called the **star point* or *neutral point*. The three line conductors are run from the three free ends (finish ends *F* in this case) and are designated as *R, Y* and *B*. This constitutes a *3-phase, 3-wire star-connected system*. The voltage between any line and the neutral point (*i.e.*, voltage across each winding) is called the *phase voltage* while the voltage between any two lines is called the *line voltage*. The currents flowing in the phases are called the *phase currents* while those flowing in the lines are called the *line currents*. Note that the phase sequence is *RYB*.

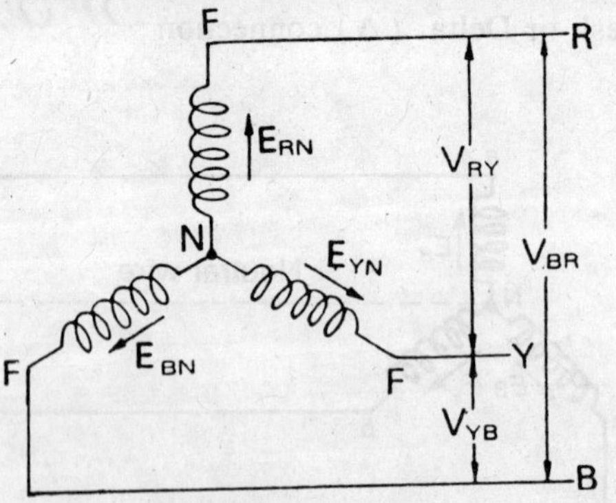

Fig. 19.6

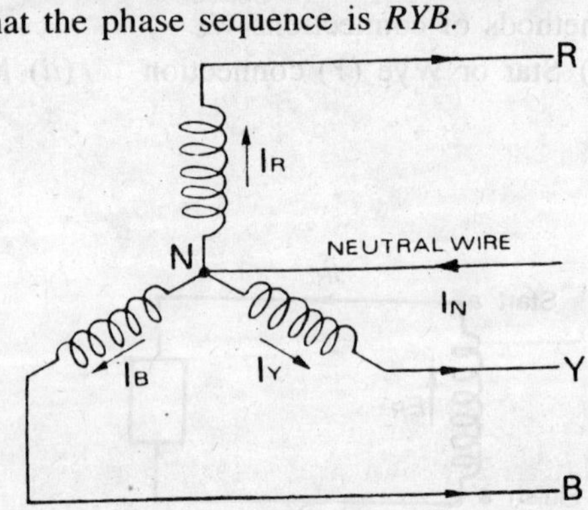

Fig. 19.7

Sometimes, a 4th wire, called neutral wire, is run from the neutral point as shown in Fig. 19.7. This gives *3-phase, 4-wire star-connected system*.

(*i*) The three phase voltages (*i.e.*, E_{RN}, E_{YN} and E_{BN}) are equal in magnitude but displaced 120° electrical from each other. The same is true for line voltages (*i.e.* V_{RY}, V_{YB} and V_{BR}). Such a supply systemic is called balanced supply system.

(*ii*) Line voltage = $\sqrt{3}$ × Phase voltage ... see Art. 19.7.

Thus *Y* connection enables us to use two voltages *viz.* phase voltage and line voltage.

(*iii*) Line current = Phase current

---

* The reader may note that the figure looks like a star or inverted *Y*. Hence, the name. But it is a misnomer. It is because star-connected windings can be represented diagramatically in a manner which does not look like a star or inverted *Y*.

(iv) For a 3-phase, 4-wire star-connected supply, the current I_N in the neutral wire is the phasor sum of the three line currents. For a balanced load, $I_N = 0$.

Note. The arrowheads alongside currents (or voltages) indicate their directions when they are assumed to be *positive* and not their actual directions at a particular instant. At no instant will *all the three line currents flow in the same direction either outwards or inwards.* This is expected because the three line currents are displaced 120° from one another. When one is positive, the other two might both be negative or one positive and one negative. Thus, at any one instant, current flows from the alternator through one of the lines to the load and returns through the other two lines. Or else current flows from the alternator through two of the lines and returns by means of the third.

19.7. Voltages and Currents in Balanced Y-Connection

Fig. 19.8 shows a balanced 3-phase *Y*-connected system in which the r.m.s. values of the e.m.f.s generated in the three phases are E_{RN}, E_{YN} and E_{BN}. It is clear from the circuit diagram (See Fig. 19.8) that p.d. between any two line terminals (*i.e.* line voltage) is the phasor difference between the potentials of these terminals *w.r.t* neutral point *i.e.*,

P.D. between lines *R* and *Y*, $*V_{RY} = E_{RN} - E_{YN}$ ... *phasor difference*

P.D. between lines *Y* and *B*, $V_{YB} = E_{YN} - E_{BN}$ ... —do—

P.D. between lines *B* and *R*, $V_{BR} = E_{BN} - E_{RN}$ ... —do—

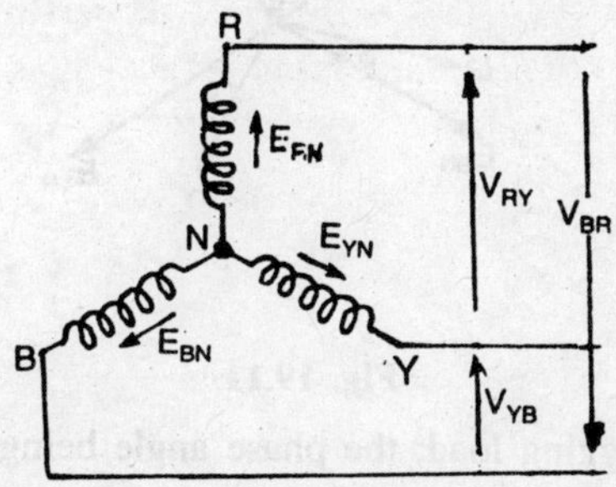

Fig. 19.8

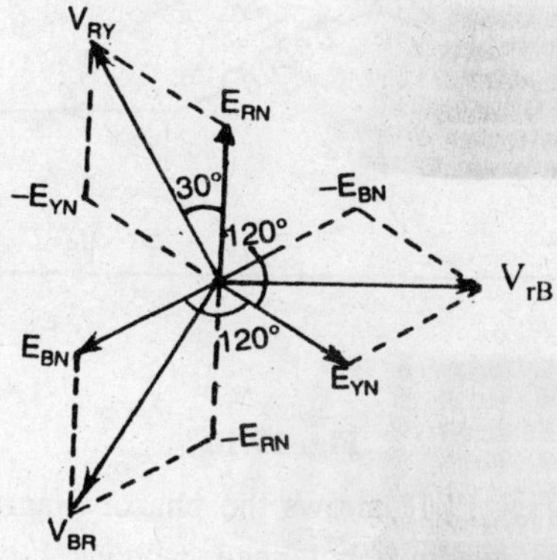

Fig. 19.9

1. Relation between line voltage and phase voltage

Considering the lines *R* and *Y*, the line voltage V_{RY} is equal to the phasor difference of E_{RN} and E_{YN}. To subtract E_{YN} from E_{RN}, reverse the phasor E_{YN} and find its phasor sum with E_{RN} as shown in the phasor diagram in Fig. 19.9 The two phasors E_{RN} and $-E_{YN}$ are equal in magnitude $(= E_{ph})$ and are 60° apart.

$$V_{RY} = 2E_{ph} \cos(60°/2) = 2E_{ph} \cos 30° = \sqrt{3}\, E_{ph}$$

Similarly, $V_{YB} = E_{YN} - E_{BN}$ ... *phasor difference*

$$= \sqrt{3}\, E_{ph}$$

and $V_{BR} = E_{BN} - E_{RN}$ ... *phasor difference*

$$= \sqrt{3}\, E_{ph}$$

---

* $V_{RY} = E_{RN} + E_{NY}$ .... phasor sum

$= E_{RN} - E_{YN}$ .... phasor difference

Between lines *R* and *Y*, there is point *N* (*i.e.*, neutral point). The voltage V_{RY} is the phasor sum of voltages from *R* to *N* and *N* to *Y*.

Hence in a balanced 3-phase *Y*-connection,

(*i*) Line voltage, $V_L = \sqrt{3}\, E_{ph}$

(*ii*) All line voltages are equal in magnitude (*i.e.*, = $\sqrt{3}\, E_{ph}$) but displaced 120° apart from one another (see the phasor diagram in Fig. 19.9).

(*iii*) Line voltages are 30° ahead of their respective phase voltages.

2. Relation between line current and phase current

In *Y*-connection, each line conductor is connected in series to a separate phase as shown in Fig. 19.10. Therefore, current in a line conductor is the same as that in the phase to which the line conductor is connected.

$\therefore$ Line current, $I_L = I_{ph}$

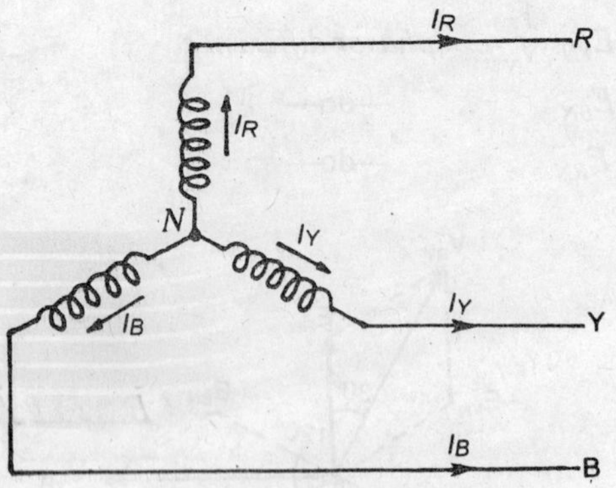

Fig. 19.10

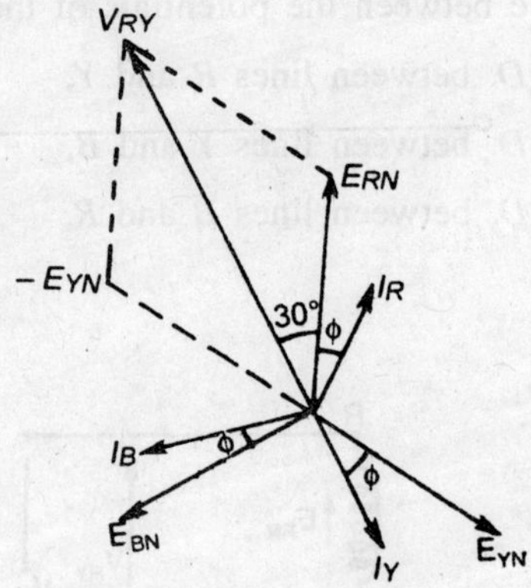

Fig. 19.11

Fig. 19.11 shows the phasor diagram for a balanced lagging load; the phase angle being ϕ. Hence in a balanced 3-phase *Y*-connection :

(*i*) Line current, $I_L = I_{ph}$

(*ii*) All the line currents are equal in magnitude (*i.e.* = I_{ph}) but displaced 120° from one another.

(*iii*) The angle between the line currents and the corresponding line voltages is $30 \pm \phi$; + if p.f. is lagging and – if it is leading.

Power

$$\text{Total power } P = 3 \times \text{Power in each phase}$$

$$= 3 \times E_{ph}\, I_{ph} \cos\phi$$

$$= 3E_{ph}\, I_{ph} \cos\phi \qquad \text{...(i)}$$

$$\text{For } Y\text{-connection}, E_{ph} = V_L/\sqrt{3};\quad I_{ph} = I_L$$

$$\therefore \quad P = 3 \times (V_L/\sqrt{3}) \times I_L \times \cos\phi$$

$$\text{or} \quad P = \sqrt{3}\, V_L\, I_L \cos\phi \qquad \text{...(ii)}$$

Either of relations (*i*) and (*ii*) can be used to determine the power. It may be noted that ϕ is the phase difference between a phase voltage and the corresponding phase current and *not* between the line current and corresponding line voltage.

Example 19.1. *Three coils, each having a resistance of 20 Ω and an inductive reactance of 15 Ω, are connected in star to a 400V, 3-phase, 50 Hz supply. Calculate (i) the line current (ii) power factor and (iii) power supplied.*

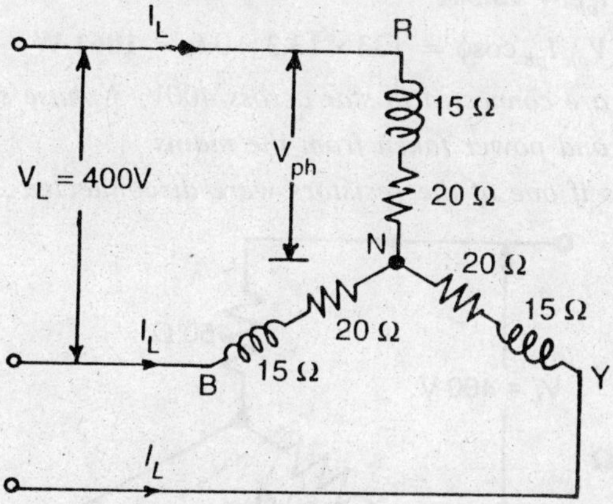

Fig. 19.12

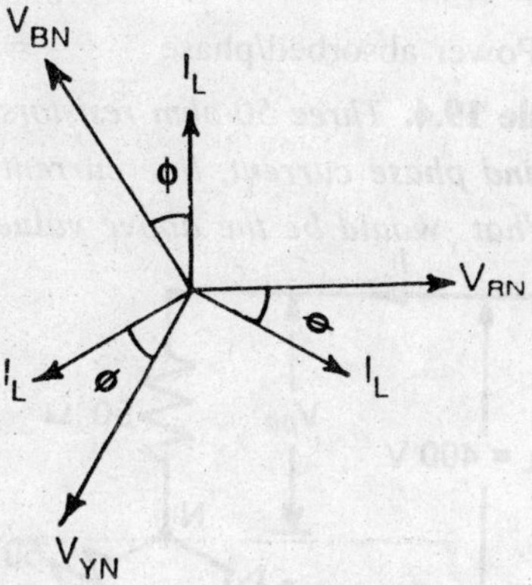

Fig. 19.13

Solution.

$$V_{ph} = V_L/\sqrt{3} = 400/\sqrt{3} = 231V$$

$$Z_{ph} = \sqrt{20^2 + 15^2} = 25\Omega$$

$$I_{ph} = V_{ph}/Z_{ph} = 231/25 = 9.24\,A$$

(i) $I_L = I_{ph} = \mathbf{9.24\,A}$

(ii) $p.f. = \cos\phi = R_{ph}/Z_{ph} = 20/25 = \mathbf{0.8}\ lag$

(iii) $P = \sqrt{3}\ V_L I_L \cos\phi = \sqrt{3} \times 400 \times 9.24 \times 0.8 = \mathbf{5121\ W}$

Alternatively, $P = 3I_{ph}^2 R_{ph} = 3 \times (9.24)^2 \times 20 = 5121\text{W}$

Example 19.2. *Calculate the active and reactive components of current in each phase of a star-connected 10,000 volts, 3-phase generator supplying 5,000 kW at a lagging p.f. 0.8. Find the new output if the current is maintained at the same value but the p.f. is raised to 0.9 lagging.*

Solution.

$$P = \sqrt{3}\ V_L I_L \cos\phi$$

or $$I_L = \frac{P}{\sqrt{3}\ V_L \cos\phi} = \frac{5000 \times 10^3}{\sqrt{3} \times 10{,}000 \times 0.8} = 360.8\text{A}$$

$$I_{ph} = I_L = 360.8\ \text{A}$$

Active component $= I_{ph}\cos\phi = 360.8 \times 0.8 = \mathbf{288.64\ A}$

Reactive component $= I_{ph}\sin\phi = 360.8 \times 0.6 = \mathbf{216.48\ A}$

Power output when p.f. is raised to 0.9 lagging

$$= \sqrt{3} \times 10{,}000 \times 360.8 \times 0.9 = \mathbf{5624 \times 10^3\ W}$$

Example 19.3. *A balanced star-connected load of impedance (6 + j8) ohms per phase is connected to a 3-phase, 230V, 50Hz supply. Find the line current and power absorbed by each phase.*

Solution.

$$Z_{ph} = \sqrt{6^2 + 8^2} = 10\,\Omega$$

$$V_{ph} = V_L/\sqrt{3} = 230/\sqrt{3} = 133\,V$$

$$\cos\phi = R_{ph}/Z_{ph} = 6/10 = 0.6 \text{ lag}$$

$$I_{ph} = V_{ph}/Z_{ph} = 133/10 = 13.3 \text{ A}$$

$\therefore$ Line current, $I_L = I_{ph} = \mathbf{13.3\ A}$

Power absorbed/phase $= V_{ph} I_{ph} \cos\phi = 133 \times 13.3 \times 0.6 = \mathbf{1061\ W}$

Example 19.4. *Three 50-ohm resistors are connected in star across 400V, 3-phase supply.*

(*i*) *Find phase current, line current and power taken from the mains.*

(*ii*) *What would be the above values if one of the resistors were disconnected ?*

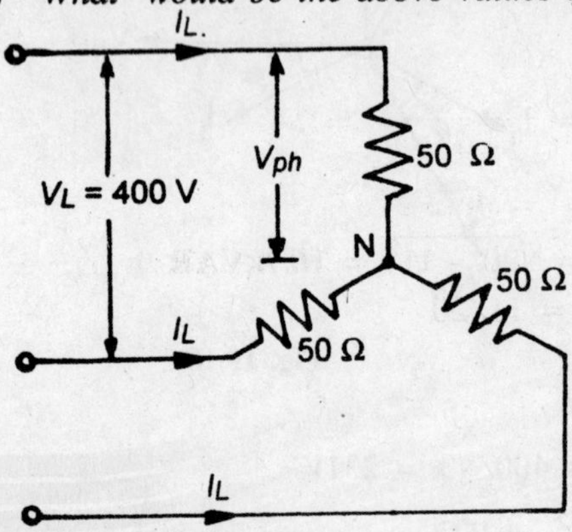

Fig.19.14

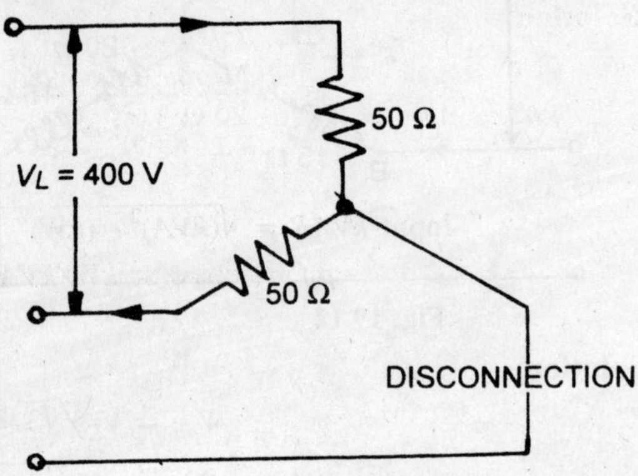

Fig. 19.15

Solution.

$$V_L = 400V;\ V_{ph} = 400/\sqrt{3} = 231V;\quad R_{ph} = 50\Omega;\quad \cos\phi = 1$$

(*i*) **When the three resistors are star-connected**. [See Fig. 19.14].

$$I_{ph} = V_{ph}/R_{ph} = 231/50 = \mathbf{4.62\ A}$$

$$\therefore \quad I_L = I_{ph} = \mathbf{4.62A}$$

Power taken, $P = \sqrt{3}\, V_L I_L \cos\phi = \sqrt{3} \times 400 \times 4.62 \times 1 = \mathbf{3200\ W}$

(*ii*) **When one of the resistors is disconnected.** When one of the resistors is disconnected [See Fig. 19.15], the remaining two resistors behave as if they were connected in series across the line voltage. In fact, the circuit behaves as a single phase circuit.

$$\therefore \quad I_{ph} = I_L = \frac{400}{50+50} = \mathbf{4\ A}$$

Power taken, $P = V_L I_L \cos\phi = 400 \times 4 \times 1 = \mathbf{1600\ W}$

Hence, by disconnecting one of the resistors, power consumption is reduced by half.

Example 19.5. *Three similar coils, connected in star, take a total power of 1.5 kW at a p.f. of 0.2 lagging from a 3-phase, 400V, 50Hz supply. Calculate the resistance and inductance of each coil.*

Solution.

$$V_{ph} = V_L/\sqrt{3} = 400/\sqrt{3} = 231V$$

$$P = \sqrt{3}\, V_L I_L \cos\phi$$

$$\therefore \quad I_L = \frac{P}{\sqrt{3}\, V_L \cos\phi} = \frac{1500}{\sqrt{3} \times 400 \times 0.2} = 10.83 \text{ A}$$

$$I_{ph} = I_L = 10.83\,A$$

$$Z_{ph} = V_{ph}/I_{ph} = 231/10.83 = 21.33\Omega$$
$$R_{ph} = Z_{ph}\cos\phi = 21.33 \times 0.2 = \mathbf{4.27\Omega}$$
$$X_{ph} = \sqrt{(21.33)^2 - (4.27)^2} = 20.9\Omega$$
$$\therefore \quad L_{ph} = X_{ph}/2\pi f = 20.9/2\pi \times 50 = \mathbf{0.0665H}$$

Example 19.6. *The load to a 3-phase supply comprises three similar coils connected in star. The line currents are 25A and the kVA and kW inputs are 20 and 11 respectively. Find (i) the phase and line voltages (ii) the kVAR input and (iii) resistance and reactance of each coil.*

Solution.

$$V_{ph} = \frac{\text{Apparent power}}{3 \times I_{ph}} = \frac{20 \times 10^3}{3 \times 25} = \mathbf{267V}$$
$$V_L = \sqrt{3}\, V_{ph} = \sqrt{3} \times 267 = \mathbf{462V}$$
$$\text{Input } kVAR = \sqrt{(kVA)^2 - (kW)^2} = \sqrt{20^2 - 11^2} = \mathbf{16.7kVAR}$$
$$p.f. = \cos\phi = kW/kVA = 11/20$$
$$Z_{ph} = V_{ph}/I_{ph} = 267/25 = 10.68\Omega$$
$$R_{ph} = Z_{ph}\cos\phi = 10.68 \times 11/20 = \mathbf{5.87\Omega}$$
$$X_{ph} = \sqrt{(10.68)^2 - (5.87)^2} = \mathbf{8.92\Omega}$$

Example 19.7. *If the phase voltage of a three-phase star connected alternator be 231V, what will be the line voltages (i) when the phases are correctly connected (ii) when the connections of one of the phases are reversed ?*

Solution.

(*i*) **When the phases are correctly connected.** When the phases are connected correctly (*i.e.* start or finish ends are joined together), the phasor diagram will be as shown in Fig. 19.9 (refer back). As proved in Art 19.7,

$$V_{RY} = V_{YB} = V_{BR} = \sqrt{3}\, E_{ph}$$

Each line voltage, $\quad V_L = \sqrt{3}\, E_{ph} = \sqrt{3} \times 231 = \mathbf{400V}$

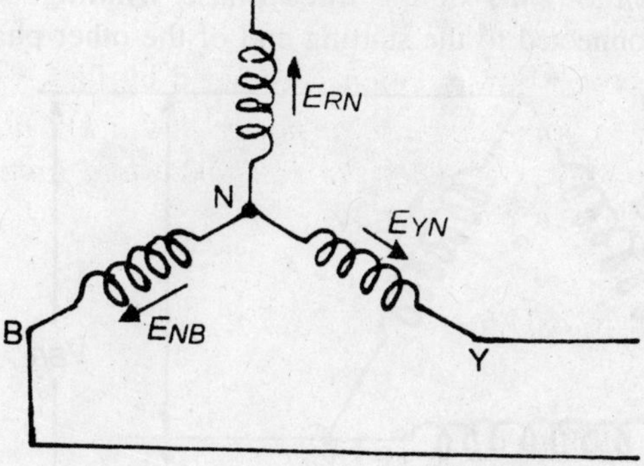

Fig. 19.16

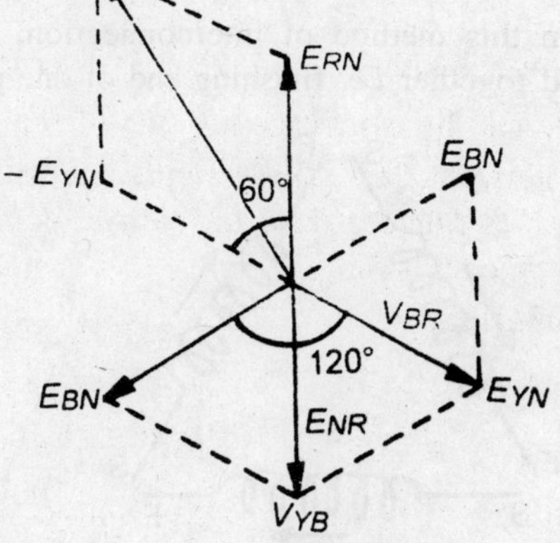

Fig. 19.17

(*ii*) **When one phase connection is reversed**. Suppose connections to phase B have been reversed. It simply means that the finished end of phase B has been connected to the start ends of the other two phases at N as shown in Fig. 19.16. Note that the voltage in phase B is now E_{NB} (instead of E_{BN}) acting outwards from N. Thus we have three phase voltages E_{RN}, E_{YN} and E_{NB} 120° apart as shown in Fig. 19.17.

$$V_{RY} = E_{RN} - E_{YN} \quad \text{... phasor difference}$$

$$= 2E_{ph}\cos(60°/2) = 2 \times 231 \times \cos 30° = \mathbf{400\ V}$$

$$V_{YB} = E_{YN} + E_{NB} \quad \text{... phasor sum}$$

$$= 2\,E_{ph}\cos(120°/2) = 2 \times 231 \times \cos 60° = \mathbf{231V}$$

$$V_{BR} = E_{BN} + E_{NR} \quad \text{... phasor sum}$$

$$= 2\,E_{ph}\cos(120°/2) = 2 \times 231 \times \cos 60° = \mathbf{231V}$$

Tutorial Problems

1. Three similar coils are star connected to a 3-phase, 400V, 50Hz supply. If the inductance and resistance of each coil are 38.2 mH and 16 Ω respectively, determine (*i*) line current (*ii*) power factor (*iii*) power consumed. **[(*i*) 11.55A (*ii*) 0.8 *lag* (*iii*) 6.4 kW]**
2. The voltage measured between the terminals of a 3-phase, 3-wire alternator is recorded as 208V. A 3-phase load consisting of three 10-ohm resistors in star is connected to the terminals of the alternator. If one of the resistors should become open-circuited, what would be the current, the voltage and the power of the remaining resistors ? **[10.4A ; 104V ; 1081.6W]**
3. A 6600 volts three-phase,star-connected alternator supplies 4000 kW at a p.f. of 0.8 lagging. Calculate the line current. If the load p.f. is raised to 0.95, the total current remaining the same, find the new output. **[437.5A ; 4750kW]**
4. Three 20 μF capacitors are star-connected across 420V, 50Hz, three-phase, three-wire supply.
   (*i*) Calculate the current in each line.
   (*ii*) If one of the capacitors is short-circuited, calculate the line currents.
   (*iii*) If one of the capacitors is open-circuited, calculate the line currents and p.d. across each of the other two capacitors.
   [(*i*) 1.525A (*ii*) 2.64A, 2.64A, 4.57A (*iii*) 1.32A, 132A, 0A; 210V]
5. Three similar coils, connected in star, take a total power of 3 kW at a p.f. of 0.8 lagging from a 3-phase, 400V, 50Hz supply. Calculate the resistance and reactance of each coil. **[33.92 Ω; 25.68 Ω]**

19.8. Delta (Δ) or Mesh Connection

In this method of interconnection, the *dissimilar ends* of the three phase windings are joined together *i.e.* finishing end of one phase is connected to the starting end of the other phase

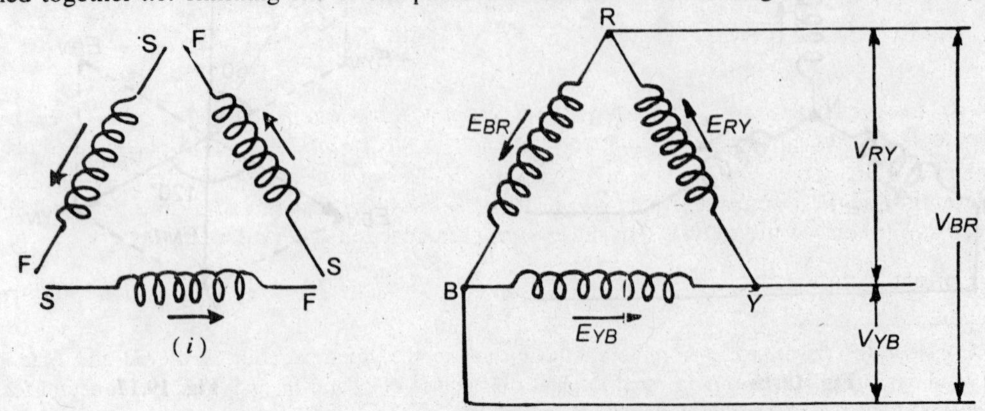

Fig. 19.18

and so on to obtain mesh or delta as shown in Fig. 19.18. The three line conductors are taken from the three junctions of the mesh or delta and are designated as *R, Y* and *B*. This is called *3- phase, 3-wire, delta-connected system.* Since no neutral exists in a Δ-connection, only 3-phase, 3-wire system can be formed.

In Fig. 19.18 (*ii*), it may appear that the three phases are short-circuited on themselves. But this is not the case. The finishing end of one phase is connected to the starting end of the other phase so that the resultant voltage round the mesh is the phasor sum of the three phase voltages. Since the three phase voltages are equal in magnitude and displaced 120° from one another, their phasor sum is zero (see Art. 19.3). Consequently, no current can flow round the mesh when the terminals are open.

Note. At no instant will all the three line currents flow in the same direction either outwards or inwards. This is expected because the three line currents are displaced 120° from one another. When one is positive, the other two might both be negative or one positive and one negative. Thus at any one instant, current flows from the alternator through one of the lines to the load and returns through the other two lines. Or else current flows from the alternator through two of the lines and returns by means of the third. It may be noted that arrows placed alongside currents (or voltages) in the diagram indicate the directions of currents (or voltages) when they are assumed to be positive and not their actual directions at a particular instant.

19.9. Voltages and Currents in Balanced Δ Connection

We shall now investigate the characteristics of a balanced Δ-connection.

(*i*) **Line voltage and phase voltage.** Since the system is balanced, the three phase voltages are equal in magnitude (say each equal to V_{ph}, the phase voltage) but displaced 120° from one another. An examination of Fig. 19.19 shows that only one phase winding is included between any pair of lines. Hence in Δ connection, the line voltage is equal to the phase voltage *i.e.*

$$V_L = V_{ph}$$

Since the phase sequence is *RYB*, the line voltage V_{RY} is 120° ahead of V_{YB} and 240° ahead of V_{BR}. Incidentally, these are also the phase voltages.

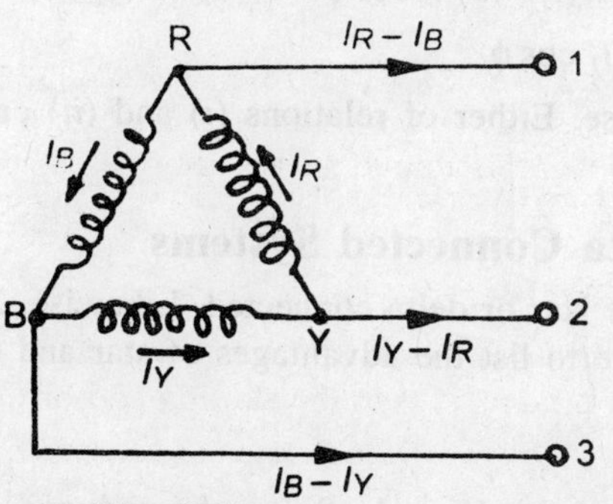

Fig. 19.19

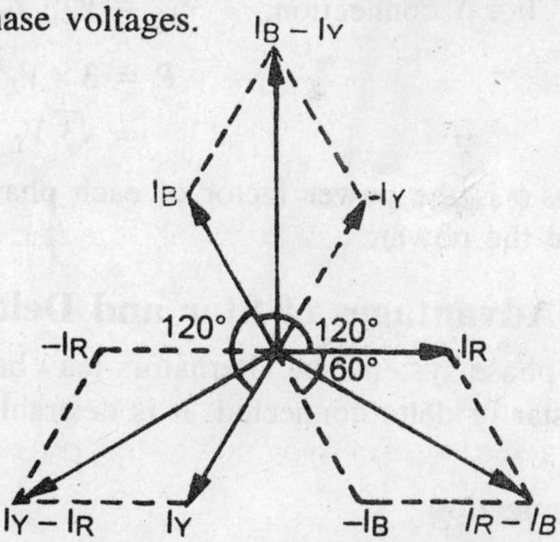

Fig. 19.20

(*ii*) **Line current and phase current.** Since the system is balanced, the three phase currents I_R, I_Y and I_B are equal in magnitude (say each equal to I_{ph}, the phase current) but displaced 120° from one another as shown in the phasor diagram in Fig. 19.20. An examination of the circuit diagram in Fig. 19.19 shows that current in any line is equal to the *phasor difference* of the currents in the two phases attached to that line. Thus :

Current in line 1, $\qquad I_1^* = I_R - I_B \qquad$ ... *phasor difference*

---

* Consider the line current I_1 in line 1 connected to the common point *R* of red and blue phase windings. It is clear that I_1 is equal to phasor difference of I_R and I_B since positive direction for I_R is towards point *R* and for I_B it is away from point *R*.

Current in line 2, $I_2 = I_Y - I_R$ ... *phasor difference*

Current in line 3, $I_3 = I_B - I_Y$ ... —do—

The current I_1 in line 1 is the phasor difference of I_R and I_B. To subtract I_B from I_R, reverse the phasor I_B and find its phasor sum with I_R as shown in Fig. 19.20. The two phasors I_R and $-I_B$ are equal in magnitude $(= I_{ph})$ and are 60° apart.

$$\therefore \quad I_1 = 2I_{ph}\cos(60°/2) = 2\,I_{ph}\cos 30° = \sqrt{3}\,I_{ph}$$

Similarly, $I_2 = I_Y - I_R$ ... *phasor difference*

$= \sqrt{3}\,I_{ph}$

and $I_3 = I_B - I_Y$ ... *phasor difference*

$= \sqrt{3}\,I_{ph}$

The three line currents I_1, I_2 and I_3 are equal in magnitude; each being equal to $\sqrt{3}\,I_{ph}$.

Hence in a balanced Δ connection :

(*a*) Line current, $I_L = \sqrt{3}\,I_{ph}$

(*b*) All the line currents are equal in magnitude $(= \sqrt{3}\,I_{ph})$ but displaced 120° from one another as seen from Fig. 19.20.

(*c*) Line currents are 30° behind the respective phase currents.

(*iii*) **Power**

$$\text{Total power, } P = 3 \times \text{Power per phase}$$

$$= 3\,V_{ph}\,I_{ph}\cos\phi \qquad \text{... } (i)$$

For Δ connection, $V_{ph} = V_L;\ I_{ph} = I_L/\sqrt{3}$

$$\therefore \quad P = 3 \times V_L \times (I_L/\sqrt{3}) \times \cos\phi$$

$$= \sqrt{3}\,V_L\,I_L\cos\phi \qquad \text{... } (ii)$$

where cos ϕ is the power factor of each phase. Either of relations (*i*) and (*ii*) can be used to determine the power.

19.10. Advantages of Star and Delta Connected Systems

In 3-phase system, the alternators may be star or delta connected. Likewise, 3-phase loads may be star or delta connected. It is desirable to list the advantages of star and delta connections.

Star Connection

(*i*) In star connection, phase voltage $V_{ph} = V_L/\sqrt{3}$. Since the induced e.m.f. in the phase winding of an alternator is directly proportional to the number of turns, a star-connected alternator will require less number of turns than a Δ-connected alternator for the same line voltage.

(*ii*) For the same line voltage, a star-connected alternator requires less insulation than a delta-connected alternator.

Due to above two reasons, 3-phase alternators are generally star-connected. One can hardly find delta-connected alternators.

(*iii*) In star connection, we can get 3-phase, 4-wire system. This permits to use two voltages *viz.*, phase voltages as well as line voltages. Remember that in star connection, $V_L = \sqrt{3}\, E_{ph}$. Single phase loads (*e.g.* lights *etc*) can be connected between any one line and the neutral wire while the 3-phase loads (*e.g.* 3-phase motors) can be put across the three lines. Such a flexibility is not available in Δ connection.

(*iv*) In star connection, the neutral point can be earthed. Such a measure offers many advantages. For example, in case of line to earth fault, the insulators have to bear $1/\sqrt{3}$ (*i.e.* 57.7%) times the line voltage. Moreover, earthing of neutral permits to use protective devices (*e.g.* relays) to protect the system in case of ground faults.

Delta Connection

(*i*) This type of connection is most suitable for rotary convertors.

(*ii*) Most of the 3-phase loads are Δ-connected rather than Y-connected. One reason for this, at least for the case of an unbalanced load, is the flexibility with which loads may be added or removed on a single phase. This is difficult (*or* impossible) to do with a Y-connected 3-wire load.

(*iii*) Most of 3-phase induction motors are delta-connected.

Example 19.8. *A balanced 3-phase, Δ-connected load has per phase impedance of (25 + j40) Ω. If 400V, 3-phase supply is connected to this load, find (i) phase current (ii) line current (iii) power supplied to the load.*

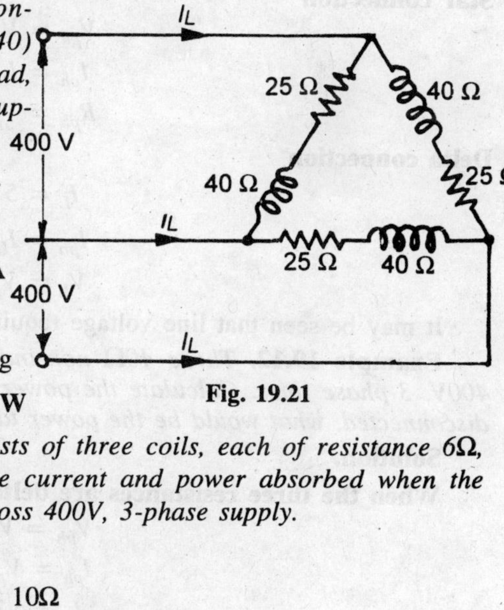

Fig. 19.21

Solution. Fig. 19.21 shows the circuit diagram.

$$Z_{ph} = \sqrt{25^2 + 40^2} = 47.17\,\Omega$$

(*i*) $I_{ph} = V_{ph}/Z_{ph} = 400/47.17 = \mathbf{8.48A}$

(*ii*) $I_L = \sqrt{3}\, I_{ph} = \sqrt{3} \times 8.48 = \mathbf{14.7A}$

(*iii*) $\cos\phi = R_{ph}/Z_{ph} = 25/47.17 = 0.53$ lag

$$P = \sqrt{3}\, V_L I_L \cos\phi = \sqrt{3} \times 400 \times 14.7 \times 0.53 = \mathbf{5397.76\ W}$$

Example 19.9. *A balanced 3-phase load consists of three coils, each of resistance 6Ω, and inductive reactance of 8Ω. Determine the line current and power absorbed when the coils are (i) star-connected (ii) delta-connected across 400V, 3-phase supply.*

Solution.

$$Z_{ph} = \sqrt{6^2 + 8^2} = 10\Omega$$

$$\cos\phi = R_{ph}/Z_{ph} = 6/10 = 0.6 \text{ lag}$$

(*i*) **Star connection**

$$V_{ph} = V_L/\sqrt{3} = 400/\sqrt{3} = 231 \text{ V}$$

$$I_{ph} = V_{ph}/Z_{ph} = 231/10 = 23.1 \text{ A}$$

∴ $$I_L = I_{ph} = \mathbf{23.1A}$$

$$P = \sqrt{3}\, V_L I_L \cos\phi = \sqrt{3} \times 400 \times 23.1 \times 0.6 = \mathbf{9602.5\ W}$$

(*ii*) **Delta connection**

$$V_{ph} = V_L = 400V$$

$$I_{ph} = V_{ph}/Z_{ph} = 400/10 = 40 \text{ A}$$

$$I_L = \sqrt{3}\, I_{ph} = \sqrt{3} \times 40 = \mathbf{69.28A}$$

$$P = \sqrt{3}\, V_L I_L \cos\phi = \sqrt{3} \times 400 \times 69.28 \times 0.6 = \mathbf{28799\ W}$$

It may be seen that when connected in Δ, the line current and power drawn are *three times* as that for star connection.

Example 19.10. *A balanced Δ-connected load takes a line current of 18A at a p.f. of 0.85 leading from a 400V, 3-phase, 50Hz supply. Calculate the resistance of each leg of the load.*

Solution.

$$V_{ph} = V_L = 400\text{ V}; \quad I_L = 18A; \quad \cos\phi = 0.85 \text{ lead}$$

$$I_{ph} = I_L/\sqrt{3} = 18/\sqrt{3} = 10.39\text{A}$$

$$Z_{ph} = V_{ph}/I_{ph} = 400/10.39 = 38.5\Omega$$

$$\therefore \quad R_{ph} = Z_{ph}\cos\phi = 38.5\times 0.85 = \mathbf{32.72\,\Omega}$$

Example 19.11. *Three similar resistors are connected in star across 400V, 3-phase supply. The line current is 5A. Calculate the value of each resistance. To what value should the line voltage be changed to obtain the same line current with the resistors connected in delta ?*

Solution.

Star connection

$$V_{ph} = V_L/\sqrt{3} = 400/\sqrt{3} = 231\text{V}$$

$$I_{ph} = I_L = 5\text{ A}$$

$$\therefore \quad R_{ph} = V_{ph}/I_{ph} = 231/5 = \mathbf{46.2\,\Omega}$$

Delta connection

$$I_L = 5\text{ A} \quad \text{... given}$$

$$I_{ph} = I_L/\sqrt{3} = 5/\sqrt{3} = 2.88\text{ A}$$

$$V_L = V_{ph} = I_{ph}R_{ph} = 2.88\times 46.2 = \mathbf{133V}$$

It may be seen that line voltage required is one-third that of star value.

Example 19.12. *Three 40Ω non-inductive resistances are connected in delta across 400V, 3-phase lines. Calculate the power taken from the mains. If one of the resistances is disconnected, what would be the power taken from the mains ?*

Solution.

When the three resistances are delta connected. [Fig. 19.22].

$$V_{ph} = V_L = 400\text{V}; \quad R = 40\,\Omega$$

$$I_{ph} = V_{ph}/R = 400/40 = 10\text{ A}$$

$$I_L = \sqrt{3}\,I_{ph} = \sqrt{3}\times 10 = 17.32\text{ A}$$

$$P = \sqrt{3}\,V_L I_L\cos\phi$$

$$= \sqrt{3}\times 400\times 17.32\times 1 = \mathbf{12000\ W}$$

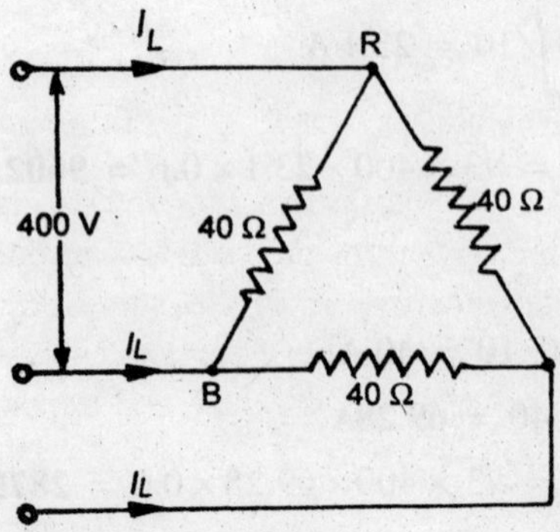

Fig.19.22

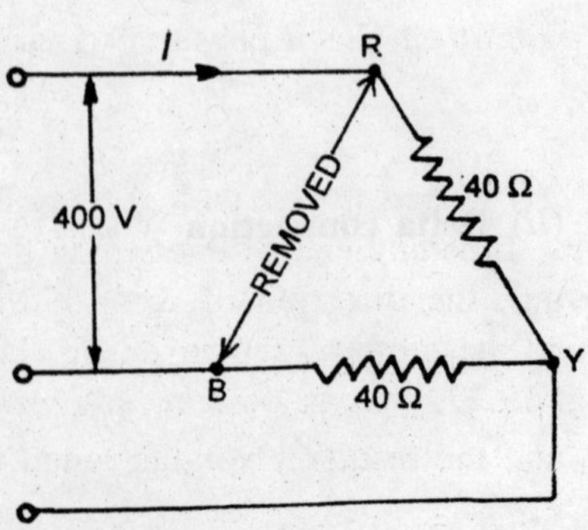

Fig. 19.23

When one resistor is removed. Fig. 19.23 shows the circuit with one resistor removed. In this case, each of the two resistors acts independently as if 400V were applied to 40 Ω.

Current in each resistor, $I = 400/40 = 10$ A

Power consumed in the two $= 2I^2R = 2 \times (10)^2 \times 40 = \mathbf{8000W}$

Hence by disconnecting one resistor, the power consumption is reduced by one-third.

Example 19.13. *Three identical resistances, each of 15Ω, are connected in delta across 400V, 3-phase supply. What value of resistance in each leg of balanced star-connected load would take the same line current ?*

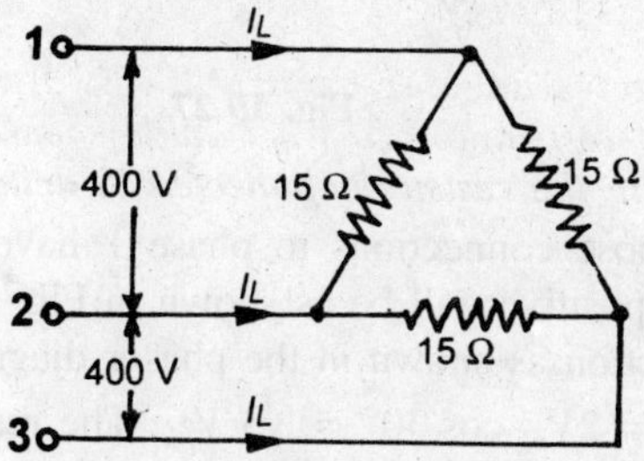

Fig. 19.24

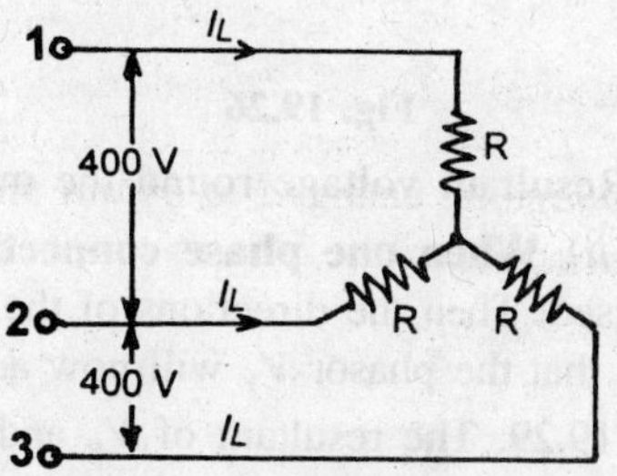

Fig. 19.25

Solution. Let R ohms be the required resistance in each leg of the star-connection (see Fig. 19.25). Since the two circuits have the same line voltage and line current, the resistance between any two corresponding terminals of the two circuits is the same. Considering the terminals 1 and 3,

$$\text{For delta connection, } R_{13} = 15 \parallel (15+15) = \frac{15 \times 30}{15+30} = 10\Omega$$

$$\text{For star connection, } R_{13} = R + R = 2R$$

$$\therefore \qquad 2R = 10 \text{ or } R = 10/2 = \mathbf{5\Omega}$$

Note. Thus Δ-connected impedances can be replaced by the equivalent Y-connected impedances using the following relation:

$$Z_Y = Z_\Delta/3$$

Using this relation, the value of R in the above problem is $15/3 = 5\Omega$

Example 19.14. *A delta connected alternator is on no load. The voltage per phase is 200V. What will be the resultant voltage round the mesh (i) when the phases are connected correctly (ii) when the connections of one of the phases are reversed ?*

Solution. Each phase voltage is 200V *i.e.*

$$V_R = V_Y = V_B = V_{ph} = 200V \qquad \text{... in magnitude}$$

(*i*) **When the phases are connected correctly.** When the phases are connected correctly (*i.e.*, finishing end of one winding is connected to the starting end of the second winding and so on), the direction of each of the phase voltages when positive is in the same direction round the mesh as shown in Fig. 19.26. The phasor diagram for this arrangement is shown in Fig. 19.27. As can be seen, the resultant of V_R and V_Y, shown as V_a, is equal and opposite to V_B and the resultant voltage round the mesh is zero *i.e.*

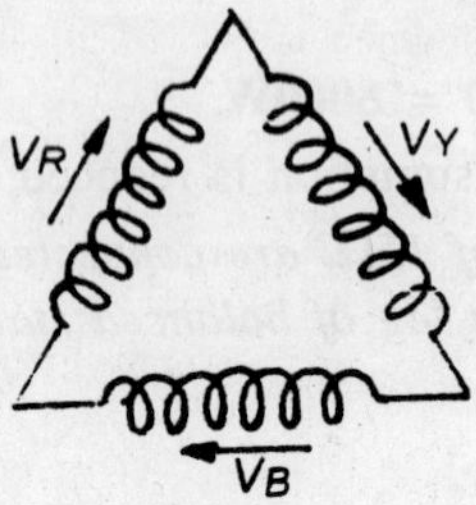

Fig. 19.26

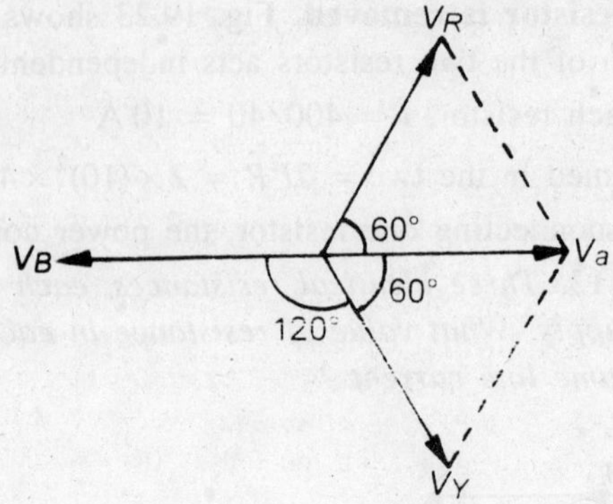

Fig. 19.27

Resultant voltage round the mesh = **0V**

(*ii*) **When one phase connection is reversed.** Suppose connections to phase Y have been reversed. Then the directions of the phase voltages when positive will be as shown in Fig. 19.28. Note that the phasor V_Y will now act in the reversed direction as shown in the phasor diagram in Fig. 19.29. The resultant of V_R and V_Y, shown as V_a, is $= 2V_{ph} \cos 30° = \sqrt{3}\, V_{ph}$. The resultant voltage round the mesh is the phasor sum of V_a and V_B. These two phasors are 90° apart.

∴ Resultant voltage round the mesh

$$= \sqrt{V_a^2 + V_B^2} = \sqrt{(\sqrt{3}\, V_{ph})^2 + (V_{ph})^2} = 2V_{ph} = 2 \times 200 = \mathbf{400V}$$

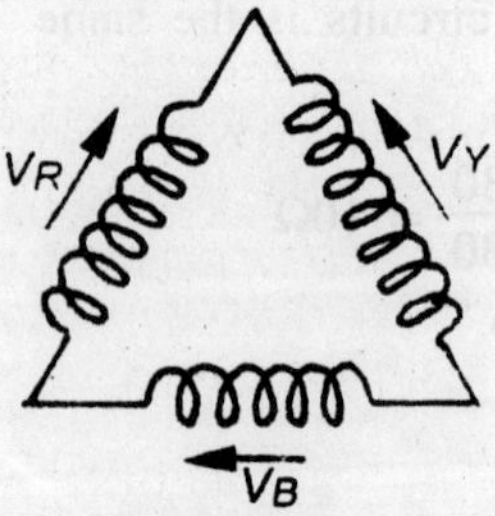

Fig. 19.28

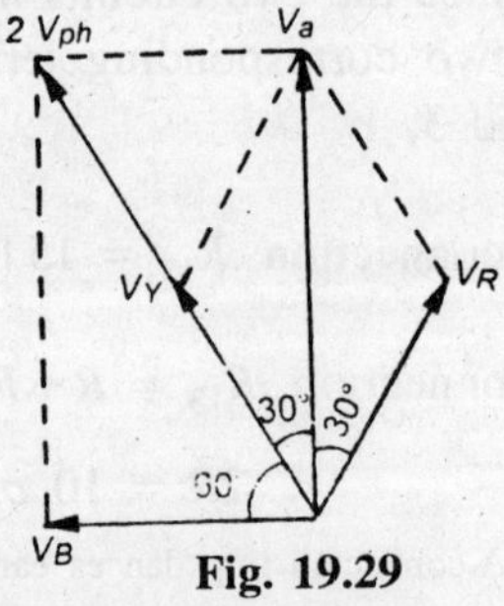

Fig. 19.29

Thus if one phase of delta connection is reversed, the resultant voltage round the mesh is equal to *twice* the phase voltage. This high voltage may cause high current to flow through the windings, even without an external load, and may damage the winding.

Tutorial Problems

1. Calculate the phase and line currents in a balanced delta connected load taking 75 kW at a power factor 0.8 from a 3-phase 440V supply. **[71.02A; 123A]**
2. Three identical inductive loads of resistance 15Ω and reactance of 40 Ω are connected in star to a 440V, 3-phase supply. Calculate (*i*) phase current (*ii*) line current and (*iii*) power absorbed. **[(*i*) 5.94A (*ii*) 5.94A (*iii*) 1589.5W]**
3. Three identical resistances, each of 18Ω, are connected in delta across 400V, 3-phase supply. What value of resistance in each leg of balanced star-connected load would take the same line current ? **[6Ω]**
4. Three similar resistors are connected in star across a 415V, 3-phase supply. The line current is 10A. Calculate (*i*) the value of each resistance (*ii*) the line voltage required to give the same line current if the resistors were delta-connected. **[(*i*) 23.96Ω (*ii*) 138.33V]**

5. A balanced 3-phase load consists of three coils, each of resistance 4Ω and inductance 0.02H. Determine the total power when the coils are (*i*) star-connected (*ii*) delta-connected to 400V, 3-phase, 50Hz supply. **[(*i*) 11.56 kW (*ii*) 34.68 kW]**
6. Three capacitors each 50 μF, are connected in delta to a 400V, 3-phase-50Hz supply. What will be the capacitance of the three capacitors such that when connected in star across the same supply, the line current remains the same ? **[150 μF]**
7. Three 6Ω non-inductive resistances are connected in (*i*) star (*ii*) delta across 400V, 50Hz, 3-phase supply. Calculate the total power drawn in each case. In the event of one of the three resistances getting open-circuited, what would be the value of the total power taken from the mains in each of the two cases ? **[*star*: 26.66 kW ; 13.33 kW *Delta*: 80kW; 53.33 kW]**
8. A 440V, 3-phase delta-connected induction motor has an output of 14.92 kW at a p.f. of 0.82 and efficiency 85%. If another star-connected load of 10 kW at 0.85 p.f. is added in parallel, find (*i*) current drawn from the supply (*ii*) total power. **[(*i*) 43.56A (*ii*) 27.6 kW]**

19.11. 3-Phase, 4-Wire Star-Connected System

3-phase, 4-wire star-connected system is widely used to distribute electric power in domestic, commercial and industrial establishments. The 3-phase loads (*e.g.* 3-phase induction motors *etc.*) are directly connected across the three lines while single-phase loads (*e.g.* lamps, fans, single-phase motors *etc.*) are connected between one of the lines and the neutral wire. Although 3-phase loads on the system are mostly balanced, these are the single-phase loads that introduce the imbalance. This is because it is rarely possible that single-phase loads on all the three phases have the same magnitude and power factor. Consequently, the line currents (I_R, I_Y and I_B) will be different in magnitude and displaced from one another by unequal angle. The current in the neutral wire will be the phasor sum of the three line currents *i.e.*.

$$\text{Current in neutral wire, } I_N = I_R + I_Y + I_B \quad \text{... phasor sum}$$

Example 19.15. *Non-reactive loads of 10kW, 8kW and 5kW are connected between the neutral and the red, yellow and blue phases respectively of a 3-phase, 4-wire system. The line voltage is 400V. Calculate (i) the current in each line and (ii) the current in the neutral wire.*

Solution.

(*i*) Phase voltage $= 400/\sqrt{3} = 231$ V

$$I_R = 10 \times 10^3/231 = \mathbf{43.3A}$$

$$I_Y = 8 \times 10^3/231 = \mathbf{34.6A}$$

$$I_B = 5 \times 10^3/231 = \mathbf{21.65A}$$

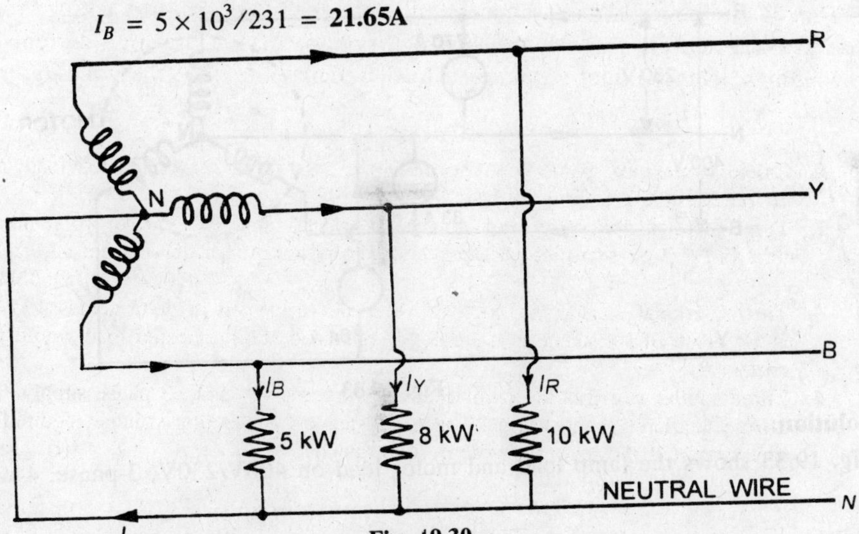

Fig. 19.30

(*ii*) The three line currents are represented by the respective phasors in Fig. 19.31. Note that the three line currents are of different magnitude but displaced 120° from one another. The current in the neutral wire will be the phasor sum of the three line currents. Resolving the three currents along *X*-axis and *Y*-axis, we have,

$$\text{Total } X\text{-component} = I_Y \cos 30° - I_B \cos 30°$$
$$= 34.6 \times 0.866 - 21.65 \times 0.866 = 11.22 \text{ A}$$
$$\text{Total } Y\text{-component} = I_R - I_Y \cos 60° - I_B \cos 60°$$
$$= 43.3 - 34.6 \times 0.5 - 21.65 \times 0.5 = 15.2 \text{ A}$$

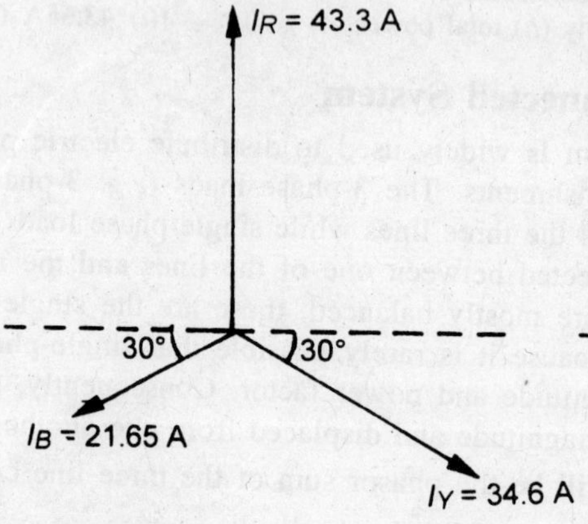

Fig. 19.31

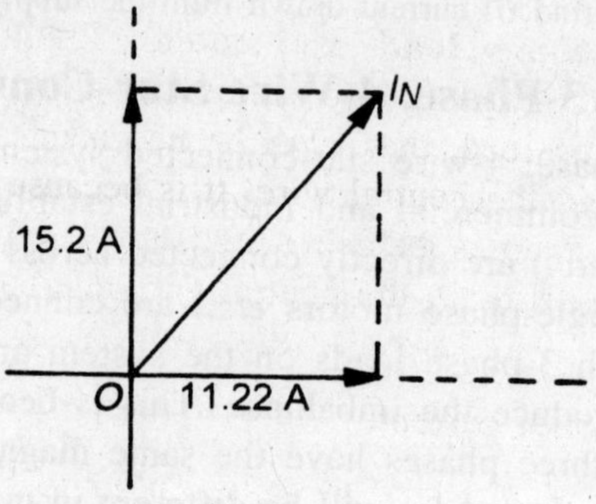

Fig. 19.32

As shown in Fig. 19.32, current in the neutral wire,

$$I_N = \sqrt{(11.22)^2 + (15.2)^2} = \mathbf{18.9 A}$$

Example 19.16. *A 3-phase, 4-wire system supplies power at 400V and lighting at 230V. If the lamps in use require 70, 84 and 33 amperes in each of the three lines, what should be the current in the neutral wire ? If a 3-phase motor is now started, taking 200A from the lines at a p.f. of 0.2 lagging, what should be the total current in each line and the neutral wire ? Find also the total power supplied to the lamps and the motor.*

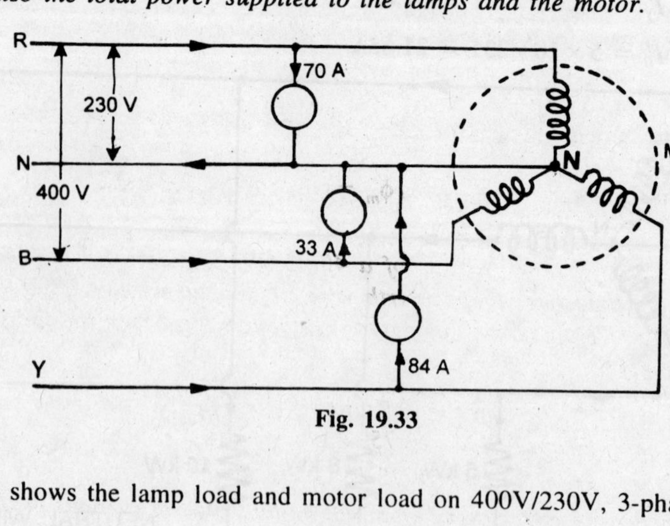

Fig. 19.33

Solution.

Fig. 19.33 shows the lamp load and motor load on 400V/230V, 3-phase, 4-wire supply.

Lamp load alone. If there is lamp load alone, the line currents in phases R, Y and B are 70A, 84A and 33A respectively. These currents will be 120° apart (assuming phase sequence RYB) as shown in Fig. 19.34.

Resultant X-component $= 84 \cos 30° - 33 \cos 30°$

$= 44.17$ A

Resultant Y-component $= 70 - 33 \cos 60° - 84 \cos 60°$

$= 11.5$ A

∴ Neutral current, $I_N = \sqrt{(44.17)^2 + (11.5)^2} = \mathbf{45.64A}$

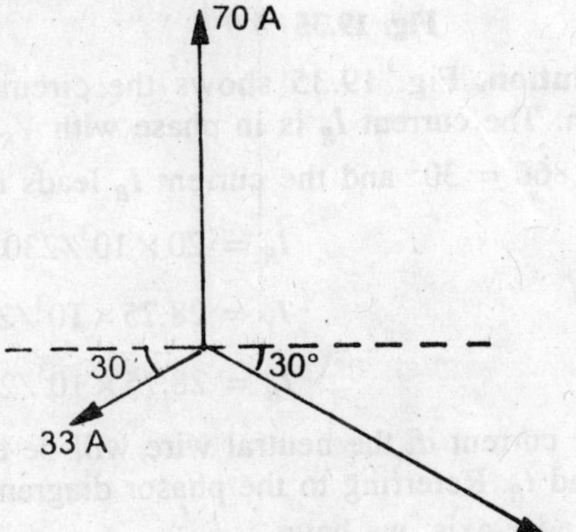

Fig. 19.34

Both lamp load and motor load. When motor load is also connected along with lighting load, there will be no change in current in the neutral wire. It is because the motor load is balanced and hence no current will flow in the neutral wire due to this load.

∴ Neutral current, $I_N = \mathbf{45.64A}$

... *same as before*

The current in each line is the phasor sum of the line currents due to lamp load and motor load.

Active component of motor current $= 200 \times \cos \phi_m = 200 \times 0.2 = 40$ A

Reactive component of motor current $= 200 \times \sin \phi_m = 200 \times 0.98 = 196$ A

$$I_R = \sqrt{(\text{sum of active comps.})^2 + (\text{reactive comp.})^2}$$

$$= \sqrt{(40+70)^2 + (196)^2} = \mathbf{224.8A}$$

$$I_Y = \sqrt{(40+84)^2 + (196)^2} = \mathbf{232A}$$

$$I_B = \sqrt{(40+33)^2 + (196)^2} = \mathbf{209.15A}$$

Power supplied

Power supplied to lamps $= 230\,(70 + 84 + 33) \times 1 = \mathbf{43010\ W}$

Power supplied to motor $= \sqrt{3}\, V_L I_L \cos \phi_m = \sqrt{3} \times 400 \times 200 \times 0.2$ $\quad (\because \cos \phi_m = 0.2)$

$= \mathbf{27712\ W}$

Example 19.17. *The three line leads of a 400/230V, 3-phase, 4-wire supply are designated as R, Y, and B respectively. The fourth wire or neutral wire is designated as N. The phase sequence is RYB. Compute the currents in the four wires when the following loads are connected to this supply :*

From R to N.: 20 kW, unity power factor

From Y to N: 28.75 kVA, 0.866 lag

From B to N: 28.75 kVA, 0.866 lead

If the load from B to N is removed, what will be the value of currents in the four wires ?

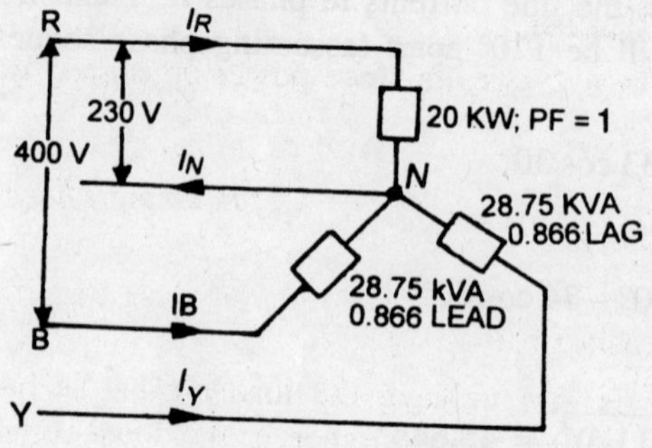

Fig. 19.35

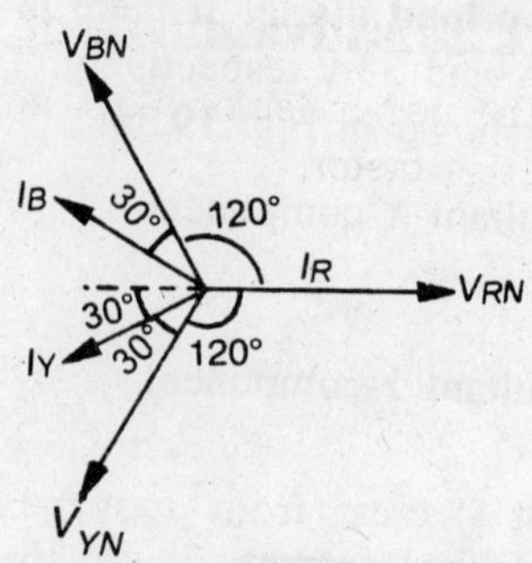

Fig. 19.36

Solution. Fig. 19.35 shows the circuit diagram whereas Fig. 19.36 shows its phasor diagram. The current I_R is in phase with V_{RN}, current I_Y lags behind its phase voltage V_{YN} by $\cos^{-1} 0.866 = 30°$ and the current I_B leads its phase voltage V_{BN} by $\cos^{-1} 0.866 = 30°$.

$$I_R = 20 \times 10^3/230 = \mathbf{86.96\ A}$$

$$I_Y = 28.75 \times 10^3/230 = \mathbf{125A}$$

$$I_B = 28.75 \times 10^3/230 = \mathbf{125A}$$

The current in the neutral wire will be equal to the phasor sum of the three line currents I_R, I_Y and I_B. Referring to the phasor diagram in Fig. 19.36 and resolving these currents along x-axis and y-axis, we have,

Resultant X-component $= 86.96 - 125\cos 30° - 125\cos 30°$

$= 86.96 - 108.25 - 108.25 = -129.54\ A$

Resultant Y-component $= 0 + 125\sin 30° - 125\sin 30° = 0$

$\therefore$ Neutral current, $I_N = \sqrt{(-129.54)^2 + (0)^2} = \mathbf{129.54A}$

When load from B to N removed. When load from B to N is removed, the various line currents are :

$I_R = \mathbf{86.96A}$ in phase with V_{RN}: $I_Y = \mathbf{125A}$ lagging V_{YN} by 30°; $I_B = \mathbf{0A}$

The current in the neutral wire is equal to the phasor sum of these three line currents. Resolving the currents along x-axis and y-axis, we have,

Resultant X-component $= 86.96 - 125\cos 30° = 86.96 - 108.25 = -21.29\ A$

Resultant Y-component $= 0 - 125\sin 30° = 0 - 125 \times 0.5 = -62.5\ A$

$\therefore$ Neutral current, $I_N = \sqrt{(-21.29)^2 + (-62.5)^2} = \mathbf{66.03A}$

Tutorial Problems

1. Non-reactive loads of 10 kW, 6 kW and 4 kW are connected between the neutral and red, yellow and blue phases respectively of a 3-phase, 4-wire 400/230V supply. Find the current in each line and in the neutral wire. **[I_R = 43.3 A; I_Y = 26 A; I_B = 17.3 A; I_N = 22.9 A]**
2. A factory has the following loads with a power factor of 0.9 lagging in each case. Red phase 40A, yellow phase 50 A and blue phase 60 A. If the supply is 400V, 3-phase, 4-wire, calculate the current in the neutral wire and the total power. **[17.3A, 31.2kW]**
3. In a 3-phase, 4-wire system, two phases have currents of 10A and 6A at lagging power factors of 0.8 and 0.6 respectively, while the third phase is open-circuited. Calculate the current in the neutral wire. **[7A]**
4. A 3-phase, 4-wire system supplies a lighting load of 40A, 30A and 20A respectively in the three phases. If the line voltage is 400V, determine the current in the neutral wire. **[17.32 A]**

19.12. Disadvantages of Low Power Factor

The power factor plays an important role in a.c. circuits since power consumed depends upon this factor.

$$I_L = \frac{F}{V_L \cos\phi} \quad \text{... for single phase supply}$$

$$I_L = \frac{P}{\sqrt{3}\, V_L \cos\phi} \quad \text{... for 3-phase supply}$$

It is clear from above that for fixed power and voltage, the load current is inversely proportional to the power factor. Smaller the power factor, higher is the load current and *vice-versa*. The large current due to poor power factor results in the following disadvantages :

(*i*) **Large kVA rating of equipment.** The electrical machinery (*e.g.* alternators, transformers, switchgear *etc.*) is always rated in *kVA.

Now, $$kVA = \frac{kW}{\cos\phi}$$

It is clear that *kVA* rating of the equipment is inversely proportional to power factor. The smaller the power factor, the larger is the *kVA* rating. Therefore, at low power factor, the *kVA* rating of the equipment has to be made more, making the equipment larger and expensive.

(*ii*) **Greater conductor size.** To transmit or distribute a given amount of power at constant voltage, the conductor will have to carry more current at low power factor. This necessitates large conductor size.

(*iii*) **Large copper losses.** The large current at low power factor causes more I^2R losses in all elements of power supply. This results in poor efficiency.

(*iv*) **Poor voltage regulation.** The large current at low lagging power factor causes greater voltage drops in alternators, transformers, transmission lines and distributors. This results in the reduced voltage available to the utilisation devices, impairing their performance *e.g.* lighting becomes dimmer, starting torque of motors is reduced *etc.* In order to keep the supply voltage within permissible limits, extra equipment (*i.e.* voltage regulators) is required.

19.13. Power Factor Improvement

The low power factor is mainly due to the fact that most of the power loads (*e.g.* induction motors *etc.*) are inductive in nature and, therefore, taking lagging currents. In order to improve the power factor, some device taking leading current should be connected in parallel with the load. One of such devices can be a capacitor. The capacitor draws a leading current and partly or completely neutralises the lagging reactive component of load current. This raises the power factor of the load.

Illustration. To illustrate the power factor improvement by a capacitor, consider a **single phase load taking lagging current I at a p.f. of $\cos\phi_1$ as shown in Fig. 19.37 (*i*). The capacitor C is connected in parallel with the load and draws current I_C which leads the supply voltage by 90°. The resulting line current I' is the phasor sum of I and I_C and its angle of lag is ϕ_2 as shown in the phasor diagram in Fig. 19.37 (*iii*). It is clear that ϕ_2 is less than ϕ_1, so that $\cos\phi_2$ is greater than $\cos\phi_1$. Hence, the power factor of the load is increased.

---

* The electrical machinery is rated in kVA because the power factor is not known when the machinery is manufactured in the factory.

** The treatment can be used for 3-phase balanced loads *e.g.* 3-phase induction motor. In a balanced 3-phase load, analysis of only one phase leads to the desired results.

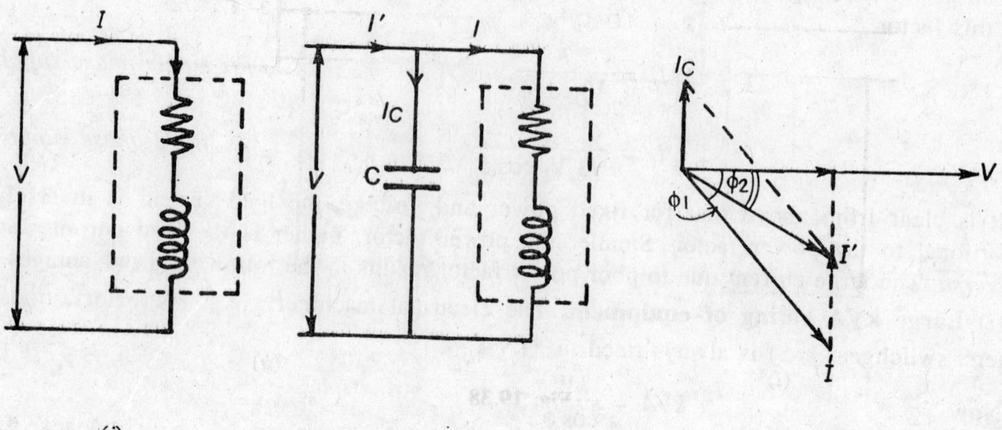

Fig. 19.37

The following point are worth noting :

(*i*) The circuit current I' after p.f. correction is less than the original circuit current I.

(*ii*) The active or wattful component remains the same before and after p.f. correction because only the lagging reactive component of load is reduced by the capacitor *i.e.*,

$$I \cos \phi_1 = I' \cos \phi_2$$

or

$$VI \cos \phi_1 = VI' \cos\phi_2 \qquad \text{(Multiply by } V)$$

Therefore, active power (*kW*) remains unchanged during p.f. improvement.

(*iii*) The original reactive component of load is $I \sin \phi_1$. However, after p.f. correction, it is reduced by I_C and becomes $I' \sin \phi_2$ *i.e.*,

$$I' \sin \phi_2 = I \sin \phi_1 - I_C$$

or

$$VI' \sin \phi_2 = VI \sin \phi_1 - VI_C$$

i.e. Net *kVAR after p.f.* correction = Lagging *kVAR* before p.f. correction– Leading *kVAR* of capacitor

(*iv*)

$$I_C = I \sin \phi_1 - I' \sin \phi_2$$

∴ Capacitance of the capacitor to improve p.f. from $\cos \phi_1$ to $\cos \phi_2$

$$= \frac{I_C}{\omega V} \qquad (\because X_C = V/I_C = 1/\omega C)$$

19.14. Power Factor Improvement Equipment

Power factor improvement may be achieved in the following three principal ways :

(*i*) **Static capacitors.** The power factor can be improved by connecting *capacitors in parallel with equipment operating at low lagging power factor. For 3-phase loads, the capacitors can be connected in delta or star as shown in Fig. 19.38.

---

* called static capacitor to distinguish from the so called *synchronous condenser* which is a synchronous motor running at no load and taking leading current.

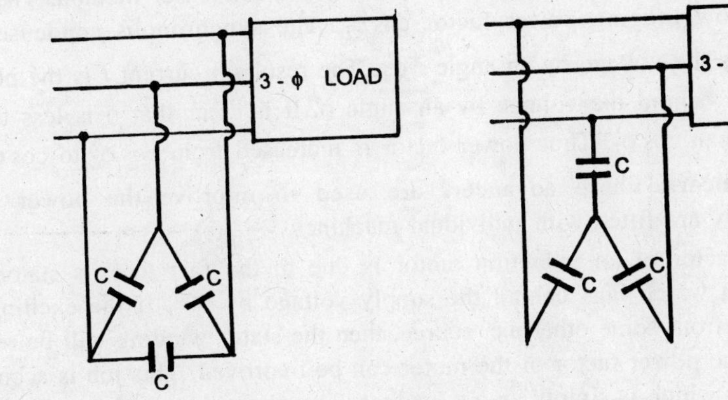

(*i*) (*ii*)

Fig. 19.38

Since static capacitors are not variable, they tend to overcompensate on light loads. As most of the loads have lagging power factors, this overcompensation is not a problem.

(*ii*) **Synchronous condensers.** A synchronous motor takes a leading current when over excited and, therefore, behaves as a capacitor. An over-excited synchronous motor running on no load is known as a *synchronous condenser*. When such a machine is connected in parallel with the load, it takes a leading current which partly neutralises the lagging reactive component of the load current. Thus, the p.f. is improved.

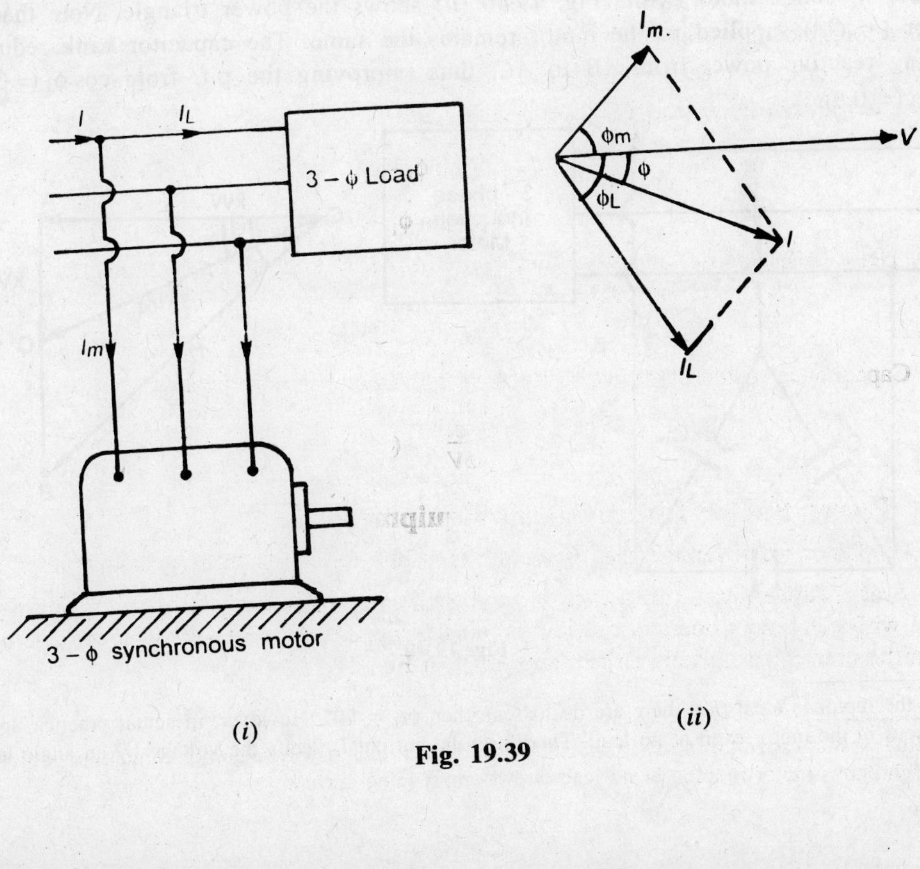

(*i*) (*ii*)

Fig. 19.39

Fig. 19.39 shows the p.f. improvement by synchronous condenser method. The 3-ϕ load takes current I_L at low lagging power factor cos ϕ_L. The synchronous condenser takes a current I_m which leads the voltage by an angle *ϕ_m. The resultant current I is the phasor sum of I_m and I_L and lags behind the voltage by an angle ϕ. It is clear that ϕ is less than ϕ_L so that cos ϕ is greater than cos ϕ_L. Thus power factor is increased from cos ϕ_L to cos ϕ.

(*iii*) **Phase advancers.** Phase advancers are used to improve the power factor of induction motors. They are fitted with individual machines.

The low power factor of an induction motor is due to the fact that its stator winding draws exciting current which lags behind the supply voltage by 90°. If the exciting ampere turns can be provided from some other a.c. source, then the stator winding will be relieved of exciting current and the power factor of the motor can be improved. This job is accomplished by the phase advancer which is simply an a.c. exciter. The phase advancer is mounted on the same shaft as the main motor and is connected in the rotor circuit of the motor. It provides exciting ampere turns to the rotor circuit at slip frequency. By providing more ampere turns than required, the induction motor can be made to operate on leading power factor like an over-excited synchronous motor.

Note. It should be noted that although power factor improvement equipment changes (*i.e.* reduces) the lagging reactive power and improves the power factor, it does not change the active power taken by the load.

Example 19.18. *A 3-phase, 5kW induction motor has a p.f. of 0.75 lagging. A bank of capacitors is connected in delta across the supply terminals and p.f. raised to 0.9 lagging. Determine the kVAR ratings of the capacitors connected in each phase.*

Solution. Fig. 19.40 (*i*) shows the delta connected capacitor bank across the terminals of 3-phase induction motor while Fig. 19.40 (*ii*) shows the power triangle. Note that active power (= OA) supplied to the motor remains the same. The capacitor bank reduces the lagging reactive power from AB to AC, thus improving the p.f. from cos ϕ_1 (= 0.75) to cos ϕ_2 (= 0.9).

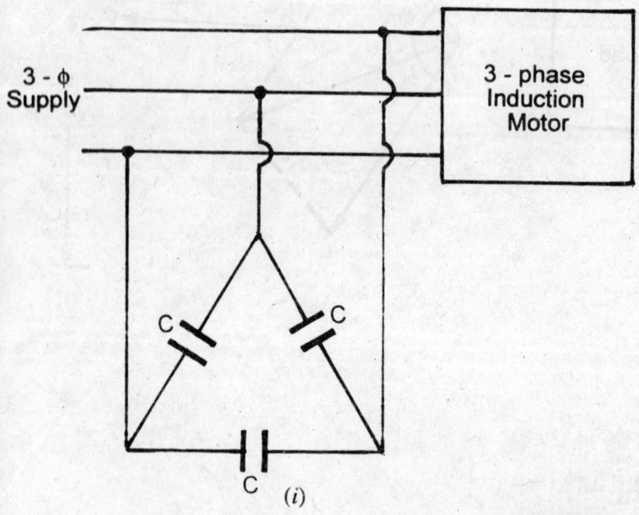

(*i*)

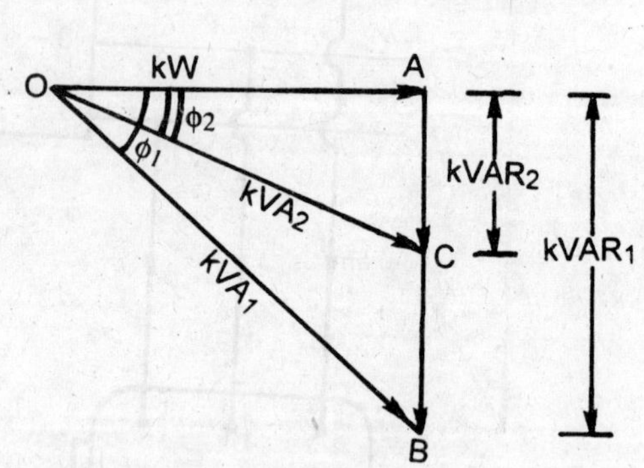

(*ii*)

Fig. 19.40

* If the motor is ideal (*i.e.* there are no losses), then ϕ_m = 90°. However, in actual practice, losses do occur in the motor even at no load. Therefore, the current I_m leads the voltage by an angle less than 90°.

Motor input (OA), $P = 5\ kW$ ... assuming 100% efficiency

Leading $kVAR$ supplied by the condenser bank

$$= BC = AB - AC = kVAR_1 - kVAR_2$$

$$= OA\ (\tan\phi_1 - \tan\phi_2) = P\ (\tan\phi_1 - \tan\phi_2)$$

Now, $\cos\phi_1 = 0.75;\ \phi_1 = \cos^{-1}(0.75) = 41°\ 24'\ ;\ \tan\phi_1 = 0.8819$

$\cos\phi_2 = 0.9;\ \phi_2 = \cos^{-1}(0.9) = 25°\ 50'\ ;\ \tan\phi_2 = 0.4843$

∴ Leading $kVAR$ supplied by the condenser bank

$$= 5\ (0.8819 - 0.4843) = 1.99\ kVAR$$

∴ Rating of the capacitor connected in each phase

$$= 1.99/3 = \mathbf{0.663\ kVAR}$$

Example 19.19. *A 3-phase, 50Hz, 400V motor develops 100H.P. (74.6 kW), the power factor being 0.75 lagging and efficiency 93%. A bank of capacitors is connected in delta across the supply terminals and power factor raised to 0.95 lagging. Each of the capacitance unit is built of 4 similar 100V capacitors. Determine the capacitance of each capacitor.*

Solution. Fig. 19.41 (*i*) shows the delta connected capacitor bank connected across the terminals of 3-phase motor while Fig. 19.41 (*ii*) shows the power triangle. Note that active power (= OA) supplied to the motor remains the same. The capacitor bank reduces the lagging reactive power from AB to AC, thus improving the p.f. from $\cos\phi_1$ (= 0.75) to $\cos\phi_2$ (= 0.95)

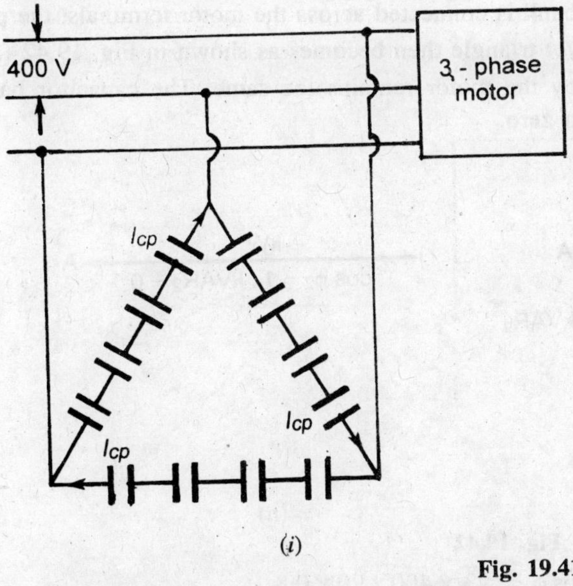

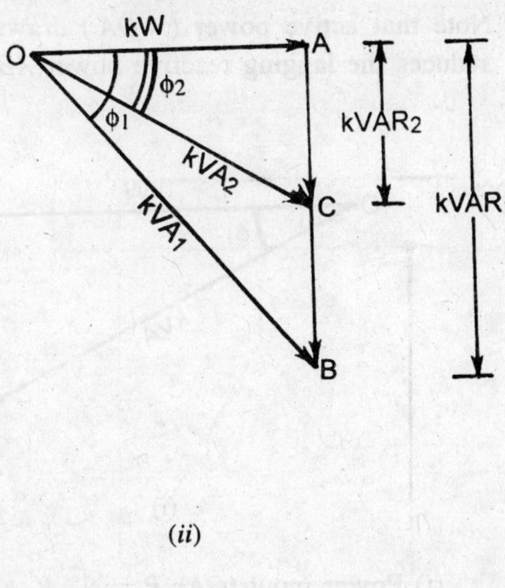

Fig. 19.41

Power input (0A), $P = 74.6/\eta = 74.6/0.93 = 80\ kW$

Leading $kVAR$ supplied by the condenser bank

$$= P\ (\tan\phi_1 - \tan\phi_2)$$

Now, $\cos\phi_1 = 0.75;\quad \phi_1 = \cos^{-1}(0.75) = 41°24';\quad \tan\phi_1 = 0.8819$

$\cos\phi_2 = 0.95;\quad \phi_2 = \cos^{-1}(0.95) = 18°12';\quad \tan\phi_2 = 0.3288$

$\therefore$ Leading $kVAR$ supplied by the condenser bank

$$= 80\,(0.8819 - 0.3288) = 44.25\,kVAR$$

Leading $kVAR$ supplied by the capacitor bank/phase

$$= 44.25/3 = 14.75\,kVAR \qquad \ldots (i)$$

Let C farad be the combined capacitance of 4 capacitors in each phase.

Phase current of capacitor, $I_{cp} = V_{ph}/X_C = 2\pi fCV_{ph}$

$$= 2\pi \times 50 \times C \times 400 = 1{,}25{,}600\,C \text{ amperes}$$

$$kVAR/\text{phase} = \frac{{}^*V_{ph}\,I_{cp}}{1000} = \frac{400 \times 1{,}25{,}600\,C}{1000} = 50240\,C \qquad \ldots (ii)$$

Equating exps (*i*) and (*ii*), we get,

$$50240\,C = 14.75$$

$$\therefore \qquad C = 14.75/50240 = 293.4 \times 10^{-6}\,F = 293.4\,\mu F$$

Since it is the combined capacitance of four equal capacitors joined in series,

$\therefore$ Capacitance of each capacitor $= 4 \times 293.4 =$ **1173.6 μF**

Example 19.20. *A 3-phase, star-connected motor, connected across a 3-phase, star-connected supply of 400V, 50Hz takes a current of 20A at 0.8 p.f. lagging. Determine :*

(*i*) *the capacitance of the capacitors per phase that are to be connected in delta across the terminals of the motor to raise the p.f. to unity.*

(*ii*) *the new value of supply line current with capacitors connected.*

Solution. Fig. 19.42 (*i*) shows the power triangle before the connection of capacitor bank. The active power drawn by the motor is OA while the lagging reactive power is AB; the p.f. being $\cos\phi_1$ (=0.8). When the capacitor bank is connected across the motor terminals, the p.f. is raised to unity *i.e* $\cos\phi_2 = 1$. The power triangle then becomes as shown in Fig. 19.42 (*ii*). Note that active power ($= OA$) drawn by the motor remains the same. The capacitor bank reduces the lagging reactive power AB to zero.

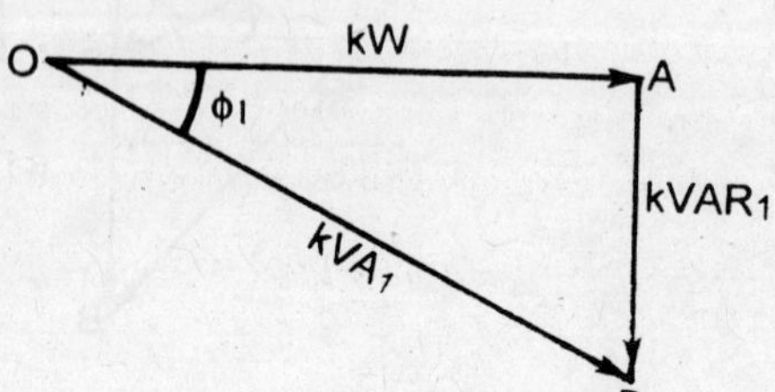

(*i*)

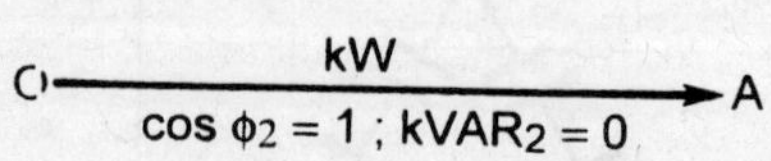

(*ii*)

Fig. 19.42

(*i*) Power input (OA), $P = \sqrt{3}\,V_L I_L \cos\phi = \sqrt{3} \times 400 \times 20 \times 0.8$

$$= 11.085 \times 10^3\text{ W} = 11.085\text{kW}$$

Leading $kVAR$ supplied by delta-connected capacitor bank

$$= AB = P\tan\phi_1 = 11.085\tan(\cos^{-1}0.8) = 8.314\;\;kVAR$$

* Since the capacitors are assumed loss-free, $kVAR = kVA$.

If I_C is the phase current in the delta-connected bank, then,

$$\text{Total } VAR = 3V_{ph}I_C = 3\times 400\times I_C = 1200\,I_C$$

$$\therefore \quad 1200\,I_C = 8314 \quad or \ I_C = 8314/1200 = 6.93A$$

Capacitive reactance/phase, $= X_C = V_{ph}/I_C = 400/6.93 = 57.72\Omega$

Capacitance/phase, $C = \dfrac{1}{2\pi f\,X_C} = \dfrac{1}{2\pi\times 50\times 57.72} = 55.17\times 10^{-6}\,F = \mathbf{55.17\mu F}$

(*ii*) Since the p.f. is raised to unity, the new line current I_L is given by ;

$$P = \sqrt{3}\,V_L I_L \qquad (\because \cos\phi_2 = 1)$$

$$\therefore \quad I_L = \frac{P}{\sqrt{3}\,V_L} = \frac{11.085\times 10^3}{\sqrt{3}\times 400} = \mathbf{16A}$$

Example 19.21. *A synchronous motor improves the power factor of a load of 200kW from 0.8 lagging to 0.9 lagging. Simultaneously the motor carries a load of 80kW. Find (i) the leading kVAR supplied by the motor (ii) kVA rating of the motor and (iii) the power factor at which the motor operates.*

Solution.

Load, $P_1 = 200\,kW$; Motor load, $P_2 = 80\,kW$

p.f. of load, $\cos\phi_1 = 0.8\ lag$

p.f. of combined load, $\cos\phi_2 = 0.9\ lag$

Combined load,

$$P = P_1 + P_2 = 200 + 80 = 280\,kW$$

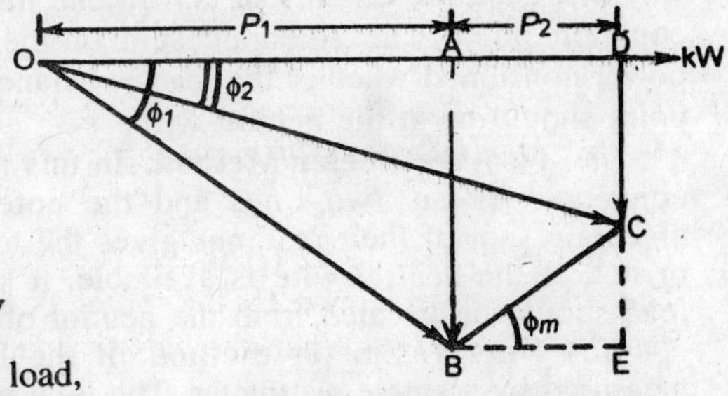

Fig. 19.43

In Fig. 19.43, $\Delta\,OAB$ is the power triangle for load, $\Delta\,ODC$ for combined load and $\Delta\,BEC$ for the motor.

(*i*) Leading $kVAR$ supplied by the motor

$$= CE = DE - DC = AB - DC \qquad (\because AB = DE)$$

$$= P_1 \tan\phi_1 - P^* \tan\phi_2$$

$$= 200\tan(\cos^{-1}0.8) - 280\tan(\cos^{-1}0.9)$$

$$= 200\times 0.75 - 280\times 0.4843 = \mathbf{14.4\,kVAR}$$

(*ii*) kVA rating of the motor $= BC = \sqrt{(BE)^2 + (EC)^2} = \sqrt{(80)^2 + (14.4)^2} = \mathbf{81.28\,kVA}$

(*iii*) p.f. of motor, $\cos\phi_m = \dfrac{\text{Motor } kW}{\text{Motor } kVA} = \dfrac{80}{81.28} = \mathbf{0.984}\ lead$

Tutorial Problems

1. A single-phase motor connected to 240V, 50Hz supply takes 20A at a p.f. of 0.75 lagging. A capacitor is shunted across the motor terminals to improve the p.f. to 0.9. Determine the capacitance of the capacitor to be used. **[59.8 μF]**
2. A 3-phase, 50Hz, 3300V star-connected induction motor develops 250H.P. (186.5 kW), the p.f. being 0.707 lagging and efficiency 0.86. Three capacitors in delta are connected across the supply terminals and the p.f. raised to 0.9 lagging. Calculate :
   (*i*) the *kVAR* rating of the capacitor bank.
   (*ii*) the capacitance of each unit. **[(*i*) 111.8 kVAR (*ii*) 10.9μF]**

---

* In right angled $\Delta\,OAB$, $AB = P_1\tan\phi_1$

In right angled $\Delta\,ODC$,

$$DC = OD\tan\phi_2 = (P_1 + P_2)\tan\phi_2 = P\tan\phi_2$$

3. A 3-phase, 50Hz, 3000V motor develops 600 H.P. (447.6 kW), the p.f. being 0.75 lagging and efficiency 0.93. A bank of capacitors is connected in delta across the supply terminals and power factor raised to 0.95 lagging. Each of the capacitance unit is built of five similar 600V capacitors. Determine the capacitance of each capacitor. **[156 μF]**
4. Three similar coils, each of resistance 20Ω and inductive reactance 15Ω are connected in star to a 400V, 3–ϕ, 50Hz supply. Calculate the line current, power supplied and the power factor. If three capacitors, each of the same capacitance, are connected in delta to the same supply so as to form a parallel circuit with the above impedance coils, calculate the capacitance of each capacitor to obtain an overall p.f. of 0.95 lagging. **[(*i*) 9.24A ; 5120W; 0.8 *lag* (*ii*) 14.31 μF]**
5. A capacitor bank taking a load of 1200kVA is necessary to improve the p.f. of a 3-phase balanced load in a factory from 0.75 lagging to 0.95 lagging. Find the factory load in kW and kVA (*i*) before connection of capacitors (*ii*) after connection of capacitors. **[(*i*) 2150 kW; 2260 kVA (*ii*) 2150 kW; 2870 kVA]**

19.15. Power Measurement in 3-Phase Circuits

The following methods are available for measuring power in a 3-phase load (star or delta connected):

(*i*) **Three-Wattmeter Method.** In this method, three wattmeters are connected in such a way that each has its current coil in one line and its potential coil between that line and some common point. The algebraic sum of the readings of the three wattmeters gives the total power consumed whether the load is balanced or not. If neutral wire is available, the common point should be at the neutral wire.

(*ii*) **Two-Wattmeter Method.** In this method, the current coils of the two wattmeters are connected in *any* two lines and the potential coil of each joined to the third line. The algebraic sum of their readings gives the total power consumed whether the load is balanced or not. If the neutral wire is available, it should carry no current. Or else the neutral of the load should be isolated from the neutral of the source.

(*iii*) **One-Wattmeter method.** If the load is balanced, the power in any phase can be measured by a single wattmeter. The total circuit power is given by multiplying the wattmeter reading by three. This method can only be used if the load is balanced. Its principal disadvantage is that it is not always possible to make the required connections.

It may be noted that Two-wattmeter method is a universal method of measuring power in a 3-phase circuit.

19.16. Three-Wattmeter Method

In this method, the three wattmeters are connected in such a way that each has its current coil in one line and its potential coil between that line and some common point *x as shown in Fig. 19.44. It can be shown mathematically that algebraic sum of their readings gives the total power consumed whether the load is balanced or not *i.e.*

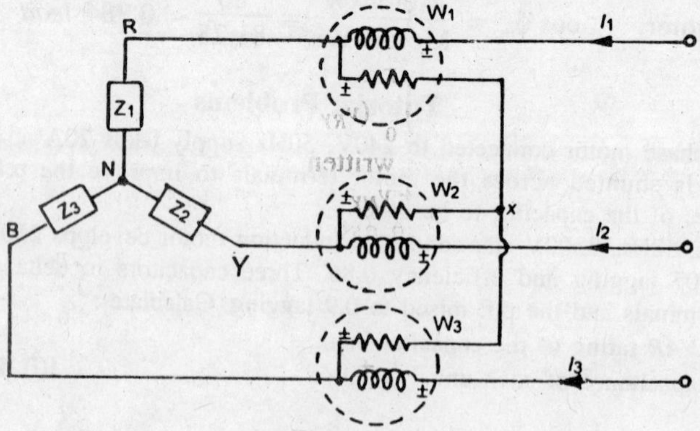

Fig. 19.44

* The point x may be specified point in the three phase system or it may be merely a point in space at which the three potential coils have a common junction.

$$\text{Total power} = W_1 + W_2 + W_3 \qquad \text{... algebraic sum}$$

A *caution* if neutral wire is available. Then common point x should be at the neutral wire.

19.17. Two-Wattmeter Method

In this *method, current coils of the two wattmeters are connected in any two lines and the potential coil of each is joined to the third line as shown in Fig. 19.45. It can be proved mathematically that algebraic sum of the readings of the two wattmeters gives the total power consumed whether the load is balanced or not *i.e.*

$$\text{Total power} = W_1 + W_2 \qquad \text{... algebraic sum}$$

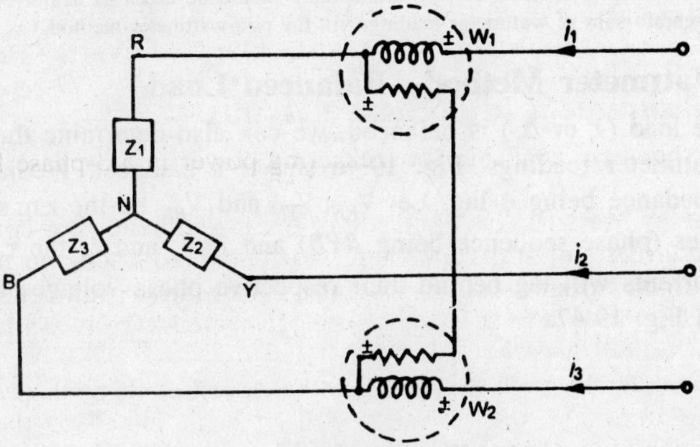

Fig. 19.45

The only precaution in this method is that the neutral wire, if available, should not carry any current. Or else the neutral of the load should be isolated from the neutral of the source. The principal advantage is that the algebraic sum of the readings of the two wattmeters indicates the total power regardless of (*i*) load imbalance (*ii*) source imbalance (*iii*) difference in wattmeters (*iv*) wave-form of the source and (*v*) phase sequence.

Proof. We shall now prove mathematically that the algebraic sum of two wattmeter readings is equal to the total power drawn by the Y load in Fig. 19.45.

Average power indicated by wattmeter W_1 is

$$W_1 = \frac{1}{T}\int_0^T v_{RY}\, i_1\, dt$$

where T is the period for all the voltage sources.

Similarly
$$W_2 = \frac{1}{T}\int_0^T v_{BY}\, i_3\, dt$$

$$\therefore \quad W_1 + W_2 = \frac{1}{T}\int_0^T (v_{RY}\, i_1 + v_{BY}\, i_3)\, dt \qquad \text{... (i)}$$

Each of the above two voltages can be written as :

$$v_{RY} = v_{RN} + v_{NY} \qquad v_{BY} = v_{BN} + v_{NY}$$

Substituting the values of v_{RY} and v_{BY} in exp (*i*)

$$W_1 + W_2 = \frac{1}{T}\int_0^T (v_{RN}\, i_1 + v_{BN}\, i_3)\, dt + \frac{1}{T}\int_0^T v_{NY}(i_1 + i_3)\, dt$$

Since $\quad i_1 + i_2 + i_3 = 0; \quad i_1 + i_3 = -i_2$

$$\therefore \quad W_1 + W_2 = \frac{1}{T}\int_0^T (v_{RN}\, i_1 + v_{BN}\, i_3)\, dt + \frac{1}{T}\int_0^T v_{NY}(-i_2)\, dt$$

---

* The two-wattmeter method is essentially the three-wattmeter method with common point shifted to one of the lines.

$$= \frac{1}{T}\int_0^T (v_{RN}\, i_1 + v_{BN}\, i_3)\, dt + \frac{1}{T}\int_0^T v_{YN}\, i_2\, dt$$

$$\therefore \quad W_1 + W_2 = \frac{1}{T}\int_0^T (v_{RN}\, i_1 + v_{YN}\, i_2 + v_{BN}\, i_3)\, dt \qquad \text{... (ii)}$$

The R.H.S. of exp (*ii*) is the sum of average powers taken by each phase of the load.

$$\therefore \quad W_1 + W_2 = \text{Total power} \qquad \text{... algebraic sum}$$

Note. If the p.f. of 3-phase balanced load is less than 0.5 (See Art. 19.19), then one of the wattmeters will read down scale. In order to obtain upscale reading, reverse the connections of either potential or current coil of the wattmeter. The reading obtained after reversal of coil connections should be taken as negative. This explains the significance of the term "algebraic sum of wattmeter readings" in the two-wattmeter method.

19.18. Two-Wattmeter Method—Balanced Load

If the 3-phase load (Y or Δ) is balanced, we can also determine the power factor of the load from the wattmeter readings. Fig. 19.46 shows a balanced Y-connected load; the p.f. angle of load impedance being ϕ lag. Let V_{RN}, V_{YN} and V_{BN} be the r.m.s. values of the three load phase voltages (phase sequence being *RYB*) and I_R, I_Y and I_B the r.m.s. values of phase currents. These currents will lag behind their respective phase voltages by ϕ as shown in the phasor diagram in Fig. 19.47.

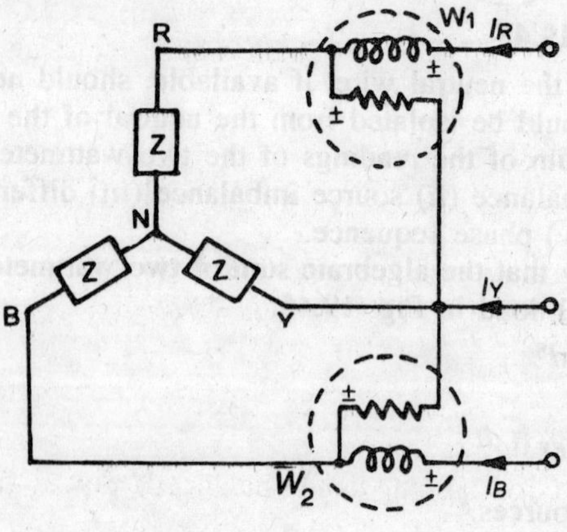

Fig. 19.46

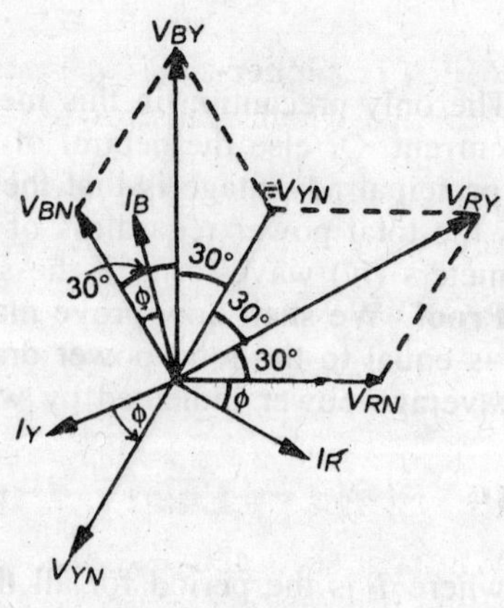

Fig. 19.47

Current through current coil of $W_1 = I_R$

P.D. across potential coil of W_1, $V_{RY} = V_{RN} - V_{YN}$ ... *phasor difference*

To obtain V_{RY}, find the phasor sum of V_{RN} and $-V_{YN}$ as shown in Fig. 19.47. It is clear from the phasor diagram that phase angle between V_{RY} and I_R is ($30° + \phi$).

$$\therefore \quad W_1 = V_{RY}\, I_R \cos(30° + \phi)$$

Current through current coil of $W_2 = I_B$

P.D. across potential coil of W_2, $V_{BY} = V_{BN} - V_{YN}$ ... *phasor difference*

To obtain V_{BY}, find the phasor sum of V_{BN} and $-V_{YN}$ as shown in Fig. 19.47. It is clear from the phasor diagram that phase angle between V_{BY} and I_B is $(30° - \phi)$.

$$\therefore \quad W_2 = V_{BY} I_B \cos(30° - \phi)$$

Since the load is balanced, $V_{RY} = V_{BY}$ = Line voltage, V_L and $I_R = I_B$ = Line current, I_L

$$\therefore \quad W_1 = V_L I_L \cos(30° + \phi)$$

$$\text{and} \quad W_2 = V_L I_L \cos(30° - \phi)$$

$$\therefore \quad W_1 + W_2 = V_L I_L [\cos(30° + \phi) + \cos(30° - \phi)]$$
$$= V_L I_L (2\cos 30° \cos\phi)]$$
$$= \sqrt{3}\, V_L I_L \cos\phi = \text{Total power in the 3-phase load}$$

Hence, the algebraic sum of the two wattmeter readings gives the total power consumed in the 3-phase load.

Power factor

$$W_2 + W_1 = \sqrt{3}\, V_L I_L \cos\phi$$

$$\text{Also} \quad {}^{*}W_2 - W_1 = V_L I_L \sin\phi$$

$$\therefore \quad \tan\phi = \sqrt{3}\,\frac{W_2 - W_1}{W_2 + W_1}$$

Thus from the two wattmeter readings, we can find ϕ and hence the load power factor $\cos\phi$.

Which is higher-reading wattmeter ? How can we tell which wattmeter reads higher (*i.e.* W_2) and which reads lower (*i.e.* W_1) ?

$$W_2 = V_L I_L \cos(30° - \phi)$$
$$W_1 = V_L I_L \cos(30° + \phi)$$

Since the value of load p.f. can vary from 0 to 1 (*i.e.* ϕ can vary from 90° to 0°), it is clear that wattmeter whose deflection is proportional to $(30° - \phi)$ *is **always positive and always the higher reading wattmeter* (*i.e.* W_2 in this case).

19.19. Effect of Load p.f. on Wattmeter Readings

We have seen above (See Fig. 19.46) that for lagging load (balanced) power factor of $\cos\phi$, the two wattmeter readings are :

$$W_2 = V_L I_L \cos(30° - \phi)$$
$$W_1 = V_L I_L \cos(30° + \phi)$$

It is clear that readings of the two wattmeters depend upon load p.f. angle ϕ.

(*i*) *When p.f. is unity (i.e.* $\phi = 0°$)

$$W_2 = V_L I_L \cos 30°; \quad W_1 = V_L I_L \cos 30°$$

Both wattmeters indicate equal and positive (*i.e.* upscale) readings.

---

* $W_2 - W_1 = V_L I_L [\cos(30° - \phi) - \cos(30° + \phi)]$
$= V_L I_L [(\cos 30° \cos\phi + \sin 30° \sin\phi) - (\cos 30° \cos\phi - \sin 30° \sin\phi)]$
$= V_L I_L (2\sin 30° \sin\phi) = V_L I_L \sin\phi$

** Except for the case when load p.f. is unity (*i.e.* $\phi = 0°$)at which the two wattmeters have equal readings.

(ii) When p.f. is 0.5 (*i.e.* $\phi = 60°$)

$$W_2 = V_L I_L \cos 30°; \quad W_1 = V_L I_L \cos 90° = 0$$

Hence, total power is measured by wattmeter W_2 alone.

(iii) When p.f. is less than 0.5 *but greater than 0 (i.e.* $90° > \phi > 60°$).

$$*W_2 = \text{positive reading}; \quad W_1 = \text{negative reading}$$

In order to obtain upscale reading on wattmeter W_1, reverse the connections of potential or current coil. *The reading obtained after reversal of coil connections should be taken as negative.*

$$\text{Total power} = W_2 + (-W_1) \quad \text{... algebraic sum}$$

$$= W_2 - W_1$$

Thus if the load p.f. is less than 0.5, then lower reading wattmeter (*i.e.* W_1 in this case) will give negative reading.

(iv) When p.f. is zero (i.e. $\phi = 90°$). Such a case will occur when the load consists of pure inductance and/or capacitance.

$$W_2 = V_L I_L \cos(30° - 90°) = V_L I_L \sin 30°$$

$$W_1 = V_L I_L \cos(30° + 90°) = -V_L I_L \sin 30°$$

Thus the two wattmeters will read equal and opposite.

$$\therefore \quad W_1 + W_2 = 0$$

The above facts are summarised below in the tabular form.

ϕ	0°	60°	more than 60°	90°
$\cos\phi$	1	0.5	< 0.5	0
W_2	positive	positive	positive	positive
W_1	positive	0	negative	negative
Conclusion	$W_1 = W_2$ Total power $= W_1 + W_2$	$W_1 = 0$ Total power $= W_2$	Total power $= W_2 - W_1$	$W_2 = -W_1$ Total power $= 0$

19.20. Leading Power Factor

In discussing two-wattmeter method for measuring power in a 3-phase balanced load, we have considered lagging load power factor *i.e.* p.f. angle ϕ is considered positive. For leading power factor, angle ϕ becomes negative. Therefore by putting the value of load p.f. angle as $-\phi$ in the readings of the wattmeters above, we have,

$$W_2 = V_L I_L \cos(30° + \phi)$$

$$W_1 = V_L I_L \cos(30° - \phi)$$

Note that effect of leading power factor is that the readings of the two wattmeters are interchanged. Now wattmeter W_1 has become the higher-reading wattmeter.

$$\therefore \quad \tan\phi = \sqrt{3}\,\frac{W_1 - W_2}{W_1 + W_2}$$

---

* The wattmeter W_2 reads positive (*i.e.*, upscale) because for the given conditions (*i.e.*, $90° > \phi > 60°$), the phase angle between voltage and current will be less than 90°. However, for wattmeter W_1, the phase angle between voltage and current will be more than 90° and hence this wattmeter gives negative (*i.e.* downscale) reading.

19.21. How to Apply p.f. Formula ?

As shown above, the value of tan ϕ (and hence the p.f. cos ϕ) can be determined from the two wattmeter readings.

$$\tan\phi = \sqrt{3}\,\frac{W_2 - W_1}{W_2 + W_1} \quad \text{... lagging p.f.}$$

$$\tan\phi = \sqrt{3}\,\frac{W_1 - W_2}{W_1 + W_2} \quad \text{... leading p.f.}$$

It may be noted that for lagging p.f., W_2 is the higher-reading wattmeter whereas it is W_1 for leading p.f. To avoid any confusion, the reader is well advised to remember the following general formula for both lagging and leading p.f.

$$\tan\phi = \sqrt{3}\,\frac{(\text{Higher reading}) - (\text{Lower reading})}{(\text{Higher reading}) + (\text{Lower reading})}$$

For example, in the two-wattmeter method if the two wattmeter readings are 12.5 kW and – 4.8 kW, then

$$\tan\phi = \sqrt{3}\,\frac{(12.5) - (-4.8)}{(12.5) + (-4.8)} = \sqrt{3}\,\frac{17.3}{7.7} = 3.89$$

$$\therefore \quad \phi = \tan^{-1} 3.89 = 75.6°$$

$$\therefore \quad \text{Power factor} = \cos\phi = \cos 75.6° = 0.2487$$

19.22. One-Wattmeter Method

If the 3-phase load (Y or Δ) is balanced, then one wattmeter is sufficient to measure the power drawn by the load. In this method, two readings of the two-wattmeter method are taken with a single wattmeter as shown in Fig. 19.48. The current coil of the wattmeter is connected in any one line and the pressure coil is connected alternately between this and the other two lines. The algebraic sum of the two readings gives the total power drawn by the balanced 3-phase load.

This method only holds for a balanced load and cannot be applied to an unbalanced load.

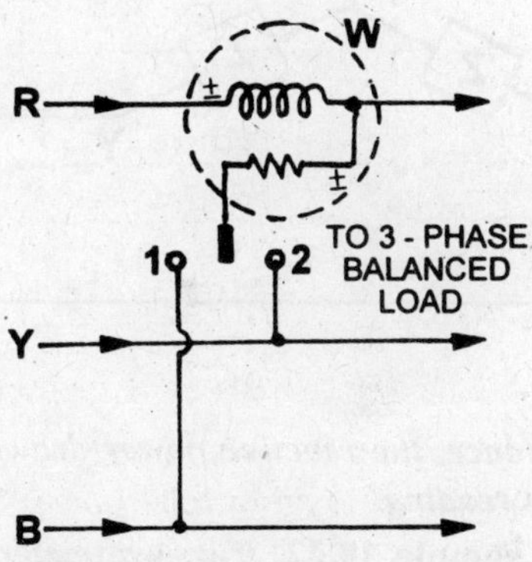

Fig. 19.48

19.23. Reactive Power with Two-Wattmeter Method

The reactive power (*i.e. VAR*) in a 3-phase balanced load can be determined from the readings of the two-wattmeter method.

$$W_2 - W_1 = V_L I_L \sin\phi \quad \text{... See Art. 19.18}$$

Multiplying throughout by $\sqrt{3}$, we get,

$$\sqrt{3}\,(W_2 - W_1) = \sqrt{3}\,V_L I_L \sin\phi$$

But $\sqrt{3}\,V_L I_L \sin\phi$ is equal to the reactive power taken by the 3-phase balanced load.

$$\therefore \quad \text{Reactive power} = \sqrt{3}\,(^{*}W_2 - W_1)\ VAR$$

* Note that W_2 is the higher reading wattmeter. Further $W_2 - W_1$ is the algebraic difference of two wattmeter readings.

Thus, for a balanced load, the reactive power is equal to $\sqrt{3}$ times the algebraic difference of the readings of the two wattmeters.

19.24. Reactive Power with One Wattmeter

A single wattmeter may be used to read the reactive power in a balanced 3-phase load. The method of connection is shown in Fig. 19.49. The current coil of the wattmeter is connected in one line (say line R) and potential coil is connected between the other two lines.

The voltage across the potential coil is V_{YB} (= line voltage V_L) and current though the current coil is I_R (= line current I_L). The phase angle between V_{YB} and I_R is $90° - \phi$ as shown in the phasor diagram in Fig. 19.49.

$\therefore$ Wattmeter reading, $W = V_{YB} I_R \cos(90° - \phi) = V_L I_L \sin\phi$

But Reactive power $= \sqrt{3}\, V_L I_L \sin\phi = \sqrt{3} \times$ wattmeter reading

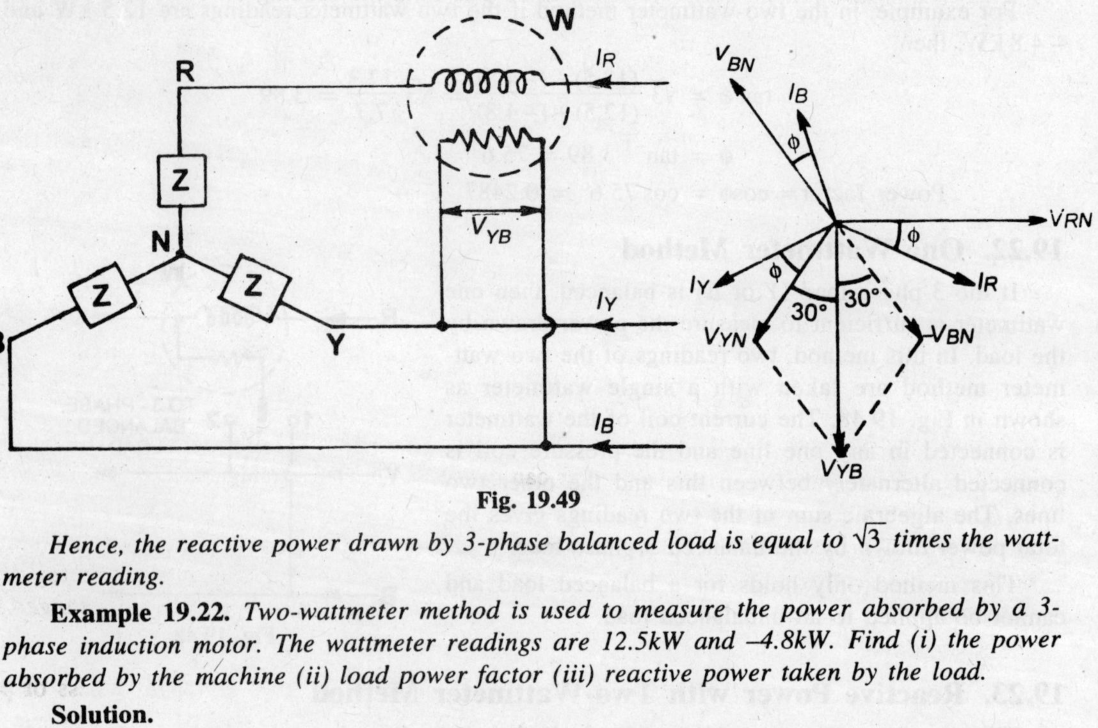

Fig. 19.49

Hence, the reactive power drawn by 3-phase balanced load is equal to $\sqrt{3}$ times the wattmeter reading.

Example 19.22. *Two-wattmeter method is used to measure the power absorbed by a 3-phase induction motor. The wattmeter readings are 12.5kW and –4.8kW. Find (i) the power absorbed by the machine (ii) load power factor (iii) reactive power taken by the load.*

Solution.

$$W_2 = 12.5\,kW\ ; \quad W_1 = -4.8\,kW$$

(*i*) Power absorbed $= W_2 + W_1 = 12.5 + (-4.8) = \mathbf{7.7\ kW}$

(*ii*) $$\tan\phi = \sqrt{3}\,\frac{W_2 - W_1}{W_2 + W_1} = \frac{12.5 - (-4.8)}{(12.5) + (-4.8)} = 3.89$$

$\therefore$ $\phi = \tan^{-1} 3.89 = 75.6°$

$\therefore$ Power factor $= \cos\phi = \cos 75.6° = \mathbf{0.2487}$ *lag*

Since the load p.f. is less than 0.5, the lower-reading wattmeter must read downscale as is evident from the data of the problem.

(*iii*) Reactive power $= \sqrt{3}\,(W_2 - W_1) = \sqrt{3}\,(12.5 + 4.8) = \mathbf{29.96\ kVAR}$

Example 19.23. *Two-wattmeter method is used to measure the power taken by a 3-phase induction motor on no load. The wattmeter readings are 375W and –50W. Calculate (i) power factor of the motor at no load (ii) phase difference of voltage and current in two wattmeters.*

Solution.

(*i*) $$\tan\phi = \sqrt{3}\,\frac{W_2 - W_1}{W_2 + W_1} = \sqrt{3}\,\frac{375-(-50)}{375+(-50)} = 2.265$$

$$\phi = \tan^{-1} 2.265 = 66.18°$$

$\therefore$ $$\text{No load p.f.} = \cos\phi = \cos 66.18° = \mathbf{0.404}\ \textit{lag}$$

(*ii*) $$\text{Phase angle in wattmeter } W_2 = 30° - \phi = 30° - 66.18° = \mathbf{-36.18°}$$

$$\text{Phase angle in wattmeter } W_1 = 30° + \phi = 30° + 66.18° = \mathbf{96.18°}$$

Example 19.24. *A 3-phase motor load has a p.f. of 0.397 lagging. Two wattmeters connected to measure power show the input as 30 kW. Find the reading on each wattmeter.*

Solution.

$$W_2 + W_1 = 30kW \qquad \text{... (i)}$$

$$\text{p.f. angle, } \phi = \cos^{-1} 0.397 = 66.6°$$

$\therefore$ $$\tan\phi = \tan 66.6° = 2.311$$

Now $$\tan\phi = \sqrt{3}\,\frac{W_2 - W_1}{W_2 + W_1}$$

or $$2.311 = \sqrt{3}\,\frac{W_2 - W_1}{30}$$

$\therefore$ $$W_2 - W_1 = 40kW \qquad \text{... (ii)}$$

Solving eqs. (*i*) and (*ii*), we get, $W_2 = \mathbf{35\ kW}$; $W_1 = \mathbf{-5\ kW}$

Since the load p.f. is less than 0.5, the lower-reading wattmeter W_1 gives downscale reading. The upscale reading of this wattmeter can be obtained by reversing the connections of its current or potential coil.

Example 19.25. *Each of the two wattmeters connected to measure the input to a 3-phase induction motor reads 10kW. If the power factor of the motor be changed to 0.866 lagging, determine the readings of the two wattmeters, the total input power remaining unchanged.*

Solution.

$$\text{Total power} = W_2 + W_1 = 20\ kW \qquad \text{.. (i) regardless of p.f.}$$

$$\text{p.f. angle, } \phi = \cos^{-1} 0.866 = 30°$$

$\therefore$ $$\tan\phi = \tan 30° = 0.577$$

Now $$\tan\phi = \sqrt{3}\,\frac{W_2 - W_1}{W_2 + W_1}$$

or $$0.577 = \sqrt{3}\,\frac{W_2 - W_1}{20}$$

$\therefore$ $$W_2 - W_1 = 6.66\ kW \qquad \text{... (ii)}$$

Solving eqs. (*i*) and (*ii*), we get, $W_2 = \mathbf{13.33}\ \textbf{\textit{kW}}$; $W_1 = \mathbf{6.67\ kW}$

Example 19.26. *Three indentical coils, each having a resistance of 10Ω and a reactance of 10 Ω are connected in (i) star (ii) delta, across 400V, 3-phase supply. Find in each case the line current and the readings on each of the two wattmeters connected to measure the power.*

Solution.

(*i*) **Star connection**

$$Z_{ph} = \sqrt{10^2 + 10^2} = 10\sqrt{2}\ \Omega$$

$$V_{ph} = V_L/\sqrt{3} = 400/\sqrt{3} = 231\text{V}$$

$$I_L = I_{ph} = 231 \div 10\sqrt{2} = \mathbf{16.33A}$$

$$\phi = \tan^{-1} X_{ph}/R_{ph} = \tan^{[-1]} 10/10 = 45°$$

$$\text{Power factor} = \cos\phi = \cos 45° = 1/\sqrt{2} = 0.707$$

$$\text{Total power} = \sqrt{3}\, V_L I_L \cos\phi = \sqrt{3} \times 400 \times 16.33 \times 1/\sqrt{2} = 8000\text{ W}$$

Let W_1 and W_2 be the wattmeter readings; W_2 being the higher-reading wattmeter. Since the load p.f. is greater than 0.5, wattmeter W_1 will also read positive.

$$W_2 + W_1 = 8000 \quad \text{... (i)}$$

Also $$\tan 45° = \sqrt{3}\,\frac{W_2 - W_1}{W_2 + W_1}$$

or $$1 = \sqrt{3}\,\frac{W_2 - W_1}{8000}$$

∴ $$W_2 - W_1 = 4619 \quad \text{... (ii)}$$

Solving eqs. (*i*) and (*ii*), we get, W_2 = **6309.5 W**; W_1 = **1690.5 W**

(*ii*) **Delta connection**

$$Z_{ph} = 10\sqrt{2} \quad \text{... same as for star}$$

$$V_{ph} = V_L = 400\text{ V}$$

$$I_{ph} = V_{ph}/Z_{ph} = 400 \div 10\sqrt{2} = 28.28\text{ A}$$

$$I_L = \sqrt{3}\, I_{ph} = \sqrt{3} \times 28.28 = \mathbf{48.98A}$$

$$\text{Power factor} = \cos\phi = \cos 45° = 1/\sqrt{2}$$

$$\text{Total power} = \sqrt{3}\, V_L I_L \cos\phi = \sqrt{3} \times 400 \times 48.98 \times 1/\sqrt{2} = 23995\text{ W}$$

Now, $$W_2 + W_1 = 23995 \quad \text{...(i)}$$

Also $$\tan 45° = \sqrt{3}\,\frac{W_2 - W_1}{W_2 + W_1}$$

or $$1 = \sqrt{3}\,\frac{W_2 - W_1}{23995}$$

∴ $$W_2 - W_1 = 13854 \quad \text{... (ii)}$$

Solving eqs. (*i*) and (*ii*), we get, W_2 = **18924.5 W**; W_1 = **5070.5 W**

Example 19.27. *The power input to a 2000V, 50Hz 3-phase motor running on full load at an efficiency of 90% is measured by two wattmeters which indicate 300kW and 100kW respectively. Find (i) the input (ii) the power factor (iii) the line current and output.*

Solution.

(*i*) Input to motor = $W_2 + W_1 = 300 + 100 =$ **400 kW**

(ii) $\tan\phi = \sqrt{3}\dfrac{W_2 - W_1}{W_2 + W_1} = \sqrt{3}\dfrac{300-100}{300+100} = 0.866$

$\therefore \quad \phi = \tan^{-1} 0.866 = 40.9°$

$\therefore \quad$ Power factor $= \cos\phi = \cos 40.9° =$ **0.756**

(iii) $400 \times 10^3 = \sqrt{3}\, V_L I_L \cos\phi$

$\therefore \quad$ Line current, $I_L = \dfrac{400 \times 10^3}{\sqrt{3} \times 2000 \times 0.756} =$ **152.74 A**

(iv) Motor output = Input × 0.9 = 400 × 0.9 = **360 kW**

Example 19.28. *The ratio of the readings of the two wattmeters connected to measure power in a 3-phase balanced load is 3 : 1. The load is known to be inductive with a lagging power factor. Calculate the power factor of the load.*

Solution.

$$W_2 / W_1 = 3; \quad W_2 \text{ being the higher reading.}$$

Now

$$\tan\phi = \sqrt{3}\frac{W_2 - W_1}{W_2 + W_1}$$

$$= \sqrt{3}\frac{(W_2/W_1) - 1}{(W_2/W_1) + 1} = \sqrt{3}\frac{3-1}{3+1} = \sqrt{3}/2$$

$$\therefore \quad \phi = \tan^{-1}(\sqrt{3}/2) = 40.9°$$

$$\therefore \quad \text{Load p.f.} = \cos\phi = \cos 40.9° = \mathbf{0.756}\ \textit{lag}$$

Tutorial Problems

1. Two wattmeters are used to measure power in a 3-phase balanced load. The wattmeter readings are 8.2 kW and 7.5 kW. Calculate (*i*) total power (*ii*) power factor and (*iii*) total reactive power. **[(*i*) 15.7kW (*ii*) 0.997 (*iii*) 1.21kVAR]**
2. A balanced 3-phase load takes 10kW at a p.f. of 0.9 lagging. Calculate the readings on each of the two wattmeters connected to read the input power. **[6398W, 3602W]**
3. A 440V, 3-phase induction motor has an output of 20.7kW at a p.f. of 0.82 and efficiency 85%. Calculate the readings on each of the two wattmeters connected to measure input. **[12.35kW; 5.25kW]**
4. The power input to a 400V, 3-phase, 50Hz induction motor is measured by two-wattmeter method. The readings of the two wattmeters are 40kW and 10kW. Calculate (*i*) the input power (*ii*) power factor (*iii*) line current. **[(*i*) 30kW (*ii*) 0.327 (*iii*) 132.42A]**
5. Three identical coils, each having a resistance of 20 Ω and a reactance of 20 Ω are connected in (*i*) star (*ii*) delta across 440V, 3-phase lines. Calculate for each method of connection the line current and readings on each of the two wattmeters connected to measure the power. **[(*i*) 8.98A; 3817.5W, 1022.5W (*ii*) 26.95A; 12452.5W, 3067.5W]**

OBJECTIVE QUESTIONS

1. In a two phase generator, the electrical displacement between the two phases or windings is .... electrical degrees.
   (*i*) 120 (*ii*) 90
   (*iii*) 180 (*iv*) none of the above
2. In a six-phase generator, the electrical displacement between different phases or windings is ....electrical degrees.
   (*i*) 60 (*ii*) 90
   (*iii*) 120 (*iv*) 45
3. The torque on the rotor of a 3-phase motor is more constant than that of a single phase motor because................
   (*i*) single phase motors are not self-starting
   (*ii*) single phase motors are small in size
   (*iii*) 3-phase power is of constant value

(*iv*) none of the above

4. For the same rating, the size of a 3-phase motor will be ....single phase motor.
   (*i*) less than that of
   (*ii*) more than that of
   (*iii*) same as that of
   (*iv*) none of the above

5. To transmit the same amount of power over a fixed distance at a given voltage, the 3-phase system requires ....the weight of copper.
   (*i*) 3 times (*ii*) 3/4th times
   (*iii*) 1.5 times (*iv*) 0.5 times

6. The phase sequence of a three-phase system is RYB. The other possible phase sequence can be ....
   (*i*) BRY (*ii*) YRB
   (*iii*) RBY (*iv*) none of the above

7. If in Fig. 19.50, the phase sequence is RYB, then, ...
   (*i*) L_1 will burn more brightly than L_2
   (*ii*) L_2 will burn more brightly than L_1
   (*iii*) both lamps will be equally bright
   (*iv*) none of the above

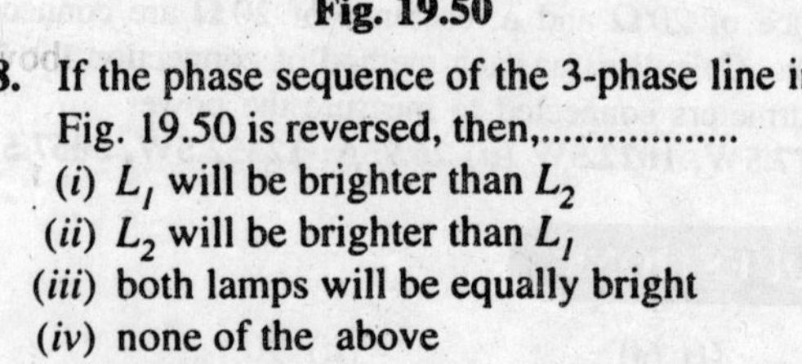

Fig. 19.50

8. If the phase sequence of the 3-phase line in Fig. 19.50 is reversed, then,................
   (*i*) L_1 will be brighter than L_2
   (*ii*) L_2 will be brighter than L_1
   (*iii*) both lamps will be equally bright
   (*iv*) none of the above

9. The advantage of star-connected supply system is that ....
   (*i*) line current is equal to phase current
   (*ii*) two voltages can be used
   (*iii*) phase sequence can be easily changed
   (*iv*) it is a simple arrangement

10. In a balanced star-connected system, line voltages are .... ahead of their respective phase voltages.
   (*i*) 30º (*ii*) 60º
   (*iii*) 120º (*iv*) none of the above

11. In a star-connected system, the relation between the line voltage V_L and phase voltage V_{ph} is ....
   (*i*) $V_L = V_{ph}$ (*ii*) $V_L = V_{ph}/\sqrt{3}$
   (*iii*) $V_L = \sqrt{3}V_{ph}$ (*iv*) none of the above

12. Fig. 19.51 shows a balanced star-connected system. The line voltage V_{RY} is given by
   (*i*) $V_{RY} = E_{RN} - E_{NY}$..............phasor sum
   (*ii*) $V_{RY} = E_{RN} - E_{YN}$.............. phasor difference
   (*iii*) $V_{RY} = E_{RN} + E_{YN}$ .... phasor sum
   (*iv*) none of the above

13. If the load connected to the 3-phase generator shown in Fig. 19.51 has a lagging p.f. of $\cos\phi$, then phase angle between V_{RY} and I_R is ....
   (*i*) $30º + \phi$ (*ii*) $30º - \phi$
   (*iii*) $60º + \phi$ (*iv*) $120º - \phi$

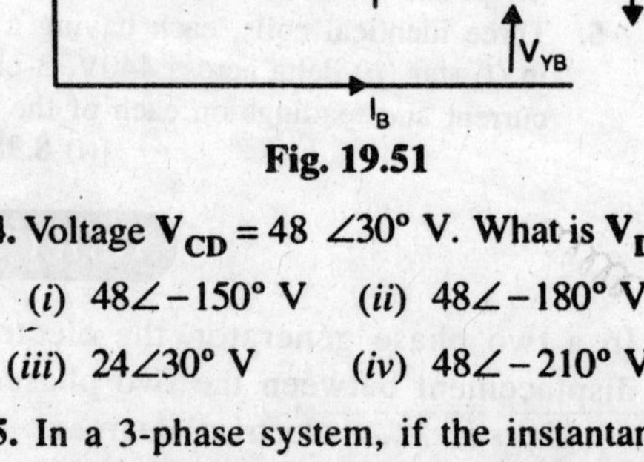

Fig. 19.51

14. Voltage $V_{CD} = 48 \angle 30º$ V. What is V_{DC}?
   (*i*) $48\angle -150º$ V (*ii*) $48\angle -180º$ V
   (*iii*) $24\angle 30º$ V (*iv*) $48\angle -210º$ V

15. In a 3-phase system, if the instantaneous values of phase *R* and *Y* are +60 V and –40 V respectively, than instantaneous voltage of phase B is ....
   (*i*) –20 V (*ii*) 40 V

(iii) 120 V (iv) none of the above

16. In a 3-phase system, $\mathbf{V_{YN}} = 100\angle-120°$ V and $\mathbf{V_{BN}} = 100\angle120°$ V. What is $\mathbf{V_{YB}}$?

(i) $170\angle90°$ V (ii) $173\angle-90°$ V

(iii) $200\angle60°$ V (iv) none of the above

17. In the above question, does $\mathbf{V_{YB}}$ lead or lag $\mathbf{V_{YN}}$?

(i) lead (ii) lag

(iii) Sometimes lead, somtimes lags

(iv) none of the above

18. In question 16, what is the phase angle between $\mathbf{V_{YB}}$ and $\mathbf{V_{YN}}$?

(i) 60° (ii) 45°

(iii) 120° (iv) 30°

19. Voltage $V_{Ee} = 15\angle0°$ V and $V_{Ff} = 12\angle180°$ V. These voltages are connected in series with *e* connected to *f*. What is V_{FE}?

(i) $18\angle-120°$ V (ii) $9\angle-60°$ V

(iii) $27\angle180°$ V (iv) $18\angle-30°$ V

20. The algebraic sum of instantaneous phase voltages in a three-phase circuit is equal to

(i) zero (ii) line voltage

(iii) phase voltage (iv) none of the above

21. If the load connected to the 3-phase generator shown in Fig. 19.52 has a leading p.f. of cosϕ, then angle between V_{RY} and I_R is

(i) $90°-\phi$ (ii) $90°+\phi$

(iii) $60°+\phi$ (iv) $30°-\phi$

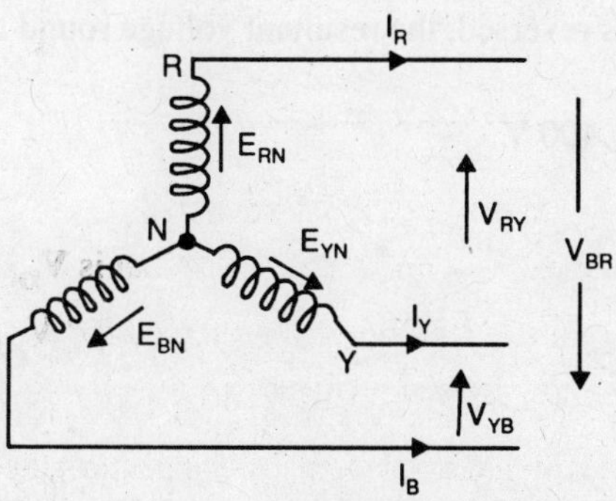

Fig. 19.52

22. Each phase voltage in Fig. 19.52 is 230 V. If connections of phase B are reversed, then, ...

(i) $V_{RY} = 230$ V (ii) $V_{RY} > 230$ V

(iii) $V_{RY} < 230$ V (iv) $V_{RY} = 0$

23. The power delivered by the 3-phase system shown in Fig. 19.52 is $\sqrt{3}\,V_L I_L \cos\phi$. Here ϕ is the phase difference between ....

(i) line voltae and corresponding line current

(ii) phase voltage and correspondig phase current

(iii) phase current and line current

(iv) none of the above

24. A 3-phase load is balanced if all the three phases have the same ....

(i) impedance (ii) power factor

(iii) impedance and power factor

(iv) none of the above

25. Three 50-ohm resistors are connected in star across 400 V, 3-phase supply. If one of the resistors is disconnected, then, line current will be ....

(i) 8 A (ii) 4 A

(iii) $8\sqrt{3}$ A (iv) $8/\sqrt{3}$ A

26. Fig.19.53 shows balanced delta-connected supply system. The current in line 1 is ....

(i) $I_R - I_B$ .... phasor difference

(ii) $I_B - I_R$ .... phasor difference

(iii) $I_Y - I_R - I_B$ .... phase difference

(iv) none of the above

27. In Fig. 19.53, line currents are .... behind the respective phase currents.

(i) 60° (ii) 30°

(iii) 120° (iv) none of the above

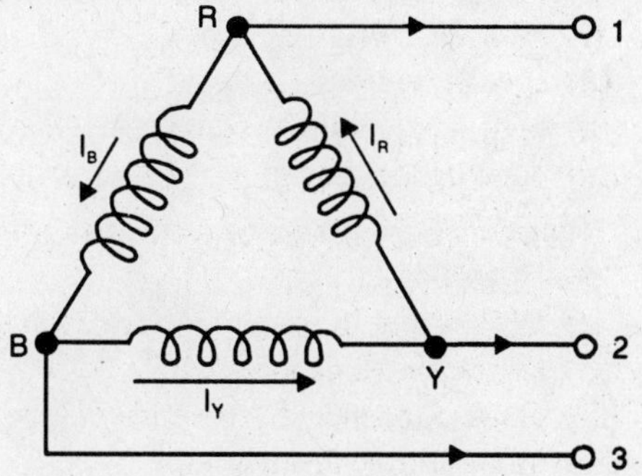

Fig. 19.53

28. The delta-connected generator shown in Fig. 19.53 has phase voltage of 200 V on no load. If connections of one of the phases is reversed, then, resultant voltage across the mesh is ....

(*i*) 200 V (*ii*) $200 \times \sqrt{3}$ V

(*iii*) 400 V (*iv*) none of the above

29. If one line conductor of a 3-phase line is cut, the load is then supplied by .... voltage.

(*i*) single phase (*ii*) two phase

(*iii*) three phase (*iv*) none of the above

30. The resistance between any two terminals of a balanced star-connected load is 12Ω. The resistance of each phase is ....

(*i*) 12Ω (*ii*) 24Ω

(*iii*) 6Ω (*iv*) none of the above

ANSWERS TO OBJECTIVE QUESTIONS

1. (*ii*)	**2.** (*i*)	**3.** (*iii*)	**4.** (*i*)	**5.** (*ii*)	**6.** (*iii*)	**7.** (*i*)	**8.** (*ii*)
9. (*ii*)	**10.** (*i*)	**11.** (*iii*)	**12.** (*ii*)	**13.** (*i*)	**14.** (*i*)	**15.** (*i*)	**16.** (*ii*)
17. (*i*)	**18.** (*iv*)	**19.** (*iii*)	**20.** (*i*)	**21.** (*iv*)	**22.** (*ii*)	**23.** (*ii*)	**24.** (*iii*)
25. (*ii*)	**26.** (*i*)	**27.** (*ii*)	**28.** (*iii*)	**29.** (i)	**30.** (*iii*)		

HINTS TO SELECTED OBJECTIVE QUESTIONS

1. In a polyphase system, the electrical displacement between phases or windings = 360/*N* where *N* is the number of phases. The only exception is 2-phase where electrical displacement between phases is 90°.

7. The phase sequence is in the order : *bright lamp-dim lamp-capacitor.* Since phase sequence is RYB, and lamp L_1 is connected in R-phase, it will burn more brightly than the lamp L_2.

15. $V_R + V_Y + V_B = 0$ *or* $(+60) + (-40) + V_B = 0$ $\therefore$ $V_B = -20$ V

16. $\mathbf{V_{YB}} = \mathbf{V_{YN}} + \mathbf{V_{NB}} = \mathbf{V_{YN}} - \mathbf{V_{BN}} = 100\angle -120° - 100\angle 120° = 173\angle -90°$ V

25. When one of the resistors is disconnected, the remaining two resistors behave as if they were connected in series across the line voltage (*i.e.,* 400V).

$$\therefore \quad I_{ph} = I_L = \frac{400}{50+50} = 4\text{A}$$

26. The current in line 1 is the phasor difference of I_R and I_B since 'positive' direction for I_R is towards point *R* and for I_B, it is away from point *R*.

28. It can be shown that if one phase of delta connection is reversed, the resultant voltage round the mesh is equal to twice the phase voltage *i.e.*

Resultant voltage round the mesh = $2\,E_{ph} = 2 \times 200 = 400$ V.

20

Transformers

Introduction

The transformer is probably one of the most useful electrical devices ever invented. It can change the magnitude of alternating voltage or current from one value to another. This useful property of transformer is mainly responsible for the widespread use of alternating currents rather than direct currents *i.e.* electric power is generated, transmitted and distributed in the form of alternating current. Transformers have no moving parts, rugged and durable in construction, thus requiring very little attention. They also have a very high efficiency — as high as 99%. In this chapter, we shall study some of the basic properties of transformers.

20.1. Transformer

A transformer is a static piece of equipment used either for raising or lowering the voltage of an a.c. supply with a corresponding decrease or increase in current. It essentially consists of two windings, the primary and secondary, wound on a common laminated magnetic core as shown in Fig. 20.1. The winding connected to the a.c. source is called *primary winding* (or primary) and the one connected to load is called *secondary winding* (or secondary). The alternating voltage V_1 whose magnitude is to be changed is applied to the primary. Depending upon the number of turns of the primary (N_1) and secondary (N_2), an alternating e.m.f. E_2 is induced in the secondary. This induced e.m.f. E_2 in the secondary causes a secondary current I_2. Consequently, terminal voltage V_2 will appear across the load. If $V_2 > V_1$, it is called a *step up-transformer*. On the other hand, if $V_2 < V_1$, it is called a *step-down transformer*.

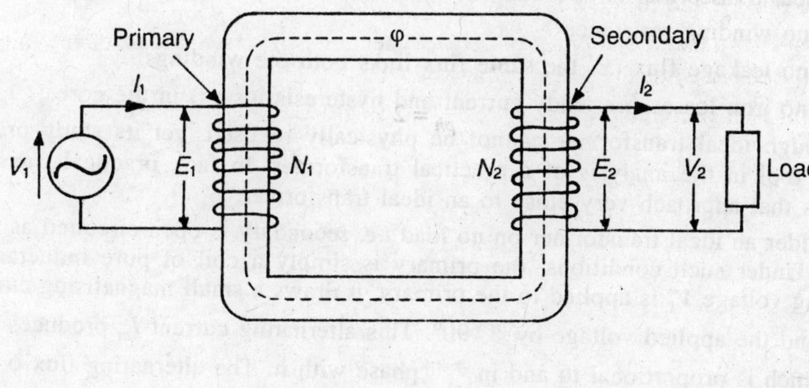

Fig. 20.1

Working. When an alternating voltage V_1 is applied to the primary, an alternating flux ϕ is set up in the core. This alternating flux links both the windings and induces e.m.f.s E_1 and E_2 in them according to Faraday's laws of electromagnetic induction. The e.m.f. E_1 is termed as primary e.m.f. and e.m.f. E_2 is termed as secondary e.m.f.

$$\text{Clearly, } E_1 = -N_1 \frac{d\phi}{dt}$$

$$\text{and} \qquad E_2 = -N_2 \frac{d\phi}{dt}$$

$$\therefore \qquad \frac{E_2}{E_1} = \frac{N_2}{N_1}$$

Note that magnitudes of E_2 and E_1 *depend upon the number of turns on the secondary and primary respectively. If $N_2 > N_1$, then $E_2 > E_1$ (or $V_2 > V_1$) and we get a step-up transformer. On the other hand, if $N_2 < N_1$, then $E_2 < E_1$ (or $V_2 < V_1$) and we get a step-down transformer. If load is connected across the secondary winding, the secondary e.m.f. E_2 will cause a current I_2 to flow through the load. Thus, a transformer enables us to transfer a.c. power from one circuit to another with a change in voltage level.

The following points may be noted carefully :

(*i*) The transformer action is based on the laws of electromagnetic induction.

(*ii*) There is no electrical connection between the primary and secondary. The a.c. power is transferred from primary to secondary through magnetic flux.

(*iii*) There is no change in frequency *i.e.* output power has the same frequency as the input power.

(*iv*) The losses that occur in a transformer are :

(*a*) core losses—eddy current and hysteresis losses.

(*b*) copper losses—in the resistance of the windings.

In practice, these losses are very small so that output power is nearly equal to the input primary power. In other words, a transformer has very high efficiency.

20.2. Theory of An Ideal Transformer

An ideal transformer is one that has

(*i*) no winding resistance

(*ii*) no leakage flux *i.e.* the same flux links both the windings

(*iii*) no iron losses (*i.e.* eddy current and hysteresis losses) in the core.

Although ideal transformer cannot be physically realised, yet its study provides a very powerful tool in the analysis of a practical transformer. In fact, practical transformers have properties that approach very close to an ideal transformer.

Consider an ideal transformer on no load *i.e.* secondary is open-circuited as shown in Fig. 20.2 (*i*). Under such conditions, the primary is simply a coil of pure inductance. When an alternating voltage V_1 is applied to the primary, it draws a small magnetising current I_m which lags behind the applied voltage by **90°. This alternating current I_m produces an alternating flux ϕ which is proportional to and in ***phase with it. The alternating flux ϕ links both the

* Since $d\phi/dt$ is common for both the windings.

** In a pure inductive circuit, current lags behind the voltage by 90°.

***Since there are no iron losses, flux ϕ is in phase with I_m.

windings and induces e.m.f. E_1 in the primary and e.m.f. E_2 in the secondary. The primary e.m.f. E_1 is, at every instant, equal to and in opposition to V_1 (Lenz's law). Both e.m.f.s E_1 and E_2 lag behind flux ϕ by 90° (See Art 20.3). However, their magnitudes depend upon the number of primary and secondary turns.

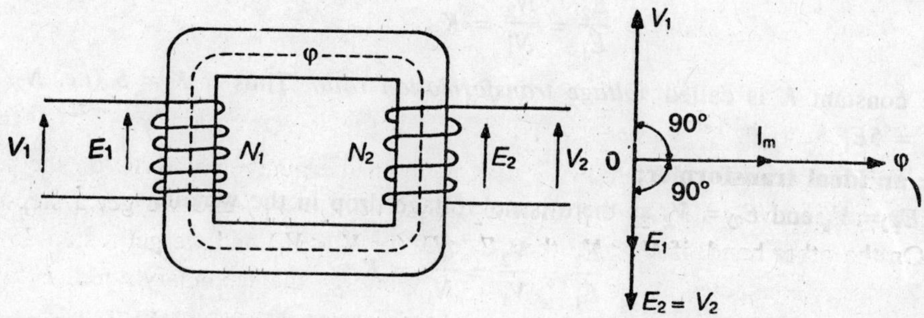

Fig. 20.2

Fig. 20.2 (*ii*) shows the phasor diagram of an ideal transformer on no load. Since flux ϕ is common to both the windings, it has been taken as the reference phasor. As shown in Art 20.3, the primary e.m.f. E_1 and secondary e.m.f. E_2 lag behind the flux ϕ by 90°. Note that E_1 and E_2 are in phase. But E_1 is equal to V_1 and 180° out of phase with it.

20.3. E.M.F. Equation of a Transformer

Consider that an alternating voltage V_1 of frequency f is applied to the primary as shown in Fig. 20.2 (*i*). The sinusoidal flux ϕ produced by the primary can be represented as:

$$\phi = \phi_m \sin \omega t$$

The instantaneous e.m.f. e_1 induced in the primary is

$$\begin{aligned} e_1 &= -N_1 \frac{d\phi}{dt} = -N_1 \frac{d}{dt}(\phi_m \sin \omega t) \\ &= -\omega N_1 \phi_m \cos \omega t \\ &= -2\pi f N_1 \phi_m \cos \omega t \\ &= 2\pi f N_1 \phi_m \sin(\omega t - 90°) \qquad \text{... } (i) \end{aligned}$$

It is clear from the above equation that maximum value of induced e.m.f. in the primary is :

$$E_{m1} = 2\pi f N_1 \phi_m$$

The r.m.s. value E_1 of the primary e.m.f. is :

$$E_1 = \frac{E_{m1}}{\sqrt{2}} = \frac{2\pi f N_1 \phi_m}{\sqrt{2}}$$

or $$E_1 = 4.44 f N_1 \phi_m$$

*Similarly $$E_2 = 4.44 f N_2 \phi_m$$

---

* The e.m.f. E_2 induced in the secondary is produced by the same flux (*i.e.* $\phi = \phi_m \sin \omega t$) that causes e.m.f. E_1. Thus the only difference in the r.m.s. values of the two is the difference in the number of turns.

In an ideal transformer, $E_1 = V_1$ and $E_2 = V_2$

Note : It is clear from exp (*i*) above that e.m.f. E_1 induced in the primary lags behind the flux ϕ by 90°. Likewise, e.m.f. E_2 induced in the secondary lags behind flux ϕ by 90°.

20.4. Voltage Transformation Ratio (*K*)

From the above equations of induced e.m.f., we have,

$$\frac{E_2}{E_1} = \frac{N_2}{N_1} = K$$

The constant K is called *voltage transformation ratio*. Thus if $K = 5$ (*i.e.* $N_2/N_1 = 5$), then $E_2 = 5E_1$

For an ideal transformer;

(*i*) $E_1 = V_1$ and $E_2 = V_2$ as there is no voltage drop in the windings.

$$\therefore \quad \frac{E_2}{E_1} = \frac{V_2}{V_1} = \frac{N_2}{N_1} = K$$

(*ii*) there are no losses. Therefore, volt-amperes input to the primary are equal to the output volt-amperes *i.e.*

$$V_1 I_1 = V_2 I_2$$

$$\text{or} \quad \frac{I_2}{I_1} = \frac{V_1}{V_2} = \frac{1}{K}$$

Hence, currents are in the inverse ratio of voltage transformation ratio. This simply means that if we raise the voltage, there is a corresponding decrease of current.

Example 20.1. *A 2000/200V, 20kVA transformer has 66 turns in the secondary. Calculate (i) primary turns (ii) primary and secondary full-load currents. Neglect the losses.*

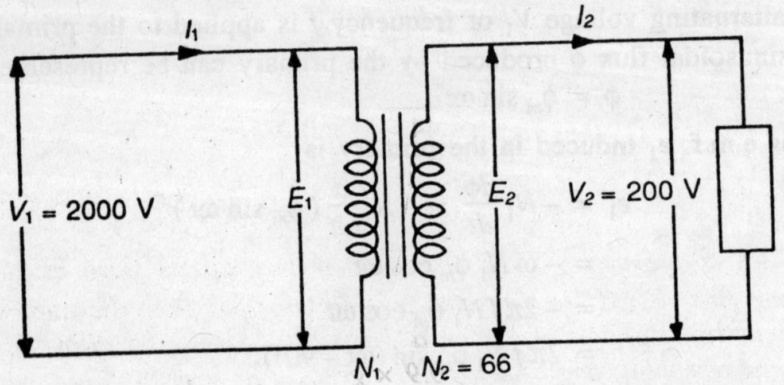

Fig. 20.3

Solution. Fig. 20.3 represents the conditions of the problem.

$$K = V_2/V_1 = 200/2000 = 1/10$$

(*i*)
$$\frac{N_2}{N_1} = K = \frac{1}{10}$$

$$\therefore \quad N_1 = N_2 \times 10 = 66 \times 10 = \mathbf{660\ turns}$$

(*ii*)
$$V_1 I_1 = V_2 I_2 = 20 \times 10^3$$

$$\therefore \quad I_2 = \frac{20 \times 10^3}{V_2} = \frac{20 \times 10^3}{200} = \mathbf{100A}$$

$$I_1 = \frac{20\times10^3}{V_1} = \frac{20\times10^3}{2000} = \mathbf{10A}$$

Alternatively, $I_1 = KI_2 = (1/10)\,100 = 10A$

Example 20.2. *An ideal 25kVA transformer has 500 turns on the primary winding and 40 turns on the secondary winding. The primary is connected to 3000V, 50Hz supply. Calculate (i) primary and secondary currents on full-load (ii) secondary e.m.f. and (iii) the maximum core flux.*

Solution.

$$K = N_2/N_1 = 40/500 = 4/50$$

(*i*) $$I_1 = \frac{\text{Volt-amperes}}{V_1} = \frac{25\times10^3}{3000} = \mathbf{8.33\ A}$$

$$I_2 = I_1/K = 8.33\times50/4 = \mathbf{104.2\ A}$$

(*ii*) $$E_2/E_1 = K$$

$\therefore$ $$E_2 = KE_1 = (4/50)\times3000 = \mathbf{240V}$$

(*iii*) $$E_1 = 4.44 f N_1 \phi_m$$

or $$3000 = 4.44\times50\times500\times\phi_m$$

$\therefore$ $$\phi_m = \frac{3000}{4.44\times50\times500} = 27\times10^{-3}\ Wb = \mathbf{27\ mWb}$$

Example 20.3. *A single phase 2200/250V, 50Hz transformer has a net core area of 36 cm^2 and a maximum flux density of 6Wb/m^2. Calculate the number of turns of primary and secondary.*

Solution.

$$E_1 = 4.44 f N_1 \phi_m$$

Here $$E_1 = 2200\ V;\ f = 50Hz;\ \phi_m = B_m\times A = 6\times36\times10^{-4} = \mathbf{0.0216\ Wb}$$

$\therefore$ $$N_1 = \frac{E_1}{4.44 f\phi_m} = \frac{2200}{4.44\times50\times0.0216} = \mathbf{459}$$

Also $$N_2 = \frac{E_2}{4.44 f\phi_m} = \frac{250}{4.44\times50\times0.0216} = \mathbf{52}$$

Example. 20.4. *A single phase 50Hz transformer has square core of 20 cm side. The permissible maximum flux density in the core is 1Wb/m^2. Calculate the number of turns per limb on the high and low voltage sides for a 3000/220V ratio. To allow for insulation of stampings, assume the net iron length to be 0.9 × gross iron length.*

Solution. Net iron length = $0.9\times20 = 18$ cm

Net *x*-sectional area, A = $18\times20 = 360\ \text{cm}^2 = 0.036\ \text{m}^2$

Max flux in core, $\phi_m = B_m\times A = 1\times0.036 = 0.036$ wb

$$E_1 = 4.44 f N_1 \phi_m$$

$\therefore$ $$N_1 = \frac{E_1}{4.44 f\phi_m} = \frac{3000}{4.44\times50\times0.036} = \mathbf{375}$$

$$N_2 = \frac{E_2}{4.44 f\phi_m} = \frac{220}{4.44\times50\times0.036} = \mathbf{27}$$

Note that one length is affected due to stamping.

Tutorial Problems

1. A 50kVA, 6600/250V transformer has 52 secondary turns. Find (*i*) the number of primary turns (*ii*) full-load primary and secondary currents. Neglect losses. **[(*i*) 1373 (*ii*) 7.58A; 200A]**
2. The net cross-sectional area of the core of 400/3000V, 50Hz transformer is 600 cm^2. If the maximum flux density in the core is 1.3 Wb/m^2, find the number of turns on the primary and secondary. **[24; 180]**
3. In a certain 50kVA transformer, the number of turns on the primary and secondary windings is 834 and 58 respectively. If primary is connected to a 3300V supply, find (*i*) secondary voltage (*ii*) the primary and secondary currents when the transformer is fully loaded. Neglect the losses. **[(*i*) 230V (*ii*) 15.2A; 218A]**
4. A single phase transformer has a ratio of 1 : 10 and a secondary winding of 1000 turns. The primary winding is connected to a 25-V sinusoidal supply. If the maximum value of flux in the core is 2.15 mWb, find (*i*) the frequency of the supply (*ii*) the number of primary turns (*iii*) the secondary voltage on open-circuit. **[(*i*) 25Hz (*ii*) 100 turns (*iii*) 250V]**
5. A single phase 50Hz transformer has 20 primary turns and 273 secondary turns. The net cross-sectional area of the core is 400 cm^2. If the primary winding is connected to 230V supply, find (*i*) peak value of flux density in the core (*ii*) voltage induced in the secondary winding. **[(*i*) 1.24 Wb/m^2 (*ii*) 3003V]**

20.5. Practical Transformer

A practical transformer differs from the ideal transformer in many respects. The practical transformer has (*i*) iron losses (*ii*) winding resistances and (*iii*) magnetic leakage, giving rise to leakage reactances.

(*i*) **Iron losses.** Since the iron core is subjected to alternating flux, there occurs eddy current and hysteresis loss in it. These two losses together are known as *iron losses* or *core losses*. The iron losses depend upon the supply frequency, maximum flux density in the core, volume of the core etc. It may be noted that magnitude of iron losses is quite small in a practical transformer.

(*ii*) **Winding resistances.** Since the windings consist of copper conductors, it immediately follows that both primary and secondary will have winding resistance. The primary resistance R_1 and secondary resistance R_2 act in series with the respective windings as shown in Fig. 20.4

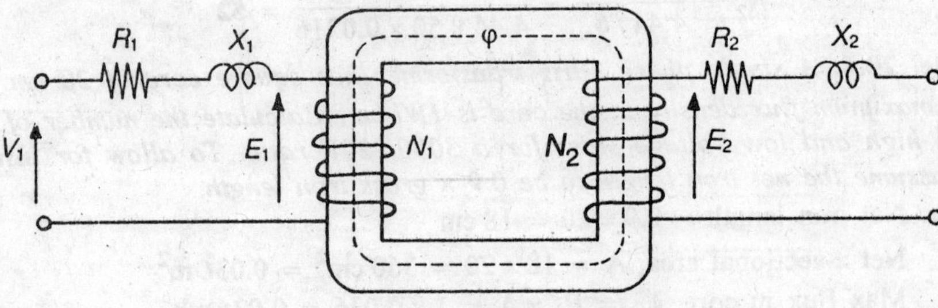

Fig. 20.4

(*iii*) **Leakage reactances.** Both primary and secondary currents produce flux. The flux ϕ which links both the windings is the useful flux and is called mutual flux. However, primary current would produce some flux ϕ_1 which would not link the secondary winding (See Fig. 20.5). Similarly, secondary current would produce some flux ϕ_2 that would not link the primary winding. The flux such as ϕ_1 or ϕ_2 which links only one winding is called *leakage flux*. The leakage flux paths are mainly through the air. The *effect of these leakage fluxes

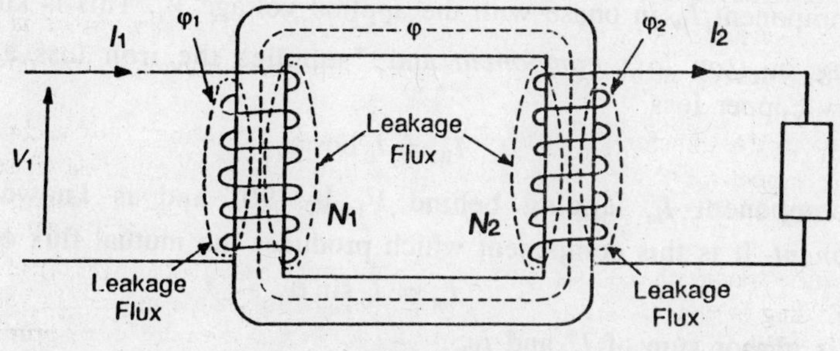

Fig. 20.5

would be the same as though inductive reactance were connected in series with each winding of transformer that had no leakage flux as shown in Fig. 20.4. In other words, the effect of primary leakage flux ϕ_1 is to introduce an inductive reactance X_1 in series with the primary winding as shown in Fig. 20.4. Similarly, the secondary leakage flux ϕ_2 introduces an inductive reactance X_2 in series with the secondary winding.

Note : Although leakage flux in a transformer is quite small (about 5% of ϕ) compared to the mutual flux ϕ, yet it cannot be ignored. It is because leakage flux paths are through air of high reluctance and hence require considerable m.m.f. It may be noted that energy is conveyed from the primary winding to the secondary winding by mutual flux ϕ which links both the windings.

20.6. Practical Transformer on NO Load

Consider a practical transformer on no load *i.e.* secondary on open-circuit as shown in Fig. 20.6 (*i*). The primary will draw a small current I_0 to supply (*i*) the iron losses and (*ii*) a very **small amount of copper loss in the primary. Hence the primary no load current I_0 is not 90° behind the applied voltage V_1 but lags it by an angle $\phi_0 < 90°$ as shown in the phasor diagram in Fig. 20.6 (*ii*).

No load input power, $W_0 = V_1 I_0 \cos \phi_0$

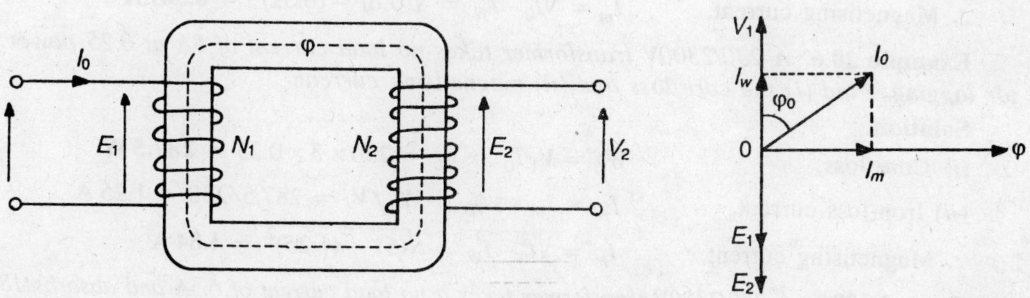

Fig. 20.6

* The primary leakage flux ϕ_1 is alternating one and hence induces back e.m.f. *e* in the primary winding.

$\therefore$ Primary leakage inductance, $L_1 = \dfrac{\text{Primary leakage flux linkages}}{\text{Primary current}} = \dfrac{N_1 \phi_1}{I_1}$

$\therefore$ Primary leakage reactance, $X_1 = 2\pi f L_1$

Also secondary leakage reactance, $X_2 = 2\pi f L_2$ where $L_2 = N_2 \phi_2 / I_2$

** At no load, copper loss in the primary is $I_0^2 R_1$. Since I_0 is quite small, this loss may be neglected.

As seen from the phasor diagram in Fig. 20.6 (*ii*), the no-load primary current I_0 can be resolved into two rectangular components *viz.*

(*i*) The component I_W in phase with the applied voltage V_1. This is known as *active* or *working* or *iron loss component* and *supplies the iron loss and a very small primary copper loss.

$$I_W = I_0 \cos \phi_0$$

(*ii*) The component I_m lagging behind V_1 by 90° and is known as *magnetising component*. It is this component which produces the mutual flux ϕ in the core.

$$I_m = I_0 \sin \phi_0$$

Clearly, I_0 is phasor sum of I_m and I_W.

$\therefore$ $$I_0 = \sqrt{I_m^2 + I_W^2}$$

No load p.f., $$\cos \phi_0 = I_W / I_0$$

It is emphasised here that no load primary copper loss (*i.e.* $I_0^2 R_1$) is very small and may be neglected. Therefore, the no load primary input power is practically equal to the iron loss in the transformer *i.e.*

No load input power, W_0 = Iron loss

Note : At no load, there is no current in the secondary so that $V_2 = E_2$. On the primary side, the drops in R_1 and X_1 due to I_0 are also very small because of the smallness of I_0. Hence, we can say that at no load, $V_1 = E_1$.

Example 20.5. *A transformer takes a current of 0.6A and absorbs 64W when primary is connected to its normal supply of 200V, 50Hz; the secondary being on open circuit. Find the magnetising and iron loss currents.*

Solution.

No load primary power, $W_0 = V_1 I_0 \cos \phi_0$

Iron loss component, $I_W = I_0 \cos \phi_0 = W_0 / V_1 = 64/200 = \mathbf{0.32\ A}$

No load current, $I_0 = \sqrt{I_m^2 + I_W^2}$

$\therefore$ Magnetising current, $I_m = \sqrt{I_0^2 - I_W^2} = \sqrt{(0.6)^2 - (0.32)^2} = \mathbf{0.507\ A}$

Example 20.6. *A 230/2300V transformer takes no load current of 5A at 0.25 power factor lagging. Find (i) the core loss and (ii) magnetising current.*

Solution.

(*i*) Core loss, $W_0 = V_1 I_0 \cos \phi_0 = 230 \times 5 \times 0.25 = \mathbf{287.5\ W}$

(*ii*) Iron-loss current, $I_W = I_0 \cos \phi_0 = W_0 / V_1 = 287.5/230 = 1.25\ A$

$\therefore$ Magnetising current, $I_m = \sqrt{I_0^2 - I_W^2} = \sqrt{(5)^2 - (1.25)^2} = \mathbf{4.84\ A}$

Example 20.7. *A 230/2300V transformer takes a no load current of 6.5A and absorbs 187 W. If the resistance of primary is 0.06 Ω, find (i) the core loss (ii) no load p.f. (iii) active component of current and (iv) magnetising current.*

Solution.

No-load loss, $W_0 = 187\ W$

Primary Cu loss due to no-load current $= I_0^2 R_1 = (6.5)^2 \times 0.06 = 2.5\ W$

---

* Remember that the component of current in phase with voltage produces the true or real power. However, component 90° out of phase with voltage produces reactive power *i.e.* flux in this case.

(*i*) Iron loss $= 187 - 2.5 = \mathbf{184.5\ W}$

(*ii*) No load p.f., $\cos\phi_0 = \dfrac{W_0}{V_1 I_0} = \dfrac{187}{230 \times 6.5} = \mathbf{0.125}\ lag$

(*iii*) Active component, $I_W = W_0 / V_1 = 187/230 = \mathbf{0.81\ A}$

(*iv*) Magnetising current, $I_m = \sqrt{I_0^2 - I_W^2} = \sqrt{(6.5)^2 - (0.81)^2} = \mathbf{6.4\ A}$

Tutorial Problems

1. A 2200/200V transformer takes a no load current of 0.6A and absorbs 400 watts. Find (*i*) the magnetising current and (*ii*) iron-loss current. **[(*i*) 0.572A (*ii*) 0.18A]**
2. A 2200/200V transformer takes 0.5A at a p.f. of 0.3 lagging at no load. Find (*i*) the magnetising current and (*ii*) iron-loss current. **[(*i*) 0.477A (*ii*) 0.15A]**
3. The no load current of a transformer is 5A at 0.3 p.f. lagging when supplied at 230V, 50Hz. The number of turns on the primary winding is 200. Calculate (*i*) the maximum value of flux in the core (*ii*) the core loss and (*iii*) the magnetising current. **[(*i*) 5.18 mwb (*ii*) 345W (*iii*) 4.77A]**

20.7. Ideal Transformer on Load

Let us connect a load Z_L across the secondary of an *ideal transformer as shown in Fig. 20.7(*i*). The secondary e.m.f. E_2 will cause a current I_2 to flow through the load.

$$I_2 \overset{**}{=} \frac{E_2}{Z_L} = \frac{V_2}{Z_L}$$

The angle at which I_2 leads or lags V_2 (or E_2) depends upon the resistance and reactance of the load. In the present case, we have considered inductive load so that current I_2 lags behind V_2 (or E_2) by ϕ_2.

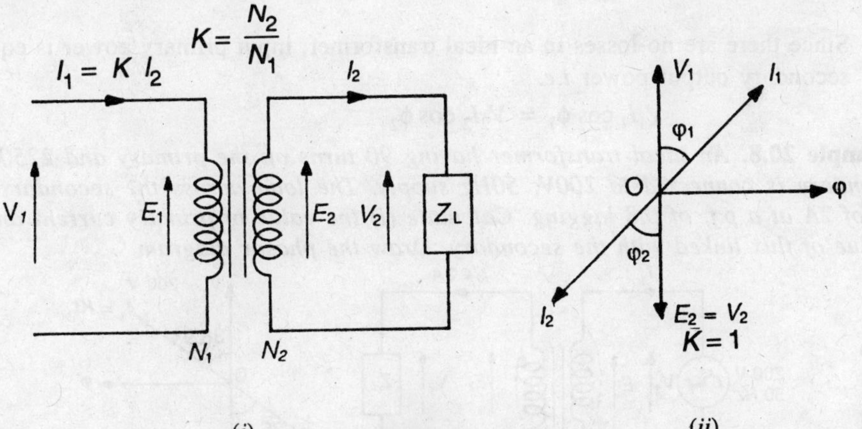

Fig. 20.7

The secondary current I_2 sets up an m.m.f. $N_2 I_2$ which produces a flux in the ***opposite direction to the flux ϕ originally set up in the primary by the magnetising

---

* Since practical transformers have full-load efficiencies well over 99%, they closely approach the ideal transformers.

** Since there is no voltage drop in an ideal transformer, $E_2 = V_2$.

***Lenz's law states that voltage induced in the secondary (*i.e.* E_2) acts in a direction so as to oppose what is causing it. This means that m.m.f. set up by the secondary current will oppose the mutual flux ϕ which is producing E_2.

current. This will change the flux in the core from the original value. However, the flux in the core should not *change from the original value. In order to fulfill this condition, the primary must develop an m.m.f. which exactly counterbalances the secondary m.m.f. $N_2 I_2$. Hence a primary current I_1 must flow such that :

$$N_1 I_1 = N_2 I_2$$

or $$I_1 = \frac{N_2}{N_1} I_2 = K I_2$$

Thus when a transformer is loaded and carries a secondary current I_2, then a current I_1 (= KI_2) must flow in the primary to maintain the m.m.f. balance. In other words, the primary must draw enough current to neutralise the demagnetising effect of secondary current so that mutual flux ϕ remains constant. Thus as the secondary current increases, the primary current I_1 (= KI_2) increases in unison and keeps the mutual flux ϕ constant. The power input, therefore, automatically increases with the output. For example if $K = 2$ and $I_2 = 2A$, then primary will draw a current $I_1 = KI_2 = 2 \times 2 = 4A$. If secondary current is increased to 4A, then primary current will become $I_1 = KI_2 = 2 \times 4 = 8A$.

Phasor diagram. Fig. 20.7 (*ii*) shows the phasor diagram of an ideal transformer on load. Note that in drawing the phasor diagram, the value of K has been assumed unity so that primary phasors are equal to secondary phasors. The secondary current I_2 lags behind V_2 (or E_2) by ϕ_2. It causes a primary current $I_1 = KI_2 = 1 \times I_2 = I_2$ which is in *antiphase* with it.

(*i*) $$\phi_1 = \phi_2$$

or $$\cos \phi_1 = \cos \phi_2$$

Thus, power factor on the primary side is equal to the power factor on the secondary side.

(*ii*) Since there are no losses in an ideal transformer, input primary power is equal to the secondary output power *i.e.*

$$V_1 I_1 \cos \phi_1 = V_2 I_2 \cos \phi_2$$

Example 20.8. *An ideal transformer having 90 turns on the primary and 2250 turns on the secondary is connected to 200V, 50Hz supply. The load across the secondary draws a current of 2A at a p.f. of 0.8 lagging. Calculate (i) the value of primary current and (ii) the peak value of flux linked with the secondary. Draw the phasor diagram.*

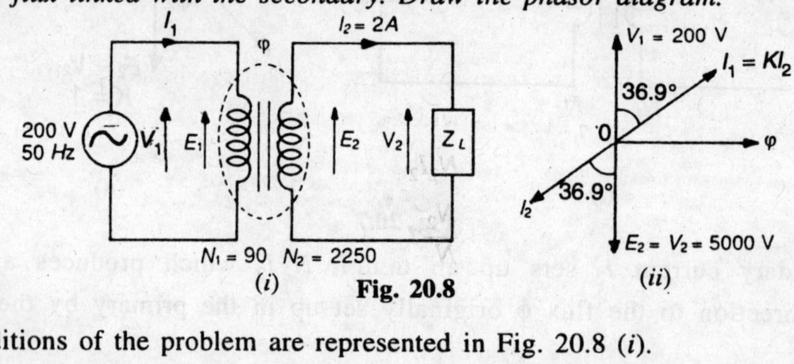

Fig. 20.8

Solution.

The conditions of the problem are represented in Fig. 20.8 (*i*).

$$K = N_2 / N_1 = 2250/90 = 25$$

* $E_1 = 4.44 f \varphi_m N_1$

* In an ideal transformer, $V_1 = E_1$. Since applied voltage V_1 is kept fixed, E_1 must remain unchanged. This is possible only if the flux remains fixed. Hence, mutual flux ϕ remains fixed whether a load is connected or not.

(*i*) $I_1 = KI_2 = 25 \times 2 = \mathbf{50\ A}$

(*ii*) $E_1 = 4.44 f N_1 \phi_m$

or $200 = 4.44 \times 50 \times 90 \times \phi_m$

$\therefore$ $\phi_m = \dfrac{200}{4.44 \times 50 \times 90} = \mathbf{0.01\ wb}$

This flux links both the windings.

Phasor diagram. Fig. 20.8 (*ii*) shows the phasor diagram of the transformer.

$$E_2 = KE_1 = 25 \times 200 = 5000\text{V}$$

$$\phi = \text{phase angle between } V_2 \text{ and } I_2$$

$$= \cos^{-1} 0.8 = 36.9°$$

The phase angle between V_1 and I_1 is also 36.9°.

Example 20.9. *An ideal transformer has 1000 turns on its primary and 500 turns on its secondary. The driving voltage on the primary side is 100V and the load resistance is 5Ω. Calculate V_2, I_1 and I_2.*

Solution.

$$K = N_2/N_1 = 500/1000 = 1/2$$

$$V_2 = KV_1 = (1/2) \times 100 = \mathbf{50\ V}$$

$$I_2 = V_2/R_2 = 50/5 = \mathbf{10\ A}$$

$$I_1 = KI_2 = (1/2) \times 10 = \mathbf{5\ A}$$

20.8. Practical Transformer on Load

We shall consider two cases (*i*) when such a transformer is assumed to have no winding resistance and leakage flux (*ii*) when the transformer has winding resistance and leakage flux.

(*i*) **No winding resistance and leakage flux**

Fig. 20.9. shows a practical transformer with the assumption that resistances and leakage reactances of the windings are negligible. With this assumption, $V_2 = E_2$ and $*V_1 = E_1$. Let us take the usual case of inductive load which causes the secondary current I_2 to lag the secondary voltage V_2 by ϕ_2. The total primary current I_1 must meet two requirements *viz.*

(*i*) It must supply the no-load current I_0 to meet the iron losses in the transformer and to provide flux in the core.

(*ii*) It must supply a current I_2' to counteract the demagnetising effect of secondary current I_2. The magnitude of I'_2 will be such that :

$$N_1 I_2' = N_2 I_2$$

or $$I'_2 = \frac{N_2}{N_1} I_2 = KI_2$$

The total primary current I_1 is the phasor sum of I_2' and I_0 *i.e.*

$$I_1 = I_2' + I_0 \qquad \text{... phasor sum}$$

where $$I_2' = -KI_2$$

Note that I_2' is 180° out of phase with I_2.

---

* However, V_1 and E_1 are 180° out of phase.

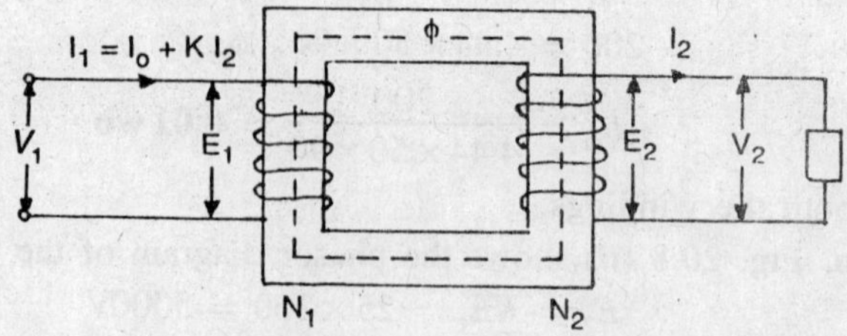

Fig. 20.9

Phasor diagram. Fig. 20.10 shows the phasor diagram for the usual case of inductive load. Both E_1 and E_2 lag behind the mutual flux ϕ by 90°. The current I_2' represents the primary current to neutralise the demagnetising effect of secondary current I_2. Now $I_2' = KI_2$ and is *antiphase* with I_2. I_0 is the no-load current of the transformer. The phasor sum of I_2' and I_0 gives the total primary current I_1. Note that in drawing the phasor diagram, the value of K is assumed to be unity so that primary phasors are equal to secondary phasors.

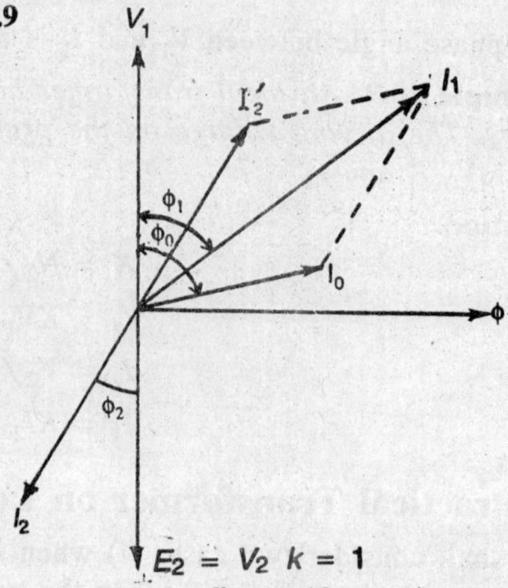

Fig. 20.10

Primary p.f. = $\cos\phi_1$; Secondary p.f. = $\cos\phi_2$

Primary input power = $V_1 I_1 \cos\phi_1$; Secondary output power = $V_2 I_2 \cos\phi_2$

Example 20.10. *A single phase transformer with a ratio of 440/110V takes no-load current of 5A at 0.2 p.f. lagging. If the secondary supplies a current of 120A at a p.f. 0.8 lagging, find the current taken by the primary.*

Solution. Fig. 20.11 shows the phasor diagram of the transformer.

$$\cos\phi_2 = 0.8 \quad \therefore \quad \phi_2 = 36.87°$$

$$\cos\phi_0 = 0.2 \quad \therefore \quad \phi_0 = 78.46°$$

Now $K = V_2/V_1 = 110/440 = 1/4$

$$I_2' = KI_2 = (1/4) \times 120 = 30 \text{ A}$$

$$I_0 = 5\text{A}$$

Angle between I_2' and I_0

$= 78.46° - 36.87° = 41.59°$

Using parallelogram method, we have,

$$I_1 = \sqrt{(I_0)^2 + (I_2')^2 + 2I_0 I_2' \cos 41.59°}$$

$$= \sqrt{(5)^2 + (30)^2 + 2 \times 5 \times 30 \times \cos 41.59°} = \mathbf{33.9\ A}$$

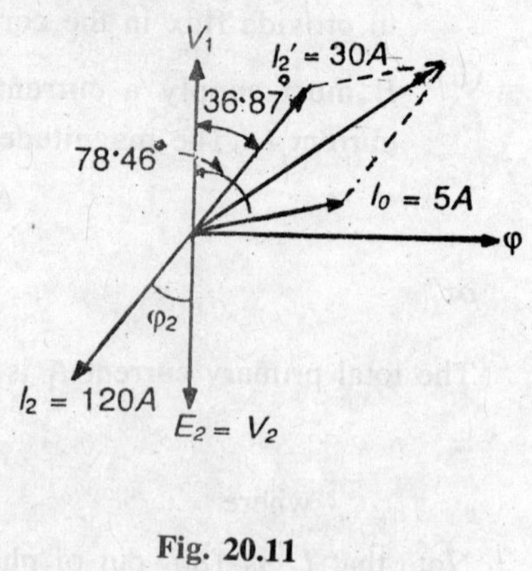

Fig. 20.11

Example 20.11. *A single phase transformer has 1000 turns on the primary and 200 turns on the secondary. The no-load current is 3A at a p.f. of 0.2 lagging. Calculate the primary current and power factor when the secondary current is 280A at a p.f. of 0.8 lagging.*

Solution. Fig 20.12 shows the phasor diagram of the transformer.

$$K = \frac{200}{1000} = \frac{1}{5}$$

$$\therefore \quad I_2' = KI_2 = \frac{1}{5} \times 280 = 56 \text{ A}$$

$$\cos\phi_2 = 0.8 \quad \therefore \quad \sin\phi_2 = 0.6$$

$$\cos\phi_0 = 0.2 \quad \therefore \quad \sin\phi_0 = 0.98$$

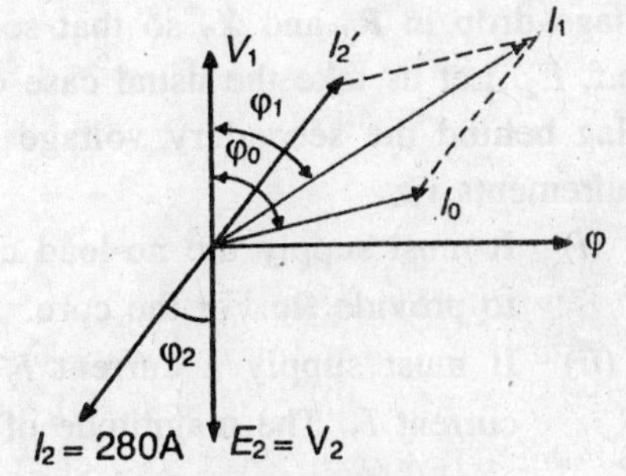

Fig. 20.12

Referring to Fig. 20.12 and resolving the currents along rectangular axes, we have,

$$I_1\cos\phi_1 = I_2'\cos\phi_2 + I_0\cos\phi_0$$

$$= (56 \times 0.8) + (3 \times 0.2) = 45.4 \text{ A}$$

$$I_1\sin\phi_1 = I_2'\sin\phi_2 + I_0\sin\phi_0$$

$$= (56 \times 0.6) + (3 \times 0.98) = 36.54 \text{ A}$$

Now,
$$I_1 = \sqrt{(I_1\cos\phi_1)^2 + (I_1\sin\phi_1)^2}$$

$$= \sqrt{(45.4)^2 + (36.54)^2} = \mathbf{58.3\ A}$$

$$\tan\phi_1 = \frac{I_1\sin\phi_1}{I_1\cos\phi_1} = \frac{36.54}{45.4} = 0.805$$

$$\therefore \quad \phi_1 = \tan^{-1} 0.805 = 38.83°$$

$\therefore$ Primary p.f. $= \cos\phi_1 = \cos 38.83° = \mathbf{0.78}$ *lag*

Example 20.12. *A single phase transformer with a ratio of 6600/600V has a load impedance of (4 + j3) ohms connected across the terminals of low voltage winding. Calculate the kW delivered to the load. Neglecting losses in the transformer, find the current taken by it from the supply lines.*

Solution.

(*i*) Load impedance, $Z = \sqrt{4^2 + 3^2} = 5\Omega$

Load p.f., $\cos\phi_2 = R/Z = 4/5 = 0.8$ lag

Load current, $I_2 = V_2/Z = 600/5 = 120$ A

Power delivered to load $= V_2 I_2 \cos\phi_2 = 600 \times 120 \times 0.8 = \mathbf{57.6 \times 10^3\ W}$

(*ii*) $K = V_2/V_1 = 600/6600 = 1/11$

Since losses are neglected, we have,

$$N_1 I_1 = N_2 I_2$$

or
$$I_1 = \left(\frac{N_2}{N_1}\right) I_2 = KI_2 = \left(\frac{1}{11}\right) \times 120 = \mathbf{10.91A}$$

Tutorial Problems

1. A 3300/240V single phase transformer on no load takes 2A at a p.f. of 0.25 lagging. Determine the primary current and power factor when the transformer is supplying a load of 60A at p.f. 0.9 lagging. **[4.43A, 1]**
2. A 4 : 1 ratio step-down transformer takes 1A at 0.15 p.f. on no load. Determine the primary current and power factor when the transformer is supplying a load of 2A at 0.8 p.f. lagging. Neglect voltage drops in the windings. **[7A ; 0.735** *lag***]**

3. A 400/200V single phase transformer is supplying a load of 50A at a p.f. of 0.866 lagging. The no-load current is 2A at 0.208 p.f. lagging. Calculate the primary current and power factor. **[26.4A; 0.838 *lag*]**

***(ii)* Transformer with resistance and leakage reactance**

Fig. 20.13 shows a practical transformer having winding resistances and leakage reactances. These are the *actual* conditions that exist in a transformer. There is voltage drop in R_1 and X_1 so that primary e.m.f. E_1 is less than the applied voltage V_1. Similarly, there is voltage drop in R_2 and X_2 so that secondary terminal voltage V_2 is less than the secondary e.m.f. E_2. Let us take the usual case of inductive load which causes the secondary current I_2 to lag behind the secondary voltage V_2 by ϕ_2. The total primary current I_1 must meet two requirements *viz.*

(*i*) It must supply the no-load current I_0 to meet the iron losses in the transformer and to provide flux in the core.

(*ii*) It must supply a current I_2' to counteract the demagnetising effect of secondary current I_2. The magnitude of I_2' will be such that :

$$N_1 I_2' = N_2 I_2$$

or

$$I_2' = \frac{N_2}{N_1} I_2 = K I_2$$

The total primary current I_1 will be the phasor sum of I_2' and I_0. *i.e.*

$$\mathbf{I}_1 = \mathbf{I}_2' + \mathbf{I}_0 \quad \text{where} \quad \mathbf{I}_2' = -\mathbf{K}\mathbf{I}_2$$

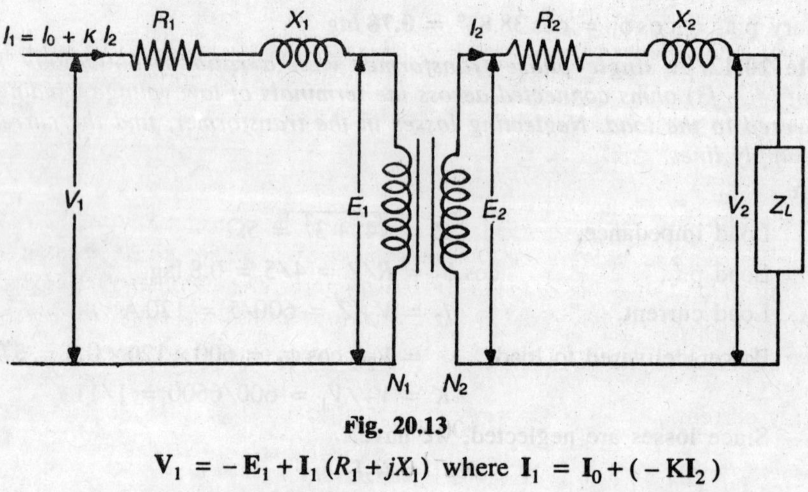

Fig. 20.13

$$\mathbf{V}_1 = -\mathbf{E}_1 + \mathbf{I}_1 (R_1 + jX_1) \text{ where } \mathbf{I}_1 = \mathbf{I}_0 + (-\mathbf{K}\mathbf{I}_2)$$

$$= -\mathbf{E}_1 + \mathbf{I}_1 \mathbf{Z}_1$$

$$\mathbf{V}_2 = \mathbf{E}_2 - \mathbf{I}_2 (R_2 + jX_2)$$

$$= \mathbf{E}_2 - \mathbf{I}_2 \mathbf{Z}_2$$

Phasor diagram. Fig. 20.14 shows the phasor diagram of a practical transformer for the usual case of inductive load. Both E_1 and E_2 lag the mutual flux ϕ by 90°. The current I_2' represents the primary current to neutralise the demagnetising effect of secondary current I_2. Now $I_2' = KI_2$ and is opposite to I_2. Also I_0 is the no-load current of the transformer. The phasor sum of I_2' and I_0 gives the total primary current I_1.

Note that counter e.m.f. that opposes the applied voltage $\mathbf{V}_1$ is $-\mathbf{E}_1$. Therefore, if we add $\mathbf{I}_1 R_1$ (in phase with $\mathbf{I}_1$) and $\mathbf{I}_1 X_1$ (90° ahead of $\mathbf{I}_1$) to $-\mathbf{E}_1$, we get the applied primary voltage $\mathbf{V}_1$. The phasor $\mathbf{E}_2$ represents the induced e.m.f. in the secondary by the mutual flux ϕ. The secondary terminal voltage $\mathbf{V}_2$ will be what is left over after subtracting $\mathbf{I}_2 R_2$ and $\mathbf{I}_2 X_2$ from $\mathbf{E}_2$.

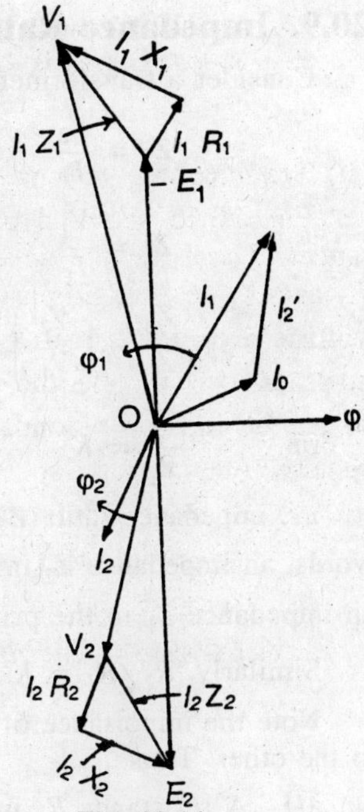

Fig. 20.14

Load power factor $= \cos \phi_2$

Primary power factor $= \cos \phi_1$

Input power to transformer, $P_1 = V_1 I_1 \cos \phi_1$

Output power of transformer, $P_2 = V_2 I_2 \cos \phi_2$

Note : The reader may draw the phasor diagram of a loaded transformer for (*i*) unity p.f. and (*ii*) leading p.f. as an exercise.

Example 20.13. *The primary of a 1000/250V transformer has a resistance of 0.15 Ω and leakage reactance of 0.8Ω. Find the primary induced e.m.f. when the primary current is 60A at 0.8 p.f. lagging.*

Solution.

Primary impedance, $\mathbf{Z}_1 = 0.15 + j\,0.8 = 0.814 \angle 79.6°\ \Omega$

Power factor angle, $\phi_1 = \cos^{-1} 0.8 = 36.9°$

Taking applied voltage as the reference phasor, we have, $\mathbf{V}_1 = 1000 \angle 0°$

Now, $-\mathbf{E}_1 = \mathbf{V}_1 - \mathbf{I}_1 \mathbf{Z}_1$

$$= 1000 \angle 0° - 60 \angle -36.9° \times 0.814 \angle 79.6°$$
$$= 1000 \angle 0° - 48.8 \angle 42.7°$$
$$= 1000 - (36 + j33)$$
$$= 964 - j33 = 964.5 \angle -2°$$

∴ Primary e.m.f., $\mathbf{E}_1 = -964 + j33 = \mathbf{964.5 \angle 178°\ V}$

Example 20.14. *The voltage on the secondary of a single phase transformer is 200V when supplying a load of 8 kW at a p.f. of 0.8 lagging. The secondary resistance is 0.04 Ω and secondary leakage reactance is 0.8 Ω. Calculate the induced e.m.f. in the secondary.*

Solution. Taking secondary voltage as the reference phasor, we have $\mathbf{V}_2 = 200 \angle 0°$.

Secondary current, $I_2 = \dfrac{8 \times 10^3}{200 \times 0.8} = 50\ \text{A}$

Power factor angle, $\phi_2 = \cos^{-1} 0.8 = 36.9°$

∴ $\mathbf{I}_2 = 50 \angle -36.9°\ \text{A}$

$$\mathbf{Z}_2 = 0.04 + j\,0.8 = 0.8 \angle 87.14°$$
$$\mathbf{E}_2 = \mathbf{V}_2 + \mathbf{I}_2 \mathbf{Z}_2$$
$$= 200 \angle 0° + 50 \angle -36.9° \times 0.8 \angle 87.14°$$
$$= 200 \angle 0° + 40 \angle 50.24°$$
$$= (200 + j\,0) + (25.58 + j30.75)$$
$$= 225.58 + j\,30.75 = \mathbf{227.67 \angle 7.8°\ volts}$$

Thus the secondary e.m.f. $\mathbf{E}_2$ leads the secondary terminal voltage $\mathbf{V}_2$ by 7.8°.

20.9. Impedance Ratio

Consider a transformer having impedance Z_2 in the secondary as shown in Fig. 20.15.

$$Z_2 = \frac{V_2}{I_2}$$

$$Z_1 = \frac{V_1}{I_1}$$

$$\therefore \quad \frac{Z_2}{Z_1} = \left(\frac{V_2}{V_1}\right) \times \left(\frac{I_1}{I_2}\right)$$

$$\text{or} \quad \frac{Z_2}{Z_1} = K^2$$

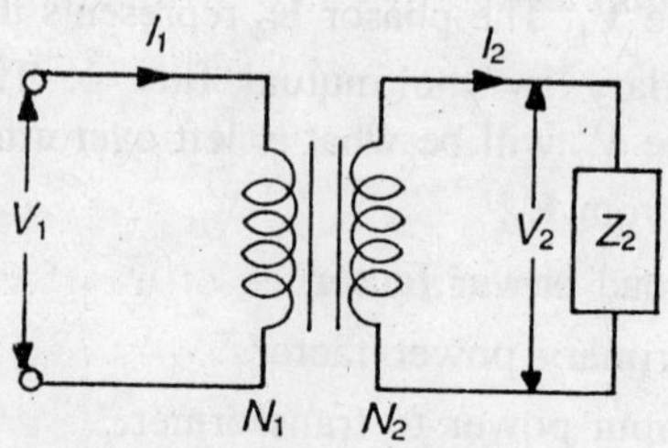

Fig. 20.15

i.e. impedance ratio (Z_2/Z_1) is equal to the square of voltage transformation ratio. In other words, an impedance Z_2 in secondary becomes Z_2/K^2 when transferred to primary. Likewise, an impedance Z_1 in the primary becomes $K^2\ Z_1$ when transferred to the secondary.

Similarly, $R_2/R_1 = K^2$ and $X_2/X_1 = K^2$

Note the importance of above relations. We can transfer the parameters from one winding to the other. Thus :

(*i*) A resistance R_1 in the primary becomes K^2R_1 when transferred to the secondary.

(*ii*) A resistance R_2 in the secondary becomes R_2/K^2 when transferred to the primary.

(*iii*) A reactance X_1 in the primary becomes K^2X_1 when transferred to the secondary.

(*iv*) A reactance X_2 in the secondary becomes X_2/K^2 when transferred to the primary.

Note : It is important to remember that :

(*i*) When transferring resistance or reactance from primary to secondary, *multiply* it by K^2.

(*ii*) When transferring resistance or reactance from secondary to primary, *divide* it by K^2.

(*iii*) When transferring voltage or current from one winding to the other, only *K is used.

20.10. Shifting Impedances in a Transformer

Fig. 20.16 shows a transformer where resistances and reactances are shown external to the windings. The resistance and reactance of one winding can be transferred to the other by appropriately using the factor K^2. This makes the analysis of the transformer a simple affair because then we have to work in one winding only.

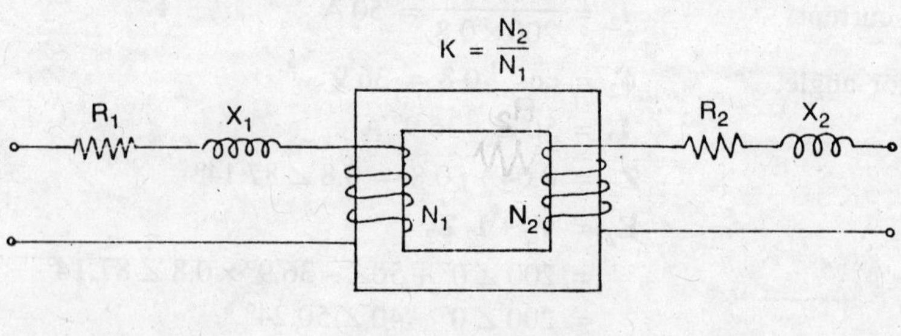

Fig. 20.16

* Thus any voltage V_1 in the primary becomes KV_1 in the secondary. On the other hand, any voltage V_2 in the secondary becomes V_2/K in the primary. Again a current I_1 in the primary becomes I_1/K in the secondary. Any current I_2 in the secondary becomes KI_2 in the primary.

(*i*) **Referred to primary.** When secondary resistance or reactance is transferred to the primary, it is **divided** by K^2. It is then called *equivalent secondary resistance or reactance referred to primary* and is denoted by R_2' or X_2'.

Equivalent resistance of transformer referred to primary

$$R_{01} = R_1 + R_2' = R_1 + R_2 / K^2$$

Equivalent reactance of transformer referred to primary

$$X_{01} = X_1 + X_2' = X_1 + X_2 / K^2$$

Equivalent impedance of transformer referred to primary

$$Z_{01} = \sqrt{R_{01}^2 + X_{01}^2}$$

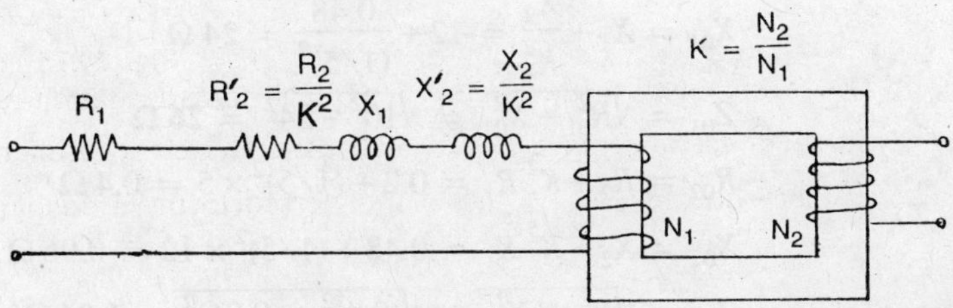

Fig. 20.17

Fig. 20.17 shows the resistance and reactance of the secondary referred to the primary. Note that secondary now has no resistance or reactance.

(*ii*) **Referred to secondary.** When primary resistance or reactance is transferred to the secondary, it is **multiplied** by K^2. It is then called *equivalent primary resistance or reactance referred to the secondary* and is denoted by R_1' or X_1'.

Equivalent resistance of transformer referred to secondary

$$R_{02} = R_2 + R_1'$$
$$= R_2 + K^2 R_1$$

Equivalent reactance of transformer referred to secondary

$$X_{02} = X_2 + X_1'$$
$$= X_2 + K^2 X_1$$

Equivalent impedance of transformer referred to secondary

$$Z_{02} = \sqrt{R_{02}^2 + X_{02}^2}$$

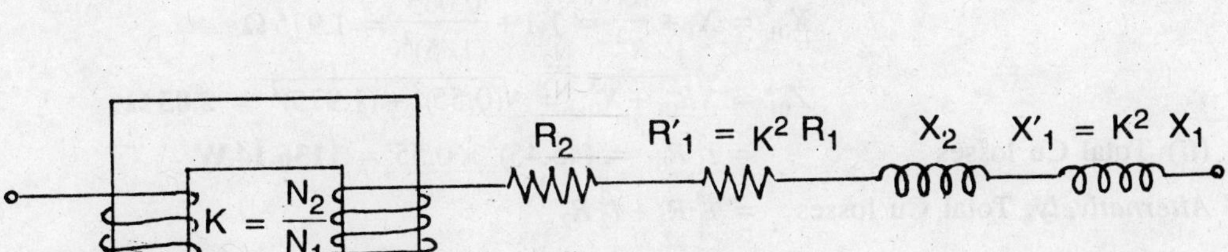

Fig. 20.18

Fig. 20.18 shows the resistance and reactance of the primary referred to the secondary. Note that primary now has no resistance or reactance.

Example 20.15. *A 10kVA, 2000/400V single phase transformer has* $R_1 = 5\,\Omega$; $X_1 = 12\,\Omega$; $R_2 = 0.2\,\Omega$ *and* $X_2 = 0.48\,\Omega$. *Determine the equivalent impedance of the transformer referred to (i) primary side (ii) secondary side.*

Solution.

$$K = 400/2000 = 1/5$$

(*i*)
$$R_{01} = R_1 + \frac{R_2}{K^2} = 5 + \frac{0.2}{(1/5)^2} = 10\,\Omega$$

$$X_{01} = X_1 + \frac{X_2}{K^2} = 12 + \frac{0.48}{(1/5)^2} = 24\,\Omega$$

$$\therefore \quad Z_{01} = \sqrt{R_{01}^2 + X_{01}^2} = \sqrt{10^2 + 24^2} = \mathbf{26\,\Omega}$$

(*ii*)
$$R_{02} = R_2 + K^2 R_1 = 0.2 + (1/5)^2 \times 5 = 0.4\,\Omega$$

$$X_{02} = X_2 + K^2 X_1 = 0.48 + (1/5)^2 \times 12 = 0.96\,\Omega$$

$$\therefore \quad Z_{02} = \sqrt{R_{02}^2 + X_{02}^2} = \sqrt{(0.4)^2 + (0.96)^2} = \mathbf{1.04\,\Omega}$$

Alternatively;
$$Z_{02} = K^2 Z_{01} = (1/5)^2 \times 26 = 1.04\,\Omega$$

Example 20.16. *A 100kVA, 2200/440V transformer has* $R_1 = 0.3\,\Omega$; $X_1 = 1.1\,\Omega$; $R_2 = 0.01\,\Omega$ *and* $X_2 = 0.035\,\Omega$. *Calculate (i) the equivalent impedance of the transformer referred to the primary and (ii) total copper losses.*

Solution.

Voltage transformation ratio, $K = 440/2200 = 1/5$

Assuming the efficiency of transformer to be 100 %,

Full-load primary current, $$I_1 = \frac{100 \times 1000}{2200} = 45.45\text{ A}$$

Full-load secondary current, $$I_2 = \frac{100 \times 1000}{440} = 227.25\text{ A}$$

Alternatively; $$I_2 = I_1 / K = 45.45 \times 5 = 227.25\text{ A}$$

(*i*)
$$R_{01} = R_1 + \frac{R_2}{K^2} = 0.3 + \frac{0.01}{(1/5)^2} = 0.55\,\Omega$$

$$X_{01} = X_1 + \frac{X_2}{K^2} = 1.1 + \frac{0.035}{(1/5)^2} = 1.975\,\Omega$$

$$\therefore \quad Z_{01} = \sqrt{R_{01}^2 + X_{01}^2} = \sqrt{(0.55)^2 + (1.975)^2} = \mathbf{2.05\,\Omega}$$

(*ii*) Total Cu losses $= I_1^2 R_{01} = (45.45)^2 \times 0.55 = \mathbf{1136.14\,W}$

Alternatively; Total Cu losses $= I_1^2 R_1 + I_2^2 R_2$

$$= (45.45)^2 \times 0.3 + (227.25)^2 \times 0.01 = 1136.14\text{W}$$

20.11. Importance of Shifting Impedances

If we shift all the impedances from one winding to the other, the transformer is eliminated and we get an equivalent electrical circuit. Various voltages and currents can be readily obtained by solving this electrical circuit.

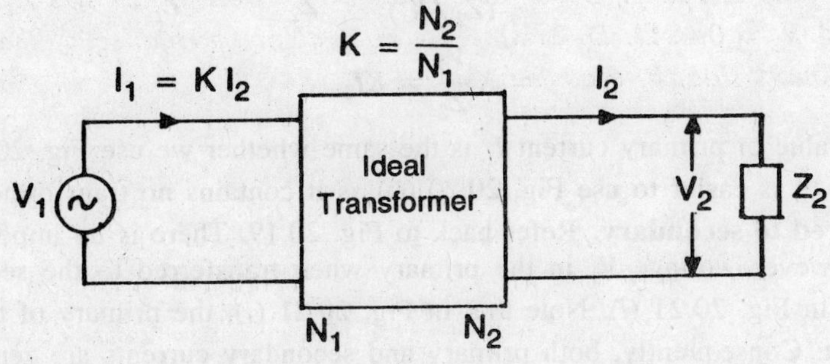

Fig. 20.19

Consider an *ideal transformer having an impedance Z_2 in the secondary as shown in Fig. 20.19.

(*i*) **Referred to primary.** When impedance Z_2 in the secondary is transferred to the primary, it becomes Z_2/K^2 as shown in Fig. 20.20 (*i*). Note that in Fig. 20.20 (*i*), the secondary of the ideal transformer is on open-circuit. Consequently, both primary and secondary currents are **zero. We can, therefore, remove the transformer, yielding the equivalent circuit shown in Fig. 20.20 (*ii*). The primary current can now be readily found out.

$$I_1 = \frac{V_1}{(Z_2/K^2)}$$

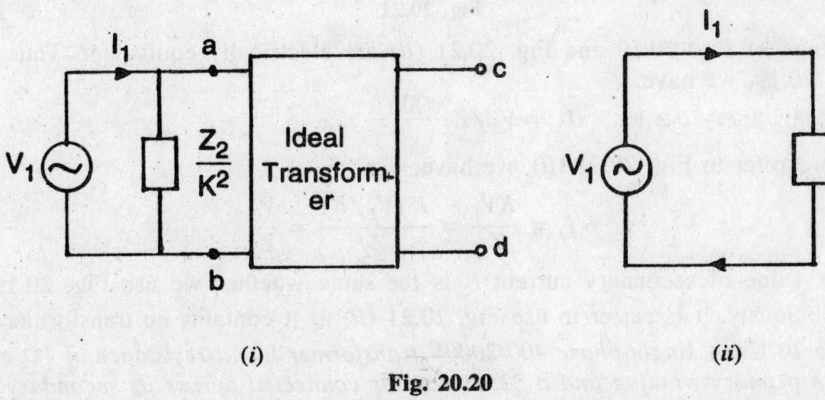

Fig. 20.20

The circuits of Fig. 20.19 and Fig. 20.20 (*ii*) are electrically equivalent. Thus referring to Fig. 20.19,

$$I_1 = KI_2$$

---

* A practical transformer consists of an ideal transformer with various resistances and reactances external to it. The treatment will, therefore, be equally valid for a practical transformer.

** The ideal transformer is located between primary terminals *ab* and secondary terminals *cd* [See Fig. 20.20 (*i*)]. Terminals *c, d* are open and hence secondary current is zero. Since secondary current is zero, primary current will also be zero *i.e.* no current will flow through terminals *a, b*. The ideal transformer can thus be removed.

Also if we refer to Fig. 20.20 (*ii*), we have,

$$I_1 = \frac{V_1}{(Z_2/K^2)} = \frac{K^2 V_1}{Z_2} = \frac{K(KV_1)}{Z_2}$$

$$= K\frac{V_2}{Z_2} = KI_2 \qquad (\because I_2 = V_2/Z_2)$$

Thus the value of primary current I_1 is the same whether we use Fig. 20.19 or Fig. 20.20 (*ii*). Obviously, it is easier to use Fig. 20.20 (*ii*) as it contains no transformer.

(*ii*) **Referred to secondary.** Refer back to Fig. 20.19. There is no impedance on the primary side. However, voltage V_1 in the primary when transferred to the secondary becomes KV_1 as shown in Fig. 20.21 (*i*). Note that in Fig. 20.21 (*i*), the primary of the transformer is on open circuit. Consequently, both primary and secondary currents are zero. As before, we can remove the transformer, yielding the equivalent circuit shown in Fig. 20.21 (*ii*). The secondary current I_2 can be readily found out as :

$$I_2 = \frac{KV_1}{Z_2}$$

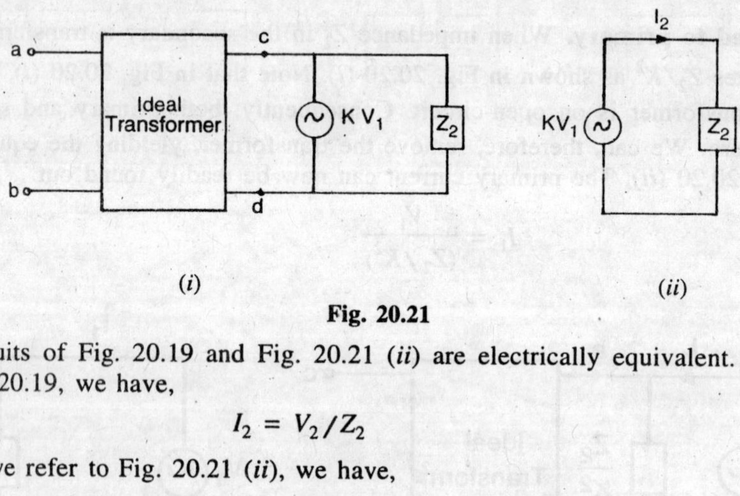

(*i*) (*ii*)

Fig. 20.21

The circuits of Fig. 20.19 and Fig. 20.21 (*ii*) are electrically equivalent. Thus referring back to Fig. 20.19, we have,

$$I_2 = V_2/Z_2$$

Also if we refer to Fig. 20.21 (*ii*), we have,

$$I_2 = \frac{KV_1}{Z_2} = \frac{K(V_2/K)}{Z_2} = \frac{V_2}{Z_2}$$

Thus the value of secondary current I_2 is the same whether we use Fig. 20.19 or Fig. 20.21 (*ii*). Obviously, it is easier to use Fig. 20.21 (*ii*) as it contains no transformer.

Example 20.17. *A single phase 400/2000V transformer has a resistance of 1Ω connected in series with primary winding and a 225Ω resistor connected across its secondary winding. Calculate the primary current when the circuit is supplied at 400V.*

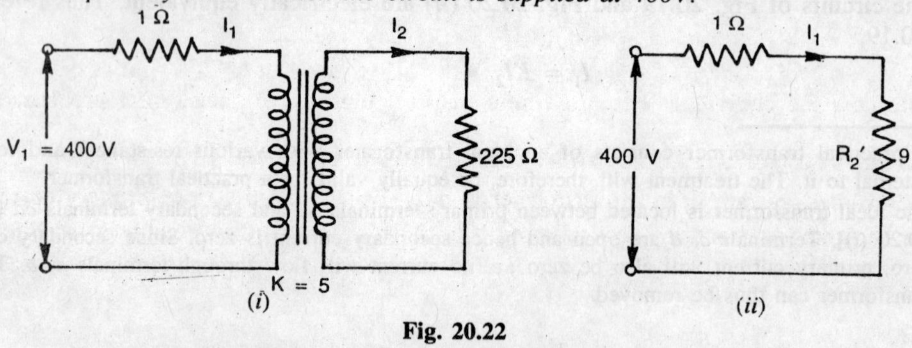

(*i*) (*ii*)

Fig. 20.22

Solution. The conditions of the problem are represented in Fig. 20.22 (*i*).

$$K = 2000/400 = 5\ ;\ R_1 = 1\Omega\ ;\ R_2 = 225\ \Omega$$

Equivalent resistance of transformer referred to primary,

$$R_{01} = R_1 + R_2'$$

$$= R_1 + \frac{R_2}{K^2} = 1 + \frac{225}{(5)^2} = 10\ \Omega$$

The equivalent circuit of the transformer referred to primary is shown in Fig. 20.22 (*ii*). This is a simple series circuit.

Primary current, $$I_1 = \frac{V_1}{R_{01}} = \frac{V_1}{R_1 + R_2'} = \frac{400}{1+9} = \mathbf{40A}$$

Example 20.18. *Calculate the secondary voltage V and primary current I in the circuit shown in Fig. 20.23 (i). The transformer has primary to secondary turn ratio of 1 :100.*

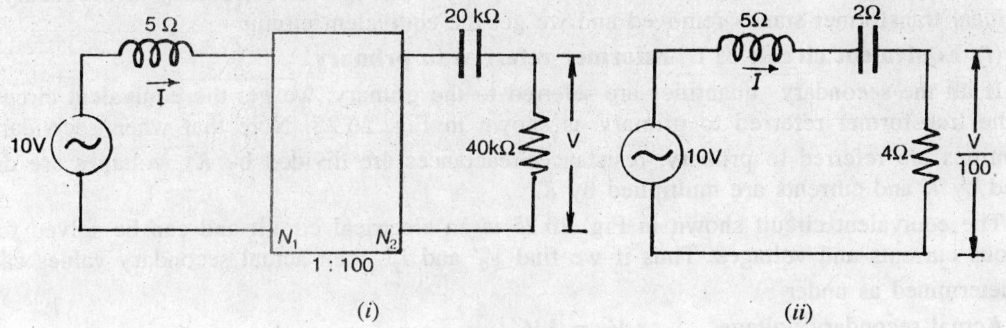

Fig. 20.23

Solution.

$$K = 100/1 = 100\ ;\quad Z_2 = (40000 - j\,20000)\Omega$$

Equivalent impedance of secondary referred to primary,

$$Z_2' = \frac{Z_2}{K^2} = \frac{1}{(100)^2}(40000 - j\,20000) = (4 - j2)\Omega$$

The equivalent circuit of the transformer referred to primary is shown in Fig. 20.23 (*ii*). Note that voltage across 40 K Ω resistance in Fig. 20.23 (*i*) is *V*. But when this resistance is transferred to the primary, the voltage across this resistance becomes *V/K* or *V*/100.

Referring to the equivalent circuit in Fig. 20.23 (*ii*), we have,

Circuit impedance, $Z = \sqrt{R^2 + (X_L - X_C)^2} = \sqrt{4^2 + (5-2)^2} = 5\Omega$

∴ Primary current, $I = 10/Z = 10/5 = \mathbf{2A}$

Now, $$\frac{V}{100} = I \times 4 = 2 \times 4$$

∴ $$V = (2 \times 4) \times 100 = \mathbf{800V}$$

Note how the equivalent circuit permits easy calculations of transformer problems.

20.12. Approximate Equivalent Circuit of a Loaded Transformer

The no-load current I_0 in a transformer is only 1 – 3% of the rated primary current and may be neglected without any serious error. The transformer can then be shown as Fig. 20.24.

This is an approximate representation because no-load current has been neglected. Note that all the circuit elements have been shown external so that the transformer is the ideal one.

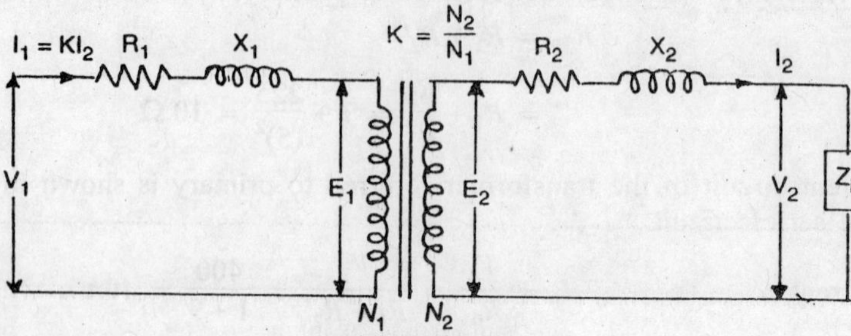

Fig. 20.24

As shown in Art 20.11, if we refer all the quantities to one side (primary or secondary), the ideal transformer stands removed and we get the equivalent circuit.

(*i*) Equivalent circuit of transformer referred to primary

If all the secondary quantities are referred to the primary, we get the equivalent circuit of the transformer referred to primary as shown in Fig. 20.25. Note that when secondary quantities are referred to primary, resistances/reactances are divided by K^2, voltages are divided by K and currents are multiplied by K.

The equivalent circuit shown in Fig. 20.25 is an electrical circuit and can be solved for various currents and voltages. Thus if we find V_2' and I_2', then actual secondary values can be determined as under :

Actual secondary voltage, $V_2 = KV_2'$

Actual secondary current, $I_2 = I_2'/K$

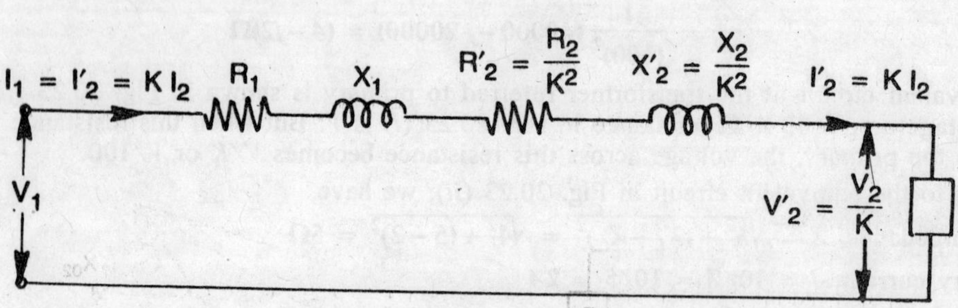

Fig. 20.25

(*ii*) Equivalent circuit of transformer referred to secondary

If all the primary quantities are referred to secondary, we get the equivalent circuit of the transformer referred to secondary as shown in Fig. 20.26. Note that when primary quantities are referred to secondary, resistances/reactances are multiplied by K^2, voltages are multiplied by K and currents are divided by K.

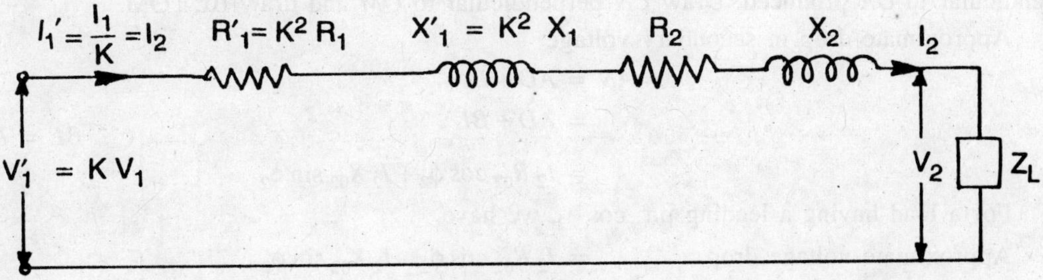

Fig. 20.26

The equivalent circuit shown in Fig. 20.26 is an electrical circuit and can be solved for various voltages and currents. Thus if we find V_1' and I_1', then actual primary values can be determined as under :

Actual primary voltage, $\quad V_1 = V_1'/K$

Actual primary current, $\quad I_1 = KI_1'$

Note. The same final answers will be obtained whether we use the equivalent circuit referred to primary or secondary. The use of a particular equivalent circuit would depend upon the conditions of the problem.

20.13. Approximate Voltage Drop in a Transformer

The approximate equivalent circuit of transformer referred to secondary is shown in Fig. 20.27. At no-load, the secondary voltage is KV_1. When a load having a lagging p.f. $\cos \phi_2$ is applied, the secondary carries a current I_2 and voltage drops occur in $(R_2 + K^2R_1)$ and $(X_2 + K^2X_1)$. Consequently, the secondary voltage falls from KV_1 to V_2. Referring to Fig. 20.27, we have,

$$\mathbf{V}_2 = K\mathbf{V}_1 - \mathbf{I}_2\,[(R_2 + K^2 R_1) + j\,(X_2 + K^2 X_1)]$$

$$= \mathbf{KV}_1 - \mathbf{I}_2\,(R_{02} + j\,X_{02})$$

$$= \mathbf{KV}_1 - \mathbf{I}_2\,\mathbf{Z}_{02}$$

Drop in secondary voltage $= \mathbf{KV}_1 - \mathbf{V}_2 = \mathbf{I}_2\,\mathbf{Z}_{02}$

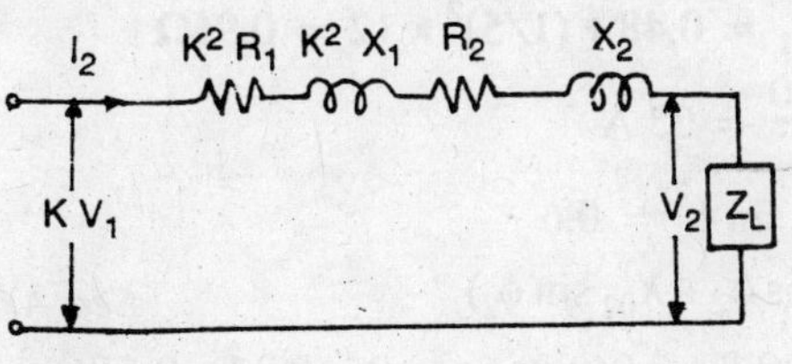

Fig. 20.27

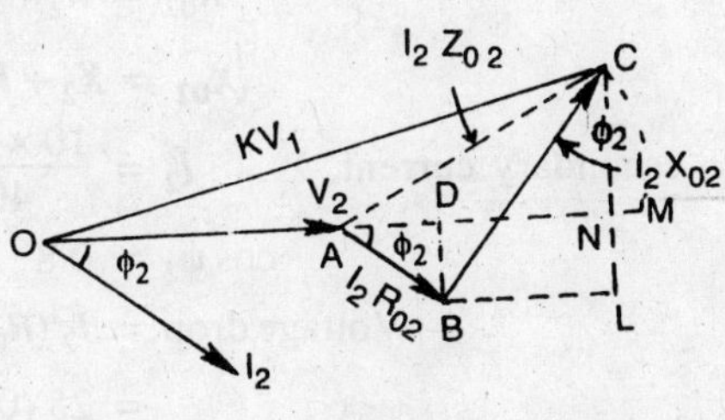

Fig. 20.28

The phasor diagram is shown in Fig. 20.28. It is clear from the phasor diagram that drop in secondary voltage is $AC = I_2 Z_{02}$. It can be found as follows. With O as centre and OC as radius, draw an arc cutting *OA* produced at *M*. Then $AC = AM \simeq AN$. From *B*, draw *BD* perpendicular to *OA* produced. Draw *CN* perpendicular to *OM* and draw BL || OM.

Approximate drop in secondary voltage

$$= AN = AD + DN$$

$$= AD + BL \qquad (\because BL = DN)$$

$$= I_2 R_{02} \cos\phi_2 + I_2 X_{02} \sin\phi_2$$

For a load having a leading p.f. $\cos\phi_2$, we have,

Approximate voltage drop $= I_2 R_{02} \cos\phi_2 - I_2 X_{02} \sin\phi_2$

Note : If the circuit is referred to primary, then it can be easily established that :

Approximate voltage drop $= I_1 R_{01}$ *$\cos\phi_2 \pm I_1 X_{01} \sin\phi_2$

20.14. Voltage Regulation

*The **voltage regulation** of a transformer is the arithmetic difference (not phasor difference) between the no-load secondary voltage ($_0V_2$) and the secondary voltage V_2 on load expressed as percentage of no-load voltage i.e.*

$$\%\text{age voltage regulation} = \frac{_0V_2 - V_2}{_0V_2} \times 100$$

where $_0V_2$ = No-load secondary voltage $= KV_1$

V_2 = Secondary voltage on load

As shown in Art 20.13, $_0V_2 - V_2 = I_2 R_{02} \cos\phi_2 \pm I_2 X_{02} \sin\phi_2$

The +ve sign is for lagging p.f. and –ve sign for leading p.f.

It may be noted that %age voltage regulation of the transformer will be the same whether primary or secondary side is considered.

Example 20.19. *A 10 kVA, 2000/400V single phase transformer has the following data:*

$$R_1 = 5\,\Omega; \quad X_1 = 12\,\Omega; \quad R_2 = 0.2\,\Omega; \quad X_2 = 0.48\,\Omega$$

Determine the secondary terminal voltage at full-load, 0.8 p.f. lagging when the primary supply voltage is 2000V.

Solution.

$$K = 400/2000 = 1/5$$

$$R_{02} = R_2 + K^2 R_1 = 0.2 + (1/5)^2 \times 5 = 0.4\Omega$$

$$\mathbf{X_{02} = X_2 + K^2 X_1 = 0.48 + (1/5)^2 \times 12 = 0.96\Omega}$$

F.L. secondary current, $I_2 = \dfrac{10 \times 1000}{400} = 25\text{ A}$

$$\cos\phi_2 = 0.8 \quad \therefore \quad \sin\phi_2 = 0.6$$

Voltage drop $= I_2 (R_{02} \cos\phi_2 + X_{02} \sin\phi_2)$ ... *See Art 20.13*

$$= 25\,(0.4 \times 0.8 + 0.96 \times 0.6) = 25\,(0.32 + 0.576) = 22.4\text{ V}$$

$\therefore$ Secondary terminal voltage at full-load,

$$V_2 = 400 - 22.4 = \mathbf{377.6\ V}$$

---

* Assuming primary p.f., $\cos\phi_1$ = secondary p.f. $\cos\phi_2$.

This assumption is quite reasonable because the two values differ by a very small amount.

Example 20.20. *The primary and secondary windings of a 40kVA, 6600/250V single phase transformer have resistances of 10 Ω and 0.02 Ω respectively. The leakage reactance of the transformer referred to the primary side is 35Ω. Calculate the percentage voltage regulation of the transformer when supplying full-load current at a p.f. of 0.8 lagging.*

Solution.

$$K = 250/6600 = 25/660$$

$$R_{01} = R_1 + \frac{R_2}{K^2} = 10 + 0.02 \times (660/25)^2 = 23.93\ \Omega$$

$$X_{01} = 35\Omega \qquad \text{... given}$$

F.L. primary current, $I_1 = \dfrac{40 \times 1000}{6600} = 6.06\ \text{A}$

$$\cos\phi_2 = 0.8 \quad \therefore \quad \sin\phi_2 = 0.6$$

$$\text{Voltage drop} = I_1 (R_{01} \cos\phi_2 + X_{01} \sin\phi_2)$$

$$= 6.06\,(23.93 \times 0.8 + 35 \times 0.6) = 243\ \text{V}$$

$\therefore$ $\%\text{ age regulation} = \dfrac{243}{6600} \times 100 = \mathbf{3.7\%}$

Tutorial Problems

1. A 500 kVA, 3800/400V, 50Hz single phase transformer has the following data :
$R_1 = 0.114\Omega$; $X_1 = 0.459\Omega$; $R_2 = 0.00102\Omega$; $X_2 = 0.00416\Omega$
Find the percentage voltage regulation of the transformer when supplying full-load current at a p.f. of 0.8 lagging. **[2.2 %]**
2. A 100kVA transformer has 400 turns on the primary and 80 turns on the secondary. The primary and secondary resistances are 0.3Ω and 0.1Ω respectively and the corresponding reactances are 1.1Ω and 0.035Ω respectively. The supply voltage is 2200V. Calculate the voltage regulation and secondary terminal voltage for full-load having a p.f. of (*i*) 0.8 lagging (*ii*) 0.8 leading. **[(*i*) 3.36%; 425.2V (*ii*) – 1.54%; 446.8 V]**
3. The primary and secondary windings of a 30kVA, 6000/230V transformer have resistances of 10 Ω and 0.016 Ω respectively. The total reactance of the transformer referred to the primary is 23 Ω. Calculate the percentage voltage regulations of the transformer when supplying full-load current at a p.f. 0.8 lagging. **[2.55 %]**

20.15. Transformer Tests

The circuit constants, efficiency and voltage regulation of a transformer can be determined by two simple tests (*i*) *open-circuit test* and (*ii*) *short-circuit test*. These tests are very convenient as they provide the required information without actually loading the transformer. Further, the power required to carry out these tests is very small as compared with full-load output of the transformer.

20.16. Open-circuit or No-load Test

In this test, the rated voltage is applied to the primary (usually *low-voltage winding) while the secondary is left open-circuited. The applied primary voltage V_1 is measured by the voltmeter, the no-load current I_0 by ammeter and no-load input power W_0 by wattmeter as shown in Fig. 20.29.

---

* The open-circuit test is always made on the low-voltage winding of the transformer. Of course, iron losses will be the same if measured on either winding. But if measurement is made on high-voltage winding, the current I_0 would be inconveniently small and the applied voltage inconveniently large.

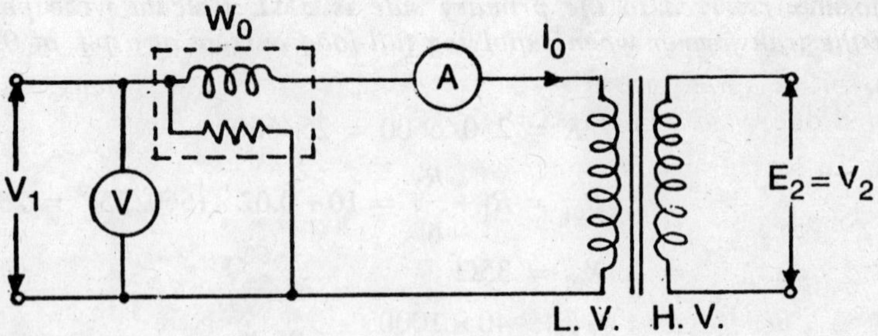

Fig. 20.29

As the normal rated voltage is applied to the primary, therefore, normal iron losses will occur in the transformer core. Hence wattmeter will record the iron losses and small copper loss in the primary. Since no-load current I_0 is very small (usually 2 – 10% of rated current), Cu losses in the primary under no load condition are negligible as compared with iron losses. *Hence, wattmeter reading practically gives the iron losses in the transformer.* It is reminded that iron losses are the same at all loads.

Iron losses, P_i = Wattmeter reading = W_0

No-load current = Ammeter reading = I_0

Applied voltage = Voltmeter reading = V_1

Input power, $W_0 = V_1 I_0 \cos \phi_0$

∴ No-load p.f., $\cos \phi_0 = W_0 / V_1 I_0$

$I_W = I_0 \cos \phi_0;\quad I_m = I_0 \sin \phi_0$

Thus open-circuit test gives P_i, I_0, $\cos \phi_0$, I_W and I_m.

20.17. Short-Circuit or Impedance Test

In this test, the secondary (usually *low-voltage winding) is short-circuited by a thick conductor and variable low voltage is applied to the primary as shown in Fig. 20.30 (*i*). The low input voltage is gradually raised till at voltage V_{SC}, full-load current I_1 flows in the primary. Then I_2 in the secondary also has full-load value since $I_1/I_2 = N_2/N_1$. Under such conditions, the copper loss in the windings is the same as that on full load.

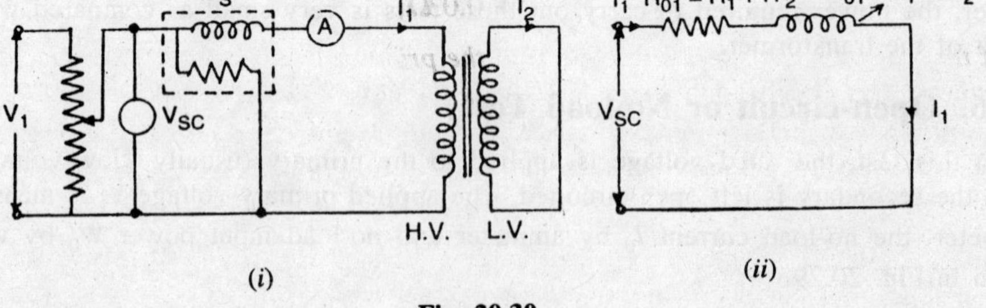

Fig. 20.30

* While making a short-circuit test, low-voltage winding is always short-circuited and measurements are made on the high-voltage winding. If measurements are made on the low-voltage winding (of course high-voltage winding being short-circuited), the current will be inconveniently high.

There is no output from the transformer under short-circuit conditions. Therefore input power is all loss and this loss is almost entirely copper loss. It is because iron loss in the core is negligibly *small since the voltage V_{SC} is very small. *Hence, the wattmeter will practically register the full-load copper losses in the transformer windings.* Fig. 20.30 (*ii*) shows the equivalent circuit of a transformer on short circuit as referred to primary ; the no-load current being neglected due to its smallness.

Full load Cu loss, P_C = Wattmeter reading = W_S

Applied voltage = Voltmeter reading = V_{SC}

F.L. primary current = Ammeter reading = I_1

$$P_C = I_1^2 R_1 + I_1^2 R'_2 = I_1^2 R_{01}$$

$$\therefore \quad R_{01} = P_C / I_1^2$$

where R_{01} is the total resistance of transformer referred to primary.

Total impedance referred to primary, $Z_{01} = V_{SC} / I_1$

Total leakage reactance referred to primary, $X_{01} = \sqrt{Z_{01}^2 - R_{01}^2}$

Short-circuit p.f., $\cos \phi_s = P_C / V_{SC} I_1$

Thus short-circuit test gives full-load Cu loss, R_{01} and X_{01}.

Note : The short circuit test will give full-load Cu loss only if the applied voltage V_{SC} is such so as to circulate full-load currents in the windings. If in a short-circuit test, current value is other than full-load value, then Cu loss will be corresponding to that current value.

20.18. Advantages of Transformer Tests

The above two simple transformer tests offer the following advantages :

(*i*) The power required to carry out these tests is very small as compared to the full-load output of the transformer. In case of open-circuit test, power required is equal to the iron loss whereas for a short-circuit test, power required is equal to full-load copper loss.

(*ii*) These tests enable us to determine the efficiency of the transformer accurately at any load and p.f. without actually loading the transformer.

(*iii*) The short-circuit test enables us to determine R_{01} and X_{01} (or R_{02} and X_{02}). We can thus find the total voltage drop in the transformer as referred to primary or secondary. This permits us to calculate voltage regulation of the transformer.

Example 20.21. *A single phase 50Hz transformer has a full-load secondary current of 500A; the primary current being one-fifth of this value. The transformer has the following parameters:*

$$R_1 = 0.6\Omega; \; R_2 = 0.03\Omega; \; X_1 = 2\Omega \text{ and } X_2 = 0.06\Omega$$

If the secondary is short-circuited, find the primary voltage required to circulate full-load current. Neglect the no-load current. What is the power factor on short circuit ?

Solution. Neglecting no-load current, the equivalent circuit of transformer on short-circuit as referred to primary will be as shown in Fig. 20.31. It is given that full-load $I_2 = 500\,A$ and full load $I_1 = 100A$. Hence $K = I_1 / I_2 = 1/5$.

---

* Iron loss is proportional to the square of flux and hence to the square of applied voltage. Since applied voltage V_{SC} under short circuit conditions is about 1/10th of normal voltage, the iron loss will be 1/100th of the normal value. Consequently, the iron loss is negligible under short-circuit conditions.

$$R_{01} = R_1 + R_2/K^2 = 0.6 + 0.03 \times 5^2 = 1.35\,\Omega$$

$$X_{01} = X_1 + X_2/K^2 = 2 + 0.06 \times 5^2 = 3.5\,\Omega$$

$$\therefore\ Z_{01} = \sqrt{R_{01}^2 + X_{01}^2} = \sqrt{(1.35)^2 + (3.5)^2} = 3.75\,\Omega$$

Primary voltage required to circulate full-load currents

$$V_{SC} = I_1 Z_{01} = 100 \times 3.75 = \mathbf{375\,V}$$

Short circuit p.f.,

$$\cos\phi_s = R_{01}/Z_{01} = 1.35/3.75 = \mathbf{0.36}\ lag$$

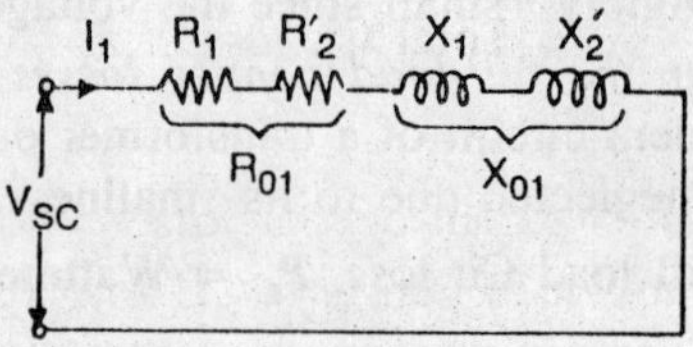

Fig. 20.31

Example 20.22. *The low-voltage winding of a 300kVA, 11000/2200V, 50Hz transformer has 190 turns and a resistance of 0.06 Ω. The high voltage side has 910 turns and a resistance of 1.6 Ω. When the low-voltage winding is short-circuited, the full-load current is obtained with 550V applied to h.v. side. Calculate the equivalent resistance and leakage reactance referred to high-voltage side. Assume full-load efficiency of 98.5%.*

Solution.

$$V_{SC} = 550\text{V};\ K = 190/910 = 19/91;\ \eta = 98.5\%$$

F.L. primary current, $$I_1 = \frac{300 \times 10^3}{11000 \times 0.985} = 27.7\text{ A}$$

$$Z_{01} = V_{SC}/I_1 = 550/27.7 = 19.8\,\Omega$$

$$R'_2 = R_2/K^2 = 0.06/(19/91)^2 = 1.38\,\Omega$$

$$R_{01} = R_1 + R'_2 = 1.6 + 1.38 = \mathbf{2.98\,\Omega}$$

$$X_{01} = \sqrt{Z_{01}^2 - R_{01}^2} = \sqrt{(19.8)^2 - (2.98)^2} = \mathbf{19.5\,\Omega}$$

20.19. Losses in a Transformer

The power losses in a transformer are of two types, namely;

1. Core or Iron losses
2. Copper losses

These losses appear in the form of heat and produce (*i*) an increase in temperature and (*ii*) a drop in efficiency.

1. Core or Iron losses (P_i). These consist of hysteresis and eddy current losses and occur in the transformer core due to the alternating flux. These can be determined by open-circuit test.

$$\text{Hysteresis loss} = k_h f B_m^{1.6}\ \text{watts/m}^3$$

$$\text{Eddy current loss} = k_e f^2 B_m^2 t^2\ \text{watts/m}^3$$

Both hysteresis and eddy current losses depend upon (*i*) maximum flux density B_m in the core and (*ii*) supply frequency *f*. Since transformers are connected to constant-frequency, constant voltage supply, both *f* and B_m are constant. Hence, core or iron losses are practically the same at all loads.

$$\text{Iron or core losses, } P_i = \text{Hysteresis loss} + \text{Eddy current loss}$$

$$= \text{Constant losses}$$

The hysteresis loss can be minimised by using steel of high silicon content whereas eddy current loss can be reduced by using core of thin laminations.

2. Copper losses. These losses occur in both the primary and secondary windings due to their ohmic resistance. These can be determined by short-circuit test.

$$\text{Total Cu losses, } P_C = I_1^2 R_1 + I_2^2 R_2$$
$$= I_1^2 R_{01} \text{ or } I_2^2 R_{02}$$

It is clear that copper losses vary as the square of load current. Thus if copper losses are 400W at a load current of 10A, then they will be $(1/2)^2 \times 400 = 100\text{W}$ at a load current of 5A.

$$\text{Total losses in a transformer} = P_i + P_C$$
$$= \text{Constant losses} + \text{Variable losses}$$

It may be noted that in a transformer, copper losses account for about 90% of the total losses.

20.20. Efficiency of a Transformer

Like any other electrical machine, the efficiency of a transformer is defined as the ratio of output power (in watts or kW) to input power (watts or kW) *i.e.*

$$\text{Efficiency} = \frac{\text{Output power}}{\text{Input power}}$$

It may appear that efficiency can be determined by directly loading the transformer and measuring the input power and output power. However, this method has the following drawbacks :

(*i*) Since the efficiency of a transformer is very high, even 1% error in each wattmeter (output and input) may give ridiculous results. This test, for instance, may give efficiency higher than 100%.

(*ii*) Since the test is performed with transformer on load, considerable amount of power is wasted. For large transformers, the cost of power alone would be considerable.

(*iii*) It is generally difficult to have a device that is capable of absorbing all of the output power.

(*iv*) The test gives no information about the proportion of various losses.

Due to these drawbacks, direct loading method is seldom used to determine the efficiency of a transformer. In practice, open-circuit and short-circuit tests are carried out to find the efficiency.

$$\text{Efficiency} = \frac{\text{Output}}{\text{Input}} = \frac{\text{Output}}{\text{Output} + \text{Losses}}$$

The losses can be determined by transformer tests.

20.21. Efficiency from Transformer Tests

F.L. Iron loss $= P_i$ ... *from open-circuit test*

F.L. Cu loss $= P_C$ ... *from short-circuit test*

Total F.L. losses $= P_i + P_C$

We can now find the full-load efficiency of the transformer at any p.f. without actually loading the transformer.

$$\text{F.L. efficiency, } \eta_{F.L.} = \frac{\text{Full–load } VA \times \text{p.f.}}{(\text{Full–load } VA \times \text{p.f.}) + P_i + P_C}$$

Also for *any load equal to $x \times$ full-load,

Corresponding total losses $= P_i + x^2 P_C$

Corresponding $\qquad \eta_x = \dfrac{(x \times \text{full-load } VA) \times \text{p.f.}}{(x \times \text{Full-load } VA \times \text{p.f.}) + P_i + x^2\ P_C}$

Note that iron loss remains the same at all loads.

20.22. Condition for Maximum Efficiency

$$\text{Output power} = V_2 I_2 \cos\phi_2$$

If R_{02} is the total resistance of the transformer referred to secondary, then,

Total Cu loss, $\qquad P_C = I_2^2 R_{02}$

Total losses $\qquad = P_i + P_C$

$$\therefore \qquad \text{Transformer } \eta = \frac{V_2 I_2 \cos\phi_2}{V_2 I_2 \cos\phi_2 + P_i + I_2^2 R_{02}}$$

$$= \frac{V_2 \cos\phi_2}{V_2 \cos\phi_2 + P_i / I_2 + I_2 R_{02}} \qquad \text{... (i)}$$

For a normal transformer, V_2 is approximately constant. Hence for a load of given p.f., efficiency depends upon load current I_2. It is clear from exp (*i*) above that numerator is constant and for the efficiency to be maximum, the denominator should be minimum *i.e.*

$$\frac{d}{dI_2}\text{(denominator)} = 0$$

or $$\frac{d}{dI_2}(V_2 \cos\phi_2 + P_i/I_2 + I_2 R_{02}) = 0$$

or $$0 - \frac{P_i}{I_2^2} + R_{02} = 0$$

or $$P_i = I_2^2 R_{02} \qquad \text{... (ii)}$$

i.e. $$\text{Iron losses} = \text{Copper losses}$$

Hence efficiency of a transformer will be maximum when copper losses are equal to constant or iron losses.

From eq. (*ii*) above, the load current I_2 corresponding to maximum efficiency is given by ;

$$I_2 = \sqrt{\frac{P_i}{R_{02}}}$$

Note. In a transformer,iron losses are constant whereas copper losses are variable. In order to obtain maximum efficiency, the load current should be such that total Cu losses become equal to iron losses.

20.23. Output kVA Corresponding to Maximum Efficiency

Let $\qquad P_C =$ copper losses at full–load kVA

$P_i =$ Iron losses

$x =$ Fraction of full–load kVA at which efficiency is maximum

* Thus if rated *VA* (*i.e.* full-load *VA*) of a transformer is 1000, then half-load *VA* = 500. Obviously, the value of $x = 1/2$.

Total Cu losses $= x^2 P_C$

$\therefore$ $x^2 P_C = P_i$ ...*for maximum efficiency*

or $$x = \sqrt{\frac{P_i}{P_C}} = \sqrt{\frac{\text{Iron loss}}{\text{F.L. Cu loss}}}$$

$\therefore$ Output kVA corresponding to maximum efficiency

$$= x \times \text{Full–load kVA} = \text{Full–load kVA} \times \sqrt{\frac{\text{Iron loss}}{\text{F.L. Cu loss}}}$$

It may be noted that the value of kVA at which the efficiency is maximum is independent of p.f. of the load.

Example 20.23. *In a 50 kVA transformer, the iron loss is 500W and full-load copper loss is 800W. Find the efficiency at full-load and half load at 0.8 p.f. lagging.*

Solution.

Full-load, 0.8 p.f.

Full-load output $= 50 \times 0.8 = 40$ kW

Total Full-load losses $= 500 + 800 = 1300\text{W} = 1.3$ kW

Full-load input $= 40 + 1.3 = 41.3$ kW

Full-load $\eta = \frac{40}{41.3} \times 100 =$ **96.85%**

Half-load, 0.8 p.f.

Output at half-load $= (50 \times 1/2) \times 0.8 = 20$ kW

Total loss at half-load $= 500 + (1/2)^2 \times 800 = 700\text{W} = 0.7$ kW

Input at half-load $= 20 + 0.7 = 20.7$ kW

Half-load $\eta = \frac{20}{20.7} \times 100 =$ **96.6%**

Example 20.24. *A 40kVA transformer has iron loss of 450W and full-load copper loss of 850W. If the power factor of the load is 0.8 lagging, calculate (i) full-load efficiency (ii) the load at which maximum efficiency occurs and (iii) the maximum efficiency.*

Solution.

(*i*) Total F.L. losses $= 450 + 850 = 1300$ W $= 1.3$ kW

F.L. output $= 40 \times 0.8 = 32$ kW

F.L. Input $= 32 + 1.3 = 33.3$ kW

F.L. efficiency $= \frac{32}{33.3} \times 100 =$ **96.1%**

(*ii*) Load for max. efficiency $= \text{F.L. kVA} \times \sqrt{\frac{\text{Iron loss}}{\text{F.L. Cu loss}}}$

$$= 40 \times \sqrt{\frac{450}{850}} = \textbf{29.1 kVA}$$

Corresponding output kW $= 29.1 \times 0.8 = 23.3$ kW

(*iii*) For maximum efficiency, Cu loss = Iron loss

$\therefore$ Total losses $= 2 \times 450 = 900$ W $= 0.9$ kW

Input $= 23.3 + 0.9 = 24.2$ kW

$$\text{Maximum efficiency} = \frac{23.3}{24.2} \times 100 = \mathbf{96.28\%}$$

Example 20.25. *The primary and secondary windings of a 50 kVA, 6600/220 V transformer have resistances of 7.8Ω and 0.0085 Ω respectively. The transformer draws no-load current of 0.328A at p.f. of 0.3 lagging. Calculate the efficiency at full-load if the p.f. of the load is 0.8 lagging.*

Solution.

$$K = 220/6600 = 1/30$$

F.L. Iron losses $= 6600 \times 0.328 \times 0.3 = 650\text{ W}$

$$R_{01} = R_1 + R_2/K^2 = 7.8 + (0.0085) \times (30)^2 = 15.45\Omega$$

F.L. primary current, $I_1 = 50 \times 10^3/6600 = 7.57\text{ A}$

F.L. Cu losses $= I_1^2 R_{01} = (7.57)^2 \times 15.45 = 885.4\text{ W}$

Total F.L. losses $= 650 + 885.4 = 1535.4\text{ W}$

F.L. output $= 50 \times 10^3 \times 0.8 = 40{,}000\text{ W}$

F.L. input $= 40{,}000 + 1535.4 = 41535.4\text{ W}$

F.L. efficiency $= \dfrac{40{,}000}{41535.4} \times 100 = \mathbf{96.3\%}$

Example 20.26. *A 440/110V transformer has a primary resistance of 0.03Ω and secondary resistance of 0.02Ω. Its iron loss at normal input is 150W. Determine the secondary current at which maximum efficiency will occur and the value of this maximum efficiency at a unity p.f. load.*

Solution.

$$K = 110/440 = 0.25$$

$$R_{02} = R_2 + K^2 R_1 = 0.02 + (0.25)^2 \times 0.03 = 0.022\ \Omega$$

Cu loss = Iron loss ... *for maximum* η

or $I_2^2 R_{02} = 150$

$$\therefore \quad I_2 = \sqrt{\frac{150}{R_{02}}} = \sqrt{\frac{150}{0.022}} = \mathbf{82.58\,A}$$

Output at unity p.f. $= 110 \times 82.58 \times 1 = 9083.8\text{ W}$

Total losses $= 150 + 150 = 300\text{ W}$ ... *for maximum* η

∴ Maximum efficiency $= \dfrac{9083.8}{(9083.8 + 300)} \times 100 = \mathbf{96.8\%}$

Example 20.27. *The efficiency of a 400kVA, single phase transformer is 98.77% when delivering full-load at 0.8 p.f. lagging and 99.13% at half-load at unit p.f. Calculate (i) iron loss and (ii) full-load copper loss.*

Solution.

Let P_i = Iron losses ... constant at all loads

P_C = F.L. copper losses

At full-load

F.L. output at 0.8 p.f. $= 400 \times 0.8 = 320\text{ kW}$

F.L. input at 0.8 p.f. $= 320/0.9877 = 324$ kW

Total F.L. losses $= 324 - 320 = 4$ kW

i.e. $P_i + P_C = 4$ kW ... (*i*)

At half-load

Output at unity p.f. $= (1/2) \times 400 \times 1 = 200$ kW

Input at unity p.f. $= 200/0.9913 = 201.8$ kW

Total losses at half-load $= 201.8 - 200 = 1.8$ kW

i.e. $P_i + 0.25 P_C = 1.8$ kW ... (*ii*)

Solving eqs (*i*) and (*ii*), P_i = **1.07 kW**; P_C = **2.93 kW**

Tutorial Problems

1. In a 100 kVA transformer, the iron loss is 1.2 kW and full-load copper loss is 2 kW. If the load p.f. is 0.8 lagging, find the efficiency at (*i*) full-load and (*ii*) half-load. **[(*i*) 96.16% (*ii*) 95.94%]**
2. The primary and secondary windings of a 500 kVA, 6600/400 V transformer have resistances of 0.42Ω and 0.0011Ω respectively and the iron loss is 2.9kW. Calculate the efficiency at (*i*) full-load and (*ii*) half-load, assuming the power factor of the load to be 0.8 lagging. **[(*i*) 98.27% (*ii*) 98.07%]**
3. A 500/250 V, 1-phase transformer has constant losses of 150 W. When delivering an output of 8 kW to a load, the transformer takes 21A at a p.f. of 0.8 lagging from 500 V supply. Calculate the variable losses of the transformer at this load.
4. A 100 kVA transformer has an iron loss of 1 kW and full-load Cu loss of 1.5kW. Calculate the kVA loading at which the efficiency is maximum and its efficiency at this loading at 0.8 p.f. lagging. **[82.3 kVA; 97.05%]**
5. A 150 kVA transformer has iron-loss of 1.4 kW and full-load copper loss of 2.8 kW. Calculate (*i*) the efficiency of transformer at full-load (*ii*) the maximum efficiency of the transformer and (*iii*) the output power at the maximum level of efficiency. Assume unity power factor **[(*i*) 97.2% (*ii*) 98.13% (*iii*) 106.05 kW]**

20.24. Why Transformer Rating in kVA ?

We have seen that Cu loss depends on current and iron loss depends upon voltage. Hence the total loss in a transformer depends upon volt-ampere (*VA*) only and not on the phase angle between voltage and current *i.e.* it is independent of load power *factor. That is why the rating of a transformer is given in kVA and not in kW. It may be further pointed out that temperature rise of the transformer is directly proportional to the apparent power (*i.e.* kVA) which flows through it. This means that a 500 kVA transformer will get just as hot supplying a 500 kVAR inductive load as a 500 kW resistive load.

20.25. All-day (or Energy) Efficiency

The ordinary or commercial efficiency of a transformer is defined as the ratio of output power to the input power *i.e.*

$$\text{Commercial efficiency} = \frac{\text{Output power}}{\text{Input power}}$$

There are certain types of transformers whose performance cannot be judged by this efficiency. For instance, distribution transformers used for supplying lighting loads have their

* Rating in $kW = kVA \times \text{load p.f.}$

The load p.f. depends upon the nature of load. Thus, kW rating of a transformer will be different at different load power factors.

primaries energised all the 24 hours in a day but the secondaries supply little or no load during the major portion of the day. It means that a constant loss occurs during the whole day but copper loss occurs only when the transformer is loaded and would depend upon the magnitude of load. Consequently, the copper loss varies considerably during the day and the commercial efficiency of such transformers will vary from a low value (or even zero) to a high value when the load is high. The performance of such transformers is judged on the basis of energy consumption during the whole day (*i.e.* 24 hours). This is known as all-day or energy efficiency.

The ratio of output in kWh to the input in kWh of a transformer over a 24-hour period is known as **all-day efficiency** *i.e.*

$$\eta_{\text{all-day}} = \frac{\text{kWh output in 24 hours}}{\text{kWh input in 24 hours}}$$

All-day efficiency is of special importance for those transformers whose primaries are never open-circuited but the secondaries carry little or no load much of the time during the day. In the design of such transformers, efforts should be made to reduce the iron losses which continuously occur during the whole day.

Note : Efficiency of a transformer means commercial efficiency unless stated otherwise.

Example 20.28. *A 5kVA distribution transformer has a full-load efficiency of 95% at which Cu loss is equal to the iron loss. The transformer is loaded in 24 hours as under.*

No load for 10 hours; one-fourth full-load for 7 hours, half full-load for 5 hours and full-load for 2 hours.

Calculate the all-day efficiency of the transformer.

Solution. Since a distribution transformer supplies mostly lighting load, it has a load p.f. of unity.

F.L. output $= 5 \times 1 = 5$ kW

F.L. input $= 5/0.95 = 5.26$ kW

Total F.L. losses $= 5.26 - 5 = 0.26$ kW

Iron loss $=$ F.L. Cu loss $= 0.26/2 = 0.13$ kW

Cu loss at 1/4 F.L. $= (1/16) \times 0.13 = 0.008$ kW

Cu loss at 1/2 F.L. $= (1/4) \times 0.13 = 0.03$ kW

Iron loss in 24 hours $= 0.13 \times 24 = 3.12$ kWh

Cu loss in 24 hours $= (0.008 \times 7) + (0.03 \times 5) + (0.13 \times 2)$

$= 0.056 + 0.15 + 0.26 = 0.466$ kWh

Total losses in 24 hours $= 3.12 + 0.466 = 3.586$ kWh

Total output in 24 hours $= (7 \times 5/4) + (5 \times 5/2) + (5 \times 2) = 31.25$ kWh

$\therefore$ All-day efficiency $= \dfrac{31.25}{31.25 + 3.586} \times 100 = \mathbf{89.7\%}$

Note that commercial efficiency of this transformer is 95% while its all-day efficiency is 89.7%.

Example 20.29. *A 40 kVA distribution transformer has iron loss of 500W and full-load copper loss of 500W. The transformer is supplying a lighting load (unity p.f.). The load cycle is as under :*

Full-load for 4 hours; half-load for 8 hours and no-load for 12 hours.

Calculate the all-day efficiency of the transformer.

Solution.

Iron loss	$= 500 \text{ W} = 0.5 \text{ kW}$
F.L. Cu loss	$= 500 \text{ W} = 0.5 \text{ kW}$
Cu loss at half-load	$= (1/4) \times 0.5 = 0.125 \text{ kW}$
Iron loss in 24 hours	$= 0.5 \times 24 = 12 \text{ kWh}$
Cu loss in 24 hours	$= (0.5 \times 4) + (0.125 \times 8) = 3 \text{ kWh}$
Total loss in 24 hours	$= 12 + 3 = 15 \text{ kWh}$
Output in 24 hours	$= (40 \times 4) + (40 \times 1/2 \times 8) = 320 \text{ kWh}$
All-day efficiency	$= \dfrac{320}{320+15} \times 100 = \mathbf{95.52\%}$

20.26. Construction of a Transformer

We usually design a power transformer so that it approaches the characteristics of an ideal transformer. To achieve this, following design features are incorporated :

(*i*) The core is made of silicon steel which has low hysteresis loss and high permeability. Further, core is laminated in order to reduce eddy current loss. These features considerably reduce the iron losses and the no-load current.

(*ii*) Instead of placing primary on one limb and secondary on the other, it is a usual practice to wind one-half of each winding on one limb. This ensures tight coupling between the two windings. Consequently, leakage flux is considerably reduced.

(*iii*) The winding resistances R_1 and R_2 are minimised to reduce $I^2 R$ loss and resulting rise in temperature and to ensure high efficiency.

20.27. Types of Transformers

Depending upon the manner in which the primary and secondary are wound on the core, transformers are of two types *viz* (*i*) core-type transformer and (*ii*) shell-type transformer.

(*i*) **Core-type transformer.** In a core-type transformer, half of the primary winding and half of the secondary winding are placed round each limb as shown in Fig. 20.32. This reduces the leakage flux. It is a usual practice to place the low-voltage winding below the high-voltage winding for mechanical considerations.

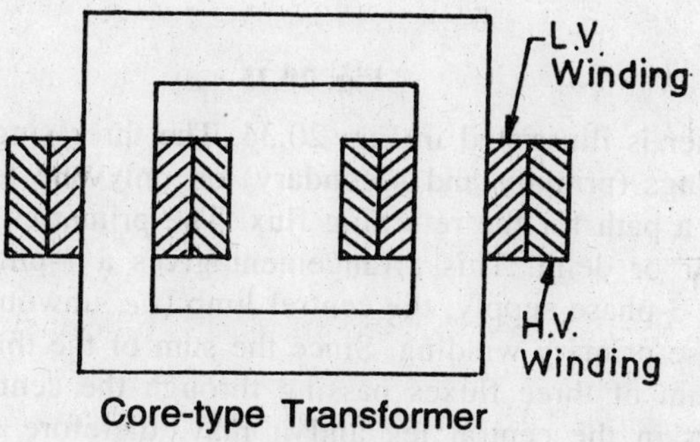

Fig. 20.32

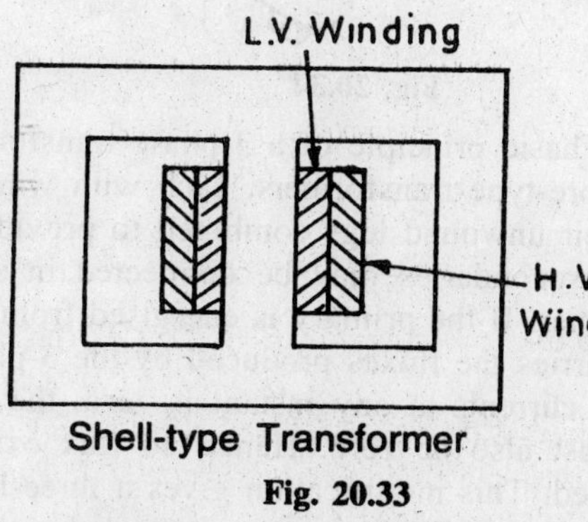

Fig. 20.33

(*ii*) **Shell-type transformer.** This method of construction involves the use of a double magnetic circuit. Both the windings are placed round the central limb (See Fig. 20.33), the other two limbs acting simply as a low-reluctance flux path.

The choice of type (whether core or shell) will not greatly affect the efficiency of the transformer. The core-type is generally more suitable for high voltage and small output while the shell-type is generally more suitable for low voltage and high output.

20.28. Cooling of Transformers

Heat is produced in a transformer by the iron losses in the core and I^2R loss in the windings. To prevent undue temperature rise, this heat is removed by cooling.

(*i*) In small transformers (below 50 kVA), natural air cooling is employed *i.e.* the heat produced is carried away by the surrounding air.

(*ii*) Medium size power or distribution transformers are generally cooled by housing them in tanks filled with oil. The oil serves a double purpose, carrying the heat from the windings to the surface of the tank and insulating the primary from the secondary.

(*iii*) For large transformers, external radiators are added to increase the cooling surface of the oil filled tank. The oil circulates around the transformer and moves through the radiators where the heat is released to surrounding air. Sometimes cooling fans blow air over the radiators to accelerate the cooling process.

20.29. Three-phase Transformer

In a three-phase system, the voltage is lowered or raised either by a bank of three single-phase transformers or by one 3-phase transformer. In either case, the windings may be connected in $Y-Y$, $\Delta-\Delta$, $Y-\Delta$ or $\Delta-Y$. For the same capacity, a 3-phase transformer weighs less, occupies less space and costs about 20% less than a bank of three single-phase transformers. Because of these advantages, 3-phase transformers are in common use, especially for large power transformations.

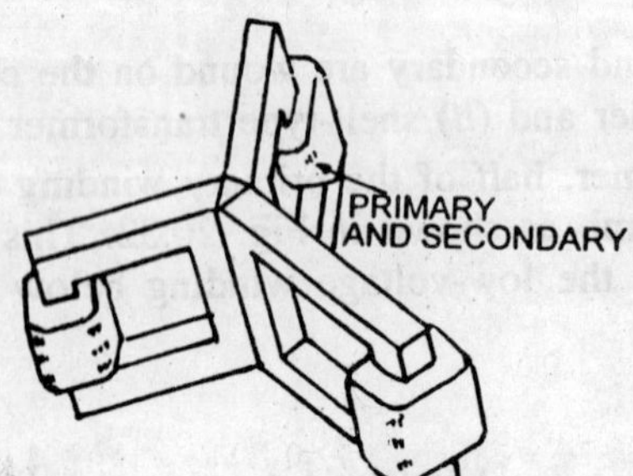

Fig. 20.34

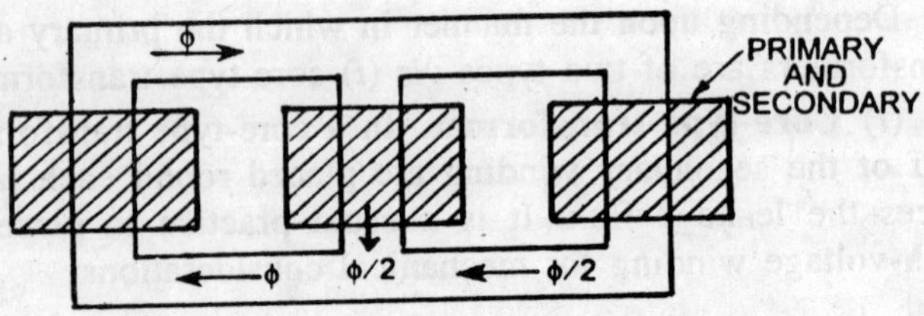

Fig. 20.35

The basic principle of a 3-phase transformer is illustrated in Fig. 20.34. The three single phase core-type transformers, each with windings (primary and secondary) on only one leg, have their unwound legs combined to provide a path for the returning flux. The primaries as well as secondaries may be connected in star or delta. This arrangement gives a 3-phase transformer. If the primary is energised from a 3-phase supply, the central limb (*i.e.* unwound limb) carries the fluxes produced by the 3-phase primary winding. Since the sum of the three primary currents at any instant is zero, the sum of three fluxes passing through the central limb must also be zero. Hence no flux exists in the central leg and it may, therefore, be eliminated. This modification gives a three-leg transformer shown in Fig. 20.35. In this case, any two legs will act as a return path for the flux in the third leg. All the connections of a 3-phase transformer are made inside the case so that only three primary leads and three secondary leads are brought out of the case.

Phase transformation ratio (K). It is the ratio of secondary phase voltage to primary phase voltage and is denoted by K.

$$\text{Phase transformation ratio, } K = \frac{\text{Secondary phase voltage}}{\text{Primary phase voltage}}$$

Example 20.30. *A 3-phase, 50Hz transformer has a delta-connected primary and star-connected secondary, the line voltages being 22000V and 400V respectively. The secondary has a star-connected balanced load at 0.8 power factor lagging. The line current on the primary side is 5A. Determine the current in each coil of the primary and in each secondary line. What is the output of the transformer in kW ?*

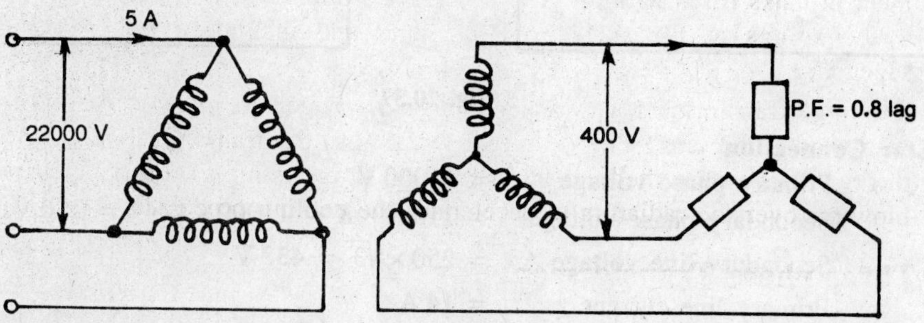

Fig. 20.36

Solution. Fig. 20.36 shows the conditions of the problem.

Primary phase voltage $= 22000$ V

Secondary phase voltage $= 400 \div \sqrt{3} = 400/\sqrt{3}$ V

$$\therefore \quad K = \frac{400/\sqrt{3}}{22000} = \frac{1}{55\sqrt{3}}$$

Primary phase current $= 5 \div \sqrt{3} = \mathbf{5/\sqrt{3}\ A}$

Secondary phase current $= 5/\sqrt{3} \div K = \dfrac{5}{\sqrt{3}} \div \dfrac{1}{55\sqrt{3}} = 275$ A

Secondary line current $= \mathbf{275\ A}$

Output power $= \sqrt{3}\, V_L I_L \cos\phi$

$= \sqrt{3} \times 400 \times 275 \times 0.8 = 152400\text{ W} = \mathbf{152.4\ kW}$

Example 20.31. *The input current to a 3-phase step down transformer connected to an 11kV supply system is 14A. Calculate the secondary line voltage and current for (i) star-star and (ii) delta-star if the phase turn ratio is 44.*

Solution.

$$K = \frac{\text{Secondary phase voltage}}{\text{Primary phase voltage}} = \frac{1}{44}$$

Star-Star Connection

Primary phase voltage $= 11000 \div \sqrt{3} = 6351$ V

Secondary phase voltage $= 6351 \times K = 6351 \times 1/44 = 144.34$ V

Secondary line voltage $= 144.34 \times \sqrt{3} = \mathbf{250\ V}$

Primary phase current $= 14$ A

Secondary phase current $= 14 \div K = 14 \times 44 = \mathbf{616\ A}$

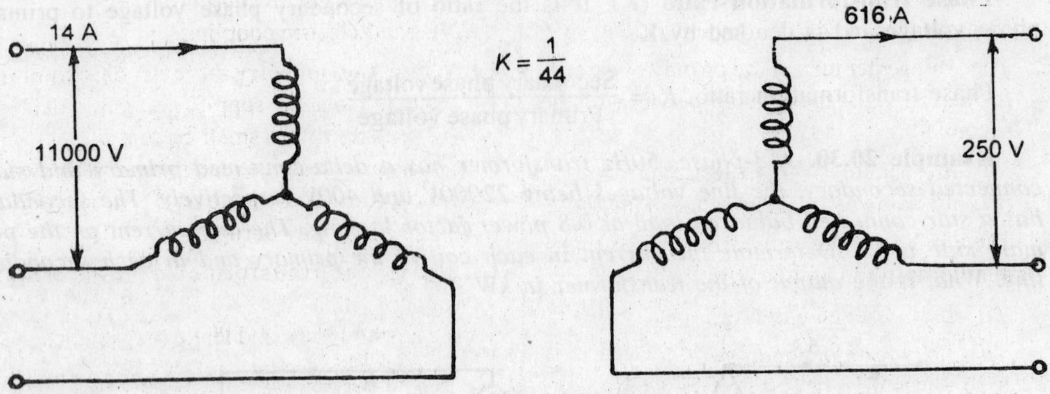

Fig. 20.37

Delta-Star Connection

Primary phase voltage	$= 11000$ V
Secondary phase voltage	$= 11000 \times K = 11000 \times 1/44 = 250$ V
Secondary line voltage	$= 250 \times \sqrt{3} =$ **433 V**
Primary line current	$= 14$ A
Primary phase current	$= 14 \div \sqrt{3} = 8.08$ A
Secondary phase current	$= 8.08 \div K = 8.08 \times 44 = 355.5$ A
Secondary line current	$=$ **355.5 A**

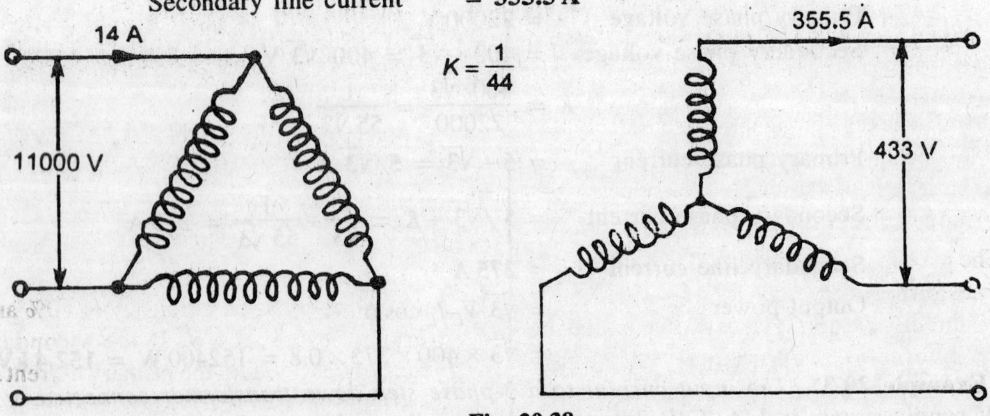

Fig. 20.38

Fig. 20.38 shows the delta-star connection.

OBJECTIVE QUESTIONS

1. A transformer will work on..................
   (*i*) a.c.only (*ii*) d.c. only
   (*iii*) a.c.as well and d.c
   (*iv*) none of the above
2. The primary ad secondary of a transformer are .... coupled.
   (*i*) electrically (*ii*) magnetically
   (*iii*) electrically and magnetically
   (*iv*) none of above
3. A transformer is an efficient device because it ....
   (*i*) is a static device
   (*ii*) uses inductive coupling
   (*iii*) uses capacitive coupling
   (*iv*) uses electric coupling
4. A transformer transfers electrical enegry from primary to secondary usually with a change in ....

(*i*) frequency (*ii*) power
(*iii*) voltage (*iv*) time period

5. The voltage per turn of the primay of a transformer is.......... the voltage per turn of the secondary.
(*i*) more than (*ii*)less than
(*iii*) the same as (*iv*) none of the above

6. The winding of the transformer with greater number of turns will be ....
(*i*) high-voltage winding
(*ii*) low -voltage winding
(*iii*) either high or low voltage winding
(*iv*) none of the above

7. A transformer does not posses .... changing property.
(*i*) impedance (*ii*) voltage
(*iii*) current (*iv*) power

8. The iron-core is used to .... of the transformer.
(*i*) increase the weight
(*ii*) provide tight magnetic coupling
(*iii*) reduce core losses
(*iv*) none of the above

9. If a transformer core has air gaps, then, ...
(*i*) refluctance of magnetic path is decreased
(*ii*) hysteresis loss is decreased
(*iii*) magnetising current is greatlyincreased
(*iv*) eddy current is increased

10. The maximum flux produced in the core of a transformer is ....
(*i*) directly proportional to supply frequency
(*ii*) inversely proportional to supply frequency
(*iii*) inversely proportional to primary voltage
(*iv*) none of the above

11. The flux in the core of a single-phase transformer is ....
(*i*) purely alternating one
(*ii*) purely rotating one
(*iii*) partly alternating and partly rotating
(*iv*) none of the above

12. A transformer is so designed that primary and secondary have ....
(*i*) high leakage reactance
(*ii*) large resistance
(*iii*) tight magnetic coupling
(*iv*) good electric coupling

13. When the primary of a transformer is conected to a d.c. supply, ...
(*i*) primary draws small current
(*ii*) primary leakage reactance is increased
(*iii*) core losses are increased
(*iv*) primary may burn out

14. An ideal transformer is one which ....
(*i*) has no losses and leakage reactance
(*ii*) does not work
(*iii*) has same number of primary and secondary turns
(*iv*) none of the above

15. A low-voltage outdoor lighting system uses a transformer that steps 120 V down to 24 V for safety. The equivalent resistance of all low-voltage lamps is 9.6 Ω. What is the current in the secondary coil? Assume the transformer is ideal and there are no losses in the line.
(*i*) 2 A (*ii*) 2.5 A
(*iii*) 4.5 A (*iv*) 1.5 A

16. In the above question, what is the current in the primary coil?
(*i*) 1.5 A (*ii*) 4.5 A
(*iii*) 2.5 A (*iv*) 0.5 A

17. In Q. 15, how much power is used?
(*i*) 30 W (*ii*) 20 W
(*iii*) 60 W (*iv*) 40 W

18. A transformer has an efficiency of 80% and works at 100 V, 4 kW. If the secondary voltage is 240 V, find the primary current?
(*i*) 40 A (*ii*) 30 A
(*iii*) 20 A (*iv*) 10 A

19. In the above question, what is the secondary current?
(*i*) 12.55 A (*ii*) 9.42 A
(*iii*) 11.56 A (*iv*) 13.33 A

20. A 2000/200 V, 20 kVA ideal transformer has 66 turns in the secondary. The number of primary turns is ....
(*i*) 440 (*ii*) 660
(*iii*) 550 (*iv*) 330

21. In the above question, what is the full-load secondary current?

(*i*) 50 A (*ii*) 30 A
(*iii*) 150 A (*iv*) 100 A

22. In Q. 20, what is the full-load primary current?
(*i*) 5 A (*ii*) 20 A
(*iii*) 10 A (*iv*) 30 A

23. The no-load ratio of a 50 Hz single phase transformer is 6000/250 V. The maximum flux in the core is 0.06 Wb. What is the number of primary turns ?
(*i*) 450 (*ii*) 900
(*iii*) 350 (*iv*) 210

24. In the above question, what is the number of secondary turns?
(*i*) 38 (*ii*) 19
(*iii*) 76 (*iv*) 104

25. A 20-turn iron-cored inductor is connected to a 100 V, 50 Hz source. The maximum flux density in the core is 1 Wb/m^2. The cross-sectional area of the core is
(*i*) 0.152 m^2 (*ii*) 0.345 m^2
(*iii*) 0.056 m^2 (*iv*) 0.0225 m^2

26. Calculate the core area requird for a 1600 k VA, 6600/440 V, 50 Hz single phase core-type power transformer? Assume a maximum flux density of 1.2 Wb/m^2 and induced voltage per turn of 30 V.
(*i*) 975 cm^2 (*ii*) 1100 cm^2
(*iii*) 1125 cm^2 (*iv*) 1224 cm^2

27. An iron-cored transformer working at a maximum flux density of 0.8 Wb/m^2 is replaced by silicon steel working at a maximum flux density of 1.2 Wb/m^2. If the total flux is to remain unchanged, what is the reduction in volume expressed as a percentage of the original volume? The frequency and voltage per turn are the same.
(*i*) 33% (*ii*) 9%
(*iii*) 22% (*iv*) 11%

28. A single phase transformer has 400 primary and 1000 secondary turns. The net crossectional area of the core is 60 cm^2. If the primary is connected to 500 V, 50 Hz source, the peak value of flux density in the core is
(*i*) 1.2 Wb/m^2 (*ii*) 0.94 Wb/m^2
(*iii*) 2.1 Wb/m^2 (*iv*) 0.25 Wb/m^2

29. In the above question, what is the voltage induced in the secondary?
(*i*) 950 V (*ii*) 1125 V
(*iii*) 840 V (*iv*) 1250 V

30. A single phase 50 Hz, core-type transformer has square cores of 20 cm side. The permissible flux density is 1 Wb/m^2. If the stacking factor is 0.9, the voltage induced per turn is
(*i*) 6 (*ii*) 11
(*iii*) 8 (*iv*) 14

ANSWERS TO OBJECTIVE QUESTIONS

1. (*i*)	**2.** (*ii*)	**3.** (*i*)	**4.** (*iii*)	**5.** (*iii*)	**6.** (*i*)	**7.** (*iv*)	**8.** (*ii*)
9. (*iii*)	**10.** (*ii*)	**11.** (*i*)	**12.** (*iii*)	**13.** (*iv*)	**14.** (*i*)	**15.** (*ii*)	**16.** (*iv*)
17. (*iii*)	**18.** (*i*)	**19.** (*iv*)	**20.** (*ii*)	**21.** (*iv*)	**22.** (*iii*)	**23.** (*i*)	**24.** (*ii*)
25. (*iv*)	**26.** (*iii*)	**27.** (*i*)	**28.** (*ii*)	**29.** (iv)	**30.** (*iii*)		

HINTS TO SELECTED OBJECTIVE QUESTIONS

1. Transformer action demands only the existence of alternating flux linking the two windings. Since alternating flux can be produced by a.c., a transformer will work only on a.c.

3. Since tranformer is a static device, the friction and windage losses are absent.

5. Since the same mutual flux links both the windings, the voltage per turn of the primary is the same as the voltage per turn of the secondary.

8. Iron core ensures tight magnetic coupling between the two windings. The co-efficient of coupling between the windings of a practical transformer may be high as 0.99.

10. The voltage equation of a transformer is given by :

$V_1 = 4.44\, N_1\, \phi_{max}\, f$

Where v_1 = applied primary voltage ; N_1 = number of primary turns

$$\therefore \quad \phi_{max} = \frac{V_1}{4.44\, N_1 f}$$

Clearly, maximum flux produced in the core is inversely proportional to supply frequency.

13. When the primary of a transformer is connected to a d.c. supply, steady flux is produced in the core. Consequently, no e.m.f. is induced in the primary. The primary winding will draw a large current due to small resistance of primary. This may cause the primary to burn out.

14. An ideal transformer is one that has (i) no winding resistance (ii) no leakage flux and (iii) no core losses. An actual transformer has some losses, winding resistance and leakage reactance. It may be noted that losses in an actual transformer are quite small. Moreover, winding resistances and leakage reactances are also small. Consequently, the assumption of ideal transformer is satisfactory for most considerations.

15. Secondary current, $I_2 = \dfrac{V_2}{R} = \dfrac{24}{9.6} = 2.5$ A

16. $V_1 I_1 = V_2 I_2$ or Primary current, $I_1 = \dfrac{V_2 I_2}{V_1} = \dfrac{24 \times 2.5}{120} = 0.5$ A

17. For lamp load, power factor is unity. Therefore $P = V_2 I_2 \cos\phi_2 = 24 \times 2.5 \times 1 = 60$ W.

Alternatively, $P = V_1 I_1 \cos\phi_1 = 120 \times 0.5 \times 1 = 60$ W

18. We shall assume power factor to be unity.

$V_1 I_1 \cos\phi_1 = 4 \times 10^3$ or Primary current, $I_1 = \dfrac{4 \times 10^3}{100 \times 1} = 40$ A

19. $V_2 I_2 \cos\phi_2 = 0.8 \times 4 \times 10^3$ or Secondary curent, $I_2 = \dfrac{0.8 \times 4 \times 10^3}{240} = 13.33$ A

20. A 2000/200 V transformer is one that has normal primary and secondary voltages of 2000 V and 200 V respectively.

$K = \dfrac{E_2}{E_1} = \dfrac{200}{2000} = \dfrac{1}{10}$. Now $\dfrac{N_2}{N_1} = K$ or $\dfrac{66}{N_1} = \dfrac{1}{10}$ $\therefore$ $N_1 = 660$

21. Secondary output = 20 kVA = 20×10^3 VA

$\therefore$ Full-load secondary current, $I_2 = \dfrac{\text{output}}{V_2} = \dfrac{20 \times 10^3}{200} = 100$ A

22. Full-load primary current, $I_1 = \dfrac{20 \times 10^3}{2000}$ 10 A

Alternatively : $I_1 = K I_2 = \dfrac{1}{10} \times 100 = 10$ A

23. $E_1 = 4.44 f N_1 \phi_m$ or $6000 = 4.44 \times 50 \times N_1 \times 0.06$ $\therefore$ $N_1 = 450$

24. $E_2 = 4.44 f N_2 \phi_m$ or $250 = 4.44 \times 50 \times N_2 \times 0.06$ $\therefore$ $N_2 = 19$

Alternatively : $N_2 = K N_1 = \frac{250}{6000} \times 450 = 19$

25. $E = 4.44 f N \phi_m$ or $100 = 4.44 \times 50 \times 20 \times \phi_m$ $\therefore$ $\phi_m = 0.0225$ Wb

Max. flux density, $B_m = \frac{\phi_m}{a}$ $\therefore$ $a = \frac{\phi_m}{B_m} = \frac{0.0225}{1} = 0.0225 \text{ m}^2$

26. $E = 4.44 f N \phi_m$ or $E / N = 4.44 f \phi_m$ or $30 = 4.44 \times 50 \times \phi_m$

$\therefore$ $\phi_m = 0.135$ Wb ; Area of *x*-section $= \frac{\phi_m}{B_m} = \frac{0.135}{1.2} = 0.1125 \text{ m}^2 = 1125 \text{ cm}^2$

27. Voltage per tun $= 4.44\ f \phi_m = 4.44 f B_m a$

Since voltage per turn and frequency are the same, $B_m a$ = constant

$\therefore\ a \propto \frac{1}{B_m}$; $a_1 \propto \frac{1}{0.8} = 1.25$ and $a_1 \propto \frac{1}{1.2} = 0.83$

$\therefore$ % reduction in core volume = % reduction in core area

$$= \frac{1.25 - 0.83}{1.25} \times 100 = 33\%$$

28. $E_1 = 4.44 f N_1 \phi_m$ or $500 = 4.44 \times 50 \times 400 \times \phi_m$ $\therefore \phi_m = 5.63 \times 10^{-3}$ WB

Maximum flux density, $B_m = \frac{\phi_m}{\text{core area}} = \frac{5.63 \times 10^{-3}}{60 \times 10^{-4}} = 0.94 \text{ Wb/m}^2$

29. Voltage induced in secondary, $E_2 = K E_1 = \frac{N_2}{N_1} \times E_1 = \frac{1000}{400} \times 500 = 1250 \text{ V}$

30. $\phi_m = B_m \times (\text{area}) = 1 \times (20 \times 20 \times 10^{-4}) \times 0.9 = 0.036$ Wb

Voltage per turn $= 4.44 f \phi_m = 4.44 \times 50 \times 0.036 = 8$

21

Three Phase Induction Motors

Introduction

The three-phase induction motors are the most widely used electric motors in industry. They run at essentially constant speed from no-load to full-load. However, the speed is frequency dependent and consequently these motors are not easily adapted to speed control. We usually prefer d.c. motors when large speed variations are required. Nevertheless, the 3-phase induction motors are simple, rugged, low-priced, easy to maintain and can be manufactured with characteristics to suit most industrial requirements. In this chapter, we shall focus our attention on the general principles of 3-phase induction motors.

21.1. Three-phase Induction Motor

Like any electric motor, a 3-phase induction motor has a stator and rotor. The stator carries a 3-phase winding (called stator winding) while the rotor carries a short circuited winding (called rotor winding). *Only the stator winding is fed from 3-phase supply. The rotor winding derives its voltage and power from the externally energised stator winding through electromagnetic induction and hence the name. The induction motor may be considered to be a transformer with a rotating secondary and it can, therefore, be described as a "transformer-type" a.c. machine in which electrical energy is converted into mechanical energy.

Advantages

(*i*) It has simple and rugged construction.

(*ii*) It is relatively cheap.

(*iii*) It requires little maintenance.

(*iv*) It has high efficiency and reasonably good power factor.

(*v*) It has self-starting torque.

Disadvantages :

(*i*) It is essentially a constant speed motor and its speed cannot be changed easily.

(*ii*) Its starting torque is inferior to d.c. shunt motor.

21.2. Construction

A 3-phase induction motor has two main parts (*i*) stator and (*ii*) rotor. The rotor is separated from the stator by a small air-gap which ranges from 0.4 mm to 4 mm, depending on the power of the motor.

1. Stator. It consists of a steel frame which encloses a hollow, cylindrical core made up of thin laminations of silicon steel to reduce hysteresis and eddy current loss. A number of evenly spaced slots are provided on the inner periphery of the laminations. The insulated

---

* Unlike a d.c. motor where both stator winding (*i.e.*, field winding) and the rotor winding (*i.e.*, armature winding) are connected to a voltage source.

conductors are placed in the stator slots and are suitably connected to form a balanced 3-phase star or delta connected circuit. The 3-phase stator winding is wound for a definite number of poles as per requirement of speed. Greater the number of poles, lesser is the speed of the motor and *vice-versa.* When 3-phase supply is given to the stator winding, a rotating magnetic field (See Art 21.3) of constant magnitude is produced. This rotating field induces currents in the rotor by electromagnetic induction.

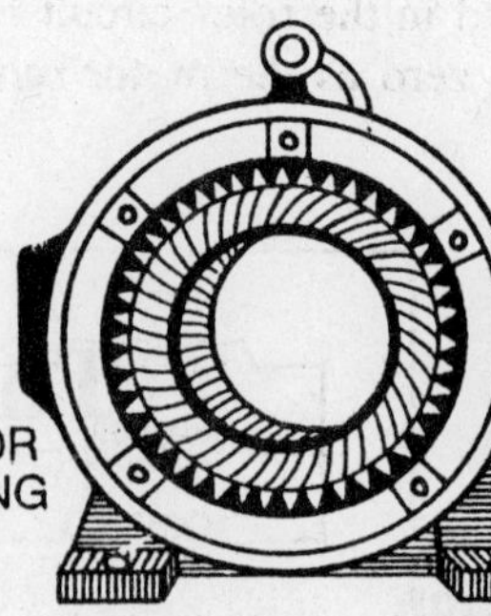

Fig. 21.1

2. Rotor. The rotor, mounted on a shaft, is a hollow laminated core having slots on its outer periphery. The winding placed in these slots (called rotor winding) may be one of the following two types :

(*i*) Squirrel cage type (*ii*) Wound type

(*i*) **Squirrel cage rotor.** It consists of a laminated cylindrical core having parallel slots on its outer periphery. One copper or aluminium bar is placed in each slot. All these bars are joined at each end by metal rings called end rings (See Fig. 21.2). This forms a permanently short-circuited winding which is indestructible. The entire construction (bars and end rings) resembles a squirrel cage and hence the name. The rotor is not connected electrically to the supply but has current induced in it by transformer action from the stator.

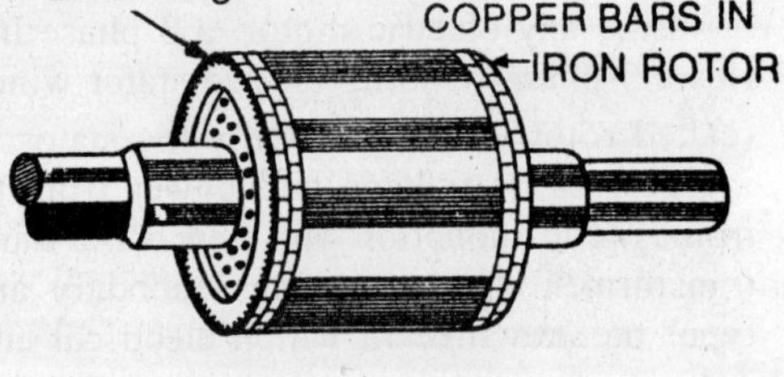

Fig. 21.2

Those induction motors which employ squirrel cage rotor are called *squirrel cage induction motors*. Most of 3-phase induction motors use squirrel cage rotor as it has a remarkably simple and robust construction enabling it to operate in the most adverse circumstances. However, it suffers from the disadvantage of a low starting torque. It is because the rotor bars are permanently short-circuited and it is not possible to add any external resistance to the rotor circuit to have a large starting torque.

(*ii*) **Wound rotor.** It consists of a laminated cylindrical core and carries a 3-phase winding, similar to the one on the stator (See Fig. 21.3). The rotor winding is uniformly distributed in the slots and is usually star-connected. The open ends of the rotor winding are brought out and joined to three insulated slip rings mounted on the rotor shaft with one brush resting on each slip ring. The three brushes are connected to a 3-phase star-connected rheostat as shown in Fig. 21.4. At starting, the external resistances

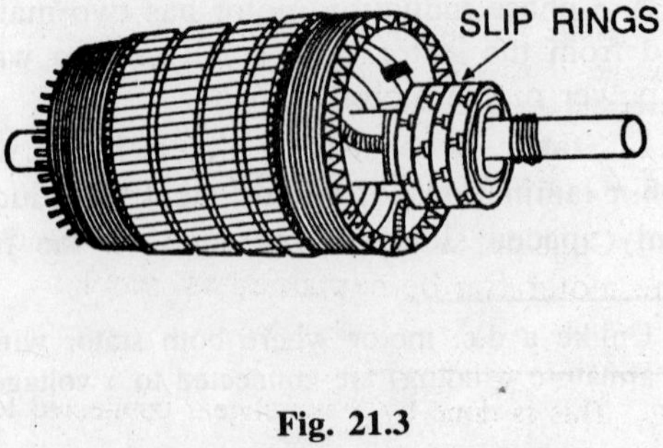

Fig. 21.3

are included in the rotor circuit to give a large starting torque. These resistances are gradually reduced to zero as the motor runs up to speed.

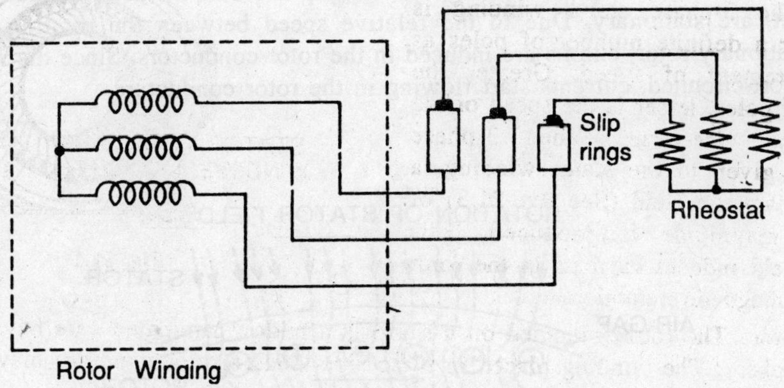

Fig. 21.4

The external resistances are used during starting period only. When the motor attains normal speed, the three brushes are *short-circuited so that the wound rotor runs like a squirrel cage rotor.

21.3. Rotating Magnetic Field

When a 3-phase winding is energised from a 3-phase supply, a rotating magnetic field is produced. This field is such that its poles do not remain in a fixed position on the stator but go on shifting their positions around the stator. For this reason, it is called a rotating field. It can be shown that magnitude of this rotating field is constant and is equal to 1.5 ϕ_m where ϕ_m is the maximum flux due to any phase.

Synchronous speed. The speed at which the revolving flux rotates is called synchronous speed (N_s). Its value depends upon the number of poles and the supply frequency and is given by ;

$$f = \frac{N_s P}{120}$$

or Synchronous speed, $N_s = 120 f/P \text{ r.p.m.}$

where P = number of poles on the stator

f = supply frequency in Hz

Thus for a 6-pole, 50 Hz motor, $N_s = 120 \times 50/6 = 1000$ r.p.m. It means that the flux rotates around the stator at a speed of 1000 r.p.m.

Note. The direction of rotating flux (*i.e.*, clockwise or anticlockwise) depends upon the phase sequence of supply voltage. If any two of the three lines to the stator winding are reversed, the phase sequence is reversed, thereby reversing the direction of rotating flux.

21.4. Principle of Operation

Consider a portion of 3-phase induction motor as shown in Fig. 21.5. The operation of the motor can be explained as under :

---

* This is done by a switchgear connected to the brushes

(*i*) When 3-phase stator winding is energised from a 3-phase supply, a rotating magnetic field is set up which rotates round the stator at synchronous speed N_s (= $120 f/P$).

(*ii*) The rotating field passes through the air-gap and cuts the rotor conductors, which as yet, are stationary. Due to the relative speed between the rotating flux and the stationary rotor, e.m.f.s are induced in the rotor conductors. Since the rotor circuit is short-circuited, currents start flowing in the rotor conductors.

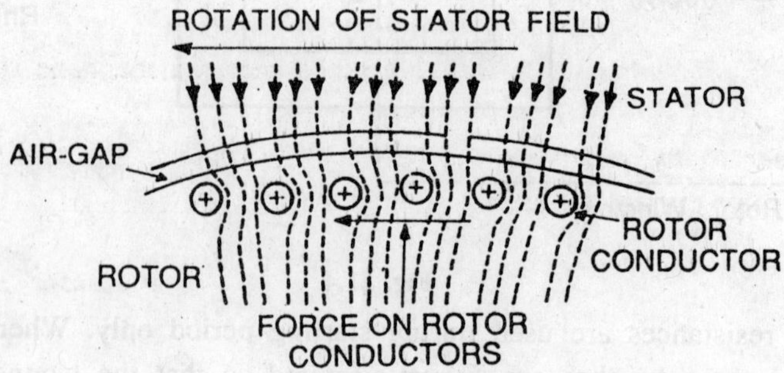

Fig. 21.5

(*iii*) The current-carrying rotor conductors are placed in the magnetic field produced by the stator. Consequently, mechanical force acts on the rotor conductors. The sum of the mechanical forces on all the rotor conductors produces a torque which tends to move the rotor in the *same direction* as the rotating field.

(*iv*) The fact that rotor is urged to follow the stator field (*i.e.*, rotor moves in the direction of stator field) can be explained by Lenz's law. According to this law, the direction of rotor currents will be such that they tend to oppose the cause producing them. Now, the cause producing the rotor currents is the relative speed between the rotating field and the stationary rotor conductors. Hence to reduce this relative speed, the rotor starts running in the same direction as that of stator field and tries to catch it.

21.5. Slip

We have seen above that rotor rapidly accelerates in the direction of rotating field. In practice, the rotor can never reach the speed of stator flux. If it did, there would be no relative speed between the stator field and rotor conductors, no induced rotor currents and, therefore, no torque to drive the rotor. The friction and windage would immediately cause the rotor to slow down. Hence, the rotor speed (N) is always less than the stator field speed (N_s). This difference in speed depends upon load on the motor.

The difference between the synchronous speed N_s of the rotating stator field and the actual rotor speed N is called ***slip.** It is usually expressed as a percentage of synchronous speed *i.e.*,

$$\%\text{age slip, } s = \frac{N_s - N}{N_s} \times 100$$

* The term slip is used because it describes what an observer riding on the stator field sees looking at the rotor—it appears to be slipping backward.

(*i*) The quantity $N_s - N$ is sometimes called slip speed.

(*ii*) When the rotor is stationary (*i.e.*, $N = 0$), slip, $s = 1$ or 100%.

(*iii*) In an induction motor, the change in slip from no-load to full-load is hardly 0.1% to 3% so that it is essentially a constant-speed motor.

21.6. Rotor Current Frequency

The frequency of a voltage or current induced due to the relative speed between a winding and a magnetic field is given by the general formula ;

$$\text{Frequency} = \frac{NP}{120}$$

where N = Relative speed between magnetic field and the winding

P = Number of poles

For a rotor speed N, the relative speed between the rotating flux and the rotor is $N_s - N$. Consequently, the rotor current frequency f' is given by ;

$$f' = \frac{(N_s - N)P}{120}$$

$$= \frac{sN_sP}{120} \qquad \left(\because s = \frac{N_s - N}{N_s}\right)$$

$$= sf \qquad (\because f = N_sP/120)$$

i.e., *Rotor current frequency = Fractional slip × Supply frequency*

(*i*) When the rotor is at standstill or stationary (*i.e.*, $s = 1$), the frequency of rotor current is the same as that of supply frequency ($f' = sf = 1 \times f = f$).

(*ii*) As the rotor picks up speed, the relative speed between the rotating flux and the rotor decreases. Consequently, the slip s and hence rotor current frequency decreases.

Note : The relative speed between the rotating field and stator winding is $N_s - 0 = N_s$. Therefore, the frequency of induced current or voltage in the stator winding is $f = N_sP/120$—the supply frequency.

Example 21.1. *A 6-pole, 3-phase induction motor is connected to 50Hz supply. If it is running at 970 r.p.m., find the slip.*

Solution.

Synchronous speed, $N_s = 120f/P = 120 \times 50/6 = 1000$ r.p.m.

$$\text{Slip, } s = \frac{N_s - N}{N_s} \times 100 = \frac{1000 - 970}{1000} \times 100 = \mathbf{3\%\ \textit{or}\ 0.03}$$

Example 21.2. *A 3-phase induction motor is wound for 4 poles and is supplied from 50 Hz system. Calculate (i) the synchronous speed (ii) the speed of the motor when slip is 4% and (iii) the rotor current frequency when the motor runs at 600 r.p.m.*

Solution.

(*i*) Synchronous speed, $N_s = 120f/P = 120 \times 50/4 =$ **1500 r.p.m.**

(*ii*) $$\text{Slip, } s = \frac{N_s - N}{N_s} \times 100$$

or $$4 = \frac{1500 - N}{1500} \times 100$$

$\therefore$ $$N = \mathbf{1440\ r.p.m.}$$

(*iii*) When $N = 600$ r.p.m., $s = \dfrac{1500 - 600}{1500} = 0.6$

$\therefore$ Rotor current frequency, $f' = sf = 0.6 \times 50 =$ **30 Hz**

Example 21.3. *A 6-pole alternator running at 1000 r.p.m. supplies an 8-pole induction motor. Find the actual speed of the motor if the slip is 2.5%.*

Solution. The frequency of 3-phase supply fed to the induction motor is determined from the speed of the alternator and its number of poles.

$$\text{Supply frequency, } f = NP/120 = 1000 \times 6/120 = 50 \text{ Hz}$$

$$\text{Synchronous speed, } N_s = 120f/P = 120 \times 50/8 = 750 \text{ r.p.m.}$$

$$\text{Slip, } s = \frac{N_s - N}{N_s} \times 100$$

or
$$2.5 = \frac{750 - N}{750} \times 100$$

$\therefore$
$$N = \mathbf{731.25 \text{ r.p.m.}}$$

Example 21.4. *A 50Hz, 4-pole, 3-phase induction motor has a rotor current of frequency 2Hz. Determine (i) the slip and (ii) speed of the motor.*

Solution.

(*i*)
$$f' = sf$$

or
$$s = f'/f = 2/50 = \mathbf{0.04} \text{ or } \mathbf{4\%}$$

(*ii*)
$$N_s = 120f/P = 120 \times 50/4 = 1500 \text{ r.p.m.}$$

$$s = \frac{N_s - N}{N_s}$$

or
$$0.04 = \frac{1500 - N}{1500}$$

$\therefore$
$$N = \mathbf{1440 \text{ r.p.m.}}$$

Tutorial Problems

1. A 2-pole, 3-phase, 50Hz induction motor is running on no load with a slip of 4%. Calculate (*i*) the synchronous speed and (*ii*) speed of the motor. **[(*i*) 3000 r.p.m. (*ii*) 2880 r.p.m.]**
2. The frequency of e.m.f. in the stator of a 4-pole, 3-phase induction motor is 50Hz and that in the rotor is 1.5Hz. Determine (*i*) the slip and (*ii*) speed of the motor. **[(*i*) 3% (*ii*) 1455 r.p.m.]**
3. A 3-phase, 50 Hz induction motor has 8 poles. If the full-load slip is 2.5%, determine (*i*) synchronous speed (*ii*) rotor speed and (*iii*) rotor frequency. **[(*i*) 750 r.p.m. (*ii*) 731 r.p.m. (*iii*) 1.25Hz]**

21.7. Effect of Slip on the Rotor Circuit

When the rotor is stationary, $s = 1$. Under these conditions, the per phase rotor e.m.f. E_2 has a frequency equal to that of supply frequency f. At any slip s, the relative speed between stator field and the rotor is decreased. Consequently, the rotor e.m.f. and frequency are reduced proportionally to sE_2 and sf respectively. At the same time, per phase rotor reactance X_2, being frequency dependent, is reduced to sX_2.

Consider a 6-pole, 3-phase, 50Hz induction motor. It has synchronous speed $N_s = 120f/P = 120 \times 50/6 = 1000$ r.p.m. At standstill, the relative speed between stator flux and rotor is 1000 r.p.m. and rotor e.m.f./phase $= E_2$ (say). If the full-load speed of the motor is 960 r.p.m., then

$$s = \frac{1000-960}{1000} = 0.04$$

(*i*) The relative speed between stator flux and the rotor is now only 40 r.p.m. Consequently, rotor e.m.f./phase is reduced to :

$$^*E_2 \times \frac{40}{1000} = 0.04\,E_2 \text{ or } s\,E_2$$

(*ii*) The frequency is also reduced in the same ratio to :

$$50 \times \frac{40}{1000} = 50 \times 0.04 \text{ or } s\,f$$

(*iii*) The per phase rotor reactance X_2 is likewise reduced to :

$$X_2 \times \frac{40}{1000} = 0.04\,X_2 \text{ or } s\,X_2$$

Thus at any slip *s*,

Rotor e.m.f./phase $= s\,E_2$

Rotor reactance/phase $= s\,X_2$

Rotor frequency $= s\,f$

where E_2, X_2 and f are the corresponding values at standstill.

Example 21.5. *A 3-phase, 6-pole induction motor is connected to a 60Hz supply. The voltage induced in the rotor bars is 4V when the rotor is at standstill. Calculate the voltage and frequency induced in the rotor bars at 300 r.p.m.*

Solution.

Synchronous speed, $N_s = 120f/P = 120 \times 60/6 = 1200$ r.p.m.

Slip, $s = \frac{N_s - N}{N_s} = \frac{1200-300}{1200} = 3/4$

Corresponding to this slip,

Induced voltage $= 4 \times s = 4 \times 3/4 =$ **3V**

Frequency $= f \times s = 60 \times 3/4 =$ **45 Hz**

21.8. Rotor Current

Fig. 21.6 shows the circuit of a 3-phase induction motor at any slip *s*. The rotor is assumed to be of wound type and star connected. Note that rotor e.m.f./phase and rotor reactance/phase are $s\,E_2$ and $s\,X_2$ respectively. The rotor resistance/phase is R_2 and is independent of frequency and, therefore, does not depend upon slip. Likewise, stator winding values R_1 and X_1 do not depend upon slip.

Since the motor represents a balanced 3-phase load, we need consider one phase only ; the conditions in the other two phases being similar.

---

* If the relative speed between stator flux and rotor is 1000 r.p.m., the rotor e.m.f./phase = E_2. If the relative speed is now 40 r.p.m., rotor e.m.f./phase is

$$= \frac{40}{1000} \times E_2$$ ... ***unitary method***

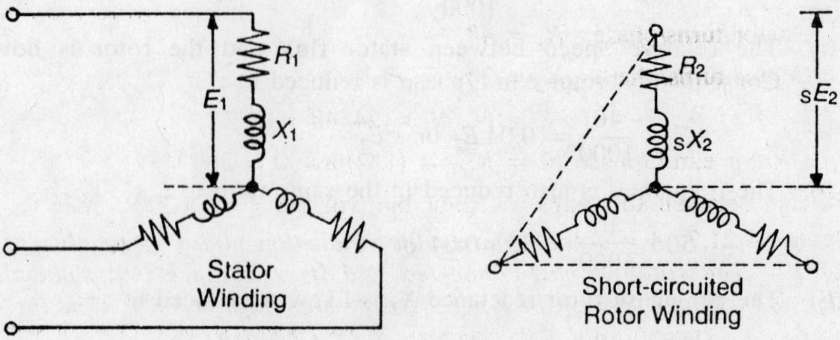

Fig. 21.6

At standstill. Fig. 21.7 (*i*) shows one phase of the rotor circuit at standstill.

$$\text{Rotor current/phase, } I_2 = \frac{E_2}{Z_2} = \frac{E_2}{\sqrt{R_2^{\;2} + X_2^{\;2}}}$$

$$\text{Rotor p.f. } \cos\phi_2 = \frac{R_2}{Z_2} = \frac{R_2}{\sqrt{R_2^{\;2} + X_2^{\;2}}}$$

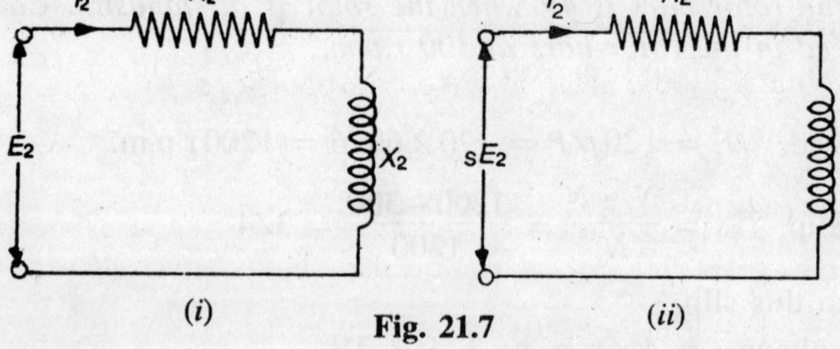

Fig. 21.7

When running at slip *s*. Fig. 21.7 (*ii*) shows one phase of the rotor circuit when the motor is running at slip *s*.

$$\text{Rotor current, } \quad I_2' = \frac{s\,E_2}{Z_2'} = \frac{s\,E_2}{\sqrt{R_2^2 + (s\,X_2)^2}}$$

$$\text{Rotor p.f., } \quad \cos\phi_2' = \frac{R_2}{Z_2'} = \frac{R_2}{\sqrt{R_2^2 + (\,s\,X_2\,)^2}}$$

Example 21.6. *A 3-phase, 400 V wound rotor motor has delta-connected stator winding and star-connected rotor winding. The stator has 48 turns/phase while the rotor has 24 turns per phase. Find the standstill or open-circuited voltage across the slip rings.*

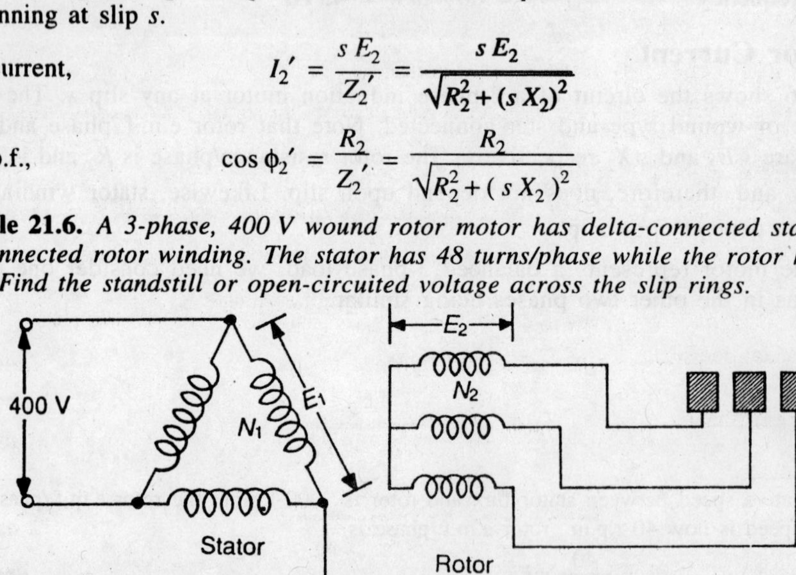

Fig. 21.8

Solution.

Stator e.m.f./phase, $E_1 = 400$ V

Stator turns/phase, $N_1 = 48$

Rotor turns/phase, $N_2 = 24$

$\therefore \quad K = N_2/N_1 = 24/48 = 1/2$

$\therefore$ Rotor e.m.f./phase $= KE_1 = (1/2) \times 400 = 200$ V

Voltage between slip rings = Rotor line voltage $= \sqrt{3} \times 200 =$ **346 V**

Example 21.7. *A 6-pole, 3-phase 50Hz induction motor is running at full-load with a slip of 4%. The rotor is star-connected and its resistance and standstill reactance are 0.25 Ω and 1.5 Ω per phase. The e.m.f. between slip rings is 100V. Find the rotor current per phase and p.f., assuming the slip rings are short-circuited.*

Solution.

Rotor e.m.f./phase at standstill, $E_2 = 100/\sqrt{3} = 57.7$ V

Rotor e.m.f./phase at full-load $= s\,E_2 = 0.04 \times 57.7 = 2.31$ V

Rotor reactance/phase at full-load $= s \times X_2 = 0.04 \times 1.5 = 0.06\ \Omega$

Rotor impedance/phase at full-load $= \sqrt{(0.25)^2 + (0.06)^2} = 0.257\ \Omega$

Full-load rotor current/phase $= 2.31/0.257 =$ **9 A**

Rotor power factor $= 0.25/0.257 =$ **0.97** ***lag***

21.9. Rotor Torque

The torque T developed by the rotor is directly proportional to :

(*i*) rotor current

(*ii*) rotor e.m.f.

(*iii*) power factor of the rotor circuit.

$$\therefore \quad T \propto E_2 I_2 \cos \phi_2$$

or

$$T = KE_2 I_2 \cos \phi_2$$

where I_2 = rotor current at standstill

E_2 = rotor e.m.f. at standstill

$\cos \phi_2$ = rotor p.f. at standstill

Note. The values of rotor e.m.f., rotor current and rotor power factor are taken for the given conditions.

21.10. Starting Torque (T_s)

Let E_2 = rotor e.m.f. per phase at standstill

X_2 = rotor reactance per phase at standstill

R_2 = rotor resistance per phase

Rotor impedance/phase, $Z_2 = \sqrt{R_2^2 + X_2^2}$ *...at standstill*

Rotor current/phase, $I_2 = \dfrac{E_2}{Z_2} = \dfrac{E_2}{\sqrt{R_2^2 + X_2^2}}$ *... at standstill*

Rotor p.f., $\cos \phi_2 = \dfrac{R_2}{Z_2} = \dfrac{R_2}{\sqrt{R_2^2 + X_2^2}}$ *... at standstill*

$\therefore$ Starting torque, $T_s = KE_2 I_2 \cos\phi_2$

$$= KE_2 \times \frac{E_2}{\sqrt{R_2^2 + X_2^2}} \times \frac{R_2}{\sqrt{R_2^2 + X_2^2}}$$

$$= \frac{K E_2^2 R_2}{R_2^2 + X_2^2}$$

Generally, the stator supply voltage V is constant so that flux per pole ϕ set up by the stator is also fixed. This in turn means that e.m.f. E_2 induced in the rotor will be constant.

$$\therefore \quad T_s = \frac{K_1 R_2}{R_2^2 + X_2^2} = \frac{K_1 R_2}{Z_2^2}$$

where K_1 is another constant.

It is clear that the magnitude of starting torque would depend upon the relative values of R_2 and X_2 *i.e.*, rotor resistance/phase and standstill rotor reactance/phase.

21.11. Condition for Maximum Starting Torque

It can be proved that starting torque will be maximum when rotor resistance/phase is equal to standstill rotor reactance/phase.

Now
$$T_s = \frac{K_1 R_2}{R_2^2 + X_2^2} \quad ...(i)$$

Differentiating eq. (*i*) *w.r.t.* R_2 and equating the result to zero,

$$\frac{dT_s}{dR_2} = K_1 \left[\frac{1}{R_2^2 + X_2^2} - \frac{R_2 (2R_2)}{(R_2^2 + X_2^2)^2} \right] = 0$$

or $R_2^2 + X_2^2 = 2R_2^2$

or $R_2 = X_2.$

Hence starting torque will be maximum when :

Rotor resistance/phase = Standstill rotor reactance/phase

Under the condition of maximum starting torque, $\phi_2 = 45°$ and rotor power factor is 0.707 lagging [See Fig. 21.9 (*ii*)].

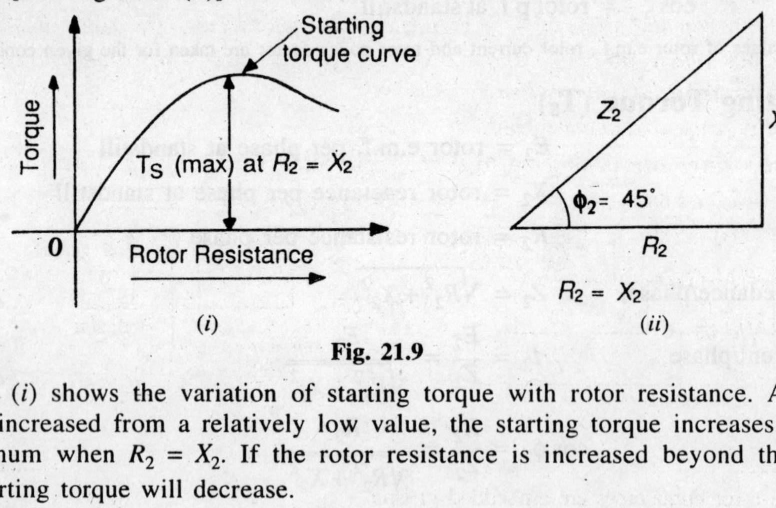

Fig. 21.9

Fig. 21.9 (*i*) shows the variation of starting torque with rotor resistance. As the rotor resistance is increased from a relatively low value, the starting torque increases until it becomes maximum when $R_2 = X_2$. If the rotor resistance is increased beyond this optimum value, the starting torque will decrease.

21.12. Effect of Change of Supply Voltage

$$T_s = \frac{K E_2^2 R_2}{R_2^2 + X_2^2}$$

Since $E_2 \propto$ supply voltage V

$$\therefore \quad T_s = \frac{K_2 V^2 R_2}{R_2^2 + X_2^2}$$

where K_2 is another constant.

$$\therefore \quad T_s \propto V^2$$

Therefore, the starting torque is very sensitive to changes in the value of supply voltage. For example, a drop of 10% in supply voltage will decrease the starting torque by about 20%.

21.13. Starting Torque of 3-phase Induction Motors

The rotor circuit of an induction motor has low resistance and high *inductance. At starting, the rotor frequency is equal to the stator frequency (*i.e.*, 50Hz) so that rotor reactance is large compared with rotor resistance. Therefore, rotor current lags the rotor e.m.f. by a large angle, the power factor is low and consequently the starting torque is small. When resistance is added to the rotor circuit, the rotor power factor is improved which results in improved starting torque. This, of course, increases the rotor impedance and therefore decreases the value of rotor current but the effect of improved power factor predominates and the starting torque is increased.

(*i*) **Squirrel-cage motors.** Since the rotor bars are permanently short-circuited, it is not possible to add any external resistance in the rotor circuit at starting. Consequently, the starting torque of such motors is low. Squirrel-cage motors have starting torque of 1.5 to 2 times the full-load value with starting current of 5 to 9 times the full-load current.

(*ii*) **Wound rotor motors.** The resistance of the rotor circuit of such motors can be increased through the addition of external resistance. By inserting the proper value of external resistance (so that $R_2 = X_2$), maximum starting torque can be obtained. As the motor accelerates, the external resistance is gradually cut out until the rotor circuit is short-circuited on itself for running conditions.

Example 21.8. *The 3-phase induction motor having star-connected rotor has an induced e.m.f. of 50 V between the slip rings at standstill on open-circuit. The rotor has a resistance and reactance per phase of 0.5 Ω and 4.5 Ω respectively. Find the rotor current per phase and the power factor at starting when*

(*i*) *the slip rings are short-circuited*

(*ii*) *the slip rings are connected to a star-connected rheostat of 4 Ω per phase.*

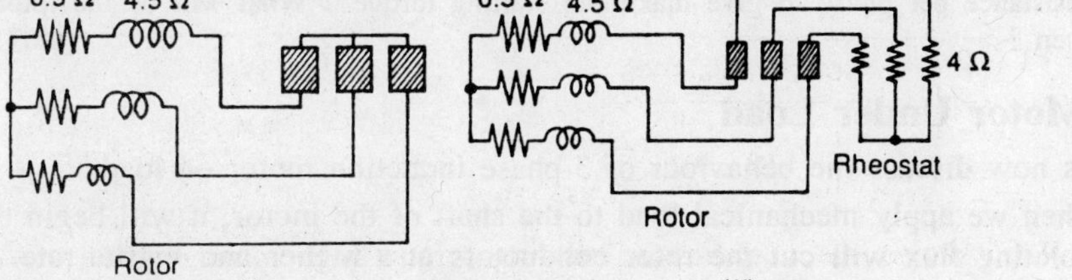

(*i*) **Fig. 21.10** (*ii*)

* Because the rotor conductors are embedded in iron

Solution.

Standstill rotor e.m.f./phase, $E_2 = 50/\sqrt{3} = 28.87$ V

When slip rings are short-circuited [See Fig. 21.10 (*i*)].

Rotor impedance/phase $= \sqrt{(0.5)^2 + (4.5)^2} = 4.53\ \Omega$

Rotor current/phase $= 28.87/4.53 =$ **6.38 A**

Rotor power factor $= 0.5/4.53 =$ **0.11** *lag*

When slip rings connected to Y-connected rheostat [See Fig. 21.10 (*ii*)].

Rotor resistance/phase $= 4 + 0.5 = 4.5\ \Omega$

Rotor impedance/phase $= \sqrt{(4.5)^2 + (4.5)^2} = 6.36\ \Omega$

Rotor current/phase $= 28.87/6.36 =$ **4.54 A**

Rotor power factor $= 4.5/6.36 =$ **0.707** *lag*

It is clear that by inserting additional resistance in the rotor circuit, the p.f. of the rotor circuit is improved, though the rotor current is decreased. However, the effect of improved p.f. predominates the decrease in rotor current. Hence, starting torque is increased.

Example 21.9. *A star-connected rotor winding of a 3-phase induction motor has a resistance of 0.2 Ω per phase and a standstill reactance of 1 Ω per phase. Find the value of external resistance per phase to be added to give maximum starting torque. What is the power factor of the rotor circuit then ?*

Solution.

Let R ohm be the external resistance per phase to be added to the rotor circuit to give maximum starting torque. For maximum starting torque,

Rotor resistance/phase = Standstill rotor reactance/phase

or $0.2 + R = 1$

$\therefore$ $R = 1 - 0.2 = \mathbf{0.8\ \Omega}$

Rotor impedance/phase $= \sqrt{1^2 + 1^2} = \sqrt{2}$

Rotor power factor $= 1/\sqrt{2} =$ **0.707** *lag*

Tutorial Problems

1. A 3-phase induction motor at standstill has a rotor voltage of 120V between the slip rings on open-circuit. The rotor winding is star-connected and has reactance of 1 Ω per phase at standstill and a resistance of 0.2 Ω per phase. Find the rotor current per phase and power factor at start. **[13.6A ; 0.98** *lag***]**
2. A 3-phase induction motor with a star-connected rotor has an induced e.m.f. of 80 V between the slip rings at standstill on open-circuit. The resistance and standstill reactance of the rotor are 0.35 Ω and 2.3 Ω per phase respectively. Calculate the rotor current per phase and the power factor at start with an external resistance of 2 Ω per phase. **[14A ; 0.71** *lag***]**
3. If the star-connected rotor winding of a 3-phase induction motor has a resistance of 0.01 Ω per phase and a standstill reactance of 0.08 Ω per phase, what must be the value of external resistance per phase to give maximum starting torque ? What will be the rotor power factor then ? **[0.07 Ω ; 0.707** *lag***]**

21.14. Motor Under Load

Let us now discuss the behaviour of 3-phase induction motor on load.

(*i*) When we apply mechanical load to the shaft of the motor, it will begin to slow down and the rotating flux will cut the rotor conductors at a higher and higher rate. The induced voltage and resulting current in rotor conductors will increase progressively, producing greater and greater torque.

(*ii*) The motor and mechanical load will soon reach a state of equilibrium when the motor torque is exactly equal to the load torque. When this state is reached, the speed will cease to drop any more and the motor will run at the new speed at a constant rate.

(*iii*) The drop in speed of the induction motor on increased load is small. It is because the rotor impedance is *low and a small decrease in speed produces a large rotor current. The increased rotor current produces a higher torque to meet the increased load on the motor. This is why induction motors are considered to be constant-speed machines. However, because they never actually turn at synchronous speed, they are sometimes called *asynchronous machines.*

Note that change in load on the induction motor is met through the adjustment of slip. When load on the motor increases, the slip increases slightly (*i.e.*, motor speed decreases slightly). This results in greater relative speed between the rotating flux and rotor conductors. Consequently, rotor current is increased, producing a higher torque to meet the increased load. Reverse happens should the load on the motor decrease.

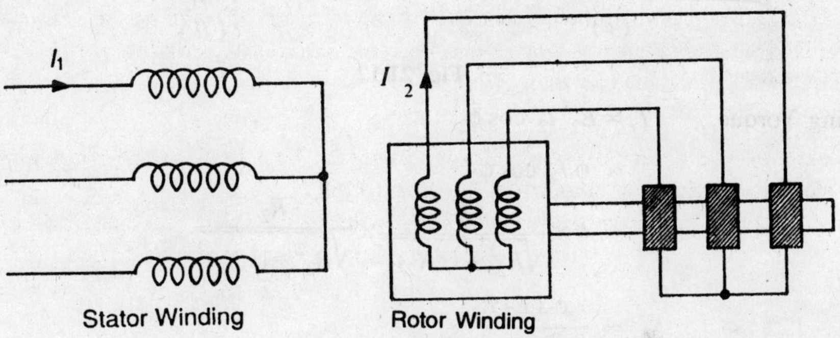

Fig. 21.11

(*iv*) With increasing load, the increased load currents I_2 are in such a direction so as to decrease the stator flux (Lenz's law), thereby decreasing the counter e.m.f. in the stator windings. The decreased counter e.m.f. allows motor stator current (I_1) to increase, thereby increasing the power input to the motor. It may be noted that action of the induction motor in adjusting its stator or primary current with changes of current in the rotor or secondary is very much similar to the changes occurring in transformer with changes in load.

21.15. Torque Under Running Conditions

Let the rotor at standstill have per phase induced e.m.f. E_2, reactance X_2 and resistance R_2. Then under running conditions at slip s,

Rotor e.m.f./phase, $E_2' = s\,E_2$

Rotor reactance/phase, $X_2' = s\,X_2$

Rotor impedance/phase, $Z_2' = \sqrt{R_2^2 + (s\,X_2)^2}$

---

* Rotor resistance is a fixed small value. The rotor frequency under running conditions is very small ($f' = sf$) and hence rotor reactance is also low. This results in low rotor impedance during running conditions.

Rotor current/phase, $I_2' = \dfrac{E_2'}{Z_2'} = \dfrac{s\,E_2}{\sqrt{R_2^2 + (s\,X_2)^2}}$

Rotor p.f., $\cos\phi_2' = \dfrac{R_2}{\sqrt{R_2^2 + (s\,X_2)^2}}$

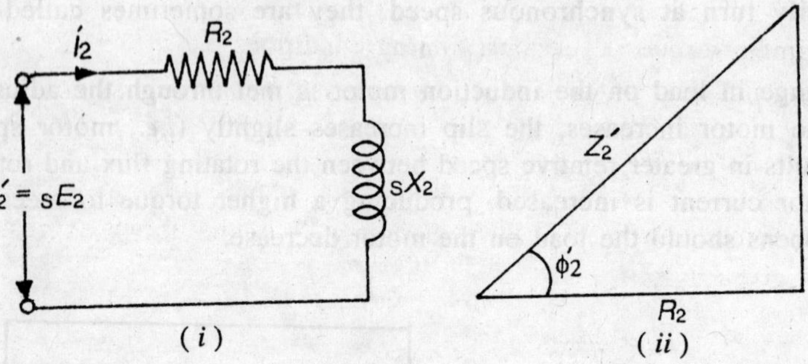

Fig. 21.12

Running Torque, $T_r \propto E_2'\,I_2' \cos\phi_2'$

$$\propto \phi\, I_2' \cos\phi_2' \qquad (\because\ E_2' \propto \phi)$$

$$\propto \phi \times \frac{s\,E_2}{\sqrt{R_2^2 + (s\,X_2)^2}} \times \frac{R_2}{\sqrt{R_2^2 + (s\,X_2)^2}}$$

$$\propto \frac{\phi\, s\, E_2\, R_2}{R_2^2 + (s\,X_2)^2}$$

$$= \frac{K\,\phi\, s\, E_2\, R_2}{R_2^2 + (s\,X_2)^2}$$

$$= \frac{K_1\, s\, E_2^2 R_2}{R_2^2 + (s\,X_2)^2} \qquad (\because\ E_2 \propto \phi)$$

If the stator supply voltage V is constant, then stator flux and hence E_2 will be constant.

$$\therefore \qquad T_r = \frac{K_2\, s\, R_2}{R_2^2 + (s\,X_2)^2}$$

where K_2 is another constant.

It may be seen that running torque is :

(*i*) directly proportional to slip *i.e.*, if slip increases (*i.e.*, motor speed decreases), the torque will increase and *vice-versa.*

(*ii*) directly proportional to square of supply voltage ($\because\ E_2 \propto V$).

21.16. Maximum Torque Under Running Conditions

$$T_r = \frac{K_2\, s\, R_2}{R_2^2 + s^2 X_2^2} \qquad \ldots(i)$$

In order to find the value of rotor resistance that gives maximum torque under running conditions, differentiate exp. (*i*) *w.r.t.* *s* and equate the result to zero *i.e.*,

$$\frac{dT_r}{ds} = \frac{[R_2 (R_2^2 + s^2 X_2^2) - 2s X_2^2 (s R_2)]}{(R_2^2 + s^2 X_2^2)^2} = 0$$

$$\text{or} \quad (R_2^2 + s^2 X_2^2) - 2s^2 X_2^2 = 0$$

$$\text{or} \quad R_2^2 = s^2 X_2^2$$

$$\text{or} \quad R_2 = s X_2$$

Thus for maximum torque (T_m) under running conditions :

Rotor resistance/phase = Fractional slip × Standstill rotor reactance/phase

Now, $$T_r \propto \frac{s R_2}{R_2^2 + s^2 X_2^2}$$ *...from exp (i) above*

For maximum torque, $R_2 = s X_2$. Putting $R_2 = s X_2$ in the above expression, the maximum torque T_m is given by ;

$$T_m \propto \frac{1}{2 X_2}$$

Slip corresponding to maximum torque, $s = R_2 / X_2$.

21.17. Torque-Slip Characteristics

As shown in Art 21.15, the motor torque under running conditions is given by ;

$$T = \frac{K_2 s R_2}{R_2^2 + s^2 X_2^2}$$

If a curve is drawn between the torque and slip for a particular value of rotor resistance R_2, the graph thus obtained is called torque-slip characteristic. Fig. 21.13 shows a family of torque-slip characteristics for a slip range from $s = 0$ to $s = 1$ for various values of rotor resistance.

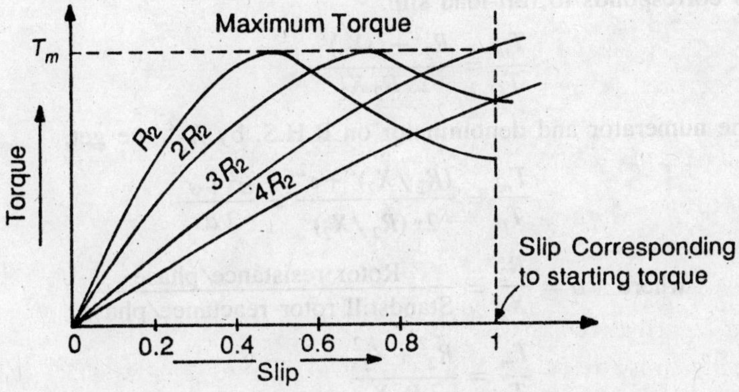

Fig. 21.13

The following points may be noted carefully :

(*i*) At $s = 0$, $T = 0$ so that torque-slip curve starts from the origin.

(*ii*) At normal speed, slip is small so that $s X_2$ is negligible as compared to R_2.

$$\therefore \quad T \propto s/R_2$$

$$\propto s \qquad \text{...as } R_2 \text{ is constant}$$

Hence, torque-slip curve is a straight line from zero slip to a slip that corresponds to full-load.

(*iii*) As slip increases beyond full-load slip, the torque increases and becomes maximum at $s = R_2/X_2$. This maximum torque in an induction motor is called *pull-out torque* or *breakdown torque*. Its value is at least twice the full-load value when the motor is operated at rated voltage and frequency.

(*iv*) When slip increases beyond that corresponding to maximum torque, the term $s^2X_2^2$ increases very rapidly so that R_2^2 may be neglected as compared to $s^2X_2^2$.

$$\therefore \quad T \propto s/s^2X_2^2$$

$$\propto 1/s \qquad \text{...as } X_2 \text{ is constant}$$

Thus the torque is now inversely proportional to slip. Hence torque-slip curve is a rectangular hyperbola.

It may be seen from the torque-slip characteristics that *addition of resistance to the rotor circuit does not change the value of maximum torque but it only changes the value of slip at which maximum torque occurs.*

21.18. Full-load, Starting and Maximum Torques

$$T_f \propto \frac{s\,R_2}{R_2^2 + (s\,X_2)^2} \qquad \text{...See Art 21.15}$$

$$T_s \propto \frac{R_2}{R_2^2 + X_2^2} \qquad \text{...See Art 21.10}$$

$$T_m \propto \frac{1}{2\,X_2} \qquad \textbf{... See Art 21.16}$$

Note that s corresponds to full-load slip.

(*i*) $\therefore$
$$\frac{T_m}{T_f} = \frac{R_2^2 + (sX_2)^2}{2s\,R_2\,X_2}$$

Dividing the numerator and denominator on R.H.S. by X_2^2, we get,

$$\frac{T_m}{T_f} = \frac{(R_2/X_2)^2 + s^2}{2s\,(R_2/X_2)} = \frac{a^2 + s^2}{2\,as}$$

$$\text{where} \quad a = \frac{R_2}{X_2} = \frac{\text{Rotor resistance/phase}}{\text{Standstill rotor reactance/phase}}$$

(*ii*)
$$\frac{T_m}{T_s} = \frac{R_2^2 + X_2^2}{2\,R_2\,X_2}$$

Dividing the numerator and denominator on R.H.S. by X_2^2, we get,

$$\frac{T_m}{T_s} = \frac{(R_2/X_2)^2 + 1}{2\,(R_2/X_2)} = \frac{a^2 + 1}{2a}$$

$$\text{where} \quad a = \frac{R_2}{X_2} = \frac{\text{Rotor resistance/phase}}{\text{Standstill rotor reactance/phase}}$$

Example 21.10. *A 50Hz, 8-pole induction motor has full-load slip of 4%. The rotor resistance and standstill reactance are 0.01 Ω and 0.1 Ω per phase respectively. Find (i) the speed at which maximum torque occurs and (ii) the ratio of maximum torque to full-load torque.*

Solution.

(*i*) Synchronous speed, $N_s = 120f/P = 120 \times 50/8 = 750\,\text{r.p.m.}$

Slip at which max. torque occurs $= R_2/X_2 = 0.01/0.1 = 0.1$

$\therefore$ Rotor speed at max. torque $= {}^*(1-0.1)\,N_s = (1-0.1)\,750 = \mathbf{675\ r.p.m.}$

(*ii*)
$$\frac{T_m}{T_f} = \frac{a^2+s^2}{2as}$$
...See Art 21.18

where s = full-load slip = 0.04

$a = R_2/X_2 = 0.01/0.1 = 0.1$

$$\therefore \quad \frac{T_m}{T_f} = \frac{(0.1)^2+(0.04)^2}{2\times 0.1\times 0.04} = \mathbf{1.45.}$$

Example 21.11. *An 8-pole, 3-phase, 50Hz induction motor has a rotor resistance of 0.025 Ω/phase and a rotor standstill reactance of 0.1 Ω/phase. At what speed is the torque maximum ? What proportion of maximum torque is the starting torque ?*

Solution.

(*i*) $N_s = 120f/P = 120 \times 50/8 = 750\,\text{r.p.m.}$

$R_2 = s\,X_2$ ...*for maximum torque*

$\therefore \quad s = R_2/X_2 = 0.025/0.1 = 0.25$

Corresponding speed, $N = (1-s)\,N_s = (1-0.25)\,750 = \mathbf{562.5\ r.p.m.}$

(*ii*)
$$\frac{T_s}{T_m} = \frac{2a}{a^2+1}$$
...See Art 21.18

$a = R_2/X_2 = 0.025/0.1 = 0.25$

$$\therefore \quad \frac{T_s}{T_m} = \frac{2\times 0.25}{(0.25)^2+1} = \mathbf{0.47}$$

21.19. Speed Regulation of Induction Motors

Like any other electrical motor, the speed regulation of an induction motor is given by ;

$$\%\text{age speed regulation} = \frac{N_0 - N_{F.L.}}{N_{F.L.}} \times 100$$

where N_0 = no-load speed of the motor

$N_{F.L.}$ = full-load speed of the motor

If the no-load speed of the motor is 800 r.p.m. and its full-load speed is 780 r.p.m., then change in speed is $800-780 = 20$ r.p.m. and percentage speed regulation $= 20 \times 100/780 = 2.56\%$.

At no load, only a small torque is required to overcome the small mechanical losses and hence motor slip is small *i.e.*, about 1%. When the motor is fully loaded, the slip increases

* $s = \dfrac{N_s - N}{N_s} = 1 - \dfrac{N}{N_s} \quad \therefore \quad N = (1-s)\,N_s$

slightly *i.e.*, motor speed decreases slightly. It is because rotor impedance is low and a small decrease in speed produces a large rotor current. The increased rotor current produces a high torque to meet the full load on the motor. For this reason, the change in speed of the motor from no-load to full-load is small *i.e.*, the speed regulation of an induction motor is low. The speed regulation of an induction motor is 3% to 5%. Although the motor speed does decrease slightly with increased load, the speed regulation is low enough that the induction motor is classed as a constant-speed motor.

21.20. Speed Control of 3-phase Induction Motors

$$N = (1 - s) N_s$$

$$= (1 - s) \frac{120 f}{P} \qquad ...(i)$$

An inspection of eq. (*i*) reveals that the speed N of an induction motor can be varied by changing (*i*) supply frequency f (*ii*) number of poles P on the stator and (*iii*) slip s. The change of frequency is generally not possible because the commercial supplies have constant frequency. Therefore, the practical methods of speed control are either to change the number of stator poles or the motor slip.

1. Squirrel cage motors. The speed of a squirrel cage motor is *changed by changing the number of stator poles. Only two or four speeds are possible by this method. Two-speed motor has one stator winding that may be switched through suitable control equipment to provide two speeds, one of which is half of the other. For instance, the winding may be connected for either 4 or 8 poles, giving synchronous speeds of 1500 and 750 r.p.m. Four-speed motors are equipped with two separate stator windings each of which provides two speeds. The disadvantages of this method are :

(*i*) It is not possible to obtain gradual continuous speed control.

(*ii*) Because of the complications in the design and switching of the interconnections of the stator winding, this method can provide a maximum of four different synchronous speeds for any one motor.

2. Wound rotor motors. The speed of wound rotor motors is **changed by changing the motor slip. This can be achieved by :

(*i*) varying the stator line voltage.

(*ii*) varying the resistance of the rotor circuit.

(*iii*) inserting and varying a foreign voltage in the rotor circuit.

21.21. Power Factor of Induction Motor

Like any other a.c. machine, the power factor of an induction motor is given by ;

$$\text{Power factor, } \cos\phi = \frac{\text{Active component of current } (I\cos\phi)}{\text{Total current } (I)}$$

The presence of air-gap between the stator and rotor of an induction motor greatly increases the reluctance of the magnetic circuit. Consequently, an induction motor draws a large magnetising current (I_m) to produce the required flux in the air-gap.

---

* The slip of an induction motor can be changed by changing the characteristics of the rotor circuit. Since the bars of a squirrel cage rotor are permanently short-circuited, the motor slip cannot be changed.

** Pole-changing speed control is generally not practicable for wound rotor motors.

(*i*) At no load, an induction motor draws a large magnetising current and a small active component to meet the no-load losses. Therefore, the induction motor takes a †high no-load current lagging the applied voltage by a large angle. Hence, the power factor of an induction motor on no load is low *i.e.*, about 0.1 lagging.

(*ii*) When an induction motor is loaded, the active component of current increases while the magnetising component remains about the same. Consequently, the power factor of the motor is increased. However, because of the large value of magnetising current, which is present regardless of load, the power factor of an induction motor even at full load seldom exceeds 0.9 lagging.

21.22. Power Stages in An Induction Motor

The input electric power fed to the stator of the motor is converted into mechanical power at the shaft of the motor. The various losses during this energy conversion are :

1. Fixed losses

(*i*) Stator iron loss (*ii*) Friction and windage loss

The rotor iron loss is negligible because the frequency of rotor currents under normal running condition is small.

2. Variable losses

(*i*) Stator copper loss (*ii*) Rotor copper loss

Fig. 21.14 shows how electric power fed to the stator of an induction motor suffers losses and finally converted into mechanical power.

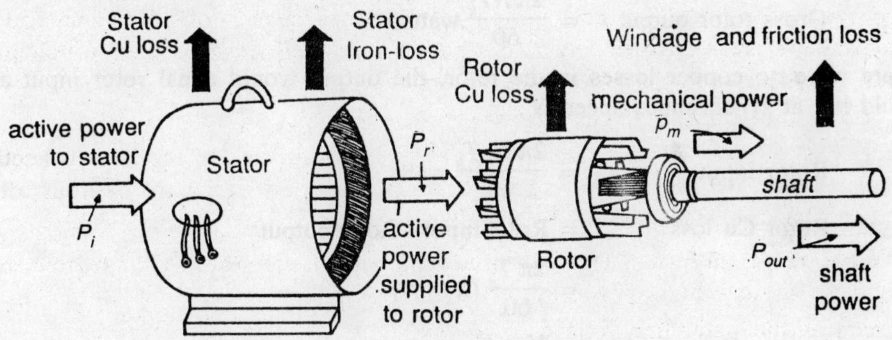

Fig. 21.14

The following points may be noted from the above diagram :

(*i*) Stator input, P_i = Stator output + Stator losses

= Stator output + Stator Iron loss + Stator Cu loss

(*ii*) Rotor input, P_r = Stator output

It is because stator output is entirely transferred to the rotor through air-gap by electromagnetic induction.

(*iii*) Mechanical power available, $P_m = P_r -$ Rotor Cu loss

This mechanical power available is the gross rotor output and will produce a gross torque T_g.

(iv) Mechanical power at shaft, $P_{out} = P_m -$ Friction and windage loss

Mechanical power available at the shaft produces a shaft torque T_{sh}.

Clearly, $P_m - P_{out}$ = Friction and windage loss

† as compared to transformer no-load current.

21.23. Induction Motor Torque

The mechanical power P available from any electric motor can be expressed as :

$$P = \frac{2\pi NT}{60} \text{ watts}$$

where N = speed of the motor in r.p.m.

T = torque developed in $N\text{-}m$

$$\therefore \quad T = \frac{60}{2\pi}\frac{P}{N} = 9.55\frac{P}{N} \; N\text{-}m$$

If the gross output of the rotor of an induction motor is P_m and its speed is N r.p.m., then gross torque T_g developed is given by ;

$$T_g = 9.55\frac{P_m}{N} \; N\text{-}m$$

Similarly,

$$T_{sh} = 9.55\frac{P_{out}}{N} \; N\text{-}m.$$

Note. Since windage and friction loss is small, $T_g = T_{sh}$. This assumption hardly leads to any significant error.

21.24. Rotor Output

If T_g newton-metre is the gross torque developed and N r.p.m. is the speed of the rotor, then

$$\text{Gross rotor output} = \frac{2\pi NT_g}{60} \text{ watts}$$

If there were no copper losses in the rotor, the output would equal rotor input and the rotor would run at synchronous speed N_s.

$$\therefore \quad \text{Rotor input} = \frac{2\pi N_s T_g}{60} \text{ watts}$$

$$\therefore \quad \text{Rotor Cu loss} = \text{Rotor input} - \text{Rotor output}$$

$$= \frac{2\pi T_g}{60}(N_s - N)$$

(i) $$\frac{\text{Rotor Cu loss}}{\text{Rotor input}} = \frac{N_s - N}{N_s} = s$$

$$\therefore \quad \text{Rotor Cu loss} = s \times \text{Rotor input}$$

(ii) Gross rotor output, $$P_m = \text{Rotor input} - \text{rotor Cu loss}$$

$$= \text{Rotor input} - s \times \text{rotor input}$$

$$\therefore \quad P_m = \text{Rotor input}\,(1 - s)$$

(iii) $$\frac{\text{Gross rotor output}}{\text{Rotor input}} = 1 - s = \frac{N}{N_s}$$

(iv) $$\frac{\text{Rotor Cu loss}}{\text{Gross rotor output}} = \frac{s}{1 - s}$$

It is clear that if the input power to rotor is P_r, then $s\,P_r$ is lost as rotor Cu loss and the remaining $(1 - s)\,P_r$ is converted into mechanical power. Consequently, induction motor operating at high slip has poor efficiency.

Note.

$$\frac{\text{Gross rotor output}}{\text{Rotor input}} = 1 - s$$

If the stator losses as well as friction and windage losses are neglected, then,

$$\text{Gross rotor output} = \text{Useful output}$$

$$\text{Rotor input} = \text{Stator input}$$

$$\therefore \quad \frac{\text{Useful output}}{\text{Stator input}} = 1 - s = \text{Efficiency}$$

Hence the approximate efficiency of an induction motor is $1 - s$. Thus if the slip of an induction motor is 0.125, then its approximate efficiency is $= 1 - 0.125 = 0.875$ or 87.5%.

Example 21.12. *The input power to a 6-pole, 3-phase, 50 Hz induction motor is 42 kW; the speed is 970 r.p.m. The stator losses are 1.2 kW and the friction and windage losses 1.8 kW. Find (i) the rotor Cu loss and (ii) the efficiency of the motor.*

Solution.

$$\text{Synchronous speed, } N_s = 120f/P = 120 \times 50/6 = 1000 \text{ r.p.m.}$$

$$\text{Slip, } \quad s = \frac{N_s - N}{N_s} = \frac{1000 - 970}{1000} = 0.03$$

(*i*)

$$\text{Stator output} = \text{Stator input} - \text{stator losses}$$

$$= 42 - 1.2 = 40.8 \text{ kW}$$

$$\text{Rotor input} = \text{Stator output} = 40.8 \text{ kW}$$

$$\text{Rotor Cu loss} = s \times \text{Rotor input} = 0.03 \times 40.8 = \mathbf{1.224 \text{ kW}}$$

(*ii*)

$$\textbf{Useful output, } P_{out} = \textbf{Rotor input} - \textbf{Rotor Cu loss} - \textbf{Windage and friction loss}$$

$$= 40.8 - 1.224 - 1.8 = 37.77 \text{ kW}$$

$$\therefore \quad \text{Motor efficiency} = \frac{37.77}{42} \times 100 = \mathbf{89.9\%}$$

Example 21.13. *A 7.46kW, 230V, 3-phase 50Hz, 6-pole squirrel cage induction motor operates at full-load slip of 4% when rated voltage and rated frequency are applied. Determine (i) speed of rotation of stator m.m.f. (ii) full-load speed (iii) frequency of rotor current under this condition and (iv) full-load torque in newton-metre.*

Solution.

(*i*) The stator m.m.f. revolves at synchronous speed.

$$N_s = 120f/P = 120 \times 50/6 = \mathbf{1000 \text{ r.p.m.}}$$

(*ii*)

$$N = (1 - s) N_s = (1 - 0.04)\, 1000 = \mathbf{960 \text{ r.p.m.}}$$

(*iii*)

$$f' = sf = 0.04 \times 50 = \mathbf{2 \text{ Hz}}$$

(*iv*) Neglecting windage and friction loss, $T_g = T_{sh}$

$$\therefore \quad T_g = 9.55 \frac{P_{out}}{N} = 9.55 \times \frac{7.46 \times 10^3}{960} = \mathbf{74.2 \text{ N-m}}$$

Example 21.14. *The full-load power input to a 3-phase induction motor is 50 kW and the slip is 3%. Neglecting stator losses, calculate (i) the full-load rotor Cu losses per phase and (ii) total mechanical power developed.*

Solution.

(*i*) Rotor input = 50 kW ...*Stator losses neglected*

$$\text{Rotor Cu loss} = s \times \text{rotor input} = 0.03 \times 50 = 1.5 \text{ kW}$$

$$\text{Rotor Cu loss/phase} = 1.5/3 = \mathbf{0.5 \text{ kW}}$$

(*ii*) Total mechanical power developed

$$P_m = 50 - 1.5 = \mathbf{48.5\ kW}$$

Example 21.15. *A 6-pole, 50 Hz, 3-phase induction motor runs at 960 r.p.m. when the torque on the shaft is 200 N-m. If the stator losses are 1500 W and friction and windage losses are 500 W, find (i) rotor Cu loss and (ii) efficiency of the motor.*

Solution.

$$N_s = 120f/P = 120 \times 50/6 = 1000 \text{ r.p.m.}$$

$$s = \frac{N_s - N}{N_s} = \frac{1000 - 960}{1000} = 0.04$$

(*i*) Rotor output, $P_{out} = \frac{2\pi N T_{sh}}{60} = \frac{2\pi \times 960 \times 200}{60} = 20096 \text{ W}$

Gross rotor output, $P_m = 20096 + 500 = 20596 \text{ W}$

$$\frac{\text{Rotor Cu loss}}{\text{Gross rotor output}} = \frac{s}{1-s}$$ ...See Art 21.24

∴ Rotor Cu loss $= \frac{0.04}{1 - 0.04} \times 20596 = \mathbf{858\ W}$

(*ii*) Rotor input = Stator output = 20596 + 858 = 21454 W

Stator input, $P_i = 21454 + 1500 = 22954 \text{ W}$

∴ Efficiency $= \frac{P_{out}}{P_i} \times 100 = \frac{20096}{22954} \times 100 = \mathbf{87.5\%}$

Tutorial Problems

1. A 3-phase, 6-pole, 50Hz induction motor develops 3.68kW at 950 r.p.m. If the stator losses are 300W, find the stator input. **[4.17kW]**
2. The power input to a 3-phase induction motor is 40 kW. The stator losses total 1 kW and the mechanical losses are 1.6 kW. Determine (*i*) shaft power (*ii*) rotor Cu loss and (*iii*) efficiency of the motor with a slip of 3.5%. [(*i*) **36kW** (*ii*) **445W** (*iii*) **90%**]
3. An 8-pole, 3-phase, 50Hz induction motor is taking 50kW and running at 725 r.p.m. Stator losses are 1.2 kW and frictional losses 1.8 kW. Find (*i*) rotor Cu loss and (*ii*) efficiency of the motor. [(*i*) **1.627 kW** (*ii*) **90.75%**]
4. A 37.3kW, 4-pole, 50Hz induction motor has a **friction** and windage torque of 22 N-m. The stator losses equal rotor Cu loss. Calculate the powei input to the stator when delivering full-load output at a speed of 1440 r.p.m. **[44kW]**
5. A 50Hz, 8-pole, 3-phase induction motor delivers an output of 3.1 kW at a slip of 0.02. The losses due to friction and windage are 100W. Find the value of gross torque in N-m. **[41.58 N-m]**

21.25. Methods of Starting Squirrel Cage Motors

The following four methods are available for starting squirrel cage motors :

(*i*) **Direct on line starting.** This method of starting is just what the name implies—the motor is started by connecting it directly to 3-phase supply. The impedance of the motor at standstill is relatively low and when it is directly connected to the supply system, the starting current will be high (4 to 10 times the full-load current) and at a low power factor. Consequently, this method of starting is suitable for relatively small (upto 7.5kW) machines.

(*ii*) **Stator resistance starting.** In this method, external resistances are connected in series with each phase of stator winding during starting. This causes voltage drop across the resistances so that voltage available across motor terminals is reduced and hence the starting current. The starting resistances are gradually cut out in steps (two or more steps) from the stator circuit as the motor picks up speed. When the motor attains rated speed, the resistances are completely cut out and full line voltage is applied to the motor.

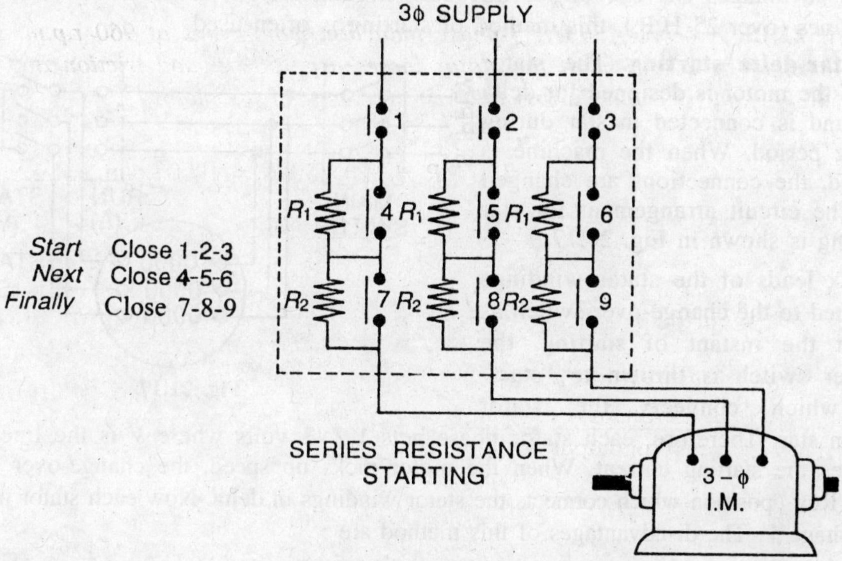

Fig. 21.15

This method suffers from two drawbacks. First, the reduced voltage applied to the motor during the starting period *lowers the starting torque and hence increases the accelerating time. Secondly, a lot of power is wasted in the starting resistances.

(iii) **Auto transformer starting.** This method also aims at connecting the induction motor to a reduced supply at starting and then connecting it to the full voltage as the motor picks up sufficient speed. Fig. 21.16 shows the circuit arrangement for auto transformer starting. The tapping on the auto transformer is so set that when it is in the circuit, 65% to 80% of line voltage is applied to the motor.

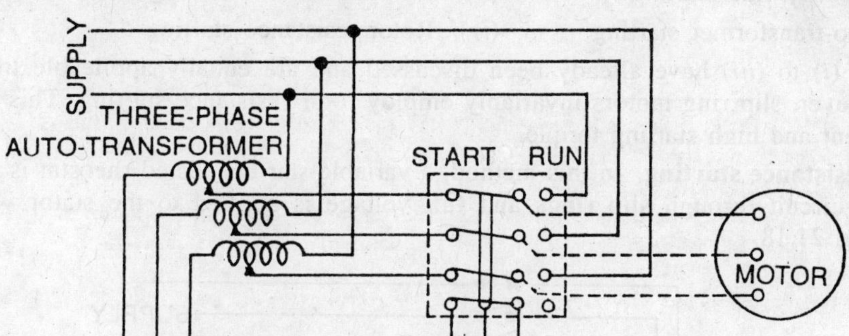

Fig. 21.16

At the instant of starting, the change-over switch is thrown to "start" position. This puts the auto transformer in the circuit and thus reduced voltage is applied to the circuit. Consequently, starting current is limited to safe value. When the motor attains about 80% of

* As shown in Art 21.12, $T_s \propto V^2$ where V is the supply voltage.

normal speed, the change-over switch is thrown to "run" position. This takes out the auto-transformer from the circuit and puts the motor to full line voltage. Auto transformer starting has several advantages *viz.* low power loss, low starting current and less radiated heat. For large machines (over 25 H.P.), this method of starting is often used.

(*iv*) **Star-delta starting.** The stator winding of the motor is designed for delta operation and is connected in star during the starting period. When the machine is up to speed, the connections are changed to delta. The circuit arrangement for star delta starting is shown in Fig. 21.17.

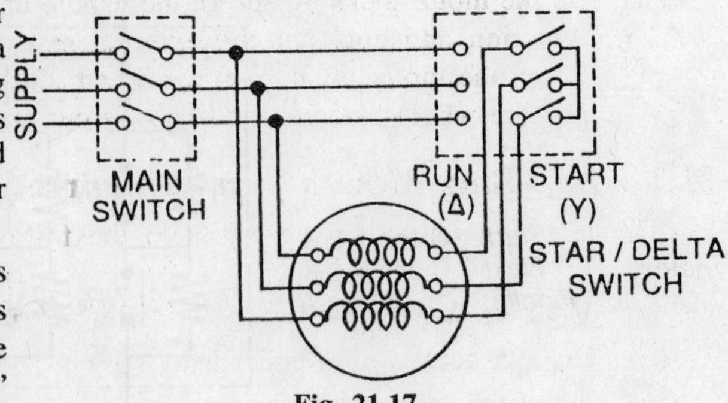

Fig. 21.17

The six leads of the stator windings are connected to the change-over switch as shown. At the instant of starting, the change-over switch is thrown to "start" position which connects the stator windings in star. Therefore, each stator phase gets $V/\sqrt{3}$ volts where V is the line voltage. This reduces the starting current. When the motor picks up speed, the change-over switch is thrown to "Run" position which connects the stator windings in delta. Now each stator phase gets full line voltage V. The disadvantages of this method are :

(*a*) With star-connection during starting, stator phase voltage is $1/\sqrt{3}$ times the line voltage. Consequently, starting torque is $(1/\sqrt{3})^2$ or 1/3 times the value it would have with Δ-connection. This is rather a large reduction in starting torque.

(*b*) The reduction in voltage is fixed.

This method of starting is used for medium-size machines (up to about 25 H.P.).

21.26. Methods of Starting Slip Ring Motors

The following four methods are available for starting slip ring motors :

(*i*) Direct-on-line starting (*ii*) Stator resistance starting

(*iii*) Auto-transformer starting (*iv*) Rotor resistance starting

Methods (*i*) to (*iii*) have already been discussed and are equally applicable to slip ring motors. However, slip ring motors invariably employ rotor resistance starting. This gives low starting current and high starting torque.

Rotor resistance starting. In this method, a variable star-connected rheostat is connected in the rotor circuit through slip rings and full voltage is applied to the stator winding as shown in Fig. 21.18.

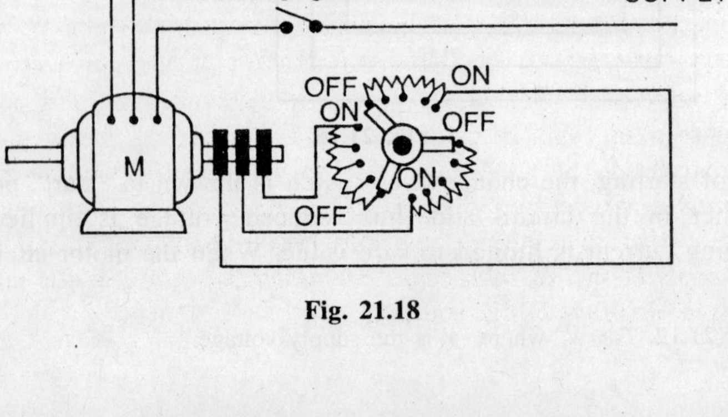

Fig. 21.18

(*i*) At starting, the handle of rheostat is set in the *OFF* position so that maximum resistance is placed in each phase of the rotor circuit. This reduces the starting current and at the same time starting torque is increased.

(*ii*) As the motor picks up speed, the handle of rheostat is gradually moved in clockwise direction and cuts out the external resistance in each phase of the rotor circuit. When the motor attains normal speed, the change-over switch is in the *ON* position and the whole external resistance is cut out from the rotor circuit.

21.27. Slip-Ring Motors Versus Squirrel Cage Motors

The slip-ring induction motors have the following advantages over the squirrel cage motors :

(*i*) High starting torque with low starting current.

(*ii*) Smooth acceleration under heavy loads.

(*iii*) No abnormal heating during starting.

(*iv*) Good running characteristics after external rotor resistances are cut out.

(*v*) Adjustable speed.

The disadvantages of slip-ring motors are :

(*i*) The initial and maintenance costs are greater than those of squirrel cage motors.

(*ii*) The speed regulation is poor when run with resistance in the rotor circuit.

21.28. Abnormal Conditions

There are a variety of abnormal conditions in the operation of a 3-phase induction motor. Some of them are :

(*i*) **Mechanical overload.** Although standard induction motors can develop twice their rated power for short periods, they should not be allowed to run continuously beyond their rated capacity. Overloads cause overheating which deteriorates the insulation and reduces its useful life. In practice, the higher motor current causes thermal overload relay to trip, bringing the motor to a stop before its temperature gets too high.

(*ii*) **Line voltage variations.** The torque of an induction motor at any speed is directly proportional to the square of the applied voltage. Thus, if the stator voltage decreases by 10%, the torque will drop by 20%. The voltage drop in the line is often produced during start-up due to the large starting current drawn from the line. On the other hand, if the line voltage is too high, the flux per pole will also be high. This increases both the iron losses and the magnetising current. As a result, the temperature increases slightly and the power factor is somewhat lowered.

(*iii*) **Single-phasing.** If one line of a 3-phase line is accidentally broken or if a fuse blows while the motor is running, the machine will continue to run as a single-phase motor. The current drawn from the remaining two lines will almost double and the motor will begin to overheat. This large current may trip the circuit breaker or may blow the fuses.

(*iv*) **Frequency variation.** The most important consequence of a frequency change ($f = NP/120$) is the resulting change in motor speed. Thus if the frequency drops by 20%, the motor speed also drops by 20%. Machine tools and other motor-driven equipment imported from countries where frequency is 60 Hz, may cause problems when they are connected to 50 Hz system. Everything runs 20% faster than normal and this may not be acceptable in some applications. In such cases, we either have to gear the motor speed down or supply an expensive auxiliary 50 Hz source.

Note. A 60 Hz motor can also operate on a 50 Hz line but its terminal voltage should be reduced to 5/6 (or 83%) of its nameplate value. The breakdown torque and starting torque are then about the same as before. Power factor, efficiency and temperature rise remain satisfactory.

OBJECTIVE QUESTIONS

1. The stator of a 3-phase induction motor produces .... magnetic field.
(*i*) steady (*ii*) rotating
(*iii*) alternating (*iv*) none of the above

2. An induction motor is preferred to a d.c. motor because it ....
(*i*) provides high starting torque
(*ii*) provides fine speed control
(*iii*) has simple and rugged construction
(*iv*) none of the above

3. A 3-phase induction motor is ....
(*i*) essentially a constant-speed motor
(*ii*) a variable speed motor
(*iii*) very costly
(*iv*) not easily maintainable

4. The air-gap between stator and rotor of a 3-phase induction motor ranges from ....
(*i*) 2 cm to 4 cm (*ii*) 0.4 mm to 4 mm
(*iii*) 1 cm to 2 cm (*iv*) 4 cm to 6 cm

5. If the frequency of 3-phase supply to the stator of a 3-phase induction motor is increased, then synchronous speed ....
(*i*) is decreased
(*ii*) is increased
(*iii*) remains unchanged
(*iv*) none of the above

6. If N_s is the speed of rotating flux and N the speed of the rotor, then, the rate at which the flux cuts the rotor conductors is directly proportional to ....
(*i*) N_s (*ii*) N
(*iii*) $N_s - N$ (*iv*) $N - N_s$

7. In a 3-phase induction motor, the rotor speed is .... the synchronous speed.
(*i*) greater than (*ii*) smaller than
(*iii*) equal to (*iv*) none of the above

8. The synchronous speed of a 3-phase induction motor having 20 poles and connected to a 50 Hz source is ....
(*i*) 600 r.p.m. (*ii*) 1000 r.p.m.
(*iii*) 1200 r.p.m. (*iv*) 300 r.p.m.

9. The relation among synchronous speed (N_s), rotor speed (N) and slip (s) is ....
(*i*) $N = (s - 1) N_s$ (*ii*) $N = (1 - s) N_s$
(*iii*) $N = (1 + s) N_s$ (*iv*) $N = sN_s$

10. When a 3-phase induction motor is at no load, the slip is ....
(*i*) 1 (*ii*) 0.5
(*iii*) practically zero (*iv*) 0.2

11. When the rotor of a 3-phase induction motor is blocked, the slip is ....
(*i*) zero (*ii*) 0.5
(*iii*) 0.1 (*iv*) 1

12. The speed of the rotating flux in Fig. 21.19 will be ....
(*i*) 1500 r.p.m. (*ii*) 1000 r.p.m.
(*iii*) 750 r.p.m. (*iv*) none of the above

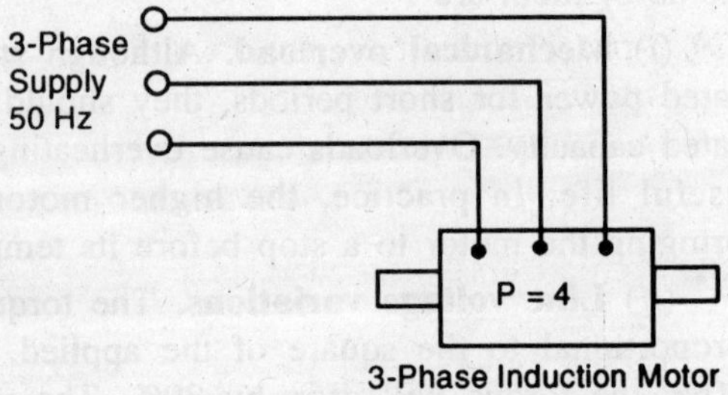

Fig. 21.19

13. If the induction motor shown in Fig. 21.19 runs at 1450 r.p.m., then, slip is ....
(*i*) 50% (*ii*) 3.3%
(*iii*) 5% (*iv*) none of the above

14. In Fig. 21.19 the rotor frequency is .... when the motor is at standstill.
(*i*) zero (*ii*) 25 Hz
(*iii*) 50 Hz (*iv*) none of the above

15. If one of the lines in Fig. 21.19 is interchanged, then, ...
(*i*) flux speed increases
(*ii*) flux speed decreases
(*iii*) flux speed remains same but direction is reversed

(*iv*) none of the above

16. If a 4-pole induction motor has a synchronous speed of 1500 r.p.m., then, supply frequency is ....

(*i*) 50 Hz (*ii*) 25 Hz
(*iii*) 60 Hz (*iv*) none of the above

17. The full-load slip of a 3-phase induction motor ranges from ....

(*i*) 10% to 20% (*ii*) 20% to 30%
(*iii*) 2% to 5 % (*iv*) none of the above

18. The direction of rotation of field in a 3-phase induction motor depends upon ....

(*i*) number of poles
(*ii*) magnitude of supply voltage
(*iii*) supply frequency
(**iv**) phase sequence of supply voltage

19. The rotor winding of a 3-phase wound rotor induction motor is generally .... connected.

(*i*) Star (*ii*) delta
(*iii*) partly star and partly delta
(*iv*) none of the above

20. 3-phase wound rotor motors are also called .... motors.

(*i*) synchronous (*ii*) slip ring
(*iii*) series (*iv*) commutator

21. The advantage of wound rotor motor is that ....

(*i*) it is inexpensive
(*ii*) it requires less maintenance
(*iii*) external resistance can be inserted in the rotor circuit
(*iv*) none of the above

22. A wound rotor motor is mainly used in application where ....

(*i*) high starting torque is required
(*ii*) speed control is required
(*iii*) less costly motor is not required
(*iv*) high rotor resistance is required during running

23. The torque characteristic of a 3-phase induction motor is similar to that of ....

(*i*) d.c. series motor (*ii*) d.c. shunt motor
(*iii*) d.c. differentially compounded motor
(*iv*) d.c. cumulatively compounded motor

24. In a wound rotor motor, the rotor winding is wound for poles .... that of stator winding.

(*i*) equal to (*ii*) greater than
(*iii*) small than (*iv*) none of the above

25. Wound rotor motors are less extensively used than squirrel cage motors because ....

(*i*) slip rings are required on the rotor circuit
(*ii*) rotor windings are generally Y-connected
(*iii*) they are costly and require greater maintenance
(*iv*) none of the above

26. A 4-pole, 50 Hz induction motor operates at 5% slip. The frequency of e.m.f. induced in the rotor will be ....

(*i*) 25 Hz (*ii*) 50 Hz
(*iii*) 2.5 Hz (*iv*) none of the above

27. The maximum voltage is induced in the rotor of a 3-phase induction motor when it ....

(*i*) runs at no load (*ii*) runs at full load
(*iii*) is blocked (*iv*) none of the above

28. The reactance of the rotor circuit of a 3-phase induction motor is maximum at ....

(*i*) no-load (*ii*) full-load
(*iii*) half full-load (*iv*) starting

29. The rotor current in a 3-phase induction motor is .... slip.

(*i*) inversely proportional to
(*ii*) directly proportional to
(*iii*) independent of
(*iv*) none of the above

30. At starting, rotor reactance of a 3-phase induction motor is .... rotor resistance.

(*i*) small as compared to
(*ii*) equal to that of
(*iii*) large as compared to
(*iv*) none of the above

ANSWERS TO OBJECTIVE QUESTIONS

1. (*ii*)	**2.** (*iii*)	**3.** (*i*)	**4.** (*ii*)	**5.** (*ii*)	**6.** (*iii*)	**7.** (*ii*)	**8.** (*iv*)
9. (*ii*)	**10.** (*iii*)	**11.** (*iv*)	**12.** (*i*)	**13.** (*ii*)	**14.** (*iii*)	**15.** (*iii*)	**16.** (*i*)
17. (*iii*)	**18.** (*iv*)	**19.** (*i*)	**20.** (*ii*)	**21.** (*iii*)	**22.** (*i*)	**23.** (*ii*)	**24.** (*i*)
25. (*iii*)	**26.** (*iii*)	**27.** (*iii*)	**28.** (*iv*)	**29.** (*ii*)	**30.** (*iii*)		

HINTS TO SELECTED OBJECTIVE QUESTIONS

3. At no load, the rotor lags behind the stator flux by only a small amount since the only torque required is that needed to overcome small no-load losses. As mechanical load is added, the rotor speed decreases. A decrease in rotor speed allows the constant speed rotating field to sweep across the rotor conductors at a faster rate, thereby inducing larger rotor current (since rotor impedance is low). This results in a large increase in torque which tends to bring the speed to the original value. Although the motor speed does decrease slightly with increased load, the speed regulation is good enough that induction motor is classed as a constant-speed motor.

5. $f = \dfrac{N_s P}{120}$

It is clear that $f \propto N_s$ (synchronous speed).

8. $N_s = \dfrac{120 f}{P} = \dfrac{120 \times 50}{20} = 300$ r.p.m.

9. $s = \dfrac{N_s - N}{N_s} = \dfrac{N_s}{N_s} - \dfrac{N}{N_s} = 1 - \dfrac{N}{N_s}$

$\therefore \quad N = (1 - s)N_s$

11. $s = \dfrac{N_s - N}{N_s}$

When the rotor is blocked (*i.e.* rotor cannot rotate), $N = 0$ so that $s = N_s/N_s = 1$

14. Rotor frequency, $f' = sf$ where f is supply frequency. At standstill, $s = 1$ so that $f' = f = 50$ Hz

16. $f = \dfrac{N_s P}{120} = \dfrac{1500 \times 4}{120} = 50$ Hz

19. So that additional external resistance may be inserted in the rotor circuit at starting to increase the starting torque and decrease the starting current. As the motor gains the speed, these external resistances are cut out of the rotor circuit.

22. At starting, the rotor frequency and reactance are high because s = 1. This means that at starting, rotor reactance is very large as compared to rotor resistance. Thus in the highly reactive rotor circuit, the rotor currents lag the rotor e.m.f. by a large angle. This results in high starting currents at low power factor which leads to a low value of starting torque.

If the rotor circuit has high resistance, rotor p.f. and hence the starting characteristics will improve. This can be easily done in a wound rotor by inserting external resistances in the rotor circuit at the time of starting.

26. $f' = sf$

where f' = rotor frequency

f = supply frequency (*i.e.* 50 Hz)

s = decimal slip

∴ $f' = 0.05 \times 50 = 2.5$ Hz

28. $X'_2 = s X_2$

where X'_2 = rotor reactance at slip s

X_2 = rotor reactance at standstill

Since $s = 1$ at starting, rotor reactance will be maximum under starting conditions.

29. $$I'_2 = \frac{E'_2}{Z'_2} = \frac{sE_2}{\sqrt{R_2^2 + (sX_2)^2}}$$

where I'_2 = rotor current per phase

E_2 = rotor e.m.f./phase at standstill

X_2 = rotor reactance/phase at standstill

R_2 = rotor resistance/phase

At normal speed, close to synchronism, sX_2 is very small and may be neglected. Since E_2 and R_2 are fixed,

$$I'_2 \propto s$$

22

Single-Phase Motors

Introduction

As the name suggests, these motors are used on single-phase supply. Single-phase motors are the most familiar of all electric motors because they are extensively used in home appliances, shops, offices *etc.* It is true that single-phase motors are less efficient substitute for 3-phase motors but 3-phase power is normally not available except in large commercial and industrial establishments. Since electric power was originally generated and distributed for lighting only, millions of homes were given single-phase supply. This lead to the development of single-phase motors. Even where 3-phase mains are present, the single-phase supply may be obtained by using one of the three lines and the neutral. In this chapter, we shall focus our attention on the construction, working and characteristics of commonly used single-phase motors.

22.1. Types of Single-Phase Motors

Single-phase motors are generally built in the fractional-horsepower range and may be classified into the following four basic types :

1. Single-phase induction motors

(*i*) split-phase type (*ii*) capacitor type (*iii*) shaded-pole type

2. A.C. series motor or universal motor

3. Repulsion motors

(*i*) Repulsion-start induction-run motor (*ii*) Repulsion-induction motor

4. Synchronous motors

(*i*) Reluctance motor (*ii*) Hysteresis motor

22.2. Single-phase Induction Motors

A single phase motor is very similar to a 3-phase squirrel cage induction motor. It has (*i*) a squirrel-cage rotor identical to a 3-phase motor and (*ii*) a single-phase winding on the stator.

Unlike a 3-phase induction motor, a single-phase induction motor is not self-starting but requires some starting means. The single-phase stator winding produces a magnetic field that pulsates in strength in a sinusoidal manner. The field polarity reverses after each half cycle *but the field does not rotate.* Consequently, the alternating flux cannot produce rotation in a stationary squirrel-cage rotor. However, if the rotor of a single-phase motor is rotated in one direction by some mechanical means, it will continue to run in the direction of rotation. As a matter of fact, the rotor quickly accelerates until it reaches a speed slightly below the synchronous speed. Once the motor is running at this speed, it will continue to rotate even though single-phase current is flowing through the stator winding. This method of starting is generally not convenient for large motors. Nor can it be employed for a motor located at some inaccessible spot.

Making self-starting. To make a single-phase induction motor self-starting, we should somehow produce a revolving stator magnetic field. This may be achieved by converting a single-phase supply into two phase supply through the use of an additional winding. When the motor attains sufficient speed, the starting means (*i.e.*, additional winding) may be removed depending upon the type of the motor. As a matter of fact, single-phase induction motors are classified and named according to the method employed to make them self-starting.

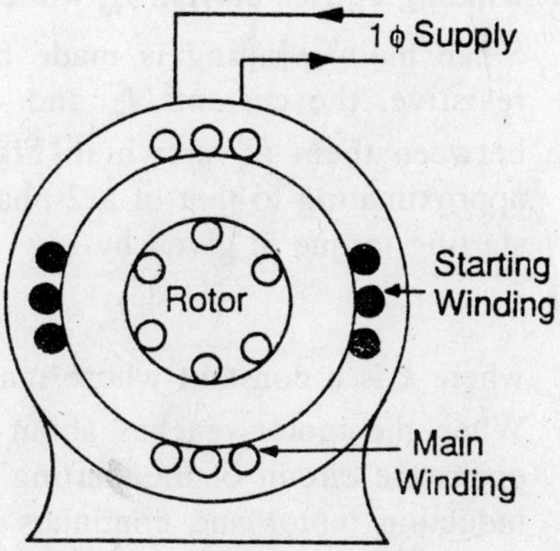

Fig. 22.1

(*i*) *Split-phase motors*—started by two-phase motor action through the use of an auxiliary or starting winding.

(*ii*) *Capacitor motors*—started by two-phase motor action through the use of an auxiliary winding and a capacitor.

(*iii*) *Shaded-pole motors*—started by the motion of the magnetic field produced by means of a shading coil around a portion of the pole structure.

22.3. Split-phase Induction Motor

The stator of a split-phase motor is provided with an auxiliary or starting winding *S* in addition to the main or running winding *M*. The starting winding is located 90° electrical from the main winding [See Fig. 22.2 (*i*)] and operates only during the brief period when the motor starts up. The two windings are so *designed that the starting winding *S* has a high resistance and relatively small reactance while the main winding *M* has relatively low resistance and large reactance as shown in the schematic connections in Fig. 22.2 (*ii*). Consequently, the currents flowing in the two windings have reasonable phase difference α (25° to 30°) as shown in the phasor diagram in Fig. 22.2 (*iii*).

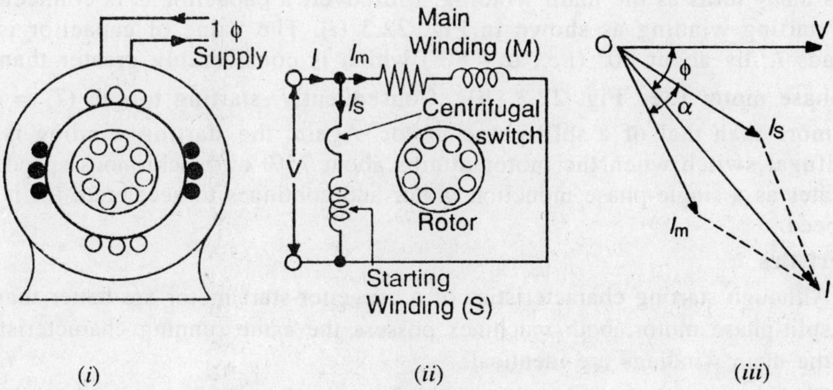

Fig. 22.2

Operation.

* This can be done by having main winding of large diameter and number of turns while starting winding of a fine wire of a small number of turns.

(*i*) When the two stator windings are energised from a single-phase supply, the main winding carries current I_m while the starting winding carries current I_s.

(*ii*) Since main winding is made highly inductive while the starting winding highly resistive, the currents I_m and I_s have a reasonable phase angle α (25° to 30°) between them as shown in Fig. 22.2 (*iii*). Consequently, a weak revolving field approximating to that of a 2-phase machine is produced which starts the motor. The starting torque is given by :

$$T_s = k\, I_m\, I_s \sin\alpha$$

where k is a constant whose magnitude depends upon the design of the motor.

(*iii*) When the motor reaches about 75% of synchronous speed, the centrifugal switch opens the circuit of the starting winding. The motor then operates as a single-phase induction motor and continues to accelerate till it reaches the normal speed. The normal speed of the motor is below the synchronous speed and depends upon the load on the motor.

Characteristics

(*i*) The starting torque is 1.5 to 2times the full-load torque and the starting current is 6 to 8 times the full-load current.

(*ii*) Due to their low cost, split-phase induction motors are most popular single-phase motors in the market.

(*iii*) Since the starting winding is made of fine wire, the current density is high and the winding heats up quickly. If the starting period exceeds 5 seconds, the winding may burn out unless the motor is protected by built-in-thermal relay. This motor is, therefore, suitable where starting periods are not frequent.

(*iv*) These motors are suitable where a moderate starting torque is required and where starting periods are infrequent *e.g.*, to drive :

(*a*) fans (*b*) washing machines (*c*) oil burners (*d*) small machine tools *etc.*

The power rating of such motors generally lies between 60W and 250W.

22.4. Capacitor-start Motor

The capacitor-start motor is identical to a split-phase motor except that the starting winding has as many turns as the main winding. Moreover, a capacitor C is connected in series with the starting winding as shown in Fig. 22.3 (*i*). The value of capacitor is so chosen that I_s leads I_m by about 80° (*i.e.*, $\alpha \simeq 80°$) which is considerably greater than 25° found in split-phase motor [See Fig. 22.3 (*ii*)]. Consequently, starting torque ($T_s = k\, I_m\, I_s \sin\alpha$) is much more than that of a split-phase motor. Again, the starting winding is opened by the centrifugal switch when the motor attains about 75% of synchronous speed. The motor then operates as a single-phase induction motor and continues to accelerate till it reaches the normal speed.

Characteristics

(*i*) Although starting characteristics of a capacitor-start motor are better than those of a split-phase motor, both machines possess the same running characteristics because the main windings are identical.

(*ii*) The phase angle between the two currents is about 80° compared to about 25° in a split-phase motor. Consequently, for the same starting torque, the current in the starting winding is only about half that in a split-phase motor. Therefore, the starting winding of a capacitor-start motor heats up less quickly and is well suited to applications involving either frequent or prolonged starting periods.

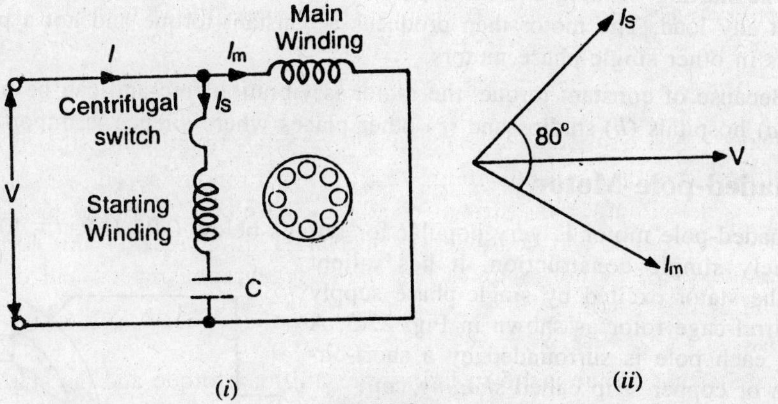

Fig. 22.3

(*iii*) Capacitor-start motors are used where high starting torque is required and where the starting period may be long *e.g.*, to drive

(*a*) compressors (*b*) large fans (*c*) pumps (*d*) high inertia loads

The power rating of such motors lies between 120W and 7.5kW.

22.5. Capacitor-start Capacitor-run Motor

This motor is identical to a capacitor-start motor except that starting winding is not opened after starting so that both the windings remain connected to the supply when running as well as at starting. Two designs are generally used.

(*i*) In one design, a single capacitor C is used for both starting and running as shown in Fig. 22.4 (*i*). This design eliminates the need of a centrifugal switch and at the same time improves the power factor and efficiency of the motor.

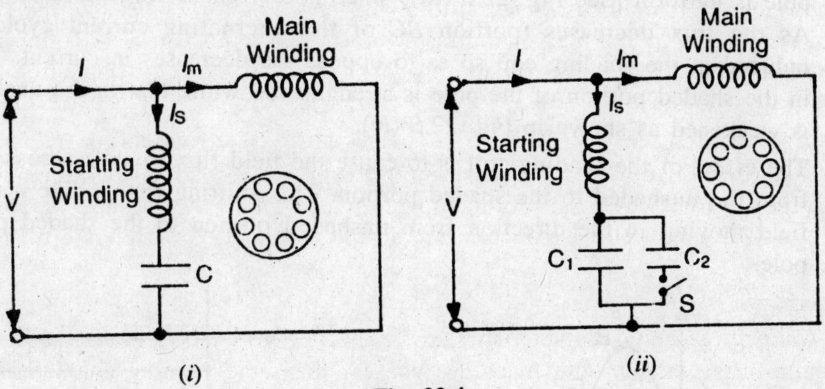

Fig. 22.4

(*ii*) In the other design, two capacitors C_1 and C_2 are used in the starting winding as shown in Fig. 22.4 (*ii*). The smaller capacitor C_1 required for optimum running conditions is permanently connected in series with the starting winding. The much larger capacitor C_2 is connected in parallel with C_1 for optimum **starting and**

remains in the circuit during starting. The starting capacitor C_2 is disconnected when the motor approaches about 75% of synchronous speed. The motor then runs as a single-phase induction motor.

Characteristics

(*i*) The starting winding and the capacitor can be designed for perfect 2-phase operation at any load. The motor then produces a constant torque and not a pulsating torque as in other single-phase motors.

(*ii*) Because of constant torque, the motor is vibration free and can be used in . (*a*) hospitals (*b*) studios and (*c*) other places where silence is important.

22.6. Shaded-pole Motor

The shaded-pole motor is very popular for ratings below 0.05 H.P. (≈ 40 W) because of its extremely simple construction. It has salient poles on the stator excited by single-phase supply and a squirrel-cage rotor as shown in Fig. 22.5. A portion of each pole is surrounded by a short-circuited turn of copper strip called *shading coil.*

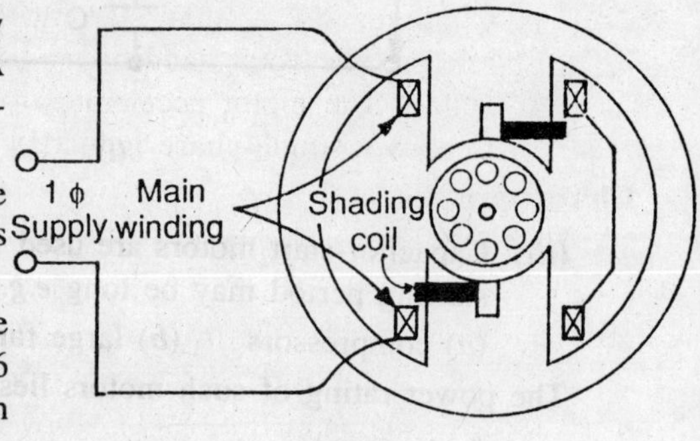

Fig. 22.5

Operation. The operation of the motor can be understood by referring to Fig. 22.6 which shows one pole of the motor with a shading coil.

(*i*) During the portion *OA* of the alternating-current cycle [See Fig. 22.6 (*i*)], the flux begins to increase and an e.m.f. is induced in the shading coil. The resulting current in the shading coil will be in such a direction (Lenz's law) so as to oppose the change in flux. Thus the flux in the shaded portion of the pole is weakened while that in the unshaded portion is strengthened as shown in Fig. 22.6 (*ii*).

(*ii*) During the portions *AB* of the alternating-current cycle, the flux has reached almost maximum value and is not changing. Consequently, the flux distribution across the pole is uniform [See Fig. 22.6 (*iii*)] since no current is flowing in the shading coil. As the flux decreases (portion *BC* of the alternating current cycle), current is induced in the shading coil so as to oppose the decreases in current. Thus the flux in the shaded portion of the pole is strengthened while that in the unshaded portion is weakened as shown in Fig. 22.6 (*iv*).

(*iii*) The effect of the shading coil is to cause the field flux to shift across the pole face from the unshaded to the shaded portion. This shifting flux is like a rotating weak field moving in the direction from unshaded portion to the shaded portion of the pole.

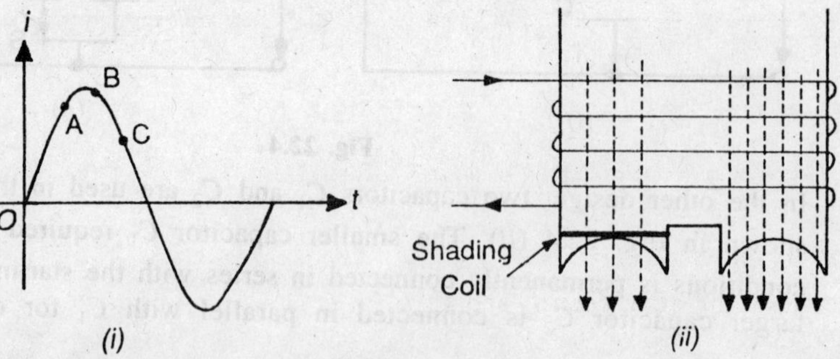

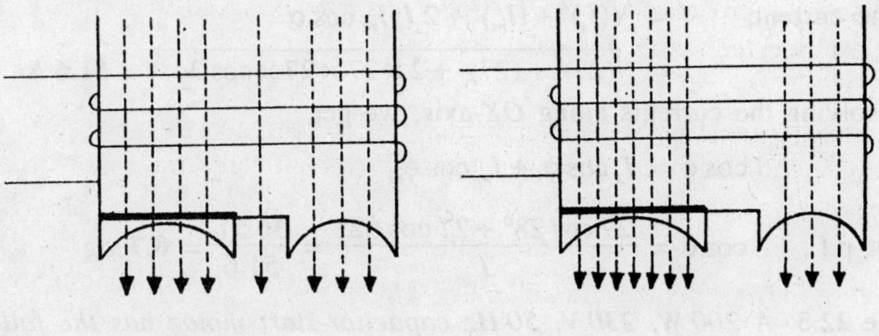

(*iii*) (*iv*)

Fig. 22.6

(*iv*) The rotor is of the squirrel-cage type and is under the influence of this moving field. Consequently, a small starting torque is developed. As soon as this torque starts to revolve the rotor, additional torque is produced by single-phase induction-motor action. The motor accelerates to a speed slightly below the synchronous speed and runs as a single-phase induction motor.

Characteristics

(*i*) The salient features of this motor are extremely simple construction and absence of centrifugal switch.

(*ii*) Since starting torque, efficiency and power factor are very low, these motors are only suitable for low power applications *e.g.*, to drive

(*a*) small fans (*b*) toys (*c*) hair dryers (*d*) desk fans *etc.*

The power rating of such motors is upto about 30W.

Example 22.1. *A 4-pole, 50Hz, single-phase induction motor is running with a slip of 3.4%. Calculate the speed of the motor.*

Solution.

Synchronous speed, $N_s = 120f/P = 120 \times 50/4 = 1500$ r.p.m.

Motor speed, $N = (1-s)\,N_s = (1-0.034)\,1500 =$ **1449 r.p.m.**

Example 22.2. *At starting, the windings of a 230V, 50Hz, split-phase induction motor have the following parameters :*

Main winding : $R = 4\Omega$; $X_L = 7.5\,\Omega$

Starting winding : $R = 7.5\,\Omega$; $X_L = 4\,\Omega$

Find (i) current I_m in the main winding (ii) current I_s in the starting winding (iii) phase angle between I_s and I_m and (iv) line current and (v) power factor of the motor.

Solution.

(*i*) $I_m = \dfrac{V}{Z_m} = \dfrac{230}{\sqrt{4^2+(7.5)^2}} = \mathbf{27A}$

$\phi_m = \tan^{-1} 7.5/4 = 62°$

(*ii*) $I_s = \dfrac{V}{Z_s} = \dfrac{230}{\sqrt{(7.5)^2+4^2}} = \mathbf{27A}$

$\phi_s = \tan^{-1} 4/7.5 = 28°$

(*iii*) Phase angle between I_s and I_m

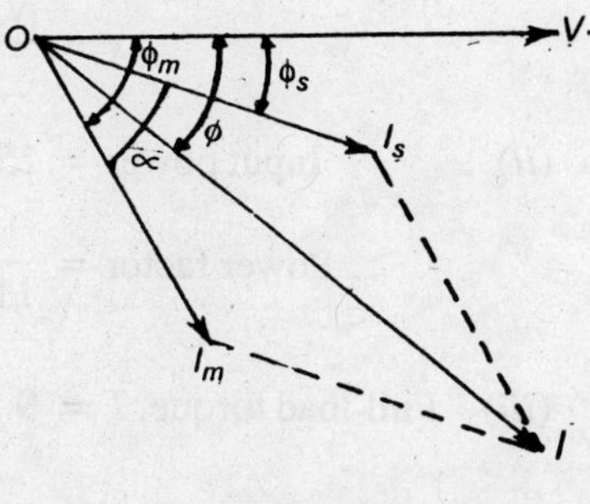

Fig. 22.7

$$\alpha = \phi_m - \phi_s = 62° - 28° = \mathbf{34°}$$

(*iv*) Line current, $I = \sqrt{(I_s)^2 + (I_m)^2 + 2\, I_s\, I_m \cos\alpha}$

$$= \sqrt{(27)^2 + (27)^2 + 2 \times 27 \times 27 \times \cos 34°} = \mathbf{51.6\ A}$$

(*v*) Resolving the currents along *OX*-axis, we get,

$$I\cos\phi = I_s \cos\phi_s + I_m \cos\phi_m$$

$\therefore$ Motor p.f., $\cos\phi = \dfrac{27\cos 28° + 27\cos 62°}{I} = \dfrac{36.51}{51.6} = \mathbf{0.7}\ \textit{lag}$

Example 22.3. *A 200 W, 230 V, 50 Hz capacitor-start motor has the following winding constants :*

Main winding : $R = 4.5\,\Omega$; $X_L = 3.7\,\Omega$

Starting winding : $R = 9.5\,\Omega$; $X_L = 3.5\,\Omega$

Find the value of starting capacitance that will result in the maximum starting torque.

Solution. Fig. 22.8 shows the phasor diagram of the motor. The current I_s in the starting winding leads the applied voltage *V* by ϕ_s while the current I_m in the main winding lags *V* by ϕ_m. The starting torque will be maximum when phase angle α between I_s and I_m is 90°.

$$\phi_m = \tan^{-1} 3.7/4.5 = 39.6°$$

$$\phi_s = 39.6° - 90° = -50.4°$$

Let *C* farad be the capacitance of the starting capacitor to give maximum starting torque.

$$\tan(-50.4°) = \frac{3.5 - X_C}{9.5}$$

or
$$-1.21 = \frac{3.5 - X_C}{9.5}$$

or
$$X_c = 15\,\Omega$$

$\therefore$
$$C = \frac{1}{2\pi f X_C} = \frac{1}{2\pi \times 50 \times 15}$$

$$= 212 \times 10^{-6}\,\text{F} = \mathbf{212\ \mu F}$$

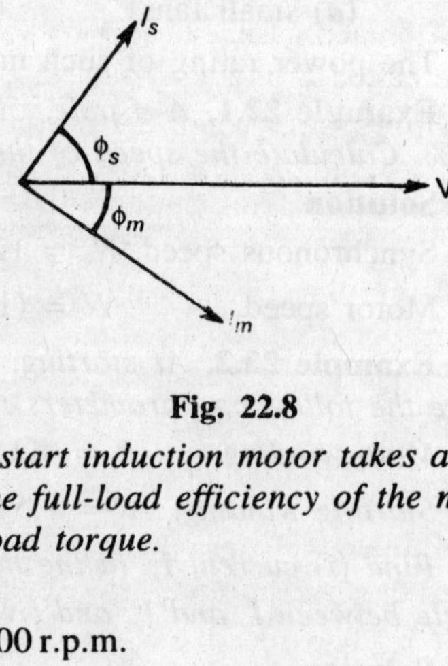

Fig. 22.8

Example 22.4. *A 4-pole, 250W, 115V, 60Hz capacitor-start induction motor takes a full-load line current of 5.3A while running at 1760 r.p.m. If the full-load efficiency of the motor is 64%, find (i) motor slip (ii) power factor and (iii) full-load torque.*

Solution.

(*i*) $N_s = 120f/P = 120 \times 60/4 = 1800$ r.p.m.

$$s = \frac{N_s - N}{N_s} = \frac{1800 - 1760}{1800} = \mathbf{0.022}$$

(*ii*) Input power $= 250/0.64 = 390.6$ W

$$\text{Power factor} = \frac{390.6}{115 \times 5.3} = \mathbf{0.64}\ \textit{lag}$$

(*iii*) Full-load torque, $T = 9.55\dfrac{P_{out}}{N} = 9.55\dfrac{250}{1760} = \mathbf{1.35\ N\text{-}m}$

22.7. A.C. Series Motor or Universal Motor

A d.c. series motor will rotate in the *same direction regardless of the polarity of the supply. One can expect that a d.c. series motor would also operate on a single-phase supply. It is then called an a.c. series motor. However, some changes must be made in a d.c. motor that is to operate satisfactorily on a.c. supply. The changes effected are :

(*i*) The entire magnetic circuit is laminated in order to reduce the eddy current loss. Hence an a.c. series motor requires a more expensive construction than a d.c. series motor.

(*ii*) The series field winding uses as few turns as possible to reduce the reactance of the field winding to a minimum. This reduces the voltage drop across the field winding.

(*iii*) A high field flux is obtained by using a low-reluctance magnetic circuit.

(*iv*) There is considerable sparking between the brushes and the commutator when the motor is used on a.c. supply. It is because the alternating flux establishes high currents in the coils short-circuited by the brushes. When the short-circuited coils break contact from the commutator, excessive sparking is produced. This can be eliminated by using high-resistance leads to connect the coils to the commutator segments.

Construction. The construction of an a.c. series motor is very similar to a d.c. series motor except that above modifications are incorporated (See Fig. 22.9). Such a motor can be operated either on a.c. or d.c. supply and the resulting torque-speed curve is about the same in each case. For this reason, it is sometimes called a *universal motor.*

Operation. When the motor is connected to an a.c. supply, the same alternating current flows through the field and armature windings. The field winding produces an alternating flux ϕ that reacts with the current flowing in the armature to produce a torque. Since both armature current and flux reverse simultaneously, the torque always acts in the same direction. It may be noted that no rotating flux is produced in this type of machines ; the principle of operation is the same as that of a d.c. series motor.

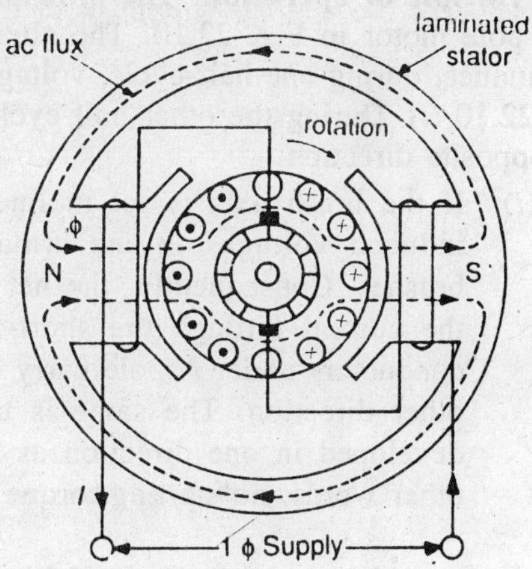

Fig. 22.9

Characteristics. The operating characteristics of an a.c. series motor are similar to those of a d.c. series motor.

(*i*) The speed increases to a high value with a decrease in load. In very small series motors, the losses are usually large enough at no load that limit the speed to a **definite value (1500—15,000 r.p.m.).

(*ii*) The motor torque is high for large armature currents, thus giving a high starting torque.

---

* In a series motor, both armature and field windings are in series. Since the armature current and flux reverse simultaneously, the torque always acts in the same direction regardless of the polarity of the supply.

** However, in a d.c. series motor, the speed at no load may rise to a dangerously high value.

(*iii*) At full-load, the power factor is about 90%. However, at starting or when carrying an overload, the power factor is lower.

Applications. The fractional horsepower a.c. series motors have high-speed (and corresponding small size) and large starting torque. They can, therefore, be used to drive

(*a*) high-speed vacuum cleaners (*b*) sewing machines
(*c*) electric shavers (*d*) drills (*e*) machine tools *etc.*

22.8. Repulsion Motor

A repulsion motor is similar to an a.c. series motor except that :

(*i*) brushes are not connected to supply but are short-circuited [See Fig. 22.10]. Consequently, currents are induced in the armature conductors by *transformer action.

(*ii*) the field structure has non-salient pole construction.

By adjusting the position of short-circuited brushes on the commutator, the starting torque can be developed in the motor.

Principle of operation. The principle involved in the starting of this motor is illustrated by 2-pole motor in Fig. 22.10. The alternating flux produced by the stator or field winding will induce, during one half-cycle, voltages in the respective armature conductors as shown in Fig. 22.10 (*i*). During the other half-cycle, the directions of the respective voltages will be in the opposite direction.

(*i*) If the brush axis is set in line with the stator-field axis [See Fig. 22.10 (*i*)], the induced voltages in the armature conductors are additive on each side of the brushes. Consequently, the net voltage between the brushes is maximum and so is the current through the short-circuited armature. It may be seen that half of the conductors under *N*-pole carry current in one direction and half carry current in the other direction. The same is true under the *S*-pole. Therefore, as much torque is developed in one direction as in the other and the armature remains stationary. In other words, no starting torque is developed.

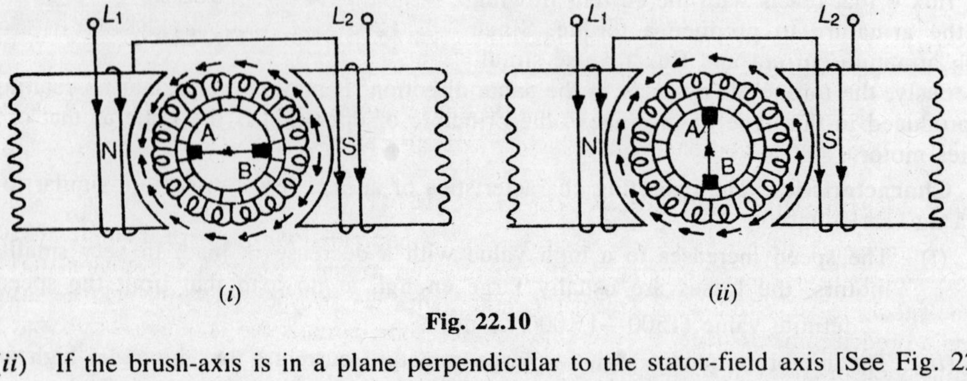

Fig. 22.10

(*ii*) If the brush-axis is in a plane perpendicular to the stator-field axis [See Fig. 22.10 (*ii*)], the voltage in each path will be neutralised. As a result, the net voltage between the brushes is zero; therefore no current flows in the armature circuit and no torque is produced.

---

* In an a.c. series motor, the field and armature windings are conductively coupled *i.e.*, the armature current is obtained by conduction from the supply. However, in a repulsion motor, the armature currents are obtained by induction *i.e.*, transformer action.

(*iii*) If the plane of the brush-axis is other than 0° or 90° to the plane of the stator-field [See Fig. 22.11 (*i*)], a resultant voltage will exist in each path, sending current through the armature. Now all the conductors under one pole carry current in one direction while all the conductors under opposite pole carry current in the opposite direction. Consequently, net torque is developed. *The direction of rotation would depend upon the direction in which the brushes are shifted from the plane parallel to the field.* Thus in Fig. 22.11 (*i*), the brushes are shifted clockwise and the net torque acts in a clockwise direction. If the brushes are shifted in the anticlockwise direction [See Fig. 22.11 (*ii*)], the currents in the armature conductors under the poles are reversed and the torque is developed in anticlockwise direction.

(*iv*) After a repulsion motor has been started in either direction, it will continue to develop torque in that direction and give series-motor type performance characteristics.

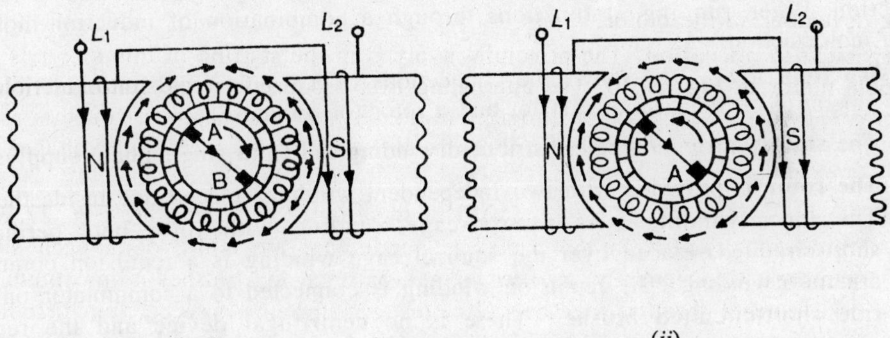

Fig. 22.11

Characteristics

(*i*) The repulsion motor has characteristics very similar to those of an a.c. series motor *i.e.*, it has a high starting torque and a high speed at no load.

(*ii*) The speed which the repulsion motor develops for any given load will depend upon the *position of the brushes.

(*iii*) In comparison with other single-phase motors, the repulsion motor has a high starting torque and relatively low starting current.

22.9. Repulsion-start Induction-run Motor

Sometimes the action of a repulsion motor is combined with that of a single-phase induction motor to produce repulsion-start induction-run motor (also called repulsion-start motor). The machine is started as a repulsion motor with a corresponding high starting torque. At some predetermined speed, a centrifugal device short-circuits the commutator so that the machine then operates as a single-phase induction motor.

The repulsion-start induction-run motor has the same general construction of a repulsion motor. The only difference is that in addition to the basic repulsion-motor construction, it is equipped with a centrifugal device fitted on the armature shaft. When the motor reaches 75% of its full running speed, the centrifugal device forces a

* Shifting the brush position has the same effect as changing the number of turns of the series winding of an a.c. series motor *i.e.*, speed of the motor is affected.

short-circuiting ring to come in contact with the inner surface of the commutator. This short-circuits all the commutator bars. The rotor then resembles squirrel-cage type and the motor runs as a single-phase induction motor. At the same time, the centrifugal device raises the brushes from the commutator which reduces the wear of the brushes and commutator as well as makes the operation quiet.

Characteristics

(*i*) The starting torque is 2.5 to 4.5 times the full-load torque and the starting current is 3.75 times the full-load value.

(*ii*) Due to their high starting torque, repulsion-motors were used to operate devices such as refrigerators, pumps, compressors *etc.*

However, they posed a serious problem of maintenance of brushes, commutator and the centrifugal device. Consequently, manufacturers have stopped making them in view of the development of capacitor motors which are small in size, reliable and low-priced.

22.10. Repulsion-Induction Motor

The repulsion-induction motor produces a high starting torque entirely due to repulsion motor action. When running, it functions through a combination of induction-motor and repulsion motor action.

Construction. Fig. 22.12 shows the connections of a 4-pole repulsion-induction motor for 230V operation. It consists of a stator and a rotor (or armature).

(*i*) The **stator** carries a single distributed winding fed from single-phase supply.

(*ii*) The **rotor** is provided with two independent windings placed one inside the other. The inner winding is a squirrel-cage winding with rotor bars permanently short-circuited. Placed over the squirrel cage winding is a repulsion commutator armature winding. The repulsion winding is connected to a commutator on which ride short-circuited brushes. There is no centrifugal device and the repulsion winding functions at all times.

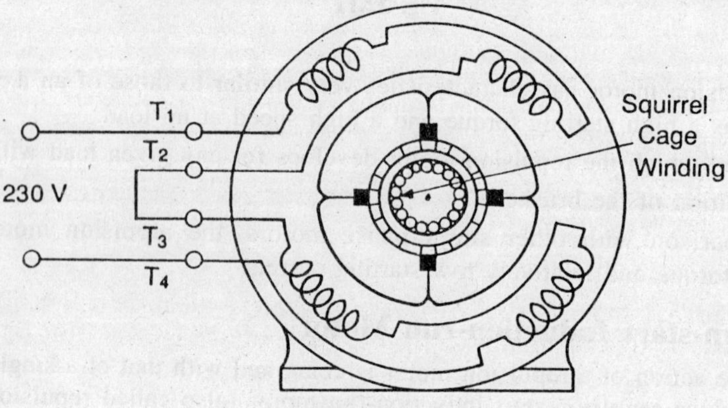

Fig. 22.12

Operation.

(*i*) When single-phase supply is given to the stator winding, the repulsion winding (*i.e.*, outer winding) is *active. Consequently, the motor starts as a repulsion motor with a corresponding high starting torque.

---

* The inner winding (*i.e.*, squirrel cage winding), being nearly surrounded by iron, has a high inductance. At starting, rotor frequency is equal to the supply frequency and the impedance of the outer winding is less than that of the inner winding. This results in a large proportion of rotor current flowing in the outer winding.

(ii) As the motor speed increases, the current shifts from the outer to inner winding due to the decreasing impedance of the inner winding with increasing speed. Consequently, at running speed, the squirrel cage winding carries the greater part of rotor current. This shifting of repulsion-motor action to induction-motor action is thus achieved without any switching arrangement.

(iii) It may be seen that the motor starts as a repulsion motor. When running, it functions through a combination of principle of induction and repulsion ; the former being predominant.

Characteristics

(i) The no-load speed of a repulsion-induction motor is somewhat above the synchronous speed because of the effect of repulsion winding. However, the speed at full-load is slightly less than the synchronous speed as in an induction motor.

(ii) The speed regulation of the motor is about 6%.

(iii) The starting torque is 2.25 to 3 times the full-load torque ; the lower value being for large motors. The starting current is 3 to 4 times the full-load current.

This type of motor is used for applications requiring a high starting torque with essentially a constant running speed. The common sizes are 0.25 to 5 H.P.

22.11. Single-phase Synchronous Motors

Very small single-phase motors have been developed which run at true synchronous speed. They do not require d.c. excitation for the rotor. Because of these characteristics, they are called *unexcited single-phase synchronous motors*. The most commonly used types are :

(i) Reluctance Motors (ii) Hysteresis motors

The efficiency and torque-developing ability of these motors is low. The output of most of the commercial motors is only a few watts.

22.12. Reluctance Motor

It is a single-phase synchronous motor which does not require d.c. excitation to the rotor. Its operation is based upon the following principle :

Whenever a piece of ferromagnetic material is located in a magnetic field, a force is exerted on the material, tending to align the material so that reluctance of the magnetic path that passes through the material is minimum.

Construction. A reluctance motor (also called synchronous reluctance motor) consists of :

(i) a **stator** carrying a single-phase winding along with an auxiliary winding to produce a synchronously-revolving magnetic field.

(ii) a **squirrel-cage rotor** having unsymmetrical magnetic construction. This is achieved by symmetrically removing some of the teeth from the squirrel-cage rotor to produce salient poles on the rotor. As shown in Fig. 22.13 (i), 4 salient poles have been produced on the rotor. The salient poles created on the rotor must be equal to the poles on the stator.

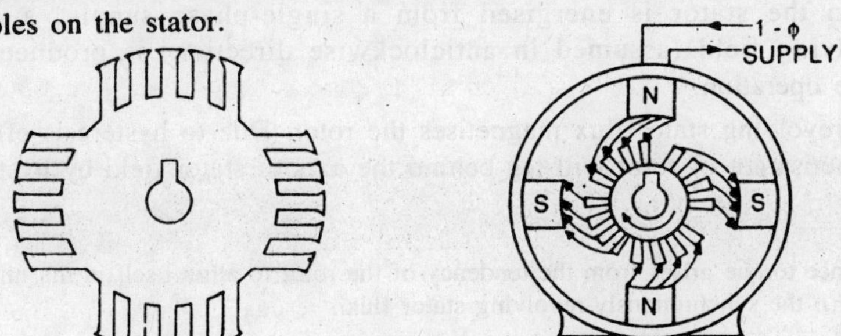

(i) Fig. 22.13 (ii)

Note that rotor salient poles offer low reluctance to the stator flux and therefore, become strongly magnetised.

Operation.

(*i*) When single-phase stator having an auxiliary winding is energised, a synchronously-revolving field is produced. The motor starts as a standard squirrel-cage induction motor and will accelerate to near its synchronous speed.

(*ii*) As the rotor approaches synchronous speed, the rotating stator flux will exert ****reluctance torque*** on the rotor poles tending to align the salient-pole axis with the axis of the rotating field. The rotor assumes a position where its salient poles lock with the poles of the revolving field [See Fig. 22.13 (*ii*)]. Consequently, the motor will continue to run at the speed of revolving flux *i.e.*, at the synchronous speed.

(*iii*) When we apply a mechanical load, the rotor poles fall slightly behind the stator poles, while continuing to turn at synchronous speed. As the load on the motor is increased, the mechanical angle between the poles increases progressively. Nevertheless, magnetic attraction keeps the rotor locked to the rotating flux. If the load is excessive, the motor may pull out of synchronism and stop.

Characteristics

(*i*) These motors have poor torque, power factor and efficiency.

(*ii*) These motors cannot accelerate high-inertia loads to synchronous speed.

(*iii*) The pull-in and pull-out torques of such motors are weak.

Despite the above drawbacks, the reluctance motor is cheaper than any other type of synchronous motor. They are widely used for constant-speed applications such as timing devices, signalling devices *etc.*

22.13. Hysteresis Motor

It is a single-phase motor whose operation depends upon the hysteresis effect *i.e.*, magnetisation produced in a ferromagnetic material lags behind the magnetising force.

Construction. It consists of :

(*i*) a **stator** designed to produce a synchronously-revolving field from a single-phase supply. This is accomplished by using permanent-split capacitor type construction. Consequently, both the windings (*i.e.*, starting as well as main winding) remain connected in the circuit during running operation as well as at starting. The value of capacitance is so adjusted as to result in a flux revolving at synchronous speed.

(*ii*) a **rotor** consisting of a smooth cylinder of magnetically hard steel, without winding on teeth.

Operation.

(*i*) When the stator is energised from a single-phase supply, a synchronously-revolving field (assumed in anticlockwise direction) is produced due to split-phase operation.

(*ii*) The revolving stator flux magnetises the rotor. Due to hysteresis effect, the axis of magnetisation of rotor will lag behind the axis of stator field by hysteresis lag angle

* The reluctance torque arises from the tendency of the rotor to align itself in the minimum reluctance position *w.r.t.* the synchronously-revolving stator flux.

α as shown in Fig. 22.14. If the rotor is stationary, the starting torque produced is given by

$$T_s \propto \phi_s \phi_r \sin \alpha$$

where ϕ_s = stator flux

ϕ_r = rotor flux

From now onwards, the rotor accelerates to synchronous speed with a *uniform torque.

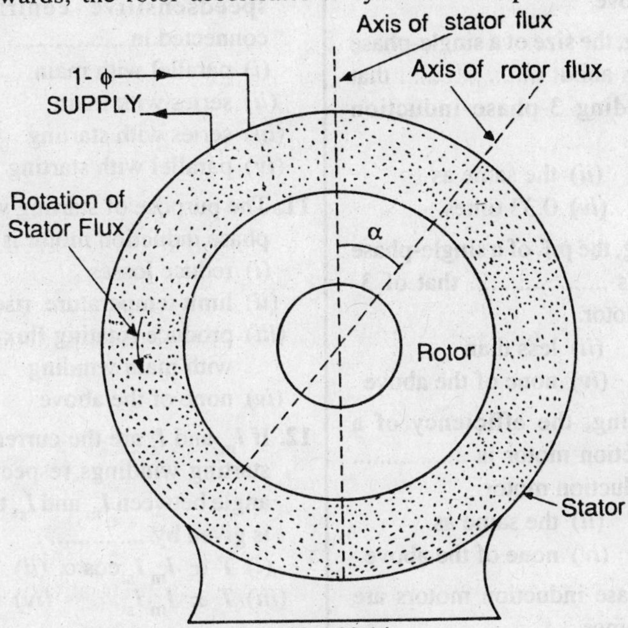

Fig. 22.14

(*iii*) After reaching synchronism, the motor continues to run at synchronous speed and adjusts its torque angle so as to develop the torque required by the load.

Characteristics

(*i*) A hysteresis motor can synchronise any load which it can accelerate, no matter how great the inertia. It is because the torque is uniform from standstill to synchronous speed.

(*ii*) Since the rotor has no teeth or salient poles or winding, a hysteresis motor is inherently quiet and produces smooth rotation of the load.

(*iii*) The rotor takes on the same number of poles as the stator field. Thus by changing the number of stator poles through pole-changing connections, we can get a set of synchronous speeds for the motor.

Applications. Due to their quiet operation and ability to drive high-inertia loads, hysteresis motors are particularly well suited for driving (*i*) electric clocks (*ii*) timing devices (*iii*) tape-decks (*iv*) turn-tables and other precision audio-equipment.

* Now ϕ_s and ϕ_r are constant. When the rotor accelerates, the lag angle α remains constant since the angle α depends merely upon the hysteresis loop of the rotor and is independent of the rate at which the loop is traversed. The motor, therefore, develops constant torque from standstill to synchronous speed.

OBJECTIVE QUESTIONS

1. A single-phase induction motor employs .... rotor.
   (i) squirrel cage (ii) wound
   (iii) either squirrel cage or wound
   (iv) none of the above
2. For the same rating, the size of a single-phase induction motor is about .... that of the corresponding 3-phase induction motor.
   (i) 3 times (ii) the same as
   (iii) 1.5 times (iv) 0.33 times
3. For the same rating, the p.f. of a single-phase induction motor is .... that of 3-phase induction motor.
   (i) the same as (ii) less than
   (iii) more than (iv) none of the above
4. For the same rating, the efficiency of a single-phase induction motor is .... that of 3-phase induction motor.
   (i) less than (ii) the same as
   (iii) more than (iv) none of the above
5. Most of single-phase induction motors are .... machines.
   (i) 2-pole (ii) 6-pole
   (iii) 8-pole (iv) 4-pole
6. The main winding and starting winding of a single-phase induction motor are connected in .... across the supply.
   (i) series (ii) parallel
   (iii) series-parallel (iv) none of the above
7. The starting winding of a single-phase induction motor has .... that of main winding.
   (i) more poles than
   (ii) less poles than
   (iii) same number of poles as
   (iv) none of the above
8. A 50 Hz, 4-pole single-phase induction motor will have a synchronous speed of ....
   (i) 1500 r.p.m. (ii) 750 r.p.m.
   (iii) 1200 r.p.m. (iv) none of the above
9. A 4-pole, 50 Hz single-phase induction motor has a slip of 5%. The speed of the motor will be ....
   (i) 1500 r.p.m. (ii) 1425 r.p.m.
   (iii) 1200 r.p.m. (iv) none of the above
10. In a single-phase induction motor, speedsensitive centrifugal switch is connected in .... winding.
    (i) parallel with main
    (ii) series with main
    (iii) series with starting
    (iv) parallel with starting
11. The purpose of starting winding in a single-phase induction motor is to ....
    (i) reduce losses
    (ii) limit temperature rise of the machine
    (iii) produce rotating flux in conjunction with main winding
    (iv) none of the above
12. If I_m and I_s are the currents in the main and starting windings respectively and α is the angle between I_m and I_s, then motor torque T is given by ....
    (i) $T \propto I_m I_s \cos\alpha$ (ii) $T \propto I_m I_s \sin\alpha$
    (iii) $T \propto I_m I_s$ (iv) none of the above

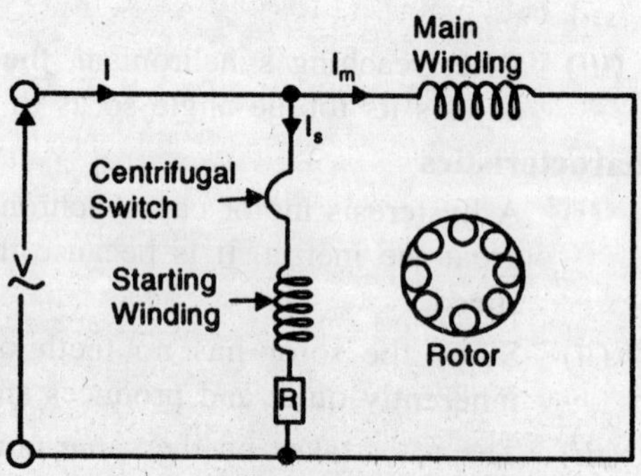

Fig. 22.15

13. In the resistance split-phase induction motors shown in Fig. 22.15, the angel α between I_m and I_s is about ....
    (i) 80° (ii) 75°
    (iii) 65° (iv) 25°
14. The starting winding of split-phase induction motor shown in Fig. 22.15 is disconnected when the motor attains about .... of synchronous speed.

(*i*) 75 % (*ii*) 50 %
(*iii*) 25 % (*iv*) 100 %

15. In the resistance split-phase induction motor shown in Fig 22.15 the main winding has ....
(*i*) high resistance and high inductance
(*ii*) high resistance and low inductance
(*iii*) low resistance and high inductance
(*iv*) low resistance and low inductance

16. At starting, the current I_s in the starting winding shown in Fig. 22.15 ....
(*i*) lags V by 90°
(*ii*) leads V by 90°
(*iii*) is nearly in phase with V
(*iv*) leads V by 75°

17. A resistance split-phase inducti used for ....
(*i*) low inertia loads
(*ii*) high inertia loads
(*iii*) very high inertia loads
(*iv*) none of the above

18. The direction of split-phase induction motor can be reversed by interchanging the connections of supply of ....
(*i*) either main or starting winding
(*ii*) both main and starting winding
(*iii*) rotor winding
(*iv*) none of the above

19. The resistance split-phase induction motor has .... starting torque.
(*i*) high (*ii*) moderate
(*iii*) very low (*iv*) none of the above

20. The resistance split-phase induction motor in used in those applications where ....
(*i*) starting is frequent
(*ii*) starting period is long
(*iii*) starting is infrequent
(*iv*) none of the above

21. The resistance split -phase induction motors are the most popular single-phase motors because of ....
(*i*) their low cost
(*ii*) their high starting torque
(*iii*) the long starting period
(*iv*) none of the above

22. In the capacitor start induction motor shown in Fig. 22.16, the angle α between I_m and I_s is ....

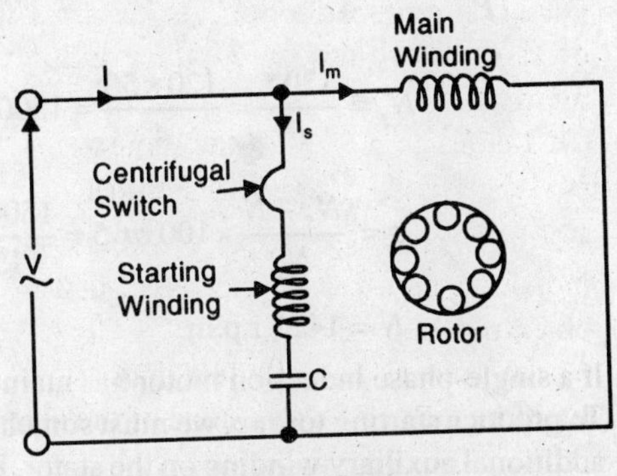

Fig. 22.16

(*i*) exactly 90° (*ii*) about 30°
(*iii*) about 40° (*iv*) about 75°

23. The value of current I_s in capacitor-start motor shown in Fig. 22.16 is .... the current I_m.
(*i*) equal to (*ii*) less than that of
(*iii*) more than that of
(*iv*) none of the above

24. The capacitor C used in the capacitor-start motor shown in Fig. 22.16 is always a .... capacitor.
(*i*) paper (*ii*) ceramic
(*iii*) mica (*iv*) electrolyte

25. At starting, the line current of a capacitor-start induction motor is .... the normal full-load current.
(*i*) 8 to 10 times (*ii*) 4 to 5 times
(*iii*) equal to (*iv*) 7 to 8 times

ANSWERS TO OBJECTIVE QUESTIONS

1. (*i*)	**2.** (*iii*)	**3.** (*ii*)	**4.** (*i*)	**5.** (*iv*)	**6.** (*ii*)	**7.** (*iii*)	**8.** (*i*)
9. (*ii*)	**10.** (*iii*)	**11.** (*iii*)	**12.** (*ii*)	**13.** (*iv*)	**14.** (*i*)	**15.** (*iii*)	**16.** (*iii*)
17. (*i*)	**18.** (*i*)	**19.** (*ii*)	**20.** (*iii*)	**21.** (*i*)	**22.** (*iv*)	**23.** (*ii*)	**24.** (*iv*)
25. (*ii*)							

HINTS TO SELECTED OBJECTIVE QUESTIONS

1. Structurally, a single phase induction motor resembles the 3-phase squirrel cage motor except that the stator winding is a single phase winding.

7. The stator of a single phase induction motor carries the main winding which creates a set of N-S poles. It also carries a smaller *starting winding* that only operates during the brief period when the motor starts up. The starting winding has the same number of poles as the main winding has.

8. $N_s = \dfrac{120f}{P} = \dfrac{120 \times 50}{4} = 1500$ r.p.m

9. $$N_s = \frac{120f}{P} = \frac{120 \times 50}{4} = 1500 \text{ r.p.m}$$

$$s = \frac{N_s - N}{N_s} \times 100 \text{ or } 5 = \frac{1500 - N}{1500} \times 100$$

$$\therefore \quad N = 1425 \text{ r.p.m.}$$

11. If a single-phase induction motor has main winding on the stator only, it will not be self-starting. To produce starting torque, we must somehow create a revolving field. This is done by placing an additional auxiliary winding on the stator. Both main and auxiliary windings (also called starting winding) are connected in parallel across the single-phase supply. The currents in the two windings have a phase difference and hence produce a weak revolving field.

15. The main winding of a resistance split-phase motor :

(*i*) is always made of relatively large wire to reduce $I^2 R$ losses.

(*ii*) has a relatively large number of turns.

Due to these reasons, the main winding has low resistance and high inductance.

16. The starting winding of a resistance split-phase induction motor has a relatively small number of turns of fine wire. Consequently, the starting winding has high resistance and low inductance. The result is that at starting, the current I_s in the starting winding is nearly in phase with the applied voltage *V*.

22. The capacitor C is so chosen that current I_s in the starting winding leads the current I_m in the main winding by about 75°.

23. In a capacitor-start motor, the staring winding has about as many turns as the main winding. Furthermore, a capacitor is connected in series with the starting winding. For this reason, the value of I_s is less than I_m at starting.

24. Because these capacitors are available in small size and are quite reliable. Moreover, they are also not expensive.

23

Alternators

Introduction

A.C. system has a number of advantages over d.c. system. These days 3-phase a.c. system is being exclusively used for generation, transmission and distribution of power. The machine which produces 3-phase power from mechanical power is called an alternator or synchronous generator. Alternators are the primary source of all the electrical energy we consume. These machines are the largest energy converters found in the world. They convert mechanical energy into a.c. energy. In this chapter, we shall discuss the construction and characteristics of alternators.

23.1. Alternator

An alternator operates on the same fundamental principle of electromagnetic induction as a d.c. generator *i.e.*, when the flux linking a conductor changes, an e.m.f. is induced in the conductor. Like a d.c. generator, an alternator also has an armature winding and a field winding. But there is one important difference between the two. In a d.c. generator, the armature winding is placed on the rotor in order to provide a way of converting alternating voltage generated in the winding to a direct voltage at the terminals through the use of a rotating commutator. The field poles are placed on the stationary part of the machine. Since no commutator is required in an alternator, it is usually more convenient and advantageous to place the field winding on the rotating part (*i.e.*, rotor) and armature winding on the stationary part (*i.e.*, stator).

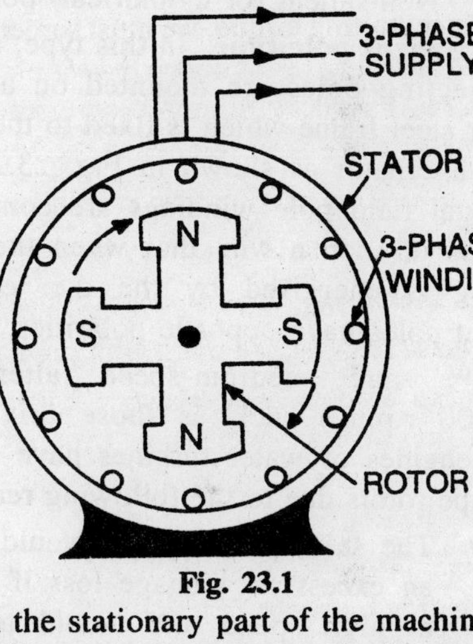

Fig. 23.1

Advantages of stationary armature. The field winding of an alternator is placed on the rotor and is connected to d.c. supply through two slip rings. The 3-phase armature winding is placed on the stator. This arrangement has the following advantages :

(*i*) It is easier to insulate stationary winding for high voltages for which the alternators are usually designed. It is because they are not subjected to centrifugal forces and also extra space is available due to the stationary arrangement of the armature.

(*ii*) The stationary 3-phase armature can be directly connected to load without going through large, unreliable slip rings and brushes.

(*iii*) Only two slip rings are required for d.c. supply to the field winding on the rotor. Since the exciting current is small, the slip rings and brush gear required are of light construction.

(*iv*) Due to simple and robust construction of the rotor, higher speed of rotating d.c field is possible. This increases the output obtainable from a machine of given dimensions.

Note : All alternators above 5 kVA employ a stationary armature (or stator) and a revolving d.c. field.

23.2. Construction of Alternator

An alternator has 3-phase winding on the stator and a d.c. field winding on the rotor.

1. Stator. It is the stationary part of the machine and is built up of sheet-steel laminations having slots on its inner periphery. A 3-phase winding is placed in these slots and serves as the armature winding of the alternator. The armature winding is always connected in *star and the neutral is connected to **ground.

2. Rotor. The rotor carries a field winding which is supplied with direct current through two slip rings by a separate d.c. source. This d.c. source (called exciter) is generally a small d.c. shunt or compound generator mounted on the shaft of the alternator. Rotor construction is of two types, namely ;

(*i*) Salient (or projecting) pole type

(*ii*) Non-salient (or cylindrical) pole type

(*i*) **Salient pole type.** In this type, salient or projecting poles are mounted on a large circular steel frame which is fixed to the shaft of the alternator as shown in Fig. 23.2. The individual field pole windings are connected in series in such a way that when the field winding is energised by the d.c. exciter, adjacent poles have opposite polarities.

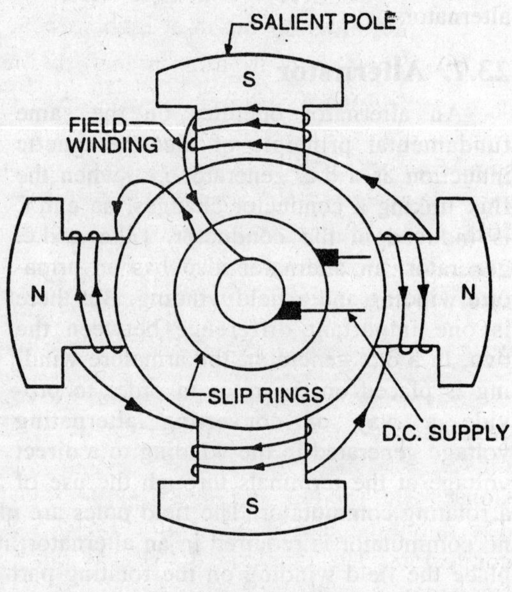

Fig. 23.2

Low and medium-speed alternators (120-400 r.p.m.) such as those driven by diesel engines or water turbines have salient pole type rotors due to the following reasons :

(*a*) The salient field poles would cause an excessive windage loss if driven at high speed and would tend to produce noise.

(*b*) Salient-pole construction cannot be made strong enough to withstand the mechanical stresses to which they may be subjected at higher speeds.

Since a frequency of 50Hz is required, we must use a ***large number of poles on the rotor of slow-speed alternators. Low-speed rotors always possess a large diameter to provide the necessary space for the poles. Consequently, salient-pole type rotors have large diameters and short axial lengths.

---

* In star connection, phase voltage $V_{ph} = V_L/\sqrt{3}$. Since the induced e.m.f. in the phase winding of an alternator is directly proportional to the number of turns, a star-connected alternator will require less number of turns than a delta-connected alternator for the same line voltage.

** This permits to use protective devices (*e.g.*, relays) to protect the system in case of ground faults.

*** $f = NP/120$

(*ii*) **Non-salient pole type.** In this type, the rotor is made of smooth solid forged-steel radial cylinder having a number of slots along the outer periphery. The field windings are embedded in these slots and are connected in series to the slip rings through which they are energised by the d.c. exciter. The regions forming the poles are usually left unslotted as shown in Fig. 23.3. It is clear that the poles formed are non-salient *i.e.*, they do not project out from the rotor surface.

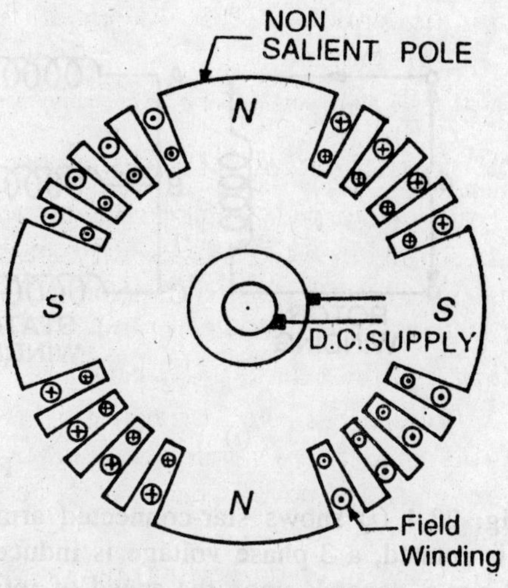

Fig. 23.3

High-speed alternators (1500 or 3000 r.p.m.) are driven by steam turbines and use non-salient type rotors due to the following reasons :

(*a*) This type of construction has mechanical robustness and gives noiseless operation at high speeds.

(*b*) The flux distribution around the periphery is nearly a sine wave and hence a better e.m.f. waveform is obtained than in the case of salient-pole type.

Since steam turbines run at high speed and a frequency of 50Hz is required, we need a small number of poles on the rotor of high-speed alternators (also called turboalternators). We can use no less than 2 poles and this fixes the highest possible speed. For a frequency of 50Hz, it is 3000 r.p.m. The next lower speed is 1500 r.p.m. for a 4-pole machine. Consequently, turboalternators possess 2 or 4 poles and have *small diameters and very long axial lengths.

23.3. Alternator Operation

The rotor winding is energised from the d.c. exciter and alternate *N* and *S* poles are developed on the rotor. When the rotor is rotated in anticlockwise direction by a prime mover, the stator or armature conductors are cut by the magnetic flux of rotor poles. Consequently, e.m.f. is induced in the armature conductors due to electromagnetic induction. The induced e.m.f. is alternating since *N* and *S* poles of rotor alternately pass the armature conductors. The direction of induced e.m.f. can be found by Fleming's right hand rule and frequency is given by ;

$$f = \frac{NP}{120}$$ ...See Art 23.4

where N = speed of rotor in r.p.m.

P = number of rotor poles

---

* The high speed of rotation produces strong centrifugal forces which impose an upper limit on the diameter. On the other hand, to build powerful alternators, we have to use massive rotors. Therefore, high-power and high-speed rotors have to be very long.

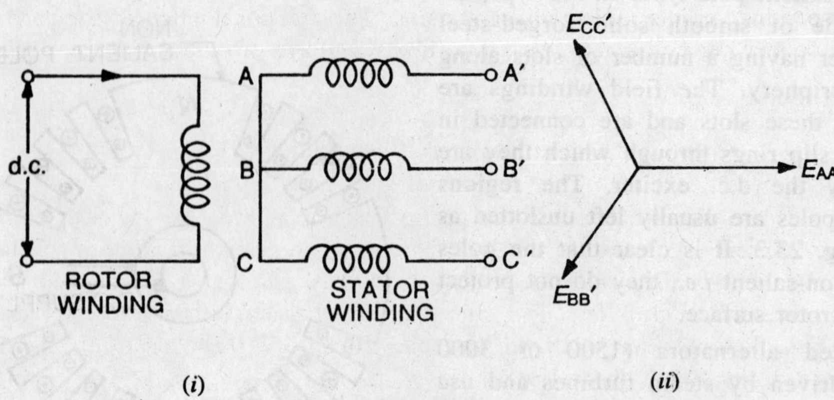

Fig. 23.4

Fig. 23.4 (*i*) shows star-connected armature winding and d.c. field winding. When the rotor is rotated, a 3-phase voltage is induced in the armature winding. The magnitude of induced e.m.f. depends upon the speed of rotation and the d.c. exciting current. The magnitude of e.m.f. in each phase of the armature winding is the same. However, they differ in phase by 120° electrical as shown in the phasor diagram in Fig. 23.4 (*ii*).

23.4. Frequency

The frequency of induced e.m.f. in the armature conductors depends upon speed and the number of poles.

Let N = rotor speed in r.p.m.

P = number of rotor poles

f = frequency of e.m.f. in Hz

Consider a stator conductor that is successively swept by the N and S poles of the rotor. If a *positive* voltage is induced when a N-pole sweeps across the conductor, a similar *negative* voltage is induced when a S-pole sweeps by. This means that *one complete cycle* of e.m.f. is generated in the conductor as a pair of poles passes it *i.e.*, one N-pole and the adjacent following S-pole. The same is true for every other armature conductor.

$\therefore$ No. of cycles/revolution = No. of pairs of poles = $P/2$

No. of revolutions/second = $N/60$

$\therefore$ No. of cycles/second = *$(P/2)\,(N/60)$ = $NP/120$

But number of cycles of e.m.f. per second is its frequency.

$$\therefore \qquad f = \frac{NP}{120}$$

It may be noted that N is the synchronous speed. For a given alternator, the number of rotor poles is fixed and, therefore, the alternator must be run at synchronous speed to give an output of desired frequency. For this reason, an alternator is sometimes called *synchronous generator.*

23.5. Pitch Factor and Distribution Factor

The armature winding of an alternator is distributed over the entire armature. The distributed winding produces nearly a sine wave form and the heating is more uniform.

* No. of cycles/second = No. of cycles/revolution × No. of revolutions/second.

Likewise, the coils of armature winding are not full-pitched *i.e.*, the two sides of a coil are not at corresponding points under adjacent poles. The fractional pitched armature winding requires less copper per coil and at the same time waveform of output voltage is improved.

(*i*) **Pitch factor** (K_p)

$$\text{Pitch factor, } K_p = \frac{\text{e.m.f. with short–pitch coil}}{\text{e.m.f. with full–pitch coil}}$$

Pole pitch is the distance between the centre lines of adjacent N and S poles measured along the circumference of armature surface. When the two sides of a coil are full pole pitch apart, it is called full-pitched coil. The e.m.fs in the coil sides of a full-pitched coil are in phase. In practice, coil pitch is less than pole pitch and hence e.m.fs in the coil sides have a phase difference. The resultant e.m.f. in the coil will be less than that of full-pitched coil. Therefore, for full-pitch coil, $K_p = 1$ while for short-pitch coil, $K_p < 1$.

(*ii*) **Distribution factor** (K_d)

$$\text{Distribution factor, } K_d = \frac{\text{e.m.f. with distributed winding}}{\text{e.m.f. with concentrated winding}}$$

Since the conductors are spread over the surface of the armature in slots, their e.m.fs differ in phase and the total e.m.f. is the vector sum and not arithmetic sum. Hence $K_d = 1$ for concentrated winding but it is less than 1 for distributed winding.

23.6. E.M.F. Equation of an Alternator

Let Z = No. of conductors or coil sides in series *per phase*

ϕ = Flux per pole in webers

P = Number of rotor poles

N = Rotor speed in r.p.m.

In one revolution (*i.e.*, 60/N second), each stator conductor is cut by $P\phi$ webers *i.e.*,

$$d\phi = P\phi\ ; \quad dt = 60/N$$

$\therefore$ Average e.m.f. induced in one stator conductor

$$= \frac{d\phi}{dt} = \frac{P\phi}{60/N} = \frac{P\phi N}{60} \text{ volts}$$

Since there are Z conductors in series per phase,

$$\therefore \text{ Average e.m.f./phase} = \frac{P\phi N}{60} \times Z$$

$$= \frac{P\phi Z}{60} \times \frac{120 f}{P} \qquad \left(\because N = \frac{120 f}{P}\right)$$

$$= 2 f \phi Z \text{ volts}$$

$$\text{R.M.S. value of e.m.f./phase} = \text{Average value/phase} \times \text{form factor}$$

$$= 2 f \phi Z \times 1.11 = 2.22 f \phi Z$$

$$\therefore \quad E_{r.m.s.}/\text{phase} = 2.22 f \phi Z \text{ volts} \qquad \text{...(i)}$$

If K_p and K_d are the pitch factor and distribution factor of the armature winding, then,

$$E_{r.m.s.}/\text{phase} = 2.22\, K_p\, K_d\, Z f \phi \text{ volts} \qquad \text{...(ii)}$$

Example 23.1. *A 3-phase, 50Hz star-connected alternator has 180 conductors per phase and flux per pole is 0.0543 wb. Find (i) e.m.f. generated per phase and (ii) e.m.f. between line terminals. Assume the winding to be full pitched and distribution factor to be 0.96.*

Solution.

(*i*) Generated e.m.f./phase, $E_{ph} = 2.22\, K_p\, K_d\, Z f \phi$

$$= 2.22 \times 1 \times 0.96 \times 180 \times 50 \times 0.0543 = \mathbf{1041.5\ V}$$

(*ii*) Line voltage, $E_L = \sqrt{3}\, E_{ph} = \sqrt{3} \times 1041.5 = \mathbf{1803.19\ V}$.

Example 23.2. *Find the number of armature conductors in series per phase required for the armature of a 3-phase, 50Hz, 10-pole alternator. The winding is star-connected to give a line voltage of 11000V. The flux per pole is 0.16 wb. Assume $K_p = 1$ and $K_d = 0.96$.*

Solution.

Generated e.m.f./phase, $E_{ph} = E_L/\sqrt{3} = 11000/\sqrt{3} = 6352$ V

Let Z be the number of conductors in series per phase.

$$E_{ph} = 2.22\, K_p\, K_d\, Z f \phi$$

$$\therefore \quad Z = \frac{E_{ph}}{2.22\, K_p\, K_d f \phi} = \frac{6352}{2.22 \times 1 \times 0.96 \times 50 \times 0.16} = \mathbf{372.5}$$

Example 23.3. *The armature of an 8-pole, 3-phase, 50Hz alternator has 18 slots and 10 conductors/slot. A flux of 0.04 wb is entering the armature from one pole. Calculate the induced e.m.f. per phase.*

Solution. Since the values of K_p and K_d are not given, they will be assumed 1.

Total no. of conductors $= 18 \times 10 = 180$

No. of conductors/phase, $Z = 180/3 = 60$

Induced e.m.f./phase, $E_{ph} = 2.22\, K_p\, K_d\, Z f \phi$

$$= 2.22 \times 1 \times 1 \times 60 \times 50 \times 0.04 = \mathbf{266.4\ V}$$

23.7. Alternator on NO Load

When the rotor is rotating and energised, the circuit diagram for an alternator with its stator circuit open (*i.e.*, no load condition) is as shown in Fig. 23.5 (*i*). Each phase generates an e.m.f. E. The per phase armature resistance and leakage reactance are R_a and X_L respectively.

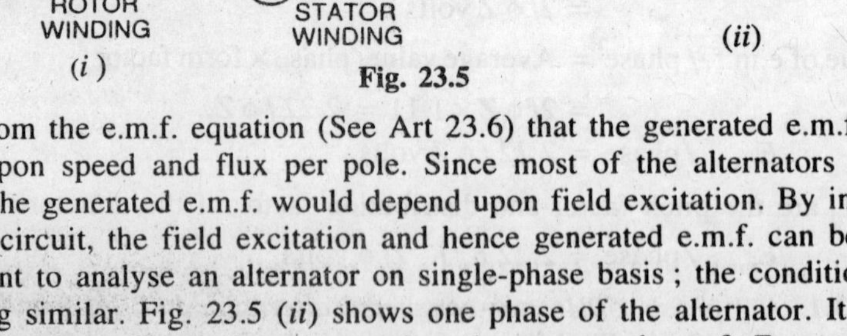

Fig. 23.5

It is clear from the e.m.f. equation (See Art 23.6) that the generated e.m.f. E in an alternator depends upon speed and flux per pole. Since most of the alternators are operated at constant speed, the generated e.m.f. would depend upon field excitation. By including a rheostat in the field circuit, the field excitation and hence generated e.m.f. can be changed. It is always convenient to analyse an alternator on single-phase basis ; the conditions in the other two phases being similar. Fig. 23.5 (*ii*) shows one phase of the alternator. It is clear that at no load, terminal voltage V per phase is equal to the generated e.m.f. E per phase *i.e.*

$$V = E \qquad \text{...at no load}$$

23.8. Alternator on Load

When load on an alternator changes, the terminal voltage V also changes. The change in V is due to the following three effects :

(i) voltage drop in armature resistance R_a

(ii) voltage drop in armature leakage reactance X_L

(iii) voltage drop due to armature reaction

The voltage drop due to armature *reaction is accounted for by assuming a fictitious reactance X_a in the armature winding. The phasor sum of X_L and X_a gives *synchronous reactance* X_s. The per phase equivalent circuit of an alternator on load can be represented as shown in Fig. 23.6.

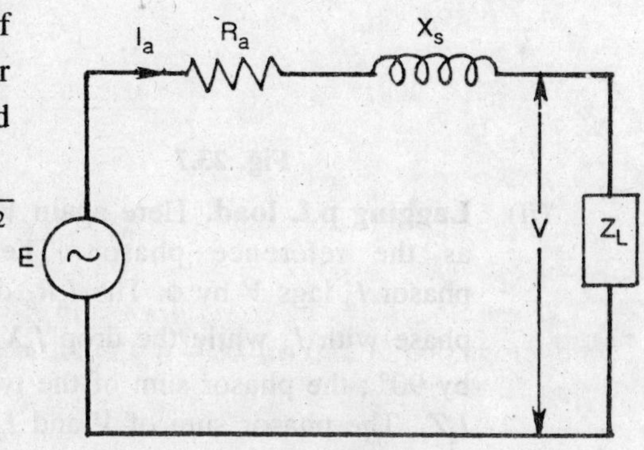

Fig. 23.6

Synchronous impedance/phase, $Z_s = \sqrt{R_a^2 + X_s^2}$

If V is the terminal voltage/phase and E is the generated e.m.f./phase, then,

$$\mathbf{E} = \mathbf{V} + \mathbf{I}_a(\mathbf{R}_a + j\,\mathbf{X}_s) = \mathbf{V} + \mathbf{I}_a\mathbf{Z}_s$$

Note : The value of X_s is 10 to 100 times greater than R_a. Consequently, we can neglect R_a without any serious error.

23.9. Phasor Diagram of a Loaded Alternator

Let E = no–load e.m.f./phase

V = terminal voltage/phase

I_a = armature current/phase

Z_s = synchronous impedance/phase

ϕ = load p.f. angle

(i) **Unity p.f. load.** Fig. 23.7 shows the phasor diagram of an alternator for unity p.f. load. Here V is taken as the reference phasor. The current phasor I_a is in phase with V. The voltage drop I_aR_a is in phase with I_a while the voltage drop I_aX_s leads I_a by 90° ; the phasor sum of the two giving I_aZ_s. The phasor sum of V and I_aZ_s gives E.

* At no load, only flux ϕ_F due to field excitation exists in the machine. However, when the alternator is loaded, the armature current also produces flux ϕ_A. The action of ϕ_A on ϕ_F is called armature reaction. The actual flux ϕ in a loaded alternator is the phasor sum of ϕ_F and ϕ_A. With lagging power factor loads which usually exist with alternators, the effect of armature reaction is to reduce the resultant flux ϕ and hence the generated e.m.f. The armature reaction effect is accounted for by assuming a fictitious reactance X_a in the armature winding.

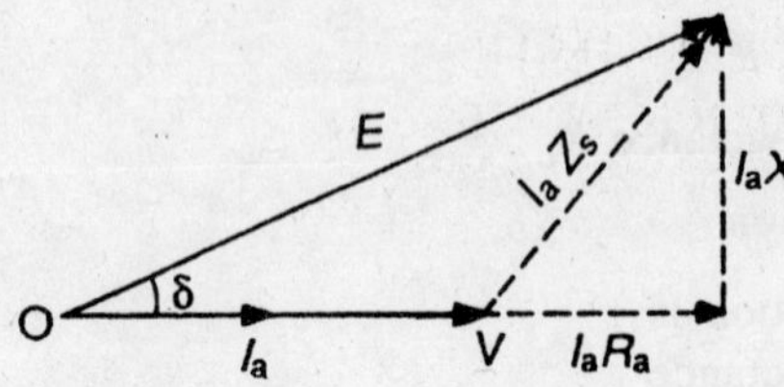

Fig. 23.7

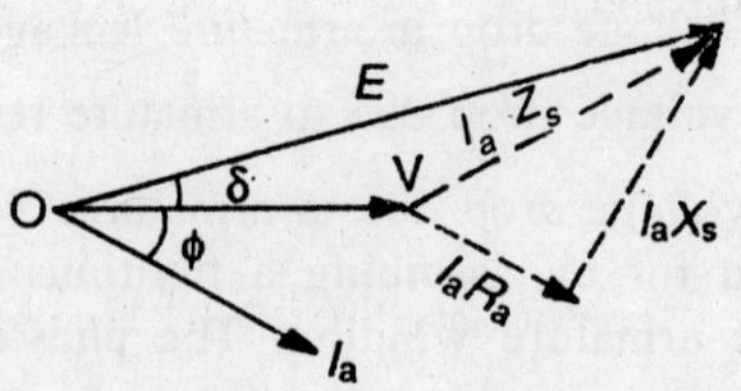

Fig. 23.8

(*ii*) **Lagging p.f. load.** Here again V is taken as the reference phasor. The current phasor I_a lags V by ϕ. The I_aR_a drop is in phase with I_a while the drop I_aX_s leads I_a by 90° ; the phasor sum of the two giving I_aZ_s. The phasor sum of V and $I_a Z_s$ gives E [See Fig. 23.8].

(*iii*) **Leading p.f. load.** Fig. 23.9 shows the phasor diagram of an alternator for leading p.f. load. As usual, V is taken as the reference phasor. The current phasor I_a leads V by ϕ. The I_aR_a drop is in phase with I_a while the drop I_aX_s leads I_a by 90° ; the phasor sum of the two giving I_aZ_s. The phasor sum of V and I_aZ_s gives E.

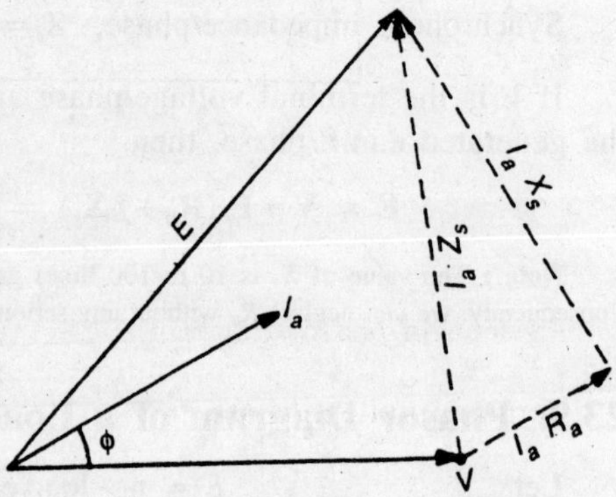

Fig. 23.9

Note : The above phasor diagrams can also be drawn by taking current as the reference phasor.

23.10. Voltage Regulation

The voltage regulation of an alternator is defined as the percentage rise in terminal voltage when full-load is removed *i.e.*,

$$\%\text{ age regulation} = \frac{E - V}{V} \times 100$$

where E = no–load voltage per phase

V = full–load voltage per phase

Its value depends upon load current and load power factor.

Example 23.4. *A 3-phase star-connected, 50Hz alternator has 96 conductors per phase and a flux/pole 0.1wb. The alternator winding has a synchronous reactance of 5 Ω/phase and negligible resistance. The distribution factor for the stator winding is 0.96.*

Calculate the terminal voltage when three non-inductive resistors, of 10 Ω/phase, are connected in star across the terminals.

Solution. Fig. 23.10 (*i*) shows the equivalent circuit for one phase while Fig. 23.10 (*ii*) shows its phasor diagram.

Generated e.m.f./phase, $E = 2.22\, K_p\, K_d\, Z f \phi$

$= 2.22 \times 1 \times 0.96 \times 96 \times 50 \times 0.1 = 1023\text{ V}$

Impedance/phase, $Z = \sqrt{5^2 + 10^2} = 11.18\ \Omega$

Current/phase, $I_a = E/Z = 1023/11.18 = 91.5\text{ A}$

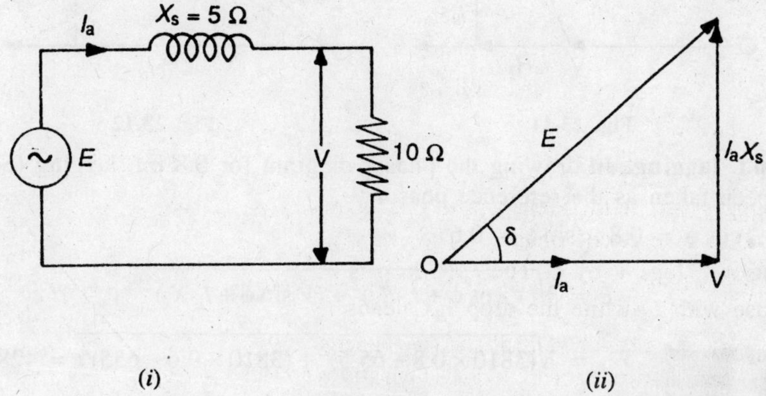

Fig. 23.10

As seen in phasor diagram in Fig. 23.10 (*ii*),

Terminal voltage/phase, $V = \sqrt{E^2 - (I_a X_s)^2}$

$= \sqrt{(1023)^2 - (91.5 \times 5)^2} = 915\text{ V}$

Terminal line voltage $= \sqrt{3} \times 915 = \mathbf{1585\ V}$

Example 23.5. *A 1500KVA, 6.6kV, 3-phase, star-connected alternator has a resistance of 0.5Ω/phase and a synchronous reactance of 5 Ω/phase. Find its voltage regulation for (i) unity p.f. (ii) 0.8 lagging p.f. and (iii) 0.8 leading p.f.*

Solution.

Line current, $I_L = \dfrac{1500 \times 10^3}{\sqrt{3} \times 6600} = 131\text{A}$

Armature current/phase, $I_a = I_L = 131\text{A}$

Voltage/phase, $V = 6600/\sqrt{3} = 3810\text{V}$

$I_a R_a = 131 \times 0.5 = 65.5\text{ V}$

$I_a X_s = 131 \times 5 = 655\text{ V}$

(*i*) **Unity p.f.** In drawing the phasor diagram for unity p.f. (See Fig. 23.11), voltage has been taken as the reference phasor.

$$E = \sqrt{(V + I_a R_a)^2 + (I_a X_s)^2}$$

$$= \sqrt{(3810 + 65.5)^2 + (655)^2} = 3930\text{ V}$$

$\therefore$ % age regulation $= \dfrac{3930 - 3810}{3810} \times 100 = \mathbf{3.15\%}$

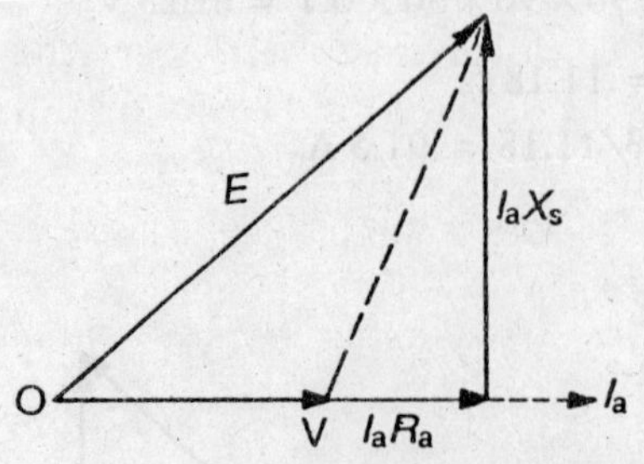

Fig. 23.11

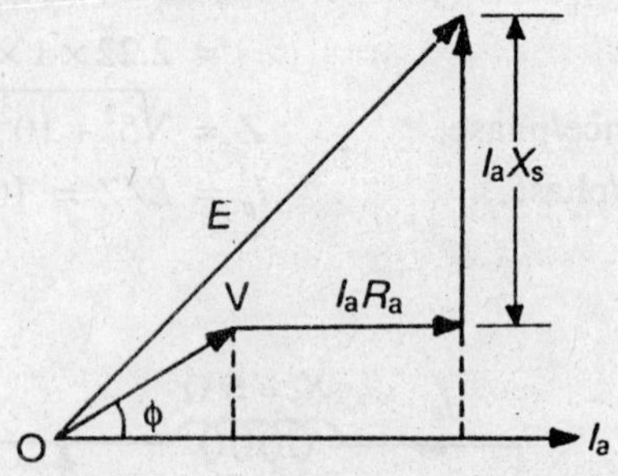

Fig. 23.12

(*ii*) **0.8 p.f. lagging.** In drawing the phasor diagram for 0.8 p.f. lagging (See Fig 23.12), current has been taken as the reference phasor.

$$\cos\phi = 0.8\,;\quad \sin\phi = 0.6$$

$$E = \sqrt{(V\cos\phi + I_a R_a)^2 + (V\sin\phi + I_a X_s)^2}$$

$$= \sqrt{(3810\times 0.8 + 65.5)^2 + (3810\times 0.6 + 655)^2} = 4283\text{ V}$$

$$\therefore \quad \%\text{ age regulation} = \frac{4283 - 3810}{3810}\times 100 = \mathbf{12.4\%}$$

(*iii*) **0.8 p.f. leading.** Fig. 23.13 shows the phasor diagram for 0.8 p.f. leading. Note that current has been taken as the reference phasor.

$$OB = OA + AB \quad = V\cos\phi + I_a R_a$$

$$BC = BD - CD \quad = V\sin\phi - I_a X_s$$

In right angled triangle *OBC*,

$$E = \sqrt{(OB)^2 + (BC)^2}$$

$$= \sqrt{(V\cos\phi + I_a R_a)^2 + (V\sin\phi - I_a X_s)^2}$$

$$= \sqrt{(3810\times 0.8 + 65.5)^2 + (3810\times 0.6 - 655)^2} = 3515\text{ V}$$

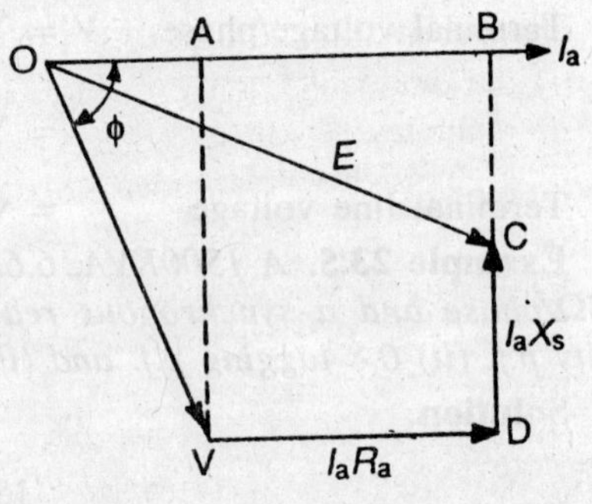

Fig. 23.13

$$\therefore\ \%\text{ age regulation} = \frac{3515 - 3810}{3810}\times 100 = \mathbf{-7.7\%}$$

Tutorial Problems

1. A 3-phase, 11kV star-connected alternator has armature resistance of 1 Ω/phase and a synchronous reactance of 20 Ω/phase. Calculate the voltage regulation for a load of 1500 kW at p.f. of (*i*) 0.8 lagging (*ii*) unity (*iii*) 0.8 leading. [(*i*) **22%** (*ii*) **4.25%** (*iii*) **–13.4%**]
2. A 3-phase, star-connected alternator is rated at 1500 kVA, 11kV. The resistance and reactance per phase are 1.5 Ω and 30 Ω respectively. To what value will the terminal voltage rise when full-load at p.f. 0.8 lagging is switched off ? **[13970V]**
3. A 3-phase alternator generates an open-circuit line voltage of 6920V. The synchronous reactance per phase is 5 Ω while resistance/phase is negligible. Find the terminal voltage if three 12 Ω resistors are connected in star across the alternator.

OBJECTIVE QUESTIONS

1. Majority of alternators in the use have ....
(*i*) revolving a.c. armature winding
(*ii*) stationary field type construction
(*iii*) revolving field type construction
(*iv*) none of the above.

2. The stator of an alternator is identical to that of a ....
(*i*) d.c. generator
(*ii*) 3-phase induction motor
(*iii*) 1-phase induction motor
(*iv*) Rosenberg generator

3. The stator of an alternator rarely uses .... slots.
(*i*) wide open type (*ii*) semi-closed type
(*iii*) closed type (*iv*) none of the above

4. The field winding of an alternator is .... excited.
(*i*) d.c. (*ii*) a.c.
(*iii*) both d.c. and a.c.(*iv*) none of the above

5. The a.c. armature winding of an alternator operates at .... the field winding.
(*i*) the same voltage as
(*ii*) much lesser voltage than
(*iii*) much higher voltage than
(*iv*) none of the above

6. The salient-pole construction for field structure of an alternator is generally used for .... machine.
(*i*) 2-pole (*ii*) 4-pole
(*iii*) 8-pole (*iv*) none of the above

7. An alternator is sometimes called .... generator.
(*i*) synchronous (*ii*) asynchronous
(*iii*) Rosenberg (*iv*) none of the above

8. A turbo-alternator uses ....
(*i*) salient-pole field structure
(*ii*) nonsalient-pole field structure
(*iii*) rotating a.c. armature winding
(*iv*) none of the above

9. The nonsalient-pole field construction is used for .... alternator.
(*i*) low-speed (*ii*) medium-speed
(*iii*) high-speed (*iv*) none of the above

10. The a. c. armature winding of an alternator is ....
(*i*) always star-connected
(*ii*) generally delta-connected
(*iii*) star-delta connected
(*iv*) none of the above

11. Low-speed alternators are driven by ....
(*i*) hydraulic turbines
(*ii*) steam engines
(*iii*) steam turbines
(*iv*) none of the above

12. High-speed alternators are driven by ....
(*i*) diesel engines (*ii*) hydraulic turbines
(*iii*) steam turbines (*iv*) none of the above

13. The air-gap in an alternator is .... in an induction machine.
(*i*) much shorter than
(*ii*) much longer than
(*ii*) about the same as
(*iv*) none of the above

14. The stator of an alternator is wound for .... on the rotor.
(*i*) more number of poles than
(*ii*) less number of poles than
(*iii*) the same number of poles as
(*iv*) none of the above

15. Turbo-alternators have rotors of ....
(*i*) small diameter and long axial length
(*ii*) large diameter and long axial length
(*iii*) large diameter and small axial length
(*iv*) same diameter and axial length

16. The rotor of a turbo-alternator is made cylindrical in order to reduce ....
(*i*) eddy current loss (*ii*) hysteresis loss
(*iii*) windage losses (*iv*) none of the above

17. The number of cycles generated in a 6-pole alternator in one revolution is ....
(*i*) 3 (*ii*) 6
(*iii*) 50 (*iv*) none of the above

18. The speed at which a 6-pole alternator should be driven to generate 50 cycles per second is ....
(*i*) 1500 r.p.m. (*ii*) 1000 r.p.m.
(*iii*) 500 r.p.m. (*iv*) none of the above

19. The frequency of e.m.f. generated in an 8-pole alternator running at 900 r.p.m. is ....

(*i*) 50 Hz (*ii*) 60 Hz
(*iii*) 120 Hz (*iv*) none of the above

20. In case of a 4-pole machine, 1 mechanical degree corresponds to .... electrical degrees.

(*i*) 2 (*ii*) 4
(*iii*) 8 (*iv*) none of the above

21. For the same rating, the size of low-speed alternator is .... that of high-speed alternator.

(*i*) about the same as (*ii*) more than
(*iii*) less than (*iv*) none of the above

22. The synchronous reactance of an alternator is due to ....

(*i*) leakage flux
(*ii*) d.c. field excitation
(*iii*) armature reaction
(*iv*) none of the above

23. The synchronous reactance of an alternator is generally .... armature resistance.

(*i*) 5 times smaller than
(*ii*) 5 times greater than
(*iii*) 10 to 100 times greater than
(*iv*) 10 times smaller than

24. The synchronous reactance of an alternator .... as the iron is saturated.

(*i*) decreases
(*ii*) increases
(*iii*) remains unchanged
(*iv*) none of the above

25. In an alternataor, the effect of armature reaction is minimum at power factor of ....

(*i*) 0.866 lagging (*ii*) 0.866 leading
(*iii*) 0.5 lagging (*iv*) unity

26. If the lagging load p.f. of an alternator is decreased, the demagnetising effect of armature reaction ....

(*i*) remains the same (*ii*) is decreased
(*iii*) is increased (*iv*) none of the above.

27. A 3-phase alternator generates an open-circuit phase voltage of 4000 V when exciting current is 50 A ; the short-circuit current for the same excitation being 800 A. The synchronous reactance per phase is ....

(*i*) 80 Ω (*ii*) 5 Ω
(*iii*) 15 Ω (*iv*) none of the above

28. A 30 MVA, 15 kV alternator will have a per phase nominal impedance of ....

(*i*) 9 Ω (*ii*) 15 Ω
(*iii*) 7.5 Ω (*iv*) none of the above

29. When load on an alternator is increased, the terminal voltage increases if the load p.f. is ....

(*i*) unity (*ii*) lagging
(*iii*) leading (*iv*) zero

30. The eficiency of the turbo-alternator .... with increase in speed.

(*i*) decreases
(*ii*) increases
(*iii*) remains unchanged
(*iv*) none of the above

ANSWERS TO OBJECTIVE QUESTIONS

1. (*iii*)	**2.** (*ii*)	**3.** (*iii*)	**4.** (*i*)	**5.** (*iii*)	**6.** (*iii*)	**7.** (*i*)	**8.** (*ii*)
9. (*iii*)	**10.** (*i*)	**11.** (*i*)	**12.** (*iii*)	**13.** (*ii*)	**14.** (*iii*)	**15.** (*i*)	**16.** (*iii*)
17. (*i*)	**18.** (*ii*)	**19.** (*ii*)	**20.** (*i*)	**21.** (*ii*)	**22.** (*iii*)	**23.** (*iii*)	**24.** (*i*)
25. (*iv*)	**26.** (*iii*)	**27.** (*ii*)	**28.** (*iii*)	**29.** (*iii*)	**30.** (*ii*)		

HINTS TO SELECTED OBJECTIVE QUESTION

3. The closed-type slots have the following disadvantages :

(*i*) They increase the inductance of the windings.

(*ii*) They are expensive in the initial cost.

(*iii*) They present a problem of end connections.

Due to these disadvantages, wholly closed slots are not used.

5. The field winding is supplied with direct current, usually at 125 V or 250 V. However, voltages as high as 33 k V are generated in the armature winding.

6. The salient pole rotors are driven by low-speed water turbines or diesel engines. It is because the salient-pole type construction is difficult to build to withstand the stresses at high speeds. Because a salient pole rotor turns a low speed (50 to 300 r.p.m.) and because a frequency of 50 Hz is required, we must place a large number of poles on the rotor ($f = NP / 120$).

8. High-speed or turbo-alternators have nonsalient-pole (*i.e.* cylindrical rotor) rotor to withstand the stresses at high speeds. Because the rotor is driven at a high speed, we require a small number of poles.

12. Steam turbines are smaller and more efficient when they turn at high speed. The same is true of alternators.

15. The turbo-alternators run at high speeds. The high speed of rotation produces strong centrifugal forces which impose an upper limit on the diameter of the rotor. In case of a rotor turning at 3600 r.p.m., the elastic limit of the steel requires the manufacturer to limit the diameter to a maximum of 1.2 m. We can double the diameter when the speed is 1800 r.p.m. but because of the transportation problems, we seldom go beyond 1.8 m.

18. $f = \frac{NP}{120} \quad \therefore \quad N = \frac{120\,f}{P} = \frac{120 \times 50}{6} = 1000 \text{ r.p.m.}$

19. $f = \frac{NP}{120} = \frac{900 \times 8}{120} = 60 \text{ Hz}$

23. The value of synchronous reactance is 10 to 100 times greater than armature resistance. Consequently, we can neglect armature resistance unless we are interested in efficiency or heating effects.

27. Synchronous reactance, $X_s = \frac{E_{ph}}{I_{sc}} = \frac{4000}{800} = 5\ \Omega$

where E_{ph} = rated open-circuit phase voltage

I_{sc} = short-circuit current per phase, using the same exciting current that was required to produce E_{ph}.

28. Nominal impedance, $Z_n = \frac{E_L^2}{S} = \frac{(15000)^2}{30 \times 10^6} = 7.5\ \Omega$

Where E_L = rated line voltage in volts

S = rated power of alternator in VA

24

Synchronous Motors

Introduction

It may be recalled that a d.c. generator can be run as a d.c. motor. In like manner, an alternator may operate as a motor by connecting its armature winding to a 3-phase supply. It is then called a synchronous motor. As the name implies, a synchronous motor runs at synchronous speed ($N_S = 120f/P$) *i.e.*, in synchronism with the revolving field produced by the 3-phase supply. The speed of rotation is, therefore, tied to the frequency of the source. Since the frequency is fixed, the motor speed stays constant irrespective of the load or voltage of 3-phase supply. However, synchronous motors are not used so much because they run at constant speed (*i.e.*, synchronous speed) but because they possess other unique electrical properties. In this chapter, we shall discuss the working and characteristics of synchronous motors.

24.1. Construction

A synchronous motor is a machine that operates at synchronous speed and converts electrical energy into mechanical energy. It is fundamentally an alternator operated as a motor. Like an alternator, a synchronous motor has the following two parts :

(*i*) a **stator** which houses 3-phase armature winding in the slots of the stator core and receives power from a 3-phase supply (See Fig. 24.1).

(*ii*) a **rotor** that has a set of salient poles excited by direct current to form alternate *N* and *S* poles. The exciting coils are connected in series to two slip rings and direct current is fed into the winding from an external exciter mounted on the rotor shaft.

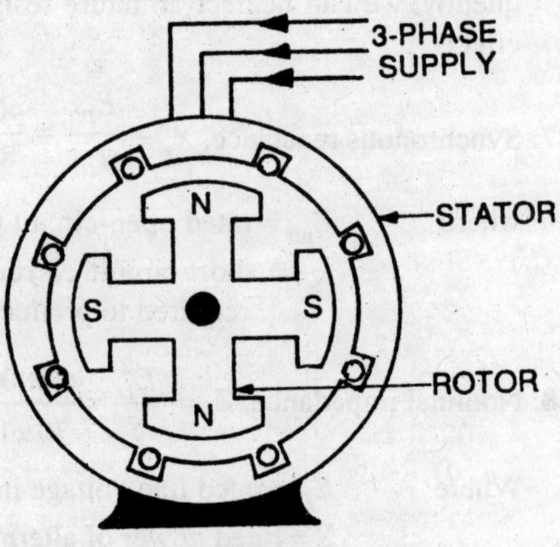

Fig. 24.1

The stator is wound for the same number of poles as the rotor poles. As in the case of an induction motor, the number of poles determines the synchronous speed of the motor :

$$\text{Synchronous speed,} \quad N_S = \frac{120f}{P}$$

f = frequency of supply in Hz

P = number of poles

An important drawback of a synchronous motor is that it is not self-starting and auxiliary means have to be used for starting it.

24.2. Operating Principle

The fact that a synchronous motor has no starting torque can be easily explained.

(*i*) Consider a 3-phase synchronous motor having two rotor poles N_R and S_R. Then the stator will also be wound for two poles N_S and S_S. The motor has direct voltage applied to the rotor winding and a 3-phase voltage applied to the stator winding. The stator winding produces a rotating field which revolves round the stator at synchronous speed N_S (= $120 f/p$). The direct (or zero frequency) current sets up a two-pole field which is stationary so long as the rotor is not turning. Thus we have a situation in which there exists a pair of revolving armature poles (*i.e.*, $N_S - S_S$) and a pair of stationary rotor poles (*i.e.*, $N_R - S_R$).

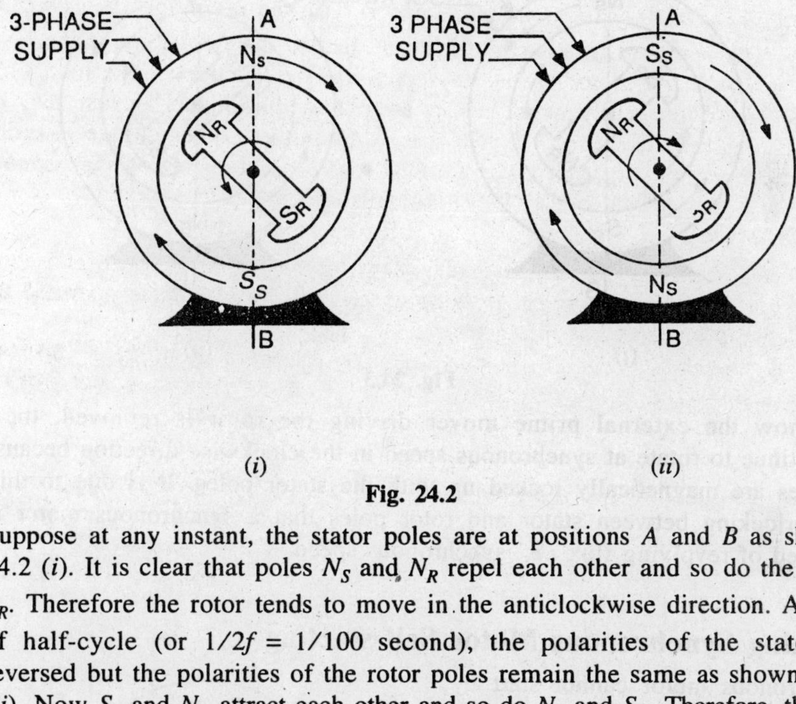

Fig. 24.2

(*ii*) Suppose at any instant, the stator poles are at positions *A* and *B* as shown in Fig. 24.2 (*i*). It is clear that poles N_S and N_R repel each other and so do the poles S_S and S_R. Therefore the rotor tends to move in the anticlockwise direction. After a period of half-cycle (or $1/2f = 1/100$ second), the polarities of the stator poles are reversed but the polarities of the rotor poles remain the same as shown in Fig. 24.2 (*ii*). Now S_S and N_R attract each other and so do N_S and S_R. Therefore, the rotor tends to move in the clockwise direction. Since the stator poles change their polarities rapidly, they tend to pull the rotor first in one direction and then after a period of half-cycle in the other. Due to high inertia of the rotor, the motor fails to start.

Hence, a synchronous motor has no self-starting torque i.e., a synchronous motor cannot start by itself.

How to get continuous unidirectional torque ? If the rotor poles are rotated by some external means at *such* a speed that they interchange their positions along with the stator poles, then the rotor will experience a continuous unidirectional torque. This can be understood from the following discussion :

(*i*) Suppose the stator field is rotating in the clockwise direction and the rotor is also rotated clockwise by some external means at such a speed that the rotor poles interchange their positions along with the stator poles.

(*ii*) Suppose at any instant the stator and rotor poles are in the position shown in Fig. 24.3 (*i*). It is clear that torque on the rotor will be clockwise. After a period of half-cycle, the stator poles reverse their polarities and at the same time rotor poles also *interchange their positions as shown in Fig. 24.3 (*ii*). The result is that again the torque on the rotor is clockwise. Hence a continuous unidirectional torque acts on the rotor and moves it in the clockwise direction. Under this condition, *poles on the rotor always face poles of opposite polarity on the stator* and a strong magnetic attraction is set up between them. This mutual attraction locks the rotor and stator together and the rotor is virtually pulled into step with the speed of revolving flux (*i.e.*, synchronous speed).

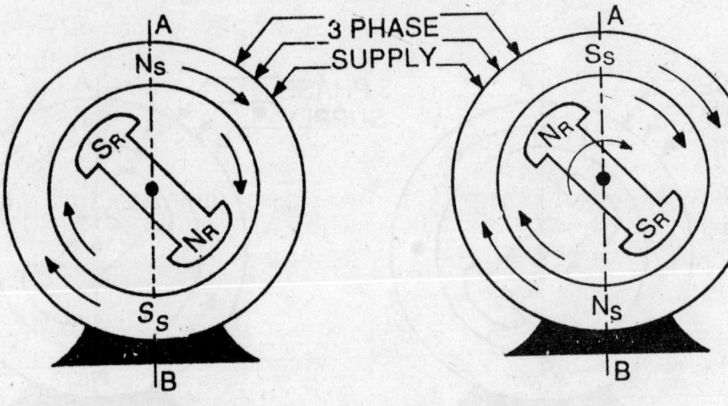

(*i*) (*ii*)

Fig. 24.3

(*iii*) If now the external prime mover driving the rotor is removed, the rotor will continue to rotate at synchronous speed in the clockwise direction because the rotor poles are magnetically locked up with the stator poles. It is due to this magnetic interlocking between stator and rotor poles that a synchronous motor runs at the speed of revolving flux *i.e.*, synchronous speed.

24.3. Making Synchronous Motor Self-starting

A synchronous motor cannot start by itself. In order to make the motor self-starting, a **squirrel cage winding (also called damper winding) is provided on the rotor. The damper winding consists of copper bars embedded in the pole faces of the salient poles of the rotor as shown in Fig. 24.4. The bars are short-circuited at the ends to form in effect a partial squirrel cage winding. The damper winding serves to start the motor.

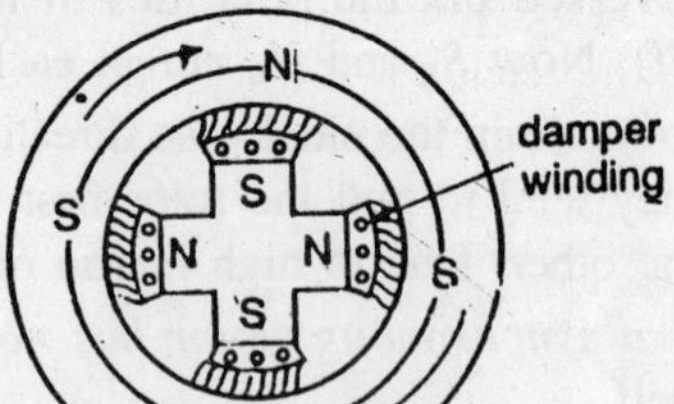

Fig. 24.4

---

* Since the rotor poles are being rotated by an external means at such a speed that they interchange their positions along with the stator poles.

** similar to that in an induction motor.

(*i*) To start with, 3-phase supply is given to the stator winding while the rotor field winding is left unenergised. The rotating stator field induces currents in the damper or squirrel cage winding and the motor starts as an induction motor.

(*ii*) As the motor approaches the synchronous speed, the rotor is excited with direct current. Now the resulting poles on the rotor face poles of opposite polarity on the stator and a strong magnetic attraction is set up between them. The rotor poles lock in with the poles of rotating flux. Consequently, the rotor revolves at the same speed as the stator field *i.e.*, at synchronous speed.

(*iii*) Because the bars of squirrel cage portion of the rotor now rotate at the same speed as the rotating stator field, these bars do not cut any flux and, therefore, have no induced currents in them. Hence squirrel cage portion of the rotor is, in effect, removed from the operation of the motor.

It may be emphasised here that due to magnetic interlocking between the stator and rotor poles, a synchronous motor can only run at synchronous speed. At any other speed, this magnetic interlocking (*i.e.*, rotor poles facing opposite polarity stator poles) ceases and the average torque becomes zero. Consequently, the motor comes to a halt with a severe disturbance on the line.

Note : It is important to excite the rotor with direct current at the right moment. For example, if the d.c. excitation is applied when *N*-pole of the stator faces *N*-pole of the rotor, the resulting magnetic repulsion will produce a violent mechanical shock. The motor will immediately slow down and the circuit breakers will trip. In practice, starters for synchronous motors are designed to detect the precise moment when excitation should be applied.

24.4. Equivalent Circuit

Unlike the induction motor, the synchronous motor is connected to two electrical systems; a d.c. source at the rotor terminals and an a.c. system at the stator terminals.

1. Under normal conditions of synchronous motor operation, no voltage is induced in the rotor by the stator field because the rotor winding is rotating at the same speed as the stator field. Only the impressed direct current is present in the rotor winding and ohmic resistance of this winding is the only opposition to it as shown in Fig. 24.5 (*i*).

2. In the stator winding, two effects are to be considered ; the effect of stator field on the stator winding and the effect of the rotor field cutting the stator conductors at synchronous speed.

(*i*) The effect of stator field on the stator (or armature) conductors is accounted for by including an inductive reactance in the armature winding. This is called *synchronous reactance* X_S. A resistance R_a must be considered to be in series with this reactance to account for the copper losses in the stator or armature winding as shown in Fig. 24.5 (*i*). This resistance combines with synchronous reactance and gives the synchronous impedance of the machine.

(*ii*) The second effect is that a voltage is generated in the stator winding by the synchronously-revolving field of the rotor as shown in Fig. 24.5 (*i*). This generated e.m.f. E_b is known as back e.m.f. and opposes the stator voltage *V*. The magnitude of

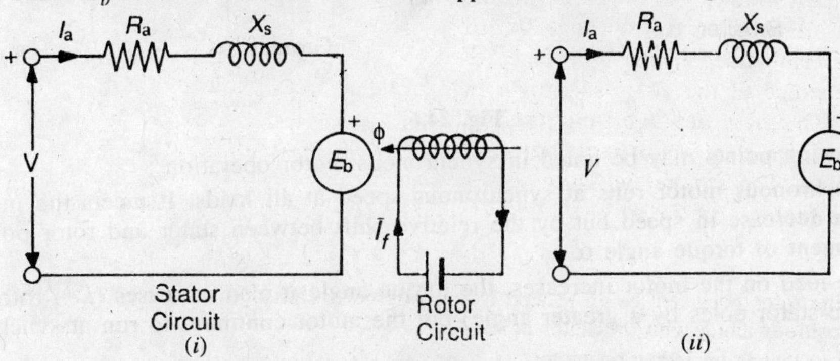

Fig. 24.5

E_b depends upon rotor speed and rotor flux ϕ per pole. Since rotor speed is constant, the value of E_b depends upon the rotor flux per pole *i.e.*, exciting rotor current I_f.

Fig 24.5 (*i*) shows the schematic diagram for one phase of a star-connected synchronous motor while Fig. 24.5 (*ii*) shows its equivalent circuit. Referring to the equivalent circuit in Fig. 24.5 (*ii*),

Net voltage/phase in stator winding,

$$E_r = V - E_b \qquad \text{...phasor difference}$$

Armature current/phase, $I_a = \dfrac{E_r}{Z_s}$

where $Z_s = \sqrt{R_a^{\,2} + X_s^{\,2}}$

This equivalent circuit helps considerably in understanding the operation of a synchronous motor.

Note : In a synchronous motor, the value of X_s is 10 to 100 times greater than R_a. Consequently, we can neglect R_a unless we are interested in efficiency or heating effects.

24.5. Motor on Load

In d.c. motors and induction motors, an addition of load causes the motor speed to decrease. The decrease in speed reduces the counter e.m.f. enough so that additional current is drawn from the source to carry the increased load at a reduced speed. This action cannot take place in a synchronous motor because it runs at a constant speed (*i.e.*, synchronous speed) at all loads.

What happens when we apply mechanical load to a synchronous motor ? The rotor poles fall slightly behind the stator poles, while continuing to run at synchronous speed. The angular displacement between stator and rotor poles (called torque angle α) causes the phase of back e.m.f. E_b to change *w.r.t.* supply voltage V. This increases the net e.m.f. E_r in the stator winding. Consequently, stator current I_a (= E_r / Z_s) increases to carry the load.

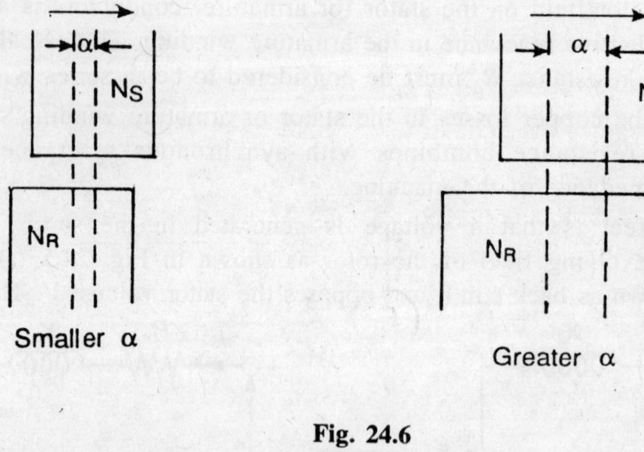

Fig. 24.6

The following points may be noted in synchronous motor operation :

(*i*) A synchronous motor runs at synchronous speed at all loads. It meets the increased load not by a decrease in speed but by the relative shift between stator and rotor poles *i.e.*, by the adjustment of torque angle α.

(*ii*) If the load on the motor increases, the torque angle α also increases (*i.e.*, rotor poles lag behind the stator poles by a greater angle) but the motor continues to run at synchronous speed. The increase in torque angle α causes a greater phase shift of back e.m.f. E_b *w.r.t.* supply voltage V. This increases the net voltage E_r in the stator winding. Consequently, armature current I_a (= E_r/Z_s) increases to meet the load demand.

(*iii*) If the load on the motor decreases, the torque angle α also decreases. This causes a smaller phase shift of E_b *w.r.t.* V. Consequently, the net voltage E_r in the stator winding decreases and so does the armature current I_a (= E_r/Z_s).

24.6. Pull-out Torque

There is a limit to the mechanical load that can be applied to a synchronous motor. As the load increases, the torque angle α also increases so that a stage is reached when the rotor is pulled out of synchronism and the motor comes to a standstill. This load torque at which the motor pulls out of synchronism is called *pull-out* or *breakdown torque.* Its value varies from 1.5 to 3.5 times the full-load torque.

When a synchronous motor pulls out of synchronism, there is a major disturbance on the line and the circuit breakers immediately trip. This protects the motor because both squirrel cage and stator winding heat up rapidly when the machine ceases to run at synchronous speed.

24.7. Motor Phasor Diagram

Consider a synchronous motor supplied with fixed excitation *i.e.*, back e.m.f. E_b is constant.

Let V = supply voltage/phase

E_b = back e.m.f./phase

Z_s = synchronous impedance/phase

(*i*) **Motor on no load.** When the motor is on no load, the torque angle α is small as shown in Fig. 24.7 (*i*). Consequently, back e.m.f. E_b lags behind the supply voltage V by a small angle *δ as shown in the phasor diagram in Fig. 24.7 (*iii*). The net voltage/phase in the stator winding is E_r.

$$\text{Armature current/phase, } I_a = E_r/Z_s$$

The armature current I_a lags behind E_r by $\theta = \tan^{-1} X_s/R_a$. Since $X_s >> R_a$, I_a lags E_r by nearly 90°. The phase angle between V and I_a is φ so that motor power factor is cos φ

$$\text{Input power/phase} = V I_a \cos \phi$$

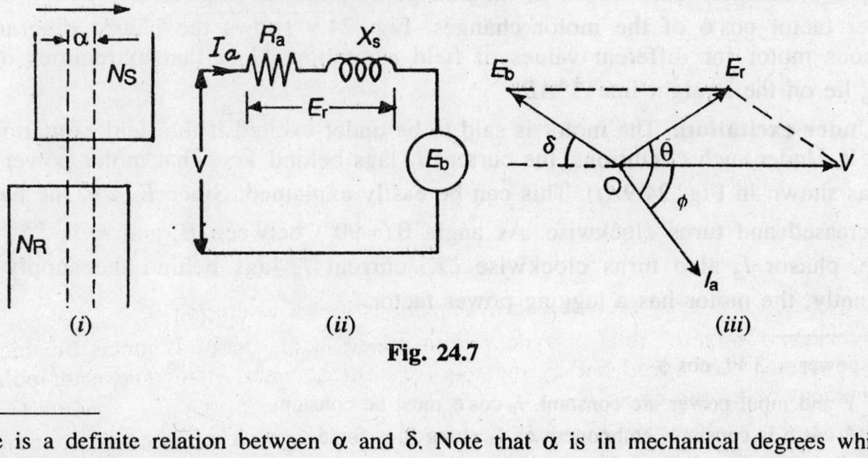

Fig. 24.7

* There is a definite relation between α and δ. Note that α is in mechanical degrees while δ is in electrical degrees.

Thus at no load, the motor takes a small power $V I_a \cos\phi$/phase from the supply to meet the no-load losses while it continues to run at synchronous speed.

(*ii*) **Motor on load.** When load is applied to the motor, the torque angle α increases as shown in Fig. 24.8 (*i*). This causes E_b (its magnitude is constant as excitation is fixed) to lag behind V by a greater angle as shown in the phasor diagram in Fig. 24.8 (*ii*). The net voltage/phase E_r in the stator winding increases. Consequently, the motor draws more armature current I_a (= E_r/Z_s) to meet the applied load.

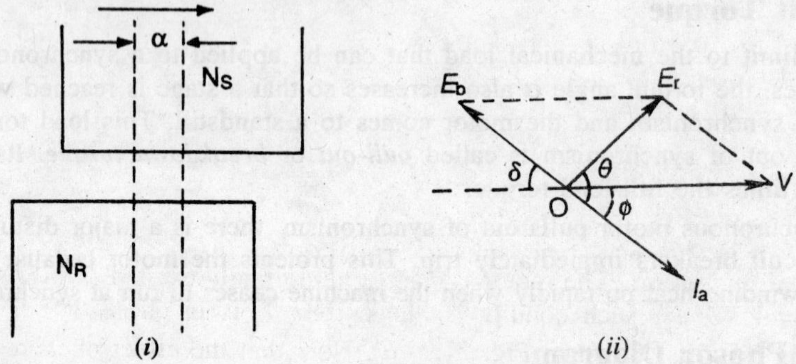

Fig. 24.8

Again I_a lags E_r by about 90° since $X_s >> R_a$. The power factor of the motor is cos φ.

Input power/phase, $P_i = V I_a \cos\phi$

Mechanical power developed by motor/phase

$$P_m = E_b \times I_a \times \text{cosine of angle between } E_b \text{ and } I_a$$
$$= E_b I_a \cos(\delta - \phi)$$

24.8. Effect of Changing Field Excitation

One of the most important features of a synchronous motor is that by changing the field excitation, it can be made to operate from lagging to leading power factor.

Consider a synchronous motor having a fixed supply voltage and driving a constant mechanical load. Since the mechanical load as well as the speed is constant, the power input to the motor (= $3 V I_a \cos\phi$) is also constant. This means that the in-phase component $I_a \cos\phi$ drawn from the supply will remain *constant. If the field excitation is changed, back e.m.f. E_b also changes. This results in the change of phase position of I_a *w.r.t.* V and hence the power factor cos φ of the motor changes. Fig. 24.9 shows the phasor diagram of the synchronous motor for different values of field excitation. Note that extremities of current phasor I_a lie on the straight line **AB.

(*i*) **Under excitation.** The motor is said to be under-excited if the field excitation is such that $E_b < V$. Under such conditions, the current I_a lags behind V so that motor power factor is lagging as shown in Fig. 24.9 (*i*). This can be easily explained. Since $E_b < V$, the net voltage E_r is decreased and turns clockwise. As angle θ (≈ 90°) between E_r and I_a is ***constant, therefore, phasor I_a also turns clockwise *i.e.*, current I_a lags behind the supply voltage. Consequently, the motor has a lagging power factor.

---

* Input power = $3 V I_a \cos\phi$

Since V and input power are constant, $I_a \cos\phi$ must be constant.

** Since $I_a \cos\phi$ is constant, component of I_a along V is fixed.

*** $\theta = \tan^{-1} X_s/R_a$. Since $X_s >> R_a$ and both are fixed, $\theta \approx 90°$ and is fixed.

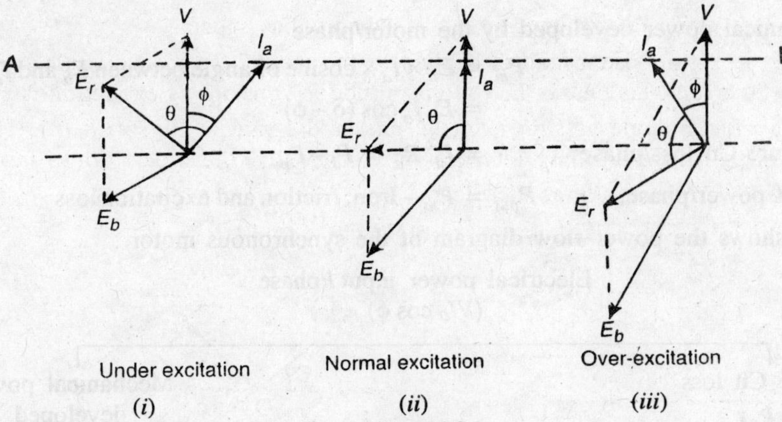

Fig. 24.9

(*ii*) **Normal excitation.** The motor is said to be normally excited if the field excitation is such that $E_b = V$. Under such conditions, the current I_a is in phase with V and the motor power factor is unity as shown in Fig. 24.9 (*ii*). Note that the effect of increasing excitation (*i.e.*, increasing E_b) is to turn the phasor E_r and hence I_a in the anticlockwise direction *i.e.*, I_a tends to be in phase with V. Consequently motor power factor is unity.

(*iii*) **Over excitation.** The motor is said to be over-excited if the field excitation is such that $E_b > V$. Under such conditions, current I_a leads V and the motor power factor is leading as shown in Fig. 24.9 (*iii*). Note that E_r and hence I_a further turn anticlockwise from the normal excitation position. Consequently, I_a leads V.

From the above discussion, it is concluded that if the synchronous motor is under-excited, it has a lagging power factor. As the excitation is increased, the power factor improves till it becomes unity at normal excitation. Under such conditions, the current drawn from the supply is minimum. If the excitation is further increased (*i.e.*, over excitation), the motor power factor becomes leading.

24.9. Power Relations

Consider an under-excited star-connected synchronous motor driving a mechanical load. Fig. 24.10 (*i*) shows the equivalent circuit for one phase while Fig. 24.10 (*ii*) shows the phasor diagram.

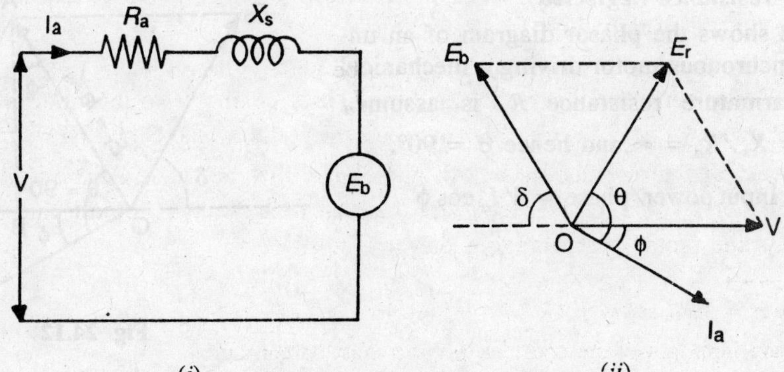

Fig. 24.10

(*i*) Input power/phase, $P_i = V I_a \cos\phi$

(*ii*) Mechanical power developed by the motor/phase

$$P_m = E_b \times I_a \times \text{cosine of angle between } E_b \text{ and } I_a$$
$$= E_b I_a \cos(\delta - \phi)$$

(*iii*) Armature Cu loss/phase $= I_a^2 R_a = P_i - P_m$

(*iv*) Output power/phase, $P_{out} = P_m -$ Iron, friction and excitation loss

Fig. 24.11 shows the power flow diagram of the synchronous motor.

Electrical power input / phase
($VI_a \cos\phi$)

Armature Cu loss
($I_a^2 R_a$)

Mechanical power developed
$P_m = E_b I_a \cos(\delta - \phi)$

Iron, friction and excitation loss

Output power
P_{out}

Fig. 24.11

24.10. Motor Torque

$$\text{Gross torque, } T_g = 9.55 \frac{P_m}{N_s} \text{ N–m}$$

where P_m = Gross motor output in watts $= E_b I_a \cos(\delta - \phi)$
N_s = synchronous speed in r.p.m.

$$\text{Shaft torque, } T_{sh} = 9.55 \frac{P_{out}}{N_s} \text{ N–m}$$

It may be seen that torque is directly proportional to the mechanical power because rotor speed (*i.e.*, N_s) is fixed.

24.11. Mechanical Power Developed by Motor

(*Armature resistance neglected*)

Fig. 24.12 shows the phasor diagram of an under-excited synchronous motor driving a mechanical load. Since armature resistance R_a is assumed *zero, $\tan\theta = X_s / R_a = \infty$ and hence $\theta = 90°$.

$$\text{Input power/phase} = V I_a \cos\phi$$

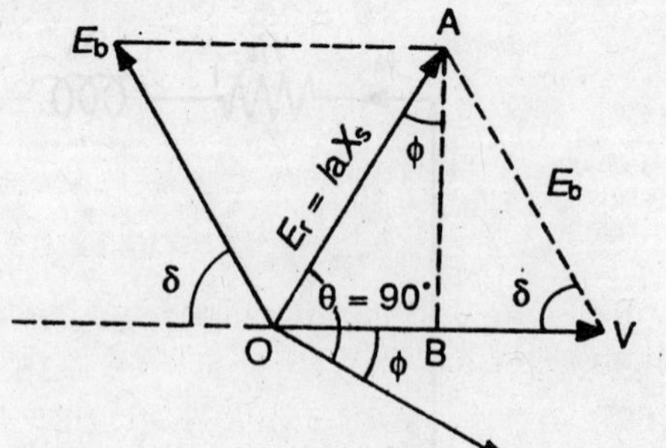

Fig. 24.12

* In a practical motor, the value of X_s is 10 to 100 times greater than R_a. Consequently, we can neglect R_a without significant error.

Since R_a is assumed zero, stator Cu loss ($I_a^2 R_a$) will be zero. Hence input power is equal to the mechanical power P_m developed by the motor.

Mech. power developed/phase, $P_m = V I_a \cos\phi$ ...(*i*)

Referring to the phasor diagram in Fig. 24.12,

$$AB = E_r \cos\phi = I_a X_s \cos\phi$$

Also $$AB = E_b \sin\delta$$

$$\therefore \quad I_a X_s \cos\phi = E_b \sin\delta$$

or $$I_a \cos\phi = \frac{E_b \sin\delta}{X_s}$$

Substituting the value of $I_a \cos\phi$ in exp. (*i*) above,

$$P_m = \frac{V E_b}{X_s}\sin\delta \quad \text{...per phase}$$

$$= \frac{3 V E_b}{X_s}\sin\delta \quad \text{...for 3-phase}$$

It is clear from the above relation that mechanical power increases with torque angle (in electrical degrees) and its maximun value is reached when $\delta = 90°$ (electrical).

$$P_{max} = \frac{V E_b}{X_s} \quad \text{...per phase}$$

Under this condition, the poles of the rotor will be mid-way between *N* and *S* poles of the stator.

Example 24.1. *A 2000 V, 3-phase, star-connected motor has resistance and synchronous reactance per phase of 0.2 Ω and 1.9 Ω respectively. Calculate the generated (back) e.m.f. per phase with an input of 800 kW at p.f. 0.8 lagging.*

Solution.

Supply voltage/phase, $V = 2000/\sqrt{3} = 1155\text{ V}$

Line current, $$I_L = \frac{800 \times 10^3}{\sqrt{3} \times 2000 \times 0.8} = 288\text{ A}$$

Armature current/phase, $I_a = I_L = 288\text{ A}$

$$\theta = \tan^{-1} 1.9/0.2 = 84°$$

$$\phi = \cos^{-1} 0.8 = 36.8°$$

$$Z_s = \sqrt{(0.2)^2 + (1.9)^2} = 1.91\ \Omega$$

$$E_r = I_a Z_s = 288 \times 1.91 = 550\text{ V}$$

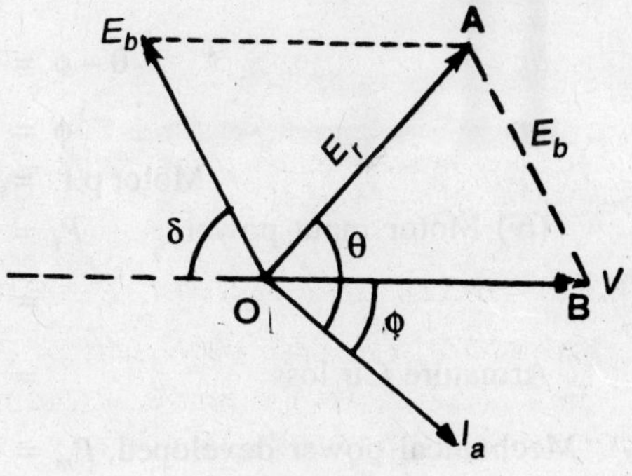

Fig. 24.13

Refer to the phasor diagram of the motor shown in Fig. 24.13. Applying rule of trigonometry to Δ *OAB*,

$$AB^2 = OA^2 + OB^2 - 2 \times OA \times OB \times \cos(\theta - \phi)$$

or $$E_b^2 = (550)^2 + (1155)^2 - 2 \times 550 \times 1155 \times \cos\mathbf{(84° - 36.8°)}$$

$$= 773296$$

$$\therefore \quad E_b = \sqrt{773296} = \mathbf{879\ V/phase}$$

Example 24.2. *A 3980 V, 50 Hz, 4-pole, star-connected synchronous motor generates back e.m.f. of 1790 V per phase. The resistance and synchronous reactance per phase are 2.2 Ω and 22 Ω respectively. The torque angle is 30° electrical. Calculate (i) the resultant armature voltage/phase (ii) armature current/phase (iii) power factor of the motor (iv) gross torque developed by the motor.*

Solution. Fig. 24.14 shows the phasor diagram of the motor. Since $V > E_b$, the motor power factor will be lagging.

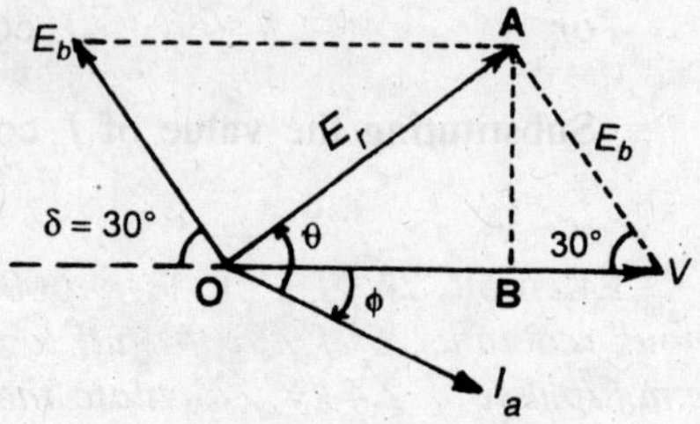

Fig. 24.14

Armature voltage/phase, $V = 3980/\sqrt{3} = 2300 \text{ V}$

Back e.m.f./phase, $E_b = 1790 \text{ V}$

$$(i)\ E_r^2 = V^2 + E_b^2 - 2\,VE_b \cos\delta$$

$$= (2300)^2 + (1790)^2 - 2 \times 2300 \times 1790 \times \cos 30°$$

$$= 1363000$$

$$\therefore\ E_r = \sqrt{1363000} = \mathbf{1167\ V}$$

$$(ii)\ Z_s = \sqrt{R_a^2 + X_s^2} = \sqrt{(2.2)^2 + (22)^2} = 22.1\ \Omega$$

$$I_a = E_r/Z_s = 1167/22.1 = \mathbf{52.8\ A}$$

$$(iii)\ \theta = \tan^{-1} X_s/R_a = \tan^{-1} 22/2.2 = 84.3°$$

In right angled ΔABO, $\angle AOB = \theta - \phi$.

$$\therefore\ \tan(\theta - \phi) = \frac{AB}{OB} = \frac{E_b \sin 30°}{V - E_b \cos 30°}$$

$$= \frac{1790 \sin 30°}{2300 - 1790 \cos 30°} = 1.193$$

or $$\theta - \phi = \tan^{-1} 1.193 = 50°$$

or $$\phi = \theta - 50° = 84.3° - 50° = 34.3°$$

$$\therefore\ \text{Motor p.f.} = \cos\phi = \cos 34.3° = \mathbf{0.826}\ \textit{lag}$$

(*iv*) Motor input power, $P_i = \sqrt{3}\, V_L I_L \cos\phi$

$$= \sqrt{3} \times 3980 \times 52.8 \times 0.826 = 300647 \text{ W}$$

Armature Cu loss $= 3\, I_a^2 R_a = 3 \times (52.8)^2 \times 2.2 = 18400 \text{ W}$

Mechanical power developed, $P_m = 300647 - 18400 = 282247 \text{ W}$

Synchronous speed, $N_s = 120 f/P = 120 \times 50/4 = 1500 \text{ r.p.m.}$

Gross torque, $$T_g = 9.55 \frac{P_m}{N_s}$$

$$= 9.55 \frac{282247}{1500} = \mathbf{1797\ N\text{–}m}$$

Example 24.3. *A 3-phase synchronous motor has 12 poles and operates from 440V, 50 Hz supply. Calculate its speed. If it takes a line current of 100 A at 0.8 power factor lead, what torque the motor will be developing ? Neglect losses.*

Solution.

Motor speed, $$N_s = \frac{120f}{P} = \frac{120 \times 50}{12} = \textbf{500 r.p.m.}$$

Input power, $$P_i = \sqrt{3}\, V_L I_L \cos\phi$$

$$= \sqrt{3} \times 440 \times 100 \times 0.8 = 60966 \text{ W}$$

As losses are neglected, output power, $P_{out} = P_i = 60966\text{W}$

∴ Motor torque, $$T = 9.55 \frac{P_{out}}{N_s}$$

$$= 9.55 \frac{60966}{500} = \textbf{1165 N–m}$$

Example 24.4. *A 3-phase, 6000 kW, 4kV, 180 r.p.m. 50Hz motor has per phase synchronous reactance of 1.2 Ω. At full load, the torque angle is 20° electrical. If the generated back e.m.f./phase is 2.4 kV, calculate the mechanical power developed. What will be the maximum mechanical power developed ?*

Solution.

$$V = 4/\sqrt{3} = 2.3 \text{ kV}$$

$$P_m = \frac{3\, V E_b}{X_s} \sin\delta$$

$$= \frac{3\,(2.3 \times 10^3) \times (2.4 \times 10^3)}{1.2} \times \sin 20°$$

$$= 4.71 \times 10^6 \text{ W} = \textbf{4.71 Mw}$$

The mechanical power developed will be maximum when δ = 90°.

∴ $$P_{max} = \frac{3\, V E_b}{X_s} = \frac{3 \times (2.3 \times 10^3) \times (2.4 \times 10^3)}{1.2}$$

$$= 13.8 \times 10^6 \text{ W} = \textbf{13.8 MW}$$

Tutorial Problems

1. A 2200 V, 3-phase, 50 Hz star-connected synchronous motor has a resistance of 0.6 Ω/phase and a synchronous reactance of 6 Ω per phase. Find the generated e.m.f. when the input is 200 kW at (*i*) a power factor of unity (*ii*) a power factor of 0.8 leading. [(*i*) **2200V** (*ii*) **2640V**]
2. A 3-phase, star-connected synchronous motor takes 20 kW at 400 V from 50 Hz supply. The synchronous reactance/phase is 4 Ω/phase and the effective resistance is negligible. If the exciting current is so adjusted that back e.m.f. is 550 V, find the power factor of the motor. [**0.872** *lead*]
3. In a 3-phase, star-connected 660 V synchronous motor, the synchronous reactance is 2 Ω/phase. When a certain load is applied, the input is 90 kW and the back e.m.f. is 890V. Neglecting resistance, calculate the power factor. [**0.824** *lead*]

24.12. Power Factor of Synchronous Motors

In an induction motor, only one winding (*i.e.*, stator winding) produces the necessary flux in the machine. The stator winding must draw reactive power from the supply to set up the flux. Consequently, induction motor must operate at lagging power factor.

But in a synchronous motor, there are two possible sources of excitation ; alternating current in the stator or *direct current in the rotor. The required flux may be produced either by stator or rotor or both.

(*i*) If the rotor exciting current is of such magnitude that it produces all the required flux, then no magnetising current or reactive power is needed in the stator. As a result, the motor will operate at unity power factor.

(*ii*) If the rotor exciting current is less (*i.e.*, motor is under-excited), the deficit in flux is made up by the stator. Consequently, the motor draws reactive power to provide for the remaining flux. Hence motor will operate at a lagging power factor.

(*iii*) If the rotor exciting current is greater (*i.e.*, motor is over-excited), the excess flux must be counterbalanced in the stator. Now the stator, instead of absorbing reactive power, actually *delivers* reactive power to the 3-phase line. The motor then behaves like a source of reactive power, as if it were a capacitor. In other words, the motor operates at a leading power factor.

To sum up, *a synchronous motor absorbs reactive power when it is under-excited and delivers reactive power to source when it is over-excited.*

24.13. Synchronous Capacitor

An over-excited synchronous motor running on no load is known as **synchronous capacitor.**

We have seen that a synchronous motor takes a leading current when over excited and, therefore, behaves as a capacitor. When such a machine is connected in parallel with induction motors or other devices that operate at lagging power factor, the leading kVAR supplied by the synchronous capacitor partly neutralises the lagging reactive kVAR of the loads. Consequently, power factor of the system is improved.

Synchronous capacitors (*i.e.*, unloaded over-excited synchronous motors) are installed in power systems solely for power factor improvement. They are economical in large sizes than the static capacitors.

24.14. Applications of Synchronous Motors

(*i*) Synchronous motors are particularly attractive for **low speeds (< 300 r.p.m.) because the power factor can always be adjusted to unity and efficiency is high.

(*ii*) Over-excited synchronous motors can be used to improve the power factor of a plant while carrying their rated loads.

(*iii*) They are used to improve the voltage regulation of transmission lines.

(*iv*) High-power electronic converters generating very low frequencies enable us to run synchronous motors at ultra-low speeds. Thus huge motors in the 10 MW range drive crushers, rotary kilns and variable-speed ball mills.

---

* The rotor produces a d.c. magnetomotive force (m.m.f.). However, as far as the stator is concerned, d.c. m.m.f. appears as an a.c. m.m.f. because the rotor is constantly turning. Moreover, rotor m.m.f. has the same frequency as the stator m.m.f. because the rotor turns at synchronous speed. The total flux in the machine is, therefore, due to the combined action of the two m.m.fs.

** At low speeds, induction motors become heavy, costly and have relatively low power factor and low efficiency.

24.15. Comparison of Synchronous and Induction Motors

S.No.	Particular	Synchronous Motor	3-phase Induction Motor
1.	*Speed*	Remains constant (*i.e.*, N_s) from no-load to full-load	Decreases with load.
2.	*Power factor*	Can be made to operate from lagging to leading power factor.	Operates at lagging power factor.
3.	*Excitation*	Requires d.c. excitation at the rotor.	No excitation for the rotor.
4.	*Economy*	Economical for speeds below 300 r.p.m.	Economical for speeds above 600 r.p.m.
5.	*Self-starting*	No self-starting torque. Auxiliary means have to be provided for starting.	Self-starting
6.	*Construction*	Complicated	Simple
7.	*Starting torque*	More	less

OBJECTIVE QUESTIONS

1. The rotor of a synchronous motor is ....

(*i*) salient-pole type
(*ii*) nonsalient-pole type
(*iii*) identical to that of a d.c. motor
(*iv*) none of the above

2. Damper winding in a synchronous motor ....

(*i*) reduces windage losses
(*ii*) serves to start the motor
(*iii*) improves p.f. of the motor
(*iv*) increases hunting of the motor

3. Small synchronous motors are started by ....

(*i*) pony motor
(*ii*) damper winding
(*iii*) variable-frequency source
(*iv*) none of the above

4. A synchronous motor runs at speeds ranging from ....

(*i*) 1800 to 3600 r.p.m.
(*ii*) 3600 to 6000 r.p.m.
(*iii*) 150 to 1800 r.p.m.
(*iv*) none of the above

5. The full-load slip of a synchronous motor is ....

(*i*) 5% (*ii*) 1%
(*iii*) 2% (*iv*) zero

6. A 50 Hz synchronous motor runs at 200 r.p.m. The number of salient poles on the rotor are ....

(*i*) 36 (*ii*) 16
(*iii*) 30 (*iv*) none of the above

7. If the supply frequency of synchronous motor is 60 cycles/second, then the rotor must revolve at ....

(*i*) 25 cycles/second
(*ii*) 60 cycles/second
(*iii*) 100 cycles/second
(*iv*) none of the above

8. A synchronous motor is a ....motor.

(*i*) variable p.f. (*ii*) variable speed
(*iii*) singly fed (*iv*) none of the above

9. The rotor of a synchronous motor is excited with direct current when the motor ....

(*i*) is at standstill (*i.e.* at starting)
(*ii*) approaches synchronous speed
(*iii*) approaches half synchronous speed
(*iv*) none of the above

10. When the synchronous motor runs at synchronous speed, the voltage induced in the damper winding is ....

(*i*) maximum (*ii*) minimum
(*iii*) zero (*iv*) none of the above

11. In order to reverse the direction of a synchronous motor, ...

(*i*) interchange any two stator lines
(*ii*) reduce d.c. field excitation to zero
(*iii*) change supply frequency
(*iv*) none of the above

12. A synchronous motor runs at only one speed (*i.e.* synchronous speed) because it ....

(*i*) has no losses
(*ii*) is a doubly fed machine
(*iii*) has a damper winding
(*iv*) none of the above

13. The speed of a synchronous motor can be changed by varying ....
(*i*) mechanical load (*ii*) field excitation
(*iii*) supply frequency
(*iv*) none of the above

14. If a synchronous motor is operating at a supply frequency of 50 cycles/second, then, rotor revolves at ....
(*i*) 25 cycles/second (*ii*)50 cycles/second
(*iii*) 100 cycles/second
(*iv*) none of the above

15. When the synchronous motor is on no load, the axes of the rotor poles .... that of stator poles.
(*i*) nearly coincide with
(*ii*) are far ahead
(*iii*) are far behind (*iv*) none of the above

16. If the mechanical angle α between the stator and rotor poles increases, then, stator current ....
(*i*) is decreased
(*ii*) remains unchanged
(*iii*) is increased (*iv*) none of the above

17. The increased load demand in a synchronous motor is met by ....
(*i*) reduction in speed
(*ii*) increase in speed
(*iii*) decrease in stator current
(*iv*) relative shift between stator and rotor poles

18. The pull-out torque of a practical synchronous motor will occur when the torque angle is about ....
(*i*) 0° (*ii*) 30°
(*iii*) 45° (*iv*) 75°

19. The pull-out torque of a synchronous motor is generally .... times its full-load torque.
(*i*) 1.25 to 3.5 (*ii*) 5 to 10
(*iii*) 6 to 9 (*iv*) 0.5

20. The pull-out torque of a synchronous motor will .... as the field excitation is increased.
(*i*) decrease (*ii*) remain unchanged
(*iii*) increase (*iv*) none of the above

21. The change of load on a synchronous motor will not result in the change of ....
(*i*) armature current (*ii*) speed
(*iii*) torque angle (*iv*) power factor

22. For a given load, the normal field excitation of synchronous motor is that which gives .... power factor.
(*i*) 0.8 leading (*ii*) 0.8 lagging
(*iii*) unity (*iv*) none of the above

23. For a given load, the armature current of a synchronous motor will be minimum for .... power factor.
(*i*) unity (*ii*) 0.8 leading
(*iii*) 0.8 lagging (*iv*) none of the above

24. For a given load, if the field excitation of a synchronous motor is increased, then, armature current..................
(*i*) decreases (*ii*) increases
(*iii*) becomes minimum
(*iv*) remains unchanged

25. An over-excited synchronous motor behaves as ....
(*i*) a resistor (*ii*) an inductor
(*iii*) a capacitor (*iv*) none of the above

ANSWERS TO OBJECTIVE QUESTIONS

1. (*i*)	**2.** (*ii*)	**3.** (*ii*)	**4.** (*iii*)	**5.** (*iv*)	**6.** (*iii*)	**7.** (*ii*)	**8.** (*i*)
9. (*ii*)	**10.** (*iii*)	**11.** (*i*)	**12.** (*ii*)	**13.** (*iii*)	**14.** (*ii*)	**15.** (*i*)	**16.** (*iii*)
17. (*iv*)	**18.** (*iv*)	**19.** (*i*)	**20.** (*iii*)	**21.** (*ii*)	**22.** (*iii*)	**23.** (*i*)	**24.** (*ii*)
25. (*iii*)							

HINTS TO SELECTED OBJECTIVE QUESTIONS

4. Most synchronous motors are rated between 150 kW (200 H. P.) and 15 MW (20,000 H. P.) and turn at speeds ranging from 150 to 1800 r.p.m. Consequently, these machines are used in heavy industry.

5. A synchronous motor runs at one speed only *i.e.* synchronous speed. Consequently, slip is zero.

7. The primary requirement is that frequency of rotation of the rotor should be such *that stator and rotor fields are stationary with respect to each other*. For this to happen, obviously, the rotor must revolve at 60 cycles/second.

10. When the motor runs at synchronous speed, there is no relative motion between rotor and revolving flux produced by stator currents. Hence e.m.f. in the damper winding placed on the rotor is zero.

12. The significant and distinguishing feature of synchronous motors in contrast to induction motors is that they are douby fed. Electrical energy is supplied both to the field and the armature windings. When this is done, torque can only be developed at one a speed-the *synchronous speed*. At any other speed, the torque is zero. In fact, it is this characteristic (*i.e.* doubly fed) which enables the synchronous motor to develop non-zero torque at only one speed (*i.e.* synchronous speed) and hence the name synchronous motor.

13. $f = \dfrac{N_s P}{120} \quad \therefore \quad N_s = \dfrac{120 f}{P}$

For a given machine, the value of P is fixed. Hence, the speed of a asynchronous motor can only be changed by varying the supply frequency.

16. When synchronous motor is on no-load, the axes of the rotor and stator poles nearly coincide. As the motor is loaded, the rotor poles fall slightly behind the stator poles, while continuing to turn at synchronous speed. The increased power must come from the a.c. source (*i.e.* stator winding). Hence, stator current increases.

18. The mechanical power developed by a synchronous motor per phase is given by :

$$P_m = \frac{V E_b}{X_s} \sin \delta$$

where P_m = mechanical power of motor per phase
E_b = phase voltage induced by excitation
V = phase voltage of a.c. source
X_s = synchronous reactance per phase
δ = torque angle (in electrical degrees between E_b and V)

The maximum torque that motor can develop is called the pull-out torque and will occur at δ = 90°. This is true for motors having smooth rotors. Most synchronous motors have salient poles; in this case the pull-out torque occurs at a torque angle of about 75° (electrical).

25

Electrical Instruments and Electrical Measurements

Introduction

Electrical energy is being used in the manufacture of many commodities. In order to ensure quality and efficiency, it is important that we should be able to measure accurately the electrical quantities involved. The instruments used to measure electrical quantities (*e.g.* current, voltage, power, energy *etc.*) are called electrical instruments. These instruments are generally named after the electrical quantity to be measured. Thus the instruments which measure current, voltage, power and energy are called ammeter, voltmeter, wattmeter and energy meter respectively. The accuracy, convenience and reliability of electrical instruments are mainly responsible for the widespread use of electrical methods of measurements. In this chapter, we shall confine our attention to the construction, working and applications of some important electrical instruments.

25.1. Types of Electrical Instruments

Electrical measuring instruments may be classified according to their functions as (*i*) indicating instruments (*ii*) integrating instruments and (*iii*) recording instruments.

(*i*) **Indicating instruments.** Those instruments which directly indicate the value of the electrical quantity *at the time* when it is being measured are called indicating instruments *e.g. ammeters, voltmeters* and *wattmeters*. In such instruments, a pointer moving over a graduated scale directly gives the value of the electrical quantity being measured. For example, when an ammeter is connected in the circuit, the pointer of the meter directly indicates the value of current flowing in the circuit at that time.

(*ii*) **Integrating instruments.** Those instruments which measure the total quantity of electricity (in ampere-hours) or electrical energy (in watt-hours) in a given time are called integrating instruments *e.g. ampere-hour meter* and *watt-hour meter*. In such instruments, there are sets of dials and pointers which register the total quantity of electricity or electrical energy supplied to the load.

(*iii*) **Recording instruments.** Those instruments which give a continuous record of the variations of the electrical quantity to be measured are called recording instruments. A recording instrument is merely an indicating instrument with a pen attached to its pointer. The pen rests lightly on a chart wrapped over a drum moving with a slow uniform speed. The motion of the drum is in a direction perpendicular to that direction of the pointer. The path traced out by the pen indicates the manner in which the quantity, being measured, has varied during the time of the record. Recording voltmeters are used in supply stations to record the voltage of the supply mains during the day. Recording ammeters are employed in supply stations for registering the current taken from the batteries.

25.2. Principles of Operation of Electrical Instruments

An electrical instrument essentially consists of a movable element and a scale to indicate or register the electrical quantity being measured. The movable element is supported on jewelled bearings and carries a pointer or sets of dials. The movement of the movable element is caused by utilising one or more of the following effects of current or voltage :

1. Magnetic effect ...... Moving-iron instruments
2. Electrodynamic effect ..... (*i*) Permanent-magnet moving coil (*ii*) Dynamometer type
3. Electromagnetic-induction ..... Induction type instruments
4. Thermal effect ...... Hot-wire instruments
5. Chemical effect ...... Electrolytic instruments
6. Electrostatic effect .... Electrostatic instruments

S.No	Type	Effect	Suitable for	Instrument
1.	Moving-iron	Magnetic effect	*d.c.* and *a.c.*	Ammeter, Voltmeter
2.	Permanent-magnet moving coil	Electrodynamic effect	*d.c.* only	Ammeter, Voltmeter
3.	Dynamometer type	Electrodynamic effect	*d.c.* and *a.c.*	Ammeter, Voltmeter, Wattmeter
4.	Induction type	Electro-magnetic induction effect	*a.c.* only	Ammeter, Voltmeter, Wattmeter, Energy meter
5.	Hot-wire	Thermal effect	*d.c.* and *a.c.*	Ammeter, Voltmeter
6.	Electrolytic meter	Chemical effect	*d.c.* only	Ampere-hour meter
7.	Electrostatic type	Electrostatic effect	*d.c.* and *a.c.*	Voltmeter only

The principles of operation of electrical instruments are given in the above table for facility of reference.

25.3. Essentials of Indicating Instruments

An indicating instrument essentially consists of moving system pivoted in jewel bearings. A pointer is attached to the moving system which indicates on a graduated scale, the value of the electrical quantity being measured. In order to ensure proper operation of indicating instruments, the following three torques are required :

(*i*) Deflecting (or operating) torque

(*ii*) Controlling (or restoring) torque

(*iii*) Damping torque

The deflecting torque is produced by utilising the various effects of electric current or voltage and causes the moving system (and hence the pointer) to move from zero position. The controlling torque is provided by spring or gravity and opposes the deflecting torque. The pointer comes to rest at a position where these two opposing torques are equal. The damping torque is provided by air friction or eddy currents. It ensures that the pointer comes to the final position without oscillations, thus enabling accurate and quick readings to be taken.

25.4. Deflecting Torque

One important requirement in indicating instruments is the arrangement for producing deflecting or operating torque (T_d) when the instrument is connected in the circuit to measure the given electrical quantity. This is achieved by utilising the various effects of electric current or voltage mentioned in Art 25.2. The deflecting torque causes the moving system (and hence the pointer attached to it) to move from zero position to indicate on a graduated scale the value of electrical quantity being measured. The actual method of producing the deflecting torque depends upon the type of instrument and shall be discussed while dealing with particular instrument.

25.5. Controlling Torque

If deflecting torque were acting alone, the pointer would continue to move indefinitely and would swing over to the maximum deflected position irrespective of the magnitude of current (or voltage or power) to be measured. This necessitates to provide some form of controlling or opposing torque (T_C). This controlling torque should oppose the deflecting torque and should increase with the deflection of the moving system. The pointer will be brought to rest at a position where the two opposing torques are equal *i.e.* $T_d = T_C$. The controlling torque performs two functions :

(*i*) It increases with the deflection of the moving system so that the final position of the pointer on the scale will be according to the magnitude of current (or voltage or power) to be measured.

(*ii*) It brings the pointer back to zero position when the deflecting torque is removed. If it were not provided, the pointer once deflected would not return to zero position on removing the deflecting torque.

The controlling torque in indicating instruments may be provided by one of the following two methods :

1. By one or more springs ... *Spring control*
2. By weight of moving parts ... *Gravity control*

1. Spring control. This is the most common method of providing controlling torque in electrical instruments. A spiral *hairspring made of some non-magnetic material like phosphor bronze is attached to the moving system of the instrument as shown in Fig. 25.1. With the deflection of the pointer, the spring is twisted in the opposite direction. This twist in the spring provides the controlling torque. Since the torsion torque of a spiral spring is proportional to the angle of twist, the controlling torque is directly proportional to the deflection of the pointer *i.e.* $T_C \propto \theta$

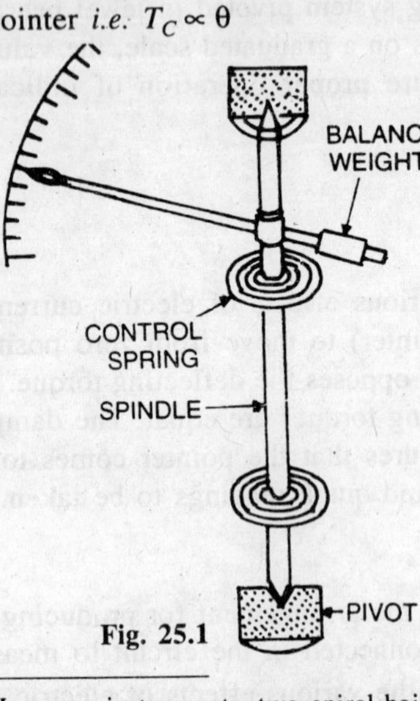

Fig. 25.1

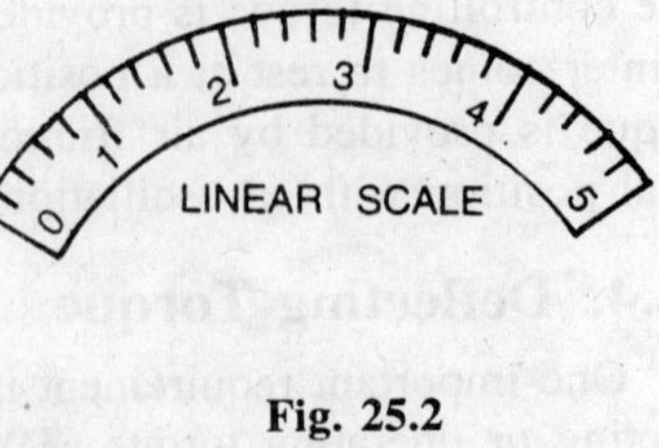

Fig. 25.2

* In some instruments, two spiral hairsprings wound in the opposite direction are used. The two springs serve the additional purpose of leading current to the moving system (i.e. operating coil).

The pointer will come to rest at a position where controlling torque T_C is equal to the deflecting torque T_d *i.e.*, $T_d = T_C$.

In an instrument where the deflecting torque is uniform, spring control provides a linear or evenly-spaced scale over the whole range. For example, in a permanent-magnet moving coil instrument, the deflecting torque is directly proportional to the current flowing through the operating coil *i.e.*,

$$T_d \propto I$$

With spring control, $T_C \propto \theta$

In the final deflected position, $T_d = T_C$

$$\therefore \qquad \theta \propto I$$

Since the deflection is directly proportional to I, scale of such an instrument will be uniform as shown in Fig. 25.2.

2. Gravity Control. In this method, a small adjustable weight W is attached to the moving system [See *Fig. 25.3 (*i*)] which provides the necessary controlling torque. In the zero position of the pointer, the control weight hangs vertically downward and therefore provides no controlling torque. However, under the action of deflecting torque, the pointer moves from zero position (from left to right) and control weight moves in the opposite direction. Due to gravity, the control weight would tend to come to original position (*i.e.*, vertical) and thus provides an opposing or controlling torque. The pointer comes to rest at a position where controlling torque is equal to the deflecting torque.

In the deflected position shown in Fig. 25.3 (*ii*), weight W can be resolved into two rectangular components *viz* $W \cos\theta$ and $W \sin\theta$. Only the component $W \sin\theta$ provides the controlling torque T_C.

$$\therefore \qquad T_C = Wl \sin\theta$$

or $\qquad T_C \propto \sin\theta$ (for fixed W and l)

Thus, the controlling torque is proportional to the sine of angle of deflection. It may be seen that the value of T_C can be varied by changing l *i.e.*, by changing the position of control weight W on the arm.

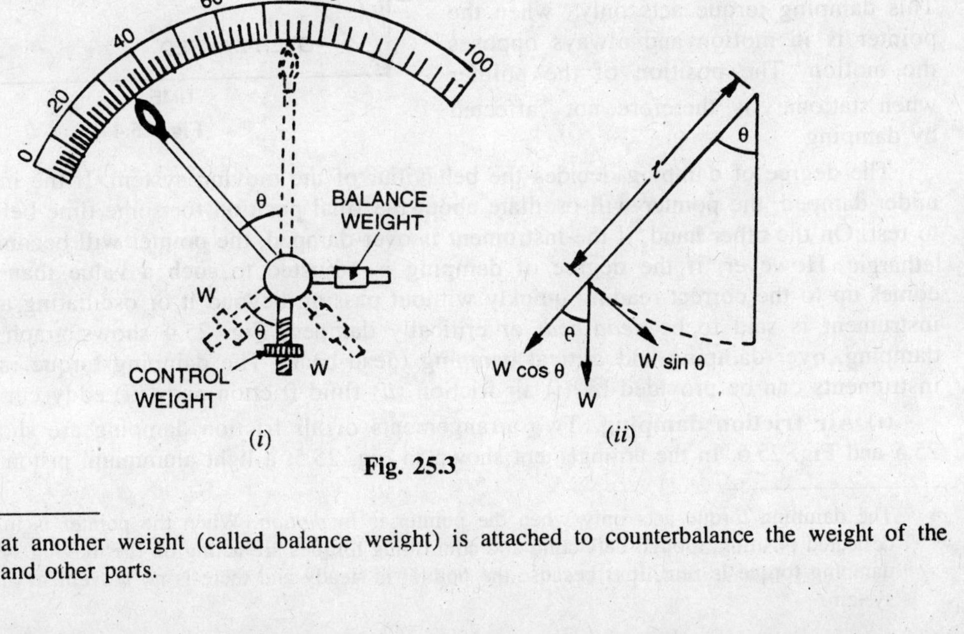

Fig. 25.3

---

* Note that another weight (called balance weight) is attached to counterbalance the weight of the pointer and other parts.

Taking the case of an instrument in which the deflecting torque is directly proportional to current (*i.e.* permanent-magnet moving coil instrument), we have for gravity control,

$$T_d \propto I \text{ and } T_C \propto \sin\theta$$

In the final deflected position, $T_d = T_C$

$$\therefore \quad I \propto \sin\theta$$

Since I is proportional to the sine of angle of deflection, gravity-controlled instruments have non-uniform scales; being crowded at the beginning as shown in Fig. 25.3 (*i*).

Advantages

(*i*) It is slightly cheaper in manufacturing costs than spring control.

(*ii*) It is unaffected by temperature variations.

(*iii*) It is not subjected to fatigue.

(*iv*) The controlling torque can be changed easily.

Disadvantages

(*i*) The instrument has to be kept in vertical position.

(*ii*) The control weight increases the weight of the moving system.

(*iii*) Gravity-controlled instruments have non-uniform scale.

25.6. Damping Torque

If the moving system is acted upon by deflecting and controlling torques alone, then the pointer, due to inertia, will oscillate about its final deflected position for quite some time before coming to rest. This is often undesirable because it makes difficult to obtain quick and accurate readings. In order to avoid these oscillations of the pointer and to bring it quickly to its final deflected position, a damping torque is provided in the indicating instruments. This damping torque acts only when the pointer is in motion and always opposes the motion. The position of the pointer when stationary is, therefore, not *affected by damping.

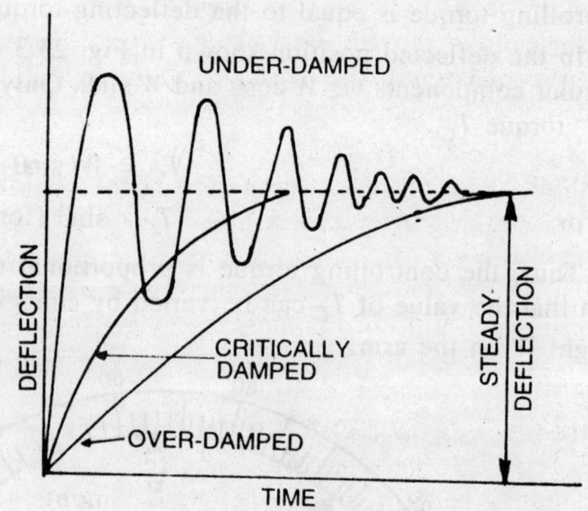

Fig. 25.4

The degree of damping decides the behaviour of the moving system. If the instrument is under-damped, the pointer will oscillate about the final position for some time before coming to rest. On the other hand, if the instrument is over-damped, the pointer will become slow and lethargic. However, if the degree of damping is adjusted to such a value that the pointer comes up to the correct reading quickly without passing beyond it or oscillating about it, the instrument is said to be *dead-beat* or critically damped. Fig. 25.4 shows graph for under-damping, over damping and critical damping (dead-beat). The damping torque in indicating instruments can be provided by (*i*) air-friction (*ii*) fluid friction and (*iii*) eddy currents.

(*i*) **Air friction damping.** Two arrangements of air friction damping are shown in Fig. 25.5 and Fig. 25.6. In the arrangement shown in Fig. 25.5, a light aluminium piston is attached

* The damping torque acts only when the pointer is in *motion*. When the pointer is in a particular deflected position, though deflecting and controlling torques are acting on the moving system but the damping torque is zero. It is because the pointer is steady and there is no movement of the moving system.

to the spindle that carries the pointer and moves with a very little clearance in a rectangular or circular air chamber closed at one end. The cushioning action of the air on the piston damps out any tendency of the pointer to oscillate about the final deflected position. This method is not favoured these days and the one shown in Fig. 25.6 (*i*) is preferred.

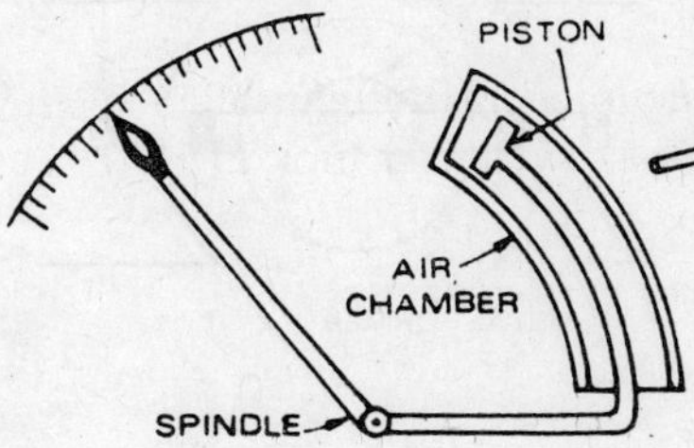

Fig. 25.5

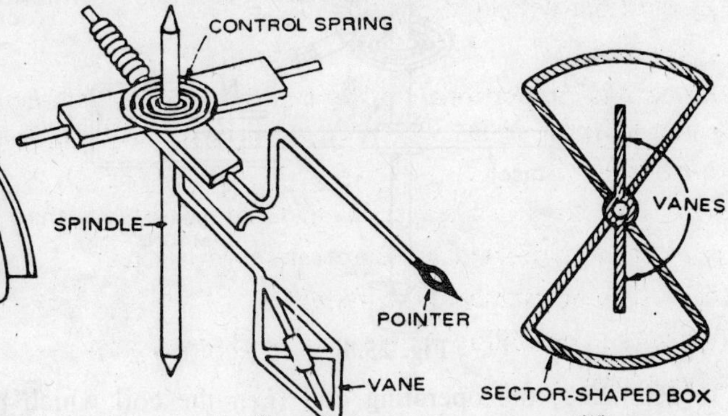

(*i*) (*ii*)

Fig. 25.6

The arrangement shown in Fig. 25.6 (*ii*) is widely used. In this method, one or two light aluminium vanes are attached to the same spindle that carries the pointer. The vanes are permitted to swing in a sector-shaped closed box that is just large enough to accommodate the vanes. As the pointer moves, the vanes swing in the box, compressing the air in front of them. The pressure of compressed air in the vanes provides the necessary damping force to reduce the tendency of the pointer to oscillate.

(*ii*) **Fluid friction damping.** In this method, discs or vanes attached to the spindle of the moving system are kept immersed in a pot containing oil of high viscosity (See Fig. 25.7). As the pointer moves, the friction between the oil and vanes opposes the motion of the pointer and thus necessary damping is provided. This method is not suitable for portable instruments because of the oil contained in the instrument. In general, fluid friction damping is not employed in indicating instruments, although one can find its use in Kelvin electrostatic voltmeter.

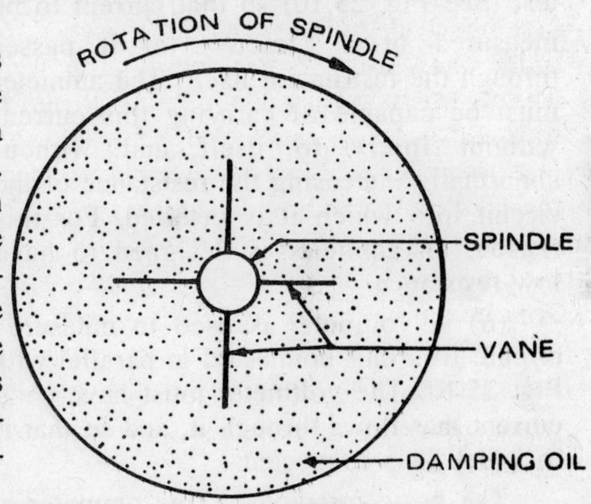

Fig. 25.7

(*iii*) **Eddy current damping.** Two methods of eddy current damping are shown in Fig. 25.8 and Fig. 25.9. In Fig. 25.8, a thin *aluminium or copper disc attached to the moving system is allowed to pass between the poles of a permanent magnet. As the pointer moves, the disc cuts across the magnetic field and **eddy currents are induced in the disc. These eddy currents react with the field of the magnet to produce a force which opposes the motion (Lenz's Law). In this way, eddy currents provide the damping torque to reduce the oscillations of the pointer.

---

* The disc must be a conductor but non-magnetic.

** The currents induced in the disc assume the form of little whirls or eddies and hence the name eddy currents.

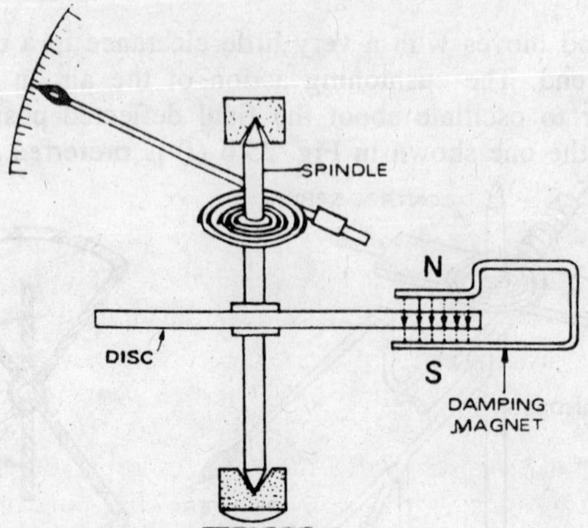

Fig. 25.8

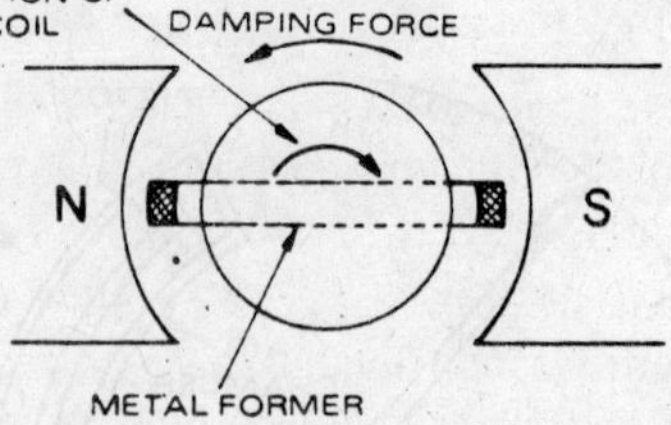

Fig. 25.9

In Fig. 25.9, the operating coil (*i.e.,* the coil which produces the deflecting torque) is wound on the aluminium former. As the coil moves in the field of the instrument, eddy currents are induced in the aluminium former to provide the necessary damping torque.

25.7. Ammeters and Voltmeters

(*i*) An ammeter is used to measure the flow of current in a circuit. It is thus connected in series with the circuit under test (See Fig. 25.10) so that current to be measured or a *fraction of it passes through the instrument itself. The ammeter must be capable of carrying this current without injury to itself and without abnormally increasing the resistance of the circuit into which it is inserted. For this reason, an ammeter is designed to have low resistance.

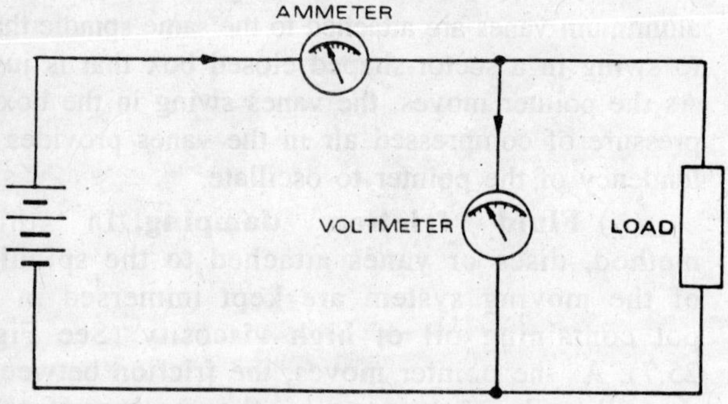

Fig. 25.10

(*ii*) A voltmeter is used to measure the potential difference between two points of a circuit. It is thus connected in parallel with the circuit or some part of the circuit as shown in Fig. 25.10. The voltmeter must have enough resistance so that it will not be injured by the current that flows through it, and so that it will not materially affect the current in the circuit to which it is connected.

The basic principle of the ammeter and of the voltmeter is the same. Both are current operated devices *i.e.,* deflecting torque is produced when current flows through their operating coils. In the ammeter, the deflecting torque is produced by the current we wish to measure, or a certain fraction of that current. In the voltmeter, the deflecting torque is produced by a current which is proportional to the potential difference we wish to measure. Thus, the same instrument can be used as an ammeter or voltmeter with proper design.

The following types of instruments are used for making voltmeters and ammeters :

(*i*) Permanent-magnet moving coil type (*ii*) Dynamometer type
(*iii*) Moving-iron type (*iv*) Hot-wire type
(*v*) Electrostatic type (for voltmeters only) (*vi*) Induction type

---

* If the circuit current is large, a shunt is used to divert a major portion of current so that only a small current flows through the instrument.

The instrument at Sr. No. (*i*) can be used for d.c. work only whereas instrument at Sr. No (*vi*) is employed for a.c. work only. However, instruments from Sr. No. (*ii*) to (*v*) can be used for both d.c. and a.c. measurements.

25.8. Permanent-Magnet Moving Coil Instruments

These instruments are used either as ammeters or voltmeters and are suitable for d.c. work only. This type of instrument is based on the principle that when a current carrying conductor is placed in a magnetic field, mechanical force acts on the conductor.

Construction. Fig. 25.11 shows the various parts of a permanent-magnet moving coil instrument. It consists of a light rectangular coil of many turns of fine wire wound on an aluminium former inside which is an iron core as shown in Fig. 25.11 (*i*). The coil is delicately pivoted upon jewel bearings and is mounted between the poles of a permanent horse-shoe magnet. Attached to these poles are two soft-iron pole pieces which concentrate the magnetic field. The current is led into and out of the coil by means of two control hair-springs, one above and the other below the coil, as shown in Fig. 25.11 (*ii*). These springs also provide the controlling torque. The damping torque is provided by eddy currents induced in the aluminium former as the coil moves from one position to another.

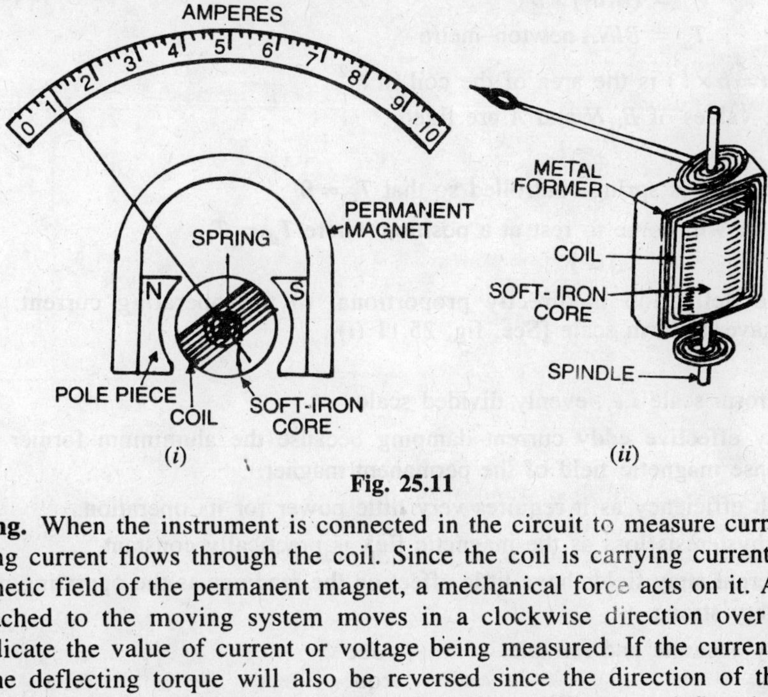

(*i*) (*ii*)

Fig. 25.11

Working. When the instrument is connected in the circuit to measure current or voltage, the operating current flows through the coil. Since the coil is carrying current and is placed in the magnetic field of the permanent magnet, a mechanical force acts on it. As a result, the pointer attached to the moving system moves in a clockwise direction over the graduated scale to indicate the value of current or voltage being measured. If the current in the coil is reversed, the deflecting torque will also be reversed since the direction of the field of the permanent magnet is the same. Consequently, the pointer will try to deflect below zero. Deflection in this direction (*i.e.*, reverse direction) is prevented by a spring "stop". Since the deflecting torque reverses with the reversal of current in the coil, such instruments can be used to measure direct current and voltage *only.

* The instrument can be used to measure alternating currents and voltages by using a rectifier. The given alternating quantity is converted to d.c. by using a bridge type rectifier. This d.c. or average value is measured by the permanent-magnet moving coil instrument. Since a.c. meters are calibrated to indicate *effective* or *r.m.s.* values, we must multiply the average value registered on this meter by 1.11 (the form factor of sinusoidal wave) to obtain the r.m.s. values. Some instruments are scaled to read r.m.s. values directly.

Deflecting torque. The magnetic field in the air gap is radial due to the presence of *soft-iron core. This means that conductors of the coil will always move at right angles to the field. When current is passed through the coil, forces act on its both sides which produce the deflecting torque. Referring to Fig. 25.12, let,

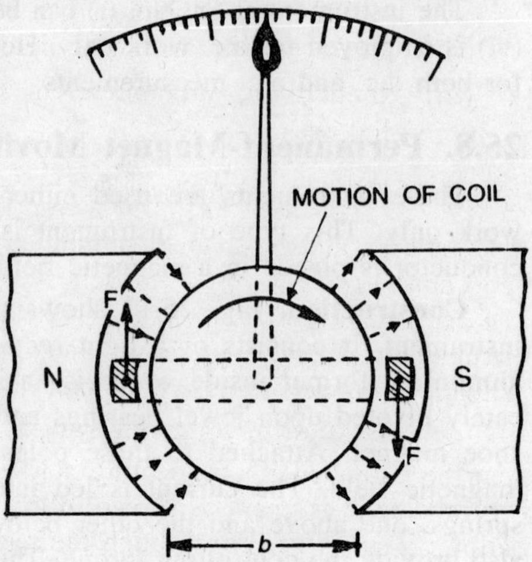

Fig. 25.12

B = flux density in wb/m^2

l = length or depth of coil in m

b = breadth of coil in m

N = No. of turns in the coil

If a current of I amperes flows in the coil, then force acting on each coil side is given by;

Force on each coil side, $F = BIlN$ newtons

Deflecting torque, T_d = Force × perpendicular distance

$$= (BIlN) \times b$$

$$\therefore \quad T_d = BINA \text{ newton–metre}$$

where A (= $b \times l$) is the area of the coil in m^2

Since the values of B, N and A are fixed,

$$\therefore \quad T_d \propto I$$

The instrument is spring-controlled so that $T_C \propto \theta$

The pointer will come to rest at a position where $T_d = T_C$

$$\therefore \quad \theta \propto I$$

Thus, the deflection is directly proportional to the operating current. Hence, such instruments have uniform scale [See. fig. 25.11 (*i*)].

Advantages

(*i*) Uniform scale *i.e.*, evenly divided scale.

(*ii*) Very effective eddy current damping because the aluminium former moves in an intense magnetic field of the permanent magnet.

(*iii*) High efficiency as it requires very little power for its operation.

(*iv*) No hysteresis loss as the magnetic flux is practically constant.

(*v*) External stray fields have little effect on the readings as the operating magnetic field is very strong.

(*vi*) Very accurate and reliable.

Disadvantages

(*i*) Such instruments cannot be used for a.c. measurements.

(*ii*) About 50% more expensive than moving-iron instruments because of their accurate design.

(*iii*) Some errors are caused due to variations (with time or temperature) either in the strength of permanent magnet or in the control springs.

Applications. Permanent-magnet moving coil instruments are acknowledged to be the best type for all d.c. measurements. They are very sensitive and maintain a high degree of accuracy over long periods. The chief applications of such instruments are :

* The soft-iron results in the decreased reluctance of the magnetic circuit and in a uniform, radial flux in the air gap.

(*i*) In the measurement of direct currents and voltages.

(*ii*) In d.c. galvanometers to detect small currents.

(*iii*) In ballistic galvanometers used mainly for measuring changes of magnetic flux linkages.

25.9 Extension of Range

In a permanent-magnet moving coil instrument, the moving coil and the spring used as coil connections have a very delicate design and can carry maximum current of about 50 mA i.e., full-scale deflection (*f.s.d.*) will occur when about 50 mA current flows through the instrument coil. However, in practice, we have to measure large currents and voltages. In such situations, some means are adopted to increase the range of the instruments.

1. Extension of ammeter range. The range of a permanent-magnet moving coil ammeter can be extended by connecting a low resistance, called **shunt*, in parallel with the moving coil of the instrument as shown in Fig. 25.13. The shunt bypasses most of the line current and allows a small current through the meter which it can handle without burning.

Let R_m = meter resistance

S = shunt resistance

I_m = full–scale deflection (*f.s.d.*) current

I = Full range current of the meter

Voltage across shunt = Voltage across the meter

or $(I - I_m)\,S = I_m R_m$

$$\therefore \quad S = \frac{I_m R_m}{I - I_m} \quad \ldots (i)$$

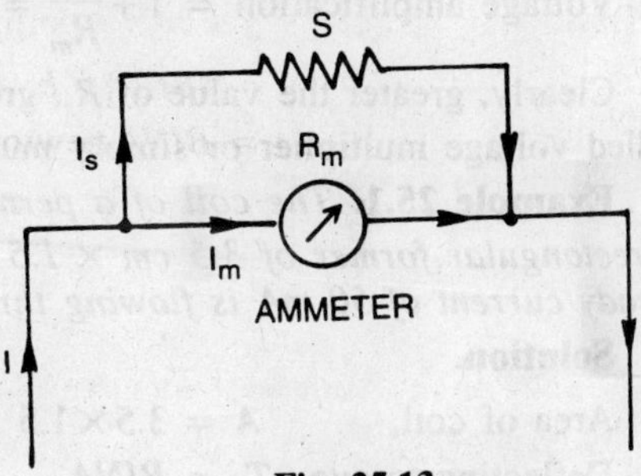

Fig. 25.13

$$\text{Multiplying power of shunt} = \frac{I}{I_m} = \frac{R_m + S}{S} \quad \text{[From eq (i)]}$$

Suppose the meter has a resistance of 5 Ω and requires 15 mA for full-scale deflection. In order that the meter may read 1A, the value of shunt is given by;

$$S = \frac{I_m R_m}{I - I_m} = \frac{0.015 \times 5}{1 - 0.015} = 0.0761\ \Omega$$

$$\text{Multiplying power} = I/I_m = 1/0.015 = 66.67$$

2. Extension of Voltmeter Range. The range of a permanent-magnet moving coil voltmeter can be increased by connecting a high resistance R_S called **multiplier**, in **series with it as shown in Fig. 25.14.

Let

R_m = meter resistance

R_s = series resistance *i.e.*, multiplier

I_m = full–scale deflection current

V = full–range voltage of the meter

---

* Shunts are made of a material such as manganin having low temperature co-efficient of resistance.

** A voltmeter is connected in parallel with the load. Consequently, current through the meter will be proportional to the voltage across the load *i.e.* voltage we wish to measure. In order to restrict the current through the meter (about 50mA), a high resistance R_s is connected in series with it.

Voltage across AB = Voltage across the meter

or $V = I_m (R_s + R_m)$

$$\therefore \quad R_s = \frac{V}{I_m} - R_m \quad \quad (i)$$

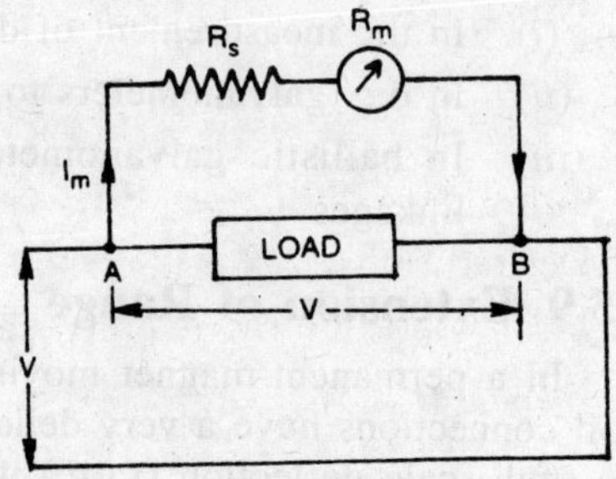

Fig. 25.14

$$\text{Voltage amplification} = \frac{\text{Voltage to be measured}}{\text{Voltage across meter}}$$

$$= \frac{V}{I_m R_m} = 1 + \frac{R_s}{R_m} \quad \text{[From eq } (i)]$$

Suppose the meter has a resistance of 5 Ω and requires 15mA for full-scale deflection. In order that the meter may read 15V, the value of series resistance R_s is given by;

$$R_s = \frac{V}{I_m} - R_m = \frac{15}{0.015} - 5 = 995\,\Omega$$

$$\text{Voltage amplification} = 1 + \frac{R_s}{R_m} = 1 + \frac{995}{5} = 200$$

Clearly, greater the value of R_s, greater is the voltage amplification. For this reason, R_s is called voltage multiplier or simply multiplier.

Example 25.1. *The coil of a permanent-magnet moving coil instrument has 20 turns on a rectangular former of 3.5 cm × 1.5 cm and swings in a uniform field of 0.18 wb/m². If a steady current of 50 mA is flowing through the coil, calculate the deflecting torque.*

Solution.

Area of coil, $A = 3.5 \times 1.5 = 5.25\,\text{cm}^2 = 5.25 \times 10^{-4}\,\text{m}^2$

Deflecting torque, $T_d = BINA$ ... *See Art. 25.8*

$$= (0.18) \times (50 \times 10^{-3}) \times (20) \times (5.25 \times 10^{-4}) = \mathbf{945 \times 10^{-7}\ Nm}$$

Example 25.2. *The meter element of a permanent-magnet moving coil instrument has a resistance of 5 ohms and requires 15mA for full-scale deflection. Calculate the resistance to be connected (i) in parallel to enable the instrument to read upto 1A (ii) in series to enable it to read upto 15V.*

Solution.

Meter resistance, $R_m = 5\Omega$

Full-scale meter current, $I_m = 15mA = 0.015$ A

As Ammeter.

Full-scale circuit current, I = 1 A

Let S ohms be the required value of the shunt [See Fig. 25.15].

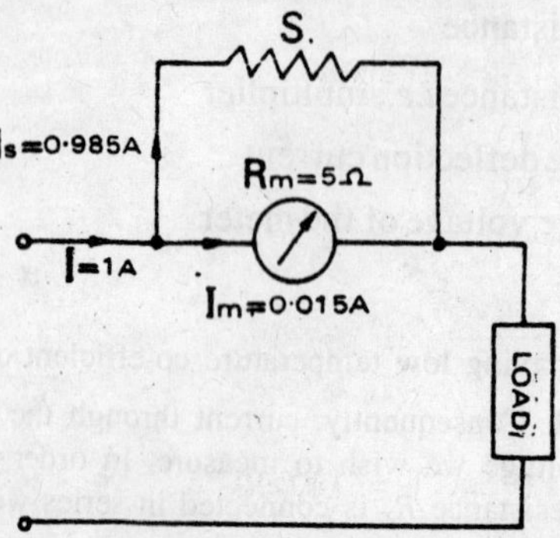

Fig. 25.15

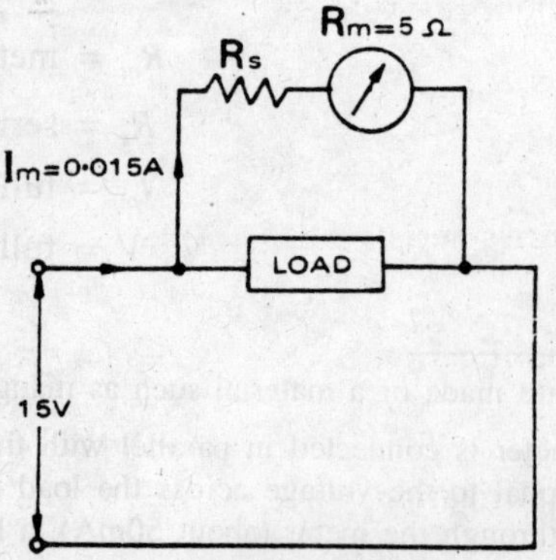

Fig. 25.16

Now $$I_m R_m = (I - I_m)\, S$$

$\therefore$ $$S = \frac{I_m R_m}{I - I_m} = \frac{0.015 \times 5}{1 - 0.015} = \mathbf{0.0761\,\Omega}$$

As Voltmeter.

Desired full-scale reading, $V = 15$ volts

Let R_s ohms be the required series resistance [See Fig. 25.16].

Now $$V = I_m\,(R_s + R_m)$$

or $$R_s = \frac{V}{I_m} - R_m = \frac{15}{0.015} - 5 = \mathbf{995\,\Omega}$$

Example 25.3. *A moving coil milliammeter with a resistance of 1.6 Ω is connected with a shunt of 0.228 Ω. What will be the current flowing through the instrument if it is connected in a circuit in which a current of 200mA is flowing ?*

Solution.

Multiplying power of shunt $= \dfrac{I}{I_m} = \dfrac{R_m + S}{S} = \dfrac{1.6 + 0.228}{0.228} = 8$

$\therefore$ $$I_m = I/8 = 200/8 = \mathbf{25\ mA}$$

Example 25.4. *What should be the resistance of the moving coil of an ammeter which requires 2.5 mA for full-scale deflection so that it may be used with a shunt having a resistance of 0.0025 Ω for a range of 0 – 10A.*

Solution.

Multiplying power of shunt $= \dfrac{I}{I_m} = \dfrac{10}{2.5 \times 10^{-3}} = 4000$

Now, $$4000 = \frac{R_m + S}{S} = \frac{R_m + 0.0025}{0.0025}$$

$\therefore$ $$R_m = \mathbf{10\,\Omega}$$

Example 25.5. *A permanent-magnet moving coil instrument gives full-scale reading of 25mA when a p.d. across its terminals is 75mV. Show how it can be used (i) as an ammeter for a range of 0 – 100A (ii) as a voltmeter for a range of 0 – 750V. Also find the multiplying power of the shunt and voltage amplification.*

Solution.

Meter resistance, $R_m = 75/25 = 3\,\Omega$

Full-scale meter current, $I_m = 25\text{ mA} = 0.025\text{ A}$

Ammeter

Required shunt resistance, $S = \dfrac{I_m R_m}{I - I_m} = \dfrac{0.025 \times 3}{100 - 0.025} = \mathbf{0.00075\,\Omega}$

Multiplying power $= I/I_m = 100/0.025 = \mathbf{4000}$

Voltmeter

Required series resistance, $R_s = \dfrac{V}{I_m} - R_m = \dfrac{750}{0.025} - 3 = \mathbf{29997\,\Omega}$

Voltage amplification, $= 1 + \dfrac{R_s}{R_m} = 1 + \dfrac{29997}{3} = \mathbf{10{,}000}$

Alternatively

Voltage amplification $= V/v = 750\text{V}/(75m\text{V}) = \mathbf{10{,}000}$

Tutorial Problems

1. A permanent magnet moving coil instrument gives full-scale deflection with 5 mA and has a resistance of 5Ω. Calculate the resistance of the necessary components in order that the instrument may be used as (*i*) a 2A ammeter (*ii*) a 100V voltmeter.
[(*i*) 0.0378 Ω in parallel (*ii*) 6662 Ω in series]
2. The coil of a permanent-magnet moving coil instrument has a resistance of 25Ω and gives full-scale deflection when a current of 2mA passes through it. Calculate the value of shunt required to convert the instrument to an ammeter having a full-scale deflection current of 1.5A.
[0.0334 Ω]
3. The coil of a permanent-magnet moving coil instrument has 50 turns and its effective length and breadth are 2.8 cm and 2 cm. The control springs have a total torque of 0.8×10^{-6} Nm per degree of deflection. Find the air-gap flux density that will give a full-scale deflection of 100° at 15 mA.
[0.19 wb/m²]
4. A permanent-magnet moving coil instrument used as a voltmeter has a coil of 50 turns with a width of 3 cm and active length of 3 cm. The gap flux density is 0.15 wb/ m². If the full-scale reading is 150V and the total resistance is 100KΩ, find the torque exerted by the springs at full-scale.
[8.1×10^{-5} Nm]
5. A permanent-magnet moving coil instrument has an air-gap flux density of 0.08 wb/m². There are 60 turns of wire on the moving coil which has an effective length in the gap of 4 cm, a width of 2.5 cm and negligible resistance. The control springs exert a torque of 49.05×10^{-6} Nm at full-scale deflection. Find the resistance to be connected in series with the coil to enable the instrument to be used as a voltmeter reading upto 500V.
[49,000 Ω]

25.10. Dynamometer Type Instruments

These instruments are the modified form of permanent-magnet type. Here magnetic field is not produced by a permanent magnet but by two air-cored fixed coils placed on either side of the moving coil. Such instruments can be used as ammeters or as voltmeters but are generally used as wattmeters. They are suitable for d.c. as well as a.c. work.

Principle. These instruments are based on the principle that mechanical force exists between the current carrying conductors.

Construction. It essentially consists of a fixed coil and a moving coil. The fixed coil is split into two equal parts (*F*, *F*) which are placed close together and parallel to each other. The moving coil (*M*) is pivoted in between the two fixed coils and carries a pointer as shown in Fig. 25.17. The current is led into and out of the moving coil by means of two spiral hair-springs which also provide the controlling torque. *Air friction damping is provided by means of the aluminium vanes that move in the sector shaped chamber at the bottom of the instrument.

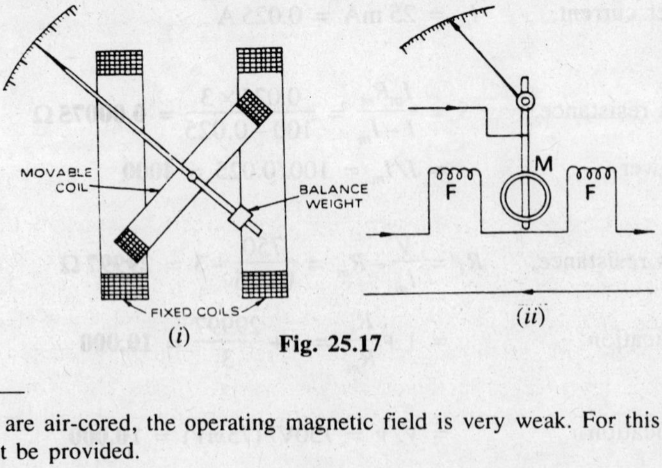

(*i*) (*ii*)

Fig. 25.17

* Since the coils are air-cored, the operating magnetic field is very weak. For this reason, eddy current damping cannot be provided.

Working. For use as an ammeter or voltmeter, the fixed coils *FF* and the moving coil *M* are so connected that the same current flows through the two coils. Due to these currents, mechanical force exists between the coils. The result is that moving coil *M* moves the pointer over the scale. The pointer comes to rest at a position where deflecting torque is equal to the controlling torque. Since the polarity of the fields produced by both fixed and moving coils is reversed by the reversal of current, the deflection of the moving system is always in the same direction regardless of the direction of current through the coils. For this reason, dynamometer instruments can be used for both d.c. and a.c. measurements.

Deflecting torque. The force of attraction or repulsion between the fixed and moving coils is directly proportional to the product of ampere-turns of fixed coils and the moving coils *i.e.,*

Deflecting torque, $$T_d \propto N_f I_f \times N_m I_m$$

Since N_m and N_f are constant,

$$\therefore \quad T_d \propto I_f I_m$$

Since the instrument is spring-controlled, the controlling torque is proportional to the angular deflection θ *i.e.,*

$$T_C \propto \theta$$

In the steady position of deflection, $T_d = T_C$

$$\therefore \quad \theta \propto I_f I_m$$

Thus deflection (θ) is directly proportional to the product of currents in the fixed coils and the moving coil.

(*i*) **As ammeter.** When the instrument is used as an ammeter, the fixed coils and the moving coil are connected in series so that the same current flows through the two coils as shown in Fig. 25.18. In that case, $I_f = I_m = I$ so that :

$$\theta \propto I^2$$

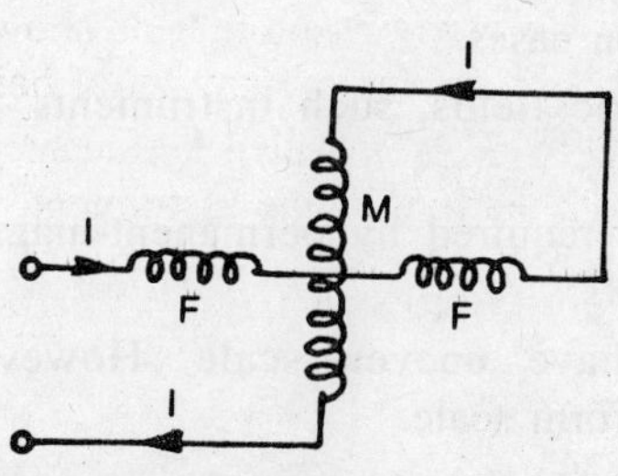

Fig. 25.18

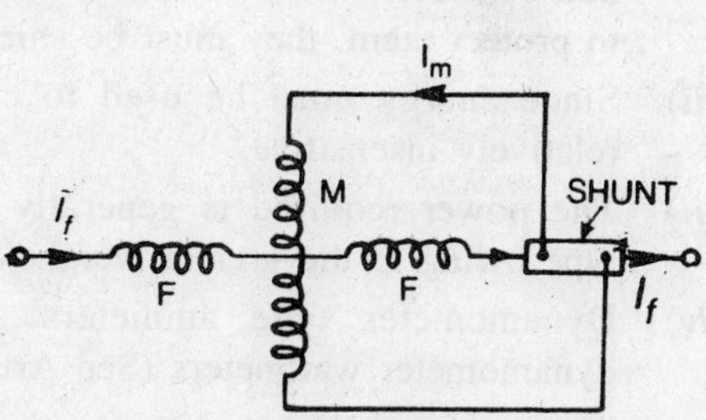

Fig. 25.19

For measuring large currents, the moving coil is shunted; the shunt being in series with the fixed coils as shown in Fig. 25.19. The fixed coils carry the main current while the moving coil carries a current proportional to the main current.

(*ii*) **As voltmeter.** When the instrument is used as a voltmeter, both fixed coils and the moving coil are connected in series together with a high resistance R_s (called multiplier) having a negligible temperature co-efficient as shown in Fig. 25.20. Therefore, current in both the coils is the same and is proportional to the voltage *V* being measured.

$$\therefore \qquad \theta \propto V^2$$

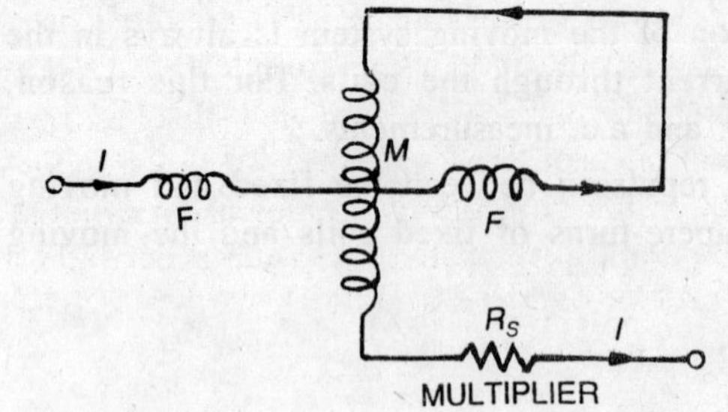

Fig. 25.20

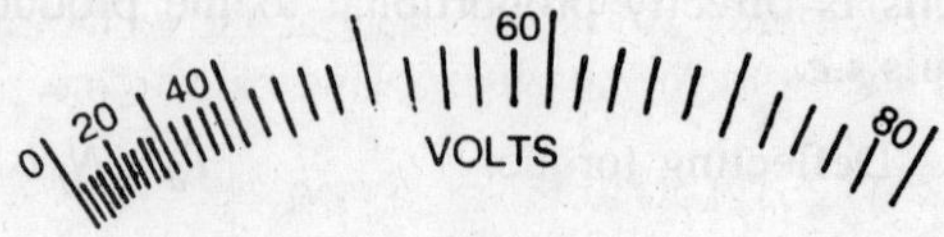

Fig. 25.21

It may be seen that whether the instrument is used as an ammeter or voltmeter, the deflection is directly proportional to the square of quantity (current or voltage) being measured. Doubling the current will make the deflection four times as large. Hence, the scale of dynamometer type instruments is *not uniform; being crowded at the beginning and open at the upper end of the scale as shown in Fig. 25.21. The obvious disadvantage of such a scale is that the divisions near the start of the scale are small and cannot be read accurately.

Advantages

(*i*) These instruments can be used for both d.c. and a.c. measurements.

(*ii*) Since air-coured coils are used, they are generally free from hysteresis and eddy current errors when used on a.c. circuits.

Disadvantages

(*i*) Since air-cored coils are used, the operating field of these instruments is so weak that considerable errors may be introduced due to stray magnetic fields and in order to protect them, they must be shielded with cast-iron cases.

(*ii*) Since energy must be used to create two magnetic fields, such instruments are relatively insensitive.

(*iii*) The power required is generally greater than that required by permanent-magnet type owing to the greater weight of the moving parts.

(*iv*) Dynamometer type ammeters and voltmeters have uneven scale. However, dynamometer wattmeters (See Art. 25.28) have uniform scale.

(*v*) They are more expensive than the permanent-magnet type instruments.

Applications. It is clear from above that dynamometer ammeters and voltmeters are inferior to the permanent-magnet moving coil instruments and are seldom used for d.c. measurements. Their chief sphere of application is in a.c. measurements and particularly when the same instruments is required to read both direct and alternating currents as in the a.c. potentiometer.

* For ammeters and voltmeters only. However, dynamometer wattmeters have uniform scale (See Art. 25.28).

25.11. Moving-Iron Ammeters and Voltmeters

This type of instrument is principally used for the measurement of alternating currents and voltages, though it can also be used for d.c. measurements. There are two types of moving-iron instruments.

(*i*) *Attraction type* in which a single soft-iron vane (or moving iron) is mounted on the spindle and is attracted towards the coil when operating current flows through it.

(*ii*) *Repulsion type* in which two soft-iron vanes are used; one fixed and attached to the stationary coil while the other is movable (*i.e.* moving iron) and mounted on the spindle of the instrument. When operating current flows through the coil, the two vanes are magnetised, developing similar polarity at the same ends. Consequently, repulsion takes place between the vanes and the movable vane causes the pointer to move over the scale.

25.12. Attraction Type

Fig. 25.22 shows the constructional details of an attraction type moving-iron instrument. It consists of a cylindrical coil or solenoid which is kept fixed. An oval-shaped soft-iron is attached to the spindle in such a way that it can move in and out of the coil. A pointer is attached to the spindle so that it is deflected with the motion of the soft-iron piece. The controlling torque is provided by one spiral spring arranged at the top of the moving element. It should be noted that in this instrument, the springs do not carry the current as the same is carried by the stationary coil. The *damping device is an aluminium vane attached to the spindle, as shown in Fig 25.22, which moves in a closed chamber. In some instruments, damping is provided by the movement of a piston inside the curved chamber [See Fig. 25.23]; the piston being attached to the spindle.

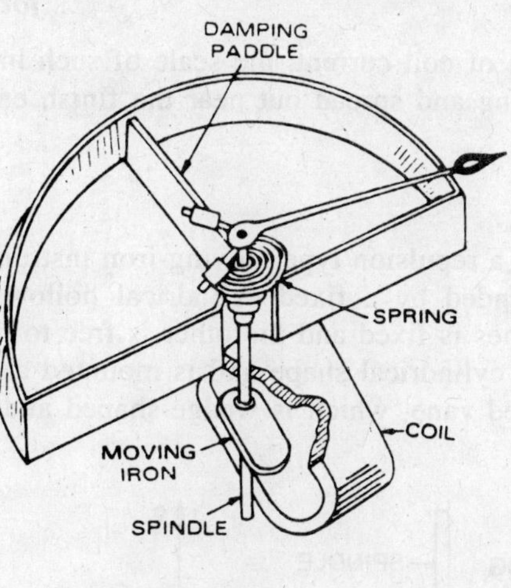

Fig. 25.22

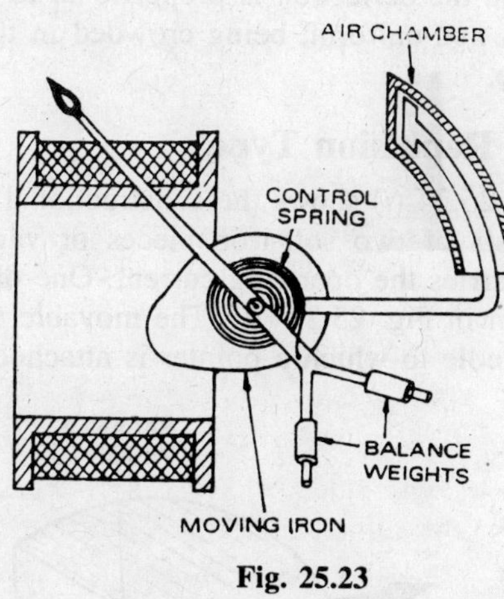

Fig. 25.23

Working. When the instrument is connected in the circuit to measure current or voltage, the operating current flowing through the coil sets up a magnetic field. In other words, the coil behaves like a magnet and therefore it attracts the soft-iron piece towards it. The result is that the pointer attached to the moving system moves from zero position. The pointer will come to rest at a position where deflecting torque is equal to the controlling torque. If current in the coil is reversed, the direction of magnetic field also reverses and so does the magnetism

* Eddy current damping is not provided in moving-iron instruments because the presence of the permanent magnet required for the purpose would affect the field due to the coil. This will, in turn, affect the reading of the instrument.

produced in the soft-iron piece. Hence, the direction of the deflecting torque remains unchanged. For this reason, such instruments can be used for both d.c. and a.c. measurements.

Deflecting torque. The force F pulling the soft-iron piece towards the coil is directly proportional to :

(*i*) field strength H produced by the coil

(*ii*) pole strength m developed in the iron piece

$$\therefore \quad F \propto mH$$

$$\propto H^2 \qquad (\because m \propto H)$$

$\therefore$ Instantaneous deflecting torque $\propto H^2$

If the permeability of iron is assumed constant, then,

$H \propto i$ where i is the instantaneous coil current.

$\therefore$ Instantaneous deflecting torque $\propto i^2$

Average deflecting torque, $T_d \propto$ mean of i^2 over a cycle

Since the instrument is spring controlled,

$$T_C \propto \theta$$

In the steady position of deflection, $T_d = T_C$

$$\therefore \quad \theta \propto \text{mean of } i^2 \text{ over a cycle}$$

$$\propto I^2 \qquad \text{.... for d.c.}$$

$$\propto I^2_{r.m.s.} \qquad \text{.... for a.c.}$$

Since the deflection is proportional to the square of coil current, the scale of such instruments is non-uniform; being crowded in the beginning and spread out near the finish end of the scale.

25.13. Repulsion Type

Fig. 25.24 (*i*) shows the constructional details of a repulsion type moving-iron instrument. It consists of two soft-iron pieces or vanes surrounded by a fixed cylindrical hollow coil which carries the operating current. One of these vanes is fixed and the other is free to move as shown in Fig. 25.24 (*ii*). The movable vane is of cylindrical shape and is mounted axially on a spindle to which a pointer is attached. The fixed vane, which is wedge-shaped and has

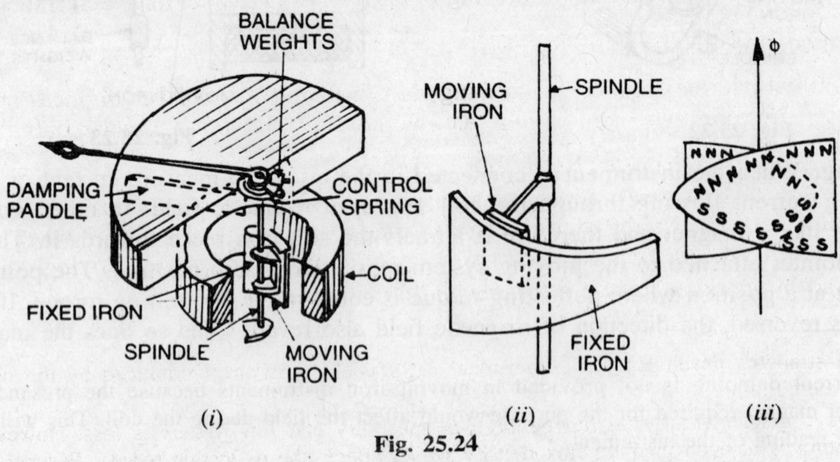

(*i*) (*ii*) (*iii*)

Fig. 25.24

a larger radius, is attached to the stationary coil. The controlling torque is provided by one spiral spring at the top of the instrument. It may be noted that in this instrument, springs do not provide the electrical connections. Damping is provided by air friction due to the motion of a piston in an air chamber.

Working. When current to be measured or current proportional to the voltage to be measured flows through the coil, a magnetic field is set up by the coil. This magnetic field magnetises the two vanes in the same direction *i.e.*, similar polarities are developed at the same ends of the vanes as shown in Fig. 25.24 (*iii*). Since the adjacent edges of the vanes are of the same polarity, the two vanes repel each other. As the fixed vane cannot move, the movable vane deflects and causes the pointer to move from zero position. The pointer will come to rest at a position where deflecting torque is equal to the controlling torque provided by the spring. If the current in the coil is reversed, the direction of deflecting torque remains unchanged. It is because reversal of the field of the coil reverses the magnetisation of both iron vanes so that they repel each other regardless of which way the current flows through the coil. For this reason, such instruments can be used for both d.c. and a.c. applications.

Deflecting torque. The deflecting torque results due to the repulsion between the similarly charged soft-iron pieces or vanes. If the two pieces develop pole strengths of m_1 and m_2 respectively, then,

Instantaneous deflecting torque $\propto m_1 m_2 \propto {}^{*}H^2$

If the permeability of iron is assumed constant, then,

$H \propto i$ where i is coil current.

Instantaneous deflecting torque $\propto i^2$

$\therefore$ Average deflecting torque, $T_d \propto$ mean of i^2 over a cycle

Since the instrument is spring-controlled,

$$T_C \propto \theta$$

In the steady position of deflection, $T_d = T_C$

$$\therefore \quad \theta \propto \text{mean of } i^2 \text{ over a cycle}$$

$$\propto I^2 \qquad \text{... for d.c.}$$

$$\propto I^2_{r.m.s.} \qquad \text{... for a.c.}$$

Thus, the deflection is proportional to the square of coil current as is the case with attraction type moving-iron instrument. Therefore, the scale of such instruments is also non-uniform; being crowded in the beginning and spread out near the finish end of the scale. However, the non-linearity of the scale can be corrected to some extent by the accurate shaping (*e.g.* using tongue-shaped vanes) and positioning of iron vanes in relation to the operating coil.

25.14. Sources of Errors

The errors which may occur in moving-iron instruments can be divided into two categories *viz.*, (*i*) errors with both d.c. and a.c. work (*ii*) errors with a.c. work only.

1. Errors with both d.c. and a.c. work

(*i*) **Error due to hysteresis.** Since the iron parts move in the magnetic field, hysteresis loss occurs in them. The effect of this error is that the readings are higher when current increases than when it decreases. This error can be reduced by employing vanes of **mumetal and by working it over a low range of flux densities.

---

* Since pole strengths developed are proportional to the field strength H produced by the operating current in the coil.

** Mumetal is a magnetic material having high permeability and low hysteresis loss. However, this material saturates at even medium flux density which affects the deflecting torque. For this reason, working flux density should have a low value.

(*ii*) **Error due to stray fields.** Since the operating field is comparatively weak (say 7×10^{-3} wb/m^2), such instruments are readily affected by stray fields. This may give rise to wrong readings. This error can be reduced by enclosing the movement in an iron case.

(*iii*) **Error due to temperature.** Change of temperature affects the instrument resistance and stiffness of the control spring.

(*iv*) **Error due to friction.** Due to the friction of moving parts, a slight error may be introduced. This can be avoided by making torque-weight ratio high.

2. Errors with a.c. work only. With the change in frequency, the impedance of the instrument coil changes. This will cause a change of current in the coil. This is particularly important in case of voltmeters since these are connected in parallel with the circuit. The indicated voltage V is given by :

$$V = I_m \sqrt{(R_m + R_s)^2 + \omega^2 L_m^2}$$

where I_m is the meter current, R_m and L_m are the resistance and inductance of the coil of the meter, and R_s is multiplier. For high frequencies, the meter gives low readings and *vice-versa*. This error can be eliminated by connecting a capacitor of suitable value in parallel with the swamp resistance r of the voltmeter. The value of capacitor is given by $C = L_m / r^2$.

25.15. Characteristics of Moving-Iron Instruments

Advantages

(*i*) These are less expensive, robust and simple in construction.

(*ii*) These can be used for both d.c. and a.c. measurements. However, when used with *d.c., they are liable to small errors due to residual magnetism.

(*iii*) These instruments have high operating torque.

(*iv*) These instruments are reasonably accurate.

Disadvantages

(*i*) These instruments have non-linear scales.

(*ii*) These instruments are not as sensitive as the permanent-magnet moving coil instruments.

(*iii*) Errors are introduced due to change in frequency in case of a.c. measurements.

Applications. The moving-iron instruments are primarily used for a.c. measurements *viz.*, alternating currents and voltages. They are not used to measure direct currents and voltages because their characteristics are inferior to permanent-magnet moving coil instruments.

25.16. Extending Range of Moving-Iron Instruments

As explained above, moving-iron instruments are used mainly on a.c. circuits. Therefore, range extension shall be discussed with reference to a.c. measurements.

1. Ammeter. Shunts are not used to extend the range of moving-iron a.c. ammeters. It is because the division of current between the operating coil and the shunt varies with frequency (since reactance of the coil depends upon frequency). In practice, the range of moving-iron a.c. ammeter is extended by one of the following two methods :

(*i*) **By changing the number of turns** of the operating coil. For example, suppose that full-scale deflection is obtained with 400 ampere-turns. For full-scale reading with 100A, the number of turns required would be = 400/100 = 4.

---

* A moving-iron instrument can be used to measure direct currents and voltages as long as the effects of a possible residual magnetism in the iron are considered. Thus a second reading should be made with the leads interchanged when the meter is used on d.c. circuits. The average of the two readings gives the true value.

Similarly, for full-scale reading with 50A, the number of turns required is = 400/50 = 8. Thus the ammeter can be arranged to have different ranges by merely having different number of turns on the coil. Since the coil carries the whole of the current to be measured, it has a few turns of thick wire. The usual ranges obtained by this method are 0–250A.

(*ii*) For ranges above 0 – 250A, a **current transformer** is used in conjunction with 0–5A a.c. ammeter as shown in Fig 25.25. The current transformer is a step-up transformer *i.e.*, number of secondary turns is more than the *primary turns. The primary of this transformer is connected in series with the load and carries the load current. The a.c. ammeter is connected across the secondary of the transformer. Since in fig 25.25, the current transformer ratio is 10 : 1, it means that the line (or load) current is equal to 10 times the reading on the a.c. meter.

$\therefore$ Load current, $I_L = 3 \times 10 = 30\,A$

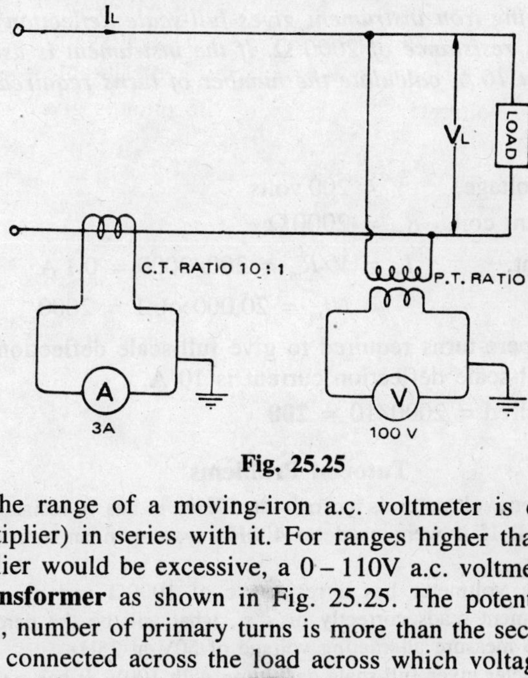

Fig. 25.25

2. Voltmeter. The range of a moving-iron a.c. voltmeter is extended by connecting a high resistance (multiplier) in series with it. For ranges higher than 0 – 750V, where power wasted in the multiplier would be excessive, a 0 – 110V a.c. voltmeter is used in conjunction with a **potential transformer** as shown in Fig. 25.25. The potential transformer is a step-down transformer *i.e.*, number of primary turns is more than the secondary turns. The primary of the transformer is connected across the load across which voltage is to be measured. The a.c. voltmeter is connected across the secondary. Since in Fig 25.25, the potential transformer ratio is 20 : 1, the load voltage is equal to 20 times the reading on the a.c. voltmeter.

$\therefore$ Load voltage, $V_L = 100 \times 20 = 2000\text{ V}$

Note that both secondaries of the instrument transformers are grounded as a safety measure.

Example 25.6. *A 15V moving-iron voltmeter has a resistance of 300Ω and an inductance of 0.12H. Assume that the voltmeter reads correctly on d.c., what will be the percentage error when the instrument is placed on 15V a.c. supply at 100Hz ?*

Solution.

Resistance of meter, $R_m = 300\,\Omega$

Inductance of meter, $L_m = 0.12\,H$

---

* Usually, the primary winding of the transformer consists of a single turn or at the most of a few turns.

On d.c. circuit.

Meter current, $I_m = V/R_m = 15/300 = 0.05$ A

Since the meter reads correctly on d.c., it means that a current of 0.05A flowing through the instrument will give a reading of 15 V.

On a.c. circuit.

Meter impedance, $Z_m = \sqrt{R^2m + (2\pi f L_{..})^2}$

$= \sqrt{(300)^2 + (2\pi \times 100 \times 0.12)^2} = 309.33\Omega$

$\therefore$ Meter current, $I'_m = V/Z_m = 15/309.33 = 0.0485$A

$\therefore$ Meter reading $= \dfrac{15}{0.05} \times 0.0485 = 14.55$ V

% age error $= \dfrac{15 - 14.55}{15} \times 100 = \mathbf{3\%}$

Example 25.7. *A moving-iron instrument gives full-scale deflection with 200 V. It has a coil of 20,000 turns and a resistance of 2000 Ω. If the instrument is used as an ammeter to give full-scale deflection at 10 A, calculate the number of turns required.*

Solution.

As voltmeter.

Full-scale deflection voltage, $V = 200$ volts

Resistance of instrument coil, $R_m = 2000\Omega$

Full-scale meter current, $I_m = V/R_m = 200/2000 = 0.1$ A

Full-scale AT required $= NI_m = 20{,}000 \times 0.1 = 2000$

As ammeter. The ampere-turns required to give full-scale deflection should be the same *i.e.* 2000 *AT*. Since the full-scale deflection current is 10 A.

$\therefore$ No. of turns required = 2000/10 = **200**

Tutorial Problems

1. A 150V moving-iron voltmeter is correct at 50Hz; it has a resistance of 2000Ω and an inductance of 0.35H. If it were used on 400Hz system in an aircraft, what would be its full-scale error ? **[8.25%]**
2. A 50V moving-iron voltmeter has a resistance of 1000Ω and an inductance of 0.765 H. Assuming the instrument reads correctly on d.c., what will be the percentage error when the instrument is used to measure alternating voltage of 50V at 25Hz. **[0.72%]**
3. A moving-iron voltmeter gives full-scale deflection with 100V. It has a coil of 10,000 turns and a resistance of 2000Ω. If the instrument is to be used as an ammeter to give full-scale deflection at 20A, calculate the number of turns required in the coil **[25]**

25.17. Hot-Wire Ammeters and Voltmeters

The operation of these instruments depends upon the expansion of a wire which is heated due to the passage of electric current. The expansion of the wire is taken up by the moving system which causes the pointer to move over a graduated scale. They are suitable for d.c. as well as a.c. measurements.

Construction. Fig. 25.26 shows the simplified diagram of a hot-wire instrument. It consists of a very thin platinum-iridium wire *AB* stretched between terminals *A* and *B*. The wire *AB* is generally called the hot-wire and carries the operating current of the instrument. A fine phosphor-bronze wire *CD* (called *tension wire*) is attached to the centre *C* of the hot-wire; the other end *D* being fixed to the base of the instrument. A fine silk thread attached to *CD* at *E* passes round a pulley on the spindle and is connected to a spring *F*. The effect of wire *CD* and the silk thread is to *magnify the expansion of the wire *AB*. A pointer is attached to the spindle which moves over the scale as the hot-wire *AB* expands. A thin light aluminium disc is attached to the pulley and moves between the poles of a permanent magnet *M*. Eddy currents produced in the disc provide the necessary damping. The instrument is generally spring controlled.

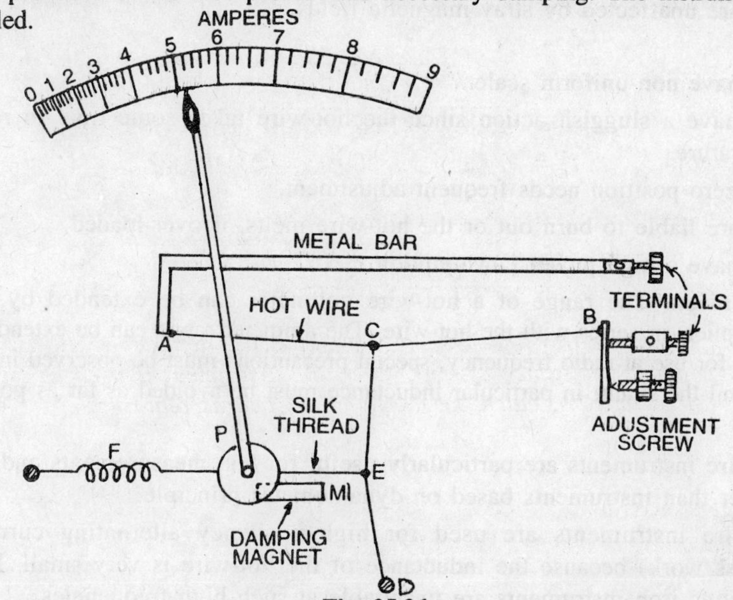

Fig. 25.26

Working. When the instrument is connected in the circuit to measure current or voltage, the operating current flows through the hot-wire *AB*. This causes the hot-wire to expand. The slack in the hot-wire is taken up by the spring *F*, causing the pulley *P* to turn and move the pointer over the scale. When the operating current ceases to flow, the wire *AB* cools and contracts, and the silk thread is pulled to the right causing the pointer to return to zero. Since the tension of wire *AB* changes with temperature, the pointer is set to zero (with adjustment screw) practically each time the instrument is to be used. As the heating effect of electric current is independent of the direction of current flow, therefore, such instruments can be used for both d.c. and a.c. work.

Deflecting torque. The expansion of the hot-wire depends upon $i^2 R$ loss where i is the instantaneous current through the hot wire *AB*.

$\therefore$ Average deflecting torque, $T_d \propto$ mean of i^2 over a cycle

Since the instrument is spring-controlled,

$$T_C \propto \theta$$

In the steady position of deflection, $T_d = T_C$

$$\therefore \quad \theta \propto \text{mean of } i^2 \text{ over a cycle}$$

$$\propto I^2 \quad \text{... for d.c.}$$

$$\propto I^2_{r.m.s.} \quad \text{... for a.c.}$$

Note that deflection is directly proportional to the square of current through the hot-wire. If current is increased two times, the deflection is increased four times. Hence such instruments have square-law scale; being crowded in the beginning and open near the end of the scale.

---

* The co-efficient of linear expansion of most metals is quite small, say 0.000015. Therefore, passage of current through the hot-wire will cause a very little expansion of the wire. With this arrangement, we measure the sag (*i.e.*, downward movement of point *C* due to the expansion of the wire *AB*) which would be much larger than the actual expansion of the wire *AB*.

Advantages

(*i*) They can be used for both d.c. and a.c. measurements.

(*ii*) Their readings are independent of waveform and frequency.

(*iii*) They are unaffected by stray magnetic fields.

Disadvantages

(*i*) They have non-uniform scale.

(*ii*) They have a sluggish action since the hot-wire takes some time to reach its final temperature.

(*iii*) Their zero position needs frequent adjustment.

(*iv*) They are liable to burn out or the hot-wire melts, if over-loaded.

(*v*) They have a high power consumption.

Range extension. The range of a hot-wire voltmeter can be extended by using a high resistance (multiplier) in series with the hot-wire. The ammeter range can be extended by using a shunt. However, for use at radio frequency, special precautions must be observed in the design of the instrument and the shunt; in particular inductance must be avoided as far as possible.

Applications

(*i*) Hot-wire instruments are particularly useful for a.c. measurements and are as a rule cheaper than instruments based on dynamometer principle.

(*ii*) Hot-wire instruments are used for high-frequency alternating currents (*e.g.* in wireless work) because the inductance of the hot-wire is very small. Dynamometer or moving-iron instruments are unsuitable at such high frequencies.

(*iii*) Hot-wire ammeters are not, however, as accurate as dynamometer ammeters, and must be calibrated by comparison with a d.c. ammeter, say of the moving-coil type.

25.18 Electrostatic Voltmeters

These instruments are based on the fact that an electric force (attraction or repulsion) exists between charged plates or objects. An electrostatic voltmeter is essentially an air condenser; one plate is fixed while the other, which is coupled to the pointer, is free to rotate on jewelled bearings. When p.d. to be measured is applied across the plates, the electric force between the plates gives rise to a deflecting torque. Under the action of deflecting torque, the movable plate moves and causes the deflection of the pointer to indicate the voltage being measured. Such instruments can be used to measure direct as well as alternating voltages.

There are three types of electrostatic voltmeters *viz.* :

(*i*) Attracted disc type — usual range from 500V to 500kV

(*ii*) Quadrant type — usual range from 250V to 10kV

(*iii*) Multicellular type — usual range from 30V to 300V

Two things are worth noting about electrostatic voltmeters. First, the deflecting torque is very small for low voltages. For this reason, they are not very sensitive to measure small voltages. Secondly, the instrument is only available for the measurement of p.d., that is to say as voltmeter. It cannot be used as an *ammeter.

25.19. Attracted Disc Type Voltmeter

Fig 25.27 shows the simplified diagram of an attracted disc electrostatic voltmeter. It consists of two mushroom-shaped plates *A* and *B*, each mounted on insulated pedestal. The plate *B* is fixed while the plate *A* (negative, for direct voltage) has a movable central portion-the

* When used as an ammeter, there will be a few millivolts voltage across the instrument. This extremely small p.d. is unsufficient to produce any deflecting torque.

attracted disc. The movable plate A is attached to a horizontal rod which is suspended by two phosphor bronze strips. When p.d. to be measured is applied across the plates, the plate A moves towards the fixed plate B and actuates the pointer *via* a pulley or link mechanism. The control force is provided by gravity and damping force by air dash pot. If the plates are too close together or if the applied voltage is too high, a spark discharge may occur. In order to prevent such a possibility, a *ballast resistor is included in the circuit. The function of this resistor is to limit the current if any sparking-over occurs. If the applied voltage reverses in polarity, there is a simultaneous change in the sign of charge on the plates so that the direction of deflecting force remains unchanged. Hence such instruments can be used for both d.c. and a.c. measurements.

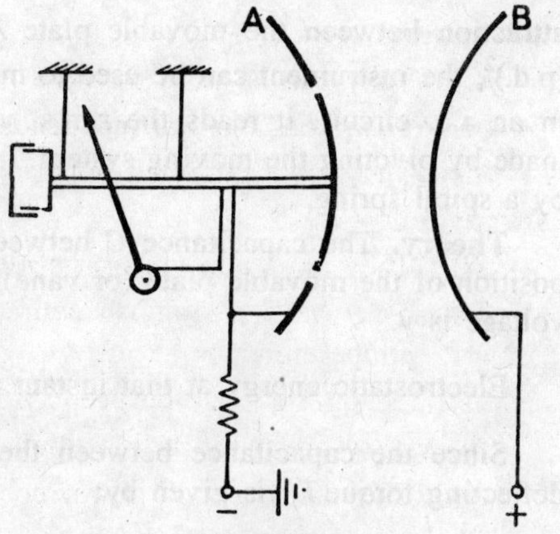

Fig. 25.27

Theory. The force of attraction F between the charged plates is given by ;

$$F = \frac{1}{2}\frac{dC}{dx}V^2$$

where x = distance between the plates

C = capacitance between the plates

V = applied voltage

Since x is always small, dC/dx is practically constant.

$$\therefore \qquad F \propto V^2$$

Obviously, the scale of the instrument will be non-uniform.

25.20. Quadrant Type Voltmeter

Fig 25.28 shows the simplified diagram of a quadrant electrostatic voltmeter. It consists of a light aluminium vane A suspended by a phosphor-bronze string mid-way between two inter-connected quadrant shaped brass plates BB. One terminal is joined to fixed plates BB (positive for direct voltage) and the other to the movable plate A (negative for direct voltage). The controlling torque is provided by the torsion of the suspension string. Damping is provided by air friction due to the motion of another vane in a partially closed box.

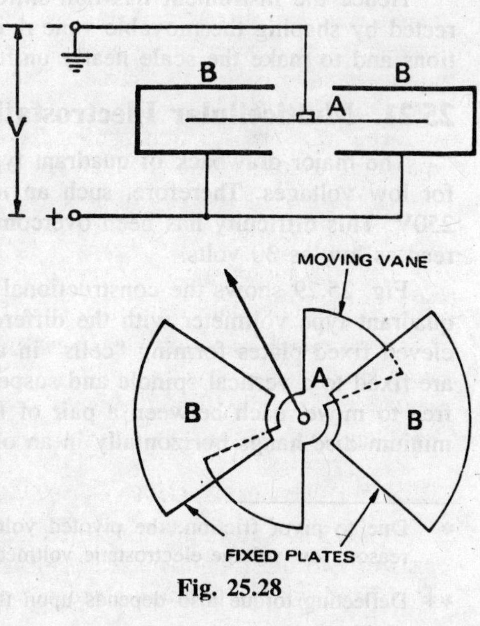

Fig. 25.28

Working. When the instrument is connected in the circuit to measure the p.d., an electric force exists between the plates. Consequently, the movable vane A moves in between the fixed plates and causes the deflection of the pointer. The pointer comes to rest at a position where deflecting torque is equal to the controlling torque. Since the force

---

* When direct voltage is being measured, there will be no voltage drop across this resistor. For alternating voltages, the drop will be negligible at low frequencies. Hence, this resistor will not cause any error in the measurement of voltage.

attraction between the movable plate A and the fixed plates BB is directly proportional to $(\text{p.d.})^2$, the instrument can be used to measure either direct or alternating voltages. When used in an a.c. circuit, it reads the r.m.s. values. More robust but *less accurate voltmeters are made by pivoting the moving system. In pivoted voltmeters, the controlling torque is provided by a spiral spring.

Theory. The capacitance C between the plates depends upon deflection θ *i.e.*, upon the position of the movable plate (or vane) A. Suppose that at any instant, the applied alternating voltage is v.

Electrostatic energy at that instant $= \frac{1}{2} C v^2$

Since the capacitance between the plates depends upon deflection θ, the instantaneous deflecting torque T'_d is given by;

$$T'_d = \frac{1}{2} \frac{dC}{d\theta} v^2$$

Average deflecting torque, T_d = Average of T'_d over a cycle

$$= \frac{1}{T} \int_0^T \frac{1}{2} \frac{dC}{d\theta} v^2 \, dt$$

$$= \frac{1}{2} \frac{dC}{d\theta} \cdot \frac{1}{T} \int_0^T v^2 \, dt$$

$$\therefore \quad T_d = \frac{1}{2} \frac{dC}{d\theta} V^2$$

where V = r.m.s. value of alternating voltage.

This equation equally applies to direct voltages. If $dC/d\theta$ were constant, then

$$T_d \propto V^2$$

Hence the instrument has non-uniform scale. The non-linearity in the scale can be corrected by shaping the movable vane A in such a way as to increase $dC/d\theta$ for small deflections and to make the scale nearly uniform for larger ones.

25.21. Multicellular Electrostatic Voltmeter

The major drawback of quadrant type voltmeter is that deflecting torque is very **small for low voltages. Therefore, such an instrument cannot measure accurately voltages below 250V. This difficulty has been overcome in a multicellular electrostatic voltmeter which can read as low as 30 volts.

Fig. 25.29 shows the constructional details of a multicellular voltmeter. It is essentially a quadrant type voltmeter with the difference that it has ten moving vanes instead of one and eleven fixed plates forming "cells" in and out of which the vanes move. The moving vanes are fixed to a vertical spindle and suspended by a phosphor-bronze wire so that the vanes are free to move, each between a pair of fixed plates. At the lower end of the spindle, an aluminium disc hangs horizontally in an oil bath and provides damping torque due to fluid fric-

* Due to pivot friction, the pivoted voltmeters are less accurate than the suspension type. For this reason, low voltage electrostatic voltmeters are always of suspension type.

** Deflecting torque also depends upon the capacitance between plates [*i.e.* $T_d = \frac{1}{2} (dC/d\theta) V^2$. In a quadrant voltmeter, the capacitance cannot be increased since the number of vanes is limited by space consideration.

tion. The controlling torque is provided by the torsion of the suspension wire as the moving system rotates. The upper end of the suspension wire is attached through a coach spring *S* to a torsion head *H*. The torsion head is provided with a tangent screw for zero adjustment. The function of the coach spring is to prevent the suspension wire from breaking when accidentally jerked. Should the moving vanes be jerked downward, then the coach spring yields sufficiently to allow the safety sleeve *E* to come into contact with the guide stop *G* before the suspension wire is over strained. The scale is horizontal if the pointer is straight but the indications can be given on a vertical scale by bending the pointer at right angles.

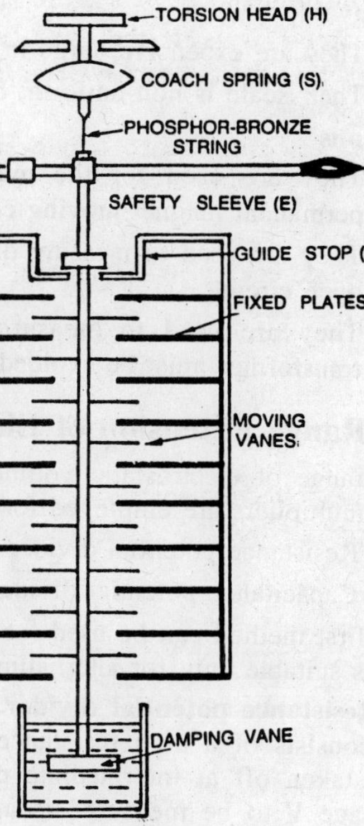

Fig. 25.29

The working principle of multicellular voltmeter is exactly similar to the quadrant type. By using a number of inter-leaved stationary and moving plates, we are able to increase the capacitance and hence the deflecting torque. Consequently, the multicellular voltmeter is much more sensitive than the quadrant type and can accurately measure low voltages.

25.22. Characteristics of Electrostatic Voltmeters

It is worthwhile to mention the advantages, disadvantages and applications of electrostatic voltmeters.

Advantages

(*i*) They can be used for both d.c. and a.c. measurements.

(*ii*) They draw *negligible power from the mains. Hence such voltmeters do not alter the condition of the circuit to which they are connected.

(*iii*) They are free from hysteresis and eddy current losses as no iron is used in their construction.

(*iv*) Their readings are independent of waveform and frequency.

(*v*) They are unaffected by stray magnetic fields, although electrostatic fields (set up, for instance, by such a simple process as rubbing the glass of the case to clean it) may cause considerable errors.

Disadvantages

(*i*) The operating force is very small for low voltages so that they are particularly suitable for the measurement of high voltages.

* With direct voltage, the instrument draws only the initial charging current. With alternating voltages, the alternating current drawn is extremely small.

(ii) Since the operating force is generally small, errors due to friction are difficult to avoid.

(iii) They are expensive, large in size and are not robust in construction.

(iv) Their scale is non-uniform; being crowded in the beginning of the scale.

Applications

(i) They are used for the measurement of very high direct voltages at which a permanent magnet moving coil instrument and the multiplier would be unsuitable.

(ii) They are used to measure direct low voltages when it is necessary to preserve an open circuit.

(iii) They are used to measure very high alternating voltages when the use of a transformer must be avoided.

25.23. Range Extension of Electrostatic Voltmeters

The range of electrostatic voltmeters can be increased by the use of multipliers. Two types of multipliers are employed for this purpose *viz.*

(i) Resistance potential divider —for ranges upto 40kV

(ii) Capacitance potential divider — for ranges upto 1000kV

The first method can be used for both direct and alternating voltages whereas the second method is suitable only for alternating voltages.

(i) **Resistance potential divider.** This divider consists of a high resistance with tappings taken off at intermediate points. The voltage V to be measured is applied across the whole of the potential divider and the electrostatic voltmeter connected across part of it (resistance r in this case) as shown in Fig. 25.30. Since the voltmeter * practically carries no current, the p.d. v across it is the same fraction of the applied voltage V as the resistance across it (*i.e.*, r) is of the whole resistance (*i.e.* R) *i.e.*,

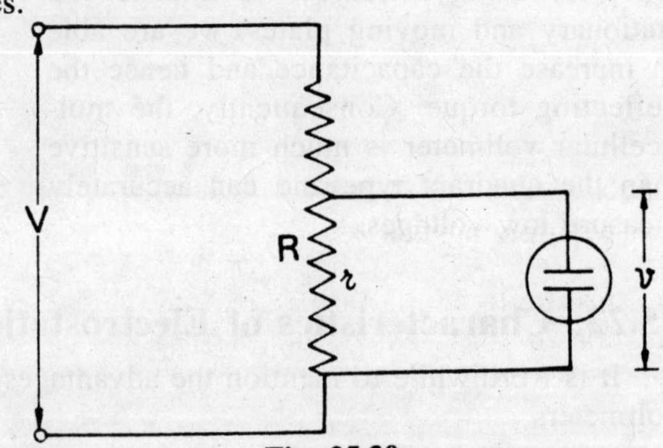

Fig. 25.30

$$\text{Multiplying factor, } \frac{V}{v} = \frac{R}{r}$$

Thus if the voltmeter is connected across 1/5 of the whole resistance (*i.e.*, $R/r = 5$), then voltage V to be measured is 5 times the reading of the voltmeter. The advantage of this method is that there is no shunting effect of the voltmeter. The drawback is that there is power loss in the resistance divider.

(ii) **Capacitance potential divider.** In this method, a single capacitor of capacitance C is connected in series with the voltmeter and the whole circuit is connected across the voltage V to be measured as shown in Fig. 25.31. Let v volts be the reading of the voltmeter. Since the voltage across a capacitor is inversely proportional to its capacitance,

$$V \propto \frac{^{**}C + C_v}{C \times C_v}$$

---

* An electrostatic voltmeter is essentially an air capacitor. For direct voltages, no current can flow through it. For alternating voltages, the current through the voltmeter is extremely small.

** Total circuit capacitance $= \dfrac{C \times C_v}{C + C_v}$ because C and C_v are in series.

and $$v \propto \frac{1}{C_v}$$

∴ Multiplying factor,

$$\frac{V}{v} = \frac{C + C_v}{C}$$

$$= 1 + \frac{C_v}{C}$$

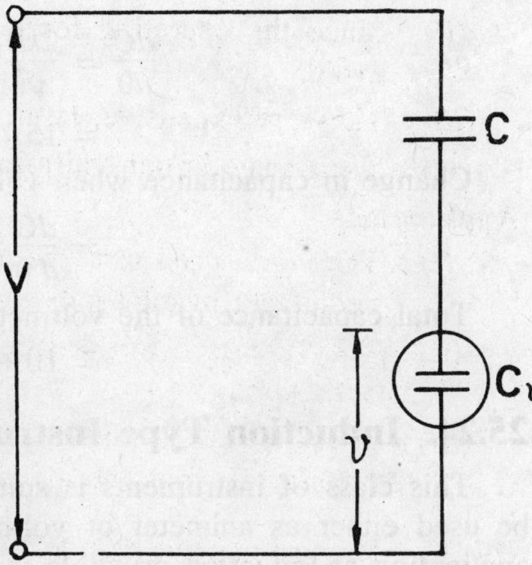

Fig. 25.31

By using capacitors of different capacitances, different voltage ranges can be obtained. This method has the advantage that the circuit consumes no power. However, the drawback is that capacitance current taken is greatly increased.

Example 25.8. *A non-inductive coil AB is connected across 440V mains and an electrostatic voltmeter of negligible capacitance is connected across a portion CD of this coil. The resistance of the coil AB is 8000Ω and that of portion CD is 2000Ω. Find (i) the reading of the voltmeter and (ii) multiplying factor.*

Solution. $$\frac{V}{v} = \frac{R}{r}$$

Here V = 440 volts, R = 8000 Ω; r = 2000 Ω ∴ $\frac{440}{v} = \frac{8000}{2000}$

or Voltmeter reading, $v = \frac{2000}{8000} \times 440 =$ **110 volts**

Multiplying factor $= \frac{V}{v} = \frac{440}{110} = 4$

Example 25.9. *An electrostatic voltmeter has a capacitance of 0.1μF and full-scale deflection of 10kV. Determine the capacitance of the capacitor to be connected in series which will make the full-scale deflection represent 60kV.*

Solution. Let *C*μ*F* be the desired capacitance of the capacitor connected in series with the voltmeter.

$$\frac{V}{v} = 1 + \frac{C_v}{C}$$ *...See Art 25.23*

Here V = 60 kV; v = 10 kV; C_v = 0.1 μF

∴ $\frac{60}{10} = 1 + \frac{0.1}{C}$ or C = **0.02μF**

Example 25.10. *An electrostatic voltmeter reading upto 1000 volts is controlled by a spring with a torsion constant of 9.81×10^{-8} Nm per degree and has a full-scale deflection of 80°. The capacitance of the voltmeter is 10μμF when it reads zero. What is the capacitance when the pointer indicates 1000V ?*

Solution.

Deflecting torque, $$T_d = \frac{1}{2}\frac{dC}{d\theta}V^2$$ *...See Art 25.20*

Controlling torque, $$T_C = k\theta$$

In the steady position of deflection, $$T_d = T_C$$

∴ $$k\theta = \frac{1}{2}\frac{dC}{d\theta}V^2$$

or $$\frac{dC}{d\theta} = \frac{2k\theta}{V^2} = \frac{2\times(9.81\times10^{-8})\times80}{(1000)^2} = 15.7\times10^{-12} \text{ F per radian}$$

$$= 15.7 \ \mu\mu \text{ F per radian} = 0.274 \ \mu\mu\text{F per degree}$$

Change in capacitance when voltmeter reads 1000V

$$= \frac{dC}{d\theta}\times\theta = 0.274\times80 = 21.92 \ \mu\mu\text{F}$$

Total capacitance of the voltmeter when it reads 1000V

$$= 10 + 21.92 = \mathbf{31.92 \ \mu\mu F}$$

25.24. Induction Type Instruments

This class of instruments is suitable only for a.c. measurements. These instruments may be used either as ammeter or voltmeter or wattmeter or energy meter. Perhaps the widest application of induction principle is in watt-hour or energy meter.

Principle. Fig. 25.32 illustrates the principle of induction type instruments. Two alternating fluxes ϕ_1 and ϕ_2 (whose magnitudes depend upon the current or voltage to be measured) having a phase difference θ pass through a metallic disc, usually of copper or aluminium. These alternating fluxes induce currents in the disc. The current produced by one flux reacts with the other flux, and *vice-versa,* to produce the deflecting torque that acts on the disc. It can be proved that net deflecting torque on the disc is given by ;

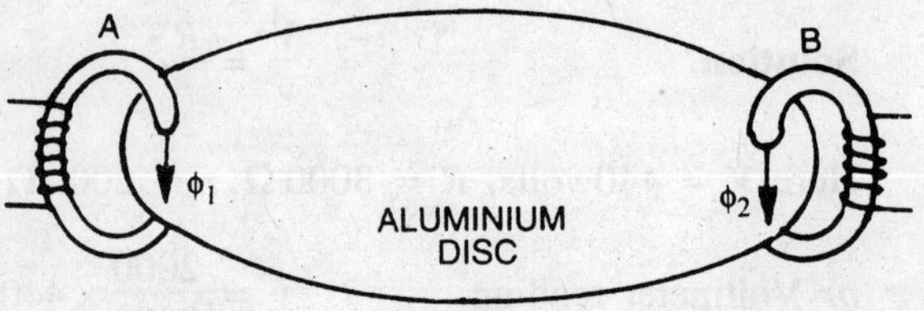

Fig. 25.32

$$T_d \propto \phi_{1m}\,\phi_{2m}\sin\theta$$

where ϕ_{1m} = maximum value of alternating flux ϕ_1

ϕ_{2m} = maximum value of alternating flux ϕ_2

Obviously, to obtain maximum deflecting torque, the angle θ (*i.e.* phase angle between ϕ_1 and ϕ_2) should be 90°.

The induction type instruments are worked on single phase. The question arises how to obtain two fluxes having a phase difference θ (= 90° as far as possible) from a single phase supply. This can be achieved in two ways *viz.*

(*i*) Splitting the phase (*i.e.* Ferrari's Principle)

(*ii*) Shaded-pole arrangement

1. Splitting the phase. In this method, two flux-producing windings are connected in parallel across a single phase supply; an inductive coil L in series with one and a resistance R in series with the other. The values of R and L are so selected that currents through the two windings [*i.e.* I_R and I_L in Fig. 25.33 (*i*)] have a phase difference of nearly 90°. The result is that we have two alternating fluxes with a relative phase shift of 90°. These fluxes pass through the aluminium disc and induce currents in it to produce the necessary driving torque.

The fluxes produced by the two currents may be represented as :

$$\phi_1 = \phi_{1m}\sin\omega t$$

$$\phi_2 = \phi_{2m}\sin(\omega t + \theta)$$

where θ is the phase angle by which ϕ_2 leads ϕ_1.

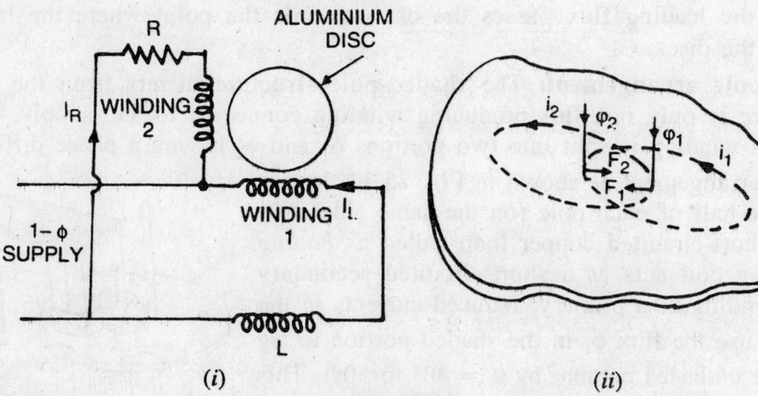

Fig. 25.33

The two fluxes ϕ_1 and ϕ_2 will induce e.m.f.s e_1 and e_2 respectively in the disc. Assuming r to be the resistance offered by the disc to each induced e.m.f., the induced currents are given by;

$$i_1 = \frac{e_1}{r} = \frac{d\,\phi_1/dt}{r} = \frac{1}{r}\frac{d}{dt}(\phi_{1m}\sin\omega t)$$

$$= \frac{\omega\,\phi_{1m}\cos\omega t}{r}$$

or $\quad i_1 \propto \phi_{1m}\cos\omega t \quad$ *... as r and ω are constant*

Similarly, $\quad i_2 \propto \phi_{2m}\cos(\omega t + \theta)$

The portion of the disc which is traversed by flux ϕ_1 and carries current i_2 experiences a force F_1 along the direction indicated. The magnitude of this force is given by ;

$$F_1 \propto \phi_1\, i_2$$

Similarly, $\quad F_2 \propto \phi_2\, i_1$

Since the direction of both fluxes and both currents are the same, these forces will be in the opposite directions. This can be easily ascertained by applying left-hand rule.

∴ Resultant force, $\quad F \propto F_2 - F_1$

$$\propto (\phi_2\, i_1 - \phi_1\, i_2)$$

$$\propto \phi_{1m}\,\phi_{2m}\,[\,\sin(\omega t + \theta)\cos\omega t - \sin\omega t\cos(\omega t + \theta)]$$

$$\propto \phi_{1m}\,\phi_{2m}\sin\theta$$

This resultant force will produce the deflecting torque T_d which is directly proportional to it.

$$\therefore \quad T_d \propto \phi_{1m}\,\phi_{2m}\sin\theta$$

(*i*) If $\theta = 0°$ (*i.e.* the two fluxes are in phase), then deflecting torque is zero. The deflecting torque will be maximum when $\theta = 90°$ *i.e.* when the alternating fluxes have a phase difference of 90°.

(*ii*) The deflecting torque is the same at every instant since ϕ_{1m}, ϕ_{2m} and θ are fixed for a given condition.

(*iii*) The direction of deflecting torque depends upon which flux is leading the other. The deflecting torque acts in such a direction so as to rotate the disc from the point where the leading flux passes the disc towards the point where the lagging flux passes the disc.

2. Shaded-pole arrangement. The shaded-pole structure differs from the split-phase type in that there is only one flux-producing winding connected to a.c. supply. The flux ϕ produced by this winding is split into two portions ϕ_1 and ϕ_2 having a phase difference of θ by shaded-pole arrangement as shown in Fig. 25.34. In this arrangement, one-half of each pole (on the same side) embraces a thick short-circuited copper loop called a *shading coil*. The shading coil acts as a short-circuited secondary and the main winding as a primary. Induced currents in the *shading coil cause the flux ϕ_1 in the shaded portion to lag the flux ϕ_2 in the unshaded portion by θ (= 40° to 50°). This displacement between the two fluxes produces the necessary deflecting torque given by;

$$T_d \propto \phi_{1m}\, \phi_{2m} \sin\theta$$

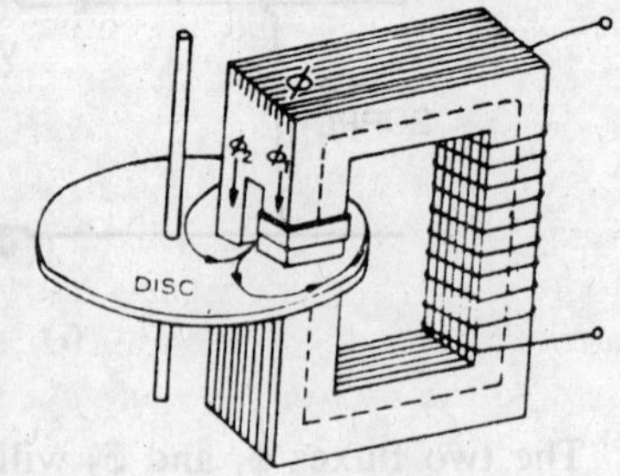

Fig. 25.34

25.25. Induction Ammeters and Voltmeters

Induction ammeters and voltmeters can be used for a.c. measurements only and can be of shaded-pole type or split-phase type.

1. Shaded-pole type. Fig. 25.35 shows the principal parts of a shaded-pole type instrument. It consists of a specially shaped aluminium disc coupled to a pointer and suspended in jewelled bearings. The disc passes through two air-gaps; the first located in an electromagnet having a shading coil and the second in a permanent magnet. The permanent magnet provides the necessary damping torque. The controlling torque is provided by a spiral spring attached to the moving system. As shown above, the deflecting torque is given by;

$$T_d \propto \phi_{1m}\, \phi_{2m} \sin\theta$$

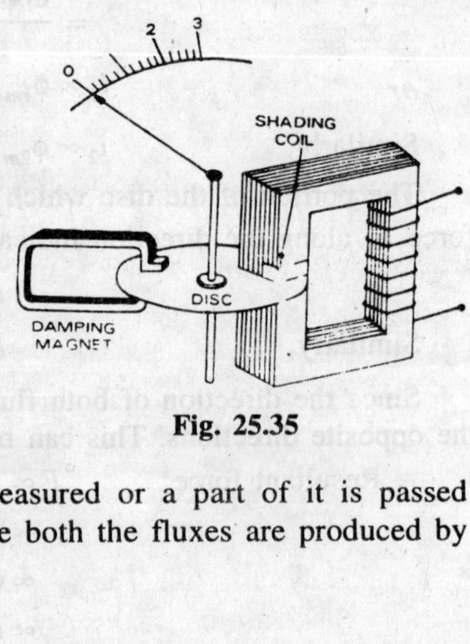

Fig. 25.35

(*i*) When used as an ammeter, the current to be measured or a part of it is passed through the operating coil of the instrument. Since both the fluxes are produced by the same alternating current I (r.m.s. value)

$$\therefore \quad T_d \propto I^2$$

As the instrument is spring controlled, $T_C \propto \theta$

$$\therefore \quad \theta \propto I^2$$

(*ii*) When used as a voltmeter, current proportional to the voltage to be measured is passed through the operating coil.

$$\therefore \quad T_d \propto V^2$$

Again the instrument is spring controlled, $T_C \propto \theta$

$$\therefore \quad \theta \propto V^2$$

---

* Note that shading coil serves the same purpose as connecting resistance and inductance in split-phase arrangement.

It is clear that shaded-pole instruments have uneven scale, being crowded in the beginning and open near the end of the scale. However, the non-linearity in the scale can be corrected to a considerable extent by modifying the shape of the disc *i.e.,* by using cam-shaped disc.

2. Split-phase type. In this method, the windings of the two electromagnets *A* and *B* are connected in parallel across a single phase supply; an inductive coil *L* in series with one and a resistance *R* in series with the other. The values of *R* and *L* are so selected that the currents through the two windings [See Fig. 25.36] have a phase difference of nearly 90°. This produces the deflecting torque on the aluminium disc. The permanent magnet provides the necessary damping torque. The controlling torque is provided by a spiral spring attached to the moving system. As shown above, the deflecting torque is given by;

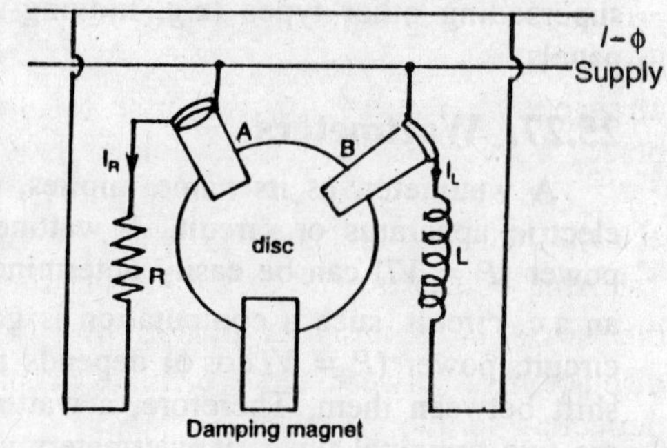

Fig. 25.36

$$T_d \propto \phi_{1m}\, \phi_{2m} \sin\theta$$

Both fluxes are proportional to current or voltage to be measured

$$\therefore \quad T_d \propto I^2 \quad \text{... for ammeter}$$

$$\propto V^2 \quad \text{... for voltmeter}$$

As the instrument is spring controlled, $T_C \propto \theta$

$$\therefore \quad \theta \propto I^2 \quad \text{... for ammeter}$$

$$\propto V^2 \quad \text{... for voltmeter}$$

Obviously, the scale of this type of instrument is also non-uniform.

25.26. Characteristics of Induction Ammeters and Voltmeters

The characteristics of induction ammeters and voltmeters are given below :

Advantages

(*i*) There is no moving iron in the instrument.

(*ii*) The moving element (*i.e.* disc) is not electrically connected to the circuit.

(*iii*) A full-scale deflection of over 250° can be obtained.

(*iv*) They provide very effective damping.

(*v*) They are not easily affected by stray magnetic fields owing to the intense concentration of instrument's own field.

Disadvantages

(*i*) They have high cost.

(*ii*) They have non-linear scales.

(*iii*) They can be used for a.c. measurements only.

(*iv*) They introduce fairly large errors due to temperature, frequency and wave-form variations.

(*v*) They consume fairly large power because of relatively large power losses in the shading coil.

Range extension. In voltmeters, the operating coil is of fine wire, a non-inductive resistance *R* being connected in series with it. In ammeters, the coil is comparatively of thick wire. A voltmeter is connected in the circuit either directly or through a potential transformer depending upon the voltage to be measured. An ammeter is connected directly in the circuit for currents upto 10A provided the circuit voltage does not exceed 650V. Beyond that it is usual to employ a current transformer.

Applications. The sources of errors in induction ammeters and voltmeters have been considerably reduced by the use of special alloys and modern design. Consequently, they are superseding other types (*e.g.* moving-iron voltmeters and ammeters) for switchboards and panels.

25.27. Wattmeters

A wattmeter, as its name implies, measures electric power given to or developed by an electric apparatus or circuit. A wattmeter is hardly ever required in a d.c. circuit because power ($P = VI$) can be easily determined from voltmeter and ammeter readings. However, in an a.c. circuit, such a computation is generally *speaking impossible. It is because in an a.c. circuit, power ($P = VI \cos \phi$) depends not only on voltage and current but also on the phase shift between them. Therefore, a wattmeter is necessary for a.c. power measurement. There are two principal types of wattmeters *viz.*,

(*i*) Dynamometer wattmeter —for both d.c. and a.c. power

(*ii*) Induction wattmeter —for a.c. power only

25.28. Dynamometer Wattmeter

A dynamometer wattmeter is almost universally used for the measurement of d.c. as well as a.c. power. It works on the dynamometer principle *i.e.* mechanical force exists between two current carrying conductors or coils.

Construction. When a dynamometer instrument is used as a wattmeter, the fixed coils are connected in series with the load and carry the load current (I_1) while the moving coil is connected across the load through a series multiplier R and carries a current (I_2) proportional to the load voltage as shown in Fig 25.37. The fixed coil (or coils) is called the *current coil* and the movable coil is known as *potential coil.* The controlling torque is provided by two spiral springs which also serve the additional purpose of leading current into and out of the moving coil. Air friction damping is provided in such instruments. A pointer is attached to the movable coil.

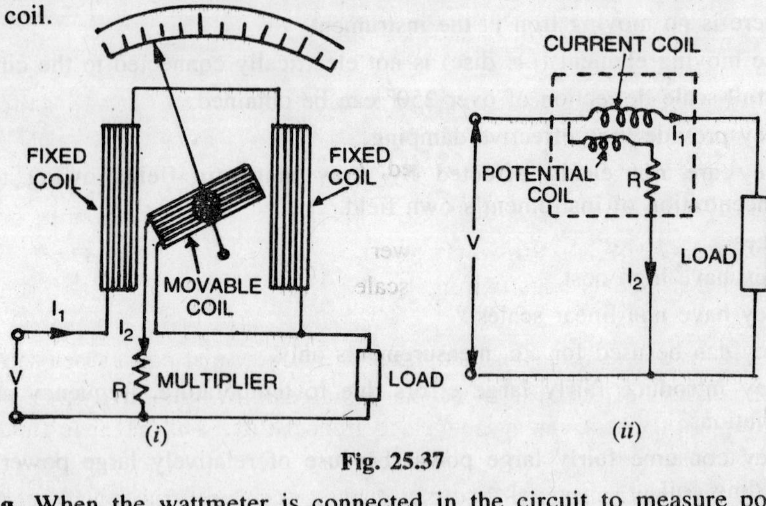

Fig. 25.37

Working. When the wattmeter is connected in the circuit to measure power (See Fig. 25.37), the current coil carries the load current and potential coil carries current proportional to the load voltage. Due to currents in the coils, mechanical force exists between them. The

* Except for the case of pure resistance when $P = VI$ ($\because \cos \phi$ is 1 for pure resistance).

result is that movable coil moves the pointer over the scale. The pointer comes to rest at a position where deflecting torque is equal to the controlling torque. Reversal of current reverses currents in both the fixed coils and the movable coil so that the direction of deflecting torque remains unchanged. Hence, such instruments can be used for the measurement of d.c. as well as a.c. power.

Deflecting torque. We shall now prove that deflecting torque is proportional to load power in a d.c. as well as a.c. circuit.

(*i*) Consider that the wattmeter is connected in a d.c. circuit to measure power as shown in Fig. 25.37 (*ii*). The power taken by the load is VI_1.

Deflecting torque, $T_d \propto I_1 I_2$

Since I_2 is directly proportional to V,

$\therefore$ Deflecting torque, $T_d \propto VI_1 \propto$ load power

(*ii*) Consider that the wattmeter is connected in an a.c. circuit to measure power. Suppose at any instant, current through the load is i and voltage across the load is v. Let the load power factor be $\cos\phi$ lagging. Then,

$$v = V_m \sin\theta$$

$$i = I_m \sin(\theta - \phi)$$

Instantaneous deflecting torque $\propto vi$

The pointer cannot follow the rapid changes in the instantaneous power owing to the large inertia of the moving system. Hence the instrument indicates the mean or average power.

$\therefore$ Average deflecting torque, $T_d \propto$ Average of vi over a cycle

$$\propto \frac{1}{2\pi}\int_0^{2\pi} V_m I_m \sin\theta \sin(\theta - \phi)\, d\theta$$

$$\propto \frac{V_m I_m}{2}\cos\phi$$

$$\propto VI\cos\phi$$

$\therefore$ $T_d \propto$ load power

Thus whether the instrument is used to measure d.c. or a.c. power, deflecting torque is proportional to load power.

Since the instrument is spring-controlled, $T_C \propto \theta$

In the steady position of deflection, $T_d = T_C$

$\therefore$ $\theta \propto$ load power

Hence such instruments have uniform scale.

Disadvantages

(*i*) At low power factors, the inductance of the potential coil causes serious errors.

(*ii*) The readings of the instrument may be affected by stray magnetic fields. In order to prevent it, the instrument is shielded from the external magnetic fields by enclosing it in a soft-iron case.

Errors. A wattmeter may not give true reading due to several sources of error such as(*i*) error due to connection of potential coil circuit (*ii*) error due to inductance of potential coil (*iii*) error due to capacitance in potential coil circuit (*iv*) error due to stray fields and (*v*) errors due to eddy currents.

Example 25.11. *A dynamometer type wattmeter with its voltage coil connected across the load side reads 192W. The load voltage is 208V and the resistance of the potential coil circuit is 3825Ω. Calculate (i) true load power and (ii) percentage error due to wattmeter connection.*

Solution. Since the voltage coil of the wattmeter is connected on the load side (See Fig. 25.38), the power consumed by it is also included in the reading of the wattmeter.

Wattmeter reading = 192 W

Power taken by potential circuit,

$$= \frac{(208)^2}{3825}$$

$$= 11.3 \text{ W}$$

(*i*) True load power = 192 − 11.3 = **180.7 W**

(*ii*) %age error $= \frac{192 - 180.7}{180.7} \times 100 =$ **6.25%**

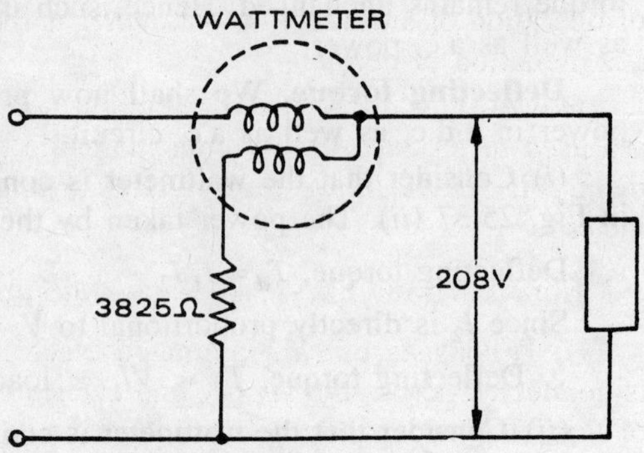

Fig. 25.38

Example 25.12. *The resistances of the two coils of a wattmeter are 0.01Ω and 1000Ω respectively and both are non-inductive. The load is taking a current of 20A at 200V and 0.8 p.f. lagging. Show the two ways in which the voltage coil can be connected and find the error in the reading of the meter in each case.*

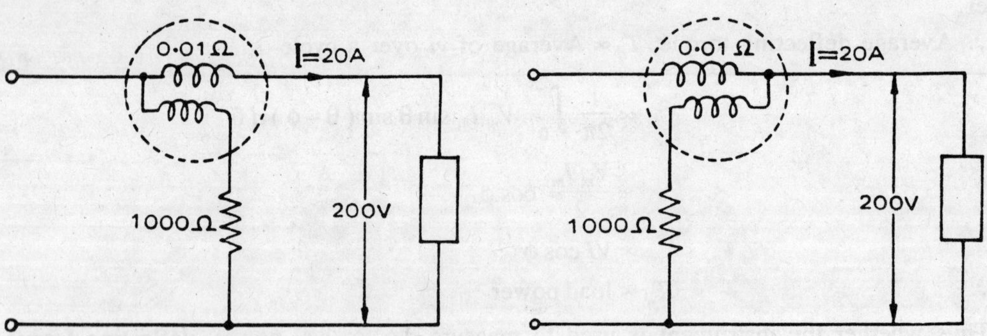

(*i*) (*ii*)

Fig. 25.39

Solution. Fig. 25.39 shows the two possible ways of connecting the voltage coil of the wattmeter.

Load power $= VI \cos \phi = 200 \times 20 \times 0.8 = 3200$ W

(*i*) Consider the connections shown in Fig. 25.39(*i*)

Power loss in current coil $= I^2 R_C = (20)^2 \times 0.01 = 4$ W

∴ Wattmeter reading $= 3200 + 4 = 3204$ W

% age error $= (4/3200) \times 100 =$ **0.125%**

(*ii*) Consider the connections shown in Fig. 25.39 (*ii*).

Power loss in voltage coil $= V^2/R_p = (200)^2/1000 = 40$ W

∴ Wattmeter reading $= 3200 + 40 = 3240$ W

% age error $= (40/3200) \times 100 =$ **1.25%**

25.29. Induction Wattmeters

The induction type wattmeter can be used to measure a.c. power only in contrast to dynamometer wattmeter which can be used to measure d.c. as well as a.c. power. The principle of operation of an induction wattmeter is the same as that of induction ammeter and voltmeter *i.e.* induction principle.

However, it differs from induction ammeter or voltmeter in so far that two separate coils are used to produce the rotating magnetic field in place of one coil with phase split arrangement. Fig 25.41 shows the physical arrangement of the various parts of an induction wattmeter.

Construction. Fig. 25.40 shows the principal parts of an induction wattmeter.

(*i*) It consists of two laminated electromagnets. One electromagnet, called *shunt magnet* is connected across the supply and carries current proportional to the supply voltage. The coil of this magnet is made highly inductive so that the current (and hence the flux produced) in it lags behind the supply voltage by 90°. The other electromagnet, called *series magnet* is connected in series with the supply and carries the load current. The coil of this magnet is made highly non-inductive so that angle of lag or lead is wholly determined by the load.

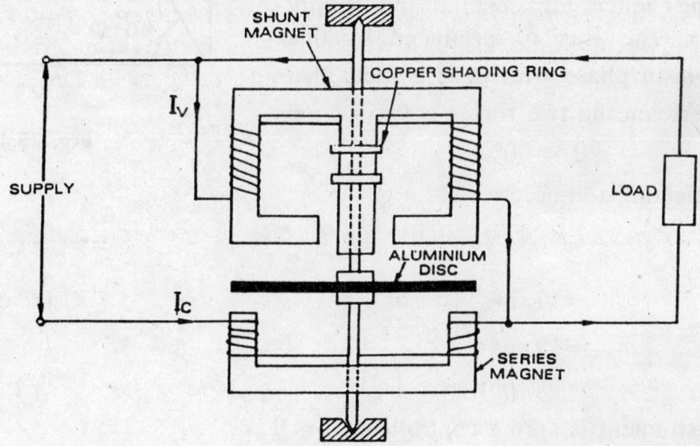

Fig. 25.40

(*ii*) A thin aluminium disc mounted on the spindie is placed between the two magnets so that it cuts the flux of both the magnets. The controlling torque is provided by spiral springs. The damping is electromagnetic and is usually provided by a permanent magnet embracing the aluminium disc (See Fig. 25.40). Two or more closed copper rings (called *shading rings*) are provided on the central limb of the shunt magnet. By adjusting the position of these rings, the shunt magnet flux can be made to lag behind the supply voltage by exactly 90°.

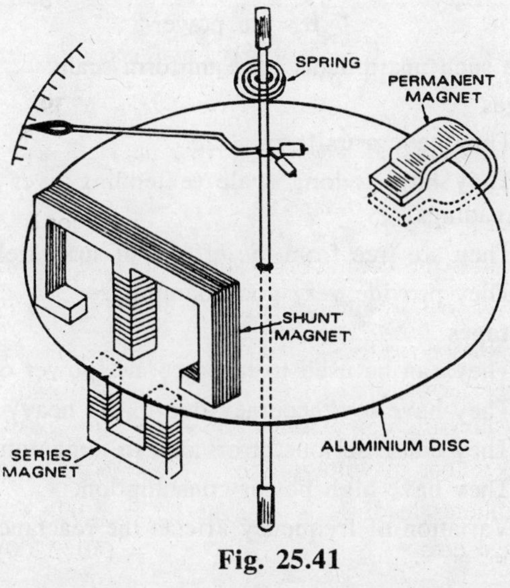

Fig. 25.41

Working. When the wattmeter is connected in the circuit (See Fig. 25.41) to measure a.c. power, the shunt magnet carries current proportional to the supply voltage and the series magnet carries the load current. The two fluxes produced by the magnets induce eddy currents in the aluminium disc. The interaction between the fluxes and eddy currents produces the deflecting torque on the disc, causing the pointer connected to the moving system to move over the scale. The pointer comes to rest at a position where deflecting torque is equal to the controlling torque.

Let V = supply voltage

I_V = currentcarried by shunt magnet

I_C = currentcarried by series magnet

$\cos \phi$ = lagging power factor of the load

The phasor diagram is shown in Fig. 25.42. The current I_V in the shunt magnet lags the supply voltage V by 90° and so does the flux ϕ_V produced by it. The current I_C in the series magnet is the load current and hence lags behind the supply voltage V by ϕ. The flux ϕ_C produced by this current (*i.e.* I_C) is in phase with it. It is clear that phase angle θ between the two fluxes is $90° - \phi$ *i.e.*

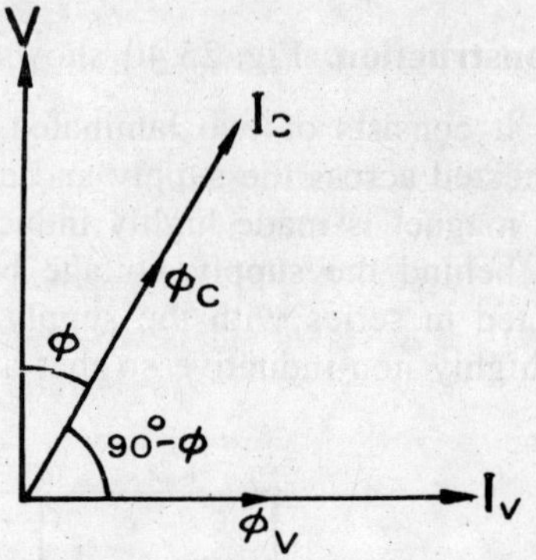

Fig, 25.42

$$\theta = 90° - \phi$$

∴ Mean deflecting torque,

$$T_d \propto \phi_V \phi_C \sin \theta \qquad \text{... See Art 25.24}$$

$$\propto VI \sin(90° - \phi) \qquad [\because \phi_V \propto V \text{ and } \phi_c \propto I]$$

$$\propto VI \cos \phi$$

$$\propto \text{a.c. power}$$

Since the instrument is spring controlled, $T_C \propto \theta$

For steady deflected position, $T_d = T_C$

$$\therefore \qquad \theta \propto \text{a.c. power}$$

Hence such instruments have uniform scale.

Advantages

(*i*) They have a uniform scale.

(*ii*) They have a long scale (extending over 300°), making possible to take accurate readings.

(*iii*) They are free from the effects of stray fields.

(*iv*) They provide very good damping.

Disadvantages

(*i*) They can be used to measure a.c. power only.

(*ii*) They have low accuracy due to the heavy moving system.

(*iii*) They cause serious errors due to temperature variations.

(*iv*) They have high power consumption.

(*v*) Variation of frequency affects the reactance of the windings.

Applications. Due to low accuracy and high power consumption, the characteristics of induction wattmeters are inferior to those of dynamometer wattmeters. For this reason, dynamometer wattmeters are almost universally used of the measurement of a.c. as well as d.c. power. However, induction wattmeters have their chief application as panel instruments where the variations in frequency are not too much.

25.30. Single Phase Induction Watthour Meter

Single phase induction watthour meters (or energy meters) are extensively used for the measurement of electrical energy in a.c. circuits. One can find such meters installed in homes.

An induction watthour meter is essentially an induction wattmeter with control spring and pointer removed but brake magnet and counting mechanism provided.

Construction. Fig 25.43 shows the various parts of a single-phase induction watthour meter.

(*i*) It consists of (*a*) two a.c. electromagnets; the series magnet and shunt magnet (*b*) an aluminium disc or rotor placed between the two electromagnets (*c*) brake magnet and (*d*) counting mechanism.

(*ii*) The shunt magnet is wound with a fine wire of many turns and is connected across the supply so that it carries current proportional to the supply voltage. Since the coil of shunt magnet is highly *inductive, the current (and hence the flux) in it lags the supply voltage by 90°.

The series magnet is wound with a heavy wire of few turns and is connected in series with the load so that is carries the load current. The coil of this magnet is highly non-inductive so that angle of lag or lead is determined wholly by the load.

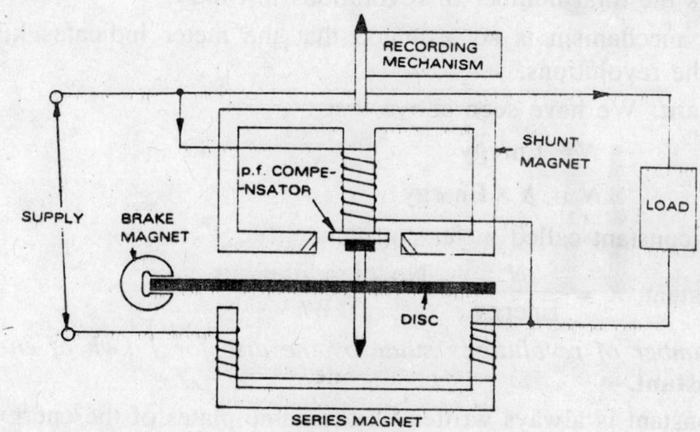

Fig. 25.43

(*iii*) A thin aluminium disc mounted on the spindle is placed between the shunt and series magnets so that it cuts the fluxes of both the magnets.

(*iv*) The braking torque is obtained by placing a permanent magnet near the rotating disc so that the disc rotates in the field established by the permanent magnet. Eddy currents induced in disc produce a braking or retarding torque that is proportional to the disc speed.

(*v*) A short-circuited copper loop (also known as power factor compensator) is provided on the central limb of the shunt magnet. By adjusting the position of this loop, the shunt magnet flux can be made to lag behind the supply voltage exactly by 90°

* The coil has a very large number of turns and the reluctance of its magnetic circuit is very small due to the presence of small air gaps. This makes the coil highly inductive.

Frictional compensation is obtained by means of two adjustable short-circuited loops placed in the leakage gaps of the shunt magnet. Geared to the rotating element is counting mechanism which indicates the energy consumed directly in kilowatthours (*KWh*).

Theory. When induction watthour meter is connected in the circuit to measure energy, the shunt magnet carries current proportional to the supply voltage and the series magnet carries the load current. Therefore, expression for the driving torque is the same as for induction wattmeter. Referring back to the phasor diagram in Fig. 25.42.

Driving torque, $T_d \propto \phi_v \phi_c \sin\theta$

$$\propto VI \sin(90° - \phi)$$

$$\propto VI \cos\phi$$

$$\propto \text{power}$$

The braking torque is due to the eddy currents induced in the aluminium disc. Since the magnitude of eddy currents is proportional to the disc speed, the braking torque will also be proportional to the disc speed *i.e.*,

Braking torque, $^*T_B \propto n$ (*i.e.* disc speed)

For steady speed of rotation, $T_d = T_B$

$\therefore$ Power $\propto n$

Multiplying both sides by t, the time for which power is supplied,

$$\text{Power} \times t \propto nt$$

or Energy $\propto N$

where $N\,(= nt)$ is the total number of revolutions in time t.

The counting mechanism is so arranged that the meter indicates kilowatthours (*kWh*) directly and not the revolutions.

Meter Constant. We have seen above that :

$$N \propto \text{Energy}$$

or $N = K \times \text{Energy}$

where K is a constant called meter constant.

$$\therefore \text{ Meter constant, } K = \frac{N}{\text{Energy}} = \frac{\text{No. of revolutions}}{kWh}$$

Hence the number of revolutions made by the disc for 1 kwh of energy consumption is called **meter constant.**

The meter constant is always written on the name plates of the energy meters installed in homes, commercial and industrial establishments. If the meter constant of an energy meter is 1500 rev/kWh, it means that for consumption of 1 kWh, the disc will make 1500 revolutions.

25.31. Errors in Induction Watthour Meters

The users of electrical energy are charged according to the readings of the energy meters installed in their premises. It is, therefore, very important that construction and de-

---

* Let ϕ = flux of permanent magnet; disc speed = n

E.M.F. induced in the disc, $e \propto \phi n$

If R is the resistance of the eddy current path, then,

$$i = e/R \propto \phi n/R$$

Braking torque, $T_B \propto \phi i \propto \phi^2 n/R$

The braking mechanism is so designed that ϕ and R are constant.

$\therefore$ $T_B \propto n$

sign of energy meters should be such as to ensure long-time accuracy *i.e.*, they should give correct readings over a period of several years under normal use conditions. Some of the common errors in energy meters and their remedial measures are discussed below :

(*i*) **Phase error.** The meter will read correctly only if the shunt magnet flux lags behind the supply voltage by exactly 90°. Since the shunt magnet coil has some resistance and is not completely reactive, the shunt magnet flux does not lag the supply voltage by exactly 90°. The result is that the meter will not read correctly at all power factors.

Adjustment. The flux in the shunt magnet can be made to lag behind the supply voltage by exactly 90° by adjusting the position of the shading coil placed round the lower part of the central limb of the shunt magnet. A current is induced in the shading coil by the shunt magnet flux and causes a further displacement of the flux. By moving the shading coil up or down the limb, the displacement between shunt magnet flux and the supply voltage can be adjusted to 90°. This adjustment is known as *lag adjustment* or *power factor adjustment.*

(*ii*) **Speed error.** Sometimes the speed of the disc of the meter is either fast or slow, resulting in the wrong recording of energy consumption.

Adjustment. The speed of the disc of the energy meter can be adjusted to the desired value by changing the position of the brake magnet. If the brake magnet is moved towards the centre of the spindle, the braking torque is reduced and the disc speed is *increased. Reverse would happen should the brake magnet be moved away from the centre of the spindle.

(*iii*) **Frictional error.** Frictional forces at the rotor bearings and in the counting mechanism cause noticeable error especially at light loads. At light loads, the torque due to friction adds considerably to the braking torque. Since friction torque is not proportional to the speed but is roughly constant, it can cause considerable error in meter reading.

Adjustment. In order to compensate for this error, it is necessary to provide a constant addition to the driving torque that is equal and opposite to the friction torque. This is produced by means of two adjustable short-circuited loops placed in the leakage gaps of the shunt magnet. These loops upset the symmetry of the leakage flux and produce a small torque to oppose the friction torque. This adjustment is known as *light-load adjustment.* The loops are adjusted so that when no current is passing through the current coil (*i.e.* exciting coil of the series magnet), the torque produced is just sufficient to overcome the friction in the system, without actually rotating the disc.

(*iv*) **Creeping.** Sometimes the disc of the meter makes slow but continuous rotation at no load *i.e.*, when potential coil is excited but with no current flowing in the load. This is called *creeping*. This error may be caused due to overcompensation for friction, excessive supply voltage, vibrations, stray magnetic fields *etc.*

Adjustment. In order to prevent this creeping, two diametrically opposite holes are drilled in the disc. This causes sufficient distortion of the field. The result is that the disc tends to remain stationary when one of the holes comes under one of the poles of the shunt magnet.

(*v*) **Temperature error.** Since watthour meters are frequently required to operate in outdoor installations and are subject to extreme temperatures, the effects of temperature and their compensation are very important. The resistance of the disc, of the potential coil and characteristics of magnetic circuit and the strength of brake magnet are affected by the changes in temperature. Therefore, great care is exercised in the design of the meter to eliminate the errors due to temperature variations.

---

* Moving the brake magnet inwards means that speed of the part of disc under the pole face of brake magnet will be less. This results in the lesser induced voltage in the disc and hence reduced braking torque.

(*vi*) **Frequency variations.** The meter is designed to give minimum error at a particular frequency (generally 50Hz). If the supply frequency changes, the reactance of the coils also changes, resulting in a small error. Fortunately, this is not of much significance because commercial frequencies are held within close limits.

(*vii*) **Voltage variations.** The shunt magnet flux will increase with an increase in voltage. The driving torque is proportional to the first power of flux whereas braking torque is proportional to the square of the flux. Therefore, if the supply voltage is higher than the normal value, the braking torque will increase much more than the driving torque and *vice-versa.* The result is that the meter has the tendency to run slow at higher than normal voltages and fast at reduced voltages. However, the effect is small for most of the meters and is not more than 0.2% to 0.3% for a voltage change of 10% from the rated value. The small error due to voltage variations can be eliminated by the proper design of the magnetic circuit of the shunt magnet.

Example 25.13. *An energy meter whose constant is 1500 revolutions per kWh makes 20 revolutions in 30 seconds. Calculate the load in kW.*

Solution.

Energy consumed when disc makes 20 revolutions

$$= (1/1500) \times 20 = 1/75 \text{ kWh}$$

Now energy consumed is equal to load in kW multiplied by time in hours *i.e.*

$$\text{Load} \times \frac{30}{3600} = \frac{1}{75}$$

$$\therefore \quad \text{Load} = \frac{1}{75} \times \frac{3600}{30} = \mathbf{1.6\ kW}$$

Example 25.14. *A 230V single phase energy meter has a constant load current of 4A passing through it for 5 hours at unity power factor. If the meter makes 1104 revolutions during this period, what is the meter constant in revolutions per kWh? If the load power factor is 0.8, what number of revolutions the disc will make in the above time ?*

Solution.

Energy supplied $= \dfrac{VI\cos\phi}{1000} \times t = \dfrac{230 \times 4 \times 1}{1000} \times 5 = 4.6 \text{ kWh}$

No. of revolutions $= 1104$

$\therefore$ Meter constant $= 1104/4.6 = \mathbf{240\ rev/kWh}$

Energy supplied when the load p.f. is 0.8,

$$= \frac{230 \times 4 \times 0.8}{1000} \times 5 = 3.68 \text{ kWh}$$

$\therefore$ No. of revolutions $= 240 \times 3.68 = \mathbf{883.2}$

Example 25.15. *A 50A, 230V energy meter on full load test makes 61 revolutions in 37 seconds. If the meter constant is 520 rev/kWh, what is the percentage error ?*

Solution.

Meter constant $= 520 \text{ rev/kWh}$

Energy supplied $= \dfrac{230 \times 50}{1000} \times \dfrac{37}{60 \times 60} = 0.1182 \text{kWh}$

No. of rev/kWh $= \dfrac{61}{0.1182} = 516.07$

$\therefore$ % age error $= \dfrac{520 - 516.07}{520} \times 100 = \mathbf{0.76\%}$ ***slow***

Example 25.16. *A 230V, 50Hz single phase energy meter has a constant of 1200 rev/kWh. Determine the speed of the disc in r.p.m. for current of 10A at a p.f. of 0.8 lagging.*

Solution.

Energy consumed by the load in 1 minute

$$= \frac{VI\cos\phi}{1000}\times t$$

$$= \frac{230\times 10\times 0.8}{1000}\times(1/60) = 0.0307\,\text{kWh}$$

Revolutions made by the disc in 1 minute

$$= 0.0307\times 1200 = 36.84$$

$\therefore$ Disc speed = **36.84 r.p.m.**

Example 25.17. *A single phase energy meter has a constant of 1200 rev/kWh. When a load of 200 watts is connected, the disc rotates at 4.2 r.p.m. If the load is on for 10 hours, how many units are recorded as error ? Also find the percentage error.*

Solution.

Actual energy consumed by the load in 10 hours

$$= (200\times 10^{-3})\times 10 = 2\,\text{kWh}$$

No. of revolutions made by the disc in 10 hours

$$= \text{speed in r.p.m}\times\text{minutes in 10 hours}$$

$$= 4.2\times 10\times 60 = 2520\ \text{revolutions}$$

Energy recorded by the energy meter

$$= 2520/1200 = 2.1\,\text{kWh}$$

The meter records 2.1 kWh whereas the actual energy consumed by load during the given period (*i.e.*, 10 hours) is 2 kWh. Therefore, the meter records **0.1kWh** *or* **0.1 unit** more.

$$\%\text{age error} = \frac{0.1}{2}\times 100 = \mathbf{5\%}$$

Example 25.18. *A 230V, 10A single phase energy meter has a meter constant of 600 rev/kWh when correctly adjusted. If the lag adjustment is disturbed so that the phase angle between shunt magnet flux and the applied voltage is 86°, calculate the error introduced at (i) unity p.f. (ii) 0.5 p.f. lagging.*

Solution. As proved in Art 25.29, the driving torque is given by ;

$$T_d \propto \phi_v\,\phi_c\sin\theta$$

where θ is the phase angle between the two fluxes.

The meter will read correctly if $^*\theta = 90° - \phi$. In other words, the meter will register correctly if phase angle between shunt magnet flux and supply voltage is 90°. If this is not so, the meter will give a wrong reading.

(*i*) **At unity p.f.**

Power factor angle, $\phi = \cos^{-1} 1 = 0°$

$\therefore\ \theta = 86° - 0° = 86°$ whereas it should be $90° - 0° = 90°$

$$\therefore\ \%\text{age error} = \frac{\sin 90° - \sin 86°}{\sin 90°}\times 100 = \mathbf{0.24\%}$$

---

* In that case $T_d \propto \phi_v\,\phi_c \sin(90° - \phi)$

$\propto VI\cos\phi$

$\propto$ True power

(*ii*) **At 0.5 p.f. lagging**

Power factor angle, $\phi = \cos^{-1} 0.5 = 60°$

$\therefore\ \theta = 86° - 60° = 26°$ whereas it should be $= 90° - 60° = 30°$

$$\therefore \text{ \%age error } = \frac{\sin 30° - \sin 26°}{\sin 30°} \times 100 = \mathbf{12.32\%}$$

The reader may note that at low power factors, the error due to shunt magnet flux not being in quadrature with supply voltage is considerable.

Tutorial Problems

1. A meter whose constant is 600 revolutions/kWh makes 5 revolutions in 20 seconds. Calculate the load in kW. **[1.5 kW]**
2. The number of revolutions per kWh for a 230V, 10A watthour meter is 900. On test at half-load, the time for 20 revolutions is found to be 69 seconds. Determine the meter error at half-load. **[0.82% *fast*]**
3. A correctly adjusted single phase 240V energy meter has a meter constant of 600 rev/kWh. Determine the speed of the disc for a current of 10A at a p.f. 0.8 lagging. **[19.2 r.p.m.]**
4. A single phase energy meter has a meter constant of 200 rev./kWh. When supplying a non-inductive load of 4.4A at normal voltage, the meter takes 3 minutes for 10 revolutions. Find the percentage error. **[1.186%]**
5. An energy meter has a meter constant of 600 rev/kWh when correctly adjusted. If the lag adjustment is altered so that the phase angle between shunt magnet flux and the supply voltage is 85°, calculate the percentage error at 0.8 p.f. lagging. **[7%]**

OBJECTIVE QUESTIONS

1. An ammeter is ....instrument.
   (*i*) an indicating (*ii*) an integrating
   (*iii*) a recording (*iv*) none of the above
2. The controlling torque of an indicating instrument .... as the deflection of the moving system increases.
   (*i*) remains unchanged
   (*ii*) decreases
   (*iii*) increases (*iv*) none of the above
3. When the pointer of an indicating instrument comes to rest in the final deflected position..................
   (*i*) only controlling torque acts
   (*ii*) only deflecting torque acts
   (*iii*) both deflecting and controlling torques act
   (*iv*) none of the above
4. When the pointer of an indicating instrument is in motion, then deflecting torque is opposed by ....
   (*i*) controlling torque only
   (*ii*) damping torque only
   (*iii*) both damping & controlling torques
   (*iv*) none of the above
5. The pointer of an indicating instrument is generally made of ....
   (*i*) copper (*ii*) aluminium
   (*iii*) silver (*iv*) soft steel
6. When the pointer of an indicating instrument is in the final deflected position ....
   (*i*) deflecting torque is zero
   (*ii*) controlling torque is zero
   (*iii*) damping torque is zero
   (*iv*) both deflecting & controlling torques are zero
7. In eddy current damping, disc or former is made of a material that is a ....
   (*i*) conductor but non-magnetic
   (*ii*) conductor but magnetic
   (*iii*) non-conductor and non-magnetic
   (*iv*) non-conductor but magnetic
8. In general, fluid friction damping is not employed in indicating instruments although one can find its use in ....
   (*i*) dynamometer wattmeter
   (*ii*) hot-wire ammeter
   (*iii*) induction type energy meter
   (*iv*) Kelvin electrostatic voltmeter
9. Permanent-magnet moving coil instrument can be used for ....
   (*i*) a.c. work only

(*ii*) d.c. work only
(*iii*) both d.c. and a.c. work
(*iv*) none of the above

10. The scale of a permanent-magnet moving coil instrument is uniform because............
(*i*) of effective eddy current damping
(*ii*) external magnetic fields have no effect
(*iii*) it is spring controlled
(*iv*) it has no hysteresis loss

11. Shunts are generally made of ....
(*i*) copper (*ii*) aluminium
(*iii*) silver (*iv*) manganin

12. The range of a permanent-magnet moving coil instrument is 0-10 A. If the full-scale deflection current of the meter is 2 mA, then multiplying power of the shunt is..............
(*i*) 2500 (*ii*) 10000
(*iii*) 5000 (*iv*) none of the above

13. A moving coil instrument having meter resistance of 5 Ω is to be used as a voltmeter of range 0-100 V. If the full-scale deflection current is 10 mA, then required series resistance is ....
(*i*) 20 Ω (*ii*) 1000 Ω
(*ii*) 9995 Ω (*iv*) none of the above

14. The multiplying power of the shunt of a milliammeter is 8. If the circuit current is 200 mA, then current through the meter is ....
(*i*) 200 mA (*ii*) 25 mA
(*iii*) 1600 mA (*iv*) none of the above

15. The material of the shunt should have ....temperature co-efficient of resistance.
(*i*) negligible (*ii*) positive
(*iii*) negative (*iv*) none of the above

16. A small swamping resistance is put in series with operating coil of a moving coil ammeter in order to compensate for the effects of ....
(*i*) temperature variation
(*ii*) external magnetic fields
(*iii*) hysteresis loss
(*iv*) none of the above

17. A moving coil voltmeter gives full-scale deflection of 100 V for a meter current of 1 mA. For 45 V reading, the meter current will be ....
(*i*) 0.45 mA (*ii*) 1.45 mA
(*iii*) 2.22 mA (*iv*) none of the above

18. Dynamometer type instruments can be used for ....
(*i*) a.c. work only (*ii*) d.c. work only
(*iii*) both d.c. and a.c. work
(*iv*) none of the above

19. A dynamometer instrument is chiefly used as a ....
(*i*) d.c. ammeter (*ii*) d.c. voltmeter
(*iii*) wattmeter (*iv*) none of the above

20. In a dynamometer type instrument, damping is provided by ....
(*i*) air friction (*ii*) eddy currents
(*iii*) fluid friction (*iv*) none of the above

21. Dynamometer type .... has uniform scale.
(*i*) ammeter (*ii*) wattmeter
(*iii*) voltmeter (*iv*) none of the above

22. The instrument in which springs provide the controlling torque as well as serve to lead current into and out of the operating coil is ....instrument.
(*i*) moving-iron (*ii*) hot-wire
(*iii*) permanent-magnet moving coil
(*iv*) none of the above

23. If current through the operating coil of a moving-iron instrument is doubled, the operating force becomes.......................
(*i*) two times (*ii*) four times
(*iii*) one-half time (*iv*) three times

24. The full-scale deflection current of a moving coil instrument is about ....
(*i*) 50 mA (*ii*) 1 A
(*iii*) 3 A (*iv*) 2 A

25. For measuring high values of alternating current with a dynamometer ammeter, we use a ....
(*i*) shunt (*ii*) multiplier
(*iii*) potential transformer
(*iv*) current transformer

ANSWERS TO OBJECTIVE QUESTIONS

1. (*i*)	**2.** (*iii*)	**3.** (*iii*)	**4.** (*iii*)	**5.** (*ii*)	**6.** (*iii*)	**7.** (*i*)	**8.** (*iv*)
9. (*ii*)	**10.** (*iii*)	**11.** (*iv*)	**12.** (*iii*)	**13.** (*iii*)	**14.** (*ii*)	**15.** (*i*)	**16.** (*i*)
17. (*i*)	**18.** (*iii*)	**19.** (*iii*)	**20.** (*i*)	**21.** (*ii*)	**22.** (*iii*)	**23.** (*ii*)	**24.** (*i*)
25. (*iv*)							

HINTS TO SELECTED OBJECTIVE QUESTIONS

5. Because aluminium is light and does not increase the weight of the moving system.

6. Damping torque acts only when the pointer is in motion and always opposes motion.

8. Most of the indicating instruments are portable. Fluid friction damping is not suitable because of the oil contained in the instrument.

10. In a permanent-magnet moving coil instrument,

$$T_d \propto I \qquad \text{and} \qquad T_c \propto \theta$$

In the final deflected position, $T_d = T_c$

$$\therefore \quad \theta \propto I$$

Hence such instruments have uniform scale.

11. Since manganin has negligible temperature co-efficient of resistance.

12. Multiplying power of shunt

$$= \frac{I}{I_m} = \frac{10\text{ A}}{2\text{ mA}} = 5000$$

Where I is the circuit current and I_m is the meter current. In the present case, I is taken equal to 10 A so that I_m will be the full-scale meter current.

13. Series resistance, $R_s = \dfrac{V}{I_m} - R_m = \dfrac{100\text{ V}}{10\text{ mA}} - 5 = 10000 - 5 = 9995\Omega$.

17. For full scale deflection (*i.e.,* 100 V), the meter current is 1 mA. For 45 V reading, the meter current will be = (1/100) × 45 = 0.45 mA.

23. In a moving-iron instrument.

$$T_d \propto I^2 \text{ .... for d.c.}$$

$$\propto I^2_{r.m.s.} \text{ .... for } a.c.$$

If current is doubled, operating torque is increased four times.

25. For measuring alternating currents larger than 5A, shunts are not practicable with a dynamometer ammeter. It is because the division of current between the shunt and the coils varies with frequency (since reactance of coils depends upon frequency). Therefore, the instrument will be accurate only at the frequency at which it is calibrated.

26

Power System

Introduction

In this modern world, the dependence on electricity is so much that it has become a part and parcel of our life. The ever increasing use of electric power for domestic, commercial and industrial purposes necessitates to provide bulk electric power economically. This is achieved with the help of suitable power producing units, known as *power plants* or electric power generating stations. The electrical energy produced at the power plant has to be supplied to the consumers. There is a large network of conductors between the power plant and the consumers. This network can be broadly divided into two parts *viz* transmission of electrical energy and distribution of electrical energy. In this chapter, we shall deal with some fundamental aspects of electric power system.

26.1. Structure of Electric Power System

The electrical system consisting of generation, transmission and distribution of electric power is called **electric power system.**

The function of an electric power system is to connect the power station to the consumers' loads by means of interconnected system of transmission and distribution networks. Therefore, an electric power system consists of three principal components : the *power station,* the *transmission lines* and the *distribution systems.* The transmission lines are the connecting link between the power station and the distribution systems. A distribution system connects all the individual loads in a given locality to the transmission lines.

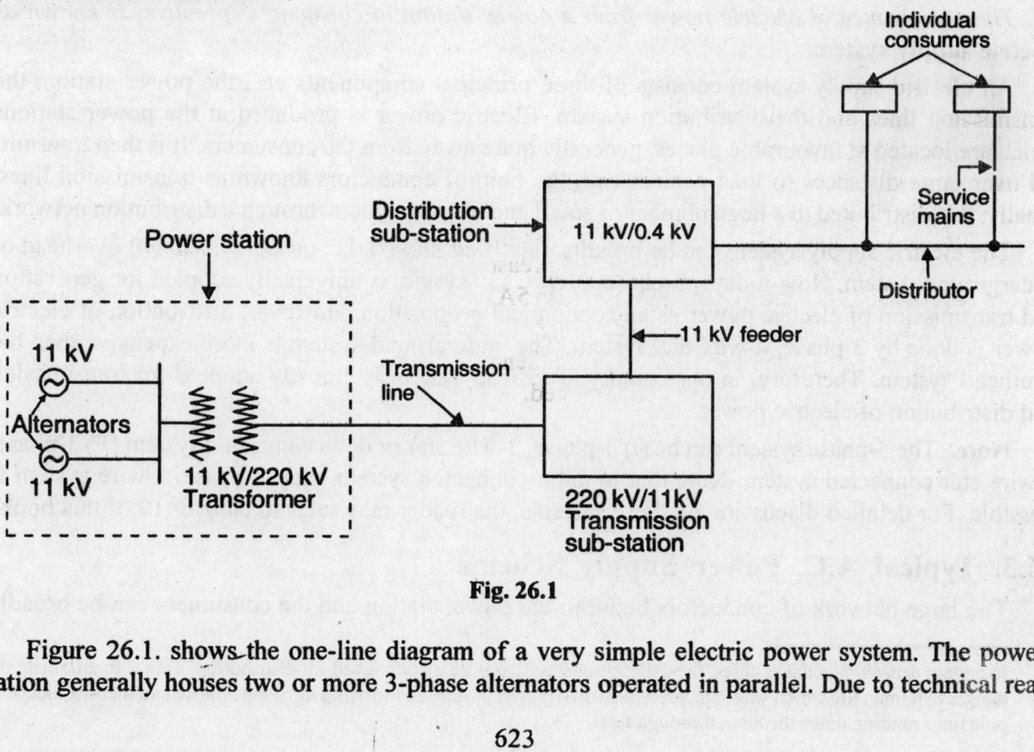

Fig. 26.1

Figure 26.1. shows the one-line diagram of a very simple electric power system. The power station generally houses two or more 3-phase alternators operated in parallel. Due to technical reasons, the generation voltage is not very high; it is usually 11 kV to 25 kV. The power stations are located at favourable places, generally quite far away from the consumers. Owing to economical reasons, electric power has to be transmitted through transmission lines at a very high voltage. Therefore, voltage at the power station is stepped to a high value (say 220 kV) by the step-up transformer. The electric power is carried by the transmission lines at this high voltage. At the outskirts of the city, this voltage is reduced considerably (say 11 kV) by a step-down transformer (also called transmission sub-station). The feeders carry the power to distribution sub-station. Feeders should not be tapped for direct supply. At the distribution sub-station, the step-down transformer reduces the voltage to 400 V (11 kV/0.4 kV transformer — 400 V line voltage and 230 V phase voltage). Distributors are used to supply power to various consumers. The following points may be noted :

(*i*) For the sake of simplicity, we show the 3-phase circuit by a single-line diagram.

(*ii*) The 3-phase alternators at the power station are designed to produce balanced voltages. Since a balanced 3-phase system is always solved as a single-phase circuit composed of one of the three lines and a neutral return , it is seldom necessary to show more than one phase and the neutral return when drawing a diagram of the circuit. Often the diagram is simplified further by omitting the neutral wire and by indicating the component parts by standard symbols rather than by their equivalent circuits. Such a simplified diagram of an electric power system is called one-line diagram.

(*iii*) In an actual electric power system, there are a number of protective devices (fuses, circuit breakers, relays etc.) and other equipment. The amount of information included in the single-line diagram depends on the purpose for which the diagram is intended. For example, here we are interested to study the general layout of electric power system. For this purpose, circuit breakers, relays etc. are unimportant and are omitted in the diagram shown in Fig. 26.1.

26.2. Electric Supply System

The conveyance of electric power from a power station to consumers' premises is known as **electric supply system.**

An electric supply system consists of three principal components *viz.*, the power station, the transmission lines and the distribution system. Electric power is produced at the power stations which are located at favourable places, generally quite away from the consumers. It is then transmitted over large distances to load centres with the help of conductors known as transmission lines. Finally, it is distributed to a large number of small and big consumers through a distribution network.

The electric supply system can be broadly classified into (*i*) d.c. or a.c. system (*ii*) overhead or underground system. Now-a-days, 3-phase, 3-wire a.c. system is universally adopted for generation and transmission of electric power as an economical proposition. However, distribution of electric power is done by 3-phase, 4-wire a.c. system. The underground system is more expensive than the overhead system. Therefore, in our country, overhead system is *mostly adopted for transmission and distribution of electric power.

Note. The 3-phase system can be (*i*) 3-phase, 3-wire star or delta connected system (*ii*) 3-phase, 4-wire star connected system. Note that in delta-connected system, only 3-phase, 3-wire system is possible. For detailed discussion on these systems, the reader may refer to chapter 19 of this book.

26.3. Typical A.C. Power Supply Scheme

The large network of conductors between the power station and the consumers can be broadly

* In certain densely populated cities, the underground system is being employed for distribution. This is to eliminate the danger to human life which would be present with overhead system and to avoid ugly appearance and inconvenience of pole lines running down the main thorough fares.

divided into two parts *viz.*, transmission system and distribution system. Each part can be further subdivided into two—primary transmission and secondary transmission and primary distribution and secondary distribution. Fig. 26.2 shows the layout of a typical a.c. power supply scheme by a single line diagram. It may be noted that it is not necessary that all power schemes include all the stages shown in the figure. For example, in a certain power scheme, there may be no secondary transmission and in another case, the scheme may be so small that there is only distribution and no transmission.

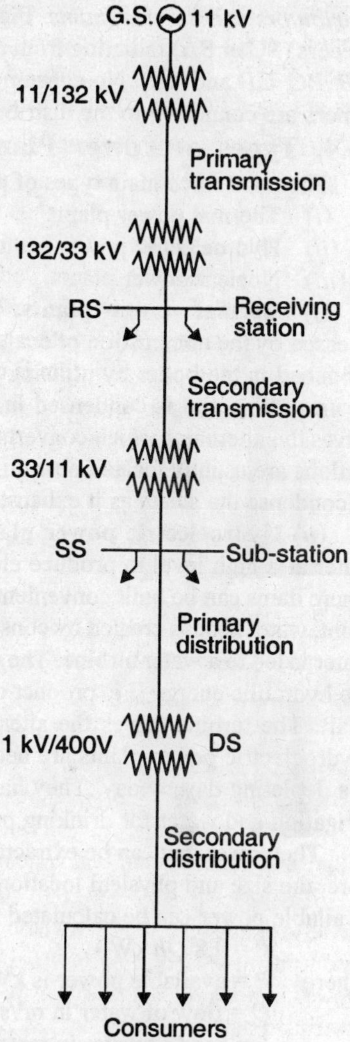

Fig. 26.2

(*i*) **Generating station.** In Fig. 26.2, G.S. represents the generating station where electric power is produced by 3-phase alternators operating in parallel. The usual generation voltage is † 11 kV. For economy in the transmission of electric power, the generation voltage (*i.e.*, 11 kV) is stepped up to 132 kV (or *more) at the generating station with the help of 3-phase transformers. The transmission of electric power at high voltages has several advantages including the saving of conductor material and high transmission efficiency. It may appear advisable to use the highest possible voltage for transmission of electric power to save conductor material and have other advantages. But there is a limit to which this voltage can be increased. It is because increase in transmission voltage introduces insulation problems as well as the cost of switchgear and transformer equipment is increased. Therefore, the choice of proper transmission voltage is essentially a question of economics. Generally the primary transmission is carried at 66 kV, 132 kV, 220 kV or 400 kV.

(*ii*) **Primary transmission.** The electric power at 132 kV is transmitted by 3-phase, 3-wire overhead system to the outskirts of the city. This forms the primary transmission.

(*iii*) **Secondary transmission.** The primary transmission line terminates at the receiving station (RS) which usually lies at the outskirts of the city. At the receiving station, the voltage is reduced to 33 kV by step-down transformers. From this station, electric power is transmitted at 33 kV by 3-phase, 3-wire overhead system to various sub-stations (SS) located at the strategic points in the city. This forms the secondary transmission.

(*iv*) **Primary distribution.** The secondary transmission line terminates at the sub-station (SS) where voltage is reduced from 33 kV to 11 kV, 3-phase, 3-wire. The 11 kV lines run along the important road sides of the city. This forms the primary distribution. It may be noted that big consumers (having demand more than 50 kW) are generally supplied power at 11 kV for further handling with their own sub-stations.

(*v*) **Secondary distribution.** The electric power from primary distribution line (11 kV) is delivered to distribution

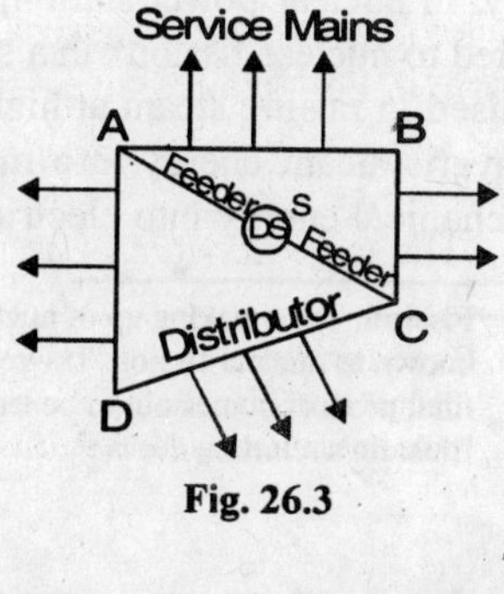

Fig. 26.3

† It may be 6.6 kV or even 33 kV in certain cases.

* Depending upon the length of transmission line and the amount of power to be transmitted.

sub-stations (DS). These sub-stations are located near the consumers' localities and step down the voltage to 400 V, 3-phase, 4-wire for secondary distribution. The voltage between any two phases is 400 V and between any phase and neutral is 230 V. The single-phase residential lighting load is connected between any one phase and neutral, whereas 3-phase, 400 V motor load is connected across 3-phase lines directly.

It may be worthwhile to mention here that secondary distribution system consists of *feeders, distributors* and *service mains.* Fig. 26.3 shows the elements of low voltage distribution system. Feeders (SC or SA) radiating from the distribution sub-station (DS) supply power to the distributors (*AB, BC, CD* and *AD*). No consumer is given direct connection from the feeders. Instead, the consumers are connected to the distributors through their service mains.

26.4. Types of Power Plants

There are three main types of power plants *viz*

(*i*) Thermal power plants
(*ii*) Hydroelectric power plants
(*iii*) Nuclear power plants

(*i*) **Thermal power plants.** Thermal power stations produce electrical energy from the heat released by the combustion of coal, oil or natural gas. For example, in a steam power plant, steam is produced in the boiler by utilising the heat of coal combustion. The steam is then expanded in the steam turbine and is condensed in a condenser to be fed into the boiler again. The steam turbine drives the alternator which converts mechanical energy of the turbine into electrical energy. Thermal stations are usually located near a river or lake because large quantities of cooling water are needed to condense the steam as it exhausts from the turbine.

(*ii*) **Hydroelectric power plants.** Hydroelectric power plants utilise the potential energy of water at a high level to produce electrical energy. These plants are generally located in hilly areas where dams can be built conveniently and large reservoirs can be obtained. In a hydroelectric power plant, water head is created by constructing a dam across a river or lake. From the bottom of the dam, water is led to a water turbine. The water turbine captures the energy in the falling water and changes the hydraulic energy (*i.e.* product of head and flow of water) into mechanical energy at the turbine shaft. The turbine drives the alternator which converts mechanical energy into electrical energy. Hydroelectric power plants are becoming very popular because the reserves of fuels (coal and oil) are depleting day by day. They have the added importance for flood control, storage of water for irrigation and water for drinking purposes.

The power that can be extracted from a waterfall depends on its height and rate of flow. Therefore, the size and physical location of hydroelectric power plant depends on these two factors. The available power can be calculated from the equation :

$$P = 9.8\, Qh \text{ kW}$$

where P = available power is kW
Q = flow of water in m^3/s
h = head of water in metres

(*iii*) **Nuclear power plants.** Nuclear power plants convert nuclear energy into electrical energy. In nuclear power station, heavy elements such as Uranium (U^{235}) or Thorium (Th^{232}) are subjected to nuclear fission* in a special apparatus known as a *reactor.* The heat energy thus released is utilised in raising steam at high temperature and pressure. The steam runs the steam turbine which converts steam energy into mechanical energy. The turbine drives the alternator which converts mechanical energy into electrical energy.

---

* **Fission.** The breaking up of nuclei of heavy atoms into two nearly equal parts with release of huge amount of energy is known as nuclear fission. The release of huge amount of energy during fission is due to *mass defect i.e.* the mass of the final product comes out to be less than the initial product. This mass defect is converted into heat energy according to Einstein's relation, $E = mc^2$.

The most important feature of a nuclear power station is that huge amount of electrical energy can be produced from a relatively small amount of nuclear fuel as compared to other conventional types of power stations. It has been found that complete fission of 1 kg of Uranium (U^{235}) can produce as much energy as can be produced by the burning of 4,500 tons of high grade coal. Although the recovery of principal nuclear fuels (*i.e.*, Uranium and Thorium) is difficult and expensive, yet the total energy content of the estimated world reserves of these fuels are considerably higher than those of conventional fuels, *viz.*, coal, oil and gas. At present, energy crisis is gripping us and, therefore, nuclear energy can be successfully employed for producing low cost electrical energy on a large scale to meet the growing commercial and industrial demands.

Example 26.1. *A large hydroelectric power station has a head of 324 m and an average flow of 1370 m³/s. The reservoir has an area of 6400 km². Calculate (i) the available hydraulic power (ii) the number of days this power could be sustained if the level of water in the reservoir were allowed to drop by 1 m.*

Solution.

(*i*) Available hydraulic power, $P = 9.8\ Q\ h$

$$= 9.8 \times 1370 \times 324 \text{ kW} = 4350 \times 10^3 \text{ kW} = \mathbf{4350\ MW}$$

(*ii*) A drop of 1 m in water level corresponds to 6400×10^6 m³ of water. Since the flow is 1370 m³/s,

∴ Time for all this water to flow through the turbine is

$$t = \frac{6400 \times 10^6}{1370} \text{s} = 4.67 \times 10^6 \text{ s} = 1298 \text{ h} = \mathbf{54\ days}$$

26.5. Variable Load on Power Plants

The function of a power station is to deliver power to a large number of consumers. However, the power demands of different consumers vary in accordance with their activities. The result of this variation in demand is that load on a power station is never constant, rather it varies from time to time. Most of the complexities of modern power plant operation arise from the inherent variability of the load demanded by the users. Unfortunately, electrical power cannot be stored and, therefore, the power station must produce power as and when demanded to meet the requirements of the consumers. On one hand, the power engineer would like that the alternators in the power station should run at their rated capacity for maximum efficiency and on the other hand, the demands of the consumers have wide variations. This makes the design of a power station highly complex.

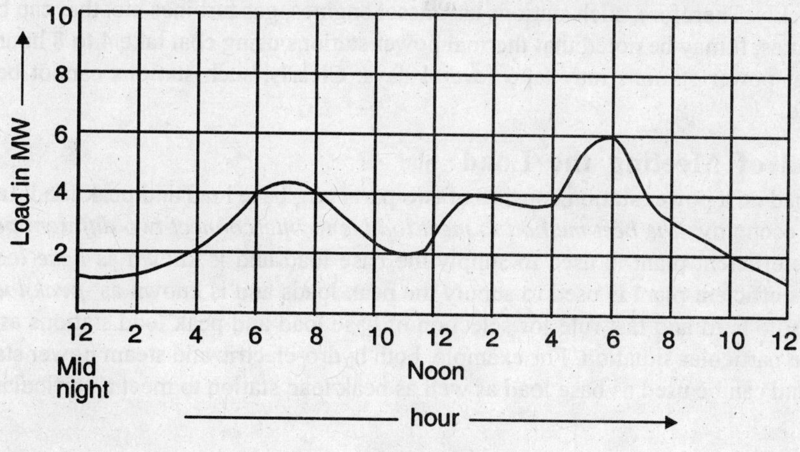

Fig. 26.4

The load on a power station is never constant; it varies from time to time. These load variations during the whole day (*i.e.*, 24 hours) are recorded half-hourly or hourly and are plotted against time on the graph. The curve thus obtained is known as *daily load curve* as it shows the variations of load *w.r.t.* time during the day. Fig. 26.4 shows a typical daily load curve of a power station. It is clear that load on the power station is varying, being maximum at 6 P.M. in this case. It may be seen that load curve indicates at a glance the general character of the load that is being imposed on the plant. Such a clear representation cannot be obtained from tabulated figures.

26.6. Base Load and Peak Load on Power Station

The changing load on the power station makes its load curve of variable nature. Fig. 26.5 shows the typical load curve of a power station. It is clear that load on the power station varies from time to time. However, a close look at the load curve reveals that load on the power station can be considered in two parts, namely;

(*i*) Base load

(*ii*) Peak load

(*i*) **Base load.** *The unvarying load which occurs almost the whole day on the station is known as* **base load.**

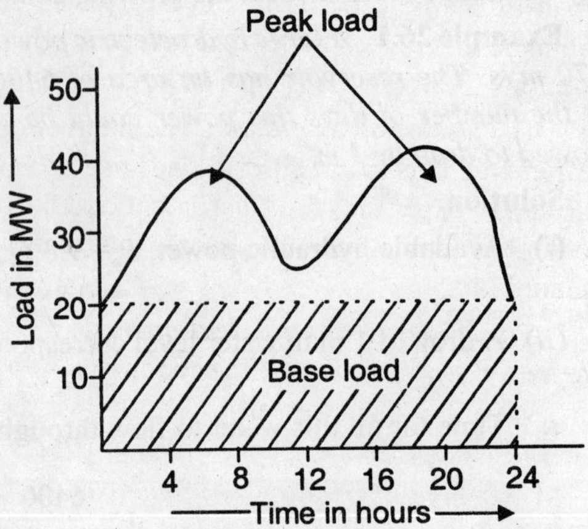

Fig. 26.5

Referring to the load curve of Fig. 26.5, it is clear that 20 MW of load has to be supplied by the station at all times of day and night *i.e.* throughout 24 hours. Therefore, 20 MW is the base load of the station. As base load on the station is almost of constant nature, therefore, it can be suitably supplied without facing the problems of variable load. Those power stations which supply base load are called *base load power stations.* Nuclear power stations and coal-fired power stations are very suitable to supply the base load.

(*ii*) **Peak load.** *The various peak demands of load over and above the base load of the station is known as* **peak load.**

Referring to the load curve of Fig. 26.5, it is clear that there are peak demands of load excluding base load. These peak demands of the station generally form a small part of the total load and may occur throughout the day. Those power stations which supply the peak load are called *peak load power stations.* These stations deliver for brief time interval in the day. Such stations must be put in service very quickly. Therefore, such stations use diesel engines, gas turbines etc. that can be started up in a few minutes. It may be noted that thermal power stations using coal take 4 to 8 hours to start up while nuclear power stations may take several days. Clearly, such stations cannot be used to supply peak load.

26.7 Method of Meeting the Load

The total load on a power station consists of two parts *viz.*, base load and peak load. In order to achieve overall economy, *the best method to meet load is to interconnect two different power stations.* The more efficient plant is used to supply the base load and is known as *base load power station.* The less efficient plant is used to supply the peak loads and is known as *peak load power station.* There is no hard and fast rule for selection of base load and peak load stations as it would depend upon the particular situation. For example, both hydro-electric and steam power stations are quite efficient and can be used as base load as well as peak load station to meet a particular requirement.

26.8 Interconnected Grid System

The connection of several generating stations in parallel is known as **interconnected grid system.**

The various problems facing the power engineers are considerably reduced by interconnecting different power stations in parallel. Although interconnection of station involves extra cost, yet considering the benefits derived from such an arrangement, it is gaining much favour these days. Some of the advantages of interconnected system are listed below :

(*i*) ***Exchange of peak loads*** : An important advantage of interconnected system is that the peak load of the power station can be exchanged. If the load curve of a power station shows a peak demand that is greater than the rated capacity of the plant, then the excess load can be shared by other stations interconnected with it.

(*ii*) ***Use of older plants*** : The interconnected system makes it possible to use the older and less efficient plants to carry peak-loads of short durations. Although such plants may be inadequate when used alone, yet they have sufficient capacity to carry short peaks of loads when interconnected with other modern plants. Therefore, interconnected system gives a direct key to the use of obsolete plants.

(*iii*) ***Ensures economical operation*** : The interconnected system makes the operation of concerned power stations quite economical. It is because sharing of load among the stations is arranged in such a way that more efficient stations work continuously throughout the year at a high load factor and the less efficient plants work for peak load hours only.

(*iv*) ***Increases diversity factor*** : The load curves of different interconnected stations are generally different. The result is that the maximum demand on the system is much reduced as compared to the sum of individual maximum demands on different stations. In other words, the diversity factor of the system is improved, thereby increasing the effective capacity of the system.

(*v*) ***Reduces plant reserve capacity*** : Every power station is required to have a standby unit for emergencies. However, when several power stations are connected in parallel, the reserve capacity of the system is much reduced. This increases the efficiency of the system.

(*vi*) ***Increases reliability of supply.*** The interconnected system increases the reliability of supply. If a major breakdown occurs in one station, continuity of supply can be maintained by other healthy stations.

26.9. Transmission of Electrical Energy

In early days, there was a little demand for electrical energy so that small power stations were built to supply lighting and heating loads. However, the widespread use of electrical energy by modern civilisation has necessitated to produce bulk electrical energy economically and efficiently. The increased demand of electrical energy can be met by building big power stations at favourable places where fuel (coal or gas) or water energy is available in abundance. This has shifted the site of power stations to places quite away from the consumers. The electric power produced at the power station is transmitted over large distances to load centres with the help of conductors known as *transmission lines.* The transmission of electrical energy forms a very important part of power system. It is because the human and material resources in transmission of electrical energy are much greater than those devoted to generation of electrical energy.

26.10. Components of a Transmission Line

As a rule, the transmission of power over long distances is carried out by overhead lines. The main components of a transmission line are (*i*) conductors (*ii*) insulators and (*iii*) supports.

(*i*) **Conductors.** The conductors carry electric power from the power station to the load centres. Conductors for high-voltage overhead lines are always bare. We generally use steel-reinforced aluminium conductors (ACSR) for this purpose. Just like the plates of a capacitor, the parallel conductors of a transmission line remain charged after the line is de-energised. To prevent fatal shocks, we must always discharge the conductors to ground before beginning work on a de-energised transmission line.

(*ii*) **Insulators.** Insulators serve to anchor the conductors and to insulate them from the ground. Insulators are usually made of porcelain but glass and other synthetic insulating materials are also used. From a mechanical stand point, the insulators must be strong enough to resist the pull due to weight of the conductors. For high-voltage (> 70 kV) transmission line, we always use suspension-type insulators. The number of insulators depends upon the voltage. For 110 kV, we generally use 4 to 7 ; for 230 kV from 13 to 16.

(*iii*) **Supports.** The supporting structure must keep the conductors at a safe height from the ground and at an adequate distance from each other. For very high voltage lines, we always use steel towers made of galvanised angle-iron, bolted together. The spacing between the conductors must be sufficient to prevent arc-over under gusty wind conditions. The spacing between the conductors has to be increased as we increase the distance between towers and as the line voltage becomes higher.

26.11. Distribution of Electrical Energy

The electrical energy produced at the generating station is conveyed to the consumers through a network of transmission and distribution systems. It is often difficult to draw a line between the transmission and distribution systems of a large power system. It is impossible to distinguish the two merely by their voltage because what was considered as a high voltage a few years ago is now considered as a low voltage. In general, distribution system is that part of power system which distributes power to the consumers for utilisation.

The transmission and distribution systems are similar to man's circulatory system. The transmission system may be compared with arteries in the human body and distribution system with capillaries. They serve the same purpose of supplying the ultimate consumer in the city with the life-giving blood of civilisation—electricity.

26.12. Distribution System

That part of power system which distributes electric power for local use is known as **distribution system.**

In general, the distribution system is the electrical system between the sub-station fed by the transmission system and the consumers meters. It generally consists of *feeders, distributors* and the *service mains.* Fig. 26.6 shows the single line diagram of a typical low tension distribution system.

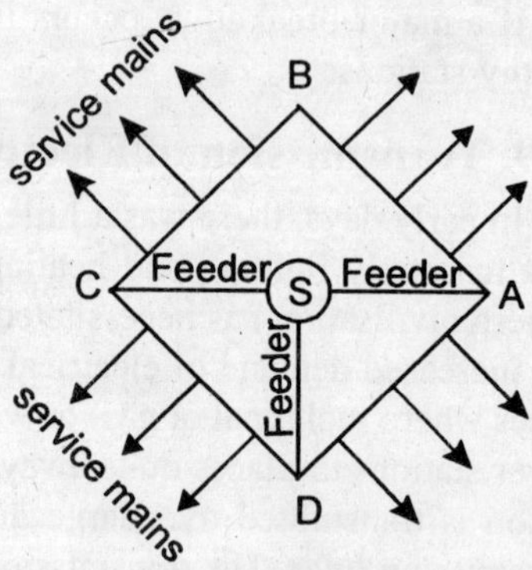

Fig. 26.6

(*i*) *Feeders.* A feeder is a conductor which connects the sub-station (or localised generating station) to the area where power is to be distributed. Generally, no tappings are taken from the feeder so that current in it remains the same throughout. The main consideration in the design of a feeder is the current carrying capacity.

(*ii*) *Distributor.* A distributor is a conductor from which tappings are taken for supply to the consumers. In Fig. 26.6, *AB, BC, CD* and *DA* are the distributors. The current through a distributor is not constant because tappings are taken at various places along its length. While designing a distributor, voltage drop along its length is the main consideration since the statutory limit of voltage variations is ± 6% of rated value at the consumers' terminals.

(*iii*) *Service mains.* A service mains is generally a small cable which connects the distributor to the consumers' terminals.

26.13. Classification of Distribution Systems

A distribution system may be classified according to ;

(*i*) *Nature of current.* According to nature of current, distribution system may be classified as (*a*) d.c. distribution system (*b*) a.c. distribution system. Now-a-days, a.c. system is universally adopted for distribution of electric power as it is simpler and more economical than direct current method.

(*ii*) *Type of construction.* According to type of construction, distribution system may be classified as (*a*) overhead system (*b*) underground system. The overhead system is generally employed for distribution as it is 5 to 10 times cheaper than the equivalent underground system. In general, the underground system is used at places where overhead construction is impracticable or prohibited by the local laws.

(*iii*) *Scheme of connection.* According to scheme of connection, the distribution system may be classified as (*a*) radial system (*b*) ring main system (*c*) inter-connected system. Each scheme has its own advantages and disadvantages.

26.14. A.C. Distribution

Now-a-days electrical energy is generated, transmitted and distributed in the form of alternating current. One important reason for the widespread use of alternating current in preference to direct current is the fact that alternating voltage can be conveniently changed in magnitude by means of a transformer. Transformer has made it possible to transmit a.c. power at high voltage and utilise it at a safe potential. High transmission and distribution voltages have greatly reduced the current in the conductors and the resulting line losses.

There is no definite line between transmission and distribution according to voltage or bulk capacity. However, in general, the a.c. distribution system is the electrical system between the step-down sub-station fed by the transmission system and the consumers' meters. The a.c. distribution system is classified into (*i*) primary distribution system and (*ii*) secondary distribution system.

(*i*) **Primary distribution system.** It is that part of a.c. distribution system which operates at voltages somewhat higher than general utilisation and handles large blocks of electrical energy than the average low-voltage consumer uses. The voltage used for primary distribution depends upon the

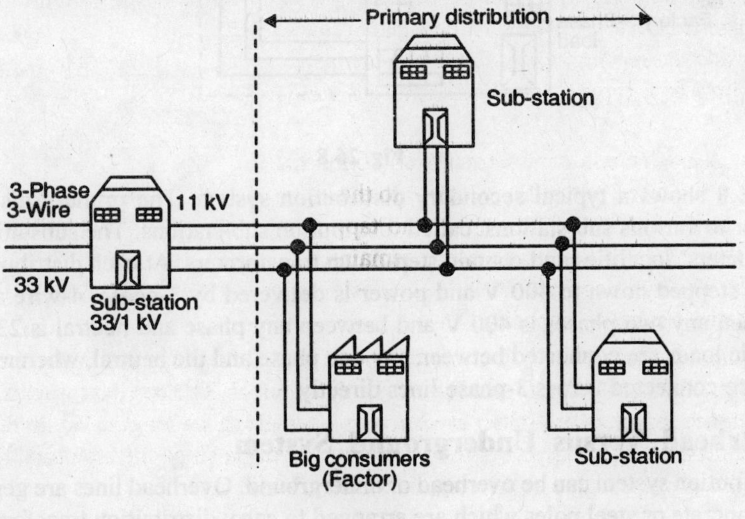

Fig. 26.7

amount of power to be conveyed and the distance of the sub-station required to be fed. The most commonly used primary distribution voltages are 11 kV, 6.6 kV and 3.3 kV. Due to economic considerations, primary distribution is carried out by 3-phase, 3-wire system.

Figure 26.7 shows a typical primary distribution system. Electric power from the generating station is transmitted at high voltage to the sub-station located in or near the city. At this sub-station, voltage is stepped down to 11 kV with the help of step-down transformer. Power is supplied to various sub-stations for distribution or to big consumers at this voltage. This forms the high voltage distribution or primary distribution.

(*ii*) **Secondary distribution system.** It is that part of a.c. distribution system which includes the range of voltages at which the ultimate consumer utilises the electrical energy delivered to him. The secondary distribution employs 400/230 V, 3-phase, 4-wire system.

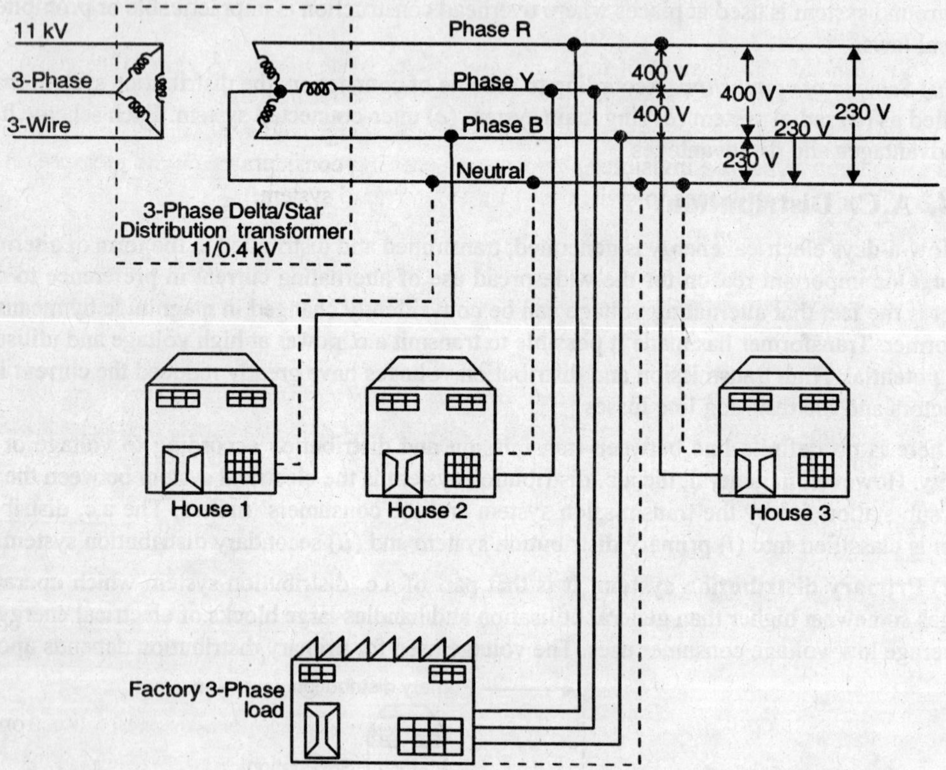

Fig. 26.8

Figure 26.8 shows a typical secondary distribution system. The primary distribution circuit delivers power to various sub-stations, called distribution sub-stations. The substations are situated near the consumers' localities and contain step-down transformers. At each distribution sub-station, the voltage is stepped down to 400 V and power is delivered by 3-phase, 4-wire a.c. system. The voltage between any two phases is 400 V and between any phase and neutral is 230 V. The single phase domestic loads are connected between any one phase and the neutral, whereas 3-phase 400 V motor loads are connected across 3-phase lines directly.

26.15. Overhead Versus Underground System

The distribution system can be overhead or underground. Overhead lines are generally mounted on wooden, concrete or steel poles which are arranged to carry distribution transformers in addition to the conductors. The underground system uses conduits, cables and manholes under the surface of streets and sidewalks. The choice between overhead and underground system depends upon a number of widely differing factors. Therefore, it is desirable to make a comparison between the two.

(i) *Public safety.* The underground system is more safe than overhead system because all distribution wiring is placed underground and there are little chances of any hazard.

(ii) *Initial cost.* The underground system is more expensive due to the high cost of trenching, conduits, cables, manholes and other special equipment. The initial cost of an underground system may be five to ten times more than that of an overhead system.

(iii) *Flexibility.* The overhead system is much more flexible than the underground system. In the latter case, manholes, duct lines etc., are permanently placed once installed and the load expansion can only be met by laying new lines. However, on an overhead system, poles, wires, transformers etc., can be easily shifted to meet the changes in load conditions.

(iv) *Faults.* The chances of faults in underground system are very rare as the cables are laid underground and are generally provided with better insulation.

(v) *Appearance.* The general appearance of an underground system is better as all the distribution lines are invisible. This factor is exerting considerable public pressure on electric supply companies to switch over to underground system.

(vi) *Fault location and repairs.* In general, there are little chances of faults in an underground sytem. However, if a fault does occur, it is difficult to locate and repair on this system. On an overhead system, the conductors are visible and easily accessible so that fault locations and rapairs can be easily made.

(vii) *Current carrying capacity and voltage drop.* An overhead distribution conductor has a considerably higher current carrying capacity than an underground cable conductor of the same material and cross-section. On the other hand, underground cable conductor has much lower inductive reactance than that of an overhead conductor because of closer spacing of conductors.

(viii) *Useful life.* The useful life of underground system is much longer than that of an overhead system. An overhead system may have a useful life of 25 years, whereas an underground system may have a useful life of more than 50 years.

(ix) *Maintenance cost.* The maintenance cost of underground system is very low as compared with that of overhead system because of less chances of faults and service interruptions from wind, ice, lightning as well as from traffic hazards.

(x) *Interference with communication circuits.* An overhead system causes electromagnetic interference with the telephone lines. The power line currents are superimposed on speech currents, resulting in the potential of the communication channel being raised to an undesirable level. However, there is no such interference with the underground system.

It is clear from the above comparison that each system has its own advantages and disadvantages. However, comparative economics (*i.e.*, annual cost of operation) is the most powerful factor influencing the choice between underground and overhead system. The greater capital cost of underground system prohibits its use for distribution. But sometimes non-economic factors (*e.g.*, general appearance, public safety etc.) exert considerable influence on choosing underground system. In general, overhead system is adopted for distribution and the use of underground system is made only where overhead construction is impracticable or prohibited by local laws.

26.16. Requirements of a Distribution System

A considerable amount of effort is necessary to maintain an electric power supply within the requirements of various types of consumers. Some of the requirements of a good distribution system are : proper voltage, availability of power on demand and reliability.

(*i*) **Proper voltage.** One important requirement of a distribution system is that voltage variations at consumer's terminals should be as low as possible. The changes in voltage are generally caused due to the variation of load on the system. Low voltage causes loss of revenue, inefficient lighting and possible burning out of motors. High voltage causes lamps to burn out permanently and may cause failure of other appliances. Therefore, a good distribution system should ensure that the voltage variations at consumers terminals are within permissible limits. The statutory limit of voltage variations is $\pm$ 6% of the rated value at the consumer's terminals. Thus, if the declared voltage is 230 V, then the highest voltage of the consumer should not exceed 244 V while the lowest voltage of the consumer should not be less than 216 V.

(*ii*) **Availability of power on demand.** Power must be available to the consumers in any amount that they may require from time to time. For example, motors may be started or shut down, lights may be turned on or off, without advance warning to the electric supply company. As electrical energy cannot be stored, therefore, the distribution system must be capable of supplying load demands of the consumers. This necessitates that operating staff must continuously study load patterns to predict in advance those major load changes that follow the known schedules.

(*iii*) **Reliability.** Modern industry is almost dependent on electric power for its operation. Homes and office buildings are lighted, heated, cooled and ventilated by electric power. This calls for reliable service. Unfortunately, electric power, like everything else that is man-made, can never be absolutely reliable. However, the reliability can be improved to a considerable extent by (*a*) interconnected system (*b*) reliable automatic control system (*c*) providing additional reserve facilities.

26.17. Protection of Power System

A great demand for electrical energy is a notable feature of modern civilization. Most of this energy is needed for lighting, heating, domestic appliances, industrial electrical machinery and electric traction. The importance of electric supply in everyday life has reached such a stage that it is desirable to protect the power system from harm during fault conditions and to ensure maximum continuity of supply. For this purpose, means must be provided to switch on or off generators, transmission lines, distributors and other equipment under both normal and abnormal conditions. This is achieved by an apparatus called *switchgear.* A switchgear essentially consists of switching and protecting devices such as switches, fuses, circuit breakers, relays etc.

During normal operation, switchgear permits to switch on or off generators, transmission lines, distributors and other electrical equipment. On the other hand, when a failure (*e.g.* short circuit) occurs on any part of power system, a heavy current flows through the equipment, threatening damage to the equipment and interruption of service to the customers. However, the switchgear detects the fault and disconnects the unhealthy section from the system. In this way, switchgear protects the system from the damage and ensures continuity of supply.

26.18. Switchgear

The apparatus used for switching, controlling and protecting the electrical circuits and equipment is known as **switchgear.**

The switchgear equipment is essentially concerned with switching and interrupting currents either under normal or abnormal operating conditions. The tumbler switch with ordinary fuse is the simplest form of switchgear and is used to control and protect lights and other equipment in homes, offices etc. For circuits of higher rating, a high-rupturing capacity (H.R.C.) fuse in conjuction with a switch may serve the purpose of controlling and protecting the circuit. However, such a switchgear cannot be used profitably on high voltage system (3.3 kV) for two reasons. Firstly, when a fuse blows, it takes sometime to replace it and consequently there is interruption of service to the customers. Secondly, the fuse cannot successfully interrupt large fault currents that result from the faults on high voltage system.

With the advancement of power system, lines and other equipments operate at high voltages and carry large currents. When a short circuit occurs on the system, heavy current flowing through the equipment may cause considerable damage. In order to interrupt such heavy fault currents, *automatic circuit breakers* (or simply circuit breakers) are useo. A circuit breaker is a switchgear which can open or close an electrical circuit under both normal and abnormal conditions. Even in instances where a fuse is adequate, as regards to breaking capacity, a circuit breaker may be preferable. It is because a circuit breaker can close circuits, as well as break them without replacement and thus has wider range of use altogether than a fuse.

26.19. Switchgear Equipment

Switchgear covers a wide range of equipment concerned with switching and interrupting currents under both normal and abnormal conditions. It includes switches, fuses, circuit breakers, relays and other equipment. A brief account of these devices is given below.

1. Switches. A switch is a device which is used to open or close an electrical circuit in a convenient way. It can be used under full-load or no-load conditions *but* it cannot interrupt the fault currents. When the contacts of a switch are opened, an *arc is produced in the air between the contacts. This is particularly true for circuits of high voltage and large current capacity. The switches may be classified into (*i*) air switches (*ii*) oil switches. The contacts of the former are opened in air and that of the latter are opened in oil.

(*i*) *Air-break switch.* It is an air switch and is designed to open a circuit *under load.* In order to quench the arc that occurs on opening such a switch, special arcing horns are provided. Arcing horns are pieces of metals between which arc is formed during opening operation. As the switch opens, these horns are spread farther and farther apart. Consequently, the arc is lengthened, cooled and interrupted. Air-break switches are generally used outdoor for circuits of medium capacity such as lines supplying an industrial load from a main transmission line or feeder.

(*ii*) *Isolator or disconnecting switch.* It is essentially a knife switch and is designed to open a circuit under *no load.* Its main purpose is to isolate one portion of the circuit from the other and is not intended to be opened while current is flowing in the line. Such switches are generally used on both sides of circuit breakers in order that repairs and replacement of circuit breakers can be made without any danger. They should never be opened until the circuit breaker in the same circuit has been opened and should always be closed before the circuit breaker is closed.

(*iii*) *Oil switches.* As the name implies, the contacts of such switches are opened under oil, usually transformer oil. The effect of oil is to cool and quench the arc that tends to form when the circuit is opened. These switches are used for circuits of high voltage and large current carrying capacities.

2. Fuses. A fuse is a short piece of wire or thin strip which melts when excessive current flows through it for sufficient time. It is inserted in series with the circuit to be protected. Under normal operating conditions, the fuse element it at a temperature below its melting point. Therefore, it carries the normal load current without overheating. However, when a short circuit or overload occurs, the current through the fuse element increases beyond its rated capacity. This raises the temperature

---

* Generally, the load contains reactive elements (inductance and capacitance). The sudden change of current in the circuit due to breaking results in the production of large induced e.m.f. (several thousand volts) which ruptures the insulation of air inbetween the contacts, causing a spark.

and the fuse element melts (or blows out), disconnecting the circuit protected by it. In this way, a fuse protects the machines and equipment from damage due to excessive currents. It is worthwhile to note that a fuse performs both detection and interruption functions.

3. Circuit breakers. A circuit breaker is an equipment which can open or close a circuit under all conditions *viz.*, no load, full load and fault conditions. It is so designed that it can be operated manually (or by remote control) under normal conditions and automatically under fault conditions. For the latter operation, a relay circuit is used with a circuit breaker. Fig. 26.9 shows the parts of an elementary circuit breaker. The tripping action is usually produced by an overload relay that can detect abnormal line conditions. Thus referring to Fig. 26.9, the relay coil is connected to the secondary of a current transformer. The primary carries the line current of the phase that has to be protected. If the line current exceeds a preset limit, the secondary current will cause the relay contacts C_1 C_2 to close. As soon as they close, the tripping coil is energised by an auxiliary d.c. source. This causes the main line contacts to open, thus interrupting the circuit.

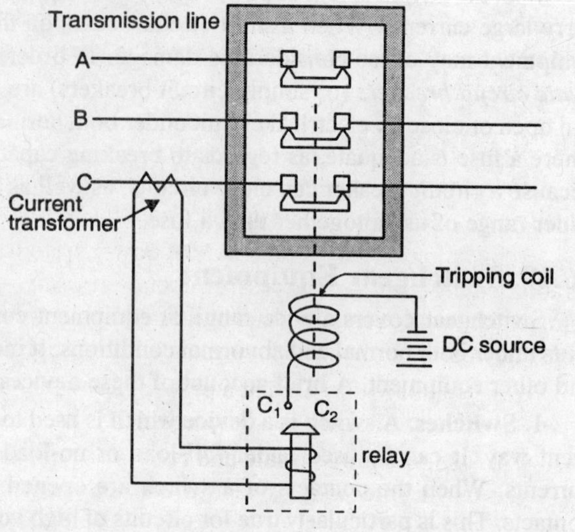

Fig. 26.9

The nameplate of a circuit breaker usually indicates (*i*) the maximum steady-state current it can carry (*ii*) the maximum interrupting current (*iii*) the maximum line voltage (*iv*) the interrupting time in cycles. The interrupting time may last 2 to 8 cycles on a 50 Hz system. High-speed interruption limits the damage to transmission lines and equipment and equally important, it improves the stability of the system.

4. Relays. A relay is a device which detects the fault and supplies information to the breaker for circuit interruption. Fig. 26.10 shows a typical relay circuit. The relay circuit can be divided into three parts *viz.*

(*i*) The primary winding of a *current transformer (C.T.) which is connected in series with the circuit to be protected. The primary winding often consists of the main conductor itself.

(*ii*) The second circuit is the secondary winding of C.T. connected to the relay operating coil.

(*iii*) The third circuit is the tripping circuit which consists of a source of supply, trip coil of circuit breaker and the relay stationary contacts.

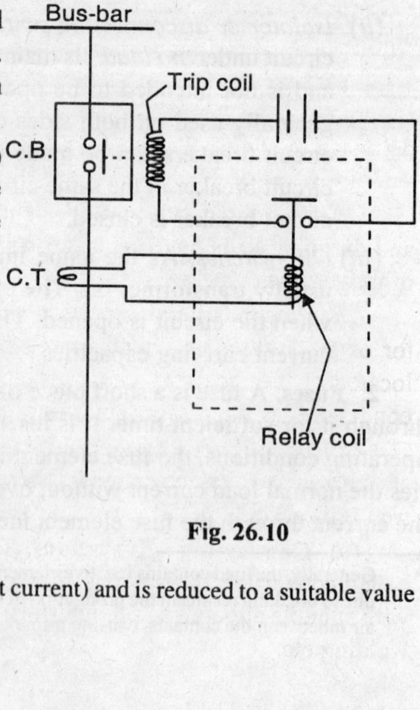

Fig. 26.10

Under normal load conditions, the e.m.f. of the secondary winding of C.T. is small and the current flowing in the

* The circuit to be protected carries a large current (normal or fault current) and is reduced to a suitable value for relay operation with the help of a current transformer.

relay operating coil is insufficient to close the relay contacts. This keeps the trip coil of the circuit breaker unenergised. Consequently, the contacts of the circuit breaker remain closed and it carries the normal load current. When a fault occurs, a large current flows through the primary of C.T. This increases the secondary e.m.f. and hence the current through the relay operating coil. The relay contacts are closed and the trip coil of the circuit breaker is energised to open the contacts of the circuit breaker.

26.20. Sub-station

The assembly of apparatus used to changs some characteristic (e.g. voltage, a.c. to d.c., frequency, p.f. etc.) of electric supply is called a **sub-station.**

Sub-stations are important part of power system. The continuity of supply depends to a considerable extent upon the successful operation of sub-stations. It is, therefore, essential to exercise utmost care while designing and building a sub-station. The following are the important points which must be kept in view while laying out a sub-station:

(*i*) It should be located at a proper site. As far as possible, it should be located at the centre of gravity of load.

(*ii*) It should provide safe and reliable arrangement. For safety, consideration must be given to the maintenance of regulation clearances, facilities for carrying out repairs and maintenance, abnormal occurrences such as possibility of explosion or fire etc. For reliability, consideration must be given for good design and construction, the provision of suitable protective gear etc.

(*iii*) It should be easily operated and maintained.

(*iv*) It should involve minimum capital cost.

26.21. Classification of Sub-stations

There are several ways of classifying sub-stations. However, the two most important ways of classifying them are according to (1) service requirement and (2) constructional features.

1. According to service requirement. A sub-station may be called upon to change voltage level or improve power factor or convert a.c. power into d.c. power etc. According to the service requirement, sub-stations may be classified into:

(*i*) **Transformer sub-stations.** Those sub-stations which change the voltage level of electric supply are called transformer sub-stations. These sub-stations receive power at some voltage and deliver it at some other voltage. Obviously, transformer will be the main component in such sub-stations. Most of the sub-stations in the power system are of this type.

(*ii*) **Switching sub-stations.** These sub-stations do not change the voltage level *i.e.* incoming and outgoing lines have the same voltage. However, they simply perform the switching operations of power lines.

(*iii*) **Power factor correction sub-stations.** Those sub-stations which improve the power factor of the system are called power factor correction sub-stations. Such sub-stations are generally located at the receiving end of transmission lines. These sub-stations generally use synchronous condensers as the power factor improvement equipment.

(*iv*) **Frequency changer sub-stations.** Those sub-stations which change the supply frequency are known as frequency changer sub-stations. Such a frequency change may be required for industrial utilisation.

(*v*) **Converting sub-stations.** Those sub-stations which change a.c. power into d.c. power are called converting sub-stations. These sub-stations receive a.c. power and convert it into d.c. power with suitable apparatus (*e.g.* ignitron) to supply for such purposes as traction, electroplating, electric welding etc.

(*vi*) **Industrial sub-stations.** Those sub-stations which supply power to individual industrial concerns are known as industrial sub-stations.

2. According to constructional features. A sub-station has many components (*e.g.* circuit breakers, switches, fuses, instruments etc.) which must be housed properly to ensure continuous and reliable service. According to constructional features, the sub-stations are classified as:

(*i*) Indoor sub-station
(*ii*) Outdoor sub-station
(*iii*) Underground sub-station
(*iv*) Pole-mounted sub-station

(*i*) **Indoor sub-stations.** For voltages upto 11 kV, the equipment of the sub-station is installed indoor because of economic considerations. However, when the atmosphere is contaminated with impurities, these sub-stations can be erected for voltages upto 66 kV.

(*ii*) **Outdoor sub-stations.** For voltages beyond 66 kV, equipment is invariably installed outdoor. It is because for such voltages, the clearances between conductors and the space required for switches, circuit breakers and other equipment becomes so great that it is not economical to install the equipment indoor.

(*iii*) **Underground sub-stations.** In thickly populated areas, the space available for equipment and building is limited and the cost of land is high. Under such situations, the sub-station is created underground.

(*iv*) **Pole-mounted sub-stations.** This is an outdoor sub-station with equipment installed overhead on *H*-pole or 4-pole structure. It is the cheapest form of sub-station for voltages not exceeding 11kV (or 33 kV in some cases). Electric power is almost distributed in localities through such sub-stations.

26.22 TARIFF

The rate at which electrical energy is supplied to a consumer is known as **tariff.**

Although tariff should include the total cost of producing and supplying electrical energy plus the profit, yet it cannot be the same for all types of consumers. It is because the cost of producing electrical energy depends to a considerable extent upon the magnitude of electrical energy consumed by the user and his load conditions. Therefore, in all fairness, due consideration has to be given to different types of consumers (*e.g.* industrial, domestic and commercial) while fixing the tariff. This makes the problem of suitable rate making highly complicated.

Objectives of tariif. Like other commodities, electrical energy is also sold at such a rate so that it not only returns the cost but also earns reasonable profit. Therefore, a tariff should include the following items:

(*i*) Recovery of cost of producing electrical energy at the power station.
(*ii*) Recovery of cost on the capital investment in transmission and distribution systems.
(*iii*) Recovery of cost of operation and maintenance of supply of electrical energy *e.g.*, metering equipment, billing etc.
(*iv*) A suitable profit on the capital investment

26.23. TYPES OF TARIFF

There are several types of tariff. However, the following are the commonly used types of tariff:

1. Simple tariff. *When there is a fixed rate per unit of energy consumed, it is called* **a simple tariff** *or* **uniform rate tariff.**

In this type of tariff, the price charged per unit is constant *i.e.*, it does not vary with increase or decrease in number of units consumed. The consumption of electrical energy at the consumer's terminals is recorded by means of an energy meter. This is the simplest of all tariffs and is readily understood by the consumers.

Disadvantages

(*i*) There is no discrimination between different types of consumers since every consumer has to pay equitably for the fixed* charges.

(*ii*) The cost per unit delivered is high.

(*iii*) It does not encourage the use of electricity.

2. Flat rate tariff. *When different types of consumers are charged at different uniform per unit rates, it is called a* **flat rate tariff.**

In this type of tariff, the consumers are grouped into different classes and each class of consumers is charged at a different uniform rate. For instance, the flat rate per kWh for lighting load may be 60 paise, whereas it may be slightly less† (say 55 paise per kWh) for power load. The different classes of consumers are made taking into account their diversity and load factors. The advantage of such a tariff is that it is more fair to different types of consumers and is quite simple in calculations.

Disadvantages

(*i*) Since the flat rate tariff varies according to the way the supply is used, separate meters are required for lighting load, power load etc. This makes the application of such a tariff expensive and complicated.

(*ii*) A particular class of consumers is charged at the same rate irrespective of the magnitude of energy consumed. However, a big consumer should be charged at a lower rate as in his case the fixed charges per unit are reduced.

3. Block rate tariff. *When a given block of energy is charged at a specified rate and the succeeding blocks of energy are charged at progressively reduced rates, it is called a* **block rate tariff.**

In block rate tariff, the energy consumption is divided into blocks and the price per unit is fixed in each block. The price per unit in the first block is the highest** and it is progressively reduced for the succeeding blocks of energy. For example, the first 30 units may be charged at the rate of 60 paise per unit; the next 25 units at the rate of 55 paise per unit and the remaining additional units may be charged at the rate of 30 paise per unit.

The advantage of such a tariff is that the consumer gets an incentive to consume more electrical energy. This increases the load factor of the system and hence the cost of generation is reduced. However, its principal defect is that it lacks a measure of the consumer's demand. This type of tariff is being used for majority of residential and small commercial consumers.

4. Two-part tariff. *When the rate of electrical energy is charged on the basis of maximum demand of the consumer and the units consumed, it is called a* **two-part tariff.**

In two-part tariff, the total charge to be made from the consumer is split into two components *viz.*, fixed charges and running charges. The fixed charges depend upon the maximum demand of the consumer while the running charges depend upon the number of units consumed by the consumer. Thus, the consumer is charged at a certain amount per kW of maximum*** demand plus a certain amount per kWh of energy consumed *i.e.*,

* The total cost of electrical energy consists of *fixed charges* and *running charges*. The greater the number of units consumed, the lesser the fixed charges per unit. Therefore, a consumer who consumes more units must pay less fixed charges per unit.

† The flat rate for power load is always less than lighting load. It is because power load is much more than the lighting load and, therefore, improves the load factor of the system to a great extent.

** Generally, fixed charges are merged into the running charges for the first and second blocks of energy so that price per unit for these blocks is high.

*** The maximum demand of consumer is generally assessed on the basis of rateable value of the premises or on the number of rooms or on the connected load.

Total charges = Rs ($b \times$ kW + $c \times$ kWh)

where, b = charge per kW of maximum demand

c = charge per kWh of energy consumed

This type of tariff is mostly applicable to industrial consumers who have appreciable maximum demand.

Advantages

(*i*) It is easily understood by the consumers.

(*ii*) It recovers the fixed charges which depend upon the maximum demand of the consumer but are independent of the units consumed.

Disadvantages

(*i*) The consumer has to pay the fixed charges irrespective of the fact whether he has consumed or not consumed the electical energy.

(*ii*) There is always error in assessing the maximum demand of the consumer.

5. Maximum demand tariff. It is similar to two-part tariff with the only difference that the maximum demand is actually measured by installing maximum demand meter in the premises of the consumer. This removes the objection of two-part tariff where the maximum demand is assessed merely on the basis of the rateable value. This type of tariff is mostly applied to big consumers. However, it is not suitable for a small consumer (*e.g.*, residential consumer) as a separate maximum demand meter is required.

6. Power factor tariff. *The tariff in which power factor of the consumer's load is taken into consideration is known as* **power factor tariff.**

In an a.c. system, power factor plays an important role. A low* power factor increases the rating of station equipment and line losses. Therefore, a consumer having low power factor must be penalised,. The following are the important types of power factor tariff:

(*i*) *kVA maximum demand tariff:* It is a modified form of two-part tariff. In this case, the fixed charges are made on the basis of maximum demand in kVA and *not* in kW. As kVA is inversely proportional to power factor, therefore, a consumer having low power factor has to contribute more towards the fixed charges. This type of tariff has the advantage that it encourages the consumers to operate their appliances and machinery at improved power factor.

(*ii*) *Sliding scale tariff:* This is also known as average power factor tariff. In this case, an average power factor, say 0.8 lagging, is taken as the reference. If the power factor of the consumer falls below this factor, suitable additional charges are made. On the other hand, if the power factor is above the reference, a discount is allowed to the consumer.

(*iii*) *kW and kVAR tariff:* In this type, both active power (kW) and reactive power (kVAR) supplied are charged separately. A consumer having low power factor will draw more reactive power and hence shall have to pay more charges.

7. Three-part tariff. *When the total charge to be made from the consumer is split into three parts viz., fixed charge, semi-fixed charge and running charge, it is known as a* **three-part tariff** *i.e.*,

---

* See page 405 on power factor improvement.

Total charge = Rs. $(a + b \times \text{kW} + c \times \text{kWh})$

where a = fixed charge made during each billing period. It includes interest and depreciation on the cost of secondary distribution and labour cost of collecting revenues,

b = charge per kW of maximum demand,

c = charge per kWh of energy consumed.

It may be seen that by adding fixed charge or consumer's charge (*i.e., a*) to two-part tariff, it becomes three-part tariff. The principal objection of this type of tariff is that the charges are split into three components. This type of tariff is generally applied to big consumers.

Example 26.2. *A consumer has a maximum demand of 200 kW at 40% load factor. If the tariff is Rs. 100 per kW of maximum demand plus 10 paise per kWh, find the overall cost per kWh.*

Solution.

Units consumed/year = Max.demand × L.F. × Hours in a year

$= (200) \times (0.4) \times 8760 = 7{,}00{,}800$ kWh

Annual charges = Annual M.D. charges + Annual energy charges

$=$ Rs $(100 \times 200 + 0.1 \times 7{,}00{,}800)$

$=$ Rs. 90,080

$\therefore$ Overall cost/kWh $= \text{Rs}\ \dfrac{90{,}080}{7{,}00{,}800} = \text{Re}\ 0.1285 =$ **12.85 paise**

Example 26.3. *The maximum demand of a consumer is 20 A at 220 V and his total energy consumption is 8760 kWh. If the energy is charged at the rate of 20 paise per unit for 500 hours use of the maximum demand per annum plus 10 paise per unit for additional units, calculate: (i) annual bill (ii) equivalent flat rate.*

Solution.

Assume the load factor and power factor to be unity.

$\therefore$ Maximum demand $= \dfrac{220 \times 20 \times 1}{1000} = 4.4$ kW

(*i*) Units consumed in 500 hrs $= 4.4 \times 500 = 2200$ kWh

Charges for 2200 kWh $=$ Rs. $0.2 \times 2200 =$ Rs 440

Remaining units $= 8760 - 2200 = 6560$ kWh

Charges for 6560 kWh $=$ Rs. $0.1 \times 6560 =$ Rs. 656

$\therefore$ Total annual bill $=$ Rs. $(440 + 656) =$ **Rs. 1096**

(*ii*) Equivalent flat rate $= \text{Rs}\ \dfrac{1096}{8760} = \text{Rs.}\ 0.125 =$ **12.5 paise**

Example 26.4. *The following two tariffs are offered:*

(a) Rs. 100 plus 15 paise per unit;

(b) A flat rate of 30 paise per unit;

At what consumption is first tariff economical?

Solution. Let x be the number of units at which charges due to both tariffs become equal.

Then,

$$100 + 0.15x = 0.3x$$

or $$0.15x = 100$$

$$\therefore \quad x = 100/0.15 = \mathbf{666.67\ units}$$

Therefore, tariff (*a*) is economical if consumption is more than 666.67 units.

Example 26.5. *A supply is offered on the basis of fixed charges of Rs. 30 per annum plus 3 paise per unit or alternatively, at the rate of 6 paise per unit for the first 400 units per annum and 5 paise per unit for all the additional units. Find the number of units taken per annum for which the cost under the two tariffs becomes the same.*

Solution. Let x (> 400) be the number of units taken per annum for which the annual charges due to both tariffs become equal.

Annual charges due to first tariff = Rs $(30 + 0.03x)$

Annual charges due to second tariff = Rs. $[(0.06 \times 400) + (x - 400) \times 0.05]$

$= \text{Rs. } (4 + 0.05x)$

As the charges in both cases are equal,

$$\therefore \quad 30 + 0.03x = 4 + 0.05x$$

or $$x = \frac{30-4}{0.05-0.03} = \mathbf{1300\ kWh\ or\ 1300\ units}$$

26.24. Power Factor Improvement

Most of the loads (*e.g.*, induction motors) operating on a.c. supply have low lagging power factors. Any decrease in the power factor results in the increase in current for the same power output. Consequently, decrease in power factor of the power system invites many disadvantages for the supply company including greater power loss and need of increased capacity of the power producing unit. Therefore, the supply company imposes penalty on individual consumers if their load power factor falls below a certain limit. Such consumers have to either improve their load power factor or to pay the penalty. For detailed discussion on power factor improvement, the reader may refer to page 405 of this book.

OBJECTIVE QUESTIONS

1. In India, the transmission of electric power is done by
   (*i*) 3-phase, 3-wire system
   (*ii*) 3-phase, 4-wire system
   (*iii*) 1- phase system
   (*iv*) none of the above

2. As a matter of economy, voltage for power transmission should be
   (*i*) low (*ii*) medium
   (*iii*) high (*iv*) none of the above

3. The primary distribution of electric power is done by
   (*i*) 3-phase, 4-wire system
   (*ii*) 3-phase, 3-wire system
   (*iii*) 1-phase system
   (*iv*) none of the above

4. When transmission voltage is increased, the line losses are
   (*i*) decreased (*ii*) increased
   (*iii*) same (*iv*) none of the above

5. If voltage is increased by n times, the size of the conductor would
   (*i*) increase by n times
   (*ii*) reduce by $1/n$ times
   (*iii*) increase by n^2 times
   (*iv*) reduce by $1/n^2$ times

6. The thermal efficiency of a steam power station is about
   (*i*) 28% (*ii*) 69%
   (*iii*) 80% (*iv*) 75%

7. The cheapest plant in operation and maintenance is

(i) steam power plant
(ii) nuclear power plant
(iii) hydroelectric plant
(iv) diesel power plant

8. Fig. 26.11 shows the daily load curve of a generating station.

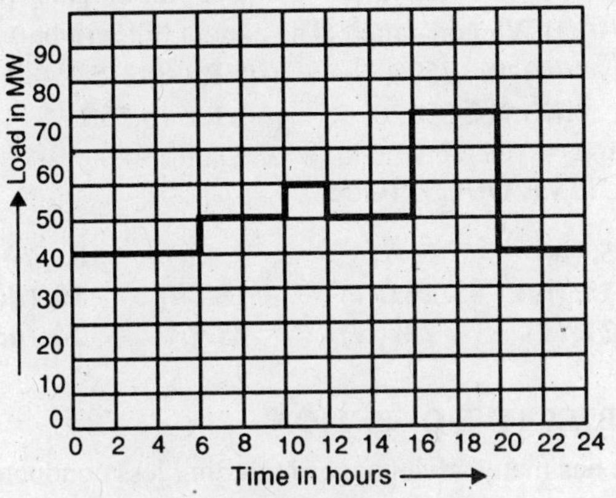

Fig. 26.11

What is the maximum demand?
(i) 40 MW (ii) 50 MW
(iii) 70 MW (iv) 55 MW

9. Referring to daily load curve in Fig. 26.11, units generated per day will be
(i) 12×10^5 kWh (ii) 6×10^6 kWh
(iii) 7.5×10^5 kWh (iv) 8.5×10^5 kWh

10. The base load power station supplies load .... in a day.
(i) 10 hours (ii) 16 hours
(iii) 24 hours (iv) 12 hours

11. The most commonly used material for insulators of overhead lines is
(i) porcelain (ii) glass
(iii) steatite (iv) none of the above

12. A 3-wire d.c. distribution system makes available
(i) one voltage (ii) two voltages
(iii) three voltages (iv) none of the above

13. A line which connects a consumer to the distributor is called
(i) feeder (ii) distributor
(iii) service mains (iv) none of above

14. An a.c. secondary distribution system employs
(i) 11 kV, 3-phase, 3-wire system
(ii) 6.6 kV, 3-phase, 3-wire system
(iii) 3.3 kV, 3-phase, 3-wire system
(iv) 400/230 V, 3-phase, 4-wire system

15. The distribution transformer is generally connected in
(i) Delta/Delta (ii) Delta/Star
(iii) Star/Star (iv) Star/Delta

16. A line which connects distributor to substation is called
(i) service mains (ii) distributor
(iii) feeder (iv) none of the above

17. A fuse is a
(i) protective device
(ii) current limiting device
(iii) curent controlling device
(iv) none of the above

18. A circuit breaker is a
(i) current controlling device
(ii) circuit interrupting device
(iii) current limiting device
(iv) none of the above

19. An isolator is designed to open a circuit under
(i) full load (ii) normal conditions
(iii) no load (iv) none of above

20. Under normal operating conditions, the contacts of the circuit breaker remain
(i) closed (ii) open
(iii) semi-closed (iv) none of the above

21. A bulk power consumer has a maximum demandof 2000 kVA at 0.8 power factor lagging. The tariff is Rs. 100 per kVA of maximum demand plus 10 paise per kWh of energy consumed. If the annual load factor is 50%, the overall cost per unit is
(i) 8.42 paise (ii) 12.85 paise
(iii) 19.25 paise (iv) 7.65 paise

22. An industrial concern has a maximum demand of 500 kVA at 0.8 p.f. lagging and an annual load factor of 60%. If electric power is supplied from a grid system at Rs. 125 per kVA of maximum demand plus 5 paise per unit, the cost per unit is
(i) 5.47 paise (ii) 16.48 paise
(iii) 2.82 paise (iv) 7.97 paise

23. A supply is offered on the basis of fixed charges of Rs. 30 per annum plus 3 paise per unit or alternatively at the rate of 6 paise per

unit for the first 400 units per annum and 5 paise per unit for all the additional units. The number of units taken per annum for which the cost under the two tariffs become equal is

(*i*) 1300 kWh (*ii*) 2600 kWh
(*iii*) 750 kWh (*iv*) 3900 kWh

24. The tariff in force is Rs. 150 per kVA of maximum demand and 8 paise per unit consumed. If the load factor is 30%, the over all cost per unit at 0.7 p.f. lagging is

(*i*) 13.7 paise (*ii*) 22.4 paise
(*iii*) 16.15 paise (*iv*) 25.6 paise

25. A consumer has a maximum demand of 100 kW. The p.f. is 0.8 lagging and the load factor is 60%. The tariff used is Rs. 75 per kVA of maximum demand plus 15 paise per kWh consumed. The annual bill will be

(*i*) Rs. 7500 (*ii*) Rs. 88215
(*iii*) Rs. 55620 (*iv*) Rs. 15,550

ANSWERS TO OBJECTIVE QUESTIONS

1. (*i*)	**2.** (*iii*)	**3.** (*ii*)	**4.** (*i*)	**5.** (*iv*)	**6.** (*i*)	**7.** (*iii*)	**8.** (*iii*)
9. (*i*)	**10.** (*iii*)	**11.** (*i*)	**12.** (*ii*)	**13.** (*iii*)	**14.** (*iv*)	**15.** (*ii*)	**16.** (*iii*)
17. (*i*)	**18.** (*ii*)	**19.** (*iii*)	**20.** (*i*)	**21.** (*ii*)	**22.** (*iv*)	**23.** (*i*)	**24.** (*iii*)
25. (*ii*)							

HINTS TO SELECTED OBJECTIVE QUESTION

2. Transmission of electric power at high voltages has many advantaages including less conductor material required, reduced line losses and decrease in percentage line drop.

4. $P = \sqrt{3}\, VI \cos\phi$ or $I = \dfrac{P}{\sqrt{3}\, VI \cos\phi}$

As transmission voltage V increases, the line current I and hence line losses are decreased.

5. Area of cross-section of conductor $\propto 1/V^2$.

6. More than 50 % of total heat of combustion is lost in the condenser.

8. It is clear from the daily load curve that maximum demand on th station is 70 MW.

9. Units generated/day = Area (in kWh) under the load curve

$$= 10^3\,[40 \times 6 + 50 \times 4 + 60 \times 2 + 50 \times 4 + 70 \times 4 + 40 \times 4]$$
$$= 10^3\,[240 + 200 + 120 + 200 + 280 + 160]\ \text{kWh}$$
$$= 12 \times 10^5\ \text{kWh}$$

12. The 3-wire d.c. system makes available two voltages *viz* V volts between any outer and néutral and 2 V volts betrween the outer.

19. Its main purpose is to isolate one portion of the circuit from the other and is not intended to be opened while current is flowing in the line. Such switches are generally used on both sides of circuit breakers in order that repairs and replacement can be made without any danger.

Appendix A

1. Energy Meters

The measurement of electrical energy is, from the viewpoint of economics, the most important of all electrical measurements. An instrument which measures electrical energy is called an energy meter or a watthour meter. Since electrical energy consumed by a load adds up as the time goes on (watt-hours = watts × hours), it is evident that watthour meter is an integrating instrument. The following three types of energy meters are most commonly used :

(*i*) Commutator motor meter .... for d.c. and a.c. circuits

(*ii*) Mercury motor meter .... for d.c. circuits

(*iii*) Induction motor meter .... for a.c. circuits only

In general, energy meters designed for d.c. circuits can be used in a.c. circuits but the reverse is not true. All energy meters are essentially *wattmeters with control spring and pointer removed but braking torque and counting mechanism provided. Since there is **no controlling torque, the moving system will rotate continuously like an electric motor and hence the name motor meters. The driving torque on the rotor will obviously be proportional to the power being supplied. A braking or retarding torque proportional to the disc speed is provided by a brake magnet. Consequently, the disc speed will be proportional to power and the number of revolutions made by the disc in a given time is a measure of the electrical energy taken by the load in that time. The counting mechanism is so designed that it counts the revolutions of the disc in terms of kilowatthours (kWh).

General Theory. The general operating principle of all watthour meters is the same. Fig. 1 (*i*) shows the general construction of a watthour meter. As indicated, the operating mechanism has a shunt circuit and a series circuit connected in the same manner as in a wattmeter. The shunt circuit carries current proportional to the supply voltage and the series circuit carries the load current. Consequently, the driving torque on the ***rotor is proportional to the power being supplied *i.e.*

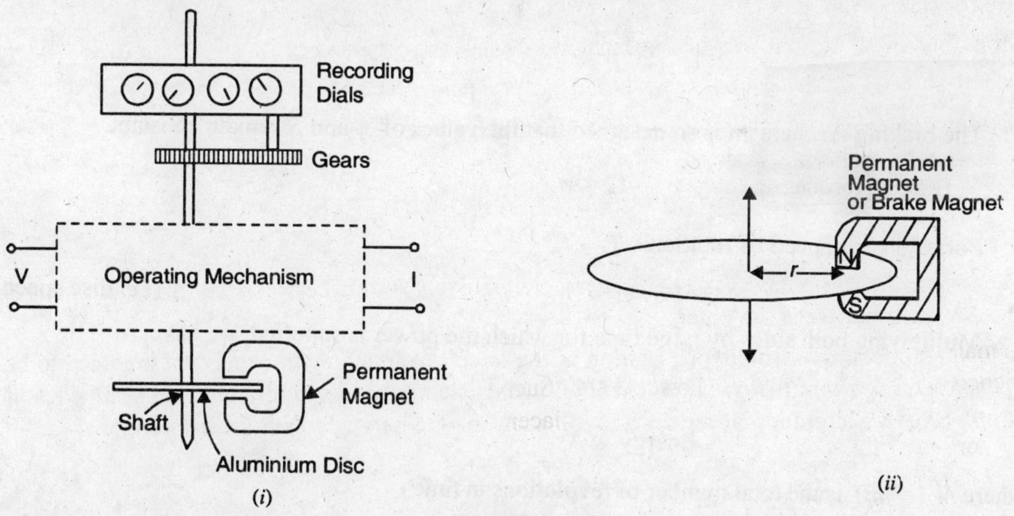

Fig. 1

* As in a wattmeter, the driving torque in a energy meter is provided by a shunt circuit (carrying current proportional to supply voltage) and a series circuit (carrying the load current).

** Unlike an indicating indicating instrument in which the controlling torque is provided which allows the moving system to rotate through a fraction of a revolution.

*** The rotor may be an aluminium disc or an armature mounted on the shaft, depending upon the type of energy meter.

Driving torque, $T_d \propto$ Power

Since there is no controlling torque, the disc will rotate continuously like an electric motor. If the speed of rotation of the rotor is made proportional to the driving torque (*i.e.* power), then its rate of rotation can be used as a measure of power and the total number of revolutions in a given time is a measure of the electrical energy taken by the load in that time. This is achieved by providing a braking torque proportional to the rotor speed. Such a braking torue is produced by rotating an aluminium disc mounted on the same spindle between the poles of a permanent magnet (called brake magnet).

Braking torque. Fig. 1 (*ii*) shows how braking torque is provided in a watthour meter. An aluminium disc is attached to the spindle and intercepts the flux of a permanent magnet (*i.e.* brake magnet). As the disc rotates, eddy currents are induced in the disc. These eddy currents react with the flux of the permanent magnet to produce a braking (or retarding) torque proportional to thedisc speed.

Let ϕ = flux of permanent magnet

n = disc speed

$\therefore$ E. M. F. induced in disc, $e \propto \phi n$

If R is the resistance of eddy current path, then,

Induced eddy current, $i = \dfrac{e}{R}$

$$\propto \frac{\phi n}{R} \qquad (\because e \propto \phi n)$$

Braking torque, $T_B \propto \phi\, i$

$$\propto \frac{\phi^2 n}{R} \qquad \ldots (i)$$

The braking mechanism is so designed that the values of ϕ and R remain constant.

$\therefore$ Braking torque, $T_B \propto n$

For a *steady speed of rotation, $T_d = T_B$

$\therefore$ Power $\propto n$ (*i.e.* disc speed)

Multiplying both sides by t, the time for which the power is supplied, we have,

$$\text{Power} \times t \propto nt$$

or Energy $\propto N$

where N $(= nt)$ is the total number of revolutions in time t.

The number of revolutions of the rotor or disc are recorded on dials which are geared to the shaft [See Fig. 1 (*i*)]. The counting mechanism is so arranged that it indicates kilowatthours (kWh) directly and not revolutions.

* The strength of the brake magnet is so designed that braking torque will balance the driving torque at a certain given speed which is usually dictated by the past practice on the speed of meters in service.

Note. The eq. (*i*) above reveals that braking torque will be proportional to disc speed only if the values of ϕ and R are constant. This implies that strength of the brake magnet should remain constant throughout the life of the meter. Also the resistance of the disc should be substantially constant.

2. Commutator Motor Meter

The first electrical energy meter was the commutator meter developed by Elihu Thomson in 1889. This meter can be used to measure d.c. as well as a.c. energy, although it is rarely used to measure a.c. energy. It is because induction watthour meter (see Art. 4) is far superior to the commutator meter for the measurement of a.c. energy.

Construction. Fig. 2 shows the various parts of a commutator motor meter. It is essentially a small motor (no iron is used in the fields or armature) with a magnetic brake. The field coils *FF* consist of a few turns of heavy copper wire and are connected in series with the load so that they carry the load current. In the field of these coils rotates an armature pivoted in jewelled bearings. The armature has a number of coils connected to the segments of a small commutator. The commutator is made of silver and the brushes are silver tipped in order to reduce friction. The armature *A* and a high resistance *R* (multiplier) are in series and connected across the supply. Therefore, armarture carries current proportional to the supply voltage. An aluminium disc mounted on the spindle (or shaft) rotates between the poles of two permanent magnets and provides the necessary braking torque. The braking torque prevents the armature from rotating too fast and from continuing to rotate after the load is disconnected.

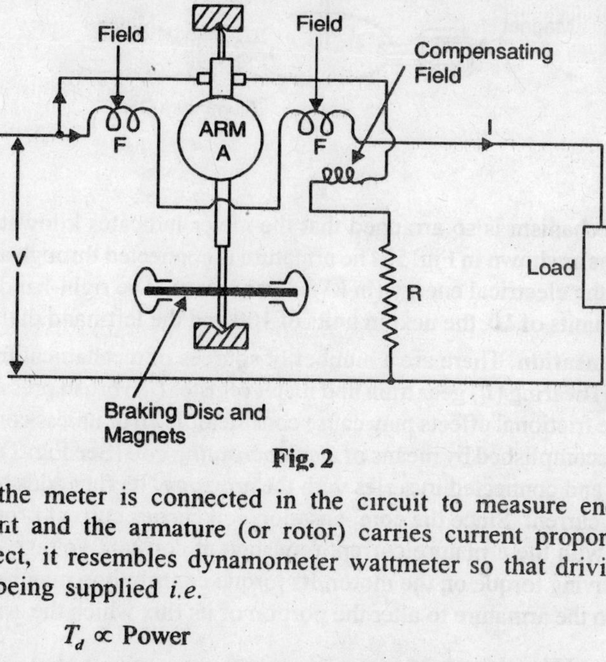

Fig. 2

Theory. When the meter is connected in the circuit to measure energy, the field coils carry the load current and the armature (or rotor) carries current proportional to the supply voltage. In this respect, it resembles dynamometer wattmeter so that driving torque is proportional to the power being supplied *i.e.*

Driving torque, $T_d \propto$ Power

The braking torque is due to the eddy currents induced in the aluminium disc. Since the magnitude of eddy currents is proportional to disc speed, the braking torque will also be proportional to the disc speed *n i. e.*

Braking torque, $T_B \propto n$ ..... See Art. 1

For steady speed of rotation, $T_d = T_B$.

$\therefore$ Power $\propto n$

Multiplying both sides by *t*, the time for which the power is supplied,

Power × time ∝ nt

or Energy ∝ N

where $N (= nt)$ is the total number of revolutions in time t.

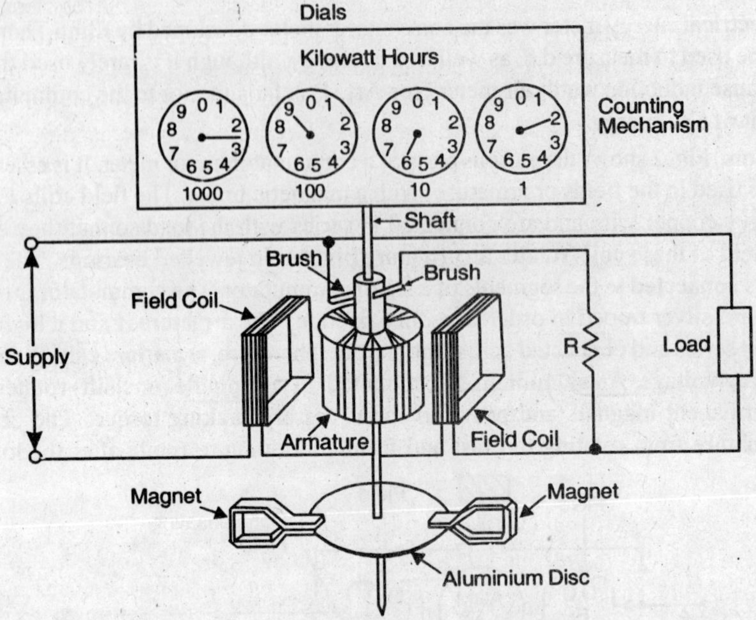

Fig. 3

The counting mechanism is so arranged that the meter indicates kilowatthours (kWh) directly and not the revolutions as shown in Fig. 3. The armature is connected through a set of gears to a series of dials that indicate the electrical energy (in kWh) consumed. The right-hand dial indicates kWh in units of 1, the next in units of 10, the next in units of 100 and the left-hand dial in units of 1000.

Friction Compensation. There are a number of sources of mechanical friction in the Thomson watt-hour meter *viz* (*i*) bearing (*ii*) gear train and meter register (*iii*) brush pressure on the commutator and (*iv*) windage. The frictional effects may cause considerable error unless compensation for friction is provided. This is accomplished by means of a *compensating coil* (See Fig. 2) placed co-axially with the two current coils and connected in series with the armature. Its flux adds to that of the field coils which carry the load current Since the compensation coil carries current proportional to the supply voltage, its reaction with the armature current (constant at constant voltage) will contribute a substantially constant driving torque on the motor. Its torque contribuiton may be adjusted by changing its position relative to the armature to alter the portion of its flux which the armature intercepts.

3. Mercury Motor Watthour Meter

The weight of the moving system of a commutator is quite large and the torque-weight ratio is low. Also the friction is large because of the brush pressure required for good commutation. Both these disadvantages are largely overcome in the mercury motor meter. This type of instrument is used for direct currents only.

Principle. The working principle of this meter is the same as that of commutator motor meter. Here the rotor is a copper disc (instead of armature) floating in a mercury chamber through which current is passed. The mercury serves the same purpose as the brushes and commutator in the commutator meter.

Construction. Figure. 4 shows the essential parts of a mercury motor watt-hour meter. It consists of a thin copper disc mounted on the spindle and floating in a mercury chamber. Below this disc is placed an electromagnet whose coils are in series with a high resistance and connected across the supply. Therefore, current in the coils of this magnet is proportional to the supply voltage. The circuit current is passed through the disc at its circumference through mercury. This current leaves the disc at a point on the circumference diametrically opposite to the entrance point. In order to ensure radial flow of current through the disc, radial slots are cut in the disc. As the disc is slotted, it cannot be used for braking purposes. Another aluminium disc is used to provide the necessary braking torque. This disc is mounted on the same spindle and rotates between the poles of two permanent magnets.

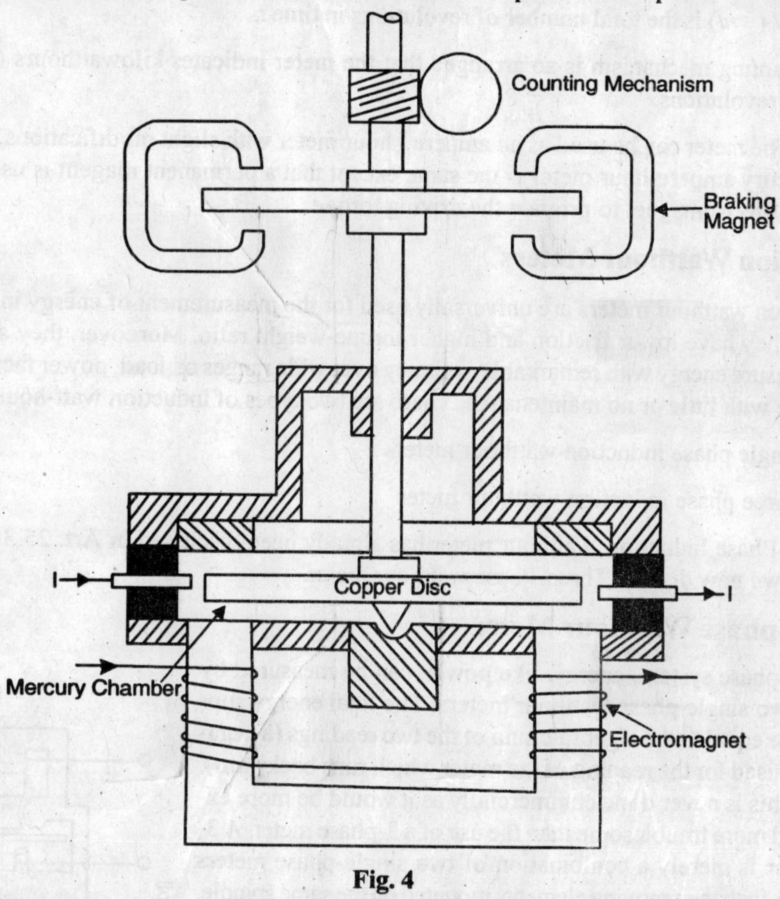

Fig. 4

Theory. When the meter is connected in the circuit to measure energy, the coils of the electromagnet carry current proportional to the supply voltage. The flux produced by the electromagnet is, therefore, proportional to the supply voltage. The reaction between the current in the disc and magnetic field due to the electromagnet produces the necessary driving torque.

$$\text{Driving torque, } T_d \propto \text{Flux of electromagnet} \times \text{Load current}$$
$$\propto VI$$
$$\propto \text{Power}$$

The braking torque is due to the eddy currents induced in the aluminium disc. Since the magnitude of eddy currents is proportional to the disc speed, the braking torque will also be proportional to the disc speed *i.e.*

Braking torque, $T_B \propto n$

For steady speed of rotation, $T_d = T_B$ (disc speed)

Power, $\propto n$

Multiplying both sides by *t*, the time for which the power is supplied,

Power $\times t \propto n t$

or Energy $\propto N$

where $N (= nt)$ is the total number of revolutions in time *t*.

The counting mechanism is so arranged that the meter indicates kilowatthours (kWh) directly and not the revolutions.

Note. The meter can be used as an ampere - hour meter with slight modifications. The construction of mercury ampere-hour meter is the same except that a permanent magent is used with disc in place of the electromagnet to provide the driving torque.

4. Induction Watthour Meters

Induction watthour meters are universally used for the measurement of energy in a.c. circuits. It is because they have lower friction and higher torque-weight ratio. Moreover, they are inexpensive and can measure energy with remarkable accuracy over wide ranges of load, power factor, voltage and temperature with little or no maintenance. There are two types of induction watt-hour meters *viz:*

(*i*) Single phase induction watthour meter

(*ii*) Three phase induction watthour meter

Single Phase Induction Watthour meter has already been discussed in Art. 25.30 (Chapter 25). Therefore, we now discuss Three-Phase watthour meter.

5. Three-phase Watthour Meter

In a 3-phase system, energy, like power, can be measured by means of two single-phase watthour meters. The total energy supplied will be equal to the algebraic sum of the two readings (a negative sign is used for the reading of the meter which runs backward). However, this is never done commercially as it would be more expensive and more troublesome than the use of a 3-phase meter. A 3-phase meter is merely a combination of two single-phase meters (See Fig. 5), with their moving elements mounted on the same spindle. The total driving torque is equal to the sum of the torques exerted by both the moving elements. Thus only one counting mechanism is required which will directly indicate the energy being supplied to the 3-phase circuit. Fig. 5 shows how a 3-phase watthour meter is connected in a 3-phase circuit to measure energy. The current coils are connected in any two lines and each potential coil is joined to the third line. In fact, the connections are similar to 2-wattmeter method used to measure power in a 3-phase circuit.

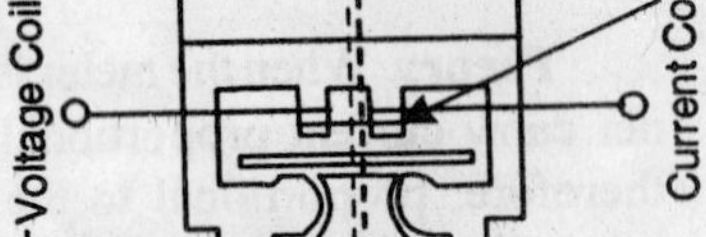

Fig. 5

It is very important that the two elements are "balanced" *i.e.* the driving torque of the two elements be exactly equal for equal amounts of power flowing through each. If this is not done, the meter will not indicate correct reading on unbalanced load. The balancing adjustment is most conveniently made with the potential coils connected in parallel and the current coils in series opposition. If the elements are *balanced, there will be no rotation of the disc for this condition. The usual lag, load and power factor adjustments are made independently for each element.

Note. In some 3-phase meters, a single disc is used and the two elements drive this disc. The disc is slotted radially to prevent interaction between eddy currents produced by one element with the flux produced by other element.

Example 1. *A 3-phase energy meter having a meter constant of 0.12 rev/kWh is used with a potential transformer of ratio 22000/110 volts and a current transformer of ratio 500/5 amperes. When connected to a load of unity p. f., the disc makes 40 revolutions in 61 seconds. If the instrument readings are 110 V and 5.25 A, find the percentage error.*

Solution.

$$\text{Actual load voltage, } V = \text{Indicated reading} \times \text{P. T. ratio}$$
$$= 110 \times (22000/110) = 22{,}000 \text{ volts}$$
$$\text{Actual load current, } I = \text{Indicated reading} \times \text{C. T. ratio}$$
$$= 5.25 \times (500/5) = 525 \text{ A}$$
$$\text{Actual energy consumption} = \sqrt{3}\ VI \cos\phi\ t$$
$$= \sqrt{3} \times (22000) \times 525 \times 1 \times (61/3600)$$
$$= 338.98 \times 10^3 \text{ Wh} = 338.98 \text{ KwH}$$
$$\text{Energy recorded by the meter} = 40/0.12 = 333.33 \text{ kWh}$$
$$\therefore \quad \%\text{age error} = \frac{338.98 - 333.33}{338.98} \times 100 = \mathbf{1.67\ \%}\ \textit{slow}$$

---

* If the disc rotates for this condition, it means that the elements are not balanced. Balancing can be easily done by adjusting the disc air gap.

Appendix B

1. Stepper Motors

Large electrical machines, d.c. or a.c., are used primarily for continuous energy conversion. However, there are many applications where continuous energy conversion is not required. For example, the printer of a computer requires that the paper moves by steps to signals received from the computer. Such applications require that motor be designed for pulse operation rather than for normal a.c. operation. The *stepper motor* (also known as step motor or stepping motor) meets this requirement. A stepper motor rotates by a specific number of degrees in response to an input electrical signal. Typical step sizes are 2°, 2.5°, 7.5° and 15° for each electrical pulse. The two most popular stepping motors are:

(*i*) Permanent-magnet (PM) stepping motor

(*ii*) Variable-reluctance (VR) stepping motor

Step angle. The angle through which the motor shaft rotates for each command pulse is called step angle. It can be shown that for any PM or VR stepping motor, the step angle is given by;

$$\text{Step angle, } \alpha = \frac{360°}{mp} \text{ degrees}$$

where m = number of stator phases

p = number of rotor teeth (or rotor poles)

For example, for 4-phase, two-pole VR stepper motor,

$$\text{Step angle, } \alpha = \frac{360°}{4 \times 2} = 45° \text{ / step}$$

2. Variable Reluctance Stepper Motor

Figure 1 shows the basic circuit for a 4 - phase, two-pole VR stepper motor. When the stator phases are excited with d.c. current in proper sequence, the resultant air gap field steps around and the rotor follows the axis of air gap field by virtue of reluctance torque. The reluctance torque is generated because of the tendency of the ferromagnetic rotor to align itself along the direction of the resultant magnetic field.

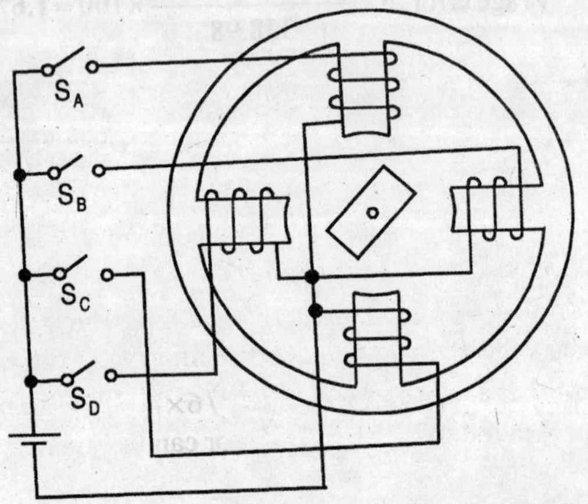

Fig. 1

Operation. The working of 4-phase, 2 - pole VR stepper moter can be understood by referring to Fig. 2. Clearly, the step angle $\alpha = 360°/4 \times 2 = 45°/$ step. The windings are energised in the sequence. A, A + B, B, B + C and so on and this sequence is repeated.

(*i*) When only winding A is excited, the rotor aligns with the axis of phase A as shown in Fig. 2 (*i*).

(*ii*) When both windings A and B are excited, the resultant m.m.f. axis moves 45° in the clockwise direction. The rotor aligns with this resultant m.m.f. [see Fig 2(*ii*)].

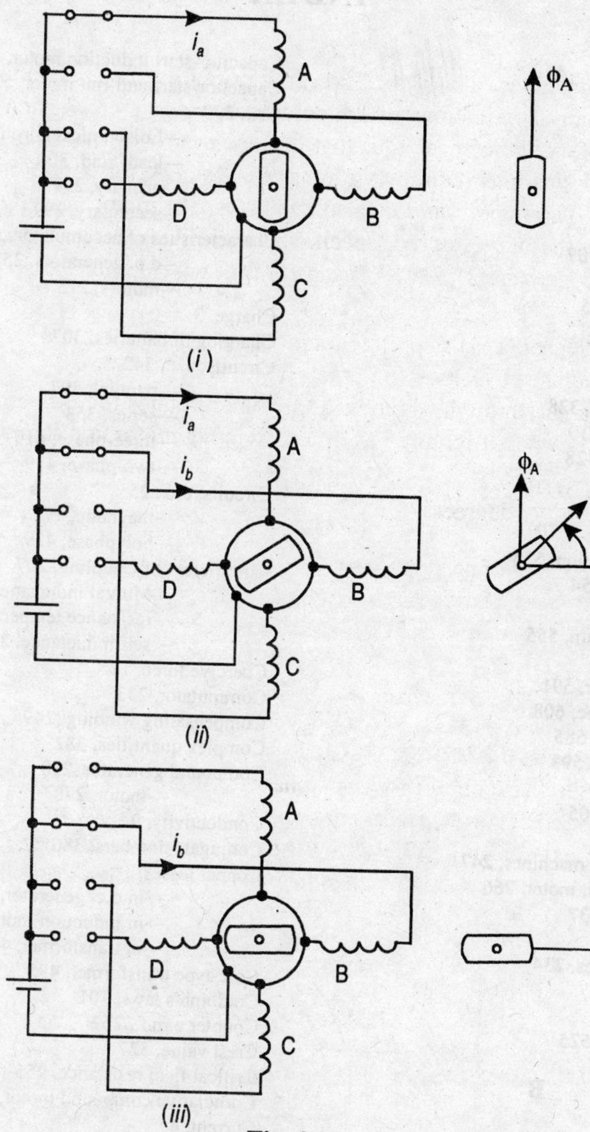

Fig. 2

(*iii*) Thus for each sequence of excitation the rotor turns through 45° clockwise.

(*iv*) In order to obtain smaller step angle, we use multipolar rotor. Thus for a four - phase, six - pole stepper motor, the step angle $\alpha = 360° / 6 \times 4 = 15° /$ step.

(*v*) The direction of rotation of the stepper motor can be reversed by reversing the sequence of switching the windings.

3. Applications of Stepper Motors

(*i*) Stepper motors are well suited for situations where either precise positioning or precise speed control or both are required in automatic systems.

(*ii*) Since stepper motors provide incremental motion, they are widely used in typewriters, line printers, tape drives, floppy disk drives etc.

(*iii*) Stepper motors are also used to perform such functions as mixing, cutting, metering etc.

(*iv*) They are also used in the manufacture of packed food stuffs, commercial end - products etc.

INDEX

IMPORTANT BOOKS ON ELECTRICAL ENGINEERING & ELECTRONICS

OBJECTIVE ELECTRICAL TECHNOLOGY

V.K. Mehta & Rohit Mehta

CONTENTS: ●·Basic Concepts ●·D.C. Circuits ● Network Theorems ● Electrical Work, Power and Energy ●·Network Theorems ●·Capacitance ● Magnetism and Electromagnetism ● Magnetic Circuits ● Electromagnetic Induction ● Chemical Effects of Electric Current ● Alternating Current ● Series A.C. Circuits ●·Phasor Algebra ●·Parallel A.C. Circuits ● Three Phase Circuits ●·Electrical Measuring Instruments ●D.C. Generators ●·D.C. Motors ● Transformers ● Three Phase Induction Motors ●·Single Phase Motors ●·Alternators ● Synchronous Motors

Code : 10 233 **2nd Edn. 2004** **ISBN : 81-219-2097-3**

ELECTRONIC PRINCIPLES
Devices & Circuits

M.L. Anand

CONTENTS: ● Basic Concepts ● Electronic Materials and Components ● Constant Voltage and Current Sources ● Electron Emission ● Vacuum Tubes ● Gas Filled Tubes ● Semiconductor Physics ● Semiconductor Diode and Rectifiers ● Special Diodes ● Transistors ● Transistor Biasing and Stabilization ● Single Stage (Small Signal) Transistor Amplifiers and Hybrid Parameters ● Multistage Transistor Amplifiers ● Audio (Class A & B) and Video Power Amplifiers ● Class C (Tuned) Amplifiers ● Vacuum Tube Amplifiers ● Feedback Theory ● Sinusoidal Oscillators ● Switching and Waveshaping Circuits Including Non-Sinusoidal Oscillators (or Multivibrators) ● Regulated D.C. Power Supply ● Field Effect Transistors (FET) ● Thyristors and UJT ● Integrated Circuits ● Operational Amplifiers (OP-AMPs) ● I.C. Timers ● Radio Communication and Reception ● Electronic Instrumentation ● Opto-electronic Devices ● Transducers ● Additional Solved and Unsolved Problems ● Appendices ● Glossary ● Index

AHW 1585 **ISBN : 81-7544-158-5**

DIGITAL SIGNAL PROCESSING

S.P. Eugene Xavier

CONTENTS: ● Introduction ● Z-Transforms ● Linear Shift invariant systems and realisation structures ● Discrete and Fast Fourier Transforms ●Finite impulse response (FIR) Filters ● Infinite impulse response (IIR) Filters ● Finite word length effects in digital filters ● Applications of DSP ● Appendix ● Bibliography ● Index

Code : 10 259 **1st Edition : 2003** **ISBN : 81-219-2305-0**

PRINCIPLES OF CONTROL SYSTEMS

S.P. Eugene Xavier & J. Joseph Cyril Babu

CONTENTS: Introduction ·●·Mathematical Tools for Control Systems Analysis ●·Mathematical Models of Physical Systems ●·Feedback Characteristics of Control Systems ●·Components of Control Systems ●·Time Domain Analysis of Control Systems ●·Stability of Control Systems ●·Root-Locus Techniques ●·Frequency Domain and Analysis of Control System ●·Design and Compensation of Control System ●·State Variable Analysis ●·Objective Type Questions with Answers ●·University Question Papers ●·Appendix ●·Bibliography ●·Index

Code : 10 184 **2nd Edn. 2004** **ISBN : 81-219-1778-6**

ELECTRONICS LABORATORY PRIMER

S. Poorna Chandra & B. Sasikala

CONTENTS: ● Network Theorems ● Electronic Device Characteristic ● Amplifiers and Oscillators ● Integrated Circuits ● Digital Electronics ● Basic Communication Circuits ● Appendix ● Index

Code : 10 290 **Rev. Edn. 2005** **ISBN : 81-219-2459-6**

SIGNALS AND SYSTEMS

S.P. Eugene Xavier

CONTENTS: ●·Classification of Signals and Systems ●·Analysis of Continuous Time Signals ●·LTI-CT Systems ●·Analysis of Discrete Time Signals ●·LTI-Discrete-Time Systems ●·Representation of Signals-Vector Space Concepts

Code : 10 199 **3rd Rev. Edn. 2003** **ISBN : 81-219-1755-7**

MODERN POWER ELECTRONICS

P.C. Sen

Code : 10 287 2nd Rev. Edn. 2005 ISBN : 81-219-2425-1

CONTENTS: ● Characteristics of Semiconductor Power Devices ● Linear Voltage Regulator ● Switching Regulators ● CUK Converter and Integrated Magnetics ● Resonant Converter ● Uninterruptible Power Supply ● Stepper and Switched Reluctance Motor Drive ● Solved Questions ● Bibliography ● Index

FUNDAMENTALS OF ELECTRICAL NETWORKS

B.R. Gupta & Vandana Singhal

CONTENTS: ● Units and Definitions ● Parameters of Electric Networks ● DC Network Analysis and Network Theorems ● Magnetic Circuits ● Self and Mutual Inductance ● Network Transients ● Alternating Current and Voltage ● Series AC Circuits ● Parallel and Series Parallel Networks ● Resonance ● AC Network Theorems ● State Variable Analysis ● Non-Sinusoidal Waves ● Polyphase Networks ● Two-Port Networks ● Filter Network ● Network Functions: Poles and Zeros ● Network Synthesis ● Mechanical Analogies ● Symmetrical Components ● Appendices ● Bibliography ● Index

Code : 10 274 Rep. Edn. 2005 ISBN : 81-219-2318-2

DIGITAL COMMUNICATIONS

S.K. Venkata Ram

CONTENTS: ● Basics of digital communication ● Digital Transmission of Analog signals ● Digital Coding of Analog_Waveforms ● Intersymbol Interference And Its Cures ● Digital Modulation Techniques ● Optimum Receivers ● Introduction to spread spectrum techniques ● Optical Fiber communications ● Satellite communication

Code : 10 247 1st Edition : 2003 ISBN : 81-219-2241-0

A TEXTBOOK OF ELECTRICAL TECHNOLOGY

B.L. Theraja & A.K. Theraja, S.G. Tarnekar,
M.K. Khedkar, R.S. Sedha

CONTENTS: Vol. I: Electric Current and Ohm's Law ●·D.C. Network Theorems ●·Work, Power and Energy ●·Electrostatics ●Capacitance ● Magnetism and Electromagnetism ●·Electromagnetic Induction ●·Magnetic Hysteresis ●·Electronical Power Sources ●·Electrical Instruments and Measurements ●·A.C. Fundamentals ●·Complex Numbers ●·Series A.C. Circuits ●·Parallel A.C. Circuits ●·A.C. Network Analysis ●·A.C. Bridges ●·A.C. Filter Networks ●·Circle Diagrams ●·Polyphase Circuits ●·Harmonics ●·Fourier Series ●·Transients ●·Symmetrical Components ●·Introduction to Electrical Energy Generation ●·**Vol II:** ● Elements of Electro-mechanical Energy Conversion ● D.C. Generators ● Armature Reaction and Commutation ● Generator Characteristics ● D.C. Motor ● Speed Control of D.C. Motors ●Testing of D.C. Machines ●Transformer ●Transformer-Three Phase ● Induction Motors● Computation And Circle Diagrams●Single-Phase Motors ● Alternators ● Synchronous Motors ● Special Machines ●·**Vol III:** D.C. Transmission and Distribution ●·A.C. Transmission ●·Distriemtion Electric Traction ● Industrial Applications of Electric Motors ●·Distribution Automation ●·Electronic Control of AC Motors ●·Electric Heating ●·Rating and Service Capacity ●·Electric Welding ●·Illumination ●·Tariffs and Economic Considerations ●·Semiconductor Physics ●·P.N. Junction Diode ●·Opto Electronic Devices ●·Special Diodes ●·**Vol IV:** ● Semiconductor Physics ● P-N Junction Diode ● Optoelectronic Devices ● Special Diodes ● D.C. Power Supplies ● Regulated Power Supply● Bipolar Junction Transistor ● Load Lines and DC Biased Circuits ●Transistor Equivalent Circuit and Models ● Single-Stage Transistor Amplifiers● Multistage and Feedback Amplifiers ● Feedback Amplifier● Field Effect Transistors● Breakdown Devices ● Sinusoidal and Non-sinusoidal Oscillators ● Modulation and Demodulation ● Integrated Circuits ● OP-AMP and its Applications ● Number Systems and Codes ● Logic Gates ● Boolean Algebra & Logic Families ● Electronic Instruments

(Also available in 4 individual parts)

Code : 10 291 24th Rev. Edn. 2005 ISBN : 81-219-0289-4

VOL. I: (Basic Electrical Engineering)

B.L. Theraja, A.K. Theraja & S.G. Tarnekar

Code : 10 292 23rd Rev. Edn. 2005 ISBN : 81-219-0290-8

VOL. II: (A.C. & D.C. Machines)

B.L. Theraja, A.K. Theraja & S.G. Tarnekar

Code : 10 293 | 23rd Rev. Edn. 2005 | ISBN : 81-219-2537-5

VOL. III: (Transmission, Distribution and Utilization)

B.L. Theraja, A.K. Theraja & M.K. Khedkar

Code : 10 004 | 23rd Rev. Edn. 2005 | ISBN : 81-219-1140-0

VOL. IV: (Electronic Devices & Circuits)

B.L. Theraja, A.K. Theraja & R.S. Sedha

Code : 10 005 | 23rd Rev. Edn. 2004 | ISBN : 81-219-1141-9

DIGITAL SIGNAL PROCESSING, TELECOMMUNICATIONS AND MULTIMEDIA TECHNOLOGY

V.K. Khanna

CONTENTS: • Communication and Information Systems Overview • Signal Processing Fundamentals • Digital Coding of Analog Waveforms • Speech and Audio Signal Processing • Image and Video Signal Processing • Digital Modulation Techniques • Networking Principles • Digital Satellite Communications • High Definition Television (HDTV) • Multimedia • Signal Processing and Communications • Bibliography • Review Exercises • Appendices • Index

Code : 10 276 | Rep. Edn. 2003 | ISBN : 81-219-2325-5

SENSORS AND TRANSDUCERS

D. Patranabis

CONTENTS: • Introduction: Definitions • Principles • Classification • Parameters • Environmental Parameters • Characterization; Mechanical and Electromechanical Sensors : Introduction • Resistive Potentiometers • Strain Gauge • Inductive Sensors • Capacitive Sensors; Thermal Sensors: Introduction • Gas Thermometric Sensors • Thermal Expansion Type Thermometric Sensors • Dielectric Constant and Refractive Index Thermosensors • Helium Low Temperature Thermometer • Nuclear Type • Magnetic Thermometer • Resistance Change Type Thermometric Sensors • Thermo emf Sensors • Junction Semiconductor Types • Thermal Radiation Sensors • Quartz Crystal Thermometric Sensors; Magnetic Sensors: Introduction • Types of Sensors Using the Principles • Magnetoresistive Sensors • Hall Effect Sensors • Inductance and Eddy Current Types • Angular/Rotary Movement Transducers • Eddy Current Sensors • Electromagnetic Flowmeter • Switching Magnetic Sensors • Squid Sensors; Radiation Sensors: Introduction • Basic Characteristics • Types of Photosensistors/Photodetectors • X-ray and Nuclear Radiation Sensors • Fibre Optic Sensors; Electroanalytical Sensors: Introduction • The Electrochemical Cell • The Cell Potential • Standard Hydrogen Electrode (SHE) • Liquid Junction and Other Potentials • Polarization • Reference Electrodes • Sensor Electrodes • Electroceramics in Gas Media • Chem FET; Smart Sensors: Introduction • Primary Sensors • Excitation • Amplification • Filters • Converters • Compensation • Information Coding/Processing • Data Communication • Review Questions • Bibliography • Index

AHW 004X | Rs. 170.00 | ISBN : 81-7544-004-X

CONCISE HANDBOOK OF ELECTRONICS AND ELECTRICAL ENGINEERING

V.K. Khanna

CONTENTS: • Mathematical and Physio-Chemical Foundations • Electromagnetics • Material Science • Network Analysis • Electrical Devices • Circuits and Systems • Telecommunication • Computers • Instrumentation • Appendices • Index

Code : 10 275 | Rep. Edn. 2003 | ISBN : 81-219-2319-0

FIBRE OPTIC COMMUNICATION

D.C. Agarwal

CONTENTS: • Principles of Fibre Optic Communication • Fibre Optical Source Devices • Mode Theory for Optical Fibres • Ray Theory for Optical Fibres • Fibre Optical Communication Components • Fibre Optical Communication Systems • Optical Fibre Measurements • Economics and Potential Applications of Optical Fibre Communication Systems • Appendix • Index

Code : 16 286 | Rep. Edn. 2004 | ISBN : 81-219-2317-4

FUNDAMENTALS OF ELECTRICAL ENGINEERING AND ELECTRONICS

B.L. Theraja

CONTENTS: Electric Current and Ohm's Law ●·Division of Current ● Network Analysis ●·Work, Power and Energy ●·Electrostatics ●·Capacitance ● Magnetism and Electromagnetism ● Electromagnetic Induction ●·Magnetic Hysteresis ●·D.C. Generators ●·Generator Characteristics ●·D.C. Motor ●·Speed Control of D.C. Motors ●·Chemical Effects of Current ●·Electrical Instruments and Measurements ●·A.C. Fundamentals ●·Series A.C. Circuits ●·Parallel A.C. Circuits ●·Complex Algebra and A.C. Circuits ●·Three Phase Circuits ●·Transformer ●·Three Phase Induction Motor ●·Single Phase Motors ●·Alternators ●·Synchronous Motor ●·Q and A on Electric Machinery ●·Semiconductor Physics ●·Semiconductor Diodes ●·Optoelectronic Devices ●·Bipolar Junction Transistors ●·Load Line and Biasing Circuits ●·Transistor Equivalent Circuits and Models ●·Transistor Amplifiers ●·Field Effect Transistors ●·Thyristors ●·Digital Electronics ●·Sinewave Oscillators ●·Analog and Digital Communication ●·Vacuum Tubes and Gas Valves ●·Electron Ballistics ●·Illumination ●·Index

Code : 10 008 **28th Edn. 2004** **ISBN : 81-219-0099-9**

CONTROL SYSTEM DESIGN
The Optimal Approach

B. Sarkar

CONTENTS: ● Introduction ● Classical Design Approach ● Modern Design Approach ● Choice of Performance Criterion ● Scaler Valued Performance Criterion ● Vector Valued Performance Criterion ● Equivalent Formulation of External Problems ● Design Approach Using a Performance Criterion ● Method of Optimization ● Appendices: A Quadratic Performance; B. Digital Simulation for Optimization; C. Evaluation of Performance Criteria

AHW 0252 **Rs. 125.00** **ISBN : 81-7544-025-2**

A TEXTBOOK OF DIGITAL ELECTRONICS

R.S. Sedha

CONTENTS: ●·Introduction ●·Number Systems and Codes ●·Logic Gates ●·Boolean Algebra and Logic Simplification ●·Flip-flops and Related Devices ●·Arithmetic Oprations and Circuits ●·Counters and Shift Registers ●·Integrated Circuit Logic Families ●·Medium-Scale Integrated Logic Circuits ●·Appendix A, B, C, D, E ●·Index

Code: 10 260 **1st Edition : 2004** **ISBN : 81-219-2378-6**

ELECTRICAL ELECTRONICS AND TELECOMMUNICATION ENGINEERING (OBJECTIVE TYPE)

B.L. Theraja

CONTENTS: Electrical Engineering: 1. Basic Electrical Engineering, 2. D.C,. Generators 3. D.C. Motors 4. Transformers 5. Three-phase Induction Motors 6. Single-phase Induction Motors, 7. Alternators 8. Synchronous Motors ● Solid State: Electronics 1. Semiconductor Physics and Diodes 2. Tranistor Fundamentals 3. Transistor Parameters and Amplifiers 4. Fets, Diacs, Triacs and Scrs 5. Oscillators 6. Television 7. Charge Coupled Devices 8. Semiconductor Memories 9. Digital Electronics 10. Computer Engineering 11. Robotics 12. Electronic Instruments ●·Telecommunications: 1. Engineering Materials 2. Applied Electronics and Circuits 3. Principles of Communication Engineering 4. Antenna and Wave Propagation 5. Electronic Measurements 6. Radio Communication Systems 7. Broadcasting and T.V. Systems 8. Microwave Communication Systems 9. Radar Engineering 10. Industral Electronics 11. Transmission Lines 12. Waveguides 13. Communication Satellites ●·Standard Symbols and Vocabulary: 1. IS: 1885 Electrotechnical Vocabulary 2. IS: 2032 Graphical Symbols for Electrical Installatins in Buildings ●·Index

Code : 10 037 **4th Edn. 2004** **ISBN : 81-219-0125-1**

BASIC ELECTRICAL ENGINEERING

V.K. Mehta & Rohit Mehta

CONTENTS: Basic Concepts ●·D.C. Circuits ●·Network Theorems (D.C.) ●·Units-Work, Power and Energy ●·Electrostatics and Capacitance ●·Magnetism and Electomagnetism ● Electomagnetic Induction ●·Batteries ● A.C. Fundamentals ● Series A.C. Circuits ●·Phasor Algebra ●·Parallel A.C. Circuits ●·Polyphase Circuits ●·Index

Code : 10 113 **3rd Edn. 2004** **ISBN : 81-219-0871-X**